D. NEARY

# Advanced Praise for *Flamethrower*

Rigg has done something most authors of "war stories" are totally incapable of doing in *Flamethrower*…he has neatly tied all three levels of conflict into a single package (i.e. tactical, operational and strategic) and done so in a magnificent manner…And in the cases of both General Kuribayashi and Corporal Williams (the two main characters of the book), as with every combatant on Iwo Jima, Rigg shows that neither was a saint nor a sinner. Each had his own flaws that have been masterfully researched and documented by the author.

—USMC 31st **Commandant, General Charles C. Krulak (1995-1999)**, author of *Operational Maneuver from the Sea* and godson of Lieutenant General Holland M. "Howlin Mad" Smith

Rigg has written a very detailed and superbly researched book in *Flamethrower* about some of our WWII great battles and some of the incredibly superb Marine Warriors who made it happen…It is a superb testimony to the efforts of our servicemen and leaders as they went about winning WWII in the Pacific.

—USMC 29th **Commandant, General Al Gray (1987-1991)**, the Al Gray Research Center at the United States Marine Corps University is named after him.

Rigg has written a compelling work of history. Ostensibly the biography of a Medal of Honor Recipient Woody Williams from the Battle for Iwo Jima, it is that and so much more. He gives us a powerful history of the Battle for the Pacific, of the very nature of war and of the powerful bonds forged by fire and fear that unite men who lived together, fought together and died together in service of our nation. It is an ode to the values of the Marine Corps and yet another well-deserved homage to the greatest generation…

—**Michael Berenbaum**, author of *The World Must Know* and *Anatomy of the Auschwitz Death Camp.* Deputy Director President's Commission on the Holocaust (1979–80), and Project Director (1988–93) and then Research Institute Director (1993-1997) of the U.S. Holocaust Memorial Museum.

Bryan Rigg has written a superb account of the Iwo Jima battle. His thorough research and riveting descriptions of the fighting make this book come alive as the stories of the individual Marines and the hell they went through are vividly recounted. Woody Williams was a hero in that battle. Medals have never been sufficient to justly reward our battlefield heroes as we have learned throughout history and there is always controversy over them in each war, but we do our best to acknowledge those who went above and beyond the call of duty, as Woody did, with the right award. His Medal of Honor is worn with due pride and remarkable humility.

—**General Anthony C. Zinni** USMC (Retired), Commander-in-Chief of Central Command (CENTCOM) (1997-2000) and author of *Before the First Shots are Fired*

*Flamethrower* is a valuable contribution to scholarship on the institutional Marine Corps, the enlisted experience in World War II, and the story of the Guam and Iwo Jima campaigns.

—**Colonel Jon T. Hoffman** USMC (Retired), author of *Chesty: The Story of Lieutenant General Lewis B. Puller* and *Once A Legend: Red Mike Edson of the Marine Raider*

This is the kind of book I love—a true story of courage and sacrifice, exhaustively researched, brilliantly conceived, and masterfully executed. Bryan Rigg is a former Marine officer, veteran of the Israeli Defense Forces, and a Cambridge Ph.D. In *Flamethrower*, he brings to bear the firepower of all three disciplines to render in vivid detail the true story of Iwo Jima Medal of Honor recipient Hershel "Woody" Williams, (an amazing character in his own right)—and, better still, Dr. Rigg sets his tale in Big Picture historical context, the Japanese side as well as the American, with the full horror of combat gut-churningly intact. *Flamethrower* belongs on the shelf next to the two great Pacific classics of WWII, E.B. Sledge's *With the Old Breed* and Robert Leckie's *Helmet for My Pillow*. Outstanding!

—**Steven Pressfield**, bestselling author of *Gates of Fire* and *The Lion's Gate* and Marine veteran

Rigg gives context to the story of one of the finest Marine Corps heroes to wear the Eagle, Globe, and Anchor. In so doing, *Flamethrower* addresses the brutality of the Pacific War in a manner that gives meaning to action—as Woody Williams kills, blasts, and blazes a trail across the most horrific landscape imaginable in the supreme iconic battle of Marine Corps history.

—**USMC Lieutenant Colonel Robert Burrell**, author of *The Ghosts of Iwo Jima*

Woody Williams was an archetypical U.S. Marine of World War II until he earned the Medal of Honor on Iwo Jima. Using Japanese and American sources, Rigg presents this event in the context of the Pacific War that at its sharp end, was a brutal death grapple between mutually loathing cultures. No less perspective and persuasive is Rigg's analysis of the post-war controversies surrounding Williams' actual deeds. Defining and controlling a narrative based on memory and repetition remains a fundamental challenge in evaluating combat experiences. This book is a page turner and is a must read for anyone who wants to learn more about World War II.

—**Dennis Showalter**, author of *Patton and Rommel*,
*If the Allies had Fallen* and *Hitler's Panzers*

The publishing marketplace is saturated with books about the Second World War, and the vast number of titles available typically make no bold or meaningful contribution to the subject's historiography. As a reflection of the either/or binary dividing the industry, it is not often that a book comes along that bridges the gap between what is scholarly and what is popular. Rigg's *Flamethrower* is one of those books. In it Dr. Rigg narrates the exceptional story of Marine Medal of Honor recipient Woody Williams through his harrowing experiences in combat during the liberation of Guam and the battle of Iwo Jima. Simultaneously, the author advances a powerful critique of the revisionist interpretations advanced by John Dower...and the late Ronal Takaki...In *Flamethrower*, it wasn't racism that produced the intense brutality that so closely characterized war with Japan, but the Bushido code and the fascistic qualities of Japanese "nationalist fanaticism" that did it. Rigg has thus authored a book that engages an ongoing intellectual discussion while it chronicles the incredible story of Woody Williams and his rendezvous with fate during the Second World War. Combining both is a majestic accomplishment that makes this book stand-out from the crowd.

—**Martin K. A. Morgan**, author of *The Americans on D-Day* and *Down to Earth: The 507th Parachute Infantry Regiment in Normandy*

*Flamethrower* is one hell'va "tour de force" of the man and the historic events that shaped Woody's character, beliefs and path to distinction. Rigg intertwines the external influences on Woody's life—upbringing, world events, the Marine Corps—to make a highly readable memoir...The book is an excellent treatise on the events that shaped the "Greatest Generation."

—**Colonel Dick Camp** (USMC), author *Last Man Standing: The 1st Marine Regiment on Peleliu, September 15-21, 1944*

Rigg's *Flamethrower* is much more than the simple biography of a hero. Yes, the focus of the book is Woody Williams, a small-town youngster from rural America who, on a grim day in Feb. 1945, at a hellhole called Iwo Jima, with his buddies dying all around him, stepped forward to perform one of the outstanding MOH actions of WWII. But *Flamethrower* goes well beyond a gripping narrative of Williams' life and his near miraculous escape from death. It is also a thoughtful examination of the immense tide of world events that swept up millions of... Americans like Williams and carried them along for better or for worse, as well as an ofttimes philosophical exploration of courage, duty, loyalty, morality, faith and the very nature of man.

—**James H. Hallas** author *Uncommon Valor on Iwo Jima* and *The Devil's Anvil: The Assault on Peleliu*

Bryan Rigg has produced a fascinating and in-depth study about the making of a modern Marine Corps hero. A former U.S. Marine himself, Rigg is at his best when peeling back the veneer of official propaganda, revealing that the motivations behind military decorations are often far more nuanced than the average reader may suspect.

—**Dwight S. Mears**, author of *The Medal of Honor: The Evolution of America's Highest Military Decoration*

*Flamethrower* is a piercing, sometimes merciless, look deep inside the American men and their valor, as well as the Japanese murderers and their atrocities that collided in the Pacific War.

—**Edwin Black**, author of *IBM and the Holocaust*

Rigg's latest book, *Flamethrower*, is a triumph on many levels. It is the story of Woody Williams who received the MOH for his actions during the battle for Iwo Jima…but in telling Williams' story, the author addresses the action on many planes. To set the stage Rigg explains in detail the fanatical, often suicidal Japanese Bushido code that guided and inspired the Japanese military during the war and contrasts it with the ethos of the U.S. Marine Corps, a code that stressed teamwork, leadership, and the value of each Marine's life. Understanding this difference is the key to appreciating the savagery our Marines faced during the Iwo campaign. Rigg…tells the story of the battle in detail, framed around the exploits of Woody, who on February 23, 1945, singlehandedly destroyed several enemy pillboxes using flamethrowers and satchel charges. It is a classic story of an ordinary man who achieved extraordinary feats under the most hell-like conditions imaginable…*Flamethrower* will take its place among the classic books about the battle of Iwo Jima. Rigg has achieved what few historians have done in the past- he has written a scholarly account of one of the key battles in U.S. history that reads like a page-turning novel.

—Retired **Captain (USN) Lee R. Mandel**, author of *Unlikely Warrior: A Pacifist Rabbi's Journey from the Pulpit to Iwo Jima*

Rigg has given us what many later-generation historians have deftly avoided: an unblinkered analysis of the militaristic ambitions and barbaric culture that drove the Empire of Japan on its deadly rampage through Asia and the Pacific. At the center of this well documented history is the dramatic story of a classic Greatest Generation hero and MOH recipient, Woody Williams. Here is a tale that will fascinate, disturb, and in the end make you wonder why this dramatic story hasn't been told before.

—**Robert Gandt** author *Angels in the Sky, Twilight Warriors*

Woody's story is not that unique among the Marines who served during that war; what makes it remarkable is the fact that enough men lived to witness his heroism, and tell his story from their viewpoint, and Rigg has chronicled one of America's best fighting men.

—**Colin D. Heaton** and **Anne Marie Lewis**, authors *Noble Warrior: The Story of Major General James E. Livingston, USMC (Ret.), Medal of Honor*

Woody was one of the great heroes of WWII, and Rigg has provided his story, as well as the accomplishments of other Marines, to the forefront.

—**USMC Major General James E. Livingston** (USMC), Medal of Honor recipient

Rigg's *Flamethrower* offers the reader a gripping account of combat in the Pacific that exposes the true face of battle from the perspective of the individual Marine, sailor, and soldier. Moreover, he expertly integrates the tactical battle with the strategic objectives and forces driving this "war without mercy." A first-rate history and an incredible story of individual courage.

—**Edward B. Westermann**, author *Hitler's Ostkrieg and the Indian Wars: Comparing Genocide and Conquest* and *Hitler's Police Battalions: Enforcing Racial War in the East*

Relying on many previously unreported facts and sources, Rigg has written a masterful account of the improbable, sustained heroism of a lone Marine Woody Williams in the most important battle of the Pacific War. It is destined to become a combat history classic.

—**Barry Beck**, Marine officer, lawyer and son of the Captain Donald Beck, Woody's company commander

Rigg's book *Flamethrower* is insightful and reaches deep on many levels into the history of WWII—in context of the island hopping campaign in the Pacific. Rigg guides us into the history, culture, psychology (and pathology) and philosophy of the two opposing sides—the fanatical Japanese army against the unbeatable U.S. Marines. Rigg explains the painful bravery of men in battle with uncanny and unflinching detail and background...The main player...is the MOH recipient, Woody Williams—the now last surviving medal of honor recipient from Iwo Jima. Rigg brings the colorful personal history of the brave Marine-flamethrower

Woody to life in a special way—from the farm, to boot camp, Guam, Iwo Jima and back again—all very compelling. One of the legacies Rigg gives us in his book, is the undeniable necessity and rightness of the atomic bombings. There was no other ethical way of ending the war with the Japanese. He explains it all very well—and to me it's the last definitive word about Iwo Jima.

—**Kenneth E. Bingham**, author *Black Hell: The Story of the 133$^{rd}$*
*Seabees that landed with the 4th Marine Div. on Iwo Jima*

Woody is an…American hero. It was courageous men like he who took the fight to the Japanese on the ground who won the war. My last base of Iwo Jima was secured by brave Marines like him…Rigg is doing our *country* a service by telling them about one of *her* bravest sons. In doing so, he also tells us about all those men who never came back to tell us about their stories. Rigg's book shows us that all those men, some of them my own comrades, are the real heroes and deserve our attention and respect.

—**Jerry Yellin**, P-51 veteran and author *The Last Fighter Pilot*

When Marines think of WWII in the Pacific, we think of a whole generation of heroism like Rigg captures in the Woody Williams' story. As a Marine who read about these battles over and over again in my career in the Corps, Rigg inspired my reading once again and adds a welcome first person account of war on foot, on bellies, and backs in the Pacific. His prose brought vivid memories of my relative who served as a Corpsmen with Marines in the Pacific. And, his battle scenes helped me recall times I was able to trace the Greatest Generation's footsteps on Iwo Jima, Okinawa, Chuuk, and Guam. *Flamethrower* brings first source research and Woody's story into a readable and historical work well worth reading and it brings great credit again to the Corps, Woody Williams and the author.

—USMC **Colonel Mant Hawkins**, Advisor to Commander US CENTCOM 2005-07,
Director Global Combatting Terrorism Network US Special Operations Command
2007-08, Director of Operations USMC Aviation and Aviation,
Ground Combat Iraq, 2004, and TOPGUN Instructor 1987-90.

Bryan Rigg chronicles the courage and sacrifice of American Marines who were determined to win the battle for Iwo Jima. He has given us a compelling and historically accurate account of many unknown aspects of the Pacific War. He also provides insight into the fanatical mind set of the Japanese defenders which caused Iwo Jima to become one of the bloodiest and most difficult battles of the Pacific War. This book is a must read for anyone who wishes for true insight into what America faced in the closing days of the war and what would have happened if Japan had not surrendered when it did. It is well researched and an important contribution to the history of WWII.

—**Anthony Eugene Bonelli,** Dean of the Meadows School at
Southern Methodist University (1978-94)


Bryan Rigg has written three interlocked stories. There is the epic of Iwo Jima, the iconic Marine Corps battle. Then there is the ferocity and sheer viciousness of the Imperial Japanese Army which made it such a frightening opponent. And, finally, the service politics and institutional pressures at work in the system of awards for valor, something seldom discussed. It is a study well worth reading.

—**Raymond Callahan**, author of *Triumph of Imphal-Kohima:
How the Indian Army Finally Stopped the Japanese Juggernaut*

# FLAMETHROWER:

Iwo Jima Medal of Honor Recipient and U.S. Marine
Woody Williams and His Controversial Award,
Japan's Holocaust and the Pacific War

## Bryan Mark Rigg

For more information about this title or to order other books
and/or electronic media, contact the publisher:

Fidelis Historia, LLC
15601 Dallas Parkway Suite 900
Addison, TX 75001
www.BryanMarkRigg.com

ISBN:
978-1-7345341-0-8 (Hardcover)
978-1-7345341-1-5 (ebook)

Printed in the United States of America

Cover and Interior design: Darlene Swanson • Van-garde Imagery

Neither the United States Marine Corps nor any
other component of the Department of Defense has
approved, endorsed, or authorized this book.

Inside covers art is the:
4th Marine Division Landing on Yellow Beach 1
painting by Donna J. Neary
Displayed in the Art Collection of the
National Museum of the Marine Corps, Triangle, VA

There aren't any great men.
There are just great challenges that ordinary men
like you and me are forced by circumstances to meet.

**—Fleet Admiral William "Bull" Halsey, Commander Third Fleet**

The only thing necessary for the triumph of evil is for good men to do nothing.

**—Edmund Burke, Anglo-Irish Statesman and philosopher (1729-1797)**

Generally speaking, the errors in religion are dangerous.

**—David Hume, Scottish Enlightenment philosopher (1711-1776)**

# IF

By: Rudyard Kipling

If you can keep your head when all about you
　　Are losing theirs and blaming it on you,
If you can trust yourself when all men doubt you,
　　But make allowance for their doubting too;
If you can wait and not be tired by waiting,
　　Or being lied about, don't deal in lies,
Or being hated, don't give way to hating,
　　And yet don't look too good, nor talk too wise:

If you can dream—and not make dreams your master;
　　If you can think—and not make thoughts your aim;
If you can meet with Triumph and Disaster
　　And treat those two impostors just the same;
If you can bear to hear the truth you've spoken
　　Twisted by knaves to make a trap for fools,
Or watch the things you gave your life to, broken,
　　And stoop and build 'em up with worn-out tools:

If you can make one heap of all your winnings
　　And risk it on one turn of pitch-and-toss,
And lose, and start again at your beginnings
　　And never breathe a word about your loss;
If you can force your heart and nerve and sinew
　　To serve your turn long after they are gone,
And so hold on when there is nothing in you
　　Except the Will which says to them: 'Hold on!'

If you can talk with crowds and keep your virtue,
    Or walk with Kings—nor lose the common touch,
If neither foes nor loving friends can hurt you,
    If all men count with you, but none too much;
If you can fill the unforgiving minute
    With sixty seconds' worth of distance run,
Yours is the Earth and everything that's in it,
And—which is more—you'll be a Man, my son!

# Table of Contents

Advanced Praise for *Flamethrower* . . . . . . . . . . . . . . . . . . . i

Dedications . . . . . . . . . . . . . . . . . . . . . . . . . . . . . xvii

Forewords by 31st USMC Commandant, General Charles C. Krulak,
29th USMC Commandant, General Alfred M. Gray Jr. and
USMC General Anthony Charles Zinni . . . . . . . . . . . . . . . xix

Index of Acronyms and Terms . . . . . . . . . . . . . . . . . . . xxvii

Note to Reader . . . . . . . . . . . . . . . . . . . . . . . . . . . xxix

Preface . . . . . . . . . . . . . . . . . . . . . . . . . . . . . . . . xli

Introduction: Background on WWII and Woody Williams . . . . . . . xlvii

Ch. 1    A Nation Rises Up—Events Before and at the Beginning of War . . . . . . 1

Ch. 2    Growing Up in West Virginia . . . . . . . . . . . . . . . . . 11

Ch. 3    Outbreak of War . . . . . . . . . . . . . . . . . . . . . . . 29

Ch. 4    Joining the Marine Corps . . . . . . . . . . . . . . . . . . . 37

Ch. 5    Background on the Japanese Military: Brave Soldiers and
         Psychopathic Rapists and Killers . . . . . . . . . . . . . . . 65

Ch. 6    Deployment—New Caledonia . . . . . . . . . . . . . . . . . 143

Ch. 7    Guadalcanal . . . . . . . . . . . . . . . . . . . . . . . . . 147

CH. 8    The Marianas . . . . . . . . . . . . . . . . . . . . . . . . 169

Ch. 9    Japanese Attack and Occupation of Guam 1941-44 . . . . . . . . 175

Ch. 10   Battle for the Marianas: Background and Preparation . . . . . . . 183

Ch. 11   Amphibious Warfare: The Marine Corps' *Forte* . . . . . . . . . 189

CH. 12    The Attack at Saipan . . . . . . . . . . . . . . . . . . . . . . . . . . 199

Ch. 13    Philippine Sea Battle: "The Great Marianas Turkey Shoot" . . . . . . 205

Ch. 14    The Attack at Saipan Continues . . . . . . . . . . . . . . . . . . 209

Ch. 15    The Attack at Tinian . . . . . . . . . . . . . . . . . . . . . . . . . 229

Ch. 16    The Liberation of Guam . . . . . . . . . . . . . . . . . . . . . . . 241

Ch. 17    "Banzai" attacks on Guam . . . . . . . . . . . . . . . . . . . . . 263

Ch. 18    Battle's End for Guam . . . . . . . . . . . . . . . . . . . . . . . . 275

Ch. 19    Events Leading to Iwo Jima . . . . . . . . . . . . . . . . . . . . 301

Ch. 20    Iwo Jima Landings Begin . . . . . . . . . . . . . . . . . . . . . . 357

Ch. 21    Woody's Landing on Iwo . . . . . . . . . . . . . . . . . . . . . . 387

Ch. 22    The Pillboxes . . . . . . . . . . . . . . . . . . . . . . . . . . . . . 409

Ch. 23    The Attack on Iwo Continues . . . . . . . . . . . . . . . . . . . 455

Ch. 24    Battle on Iwo Winds Down . . . . . . . . . . . . . . . . . . . . . 525

Ch. 25    Justification for Iwo Jima . . . . . . . . . . . . . . . . . . . . . . 553

CH. 26    After Iwo Jima . . . . . . . . . . . . . . . . . . . . . . . . . . . . 557

Ch. 27    Receiving the Medal of Honor . . . . . . . . . . . . . . . . . . . 567

Ch. 28    Observations About Military Awards and Woody's MOH . . . . . . . 635

Ch. 29    Life After the War . . . . . . . . . . . . . . . . . . . . . . . . . . 679

          Conclusion . . . . . . . . . . . . . . . . . . . . . . . . . . . . . . 693

          Afterword . . . . . . . . . . . . . . . . . . . . . . . . . . . . . . . 771

          Acknowledgements . . . . . . . . . . . . . . . . . . . . . . . . . 777

          Bibliography . . . . . . . . . . . . . . . . . . . . . . . . . . . . . 787

          Appendix #1 (Legal document of Rigg's motion to dismiss
          Woody's frivolous lawsuit) . . . . . . . . . . . . . . . . . . . . . 811

          Endnotes . . . . . . . . . . . . . . . . . . . . . . . . . . . . . . . 925

          Index . . . . . . . . . . . . . . . . . . . . . . . . . . . . . . . . . 943

# Dedications

To MY SONS, IAN AND Justin, and daughter, Sophia. May they never have to face another Iwo Jima. However, if they have to face one, may they be the first to volunteer for such a mission.

To Corporal Vernon Waters and 2nd Lieutenant Howard Chambers and all those Marines who sacrificed their lives or health in defense of our country and freedoms fighting the Japanese during World War II.

To the proclamation Woody Williams gave to me at the beginning of this research in 2015 to be absolutely "F-A-C-T-U-A-L" with my findings.

To my Texas Christian University Starpoint teacher Mary Stewart, Fort Worth Christian teacher Donna Reynolds and football coach Cam Prock, my Phillips Exeter Academy instructors Harvard V. Knowles, Andrew Polychronis and David R. Weber, football coach Edward Frey and college counselor Thomas Hassan, my Yale University professors James B. Crowley, Liselotte M. Davis, Paula E. Hyman, Paul Kennedy, Ramsay MacMullen, Jeffrey L. Sammons, Steven B. Smith and Henry Ashby Turner Jr.; my Cambridge University professor and master's and PhD advisor Jonathan Steinberg; my mentor at the *Bundesarchiv-Militärarchiv* in Freiburg i. Br. Günther Montfort; and historians and my proof-readers/reviewers 31st USMC Commandant, General Charles C. Krulak, Michael Berenbaum, General Anthony Zinni (USMC), Colonel Robert S. Burrell (USMC), Richard B. Frank and Colonel Jon T. Hoffman (USMCR). These great mentors, scholars and leaders taught me the value of research, critical thought and to never waiver in the pursuit of truth.

## Books also by the author Bryan Mark Rigg:

*Hitler's Jewish Soldiers: The Untold Story of Nazi Racial Laws and Men of Jewish Descent in the Germany Military*

*Rescued From the Reich: How one of Hitler's Soldiers Saved the Lubavitcher Rebbe*

*Lives of Hitler's Jewish Soldiers: Untold Tales of Men of Jewish Descent Who Fought for the Third Reich*

*The Rabbi Saved by Hitler's Jewish Soldiers: Rebbe Joseph Isaac Schneersohn and his Astonishing Rescue*

# Forewords
by 31st USMC Commandant,
General Charles C. Krulak, 29th USMC Commandant,
General Alfred M. Gray Jr. and USMC General Anthony Charles Zinni

## Foreword by USMC 31st Commandant (1995-1999), General Charles C. Krulak

In warfare, it is understood that there are three levels of conflict: Strategic, Operational, and Tactical.

The Strategic Level normally involves decisions that are made at the highest levels. During World War II, this was best exemplified by Roosevelt's, Churchill's, and Stalin's decisions to prioritize victory in Europe over victory against Japan. In making such a decision, each leader was provided input by many. That input was not simply about number of forces available, amount of war fighting equipment available, logistics available, etc. It also included other important factors including ongoing suffering of both combatants and non-combatants, focus and impact of the media, cultural and religious make-up of the enemy and other less measurable aspects. Taking these various metrics into consideration, in the simplest of terms, the European Theater became the *main focus* for the Allies and the Asian Theater became *an economy of force* arena.

Once this Strategic Decision was made, the Operational Level of conflict came into focus. Because the fight against Japan was to be an economy of force arena and the number of forces, war fighting equipment and logistics would be limited, it was decided that the operational execution would be two pronged…a central drive under Admiral Nimitz and a southern drive under General MacArthur. Limited capability would be allocated depending on the priority of action decided upon by these two commanders.

Once again using the Pacific Theater, the Tactical Level can best be described by individual battles…the Battle for Iwo Jima being a perfect example. Here it was that the individual Marine, sailor, soldier, and airman combined forces to impose their will on a tenacious enemy. It was a titanic struggle. The acts of individual heroism and selfless sacrifice were a 24/7 occurrence for friend and foe alike.

In the book you are about to read, *Flamethrower*, the author does a brilliant job of connecting all three levels of conflict and their impact on each other.

You cannot understand the tenacity, the fanaticism, and the savagery of the Japanese soldier without understanding their culture, the impact of religion, and the deep-seated value system of the *Samurai*. All of these are best seen in the person of General Tadamichi Kuribayashi, the architect and commanding general of the defense of Iwo Jima.

Likewise, the bravery of the Marines and sailors who assaulted across Iwo Jima's black sand beaches, who fought through the mine fields and the interlocking bands of fire from hidden pillboxes and bunkers, who endured constant artillery and mortar fire, and finally, who secured the island is hard to understand without knowledge of what *esprit de corps* and brotherhood means to Marines. All of this is best seen in the person of young Corporal Hershel Woodrow "Woody" Williams.

In the case of both General Kuribayashi and Corporal Williams (the two main characters of the book), as with every combatant on the island, neither was a saint nor a sinner. Each had his own flaws that have been masterfully researched and documented by the author. In reviewing the history of General Kuribayashi, we learn that he commanded troops during the China campaign and his troops carried out brutal acts of rape, beheadings, and torture. In reviewing some of the claims made by Corporal Williams, questions about the acts that resulted in the awarding of his Medal of Honor are raised. Additionally, the approval process that resulted in his receiving the award are brought into question. But in each instance, these flaws and questions were counterbalanced by documented acts of bravery and selflessness, and by their devotion to their fellow warriors, to their country, and to their countrymen and women. We must not allow these flaws or questions to totally obscure all they did on that island.

At the end of the day, the focus of the reader will be inextricably drawn to the island of Iwo Jima itself and the battle that took place there from 19

February-26 March 1945. It was a horrific battle. During the 36-day campaign to take the island, a Marine fell to Japanese fire every two minutes…every two minutes for 36 days. Over 26,000 Americans were killed or wounded on Iwo Jima. It was the only battle in the history of the Marine Corps where Marines suffered more casualties than the enemy. Having been to the island on four occasions, I can attest to the fact that it still bears the scars of that titanic struggle. It is a place heavy with history and long on memories. It remains a haunting place.

Iwo Jima provides the stage for this remarkable book. It is on this stage that numerous actors play out their roles. Although the author draws on his prodigious research to help us understand many of the actors, it is on the stage itself where those actors are truly revealed as they face the horrors of close combat. Like a chess game, the leaders at all levels, on both sides, must think several moves ahead. Executing those moves are individuals who are asked to do the nearly impossible. Whether, like the defenders, fighting from pillboxes and caves with little water, food, or supplies and knowing that their lives were forfeit or, like the attackers, enduring withering fire, massive casualties, and measuring success by the number of yards cleared in a day, you will know and understand the true meaning of sacrifice.

As you read this book, hopefully you will come to believe that the words engraved on the granite base of the Marine Corps War Memorial at the Arlington National Cemetery might not only be intended to honor Marines, but to recognize all who fought on that island, for in truth:

*"Uncommon Valor Was A Common Virtue."*

**Charles C. Krulak**
General, U.S. Marine Corps (Ret.)
31st Commandant of the Marine Corps
Godson of Lieutenant General Holland M. "Howlin Mad" Smith
December 2019

## Foreword by 29th USMC Commandant (1987-1991), General Alfred M. "Al" Gray

I met Bryan Mark Rigg, then a USMC 2nd Lieutenant at The Basic School, back in 2001. He had unfortunately had an injury and was transitioning to a teaching job at American Military University where I served as a board member. I encouraged the young officer in the new life he had and that he should take pride that, although his military career was cut short, he still had earned the title *Marine*. In the last 19 years, he has written several books on WWII and recently, with *Flamethrower*, he has turned his attention to the Marine Corps during that iconic struggle. He has returned home, so to speak, by giving back to the organization which has shaped his identity.

Rigg has written an extremely detailed and superbly researched book about some of our WWII great battles like Iwo Jima and documented some of our incredibly superb Marine Warriors who brought victory against Japan. Using never-before-seen documents, he actually tells us many new facts about Guam and Iwo Jima and the men who fought there that will hopefully educate the next generation of historians and Marines.

In *Flamethrower*, Rigg has focused on two men in particular, General Tadamichi Kuribayashi (the IJA Iwo Jima commander) and Medal of Honor recipient Woody Williams. And then, in general, he has focused on countless Marines whose stories until now have remained unknown and who all prove my motto: "*Every Marine is, first and foremost, a rifleman. All other conditions are secondary.*" Analyzing documents in archives throughout the U.S., in Europe and Asia, Rigg has tied together short biographies of numerous American heroes to tell the larger story of the Pacific War. And in exploring these documents, Rigg has found many controversies, especially about Kuribayashi (a mass-murderer) and the war-hero Woody Williams himself, a man I have had the privileged to meet several times and who I find to be rather humble and a good representative for the Medal of Honor Society. When encountering these controversies though, instead of shying away from them or burying them (which some encouraged him to do), Rigg, like a true Marine and thorough historian, faced them head on.

Medals throughout time rarely give the appropriate recognition to a warrior who puts his life on the line for his country and comrades. We know many Marines throughout the ages have not got the appropriate recognition they deserved. The majority of those who fought on Iwo who deserved a Silver Star or

Navy Cross or Medal of Honor never did because the bureaucracy failed them or their officers had died or their witnesses had perished—in other words, there was no one to document their deeds. And in a few cases, we also see the reverse that some people got recognized at a level that maybe they should not have been rewarded at. In *Flamethrower*, one will learn about the medal process from World War II like I have never seen. And in doing so, one will see that Woody's Medal of Honor process was confusing and full of problems, some stemming from politics, and some stemming, unfortunately, from Woody himself. But we readers and even Woody, as a true Marine, should be able to handle the factual truths contained within this book. With this being said, we must realize anyone who fought on Iwo deserves respect for what he had to undergo to defeat the Japanese and bring victory to the Corps in the toughest battle Leathernecks have fought. And Rigg uses both Kuribayashi and Woody's stories to bring to life commanders like Lieutenant General Graves B. Erskine and Captain Donald Beck so they can find their rightful place in history—Erskine especially has been ignored for too long.

*Flamethrower* also goes into the psychological and sociological background of the Japanese soldiers Marines faced. I have never read such a thoroughly investigated study about the enemy we encountered in the Pacific. When one understands who we fought and defeated, one just remains thankful for what our servicemen accomplished. Few truly understand how tenacious the average Japanese soldier was and how this enemy was the only one Marines have faced, especially at the war's beginning, who really bloodied our noses and knocked us down. But we got up, as we always do, and fought again, and eventually we brought down the Japanese soldier with powerful riflemen who were better trained, equipped and motivated than their adversary.

In the end, you are about to read one of the best documented books about what warfare was like in the Pacific War against Hirohito's Japanese forces. *Flamethrower* is a superb testimony to the efforts of our servicemen and leaders as they went about winning WWII to obtain the Victory over Japan or V-J Day.

**Alfred M. "Al" Gray Jr.**
General, U.S. Marine Corps (Ret.)
29th Commandant of the United States Marine Corps (1987-1991)
December 2019

## Foreword by General Anthony Zinni, former Commander of CENTCOM (1997-2000)

Iwo Jima is a small (8 square miles) Japanese island that, until the ferocious battle fought there in World War II, was a tiny inconsequential land speck in the vast Pacific Ocean. In 1945, as American forces were closing in on the Japanese main islands, this little sulfur smelling volcanic rock took on strategic importance since it would be the first piece of Japanese land taken and would permit fighter escort of long range bombers being launched from bases hitherto too distant for such support. The battle would become legendary because of the ferocity of the fighting and the iconic photo taken by Joe Rosenthal of Marines raising the American flag on Mount Suribachi at the southern tip of the island. Over 100 thousand American and Japanese combatants squeezed onto that island and fought a horrendous 36-day battle that was marked by vicious hand-to-hand fighting. There was no place for grand operational maneuver or brilliant tactics. It was a fight between small units and individuals being ground out over desolate rocky terrain with success measured in the few yards taken. Twenty-seven Medals of Honor and numerous other awards for heroism would be earned, far surpassing any other battles of the war. The US would suffer great numbers of casualties and the Japanese would have few survivors.

Like all other Marines of my vintage, Iwo Jima was revered as a sacred piece of Marine Corps history and heritage. Survivors of the battle were still in our ranks and highly respected when I joined. We had grown up with John Wayne in the movie *Sands of Iwo Jima* and legends like Manila John Basilone who died leading his machine gunners off the invasion beaches were the "saints" of our Corps. Marines cherish and celebrate their history and we certainly knew the story of Iwo Jima from the beginning of our service as we did all the battles fought in our centuries of existence. Iwo, however, would take on a more personal context for me.

In 1987, I took command of the 9th Marine Regiment (9th Marines) based on Okinawa. The regiment had been in the center of the line on Iwo Jima and it was part of the unit's proud history. Our division chief-of-staff at the time requested that each of the colonels in the command take a World War II battle in the Pacific for research and study to include visiting the battlefields. Then we were instructed to conduct classes on the battle for members of the command, walk the battlefields with them, describe the actions, and discuss the lessons to be learned. I was given Iwo Jima as my battle study. In the course of the next two years, I spent quite a bit of time on Iwo and in researching the battle to include discussing the fight with Japanese officers who were also students of the war and an officer who was the last one off the island before the battle. The more that I dug into the battle and researched the records, accounts, and documents, the more personally invested I became. Walking the unique black beaches and watching the long buried amphibious craft coming up out of the sand as the island was rising in the area of the landing beaches, the more that I had a sense of connection to those warriors who came ashore that February day in 1945. As I walked the island, I could literally scrape my foot in the sand and kick up spent rounds and shell casings. I realized that I was stepping over where men had bled and fought with each step. The sealed cave and tunnel entrances that still held the remains of thousands of Japanese soldiers made it seem eerily like a hallowed cemetery.

There are vast battlefields where you can see how armies moved and swept over great distances responding to the commands of generals and colonels. On Iwo, however, you had a sense of close-in fighting with bayonets, rifles, machineguns, flame throwers, and demo charges. Inch by destructive inch enemy lines, masterfully dug in, had to be reduced by sergeants, corporals, and privates. Courage, fear, desperation, and pain was a constant for one month and one week until the last line was breached and taken. Bryan Rigg has captured this true nature of the battle for Iwo Jima and the story of those men who fought it in *Flamethrower*. His is a story of warriors thrust into hell and should give all of us who enjoy our freedoms an appreciation for the men of the Greatest Generation and the battles they fought.

**Anthony Charles Zinni**
General, United States Marine Corps (Ret.)
Former Commander-in-Chief of CENTCOM (1997-2000)
December 2019

# Index of Acronyms and Terms

AA—Antiaircraft guns

Amtracs—Amphibious tractor carrying personnel

AWOL—Absent without leave

*Banzai*—Suicidal charging wave of attacking Japanese soldiers

BAR—Browning Automatic Rifle

CCC—Civilian Conservation Corps

CP—Command Post

D-Day—Invasion Day

DI—Drill Instructor

DSM—Distinguished Service Medal

FDR—Franklin Delano Roosevelt

FMFPAC—Fleet Marine Force, Pacific

IJA—Imperial Japanese Army

IJAA—Iwo Jima American Association

IJN—Imperial Japanese Navy

JAG—Judge Advocate General

Jap—Derogatory nickname for Japanese

Jarhead—Another term for Marines due to their heads looking like "jars" since their haircuts are so short

KIA—Killed-in-action

LCI—Landing Craft, Infantry (158-foot vessels)

LCI (G)—Landing Craft, Infantry, rocket-armed gunboat

LCVP—Landing Craft, Vehicle, Personnel, *aka* Higgins Boat

Leatherneck—Another name for Marine

LSM—Landing ships, medium transports

LST—Landing ship, tank (400-foot long ship for landing supplies and vehicles (jeeps, trucks and tanks) on a beach through its bow butterfly doors)

LVT—Landing Vehicle Tracked (an amphibious tractor also known as an amtrac)

LVT (As)—Landing Vehicle Tracked, armored

MarDiv—Marine Division

MCRD—Marine Corps Recruit Depot, San Diego

MIA—Missing-in-action

MOH—Medal of Honor

Navy Corpsman—The naval equivalent to an army medic

NCO—Non-commissioned officer

Nip—Derogatory nickname for Japanese

Nippon—Japanese word for Japan

PFC—Private First Class

POW—Prisoner-of-war

PTSD—Post-traumatic Stress Disorder

Rock and Shoals—Naval Legal codes

UDT—Underwater demolition team

ULTRA—this was the British code name for the intelligence unit that broke German codes; the U.S. code word for its intelligence unit that broke Japanese codes was Magic.

USMC—United States Marine Corps

VE Day—Victory in Europe

VJ Day—Victory over Japan

W-Day—Invasion Day

WIA—Wounded-in-Action

WWI—World War I

WWII—World War II

# Note to Reader

Goodness without knowledge is weak…yet knowledge without goodness is dangerous, and that both united form the noblest character, and lay the surest foundation of usefulness to mankind.
—**John Phillips**, Founder of Phillips Exeter Academy 1791[1]

History is dangerous because the truth can be uncomfortable. War probably brings out this reality more than any other human activity, and World War II is replete with uncomfortable truths.
—**Alexandra Richie**[2]

"In war, truth is the first casualty."
—**Aeschylus**, 5th century B.C.E Greek tragedian[3]

WHEN I STARTED THIS RESEARCH into the Pacific War in WWII in 2015, I wanted to preserve both U.S. Marine Corps and U.S. history and honor one particular war hero from the battle of Iwo Jima, U.S. Marine and Medal of Honor Recipient Woody Williams. I also wanted to study General Tadamichi Kuribayashi, who commanded Iwo Jima and who was reputed to be one of the toughest Imperial Japanese Army (IJA) leaders any WWII Allied command encountered. But along the way, I uncovered troubling facts about both warriors that made me struggle with a different sort of battle that many scholars encounter: The pursuit of evidence against interest.

Initially, I expected and wanted to present Woody without flaws, and Kuribayashi as an ethical and ingenious leader. However, I discovered Woody's autobiographical reporting to have many discrepancies, and that Kuribayashi committed horrible and often unmentioned atrocities. If I could have presented Woody without failings, there would have been joint Marine events with Woody accompanied by much fanfare. And if I could have written about Kuribayashi without his criminal past, the Japanese government and society most likely would have supported my work "praising" *their* hero as they did when Clint Eastwood's film *Letters from Iwo Jima* came out in 2006 glorifying Kuribayashi. My mentors USMC Commandant, General Charles C. Krulak, USMC General Anthony C. Zinni, historian Richard Frank, historian and USMC Colonel Jon T. Hoffman and former Marine Corps Historical Division Director Charles P. Neimeyer said it was troubling to read what I have documented, but they recognized the importance of my writing the truth about the facts my research had uncovered. If a historian fails to reveal *material facts* about an event or person, he will fail to do justice to history. And if he fails to report the truth, he will continue to distort history and ultimately leave "the reader dissatisfied" and "irritate his intelligence."[4]

First, let me explain some of the controversies with Kuribayashi. His grandson, the prominent Japanese Liberal Democratic Party (LDP) politician, Yoshitaka Shindo, was interviewed for this book. Once he started to see I was uncovering his grandfather's criminal past, he tried to discredit me with some of my U.S. contacts and take back what he told me, apparently, in a moment of weakness during the interview I conducted with him in April 2018.[5] Shindo even made a nuanced threat to the Iwo Jima American Association to close Iwo to American tourists in 2019 if its members did not force me to adhere to his demands (a violation of the 1968 agreement between the U.S. and Japan).[6] I later learned I should not have been surprised by Shindo's behavior because he is not only a right-wing nationalist, but also a member of *Nippon Kaigi* (a sort of "Holocaust" denial group) that disavows Japan committed WWII atrocities.[7] This book and perhaps the potential political fallout seems to have him worried.

Shindo may have also taken offense when I challenged him on not doing enough to conduct DNA-testing on the human remains found today on Iwo Jima since there are hundreds of MIA (*Missing-in-Action*) Marines whose bones still must be in this battlefield. He is one of the most prominent Japanese politicians who supports Japan's efforts to gather the fallen and cremate them adhering to Japanese religious death rituals. When Iwo Jima American Association Board member Bonnie Haynes was asked about why Shindo is not helping the Marine Corps find its dead, she replied: "He really doesn't care about our dead—are you kidding yourself to think otherwise?"[8] Shindo admitted *no DNA-testing* is being done on the skeletons found and did not take kindly when I told him that this was wrong. Apparently, when bones are found on Iwo, they are put together in a pile, set on fire and offered up to Shinto gods. Unfortunately, there probably have been several Marines whose remains have been improperly burned along with the enemy they had battled. If Shindo was offended by these questions, that may also explain why he tried to hinder my research. (see Gallery 1, Photo 1)

Next, when conflicting information surfaced about Woody and the narrative of his heroic acts that lead President Truman awarding him his Medal of Honor, he and his family attempted to prevent the publication of my book first by threats and then by litigation.[9] Woody tried to unconstitutionally silence me with *prior restraint* by filing a Federal lawsuit against me.[10] My attorneys responded to his intimidations thusly: "Your threats to tortiously interfere with Mr. Rigg's book contract are…unlawful" and "violate the First Amendment [of the U.S. Constitution]."[11] (See Appendix 1) Woody would be wise to adhere to Thomas Jefferson's mandate which says: "Our Liberty depends on freedom of the press. And that cannot be limited without being lost."[12] Moreover, after three years of thousands of hours of detailed research, travel to Iwo and other locations in order to write this book, Woody attempted to force me to sign an agreement that gave him censoring rights.[13] And lastly, he also demanded to be *personally* financially compensated for his story rather than using some book proceeds to support his charitable organization, *The Gold Star Families Memorial Foundation*, which was something originally offered. Although we discussed the possibilities of sharing royalties, Woody's terms became unacceptable. Moreover, it became clear as I started to see problems with Woody's self-reporting, that had I allowed him to have final say on this book, much misinformation would have been placed in this work and I do not write such history.[14]

Due to Woody's behavior and his threats toward my first publisher, Regnery Books, it declined to publish this book, fearing legal retribution. Although my editor Alex Novak wanted to continue with the work and loved my research, Regnery's staff said our marketing strategy was now "shot to hell;" a joint book tour with Woody would now never happen.[15] I then took my manuscript to the

University Press of Kansas, which has published three of my books. Although the university's legal department found nothing wrong with the scholarship—my factual assertions are backed by research, and I was documenting the story of Woody as a *public figure*—the press became uncomfortable with the idea of publishing a work which Woody had threatened litigation. Although my editor Joyce Harrison had communicated her commitment to publishing the book, and had gathered positive readers' reports from prominent historians (Richard Frank, Charles Neimeyer and Jon T. Hoffman), staff members put pressure on her to rescind their offer unless I redacted sections of my book. Rather than supporting my research and the freedom of speech, Kansas chose to avoid controversy. I then went to David Reisch at Stackpole Books (an imprint of Rowman & Littlefield), and despite the problems with Woody, Reisch decided to work with me on this book. But, then, once again, Woody continued his legal bombardment against me and his grandsons and his legal team continued to attack me and my now *third press*. My findings were being fought over in a federal court and the conflict between Woody and I made the rounds on web pages and in newspapers.[16] When Woody directly threaten to take on Stackpole Books, Reisch's boss, Judith Schnell got nervous. She, like University Press of Kansas, cowered, and instead of backing up the history, she wanted me to *actually* mislead the public and write a piece that would stop the legal action (i.e. not reveal the truth).[17] As a historian, I could not do so. I then went to the self-publishing arm of Simon & Schuster called Archway. At first, they were very excited about publishing the book and my director Joseph Skaggs pushed hard for the book, but then his boss, Billy Elliot, also got scared and decided while there is pending litigation, that he did not want to expose his company to litigation since Woody and his grandsons continued to threaten any entity that would support me publishing a work of history.[18] As a result, through counsel with numerous authors and lawyers, I decided to set up my own publishing company *Fidelis Historia* in order to get this book out and especially get it out for the 75th anniversary of the battle of Iwo Jima, 19 February 2020. In the end, I am thankful to Regnery, Kansas, Stackpole, Archway/Simon & Schuster because they all helped me go through the review process, the reader reports and even copy-editing. What normally a historian goes through to make a book kosher and presentable to the public, I got to go through *four* times. So, in effect, I must be thankful to all these presses for making this book one of my most reviewed and polished works I have ever done. In the end, my lawyer, Geoffrey Harper, and my mentor and world-famous Holocaust historian, Michael Berenbaum expressed what countless others have echoed when telling me they are sadden that all these presses showed little courage in presenting historical truth.[19]

In order to deal with these legal minefields, I hired the prestigious law firms Winston & Strawn, LLP in Texas and Nelson Mullins Riley & Scarborough, LLP in West Virginia and my lawyers Geoffrey Harper, Steven Stoghill and Tom Hancock have successfully guided me in my pursuit to present an accurate and faithful historical story of WWII in the Far East. My counsel *fully* supports this book since it documents the *truth*. I have followed the mandate of my cantankerous Yale University Professor Henry Ashby Turner Jr. gave: Follow the evidence wherever it goes and present conclusions based on *empiricism*.

In the end, it has been a struggle to gather the truth and publish this book, but I feel this work has faithfully portrayed the Japanese and Marines whose stories are told here. One would think in the land of James Madison and Thomas Jefferson who helped form, write and give commentary on the *First Amendment* to the U.S. Constitution guaranteeing the freedom of speech and the press, that such problems just described would neither exist today in America nor in Japan where General of the Army Douglas MacArthur and his staff helped to embody the same values in shaping Japan's then new post-war government when it adopted its own Constitution which emulates our Founding Fathers' ethics.[20] Ultimately, as Marine Corps historians Philip A. Crowl and Jeter A. Isely at Princeton University wrote, "History is of value only if it helps to solve pending and future problems."[21] But if one does not reveal the unvarnished truth of these "problems" because to do so might offend someone, then future problems cannot be solved. According to philosopher Theodor Adorno, in order to "master the past (*Aufarbeitung der Vergangenheit*),"[22] one not only has to learn the facts, but also he must fight to know the facts despite those who would prefer to falsify or delete them.

As one can see, this book has been written with much soul-searching, heartache and many a sleepless night. I wrote with a burden on my shoulders. In due course, I returned to the Marine Corps Code of *Honor, Courage* and *Commitment* and I have relied on this Code to present an honest history of *all the* battles and persons described within these pages. So it is both with great pleasure and honor, but also with much circumspection and sadness, that I present the following book to you, dear reader. I hope you will agree that it is better to deal with uncomfortable truths rather than beautiful lies if we want to learn from history to build a better society. Moreover, we should rise above threats to censor the truth. Thomas Jefferson said, "There is not a truth existing which I fear...or would wish unknown to the whole world."[23] This work has followed Jefferson's mandate of facing the facts without fear. Due to the controversies within these pages, I went to numerous experts, both historical and military, to verify that, what I was reading and discovering was correctly presented and analyzed. Consequently, this is the reason why I have requested of three of the most famous, celebrated Marine generals in the Corps' history in Krulak, Gray and Zinni to proof-read and then write forewords for this book. I also requested of two of the most distinguished Marine Corps historians, Colonel Jon T. Hoffman and Lieutenant Colonel Robert Burrell, to read the work multiple times and review the documents to make sure I get this history right (and believe me, Hoffman and Burrell gave me detailed, lengthy and tough feedback on different drafts). The reading knowledge, military experiences and martial insights all these men brought to the work were *invaluable*. These men, in their support, showed they truly live by the Marine Corps code of *Semper Fidelis* (always faithful).

This book comprises, on the one hand, biographies of Woody Williams, several of his comrades and General Kuribayashi, and on the other hand, pure history exploring Marine Corps combat, battlefield accounts and analysis of and lessons from the Pacific War. The beginning chapters give the reader some historical background to understand the events and people explored later. After these chapters, the follow on chapters largely flow chronologically except for a few exceptions when historical background is yet again necessary to understand what is about to transpire.

The Preface, Introduction and Chapter 1 explore how this study began and provide information on WWII to supply the reader with enough material to appreciate the events explored throughout the book. Without introducing several of the characters, concepts and events from WWII in these chapters, one could not understand many facts and narratives later investigated.

In Chapter 2, since Woody has become one of the most famous enlisted

Marines of WWII, his biography is used to explore other brave servicemen who otherwise would be lost to history. This is one of the first thoroughly researched biographies of an enlisted Marine from the Second World War. However, in conducting this study, it became apparent that using his biography would be more beneficial for the public in telling the stories of other Marines and larger issues of American history and WWII rather than focusing on *one single man*. As a result, in Chapter 2, Woody's upbringing is explored to tell the story of what was going on politically and sociologically in the world in the 1920s and 1930s to inform one of what had taken place for a few decades before the U.S. found itself at war in 1941.

Chapters 3-4 delve into how the war's beginning was viewed by Americans using Woody's life, yet again, to examine this issue. Moreover, these chapters discuss how Woody made it into the Corps and then, once there, explore the larger issues of how the Marines trained and indoctrinated their men for war.

Chapter 5 explains the enemy Woody and his fellow Marines would face during WWII. This lengthy chapter gives detailed evidence on Japan's radical society and military in order to help the reader understand the enemy America was fighting in the Pacific. Few Americans understand just how fanatical and grotesque the average Japanese soldier was back then. At the chapter's conclusion, one of the Marines' arch enemy commanders, General Kuribayashi, is examined. None of the literature explored for this book, both in Japanese and English, investigates how horrible Kuribayashi was while stationed in China. Readers need to know that the commander the U.S. defeated on Iwo was one of the most notorious leaders the U.S. vanquished in WWII, akin to a Heydrich, Himmler and even Hitler and their ilk. In short, Kuribayashi was a mass-murderer and Marines need to take special pride that they took down this villain.

Chapters 6-8 help one understand Woody's deployment and why he was on the islands he first landed on. They help shape the reader's understanding of the strategy that transpired in the Pacific in order to defeat the Japanese, i.e., island hopping.

Chapter 9 backtracks in time to help one appreciate the importance of Guam for Nimitz's strategy during WWII. It deals with this island's history to help the reader realize this was American territory and that the Japanese brutalized U.S. nationals for almost three years killing thousands of its population after taking it in December 1941 until the U.S. took it back in August 1944. Few in America know about Guam, its history and that a battle took place there in 1944. As a result, providing information about this island in the context of WWII is critical to understand the significance it had for America during the war.

Chapters 10-11 explain the importance of taking the Marianas by exploring the battles to come at Saipan, Tinian and Guam. Amphibious warfare is studied because without it being developed by the Marine Corps, assaults from the sea as seen at Normandy, North Africa, Saipan, Guam and at other battles would have been virtually impossible making the defeat of Nazi Germany and Imperial Japan much harder to have accomplished.

Chapters 12-15 explore the first phase of taking the Marianas studying the battles of Saipan and Tinian, and the naval battle of the Philippine Sea known as the "The Great Marianas Turkey Shoot." Since Woody's division, the 3d Marine Division (3d MarDiv), was a reserve outfit to be used in case things deteriorated on Saipan and the taking of Saipan and Tinian were crucial in liberating Guam where the division fought, these events are explored.

Chapters 16-17 describe Guam's liberation. In these chapters, besides studying this conflict in detail, one starts to get a feel for the other brave men in Woody's unit. In the hope of somewhat replicating *Band of Brothers*, one learns what Woody and his fellow Marines had to do to take this island by exploring the brotherhood necessary for the unit to perform well. Also, one of the largest *Banzai* attacks of the war happened on Guam, and since Woody's unit was in the middle of it, chapter 17 describes a facet of Japanese warfare that is always shocking in its retelling. In exploring this battle, one will learn Guam was the key to fully securing the Marianas and setting up the platform for the atomic bombs.

Chapter 19 explores global events leading up to the battle of Iwo Jima. Moreover, it focuses on how both the American and Japanese prepared for and viewed the upcoming combat on this island. It looks at Woody's training and Nimitz's strategy behind wanting to seize Iwo. Moreover, never before seen facts about Kuribayashi and his Iwo defense are detailed for the first time. Without this information, a true understanding of the battle in its entirety would be lost.

Chapters 20-21 document the actual Iwo Jima battle during the first days of the invasion. This information is necessary to understand why Woody's 3d MarDiv was deployed on Iwo since many felt it would only be a floating reserve like at Saipan. In short, these chapters give in depth information about the events leading up to Iwo and then the battlefield dynamics once the swords were drawn.

Chapter 22 chronicles Woody's Medal of Honor (MOH) actions and what actually took place. Since Woody's MOH exploits are some of the vaguest and undocumented ones of any MOH recipient from Iwo analyzed for this study, a detailed report of what can be documented versus what was thought to have happened is necessary to follow the controversy that plagues Woody's MOH to this day.

Chapters 23-25 examine the rest of the battle for Iwo after Woody's MOH acts, focusing on the 3d MarDiv. In exploring this division, one learns about Major General Graves B. Erskine and his leadership. Moreover, one will grasp what it took to conquer this island to secure this forward Pacific base. These chapters further investigate the reasons for taking the island. And throughout these chapters, more controversial stories are scrutinized about Woody's behavior.

Chapters 26-27 survey what happened to Woody after Iwo and how and why he received the MOH. Moreover, Chapter 27 gives an in depth case study of how medals were conferred during WWII and how a chain-of-command arrived at awarding a Medal of Honor. Moreover, this chapter studies the numerous problems with Woody's MOH with never before seen documentation and analysis. Woody's Medal of Honor, shockingly, never received the endorsement of Fleet Marine Force commander, Lieutenant General Roy S. Geiger, and his board, Pacific Fleet commander, Admiral Chester Nimitz, and his board or the Commandant of the Marine Corps, General Alexander A. Vandegrift due to the lack of evidence for Woody's supposed actions. And even more startlingly, Woody's platoon leader, 2nd Lieutenant Howard Chambers, refused to endorse Woody for the *Medal*, the very Marine who supposedly witnessed everything Woody had done. This research shows that these men had reason to worry about the Medal of Honor going to Woody because it appears much of the information lower level officers relied on to endorse Woody's MOH package actually came from Woody's own reporting about what he did which violates the regulations. This chapter also explores that Woody probably received the MOH due to political pressure and not because the Marine Corps felt he warranted it which, in the end, as already mentioned, the Corps at its highest levels *never* supported his MOH. And most surprisingly, this work shows that Woody probably knows this and has done a lot throughout the years to prevent people from knowing what role he actually played, albeit unknowingly at the time, in he getting the highest medal this nation can bestow on a serviceman because of how he inflated his deeds in battle.

Chapter 28 dissects the problems and issues surrounding the bestowing of valorous medals especially during and after WWII. Several case studies are given to explore the issues surrounding Woody's MOH and why there are so many more quandaries with his compared to others. In analyzing this history of medal awarding, this study supports Commandant Krulak's assessment that Woody should not have received the Medal of Honor for the package that was put in for him in 1945.

Chapter 29 concludes with what Woody did after the war and how the MOH changed his identity. Moreover, it shows how he dealt with his combat trauma and what helped him live a productive life.

Last, the Conclusion and Afterword review the lessons learned from this research. These chapters explore both macro- and micro-level issues this book raises. On the macro-level, it examines why America won the war and justifies the use of the atomic bombs in bringing down Hirohito's regime. They also explore how the U.S. evolved its strategy to win the war, why Hirohito decided to end the war and was not tried as a war criminal, and how the U.S. created a democracy in Japan and secured world peace. Moreover, these two chapters review Japan's crimes and Kuribayashi's behavior and why the Japanese have not been truthful with their wartime past. On the micro-level, Woody's life is investigated and the controversies detailed and analyzed as to why there are so many discrepancies about his narrative. At the Conclusion's end, the book finishes with philosophical lessons this work presents of trying to answer the question of how we create a more ethical society. One may find the lengthy Conclusion somewhat unorthodox, and he or she would be correct, as most conclusions are short and pithy. In contrast, it is felt necessary to write a long conclusion to try to tie the important lessons of this book together that explores strategic, operational, tactical, sociological, philosophical and biographical information. As a result, it is hoped that this organizational tactic at the end of the book will help one understand more about two of the most pivotal events in world history; namely the Pacific War and Japan's mass-murder.

In the end, until now, no one has biographically explored a WWII enlisted Marine and his combat unit like in this book. Moreover, this is the first meticulous biography of Woody Williams. And in exploring his life, many never before seen facts, events and people from the Pacific War will be presented. This book probably offers one of the most detailed descriptions of the *Banzai* on Guam; thorough assessment of how many Japanese civilians died on Saipan; described narrative of the brutal behavior of Japan on Guam; specified analysis for "Japan's Holocaust"; thorough description of General Kuribayashi's life as a military man and list of his atrocities; documented assessment of Iwo's engineering community and its accomplishments; analyzed appraisal of the 3d MarDiv's role on Iwo; described evaluation of General Erskine's part in developing amphibious warfare and his command of the 3d MarDiv at Iwo; comprehensive exploration of the men and departments involved with awarding Marine and navy MOHs for WWII servicemen; meticulous study of Marine Corps flamethrower tactics and usage during the Second World War; and, last, but not least, one of the most controversial Medal of Honor stories to come out of WWII.

After reading this book, it is hoped the reader will develop a profound respect for what America and the Marine Corps did during WWII. Moreover, this study explores the lives of many Marines so that one may learn more about what is necessary to win when the U.S. sends its citizens into harm's way. And in the end, hopefully one will come away not only learning more about the Pacific War and "Japan's Holocaust," but also with a renewed respect for the men and women who helped defeat one of the most treacherous and morally bankrupt regimes known to mankind, Imperial Japan.

Bryan Mark Rigg, PhD,
*Dallas, Texas December 2019*

# Preface

"History with its flickering lamp stumbles along the trail of the past,
trying to reconstruct its scenes, to revive its echoes, and kindle with
pale gleams the passion of former days."
—Winston Churchill[24]

EARLY MORNING OF 21 MARCH 2015, a group of United States World War II veterans gathered at the Outrigger Hotel in Tamuning, Guam. Many joked with their children and grandchildren who had accompanied them for this historic 70th commemoration of the Iwo Jima battle. Over 30 veterans with family members returned to this famous battlefield on Japanese soil. P-51 fighter pilot Jerry Yellin (19 missions),[25] B-29 flight engineer and former Japanese POW Fiske Hanley (seven missions),[26] Water Purification Engineer Ed Graham, Marine Corps riflemen Darol E. "Lefty" Lee, Medal of Honor recipient Hershel Woodrow "Woody" Williams, John R. Coltrane, Billie Griggs, James "Jim" Skinner, T. Fred Harvey,[27] Leighton R. Willhite, William "Bill" Pasewark Sr. and John Lauriello, company commander Lawrence Snowden, Navajo Code Talker Samuel Tom "Sam" Holiday,[28] navy Corpsman Leo Tuck, Sailor and AA Gunner Harold Andrews from *USS Gunston Hall* (LSD-5), as well as Imperial Japanese Navy (IJN) Sailor Tsuruji Akikusa and many others traveled back to the island and in time to recall events that shaped U.S. history and their lives. Over 400 people flew to Iwo on three United Airlines chartered planes. (see Gallery 1, Photos 2-5)

Many veterans moved slowly often with help from a grandson or son nearby. A few were in wheelchairs. If these men were walking around any city today, people might pass them by, paying little mind to someone's old granddad who had lived his life and was nearing the end. Few, if any, would know that these men, when in the bloom of youth, had gone to some of the most hellish places on earth to defend America and fight for her freedom. They became heroes when they fought courageously for their nation. When in their teens and twenties, they were strong, intrepid young warriors. Many were Marines, and they prided themselves on being an elite fighting force; they were an important part of what made the Marines legendary. Even now, these older men, when they get around each other or current Marines, will puff up their chests and bellow "*Semper Fi*" or "*Oorah*." They speak with pride of having served and still find it difficult to explain how they survived this battle. Most were returning to Iwo for the first time in seventy years with their last memories of this place filled with carnage, combat and killing. Many lost buddies there, and a few had a son or grandson carrying their comrades' name.[29] Some children were there, now in their seventies, paying respects at the place that had taken their fathers' lives asking for answers to unanswerable questions getting closer to accepting their losses.[30] These children, themselves now older men and women, needed to see where their fathers had taken their last breaths. Other children were there whose fathers had survived, but now were deceased. They just wanted to pay their respects and learn more about what their fathers had told them throughout their lives like Laura Leppert, president of *Daughters of World War II*, whose father George Broderick fought in the 5th Marine Division (5th MarDiv) and suffered wounds fighting on this island.[31]

Dressed in our Sunday best, we took our time boarding the airplanes. Since there are still thousands of missing-in-action (MIA) IJA soldiers on the island, the Japanese treat the occasion as if attending a funeral requiring a strict dress code. My 12-year-old son Justin and I wore dark navy blue suits, dress shirts and red and blue ties (but since we would be exploring the island, I also wore my black leather Marine combat boots polished to a high gloss as if I were undergoing an inspection). Veterans Jim Skinner wore a replica set of WWII Marine dungarees, Jerry Yellin wore his U.S. Army Air Corps officer outfit, John R. Coltrane wore his Dress Blues and Navajo Code Talker Sam Holiday wore a blend of a Marine Corps League and formal Navajo Tribal Gold and Red uniform. Prior to this, we attended educational seminars and toured Guam's battlefield, where the inhabitants suffered through almost three years of Japanese occupation until the United States liberated them in July 1944. We visited Andersen U.S. Air Force

Base where some of the B-29 Superfortress bombing campaigns were launched against Japan that helped end the war. We met with the governor of Guam, Eddie Calvo; the Marine Corps Commandant and later Chairman of the Joint Chiefs-of-Staff, General Joseph Dunford; Air Force Commander of Andersen Brigadier General Andrew Toth; and Commander of the Joint Region of the Marianas, Rear Admiral Babette "Bette" Bolivar. The return to Iwo Jima was a prospect that had most veterans sitting in quiet contemplation on takeoff. Once airborne though, people rose from their seats and chatted with each other as if at a relaxed social gathering. Standard airplane etiquette flew out the window; even the flight attendants mingled with the crowd. (see Gallery 1, Photos 6-8)

As the planes flew over the Pacific Ocean, it was apparent why, during WWII if a plane broke down and had to ditch, it and its crew were often lost to history. The Pacific Ocean covers over 30% of the earth's surface and that is a whole lot of ocean. "You could drop the entire landmass of the earth into the Pacific and still leave a vast sea "shroud to roll," in Herman Melville's words, "'as it rolled five thousand years ago.'"[32] We flew 6,000 miles from Los Angeles to Iwo Jima. During WWII, this distance had to be traversed by ship and there was no straight line to anything, since a vessel had to go island to island for supplies. U.S. fleets had to sail 8,000 miles to get to Japan's doorstep ("more than twice the distance from New York City to San Francisco").[33] Few nations at the time had fleets with sufficient range to attempt such a crossing with multiple refuelings. What used to take months to travel, jet airplanes can now do in days, sometimes mere hours. The logistics were still daunting, requiring my son, Justin, and me to first travel from Dallas to Los Angeles where we met the veterans with whom we would make this journey. The next day, after an airport fire department water-cannon salute for the veterans soaked our airplane, we left Los Angeles airport for Hawaii, then Guam and subsequently flew across the seemingly interminable Pacific *en route* to Iwo Jima. The journey took days, four airplanes and almost twenty hours of flight time.

The day we flew to Iwo Jima, our planes circled the island a few times and the anxious men pressed their faces against the windows, their memories flooding back as they looked at Mount Suribachi's rugged hump rising from the sea. The landing beach, where many died, was still long, black, smooth and steep. Sunken ships in shallow waters, pillboxes and machinegun placements dotted the greenish-brown landscape. The men described sounds and smells as memories flooded back. Iwo's craggy landscape and rolling beaches were surrounded by dark seas and clear skies wielding a scene worthy for a painter, a "mystery of an unknown earth."[34] (see Gallery 1, Photo 9)

We landed on Iwo Jima and slowly pulled up in line near the control tower that had the new name of the island *Iwo To* printed on its structure. According to some, the island was renamed because the Japanese do not like being reminded of their defeat especially after Clint Eastwood's 2006 films about the battle.[35] A large Japanese flag flew over the airport and several military planes with the red circle on their fuselages lined the runway. It was the postwar flag of America's democratic ally, not the rising sun flag (*Hinomaru*) of the aggressive, fascist Empire of Japan that once flew over this island and much of the Pacific. Some veterans expressed their disgust that the U.S., due to Vietnam War politics, gave this land back to the Japanese in 1968. Veteran Jim Skinner said: "We spilt so much blood for this island and Japan was such an evil nation, we should've never given it back…This is hallowed ground for America and for the Corps."[36] Veterans around him nodded in agreement. As modern Marines, who were there to be part of a joint military ceremony, moved mobile stairways to our jets, most did not know that the airport was built over an underground network of tunnels and bunkers the Japanese had constructed during WWII and that hundreds if not thousands of Japanese remains were still there, buried underneath the runways and airport buildings. We were "walking on the dead."[37] At least half of the Japanese force that fought at this battle is still buried on the island along with the remains of 300-400 Marines who were never recovered, so every step we took on Iwo would be on a graveyard. (see Gallery 1, Photos 10-11)

After disembarking, the ocean air and unmistakable sulfur fumes of the volcanic island entered our nostrils, reminding us of rotten eggs. The scent evoked the horrible memories of war for many of the veterans. Most talked of 1945—their thoughts returned to a lost era. The last time their feet had crushed Iwo sand and ash, they were young men who had fought in one of history's bloodiest battles. They told my son Justin and me incredible stories, but one stood out. The veteran does not look impressive now standing 5'4" with a bald, skeletal head and large, arthritic hands, yet he still has energetic movements and a strong gait. Often a dark blue overseas cap rests on his head denoting he is a Medal of Honor recipient with that medal resting on his chest tied loosely around his neck by the sky blue ribbon with tiny white stars associated with that decoration. Adorned with such paraphernalia, he walks with his head high. It is not often one meets a person wearing a large, upside down gold five-point-star around his shoulders. His striking blue eyes sparkle when he gives those around him answers to their inquiries. His name is Woody Williams and these days he is recognized as a brave warrior who earned the Medal of Honor, but few know the true story of this Marine. Learning about his life will help us explore some unknown facts and learn about other amazing men and stories of the Pacific War. (see Gallery 1, Photo 12)

Here is an artist's rendition of the Medal of Honor from the Smithsonian. It is the highest medal the U.S. military can bestow on a serviceman or woman for valor in combat. The medal's images for naval service have the Union personified as the goddess Minerva carrying a shield who faces and repulses Discord that symbolized the rebellious South since Congress created the MOH during the Civil War. Discord is fleeing Minerva and represents the "foul spirit of secession and rebellion." He is placed in a "crouching attitude holding in his hands serpents, which with forked tongues are striking at [Minerva]." The medal's shape is "'a star of five rays' suspended 'by the flukes of an anchor.'" Thirty-four stars surround the celestial war representing States of the Union including the secessionist southern states.

# Introduction: Background on WWII and Woody Williams

"The more we…prepare in peace the less we…suffer in war."
—Admiral Husband E. Kimmel[38]

*Si vis pacem, para bellum—*
"If you want peace, prepare for war."[39]

## Background on World War II

In 1941 the world was embroiled in a war started by Nazi Germany, Fascist Italy and Imperial Japan. Hitler and Mussolini had taken over large portions of Europe and North Africa, while Hirohito had seized Manchuria, large portions of China and other Asian lands. Great Britain, China, Australia, the Soviet Union and other Allied countries fought for survival against these fascist regimes and it was doubtful they would prevail.

The America First Committee and other isolationists insisted the U.S. stay out of the conflict in the 1930s, blocked efforts to strengthen the armed forces and obstructed aid President Franklin Delano Roosevelt (FDR) proposed to bolster Great Britain's military. Isolationists criticized interventionists who said America was threatened. Many argued the U.S. needed to finish its recovery from the Great Depression, focus on its own economy and not fight other nations' wars. They insisted that, so long as it did not pose a threat to the fascists,

America could sit out the conflict. In early 1941, Britain was suffering bombings, was running out of money to pay for war materiels on a cash and carry basis and was pleading with FDR for help. Roosevelt responded with a proposal called Lend-Lease, meaning the U.S. would retool its factories from producing civilian goods to war materiel and become the "arsenal of democracy." Republican Wendell Wilkie, who had just lost to FDR in the presidential election, testified in favor of Lend-Lease. Yet, on the other side of the fence, Medal of Honor recipient and aviator Charles Lindbergh, a leader of and spokesman for the America First Committee, testified against the bill saying it was a foregone conclusion Germany would win regardless of America's efforts. He further claimed there was no difference between Nazi Germany and Great Britain as Americans believed, and that Roosevelt was a greater threat to America than Hitler.[40] Despite these sentiments, Congress passed Lend-Lease over such opposition.

When Japanese planes sank the U.S. gunboat *USS Panay*, killing three sailors and injuring 48 others, and three Standard Oil tankers, murdering evacuees from Nanking and passengers on the Yangtze River in 1937 during the Second Sino-Japanese War (in China it's called the War Against Japanese Aggression or the War of Resistance Against Japan), the isolationists encouraged America to do nothing. Since a majority of Americans put pressure on officials desiring to avoid conflict after the WWI carnage, the government complied. Tokyo apologized and claimed it did not know *Panay* was American, while at home Japanese politicians jeered at American weakness. Admiral Isoroku Yamamoto indeed "tearfully" presented himself to the U.S. Embassy, offered an apology and a "cash indemnity of $2.2 million [$39,479,471.43 in 2019]," but he was threatened with assassination for doing so and it appeared he was an exception to the rule since most in Japan welcomed the attack on the Americans. In defiance of their government some military leaders had planned the attack in an attempt to draw America into war to remove the "white man's" presence from Asia, but it did not work.[41] Military advisor to China, Colonel Joseph "Vinegar Joe" Stilwell simply called the Japanese "bastards" who knew full well what they were up to. He hated the Japanese for what they were doing to China, and the world for that matter, and called them "arrogant, cynical, truculent, ruthless, brutal, stupid, treacherous, lying, unscrupulous, unmoral, unbalanced and hysterical."[42] Later 25th USMC Commandant, Robert E. Cushman Jr., who was a lieutenant in

Shanghai when all this happened, supported Stilwell's assessment when he proclaimed, "[The Japanese] knocked off *Panay* and in effect, told us to go to hell."[43] Simultaneously to the assault on the *Panay*, Japanese pilots and shore batteries attacked the British gunboats *HMS Ladybird, HMS Bee, HMS Cricket* and *HMS Scarab*. The attack killed one and injured the commander and several sailors on *Ladybird*, yet England, like the U.S., did not respond with violence.[44]

During the Rape of Nanking, Japanese planes attacked and sank the U.S. Naval gunboat, *USS Panay* on 12 December 1937. National Archives, College Park.

Not to be outdone by ground troops and air forces, the Japanese Navy (IJN) dispatched its own gunboats (17 ships in the 11th Battle Fleet) on the Yangtze River hugging the city of Nanking to its north to prevent any attempt of Chinese citizens or troops from fleeing northwest. Japanese scholar Yutaka Yoshida wrote the IJN attacked small boats and even some people floating on doors of houses fleeing the city "either strafing them with machinegun fire or taking pot shots at them with small arms. Rather than a battle this was more like a game of butchery."[45]

The U.S. and Britain eventually cut off shipments of raw materials to Japan due to these actions, but they stopped short of military retaliation. In April

1940, the Nazi military attaché in Washington, Lieutenant General Friedrich von Boetticher, noted that Germans were shocked the U.S. took no military action against Japan for its aggression against America's friend Chiang Kai-Shek (leader of Nationalist and "democratic" China), its economic disruption and its military presence in China.[46] But as historian Barbara Tuchman wrote about the *Panay* incident, "when there is no will to war, war does not happen."[47] Nevertheless, the Marines in China were thankful the U.S. did not declare war against Japan after the *Panay* because, as 2nd Lieutenant Cushman Jr. declared, "there were 15,000 of us and about 150,000 Japanese within hand grenade range."[48] In other words, had America declared war, these U.S. servicemen stationed in China would have been slaughtered.

Germans quietly subsidized and encouraged America's isolationist movement. Several strains of isolationists existed and while many such as the Communists (during the non-aggression pact between Hitler and Stalin 1939-41) or socialists wanted to avoid war, others started to support military build-up as Japan continued its war of aggression in the late 1930s and while Hitler overran European democracies from 1938 to 1940.[49] Even with this build-up and America's embargos, the Japanese refused to change their policies and explained to their Nazi allies that, despite U.S. measures, they could continue to wage war in China unhindered for years.[50] The support FDR needed to stop Japan and Germany only came when America was attacked.

This cartoon aptly depicts the anti-war feeling of the time: It shows a scantily clad female skeleton with "War" painted on her chest, luring a young man upstairs to her bedroom. On the back of the young man's jacket is written "Any European Youth." The deadly temptress says, "Come on in. I'll treat you right. I used to know your Daddy." New York Daily/Getty Images

Chancellor Adolf Hitler declares war against the United States on 11 December 1941 at the *Reichstag* in Berlin, Germany. National Archives, College Park.

On 7 December 1941, IJN naval forces of 350 bombers, torpedo planes and fighters made a surprise attack on the U.S. Pacific Fleet in Hawaii, sinking four battleships, destroying hundreds of planes on the ground, killing 2,403 Americans and wounding 1,100.[51] Japanese forces also attacked the U.S. possessions of the Philippines, Guam and Wake Island, British Malaya and Burma, and the Dutch East Indies. Great Britain, the Netherlands, Australia and America were *now* at war with Japan whether they liked it or not. Germany declared war on the U.S. on 11 December 1941 and Hitler exclaimed, "Now it's impossible for us to lose the war… We now have an ally [Japan] who has never been vanquished in three thousand years."[52] German navy Commander-in-Chief Admiral Erich Raeder telegrammed IJN Minister Admiral Shigetarō Shimada congratulating Japan and pledged support.[53] In six months, Japan conquered an area larger than the European theater of war. The Allies saw "disaster pile on disaster" and the Japanese seemed untouchable.[54] The "tiny people" of Japan were inflicting gigantic humiliation and defeats on Allied forces everywhere. Commander-in-Chief of the IJN Combined Fleet, Admiral Isoroku Yamamoto, predicted he could "run wild" in the Pacific for six months after the surprise attack, but he could not foresee what would happen thereafter.[55] For the time being though, the 5'3" tall admiral could stand elevated over the many enemies his navy devastated. Other Japanese officers expressed to the Nazi naval attaché in Tokyo, Rear Admiral Paul Wenneker, that war would last just a few years until the Empire actually would prevail.[56] World events seemed to support the expectations of these officers. In 1942, victory seemed remote for the Allies. Much would have to be done to defeat Japan.

This book describes many courageous Americans who liberated the world from totalitarian aggressors, specifically focusing on the U.S. Marine Corps (USMC), which played a crucial role in the Pacific War. This book analyzes Medal of Honor recipient Hershel Woodrow "Woody" Williams and his actions during the conflict. Before exploring Woody's life, it is important to understand facts about the Medal of Honor, because without it, most would never have heard of him.

Fleet Admiral Isoroku Yamamoto. He was the mastermind of the attack at Pearl Harbor and probably was the most brilliant Japanese naval officer of WWII. National Archives, College Park.

## The Medal of Honor

The Medal of Honor is the highest U.S. military decoration for valor. The medal "recognizes individual gallantry at the risk of life, above and beyond the call of duty." In the majority of cases, the president presents the MOH to the recipient. Since Congress created it in 1861, fewer than 3,500 servicemen out of 45 million who have served during wartime have received it.[57] During WWII, 464 MOHs were awarded, 266 of them posthumously (57.3%).[58] Those who receive the MOH join a fraternity of heroes that each service memorializes.

## Where Woody Fought and Earned the MOH

Woody was in the reserve division for the battle of Saipan and fought in two major Pacific battles: Guam in the Marianas Islands and Iwo Jima in the Volcano Islands (Bonin Islands). Both battles will be explored, putting them in context of the war and how their conquest played a role in defeating Japan. Many unknown facts about Iwo will be examined because this was one of the most important Pacific battles.

Before the war, Iwo Jima, part of the Tokyo Prefecture, was an obscure, sparsely populated, volcanic island located in the vast Pacific, a hunk of rock that Americans and most others had no reason to care about, much less visit. It was a sulfur-mining community, housing a few hundred families who inhabited and quarried the island. All of that changed in WWII when it became a Japanese military base. It was here in 1945 that some of the best-led and prepared Japanese servicemen clashed with some of America's finest Marines. Seventy years ago, Iwo became one of the most "written-about battles" of WWII.[59] It was a stepping stone on the road to mainland Japan and was the first piece of Japanese territory attacked by America. With time, the memory of this conflict has faded. Today through film, many might recognize the name of Iwo Jima, but have only a vague knowledge of its location or what happened there to make it famous.

For the thirty veterans who returned to the island in 2015, this battle was the most horrific experience of their lives. The harrowing events of their first trip to Iwo were seared into their memories. They knew the taste of adrenaline, fear and courage that only battle can make one experience. They understood that the saying "war is hell" was not an exaggeration, but a sincere attempt to describe combat. These men knew what it was like to push oneself to the limits of endurance and then further. They witnessed the loss of friends, shot or blown up. Casualty figures were not abstract numbers, but men they knew. In fact, many were casualties themselves. They could not forget the sight and sound of wounded men crying out for "Corpsmen" or the smells of death. They knew what it was like to be gripped by anxiety and to fight on.

As already mentioned, one veteran who returned to Iwo in 2015 was Woody Williams. Back then his buddies called him "Willie." Others called him *Zippo head* or hot-head because of his temper.[60] He returned to Iwo as the last living MOH recipient from that battle and WWII Marine MOH holder. Of the 27 men who received this *Medal* during Iwo, the most in any U.S. battle, only 13 made it off the island alive. It is remarkable Woody survived doing all he did there and returned to civilian life living into his nineties.

## Background on Woody Williams

Never a big, strong guy, according to Woody, he did not initially pass the physical to get into the Marines in the spring of 1942. However, eventually the Marines lowered their standards according to Woody and he was accepted. As a Marine, Woody was part of the reserve for Operation *Forager* at the Battle of Saipan, and fought to liberate Guam, surviving one of the largest *Banzais* (suicidal mass wave of charging Japanese) of the war.[61] On Iwo, he continued fighting for weeks after he earned the MOH while his company continued to suffer high casualties.[62]

Later in life, Woody survived a heart attack (1985), lip cancer (2003), and colon cancer twice (2005 and 2012). To eradicate the colon cancer, surgeons removed ten inches of intestine the first time and a baseball-sized-tumor the second time. Describing what the doctors took out, Woody exclaimed, "And they didn't even pay me for it."[63] At 96, he has defied the odds of living past the 76.4 years life expectancy of a Caucasian male.[64] Woody is alert and often greets others with a ready smile and a sunny disposition. To show how active and strong Woody is, in February 2017, I made a research trip to West Virginia to spend time with him. He drove the 40 miles from his town Ona to Charleston Yeager Airport to pick me up in his Marine Corps stickered Cadillac Deville. Along the way, the plastic ground effect or spoiler under the car's front bumper got partially dislodged and the left side was dragging on the road making noise. As we pulled into his driveway, he first got out, walked up the small hill his house is built on, urinated on his yard facing away from me and then came back and said he was going to fix his car. That night, not having the proper tools to repair it, Woody drove his automobile on his car lifts in his shop, took a chainsaw in hand, and then proceeded to cut the dangling hard plastic off the car's front—presto, he fixed the problem and could drive it without noise. As recent as 4 February 2018, he walked to the middle of the field with a strong gait and happy grin to toss the coin to start Super Bowl LII and on 11 November 2019, he was one of five Grand Marshals for the Veterans Day Parade in New York City and welcomed his role with energetic waves, smiles and numerous handshakes (President Trump even publically mentioned him at the opening of the event). He likes the fame, and compared to many MOH recipients, has handled it well. Furthermore, to some extent, he deserves the attention for the service he gave to his country. He looks at each day as a blessing. The primary thing that helps him live such a vigorous life, having overcome hardship, death and cancer is the remembrance that he survived Iwo Jima: Everything after that "was gravy."[65]

Woody's list of military accomplishments is impressive. A nation depends on men like him when it sends its citizens into battle to defend its freedom. Iwo was one of the most important battles the U.S. experienced, and having valiant Marines like Woody meant the difference between victory and defeat. His story is one of many and it was not easy for most to become skillful and determined warriors killing "enemy combatants." But many thousands had to do so and Woody's story is theirs as well.

## Collecting Material about Woody

I have spent hours with Woody and other veterans discussing their combat experiences. I walked Guam and Iwo Jima's battlefields with Woody in 2015 and 2018. On Guam, we stood on the beach where he landed and explored the ridge where he fended off the *Banzai* in July 1944. On Iwo, we explored the island and I walked the area where he took out pillboxes in February 1945. Historian of Guam's Pacific Battlefield National Park, James Oelke Farley led me around these battlefields and offered insightful historical context on what we saw.

To understand Woody, I explored his roots, visiting him in his home state of West Virginia where he grew up, talking to his acquaintances. In October 2017, we also journeyed to the Marine Corps Recruit Depot (MCRD) in San Diego where Woody completed boot camp on 21 August 1943 and became a *Marine*. And to verify what Woody and others told me and to document the historical events they took part in, I traveled to eighteen major archives, museums and research centers throughout China, Japan, Guam, Germany and the United States and reviewed tens of thousands of documents.

I hope this account will help military historians and Marine Corps enthusiasts understand more about the Pacific War and better appreciate the achievements of those who fought there, especially at Iwo where America sacrificed so much. Some were recognized with various medals while the heroism of most is known only to God. Woody often says that he is just the "caretaker of the medal" and that it represents those *real heroes* who never returned.[66] In WWII, it can be said, echoing Admiral Chester Nimitz, that *uncommon valor was a common virtue* for most Marines.[67]

# Ch. 1: A Nation Rises Up— Events Before and at the Beginning of War

"Do not stand idly by while your neighbor's blood is being shed."
—Talmud [Jewish Oral Bible][68]

IT IS DIFFICULT TODAY TO understand the significance of Iwo Jima for the U.S. in 1945. After suffering the humiliation at Pearl Harbor, Americans were shocked, fearful and angry. Before 7 December, military and government leaders recognized the threat of a Japanese attack somewhere—perhaps in the Philippines where General MacArthur was preparing a defensive force. Manila was only 628 miles from Japanese-held Taiwan and 1,846 miles from Yokohama in Tokyo Bay. Pearl Harbor is 4,000 miles from Yokohama and the possibility of a surprise attack occurring at such a distance was regarded as remote.[69] After the devastating attack, stunned Americans asked, "Where would the Japanese strike next?" Public images of the Japanese changed from little yellow men who were no match for Western countries to fearsome warriors.[70] "Could Americans stop them?" "Would the Japanese Empire try to conquer California?" Despite the initial losses, there were steely confidence and determination among America's military and civilian leaders to defeat the enemy. Admiral William "Bull" Halsey said after Pearl: "Before we're through with them, the Japanese language will be spoken only in hell."[71]

Japanese attack on the *USS Arizona*, Pearl Harbor, Hawaii, 7 December 1941. This ship experienced one of the most violent explosions of any ship during this battle. She lost 1,177 of her sailors and Marines in a matter of seconds. National Archives, College Park.

Japanese attack on the *USS Oklahoma*, Pearl Harbor, Hawaii, 7 December 1941. This battleship flipped over and sunk trapping many of its men below desk where they drowned. She lost 429 sailors and Marines during this battle. National Archives, College Park.

Aerial view of the Japanese attack on battleship row at Pearl Harbor, 7 December 1941. National Archives, College Park.

Across the ocean, Japanese commanders rejoiced at their victories and gave parties and toasts to their success.[72] The Lord Keeper of the Privy Seal Kōichi Kido, Hirohito's closest advisor, felt foreboding at the surprise attack, but on his way to the palace the morning when the Pacific War began, he bowed toward the sun and offered "a prayer to the gods." He felt awe and reverence that their deities had given them initial victories and he hoped the recent actions would lead Japan into a glorious future. Nearby NHK (Japanese Broadcasting Corporation) set up hundreds of loudspeakers throughout the capital and blared news of the surprise attack against the Americans. "People stopped in their tracks, startled; then, as martial music blared out, many began clapping as if it were a ball game."[73] The Japanese embraced war.

On 7 December 1941, the years' long political battle President Franklin D. Roosevelt had fought against isolationists like the America First Group ended with the attack on Pearl Harbor. The U.S. was fully in the war after the "day that will live in Infamy."[74] From 1939 to 1941, isolationists like Charles A. Lindbergh, socialist Norman Thomas, Colonel Robert McCormick of the *Chicago Tribune*, Senators Burton Wheeler of Montana and William A. Borah of Idaho and many others were determined to keep America out of the war and opposed helping those already at war like Great Britain, France and China. They also blocked efforts to increase American forces. Isolationist efforts kept the armed forces small, under-equipped and weak so as to pose no threat to the aggressors Hitler, Mussolini and Hirohito who were armed to the teeth. As war raged across Asia and Europe, isolationists responded to the growing danger with indifference and the belief the U.S. could stay out of the war by showing a "desire for peace and noninvolvement."[75] Marine Corps chaplain, Rabbi Roland Gittelsohn, believed the U.S.' most "reliable defenses were 'Captain Atlantic' and 'Corporal Pacific.'"[76] Many felt protected by "Fortress America."[77] Ignoring the greed, cruelty and bloodlust of German, Italian and Japanese militarism as well as Soviet aggression in the Baltic States and Finland, isolationists insisted no one would attack America so long as it posed no threat. Admiral Yamamoto's Fleet blew that argument sky high when his carrier planes sank several American ships at Pearl Harbor. Soon thereafter, Hitler and Mussolini declared war against the U.S.[78]

Hitler thundered to the *Reichstag*: "We will always strike first!" *Der Führer* then called Roosevelt "mad" and claimed he:

> incited war, then falsifies the causes, then odiously wraps himself in a cloak of Christian hypocrisy and slowly but surely leads mankind to war, not without calling God to witness the honesty of his attack…The fact that the Japanese government, which has been negotiating for years with this man, has at last become tired of being mocked by him in such an unworthy way, fills us all, the German people, and I think, all other decent people in the world, with deep satisfaction…under these circumstances, brought about by President Roosevelt, Germany too considers herself to be at war with the United States, as from today.[79]

Later on that day, 11 December, Germany, Italy and Japan signed yet another tripartite pact affirming their "unshakeable determination not to lay down arms" until the U.S. and England were soundly defeated.[80]

Chinese political cartoon showing the Axis relationship exemplified by the Tripartite Pact of 27 September 1940 between Fascist Italy (Mussolini), Nazi Germany (Hitler) and Imperial Japan (Tojo). Shanghai Historical Museum

Suddenly, Americans regardless of political affiliation realized "there was no place to hide in the modern world."[81] All faced a global war now. A 3d Marine Division (MarDiv) report noted, "Pearl Harbor woke us up to the realization that we had been playing the part of the ostrich with our head in the sand for... years."[82] The 4,000 miles it took the Japanese fleet to attack Pearl Harbor only required a few weeks and it made Americans unite behind FDR "to seek revenge" and become "aware of the full extent of the Axis threat to their well-being and lead them to end the long retreat from responsibility."[83] This realization led to a renewed gusto for freedom and an urge to fight the ever-present evil. Suddenly supporting allies in the fight had relevance. On the day of the Pearl Harbor attack, Secretary of State Cordell Hull received Japanese Ambassadors Saburō Kurusu and Kichisaburō Nomura and refused to shake their hands. They handed Hull a note prepared by their government concerning future good relations between the two powers with the instruction that it was supposed to be given at 1pm (it was then 2:20pm). Reeling from the deception and surprise attack, Hull looked over the formal and polite document and losing his patience, asserted, "In all my fifty years of public service I have never seen a document that was more crowded with infamous falsehoods and distortions—infamous falsehoods and distortions on a scale so huge that I never imagined until today that any government on this planet was capable of uttering them."[84] The ambassadors tried to say something in response, but Hull hushed them and dismissed them with a "curt nod toward the door." As they left his office, he muttered "Scoundrels and pissants!"[85] A few days later, Roosevelt addressed Americans on 9 December via radio setting the tone:

> We are now in this war. We are all in it—all the way. Every single man, woman and child is a partner in the most tremendous undertaking of our American history. We must share together the bad news and the good news, the defeats and the victories—the changing fortunes of war.[86]

As Americans rushed to recruiting stations, the Roosevelt administration met with industrial leaders to plan a massive shift to put the economy on a war footing. They formulated long-term plans for war on multiple fronts which predominantly required ships. Quickly, America achieved naval supremacy over its enemies soon after the war started with its shipyards working around the clock. Former Commander of Naval Operations, Admiral William D. Leahy, feeling this was going to happen, said on 7 December 1941, "The war formally declared today will in my certain opinion result in the destruction of Japan as a first class sea power regardless of how much time and treasure are required to accomplish that."[87] Unfortunately, Leahy's prophetic proclamations would take more treasure and blood than most could image, but the will and desire was there to destroy that Japanese naval power. And Leahy's predictions would slowly but surely start coming true especially with America's ship building program. For instance, the U.S. Navy went from having 352 vessels in 1942 to 8,800 by 1945, training a vastly enlarged force to support and man them (if one includes all the smaller boats, like the landing-crafts classified as Higgins boats and amtracs, then the vessels the U.S. produced comes to 88,000!).[88]

Admiral Chester Nimitz, commander of the Pacific Fleet. After having witnessed what the Marines did at Iwo Jima, he claimed "uncommon valor was a common virtue" for all of them. National Archives, College Park.

The navy had a special place in FDR's heart. In 1913, President Woodrow Wilson had appointed then 31-year-old Roosevelt the Navy's Assistant Secretary. Roosevelt oversaw its mobilization for WWI. He became acquainted with Pacific War admirals: King, Nimitz, Halsey, Leahy, Spruance, Turner and Fletcher as young officers. When WWII started, FDR recalled Admiral Ernest J. King to run the navy and selected Admiral Chester Nimitz to command the Pacific Fleet. Former Chief-of-Naval Operations and former ambassador to Vichy France, Admiral William Leahy, was re-instated and effectively became the first Chairman of the Joint Chiefs-of-Staff. He, in particular, had developed an excellent relationship with FDR. FDR set up a map room in the White House based on Churchill's and tracked events via regular meetings with Admiral Leahy and his Army Chief-of-Staff, General George C. Marshall.[89]

The armed forces went from around 500,000 at the start of 1941 to 12 million by 1945. By 1946, over 16.3 million had served or were then in uniform. The U.S. put one-sixth of its male population in uniform.[90] Soon after the disaster in Hawaii, the U.S. deployed the navy conducting raids in the Marshall Islands (February 1942), and at the battles of the Coral Sea (May 1942) and Midway Island (June 1942). The Marshall Islands raids destroyed several buildings and supplies on the islands, and damaged and sank some ships. Coral Sea was considered a draw, but America took the fight to the enemy off of Papua, New Guinea and the Solomon Islands. In fact, when Yamamoto brought his fleet back six months later to secure control of the Pacific by taking Midway and destroying the aircraft carriers which had escaped damage at Pearl Harbor, it was the U.S. Navy's Pacific Fleet under the command of Admirals Nimitz, Frank Fletcher and Raymond Spruance which surprised and defeated the Japanese. American pilots sank four fleet carriers and one heavy cruiser, while losing only a fleet carrier and a destroyer.[91] The battle of Midway crippled the Japanese fleet. Yamamoto recognized the turning of the war, saying "our one great hope for a negotiated peace" was lost with this defeat.[92] Although Yamamoto was an operational genius, he ultimately failed his nation strategically.[93] So, upon entering the war, the U.S. hit back, and hit back hard. After Midway, the long, hard slog across the Pacific began. The U.S. seized the initiative taking the fight to the Japanese, beginning with the Doolittle Raid that bombed Tokyo in April 1942 and the invasion of Guadalcanal in August 1942.[94]

Everywhere the U.S. went after the fall of the Philippines, it won. But these were hard won victories against a formidable enemy. There was hatred enough for both the Nazis and Japanese, but Americans' animosity toward the Japanese was visceral. Hitler's Germany was indeed more dangerous at the beginning, but for Americans in general, "Japan [was] the more despised antagonist."[95] In contrast to the American public's ignorance of Hitler's extermination of the Jews which began in earnest in 1941, American news media had reported Japanese atrocities in Asia since 1931. Once Japan made itself an enemy, Americans became hyper-focused on defeating this antagonist they had for years viewed with loathing.[96] It is also important not to overlook that Japan attacked Americans on American soil while the Nazis had only occasionally sunk a few ships far from the U.S. mainland. Yamamoto was right to think after Midway that it was "just a matter of time before the Americans could gain the ultimate victory."[97] However, there would be much combat, suffering and death before the U.S. obtained that victory.

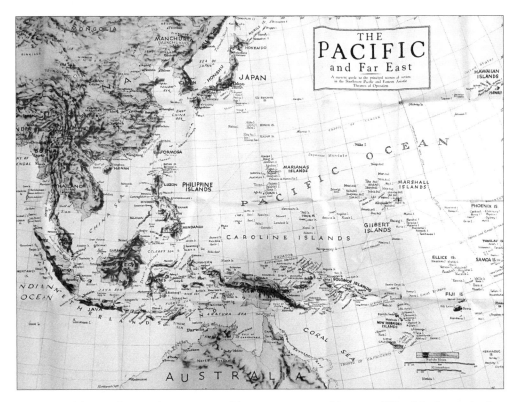

The map of the Pacific War during WWII. This map was carried by one of Woody's platoon leaders, 2nd Lieutenant Richard Tischler. Author's Collection and the Tischler Family Archive

# Ch. 2: Growing Up in West Virginia

To remember is to create links between past and present,
between past and future. To remember is to affirm man's faith
in humanity and to convey meaning to our fleeting endeavors.
—Holocaust survivor and writer Elie Wiesel[98]

WOODY WAS A FARM BOY from central West Virginia, a haven for independent thinkers. Since before the Civil War, the state's economy had depended on coal mining. When Virginia seceded from the Union in 1861, a loyalist government formed in the western part of the state. Defying the Confederate army to subdue this rebellion within a rebellion, West Virginia became a state in 1863. In later years, Woody's state boasted leaders in the women's suffrage and prohibition movements, a pioneer of rock and roll, and a woman whose decoration of war graves inspired Memorial Day. It was a state of strong-willed people. Some of Woody's ancestors have lived in West Virginia or Virginia for six generations from the early 1700s, and although his family did not have a strong military tradition, his great-great-great grandfather George Jacob Helsely Jr. fought in the Revolutionary War.[99]

Woody was born to a modest family of dairy farmers on 2 October 1923, in a tiny town called Quiet Dell, five years after *The War to End All Wars* (WWI). It was the roaring 1920s, a time of peace and prosperity throughout America. He was born prematurely, weighing only three pounds according to him, so the doctor and family members doubted he would live. Woody said, "I was so tiny my mother was either very tired since she had given birth to so many or my father had run out of juice."[100] Customary in the early 20th century, no doctor was present at Woody's humble birth. His mother had already given birth many times and was no stranger to labor pains and delivery, so a neighborhood midwife sufficed. Born prematurely, he faced many obstacles including an absence of vital prerequisites like oxygen and heat needed to sustain a new, fragile life. There were no incubators for premature babies although he was often placed in his father's shoe box next to the wood burning stove to keep warm. Children born too early back then usually died. But West Virginia's hill country people were resourceful and by chance, Woody was born under a lucky star. For weeks, many wondered if he would pull through those long fall nights. Yet after several months, he put on weight and remarkably survived.

Although Woody has given numerous interviews that he was three pounds at birth, this is highly unlikely because that meant he had been barely 30 weeks in the womb (normal, healthy gestation requires 39-40 weeks). According to former Air Force Medical Doctor David Alkek, Woody is probably incorrect he was so small. "Babies weighing that much, especially in the 1920s, died... Even babies today who are that small have a difficult time surviving and would have to spend at least one month in the ICU to have a fighting chance."[101] Certificate Registered Nurse Anesthetist (CRNA) Ann R. Mandel who is also a Labor and Delivery and Surgical ICU nurse declares:

> There's no way he was three pounds and survived back in 1923...
> He would've been ten weeks premature at that weight and around
> sixteen inches long. The only way you could feed a child that
> premature was through tubes because they don't suckle and they
> didn't have that technology back then to do so. And the lungs
> were very underdeveloped at this stage and would've needed help
> with the aid of machines which they didn't have back then. Re-
> member, President Kennedy's premature son, Patrick Bouvier
> Kennedy, was born at 34 weeks and weighed four pounds and ten
> ounces at birth and he died two days later of Hyaline Membrane
> Disease and that was in the 1960s.[102]

If truth be told, Woody was most likely at least 36 weeks in the womb and probably over five pounds at birth—still fragile, but capable of pulling through at that

time in West Virginia. Either the story his family gave him or the folklore he has developed himself is, according to Nurse Mandel, "highly unlikely."[103]

Woody spent his childhood in a rural area of the Mountain State. His family made a modest living in good times. That changed when Woody entered elementary school. The October 1929 stock market crash led to the Great Depression and failure of thousands of businesses. It was "the most devastating economic blow the nation ever suffered."[104] Unemployment was as high as one in four numbering 12 million by 1932.[105] Within a few years of the crash, federal "tax revenues dropped by more than half" and military programs got slashed: The U.S. Navy, for example, "stopped the construction of two aircraft carriers, three cruisers, one destroyer, and six submarines" putting over 5,000 industrial workers on the street.[106]

President Franklin Delano Roosevelt's First Inauguration Speech, 4 March 1933. He encouraged his fellow citizens during the Great Depression saying: "First of all, let me assert my firm belief that the only thing we have to fear is fear itself—nameless, unreasoning, unjustified terror." National Archives, College Park.

Government response was initially ineffective, and there was no safety net for the average citizen. Private charities were overwhelmed. The Depression spread to Europe and became a worldwide catastrophe. In Germany, the Depression compounded the economic hangover from WWI and helped Hitler take power as Chancellor in 1933. The presidential election of Franklin D. Roosevelt, who also took office in 1933, brought hope to Americans and a torrent of government programs to restore confidence in banks and markets, reopen factories, rebuild the economy, and get people back to work.[107] He inspired a nation with his speech on 4 March 1933 when he took the oath of office: "First of all, let me assert my firm belief that the only thing we have to fear is fear itself—nameless, unreasoning, unjustified terror."[108] Roosevelt helped the nation cope with pain and gave confidence that the future would be better than the current present.

Woody's home. Circa 1927. Woody is with his brother June by the front porch stairs and his other brother Gerald is in the corner of the photo. Woody's room is upstairs and to the right. Marine Corps Historical Division

Woody's youth was not unique; he was a country boy from the farmland at a time before heavy mechanization when a high percentage of Americans were farmers. His family worked hard and lived off the land. Yet farm life laid the foundation for a strong work ethic, a healthy body and a patriotic spirit.

The Williams family certainly knew misfortune and hardship. No stranger to heartache, Woody's mother likely prepared herself for the tragic words she had often heard: "Your baby did not make it." According to Woody, out of the eleven children to whom his mother gave birth, Woody says six died (in earlier statements, he has sometimes said only four or five). He claims they died of complications during childbirth or the Spanish flu epidemic of 1918-19.[109] Freshly-dug

Woody (center) with his brothers Gerald and June (the one with the hat). Circa 1927. Marine Corps Historical Division

graves were a daily sight in most American towns as the country and world struggled with the outbreak. Most societies were unsuccessful in containing this pestilence that would claim 50 million lives.[110] So when Woody was born on 2 October 1923, he was born into a family that had seen much sadness. Nevertheless, Woody says his mother and father were strong and never seemed to get dispirited.

Woody's full name is Hershel Woodrow Williams. His first name came from doctor Hershel Yost who visited his mother to check on him three days after Woody's birth, and his middle name came from the U.S. president his father admired, Woodrow Wilson (1913-21). Woody's mother called him Hershel, but everyone else called him Woody.[111]

As a boy, Woody rose with the sun and worked all day. The family's dairy farm had 36 head of cattle, all of which had to be milked, fed and cared for daily.[112] Without electricity on the farm, all work was done by hand. Milk and butter had to be delivered on a route to customers. Their icebox could only hold a little food. Someone had to drive to the ice house in town, pick up a block, and lash it to the bumper of the Ford Model T or Model A pickup truck to carry it home. Flat tires frequently happened because the roads were unimproved and rough. Their car always carried a spare tire and inner tube with a pump and jack.

Being without electricity meant they did not have time-saving devices like a washing machine, dryer or iron. As with all mothers of his time, Woody's spent hours each day washing, wringing and hanging out clothes. They were fortunate to have a spring up the hill from their house, so they could have fresh water carried down to the house by pipe, instead of having to carry the water in buckets.

Woody was born before television and during the advent of radio stations. As radio spread in the 1920s, it became popular entertainment for homes with electricity. For those without it, the best they could do was listen on a little crystal radio set with headphones or a wind-up record player. Sometimes Woody hooked up the radio to a car battery to listen to it. Quiet Dell was so small it did not have a church. For those things, people went to Fairmont with its population of 23,000.

Woody walked a mile to the elementary schoolhouse. By high school, since it was in Fairmont seven miles away, Woody occasionally caught a ride in the morning as milk was being delivered, but as he got older he had to hitchhike or walk back home in the afternoon. He used gas lamps to do his homework.

In Woody's day, parents rarely talked about feelings. Few remembered their parents hugging them, telling them they loved them or putting resources into education. Children were taught the family had to work together to get food on the table and that was their main focus.

Today many American parents feel their children need therapy to be normal, but on the farm the best therapy was to work with animals, fight Mother Nature's harsh realities, and be thankful they had a roof over their heads. Many did not have the luxury to worry about how they felt. The goal was to raise and grow food and if the family had excess, they would sell it at market or to a neighbor. Money was used to buy seeds, farm equipment or tools, clothes and other items the family could not grow or make themselves. To supplement their meat, Woody's brothers hunted rabbits, squirrels and ground hogs with a 12-gauge shotgun.

Woody was content in this environment, especially considering that the bucolic was all that he knew. He did not give the morrow much thought, rising each day focused on chores. Long hot summers and cold harsh winters taught Woody and his siblings to live in nature and adapt to its surroundings. Everyone paid attention to the weather because it indicated how to manage the animals and crops as their survival meant one's own survival. Life was one of paying attention to others' needs as well as the maintenance of equipment and the workable land. Time had to be made for school and church, but labor consumed people's efforts. Woody settled into a routine—and when one chore was done, there was always another: Canning vegetables, feeding hogs, repairing the barn, and so on.

Back then, the man of the house had a crucial role in the heavy work like planting, weeding and harvesting crops, tending animals, mending fences and repairing the barn and other buildings. As a consequence, on 5 December 1934, when Woody was 11, the family suffered great tragedy when his father Lloyd Dennis died of a "heart attack [at the age of 46]. He absolutely worked himself to death."[113] Life was upended for Woody with his father suddenly gone because Lloyd always awakened everyone and assigned the chores. Woody was young and does not remember the family staying grief stricken. His mother Lurenna felt distraught when it happened, but she and the children were accustomed to life's hardships.

Lloyd was buried in the E.T. Vincent graveyard next to three of his deceased children. The remaining family members undertook more responsibilities to ensure their survival. Woody learned early that those you love can and will die right before your eyes; however, instead of being paralyzed by it, he learned to move forward with life. Such events taught people to live in the moment, prepare for the future and not waste time looking back. Harsh as it was, it was a lesson that would serve Woody well a short nine years down the road.

Running a dairy farm and raising five kids was toilsome. Now the same chores had to be done without their father. Fortunately, the boys in the family were old enough to do the milking, churning butter, feeding cattle and mucking out the barn.

Unlike many families during the Great Depression, Woody's family did not go hungry. Milk, cream and butter were dietary staples and always in demand, and the Williams family grew their own food. West Virginia never suffered from drought like Oklahoma and other states did during the Dust Bowl. The economy improved after the government's New Deal began in 1933, but it was a gradual improvement that occasionally backtracked from year to year, not a sudden return to boom times.

Woody developed a profound respect for his mother Lurenna ("Rennie"), who endured the family's trials with dignity. She never complained even when distressed, but presented a strong face and did not retreat into a cocoon of sadness.

Although not well educated, she kept detailed account records of the farm and made sure their home was orderly. She educated herself by reading *McGuffey's Readers*, a series of graded primers commonly used for elementary school from first to sixth grades. Rennie made sure the family worked as a unit and ate the evening meal together. During these meals, Woody remembered two things his mother emphasized: "Don't lose your temper" and "Never use foul language." Even today, Woody rarely curses and dislikes seeing it in print or hearing it although he spent years in the Corps and the Veterans Administration (VA) where profanity was commonplace (however, Woody did curse like a sailor when a Marine).[114]

Woody and his siblings obtained an education despite their hardships, and they walked to the schoolhouse both ways, in the snow. Sometimes the kids walked a mile to and from school barefoot during the summer—in the winter, they always had shoes on. They had a basic understanding of math, history, government and English by the time they became teenagers—something becoming widespread in their generation even in West Virginian rural areas.

One of the most fundamental principles instilled in Woody's generation was the commandment "Thou shalt not kill" or "Thou shalt not commit murder" (Exodus 20:13 and Deuteronomy 5:17).[115] Many called to serve their country worried if they were violating God's law when they did so. Was it right to fire at enemy soldiers, sink an enemy ship, or shoot down an enemy plane? Were they permitted to support others who did the killing? Did it depend on whether the enemy soldier had tried or was at that instant trying to kill you or your friends? If killing in self-defense or in defense of others was not prohibited, what were the limits of defensive action? Some of the Ten Commandments had long been a cornerstone of law in most civilized Western nations and these issues were highlighted in the 1941 film *Sergeant York* starring Gary Cooper. Woody described the dilemma:

> We were raised with the belief that you didn't kill others, period. Then we were taken into boot camp and taught to kill. Our drill instructors taught us in order to win the war and survive, we had to bring overwhelming violence to the enemy and believe in the truism: Kill or be killed. Then later, when we did kill, we had a hard time reconciling this with how our mothers and churches raised us. Regardless of how effective we were on the battlefield, I feel most of my brother Marines never got over having to slaughter the Japanese.[116]

In some respects, the typical Marine was at a disadvantage because he "comes from a civilization in which the taking of life is prohibited and unacceptable."[117] Boys of the 1920s and 1930s were indeed raised in an environment that taught them physical toughness, but in learning how to "man-up" when defending yourself in a fight or argument, they also were taught to always value other human beings. This lesson was given by their parents, teachers or clergy reading verses or stories from the Bible that included the Golden Rule, the Good Samaritan, the Prodigal Son, the Sermon on the Mount and others.[118]

Woody talks about the upbringing of American boys during this time that included a high regard for human life. This sentiment plus the horror that the carnage of WWI had failed to bring lasting peace to the world gave rise to the isolationist movement in the United States. Americans "retreated even deeper" within their country, "searching for an isolationist policy that would spare them the agony of another great war."[119] Issues about how and why a military should be used globally were often discussed among U.S. leaders at this time.

When Woody talks about the necessity of a strong military and willingness to use it, he struggles with the teachings of his youth that held that everyone is made in God's image (Genesis 1:27).[120] Such conflicts continue to plague Woody as he reflects on his upbringing, military career and the U.S.' role in world affairs. But philosophy and morality were rarely discussed at Woody's house. As in most rural homes, the environment was rudimentary, leading to little discussion about larger philosophical issues. Unlike today, with television, radio and constant blog feeds raising issues of genocide, murder, war and politics hourly, life in the 1930s focused on family, earning a living and recovering from the Great Depression. Issues about how and why one should kill only became a heated debate when the U.S. entered the war full throttle in December 1941.

Woody's home had no plumbing, electricity, radios, televisions or locks. The first commercial AM radio station, KDKA in Pittsburgh, didn't begin broadcasts until 1920. It was popular, boosting the sales of radios and the openings of radio stations across the country. The use of many inventions (light bulbs, electric irons, washing machines, electric fans and movie projectors) were only for those with electricity.

During summer, they opened their windows at night due to the lack of air-conditioning and during winter, they heated each bed with a brick from the fireplace.[121] Adapting to the weather was something they, as most of their ancestors, had done and they knew they could acclimate to the weather.

Woody was conditioned as a boy to accept and adjust to his environment. Farm life toughens people. Places they were sent during the war tested their ability to adapt under extreme conditions and having such upbringings helped them adjust to the different meteorological conditions they would experience.[122]

Woody on his Uncle Clint's horse *Gail*. Circa 1940. Conducting some local competitions between other families' horses in town, Woody brags that he beat a lot of their horses in races with *Gail*. Marine Corps Historical Division

During long winter nights in the 1920s in West Virginia, they used oil lamps. In the 1930s, they used gas lights; however, Mama Williams only used them if necessary. They used an outhouse for the toilet and since they did not have the luxury of toilet paper, they used plants, newspapers or Sears Roebuck catalogues to wipe themselves. When the hole in the ground filled, they dug another nearby and moved the outhouse covering over it and then covered the old hole with dirt.

They bathed once a week in a tub of water that all siblings shared with water heated in the fireplace. Woody does not remember smelling bad; however, he realizes current American society would find it unbecoming. People were accustomed to washing their hands and faces. Most did not floss or brush their teeth, but since food was organic, Woody claims their teeth did not decay as much as children's do today because of the modern consumption of sugar. In fact, Woody owned his first toothbrush at the age of 19 when the Marines issued him one. When the family had a bad tooth, they pulled it out with a string tied to a door or used iron pliers to take it out. They were primitively self-reliant compared to today's standards.

News traveled slowly during those days when Woody worked the farm. There was a relatively "wealthy" uncle living nearby who owned a radio, whom Woody visited. It and the newspaper provided news. However, the family did not spend much time thinking about politics and world affairs. They worked to survive.[123]

Besides the harsh realities of living off the land without a father, Woody coped with an older brother who he felt was a bully and "down right mean." Lloyd Dennis Williams Jr. ("June") was six years older than Woody and picked on his siblings. After June bullied Woody one day, Woody hid in the barn and grabbed a singletree (long piece of wood for attaching a horse to a wagon) and waited behind the door for his brother. June had been harsh that day and Woody was scared June would hurt him. "If he came in that barn, I was going to hit him as hard as I could with the singletree. Fortunately for both of us, he didn't come in."[124] However, once when June was drunk, Woody snuck up from behind and hammered him across the head with a "huge stick of wood. Woody thought he had killed…[June] at first."[125] Woody took advantage of the opportunity and enacted his revenge. Usually though, Woody avoided June like the plague and spent time with another older brother, Gerald, who was Woody's best friend.

Not all of Woody's life was hard. He found the timeless comforts of coming of age swimming in rivers or lakes with his friends and playing cowboys and Indians or baseball. As a teenager, he often went to the dance halls on the weekend with local girls. This was also a place the boys showed off their strength to prove who was the toughest. "Almost every night, there was a fight outside the dance hall. When this happened, someone would yell…'Fight' and we would all exit to watch who was going to fisticuff it with each other." This was especially true if someone was called a "liar" or a "son-of-a-Bitch." "If that happened," Woody said, "then those were fighting words." When asked if he fought much, Woody said, "Yes, I had my fair share. However, I usually stayed away from fighting and remained...reserved in most situations."[126] Although he says he was shy, he liked the ladies and the occasional drink. He would enjoy singing, "Up in Lehigh Valley/ Me and a gal named Sue/ A-pimping for a whorehouse/ And a damn good one too!"[127] There was a local girl he and his brothers nicknamed "Mule" and she was the "village slut" having sex with many of the local boys, including his brother June. When asked if she taught him about the "birds and the bees," Woody gave a sly smile but then answered "No!"[128] He enjoyed performing musically for others, learned how to play the guitar and sang the songs *Don't Fence Me In* made famous by Bing Crosby in 1943, the *Yellow Rose of Texas*, *John Henry*, and *Lonesome Valley*. Such fun was rare, but a welcome break from chores and school.[129]

Today, courses like Outward Bound teach survival skills and make kids tougher, but during Woody's childhood, every year was a survival course. Regardless, amidst the hard work, he learned to enjoy the sounds of birds, shapes of the clouds, clashes of thunderstorms and the taste of well water. It was a life that seemed more real when viewed today behind the wall of electrical devices, cars and heavily-lit homes. His goals were modest; he wanted to be a cowboy like his heroes, movie stars Ken Maynard, Tom Mix and Hoot Gibson.[130] Sometimes, his extracurricular activities turned dangerous. As a 15-year old, he was thrown from a convertible after it hit a telephone pole, and his momentum carried him through a thorny hedge and dumped him in the neighbor's yard. He had a damaged shoulder, cuts and a thorn broken off next to his spine. A drunk driver and passengers did not help the situation.[131] Before the war, Woody's guardian angel was already hard at work.

By age 15, Woody's older brother had become unbearable. Since June ran the homestead and was difficult, Woody "couldn't take living at the farm any longer," so he left. He worked on other farms earning .10 cents a day and having a place to "sleep and board": "I learned early in life that one had to be responsible for their own lives, and with enough will and grit they could do what had to be done."[132]

From an early age, Woody knew how to live off the land, dress an animal, eat everything on his plate, work to help family members and most importantly, know that if one did not do his part, others suffered. There was no "I" in Woody's upbringing and this conditioning is why Woody can claim he is one of the greatest American generations. Unlike today, it was a generation not of entitlement, but one of discipline and toughness.

Many suffered during the Great Depression of the 1930s. From the beginning of his presidency in 1933, Roosevelt persuaded Congress to enact the New Deal programs to jumpstart the economy, but as the economy revived, many still struggled, and some just hoped to survive the next day. Events that erupted on the world stage in the 1930s were beyond most people's imaginations. Authoritarian, aggressive, and militaristic regimes ruled in Japan, Germany, Italy and the U.S.S.R, threatening once again to unleash the dogs of war. America entered WWI with the Allies in 1917 for the last year and a half of the conflict. While American involvement in WWI had been decisive, it affected only the last part of the war and was small as far as society was concerned compared to the impact on European belligerents. WWII, in contrast, was a massive struggle for Americans and Western civilization which dominated society and was everyone's daily focus.

For many Americans, before Pearl Harbor, the world beyond their borders was something distant and it did not involve or threaten them. They were spectators watching other, less fortunate countries hash out their differences. They echoed the sentiment of many who did not want to get involved with other nations' problems. They saw no need to look for trouble in distant lands when there were plenty of hardships at home. Japanese mistreatment of people in China and Germany's takeover of one European country after the other did not create enough concern by the average American to force the U.S. government to get involved militarily. So, it seemed, there was not a power on earth to stop the world across the oceans from spiraling out of control by 1941. Friendly democracies including Austria, Czechoslovakia, the Netherlands, Belgium, Luxembourg and France were threatened, invaded and conquered by Nazi Germany. The British Expeditionary Force barely escaped annihilation at Dunkirk as it was pursued relentlessly by the *Wehrmacht*. Yet Woody doesn't remember learning about these conflicts as they happened. Some isolationists realized in 1939 and 1940 that the U.S. was going to have to fight Hitler and they changed their views (except Communists, whose views were dictated in Moscow and reversed from hating to liking to hating the Nazis again between 1939 and 1941),[133] but that change was

Woody and his mother Lurenna at home. Woody had joined the Civilian Conservation Corps and was in their uniform. Circa 1941. Marine Corps Historical Division.

not as fast as Great Britain, France, Canada, Australia and others in the free world would have liked until Pearl Harbor.

The America of December 1941, which was suddenly in a global war, had endured a lot in the preceding decades. The Depression conditioned many to focus locally in order to survive daily hardships. The massive Dust Bowl drought ravaged America's farms and ranches as illustrated in John Steinbeck's *The Grapes of Wrath* (1939) and taught many how to live on very little. Roosevelt worked diligently to revitalize America with New Deal infrastructure and social welfare programs that helped to rebuild the U.S.' economy, but it was a slow process. Recovery from the Great Depression would not completely materialize until the outbreak of the war, when mobilization of the war economy forced a rapid uptick in industrial production that brought back employment.

One of the New Deal programs for which Woody enlisted in July 1940 was the Civilian Conservation Corps (CCC), in which more than 3,000,000 served from 1933 until 1941. His older brother June had left the farm by then and served in the CCC and had motivated Woody to join "because the pay was good." Also, Woody was tired of working on other people's farms and knew the money was better in the CCC. Reserve army officers were put in charge of CCC camps and military discipline was enforced.[134] This organization provided Woody an escape since it seems he was headed down the wrong path. In a 1956 interview Woody gave, journalist George Lawless wrote, "his reckless behavior and young devil-may-care attitude made him a thorn in the side of authorities… He quit high school in his freshman year—and school officials breathed a sigh of relief."[135] Woody even admitted he was "headed for reform school."[136] His life was not going in the right direction—he was going to be an uneducated young man with little prospects except working on his family's farm. Not wanting to go back to school, Woody decided to join the CCC. Woody was first sent to Morgantown, West Virginia where he entered a camp, was issued a uniform and work clothes, and indoctrinated about discipline and expectations.

Civilian Conservation Corps picture of young men in West Virginia. This was a New Deal program of FDR's to get young men working to build out the infrastructure of the United States. National Archives, College Park.

Soon thereafter, he worked at Cooper's Rock State Park, West Virginia running a small "jackhammer drilling holes in rocks so the rocks could be blasted loose and used to build walls." He laid stone walls around cliffs to keep people from falling into gullies.[137] Once he broke up the rock, demolition men blew up parts of the cliff to get the form they wanted. In October 1941, he, along with 200 other men, was transferred north and built jack fences in Whitehall, Montana, along government property.[138] In the Rocky Mountain forest, they cut down pine trees to make posts and fences. Woody explained: "The Jack Fence required two posts notched to fit together and forming an upside down V so that five strands of barbed wire could be nailed to each side to keep animals from being able to go through it or over it. The fence was about 8 feet tall."[139]

They built miles of fences, working from morning until nightfall supervised by military personnel and civilians skilled as woodsmen. Cutting long posts by hand from downed trees, hauling them onto trucks, unloading them and then using them to build fences was backbreaking work, but for these young men, the hard labor was inadvertently conditioning them for the military that most would join with the outbreak of the war. Woody also drove a two-ton dump truck hauling dirt and coal and learned auto mechanics.[140]

A group of young men about to go out on a work detail in the Civilian Conservation Corps. This was like a group that Woody served with in 1941. National Archives, College Park.

While in the CCC, he was doing his best to find a better-paying job to send money home. And the CCC made sure its men did help their families. Out of the eventual $30 he earned each month, he was required to send half home. This helped his mother run the farm. Even giving up some of his pay, he still felt rich.[141] Historian Robert A. Divine wrote CCC members contributed "both to their families' incomes and to the nation's welfare."[142] The CCC "was designed to get idle young men off street corners, and teach them job skills and a work ethic by getting them involved in shoring up the nation's natural environment." These men planted millions of trees, released fish in lakes and rivers and built thousands of miles of canals and roads. This organization indirectly prepared "some three million boys, many of whom would flood into the armed services" for war when it came.[143]

The U.S. Army considered the CCC experience valuable and made many in the program non-commissioned officers when the wartime draft began.[144] The CCC taught them the value of teamwork, physical fitness and loyalty.[145] While in the CCC, Woody's world changed in December 1941 as most Americans were forced to look beyond their borders and engage totalitarian regimes in war.

# Ch. 3: Outbreak of War

"Destiny hangs by a slender thread."
—General Holland M. "Howlin Mad" Smith[146]

"It is easier to start a war than to end one."
—Japanese Ambassador Saburō Kurusu[147]

ONE WEEKEND IN MONTANA, WHILE cleaning their barracks on 7 December 1941, Woody and his CCC comrades were challenged to look globally when Japan's surprise attack forced people to consider America's geopolitical role. The isolationist movement disappeared and the Japanese instantly became "the most hated enemy in American history."[148] Montana's Senator Burton K. Wheeler, an isolationist, changed immediately: "The only thing to do now is to lick the hell out of [the Japanese]."[149]

"The Japanese gave the average American a cause he could understand and believe to be worth fighting for."[150] "The sense of surprise, disbelief, and disgust at the 'sneak attack' was genuine."[151] Following Germany and Italy's declarations of war against America on 11 December came others from Romania, Hungary and Bulgaria. Historian Gerhard Weinberg notes, "The whole world was indeed [now] aflame."[152] Americans flocked to recruiting stations, although most could not have found Japan and Germany on a map. Woody recalls that most CCC colleagues did not even know the location of Hawaii. Even "well informed" Americans did not know where Pearl Harbor was.[153]

Admiral Yamamoto, who planned the attack, hoped the destruction of much of the Pacific Fleet would discourage America from intervening in Asia, causing her to sue for peace and leave Japan to rule the Far East. Japan deeply resented the "White Race" ruling Asians. For instance, the pan-Buddhist organization known as the Myō Kai issued the following proclamation: "Revering the imperial policy of preserving the Orient, the subjects of imperial Japan bear the humanitarian destiny of one billion people of color [the Asians]… We believe it is time to effect a major change in the course of human history, which has been centered on Caucasians."[154] Also, Japanese leaders felt they were better placed strategically than the U.S. or Europe to fight the rise of Communism in Asia and wanted a free hand in attacking these forces. As Japan's Prime Minister Fumimaro Konoe said in 1938:

> The new order has for its foundation a tripartite relationship of mutual aid and co-ordination between Japan, Manchukuo, and China in political, economic, cultural, and other fields. Its object is to secure international justice, to perfect the joint defense against Communism, and to create a new culture and realize a close economic cohesion stabilization of East Asia…[155]

The Empire's leaders often used flowery words to conceal aggressive, militaristic behavior hiding behind a "mask of liberation."[156] (see Gallery 1, Photos 13-14)

In reality, the Japanese were not seeking to end imperialism in Asia, they were simply determined to take the place of the Europeans and Americans as masters. When Marine Lieutenant Colonel Graves Erskine was stationed in Beijing from 1935-37, he had several meetings with Lieutenant General Kenji Doihara called the "Lawrence of Manchuria" since he was instrumental in the 18 September 1931 invasion of Manchuria (Japan had detonated a bomb at the South Manchuria Railway (Mukden Incident) and used that ruse to attack and conquer northern China and establish *Manchukuo*).[157] One day, Erskine asked Doihara, "Just what are you doing over here [in China]?" Doihara answered, "I think old lady China sick old lady. I speak in parable about China. She do not need Uncle Sam, but what she need is Doctor Japan." That "doctor" was not there to heal and help, but to find ways to keep the "old lady" so sick that she would be a slave and unable to offer resistance.[158] Ironically, Doihara was an opium addict and the mastermind of the Chinese drug trade which kept some of "old lady" China sick on opium (the League of Nations pointed out in 1937 that 90% of all "illicit drugs in the world" (primarily cocaine and opium) were of Japanese origin, supplied specifically by the IJA and its government).[159] Besides controlling the drug market, Japan also wanted a buffer state, that was China, between it and the Soviet Union which Japan felt it would eventually have to fight.[160] Japan's policies in former white colonial areas now under men like Doihara and Hirohito, according to historian Saburō Ienaga, "were even more atrocious than those of the former governing countries."[161] Japanese felt slighted at not being allowed to control China and elsewhere to the extent of European powers after WWI, and were upset the major powers had blocked a clause against racism in the Treaty of Versailles. Prime Minister of Japan, General Hideki Tojo, believed Americans had no stomach to fight and would back down at the first confrontation with his forces.[162] Many Japanese, due to propaganda, believed that if they fought the Americans, they would prevail because as a democracy, Americans "can't unite for a common purpose. One blow against them, and they'll fall to pieces."[163]

Clearly, history has since shown that the Japanese underestimated both the strength of the American character and how swiftly and effectively the U.S. would respond to an attack on its democracy. America would not make peace from a position of weakness.[164] More importantly, forcing the entry of the U.S. into the war ensured the survival of the beleaguered U.K. as well as the defeat of Nazi Germany, Fascist Italy and other Axis Powers. When Winston Churchill received news of the Japanese attack, he exclaimed as of that time the Allies had won the war—it may take time, but with the U.S. on England's side, they would defeat Hitler and Mussolini and grind Japan "to powder."[165] Japan created the opposite response it had intended: America would not leave Asia to Japan and would never ignore an attack on her sovereignty.

When war erupted, Woody was confounded. Many CCC comrades joined the army, but he did not want that service. He stayed on with the CCC until March 1942, when he requested his discharge. He returned home thinking he would "go to war right away," but to his surprise, he was not drafted. Since he did not volunteer for service after he left the CCC, a friend convinced Woody to work with him in a factory.[166]

Japan's Prime Minister General Hideki Tojo (1941-1944). He sent armies overseas knowing full well they were murdering and raping everywhere they went. He did not care. National Archives, College Park.

When he returned to Quiet Dell, besides helping his mother, Woody also took a hot and dangerous job in a steel mill in Sharon, Pennsylvania, manufacturing anchor chains. After three months there, he started having sharp chest pains and spitting up "horrible stuff" and blood. When the foreman refused to give him a different job, he quit, and returned to Fairmont to drive taxis for Conn Transportation Company. Woody had not received a driver's license yet, but a friend somehow got him one and he officially became a "chauffeur."[167]

Besides driving people around, he also delivered Western Union telegrams. As America deployed its forces, the U.S. lost servicemen killed-in-action (KIAs). The understaffed military notified families by sending Western Union telegrams. Woody had the unenviable job of delivering those telegrams to distraught families.[168]

To compound his discomfort, Woody was required to knock on the door, wait for someone to answer, hand them the telegram and wait for a family member to sign for it. Often, people read the telegram to know what they were signing. Most, upon learning their loved one had died, broke down. Several looked at Woody and asked, "Why did this happen to my husband?" "Why did this happen to my son?" Woody had no answer and was emotionally overwhelmed. These experiences gave him an appreciation for the fragility of life and randomness of death.[169]

On the other hand, he enjoyed working as a taxi driver as he enjoyed the people and the pay was good. One day, a lovely young lady, Ruby Meredith, missed her bus home to Meadowdale and flagged Woody to drive her home. He found her attractive, but they only exchanged a few pleasantries. He probably also was not as outgoing with her because he was currently dating Eleanor Genevieve Pyles. Woody was also somewhat bashful so he sometimes did not strike up conversations with strangers easily, but her curly auburn hair, full lips, beautiful brown eyes and slim figure caught his attention. He dropped her off, sorry that he would probably not see her again. A few weeks later, however, she needed a ride again. This time, he invited her to sit in the front seat and they talked the whole way to her home. Soon thereafter, he probably broke up with Eleanor and focused on Ruby (this was probably wise since Eleanor was either divorced and/or still married and had a young son).[170] Thereafter, Woody and Ruby started to date and fell deeply in love.

Either in late 1941 or in the spring of 1942, Woody explains he showed up at the Marine Corps' recruiting station, but was told he had a problem. One had to be 5'8" to join, but he claims he was 5'6" (he often fibbed and said he was

5'6" when in reality he was 5'5¼").[171] He was upset when the recruiter rejected him. Although 18, he still was not drafted yet. He could have gone to another service, but he had his heart set on the Marines.[172] However, it seems his understanding of the requirements was incorrect. Prior to Pearl Harbor, the standard for the military was to be at least 5'5", so he was tall enough to serve.[173] If he avoided volunteering, then he probably wanted to avoid the service as long as he could and wait for his "call up."[174] This discrepancy of how Woody remembers entering the Corps is reconciled by historian Jon T. Hoffman: "I suspect Woody is trying to provide reasons that were beyond his control why he waited so long to enlist, when the reality is probably something else. I'd be willing to bet he 'joined' when he was drafted."[175] No records exist of him volunteering and the only document we have about Woody's manner of enlistment is his registration draft card from 30 June 1942. In his card, one sees Woody struggled with his height again fibbing this time by two inches by writing "5'7".[176] Furthermore, it appears Woody's height also should not have been a problem by looking at other Marines' enlistment files. In August 1942, later MOH recipient Jack Lucas was accepted in the Marine Corps with the same height Woody had: 5'5¼".[177] And in September 1942, another later MOH recipient Tony Stein was accepted in the Corps with the height Woody had been telling people he had; namely, 5'6".[178] And Don Graves, Iwo Jima Veteran and fellow flamethrower operator, was accepted in the Marines in August 1942 and he was even shorter than Woody by ¼ inches standing at a flat 5'5" (65 inches). [179] Graves even claimed there were a few guys in his unit shorter than him, like Robert C. Filip in Woody's regiment who stood 5'4½".[180]

Woody states that a few months later after he had been rejected by the Corps, the Marines changed their "height requirement" and he rushed to the recruiting office to "volunteer." Woody sarcastically said that finally the Corps would take "little runts like me."[181] Influenced by Marines he had met as a child, he did not want to serve in any other service. According to Hoffman and the registration card, Woody was, however, *drafted* and did not volunteer. Yet, once drafted, he picked the Corps. So why did Woody wait so long to enter the military? According to Hoffman: "Those who waited [like Woody]…probably were hoping they wouldn't have to go at all."[182] Woody's service record supports the fact that he enlisted at the local board #2 at Fairmont, West Virginia after his draft card was pulled.[183] According to Hoffman, coupling this information with "the executive order that…put an end to voluntary enlistments [by December 1942] (except for 17 year olds…) and the more reliable info on height requirements, I'd say Woody was drafted."[184] At the time, Woody thought he was called up to protect his nation from *invasion*. After Pearl Harbor, he worried that Japan would invade the mainland, and many felt Hitler might come knocking on America's door after all of Europe fell to him. "Had I been told then," Woody said, "that I was going to be deployed halfway around the world to fight on…bizarre islands, I wouldn't have been as anxious to go. I just wanted to protect my country."[185]

That Japan would soon hit our shores was something many feared. The seemingly unstoppable Japanese military had taken Guam, Wake Island, the Philippines, the Dutch East Indies, Malaya, Burma, French Indochina, parts of China and much of Asia with incredible speed, and appeared poised to take Hawaii before invading the mainland.[186] On a personal note, when my grandmother Leona Rigg née Parr in Philadelphia heard about the Pearl Harbor attack, she became hysterical, talking incessantly about the Japanese killing Americans on our soil. My grandfather, Mark Rigg, sarcastically told her, "Leona, calm down, I'm sure they won't be here for a few more weeks." My grandmother was not alone. "In neighborhoods all across the country, people stepped out of their homes and looked skyward, as if expecting a fleet of Japanese planes to appear suddenly overhead."[187] Radio reports and newspaper articles added to the hysteria "documenting" "air raids over American cities."[188] Nonetheless, Leona and many of her fellow citizens had good reason to worry. The Nazis sought to gain favor with Latin American countries to give the German military a stepping stone to invade America and, in particular, made headway with Argentine fascist leaders. Hitler's henchmen even devised plans for invading the U.S. Meanwhile, Japanese submarines shelled the Californian coast in February and Oregon in June 1942.[189] If they could sink the U.S. Pacific Fleet in two "smaller" attacks, what could they do if they decided to attack the West Coast in force which did not have the full complement of battleships to protect it?[190] For many Americans, fear for both the nation and themselves was well-founded.

# Ch. 4: Joining the Marine Corps

"War makes…heavy demands on the soldier's strength and nerves.
For this reason, make heavy demands on your men."[191]
—USMC Training Doctrine


"We must lust for battle; our object in life must be to kill;
we must scheme and plan night and day to kill."[192]
—General Lesley J. McNair, director of
U.S. ground forces' training program 1942


President Ronald Reagan said, "Many people live their
whole lives wondering if they made a difference in the world.
Marines don't have that problem."[193]


WOODY'S MOTHER WAS UNHAPPY HE was entering the Marines although, being 19.5 years old, he was past the draft age (in some reports he claims he joined when he was 18, but the records do not bear this out).[194] She wanted him to remain on the farm, but Woody was on the road to becoming a Leatherneck, as Marines were often called.[195] After he was drafted, he was granted his request to join the Marines and left in May of 1943 for the Marine Corps Recruit Depot (MCRD) in San Diego. The documents note he was 65.25 inches tall (5'5¼") and weighed 165 pounds at enlistment although Woody continued to fudge his height claiming 5'6".[196] According to the National Heart, Lung and Blood Institute, Woody's body mass index (BMI) showed he was overweight nearly tipping the scale towards obese. When asked about how "plump" Woody was, he replied, "It was all that good farm cooking Mom did."[197] It seems this weight also bothered him like his height, and, he later would also fib about it telling journalists that he was actually 135 pounds when he started with the Corps.[198] Soon though, regardless of his weight, Woody's new lifestyle with the Corps would transform him from an average, portly citizen to a harden, combat-capable Marine. Because he lived east of the Mississippi River, he normally would have been assigned to boot camp at Parris Island, South Carolina, but Parris Island was at capacity, so Woody reported to San Diego. He was what Parris Island recruits jokingly referred to as a "Hollywood Marine," not only because of the geographic location of San Diego to tinseltown, but also because Marines there actually often "made war movies as part of their training."[199] Before going, Woody and future Marines took the following oath:

> I will bear true faith and allegiance to the United States of America; that I will serve them honestly and faithfully against all their enemies whomsoever; and that I will obey the orders of the President of the United States, and the orders of the officers appointed over me, according to the Rules and Articles for the Government of the Army, Navy and Marine Corps of the United States.[200]

Woody and fellow recruits left on 27 May 1943 with the rank of "Private" after taking their oaths, "having, in short" put their lives "in hock to the most fearsome and hazardous of the country's armed forces."[201] Woody traveled by train with six other West Virginia boys that picked up recruits from various stops across the nation during a five-day transit to "The Golden State."[202] Along the way, Woody admired America's vastness and wondered if he would make the grade to become a Marine. He pondered on the tests he would face, not only in the Marines, but also against the enemy. The day he stepped off the train near downtown San Diego was the first time he had seen an ocean.

Marine Corps Boot Camp Receiving Barracks, San Diego, circa 1944. It was boldly noted on the building, "If you want an Island, Tell it to the Marines." Then under this sign, the Marines had listed the islands conquered of "Guadalcanal, Makin, Bougainville, New Britain, Tarawa, Kwajalein, Eniwetok, Saipan, Tinian, Guam, Palau and the Philippines." Marine Corps Recruit Depot Archive, San Diego

A bus took the recruits from the train station to MCRD where there was a welcoming party. A Master Sergeant standing over 6 feet tall and weighing 250 pounds yelled for them to get off the bus. "He scared the living daylights out of me. I thought he was going to eat me alive," Woody said. As they rushed off the bus, a group who had just finished the grueling months of boot camp shouted the traditional MCRD greeting: "You'll be sorry!" At the in-processing, Woody and the others were told that in the civilian world, they had freedom to do what they liked. Sergeant Lawrence Henry Pepin further explained: "All of this was and always will be due to our great American way of living and our democratic form of government of which we will allow no other foreign sources to interfere."[203] However, of that moment, they belonged to the Corps and would learn how to kill and "carry out orders efficiently and obediently" and would have to sacrifice their individuality to survive both in the Corps and war. Pepin taught them how to address their instructors and what they should eat—he even instructed them

The cattle-car truck that brought recruits from the train station to the boot camp at San Diego 1943. Marine Corps Recruit Depot Archive, San Diego

to tell their parents not to send them candy, cookies or gum because "sweets will make you sick."[204] A platoon sergeant in perfectly pressed khaki slacks, shirt and khaki tie along with a campaign hat (resembling the hat Smokey Bear favored) barked orders. This was their drill instructor ("DI"), the man who would spend almost every waking hour with Woody and the platoon for the next few months. It was his job to turn civilians into Marines who would fight, accomplish their mission, and survive to return home when their work was done. He exclaimed, "If any of you idiots think you don't need to follow my orders, just step right out here and I'll beat your ass right now. Your soul may belong to Jesus, but your ass belongs to me and the Marines. You people are recruits. You're not Marines."[205] Recruits were more often called "Shitbirds" and "Boots" than "Recruits" by their DIs.[206]

With a booming voice, the platoon sergeant ordered the recruits to line up in formation. They were taught how to stand at attention. "Stand up straight, keep your heels together, toes at a 45-degree angle, your arms by your side, your thumbs along the seam of your trousers, shoulders back and eyes straight ahead. Don't look at me! No talking in ranks unless I ask you a question." Most had never been ordered around like that or stood in formation. Woody, on the other hand, had plenty of experience with military formalities from his CCC days giving him an edge over others.

There was no time to waste. Marines were needed for war. Everywhere they went they marched in formation at a fast pace ("quick time") with a DI counting cadence, or jogged ("double time") at a precise number of steps per minute. As soon as he got to the MCRD, his platoon was taken to get their immunizations. Walking single file, medical personnel inserted syringes in both arms simultaneously. No one knew where they might be deployed after boot camp, and the variety of inoculations would protect them from a number of illnesses. Eyes and teeth were also thoroughly examined. On average, 30% of a platoon got glasses (for many, the first and/or best they had ever had) and 80% received some form of dental repair and cleaning.[207]

At one stop, they were issued their uniforms including dungarees, caps ("covers"), semi-high top shoes ("boondockers"), socks, underwear ("skivvies") and web belts with brass buckles that were required to be kept highly polished. And there was a duffle bag ("sea bag"). Their helmets were called "Iron Kelly" or "Tin Can Derby." Each recruit was given a boot camp haircut. It took less than a minute in the barber chair for a burr cut to about an eighth of an inch.[208]

Recruits arrived at MCRD San Diego from all backgrounds, social classes, ethnic groups, religions and parts of the country; from big cities, small towns and rural areas. They looked and talked differently. Their ages ranged from 17 to mid-30s. Some had a high school education or less and others had started or finished college. There were bachelors and family men. Some were physically fit and others needed exercise (which the Corps provided in abundance). Their attitudes towards authority and each other varied widely.

What united them was a righteous anger at fascist warmongers who had started a war, and a determination to do whatever was needed to protect their country. For them to work and fight together at the highest level of efficiency, radical reshaping had to occur, and with the nation at war, this had to be accomplished quickly.

The first two weeks, the DIs tore down the recruits' civilian facades so that they could build them back up into Marines. They converted boys into men "by destroying the identity they brought with them."[209] They spoke in the third person and did away with "I," "my" or "me." If a trainee wanted something, he asked, "This Recruit requests to go to the head" or when admonished, the answer would be, "This Recruit was wrong. This Recruit will do better." Speaking this way broke down the individuals ("egoectomy")[210] in order to build them into a collective group called Marines. They had to learn to talk like Marines using the right vocabulary. For instance, a rumor was called *scuttlebutt* because that was the name of the water fountain or container on old ships of the British navy where people met and thus rumors spread. Gossiping was *shooting the breeze*, information was *dope*, news was the *scoop* or *the word*, straightening up was *square away*, and manipulating people was called *working one's bolt*.[211] The Marines had nick-names for the other services referring to U.S. Army soldiers as "Dogfaces" and Army Air Force pilots as "Doggies" or "Airedales," while sailors were "Swabbys" or "Squids."[212]

Because Marines often fought from ships or were sent by ship to secure beachheads and facilities ashore for the navy, it was vital Marines and sailors spoke the same language. Recruits did not have walls in their barracks, they had bulkheads. They did not have stairs, they had ladders. They did not mop the floor, they swabbed the deck. Upstairs became topside. The bathroom or latrine was called the "head" (the nautical term for a toilet).[213]

They learned standardized names for letters from A to Z and ways to say numbers so that soldiers, sailors, pilots and Marines could communicate clearly and quickly via a bad radio connection or in an environment filled with explosions, gunfire, and yelling. Companies, which are subdivisions in a battalion, were designated by letters: A, B, C, D, E, F, G and so on. As such, companies were verbalized as Able, Baker, Charlie, Dog, Easy, Fox and George companies. If they did not learn to communicate with the same terms as sailors with whom they worked, confusion could result, and confusion on a ship can get people killed, and quickly.

They *never, ever* referred to their weapon as a gun. It was a rifle. When someone called it a gun, the DI often had that person march around the barracks yelling, "This is my rifle," and then hold up his M1 rifle, and then grab his genitalia with his other hand and yell, "and this is my gun." "This is for fighting," he would bark holding up the M1 again, "and this is for fun," and then grab himself again.[214] This happened to Woody, so he had to run between two barracks for several minutes, nude, holding himself and his rifle while yelling out the ditty. "After that, I never, ever called my rifle a gun again," Woody explained.[215]

They jogged, hiked, ran obstacle courses, went through bayonet drills on dummies, learned hand-to-hand combat with KA-BAR knives, practiced throwing live grenades, performed gas mask drills and did calisthenics together. When not otherwise occupied, they marched, marched and marched. Their beds ("racks") had to be made exactly as instructed and the barracks kept spotless. DIs performed white glove inspections so that the recruits would clean so well nothing could be found wrong.

Physical and rifle exercises at Marine Corps boot camp, San Diego 1943. Marine Corps Recruit Depot Archive, San Diego

Marine Corps boot camp, San Diego, Obstacle Course, 1943. Marine Corps Recruit Depot Archive, San Diego

Shoes and boots had to be cleaned and cared for, and brass polished until it gleamed. The DIs yelled at recruits face to face and demanded they do things better and faster, repeating tasks until they got them right; anything else was unacceptable. Days started with a bugle calling reveille before the sun rose and ended late at night. Every evening, when Woody's head hit his pillow, he immediately fell asleep. The first two weeks, they wore their own footwear because they did not have enough boots in supply. The endless marching tore up shoes and Woody had holes in each one. For days, he put cardboard in his shoes protecting his feet as best he could from the marching.[216]

Adjusting to barracks life took time, including the lack of privacy using the restroom and showering in the open. Getting used to different personalities also took time. Food was nourishing and plentiful (on average 3,300 calorie diet per day), which was fortunate as they burned hundreds of calories, but they were always rushed in and out of the dining ("chow") hall.[217]

One day, during rare down time, Woody climbed into his bunk and fell asleep. Another Williams in his platoon decided to play a joke. While he was asleep, he tied Woody's shoe laces around the bed's springs. When the DI ran in and yelled for the recruits to form up outside, Woody could not get up.

As a result, the DI punished him. Woody was furious with the other Williams, so later, after hours, he challenged him to a fight. "I was going to show him I was tougher," Woody said. They started to slug each other and the platoon watched. When the DI noticed this, he stopped the fight, gave them boxing gloves, and then told them to "beat the Hell out of each other." Woody tried to match if not beat the other Marine, but after he was knocked down several times, he gave up.[218] That cleared the air and both Williams boys became friends.

Marines are a unique breed. They strive to be an elite fighting force, superior to other services. It is a "state of mind," and as Lieutenant General Victor H. Krulak said, "Among Marines there is a fierce loyalty to the Corps that persists long after the uniform is in mothballs… Woven through that sense of belonging, like a steel thread, is an elitist spirit. Marines are convinced that, being few in number, they are selective, better, and above all, different."[219]

There is a reason that is not bravado, but in the crested reality of training of the mind and body during boot camp. From the Corps' inception on 10 November 1775, Marines have taken pride in their accomplishments and appearance, including their distinctive uniforms. Woody admits the Dress Blues sold him on the Marines besides the tall-tales the Marines in his hometown told him.[220] By 1941, the Marines had proven themselves to be one of the best fighting organizations in history as a result of defeating the Barbary Pirates at Tripoli, Libya (1804), the Mexicans at Montezuma, Mexico (1846) and the Germans at Belleau Wood, France (1918).

The making of a Marine starts in a three-month course known as boot camp. In WWII it sometimes was truncated to between 7-10 weeks, although Woody attended it between May and August 1943 (starting with his train ride, his training extended from 26 March to 21 August).[221] The normal day during this training required rising at 5am (0500) followed by drilling on the parade ground, running the obstacle course, training with weapons, practicing small unit tactics, prepping uniforms, conducting physical fitness and then finally turning off the lights at 10pm (2200)—only to repeat the whole cycle the next day. Woody's typical boot camp day illustrates the metamorphosis of becoming a Marine.

Reveille sounded at 0500, pronounced "oh-five-hundred." The military uses a 24-hour clock to avoid confusion and miscommunication between 8:00am (0800) and 8:00pm (2000). Everyone in the barracks scrambled out of his bunk, ran to the head, shaved, made his bed, dressed and then assembled in formation outside. Hundreds of recruits filled the parade ground. In Woody's day, they went by platoons of 63 each with six squads.[222] At boot camp, everything is done with the recruit's assigned platoon.

Once a platoon completed roll call (verifying all present and accounted for), it marched to the chow hall for breakfast at 0615.[223] After breakfast, the platoon returned to the barracks to study weapons, history, or small-unit tactics and then mustered at the obstacle course to run through it. They crawled under barbed wire for many yards in sand pits, participated in pugil stick brawling to learn bayonet fighting, carried each other to practice the fireman carry, or ran through the sand. They did this with intimidating, demanding DIs circling them loudly barking orders like vicious dogs. To complicate it for those not accustomed to working as a team, the platoon performed together or was punished. If someone did not do an obstacle correctly, the group was disciplined with extra push-ups, holding their weapons out at arm's length, doing squats for several minutes or leg lifts for what seemed like an eternity. It was imperative that all help each other and that no one cheat. If there was cheating, the violator would bring down "God's" wrath upon himself and his comrades. Collective responsibility was drilled into everyone. If a "recruit should faint from exhaustion during a forced march, the rest of the unit is trained to run in circles around the body while he is tended to until he comes to."[224] Everyone was disciplined to look out for everyone and everything.

Marine Corps boot camp, San Diego. Obstacle Course, circa 1943. Marine Corps Recruit Depot Archive, San Diego

Rules ran the gamut, from how to handle a weapon to how to process trash. One day Woody violated one of these. Thinking no one looking, Woody threw his cigarette butt on the ground. It was a *huge* violation to litter and one of his DIs saw him. As a result, he pulled Woody into an office and gave him a needle and a line of thread explaining, "Look, Recruit. You violated a rule here. As a result, you must get me one hundred cigarette butts and bring them back to me on this thread or I'll make your life a living Hell. Dismissed." Woody now, in addition to everything else he had to do, had to collect butts during the few minutes of spare time he had. They had paper sheets and bags of tobacco to roll their cigarettes called "Blanket and Freckles." He told his buddies to give him their butts once they finished smoking (everyone smoked then and cigarettes were part of every sea ration kit).[225] For days, Woody circulated with what looked like a Christmas popcorn garland begging people to help him out. He stalked people smoking and asked them for their butts. He knew if he did not succeed, he would be thrashed with extra fire watch at night, push-ups, or holding his rifle at arms' length until his muscles burned. Within the allotted time, he succeeded in stringing the hundred butts and displayed the garland to his DI. The DI inspected every butt, and when he found one that had some tobacco still in it, he ripped it from the string and told Woody to go back and bring him butts. "I said to bring the butts in, not butts with tobacco." So Woody, once again, scavenged the grounds for butts and asked his buddies for their finished cigarettes. And before putting them on the string, he removed every tobacco shaving. He finished this mundane punishment and never again threw a butt on the ground. Paying attention to details was important and doing these exercises conditioned men to be situationally aware and obey orders.[226]

After one and a half years of being at war, some of the Non-Commissioned Officers (NCOs) training Woody and fellow recruits had seen battle at Tulagi, Gavutu-Tanambogo, Guadalcanal and Makin Island. They knew what Woody and his soon-to-be fellow Marines would face, and returned to the States to teach recruits lessons they had learned about enemy tactics, weaknesses and strengths. This policy of rotating men out of combat to instruct recruits improved the capabilities of U.S. warriors.

Recruits were told that Japanese soldiers were tough adversaries, but that American servicemen could defeat them. They were taught that the hatred the Japanese had for them was more intense than they could imagine. Journalist Christopher Hitchens called the Japanese "Zen-obedient zombies" bent on world domination who viewed their country's "membership of the Nazi/Fascist Axis as a manifestation of liberation theology."[227]

Their DIs told them that most Japanese viewed Americans' kindness and sense of fairness as an exploitable weakness. Feigning surrender or asking for medical help once overrun and wounded, they would pull out a grenade and kill themselves and the Americans nearby.[228] Japanese booby-trapped everything, including attaching blocks of TNT to cases of scotch whiskey and even filling coconuts "with powder charges."[229] They rigged their wounded and dead with explosives. English-speaking soldiers dressed up like Marines and called for help only to kill the Corpsman once he arrived.[230] As a result, Marines and GIs became reluctant to take prisoners, preferring to shoot first and ask questions later about the "suicide bomber."[231] The enemy seemed surreal and otherworldly. The battle-hardened commanding general on Guadalcanal and later 18th USMC

Commandant Alexander Archer Vandegrift, remarked he had "never heard or read about this kind of fighting."[232]

Although the DIs described this Japanese behavior, until the Marines experienced it in battle, they could not comprehend it. The prospect of fighting a nation of 73 million who were willing to die for a religious and patriotic cause was frightening. The prospect of a nation with one of the most modern and sophisticated engineering and military communities in the world bringing its might and fanatical resolve against the U.S. was terrifying. To bring down Japan, boot camp conditioned men to know how to kill their enemies.

The DIs' job was daunting, turning Roosevelt's war policy into action. NCOs got recruits physically fit and knowledgeable of the Corps (no small task), and conditioned to wanting to kill their enemies. The majority of Marines fought with Nimitz's forces in the central Pacific and had to conquer numerous islands to secure the stepping stones to reach *Nippon*'s shores.

This was a huge undertaking. Besides training recruits to fight as a unit and operate various weapons effectively, the military had to condition men unlike they had been raised: To be peaceful, law-abiding members of a free society. In a democratic, civil society, attacking and killing strangers was unacceptable and, if not in self-defense, criminal. Schools, the Boy Scouts and churches of pre-war America had prepared the nation's youth to be productive citizens, good family members, and ethical people.

The armed forces had mere months to prepare the recruits, volunteers and draftees to survive and succeed in a radically different environment. Marines in 1943 instructed men how to deal with an enemy whose ideals were the polar opposite of what they had been taught. In America, people valued a free press, liberty and independent thinking. In Japan, these values were dangerous weaknesses the government suppressed. The vast majority of Americans adhered to a code for humane treatment of enemy civilians and POWs. The Japanese government had signed the 1929 Geneva Convention concerning the treatment of POWs "but never ratified it because of strong opposition from the Japanese military."[233] To the Japanese, people not part of their national and racial group had no rights or value except as colonial subjects or slaves.[234]

As for POWs, they were beneath contempt for Japan's military and it degraded, beat and killed them. After surrendering to the Japanese at Wake Island, several prisoners were taken from the ship's hold and brought top deck. The navy lieutenant in charge, Toshio Saito, announced that because the Marines had inflicted so much death on the Japanese, they were to be executed. The Japanese beheaded and mutilated the Americans and then dumped them overboard. At least here, the Japanese killed them before throwing them into the sea. After Singapore fell, in one of many rounds of butchery, the Japanese murdered 6,000 Chinese, often tying their hands behind their backs before throwing them off a ship into the ocean. There were isolated cases of crucifixion, displaying Japan's mockery of Christianity. Hearing of such events, Marines were warned *never* to surrender.[235] The Marines had to re-train the recruits to view their enemy as brutal aggressors who deserved death. This enemy if given an inch of kindness would repay it with a mile of treachery.

During training, DIs yelled, "If you show weakness, the Japanese will cut off your head and shit down your neck. Understand?" In response, the men shouted, "Sir, Yes Sir." Or, "If you don't kill your enemy, he'll kill you. He wants to cut off your head and show it to the Emperor as a trophy. Don't give him that pleasure!"[236] "Sir, Yes Sir." "Never give up. If you stick together and support one another, you'll win. If you don't support one another, you'll die."[237]

Although they took prisoners (including those who surrendered in the Philippines, Guam and Wake Island), Japan treated them cruelly. IJA soldier Hino Ashihei observed: "As I watch large numbers of the surrendered soldiers, I feel like I am watching filthy water running from the sewage of a nation which derives from impure origins and has lost its pride of race."[238] On the Bataan Death March, a Japanese guard tried to get a POW's wedding band. When it would not come off, the Japanese took a machete, cut off the hand and mutilated the severed hand to get the ring. While bleeding profusely, the prisoner was forced to march on. In shock, he took another fifty steps before collapsing. Guards then ran him through with bayonets. Trucks passing by marching POWs contained sword wielding Japanese soldiers who yelled "duck" and then swiped at the column decapitating Americans who did not stoop quick enough.[239] Sentries made Filipinos bury an American POW alive. A fellow American, horrified, "saw a hand feebly, hopelessly, claw in the air above the grave."[240] Around 37% of POWs died under the Japanese.[241]

Many POWs were worked to death as laborers. Some Japanese even ate some

POWs believing they would gain strength from devouring parts of their livers. Men under Iwo Jima's commanding general, Kuribayashi, conducted such rituals. Future President George H.W. Bush nearly met that fate as a pilot who crash-landed in the ocean near Chichi Jima (under Kuribayashi's command). Unlike other airmen who went down in this area and were murdered, Bush was rescued in the nick of time by the submarine *USS Finbak*. Americans who would have liked to have taken Japanese prisoners to question rarely had the opportunity. The majority of Japanese who quit fighting were unable to fight or were dead.[242]

Woody's instructors cautioned, "They'll kill [all of] you if they can, right after they cut off your balls and shove'em down your throat. Understand?" "Sir, Yes Sir!" Such speech was not hyperbolic; Marines on Guadalcanal were horrified to find comrades, taken alive by the Japanese, murdered and holding their own decapitated heads with their severed penises in their mouths on the paths to greet the subsequent waves of attacking Marines.[243] One unlucky soul who had the barbaric ritual of having his severed member stuffed in his mouth done to him actually was not dead but only unconscious when the Japanese took down his pants, and sawed off his penis and testicles with a knife. They then shoved his bloody manhood into his mouth and left. When he came to, some fellow Americans had retaken his position. Spewing out his own penis and testicles, with blood and his own sperm running down his face and chin, he begged for his countrymen to kill him.[244] Nonetheless, instead of causing Americans to flee in terror, such acts strengthened their determination to avenge their comrades. (see Gallery 1, Photo 15)

From August 1942 until February 1943, Marines at Guadalcanal fought thousands of tough, clever, well-armed, and relentless Japanese infantry supported by IJN forces and air squadrons, and they had to kill tens of thousands to win. It was America's first major victory over the Japanese on land. After reports leaked of the treatment of those taken prisoner after the fall of the Philippines in 1942, where the Japanese executed 15,000 prisoners, Americans learned they could not expect humane treatment if they surrendered. Marine Major General Vandegrift commanding the 1st MarDiv on Guadalcanal told his men, "This will be no Bataan."[245] The knowledge of Japanese cruelty against POWs including beheading, burying alive, torturing, beating and enslaving them changed the way Americans conducted war. In fact, Japanese fanaticism made Marines more zealous. Soon after being taken prisoner by the Japanese in the Philippines, most American POWs admitted they would have *never* "surrendered had they known the fate in store for them."[246] In one account, the Japanese herded 150 U.S. Bataan soldiers, sailors and Marine POWs into an air raid shelter at Puerto Princesa prison on Palawan Island, doused them with gasoline and then burned them alive.[247] Vandegrift told his Marines they were to *never, ever* surrender. Some Americans felt shame many were taken prisoner by inferior forces, such as our Allied British comrades at Singapore when they greatly outnumbered the attacking armies. Upon hearing Allied claims of Japanese atrocities, the spokesman for Tokyo radio stated:

> If American and British leaders are so ready to raise a hue and cry over the 'maltreatment' of their War Prisoners, why don't they teach their men to stand up and fight to the finish? The way the Americans threw up their hands at Corregidor and the way the British gave up Singapore—on the heels of loud mouthed assertions that they would fight to the finish – surely shows that the men must have carried on their backs a pretty wide streak of yellow.[248]

At Guadalcanal, there would be no "streaks of yellow," and Vandegrift would do his best to make sure the Japanese did not have opportunities to commit such atrocities again.

Under horrific conditions, including repeated ground assaults, heavy naval gunfire, bombing, and strafing attacks by enemy planes and facing limited resupply, Vandegrift led his men to victory at Guadalcanal, killing 23,000 Japanese and seizing the island for an air base protecting the lifeline to Australia and New Zealand. Vandegrift received the Medal of Honor, a Navy Cross and a Distinguished Service Medal for his heroic leadership on Guadalcanal,[249] and he and many others disseminated the lessons they learned in 1942 and 1943 to future boot camp classes like Woody's.[250] (see Gallery 1, Photos 16-19)

Over 150 U.S. soldiers, sailors and Marines captured by the Japanese in 1942 at Bataan were herded into an air raid shelter at Puerto Princesa prison and burned alive. 14 December 1944. National Archives, College Park.

Woody's DIs conditioned the Marines to fight effectively as a unit under adverse conditions against a formidable enemy. This physical and psychological education began while the Marines ran through obstacle courses and conducted field exercises and drill. After exercises, Woody's platoon was covered in dirt, sand and sweat, with every muscle screaming in agony. They returned to the barracks, showered, put on a new set of clean dungarees, mustered at the parade grounds and marched off to the chow hall at 1200. Posters everywhere declared "Remember December 7th." The training manual given to recruits during Woody's time went further than this generic battle cry, saying in fighting the "enemies of democracy," the "Corps today has one battle cry. It is not 'Remember Pearl Harbor,' but 'Remember Wake Island.'" At this battle in December 1941, the Marines held out heroically receiving praise from Roosevelt for defending the garrison against "overwhelming superiority of enemy air, sea and land attacks... These units are commended for...devotion to duty and splendid conduct at...

battle stations under most adverse conditions." Almost 500 Marines and a few sailors on the island killed over 650 Japanese, sank two destroyers and held off the IJN from 8 December until 23 December when the Americans surrendered. Woody's instructors also instilled that they must kill the attackers who killed their brothers at Wake and get the land once held by Marines back to the U.S. Woody's learning booklet noted: "The Marines will not rest until the crumbled and burning walls of Tokio [sic] echo that cry [Remember Wake Island] as victorious Marines sweep through the city with bayonet and grenade."[251] Woody and his buddies were told of the battles of Tulagi, Guadalcanal and Tarawa and how their brothers had won glory for the Corps being the first to fight yet again.[252] The report sounds like a high school football pep-talk, but since most recruits were teenagers, this language spoke to their sensibilities:

> You men have been called upon by our country to take part in ending this, History's greatest Catastrophe, WWII. You men should feel quite proud that you were selected to do your part in our Corps and I'm sure that you will do your part faithfully and that you will do your utmost to uphold the traditions and the glorious spirit which has been left by those men who have fallen in battle so far in this WWII. I'm sure that when you get over there and see the small wooden crosses designating those who have already fallen that their fighting has not been in vain and that you will carry on for them.[253]

This lesson emphasized the importance to Woody and his comrades of the struggle they faced. It imparted the responsibility they had to build upon the sacrifice already given in the lost lives of those who had gone before them. As Marines, their goal was simple; end this war quickly.

Woody's instructors gave them a proper way to behave telling them such platitudes, "Don't be a continual griper. Remember this little phrase; 'I complained because I had no shoes until I met a man who had no feet'" or "'The very foundation of good breeding is courtesy.' Courtesy is culture. It is the sense of consideration for others" or "The greatest are the humblest and conceit is the essence of ignorance."[254] All were encouraged to read "good books...[D]evote two or three hours a week to good books [and] within three or four years you will have gained a liberal education."[255] During boot camp, they received lessons in humanities, philosophy and psychology. For many who had not finished high school like Woody, this was the best, albeit elementary, education received in their lives.

Throughout recruit training, besides getting in shape and learning the Corps' creeds, Woody learned arithmetic, general orders, insignia, combat signals, first aid, chemical warfare, nomenclature of weapons, rifle rules, close order drill, hand grenade use, scouting and patrolling, manners and etiquette, Chinese and Japanese insignias, jungle warfare, and swimming just to name a few. They were graded and were told "He who is best educated is most useful."[256]

The men received lectures on sex education and venereal diseases. The teacher, combat veteran from Guadalcanal and Tulagi, Purple Heart recipient (a medal received when one was wounded or died in combat), scout-sniper and Drill Instructor Sergeant Pepin, gave a heavy dose of morality saying if they contracted a disease and later gave it to a woman, their wife, whom they loved, then they could be harming her health later if they were irresponsible with their sexual exploits now.[257] "Doctors claim that 80% of the operations for removal of ovaries and other organs are due to a husband having had gonorrhea at some previous time. A case of gonorrhea may thus cause untold trouble for an innocent woman."[258] A detailed description was given of the diseases one could contract from prostitutes and women who were "unclean." Pepin continued:

> It is to be realized that you are young and human, and being human, you may occasionally fall from grace; therefore [you must] take the proper precautions to prevent serious consequences if you do happen to slip sometime. There are certain things which will assist in preventing venereal disease. First: if by any chance you should be playing around with a woman that you think is producing only for you, remember that the woman who is producing for you is apt to be producing for someone else when you are not around. Therefore, take precautions the same as you would if you were playing around with a whore…the best precaution is the use of the condom, commonly known as 'rubber.'[259]

Pepin then described how to clean the penis properly with soap and "sanitube" medical cream. He gave a parting shot explaining this information was given not to be abused:

> Do not think that these preventive precautions have been told you in order…to indulge your lives without restraining…Stay away from free and easy women. Stick to decent ones…Statistics show that all women who are easy contract venereal disease some time or other. In some localities 90% of these women already have venereal disease in some form.[260]

Pepin was overly obsessed with talking about sex, but since he was a closet homosexual, maybe talking about erotic matters in front of many good-looking young men allowed him some inner satisfaction or maybe because he could not openly be gay, he was sexually frustrated and this expressed itself with him talking about sex more. Or maybe he was just a typical young male struggling like everyone else with sexual desire and bio-chemical inputs.[261]

Nonetheless, many of the lessons he was trying to get across were indeed important. Since these men were going to train or travel for months without having a woman's comfort, it was imperative the Corps instruct them so that when they finally did have access to women on leave, they do it responsibly. More importantly, this information was useful sociologically because it would be the first time many received sex education, having come from rural and conservative backgrounds. "They showed us movies with pictures of what these diseases looked like on men's bodies," Woody said. "Moreover, they showed us what diseases like genital warts, syphilis and other diseases on a woman looked like and I wasn't about to stick my pecker into any of those," Woody claimed.[262] Those instructions made a huge impression. The Corps was interested in this matter to ensure that when called upon, Marines could fight without an "itch" in their pants or a "burning when they pissed."[263] The 4th Marines in Shanghai, a unit of over 1,000 men, had a venereal disease rate of 35% in 1940, so the Corps needed to educate its men on these issues.[264] But even with such education, boys will often be boys, and 3d MarDiv correspondent Sergeant Frederick K. "Dick" Dashiell wrote that even on Hawaii, "there are regular houses of prostitution—or were until re-

Gunnery Sergeant (known as "Gunny") Lawrence H. Pepin. He was one of Woody's Drill Sergeants at boot camp and a very educated man. He largely wrote up the training manual that guided most DIs for boot camp during the time Woody was there. Although highly decorated, he would later be dishonorably discharged from the Marine Corps for homosexuality. St. Louis Personnel Records Center.

cently. These were licensed and you could see sailors, mostly, lined up a block away waiting entrance. Kind of a sickly sight, like buying passion over the counter..."[265] (see Gallery 1, Photos 20-21)

After morning classes and noon chow, Woody and his platoon practiced close order drill marching on the parade ground and did classroom work on weapons, history and tactics. They would also take their dirty clothes, wash them by hand and hang them out to dry. Every Quonset hut where they were bivouacked had a wood burning stove that had to be cleaned out to be used that evening to keep warm.[266] At 1700 they gathered for the evening meal, then returned to the barracks to clean weapons, organize equipment, iron clothes and get ready for the next day. Everywhere they went, they carried their rifles.[267] At 2200, taps sounded, lights turned off, and everyone went to bed except for a few who would post watch for two hours and then be relieved by two or three others at 2400. They cycled through the platoon every night and everyone took a turn at standing watch over the unit.

Woody was taught that Marines were better than other services being one of the "first body of fighting troops organized in the United States." The report claimed Marines were better than all the services in "rifle and athletic tournaments" holding "all world records with the rifle." "It has a glorious tradition of never having surrendered. In no fight, when in action under the order of its own officer has it retreated or given up an inch of ground. From the revolutionary war to this day its service have been glorious... Its history is such that it is recognized by both the people of this country and abroad as among the crack troops of the world."[268] While this bravado instilled pride in the organization and although the report did cite notable battles like Belleau Wood for its claims, it was also clearly false. Although rarely do Marines retreat or surrender, it happened at Wake Island, Guam and the Philippines, just to name a few in WWII.[269] The report's author would have been wise to adhere to what historian Allan Millett later wrote: "In the continual struggle to match performance with elitist rhetoric, in the daily challenge to separate organizational mythology from relevant military doctrine, the Corps must understand its own past without excessive self-congratulation."[270] However, in Woody's time, most lessons were packed full of bravado and self-congratulation. This type of pedagogic methodology might not have been full of critical analysis, but it sure helped with the self-esteem and confidence for these Marine recruits.

Woody described fellow Marines as simple boys from the heartland who really did not understand what they were getting into. They wanted to serve their nation and hated the Japanese for harming America. Many had a difficult time harmonizing their peacetime values with the warrior ethos necessary to fight the Japanese. Marines would have to fight on the beaches, on hot, desolate islands, on rocky atolls, rugged shores, on hills, in open fields and in dense jungles against entrenched troops and, since Japanese never surrendered, they had to kill or incapacitate them with the most effective means—bullets, flamethrowers, grenades, mortars, artillery and sometimes knives in hand-to-hand combat.

Some recruits would not make it through the 7-10 weeks of boot camp's constant physical activity and mental strain.[271] When a DI doubted a recruit could handle combat, he would be tough with him. If a recruit did not become resilient, he would suffer. The DIs were not yelling, criticizing and pressuring the recruits for fun. DIs and the officers who supervised them focused on turning recruits into strong, capable Marines who would be reliable in combat. Being too soft on recruits produced weaker Marines, more likely to become casualties or die in combat causing injury or death to other Marines. Recruits learned about discipline and obedience "in an atmosphere of constant apprehension and fear, of constant challenge to physical capabilities and mental stress." Often DIs disciplined the recruits who needed it the most: i.e. the weak and the defiant. Occasionally this led to a recruit killing himself because he "couldn't comprehend or endure the mind-boggling physical and mental pressures."[272]

Woody took boot camp in stride, kept his mouth shut and did what he was told, so his instructors left him alone. He felt boot camp was easier than working a farm, because at boot camp, they got Sunday mornings off to go to church, which was encouraged. Sergeant Pepin wrote, "Remember that there are no atheists in foxholes. God doesn't want you to wait until you are over there to get close to him, no, he wants you all the time. Remember, 'A strong mind is a clean mind and a clean mind is through Love of God.'"[273] Ironically, for six days during the week at boot camp, Marines were conditioned to kill. Then on Sunday, they worshiped listening to sermons about God's love and how they should respect their fellow man.

Religion played a powerful role in the U.S. and in Japan. Both nations believed God (or the gods) were on *their* side. Both believed the heavenly host supported them, and this gave their service meaning. For Americans, this belief gave them confidence they were moral and if they died would enter God's heavenly halls. However, Americans did not want to die for religious beliefs. The beliefs just reinforced the conviction that if they died, they would inherit salvation. Conversely, the Japanese believed the gods would reward them if they died in battle, and it was the highest calling to search out death. Americans had a Judeo-Christian religious ethic that supported the goal of preserving a serviceman's life whenever possible, while the Japanese had a Shinto religious ethic that glorified death in the Emperor's service.

For Woody, religion did not play a role in his life and he did not believe in the Judeo-Christian version of God, but he was not an atheist. He felt there was a higher power, prayed to that "force" and felt it protected him. Uninvolved in any church activities growing up, he claimed, "You only know what you're taught and we weren't taught anything about God." His mother kept a large Bible in which family births and deaths were recorded. However, she kept the Bible off-limits for reading it, and she had a limited ability to read or write herself. Woody was so ignorant about Protestant, Catholic or Jewish affiliations that when he filled out his paperwork, he did not know what to write for religion. A DI told him he had to write something, so he copied his neighbor's application and wrote "C"—whatever that meant. As a result, throughout boot camp, he attended Catholic mass every Sunday although he had "no clue" what was said since the services were in Latin.[274]

Although Woody had no penchant for learning a new religion, he did take up an affinity for another new concept: riflery. The high point of boot camp was the three weeks spent on the rifle range, with every minute devoted to studying, cleaning, disassembling and assembling, and firing their rifles.[275] No matter what military occupational specialty to which a Marine is eventually assigned, every Marine is first and foremost a *Rifleman*. Marines could replace warped machine-gun barrels in the dark because they had done it a thousand times blindfolded. On the last day at the range, the recruits fired for record. They aimed at targets and did all they could not to get "Maggie's Drawers," meaning a "red flag indicating a missed shot."[276] All earned a classification as a Marksman, Sharpshooter, or Expert to be displayed in a badge on their uniforms with Expert being the highest classification. They were indoctrinated with "*Never* point a rifle at anything you don't intend to shoot. *Check* your rifle *each* time you pick it up to be sure it isn't loaded. Many *accidents* have occurred with 'unloaded' rifles."[277] They cleaned the rifle daily, "treated it gently," never used it as a leaning post, avoided touching the steel, and always placed it "in bed every night with oil."[278] In other words, they were taught to view their rifle like a lover and trusted friend as illustrated in the "Marine Rifle Creed":

This is my rifle. There are many like it, but this one is mine.

My rifle is my best friend. It is my life. I must master it as I must master my life.

My rifle, without me, is useless. Without my rifle, I am useless. I must fire my rifle true, I must shoot straighter than any enemy who is trying to kill me. I must shoot him before he shoots me. I will…

My rifle and myself know that what counts in this war is not the rounds we fire, the noise of our bursts, nor the smoke we make. We know that it is the hits that count. We will hit…

My rifle is human, even as I, because it is my life. Thus, I will learn it as a brother. I will learn its weaknesses, its strength, its parts, its accessories, its sights, and its barrel. I will ever guard it against the ravages of weather and damage as I will guard my legs, my arms, my eyes and my heart against damage. I will keep my rifle clean and ready, even as I am clean and ready. We will become part of each other. We will…

Before God I swear this creed. My rifle and I are the defenders of my country. We are the masters of our enemy. We are the saviors of my life.

So be it, until victory is America's and there is no enemy, but Peace![279]

Woody remembered this creed and said they learned it by heart, were tested on it and believed its tenets.[280] They knew it like Christians know the Lord's Prayer. They were told:

In WWI the Germans were astonished at the [Marines'] ability to shoot so deadly at ranges of 600 and 1000 yards…You men must keep up the tradition of the Corps and its men before you, who have made it so by their expert shooting and fighting. To be a 100% Marine you may attain 75% of it by being a master of your rifle.[281]

Boot camp taught Woody that if you "master your rifle, you will master your enemy."[282] This was not hyperbolic. By boot camp's conclusion, a Marine knew the ins-and-outs of his air cooled and gas operated M1. He knew the barrel was 24 inches long with 2.5 turns inside, the chamber pressure was 51,000 lbs. per square inch, and the muzzle velocity was 2,680 feet per second.

Having completed boot camp, the recruits had visibly changed. The DIs had done their job. Unlike the raw, young recruits who arrived at MCRD, many out of shape (like Woody) and undisciplined boys in wrinkled dungaree uniforms, sloppy, marching out of step, unsure of themselves and ignorant of the organization that they had joined, a new group had emerged. At graduation parade, there appeared squared-away young men who had earned the title "Marine" and the right to wear the eagle, globe and anchor insignia of the Corps. Months of vigorous exercise outdoors had made them tan and fit. Their uniforms were neatly pressed and ironed with sharp creases in the trousers. Their covers were starched, uniforms pressed and brass belt buckles gleamed. They held themselves proudly erect and marched in formation with their platoon and company, responding to each command with precision and quiet strength. In order to earn the title of "Marine," Woody and his fellow Leathernecks had just, on average, ran 90 miles, marched 250 miles, scampered 10 times through the obstacle course, completed 70 hours of calisthenics, swam 16 hours in the pool and conducted 90 hours of field training. On average, every recruit in Woody's platoon lost eight pounds of fat and gained 12 "pounds of muscle."[283] These were the measurable accomplishments—what was even more transformative was how these young men thought. They were now warriors with confidence that they could experience "violence and bloodshed" and still defeat their enemy. This confidence could not be measured, but it was definitely observed.[284]

Woody felt this change in his bones. He gained a deep respect for the Corps and was proud to be a Marine. Before boot camp, he had little knowledge about the Corps. Those he surrounded himself with gave him a sense of pride. They were the elite, "and in many respects," Woody said, "we were."[285] This was not misplaced swagger. They had some of the hardest military training to be had in the world at that time. It was a complete "transformation of my thinking," Woody declared. As the booklet Woody studied noted:

> The question may arise, 'Why is it that the Marines are always successful in whatever they undertake?' The answer is discipline and pride in their organization, pride in their uniform and feeling that we can do anything—if necessary we will do the impossible. The spirit that cultivates and the spirit which makes your organization outstanding is the spirit that never says it cannot be done.[286]

Even today, such learning and "education" give Woody pride, and most of the boys at MCRD wholeheartedly bought into the belief they were indeed the best of the best.

Marine Corps boot camp, San Diego. 21 April 1941. Marine Corps Recruit Depot Archive, San Diego

Marines marching at boot camp, San Diego, circa 1943. Marine Corps Recruit Depot Archive, San Diego

During training, Woody's focus was infantry. At first, he worried he would never measure up to the others who seemed smarter, bigger and better than he, but by graduation, he felt as capable as most of his comrades (according to Woody, he graduated in the top ten of his class).[287] This did wonders for his self-esteem and he then knew he could meet any challenge the war might bring.[288] An NCO at boot camp explained, "Now you have become one of us and we expect a willing obedience much more rigid and exact than you have ever encountered before. Almost two hundred years of Marine traditions are now yours to add to. Marines WIN. Marines never retreat; that is what we demand and you must give us the best you have…"[289]

Woody had now earned the title of Marine, and as anyone who has gone through the Corps' crucible to earn it, he bears it with a pride unknown to most militaries and unfamiliar to the other services. One could call the Marine Corps a cult.

One of the first pictures taken of Woody as a Marine. 1 July 1943. He was overweight at 165 pounds when he entered the Corps and one sees he was somewhat pudgy even here after training for one month at boot camp. When asked about this, Woody explained, "It was all that good farm cooking Mom did." St. Louis Personnel Records Center

> Recruits are imbued from the outset with the conviction that they are members of a select *corps d'élite:* that it is unthinkable for any Marine ever to let down another Marine in combat, and that failure to live up to the traditions of the Corps amounts virtually to treason. Observing this mode of thought carried into action has led more than one outsider to remark that these men seemed to be fighting less for the U.S. than for the U.S.M.C.[290]

Some believe the Corps is a way of life and many Marines would agree with this sentiment. Woody's boot camp instructional manual echoes these beliefs:

> Spirit[—]We know it as Marine Corps Spirit. It has made our Corps famous. It is that which makes a man always want to be at his best. It will carry him on to success and victory when everything else failed. It is not built on fear of correction or punishment. It is something in the man himself which will carry through when his leader has fallen and his friends and comrades are dropping by his side. It will carry him on alone, unafraid until death.[291]

After Woody finished boot camp, he belonged to the Corps and he had never been so proud in his entire life. He had been selected to be part of the chosen few. "I felt like I was 12-feet tall and could whip anybody alive," Woody said.[292] He would need to feel this way to fight the Japanese.

# Ch. 5: Background on the Japanese Military: Brave Soldiers and Psychopathic Rapists and Killers

"Those who can make you believe absurdities
can make you commit atrocities."
—Voltaire[293]


"A Lesson From the Holocaust:
Never Stop Telling the Terrible Stories."
—*Chicago Tribune*[294]


## Preamble

After entering a large museum in one of the world's most ancient cities, Nanking, the former Chinese Empire's capital, my 18-year-old daughter Sophia and I walked over a glass walkway allowing us to look down ten feet of earth and observe an ancient footpath. Lights highlighted the ground under the glass while the room we were in was dark. To our right hung numerous photographs on a black wall showing Japanese slaughtering Chinese citizens in December 1937 revealed by individual lights, the types one sees on art museum paintings. Dead babies, severed heads, piles of bodies on Yangtze River banks, and helpless POWs were all documented there from camera shots taken by the perpetrators themselves, IJA soldiers who had conquered Nanking.[295] The photo collage documented an orgy of slaughter. Walking along the pathway, there was a rectangular centerpiece in the hall ringed by three foot walls. Many, on gazing over the lips of the precipice, shook their heads. Some put their hands together and offered prayers to the heavens. Others whispered to companions, "Bastards," "I hate the Japanese," "How can Japan deny this?" or "America should've dropped more bombs on the assholes." People here had eyes both grave and focused. The air trembled with their energy.

Skeletons of victims of Japanese atrocities at Nanking at the Nanking Massacre Museum's third hall from December 1937. 11 May 2019. Author's Collection.

My daughter, standing 5'11" with blond hair and blue eyes stood out in the crowd of mostly local *Han* Chinese. Everyone was silent except for those walking away from the centerpiece. Sophia looked over the wall first and then she glanced at me with eyes wide and filled with sadness. "In truth I found myself upon the brink of an abyss, the melancholy valley containing thundering, unending wailings."[296] As I looked over, I beheld a mass grave. Many skulls showed signs of having either been crushed by a heavy object or shot by a bullet. Other skulls had been removed from their bodies, settling several feet away from the shoulders. Sophia and I did not know this crime scene was not even close to being the museum's biggest. One much larger awaited us in another building with hundreds of bodies there; old ladies, POWs and children, all of whom gave evidence the Japanese had killed them with guns, swords and clubs. We walked on a massive unplanned cemetery where everyone had met an untimely and murderous end. The Japanese had "stained the world with blood"[297] and jars of human remains mixed with the earth that had absorbed their flesh lined the exhibit. (see Gallery 1, Photos 22-25)

My daughter and I were visiting one of the largest crime scenes in history. Sophia was there because she is fluent in Chinese and was helping me conduct research. On this day, our journeys took us to *The Memorial Hall of the Victims in Nanjing Massacre by Japanese Invaders* in Nanking, China. I had been to the Holocaust Museum in Washington D.C., Yad Vashem in Jerusalem, Israel, the large Holocaust Memorial in Berlin, Germany and the Nazi death camp Auschwitz in Krakow, Poland, but nothing had prepared me for this massive complex built to commemorate *The Rape of Nanking*. All of the European sites show evidence of mass murder with Auschwitz being the most dramatic with large rooms full of shoes, shaven female hair and suitcases of the deceased. However, here, we observed hundreds of skeletons, often with their empty black eye sockets starring at us as if pleading to never forget. The gloomy air released their messages from their shattered bones relying on the living to recount their stories. The sheer size of the extermination reminded me of Dante's phrase: "I should never have believed that death could have unmade so many souls."[298] How could an apparently "cultured" civilization like Japan unleash soldiers who could do such horrible crimes?

## Background on Japan's Criminal Behavior

Japan under Hirohito from 1925 to 1945[299] produced horrors on the scale of Hitler's evils. Hirohito's and Hitler's rule shock most, especially those who have only known the democracies of today's Japan and Germany. WWII behavior of both countries was deplorable.[300] For Japan, its legacy of 1927-45 was one of atrocities, war crimes and "suicidal disgrace."[301]

Dating to the late 19th Century, Japan was instilled with an "emperor-centered ideology." This belief was used by the Imperial army "to validate [its actions in the Emperor's name and] [it] as a special institution in the Japanese polity" to unite its men for whatever campaign or political endeavor it happened to be focusing on.[302] Such a focus by the IJA allowed Hirohito to maintain more control over the military, and by extension, over his subjects and respect than the Nazi dictator. And the heavy focus on the Emperor was fully supported by Shinto and Zen-*Bushido* Buddhist doctrines. And these doctrines also made the Japanese feel superior and justified in murdering anyone not Japanese for the slightest of reasons.[303] No one was permitted to question the Imperial institution that permitted anything in the Emperor's name. Unlike in Germany, which had known enough democracy and enlightened thinking for many to recognize the evil in Nazi rule, few in Japan ever questioned the Emperor—he was always good, right and infallible. No one tried to kill Hirohito during WWII unlike some Germans tried with Hitler. The Meiji Constitution had even declared the Emperor "sacred and inviolable" and "free from all worldly responsibilities."[304]

> To anyone reading the Japanese [press] of the period, it could easily have appeared that the whole of Japanese life centered on the slightest whim of the Emperor. Dire punishments threatened anyone who dared to suggest that everyone and everything in the Imperial realm was not subject to his divine judgment.[305]

Emperor Hirohito reviewing troops prior to the outbreak of war. Notice he was the only one on a white horse which he had named *Shirayuki* (White Snow). National Archives, College Park.

The superficially timid, bookish and soft-spoken Hirohito was an absolute ruler. Hirohito (*Shōwa* Emperor) was Emperor Meiji's first grandson, born on 29 April 1901.[306] At eleven, he became the crown prince and received the ranks of army second lieutenant and navy ensign.[307] In 1926, he became Emperor. The Japanese believed he was a god in human form (*arabitogami*) and the nation's "high Shinto Priest."[308] Zen Buddhist leaders believed Hirohito was the long-term "protector" of their religion and a "Gold Wheel-Turning King" (*konrin jōō*), "one of the four manifeststations of the ideal Buddhist monarch or *cakravartin-raja*." They believed Hirohito was a "*Tathagata* [fully enlightened being] of the secular world."[309] He was also head of the armed forces and "a real war leader," "exercising his constitutional prerogatives of supreme command" (*daigensui*).[310] He stood 5'5", weighed 150 pounds, wore glasses and carried a slight build. He did not look like a prototypical blood-thirsty dictator. Moreover, he was a passive, sensitive man who enjoyed marine biology, studying fungi and reading science books. Such subjects were unusual for a brutal dictator's curriculum. The "emperor, as a source of law, transcended the constitution, whose purpose was not to place limits on his powers, but the… opposite—to protect him and provide a mechanism enabling him to exercise authority unimpeded by limits."[311] In contrast to his demeanor, Hirohito often met with his commanders to dictate strategy. He ordered "his men to commit acts in violation of laws and customs of war."[312] Few WWII leaders had such control over the halls of power.[313] He was land driven, empire-hungry, resource-coveting and indifferent to the suffering of conquered peoples and his own citizens. He seized more

Emperor Hirohito in military dress uniform. Although he never trained as a soldier or officer, he was the highest military authority in the land during the war. He liked playing soldier. National Archives, College Park.

land, conquered more countries and killed more people than the Nazi dictator. Like Hitler, Hirohito rarely allowed his forces to retreat. It became a "hallmark" for soldiers to fight in place until they died. Hirohito's armed forces accomplished some amazing feats, but the military was marred by rape, atrocities, suicidal *Banzai*s and dramatic defeats. Hirohito ruled a Japan with an "army run amok, led by fanatics whose blind devotion to the emperor encouraged barbaric behavior."[314] To understand these facts, a short discourse on modern Japan is necessary.

The homogeneous island nation of Japan had lived in self-imposed isolation as a feudal state ruled by *shoguns* (hereditary Commanders-in-Chief) from the *Samurai* warrior class for two and a half centuries before Commodore Matthew Perry and the U.S. Navy's "Black Ships" forced it in 1854 to open up to world trade, thereby shattering "the 'closed door' policy of the Tokugawa government."[315] Social upheaval followed Perry's arrival. The Emperor had long been a cultural and religious figure, apart from and nominally superior to the warrior class.[316] For centuries, the *shoguns* held power until the advent of Western influence. Some Japanese leaders had heard of the American and French Revolutions and the expansion of colonial empires. But the Opium Wars of the 1840's and 1850's and the granting of extensive privileges for European powers in China shocked Japan's elite. "[T]o see China humiliated, to see its major ports occupied and administered by barbarians from the West made it obvious…that a great shadow had begun to fall over East Asia, one that Japan could not escape."[317] Japan became hyper-focused on ensuring Western powers would not do to it what they did to China.[318]

In 1868, the Meiji Restoration returned Japan to imperial rule and later adopted a constitution for the nation (Mutsuhito took the name Meiji on becoming Emperor in 1868). Unlike the U.S. Constitution, a group of elites drafted Japan's in secret and presented it to the people as a gift from the Emperor.[319] Meiji's constitution was done to impress the West which Japan aspired to join and prove it had modern institutions. It wanted respect from powerful countries, treatment as an equal and protection from Western imperialism. Yet, Japan's constitution was not meant to make her institutions viable or citizens' rights real. For instance, the constitution guaranteed religious freedom, a provision of great interest in the West. But in practice, regulations, law and social pressure suppressed non-Japanese faiths.[320] State Shinto observances, complete with priests, spirits, gods, shrines and myths, could not be questioned and were exercises of civic duty.

There were elections in which, at first, a small percentage of men could vote, but the military was not accountable to elected officials, answering only to the Emperor. Although male suffrage increased before the war, the military remained beyond the voters' control of their elected officials; in fact, the army often controlled elections and law issuance. New institutions altered to suit Japan were imported from the West, but Japan's elite chose not to adopt Western liberal institutions. Japan strongly strove to secure independence from other nations to avoid the risk of becoming one of Europe's dominions. Following Europe's example, Japan started building its empire by seizing Okinawa and other Ryukyu islands from China in 1871.[321]

In 1871, Japan needed a new focus to develop national feeling; to inculcate patriotism and transfer loyalties previously centered on family, region, or one's local lord to that of a national level like Europe was doing.[322] Glorification of the Emperor was the answer; the constitution declared Japan had been "ruled over and governed by a line of emperors unbroken for ages eternal" and the Emperor "symbolized the unity of the...people (*ue no zettai*)."[323] The nation suffered during the transition from a feudal state to a progressive society. It was plagued by civil wars pitting the modern military against conventional minded-*Samurai* who wanted to keep Japan "traditional." Thousands died as Japan was dragged "kicking and screaming" into the modern age.[324]

The country underwent rapid industrialization and urbanization during which time it created a large, modern, conscript military answerable only to the Emperor, considered a direct descendent of the Sun Goddess, *Amaterasu Omikami* (one of the most powerful Shinto gods). This was why their servicemen were called *Soldiers of the Sun*. The sun goddess's creeds and traditions stemmed from a pre-Bronze age culture that supported the belief that they were the special objects of the divine—our solar system's star, *the Sun*. The belief that the main physical celestial sphere that dominates the sky was the sole progenitor of the Japanese shows their conviction that Japan was the center of the universe and everything in life was arranged with them in mind. This pathetic solipsism created an extreme self-centeredness and self-regard that helped create, along with other fascist ideologies, a powerful and unforgiving citizenry willing and able to commit some of the most grotesque atrocities any modern power has ever performed precisely because they believed they had divine permission to do so. Their delusions justified their goal "to acquire and wield power over others." Their beliefs lent themselves to convictions that everyone non-Japanese was the gods' unfavored children. And during the 20th century, they were the only modern nation still ruled by a god who, by the very justification for his infallibility and power, espoused totalitarian convictions supported by a state requiring all to worship him.[325] These forms of "arrogance and servility, both of which flow directly from the power of the passions," were evident throughout Japanese culture and defined their beliefs.[326] To ensure these Shinto creeds took hold, between 1870 and 1884, 10,000 evangelists were "employed in a massive campaign to promulgate" this renewed focus on the Emperor and sacred country in a Mormon-like operation.[327]

The Emperor, in Shinto belief, was a divine intermediary between the people and the gods. From 1890, an Imperial rescript (*tokuhō*—like a Catholic *Encyclical*) required school children in assemblies to memorize and recite an oath to "offer yourselves courageously to the State" and "thus guard and maintain the prosperity of Our Imperial Throne coequal with heaven and earth" as well as affirm the Emperor's teachings, decisions, and decrees as infallible.[328] Shinto priests developed rites to pay homage to the rescript and the imperial photo in ceremonies. "Every aspect of life took on a military coloring. Boys went off to school in [army] uniforms…Instructors often drilled the children indoctrinating them with military values. Every day, boys would face a picture of the Emperor and be asked 'What is your dearest ambition?'" and then in unison, answered, "To die for the Emperor!" Seeing his school catch fire, a boy rushed in to save the Emperor's photo trying to prevent the flames from destroying god's image, but died in the blaze. Instead of viewing this act as a tragedy, it was hailed as a "happy manifestation" "of love for the Emperor." "These pathetic martyrdoms swelled the hearts of the people rather than depressed them." Children were taught "myths" as truth that the Emperor discussed decisions with the Shinto gods, "in pitch-black darkness," and then after receiving wisdom from on high, declared it to the nation. Japanese faith had become "nothing now but credulity and prejudices" and "absurd mysteries." This created unquestioned obedience to the Emperor as millions were intoxicated with such legends. The Emperor was recast as a military figure who appeared in uniform and was owed incontestable loyalty.[329] Admiral Kanji Katō, occupying "all the major educational and 'command' posts in the prewar" navy, wrote in 1933:

> Whence comes the strength of our army, that during the three thousand years since the founding of our country has protected us and has not yielded an inch of land to the enemy? 'For generations, the Emperor has led our Imperial Forces.'…It is the pride of

the Imperial Forces to submit to…the Emperor…around whom everything centers…The soldier dies for the Emperor…When he is about to die, the last words that break forth from his lips are '*Tennō Heika, Banzai!*' (Our Emperor, lives 10,000 years). We die gladly. We are happy to float in his spiritual grace…Under the [Emperor's] supreme command there is no limit to our power.[330]

Everything in society took on a religious tent-revival-like atmosphere. The military focused on the Emperor as the "foundation of our national strength to explain" Japan's "invincibility, equaled by no country in the world."[331] Unquestioning allegiance to the Emperor instilled during military training permeated into civilian life when soldiers left active duty. It was a vicious circle teaching totalitarian, xenophobic and exclusionary "virtues" as being the highest forms of behavior to unify the country using Shinto beliefs as a foundation.[332] Although many would also call themselves Zen Buddhists, to be Japanese was by default to also be Shintoist.[333] "State Shinto, the government's artificial construct, was purposely designed as a cult of national morality and patriotism, to which followers of all religions [Buddhists, Taoists, Christians and Confucists] must subscribe."[334] And these Shintoist beliefs created a fascist and imperialistic nation. By WWII, no Japanese dared question the Emperor.[335] Hirohito enjoyed maintaining his people "in a condition of political tutelage and immaturity."[336] The government's propaganda was grounded in elite religious teachings and created an exaggerated metaphysical and epistemological solipsism making Japanese feel they were the best humanity had to offer led by the best the gods could create, the Emperor. As Japan modernized, these beliefs expressed themselves in armed aggression. Having one dominant sect did not lead to a culture of moral restraint and thus, Japan's society embraced violence as an acceptable means of achieving its theological mandate to rule the world.[337] Thus, the political beliefs of Japan were not liberated from religion, but rather, were theocratic and not subordinate to a secular state like in the U.S.[338]

One would think Buddhism would have prevented Japan from becoming fascist, but that was not so. Zen Buddhism at this time also took on Shinto-like characteristics and made itself subservient to the Emperor, nation and military. Although today, most Buddhists would call themselves pacifists, under fascist

Japan, their beliefs inverted themselves.[339] For Zen Buddhists, "warfare and killing were described as manifestations of Buddhist compassion." Historian and Buddhist Priest Brian Victoria continued: "The 'selflessness' of Zen meant absolute and unquestioning submission to the will and dictates of the Emperor. And the purpose of religion was to preserve the state and punish any country or person who dared interfere with its right of self-aggrandizement."[340] As Japan continued to become more imperialistic and fascistic, it had its two major religions of Buddhism and Shintoism in its back pocket. Buddhism "was indeed one, if not the only, organization capable of offering effective resistance to state policy," but instead, it offered its unquestioning loyalty.[341] Zen Buddhism became fascist claiming it was the best Buddhism:

> Buddhism in India collapsed due to…Indian culture, Buddhism in China collapsed because it ran directly contrary to the history of nature of the Chinese state, and was therefore only able to produce a few mountain temples. On the other hand, thanks to the rich cultivation Japanese Buddhism received on Japanese soil, it gradually developed into…[what] Buddhist teaching was aiming toward.[342]

Many Japanese believed Japan was "the only Buddhist country" in existence because Buddhism "didn't simply spread to Japan but was actually created there."[343] And spreading Japanese Buddhism by the sword, killing "unruly heathens" (*jama gedō*), was supported by Zen Buddhists: "discharging one's duty to the state on the battlefield is a religious act."[344] "Buddhism doesn't merely approve of wars that are in accord with its values; it vigorously supports such wars to the point of being a war enthusiast."[345] And since Japanese "religion", a synthesis of Shinto and Buddhist ideologies, was martial to its core (*Bushido*—"the Way of the Warrior"),[346] it was just a matter of time before Japan's brainwashed legions swarmed out beyond the nation's borders to enact *their* "God-given right" to rule everyone.

Eager to use the modernized military into which much effort and treasure had been expended, Japan conducted actions against China, Korea and the Russian Empire, gaining territory in Manchuria, Korea and Sakhalin. In the 1894-95 First Sino-Japanese War, Japan crushed China and gained Taiwan/Formosa.[347] As would be on display in WWII, the Japanese committed crimes everywhere. Some IJA soldiers bayoneted children through their "anuses [and] then [held them up in the air]…for fun."[348] Allying with Great Britain, Japan obtained a modern fleet which included European warships the British taught the Japanese to operate. Japan with its military of "leather-skinned dwarves" as the Russians called them, shocked the world in the Russo-Japanese War with a naval victory over the Tsar's navy at Tsushima (the strait between Japan and Korea in the Sea of Japan) in May 1905 in the first major sea battle with steam-powered vessels.[349] With Krupp siege guns, the Japanese military also defeated Russia at its Port Arthur base on the Manchurian coast. It was the first modern victory of an Asian country over a Western power, boosting the confidence of Japanese militarists.[350] The alliance with Britain in WWI gave Japan an opportunity to seize Germany's eastern empire, including islands in the Carolines, Marshalls and northern Marianas, and concessions in northern China. Its new imperialistic, totalitarian system was paying dividends.

This Japanese expansion worried the future Chairman of the Joint Chiefs-of-Staff, Fleet Admiral William D. Leahy, and many in the United States Marine Corps. As a Lieutenant Commander and the navigator on the armored cruiser *USS California*, Leahy had a chance to visit Japan in 1910. Ever since Japan won its war against Russia in 1905, Leahy had been following its society and military viewing it as a future enemy. Leahy would write after his time in country and after meeting many IJN officers and even the victor of Tsushima, Admiral Heihachirō Tōgō himself, that Japan was ruled "by a people with a high order of military talent and an intense nationalism, it will be a menace to our institutions."[351] Having such exposure to Japan's militaristic and aggressive society influenced Leahy greatly in the 1930s and 1940s as he, more than any other officer in America, shaped the United States' strategic handling and diplomatic thought concerning the enemy of Japan.[352]

Many Marines, like Leahy, long before World War II started, also viewed Japan as America's nemesis and took action to make sure Japan was part of the United States Marine Corps' war doctrine viewing this nation as a serious threat. After WWI, the brilliant and prophetic USMC Major Earl H. Ellis thought about what country in the Pacific could challenge America and her democracy. He saw what was going on in Japan and quickly warned his superiors of the danger of a future war with the Empire of Japan writing up a divinatory study entitled *Advanced Base Operations in Micronesia* (1920-21). With the support of his superiors, Ellis spent years analyzing Japan, the Pacific and Asia and amphibious warfare to prepare for the possibility of future hostilities. His study and plans convinced some of those officers higher up his chain of command, like Commandant John Archer Lejeune, to focus on amphibious warfare for the Pacific helping the Marine Corps prepare for WWII. In fact, his materials and study provided "the framework for the American strategy for a Pacific war, adopted by the Joint Board of the Army and Navy in 1924 as the 'Orange Plan.'" He mysteriously died in 1923 on Japanese-held Palau, many thinking the Japanese government had sent an assassin to do away with the Marine Corps' Messiah who had seen, correctly, the future that Japan wanted and from which America needed to defend itself.[353]

Leahy and Ellis were right to fear the Japanese as their society turned more radical throughout the early 1900s. In 1905, an era of political violence commenced and even more fascist ideology appeared in Japan.[354] A die-hard Shinto ideologist was a confidante of the Empress and lectured her son, Hirohito, who was acting as regent for his insane father.[355] The lectures which influenced his thought were published as imperial edicts upon Hirohito's coronation in 1926, thereby guiding Japan's political climate which focused mainly on military might.[356]

By the 1920s, Japan was the only state in Asia with a "modern army, navy and air force—in fact among the best in the world."[357] In 1922, it built the first aircraft carrier *Hōshō*. In 1927, it deployed troops to northern China in Shandong Province and then started a full-scale war in 1931 in Manchuria, using it as its *Lebensraum* sending hundreds of thousands of colonists there to populate it. It resigned from the League of Nations in March 1933 at about the same time as Nazi Germany did.[358] Not content with these Imperial conquests, in July 1937 Japan launched its war against China taking over the equivalent of half the U.S. in land mass.[359] It started with the Marco Polo Bridge incident outside Beijing on 7 July when Japanese and Chinese troops clashed with one another. Japan used this event to commence war primarily against Chiang Kai-Shek's Nationalist forces (*Kuomintang*) with Hirohito's blessing ordering his men to "chastise Chink forces."[360] Nationalist forces were Japan's main enemy on China's mainland (trained and supplied by Germany from 1934 to 1938 and then the U.S. thereafter).[361] Concurrently, China was undergoing a major Civil War with Chiang Kai-Shek's armies battling Mao Zedong's Communist forces to the north and both these groups fighting numerous warlords throughout the country.[362] It was a Civil War within a Civil War, fighting in different forms the outside aggressor of Japan when they were not killing each other. China was a chaotic land unable to expel the Japanese invaders largely due to the nation's inability to unite in a common cause under one leader. Furthermore, Japan saw China's instability as a threat to her protection against Communist Russia and the Allies, as well as an opportunity to take advantage of China's instability to seize territory and resources. In 1939, Prime Minister Hiranuma addressed the Japanese legislative assembly, the Diet, claiming he hoped the Chinese would co-operate with them—that is, become Japan's slaves. He continued, "As for those who fail to understand we have no other alternative but to exterminate them."[363] Japan conducted this war with such "ruthlessness and brutality," it shocked the "great powers" of the day.[364] Even Germany's Supreme Army Command in 1938, after observing Japan's combat, wrote that the IJA's bravery, loyalty to the fatherland, and lack of fear for death baffled it and could not be "measured with European sensibilities."[365] That Japanese soldiers made the Nazi army stand back in awe for its brutality makes one take note of IJA extremism born from "a fanatical nation at war."[366]

When the breakdown of their ritualistic and strict environment occurred as the Japanese invaded other countries, the "collapse of the subliminal restraints against aggressive impulses [happened] with catastrophic consequences. Frustration...pent up by the regimented rituals of social interaction within Japan exploded with...savagery against enemies abroad."[367] For decades, cadets were taught, "above all, the ethics of the warrior, the cult of *bushido*"[368] which preached a fanatical dedication to fulfilling orders and destroying the enemy. By WWII, warrior codes had changed into religious death cults. Army regulations of 1912 first laid out the Spirit Warrior doctrine. The Imperial rescript to servicemen required absolute, unthinking obedience to orders. Because officers had a direct connection to the Emperor, their orders came with divine authority making them infallible. Unlike U.S. servicemen, the Japanese had no concept of unlawful orders.[369] The IJA taught that to truly live, one must be willing to die for the Emperor and nation. If death came, it should be welcomed. Fighter ace Sub-Lieutenant Saburō Sakai claimed they lived by *Hagakure* (a popular *Bushido* text); to "live in such a way...[to] always be prepared to die."[370] In short, more was "expected of the Japanese soldier than any soldier" of WWII.[371] Their collective "valor [was] on a scale never surpassed."[372]

In the city of Newchwang, Manchuria on 13 Nov. 1933, Japanese had executed many "bandits" in public (notice the two slaughtered children in the middle of the photo).
National Archives, College Park.

One reason for this was that for decades, Japanese preached their "Aryan-like" myth—they, the *Yamato* race (the Sun clan), were the Sun Goddess' children and better than everyone else. Pseudo-scientist Hashimoto wrote, "there's no nation that can compare with our national blood solidarity which makes possible a unification like ours with the Emperor in the centre."[373] The Japanese preached their master race theory in schools and the government for decades, brainwashing generations. State-run Shinto religion inundated society and preached the "superiority of the Japanese as a race [*Nihonjin*], their divine mission in the world, and their supreme devotion to the emperor...Japan was pictured as an invincible nation." Schools taught that the "imperial throne was as ancient as the very origins of heaven and earth...The state supported financially this intensely nationalistic cult...[which produced] the blind loyalty and devotion Japan's leaders required."[374] Politician Nakajima Chikuhei claimed in 1940 it was their duty to control others since they "were pure-blooded [and] had descended from the gods. The Greater East Asia War was thus no ordinary conflict but rather a divine mission."[375] Leaders like War Minister General Sadao Araki (1931-34) devoted themselves to "the spiritual side of warfare" claiming the divine mission to rule made them superior and assured victory.[376] Lieutenant Colonel Gorō Sugimoto, a Buddhist philosopher, wrote the bestselling book *Great Duty* (*Taigi*) that recruits, officers and school children admired.[377] In part, he wrote what many believed, combining Shinto, Buddhist and fascist convictions into one ideology:

> The Emperor is identical to the...[Sun] Goddess *Amaterasu*. He is the...only God of the universe...[T]he many components [of a country] including such things as its laws and constitution, its religion, ethics, learning, and art, are expedient means by which to promote unity with the Emperor....Stop such foolishness as respecting Confucius, revering Christ, or believing in [Buddha]! Believe in the Emperor, the embodiment of Supreme Truth, the one God of the universe! Revere the Emperor for all eternity! Imperial subjects...should not seek their own personal salvation. Rather their goal should be the expansion of Imperial power... The Imperial way is...the fundamental principle for the guidance of the world...The Emperor's way is what has been taught by all the saints...The wars of the Empire are...holy...They are the [Buddhist] practice (*gyō*) of great compassion (*daijihshin*). Therefore the Imperial military must consist of holy officers and...soldiers....Warriors who sacrifice their lives for the Emperor will not die. They will live forever. Truly, they should be called gods and Buddhas for whom there is no life or death.[378]

Sugimoto preached a xenophobic, radical, religiously racist, Japanese-centric ideology that was common among IJA personnel in particular and *Nippon's* society in general. Sugimoto believed by surrendering oneself to the Emperor, a soldier could secure salvation, eternal life, deification and supreme teacherhood especially if he died for the Empire's expansion and preservation. As is often the case with religious gibberish, Sugimoto unknowingly illustrated the contradictions inherent in the dogma when he wrote that people should only believe in the Emperor, not Buddha, but later pronounced they should aspire to become gods and Buddhas. He declared Hirohito the one true "God of the universe," being equal to the Sun Goddess although by definition, since Shinto ideology claims *Amaterasu* gave birth to Hirohito's line, the Imperial household was in an inferior position. Yet, apparently Sugimoto was convincing with his illogical arguments, something one often sees with charismatic men. When one reads Hitler's *Mein Kampf* or listens to his speeches, one also reads about pseudo-religious allusions and illogical concepts (he claimed Jesus was "Aryan" and a fighter against Jews, who wanted the *Volk* to rule all mankind).[379] The world was rife with fanatical visions of what a future humanity should look like and what gods or religiously xenophobic beliefs a society should embrace. Sugimoto spent his life on campaigns for the Emperor entering the IJA in 1921. When the Second Sino-Japanese war broke out, he died in battle on 14 September 1937 in Shanxi Province. In his conviction, he became a god and Buddha, gazing toward the Imperial throne as he took his last breath.[380]

The hyperbolic language espousing the destiny of omnipresent rule as embraced by leaders like Hashimoto, Chikuhei, Araki and Sugimoto, buoyed the Japanese national self-concept, yet there was a yawning gap between their religious pretensions, their historical claims and their elect status on the one hand, and the reality that they had only a fraction of the resources possessed by their enemies.[381] As a tiny minority in the world of nations with an island land poor in raw materials that had just gone through an industrial revolution, they lagged economically behind other "inferior" nations. Able, industrious, martial and religious, Japanese felt frustrated by their manifest inability to solve their political problems and to control Asia. They were reliant on more powerful Allied countries who could cut off commodities, especially oil, without notice, crippling their economy. Was there not something monstrously wrong with a nation and its religion of Shinto and Zen Buddhism that claimed its values, people, gods and society were the best humanity had to offer, yet the world around them was disproving these claims with the U.S., Britain and Germany being far superior in empire building than the Nipponese?

When exposed to the outside world, many Japanese developed Napoleon complexes and "race inferior" hang-ups because men in Western nations were, on average, five to six inches taller than the average Japanese and appeared to have the respect of non-White nations throughout the globe because they were White and more powerful. Superior Western technology in cities and in arms also played a role in Japanese feelings of inadequacy. Japanese must have wondered why, if they were the gods' chosen people, were they so much smaller and physically and technologically weaker than other ethnicities, especially Caucasians. Japan had a social structure that had a "tendency to exclude

TOJO WANNA CRACKER ?

Anti-Japanese U.S. propaganda depicting Prime Minister Tojo as a small, monkey-like human. Depictions of the Japanese as small and weak was a common theme in most political posters at this time. National Archives, College Park

and denigrate others," which is usually a sign of deep insecurities.[382] Men of smaller stature can have the "small man syndrome" which makes them prone "to seek power, war and conquest to make up for their physical shortcomings."[383] Hitler (5'8"), Goebbels (5'5"), Mussolini (5'7") and Stalin (5'5") all had this "syndrome;" and most likely many small Japanese leaders did too, including specifically Hirohito (5'5"), Tojo (5'4") and Yamamoto (5'3").[384]

Japanese culture was also fascinated with "White people (fair skinned)" and desired being like them.[385] Worse still, even if there were individuals among the body politic who did not necessarily feel physically inferior with their height and pigmentation, the entire Western world depicted Japanese as "little yellow men" and "monkeys" or, as Rudyard Kipling claimed, "little anatomies," rather than the future world leaders to which Japan aspired irritating many Japanese diplomats, politicians and flag officers. They hungered to be like the Europeans and Americans who were more powerful physically, politically and militarily, but they also despised them for their dominance in Asia and their ridicule for Asian races.[386] If the Japanese were perfect, why did they feel these shortcomings or why did the world view them as inferior? To what extent was the disjunction between Japanese aspiration and performance the responsibility of a fallacious religious analysis? The disconnect between their cult-like indoctrination and their personal views kept the Japanese world in a perpetual belligerence that would later erupt into a global conflict.

In some ways Japanese society, Shintoism and Zen Buddhism were highly unstable, and certainly self-destructive. Japan's history, myths and religion had huge gaps and plenty of plagiarisms. It borrowed many of its beliefs from China and India, especially Buddhism, Taoism and Confucianism, and adapted the conviction in the Divine Origin of the Emperor from the political theories of China's Chou kings (1100 to 300 B.C.E). It took its "system of...writing using Chinese-derived characters" or ideographs (*Kanji*) and much of its legal theory from China, and its ethnic origins were largely Mongolian, Chinese and Korean.[387] China and Korea were Japan's forebears genetically, culturally, spiritually and linguistically.[388] Like many cultures with fanatical beliefs about a nation's origins being "pure," the Japanese exaggerated their beginnings and destiny. The more extreme Japan became in its mission to rule *everyone and everywhere* because they were superior, the more extraordinary, fanciful, mythical and fallacious its claims became about its people and their role in the world.[389] In 1934, Ryūzan Shimizu, president of Nichiren Buddhist affiliated Risshō University explained Japan's mission: "The underlying principle of the spirit of Japan is the enlightenment of the world with truth...we [must] lead all nations of the world into righteousness and establish heaven on earth...where all men shall be Buddhist saints. This is the true ideal of the spirit of Japan."[390] Zen Master Sōgaku Daiun Harada wrote in 1934:

> The spirit of Japan is the Great Way of the [Shinto] gods. It is the
> substance of the universe, the essence of Truth. The Japanese peo-
> ple are a chosen people whose mission is to control the world...

> fascist politics should be implemented…All of the people of this
> country should do Zen. That is to say, they should all awake to
> the Great Way of the Gods. This is Mahayana Zen.[391]

Despite the logical inconsistencies of their efforts, they nonetheless strove to elucidate and ultimately create that new world order. Instead of using argument, education, and intelligence to convince the world of their superiority, the Japanese used brute force.

In implementing this mission of ruling the world and training this "super race," the military became a key institution, brutalizing its men with abuse and draconian punishments, instilling belief in whatever the nation ordained. Leaders beat their subordinates, and they in turn harmed those under them. They were not allowed to question, criticize or strike back against superiors.[392] Fighter ace Sakai wrote, "We were automatons who obeyed without thinking."[393] The practice of "harsh discipline" was known as *tekken seisai* (iron fist) and *ai-no-muchi* (whip of love).[394] This bully-like behavior took the form of blood, broken bones and incredible humiliation for subordinates and much worse for enemy soldiers and civilians. The "loyalty" Japan's leaders required was taught through militarism, racism and unquestioning obedience by strict press censorship, religious indoctrination, and intimidation.[395] This training created bizarre behavior; by way of illustration, when an IJA officer "made a slip of the tongue" while reading an Emperor rescript to his troops, he fell on his sword. He made a mistake and instead of people declaring this a pathetic display of loyalty, they praised him for it.[396] Although the Japanese were taught they were exceptional, their military superiors treated them as inferior. As often the case with cultures that try to make their citizenry more than they are capable of becoming, Japan went to extremes to create an enemy and demonize him to keep control over its citizens while trying to also make them feel superior at the same time.

Japanese fascism preached that everything Japanese was superior and anything non-Japanese barbaric. *Gaizin* (or non-Japanese) were viewed as non-human. For decades, Japanese citizens had been indoctrinated with "evidence" that Japan's enemies, like Caucasians and Chinese "were inferior, even subhuman creatures for whom no respect was possible."[397] Uno Shintarō, who beheaded prisoners, wrote, "we never really considered the Chinese humans. When you're winning, the losers look…miserable. We concluded that the Yamato race was superior."[398] George Orwell wrote that for centuries, the Japanese preached "a racial theory…more extreme" than the Nazis, claiming "their own race to be divine and all others hereditarily inferior."[399] James Young, a former Japanese prisoner, wrote an insightful 1940 book *Our Enemy* in which he supported Orwell's conclusions, writing about the Nanking massacre: "[The Japanese] duplicat[ed] all Nazi brutalities, exceed[ed] them, and add[ed] a whole satanic range of their own peculiar cruelties."[400] Japanese also disparaged foreign religions, viewing their adherents as heathens.[401]

This superiority dogma created an interesting dilemma for Fascist Japan—it had to compete with its allies for the claim of genetic, eugenic and cultural greatness. Germans claimed Aryan superiority which by definition placed the Japanese in an inferior position. Yet Japan also had a master race theory which put the Germans in a lower position—there can't be two chosen people: The *Yamato* race was not the *Aryan* race. These race theories were irreconcilable, or as a Japanese journalist noted, were "like trying to mix oil and water."[402] To get around this problem with their Asian ally, the Nazis devised a plan. Justifying the alliance with the Japanese, who fit into the Nazi category of "Asiatic barbarians" and whom Hitler had degraded in *Mein Kampf,* called for a creative rationalization (Japanese translations edited these sections out of the *Führer*'s work).[403] Hitler's Armament Minister Albert Speer called Germany's alliance with Japan "from the racist point of view a dubious affair."[404] Party members disapproved of the association with "barbarian" midgets, and the Japanese did not appreciate being called inferior "non-Aryans." Particularly because of the importance of the 1936 Anti-Comintern Pact, creating a unified front against Communism, Hitler needed to ease the racial tension both internally and externally. Consequently, Hitler "officially" labeled the Japanese "honorary Aryans" because they possessed Germanic qualities calling them the "Prussians of the East." Yet in private, he felt they posed a danger to Whites. Interestingly, the Japanese continued to call Germans *Gaizin*, but the mental and verbal gymnastics seemed to allow Japan to save face and for Germany to justify its alliance with "inferior" humans.[405] When the Tripartite Pact of 27 September 1940 was signed, Hitler gave a dig at the Japanese in the preamble that ultimately the "strongest" race (i.e. Aryans) will survive the current global conflict. In response, Prime Minister Konoye,

Foreign Minister Matsuoka and War Minister Tojo skirted the issue Hitler had raised saying if Japan should fail in "her 'grand mission of spreading the Imperial Way,'" then she did not deserve to exist.[406] In the end, these racist regimes demonized those not allied with them as inferior and viewed the other Axis nations as racially acceptable at least in official documents for public consumption. And while Germany primarily focused on the Jews as being the worst of humanity, Japan labeled anything non-Japanese as unworthy of life and proved it with mass murder everywhere it conquered.

Japanese soldiers bury Chinese citizens alive in Nanking, December 1937. They usually made the victims dig their own graves first before they threw the dirt on top of them suffocating them to death. Nanking Massacre Memorial.

Consequently, as a nation, the Japanese accepted killing the "other," and thought of world conquest as a given, thus removing any obstacles to committing atrocities. Japanese treated conquered peoples as expendable resources used for

Japan's benefit without regard for their welfare, slaughtering millions. Thus, this study will call this slaughter "Japan's Holocaust."[407] Japan's culture conditioned citizens for mass slaughter. For instance, in the 1930s in one school, a boy cried about dissecting a frog. His teacher hit him asking, "Why are you crying about one lousy frog? When you grow up you'll have to kill a hundred, two hundred Chinks."[408] The government, military, and schools inculcated the populace with the message that Japan would rule everywhere declaring *Hakko Ichiu* ("eight corners of the world under one roof").[409] Their minds were

Chinese poster depicting the defense of the motherland against Japanese aggression. The Chinese often depicted the invading Japanese as murdering demons with pig faces. In bold Chinese characters, it is written: "Everyone Fight Back!!!" Shanghai Historical Museum

"stuffed with martial dreams of grandeur [and a] political philosophy [that was] a pungent admixture of National Socialism, Fascism and medieval superstition."[410] And claiming *Hakko Ichiu* showed that the Japanese wanted all to adhere to their beliefs about Hirohito and the *Yamato* race. Japan failed to realize that "to force men to say they believe what they do not, is vain, absurd, and without honour."[411] However, most Japanese did not care that such measures they enforced on others were immoral so they embraced a movement that imposed their will, culture and religion on nations they ruled. Their Shinto and Buddhist beliefs embraced atrocities. Buddhists under Hirohito claimed their belief "does not reject the killing of masses of people that takes place in war, for it sees such warfare as an inevitable part of creating an ever stronger and more sublime compassion."[412] Such was the reality of Japan's religious climate.

## Japanese Xenophobic and Imperialistic Goals in Practice

From 1927 until mid-1942, Japan conquered areas throughout Asia and the Pacific and Indian Oceans in its "reckless expansionist policy."[413] Years prior to Pearl Harbor, Japan had embarked on a program of aggression taking Formosa (Taiwan) in 1895, Korea in 1910, German islands and concessions in China in WWI from 1914 to 1918, small parts of northern China in 1927,[414] Manchuria (*Manchukuo*) in 1931,[415] and large sections of eastern China (beginning in the fall of 1937).[416] In September 1940, it allied with Nazi Germany and Fascist Italy in the Tripartite Pact.[417] Hirohito wrote that these countries benefited Japan because they "share the same intentions as ourselves."[418] From defenseless Vichy France, Hirohito's Japan wrested control of Indochina (1940-41) edging closer to the Dutch East Indies' oil reserves.[419] In December 1941, emboldened by his territorial conquests, Hirohito—as the Armed Forces' Commander-in-Chief—took his aggression to another level, by going to war against Great Britain, the Netherlands, the United States, Australia and New Zealand. The destruction of America's Pacific battleships left Japan virtually free to pursue the conquest of other Asian and Pacific islands. For months, America received only bad news from the sole front on which her troops were engaged as Japan's military enjoyed one success after another. Not only was America's Pacific fleet licking its wounds, but the British navy also suffered a grievous loss when Japanese planes sank the battleship *HMS Prince of Wales* and battlecruiser *HMS Repulse* off Singapore's coast on 10 December taking 840 sailors to the bottom of the sea. Churchill, now feeling what Roosevelt had a few days before, said,

> In all the war, I never received a more direct shock [with the sinkings of the *Prince of Wales* and *Repulse*]…As I turned over and twisted in bed the full horror of the news sank in…There were no British or American ships in the Indian Ocean or the Pacific except the American survivors of Pearl Harbor, who were hastening back to California. Across the vast expanse of waters, Japan was supreme, and we everywhere were weak and naked.[420]

Japanese soldiers hung up the heads they had cut off of "bandits" in Kharbine, Manchuria, 19 November 1937. 19 November 1937 (the Chinese characters on the sign translate as giving travelers directions to the "Airline Association"-- more evidence that the heads were displayed on a large street sign and very public place near the airport). National Archives, College Park

While there were months of tremendous U.S. and Filipino heroism in defense of the Philippines and successful missions of the Flying Tigers (American fighter pilots provided to China), these were rare "success" stories in the early months of the war. America was unprepared for the conflict thrust upon her on both sides of the world, and at first, "the chances for an Allied victory" seemed remote.[421] During the first six months after the surprise attack on the U.S., Japan won a series of victories over unprepared British, Dutch, Australian and American forces, giving it control over an enormous area newly acquired or already held, including the Philippines, Guam, Burma, Malaysia, Singapore, Hong Kong and the Dutch East Indies (Indonesia). Japanese victories extended into parts of Thailand, New Guinea and numerous Pacific islands. With the invasion and occupation of these objectives, which were ripe for exploitation, Japan showed her true *casus belli*.

Hirohito helped plan these aggressive acts and got "caught up in the fever of territorial expansion and war."[422] Japan dealt harshly with anybody who tried to resist its expansion. Often in the areas Japan took over, one witnessed the decapitated heads of those who had resisted *Nippon*'s invasion. The "Japs" hung severed skulls from telephone poles, stuck them on fence lines or placed them on roads for all to see what happened to those who refused their rule.[423]

A few weeks after Pearl Harbor, the Japanese were knocking on Australia's doors and bombing its northern city of Darwin, which had little defense because most of the country's troops were fighting Field Marshal Erwin Rommel's *Afrika Korps* in North Africa. "In the early days of the War, the Japanese were an invincible force in the Pacific, moving swiftly and easily from victory to victory—and island to island."[424] Hirohito was one of the driving forces for Japan's aggression and his empire "dwarfed Hitler's" in size. "The Rising Sun was blinding" and anyone who looked at an atlas in 1942 concluded the Allies were losing everywhere and that the Japanese reigned supreme in Asia.[425] It "appeared that [Japanese] policies were also being blessed by the Shinto gods" with their conquests.[426] They had seized territories amounting to one seventh of the earth's circumference (three times larger than the U.S. and Europe combined) in developing their "Co-Prosperity Sphere," with their "super race" ruling 500 million people.[427] Lieutenant General Kuribayashi's grandson Yoshitaka Shindo, a politician in Japan's Liberal Democratic Party (a right-wing, nationalistic, conservative party actually), observed:

> Just like America seized lands from the Indians and Mexicans or ruling people like in Guam and the Philippines, we were building a nation. It was our Manifest Destiny. It was our time to spread our borders. To prevent colonization of ourselves could only be done, we felt, by spreading our physical influence throughout Asia. Although the Japan of that time did this with the wrong means by killing people and without the freedom of speech and democracy. I...tell you this to help you understand why the Japanese felt the way they did back then.[428]

Japan's Pacific War destroyed "White" colonial rule in Asia under the guise of Pan-Asianism. In reality, Japan was supreme ruler.[429] The western Pacific had "become a Japanese lake."[430] For months after Pearl Harbor, Prime Minister and General Hideki Tojo (appointed by Hirohito)[431] could stand behind his declaration to citizens given on 8 December 1941:

> The key to victory lies in a 'faith in victory.' For 2600 years since it was founded, our Empire has never known a defeat. This record alone is enough to produce a conviction in our ability to crush any enemy no matter how strong. Let us pledge ourselves that we will never stain our glorious history, but will go forward...[432]

Tojo was not slowed by the facts that to "crush" America, he needed more aircraft carriers and field armies than he and his Axis allies had. The best he could hope for was to "deter or convince" the U.S. from further fighting in Asia and the Pacific, and that a Nazi regime would emerge victorious over Britain and the U.S.S.R, leaving America isolated and preoccupied with survival. In a twist of logic, once Tojo, Hirohito and others started war, an "irrational belief in victory...became the strategy."[433] Issues about securing victory were removed from most civilians' consciousness and they felt pride at Japan's success.[434] The German naval attaché in Tokyo, Rear Admiral Paul Wenneker, recounted how his contact Captain Nakamura depicted the war with the U.S. as ushering the empire into its greatest epoch and that both nations should work together to achieve victory.[435]

The risky, surprise attack on Pearl Harbor was made without a follow-up plan.[436] In some respects, it was conducted in response to the Nazi successes throughout Europe as well as Hitler's current invasion of Russia and his battle against Britain, all of which motivated the Japanese to act quickly in Asia before the spoils of war went to others.[437] Newsreels of Hitler's successes dating to 1939 played throughout Japan and "reinforce[ed] this sense of urgency."[438] British ambassador to Japan Sir Robert Craigie, who detested the militaristic Japanese and felt nauseated by having to continue to be "polite to the little blighters," wrote, "How, they urged, could Japan expect Hitler to divide the spoils with them unless she had been actively associated in the spoliation."[439] Foreign Minister Yōsuke Matsuoka verified Craigie's observation claiming, "When Germany wipes out the Soviet Union, we can't simply share in the spoils of victory unless we've done something...We must either shed our blood or embark on diplomacy. And it's better to shed blood."[440]

Japanese war hawks also wanted more dividends on the investment of the 400,000 men it had sacrificed in China, an adventure "proven to be a disaster," and secure their Asian gains by knocking the Americans back into the Western Hemisphere. Furthermore, the oil embargo the U.S. enforced in July/August 1941 against Japan could cripple the Empire's economy, motivating it to attack America and seize the oil-rich Dutch East Indies. Regrettably, U.S. Secretary of State, Cordell Hull, at the same time told the Japanese, whom he despised for their false pretenses of politeness, smiles, bowing and "hissing," that if they wanted peace they "would have to abandon the Tripartite Pact" and remove their troops from several areas in China. These demands as well as the restrictions of the sale of petroleum to Japan were unacceptable to Hirohito's government. While we don't know the most important motivation for the attack, some or all of these reasons prompted the Japanese to stealthily move an armada from Tokyo to Hawaii and to smash the U.S. fleet on a Sunday morning when America's guard was down.[441] The Japanese had failed to learn, according to military philosopher Carl von Clausewitz that, "an enemy who violates a frontier will be made to pay a penalty in blood."[442] It was already paying a high price "in blood" in China fighting over a 2,800 mile front against three million Chinese troops,[443] and now it dramatically increased that debt "in blood" by violating the Allies' frontiers. Just as Hitler was paying a high price for invading Russia while also trying to destroy England, now Japan left its rear unsecured while it opened a gigantic front to the east against one of the most powerful countries in the world. While Japan launched its attacks against the West, Chinese Nationalists had completed training of "another million of [its] soldiers to replenish the 'wall of steel and flesh'" which had been "checking and throwing back Japan's repeated attacks."[444] If Japan could not defeat a rag-tag group of Chinese soldiers with inferior weapons and supplies, how was it going to tackle the first-world nations of the U.S., Great Britain and her Commonwealth territories?[445] Moreover, if they waged a war against China "without just cause or cogent reason," why in the world open another front with the Allies with no definitive end goal defined or military means to destroy the newly created hostile forces?[446] The irrational levels of action the Japanese government took its nation to were incomprehensible. Their fanaticism had blinded them to *Realpolitik*.

The photograph claims it is of Chunking, China showing civilians killed when a Japanese bomb hit their shelter. However, since many of the women are nude and on stairs way above ground, it is felt this is of Nanking and shows evidence of rape and murder. December 1937. National Archives, College Park.

Nonetheless, the Japanese felt confident about the future. After Pearl Harbor, Hirohito issued the Declaration of War as an Imperial rescript. It blamed the U.S. and England for causing war by supporting China and undermining Japan's efforts to create an Asian Co-Prosperity Sphere. In other words, the war started because the Allies had disturbed "the peace in East Asia" by spreading their unbridled ambition throughout the Orient. The Emperor wanted peace, but the Allies forced him to strike first. "Our Empire, for its existence and self-defense, has no other recourse but to appeal to arms and to crush every obstacle in its path."[447] Soon overwhelming military successes greeted Japan, thereby reinforcing its citizens' belief they were the gods' chosen people. On the day of the attack, Hirohito wore his naval uniform and was in a "splendid mood," gloating over the victories. Hirohito praised frontline units and sent rescripts to commanders "which carried far more honor and prestige than did presidential citations for American commanders."[448] By 12 December, Hirohito thanked "the gods" for Japan's victories and "asked for their protection" as he led the "nation in this time of…national emergency."[449]

Japanese bombing of South Station in Shanghai on 28 August 1937. H. S. "Newsreel" Wong took this photograph of the baby. His mother lay dead nearby. The baby was one of the only survivors of the attack on the station. National Archives, College Park.

Instead of prosperity, Japan's conquests brought hardship, disease, starvation, slavery and slaughter to enemy soldiers, prisoners and civilians, as had already been proven with Japan's actions in Manchuria and China throughout the 1930s. Rape, brutality and torture were encouraged and publicized,[450] becoming common occurrences when a Japanese army was in control. This was illustrated with the "Rape of Nanking" where from July 1937 to March 1938 the Japanese slaughtered around 300,000 Chinese and conducted a minimum of 20,000 rapes fighting from Shanghai to Nanking.[451]

The majority of the slaughter happened after the Chinese army of 150,000 (the Nanking

A Japanese sailor stands proudly with the head of a Chinese man he has just cut off. National Archives, College Park.

Garrison Force) disintegrated under the better equipped and led Japanese by mid-December. The majority of Chinese soldiers had laid down their arms (approximately 90,000), 10,000 had died in battle defending Nanking and 47,000 had escaped west.[452] Many have asked why the Chinese did not fight more ferociously at Nanking. Chiang Kai-Shek's men had fought bravely around Shanghai for months, inflicting thousands of casualties on the IJA in protracted urban warfare, but when it retreated in November to Nanking, it was an exhausted force and its leaders had not prepared the capital for defensive battle. The Chinese commander, General Shengzhi Tang, abandoned Nanking and the majority of his troops. He was ordered to leave by Chiang Kai-Shek (who along with his wife also had abandoned the city), but one would think he and the *Generalissimo* would have looked out for their troops before leaving. Without effective leadership, the Chinese soldiers showed little initiative and caused much "military confusion."[453] The U.S. military advisor in China, Colonel "Vinegar Joe" Stilwell, observed that Chiang Kai-Shek was "an ignorant, illiterate peasant son-of-a-bitch" and that "The Chinese soldiers [are] excellent material, wasted and betrayed by stupid leadership."[454] As a result, the Chinese at Nanking were quickly defeated.

IJA soldiers walk around dead bodies of Chinese POWs the Japanese have killed and tried to burn to cover up their crimes in December 1937. These actions were haphazardly done as shown by the half-burnt victims. Nanking Massacre Memorial.

Capturing so many prisoners-of-war, the IJA resorted to wholesale butchery. Military correspondent Yukio Omata provided testimony to the POW executions:

Those in the first row were beheaded, those in the second row were forced to dump the severed bodies into the river before they themselves were beheaded. The killing went on non-stop, from morning until night, but they were only able to kill 2,000 persons in this way. The next day, tired of killing in this fashion, they set up machineguns. Two of them raked a cross-fire at the lined-up prisoners. Rat-atat-tat. Triggers were pulled. The prisoners fled into the water, but no one was able to make it to the other shore.[455]

Executed Chinese Nationalist POW troops by IJA soldiers at Nanking, December 1937. These men were murdered in a clear violation of the Hague Convention of 1907 which Japan had signed. Second Historical Chinese Archives.

Piles of dead, burnt bodies of Chinese Nationalist troops murdered by IJA soldiers at Nanking, December 1937. The IJA soldier who made this photo album wrote on this page: "Misery: One of the miseries are the abandoned corpses in the outer moat of the Nanking City Wall." Nanking Massacre Memorial.

It is estimated that out of the 300,000 murdered during the campaign from Shanghai to Nanking in the winter of 1937-38, 80,000-90,000 were defenseless National Chinese POWs—"it was a slaughter, a massacre."[456] Although the one treaty the Japanese had signed and ratified, the Hague Convention of 1907, forbade the killing of enemy soldiers, "who having laid down [their] arms, or having no longer means of defense, have surrendered at discretion,"[457] the Japanese ignored this accord and killed without remorse. Like with most moral codes, signed and unsigned, the Japanese ignored almost all manner of humane behavior while engaged in war. In fact, IJA commander at Nanking, General Iwane Matsui, "lacked any idea that the killing of POWs was a violation of international law and a grave crime against humanity."[458] And while the killing of defenseless POWs took place under Matsui's 10th Army, rapes were occurring everywhere.[459] Women beseeched their captives for mercy, as a "starving man" would "for food and drink,"[460] but that

Bodies of Chinese massacred by Japanese troops along a river in Nanking (Murase Moriyasa's photo). Burnt logs were interspersed among the dead, charred bodies. December 1937. Second Historical Archives, Nanking, China.

mercy rarely came. A graphic interview given by IJA soldier Shirō Azuma who committed such crimes was poignant:

> At first we used some kinky words like *Bikankan*. *Bi* means "[Pussy]," *Kankan* means "look." *Bikankan* means, "Let's see a woman open up her legs." Chinese women didn't wear underpants. Instead, they wore trousers tied with a string. There was no belt. As we pulled the string, the buttocks were exposed. We "*Bikankan*." We looked. After a while we would say something like, "It's my

day to take a bath," and we took turns raping them. It would be all right if we only raped them. I shouldn't say all right. But we always stabbed and killed them. *Because dead bodies don't talk.*[461]

Chinese cadavers killed by IJA personnel piled on the shore of the Yangtze River, Nanking (Murase Moriyasu). December 1937. Second Historical Archives, Nanking, China.

Yale University historian Jonathan Spence wrote that Nanking's "period of terror and destruction" ranks "among the worst in...modern warfare... [It was] a storm of violence and cruelty that has few parallels."[462] Today, in the Chinese consciousness, it is as significant for them as Auschwitz is for Jews. The 200 miles between Shanghai and Nanking in the fall and winter of 1937 was a flotsam of rape and dead bodies, a literal "carnival of death."[463] Horrible executions and gratuitous murder were observed at Soochow, a mid-point between Shanghai and Nanking and its devastation mirrored countless other towns. Wild dogs grew fat on the corpses throughout the city.[464] Similar acts happened in cities surrounding Nanking such as Mufushan, Wuhu, Nit'ang, Wuxi and Suzhou, and unlike in Nanking, there were no International Safety Zones to protect people from IJA soldiers.[465] The pattern of behavior not only in China, but also in Pacific Island nations demonstrates that "these murders followed such a similar pattern over such a wide range of territory and covered such a long period of time, and so many were committed after protests had been registered by neutral nations that...only positive orders from above [like Tojo, Hirohito, Matsui and Kuribayashi] made them possible."[466]

Bodies of the victims of Nanking on a riverbank, December 1937 (Murase Moriyasu). Second Historical Archives, Nanking, China.

Burial mounds of massacred citizens of Nanking, China by the Imperial Japanese Army, December 1937. Second Historical Chinese Archives.

Decapitated head of a Chinese man by an IJA soldier—the Japanese played a "joke" on the corpse and put a cigarette butt in his mouth. Nanking Massacre Memorial

A Japanese soldier beheads a Chinese citizen at Nanking, December 1937. This bearded soldier is in many photos at this scene. Notice the huge crowd of Japanese spectators. Second Historical Chinese Archives, Nanking.

A Japanese soldier beheads a Chinese citizen at Nanking, December 1937. Note the bearded soldier once again in this photo standing next to many IJA soldiers with cameras. Taking photos of crimes was common for the Imperial Japanese Army. Second Historical Chinese Archives, Nanking.

Japanese soldiers conducting another beheading at Nanking China (notice the bearded man again in the background). December 1937. National Archives, College Park.

Japanese officer at Nanking proudly displays the head he has just cut off of a Chinese citizen. Notice how he is smiling holding the severed head and his bloody sword. Second Historical Archives, Nanking, China and the Alliance for Preserving the Truth of Sino-Japanese War

As a result of troops being sent to conquer Nanking without supplies, they had to source their own food. Nevertheless, in doing so they did not have to destroy thousands of homesteads, but they did so anyway.[467] Victims called the IJA *Huang-chün*, an "army of locusts."[468] Japan used this campaign to create fear and submission. Civilians were massacred openly to entertain soldiers and encourage others to similar conduct. Sub-Lieutenants Tsuyoshi Noda and Toshiaki Mukai held a competition widely reported in newspapers about how many heads of civilians they had cut off.[469] Japanese acted "beyond the bounds of permitted aspirations."[470] *Manchester Guardian* journalist Harold John Timperley summed up these crimes writing:

> [S]hould anyone believe that the Japanese army is in this country to make life better…for the Chinese, then let him travel over the area between Shanghai and Nanking…and witness the unbelievable…destruction. This area, six months ago, was the most densely populated portion of the earth…and…prosperous sec-

tion of China. Today the traveler will see only cities bombed and pillaged; towns and villages reduced to shambles; farms desolated...The livestock has been either killed or stolen, and every sort of destruction that a brutal army...can inflict has been done... Countless numbers have been killed; others have been maimed for life; yet others are huddled in refugee camps; or hiding in mountain caves, afraid to return to their desolate farms, their empty shops and ruined businesses. Those who would dare return are not permitted to do so by the war-mad Japanese army. It is shameful...that the Japanese who control communication lines, are proclaiming to the world that they are inviting Chinese back to their ancestral homes to live in peace and plenty.[471]

The audacity that Japan thought it could commit such atrocities without incurring the world's disdain showed it had a unique arrogance that it played by a different set of rules than the rest of humanity. The martial culture practiced in China from which most of the Japanese troops came by the time they started fighting U.S. Marines like Woody had hardened them into uncaring, unsympathetic and indifferent men—potent and dangerous combinations for warriors to have when heavily armed and directed to kill the enemy at all costs.

Besides seizing more territory and resources for Japan, another IJA reason for taking Nanking was that it was Chiang Kai-Shek's *Kuomintang* capital. Taking this city was not only a military victory, but also a triumph politically and psychologically, showing the world Japan waged a "war of annihilation."[472] For destroying and raping Nanking, Hirohito decorated his generals with medals.[473]

Interestingly, within all ranks, Zen Buddhist monks and priests encouraged and fought with IJA troops. Moreover, wherever they conquered, they exported their form of Buddhism and destroyed the Chinese Buddhism where they could. For instance, IJA Buddhist monks invaded the Nanking temple estates and preached their form of theology, extolling Japan's holy war and eradicating Chinese Buddhism which it denounced as inferior and heretical. Nanking not only underwent a slaughter of its citizens, but displacement of its culture as well.[474]

Japanese troops celebrate their victory at Nanking 31 December 1937. National Archives, College Park.

All these Nanking atrocities happened with the encouragement, approval and knowledge of Japanese commanders: General Iwane Matsui (Commander 10th Army), Lieutenant General Hisao Tani (6th Division Commander), Lieutenant General Kesago Nakajima (16th Division Commander) and Lieutenant General Prince Yasuhiko Asaka (Hirohito's 50-year-old granduncle), an ultranationalist on the Supreme War Council.[475] After battle started, Lieutenant General Nakajima lent his sword to help a subordinate fulfill such criminal orders: "Takayama Kenshi visited me at noon today [13 December]. There happened to be seven captives, so I asked him to try to behead them with *my sword*. I did not expect him to do such a good job—he cut off two heads."[476] And, as mentioned before, when IJA soldiers were not executing Chinese men, they raped women everywhere. It became their obsession.

General Iwane Matsui's victory march through Nanking on 17 December 1937. Although Matsui knew his men were raping women and killing thousands of POWs and citizens, he did nothing to stop their barbarity. Nanking Massacre Memorial

At the Nanking War Trials in 1947, the military judges brought skulls of victims from Nanking to bear witness to the proceedings against Lieutenant General Hisao Tani (6th Division Commander at Nanking). Nanking Massacre Memorial.

General Iwane Matsui stands trial at the Tokyo War Trials in 1948. He was found guilty and executed by hanging. His soul has been enshrined at the Yasukuni Shinto Shrine, Tokyo Japan by the Japanese for future generations to honor. Nanking Massacre Memorial.

For example, "Chinese witnesses saw Japanese rape girls under ten years of age…and then slash them in half by sword. In some cases, the Japanese sliced open the vaginas of preteen girls…to ravish them more effectively."[477] Many girls died from internal bleeding.[478] Australian journalist Timperley called the Japanese "lust-mad."[479] In some cases, soldiers would enter a village, round up the men, escort them outside town and place them under armed guard. Then, IJA soldiers took turns returning to the village where the women were and raped them. At one village, Chinese men sat on the side of the road "trembling" as they heard the screams from their wives, daughters and sisters but could do nothing.[480] Nearby in Nanking, two husbands who ran back for their wives to save them were shot by their guards.[481] Women were taken to IJA camps and never heard of again.[482] Often, when the IJA was in control of a Chinese city, women's screams echoed throughout the streets not only during the nights, but also during the days.[483]

The coordination and network that evolved within such platoons, companies and battalions lay bare a diabolical military structure that aimed at ripping apart everything that holds a decent society together. Instead of winning Chinese hearts and minds in their pursuit of a Greater East Asia Co-Prosperity Sphere, the Japanese soldiers spawned guerilla networks costing them more casualties; a generational hatred which would take decades or longer to repair; the breakdown of infrastructure that could have fed them and their horses (they were deep in enemy territory without adequate logistical support); the devastation of farms that could have helped them by stealing and eating the livestock—without proper husbandry of animals, herds of cows, passels of hogs and strings of ponies dwindle fast, something the IJA was doing at an alarming rate; and a military environment lacking in training and repairing of weapons in order to prepare for the next battle since the China war was far from being over.[484] "These activities are not the normal manifestations of a nation of the 20th century imbued with altruistic and humanitarian ideals of helpfulness toward a weaker neighboring people. They are the actions of a nation still steeped in the traditions and barbaric conceptions of a nationalistic aggrandizement."[485]

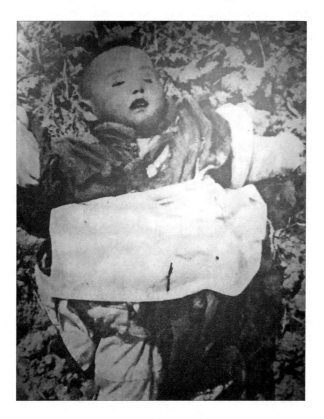

A slaughtered baby by an IJA soldier at Nanking, December 1937. Often Japanese soldiers enjoyed throwing children in the air and catching them with their bayonets as they fell back down to earth. Nanking Massacre Memorial.

Japanese "went beyond rape" and sliced off women's breasts, cut open pregnant women's bellies and decapitated the "newborn" infants in front of their bleeding mothers, killed small children and buried people alive. As one IJA soldier was raping a woman, her five-month-old baby nearby started to wail. The man stopped, went to the child and smothered it, and then returned to the woman and finished raping her. After killing a nursing mother, soldiers failed to kill her small baby boy. He later was able to struggle up to his nude mother's breast to suck what milk remained in her breast and was found "glued to his mother's corpse by frozen cubes of milk, tears and snot" (it was after all in the dead of winter). IJA soldiers castrated prisoners before using them for bayonet practice, or doused them with gasoline and lit them afire. Soldiers "impaled babies on bayonets and threw some into pots of boiling water." They hung people by their tongues on iron hooks or buried them to their midsections and released dogs on them. One 80-year-old woman died when a soldier skewered "her vulva with his bayonet." Other elderly women were "lucky" and just got raped. Obviously, these accounts illustrate that IJA soldiers were also sick misogynistic gerantophiles. Some Nazis in the International Safety Zone like John Rabe *wrote Hitler* to help stop these atrocities, calling the IJA "bestial machinery."[486] He also wrote the Japanese Embassy along with others for help detailing IJA crimes (the Embassy did nothing).[487] Women were not the only sexual victims; Chinese men and teenage boys were often sodomized and abused too. Japanese "hunted" citizens "like rabbits."[488] IJA soldier Hakudo Nagatomi wrote:

> I beheaded people, starved them to death, burned them, and buried them alive, over two hundred in all. It is terrible that I could turn into an animal and do these things. There are really no words to explain what I was doing. I was truly a devil.[489]

Uno Shintarō described cutting off heads as a sexual experience and felt "refreshed" after the killings, having first dragged the prisoner from the stockade, forced him to dig his grave, and then pushed him into a kneeling position by the hole to be his final resting place. Shintarō gripped his sword and swung down, hard to sling the head from the body. Other soldiers took a baby from its mother, placed it on the ground and hammered a nail through its forehead. In fact, the

John Rabe, Nazi Party member and hero of Nanking. He saved thousands by setting up an international security zone to house citizens fleeing the Japanese military in December 1937. He even asked Hitler for help. Nanking Massacre Memorial.

killings would have been more widespread had the expatriate community, led by Nazi Party Member John Rabe, who chaired the "International Committee for the Nanking Safety Zone," along with help from Americans and Professors Miner Searle Bates and Rev. Charles H. Riggs, not set up a precinct to protect civilians. Other Americans saving lives were Rev. John Magee of the American Church Mission and Professor Minnie Vautrin of Ginling College. These expatriates saved 200,000 from meeting the same fate as the 300,000 murdered by Matsui's forces by defending a Japanese-free zone and using international relations and pressure to keep IJA forces at bay. Although repeatedly badgered to provide them with women to rape, the expatriates refused to give into their demands, although they could not protect everyone and some were violated. Outside that zone, most Chinese suffered horribly.[490] (see Gallery 1, Photo 26)

Babies slaughtered by Japanese soldiers at Nanking December 1937. Japanese soldiers seem to have no moral compass when it came to killing innocent children. Second Historical Archives, Nanking, China and the Alliance for Preserving the Truth of Sino-Japanese War

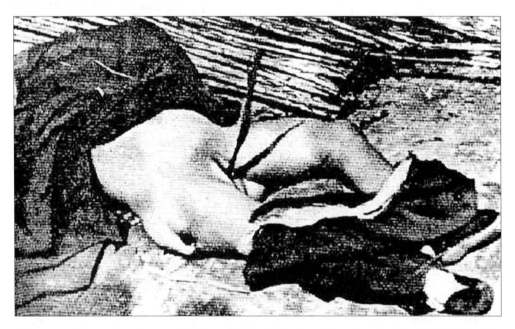

Japanese soldiers often derogated Chinese women's bodies after they raped and killed them at Nanking and in other cities by sticking branches into their vaginas for all to see. Second Historical Archives, Nanking, China and the Alliance for Preserving the Truth of Sino-Japanese War

Japanese soldiers collected the heads of Chinese citizens they had cut off during the Rape of Nanking and displayed them in a group setting December 1937. Second Historical Archives, Nanking, China and the Alliance for Preserving the Truth of Sino-Japanese War

Japanese soldiers used Chinese prisoners for bayonet practice. Often IJA command authority used such exercises to initiate soldiers into their units. Second Historical Archives, Nanking, China.

Corpses of raped, killed and desecrated women by Japanese soldiers in the streets of Nanking in December 1937. Notice that the Japanese IJA personnel took the time after they killed these women to shove sticks into their vaginas. Universal History Archive/Universal Images Group and Getty Images

As mentioned, Nanking was not unique as hundreds of cities suffered from Japanese barbarity. To the north, Beijing fell at the end of July 1937 and Baoding (Paoting) collapsed to the IJA on 24 September. After the city's capture, 30,000 troops "burst out in a week's rampage of murder, rape and pillage." Also, just as they would later burn down a third of Nanking's city, they also burned Boading's main buildings and all the schoolbooks they could find in a week-long bonfire. They also set afire Hopei Medical College's library and laboratory equipment. "A decade's records of crop statistics at the Agriculture Institute, the basis of its program for improved farming methods, were also deliberately destroyed." Japan was making war on an entire society.[491] Everywhere Japan attacked, it met with success. American military advisor, Colonel Stilwell, had to deal with IJA officers and found them insufferable: "Beijing under the control of the 'arrogant little bastards' was hard to bear."[492]

Japanese bayonet prisoners at the Rape of Nanking. December 1937. Nanking Massacre Memorial.

Throughout Japanese-conquered lands, there were numerous cases of cannibalism of their enemies, even frying up entrails, penises and testicles. Bestial acts were the sons of Japan's *modus operandi*.[493] Many Japanese soldiers seemed to relish in their demonic exploits: "No law could curb their reckless debauches, no ray of wisdom penetrate their blindness."[494] Japan had produced a nation of Hannibal Lecters from the movie *Silence of the Lambs* (1991).[495] Joseph Conrad's *Heart of Darkness* could have described them:

> They were conquerors, and for that you want only brute force— nothing to boast of, when you have it, since your strength is just an accident arising from the weakness of others. They grabbed what they could get for the sake of what was to be got. It was just robbery with violence, aggravated murder on a great scale,

and men going at it blind—as is very proper for those who tackle a darkness. The conquest of the earth, which mostly means the taking it away from those who have a different complexion or slightly flatter noses than ourselves, is not a pretty thing when you look into it too much.[496]

In taking over nations, the Japanese often encountered what they thought were "weaker" communities. Vast tracks of China were rural and any armed resistance there depended on the U.S. and Soviet Union to provide it with supplies. But citizens of the Philippines and Guam had neither the training, weapons nor infrastructure to repel the Japanese. Many occupied islands had communities of unsophisticated tribesmen like in Papua New Guinea. As a result, Japan easily used brute force. Japanese conducted their "robbery with violence" with an un-satiated appetite. Often initiations into the IJA fraternity were to have the officers behead and the enlisted men bayonet prisoners. The military turned men into "murdering demons."[497] It is difficult to fathom the "darkness" Japan had to overcome to do what it did. Looking at Japan's actions, it was "not a pretty thing." One man could have stopped this: Hirohito. Instead, he, "being knowledgeable about political and military affairs… participated in the making of national policy and issued the orders of the imperial headquarters to field commanders and admirals. He played an active role in shaping the barbaric Japanese war strategy."[498]

Chinese political poster depicting Japanese as murderers and demons killing innocent civilians. On the Chinese left arm holding the rifle defying the Japanese monster, the Communist Party name is written and on the right arm, the name of the Nationalists is written. This poster is asking both parties to unite to defend China against the Japanese invaders. Shanghai Historical Museum

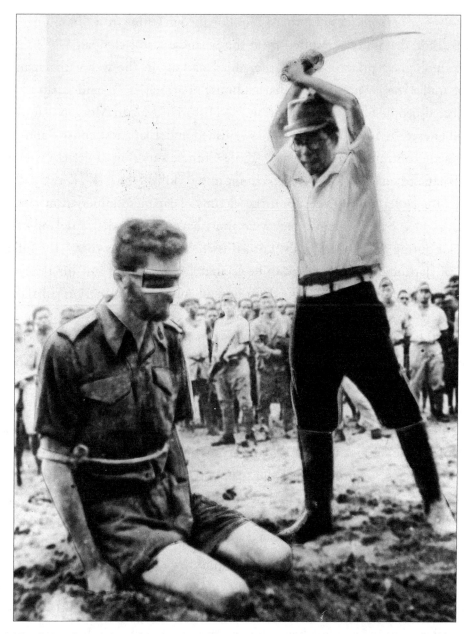

On 4 October 1943, Australian Commando Leonard Siffleet was beheaded at Aitape Beach in Papau New Guinea along with two Ambonese, H. Pattiwal and M. Reharing. Japanese soldier Yasuno Chikao did the killing. National Archives, College Park.

Japan committed mass murder killing over 20 million from 1931 through 1945.[499] This gave it fifteen years to exterminate "inferior peoples." The National Museum of the Pacific War writes the Japanese slaughtered 22,617,242.[500]

*Encyclopedia of Genocide* notes that 15 million died in China from "bombing, starvation, and disease that resulted from the Japanese terror campaign."[501] "China paid the highest price of any of the Allied nations in the war with Japan."[502] "Asia under the Japanese was a charnel house of atrocities,"[503] and Japan treated subjected people with "beastialization."[504] Historian Martin Morgan claims the number exceeds 40 million since cities vanished under Japanese control and mass graves are continually discovered.[505] Those Chinese suffering under IJA rule said the Japanese had a motto: *sanko seisaku* meaning, "kill all, burn all, destroy all."[506]

"The suffering inflicted upon the victims of Japanese imperialism remains incalculable," but it is larger than what most have calculated.[507] England's chief advisor during the postwar trials, Lord Russell of Liverpool, wrote, "The full extent of [Japanese] crimes will never be accurately known, but in China they will never be forgotten."[508] It is difficult to find an "obvious explanation" for Japan's slaughter. Japanese targeted Chinese for mass murder "regardless of sex or age" and simply marked them "out as victims" to be dispensed without any justification except that they were poor, defenseless and non-Japanese.[509] Japan disputes the numbers it slaughtered. However, Japan knows how many it lost in the war (3 million),[510] but does not *care to know* the numbers other nations give of those who died under its rule. Whereas it expects all to believe it lost a certain number, it denies other nations who have made claims against it telling China, Vietnam, the Philippines, Guam, etc. that they cannot calculate their dead.[511] There may be problems with the documentation of the barbarism, but the summary assessments made by nations after WWII cannot be denied—Japan murdered millions. However, one must explore the reasons for the lack of documentation of Hirohito's Holocaust as for Hitler's genocide.

Whereas the Nazis meticulously recorded how they "processed death," the Japanese sloppily maintained records. "One argument could be that they were so disinterested in the actions their soldiers committed, and with documenting the murder of civilians to be totally unimportant they just ignored the process."[512] Moreover, the infrastructure of the societies Germany destroyed had government records on civilians. The Chinese system of documenting people was primitive, making it arduous to verify how many were living in an area before and after the war. Although there are holes in authenticating the extent of Nazi destruction, there are historians, especially *German* ones, who tenaciously ferret out the truth.

In Japan, the government and much of its academic community avoids the past, and few are willing to document it.[513]

Japanese officials and many historians have "systematically kept all mention of their atrocities out of the nation's history textbooks." Journalist Honda Katsuichi wrote, "Unlike the Germans and Italians, the Japanese have not made their own full accounting of their prewar actions."[514] Historian Saburō Ienaga wrote that falsifying textbooks "prohibits the completion of people's developments as human beings."[515] Unlike Germany, Japan has neither atoned for its crimes by paying reparations nor at the very least, admitted its atrocities.[516] In 1994, Justice Minister and former Army Chief-of-Staff, Shigeto Nagano declared the Rape of Nanking a fabrication and claimed that Japan during WWII was "'liberating' Asian countries from Western colonial powers."[517] This would be akin to a German politician saying Auschwitz did not happen and Hitler was "liberating" Russia from communism. Tokyo's *Yushukan* War Museum mentions nothing of the atrocities. It just writes that General Matsui in 1937-38 controlled his men and warned them not to commit unlawful acts. It further claims he destroyed the Chinese at Nanking and then notes: "The Chinese soldiers disguised in civilian clothes were severely prosecuted."[518] They were not "prosecuted," they were slaughtered! Commenting on Japan's falsifications, historian Yuki Tanaka writes that this dishonorable behavior "naturally hinders full recognition of responsibility for Japan's abhorrent military acts and the losses its Asian neighbors suffered as a result of war and colonialism."[519] Holocaust survivor Primo Levi said, "Those who deny Auschwitz would be ready to remake it,"[520] and one could say Japanese leaders who falsify history, and deny Hirohito's atrocities in state-produced textbooks are people who could once again "remake Auschwitz."

There are some who are honest with the past including Japanese historians Keiichi Tsuneishi, Yoshiaki Yoshimi, Hirofumi Hayashi, Yutaka Yoshida and Yuki Tanaka to name some, but they represent a minority among Japan's political and academic leaders. There are a few brave souls in the government, but not many. For example, Japanese Consul-General in Houston, Texas, Tetsuro Amano, said in January 2018 that Japan should never forget the victims of Hirohito's Japan and that apologies and economic activity should back up that remembrance.[521] However, although there have been some reparations, they pale in comparison to Germany's. Although Japan slaughtered at least twice as many as the Nazis (Hitler murdered over 11 million, most notably Jews [6 million] and Russian POWs [3.7 million]), Japan has paid a little over 1% of what Germany has ($1 billion compared to *Deutschland's $90 billion*).[522] As a result, many claim "Japanese are both morally and monetarily deficient [to Germans] by comparison."[523] Most Japanese have little knowledge about or remorse for Japan's atrocities, and with the political stonewalling since the war, it is unlikely they ever will.[524]

Although more died during the Rape of Nanking and the rampage Japan conducted in China after Doolittle's raid than the Japanese who died from the atomic bombs (Hiroshima claimed 140,000 and Nagasaki 70,000), some Japanese maintain the illusion of victimhood because they were bombed with nuclear weapons (a strange assertion knowing they started the Pacific War).[525] Some demand apologies from America for these attacks, while the Japanese only give half-hearted apologies to their victims.

Such apologies are "half-hearted" because they are not backed up with reparations or explorations of these crimes in state-sponsored textbooks.[526] As recent as 2018, after San Francisco erected a memorial honoring sex slaves (i.e. "Comfort Women") abused by the Japanese military called the *Column of Strength,* the mayor of the Japanese sister city in Osaka Hirofumi Yoshimura cut ties and called the memorial "Japan-bashing." The previous mayor Tōru Hashimoto first denied the existence of "Comfort Women" in 2013 and then revised his statement saying they existed but were necessary so soldiers could "rest."[527] The indifference of Yoshimura and Hashimoto is commonplace among Japan's political leaders. Even the knowledgeable Air Force officer and Senior Fellow at their National Institute for Defense Studies, Colonel Yukio Yasunaga, felt "Comfort Women" followed the Japanese armies in China "and were paid for their services. All the talk today about them is political."[528] Yasunaga's sentiments are commonplace. Japan's WWII sex slaves or "Comfort Women [*ianfu*]," as they were euphemistically called, have rarely been acknowledged by Japan's government. Historians believe at least 200,000 women were abused. Since the average age of these slaves was 15, one could argue that the Japanese were not only rapists, but pedophiles (although often IJA soldiers did not have to wait for teenager "Comfort Women" to become pedophiles often raping young girls between the ages of 10 to 15, killing many due to internal hemorrhaging).[529] In some of their rape brothels, the Japanese gave women/girls a daily quota: "twenty enlisted men in the morning, two NCO's in the afternoon, and the senior officers at night."[530] Jeanne Ruff-O'Hearne, a Dutch "Comfort Woman," voiced what Japanese did to her echoing what happened to thousands: "During that time the Japanese had abused me and humiliated me. They had ruined my young life. They had stripped me of everything, my self-esteem, my dignity, my freedom, my possessions, my family."[531]

When the World Heritage Foundation UNESCO (United Nations Educational, Scientific and Cultural Organization) listed the Rape of Nanking documents as part of humanity's heritage, Japan halted its funding for UNESCO in 2016 to a tune of $40 million (since Japan is one of the biggest funders, this will hinder it).[532] Moreover, unlike Germany, Japan never investigated nor worked with foreign nations to prosecute its war criminals.

Whereas Germany condemns Hitler, Japan has never blamed Hirohito for leading his country into fascism, mass murder and a disastrous war. Japan allowed him to hold his throne until 1989 when he, a fallen god and war criminal, died of cancer. For years, he enjoyed a life of leisure often playing "rounds of golf in the postwar era, blithely oblivious to his sordid wartime past."[533] Since authority to act as the Japanese did derived from him, it represents an injustice he was never charged with crimes.[534] The judges should have executed Hirohito.[535] Woody says:

> The man wasted two years of my life and was responsible for the misery and death of millions. You'd think there would've been some consequences for such behavior. What culture allows a mass murderer to get away with it?[536]

In the 44 years after WWII when his rule continued, albeit without the power he had before and during the war, Hirohito *never* discussed the conflict publicly (his diary has not been released).[537] The wartime generation worshiped him, but when the war ended, the ex-god never accepted responsibility for Japan's crimes. "Despite the fact that the war was fought under his command," he refused to accept "responsibility for the...acts which occurred during it" in his *holy name*.[538] Many in Japan today refuse to reject Hirohito. When asked to denounce Hirohito in 2018, Japanese Deputy Consul General for Houston Ryuji Iwasaki, presenting a sphinxlike expression, refused.[539] When the mayor of Nagasaki,

Hitoshi Motoshima claimed Hirohito was to blame for the war, a right-wing nationalist shot him in 1990. He fortunately survived. MacArthur left Hirohito as Emperor after 1945 to secure the peace, and to liberalize and democratize Japan. The Allies controlled the war trials, and the responsibility of not holding Hirohito accountable lay in MacArthur's hands.[540] General Curtis LeMay explained MacArthur's position saying the Allies left Hirohito alone "because of his anti-communist posture and willingness to work with the U.S. during the Cold War [which President] Truman saw as valuable…Personally, I would've strung the bastard up by his balls."[541] (see Gallery 1, Photo 27)

Japan's recent commemorations of peace say nothing about its crimes. Every 15 August, on Japan's anniversary of surrender, the nation recalls "the war" "with a government-sponsored 'Day Commemorating the End of the War.'" Although not a national holiday, Tokyo's *Niho Budôkan* hall, "normally the site of concerts, professional wrestling matches and martial-arts events," is used where national leaders and honored guests participate in ceremonies to "Mourn the War Dead." In a sea of Japanese flags and chrysanthemum banners, leaders such as the Prime Minister, government and local officials as well as family members of famous military dead are among the VIP guests. The Emperor reads a statement covered by the press and then a moment of silence is observed. "No apologies or regrets are offered to the millions…who survived the depredations of the Japanese, and no one seems to consider anything to be amiss in that."[542] Although apologies are demanded of the Japanese, they rarely comply. The "disingenuous" Hirohito "lacked all consciousness of personal responsibility for what Japan had done…and never once admitted guilt for the war of aggression that…cost so many lives."[543]

Yet Japan demands apologies from others. As recently as May 2016, when U.S. President Obama visited Hiroshima, Terumi Tanaka, secretary general of Japan's Confederation of A-Bomb and H-Bomb Sufferers, urged Obama to apologize to the "victims," claiming the weapon "inhumane and against international law."[544] Tanaka attempted to convince the world to view the Japanese as WWII victims instead of as the assailants which caused the use of the bombs in the first place.[545] Historian John W. Dower writes about the Japanese today that, "a collective sense of Japanese victimization in that terrible war generally prevails over recollection of how grievously Imperial Japan victimized others."[546] On the other side of the equation, some veterans' groups advised Obama not to visit

Hiroshima because it would be an "implicit apology" insulting U.S. veterans. Retired Rear Admiral Lloyd "Joe" Vasey, a WWII submarine officer, wrote:

> Any presidential action or policy that even appears as an implicit apology for the use of the atomic bomb would be a gross insult to us and our valiant comrades who fought and sacrificed...to win the war and bring us a peace that liberated Asia.[547]

Obama made clear there would be no apology but rather a reflection on the bombs' destructiveness and renewed commitment to avoid nuclear war.[548] Obama's visit was followed by one of Japan's Prime Minister Abe to Pearl Harbor with an apology to the American spirits killed there. Although WWII events "seem far behind us, in many ways they continue to structure mentalities in the contemporary world."[549]

## Lieutenant General Tadamichi Kuribayashi: One of Japan's Most Evil Commanders

Many Japanese against whom Woody battled at Iwo were veterans of Japan's brutal, criminal combat. One was Iwo Jima garrison's commander, Lieutenant General Tadamichi Kuribayashi, who, had he survived the war, would have been probably charged with *Crimes against Humanity* and *Peace*.[550] In April 1939, he toured Manchuria and China as Divisional Chief of the "Remount Administrative Section, Military Service Bureau of the War Ministry" gathering data about IJA's logistics needs.[551] He must have observed that the Manchurian operation was "a mad carnival of debauchery carried out by gangs of ignorant bullies."[552] Enemy combatants there were labelled "bandits" and almost universally "exterminat[ed]" when taken prisoner.[553] Moreover, sometimes the "bandits'" villages also were "liquidated" as evidenced by the Pingdingshan Incident in September 1932 when Japanese troops machined-gunned and shot 3,000 citizens, including men, women and children, because everyone was assumed to have supported the "guerillas."[554]

From September 1941 until June 1943, Kuribayashi served as the 23rd Army's Chief-of-Staff during the battle and occupation of Hong Kong where atrocities happened. During the conflict and after conquering the city, the Japanese raped thousands: "Over 10,000 Chinese [and foreign] women, from the early teens to the sixties, are reckoned to have been raped or gang-raped by the Japanese... during the sack of Hong Kong."[555] Catholic Nuns were a special target.[556] One tall, stately 6'0" Australian nurse was raped repeatedly, but after the war, she defiantly reported: "It didn't hurt that much since all the little buggers had small pricks."[557] Was this nurse telling the truth or devising a story of defiance to help cope with her trauma? Either way, she was a victim of horrific behavior. And her flippant comments about Japanese manhood belie the fact many rape victims, even probably she herself, suffered psychic wounds. Some who survived these deplorable acts "walked and talked like zombies" afterwards, having suffering severe trauma.[558] Others throughout China "ended their own lives" while others "went insane."[559]

Later, at St. Stephen's College, seven nurses were "dragged, crying and writhing," to a room full of already butchered men where IJA soldiers held them down on top of a pile of dead, ripped off their garments and then gang-raped them. After several finished raping a nurse, a few would step forward and bayonet her to death. Blood curdling screams echoed throughout the halls as the men either thrusted their penises or their blades into the women. They would place the next victim on the burial mound next to her freshly killed colleague and then one soldier after another would sexually violate her until someone made the decision she was no longer useful "to fuck," and would dispatch her to the "undiscovered country."[560] The gutters at the college "ran with blood."[561] One "Jap" later remarked these victims "cried like a lot of pigs."[562]

One POW, a British Lieutenant, asked a Japanese officer to see his wife who was one of the nurses. Sadistically, the officer took the unsuspecting prisoner outside the building to a stack of bodies. Seeing his nude wife brutally violated, he tried to attack his captors, but they restrained him and threw him back into the storeroom where the other POWs were incarcerated. Returning to this cell, he broke down and started babbling to himself incoherently.[563] Soon after these atrocities, the living POWs had to take over a hundred bodies of their murdered comrades and nurses, pile them up outside in the courtyard and then set them afire in a bonfire ordered by their captors. As the bodies burned, guards laughed nearby.[564]

After hostilities ceased in one area now controlled by the Japanese, a British businessman tried to have a dinner with his wife and teenage daughter when IJA troops barged into his home, tied him to a chair and then raped his wife in front of him and their child.[565] Japan had produced a generation of men who were psychopathic. They had mental disorders and expressed abnormal and violent social behavior. Such were the men under Kuribayashi's command. Such was Kuribayashi who never protested these crimes.

These atrocities elicit the question, why did the Japanese rape so much? Psychologists offer one explanation for Japanese's behavior of rape during WWII. For generations, Japan viewed women as inferior (a view not necessarily unique to their culture). However, societal norms during Hirohito's reign dictated women primarily raise children and they often did so using corporal discipline. Boys growing up in Japan were taught to view women as weak, stupid and emotional and they, subconsciously, resented the fact the inferior half of society had to raise them. Ironically, Japan subjugated the half of its population responsible for raising 100% of the next generation. Rapists' childhoods are "typified by neglect alternating with harsh punishment and deprivation" which most Japanese children experienced.[566] Also, rapists are "likely to have been raised by a...single...parent" all of which many Japanese experienced because absentee fathers were common.[567] Japanese were raised in a culture marked by despair, force and violence, often witnessing their fathers physically abuse their mothers. They grew up steeped in the idea that "physical aggression is acceptable and necessary."[568] Throughout WWII, Japanese servicemen demonstrated "an aggression toward women syndrome."[569] When Japanese were released on foreign nations, their pent-up resentment against women expressed itself in violent forms; as they conquered foreign lands, they raped many woman they encountered viewing them "as adversaries to be attacked and subdued."[570]

Japan spawned a nation of men who took their psychotic drive to conquer to a whole different level of diabolical. They were what psychologists Robert Prentky and Ann Burgess called *enraged* and *sadistic* rapists who "hate their victims" and want to "humiliate and dominate women."[571] They often made "unusual and painful sexual demands and many seem to be acting out a bizarre fantasy." Frequently they "fatally injure their victims"; the Japanese did this by the millions.[572]

In addition to Kuribayashi's troops raping women in Hong Kong, they also murdered civilians (4,000 killed during the battle and 50,000 during the occupation).[573] As a British eye-witness noted, the attack on Hong Kong was accompanied with a strong "taste of Nanking."[574] The Allied troops described how horrifying it was the first night they spent on Hong Kong Island on 12 December hearing the screams on the mainland as Japanese troops raped and murdered all night long.[575]

British Foreign Secretary Anthony Eden informed the House of Commons the Japanese executed 50 defenseless British POWs.[576] In fact, the Japanese executed most of the wounded they found and POWs numbering in the hundreds, outright crimes against the principle of *hors de combat*.[577] Others were mutilated: "soldiers cut out one man's tongue and chopped off another man's ears."[578] While the screams of these victims rang out in the Red Cross Hospital, IJA soldiers slashed out others' eyes and sliced off limbs and noses. Seventy defenseless Commonwealth soldiers were slaughtered at this hospital, many dying by blade thrusts while they lay sick in their beds.[579]

During the occupation, the Japanese raped, killed and resettled tens of thousands of natives (Hong Kong's population dwindled from 1.6 million in 1941 to just 600,000 in 1945 due to forced evacuations, resettlement and starvation).[580] Right after the IJA's takeover of Hong Kong, the 23rd Army "chopped off a few hundred heads a day for some time."[581] "Guidelines adopted by the 23rd Army" and approved by Kuribayashi provided for the "immediate suppression of hostile influences"; namely, all Communist or Nationalist elements were to be "imprisoned or liquidated."[582] By New Year's Day, a former playing field in Hong Kong was "piled high with the corpses of Chinese who had been bayoneted or shot."[583] The *Kempeitai* in Kowloon at King's Park were relentless, using several citizens "for shooting and bayonet practice and even beheading some."[584] By the end of January 1942, corpses littered Hong Kong's coast line.[585] Unlike at Nanking, there was no International Safety Zone to protect helpless citizens because Japan would not have honored it since it was at war with the world by then. After taking Hong Kong, the 23rd Army proved to be a disastrous city administrator. "The colony lay prostrate in the absence of any firm order. The streets were a mess of corpses and shell-holes and severed lines, and the harbor was clogged with the sunken hulks of two hundred vessels."[586]

Troops under General Kuribayashi, Chief-of-Staff of the 23rd Army, celebrate after conquering the large Hong Kong Island belonging to the colony. During the battle and right after, such troops would rape a minimum of 10,000 women and murder 50,000 civilians. 25 December 1941. Monadori and Getty Images

As mentioned, Kuribayashi never protested these crimes or questioned his superior about them. As an obsessive man of detail and micromanager of troops, Kuribayashi had to not only have known of these crimes, but also had to have supported and/or ordered them (at best, repeatedly turned a blind eye to them).[587] He was not one to shy away from challenging commanders when he thought them wrong.[588] Kuribayashi was stationed during part of the battle at Kowloon where "the murder of civilians, looting, plunder, serious abuse, and sexual violence" became routine.[589] The IJA "systematically looted the city because it stored useful resources such as food, motor vehicles, medical and scientific equipment, but it was in a rather 'orderly manner' because it was organized by the 23rd Army."[590] The pilfering of the city damaged the remaining population later when food and medical supplies were scarce. Kuribayashi's commanding officer, 23rd Army commander General Takashi Sakai, was executed at the Chinese War Crime Trials in Nanking in 1946 for his behavior.[591] The Tribunal wrote:

"All the evidence goes to show that [Sakai] knew of the atrocities committed by his subordinates and deliberately let loose savagery upon civilians and prisoners of war."[592] The sentence further read:

> The defendant, Takashi Sakai, having been found guilty of partic-
> ipating in the war of aggression and having been found guilty of
> inciting or permitting his subordinates [Kuribayashi was his high-
> est ranking subordinate] to murder prisoners-of-war, wounded
> soldiers and non-combatants; to rape, plunder, deport civilians;
> to indulge in cruel punishment and torture; and to cause destruc-
> tion of property, is hereby sentenced to DEATH.[593]

One example out of hundreds to illustrate how grotesque Sakai was happened when he allowed two subordinates to rape two women, mutilate their bodies and then feed their butchered body parts to dogs in a Jeffrey Dahmer-like manner.[594] Such were the men loyal to Sakai and Kuribayashi.

After leading a march through Victoria's centre on 28 December, Sakai and Kuribayashi gave their victorious troops a "three-day 'holiday'" during which these commanders allowed them to rape, pillage and plunder the city.[595] This should not be a surprise because this was what Sakai had been doing for the past ten years in China, most notably the butchering of people at the battle of Jinan where he slaughtered thousands of civilians in May 1928—his men had been conditioned to do everything they were doing. In other words, this behavior was not surprising but *normal Japanese behavior.*[596] Out of 123 to stand trial at Hong Kong after the war, 21 were executed and many were junior in rank and responsibility to Kuribayashi and under his command.[597] Had Kuribayashi survived the war, he probably would have been right next to them during their trials or next to Sakai at the Nanking Trial since Kuribayashi implemented Sakai's orders. Historian Dennis Showalter said, "Kuribayashi's poor leadership and support of the crimes in his region of operations in China would've definitely put him in the cross hairs of the war trials after the war."[598] Historian Philip Snow writes of Kuribayashi's failed leadership: "Few officers [at Hong Kong] seem to have made any notable effort to hold the men back, and one or two were even prodding them to commit their excesses."[599] Moreover, men under Kuribayashi, like commanding officers Ryossaburo Tanaka of the 299th Regiment and Takeo Ito of the 28th Infantry Unit of the 38th Division, slaughtered defenseless POWs and were convicted of war crimes.[600] Tanaka issued orders to kill all "prisoners of war."[601]

Even POW camps suffered the deaths of many who "need not have died" under Kuribayashi's subordinates like the commandant of all camps, Colonel Isao Tokunaga, acting under Kuribayashi's orders.[602] Despite Kuribayashi visiting these camps, the conditions there did not improve and the senseless death continued.[603]

General Takashi Sakai with his Chief-of-Staff, Major General Kuribayashi (directly behind Sakai's lead horse in IJA uniform), marching into Hong Kong on 28 December 1941 during their victory parade. The Asahi Shimbun and Getty Images

According to Hong Kong's War Crimes Judge and Prosecutor Major Murray Ormsby, the burden of proof to convict a person of war crimes was to prove their subordinates acted against the laws of war and committed atrocities which Kuribayashi's subordinates did in huge numbers.[604] USMC Colonel Mant Hawkins, Advisor to the Commander of Centcom 2005-07 and Director of Global Combatting Terrorism Network Special Operations Command 2007-08, said, "As Chief-of-Staff, Kuribayashi knew about everything Sakai and their men did. That no action was taken to curtail the atrocities is a testament to Kuribayashi's fascist mind-set and brutal uncaring."[605] Presiding Judge and Major General Russel Reynolds, in rendering a judgement about General Tomoyuki Yamashita at his trial for the Rape of Manila, could have been writing about Kuribayashi when he wrote: "Where murder and rape and vicious, revengeful actions are widespread offenses, and there is no effective attempt by a commander to discover and control the criminal acts, such a commander may be held responsible, even criminally liable, for the lawless acts of his troops."[606] Taken to the next level, just as General Yamashita's Chief-of-Staff, Lieutenant General Akira Mutō, was sentenced to death for the Rape of Manila, so too would Kuribayashi probably have met a similar fate as Sakai's Chief-of-Staff for the Rape of Hong Kong for failing to protect POWs and civilians had he survived the war. As General of the Army MacArthur said, "a soldier's duty included the protection of the weak and unarmed," a duty Kuribayashi shirked.[607] Furthermore, MacArthur, although describing Yamashita's leadership failures during Manila, could have been describing Kuribayashi's actions writing that men like him were "a blot upon the military profession [and] a stain upon civilization."[608]

A massacre of civilians in Canton in 1938 by the Japanese military. Many of the men who committed these atrocities were later under General Kuribayashi's command. Three Lions and Getty Images

After Hong Kong's fall, Kuribayashi continued his activities as Chief-of-Staff for the 23rd Army which controlled Guangdong province, the southernmost of the mainland provinces which channels most of South China's trade. It has one of the longest coastlines of any province and encompasses the cities of Hong Kong, Shenzhen, Guangzhou (also known as Canton) where numerous crimes transpired of "Comfort Women," rape, murder, human medical experiments and pillage, all under Kuribayashi's command. Unit 8604, like Unit 731, conducted horrible medical experiments on civilians in Canton with the support of Kuribayashi's command.[609] Since Kuribayashi's 23rd Army was in control of Guangdong Province for "security maintenance…maintenance of facilities such as airfields and economic measures (finance, currency, distribution of goods, [and total] control," he had to know of these atrocities.[610] Moreover, the 23rd Army was the main hub for the IJA's "military training…and logistics base" in Guangdong which committed these crimes just mentioned above.[611] Historian Robert Burrell writes: "Kuribayashi provided prominent leadership in…these horrific arenas where Japanese brutality was the norm."[612]

It does seem the 23rd Army under Kuribayashi and Sakai's leadership had meetings about the "rape-epidemic" to control it. Months into the occupation, they did arrest some soldiers newly guilty of such crimes (there seems to have been little or no effort to punish those who had already committed such transgressions). To avert future rapes, the 23rd Army established brothels and looked for "non-family women to staff them. Posters appeared on the streets advertising for 'comfort women,' some hundreds of whom were recruited…[from] Guangdong province."[613] Kuribayashi helped establish these prostitution houses, not because of his concern for the "Chinese victims of rape by [his] soldiers but because of [his] fear of creating antagonism among the Chinese civilians."[614] Also, localizing the women being "used," the IJA medical personnel could reduce venereal diseases, something authorities worried about because such diseases "undermine the strength of their men (and hence their fighting ability)."[615] Since they needed more than these hundreds of local women to "service" all the men, Kuribayashi and his staff imported an additional 1,700 Japanese prostitutes from Canton treating women like "military supplies."[616] And while senior officers like Kuribayashi visited "Comfort Women" alongside their men, it is interesting that Kuribayashi brought in Japanese prostitutes. These women "were in a different position from the comfort women" in that they mainly serviced "high-ranking officers" and were given better food and accommodations.[617] It appears Kuribayashi was not only taking care of his flag- and field-grade officers, but he was probably also taking care of his needs. Kuribayashi and his command helped organize and take part in what would become "the largest and most elaborate system of trafficking in women in the history of mankind and one of the most brutal."[618]

Kuribayashi and Sakai also made sure, from the moment the "White" troops surrendered, to inflict the "maximum humiliation on the 10,000-odd British and Canadian soldiers who had fallen into Japanese hands."[619] They were beaten, forced to bow to Chinese citizens, required to collect trash in the streets and garrisoned in some of the worst parts of the city. "[E]very opportunity was taken to drive home the message that the ethnic tables had been turned."[620] They were put under Colonel Tokunaga, who was "gross, cruel and sadistic" and subjected his prisoners to starvation diets and soon, all were racked by "wound sepsis, dysentery and diphtheria... [and] deficiency diseases such as pellagra and beriberi."[621] All this occurred while Kuribayashi continued to labor away at Sakai's staff duties growing a nice, fat belly.[622]

Turning their attention away from the POWs, the Japanese leadership under Sakai and Kuribayashi made Japanese the colony's official language, renamed several areas using Japanese names and enforced religious festivals honoring Hirohito and the Imperial forces: The "local citizens were expected to play a full part in them" or they would be punished.[623] As the Hong Kong landscape was Nipponized, a massive monument was constructed called *The Tower of Triumph* on a "spur of Mount Cameron overlooking the Wanchai district" designed to honor the Japanese dead who fought to conquer Hong Kong and glorify "Japanese power." Sakai and Kuribayashi's engineers embedded a *Samurai* sword in the foundation solemnized "by a bevy of Shinto priests dressed in white." Chinese leaders were forced to attend the ceremony, donate money and display "their conversion to Japanese values."[624]

Like with most Japanese occupational duties, Sakai and Kuribayashi's men continued to kill Chinese citizens for the slightest of offenses.[625] And as seen throughout the new Empire, Japan was exterminating the cultures and traditions of the people it controlled in order to make them Japanese. During the battle and months afterwards, mass chaos and crimes ruled the day, but by the summer of 1942, order seemed to have been somewhat restored to Hong Kong and an organized return to a functioning city resumed. Nonetheless, the crimes of the 23rd Army will forever haunt Kuribayashi's legacy.[626]

Kuribayashi also forced conscripts from Korea in the 23rd Army to take care of horses in Guangdong and 1,600 Korean enslaved laborers (*Senjin* or *Chōkō*) on Iwo to build defenses.[627] So whether it was raping Chinese women, killing POWs, pillaging Chinese cities' resources, slaughtering civilians, dealing in human trafficking or enslaving Korean men, the list of misdeeds Kuribayashi was *directly* responsible for was legion. Kuribayashi knew about his country's crimes because he helped commit them.

When Kuribayashi's grandson, Liberal Democratic Party (LDP) politician Yoshitaka Shindo, was asked about these matters, he said, "I must be honest. I don't know about what my grandfather did in Hong Kong. I only really learned about my grandfather and what he did at Iwo Jima when Clint Eastwood made his film...There were difficult things during the war. If you find things based on historical fact, then write them."[628] Such integrity with Japanese politicians is rare, especially ones who, like him, are members of the LDP and of the revisionist group *Nippon Kaigi* (Japan Conference). Nonetheless, for a man who uses his grandfather to further his political career, placing images of his grandfather in uniform on political posters next to pictures of himself, one might find it strange Shindo has not done more research to be careful in using pictures of a man who led an army of rapists, murderers and war criminals.

In June 1944, Kuribayashi assumed command of the newly formed 109th IJA Division (The Courage Division) and moved it to Iwo Jima from Chichi Jima nearby thinking it was the island America would attack.[629] This division had troops who had also committed crimes. Major General Sadasue Senda commanded 4,600 men in the 2nd Mixed Brigade in the middle of Iwo and had a distinguished career. From 1936 to 1939, he was assigned to border protection on Korea's northern frontier against the Russians. He fought at the Battle of Lake Khasan against the Soviets (29 July—11 August 1938) leading the 76th Infantry Border Garrison Regiment (1,500 men). During this time, his troops committed crimes against Koreans, especially women. In May 1940, he took command of the 44th Infantry Regiment based in Manchuria and fought guerrilla warfare. Tens of thousands of Chinese died under his command. Marine Correspondent and Iwo veteran Bill D. Ross found evidence Senda "helped slaughter hundreds of thousands" while stationed in China. On 1 March 1944, he was promoted to major general and in November, assigned to Kuribayashi's 109th Division. Then on 16 December 1944, he took command of the 2nd Mixed Brigade which took him to the heart of Iwo.[630] The 26th Tank Regiment at

Iwo had also been "seasoned in Manchuria," stationed in Korea and commanded by the famous Olympic 1932 Gold Medalist Colonel Baron Takeichi Nishi.[631] Imperial Headquarters had as early as August 1937 issued a directive that it was not "bound in fighting Chinese to adhere to international law."[632] Unlawful orders had become lawful and Kuribayashi, Senda and Nishi supported these crimes.

## Most Japanese Soldiers were Guilty of War Crimes

It is fair to surmise that if every Japanese combatant stationed throughout Asia either indirectly or directly committed or supported the crimes against the Asians as seen from 1931-45, then the majority of the approximately 2 million soldiers stationed in Asia during these years probably participated in or at least obliquely supported the killings.[633] In January 1938, Imperial Headquarters in the name of Field Marshal Prince Kan'in Kotohito, Hirohito's uncle, wrote General Matsui, "If we look at actual conditions in the army, we must admit that much is less than blemish-free. Invidious incidents, especially as to troop discipline and morality, have occurred with increasing frequency of late."[634] By August 1938, Lieutenant General Yasuji Okamura, who replaced Matsui, declared, "It is true that tens of thousands of acts of violence, such as looting and rape, took place against civilians during the assault on Nanking. Second, front-line troops indulged in the evil practice of executing POWs on the pretext of [lacking] rations."[635] And Head of the War Ministry, Lieutenant General Seishirō Itagaki, had a "Top Secret" memorandum circulated throughout the IJA in 1939 stating, "If the army men who participated in the war [throughout all of China] were investigated individually, they would probably all be guilty of murder, robbery or rape."[636] Kotohito, Okamura and Itagaki's statements display that the vast majority, if not all, of the troops stationed in China committed atrocities. Unlike the Nazis, the Japanese did not utilize factories of industrialized death with gas chambers, poison chemicals and crematoria to kill and dispose of their victims. They used knives, bayonets, swords and bullets, which required more personnel than the Nazis used in their slaughters. The Japanese stationed on Iwo had knowledge of what their nation did from hearing accounts and/or because they had done the killing themselves.[637] This did not make the Japanese Holocaust any less gruesome than the Nazi genocide; it did, however, leave no large-scale evidence. Everyone in the IJA who served in

**Photo 1:** Author with LDP politician Yoshitaka Shindo at his office in Tokyo after interviewing him about his grandfather General Kuribayashi, the Japanese commander of Iwo Jima. He is one of the leading advocates for recovering Japanese war dead from Iwo and is very proud of his grandfather's service using the general's image in his political posters. Unfortunately, he is a member of the neo-fascist group *Nippon Kaigi* which denies Japan's wartime atrocities. He also often worships at the Yasukuni Shrine where war criminals "souls" are interred causing outrage throughout Asia. 9 April 2018. Author's Collection

**Photo 2:** Justin Rigg with Marine and Iwo veteran T. Fred Harvey. He earned the Silver Star and Purple Heart for his actions on Iwo Jima. Outrigger Hotel, Guam. 20 March 2015. Author's Collection.

**Photo 3:** Justin Rigg with Marine and Iwo veteran John Lauriello at the Outrigger Hotel on Guam. Lauriello landed on Iwo the first day of the invasion on 19 February and was on the island for the entire campaign. "Hell, I still don't know how I survived," he said in 2015. 22 March 2015. Author's Collection

**Photo 4:** Justin Rigg with B-29 flight engineer and former Japanese POW Fiske Hanley. 22 March 2015. Outrigger Hotel, Guam. Author's Collection.

**Photo 5:** Author Bryan Rigg and son Justin with Marine Corps 36th Commandant and later chairman of the Joint Chiefs-of-Staff, General Joseph Dunford. 22 March 2015. Outrigger Hotel, Guam. Author's Collection.

**Photo 6:** Author Bryan Rigg and son Justin with Japanese Iwo veteran Imperial Japanese Navy Sailor Tsuruji Akikusa. 22 March 2015. Outrigger Hotel, Guam. Author's Collection

**Photo 7:** Author Bryan Rigg with Jerry Yellin who was a P-51 fighter pilot during WWII and Iwo Jima veteran. He flew 19 missions during the war. 21 March 2015. Guam's Antonio B. Won Pat International Airport. Author's Collection.

**Photo 8:** Justin Rigg proudly stands with the Navajo Code-Talker and Iwo Jima veteran Samuel Tom "Sam" Holiday on top of Mt. Suribachi, 21 March 2015. He said about his visit to Iwo, "The first time I came as a warrior. The second time a man of peace." Author's Collection

**Photo 9:** This picture is taken from our airplane of Iwo Jima from the south looking at Mt. Suribachi. We circled the island with a plane full of Iwo veterans three times before we landed for the 70th commemoration of that battle. As we neared the island, the veterans became quiet as they gazed out their windows all flooded with the memories of a place many had not seen in seven decades. The last time they had seen it, they were young and strong Marines who had just finished defeating the Japanese. 22 March 2015. Author's collection

**Photo 10:** This picture was taken at Iwo Jima on 21 March 2015 on top of the dormant volcano Mt. Suribachi. Iwo veteran Jim Skinner is in 1945 era dungarees (left), Bruce Hammond, the son of Iwo veteran Ivan Hammond is in the middle and Woody Williams, a Medal of Honor recipient from Iwo, is to the far right wearing his medal. Dave Shively.

**Photo 13:** Anti-American Japanese propaganda. It reads: "American and Britain say they will bring peace to Asia but they are really demons." Notice that the demons in this poster are in the likeness of Roosevelt and Churchill. The blue demon is FDR and the white demon is Churchill. Author's Collection

**Photo 11:** Japanese grave marker on Iwo Jima noting where many IJA soldiers gave their lives fighting the Marines. Notice that the Japanese leave food and water at their graves for the dead. 24 March 2018. Author's Collection.

**Photo 14:** Anti-American Japanese propaganda. It reads: "We are now standing toe-to-toe with America and Britain—Chop Chop Chop—Demon Busters." Author's Collection.

**Photo 12:** Veterans' Question & Answer session at Guam's Museum of the Pacific War. Medal of Honor recipient Woody Williams (USMC) is standing with an overseas cap (back). To Woody's left with his head looking down and wearing a blue hat is Iwo veteran Ivan Hammond (USMC) (far back). To Woody's right is Iwo Jima veteran Norman Baker (USMC). P-51 Pilot and Iwo veteran Jerry Yellin (USAAF) is in a dark sweater (back). To Yellin's right is Iwo veteran Jack Colby (USMC). Water Purification Engineer and Iwo Jima veteran Ed Graham (USMC) is by the microphone (front) and he is being interviewed by radio personality David Webb. Corporal and Iwo veteran Billie Griggs (USMC) is sitting with a cane (front). To the left of Griggs in a white shirt is Iwo veteran William "Bill" Pasewark Sr. (USMC). And to the right of Griggs, in the middle of the front row, is Corporal and Iwo veteran Leighton R. Willhite (USMC). Right above Willhite's head, one can barely see a veteran with a baseball cap and sunglasses. That is Navajo Code-Talker and Iwo Jima veteran Samuel Tom "Sam" Holiday (USMC). The man in the orange jacket is unknown. 22 March 2015. Author's Collection

Photo 15: War propaganda poster admonishing Americans to remember Japanese atrocities from Bataan. National Archives, College Park

Photo 16: Revenge poster telling military personnel to remember Pearl Harbor. National Archives, College Park.

Photo 17: Remember Pearl Harbor poster that hung at many U.S. military bases throughout WWII. National Archives, College Park.

Photo 18: U.S. propaganda poster "Avenge Pearl Harbor." National Archives, College Park.

Photo 19: War propaganda poster showing Japanese atrocities against American servicemen declaring "This isn't war. It's **MURDER**...*Make 'em pay.*" National Archives, College Park

Photo 21.

Photo 20: Here are two examples of posters the U.S. Armed Forces put up on military bases throughout the world to warn American servicemen of the dangers of venereal diseases. As Gunnery Sergeant Pepin taught Marines like in Woody's boot camp class, it was important they protected their health and if they did engage in sexual activity, they were to do so using a condom. Woody even explained that while at boot camp, the recruits were shown movies of what genital warts and other diseases looked like on female genitalia, and Woody declared after watching these educational programs, he was not going to "stick my pecker into any of those." National Archives, College Park

**Photo 22:** Nanking Memorial second hall's mass grave of butchered Chinese by the Japanese, December 1937. 11 May 2019. Author's Collection

**Photo 23:** Nanking Memorial hall's first mass killing pit of massacred Chinese POWs by the IJA, December 1937. This burial ground is in the main museum building. The museum is built on a massive graveyard of thousands of dead. 11 May 2019. Author's Collection

**Photo 24:** Author Bryan Rigg outside the Nanking Massacre Memorial Museum at Nanking, China. This 36-foot statue of a mother holding her murdered child by Japanese soldiers greets visitors at the entrance gate. 11 May 2019. Author's Collection.

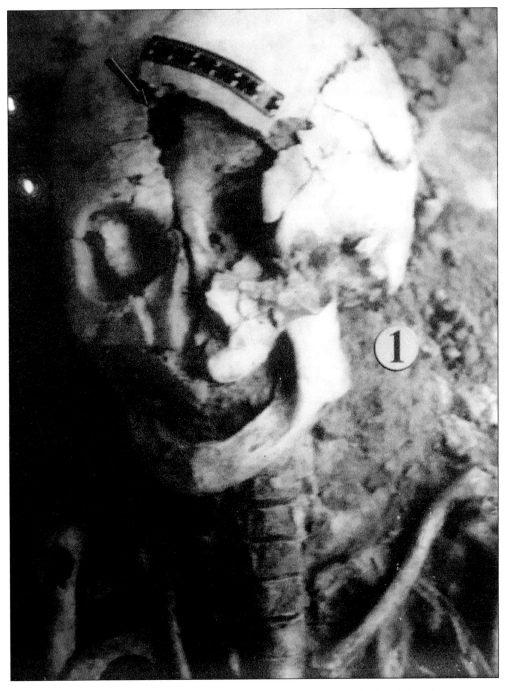

**Photo 25:** A skull of a 60-year-old woman with a bullet hole to her forehead done by a Japanese soldier at Nanking in December 1937. Nanking Massacre Memorial.

金永
陵生

明妮 · 魏特琳
Minnie Vautrin
1886—1941

**Photo 26:** Author and daughter Sophia Rigg at a monument for American Professor Minnie Vautrin of Ginling College on the grounds of the University of Nanking. During the "Rape of Nanking," she helped save the lives of hundreds of people and prevented numerous people from being raped by hiding them from the IJA soldiers. She bravely confronted the Japanese and refused to be cowered by their threats and demands. She grew up in Illinois, obtained a Master's in Education from Columbia University and taught for almost three decades in China. After Japan conquered Nanking, she returned to the United States. She would commit suicide in 1941 tormented by the memories of what she went through under the Japanese and her inability to return to China and help the innocent people she knew were suffering under the Empire of the Sun. The Chinese government would recognize her for her heroics after the war awarding her the Emblem of the Blue Jade for saving lives and being an up-stander. 11 May 2019. Author's collection.

**Photo 27:** Author Bryan Rigg at Hirohito's grave at Musashi Imperial Graveyard, Hachiōji, Tokyo on 9 April 2018. Many citizens of Japan come to his tomb here and offer prayers to him. Hirohito was never brought up on war crimes although he ordered many of the military actions and atrocities of WWII. He died peacefully in his bed in 1989. Author's Collection

**Photo 28:** Yasukuni Shinto Shrine, Tokyo Japan. This war shrine is one of the most "holy" places for the Shinto religion. In addition to the 1,068 "souls" of criminals registered there, there are 2,465,138 soldiers' "souls" interred there as well, most of whom fought for the fascist regime of WWII. Many in the world do not understand why Japanese politicians pay their respects at this shrine like Prime Minister Shinzo Abe or the grandson of General Kuribayashi LDP politician Yoshitaka Shindo. For many in the world, this shrine is akin to Germans "erecting a cathedral to Hitler in the middle of Berlin" or a monument to the Deathhead-SS. 9 April 2018. Author's Collection

**Photo 29:** Emperor Hirohito leaves the Yasukuni Shinto Shrine in Tokyo, Japan. This shrine honors Japan's war dead, but it has generated much controversy since many war criminals' "souls" from WWII are enshrined there. Yushukan War Museum

**Photo 30:** Author Bryan Rigg at the Yasukuni Shinto Shrine, Tokyo Japan. This is where Japanese politicians and citizens honor their war dead, some of whom were war criminals. This would be like Germans honoring graves of SS officers. 9 April 2018. Author's Collection

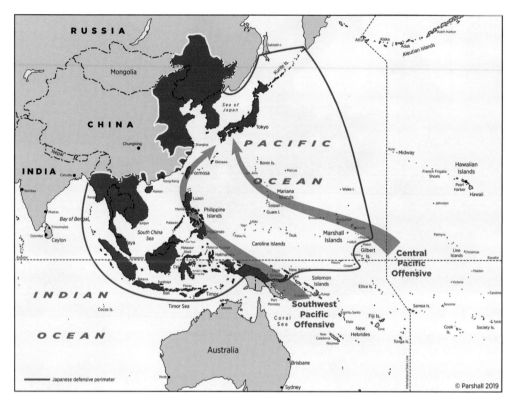

**Photo 31:** The two pronged attack through the Pacific with General MacArthur's thrust on the left and Admiral Nimitz's in the middle. John Parshall

**Photo 32:** Anti-Japanese and Hitler Poster. Author's Collection.

**Photo 33:** Anti-Japanese War Poster. Author's Collection.

China had blood on their hands. The Japanese military more broadly participated in individual acts of crimes against humanity than the Nazi military.

## Economic Spoils for the Empire of Japan

The Japanese also pillaged and plundered reaping a harvest of spoils in raw materials of rubber, steel, foodstuffs and oil. By 1942, Japan seized "all of the world's supply of quinine, most of its rubber and the greatest sources of oil in the Far East."[638] And slave labor was used throughout the Empire to benefit the Japanese economy. For example, under Kuribayashi's command in Hong Kong, the Kinkaseki Mining Company utilized POW "labor in inhuman, slave-like conditions."[639] In fact, 25% of all miners there were Korean slave laborers.[640] Therefore, Japan's business community knew where their increased profits came from.

Japanese soldiers clean their swords at Nankin, China, 22 September 1937. National Archives, College Park

## Knowledge of Japan's Crimes Around the World

Did Marines and Americans like Woody know about Japan's crimes? Throughout the 1930s and 1940s, many people knew of Japan's crimes by watching news-reels at the theater and by reading newspapers. They especially learned about the Rape of Nanking.[641] Learning about the atrocities, Roosevelt gave a speech on 5 October 1937 calling Japan a "warmaker" and a carrier "of dreaded disease" and called on the world to quarantine it.[642] So from the U.S. President to a lowly Marine like Woody, many knew about Japan's atrocities. Sergeant John Warner in Woody's unit, a China Marine assigned to the Peking (now Beijing) and Shanghai units, described how Japanese raped and murdered Chinese civilians.[643] Supporting Warner's testimony, General Clifton B. Cates, 4th MarDiv commander on Iwo and later the 19th Commandant of the Marine Corps (1948-1951), said the Japanese "were as cruel as could be. We…saw that in Shanghai… [saw] them shoot civilians and run a bayonet through them." Cates witnessed "beheadings" saying the Japanese mind is not the same as an American.[644] U.S. Office of Strategic Services called the Japanese headquarters of the Gendarmerie in China a "holy terror."[645] Female nurses on Hawaii, fearing an imminent invasion in 1941 after Pearl Harbor and knowing how the Japanese treated women in foreign lands, carried "pocketknives" resolved "to slash their wrists if the hospital was taken."[646] FDR claimed the Pearl Harbor attack provided the "climax of a decade of international immorality."[647] When the Japanese committed evil, they did so with "joyous zeal"[648] and often, with much fanfare making the news of their exploits travel far and wide.

## Do the Japanese Acknowledge Their Crimes?

As already detailed, few Japanese admit they did anything wrong. Whereas Germany has gone to great lengths to erect monuments to honor the victims of the Third *Reich* domestically, Japan has only erected monuments to its war dead in its country.[649] Their lack of remorse continues in overt ways: Pilgrims *today* go to the graves of notorious criminals like Prime Minister and General Hideki Tojo and Emperor Hirohito to worship their spirits. It is analogous to Germans laying wreaths at Hitler's grave (if there was one), yet this is what some do— even elected leaders like Prime Minister Shinzo Abe and his cabinet member

Minister of Internal Affairs and Communication Yoshitaka Shindo. This is most notable at Tokyo's Yasukuni Shinto shrine where 1,068 war criminals' "souls" are enshrined including Tojo, Prime Minister Kōki Hirota and General Matsui.[650] Before Hirohito died, he often worshiped there honoring those under his command who committed *his* atrocities.[651] This "controversial" place is not holy, but a "place of horror."[652] This shrine would be akin to Germans "erecting a cathedral to Hitler in the middle of Berlin" or a monument to the Deathhead-*SS*.[653] Japanese society openly parades its crimes and publicly honors its butchers.

## Short Review of the Yasukuni Shrine (Galley 1, Photos 28-30)

The Yasukuni Shrine is 150 years old. In addition to the 1,068 "souls" of criminals registered there, there are 2,465,138 soldiers' "souls" interred there as well, many of whom fought for the fascist regime of WWII.[654] Eight million worshipers visit it annually.[655] It not only honors those who have served the nation, but is also viewed as a place of good luck where people go to ask the Shinto gods "for help in passing college entrance exams, to get a good job, for healing from an illness, etc..."[656] So although war criminals are enshrined there, many others were decent citizens thus causing confusion about the shrine's meaning. Kuribayashi often went there and even took his children. His grandson, LDP politician Shindo, goes there regularly:

> This is the final resting place for my other grandfather and my uncle's souls. When I go there, I go there to visit them. I've brought my children there. I like the cherry blossoms and the place is...familiar to me. That other nations get upset that people go there is political. In short, we go there to console the souls of our ancestors.[657]

Since the entire nation of Japan mobilized for war during WWII, this shrine has a heavy emphasis on the years from 1931-45 when Japan offered up millions of souls to this shrine. In other words, many in Japan by the nature of their belief system feel attached to this shrine which causes confusion and controversy. For example, when Shindo visited it recently, China's government denounced him.[658] But Shindo snubbed the Chinese believing it is his religious right to honor the "spirits" of the war dead in general and of his family in particular. He is following in the footsteps, literally, of his grandfather General Kuribayashi, when he worships there. However, his actions, and those of thousands of other Japanese, sanction the continued christening of declaring men enshrined there as sacred because all souls in the shrine are apotheosized. Steeped in ancestor worship, Japan's citizenry refuses to question or interrogate one's past once his soul becomes "sacred" and takes up residence in the shrine. As a consequence, the shrine impedes mankind's struggle to honestly document the past because it prevents many Japanese, by the nature of its mandates, from critically questioning a person's past. Honorable Justice Liu Daqun, Judge of the Appeals Chamber of the International Criminal Tribunals for the Former Yugoslavia and Rwanda Genocide Trials said, "Those visits [to Yasukuni] have prompted many to allege that the country's official stance is one of defiance," and many of Shindo's and Abe's acts, claims and associations attest to this opinion.[659]

## The Japanese Soldier

The Japanese soldiers Americans fought in the Pacific were some of the most ruthless any modern nation has produced. Once on Guam or Iwo, Americans like Woody were up against experienced, well-trained, well-equipped and skillful adversaries. Japanese were committed fighters, willing to endure tremendous hardship and die for their Emperor. Woody said, "I don't know what motivated them, but they were the most committed and fanatical people I've ever…experienced…[T]hey were so crazy and possessed I sure hope in the next war they're on our flanks instead of in front of our lines."[660] These super-human deeds came in part from their belief in the Emperor who gave them "'fantasies of a strong father' in a more abstract and irrational way" than Hitler did for Germans.[661] Hirohito's subjects showed they were "able to…die for the sake of [their] ideas and values" and for him.[662]

Interestingly, their code of no surrendering motivated them to commit what Westerners would consider atrocities against their own. Japan's military treated surrender like a crime and those who were captured and returned to Japan were encouraged to kill themselves at home and many complied.[663] Before committing *seppuku* (ritualistic suicide by disembowelment, also known as *hara-kiri*), these POWs were already considered dead. "They could not return to their families," and even if they did, their families would reject them.[664] This would be like Americans telling POW survivors of the Bataan Death March to kill themselves after they returned to the U.S.

Japanese soldiers commit suicide (*jiketsu*) in a hospital in Luzon rather than surrender to the Americans, July 1945. National Archives, College Park

Moreover, if Japanese were unable to fight, wounded or threatened with capture, their own soldiers often killed them instead of leaving them to be taken prisoner. For example, "a tiny fraction" of the 8,212 Japanese wounded during the Luzon battle survived the war. Before leaving them in the field hospitals, their fellow soldiers murdered them or encouraged them to commit suicide. When the Allies found them, many "had killed themselves with grenades; some had been beheaded by their officers and others had been shot."[665] One commander ordered: "Concerning those wounded: Men who are slightly wounded will participate in this battle. When men wounded are not able to participate in battle their leaders will see to it they end their lives." At one Luzon hospital, Americans found 1,810 dead who either committed *seppuku* or *jiketsu* (literally self-determination, but in this context, suicide other than disembowelment) or were killed by comrades.[666] An IJA doctor volunteered to shoot patients in an act of "sacred murder" so they would not become prisoners, and often medics injected opium and corrosive sublimate (mercury chloride) into the veins of the injured. One soldier even killed his brother who could no longer fight.[667] On average, the Japanese had a fatality rate of 98% when in combat zones which has never been surpassed in modern warfare.[668] "While other armies merely talked of fighting to the last man, Japanese soldiers took the phrase literally, and did so."[669]

Japanese soldiers about to bayonet children in the Philippines, 1945. National Archives, College Park.

The different ways the U.S. and Japan conducted war illustrated the clash of two irreconcilable cultures. Japanese filmmakers sponsored a newsreel before Iwo, describing the conflict "as a suitable place to slaughter the American devils" and do *their* gods' work knowing they would all die doing so. The Japanese believed "to extirpate evil required the mass death of the Other, just as purifying the Self required acceptance of mass self-slaughter."[670] Long before America's fleet reached Iwo, Kuribayashi notified his troops they would never return home.[671] He "introduced the Kamikaze suicide doctrine…Men took oaths to undertake suicide missions if…necessary… Scarves with the Kamikaze symbol [神風]…

were distributed" to his troops.[672] It was as if Kuribayashi had inscrolled above his garrison's gates: "Abandon Every Hope, Who Enter Here."[673]

In contrast to Japan's philosophy of "kill all, burn all, destroy all," Roosevelt convinced the British, Russians, and Chinese to sign the "Declaration of the United Nations" in 1942, a statement "that pledged the Allies to pursue total victory (no separate peace treaties)... in order 'to defend life, liberty, and religious freedom, and to preserve human rights and justice.'"[674] Many would die to redeem this pledge. In waging unprovoked attacks, Japan brought the Allies' wrath upon them. By 1945, it was drowning in a flood of enemies.

The U.S. and its allies saved Asia from untold suffering by not allowing Japan's rule to go unchallenged. The Allied Tribunal, the Asian Nuremburg Trials, noted Japanese leaders:

> ...engaged in a conspiracy to commit atrocities that led to the 'wholesale destruction of human life, not alone on the field of battle...but in homes, hospitals, and orphanages, in factories and fields; and the victims would be the young and the old, the well and the infirm—men, women and children alike.'[675]

By insisting upon unconditional surrender, the Allies freed conquered peoples and safeguarded those who were threatened. In order to accomplish this mission, the Pacific War would turn into "one of the most brutal ever fought, on both sides."[676]

The Allies defeated an evil regime and stopped Japan's Holocaust. Although Japan did not set out to commit ethnic cleansing, it often treated the non-Japanese in conquered lands like the Nazis treated Jews, gypsies, communists and homosexuals. The Japanese killed their victims up close and inefficiently, yet with *gusto*. It took them *longer* to accomplish their killing by gang-rapping women, lopping off heads, bayonetting POWs, starving citizens and burying people alive than by using Nazi-like extermination camps. Germans occluded their crimes, only allowing a "select" few to participate in the extermination whereas the Japanese used military personnel and did not try to separate the people killed from those allowed to live. They also did not bury them in mass graves or cremate them— they just left the slaughtered to rot in the sun or threw them in a river or ocean.

Japanese troops committed extermination from the front lines to the rear echelons. Japan's Rape of Nanking killed 300,000 using 200,000 soldiers of the Shanghai Expeditionary Army slaughtering on average around 50,000 monthly during its five to six month campaign from 1937-38 in this one region alone (about 4,000 square miles or the size of Connecticut).[677] Newspapers and newsreel makers throughout the world reported the atrocities turning public opinion against Japan.[678] Neither Hirohito nor his subordinates stopped this slaughter: "War crimes may afflict all armies, but the scope of Japan's atrocities was so excessive and the punishments so disproportionate that no appeal to moral equivalency can excuse their barbarity."[679] Woody and millions of Allied servicemen defeated one of the most horrible regimes known to man and saved untold millions from further bloodbaths. This Holocaust "will forever haunt the old" Japanese military.[680] Paul Berman, author of *Terror and Liberalism*, wrote a "primer on totalitarianism" stating that such "pathological" regimes like Imperial Japan contain "a genocidal, and even suicidal, dimension…that get drunk on slaughter."[681] XIV U.S. Army Corps, after investigating Japanese atrocities at Manila, summed up what the Empire of the Sun was: "Japan…is truly an enemy of the civilized world."[682] Everywhere Japanese soldiers went, they built a "nest of wickedness."[683] "The world has long since called them blind, a people presumptuous, avaricious, envious; be sure to cleanse yourself of their foul ways."[684] By 1944-45, America and her Allies cleansed the world of Japan's filth and brought its "mad ambition" for world domination to an end.[685] And Woody and thousands of his fellow Marines would be at the point of the spear to take out some of the main bodies of criminal soldiers Hirohito had unleashed on the world.

# Ch. 6: Deployment—New Caledonia

"If a King is plagued by bandits, he must first find out
where their camp is before he can attack them."
—Buddha[686]

AFTER BOOT CAMP, THE MARINES went to additional training for their assigned disciplines, such as infantry, artillery, engineering, communications or other military specialties. On 21 August 1943, Woody trained as infantry working with tanks at Camp Jacques Farm in Southern California. Tanks needed infantry protection from enemy soldiers who might sneak up and disable them with an explosive placed in a vulnerable spot or a grenade dropped down a hatch. Symbiotically, Marines moved with tanks on open terrain or roads, using the tanks for cover, and utilizing them to machinegun or blast enemy strong points. Tanks were also employed to open paths in dense jungle that infantry would have otherwise had difficulty hacking through.

On 7 September, Woody switched to an infantry training school. During these two months, trainees learned how to work together as units with hand signals and formations in the field. Marines learned to fight, sleep and eat over terrain features they would attack. It was challenging, but the Marines were further molding these men into "fighting machines." "After that time," Woody said, "I knew how to work with my comrades and talk to them without uttering a word all with the purpose of attacking the enemy without him taking advantage of us or learning we were there."[687] On 5 November 1943, Woody completed the eight-week school.[688]

Camp Jacques Farm in Southern California in 1943 where Woody trained to work with tanks. Marine Recruit Depot Archive, San Diego

A few weeks before leaving for the South Pacific, Woody got sick with fever. In one of their last maneuvers, the trainees conducted a ten mile "hump" with full packs and gear. Marines often do 10- to 20-mile-hikes in formation with battle gear to build stamina, condition bodies and foster unit cohesion. However, on the day of the hike, Woody awoke with dizziness and chills. Although sick, he refused to report to sick bay. He hiked several miles, but when the group took a break at noon, he passed out. "I wasn't about to quit," Woody says. He was taken to the hospital where he was diagnosed with a 104° temperature, racked with chills and suffering from strep throat. Doctors wanted to remove his tonsils. However, the day the operation was set would have prevented him from leaving for the Pacific with his friends. He did not want to be left behind, so he convinced the doctors to release him with the promise that once aboard ship, he would report to sick bay and deliver his medical folder in order to have his tonsils removed.[689]

On 3 December 1943, Woody shipped out on the *M.S. Weltevreden*. He learned their destination was New Caledonia (a South Pacific island 2,000 miles east of Australia) where they would prepare to go to Bougainville in the Solomon

Islands north of Guadalcanal as part of the 32$^{nd}$ Replacement Battalion for those who had been fighting there since November. When on the ship, instead of reporting for surgery, Woody destroyed his "Doctor's orders." "They just got lost," Woody says with a sly smile. He was not about to go under the knife and let them take any of his body parts.[690] Once at New Caledonia, they were going to only be there for a few days before sailing to Guadalcanal to get further outfitted for Bougainville.

During this time, Woody's military career almost came to an end. During free time, Woody and his comrades explored the island, swam and sunbathed on the sandy beaches. One day, they crossed a lagoon to investigate the opposite beach when one of his buddies said, "Hey Woody, how about you climb one of these coconut trees and bring us down one?" That did not look hard. Without thinking, Woody climbed the tree and tried to get a coconut. "I thought if I could twist the coconut, it would come down. What happened next surprised me. When I twisted it, it snapped back and the force of it on my hand made me lose my grip on the…trunk. I was probably thirty feet in the air and when I lost my grip, I slid down the whole…trunk removing large sections of my skin on my heels, my chest, my arms and my face." Woody hit the ground bleeding and in pain. As a further introduction to his first real pain in the Corps, his buddies carried him back through the lagoon where saltwater burned his lacerations.[691]

Returning to base, they took him to the first aid station. The nurses and doctor called him an idiot and dressed his wounds. He could not move. Regardless of what he had done in his free time, he feared being court-martialed for the offense of not showing up for duty. Because Woody could not walk on his scraped feet, one of his buddies answered for him during roll call. The sergeant taking roll never noticed Woody's absence. After a week, he returned to duty. None of his superiors knew he spent the previous week convalescing in the medical hut. Luckily for Woody, the men did not have duties that required everyone to be present.

The last week before leaving New Caledonia, Woody tried out for a specialty fighting group, the *Raiders*. Everyone considered the *Raiders* to be the toughest in the Corps and Woody wanted to join. They were created in 1942 to be the first U.S. special operations forces to see combat in WWII. They had become famous for their heroics at Guadalcanal on Edson's Ridge, fighting waves of attacking Japanese who tried but could not break the defense.[692] Although Woody had scabs over his body, he signed up for the test—a 10-mile-run. He did his best, but he failed to make the time. It was just as well since, as warfare in the Pacific required large-scale operations, the *Raiders* disbanded in January 1944 and their personnel were reassigned to other units.[693]

After New Caledonia, they boarded ships for Guadalcanal on 22 January to get outfitted for Bougainville, the largest of the Solomon Islands (Woody was aboard the *USS Rixey*). This was part of Operation *Cartwheel* devised by the Joint Chiefs-of-Staff (American supreme command) to have General MacArthur and Admiral Halsey advance along the north coast of New Guinea and the Solomon and Bismarck Islands. Meanwhile, Admiral Nimitz would advance across the Central Pacific starting in the Gilbert Islands in November 1943. Halsey initiated the leap-frogging strategy in August 1943 of going around strongly held islands or defenses to strike at lightly held positions. This approach increased the effectiveness of the two pronged advance through the theatres commanded by MacArthur and Nimitz. (see Gallery 1, Photo 31)

In preparation for Bougainville, MacArthur and Halsey were in the process of skirting the Japanese stronghold of Rabaul where 100,000 troops were stationed.[694] Instead of fighting a bloody battle for Rabaul, MacArthur attacked weaker island positions, moving north toward the Philippines and Japan. Meanwhile, Lieutenant General George Kenney's Fifth Air Force pounded Rabaul destroying enemy forces. This strategy saved countless American lives. Guadalcanal and Bougainville were the southern part of these operations and those two battles started to push Japan back into its own borders. They provided the southern foundation the Marines, U.S. Army and U.S. Navy needed to project their power northward toward Japan's homeland.

# Ch. 7: Guadalcanal

"Kill Japs, kill Japs, keep on killing Japs."
—Admiral William F. Halsey[695]

AFTER CAPTURING THE BEACHHEAD AND airfield sites and surviving an earthquake on Bougainville,[696] the 3d MarDiv[697] was moved back to Guadalcanal, where Woody was already stationed having arrived there on 26 January 1944. His replacement unit would not be deployed on Bougainville, and remained on Guadalcanal awaiting orders. At Guadalcanal, which bore scars of intensive fighting, Woody continued his training for the next battle—the liberation of Guam. Woody was assigned to the 21st Regiment ("21st Marines") of the 3d MarDiv. He was attached to headquarters of C Company ("Charlie Company") in the 1st Battalion. An infantry regiment consisted of three battalions, each with around 33 officers, 1,000 men, two navy surgeons and 40 medical Corpsmen.[698] Before an operation, a regiment could be augmented. For example, Woody's 21st Marines could number 4,937 with tagged on units like it did before Guam. As augmented, it was called a Regimental Combat Team (RCT).[699] However, right before Iwo, it numbered 3,006 returning back to a traditional regiment.[700]

Woody's regiment was part of a newly formed division, the 3d MarDiv, "The Fighting Third," which was activated on 16 September 1942 at Camp Elliott in San Diego as the Corps was rapidly expanding. The 3d MarDiv's symbol was the caltrop, a three pronged figure anti-personnel/anti-horse weapon. In ancient days, it prevented full-frontal attacks. It consisted of an "iron ball and four metal spikes. It was designed so that when three spikes" touched the ground, another pointed straight up. As a result, the division's motto was "Don't Step on Me."[701]

While on Guadalcanal, besides training and learning about the next island to invade, Woody's platoon conducted patrols to hunt rogue Japanese remnants. It was hard for the new men to overcome their natural aversion to killing people but overcome it they did, especially after experiencing Japanese fanaticism. They focused on completing their assignments and returning home alive back to their families. Third MarDiv correspondent Dick Dashiell wrote his wife Vivian:

> There is not much personal to write…, really, except the old, but very, very true words, I love you. I often wonder how many times your walk, your talk, your smile, your sweet figure and all have flashed through my mind or lingered there for long, long minutes. Millions, I know, since that August afternoon on the Charlotte station platform. My love for you is a constant, growing thing that not even the finest poet could begin to describe. My whole life and future are anchored in you and Freddy [three year old son]—and with how much happiness I contemplate it. How much the memory of you two comforts me is something you'll never know.[702]

These men's homesickness was palpable and many struggled with it.

Another problem on the island was mosquitoes. Omnipresent and hungry, they swarmed the men, leaving behind itchy welts. While sitting down or conducting patrols, their skin continued to demand scratching and often the bite areas turned bloody. The men used a thick, oily liquid repellant called *Skat* which helped, but many nonetheless contracted malaria from their bites.

The natives on the island were also foreign to the Americans. Woody's company commander, Captain Donald Beck, described them:

> The natives here look like all south sea island natives. They are small and not quite as dark as our American Negro. Many…have reddish hair. I have never seen any of the women, some who have seen them say they are sad sights. They wear only a cloth around their lower extremities and therefore leaving their breasts exposed, and hanging nearly to their knees. I have talked with several of the men. They speak pigeon English and are of Christian faith.

An example of pigeon English is 'him go', meaning you are giving him an order to leave. They have a great passion for trinkets and jewelry. When they saw my ring and watch they were greatly attracted. They wear all sorts of gadgets in their ears. I saw one with a safety pin in each ear. This I assure you amused me a great deal. They have a great admiration for the Marines. They are lukewarm toward the [Army] soldiers and dislike the British.[703]

This description of Guadalcanal shows how primordial the people of the islands were the Marines were conquering. Although Beck had difficulty understanding their culture, he and his men had friendly interactions with the *Guale*. In Beck's opinion, the Marines were doing better than GIs and the previous rulers, the British, in fostering good relations. Missionaries from Spain and Britain had worked their magic with their proselytizing efforts, and the majority were Christian. However, although there were several natives on the island, the Marines carved out tracts of lands where they could train and conduct amphibious landings. As for Beck, he had some interaction with the locals, but mostly spent his time fighting the insects and conducting training.

At first, on Guadalcanal, Woody was a Browning Automatic Rifleman (*aka* BAR man).[704] The BAR was a light machinegun which usually had a bipod, a V-shaped device that held up and steadied the barrel. It fired .30 caliber rounds from a twenty-round magazine—it was a powerful albeit heavy weapon to carry, weighing in at 20 pounds. BAR men often carried 13 to 14 magazines (1.5 pounds each).

After a few weeks in this role, Woody was assigned a new duty: a special weapons expert. This attached him to company headquarters to ensure supplies for special weapons personnel were ready to be deployed quickly. Because of his short stature, Woody was chosen to be a flamethrower. A flamethrower operator had a support team whose job was to suppress enemy fire aimed at the flamethrower and follow up with explosives for the target.

Flamethrower operator was an occupational specialty that did not allow many survivors. Japanese feared backpack and tank flamethrowers and directed fire at them. One man in Woody's Iwo unit observed: "Our flamethrower moved up towards a pillbox, but was shot through the neck and died almost immediately.

Our lieutenant started to strap the flamethrower gun on himself, but was hit by a spray of bullets right through the stomach [and bled out.]"[705] Unlike tankers, flamethrowers had no armor for protection, only their cotton dungarees. Woody "heard" the average flamethrower's lifespan in combat was five minutes.[706] On Iwo Jima, one battalion (1,000 men) in Woody's 21st Regiment (3,000 men) suffered over 90% casualties among its flamethrowers and 3d MarDiv commander, Major General Graves B. Erskine, claimed the casualties, especially at the campaign's beginning, were "very high."[707] Veteran Eugene Sledge wrote: "Carrying tanks with about seventy pounds of flammable jellied gasoline through enemy fire over rugged terrain in hot weather to squirt flames into the mouth of a cave or pillbox was an assignment that few survived but all carried out with magnificent courage."[708] Considering the occupational hazards in the Pacific War, being a flamethrower was probably the most dangerous. (see Gallery 2, Photo 1)

This duty of being a flamethrower required Woody to also have TNT, satchel charges, and pole charges in addition to the flamethrowers to be ready to go at any time (this also earned him the name of "demolition man"). Satchel charges were bags of explosives that were to be thrown into a pillbox, cave or other area to destroy it and a pole charge was C-2 explosive attached at the end of a 8- to 10-foot long 2-by-2 wood pole used to place an explosive into a bunker's small opening or enclosure to blow up the structure.[709] C-2 is a moldable, "Play Doh"-like substance detonated by electrical impulse or a fuse.[710] They came in bundles of eight small boxes, with each box about the size of a stick of butter. Four of these could bring down a two-story building.

The flamethrower was the most unique of Woody's weapons. This was a new weapon and his team worked extensively on how best to deploy it. Starting in 1915, the Germans had used the flamethrower with deadly effect in the trenches.[711] The Japanese had used it on Wake Island, Hong Kong, Bataan, the Philippines and Guadalcanal during the battles against Allied forces from 1941 through 1942.[712] Americans had first used them in Papua New Guinea in December 1942, but their experience with them was a failure while using the poorly designed M1 model.[713] The flamethrower Marines used in 1944 and 1945, especially the improved M2-2, had two tanks about the size of scuba tanks, full of flammable fuel (4.5 gallons) plus a slightly smaller center tank of compressed air or nitrogen (nitrogen was preferred because it was not flammable).

All three tanks were welded together onto a backpack with shoulder straps. The compressed air canister pushed air through the hose connected to the three foot wand or gun which delivered burning fuel on a target. When fired with full force, the weapon blew fuel at a "rate of one-half gallon per second." It was heavy and awkward machinery, but deadly.[714]

On Iwo, "The flamethrowers suffered particularly heavy casualties. The Japs feared and hated their fire. When they saw one of our flamethrowers advancing, they concentrated their weapons on him. Our men swore the Japs hurled their mortar shells not at our units, but at our individual flamethrowers."[715] The oil in the diesel produced a black smoke as it burned so a flamethrower had to constantly move or the enemy would zero in on him since his battlefield signature was large. Sometimes it was not flamethrower smoke, but its sound that elicited counter fire: "When the pressure is released [from compressed nitrogen], there is a sharp hiss that immediately discloses to the enemy your exact position. 'Then,' [a veteran explained], 'you start getting hell from [him].'"[716] The weapon itself was also a hazard:

> [The] mixture of air under pressure into gasoline produces a highly flammable mixture and… gun hoses and connections have a tendency to leak under severe combat conditions. A tank, being of metal construction, containing many moving metal parts and an elaborate electrical system, is a constant source of sparks. If compressed air is used, one spark near a leaking hose would cause a violent explosion.[717]

Often in battle, these instruments exploded due to malfunctions and enemy fire (likely hitting the tubing when the flame ignited). In other words, when in a foxhole, one tended not to share it with a flamethrower. Nonetheless, without the flamethrower, war in the Pacific would have been considerably slower and more challenging.

When Woody received his flamethrower in 1943, the weapon was so new it did not come with a manual. Woody and his NCOs worked hard to discover the ideal mix of diesel and gasoline to produce the most effective flame. They also experimented using kerosene, motor oil and a napalm-like jelly called "phosphorous gel" or "phosgel," none of which worked as well as the diesel and gas mixture. If one used only diesel, the fluid was too heavy and the flame traveled only a few feet. Such a weapon had no effect on the battlefield unless one was to pour his flame into an opening in the earth. If one just used gasoline (130-octane aviation gas), then the flame lacked body and sprayed quickly into the air and was swiftly extinguished. If mixed in a 1:1 ratio, however, these two fuels gave enough body and heat (power density) to send fire twenty to thirty yards to the operator's front, unleashing hell on the battlefield.[718]

Woody's regimental report from the 2nd Battalion noted the men "preferred a mixture of gasoline and oil rather than the napalm mixture because it would carry farther."[719] Once engaged, an operator had 7-9 seconds of fuel with the trigger pulled before the tanks emptied.[720] It was a terrifying firearm: Former Marine Commandant and 4th MarDiv commander on Iwo, Major General Cates, said he could not think of a more inhumane weapon.[721] Marine Eugene Sledge wrote:

> I shouldered the heavy tanks, held the nozzle in both hands, pointed at the stump about twenty-five yards away, and pressed the trigger. With a whoosh, a stream of…flame squirted out, and the nozzle bucked. The napalm hit the stump with a loud splattering noise. I felt the heat on my face. A cloud of black smoke rushed upward. The thought of turning loose hellfire from a hose nozzle as easily as I'd water a lawn back home sobered me. To shoot the enemy with bullets or kill him with shrapnel was one of the grim necessities of war, but to fry him to death was too gruesome to contemplate. I was to learn soon, however, that the Japanese couldn't be routed from their island defenses without it.[722]

The men practiced using this weapon for hours, learning the best methods. The firing apparatus had two triggers. There was one on the wand's front where there was a cylinder with phosphorus matches to activate the flame and another at the back to send out the fuel using compressed air. Each flamethrower had five matches to the cylinder so they were used sparingly. A match stayed lit for ten seconds. The supply of "fuel" lasted 7-9 seconds when engaged, so the operator pumped the trigger in bursts to conserve fuel and direct the fire.

Whether taking out personnel in the open, in pillboxes or in trenches, one just had to ignite the first thrust of fuel and then the targeted area would already be on fire, thereby providing the flame the fuel needed to ignite subsequent bursts, so one just had to continue to hit the back trigger to release more fuel on the burning area. Releasing the fuel in bursts was the most effective method to send out more fuel to eliminate a target. Once the flame left the nozzle, it rolled out to approximately twenty feet in diameter as it created a huge fireball of force.[723] Learning how to properly operate a flamethrower was tough and Woody says controlling the flame was a skill only obtained with several weeks of training. He stated, "I can't tell you the number of times I singed off my eyebrows and the hair on my arms learning to use that rig."[724]

It was crucial for the flamethrower to fire his weapon from the correct stance. The propellant for the fuel had to be powerful enough to reach a target without getting too close. That gave the weapon a kick when compressed air was released from the center tank, forcing the barrel up as it fired. Unless the operator was prepared to hold down the flamethrower as it fired, and lean forward to absorb the kick, the barrel could wind up pointing to the sky. Not only would he miss the target, the fire would rain down, incinerating him. That happened in one instance when a trained flamethrower was brought down and an inexperienced Marine picked up the weapon and fired it. He did not lean forward, was saturated with flame and was killed in an instant.[725]

Many expressed concern about being a flamethrower operator for a variety of reasons. Carrying so much gear made them an easy target. They looked like hunchbacks. And since they were carrying metal tanks with highly explosive fuel next to a tank with compressed air, they worried their hardware would not stand up to shrapnel or bullets. This would make them walking bombs.

Hearing these concerns, one of Woody's NCOs tested the flamethrower's ruggedness. So they shot up some tanks with high caliber weapons to see what would happen. Much to everyone's relief, even .30 caliber rounds from the BAR rifle would not penetrate the tanks (the rubber tubing connecting the wand with those tanks was another matter). In short, although the tanks were tough, they still knew they would become targets and have difficulty finding cover carrying the metal backpack while engaging the enemy.

Despite this vulnerability, they recognized the flamethrower would be one of the most effective weapons to use against emplacements: Bullets cannot strike targets inside pillboxes or trenches like a fast-flowing stream of liquid, flaming gas. The flamethrower became one of the best weapons against strongly-fortified enemy positions, even more so than tanks, naval gunfire, cannons, claymores, mines and satchel charges.

Nonetheless, many have asked why battle tanks or cannons weren't used to take out enemy positions. The problem with tanks was they were unable to operate on some hilly terrain or on the soil on some of the islands, or where the enemy had constructed tank traps, and there were not enough of them to handle the multitude of fortified positions the enemy had created. Naval gunfire was great, but battleships were so far away it was difficult for them to make a pinpoint hit on a target. Moreover, it sometimes took hours to get the requested naval gunfire or close air support, and in the meantime Marines were vulnerable to enemy fire. Once the Marines landed, great skill in targeting was required. A similar caution applied to artillery, which fired smaller shells from closer distances. There were a lot of enemy fortifications that, as a practical matter, only a Marine with a flamethrower could effectively neutralize.

It was remarkable how far the U.S. had come with flamethrower technology in the short amount of time. After the Nazis had effectively used flamethrowers (Model 1935 *Flammenwerfer*) and demolitions in May 1940 against the Belgian fort of Eben Emael and secondary positions around the fortification such as machinegun nests, the Chemical Warfare Service and the Army Engineer Test Board collaborated in the development of experimental flamethrowers (E1 and E1R1s).[726] By March 1942, the model M1 started to arrive in the Pacific. It was poorly designed and suffered from defective battery ignition systems, broken fuel lines, leaky valves and frequent loss of pressure.[727] Improvements were made and the first significant use of flamethrowers by the U.S. Army and the USMC was the M1-A1 in late 1942 and 1943.[728] Yet, these versions were inferior and often the battery operated ignition system failed to ignite the fuel exposing them to counter fire.[729] Often, they used "tracer bullets, white phosphorus shells, or hand grenades" to set "target areas on fire."[730] The M2 and M2-2 which Woody used were vastly improved when they arrived in mid-1944. The M2-2 had a more rugged design and an improved ignition system that used a housed match lighting mechanism in the wand's nose.[731] This ignition system was a "waterproofed revolving cylinder which worked much like a revolver" with the matches inside.[732] By the time Woody hit Guam and Iwo, he wielded the most advanced flamethrower in the world.

On 9 January 1944, Woody received his first promotion to temporary Private First Class receiving the responsibilities of a PFC and being addressed with that rank, but not paid at that rank. Then by 21 January, he became officially a PFC with the pay of that grade. On 16 March 1944, Woody wrote his girlfriend about completing "flame throwing" school. He told her that he had nicknamed his weapon "Ruby" after her.[733] During training, barely a week went by that Woody did not receive letters from Ruby or write Ruby himself. Although their courtship had been brief, he felt she was unique and was smitten. He wrote her on 12 April 1944:

> I received the long, sweet and wonderful seven page letter from you this evening. Thank you a lot for that letter, it's the sweetest letter I ever read. It made me very lonesome, and also very happy. I want you so bad Ruby. It seemed as if you were talking to me in person instead of reading the letter. Ruby, I am so very proud of you and your love.[734]

In this letter, he declared his full-on commitment to her and apologized he was unable to get her an engagement ring before leaving overseas. "I am very proud to be engaged to you…When I can call you my wife and put the ring on your finger, I will be the happiest man alive in the world."[735] Although Woody had not given her a ring, he had asked "the question" before he left for the Marines. Even so, he still called her his "girlfriend" and never used the term fiancée. Woody explains, "I was head over heels in love and adored this woman and even though I was going to marry her, I only called her girlfriend, never my *fee-annn-cay*. That was just not a term we used."[736] Regardless of the idiom he used for Ruby, she was the most important person in his life. Often, the emotions of leaving for battle got the best of young lovers and they accelerated their relationships.

Woody's relationship strengthened because Ruby had proven a loyal girlfriend and he had grown more fond of her while away although he was also writing his ex-girlfriend Eleanor Pyles too. Maybe he hedged since he was going to be gone a long time and perhaps Ruby would find someone else so he kept someone in the wings. Woody's old girlfriend wrote him that she had seen Ruby in town with another guy. This upset Woody and on 12 April 1944, he wrote, "Ruby I never doubted your love for a single second, but as you say when I was told that they saw you with someone, I couldn't help but wonder a little bit." He admitted he got this information from his ex-girlfriend and then had to explain why he was writing her in the first place: "The only reasons I was writing [my old girlfriend] was because I was sorry for what I did [apparently Woody had some cross-over with the women and broke up with Eleanor to be with Ruby.]" Once Ruby found out about this "other woman," Woody justified the relationship in writing to Ruby as being, for him, now platonic and on "friendly terms, but she won't accept that. When I quit dating her, I told her we were through, that we just didn't click." However, Woody knew Ruby did not believe this and she questioned his dedication. As a result, Woody declared his love for Ruby to prove she was the one. Ruby was his first choice and she caught him with "his hand in the cookie jar" and probably convinced him to drop the "backup" and focus on her since she truly cared for him. Woody told Ruby he would write his ex one last time. He wanted to tell her he didn't believe her claims about Ruby and not to write again. Then he pledged: "Always remember Ruby that I love you and trust you."[737] Right as Eleanor was causing all this drama with Woody and his new girlfriend, she decided, like Woody, to also join the Marines and

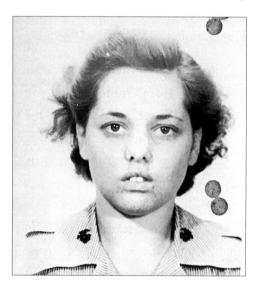

Private Eleanor G. Pyles. This was Woody's girlfriend before Ruby although she was also married at the time. It seems Woody dated both Eleanor and Ruby at the same time, or at least was writing both simultaneously while in the Marine Corps, causing some trouble for him when Ruby found this out. Soon after Ruby discovered this information, Woody cut off all ties with Pyles and then started to just focus on Ruby. Pyles entered the Marines one year after him, but soon left in disgrace having secretly married again and become pregnant. St. Louis Personnel Records Center enlisted almost one year after him on 2 February 1944 serving at Camp Lejeune and Cherry Point, North Carolina.[738] So, soon after Woody wrote Ruby about Eleanor in April 1944, Woody would not even have had his spy in Eleanor even if he had wanted to keep her checking up on Ruby's extracurricular activities since she had left to become a Leatherneck as well.

In contrast to the acceleration of relationships like Woody's, sometimes distance destroyed a couple. Because Marines rarely had options to date on islands they occupied, it was often the "girlfriend" state-side who found another lover and wrote a "Dear John Letter" or a break-up letter. One of Woody's comrades received such a letter and he was so depressed he stuck his rifle barrel in his mouth and blew his brains out. As a result, Woody's superiors called a meeting:

> Now look boys, there're millions of women…in the world. If one
> decides not to sleep with you anymore or love you any longer, go
> find another…Blowing your brains out over a girl is moronic and
> a waste of a bullet. Don't get depressed over a Dear John letter
> because many of you will receive one before this war is over.[739]

Woody admitted that had Ruby sent him such a letter, he would have been depressed. However, he "picked a good woman" back home and she remained loyal.[740]

Overseas duty put strains on many couples. Woody's commander from Iwo, Donald Beck, had an exchange with his wife Ruth that showed they both wondered what the other did in their off hours. Beck's letters discussed that the local Guam girls took up with men married or not: "Some are quite attractive and many…men and officers date with them a lot." [741] He expressed that he had danced with them, but nothing else. His son Mark notes, "Don would have been better off leaving the comments about the local women out of [his letters]." [742] It didn't help in one letter after he had received a new photograph of her, that Beck observed, "It appears that you have gained some weight [men sometimes just do not think things out with women]." [743] Ruth addressed many of his comments in a letter and Beck, in return, explained:

> I hope you have never worried about my behavior. I'm sure you
> haven't, but since you mentioned it I must assure you that it has
> never been anything but excellent. You haven't been stepping out

with any of those old broken down boys have you? Better not because Barry [their one year old son] is my chief spy and he will tell me all about it. Your comments were fine in regard to my morals, and I hope they may always be up to your expectations.[744]

As one can see, separation stressed relationships. From Woody explaining why he wrote his ex-girlfriend obtaining "info" about his "girlfriend" to Beck explaining to his wife she had nothing to fear from the beautiful tropical girls he was with while questioning her loyalty, these men continually navigated the problems of jealousy and communication. The pressures of not being with a lover made things trying for even the healthiest of relationships.

But like with Woody, when Beck was defending himself for his actions on Guam, he was effusive about his love writing things like "Darling, this is all for now. I have certainly missed you of late. I would just like to take you in my arms and hug & kiss you forever."[745] So, men like Beck and Woody tried to balance out the range of their emotions with words of affirmation.

While on Guadalcanal, little news reached the men for fear that any information received might be obtained by the enemy. Each letter written had to be submitted unsealed so a censor could ensure it would not reveal valuable information. Even if they were allowed to disclose their location, they and their loved ones had little comprehension of the geography or importance of the islands on which they fought.[746] Woody wrote Ruby in 1944, "I am on a ship and am in the Pacific somewhere, but that's all I can tell you."[747] For a short time, Woody wrote his girlfriend with a "secret understanding" that he would start each paragraph with the first letter of the name of the island or place he was so she could track him on a map, explaining all this in a letter on 17 November 1943 while stateside, and thus, not yet subjected to censors.[748]

Eventually, the censors caught on, because in one letter, they scissored out the beginning words of his paragraphs. One told Ruby in a note at the bottom of her letter: "Tell Hershel to stop using code and his letters won't be cut up. Censor."[749] Even if Ruby knew Woody's whereabouts, both had no understanding of where he was or what his unit was about to do. One veteran explained, "Our vision of the war was largely tunnel vision. To…us the most important place in the world was his foxhole. The impact of MacArthur's and Nimitz's twin offensive was lost on us. Most Marines were as ignorant of geography as their families at home."[750] (see Gallery 2, Photo 2)

Marines knew nothing about the progress of worldwide campaigns. The war seemed like it would last forever, and they did not understand the complexity of or the time required to complete their missions. Some optimistic souls chimed "Golden Gate in '48," hopeful they could win the war and sail home entering San Francisco by the Golden Gate Bridge by 1948.[751] No one knew how long it would take to defeat Germany and Japan.

Fifth MarDiv chaplain Rabbi Roland Gittelsohn who saw war from the troop ships and the battlefield lamented the Marines' lack of geographical and political awareness. They were fighting for freedom and democracy, but "the inability to realize this end vision was America's greatest failure and potentially its most disastrous defeat." The chaplain believed the Marines "were probably the least politically aware since all servicemen in that branch—regardless of their ultimate specialty—were considered infantrymen." Since Marines were absorbed with the daily tasks of warriors in combat, "the big picture goals for which they were fighting would sometimes be overlooked."[752] Woody supports this analysis, "I was there simply to defeat the enemy who attacked us."[753] The Marine Corps created excellent warriors to defeat Japan. Had the chaplain seen the war at the top levels, he would have found a strategic vision and determination to not only win the war, but also to create conditions and institutions which would democratize and liberalize the fascist enemies to prevent WWIII.

Every day was one of preparing for the next campaign so the big picture may have been lost on them. They kept their reasons for fighting simple often just saying, they were there to defeat Japan. The South Pacific's oppressive heat clung to them, and white salt crystals ringed their clothes from perspiration. Not only did they train physically, but they also conducted classroom instruction as well. They learned tactics, terrain navigation, and radio communication. They practiced assaulting beaches using the landing craft to which they would be attached when the assaults began. Moreover, their division's intelligence branch, under the direction of Major General Turnage, focused on the psychology of how to view the enemy. The inability of conditioning hatred towards their enemy before combat worried many commanders. Waiting to experience war's horrors to develop the right mindset cost several their lives before they could acclimate to combat. As a result, a training brief made the rounds throughout the 3d MarDiv that encouraged hatred.[754]

This brief described how commanders observed when Americans arrived at the front, "they did not hate their enemy." Without hatred, the brief argued, the men would "fear" the enemy giving the enemy the upper hand. It was evident from combat that IJA troops did not fear but hated Americans. Creating similar hatred in American boys was difficult since Americans were "different. [They are] fair minded and [think] that the enemy will be fair too."[755] The report continued:

> Until John Doe [average American] learns to hate he will be no good. As long as he regards his opponent as a good fellow, a man who, after all, does not really want to fight and kill him, John Doe will go into combat carelessly and not aggressively. He won't go and look for the enemy; he won't want to kill, to destroy; to win. When the enemy proves to be stronger, John Doe will not hold out and counterattack; try to beat the enemy with the last ounce of energy, beat him by his stronger will…Hate is like gin. It takes awhile, and then, suddenly, it hits you. After you have seen your buddies killed; after you see bodies, or what's left of them, piled up for burial; when you realize that they are after you too; when it finally connects in your mind that moral code does not exist in this war, then you will begin to hate, and want to retaliate. A [Marine] has to develop the primitive instinct to kill anybody who threatens him or his own. Beyond that he must learn to kill before the other can get at him. Until he hates the enemy with every instinct and every muscle, he will only be afraid…This is primitive psychology, a cruel and inhuman one. But war is all that. Hate must become first nature to a soldier and make him want to use every trick.[756]

Obviously, using hatred was one way to fight the Japanese. As elucidated earlier, the Empire of the Sun had conditioned its population to hate foreigners and the corresponding lack of empathy made it easy for Japanese to slaughter other Asians and Americans. This "deep sense of superiority" fostered a widespread mentality among Japanese that committing atrocities was permissible "without a prevailing feeling that these were morally reprehensible."[757] American boys were at a disadvantage in war because they did not have such mindsets. These 3d MarDiv officers who drafted this report believed Americans had to not only hate the Japanese, but also in order to win, they had to hate more than the Japanese hated them. This document showed that fascist regimes' escalation of fanaticism created a resolve in the American soul to fight *and hate* their enemies.

The report admitted that the Marines would not understand and embrace hatred until they had confronted the enemy, experienced combat and witnessed buddies getting ripped apart. Conditioning men that it was acceptable in war to hate and kill, they re-engineered the Judeo-Christian education many arrived to the Corps with, namely, "love your enemies" (Matthew 5:44) and "thou shalt not kill" (Exodus 20:13).[758] The report declared the will to kill the enemy must become a passion and to feed it, hate was a necessity. The new "morality" contradicted the Bible and men were taught that the higher good was to kill their enemy, not love him.

The draftees of this document noted how strange hatred training was. The Japanese and Nazis would not qualify hatred of the "other," but American authorities who sanctioned this document needed to justify what they preached. Calling this training "cruel and inhuman" was not hyperbolic on the part of those who assembled the analysis. However, becoming hating warriors was a necessary evil if the Marines were to defeat the most brutal enemies Americans had faced since confronting certain Native American tribes on the frontier.[759] As U.S. Ambassador to Japan, Joseph C. Grew, noted, "To shape our foreign policy on the unsound theory that the other nations are guided and bound by our present standards of international ethics would be to court sure disaster."[760]

How did the Marines teach hatred? Woody does not remember, but it was nonetheless disseminated and his commanders intentionally sanctioned the new language necessary for such training. When they conducted bayonet practice, the dummies were made to look like Tojo. At bases throughout the world, posters showed Japanese raping and destroying societies. Images of Japanese looking like rats, vampires or ghouls hung everywhere. When conducting classroom work, pictures of Japanese slicing off heads of innocent Allied POWs were shown. One poster showed the brutal Bataan Death March of American POWs with the declaration: "Stay on the job until every murdering Jap is wiped out!" Dehumanizing the enemy was widespread, describing the Japanese as apes, yellow monsters or bucktoothed, smiling, grotesque killers. This indoctrination gave permission to hate, something not part of the curriculum of American schools. The nickname of "Jap" used for Japanese had a negative connotation, and American servicemen also used terms like "Ring Tails, "Tojos," "Rice Bags," "Nippos" and "Nips" (taken from *Nippon,* Japanese for Japan) to define their enemy.[761] (see Gallery 1, Photos 32-33 and Gallery 2, Photos 3-4)

One training manual discussed how Marines could differentiate their Japanese enemies from other Asians. This racist document described the Japanese as having long torsos, but short legs, hardly any facial hair and "the eye of the Japanese is set at a slant to the nose." "Japanese teeth tend to 'buck' or protrude, and the nose lacks a distinct bridge." "Practically all Japanese are pigeon-toed and many are bow-legged, so that in walking they usually shuffle along."[762] Two army officers wrote in 1942 that the average Japanese soldier was 5'3" and 117 pounds and basically a "runt."[763] We know it was nothing physical that caused the Japanese to behave barbarically, it was the way they were indoctrinated in their society, brutalized in the service and led by their superiors all the way to the highest echelons that resulted in the militarism, aggression and atrocities. Every ethnicity has unique phenotypical and physiological traits, but there were many terms here to denote contempt.

And besides American leaders uniting their warriors by focusing them against a common, demonized enemy, they also created songs, creeds and poems to build *esprit de corps*. For example, Woody's 21st regiment came up with its theme song which read:

MEN OF THE TWENTY FIRST
We come from different states, We Joined the Corps on varied dates,
We laugh at fear and scorn the fates, We aim to knock down Tojo's gates,
WE'RE MEN OF THE TWENTY FIRST
We left our wives and sweethearts home, We fight for freedom not alone,
We hope to sail across the foam, We care not where by chance we roam,
WE'RE MEN OF THE TWENTY FIRST
We work and play through day and night, We fight to live and live to fight,

We never of our goal lose sight, We pledge ourselves with all our might,
WE'RE MEN OF THE TWENTY FIRST
We know the day is coming near, We look to it without a fear,
We realize well why we are here, We feel that we have not a peer,
WE'RE MEN OF THE TWENTY FIRST
We know that when our battle's won, We'll see the end of Nippon's sun,
We'll say "God Speed, our task is done, We're one for all and all for one",
WE'RE MEN OF THE TWENTY FIRST[764]

The song was one of unity and vision building a brotherhood focused on destroying Japan. It was a pledge to stick together through tribulations and work towards creating a lasting peace. The regiment's song never wavered from knowing the U.S would win and held the optimism that victory was coming.

This song discussed that despite being from all over the nation, they worked toward one common goal. Captain Beck often talked about his men, their diverse backgrounds and the different states from which they came writing that some of them came from the *Lone Star* State of Texas, the *Live Free or Die* State of New Hampshire or his state, the *Wonder* State of South Dakota, "it's a wonder it's a state."[765] Although from diverse backgrounds, the love of country and the shared determination that they wanted to defeat Japan united them into a strong fraternity.

Captain Donald Macaulay Beck stood 5'10" and weighed 160 pounds. He was a quiet man with an athletic build and dark hair. His attractive face had clear eyes and an aquiline nose framed by a square jaw and prominent forehead. He was always calculating and thinking ahead. He was born on 22 January 1917 in Gregory, South Dakota, on a ranch and grew up taking care of horses. The family lost the ranch in the Great Depression and he, like Woody, joined the Civilian Conservation Corps in 1934 and worked there for three years. He finished college at the University of South Dakota where he majored in History and Political Science, ran track and was a member of Sigma Alpha Epsilon fraternity. He entered the Corps soon after graduating college in May 1941.[766] He was a well-educated, dignified and intelligent man. (see Gallery 2, Photo 5)

When Beck's regiment was not training in the field, learning in the classroom or creating songs or bonding as a sacred brotherhood, some conducted patrols in the jungle. Guadalcanal had been the Marines' first major military success. The 1st

MarDiv established Henderson Field like an unsinkable aircraft carrier protecting the American lifeline to Australia and projecting air power against the Japanese. Although the Japanese navy evacuated 13,000 of an original 36,000 troops back in February 1943 after abandoning hope of winning control of the island, there were still many Japanese on the island. These desperate remnants of the defeated forces would not surrender, and they harassed the Americans. So Woody's fellow Marines had to search for and kill these diehard elements.[767]

This was a mission for riflemen, not flamethrowers. Since Woody was attached to headquarters of C Company, he remained at base and continued to train and ensure the demolitions were always prepared. Woody's comrade, Lefty Lee, and others in his unit who were strictly "grunts" (riflemen), conducted daily patrols and, according to Lefty, sometimes found Japanese. Lefty said one day, they were ambushed as they hiked over a jungle hill. Bullets started flying everywhere and they returned fire, closed with and killed the Japanese.[768]

Sometimes, before they found the enemy, they came across other Marines who were not so lucky. One group had been waylaid, and as Woody's platoon discovered their dead comrades, his platoon observed their mutilated bodies. "Although these Japanese knew they had lost…, they still behaved in such a way that would ensure we gave them no quarter," Lefty said. "The bodies of our buddies were sliced up like someone had taken a butcher's knife to them. Their clothes had been removed, their legs spread and their genitals cut off. Their dead, gaping jaws had been opened and their members had been shoved in their mouths."[769] Sometimes the Japanese even took the time to drape placards around the dead reading, "It took them a long time to die."[770] On Bougainville, where Lefty also served, Marines retrieved fallen comrades only to find the Americans staked to the ground "through their arms, chests, and legs." One Marine sent to get these bodies was a brother of one of the dead: "It's tough seeing your brother like that," said veteran Bill Faulkner.[771]

If the Japanese thought atrocities would terrify the Marines into running away, they were wrong. Such sights turned some Marines vicious. Lefty said:

> One guy we had…would tie the ears of Japanese he had killed
> around his neck. He volunteered for every mission to take point,
> a hazardous position…If he'd been stateside, he would've been

put in a mental ward. However, in combat, he was brutal and hated the Japanese with every fiber of his being.[772]

Abuse begets abuse. Still, this Marine's behavior in Western society was abnormal and prohibited whereas Japanese behavior of mutilating enemies was institutionalized and approved by their leaders. Woody said:

> You knew you didn't want to become a prisoner. We knew that the Japanese thought it the greatest disgrace to surrender. I'm sure I might've done so had I had no other choice, but in general, we weren't going to give up…The few [Japs] who surrendered…usually did so because they were too weak and malnourished to offer resistance. They surrendered because their bodies had given out. Many…Japs we found were mere skeletons. Yes, we knew they slaughtered our men, but in general, when we could, we showed compassion. We did have a few who had necklaces of gold teeth they had taken from dead Japanese, but this was only done once the Japanese were dead and they did not mutilate the bodies. Everyone had ways of getting their souvenirs…Many Marines just enjoyed getting Japanese flags, swords and gold teeth.[773]

One of the few souvenirs Woody's fellow flamethrower comrade, Corporal Joseph Anthony Rybakiewicz, brought back home was indeed one gold tooth he had taken from an enemy soldier he personally had killed plus his medical kit and photo of his wife and two kids.[774] The taking of gold teeth was so popular, that one of Lefty's comrades, Al Tommasi, asked if he could take his gold tooth if he ever died in combat to which Lefty laughed and said, "sure… just make sure I'm dead when you do so."[775] A small minority of servicemen even took Japanese skulls to hang on jeeps. The war so brutalized some, it made them do things they would never do in an ordered society.[776] One man was so grateful for navy nurse Kathryn Van Wagner's kindness and care flying out of the Hell of Iwo Jima that he gave her a green bag, looking at her with compassion and sad eyes and saying, "I want you to have this." Van Wagner opened the bag up and shockingly found gold teeth and "bloody, dried-up Japanese ears." She gave them back and said she "couldn't imagine using them." He was sorry this was all he could offer her but this was all he had.[777] "As early as September 1942, Admiral Nimitz ordered that 'no part of the enemy's body may be used as a souvenir' and warned that violators would face 'stern disciplinary action.'"[778] The Joint Chiefs-of-Staff issued directives to stamp out the practice. Japanese propaganda relayed these stories to their servicemen and several had these images of the Marines in their heads creating more hatred, because the Japanese abhorred "the desecration of the dead" although they desecrated bodies themselves.[779] War often exposes mankind's hypocrisy.

The war brutalized many and they started to treat body parts of the dead enemy as trophies. Here, U.S. military personnel have taken a Japanese soldier's head and impaled it on their tank on Guadalcanal. National Archives, College Park.

And although some Marines and GI's killed enemy combatants who had surrendered or who were defenseless, violating the Hague and Geneva Conventions, these were rare occurrences. Jim Skinner said while going through a tunnel in

Iwo, he came around a corner where he met a Japanese with his hands up and he was so surprised that a "Nip" was trying to surrender because he had only experienced Japanese feigning surrender to kill Americans that he just blew the "Jap" away. "I shouldn't have done so, but I just didn't know what else to do."[780] Gunnery Sergeant (known as "Gunny") Keith A. Renstrom had his men finish off several Japanese soldiers who had botched their suicide attempts or were severely wounded in battle on Tinian and Saipan by shooting them in the back of their heads.[781] Army Lieutenant Sigmund L. Liberman, a Jewish-American army officer who had fought in the Battle of the Bulge and had received the Purple Heart and Bronze Star admitted that after he and his unit had liberated the Mittlebau-Dora concentration camp, he was so enraged that the next time they took *Wehrmacht* prisoners, he made sure the "12 men," on their way back to a drop-off station, "simply did not make it." When asked why he killed unarmed prisoners, Liberman said after seeing so many of his Jewish brothers and sisters dead at the camp, he wanted revenge.[782] Also enraged by the crimes they had witnessed, Lieutenant William P. Walsh and Private Albert C. Pruitt did the same thing Liberman had done, killing four *SS* guards "disguised" as medics on the Dachau death train.[783] Although these instances with Skinner, Renstrom, Liberman, Walsh and Pruitt were rare and illegal, they did happen. However, compared to the Japanese who killed POWs and Allied wounded as a matter of course, the killing of IJA and Nazi prisoners and injured soldiers by Americans was sporadic and atypical.

As for Guadalcanal, though jungle combat patrols were dangerous, they were an effective way to accustom Marines to combat. While on Guadalcanal, few had an inkling of what lay ahead. They were stationed there until their next engagement, Operation *Forager* in the Marianas. Unbeknownst to them, their 3d MarDiv was originally going to invade Kavieng, New Ireland, but Nimitz's rapid advance through the Pacific changed his plans, and his sights turned to securing airfields for bomber bases in the Marianas. Some 3rdMarDiv's units were already underway to Kavieng when the campaign was cancelled and returned to Guadalcanal. Most never knew they were spared from taking another "damn island." Kavieng was bypassed and the Japanese there were left to starve. "At least 148,000 Japanese" soldiers died nearby on New Guinea during the Pacific war, not due to combat, but because they had run out of food and water.[784] Historian Edward Drea wrote that the majority of Japanese fatalities occurred due to logistics failures, not combat. The IJA's "incompetence killed more Japanese soldiers than did the Allies."[785]

# CH. 8: The Marianas

## (See Gallery 2, Photo 6)

"Sail On, O Union, strong and great!/ Humanity with all its fears,/
With all the hopes of future years,/ Is hanging breathless on thy fate!"
—Henry Wadsworth Longfellow[786]

A MAJOR OPERATION DIRECTED TOWARD ending the war was the quest to control the Mariana Islands, one of the main "strongholds in the Japanese defensive chain."[787] The major objectives in the Marianas were Saipan, Tinian and Guam. American control of the Marianas would permit long-range heavy bombers to make regular, round-the-clock raids on Japan and provide deep-water ports.[788]

Originally, the Marianas were not a prominent part of the U.S. pre-war plans. But Admiral Ernest King, who simultaneously held the two highest navy offices (Commander-in-Chief, U.S. Fleet and Chief-of-Naval Operations) knew the invasion of these islands would probably "draw out the remnants of the Japanese Imperial Fleet for a decisive naval battle" the U.S. Navy would win.[789] He felt the Marianas were a key to winning the war, viewing them as a stepping stone heading north to Japan from Guadalcanal and Australia.[790] According to historian Phillips Payson O'Brien, Chairman of the Joint Chiefs-of-Staff, Admiral William D. Leahy, King's boss, was the driving force behind this overall strategy saying, "The Marianas had to be taken as they were on 'a direct line with Japan' and could best be supplied by American air- and sea-power."[791] General Hap Arnold, Commanding General of the Army Air Forces, arrived at a similar conclusion. One of the most expensive weapon systems developed during the war was the B-29 Boeing Superfortress four-engine heavy bomber and it needed islands for its airfields like the ones in the Marianas (billions of dollars were thrown into this bomber project making it even more expensive than the Manhattan Project). The B-29 was called the "battleship of the skies," weighed 65 tons, had a wingspan of 141 feet and was almost 100 feet long. It could carry a 20,000-pound payload and its flying range was 6,000 miles. "Twelve .50 caliber machine guns and one 20mm cannon" defended its crew. Additionally, its "pressurized and heated cabin allowed crewmen to fly distances in relative comfort at high altitudes without the need for oxygen masks or heavy suits."[792] Arnold placed his hopes behind this airplane to bring down Japan.[793] In 1944, the B-29 could reach Tokyo and the Kanto Plain's industrial region from the Marianas, an island chain that could support 1,000 of these bombers. As historian Allan Millett put it, supporting General Arnold's conclusion, having Saipan and other bases in the Marianas provided the "advanced bases for the strategic bombardment of" Japan. Thus, on 12 March 1944, the Joint Chiefs-of-Staff approved the invasion of Guam, Saipan and Tinian with a target date for D-Day (Invasion Day) of June 15.[794] Often D-Day is associated with the 6 June 1944 Normandy invasion, but D-Day, A-Day or W-Day were terms used for the landing day for many Allied invasions.[795]

In March 1944, U.S. commanders met in Hawaii and planned Operation *Forager* (Marianas' campaign) and Operation *Stevedore* (Guam's liberation).[796] Lieutenant General Smith (V Amphibious Corps focused on Saipan and Tinian) and Major General Roy Geiger (III Amphibious Corps focused on Guam) shared quarters as they planned operations. Geiger looked like a tackle football nose guard or middle linebacker and was a "gruff, plain-talking" Marine with silver hair and ice-cold, cobalt eyes. He was known for his ability to hyper-focus and get his staff to over-plan for the missions at hand.[797] Smith described Geiger "as a heavyset, bear-like and totally fearless man...who could only have happened in the...Corps." Rear Admiral Richard "Close-in" Conolly worked with these Marine generals and commanded the navy's Southern Attack Force (Task Force 53).[798] Conolly said of Geiger, "That's the best relationship I ever had, anywhere, anytime, during the war." This was a team of top-notch leaders. They planned in detail, and after winning the Battle of Saipan, Geiger, Conolly and staff conferred and modified future plans for taking Guam based on the latest intelligence and hard-won lessons.[799] Seizing the Marianas would not only give America an advance base for the B-29s to hit Japan with devastating force, but it would also cut off strategic lines of communications and supply for the IJN, provide a forward base for submarines to operate in enemy waters and set up a staging area for the fleet. Guam, 210 square miles in size, was the primary objective because of its excellent airfields and one of the best deep-water Mariana harbors.[800] Accomplishing these goals required a lot of fighting.

Marines had to conduct extensive training to ensure that when the Marianas were attacked, they could be taken. They trained daily, learning how to operate with one another and handle their weapons. Their instructors, often through surprise, gave them every situation known on the battlefield. What happens when your platoon sergeant dies? What happens when your heavy machinegun fails to fire? What if it was missing a part? What happens if one of your buddies gets his artery sliced open? How do you stop the bleeding? How do you destroy pillboxes?

"It was like drinking from a fire hose," Woody said describing everything they learned. He continued to load and unload from ships to Higgins boats (LCVP for Landing Craft, Vehicle, Personnel), assault beaches and learned to handle ocean-to-shore engagements. He was amazed by the beautiful beaches, blue waters, raw smells of sea water, tropical islands and fish. After living in land-locked West Virginia, he felt he had landed in a foreign world as he traveled throughout the Pacific. "Without the war, I would've never had experienced the type of travel…such ventures before the war was only for rich people or government officials. I was getting a first-class world cruise on the taxpayer's dime."[801]

Most who trained with Woody had never seen combat, but many who taught them had. The eternal quest in education is to make the theoretical count practically. Stateside, Woody's girlfriend Ruby worried he was in harm's way, a reality all deployed Americans faced. Knowing he would soon "see the elephant,"[802] he wrote Ruby:

> I don't know wheather I have been putting it in my letters or not. Ruby, don't worry about me when I go oversea's cause I will get along okey. You see Ruby we get about 3 more Months of training over there before we go into actual combat. Ruby I love you and I want you to be happy. I know you won't and can't be exactly happy, but do the best you can [sic.].[803]

Woody's commander at Iwo, Donald Beck expressed the same sentiments with his wife Ruth: "I hope you don't worry about me, for I would feel badly if I thought you did. There is nothing to worry about. If you don't hear from me for some time that is still no need to worry. I may be out here for a long time and if you worry you won't be as pretty when I come back."[804] While Woody and Beck did their best to learn about war, train with others and calm fears of fighting the Japanese, they occasionally focused on consoling their women back home telling them they had nothing to fret about. This reality of social dynamics is often overlooked during war. The human cost to loved ones as they send their men and women off to war is often underappreciated. Letter writing was not always as skillful as just shown. Months later, Beck wrote his wife that he had survived Bougainville, but lost a lot of men, especially his "heavy machinegun officer [who] was killed right beside me."[805] This letter about losing men especially right next to him contradicted his earlier letter of consolation.

As Woody, Beck and others were about to meet the enemy on some far-off Pacific island, they had mothers, sisters, brothers, girlfriends and wives concerned for their welfare. If they died, their deaths would devastate those left behind. It happened worldwide: American, British, Japanese, German, Russian, etc. The communal plea of "please make it home" from family members in every corner of the globe was repeatedly spoken. Their gods and words may have varied, but the prayer for safety was universal. The cost of war is much more than body counts. Marine veteran T. Fred Harvey remembered his mother, a Comanche Indian and not one anyone messed with, telling him before he left: "Sonny, now you listen to me. There are four things you must mind me about. When the war is over, you better come home to us. You don't come back though as a drunk and you better not come back a *Coward*. Oh, and don't be coming back with any of those tattoos."[806] Harvey's Mom was an example of millions of mothers worldwide as they coped with sending their sons to Russia, to China, to Germany or to the Pacific to places they knew boys would die.

Globally the emotional strain was palpable—in print and newsreels, but mostly in living rooms. Mothers ached for the safety of their children. Spouses hoped for the long, shared lives promised when they spoke their vows. Lovers craved the touch of their beloved. Daughters missed their fathers and the physical and emotional security they provided their families. In this emotional crisis, the lines for the need for war and the need for each other became blurred as the heartache of war trudged on. But there was that spark of hope beyond reason which remained a light in the darkest nights. A letter, like the ones Woody wrote Ruby or Beck for Ruth, was a glimmer of hope to which they, and millions of other women, held fast.

After the Marianas campaign, thousands of citizens would never receive another letter, kiss or hug from their loved one because they had given their last full measure taking the islands necessary to ensure victory. Woody's letter to Ruby was an example of thousands being written: "Don't worry about me. I'll be Ok. I'll come back." Woody knew some would not come back. He knew Ruby lost an uncle in WWI. In the end, Woody and most of his comrades did come back to their loved ones, but there would be a minority in the hundreds of thousands who would die. So Woody prepared for combat and consoled his loved ones the best he could.

# Ch. 9: Japanese Attack and Occupation of Guam 1941-44

"Every morning be sure…to think of yourself as dead."
—18th Century *Samurai* Death Code[807]

THE UNITED STATES ACQUIRED GUAM as a result of the Spanish American War in 1898. Guam, the largest and most populous island in the Marianas in the North Pacific and in Micronesia, was located on the route from North America to the Asian continent and the Philippines and Japan.[808] Coconut groves flourished under clear skies and aqua ocean waters lapped white sandy beaches that surrounded the island. Jungle growth covered the interior. Guam and its sister island Saipan "were big, rugged islands, dominated by steep peaks, yawning gorges, undulating tablelands," fields of sugarcane, and thick vegetation.[809]

Guam was a coaling and supply station for ships crossing the Pacific and later served as a stop for planes like Pan American Airways clippers. It was populated by 22,000 indigenous Chamorro farmers and fishermen and their relatives, for whom family, church and village lifestyle were vital. The later first Chairman of the Joint Chiefs-of-Staff, Fleet Admiral William D. Leahy, described the people before the war: "They have little and want nothing. Living in little grass huts one story above the pigs and chickens; wearing what clothes they can find, or none at all, and subsisting on whatever fruit happens to grow near, they appear to be happy and contented beyond the usual lot of humans."[810] Most were Catholic with a Protestant minority and almost all were, simple, good-natured and kind people. Although not U.S. citizens during WWII, they had many of the rights citizens enjoyed. They spoke English, were permitted to travel to the U.S. mainland and were given U.S. identity documents.

Chamorro people on Guam were focused on family, fishing and farming. They were good-natured and a friendly people totally unprepared for the brutality coming their way when the Japanese took over their island in December 1941. MARC

A number of Chamorro served in the military, with 12 dying on ships during the attack on Pearl Harbor.[811] A small U.S. government presence existed on Guam, which before WWII included 153 Marines, 271 sailors, five nurses, 134 civilian construction workers, 247 insular Guard members (Chamorros under Marine NCOs) and a few civilian administrators. The governor was a U.S. Navy captain.[812] The garrison had a decommissioned oil storage ship, a lightly armed mine sweeper and two patrol boats.[813]

In Asia, the attack on Hawaii happened on 8 December 1941 since Pearl Harbor lies over the International Date Line. On this date, Chamorros were celebrating the Feast of the Immaculate Conception as Japanese planes attacked.[814]

Panicked parishioners exclaimed *"Asaina Yu'os! Santa Maria!"* ("Lord God! Holy Mary!"). As the capital Agaña (Hagåtña) erupted in flames as bombs rained down, frightened people exited to the hills inland. By sunset, the city was empty. Two days later, 6,000 Japanese sloshed ashore and overwhelmed Guam's garrison.[815] The small, lightly-armed American detachment put up a defense, but in the end, it was "manifestly hopeless."[816] The most powerful weapon the U.S. had, the minesweeper *USS Penguin*, engaged planes with antiaircraft fire, but when its loss was certain, it was scuttled to prevent capture. When it went down, the only guns on the island larger than .30 caliber went to the ocean's bottom. The garrison's commander, USN Captain George J. McMillin surrendered on 10 December. The Japanese commander, Major General Tomitarō Horii, a seasoned veteran of the Sino-Japanese War, then commenced with the occupation.[817]

Japanese invasion of Guam December 1941. War in the Pacific National Park Museum

To hide his true intentions, Horii issued a proclamation that "welcomed" the 22,000 Chamorros to "The Greater East Asia Co-Prosperity Sphere" (*Dai Tō Kyōei Ken*) in its pursuit to do away with the "White Man" in Asia and create a Pan-Asian realm. He stated:

> It is for the purpose of restoring liberty and rescuing the whole
> Asiatic people and creating the permanent peace in Asia. Thus
> our intention is to establish the New Order in the World…You
> all good citizens need not worry anything under the regulations
> of our…authorities and my [authority] [sic] enjoy your daily life
> as we guarantee your lives and never distress nor plunder your
> property.[818]

These soon proved to be empty promises.

*Kwantung* army troops who had occupied Manchuria garrisoned Guam. The Japanese seized farms and their rule entailed arrests, beatings, interrogations and torture. They forced civilians to grow crops, build fortifications, lay mines, dig caves and provide women for their pleasure. Those who refused were executed.[819]

The Japanese occupiers changed Guam's name to *Omiyajima*, Great Shrine Island, and the capital Agaña to *Akashi*, Red City, in honor of the rising sun, and they taught citizens Japanese culture. They also gave other towns and landmarks Japanese names, forced the population to learn them and predicted their new civilization would last ten thousand years (they would be off by a little more than 9,996 years).[820] Japanese was taught and English was banned and those caught speaking it were severely punished.

Captured U.S. servicemen were sent to POW camps. Many died *en route* to Japan in "hell ships" in which prisoners were packed together for weeks in hulls without nourishment. Many went mad or starved to death. A doctor with a stash of morphine committed suicide. "[C]razed by thirst, [some] went berserk; they slashed at the throats and wrists of companions to suck blood." Others licked and slurped up sewage from "open drains."[821] Many who survived the journey would not survive the camps.

Approximately 14,000 Japanese troops seized government buildings and private homes on Guam. Over 2,000 residents were evicted from their residences and some were placed in concentration camps such as Manenggon in 1942. There were numerous reports of rape.[822] In studying the Rape of Nanking, the Rape of Hong Kong and the Rape of Manila, and after going through the numerous War Crime affidavits on Guam with Park Ranger Kina Doreen Lewis, this study concludes that the vast majority of women and young girls on Guam were raped by Japanese soldiers since they often, in occupied lands, went "girl hunting" *all the time*. Often the reports looked at just talked about violence and abuse, but did not use the term rape although it was clear this was what in reality happened. Since the garrison outnumbered the female population by at least 2:1 and given the penchant for IJA and IJN men to rape anybody they wanted, the women on Guam had no chance to go unscathed for the almost three years they suffered under Japanese rule. "Rape is immeasurably traumatic experience for the females involved; it leaves lifelong emotional scars. Given the shame-inducing nature of this crime, victims naturally wish to keep it secret" especially for such a Catholic community like Guam.[823] "Comfort Women" were indeed brought in to join the already abused native girls—authorities established a brothel with Japanese, Korean and Chamorro women to be raped, sometimes numerous times a day.[824] The Japanese even conducted medical experiments on civilians.[825] Japan did all it could to demoralize, debase and destroy the society of the Chamorro people. As one IJA soldier claimed of his time in China, but it might as well as been from Guam: "The only skills I picked up after half a year in combat were how to rape and loot."[826]

Although possession of radios was forbidden, a number of Chamorros secretly had them. They picked up news from a station broadcasting from San Francisco's Fairmont Hotel. Despite the occupiers' efforts, the people later would get word American forces were defeating the Japanese everywhere.[827] At first, radio reports were horrible, but within six months, Guam's citizens knew America was on the offensive.

Meanwhile, the occupiers wanted to increase agriculture to support a garrison of 30,000 troops. So forced labor increased. Food was seized from islanders who received back paltry rations. By early 1944, women were used to tend crops. Men were forced to construct tank traps, fortifications, and dummy fortifications, and transport food and ammunition. If they did not perform well, their masters beat and sometimes executed them. By Operation *Stevedore*, Japanese had massacred hundreds. In forced marches starting on 10 July 1944, like to Manenggon concentration camp, some were beaten to death. The Japanese moved 80% of the population to seven camps, areas of utter deprivation: "there was very little food or medicine, no potable water, no sanitary facilities, and…only makeshift or temporary shelter from the torrential rains."[828]

Photo of Chamorro skulls. The Japanese executed these victims by decapitation during the occupation. Guam Public Library System

Some Chamorros were ordered to dig graves before being beheaded. Edward L.G. Aguon watched soldiers execute four people. The soldiers tied them to a cotton tree. Then, the "Nips" slapped and beat them playing with them as a cat does a mouse. Thereafter, the executioners bayoneted them even after "it was obvious [they] were…dead." "[T]he most painful thing I remember… [was] to see the looks on their faces when the final stab of the bayonet pierced their flesh; to hear their cries, as their last breath left their bodies." Executions increased in mid-1944, and included teenagers discovered in the jungle looking for food; they were tied to trees and decapitated.[829] The Japanese executed one man when they found out his son was a U.S. sailor.[830] Toward the end of Japanese rule on Guam, public executions became "frighteningly common."[831]

Japanese beheading of Father Jesus Baza Duenas on Guam on 12 July 1944. Guam Public Library.

This photo shows Japanese atrocities with them beheading three men on Guam. This happened shortly after they took over the island in December 1941. National Archives, College Park.

The Japanese imported Korean forced laborers to work next to Chamorro slaves. Many who failed to perform were whipped, beaten or slaughtered. Several, especially native children, died from malnutrition. The population decreased by 10% during the occupation, claiming 2,000 lives.[832] Japanese authorities on the island promulgated that if America returned to Guam, it would find an island of rotting corpses full of "flies."[833] Well, in the years they controlled Guam, they were indeed accomplishing this reality proving "Japanese civilization had come to Guam."[834]

As the war spiraled into defeat for the Japanese, their terrible behavior intensified. "As the Marianas-based Japanese prepared to defend the islands to the death, Chamorro lives became expendable. Atrocities increased in both frequency and ferocity."[835] This was especially the case for those Japanese who had been stationed in China who had "slaughtered Chinese peasants."[836] Once the population was placed in the concentration camps, hundreds died of disease and malnutrition, "in conditions of indescribable squalor."[837] The natives anxiously awaited the Americans' return and hoped they would kill every one of the "yellow devils."[838]

# Ch. 10: Battle for the Marianas: Background and Preparation

"The factor most responsible for the
miseries of mankind is man himself."
—Mikiso Hane[839]

The U.S. amassed 600 ships from the West Coast and locations in the Pacific, carrying over 300,000 men including 127,571 assault troops (2nd, 3rd, and 4th MarDivs, 1st Marine Brigade and the Army's 27th Infantry Division) for Operation *Forager*, the invasion to take the Marianas (about the size of the force invading North Africa).[840] It would be a "classic example of operational maneuver from the sea."[841] This operation "into the heart of the Japanese defenses threatened their north-south lines of communications. Allied possession of the Marianas isolated the Carolines to the south and endangered Japanese sea lines of communication to Rabaul in New Guinea and Truk." Control of the Carolines and Marianas also protected the right flank of MacArthur's upcoming Philippines invasion. Possession of the Marianas exposed many of the remaining Japanese positions and opened more Allied operational options than the Japanese could defend against: "South to the Carolines and Truk, southwest to the Palaus, west to the Philippines, northwest to Okinawa, or north to the Volcanoes and Bonins [including Iwo]."[842] Taking the Marianas was like cutting out the train hub that all traffic had to go through. Hundreds of ships took the men on Guadalcanal where Woody's 3d MarDiv was stationed and sailed toward the Marianas.

As the Saipan Battle started, another division, the Army's 77th Infantry Division in Hawaii, was transported to join the fight for Guam. Some ships in the invasion fleet would be at sea for months, during which time the navy had to keep them supplied. In an average supply ship there were enough rations to feed 90,000 for a month. "Reefer" vessels (refrigerated cargo ships) had to be loaded to provide a balance of fresh food to one ship after another.[843]

The navy provided fresh provisions for crews of combatant ships five days out of six, and for sailors and Marines ashore, one day in three. Six fleet oilers and 40 chartered tankers carried 4,495,156 barrels of oil and eight million gallons of aviation gas to forward areas in July 1944. No ship or plane missed action for want of fuel.[844] The Task Force 58 included fifteen aircraft carriers, divided into four task groups.[845]

This was an amazing feat coming two and a half years after Japan devastated the Pacific Fleet, and it was done simultaneously with the Allied invasion of Normandy, half way around the world. The logistical challenges, in the day of pen and paper, were immense. The Americans could take vengeful pleasure knowing that as the U.S. fleet appeared off the Marianas, the fleet commander who attacked Pearl Harbor, Vice Admiral Chūichi Nagumo, was "trapped" on Saipan and "frantically wired Tokyo" begging for help.[846] Some battleships his forces had sunk at Pearl Harbor, "much to his surprise," joined the fight at Saipan and bombarded numerous targets.[847]

The U.S. had indeed recovered, if not strengthened considerably beyond recovery, unimaginably quickly. Historian Lynne Olson wrote that when war started the U.S. could not even repulse Mexican bandits crossing the Rio Grande. In 1939, America had the world's 17th largest army, sandwiched numerically between Bulgaria and Portugal, not a proud position nor among respected company.[848] The Nazi military attaché in Washington D.C., Lieutenant General von Boetticher, predicted on 4 October 1940 that the U.S. did not come close to having the military to stop a Japanese campaign in Asia and, at that time, history proved him right.[849] The U.S. had around 500,000 men in its armed forces at the beginning of 1941 and there were not enough weapons for this force.[850] In contrast, the Germans had 5.8 million by April 1940 in its *Wehrmacht*, Italy had 3 million in its armed forces and Japan had 2.3 million by November 1941 in its military.[851] America was heavily outnumbered. Things had to radically change if the U.S. was to avoid a disastrous defeat and rally to the aid of the remaining free countries.

Admiral Chūichi Nagumo, Chief of the Naval General Staff and the fleet commander of the Pearl Harbor attack force. He would later blow his brains out while trapped on Saipan and facing utter defeat. National Archives, College Park.

As Axis countries overran many nations, Hirohito believed in victory against the U.S. and "exercised his authority…to start war against the United States."[852] He planned to attack what would become "the most powerful country in the world."[853] Admiral Yamamoto realistically noted that just taking some American Pacific possessions would not be enough to defeat the U.S., but the

Japanese would have to march to Washington D.C. to dictate "the terms of peace in the White House" which he knew was beyond Japan's capabilities. He feared politicians in their "armchairs" failed to understand this reality.[854] He believed Japan could not actually fully defeat the U.S.[855] For expressing such convictions, although he was a highly respected and accomplished military leader, many ultranationalists wanted to kill him for even mentioning such defeatist beliefs. As a result, Prime Minister and Admiral Mitsumasa Yonai sent Yamamoto to sea in August 1939 as commander of the Combined Fleet to escape a madman's blade or a pistol shot in the back.[856] Iwo Jima's commander, Kuribayashi, also spent years in the U.S. as Yamamoto did, and privately wrote that America "is the last country in the world Japan should fight."[857] Yet it would take years for Yamamoto's and Kuribayashi's fears to be realized.

Although America was a giant in manpower and manufacturing, it took months to mobilize its military and industry. While that U.S. giant grew into its full size, Germany, Italy and Japan, with the help from other Axis powers, pummeled America's beleaguered allies, destroying military and civilian shipping, slaughtering millions of soldiers and civilians, bombing cities, and capturing and fortifying vast amounts of territory that would take years and the lives of millions to retake. But America would rise to the occasion. After the first year at war, the U.S. increased its arms production equal "to the total of all three enemy powers put together."[858]

"It's a truism that governments seldom operate efficiently until they're squeezed by violence or the threat of it."[859] Not only did the U.S. out-produce the Axis countries for its own military, it also provided the supplies Britain and China needed to survive Hitler's and Hirohito's onslaughts, while "supplying the Soviet Union with fully two-thirds of their motor vehicles and one-half of their planes...and supplying thirteen million Soviet soldiers with their winter-boots," uniforms and blankets to fight Hitler.[860] The U.S. sent food to the U.S.S.R that saved millions.[861] Stalin toasted at the 1943 Tehran Conference: "To American Production, without which this war would have been lost."[862] Without the U.S., Germany might very well have conquered the Soviet Union. America also kept Britain in the fight with supplies provided between the defeat of France in 1940 until the end of 1941.[863]

Without the U.S., Japan would have generally tormented Asia in its pursuit of its Pan-Asian movement and in particular, dismembered China. By 1942, the U.S. provided some supplies to China and sent millions of men and their machines and weapons against the Empire. By the war's end, Japan had deployed 16 carriers to fight the U.S., whereas the United States had deployed over 140 carriers of all classes. Japan outnumbered America with aircraft carriers at Pearl Harbor, and its pilots and torpedoes were superior, but by war's end, the U.S. had the capability to bring 15 to 20 times the fleet to Japan's shores compared to what Japan brought to Pearl Harbor in 1941.[864]

Even with these facts, as its battleships lay burning in Pearl Harbor, the task facing America in 1941 was daunting. By early 1942, Japan ruled its domain with little opposition. That Japan had conquered large portions of Asia and around half the Pacific seemed to prove they were, at this stage, invincible. In order to fight Japan, the U.S. had to develop its capabilities of projecting war from the sea. And one of the main services to help America with this herculean task was the U.S. Marine Corps.

# Ch. 11: Amphibious Warfare: The Marine Corps' *Forte*

"We and all others who believe in freedom as deeply as we do,
would rather die on our feet than live on our knees."
—President Franklin Delano Roosevelt

BY SUMMER 1942 "TOKYO HAD overstretched itself and put most troops in the wrong places, but it did not recognize that. It had achieved enormous territorial gains, its homelands were intact, and the booty from its conquests was pouring back home."[865] The loss of four carriers at Midway, MacArthur's progress on Papua and battles around Guadalcanal seemed distant and unimportant to the Imperial Headquarters' generals who were taught the Prussian strategy of controlling the mainland.[866] Yet, within the country, the number of families who had sacrificed a son in war continued to increase as a reminder Japan was not doing well. By that time, thousands of little wooden boxes wrapped in cotton cloth had arrived to bereaved families with what they were told were the cremated remains of their fallen men.[867] Although people congratulated families who had sacrificed a son to the nation, many questioned their leaders' decisions.[868] As comfort, many were told by their priests that "loyal, brave, noble, and heroic spirits of those officers and men who have died shouting, 'May the Emperor live for ten thousand years!' will be reborn right there in this country. It is only natural that this should occur."[869] Nevertheless, the promise of rebirth did not fully console many families as they started to question whether the war with the world was worth all the sacrifice. Control of the mainland started to waiver as the fortunes of war turned.

Japan's outer defense perimeter began to shrink throughout 1942 and 1943. By 1944, cracks in Japan's armor appeared, although no acknowledgement of it surfaced in the tightly-controlled Japanese press. Instead, as in tyrannical regimes throughout history, the press wrote about fantastic victories and that the military was cleverly leading the Americans into a trap to destroy them.[870] In 1942, America won battles at Midway and Guadalcanal. In 1943, America killed Admiral Yamamoto in an air interception raid, taking out one of Japan's most brilliant military minds. American forces then followed by conquering the rest of the Solomon Islands and with Australian troops, made advances in Papua New Guinea. In 1944, the U.S. was finally able to conduct large-scale operations against Japan itself.[871] It was hoped that once Guam, Saipan and Tinian fell, the U.S. could heavily target Japan's home islands from the air. Hirohito understood this fact and issued a proclamation to troops on Saipan that they must be victorious so the bombing of Tokyo would not increase.[872] But in order to get the bases for these planes, one had to secure islands from the sea, a laborious procedure technologically and logistically.

To accomplish this feat, the U.S. entrusted much of the preparation for their Pacific campaign to one of the most hot-tempered, tough, perfectionist Marine generals in history, Lieutenant General Holland M. "Howlin Mad" Smith. He was one of the fathers of Marine Corps amphibious warfare, one of the most "perilous" forms of operations.[873] His ideas and foresight with the Higgins boats and ship-to-shore tactics were critical for victory.[874]

Born on 20 April 1882, Smith grew up in the Deep South in Alabama, was the grandson of two Confederate veterans and knew how to hunt and live off the land. He attended the University of Auburn (at the time, called Alabama Polytechnic Institute) and after graduating with a Bachler of Science in 1901, he entered Law School at the University of Alabama graduating in 1903. With such education, he defied, like most Marines, the stereotype that many feel Jarheads embrace; namely, knuckle-dragging, low-brow Neanderthals without any brains. During this time, he had become a first sergeant in the Alabama National Guard and liked soldiering. Instead of joining the Protestant clergy like his mother hoped or the family law firm like his father desired, he dedicated his life to the military after practicing law for only one year. He began his service in the Corps in 1905 and served against rebels in the Dominican Republic in 1916-

17. During WWI, Captain Smith and other Marines were shipped to France to fight the "*Krauts.*" Later, Smith fought at Chateau Thiery, St. Mihiel and Belleau Wood in 1918.[875] For his actions in combat, he received "the *Croix de Guerre* [French "War Cross" for valor]...by Brigadier General [James] Harbord for services in action against the enemy."[876] His citation read:

> During the operations of the 4th Brigade in the Belleau Wood and vicinity he displayed the finest courage and remarkable ability in performing duties to the utmost importance, maintaining liaison with all the units and collecting information concerning the enemy under extreme artillery and machine gun fire.[877]

In Captain Smith's first 13 years of service, he had experienced much of the world ending with fierce combat against the Germans on the Western Front. Since his WWI combat experiences were some of the most dramatic in his life up to this point, he began to study tactical, operational and strategic lessons from the Great War, especially since he was appointed as a General Staff Officer in December 1918 and then the Assistant to the G-3 (Operations) officer of the U.S. 3rd Army.[878] He drew many insights from his studies, thinking always about how he could use them to strengthen "*his*" Corps. Before returning to the States in March 1919, he obtained the rank of major.

During America's involvement during World War I from 1917-1918, at the insistence of the army (particularly General John J. Pershing), Marine units fought mainly in the 2nd Infantry Division and under army command, and these army commanders were not always fair to Marines in Smith's opinion.[879] As a result, he felt in the next war, Marines should do all they could to fight under Marine command, and he, and many other Leathernecks, made sure this would happen.

Also, after WWI, Smith, and other Marines took an acute interest in the major amphibious operations of the Great War, led by the First Lord of the Admiralty Winston Churchill. These amphibious landings against the Ottomans at Gallipoli from 1915 to 1916 were disasters. The poorly-equipped, -supported and -trained Commonwealth attackers suffered 97,000 wounded and 44,000 dead in a losing effort. The Marines learned from studying Gallipoli how not to attack from the sea. Consequently, after 1918, Captain Smith and many other

Marines took a "keen, active and progressive interest in amphibious operations" to justify themselves as a separate entity. "The result of their efforts was to be seen in a combination of convincing warfare doctrine, improved technological and logistical assets, and well-trained specialist units that [would] forever identified the Corps with massive and effective assault from the sea."[880] Although Marines had practiced this throughout the Corps' existence, conducting 160 assault landings before WWII, these operations needed new equipment and tactics.[881] While the Corps focused on amphibious operations, it had a long way to go to perfect techniques to assault heavily fortified enemy beaches. Few nations had focused on this at the time making Marines unique.[882] And by the Marine Corps focusing on such an important and distinctive form of sea maneuvers, it could also ensure autonomy away from the U.S. Army.

Before Pearl Harbor, under now Brigadier General Smith's leadership, the Marines conducted several landings on the East Coast from destroyers. Those early landings were initially fiascos, but the Marines learned valuable lessons, especially with problems with ship-to-shore troop deployment.[883] Andrew Jackson Higgins, an inventor from New Orleans and a friend of "Howlin Mad" Smith, had the solution for conducting ship-to-shore operations that were fast, efficient and simple. His solution was the Higgins boat, an amphibious craft designed to move men and supplies from large ships to the beach. It was 36 feet long with a crew of four. It could carry 36 and had a forward ramp allowing troops to run out of the boat instead of climbing over its sides, a development strongly influenced by Smith's subordinate, 1st Lieutenant Victor H. Krulak.[884] In fact, USMC Commandant and General, Charles C. Krulak (the godson of General "Howlin Mad" Smith), commenting on his father's work with Higgins, said, "It was my father who brought to light the importance of the bow launch landing craft and the ability to back off the beach."[885] Krulak's reports about similar boats he observed the Japanese use outside of Shanghai in 1937 submitted to Smith actually were the deciding factor in educating the by then "Full-Bird" Colonel Smith about the importance of Higgins' boat design and how it could shape amphibious warfare.[886] With Krulak's and Higgins' prodding and education, Colonel Smith realized that the Higgins boat was critical to the Marines' future success so he, in turn, educated and then tried to convince Admiral Ernest King the Marines needed those boats in discussions throughout the late 1930s and early

1940s. Eventually, Smith won King over and the Higgins boat was developed for amphibious war. America would build 23,358 of these crafts and they carried the bulk of the burden for successful amphibious operations around the globe.[887] Smith said without this boat, "landings on Japanese-held beaches in large numbers would have been unthinkable."[888] So, without the Higgins crafts and Smith's push for them, island invasions would have been difficult if not impossible. The Marines brought this craft to the U.S. Armed Forces which was one of the most important pieces of technology used in WWII.

So by the time of Guadalcanal, the Marines had advanced greatly in building out the force for their missions. This was an amazing fact knowing the Corps' size had shrank to 15,000 during the inter-war period due to the isolationist pressure to limit it to a basic defense role since expeditionary forces were regarded as too "interventionist."[889] In 1940, seeing war on the horizon, the Corps was able to expand to 1,410 officers and 25,070 enlisted men and continued to grow throughout the next year.[890] By the time the U.S. entered the war, the Marines numbered 66,319 and were in division strength for the first time in history to Smith's great satisfaction. As war descended on Europe and Asia, the 1st MarDiv was activated 1 February 1941 to operate with the Atlantic Fleet and the 2nd MarDiv was activated to operate with the Pacific Fleet. By war's end, six Marine divisions were fighting in the Pacific. The Corps aviation wing expanded from 13 squadrons in 1941 to 87 by 1943 (a squadron had on average circa 12-20 planes). The Corps grew to 669,100 by 1945 making it the largest it would ever become.[891] And Smith's vision for "his Corps" was to make everyone of "his Marines" an "amphibious warrior," and history would prove that he was quite successful in doing so.

Returning to General Smith's biography, by 1942, he had served throughout the world from the Philippines, to Haiti and Cuba in the West Indies, not to mention at home in Washington, D.C., Norfolk, Virginia and at the Naval War College in Newport, Rhode Island. He knew the world geographically, politically and militarily. More importantly, during the crucial years from 1935 to Pearl Harbor, Smith led the Corps' development of amphibious warfare in *all its forms*, not just with the Higgins boats.

Smith looked nothing like a typical Marine, standing 5'9" and wearing a rather round, unfit figure. His face resembled something halfway between a New

York lawyer and a shaven Santa Claus with pudgy cheeks, a fat nose and squinting eyes peering through round spectacles. But behind this façade of unwarrior-like appearances raged a focused energy and desire to perfect the Marines' ability to attack from the sea with overwhelming power. Colonel Smith was Director of the Division of Operations and Training from 1937-39, after which he was Assistant Commandant. In 1939, he commanded the 1st Marine Brigade which became the 1ˢᵗ MarDiv in 1941. After hostilities commenced, now Major General Smith trained the army and Marine Corps in amphibious warfare, having taken command in August 1942 of the Pacific Fleet's department of the Amphibious Corps. Because of the Marines' innovative tactics of sea assaults and their knowledge of both the machinery and how to implement such combat, Smith truly made amphibious warfare the Corps' "main *raison d'être.*"[892]

Consequently, the Corps, under Smith's leadership, understood its role as the principal force for the *entire United States military* responsible for the complex specialty of amphibious warfare—the very justification for its existence. Through Smith and other Marine officers' efforts, they demonstrated to the navy that amphibious landings were critical. This Marine doctrine greatly impacted America's way of conducting war. The Marine doctrine was adopted by the U.S. Army, which republished it in manuals, having never developed an amphibious warfare doctrine of its own. The Marines trained army officers as well, and they spread out around the world conducting landings in Alaska, North Africa, Italy, France, and the Philippines to name a few. Commandant and General Alexander A. Vandegrift said after the war,

> Despite its outstanding record as a combat force in the past war, the Marine Corps' far greater contribution to victory was doctrinal: that is, the fact that the basic amphibious doctrines which carried Allied troops over every beachhead of World War II had been largely shaped—often in the face of uninterested or doubting military orthodoxy—by U.S. Marines.[893]

For instance, the later Lieutenant General Victor H. Krulak, as a member of Smith's staff in the late 1930s and early 1940s, not only helped in the development of the Higgins Boats, but also with amtracs as well. These landing crafts, called "alligators," were floating tanks or heavily armoured treaded vehicles that could land on beaches and *drive* inland after being dropped into the sea via a ship off shore. These two "boats" carried the bulk of the burden for amphibious warfare. You could not have one without the other when conducting landings because amtracs protected Higgins boats, and Higgins boats brought the riflemen necessary to provide amtracs the support they needed once full ground combat commenced (which, when hitting an enemy beach, happened almost immediately). After the war, Krulak quoted a study that was even more to the point than Vandegrift's assessment above about the Marine's main mission throughout the 1920s and 1930s, writing,

> Had the Marine Corps not so devoted itself [to amphibious warfare], there would have been no amphibious doctrine for the Army to follow when the threat of war appeared and the Army, when it evidenced its first sustained interest in the amphibious problem in 1940, would have found itself twenty years late.[894]

In plain English, Krulak was making the point that the U.S. Army would have been in a disastrous position in projecting its power against Hitler and Hirohito had it not been for the Marine Corps (D-Day at Normandy would have been delayed by months if not years without the Corps). It was men like Smith and members on his staff like 1st Lieutenant Victor H. Krulak, his Chief-of-Staff Colonel Graves B. Erskine and many other Marines who perfected useable and strategically deadly procedures for large naval and land forces to attack from the sea.

Although Smith was this type of warfare's visionary, it was his Chief-of-Staff Colonel Erskine who developed much of the guts of how one should conduct it. Erskine explored the details of how Marines should be berthed on a ship before an assault, how amphibious tractors (amtracs) should be used ahead of Higgins boat landings, and how the men should deploy on the beach once leaving their crafts. He developed the procedures on how to ascertain when the beachhead was secured enough by the amtracs and Higgins-borne troops ashore in order to then bring to the beaches the *LCIs* (Landing Craft, Infantry—158-foot vessels) that could deploy up to 200 troops and all their equipment and *LSTs* (Landing Ship, Tank—382 foot vessels) that could land tanks, jeeps, artillery batteries and trucks from their bow-butterfly doors to provide a second wave of overwhelming power against the enemy. Erskine was Smith's "right-hand-man." Later, he gave interesting commentary about Smith saying they often fought and sometimes Smith would not speak to him for days because Erskine did things without his knowledge (they would not get done if Erskine had not behaved in this manner). As Chief-of-Staff, Erskine ran "the outfit" behind the scenes allowing Smith to keep amphibious operations in Marine hands by looking "after the politics." If Smith was one of the fathers of amphibious warfare, then Erskine was his brain-child to develop that warfare.[895]

Graves Blanchard Erskine (the "Big E" or "Blood and Guts")[896] was born on 28 June 1897 in Columbia, Louisiana and, like Smith, also descended from strong Southern roots—his maternal grandfather had fought for the Confederacy. Before WWI, he served in the Louisiana State Guard as a sergeant while studying for his bachelor's, which he received from Louisiana State University in 1917. In that year, he entered the Marines and in 1918, he deployed to France and fought at Belleau Wood, Chateau-Thierry and St. Mihiel. In one battle, he took shrapnel to his hip and leg, but refused to leave his unit. However, his luck ran out on 15 September, when a bullet shattered his lower right leg—one bone was sticking six inches out of his body. He received the Silver Star (third highest medal for valor)

Lieutenant General Graves B. Erskine. He helped develop many tactics of amphibious warfare used by the Marines during WWII. He also was commander of the 3d MarDiv on Iwo Jima which was Woody's division. According to USMC General Anthony Zinni, Erskine was the most brilliant Marine general of WWII. St. Louis Personnel Records Center

and a Purple Heart. After recovering, he rose through the ranks. He spent time overseas in Haiti, Nicaragua, Japan and China (American Legation in Beijing 1935-37). He was recommended for a Navy Distinguished Service Medal for his role planning the assaults at Saipan and Tinian. In the end, the award was down-graded to a Legion of Merit, which he received from Nimitz.[897] Nonetheless, this was one piece of evidence out of many proving Erskine's vital roles in helping Smith and others develop and then utilize amphibious warfare.

As one can see, as Smith's Chief-of-staff, Erskine had one of the most important jobs in the Corps. When Brigadier General Mike Edson took over Erskine's role as Chief-of-Staff of Amphibious Warfare Operations once Erskine took over the 3d MarDiv in late 1944, Edson wrote of this billet that it "was easily the fourth, and maybe the third, most important post in the entire Marine Corps and the success or failure of the Corps [resided in this position]."[898] In short, Erskine was largely responsible for developing the operational warfare for the new and more powerful Marine Corps that destroyed Japanese-held islands throughout WWII—his fingerprints, if not is whole stamp of approval, were on most of all the major island campaigns.

Unlike Smith, Erskine's handsome and muscled face and body looked like a poster-child for the Corps. Standing at a lean 6'0", he loved physical fitness and prided himself on his Scottish and Southern martial heritage. He had a sharp tongue and behind his clear, green eyes lurked a mind full of curiosity ripe with critical analysis. He was a perfect yin/yang partner for Smith in developing ship-to-shore operations.

By the Marianas campaign, no organization in the world could conduct am-phibious warfare like the Marines thanks to men like Smith, Erskine and Krulak. It is difficult to point to one particular historical event that made the difference in the war; however, amphibious warfare was one of the major reasons Japan fell. "The war ended when it did, where it did and as it did as a result of the seizure of successive advanced bases across the Central Pacific," bases that brought the U.S. closer to Japan's mainland. Amphibious warfare brought the death grip around Japan accomplishing its "basic mission on the most vital front."[899] There was no other way to take those islands without hitting the beach fast and unloading quickly which the Higgins boat and a host of other craft like *LCIs*, *LSTs* and amtracs allowed the Marines to do. In the end, as historians Philip A. Crowl and Jeter A. Isely concluded: "The most important contribution of the United States Marines to the history of modern warfare rests in their having perfected the doctrine and techniques of amphibious warfare to such a degree as to be able to cross and secure a very energetically defended beach."[900] In short, one could argue that the only way anyone in 1942 or 1943 was going to start to destroy the atrocious Hirohito regime was to assault his islands, and the Marine Corps had developed such operations that would shock Japanese leaders with how effective it was in bringing the full force and strength of American fighting men deep into the Empire of the Rising Sun.

# CH. 12: The Attack at Saipan

"Justice is only superficial courtesy among nations
and the ultimate resort is military power alone."
—Japanese Minister Kentarō Kaneko, Privy Council, 1930[901]

GUAM'S SISTER ISLAND OF SAIPAN, about 100 miles north of Guam, had been purchased from Spain by Germany in 1899, and then granted to Japan by the League of Nations after WWI when Japan was allied with the Allies. Thereafter, Japan colonized Saipan, and although there were Chamorros and Koreans there, 90% of the 30,000 civilians were Japanese by 1941.[902]

Marching toward Japan's capital, the U.S. decided Saipan must be taken first in the Marianas. The U.S. attacked Saipan before Guam because it had the best existing airfields and lay 100 miles closer to Japan's primary islands. D-Day for Saipan was 15 June 1944; W-Day, the attack on Guam was to take place a few days thereafter. Woody and the 3d MarDiv were slated to liberate Guam, but during the Battle of Saipan, the 3d MarDiv was designated a floating reserve only to go ashore as reinforcements if the fighting forces needed them. Because the Japanese First Mobile Fleet moved to attack the American fleet supporting the landings and the Saipan resistance was tougher than expected, W-Day was postponed until 21 July. Over 60,000 Marines and soldiers from the V Amphibious Corps under Vice Admiral Kelly Turner and Lieutenant General Smith (2nd and 4th MarDivs and the Army 27th Infantry Division) fought 30,000 Japanese on Saipan. As mentioned earlier, Smith's trusted Chief-of-Staff for the landing force, now Brigadier General Erskine, planned the invasion. The commanders felt they would defeat the enemy here quickly especially since they thought only 12,000-15,000 Japanese defended the island.[903]

In setting the tone for the battle, one day before the Marines landed, on 14 June, the lone American prisoner on the island, a tall pilot who had been shot down, was taken from his cell and the Chief of Police Nitta drew his saber and began hacking at his neck, arms and back. The "Yank" fell down, bleeding profusely. Although not yet dead, Nitta left him there to bleed out.[904] With this death, America suffered the first death of the Saipan battle and Nitta had freed a few men from prison duty so that they could fight the Marines on the beach.

Right before the landings began on 15 June, ships moved within 2,500 yards off the shore and pounded enemy positions.[905] IJA Corporal Takayoshi Igata was so frightened during the bombardment "that his testicles shrunk," he later confessed.[906] However, the naval bombardment of Saipan in preparation for the landing had little effect. The terrain was hilly and rocky, and the Japanese were hidden in fortifications in coral-limestone caves, protected from the shelling of ships and aerial bombing. However, the U.S. would have one lucky strike. A stray shell did plummet into the commanding general Yoshitsugu Saitō's staff during a meeting right outside their command post. After the dust cleared, Saitō "was still sitting unhurt and silent, with his sword stuck in the ground between his spread legs." However, the explosion killed half of the gathering, thus weakening the Japanese chain of command. Outside of that single shell hit, most of his soldiers survived the U.S. Navy's bombardment.[907] Once on the beach, Japanese artillery strafed attacking Marines indicating that many of the enemy positions had not been neutralized. Lieutenant General Smith knew this battle was going to be tougher than earlier ones claiming: "We learned how to pulverize atolls, but now we are up against mountains and caves where the Japs can dig in…A week from today there will be a lot of dead Marines."[908]

Saipan was the first Pacific battlefield where the Americans also encountered a large enemy civilian population, possibly as high as 27,000. This was problematic because it was mixed in with both IJA troops and a friendly population of 3,350 Chamorros and Carolinians making the bombardment of targets somewhat restricted.[909] A pamphlet given to the 2nd MarDiv cautioned:

> We must…be…sure that a civilian is fighting us or harming our installations before we shoot him. International law…demands that civilians who do not fight back …must, whenever possible be

taken alive and must not be injured or have their possessions taken from them except after a due trial by competent authority.[910]

Unlike the IJA's penchant for indiscriminately attacking civilians, even the most hardened American infantryman had no wish to harm women and children. In short, the presence of more defenders than expected and the complications of dealing with civilians made the battle longer than originally thought.

Before assaulting Saipan, some wondered why they should seize it since it seemed unnecessary to occupy if Guam was taken. Woody explained: "We should've just left them there to rot."[911] One medical officer briefed his men about Saipan, telling them that dense, hard-to-navigate jungle covered it amongst steep hills and mountains. Hidden batteries, coastal guns and pillboxes covered the island. Sharks, barracudas, and poisonous sea snakes lurked in the waters surrounding the island and the population was hostile. Once ashore, "leprosy, typhus, filariasis, typhoid and dysentery as well as snakes and giant lizards" awaited them. After the officer finished, there "was a long silence." A private then asked, "Sir, why don't we just let the Japs keep the island?"[912] Unfortunately, leaving Japan with air bases and naval stations near Guam was not a prudent option. They would have to take Saipan and nearby Tinian to provide bases for U.S. airpower and seapower while denying the same for the Japanese.

After conducting a feint invasion to the northwest above Saipan's capital of Garapan, the Marines hit the southwestern beach below the city on the morning of 15 June with a lead wave of 96 amtracs loaded with Marines accompanied by 68 armored amtracs, each with a cannon (37mm) or a howitzer (75mm) and a machinegun. In the first twenty minutes of battle, 8,000 Marines and 700 amtracs hit the beach, conducting Erskine's "amphibious blitzkrieg" since they used their tracks to not only go through the water, but also over coral reefs, beaches and land to create a strong perimeter for follow on ground troops. Several casualties happened along the way when Marines became curious and peeked over the sides of landing crafts and hit by shrapnel and bullets. Some amtracs were destroyed before reaching shore taking direct shell hits. Nonetheless, the majority hit the beach, unloaded troops and drove inland creating a moving "iron wall" to cover the following waves of Marines in Higgins boats by laying down artillery and machinegun fire. All the while, planes dropped bombs and strafed positions. Naval gunfire located caves and gun emplacements around the island and fired on them. By nightfall, 20,000 Marines had landed. Opposition was fierce from the start, however. By the end of the first day, 2,000 men, thirty percent of both 6th and 8th Regiments, had been killed or wounded.[913] That evening, the Japanese hit the American left flank with almost a thousand men and a few tanks from Garapan. The counterattack was more than U.S. commanders had anticipated, but their men were able to push the assault back into Garapan. The city was leveled. Many commanders wished they had bombed the capital before the invasion since it housed the attackers creating unnecessary urban warfare taking place that night. (see Gallery 2, Photo 7)

Lieutenant General Yoshitsugu Saitō commanded Saipan with the core organization being the 43rd Division. He was a veteran of the Sino-Japanese War as the 5th IJA Division's Chief-of-Staff in 1938 at Canton and chief of cavalry operations for the *Kwantung* Army in 1939 in Manchuria. Japanese atrocities occurred where he served, especially in the activities of Unit 731: Thousands were killed "in grisly experiments involving tolerance for freezing temperatures, poison gas, starvation, x-rays, boiling water, and pressure so extreme that their eyes came out of their sockets."[914]

During a goodwill tour of Japan in 1937, Brigadier General Erskine spent time with Saitō during a party. This evening, Erskine listened to an arrogant Saitō claiming they would take over America and govern her like a colony, but they would make sure Erskine received preferential treatment. The lack of respect shown Erskine made him "pretty goddamn furious at this point." Saitō then asked, "Now if this all fails [their takeover], what are you going to do for us?" Erskine could not think of "anything decent to say," so he replied, "I'll give you a military funeral." That did not stop the Japanese from issuing further insults in declaring that they copied their navy after the British and their army after the Germans but had copied nothing from the Americans, especially their Navy, which "was no damn good."[915] Now in 1944, Erskine took pleasure knowing that the amphibious assault against Saitō used that "no damn good" navy and knowing too that Japan's plans for world domination were failing. Erskine hoped to soon give Saitō his funeral.

Even in the face of the large enemy fleet sitting off his shores, Saitō probably did not worry as much as he should have about his potential funeral at this stage. He undoubtedly should have, knowing that Japan had failed to stop the Marine invasions at Guadalcanal, Tarawa and Kwajalein just to name three IJA defeats. Nonetheless, Operations Section Chief, Colonel Takushiro Hattori, of the Senior Officers' Section of the IJA's General Staff Office, had given commanders in the Marianas like Saitō reason to feel confident, declaring on 9 March 1944 that the Imperial defenses in the Marianas were "impregnable."[916] Several Japanese senior officers called the chain of islands forming Saipan, Tinian and Guam, the "Tojo Line,"[917] and truly believed it was invincible. They confidently judged that their weapons, tactics and manpower there would check the tide of American's amphibious warfare capabilities. These staff officers, working in Ichigaya near Tokyo, would have been wise to have studied more carefully the French experience with their Maginot Line in 1940 and how ineffective it was in stopping Hitler's *Blitzkrieg*. Time would only tell whether or not Tojo's Line could maintain its integrity or snap in numerous places like what happened to the Allied soldiers standing confidently behind their "impregnable wall" along hundreds of miles along the French border in May 1940. Indeed, the first couple of days of battle of Saipan convinced many American commanders that the "Saipan wall" they faced in breaching was more daunting than expected, but they still felt they would eventually succeed in breaking the battlements there.

The unexpected delays in taking Saipan as quickly as hoped postponed Guam's invasion, as air support for Guam was tied up at Saipan. Moreover, based on Saipan's unexpected defense, commanders decided the Marines attacking Guam needed more support. That support required the transportation of the 77th Infantry Division from Hawaii, which took weeks. A submarine that sighted an approaching Japanese fleet further delayed Guam's landing when the navy diverted its efforts from Guam to participate in the coming Battle of the Philippine Sea.

The Marines had simply underestimated the number and tenacity of Saipan's defenders. Moreover, here, the Japanese "appeared to be changing their style of fighting. Instead of trying to throw the Marines back into the sea at the shoreline, they began to burrow into the ridges and caves" to exact as many casualties as possible.[918] Hirohito exhorted his men: "Although the frontline officers and troops are fighting splendidly, if Saipan is lost, air raids on Tokyo will take place often; therefore you will hold Saipan."[919] For the common soldier, this was powerful. The Emperor told his men Japan's safety depended on them, and they would do their best to rise to the occasion.

A destroyed Japanese tank on Saipan. On 16 June 1944, Japanese launched the largest tank battle of the Pacific War on Saipan. Connie Armstrong.

For example, on the night of 16 June, the Japanese launched the largest tank battle in the Pacific War, hitting the American lines with thirty armored vehicles, hoping to destroy the "Yanks" on their beachhead. After hours of fighting, the Americans, however, destroyed almost all of them. These Japanese tankettes could inflict harm on infantry, but when going against the Sherman tanks, which had 75mm cannons, they were outgunned using their 37mm, 47mm and 57mm guns and weakly protected sporting only thin armor plating. Japanese fought with resolve inflicting more casualties than U.S. commanders imagined they would, but the Leathernecks still held their beachhead. Even though they took heavy casualties doing so, Woody's reserve division was still not required. While war raged on Saipan, Woody and his Marines conducted calisthenics, jumping jacks, windmills and muscle-building exercises like push-ups, sit-ups and leg squats on their ship's deck. Afterwards, they broke down weapons, cleaned and reassembled them and waited.

Saipan's invasion showed the U.S. fleet's location to the Japanese so they sent a large naval force to attack the American Fifth Fleet. Many U.S. ships were idle and Japan hoped that with the proper weather, element of surprise and resolve, they could inflict massive losses on the U.S. Navy and send the Americans in retreat, reversing the tide of war. Saitō believed his navy would deliver a heavy blow against the Americans as his men battled Marines on shore and held the support vessels hostage off the coast. Maybe the "Tojo Line" would hold together after all. Suddenly, the U.S. Navy found itself supporting one of the largest invasions of the war, while also fighting one of its biggest sea battles in history. It was unclear if they could pull off a victory.

# Ch. 13: Philippine Sea Battle: "The Great Marianas Turkey Shoot"

War, once declared, must be waged offensively, aggressively.
The enemy must not be fended off; but smitten down. You may
then spare him every exaction, relinquish every gain, but until
then he must be struck incessantly and remorselessly.
—Rear Admiral Alfred Thayer Mahan

SINCE THE PACIFIC WAR'S BEGINNING, Hirohito and the IJN had been intent on staging one decisive battle to destroy the Americans in order to force the U.S. to sue for peace. They believed they could recreate their stunning 1905 victory at Tsushima. By 1944 Japan had the two most powerful battleships in the world (*Yamato* and *Musashi*) and could assemble a large force with fleet carriers and supporting ships. As the war dragged on, it faced a constant shortage of fuel, limiting its ability to maneuver its fleet far from the occupied Dutch East Indies where the bulk of its oil was. Thanks to American submariners, mines and planes, the Japanese lacked tankers to run both their economy and fleet.[920] Moreover, the U.S. Navy sank troop ships carrying around 100,000 soldiers headed for the islands the Americans assaulted. To put this in perspective, a little over 100,000 garrisoned the islands from Iwo Jima down to Tarawa. In other words, the U.S. Navy prevented the garrisons that the U.S. forces attacked throughout 1943-45 from being twice as large as they ended up being.[921] As a result, the U.S. troops fighting the Japanese on Saipan were lucky they *only* faced 30,000 as the U.S. Navy now steamed away to meet the Japanese naval threat.

At that time, the IJN assembled the First Mobile Fleet under the command of Vice Admiral Jisaburō Ozawa.[922] It sailed to the central Pacific to seek the American fleet for a surprise attack that, according to the Japanese high command, would change the war. It was an awesome naval force with five fleet aircraft carriers, four light carriers and most of its major surface ships including battleships and cruisers.[923] It could launch carrier planes at a distance of 300 miles or more from their target and then land them on Guam where the planes could refuel and reload ammunition.[924] Due to fuel shortage, the First Mobile Fleet could strike the American fleet in the Marianas, but then had to withdraw to Japan's Inland Sea.[925] The Japanese were also counting on large numbers of their land-based planes (First Air Fleet) to join the battle.

The American Fifth Fleet, with its battleships, carriers, cruisers, destroyers, troop transports and supporting ships numbering in the hundreds, was commanded by Admiral Raymond Spruance, who, with Admiral Frank Jack Fletcher, had vanquished the Japanese fleet at Midway. Constant movement and concealment in heavy weather were the best protection for aircraft carriers and other warships. A fleet lay most vulnerable to attack by a Japanese carrier, submarine and battleship force while tied down supporting landings.

The First Japanese Mobile Fleet knew approximately where this American fleet was, and that most of it could not leave to hide or maneuver without endangering the 70,000 Marines, soldiers and sailors fighting on Saipan. The Japanese could maneuver freely while the Americans could not. Hirohito's message relayed to the First Mobile Fleet was: "This operation has immense bearing on the fate of the Empire. It is hoped that the forces will exert their utmost and achieve as magnificent results as in the Battle of Tsushima."[926] With luck, they would surprise the Americans, inflicting major damage before the Americans could defend themselves.

Such an attack concerned Admiral Ernest King and Admiral Nimitz. The advantage for aircraft carriers was to strike the first blow with their planes instead of reacting to planes attacking them. American subs had spotted and tracked the First Mobile Fleet as it made its way toward the Marianas and as it approached, search planes from Spruance's carriers reported contact with Japanese planes. Meanwhile, two U.S. submarines sighted IJN ships from two groups. Admiral Marc Andrew "Pete" Mitscher, his Chief-of-Staff Captain Arleigh Burke, and others favored launching an aerial attack on the First Mobile Fleet, but Spruance's orders were to cover the Saipan beachhead and transport fleet at all costs so he chose to stay put in the Marianas for a little longer, much to the relief of Lieutenant General Smith. Spruance thus postponed Guam's invasion even more.[927] But in the end, to meet the attack, some ships were diverted from Saipan to confront the force bearing down on them. Smith, who was in charge at Saipan, wrote, "It must have been amazing and cheering to the Japanese to see the American fleet disappear over the horizon. One day hundreds of ships of all sizes and purposes filled miles and miles of anchorage and the next they were gone, leaving the invasion troops hanging on to a thin beach strip, unprotected from the sea."[928] Saitō must have smiled as he gazed out of his Command Post (CP) as the IJN worked its magic and gave his troops a needed break from the naval bombardment. Now he had his chance to knock the Marines back into the sea without proper shipping support. Spirits ran high among Saitō's staff.

Conversely, many Marines were shocked to wake up to see their entire navy had vanished over the horizon. "The ground troops were stunned." PFC Albert J. Harris recalled, thinking to himself, "My God, are we doing that bad [that] they left us here? I knew we weren't doing very good but…They'd all pulled out. I didn't want anybody to leave me there."[929] For the time being, the Marines were indeed on their own and had to make do, just like their comrades had done on Guadalcanal almost two years before.[930]

On 19 June 1944, the day broke with fair weather, warm winds at 9-12 knots, mild seas and good visibility around the Marianas, making it easy to spot and intercept planes; from the USS Lexington carrier's superstructure, one could see 40 miles in any direction.[931] Ozawa's First Mobile Fleet carriers launched four air attacks south toward the Marianas against the Fifth Fleet.[932] Would this Philippine Sea battle be the decisive battle that would change the war?

Unlike in 1941 when the experienced Japanese airmen were flying Zero "Zeke" fighters superior to the aircraft of the less-experienced American pilots, the opposite was now true. The Zeros had not been substantially upgraded and their weaknesses, such as their lack of armored cockpits and self-sealing fuel tanks, were known to American pilots who had new tactics and vastly improved planes, like the Grumman F6F Hellcat, using both against the enemy with devastating effect.

In the Battle of Philippine Sea, American pilots from the Fifth Fleet under the command of Admiral Spruance destroyed 476 Japanese sea and land-launched planes at a cost of just 50 of their own. The enemy planes had come from the garrisons on nearby islands and carriers. It was so one-sided that American pilots nicknamed it, "The Great Marianas Turkey Shoot."[933] It was the largest carrier-to-carrier battle in history with 200 ships and 100,000 men engaged.[934]

In every category of plane and ships, except for cruisers, the Japanese were inferior.[935] Because the Japanese had to deal with the thrust up from the southwest Pacific by MacArthur's forces as it dealt with Nimitz's central Pacific offensive, there was uncertainty as to where the Americans would strike. Japanese air assets were thinly spread over a wide area and many were unavailable for battle. American raids conducted in the area before the battle destroyed much of the First Air Fleet and ground support before the Mariana campaign started, so of the 1,750 land-based planes Ozawa counted on, only 435 were air worthy and many pilots were undertrained.[936] During battle, *USS Essex* Helldivers put the airfield on Guam out of commission, so when the Japanese carrier planes tried to land there, most crashed.[937] The Japanese not only lost planes and carriers, but also about 445 aviators.[938]

During the three-day battle, U.S. submarines and planes sank three out of nine Japanese carriers killing 3,169 enemy sailors. The Americans stopped the Japanese naval force and sent them in retreat ending "once and for all any naval or air threat to the Marianas invasion."[939] During battle, Woody and the 3d MarDiv remained afloat. The danger that the First Mobile Fleet would sink Woody's ship had passed. The Japanese defeat in the Battle of the Philippine Sea ended their hope of holding Saipan. The defenders were doomed—they "resembled fish caught in a casting net."[940] As U.S. Navy's ships returned and their planes once again darkened the sky over Saipan, Saitō knew he and his men were going to die. The Tojo Line started to slowly but surely now crumble.

At the same time the U.S. Navy and Marines were winning "The Great Marianas Turkey Shoot" and the Battle of Saipan, half a world away Allied soldiers under General Dwight D. Eisenhower made their amphibious landing on Nazi-held Normandy beaches and penetrated Hitler's "Fortress Europe." British General Bill Slim's 14th Army at Imphal and Kohima had defeated the Japanese army in its attempted India invasion. The Red Army offensive in Belarus, Operation *Bagration*, was destroying Hitler's Army Group Center as it and a second offensive in Ukraine pushed west toward Berlin. On about all fronts, the Allies repelled the fascist powers from their conquered lands pushing them to their pre-war borders. Many enemy walls were falling down all over the world.

# Ch. 14: The Attack at Saipan Continues

"If the commander knows his enemy, where he is disposed,
and what he can do, the tremendous power at the commander's
disposal enables him to use his troops and the ground so
as to surprise and overwhelm the enemy."
—3d MarDiv Intelligence Report[941]

AFTER THE PHILIPPINE SEA BATTLE, the Americans focused on finishing off the Japanese on Saipan and other Mariana islands. IJA soldiers received the undivided attention of the Marine Corps, the U.S. Army, and the fleets of ships supporting them. The defenders' lifespan was no longer measured in years, but days. Lieutenant General Saitō reported to Tokyo on 25 June: "Please apologize deeply to the Emperor that we cannot do better than we are…There is no hope for victory in places where we do not have control of the air and we are still hoping here for aerial reinforcements…Praying for the good health of the Emperor, we all cry 'Banzai!'"[942] The support for which he hoped would never come.

On 6 July, over two weeks after the Japanese defeat in the Battle of the Philippine Sea, Saitō ordered a *gyokusai* ("honorable death/defeat"): "Whether we attack or…stay where we are, there is only death. However, in death there is life. I will advance with you to deliver another blow to the American devils and leave my bones on Saipan as a fortress of the Pacific."[943] He was correct in that they were all going to perish. However, even in defeat he characterized it as a victory claiming that by fighting on the gods' side they would gain eternal life, especially since they fought evil Americans. Japanese troops tenaciously brawled with

U.S. forces hoping to win the spiritual battle, because in the physical world, they were losing everywhere. U.S. forces had corralled them to the isle's far north. The enemy was now crowded around hundred-foot cliffs that plummeted from a plateau surrounding Mount Marpi. This mountain was

*Banzai* Charge Saipan, 7 July 1944. National Archives, College Park

bolstered on almost all sides by imposing precipices dotted with caves and the shell blasts from U.S. warships. It was an onerous place for the Japanese to offer effective defense.

Many Japanese hid in caves and fought from places of concealment. The Marines used flamethrowers and demolition charges burning opposition forces alive and/or sealing them forever in caves by blowing them shut with explosives.[944] The Marines continued to adapt and improvise during the battle, reacting to problems in real time and bringing the battle relentlessly to the enemy. The main person behind changing tactics in order to defeat the enemy as quickly and effectively as possible on Saipan was Lieutenant General Smith's Chief-of-Staff, Brigadier General Erskine, who was "the neck of the funnel of information" about the battle picking which information to act on and which not to. He shaped the whole campaign and its success.[945]

In a last-ditch effort to kill Marines, many Japanese died in *Banzai*s, the last of which happened on 7 July when over 3,000 (some reports claim 4,300 if one includes the second wave) charged screaming out of Paradise Valley at American lines, shattering two Army battalions along the west coast south of Marpi Point.[946] Japanese medics killed soldiers too sick to participate.[947] One soldier who joined in this charge wrote a diary entry after bowing north to the Emperor saying a silent goodbye to his family: "I, with my scarified body, will become the whitecaps of the Pacific and will stay on this island until the friendly forces come to reclaim the soil of the Emperor."[948] He joined the ranks, shouted "*Banzai*" and rushed toward the Americans. The men looked like "spiritless sheep being led to the slaughter" with their officers taking on the roles of being "guides to the Gates of Hell."[949] The mass of men coming down the hills at the Americans resembled a "stampede staged in the old Wild West movies."[950] Eventually, after running over the initial lines of the Americans, Marine and army gunfire mowed them down, killing almost all of them. They had done their best to obey the Emperor, fighting without regard for their lives, but their tactics proved ineffective against the well-armed, trained and disciplined Marines and soldiers. "The carnage [was] beyond belief. Burial parties needed days to deal with the great number of dead."[951] Behind the *Banzai*, civilians moved north with retreating troops stopping at the shore with no place to escape. Fearing the Marines, Japanese citizens started to take their own lives.

The Japanese dead from the large *Banzai* attack on Saipan from 7 July 1945. National Archives, College Park

Many Marines did all they could to help civilians. Even after experiencing enemy cruelty and fanaticism, American servicemen displayed kindness and did all they could to prevent the people from harming themselves. As soon as the Marines ascertained what was occurring at Marpi Point when civilians started killing themselves, they brought up interpreters on megaphones and loudspeakers, telling them they would not hurt them.[952] Scout planes dropped fliers notifying civilians of U.S. servicemen's good intentions. (See Gallery 2, Photo 8)

Seabees had to bulldoze the hundreds of dead Japanese soldiers into a mass grave from the *Banzai* attack on Saipan from 7 July 1945. National Archives, College Park

Marine Jim Reed witnessed a father throw his children, wife and then him-self off a cliff. After observing such ghastly scenes, Reed wanted to help. When he encountered some children and women, he gathered them up before they jumped. Avoiding contact was difficult and many Marines interacted with the non-combatants. In other island battles, the natives melted into the hills before-hand or sailed to nearby islands and returned after fighting stopped, but there were still over 20,000 in the Japanese-controlled area on Saipan.[953]

The civilians had a difficult time avoiding the conflict and behaved in bi-zarre ways when facing defeat. During the campaign's final days, Marines wit-nessed hundreds if not thousands of civilians kill themselves to avoid falling into American hands. Lieutenant General Saitō had ordered civilians to commit sui-cide in the event of defeat. For years, Japan had fed its citizens the propaganda de-scribing Americans wanting to cut the "testicles" off of Japanese and as "sadistic, redheaded, hairy monsters who committed unspeakable atrocities before putting all Nipponese, including women and infants, to the sword."[954] One document claimed Americans were "barbarous and execute all prisoners" and kill some "by cutting them up and crushing them with steam rollers."[955] Japanese "were taught to despise Marines, who purportedly had to murder their own parents to qualify for enlistment."[956] Many Japanese believed these lies.[957] A few

As the battle on Saipan ended, Japanese citizens killed themselves *en masse*. Here is a pile of citizens and soldiers who had just committed suicide. Connie Armstrong

civilians did take up arms against the Marines, but not many.[958] Believing the propaganda, many civilians trapped at Marpi Point decided to kill themselves. Mothers threw their infants off cliffs and then jumped. A group of a hundred passed grenades out to one another, from young child to elderly adult; then they held the grenades next to their stomachs, pushed the pins and blew themselves up. Women, men, boys and girls slit each other's wrists and bled out. Others took cyanide.[959] The 4th MarDiv reported: "Enemy soldiers, and civilians killed their families and themselves rather than surrender."[960]

Later, a patrol boat's commander said as he motored along the coast by Marpi cliff by the ocean, his craft's progress "was slow and tedious because of the hundreds of corpses floating in the water." One of the dead was that of a nude woman who had killed herself while giving birth: "The baby's head had entered the world but that was all of him."[961] There were piles of bodies floating in the surf and along the rock jetties, most of them dead, but many not and sickening, gut wrenching moans echoed throughout the wind and canyon walls as those still alive dealt with compound fractures, lacerations and internal bleeding.[962]

As civilians stood at Marpi Point debating whether or not to jump, a Japanese sniper shot those reluctant to commit suicide. While two parents agonized over whether they should throw themselves and their small children over the cliff, the sniper killed the parents. This revealed his position to Marines nearby who zeroed in on him. Realizing his position had been given up, he defiantly walked out of his cave and "crumpled under a hundred American bullets."[963] In another cave, IJA soldiers hid with women and children. When some infants started crying and threatened to reveal their hideout, a sergeant said, "Kill them your-self or I'll order my men to do it." With that order, "several mothers killed their own children."[964] A 4th MarDiv report observed that the Japanese "reputation for butchery did stand up. This was…illustrated in the days at Marpi Point where great numbers of civilians were slaughtered by maddened soldiers."[965] Marines became disgusted with such behavior.

Marine Gunnery Sergeant Keith A. Renstrom witnessed a family of seven near the lip of a cliff struggling with the compulsion whether to jump or not. The seven members consisted of a father, a mother clutching her young infant, a young boy and girl of about five years or six years of age as well as two elder sib-lings, both of whom appeared to be in their young teens, and both dressed in ex-quisite kimonos. The family stood together, their forms pronounced at the cliff's edge with the seamless back drop of the ocean behind them. With their hair whipping around their faces wildly, they stared up at the position of Renstrom and his Marines. Suddenly, with a flash of movement, the father reached out and snatched the tiny infant bundle from his wife's protective arms and hurled the baby up into the air and over the jagged rock face. Renstrom watched as the child rolled down the face of the cliff until it was stopped near the bottom, where it got hung on the sharp coral edges by its blanket and clothing. As the baby dangled, still appearing to be moving its tiny arms and legs, the men glared in horror as the waves then came relentlessly to pummel its little body against the jagged wall. Its life seemed to ebb away from it on each crest of a whitecap. The Marines returned their attention back to the remaining family members in time to see the two eldest children solemnly bow to each other and, without hesitation, turn and throw themselves off the ledge. The father then grabbed the young son and attempted to throw him over the ledge, but the little boy, having just bore witness to the horrifying death of his siblings, fought against his father with all the fierce determination his small body could wage against the size of a determined adult. The father was able to disengage with the boy and then he heaved him over the cliff to his death. The Marines could hear the boy yelling and witnessed him fighting at the air, but gravity tore at him and brought him to a crunching thud when he hit the rocks below. Then the father swiftly and me-chanically did the same with his little girl, throwing her mercilessly to her death. He then turned to his wife. High on the cliff with the shattered bodies of their children below, they began to argue. Running out of patience with his wife as she hesitated in the face of her own death, he leapt forward and shoved her over the jagged ledge to join her dead children. Taking in the scene of his family now below him, he turned to Renstrom and his Marines and yelled out "*Banzai! Banzai!*" and then followed this declaration up with his own jump.

As the men watched his descent, much to their chagrin they saw that instead of falling directly onto the rocks below like the others, in a dark twist of comedic fate, he was instead caught and spared his intended show of Japanese stoicism and landed in an embarrassment of failure cradled by the crest of an untimely wave. Thus, he did not die instantly like his wife and, at least, four of his children. The Marines stood transfixed, watching as the man writhed in pain bobbing along the surface of the turbulent

Gunnery Sergeant Keith A. Renstrom. He received the Bronze Star with V for his actions on Tinian. On Saipan, he witnessed the grotesque display of mass suicide by Japanese citizens. During the reunion of 2018 to Iwo Jima, he became well known to everyone for declaring his bravery stemmed from believing in the motto "Death before Dishonor." He was a unique Marine NCO who never drank alcohol or slept around because he was a devote Mormon (Renstrom liked to refer to himself as a member of the Church of Jesus Christ of Latter Day Saints, or LDS—not Mormon). St. Louis Personnel Records Center.

breaking waves of the rocky ocean landing having only injured himself. Obviously in extreme agony, but clearly still alive, they watched as the current began carrying him further away out to sea. After a short internal debate, Renstrom raised his Tommy-gun (Thompson submachine gun) and fired three shots, all of them missing his intended bobbing target. Quietly, a Marine nearby, visibly impacted as they all were by the senseless and nauseating scene that had just played out in front of them spoke, "Gunny, let me take care of this." Slowly, he raised his M1 rifle and shot the father in the head. As the blood rushed out of the father's skull, his body slowly disappeared into the deep, blue sea. This was all to become

Renstrom's biggest regret that followed him throughout the rest of his life, in that he had not recognized what these people were about to do to themselves in order to stop the pointless slaughtering of themselves. The shock of what he was witnessing had paralyzed him.[966] (See Gallery 2, Photo 9)

Japanese citizens start to jump off of cliffs on the north side of Saipan from Marpi Point. July 1944. National Archives, College Park.

Government-controlled news reports in Japan glorified this suicidal fanaticism. One newspaper extolled mothers who killed their children and themselves as the flowers of womanhood.[967] "This was obviously crude propaganda, but to its underlying purpose was chilling to convince…civilians that they too were expected to fight to the bitter end to protect the homeland. The army had imposed its standard of no surrender onto the civilian population to legitimize the notion of death before dishonor and collective suicide for all Japanese."[968] This "carnival of death" and "frenzied extinction" at Marpi Point shocked "battle-hardened Marines."[969] Woody said,

> The thought that my mother would kill me and my [siblings] to
> prevent us from falling into enemy hands was an absolutely foreign
> idea. And that she would kill us and herself together in an orgy
> of death was stupid. What culture creates mothers who think of

their nation and children in such terms? That Japan had mothers who would send off her sons to commit mass slaughter and then slaughter their younger [siblings] in defeat bespeaks of a failed culture...Thank God we brought them morals and democracy.[970]

V Amphibious Corps reported after the campaign that 27,000 Japanese and Korean citizens lived on the island when Marines landed. After the conflict, only 9,091 Japanese and 1,158 Koreans were found. This leaves 16,751 who killed themselves, were murdered by other Japanese or were caught in the battle's maelstrom (62%).[971] Historians Haruko and Theodore Cook write, "Japanese women, children and other noncombatants were driven into the northern corner of the island, where most committed mass suicide."[972] The Commandant of Hong Kong camps, Colonel Isao Tokunaga, under General Kuribayashi's command, explained after the war:

> According to a Japanese, not only military personnel but women and children will think it better to die than to become a POW. This principle was strongly taught [to] women and children and in this war, in the Pacific area, many women and children died rather than be POWs.[973]

Unfortunately Tokunaga knew his fellow citizens all too well. And often the citizens were to be exploited for military goals. The 4th MarDiv also reported numerous Japanese soldiers had used civilians as human shields, disregarding their citizens' welfare and killing innocent women and children.[974] The report noticed a "peculiar trait" when the soldiers knew civilians were surrendering. They set up snipers to hit the roads coming and going to the evacuation points, putting their own citizens in harm's way and killing any Americans who were helping civilians.[975]

As just mentioned, due to suicide, enemy disregard for the safety of their own citizens, and combat, 16,000 Japanese and Korean citizens died on Saipan (most of the 16,000 were Japanese). As in many campaigns, almost all Japanese troops fought to the death. The 4th MarDiv concluded, "No quarter can be given to the Jap soldier. He will employ every trick or ruse possible that will profitably yield several American lives for the sacrifice of his own."[976]

A Marine helps save a Japanese infant on Saipan. Unlike the Japanese who took their enemy's babies and often threw them into the air and impaled them on their bayonets, Marines did alot to help Japanese civilians on the islands they captured. National Archives, College Park.

Japanese behavior perplexed Americans. One Marine labelled it: "These Nips Are Nuts." Another Marine wrote the "Japs" were "plain crazy, sick in the head, that's all."[977] Marines, witnessing this fanaticism, started calling one another "Asiatic" if one acted oddly, meaning they were behaving like the wacky "Nips."[978] Japan had created a collective death cult: "Suicide became ritualized, and formally institutionalized, in the army's ethos as a laudable goal and a testament to the unique Japanese spirit."[979]

But there were cases where some of Saipan's citizens started to realize Americans would not hurt, but help them. Reed showed he had water and chocolate for them. He drank a little from his canteen and then shared it. He then took a little bite from his candy bar and handed some to kids. Before he knew it, he had a crowd around him. They gathered the kids and adults in a truck and drove them to an area away from Marpi Point where Marines took care of the civilians.[980]

Lieutenant General Yoshitsugu Saitō's funeral on Saipan. Major General Erskine made sure he got this ceremony due to an old promise he had made to the Japanese general when they both were stationed in China in 1937 together. July 1944. National Archives, College Park.

As they drove between the suicide cliffs and the beachhead, rogue Japanese units fired at them. Marines did their best to protect themselves, counterattack and kill these IJA soldiers. Reed remembered one girl of about four-years-old with a little teapot. At first, she was terrified. But slowly, through acts of kindness, she started to trust him. Before you knew it, he was teaching her how to play "patty-cake" and giving her water and candy. When it came time for him to save more civilians, she jumped into his arms. When he tried to disengage, she refused to let go. Finally, he freed himself from her embrace and with tears in his eyes, said goodbye. She stood waving as he drove away. He told his buddy, "I wish I could take her home with me."[981]

Gunny Renstrom also took care of a little girl. She too was most likely also an orphan like the child Reed showed kindness to, but instead of helping his self-appointed charge for a few hours, Renstrom took care of her for 12-14 hours and overnight. She was "just-cute-as-she-could-be," and was wet and shivering since it had been raining, and on looking at all the Marines, she saw something in Renstrom and walked up to him and nestled next to his leg under his poncho. Soon thereafter, he sat down under a cranny of a large rock, with the sweet girl grabbing his leg and remaining under the drape of the poncho, sitting her body next to the entire length of his hip and leg and placing her head under his arm. Her eyes were wide with wonder and fear as she gazed at the numerous, large men with rifles and machineguns staring at her. Renstrom slung his Tommy-gun around his shoulders, took his canteen, unscrewed its lid and gave it to her. She had to hold the bottle with both of her teeny hands, and she greedily tipped it as she drank deep from its neck. Water flowed over her little mouth dribbling down both sides of her face as she gulped as much as she could. After she finished, she handed the canteen back to the Gunny and smiled, her white teeth gleaming in the dull light of day that penetrated the heavy cloud cover. Renstrom also fed her some of his food. "She ate two full rations," and Renstrom was shocked such a "little thing" could eat so much. After her "feast," she curled back up closely to the Gunny, still under the poncho, and pulled her loose camouflage-drape over her shoulders and slowly drifted off to sleep as evening approached. Marines nearby, smiled, shook their heads and laughed inside as they watched their fearless-leader turn "into putty" as he doted on this child. Gunnys were to be feared and they definitely never showed weakness in front of their men, but these Marine-laws had disappeared when this innocent, frightened and tender girl entered into Renstrom's life.[982]

As the day turned into dusk, the salty, filthy, heavily-armed Renstrom stood watch over her little body, occasionally worried that she would give him lice ("cooties"), but then rejecting the desire to stay away from her thinking she was like one of his younger siblings who needed help. Often, throughout the night, he would bend down and tuck the poncho around her slender torso and skinny legs. A few times, he patted her head and brushed her coal-black hair away from her face. As he cared for her, suddenly a Japanese patrol hit his outfit. His men returned fire and they started to kill the enemy right and left. Renstrom grabbed his Tommy-gun, but did not fire it for fear of hitting his own men. It was fierce hand-to-hand combat and he sat there with the weapon at the ready, right by the girl, watching the shadows of fighting around him. After the engagement ended, he noticed his small friend "never even woke up. She slept through the whole engagement."[983]

Saipan civilians helped by a Marine—very different than what the Japanese had done with citizens under their control. National Archives, College Park

The next morning, Renstrom estimated that there were around 15-20 dead Japanese around their position. A few of his men, finding some of the Japanese still alive, shot these enemy soldiers in the back of their heads and "put them out of their misery." Single-shots from rifles echoed across the landscape. When "the beautiful girl" awoke, she smiled, stood up and then grabbed Renstrom's hand. He then walked some distance away from his platoon, heading toward other

Marines who were gathering civilians to take them to the camps. The little girl looked up at Renstrom with adoring eyes as they meandered through his Marines and around dead enemy bodies, hand-in-hand.

On reaching the gathering point, he pointed to the men who now would care for her. Perceiving that the Gunny was saying goodbye, her tiny hand squeezed his in a death-grip. "Sweetheart, these men are good men," he said as he pointed to his chest and then pointed to them, trying to reassure her that they would care for her. Since Renstrom outranked most there, he made sure those collecting civilians would protect "his little girl" because he did not want to chance her going back to some crazed Japanese adult who would launch her and himself off a cliff. Prying her wee-fingers away from his palm with his free hand, he hugged her and sent her on her way. When the girl left him to join a group of other children under Marine sentries, she waved to him walking backwards with pintsize tears dripping from her eyes, hitting her top lip and rolling down her chin in long, wet lines. This harden Gunny and grizzled warrior raised his hand and waved back at her, and then wiped the tears away from his eyes as his other hand held his Tommy-gun pointing at the ground. As she walked away, her head continued to swivel back and forth looking for Renstrom to assure her she would be alright. His smiles confirmed to her that she would be safe. "That's one child of God a fanatical Jap won't kill," he thought to himself. For this devout Mormon, the carnage of war and destruction of Saipan's society were events his conservative, religious upbringing in Utah had not prepared him for and he sat there and cried for several minutes as the girl disappeared in the distance, skipping along with the other children into her unknown future. After a few minutes, Renstrom checked the grenades hanging on his webbing, his ammunition and his machinegun and then returned to his men. He immediately started to bark orders and made sure his Leathernecks were sharp and ready for the next patrol they were about to conduct. The men moved quickly under his instructions: The Gunny had returned.[984]

The actions of Saipan's civilians was a harbinger of things to come as the U.S. inched its way toward the Empire's heart.[985] According to journalist Robert Sherrod, the Japanese civilians' behavior and the Saipan garrison's fight to the death had been glorified in Japan and "were intended to make the U.S. think it would be that way all across Japan." Events on Saipan did register in U.S. leaders' minds.[986]

As American forces neared Japan, the enemy had much larger armies to bring to the war zone willing to fight to the death. Typical was Saitō at Saipan who ordered his charges to swear allegiance to the ethical tenet *Senjinkum* (Battle Ethics): "We must utilize this opportunity to exalt true Japanese manhood...I will never suffer the disgrace of being taken alive."[987] The actions at Marpi Point illustrates the results of such orders on civilians and IJA personnel conditioned to follow his lead. As the fighting on Saipan dwindled by 6 July, Vice Admiral Chūichi Nagumo, who had led the Pearl Harbor raid, committed *hari-kari* shooting his brains out.[988]

Japanese defeats at the Philippine Sea and Saipan showed that the U.S. Navy had rebounded in two-and-a-half years and was able to project sustained power in the Pacific, to destroy Japanese naval capabilities, to capture well-defended islands and to kill famed enemy commanders. After the battle of the Marianas and that of Leyte Gulf, the Japanese navy ceased to be a serious threat and victory lay within the U.S.' grasp.[989] The delay of Guam's invasion by the Saipan and Philippine Sea battles allowed the U.S. Navy to pummel Guam with extra gunfire, of "a scale and length of time never before seen" in WWII.[990]

Meanwhile, Woody was unaware of current events. His world was circumscribed by his ship's hull where he sang songs and played the guitar, prepared his equipment in case of deployment and wrote letters to Ruby. It took weeks for reports to filter into the newspapers and for those reports to make it into the letters from home to front line troops. Woody, and most Marines, lived in a world of geographical, geopolitical and strategic ignorance.

Several Marines help a small Japanese girl caught in the maelstrom of battle. Gunny Keith Renstrom helped a girl like this while he was on Saipan. July 1944. National Archives, College Park.

By 9 July 1944, the Marines under Lieutenant General Smith declared Saipan secure. Sporadic fighting continued, but the Americans had won the island. America suffered 17,537 casualties taking Saipan, with 3,426 killed.[991] On 10 July, General Saitō sat down, declared "*Tenno Haikai! Banzai!*" and then disemboweled himself with his sword. As he lay over his bleeding belly with his sword shoved into his guts, his orderly shot Saitō in the head to finish him. Later, with Smith's blessing, Brigadier General Erskine found his corpse and gave Saitō a memorial with military honors. Erskine fulfilled his promise given this IJA commander in 1937 and must have taken pleasure with the outcome. While he conducted this ceremony draping Saitō's body with a Japanese flag and burying him with firing three volleys, many thought Erskine "completely nuts."[992] Little did they know the background that while Erskine rendered the courtesies Saitō's rank deserved, he also did this as a snub to Saitō in consideration of the way that Erskine had been treated by the arrogant man in 1937.

A Marine helps a Japanese mother and her children on Saipan. Many Marines did all they could to aid innocent civilians on Saipan and prevent them from killing their children and themselves at the end of battle, something their leaders had ordered them to do. July 1944. Natonal Archives, College Park

Many Japanese admirals, generals and politicians as well as the Emperor knew that once Saipan fell the homeland lay open to attack. They understood precisely what American control of Saipan, Tinian and Guam meant: Unmitigated disaster. Not only would the U.S. have key naval bases and staging areas, but also hundreds of new B-29s would soon rain destruction onto the homeland itself. On 19 July, after Saipan fell, Tojo and his entire cabinet resigned. One of the biggest war hawks had failed miserably.[993] While this was a step toward ending the war, Hirohito held out for a "crushing blow on the enemy someplace" so he could negotiate an acceptable peace.[994] He inquired about re-taking Saipan, but his advisors counseled against it.[995] Only with Saipan's fall did "Japanese with insight [know] that Japan had to prepare for the worst."[996] As the government shakeup was happening, the Japanese continued to fight unceasingly on the islands they still defended, and in China and Burma. The Allies, meanwhile, turned Saipan into a major bomber base. By November, hundreds of sorties had flown from this island alone against Japan.[997] General Curtis LeMay said: "Taking Saipan was perhaps the most critical decision made [at that time], and it was the right one. That island was needed for the bombers, and also required as a direct interdiction location to interrupt Japanese naval traffic should there be any attempt to reinforce other islands in the chain."[998]

LeMay further stated, "Once we took Saipan, those bastards knew the shit was going to hit the fan. They knew that we were going to bomb the goddamned hell out of them, and then invade."[999] LeMay, as well as millions of Americans, took pleasure in the vengeance America was dishing out. The U.S. decided it "must not only punish, but punish with impunity. A wrong is unredressed when retribution overtakes its redresser. It is equally unredressed when the avenger fails to make himself felt as such to him who has done the wrong."[1000] The U.S. was repaying the devastating blow at Pearl Harbor with punishment akin to a boxer's body blows left and right to the face, ribs and stomach. Japan was starting to wobble and there was no referee to blow the whistle. By July 1945, 1,200 bombing sorties from the Marianas would attack Japan weekly.[1001]

After securing the island, the Marines cared for civilians. In contrast to how the Japanese treated conquered peoples, the U.S. collected non-combatants and provided them "food, water, first aid and transportation" to internment camps where shelter was given.[1002] Women were respected and not raped. When the Americans discovered at the enemy camp that the Japanese did not care for orphaned babies, allowing 21 to die, they intervened. Interestingly enough, Chamorros had been caring for these orphans for days before without incident (an amazing testament to how they treated enemies), but the children were neglected when turned over to their own countrymen. In fact, one woman was witnessed taking an orphan to the edge of the camp where she "deposit[ed]" the child "over the fence and wandered away." Shocked and disgusted, the Marines set up an orphanage and cared for the children themselves.[1003] The Americans set up three different groups; they established one camp with 9,091 Japanese, one with 1,158 Koreans, and another with 2,258 Chamorros and 782 Carolinians (the Chamorros and Carolinians stayed together) for a total 13,289. All were cared for equally, both enemy and friend.[1004] The Chamorros stated to U.S. authorities that they were glad the "Americans were there and that they hoped the Japanese would never return to govern them."[1005] The Marines had made sure this wish would probably come true.

After hostilities ceased, Gunny Renstrom wrote his father on 17 July 1944 summing up the situation for the Marines who had just fought at Saipan:

> There is no glory in war, Dad! The only thing that you want is your life. When you see some of your best men die at the hands of the enemy, it makes you wonder just why do we have war anyway. Then you stop and think of the things the enemy has done to you and the men around you. It makes you mad! And the only thing that is left to do is try and get even. Well F Company made the enemy pay plenty for the ones they got of ours. It took us a little longer than we expected but as always the Marines came through again [Renstrom's company accounted for killing 600 IJA soldiers with possibly five of them being eliminated by his hand].[1006]

Renstrom knew this battle was tougher than they had expected. However, he took satisfaction that his men made the Japanese pay more in KIAs than his unit suffered. Nonetheless, the whole ordeal of making war and killing the Japanese made him question why mankind engaged in such activity as warfare. However, as a Marine, his job was to fight and defeat the enemy and he would have to do more here very soon as his unit in a matter of a few days would engage in another island assault right next to Saipan: Tinian. And instead of him being a part of a unit in this battle, he would find himself leading an entire platoon in battle due to his officer becoming a casualty.

# Ch. 15: The Attack at Tinian

"Tinian was the perfect amphibious operation in the Pacific War."
—General Holland M. Smith[1007]

AFTER SAIPAN WAS SECURED AND what was left of the Japanese First Mobile Fleet retreated to Japan's Inland Sea, Guam and Tinian moved up on the list. Tinian was a smaller, flatter island that already had two 4,700 foot long runways and another three under construction.[1008] To use either island as an air base, the Americans had to control both. Major General Harry "The Dutchman" Schmidt commanded the Tinian landing force of two divisions from Saipan and used Smith's Chief-of-Staff, Brigadier General Erskine, to plan his campaign.[1009] Most who attacked Tinian were based on Saipan and could see Tinian's northern shores. (See Gallery 2, Photo 10).

On 24 July 1944, the invasion began, code name Jig-Day. After a diversionary landing on the southwest coast off Tinian Town, an amphibious force of 39,000 Marines of the 2nd and 4th MarDivs invaded Tinian on the northwest coast, three and a half miles from Saipan. Over 9,000 Japanese defended the island, many of whom were hardened veterans from Manchuria. In a brilliant landing, the Marines achieved "complete tactical surprise" from staging areas at Saipan's most southern end arriving on two tiny beaches at the northern tip that the Japanese deemed impossible for major landings. Following Erskine's instructions, the Marines placed their artillery south of Saipan's Aslito airfield and covered the landing forces with curtains of fire using 156 field pieces. Also, from the airfield, planes flew sorties to support attacking Marines. For the first time in combat, those planes used napalm bombs (fire bombs "consisting of a jettisonable tank to be carried under the belly of a fighter plane and containing a mixture of gasoline with about six per cent napalm jelly.") Tinian was smaller and not as rugged as Saipan or Guam, but it still was daunting to conquer. Thousands of Marines engaged in fierce fighting, sealing off caves, shooting Japanese soldiers, and fending off *Banzais*.[1010]

Several hours after the landings on 24 July, Gunnery Sergeant Keith Renstrom had been placed in another unit from the one he had served in at Saipan. Late in the day, he found himself now in charge of the 2nd platoon of F Company, 2nd Battalion, 25th Marines, 4th MarDiv after its platoon leader and sergeant had been wounded in action. This platoon was at the point of the spear leading the forward elements of the attack on Tinian down the center of the island along one of the island's main roads during the morning and afternoon. The Gunny's platoon had made excellent progress and was a few miles or so away from Tinian Town, heading south on the right side of the isle as the afternoon turned into evening.[1011]

When he took over these Marines, he found they had made some poor decisions. He did not make a good first-impression when he immediately ordered them to remove themselves from their hard-to-dig foxholes because they had placed them right at the edge of a thick sugarcane field with stalks taller than six-feet. Barking orders, Renstrom told them to move to the defensive line hundreds of yards behind them. Cursing the Gunny, his Jarheads grudgingly obeyed him and pulled back behind some trees near a road they had been using before they had come to the clearing and fields. From these trees, they had some form of concealment and they had now put considerable distance between them and any force that might suddenly appear out of the sugarcane field. Although Renstrom reasoned with them about this decision, these Marines still felt he had made them do unwanted and unnecessary work to re-establish their lines since they had the Japanese on the run. Renstrom made sure his new charges stopped their gripping and followed his orders.[1012]

Besides the men being irritated by Renstrom making them move their foxholes, they also were somewhat leery of having a "goody-two-shoes" Mormon leading them. Yes, they had heard he had fought bravely on Saipan, but was a "deeply-religious" Mormon really able to fight? As one of his colleagues wrote of him: "When I first met my Gunnery Sergeant Renstrom, I couldn't believe someone like him existed, in the Marine Corps anyway. A Gunny who didn't smoke, drink, swear or chase women – it never happened!"[1013] Time would only tell whether or not they were in good hands and if an Almighty Creator had sent them a man worthy of their loyalty and trust.

Renstrom's men hunkered down into their new positions and as evening fell, the front seemed uncannily silent. Renstrom explicitly told them that if they had any visitors during the night or morning to wait for the sound of his Tommy-gun before they fired on anything. Later that night, around 2300, some of the men on watch noticed some movement and then ensured everyone woke up. From their hidden positions, they saw a small squad of six Japanese appear suddenly walking down the road. His men listened for the Gunny's Tommy-gun for guidance. The Gunny did not fire upon the enemy yet. He knew these IJA soldiers were forward elements only, and were part of a larger unit. The six enemy walked right up to Renstrom's platoon's position, but they did not see any of his men and they continued down the road toward another platoon in the rear. After several minutes, that rear echelon platoon opened up on the six enemy soldiers. Renstrom and his men heard the screams and witnessed the small arms fire tear into the squad. As they gazed north listening to the battle unfold, they noticed a lone IJA survivor from this attack running back to their position. The enemy soldier literally stopped a few yards right in front of the Gunny's foxhole: "I could have almost have reached out and touched him." All the Gunny's men listened for him for direction. Should they kill the last one or not? Definitely not. Renstrom wanted this Japanese soldier to survive and go back to his larger group and tell them what he knew the Japanese man believed—the Marines lines were behind where Renstrom was. The enemy solder, out of breath, panted for a few seconds, found a pool of water on the road, kneeled down and drank from it, stood up and then continued running away, disappearing down the road. "I was so proud of my men for maintaining fire discipline. No one shot at the lone survivor. Always let one Jap live to go back and sow confusion and fear was our motto," Renstrom said with a grin. And in this case, as the IJA soldier disappeared over the horizon, he was doing exactly what Renstrom wanted him to do. The Gunny smiled.[1014]

After several hours of nothing happening, the dull roar of engines started to pierce the air. Renstrom knew exactly what was now coming at them: Tanks. There is nothing more disheartening for infantry than to hear the rumble of iron monsters coming at them when they do not have their own tanks in support. Disregarding such obvious danger, Renstrom quickly jumped into action and directed his men on how they could successfully destroy this enemy force.[1015] Renstrom instructed his Leathernecks to let the tanks pass their position to be attacked by the other platoon in the rear who he had informed on the radio that an armor attack was on its way:

> Moving north along the coastal road, the enemy force consist- ed of...six light tanks with infantrymen riding and following on foot. First warning of the enemy move came when Marine listen- ing posts stationed along the road a short distance forward of the lines reported enemy tanks rumbling in from the south.[1016]

Renstrom knew a lot of Japanese soldiers would be trailing the reconnaissance force of armored fighting vehicles and he wanted to attack this group and not the tanks.[1017]

When the advance guard of six IJA tanks suddenly appeared, driving north on the road, Renstrom's men stared at the tanks as they approached their posi- tions. They rolled nearer and nearer to their lines. Suddenly, all six tanks near Renstrom's foxhole stopped. Mounting 37mm cannons and 7.7cm machine- guns and parked a few yards away from his men, the Gunny knew they were outgunned. "That terrified me to death," the Gunny said. "Had they seen us?" Renstrom thought. He grabbed one of his grenades and held if firmly. Suddenly, a hatch flipped open on the lead tank and an officer jumped out oblivious that over 30 Marines had him triangulated in a perfect kill zone. The officer looked around, lite up a cigarette, talked to a few of his men and then gazed down the long road. He finished his cigarette and threw it away and jumped back into his tank and slammed the hatch shut with a metallic clang.[1018]

The tanks then lumbered by, one-by-one, along the road with none of them noticing the camouflaged Marines in hiding, covered by a nice blanket of darkness. As the tanks left them, the Marines peered back out toward the sugarcane field. They took in deep breaths of air tainted with the scent of gasoline exhaust. They could hear the reeds of the plants slowly rustle as the wind blew through the land. For now, things seemed quiet, but everyone knew that would not last long.[1019]

After several minutes of uneasy quiet, from down the road, they gradually saw a column of IJA soldiers coming toward them. The pounding of boots against the road and the jangling of metal weapons rubbing against combat gear ricocheted throughout the night air. Three-man machinegun teams carried their heavy weapons surrounded by numerous men armed with rifles, bayonets, knee-fired grenade launchers and grenades. Like clockwork, Renstrom knew these troops would be coming having studied Japanese tactics when he had been on ship just a few weeks before reading books in the vessel's library about the IJA. Renstrom guessed there were around 35 men approaching them. Yard by yard, they came closer and closer. Men thought to themselves, "Why is the Gunny waiting so long?" Renstrom behaved as if he really was waiting for the sign of "seeing the whites of his enemy's eyes" before he attacked. When the enemy was about ten yards away from his men, Renstrom gave his Marines the signal to "open up" by firing his Tommy-gun. As he squeezed his trigger, the field erupted with thousands of rounds being fired in a matter of seconds. Renstrom's men waylaid the attacking soldiers, killing all of them largely using their general-purpose, large machineguns to do most of the heavy lifting—since they were on flat land, the enemy had no cover and did not have a chance to survive the onslaught. As the entire enemy force lay on the ground, Renstrom's machineguns still fired into the IJA soldiers along the whole front. The dull, sickening thuds of slugs penetrating flesh and bone were interspersed with the sharp, ding of other bullets bouncing off the coral rock surface or the dirt road to trail off either into the distance or hit enemy bodies on the rebound.[1020] As the Marine Corps official history noted: "At this juncture the…machine guns…levelled a heavy volume of enfilading fire into the area…This fire, in the words of the battalion executive officer, 'literally tore the Japanese…to pieces.'"[1021]

As Gunny Renstrom's men annihilated this Japanese unit, over the horizon, the other platoon now engaged the enemy tanks with grenades, bazookas and satchel charges supported by "75mm half-tracks, and 37mm guns." Ships near the shore "began firing illuminating shells over the area, virtually turning night into day." Explosions echoed all throughout the land.[1022] Lieutenant Jim Lucas, a reporter, witnessed this attack and later wrote:

> The three lead tanks broke through our wall of fire. One began to glow blood-red, turned crazily on its tracks and careened into a ditch. A second, mortally wounded, turned its machine guns on

its tormentors, firing into the ditches in a last desperate effort to fight its way free. One hundred yards more and it stopped dead in its tracks. The third tried frantically to turn and then retreat, but our men closed in, literally blasting it apart . . . . Bazookas knocked out a fourth tank with a direct hit which killed the driver. The rest of the crew piled out of the turret screaming. The fifth tank, completely surrounded, attempted to flee. Bazookas made short work of it. Another hit set it afire and its crew was cremated.[1023]

Out of the clouds of dust and chaos, a loan tank had disengaged and now was retreating south, full-speed back toward Renstrom's platoon. Then, once again, the tank stopped right in front of the Gunny. Without hesitation, Renstrom's bazooka man shot a round at the treads, knocking the tank out. Smoke started to spread throughout the tankette's cabin. As its officer exited the tank, Renstrom unleashed a spread of bullets killing the commander immediately in the middle of the road. It was an easy kill since he literally stood a few feet from the officer and could actually identify his insignia from the short distance: He had just taken out an IJA captain. Looking around, the Gunny did not see any more threats coming at them, but he felt uneasy and it was still dark. Unsure whether there were more Japanese or not coming for them, he had his men reconsolidate their lines, check their guns and re-load all their weapons. And then they waited. After at least 40 minutes, things started changing on the horizon to the south of Renstrom's platoon.[1024]

Sluggishly, eerie forms started to emerge between the waving, long stalks of sugarcane until dozens of follow-on Japanese infantry emerged in their skirmish line walking into the open field past the line of abandoned foxholes. Instead of the enemy coming at them in a column, slowly and methodically, the figures emerging from the field, suddenly started to run and scream *"Banzai!"* in a phalanx attack. The men could see some soldiers wielding swords flashing in the weak morning light and others holding their rifles low in a fixed, bayonet charge. Before the Gunny knew it, around another 35 Japanese, full of rage, rushed toward him and his men. When they closed within 30 yards, Renstrom gave his men the sign, once again, by firing his distinctive Tommy-gun and then his Leathernecks unleashed hellfire against the wave of suicidal attackers. The morning climaxed with the sounds of men shouting, shrieking and expiring. Thousands of rounds were expended until the field fell quiet and no more Japanese were standing. After Renstrom's men let up, they gazed down at their watches. It was now around 0530. The groans of the injured and dying echoed across the field and then suddenly, grenade explosions here and there lit up the combat zone. The wounded Japanese were killing themselves—committing *jik-etsu*. The yells of *"Banzai"* continued to ring out from the enemy's battleground punctuated by a grenade blast. Renstrom said, "That made our job later easier and I was glad they were killing themselves."[1025]

Sometimes, the Marines would hear an explosion and then see "Jap bodies...fly ten to fifteen feet in the air." Leathernecks nearby shook their heads at each other as they witnessed these IJA soldiers flipping and flying into the air, often in pieces in the light of dawn. Later, after the Marines mopped-up the area, they found that many of the Japanese had been carrying magnetic mines. "The Japs who were wounded and unable to flee were placing the tank mines under their bodies and tapping the detonators" and sending themselves into thunderous spouts of blood, guts and body parts. Some of those around, as if they were whaling sailors, yelled "there she blows."[1026]

As the sun rose around 0550, Renstrom and his men observed the utter destruction they had dished out against the enemy. They had not lost one of their men but they had killed at least 70 IJA soldiers in their two attacks. As they left their positions, Renstrom's men, when they found Japanese who had botched their suicide attempts and were still alive, shot them in the back of their heads and helped these *Samurai* get their battle deaths. They also did not want to chance that any "Jap" could kill any follow on Marines—they had learned the hard way never to leave an alive enemy in your rear. After they made sure all the IJA soldiers were dead, Renstrom's men then slowly, but surely, gathered the dead for burial, and the Japanese weapons for the logistics groups behind the lines to either destroy or use.[1027] This one Gunny had just organized the total destruction of an armored-reinforced-infantry company comprising of six tanks and 267 enemy soldiers.[1028] Now, all his men, after witnessing his leadership and intelligence in action and experiencing this incredible victory, would never question him again if he asked them to move their foxholes or make fun of him for his conservative, religious Mormon-behavior. They would follow him to Hell and back and as he walked amongst them, one could see they believed in him. In the Corps, most superiors got respect, but few would get reverence and Renstrom had that too from his Jarheads. Although put in for a Navy Cross, it was downgraded and Renstrom would receive the Bronze Star with V for his actions this day eventually in 1949.[1029]

This is a picture of the sugarcane field where the Japanese *Banzai* came from that attacked Renstrom's platoon. The Gunny's men destroyed this attack and did not lose one man. Many of the killed Japanese can be seen in this photo. 25 July 1945. National Archives, College Park

After thwarting these Japanese attacks and cleaning up the area, the Gunny had his Marines continue forward to engage the enemy. All these new Marines looked to Gunny now as a prophet and warrior without parallel. He had absolute control over them and everyone knew who was in charge, both in rank and in spirit. Unfortunately though, Renstrom would shortly thereafter be shot through the right leg between the knee and hip, and removed from combat late in the day on 25 July. He only lasted two days during this battle, but had produced one of the best examples of superior small unit tactics on the island and many others would display such actions for several more days until the island was conquered. (see Gallery 2, Photo 11)

This is a picture of the Japanese tankette that Renstrom's bazooka man took out during the battle on 25 July 1945. National Archives, College Park

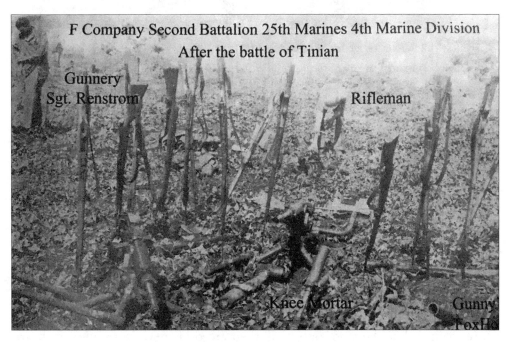

This is a collection of the weapons Renstrom's platoon captured after they removed the Japanese dead from the field of battle for burial. Renstrom's coordinated ambush killed around 70 Japanese and destroyed six Japanese tanks. Renstrom is in this photo in the top corner. Renstrom was initially put in for a Navy Cross, but probably due to a jealous senior commander, it got downgraded to a Bronze Star with V. National Archives, College Park and Renstrom Family Archives

Japanese children on Tinian being helped by Marines. Whenever American forces could, they did their best to help Japanese citizens and defenseless children. Iwo Jima and Tinian Marine veteran Connie Armstrong

A few days later, the Marines slaughtered a mass *Banzai* on 31 July taking out a large percentage of the Japanese garrison prompting an officer to say, "You don't need tanks. You need undertakers. I never saw so many dead Japs."[1030] In eight days, the Marines killed almost all the Japanese, but suffered few casualties: 290 killed, 1,515 wounded and 24 missing-in-action.[1031] The kill-ratio was thirteen to one in favor of the Marines.[1032] Admiral Spruance described this operation as "probably the most brilliantly conceived and executed amphibious operation" in WWII.[1033]

Similar to Saipan, many of the 13,000 Japanese civilians and 2,700 Korean laborers on Tinian committed suicide or were murdered by their own soldiers: Once again, IJA soldiers not only took their creed of "never surrender" to heart, but made their civilians do so by convincing them to kill themselves as well. Like at Marpi Point on Saipan, 2,500 adults threw small children and themselves off Marpo Point at the southeastern coast of Tinian.[1034] The 23rd Marines issued the following report on 3 August:

Japanese children being helped by Marines on Tinian. The Japanese citizens of Tinian and Saipan who survived the war were confused and baffled by how kind Americans were to them. Connie Armstrong

> Several freak incidents occurred during the day: (1) Jap children thrown [by their parents] over cliff into ocean; (2) [Japanese] military grouped civilians in numbers of 15 to 20 and attached explosive charges to them, blowing them to bits; (3) Both military and civilians lined up on the cliff and hurled themselves into the ocean; (4) Many civilians pushed over cliff by [Japanese] soldiers.[1035]

Between the mass suicides, murders by their own military personnel and deaths due to getting caught in the crossfire, it was estimated that a total of 4,000 citizens out of an original 13,000 died in this battle (31%).[1036]

And like at Saipan, the Tinian commanders, IJN Captain Goichi Oya and IJA Colonel Kiyochi Ogata, committed suicide. Furthermore, Imperial servicemen, often after helping citizens take their own lives, killed themselves by either jumping off cliffs or shooting themselves in the head.[1037] Trying to make sense of it, Woody said,

Think about it, if the most horrible thing you can do in your culture is surrender and the most honorable thing you can do for your country and family is commit suicide, most often, if this is how you were brought up, you're naturally going to kill yourself. It all comes to how you were raised.[1038]

The battle officially came to an end on 1 August 1944.

The Seabees completed some of their most impressive work on Tinian, moving "more than 11 million cubic yards of mud, rock and coral to build the world's largest bomber base" which boasted six runways, each 1.5 miles long. It would prove an ideal place to launch sorties of B-29s carrying atomic bombs.[1039] The Seabees, many being from New York City, gridded out the island as if it were Manhattan with the same street names and popular landmarks like Broadway and Central Park (the island is the same form and slightly bigger than Manhattan). An island that has only around 2,500 citizens today ballooned out to supporting 75,000 servicemen after the U.S. seized it. From Tinian's flat terrain, a B-29 named *Enola Gay* would take off one year later from the northern airfield, bearing in its belly the atomic bomb *Little Boy* destined for Hiroshima that hopefully would mark the beginning of the war's end.

# Ch. 16: The Liberation of Guam

"Life is seldom gentle, and war never."[1040]

(see Gallery 2, Photo 13)

IN MID-JULY 1944, THE U.S. prepared to recover Guam, American territory Japan had captured in 1941. The seizure of Guam was the first time since the War of 1812 that American territory was occupied by a foreign power (unless one includes what the Confederates took in the Civil War).[1041] For the American high command, reclaiming Guam was strategically necessary and freeing its people an added benefit for humanitarian and political purposes. The Marines' success in other combat missions before Guam gave most Leathernecks confidence they would win, but it would be a challenge.

The Japanese organized Guam's defense around the 29th Infantry Division, an outfit seasoned in Manchuria, commanded by Lieutenant General Takeshi Takashina. The Naval Guard was increased to 3,000 and there were 5,100 men in the IJA 6th Expeditionary Force. Including naval, ground and airforce personnel, there were around 18,500 Japanese planning defenses, constructing fortifications and lying in wait for the Americans.[1042] Early in 1944, the Japanese knew the Americans were coming, so they closed the schools and had every man, woman and child construct defenses around the clock.[1043] In addition to these laborers, probably over 1,000 Korean forced laborers were among Takashina's ranks and they fought alongside his men—most did not want to be there, but were forced by the Empire to serve.[1044]

Korea had been a Japanese colony for three decades since 1910 though

Koreans were defined as Japanese citizens. In general, Koreans were treated as second-class citizens under Japanese rule. Many Koreans were in the IJA; some were high ranking officers such as Lieutenant General Sa-ick Hong, the commandant of Manila's notorious POW camp. Future Korean president, Chung-hee Park graduated from the Japanese Military Academy and fought for Japan in Manchuria. Some *Kamikaze* pilots were brainwashed Korean youth. Since Japan controlled Korea for so long, there were many problems with viewing all Koreans in the IJA as second-class citizens although in general they were viewed as inferior.[1045]

The *USS* Pennsylvania battleship (BB-38) fires on Japanese targets on Guam, July 1944. National Archives, College Park

Using Chamorro and Korean laborers, Japan had almost three years subsequent to capturing Guam to build defenses. Thinking they would never be attacked, the Japanese only strengthened Guam's defenses after losing Guadalcanal in February 1943.[1046] Many coastal defense guns, with antiaircraft units and fixed emplacements, awaited the Marines. Postponement of W-day (Guam's invasion) by the five weeks necessary for the Americans to defeat the Japanese on Saipan and at the Battle of the Philippine Sea accorded the defenders more time to construct underwater beach defenses.[1047] However, the delay allowed for the most extensive U.S. naval bombardment and air bombing in the Pacific War right before an enemy island was invaded. Returning to Guam 50 years later, a Japanese officer said the vast American invasion fleet offshore had "paved the sea."[1048]

Commanding officers for the Marianas campaign on Guam 11 Aug. 1944. From Left to Right--Major General Roy S. Geiger (IIIAC), Admiral Raymond Spruance (Fifth Fleet Commander), Lieutenant General Holland M. "Howlin Mad" Smith (VAC), Admiral Chester Nimitz (Pacific Fleet Commander) and Marine Commandant Lieutenant General Alexander Archer Vandegrift. National Archives, College Park.

Major General Roy Geiger demanded that naval gunfire bombard Guam weeks before the invasion. Admiral Spruance, intent on sparing Chamorro lives, required that someone devise a plan to systematically bombard enemy positions, beaches and fortifications. Rear Admiral Conolly accepted that responsibility, directed the bombardment and commanded the amphibious assault.[1049] The navy's big guns bombarded the island for thirteen days with high-explosive shells, many of which weighed over 1500 pounds (14-inch shells) or 2700 pounds (16-inch shells). These shells created blast craters thirty to forty feet deep and a shrapnel range of a few hundred yards. Before the landing, 6,258 rounds of 16-inch and 14-inch shells were fired by the battleships and 3,862 rounds of 8-inch shells (335 pounds each) and 2,430 rounds of 6-inch shells (100 pounds each) were fired by cruisers. Various ships fired 16,214 rounds of 5-inch shells (50 pounds each).[1050]

While the U.S. Navy subjected the Japanese garrison on Guam to the "heaviest preparatory bombardment yet delivered" in the Pacific, a few lucky strikes "destroyed the water mains to Orote, and the ones to Agana and Agat" on 11 July. From that day until the landings on 21 July, the vast majority of the Japanese soldiers were without water except for the little rain water they could collect which "barely sufficed for drinking purposes." Consequently, many IJA troops had to drink from mud puddles. Tortured by parching thirst, these men threw away logic and drank water that would ensure sickness. "Skin diseases flourished, and diarrhea and dysentery increased greatly" amongst the enemy which also "softened up the landing zone" to some degree.[1051] It is easier to fight an enemy who has to defecate on a regular basis and is sick to his stomach than one who is well fed and healthy.

While the extended preparatory naval gunfire and air attacks degraded the defenses, the bombardment signaled to the Japanese where the Marines would land, which was different than the Japanese expected. With the bombardment placement, the Japanese now knew the landing zones were (1) on the northern part of Guam at Asan and (2) south of Guam's Orote peninsula at Agat. This allowed the Japanese to change their defensive strategy.[1052]

The navy took 15 days to find enough ships to move the 77th Infantry Division 2,300 miles from Honolulu to Guam to reinforce the Marines. Woody and his buddies boarded the USS Wayne on 3 June 1944 and headed toward the front. They sailed to many islands to get to the staging areas, arriving at Kwajalein Atoll in the Marshalls Islands on 8 June and departing on 13 June. Their ships kept "'cutting holes in the sea' within sailing distance of Saipan" in case Lieutenant General Smith needed them in Saipan's battle. Every night men waited to see which way the convoy sailed to ascertain if they were going to Saipan or hitting Guam. After several days, they were released from their "sitting reserve" duty and headed for Eniwetok Atoll also in the Marshalls which is 1,200 miles east of Guam, arriving there on 29 June. When on Eniwetok, they "were given a respite and a chance to go ashore to lose their 'sea legs.'" They enjoyed a few days on land, drinking warm beer and sleeping outside instead of in crowded quarters below deck. However, it was not a tropical vacation. Correspondent Dick Dashiell wrote that the men were there for three days and "sweltered and simmered in the sun."[1053] After this short respite, such as it was, Woody, Lefty Lee and their division left Eniwetok for Guam on 6 July.[1054]

Along the way, the fleet looked for enemy planes and ships. Before they arrived at Guam, six Japanese torpedo bombers attacked several *LCI* ships (Landing Craft, Infantry), 158 foot vessels with a crew of 24 and approximately 200 Marines each. A torpedo hit one *LCI* just "forward of its bridge" and blew off the bow. After removing the survivors, the ship was sunk. Nearby ships shot down three planes. Clearly, the journey to Guam had its perils.[1055]

The fleet invasion force heading to Guam, July 1944. National Archives, College Park.

Among the hundreds of ships in the Marianas were 15 aircraft carriers, five escort carriers, six battleships, nine cruisers and 57 destroyers. American "seabasing" had no equal and the navy pounded Gaum's shores with a ring of naval guns that dwarfed anything the Japanese could project. When the U.S. Navy fired at targets the entire landing area erupted into a "cloud of dust and debris, fire and smoke, screaming wounded and silent dead."[1056] It was an amazing act of coordination since hundreds of ships had to be drawn from numerous bases.

Days before the invasion, navy underwater demolition teams (UDTs or Frogmen) cleared the waterways for the amtracs and tanks. Starting 14 July, they "destroyed 900 log cribs and wire cages filled with coral." They left a large sign on the Asan reef reading "Welcome Marines! USO this way."[1057] Conolly wrote "positively, landings could not have been made on either Agat or Asan beaches nor any other suitable beaches without these...clearance operations."[1058]

Guam is 30 miles long and four to eight miles wide. "Ringed by coral reefs and rocky cliffs, Guam was a tough prospect for any amphibious attacker."[1059] It was larger and more rugged than Saipan, giving the enemy extensive cover and more room to maneuver. Commandant Vandegrift later commented that it was "some of the most rugged country I have ever seen."[1060] On invasion day the wind was minimal and the sea calm. Woody's ship had kept a safe distance away during the preliminary naval bombardment, but now it moved into position for the invasion.

The night before the attack, Marines laid out their weapons, ammo, first aid kits and canteens; sharpened Ka-Bar knives, bayonets and entrenching tools; and gathered up C rations for several days, extra socks, uniforms and ponchos, checking that everything was squared away. Valuables like letters, rings, money, and other personal items were entrusted to the ship's crew for safe-keeping.

Before dawn on W-day the Marines were served the traditional pre-landing chow of steak and eggs. For some it would be their last meal; as Marine veteran Eugene Sledge wrote, "the usual feast before the slaughter." Having a government life insurance policy was not mandatory, but almost everyone did.[1061]

On 21 July, at 0600 hours, assault ships reached assigned areas. A typhoon threatened the landing, but it veered off and its heavy rain would not strike Guam until 29 July. "W-Day dawned clear with slight overcast, light wind and calm sea, perfect for landing..."[1062] Rear Admiral Conolly gave the order that launched an amphibious assault: "Land the landing force." What unfolded next was a choreographed execution of a thousand moving dancers. Chaplains offered prayers over the loudspeakers. The transport carrying Woody lowered landing craft suspended from their davits on both sides. Woody and his brother Marines geared up, checked each other and lined up to load their crafts. Clinking gear of metal against metal echoed throughout the passageways as the men moved in unison like a long line of ants to get into their vessels. Woody's "reserve battalion" used mostly Higgins boats.[1063] After the boats filled, the coxswains circled next to the ship until other landing craft were loaded with their Marines. This slow hold pattern was dramatically changed when the boats rushed to shore and delivered their warriors in a phalanx formation. The boats lined up according to their colored flags denoting their beach destinations. They powered through the sea with their Marines, landing them quickly. Empty boats returned to the ships to ferry more Marines and supplies to the growing combat zone.

Aerial view of W-Day for Guam when the landing force hit the island, 21 July 1944. Over 30,000 men would hit Guam on the day of invasion. National Archives, College Park

Marines jumping off an amtrac and other Marines climbing down the ropes into their Higgins boats to assault Guam, 21 July 1944. National Archives, College Park. Photo Credit Corporal Carry.

Two hours before the Marines hit the beach, the fleet pounded Japanese positions. Following the naval bombardment, planes sortied out and conducted "grass-cutting, strafing and bombing sweeps" along the beaches.[1064] Naval gunfire support did not end with the landing. Every battalion had a fire control party to call in gunfire to support it, but some radios became soaked going ashore and casualties were heavy.[1065] Ship support fire came, but not as regularly as the Marines desired due to lack of communications caused by casualties and radio damage. (See Gallery 2, Photos 14-15)

The success of pre-landing bombardment was difficult to assess after battle, but the Marines got an evaluation from Japanese Colonel Hideyuki Takeda after the war (one of the few Japanese officers who did not die in battle or kill himself). He said coast defense emplacements in the open and about half under cover were "completely demolished before the landings," including many 200mm guns on points that overlooked beaches and half of the pillboxes in the inshore area of the landing beaches. Power plants concealed in caves and communications equipment remained undamaged, and antiaircraft guns were operational until the last. Even so, the Japanese suffered more damage than they thought possible. Before the invasion started in earnest, Takeda troops' morale suffered due to this naval gunfire as it not only took out many targets, but also sounded the beginning of the death knell for the last major island left in Tojo's Line.[1066] When the Marines hit the beach on 21 July, one Japanese officer, amazed by the materiel and number of enemy, thought, "This is the day I will die."[1067] Many IJA soldiers looked at each other and shouted: "We'll meet at Yasukuni!"[1068]

Chamorro teenagers in an outrigger canoe slipped out to the fleet to help the navy pinpoint Japanese strongpoints. Many Chamorros would find the Marines during the battle and provide "Intel." Some stayed as scouts. As more were liberated, volunteers received uniforms and weapons and were placed in units working with the Marines to hunt down the remaining Japanese after their line of defense was broken and they had retreated from the beachhead.[1069]

When the Marines climbed down the cargo nets into their landing craft, they looked at each other in silence. A quiet fear shone in their eyes and after hearing about Saipan's carnage, they knew taking Guam would be no "cakewalk." Captain Beck wrote: "Marines certainly took a terrific loss at Saipan. There were almost three thousand killed and several thousand wounded and the fighting still goes on although they now have the Japs where they want them…I'm glad I wasn't there."[1070] Beck knew he was about to deliver to the Japanese forces on Guam what his brother Marines had just delivered to Saipan. Most were not pessimistic about their futures. If asked, most would say they would survive. And the majority would say it would be some of their comrades, not them, who would die taking the island, or, in slang, the ones who "buy the farm" (a reference to life insurance benefits going to the deceased's family). Most maintained this mentality or they would not go to war. Marine Private and Correspondent Allen R. Matthews said he "could picture himself grieving over the death of a friend— never a friend grieving over him."[1071]

When Woody got into the landing craft that would transport him to combat, he, like most Marines, felt unsure of what would happen when he landed or what his response would be, but he did not think he would die. He had talked with others about warfare, but had not seen any war movies or photos of battles. On the ship, eating his traditional steak and eggs breakfast, or waiting for the landing craft to reach the shore, he was more afraid of letting his buddies down than he was of death. He did not want to die, though he knew he might and that thought prevented him from getting much sleep the night before. He did not take religion seriously, but with W-Day so close he felt he should pray. In the face of death, Marines seemed to have a "delusion of reprieve" that entering this battle would not result in their deaths when all that surrounded them seemed to foretell something different.[1072]

Woody was nervous at first. But once the bullets started to fly, "you don't care anymore. You don't give a darn whether you get hit or not. You just focus on doing your job and fight. There're times when you are so exhausted your self-instinct becomes lesser and you take more chances." But Woody had to hit the beach before these emotions kicked in.[1073]

Woody and fellow Marines were about to do one of the most hazardous things any military could: Assault a strongly-defended beach from the sea against a dangerous, well-equipped and determined enemy. Young and healthy, these men willingly went into harm's way, to be in the vanguard of those taking the fight to the enemy. Unlike their enemy, these Americans came not to subjugate, rape, pillage and plunder. They were there to liberate and to defend their country. They would not let fascist aggressors rule the world. They were Marines and "there to fight."[1074]

As Woody descended the rope nets into his craft, Major General Roy Geiger, commanding the III Amphibious Corps, said over a loudspeaker (See Gallery 2, Photo 16):

> The eyes of the nation watch you as you go into battle to liberate this former American bastion from the enemy. The honor which has been bestowed upon you is a signal one… Make no mistake, it will be a tough, bitter fight against a wily, stubborn foe who will doggedly defend Guam against this invasion. May the glorious traditions of the Marine *espirit de Corps* spur you to victory.[1075]

His message was followed by the Marine Corps hymn blaring away on the horns. Ahead of them, *LCI-(G)s* (Landing Craft Infantry, rocket-armed gunboats) followed by LVT-As (Landing Vehicle Tracked, armored *aka* amtracs) firing 37mm guns launched to clean up the beach. *LCIs* carried up to seven platoons and LVTs carried 20 men. Troop-carrying LVTs were 26 feet long and often carried two pintle-mounted .50 caliber Browning machineguns. Armored LVT variants mounted 37mm guns or 75mm howitzers. Some of them had a flamethrower cannon that could spray flame 150 yards. Woody's boat was one of 360 troop-carrying landing craft; his platoon used a Higgins boat[1076] and hundreds of them ferried the Marines to shore. He moved into his assigned position as the rest of his squad made their way into the craft. Lowered into the water, the Higgins boat circled lazily on the ocean, bobbing up and down out of range of the shore guns, waiting for the other boats so they could all head to battle.

Seeing some looking green, Woody's gunnery sergeant threatened to do "bodily harm" to anyone who did not spew their stomach contents overboard because so many sat on the floorboard of the boat and this would have been unpleasant for all. But due to anxiety, or sea sickness, some vomited into the bottom of the boats or on unlucky comrades. Most pulled themselves over the boats' gunwales and heaved into the ocean, but not all were so lucky. Others lost control of their bowels and bladders and soiled their pants. Some thought about the coming battle, their loved ones or God, and others, numb, gazed at the distant beach they were about to hit. All would do their best to kill as many "Japs" as they could. The Japanese commander had ordered his men to kill at least seven Americans before meeting their deaths.[1077] Murderous thoughts dominated men's minds this day.

American commanders had chosen two landing sites: The first was Asan beach between Adelup Point and Asan Point, north of Apra Harbor and South of Agaña, the island's capital;[1078] the second was at Agat, south of the Orote Peninsula with its major air strip located between the North and South landing areas. Woody's landing beach was 2,000 yards long between Asan Point (to the south) and Adelup Point (to the north)—the 3d MarDiv described the landing beach as lying between "a pair of devil's horns." The 21st Marines would hit Asan beach's middle, codenamed Green Beach.[1079] Woody's battalion had increased from 1,000 men to 1,500, becoming a Battalion Landing Team (BLT) with engineers, tanks and an artillery battery attached to it, designed to make the regiment flexible with combined arms for the task of "storming a beach."[1080]

A mantra many said to each other before entering battle was, "Somebody's going to get hurt."[1081] Others yelled "*Gung Ho*," coming from the China Marines meaning, "All Together" or "All for One and One for All." For most, it simply meant one was highly motivated.[1082] Woody liked this term the most.

Each landing vehicle had to deliver its passengers to the proper area so the men could join with their companies and companies with their battalions to move forward together. In Operation *Stevedore*, between every pair of beaches, a patrol craft marked the boundary, and between Waves 1 and 2, there were ramped landing craft for salvage. Landing craft flew large flags with color combinations and symbols like stripes and balls indicating the beach to which it was assigned. Five minutes before H-hour when the first wave of Higgins boats left the line of departure, a "meat-ball" signal pennant was hoisted on each ship's yardarm, directing the craft. Battleships, cruisers and destroyers fired at the beach defenses, halting as the landing craft got close to permit carrier planes to swoop in to strafe, rocket and bomb targets.

When the Marines' boats stopped circling and started for the island, they would all turn straight for the beach and gain speed to 12 knots, evidenced by their bows kicking up foamed seawater that trailed back along their hulls. The crafts' engines roared as they headed to the thunder of battle that awaited beyond the shoreline. As they went into range of enemy guns, they followed behind the first wave of LVTs (amtracs) that had drawn much fire. Shells threw up plumes of water often harboring shrapnel as the Japanese tried to sight in the Americans. Some LVTs took direct hits. Twenty-one were sunk before they reached the beach, while others suffered damage. Some of the Higgins boats following the LVTs ran aground before they hit the beaches. As a result, those Marines had to jump out into waist-high water and wade ashore. Hydrographic intelligence was poor in some areas. Woody and his brother Marines followed in Higgins boats and they had difficulty reaching the beach due to the obstacles and reef. Lefty Lee and some of his unit had to trudge 300 yards in the shallow ocean to get to the beach with snipers shooting at them and mortars dropping down.[1083] Moving through the ocean without any cover made them easy targets. Several dropped into the water turning it red as Marines were hit and bled out. If an LVT or LCVP dumped its passengers into the water where it was too deep to walk or fell into a water hole, then most, loaded up with their packs, ammunition belts and weapons, drowned. Such a death was terrifying, quick and cruel. Sharks prowled everywhere, feasting on cadavers abandoned to the sea's depths. Thanks to the underwater demolition squads, most of the courageous Marines steering the LVTs and navy coxswains' driving the Higgins boats were able to accomplish their missions and provide heavy machinegun and cannon support, and deliver their riflemen to the fight, landing on *cleared* beaches.[1084]

All 3d MarDiv's three regiments landed abreast on a 2,000 yard front. Major General Geiger had placed Major General Allen Turnage in command of the division, a man with battle experience from WWI and Haiti. He commanded Marines in North China in 1939 and successfully led the 3d MarDiv at Bougainville, earning a Navy Cross.[1085] As the 3d MarDiv departed for hostile shores, it seemed like it was in good hands with Turnage, although Smith would soon be disappointed in him.[1086] As the Higgins boats rushed toward the island, the "air observer flying over Asan Beach" radioed back to his command: "First wave headed for beach…The Rockets are landing and giving [the Japanese] Hell. Good effect on Beach."[1087]

The first LVT wave landed at 0829 hours, the second wave of Higgins boats about 15 minutes later, and by early afternoon 20,000 men with weapons and vehicles were on land. The beachhead was a concave, semicircle about 1,200 yards deep, heavily cratered and almost completely covered in dry rice paddies. Beyond stood an arc of steep hills covered with razor-sharp kunai grass or dense jungle, ending on the left with Chonito Cliff reaching to the sea. In the center were the two branches of the Asan River flowing through narrow steep gorges. The enemy held the high ground and could move easily by road supported with abundant supplies dispersed throughout the jungle. This area resembled "a miniature Salerno."[1088] "High ground offers three strategic assets: *greater tactical strength, protection from access, and a wider view.*"[1089] Lieutenant General Takashina had personally overseen the construction of the fortifications above the beaches.[1090] These had undergone extensive shelling for 13 days before the landing. Yet there were plenty of enemy caves in the slopes of the hills, and the Japanese ordered surviving troops, especially those in units consisting of pack artillery and mortars, to prepare defensive positions on the reverse slopes and crests. "The beachhead was so covered with troops that almost every projectile dropped onto it by the enemy inflicted casualties."[1091] Takashina watched it all and could safely hide in his CP in Mangan Quarry deep inside the opposite side of the Fonte Hill, a hill that would later become known as *Banzai* Ridge. The Marines needed to displace Takashina and his troops and seize the high ground before them.

Meanwhile, the Marines set up artillery in the 21st Marines area where the Asan River flowed to the sea so they could put accurate fire on the hills while aerial spotters directed naval gunfire. Japanese airforces had been neutralized, so if Marines saw planes, they knew they were friendly. To mark the forward lines once they had moved off the beachheads, the Marines carried brightly-colored panels to mark separation between friendly and unfriendly forces, showing pilots where to attack.[1092] (see Gallery 2, Photo 12)

After landing on Green Beach, Woody and his comrades rushed beyond the shore, laid down, and returned fire. They ran for 10-15 yards, hit the deck, got up when the sergeant directed and continued forward. Woody jokes he was saved because the "Japs" were such "bad shots" in the area he had landed.[1093]

Over 30,000 men stormed the landing areas that first day. Unlike the 1941 attack on Guam by the Japanese who overpowered Guam's defenders, now the Japanese were outnumbered and cut off from air support, reinforcement, resupply or evacuation. Although the Japanese fought tenaciously, there was no chance they could hold the island from the massive, well-planned assault. Prime Minister General Hideki Tojo told the Guam commanders, "Because the fate of the Japanese empire depends on the result of your operation, inspire the spirit of officers and men and to the very end continue to destroy the enemy gallantly and persistently; thus alleviate the anxiety of the Emperor."[1094] An optimistic Major General Kiyoshi

Shigematsu, the seasoned China veteran who commanded the 48th Independent Mixed Brigade, told his men: "The enemy, overconfident because of his successful landing on Saipan, is planning a reckless and insufficiently prepared landing on Guam. We have an excellent opportunity to annihilate him on the beaches."[1095] He died in battle a few days later. The Japanese would do nothing to alleviate the Emperor's anxiety. Can gods feel anxiety? Apparently, Shinto ones do.

Japanese commander on Guam, Lieutenant General Takeshi Takashina (in short sleeves) with Colonel Tsunetaro Suenaga, commander of the 38th Regiment. Both did not care about the citizens on Guam and allowed their men to rape and murder everywhere. National Archives, College Park

When Marines hit the beach, the roar of shells filled the air. Snaps and hisses echoed overhead as bullets raced around them. Marine howitzers returned fire.[1096] In a few days, they massed enough to place 26,000 shells on any position in a few hours.[1097] A Japanese veteran reported, "On this island [Guam] no matter where one goes the shells follow."[1098] But the Japanese returned fire as well and on landing, bodies started to fall and dead and wounded lined the area of advance. Blockhouses, trenches, and machinegun installations "spit hate and death at the Marines."[1099]

After Marines hit the beach, no barriers were there to protect them, so they sprang into action by moving forward to close with the enemy. Progress was slow at first, and in some areas Marines hitting the ground scraped their elbows and knees against the coral rock. Every Marine who advanced from the landing zones did so with skinned arms and legs.[1100]

They pushed forward, met up with a pillbox or group of Japanese and attacked. It was a slow process requiring each platoon to be flexible with weapons and tactics. If Marines faced the Japanese in ravines, they closed using machineguns and rifles and killed each one individually. When they came upon a gun emplacement or pillbox, they hit it with grenades, bazookas, flamethrowers or explosives and took out the small fortress before engaging the next obstacle.[1101] At first, there was no end in sight. One trial followed another.

Air superiority is crucial for any amphibious assault.[1102] Marines are vulnerable if they land on a beach without air cover. With control of the sky, as soon as a Japanese cannon opened up, air support did its best to silence it.

The battle was chaotic and nerve wracking. "There's no way to explain one's feeling when you know that there're those who want to kill you, waiting to do just that," Woody said.[1103] But Woody never allowed himself to think he was going to die—ever.[1104] Viktor Frankl wrote:

> Everything can be taken from a man but one thing: the last of the human freedoms—to choose one's attitude in any given set of circumstances, to choose one's own way. And there were always choices to make. Every day, every hour, offered the opportunity to make a decision, a decision which determined whether you would or would not submit to those powers which threatened to rob you of your very self, your inner freedom.[1105]

When thoughts of death entered his mind, Woody refused to let them overpower him. He refused to believe death would visit him. He knew thoughts of panic would not allow him to survive, protect his buddies, and most importantly, see his girlfriend again. He chose to believe he would live and this gave him "inner freedom" to function with a calm spirit and strong determination.

Planes continued to hit targets as thousands of Marines flooded inland toward the ridgeline where the Japanese had dug in. Along the way, Marines fought emplacements. Huge shells slammed into Americans, sending pieces of flesh, bone and bodies everywhere. In the chaos, the Marine command structure down to the platoon level took over.

To supply the Marines with ammunition, food, water and more troops, the Seabees worked behind the advancing units to build roads from the beach from which supplies and replacements flowed. They manned 358 tractors during battle and built infrastructure. Every battle in the Pacific and European theatres was a military, engineering and logistical achievement.

The ring of ridges in front of them was the toughest obstacle for the Marines to attack that first day. The terrain was awful; a jumble of intertwined hills, ridges and defiles that provided countless opportunities to bring fire on Marines from all directions. Maintaining unit cohesion was demanding. On the left, the enemy manned guns and mortars above the beach and inflicted heavy casualties on the Marines. Marines could not go around or up the 300-foot Chonito Cliff and Delup Point, so they bypassed them and hit them from the rear. Using tanks and flamethrowers, they secured these areas by noon. By the first day's end, Woody's 3d MarDiv had suffered 105 killed, 526 wounded and 56 missing.[1106] Marines headed for the caves in the high ridges which harbored harassing Japanese.[1107] Reaching the first ridgeline, Woody strapped on a flamethrower, and with his assistant, Vernon Waters, approached the ridge, and used it "a few times on caves going up the mountain but there was…little need for [it]." They eventually abandoned the flamethrower in favor of rifles and continued forward.[1108] (See Gallery 2, Photo 18)

Usually a flamethrower worked with another as a two-man team. Woody said Waters "carried my pack, rifle, bedroll and grenades" because Woody could not carry anything else but the flamethrower. If a flamethrower sat down, "the weight of the tanks would make him fall onto his back like an upended turtle" and require help to get on his feet again.[1109]

Vernon and Woody were an unlikely pair, and stood out together. Vernon was 6'4¼", weighed around 200 pounds and had a large head, hands and feet—"He was a gentle giant," Woody said. He had a powerful physique with chiseled muscles looking ever bite as though he were a "Greek god." According to Woody, they had been inseparable since boot camp.[1110] Next to him, Woody looked like a midget, standing only 5'5¼". The West Virginian, short and rather seedy-looking Woody looked out of place next to the high-school educated, erect and Hollywood-handsome Waters. Now on Guam, this odd couple worked to secure the island, shoulder to shoulder.[1111]

The Japanese resistance in Woody's area was not heavy the first day most likely due to the fact that he was in reserve and arrived in a later wave than the initial attack. He and Vernon continued on with their attacks mostly using grenades and rifles. Others nearby met with fierce resistance and suffered wounds or death. The 3d MarDiv reported: "Against vicious machine gun, mortar and rifle fire, [the Marines] fought and clawed their way up the ridge."[1112] During the day, the 1st Battalion, of which Woody, Vernon and Lefty were a part and which had landed last on the beach as a reserve, was moved into the "assembly area" at the hill "just to [the] south of [the] cemetery" and moved forward to attack. The enemy zeroed in on this area with mortars and "casualties were received."[1113]

As Lefty made his way to the ridgeline, he moved around the rice paddies and used the berms and drainage ditches for concealment. Soon his unit became pinned down at the hill's bottom. Suddenly, Lefty heard a huge *zoom boom* as a mortar hit lifting him off the ground and slamming him down. "Luckily it hit a muddy area and buried itself into the muck before it exploded, shielding me somewhat from the blast, but I still got hurt," Lefty said. He spit blood and wheezed. He could feel air flowing in and out of his left lung, and blood oozed onto his dungarees from a quarter-sized sucking chest wound. He felt dizzy. A Corpsman ran to his aid and ascertained his wound was non-lethal. Lefty still had difficulty breathing and felt disconcerted looking down at his chest and seeing the air wheeze in and out of his torso through bloody bubbles. In an act of improvisation, the Corpsman took a piece of cellophane from his gas mask and plugged his wound. Suddenly, Lefty recalls, "I could breathe and my life started coming back into my body." The "Doc" told him to return to the beach and seek aid. Lefty abandoned his gear and started back across the rice paddies and berms erupting with explosions and small arms fire.[1114] (See Gallery 2, Photo 17)

As he neared the beach, the strains of fatigue tore at his legs and he continued to lose blood and adjust to the aftershock of being blown up. Noticing a small mound, he decided to rest on it. As he stepped on it, it erupted in a gaseous explosion of sickening decay. Maggots, intestines and rot covered him as he sank into the earth that had ballooned up and covered him with grotesque decomposition. He had stepped onto a shallow grave of a water buffalo (carabao). The decomposing carcass had filled with so much gas that as soon as Lefty placed his weight on the animal, it burst, spewing a volcanic flood of decay. "I'd seen a lot in my short life up to that time," Lefty said. "I'd fought the Japs at Guadalcanal and Bougainville, horrible islands of jungle and rot...I saw a lot of death... However, I never lost my stomach until...then."[1115]

Corporal Darol Eugene "Lefty" Lee. He was on the reunion to Iwo Jima in March 2015. He made up an entire story of witnessing Woody's MOH actions on 23 February 1945. He actually was removed from Iwo on 21 February for psychoneurosis. According to his medical report, he started having problems psychologically on Guam months before he hit Iwo's beach due to his combat experiences. St. Louis Personnel Records Center

As Lefty found the first aid tent, he entered looking like a Halloween ghoul. Everyone gazed at him with horror and covered their noses. Although staff cared for many mangled bodies and heard the dying and wounded scream, Lefty presented them with a sight they had never seen. Quickly, a doctor and aides took him aside. They stripped, cleaned and then triaged him. After he got stabilized, he was transported to the hospital ship *USS Solace*. While there, they dressed his wound, covered it with plastic and then made him shower.[1116]

While Lefty recovered, Woody engaged the ridge. While moving forward, Woody came upon the area where Lefty had been hit. Blood-soaked ground mixed with turned up earth gave evidence where shells and mortars had hit and blown up Marines. Nearby, Woody noticed a helmet upside down. Looking inside of it, he noticed a wad of letters in the webbing. He pulled them out thinking of the man and how his relatives would want those letters. The name addressed on these letters was Darol E. Lee ("Lefty"). At this time, Woody did not know Lefty, but he thought this comrade dead and he would send the letters to the return address if he himself survived.[1117]

While Woody secured Lefty's letters, Lefty waited in the ship's sick bay to consult with medical staff. The war went on without him for a few days.[1118] However, for others, if they could return to battle, they did so. Many did not want to be away from their buddies.

Bravo Company commander of the First Battalion Donald Beck described PFC Leo Simon who, although wounded in the head on 22 July, requested permission to leave the field hospital and return to his lines on 23 July. Although he could no longer wear a helmet because of his bandages, he went back in the lines carrying his heavy machinegun arriving in time to cover his company's attack. That night, Simon was wounded again during a Japanese counterattack, but held his ground and killed "a large number" of the enemy. Later he was wounded again attempting to rescue an injured comrade. Nonetheless, he got his wounded friend and himself to safety.[1119] Although many Marines were wounded, their units moved toward the complex mosaic of ridges above them where IJA troops awaited them.[1120]

On the night of 22 July, the enemy hit Woody's battalion with mortars and a bayonet attack.[1121] At 2200, word came back to headquarters that Woody's Charlie Company "had been cut off and needed medical supplies and ammunition."[1122] Since the Company's Corpsman had most likely been killed and could not be found, Senior Corpsman Carroll M. Garnett was sent to the lines and started taking care of the wounded by crawling along the lines and tending to those with injuries while the Japanese continued their attack. Woody's comrade "Smitty" had suffered a horrible bayonet wound across his belly and his intestines "were protruding." Garnett cleaned and then dressed the wound as best he could and then gathered the wounded in one area so that the litter bearers could take them back to the rear for further medical treatment.[1123] The Japanese onslaught was eventually repulsed during the early hours of the morning of 23 July.[1124] Suffering no wounds, Woody remarkably had survived another day on Guam.

Then during 23 July, Chamorro scouts told 1st Battalion, 21st Marines the

Japanese were massing their forces at Mt. Tenjo to the south. At 2200, the enemy attacked Woody's lines again, but then withdrew after thirty minutes.[1125] The night of 24 July, the Japanese again attacked at sundown at 2230 continuing their assault until 0545 on 25 July. They penetrated 1st Battalion's lines in two places, but the "enemy forces were thrown back after an hour" of hard fighting.[1126] Unfortunately, fratricide once again sidetracked the Americas as one battalion of the 21st Marines was bombed "by friendly planes."[1127] In describing such unfortunate events, the 3d MarDiv's operations officer "dryly remarked": "[P]ilot error, resulting in strafing or bombing of our own troops, did not improve the troops' confidence in close air support."[1128] Dealing with your dead buddies was tough, but dealing with them when they were killed by your own men was tragic. Nonetheless, fratricide is "an untoward but inevitable aspect of warfare."[1129]

During the battles, three categories of men handled the dead and wounded. The first category was Corpsmen, who provided care for the wounded on the spot or in aid stations like Garnett. The second category was litter bearers who carried wounded who could not walk back to an aid station. The job was dangerous in that litter bearers had to stand up on the battlefield and move slowly and were thus obvious targets. The third category was body bearers, whose duty was to remove the battlefield dead. Commanders wanted the fallen removed from the field and properly buried. Removal was necessary to keep morale from degrading and to prevent disease. Also, Marines have a mission to never leave a man behind, dead or alive.

Sometimes removal of the dead could present difficulties. During a mortar attack, one of Woody's buddies on Guam had been knocked unconscious and his left arm mangled. Litter bearers passed by thinking him dead. Soon body bearers put him on a stretcher and carried him back to the large tent morgue. They placed him in line with other dead and left. Later, he regained consciousness and wanted to scream; his arm was causing him incredible anguish.

As he lay there suffering, he noticed many remained silent, and not realizing they were dead, thought they were bravely "sucking it up and refusing to yell." Thinking they had scorned the pain, he stopped whining and felt like he too could keep calm until the orderlies came. Due to the heavy casualties, the body bearers returned to the tent and started moving men out to make room for the cadavers they had brought (this action by these men seem somewhat illogical, but this is how Woody recounted the story). After they removed several, he yelled, "What about me guys?" The body bearers jumped out of their skin hearing "Lazarus" speak. They ran to him, lifted him up and moved him several hundred yards to the first aid tent. There, the corpsmen stabilized him and then he was transported to a hospital ship where the doctors amputated his arm. His fighting days were over, but at least he had returned to the world of the living and not accidentally been buried alive.[1130]

Meanwhile, Marine commanders selected the high ground overlooking Agaña Bay as their main objective. Now called Nimitz Hill, during the battle the Marines called it *Banzai* Ridge.[1131] From the ocean, one goes a mile inland and over multiple layered 1,000-foot rise in elevation to reach the top. The bulk of the Japanese lay scattered throughout the hilly summit. (See Gallery 2, Photo 19)

Yard by yard, the Marines advanced up the heights, working around its groves and gorges. It took days to inch their way up the terrain. In many areas, it was so steep and slippery Marines needed both hands to climb. They slung their rifles over their shoulders as they moved up. Along the way, the Japanese rolled grenades down on them with harrowing frequency. Woody's battalion had been positioned on the front lines on Fonte Ridge, yet another name for *Banzai* Ridge. They had to clamber up a steep bluff to get there from their line of departure. The precipitous ascent required engineers to construct a lift of pulleys and lines to haul up food, water and ammunition as well as to bring down wounded. When at the top, Woody and his men entrenched themselves in foxholes and remained there for a few days, fighting off harassing attacks.[1132]

Reviewing the front lines with field glasses, Woody's battalion's executive officer, Lieutenant Colonel Ronald R. Van Stockum, noticed a squad opposite their position in a ravine, moving quickly and hunched over. At first Van Stockum thought they were Marines. But then he saw small branches garlanded their helmets instead of the camouflaged cloth Marines wore—he was observing IJA soldiers. Days later, he appreciated they were there to ascertain the best place to make Marines pay for their invasion.[1133] For a few days, nothing seemed out of the ordinary. When the sun set on 25 July, the Japanese lines, still strong, did not seem different from previous nights. Little did Woody and his comrades know that in a few hours, they would experience one of the war's biggest *Banzais*.[1134] Many in his unit would die and it was uncertain whether Woody would live another day.

# Ch. 17: "*Banzai*" Attacks on Guam

"I observed with assumed innocence that
no man was safe from trouble in this world."
—Joseph Conrad, *Heart of Darkness*[1135]

ON 25 JULY, EYES WERE open and minds sharp once the Marines entrenched themselves on the heights above the beach. Woody was upfront in an outpost as night fell. Many dangers came with evening because the Japanese proved both adroit and active in night fighting. They possibly made some attempts that night to sneak up and infiltrate the American lines to kill sleeping Marines in their foxholes, their usual *modus operandi*. But by and large, they probably only conducted reconnaissance probes to maintain an element of surprise for the counterattack they were about to launch.

To protect the men from such infiltrations, Marines had strict passwords for moving around and changed them daily. If someone approached another, one would have to either use the names of baseball teams or cars to recognize the other as "friendly." If one said "Yankees," then one would have to answer "Indians." If one said "Ford," the other would have to answer "Chevrolet" and so on.[1136] If one did not give the correct answer, then he was often shot. Sometimes the last words spoken by a Japanese approaching Woody's unit was "Me Marine."[1137]

One evening, Woody needed to urinate so he crawled out of his foxhole to do his business. When he returned, he was commanded to halt and give the password. This evening, the password was a baseball team, and Woody did not know the teams so he had forgotten it. But he yelled out, "I don't know, but I'm Willie [the nickname his buddies called him]. Don't shoot me." Luckily, he was allowed to return to his foxhole.[1138] Sometimes during the battle's chaos, when a Marine did leave his foxhole and could not be identified, he was shot by his own troops. Van Stockum had a similar situation happen to a friend who, upon getting out of his foxhole to urinate was mistaken for a "Jap" because he wore an IJA-issued raincoat and shot in his privates.[1139] Nonetheless, all these safety measures were enough to protect most from becoming casualties.

When not policing their lines, they continued to probe for enemy weaknesses in the front often doing their best to deploy their armor. Right up until the day before the Japanese attack, Woody and his buddies used tanks that had navigated the difficult terrain to blast enemy positions. The tanks were cumbersome and could not be used freely due to the difficulty in attacking higher ground.[1140] Until they seized the high ground, they could not use tanks with freedom of movement. Before evening fell on the 25th, most tanks had pulled back into defensive positions.

During the night on the ridgeline under pelting rain, the Marines experienced a serious counterattack against the northern beachhead, the same beach where Woody had landed. It had been expertly prepared. Toward midnight, the enemy began probing for weaknesses in Marine lines. The 3d MarDiv reported "the probing attacks, at first, were so small and unrelated that it was not realized that this was the prelude to the enemy's supreme attempt to drive our forces into the sea."[1141]

Before the *Banzai*, the Japanese assigned to the attack had gotten drunk on sake (Japanese rice wine).[1142] "The attitude of the enemy…had indicated many of them were drunk—insanely so. A number of canteens containing liquor were found on the enemy dead, and empty sake bottles were strewn in front of the regiment's lines."[1143] Japanese on Guam often fought in a drunken state since the island housed the alcohol depot for Imperial Pacific forces.[1144]

What may have started as another evening in the trenches, shifted to heavy mortar fire onto the Marines' left flank of Fonte Ridge accompanied by a massive *Banzai* that hit them around midnight. The Japanese massed their forces, whipped up their troops to an emotional frenzy, stealthily probed the American lines for weaknesses, and then, led by sword-wielding officers, conducted one of the largest charges of the Pacific War screaming, throwing grenades and shooting. This assault was one of many and would worsen through the night. This was Takashina's audacious move to push the Marines back into the ocean. After the initial hit, the

Japanese tried to capitalize upon the advantage of the surprise and at several points they pushed through Marine lines. After the initial shock, the American lines suffered several openings and small bands of enemy crept down the Asan River valley into the 21ˢᵗ Regiment's area, moving toward the 3d MarDiv artillery's CP. Several hundred reached the rice paddies and fought hand to hand with Marines. One section of the line received seven attacks during that night alone in the Japanese effort to break through.[1145] Since flares failed to illuminate the infiltrating Japanese, ships broke protocol and shone their searchlights over the battle lines.[1146]

Japanese soldiers on Guam 1944. This photograph was found in one of the many caves on Guam after the war. Such were the men Woody fought from July until August 1944. Pacific War Museum Guam

The Japanese attacked with an initial wave of at least 4,100 men on the 9,000 yard front. Due to the massing of forces and the obvious direction of their attack, the Marines killed more of the Japanese than the Japanese did Americans. Marines quickly learned this night attack was indeed "more than a mere reconnaissance in force."[1147] Reinforcements rapidly plugged the holes in the broken lines. (See Gallery 2, Photo 20)

The night of the attack, after finishing his guard duty and beating back small attacks, Woody stated a comrade named Clevenger replaced him. Woody then returned to the rear. Around 2330 the 1st Battalion, 21st Marines' CP at the bottom of Fonte Ridge heard a "cacophony of machine-gun fire and explosive bursts" coming from atop the cliff. Reports poured over the airways, many of them mixed and confused. One forward observer stumbled down the cliff to the CP declaring the enemy had attacked in force and "all hell had broken loose."[1148] Woody's neighboring 3rd Battalion reported, "After probing in the dark, Nips launched a terrific attack in force."[1149] Men started to fall everywhere. See Gallery 2, Photo 22)

Mack Drake of B Company fought off the Japanese for three hours as the enemy attacked and then overran his lines. Although "wounded in the face and arm, he refused to be evacuated." He remained fighting until before dawn when his squad leader ordered him to seek medical attention. After receiving first aid, he took part in the counterattack at dawn. His actions accounted for "10 enemy killed."[1150] Woody's best friend from boot camp, PFC Ellery B. "Bud" Crabbe, serving in a pioneer unit, was moved up to counter the assault with a water-cooled .30 machine gun. As Japanese descended on his position, Bud and his comrade who fed the machinegun its ammunition belt did their best to spray the enemy. Someone told Bud later his gunnery took out 20. According to one study, over 400 Japanese dead lay in front of the 1st Battalion of the 21st Marines' lines, the lines Woody, Drake and Bud held.[1151] Drake and Bud's experience mirrored John Basilone's on Guadalcanal. Basilone killed so many Japanese during their assault that their

PFC Ellery B. "Bud" Crabbe (Woody's best friend from Boot Camp). Bud would fight bravely on Guam during the night of the *Banzai* attack, 25-26 July 1944. St. Louis Personnel Records Center

bodies piled up upon each other almost six feet high so that subsequent enemy in the rear had to crawl over the dead bodies of their comrades to continue the attack against John. Another machine gunner, like Basilone, killed 80 Japanese on Guam and "stacked them up as a barricade against the next charge."[1152]

Despite the large number of Japanese casualties, the enemy almost annihilated Woody's company. Woody said, "You wondered why and how they kept coming, and if there would ever be any end to them. We just had to keep firing and killing them until we had killed enough of them to break their main resistance."[1153] Yet, many in Woody's lines fell and the front's integrity broke down. Woody's regimental journal noted the Japanese started breaking through the Marine's first line of defense by 0100.[1154]

Upon receiving alarming reports about this attack, Woody's command tried to get his C Company and the other two companies "A" and "B" fragmentation hand grenades from the ammunition dump in the rear. In anticipation of such an attack, they should have already had them. To the consternation of many, after a "painfully long delay," the wrong grenades arrived—they were smoke grenades and useless. Things were getting dire on the front lines and many were dying. Woody frantically fired away at charging Japanese as flares illuminated the threats.[1155] Guns were blazing everywhere. Sometimes, men shot in the direction of yelling Japanese and the waves of shadows. Japanese screams were heard giving orders or shrieking in pain mixed with the sound of Marines yelling to plug holes, hollering for Corpsmen and shouting out their fierce wish to kill the "yellow bastards." Woody observed: "No one can image how the noise of battle creates the most bizarre combination of sounds that no one has ever heard other human beings make."[1156] "*Banzai!* Where are you? Over here! Corpsman, I'm hit! Kaboom. To your right, watch out! Rat-a-tat-a-tat-tat-tat-tat. *Banzai!* Ping. Ahhhhhhhhhhhhhh! Help me! Holy Fuck! Fuck you! Ching. Ching. Ching. Stupid Nip! Brrrrrrrrrrrr. *Banzai! San Nen Kire* (Cut a thousand men!), *Vompp. You Marine Die!!*" One could say Woody "heard all things in heaven and in the earth…[and] heard many things in hell."[1157] In addition to the sounds, the smell of cordite, sweat, decay and smoke permeated the air while the taste of adrenaline seeped into the mouth making their tongues and lips dry.

Woody's battalion commander, a real "go-getter", Lieutenant Colonel Marlowe C. Williams, did not know what was going on as reports flooded in that men were dying, positions were over-run and ammunition was needed. Looking at his XO, Lieutenant Colonel Van Stockum, Williams said, "Van get up there and see what's going on." Van Stockum said that with this order, "the most hazardous mission of my Marine career" commenced. He grabbed his radio operator and started to scale the cliff where the battle raged.[1158]

Lieutenant Colonel Marlowe C. Williams, commander of the 1st Battalion, 21st Marines, 3d MarDiv. Captain Beck described him as a real "go-getter." He received the Silver Star for his actions on Guam. He suffered wounds at the very beginning of Iwo and was removed from combat. Woody served in Lieutenant Colonel Williams' battalion both on Guam and Iwo. St. Louis Personnel Records Center (In this picture, Williams is now a Colonel).

Artillery rounds rained overhead sent by the batteries behind Van Stockum trying to stop the *Banzai*. Simultaneously, Japanese mortars cascaded onto the CP and the front lines. In the confusion, Van Stockum looked around to locate his radioman, but he had gotten lost in the mayhem and was nowhere to be found. Van Stockum continued forward.[1159]

Just before dawn at 0400 the *Banzai* pushed through the Marines' rear echelons—it was nothing like what the Marines had already seen. One Leatherneck, Charles Maecham, described the thick enemy crowd charging them, illuminated by flares, as resembling "brown maggots wriggling through the grass."[1160] In some places the Japanese infiltrated so quickly the fighting descended to hand-to-hand combat. Allen Shively, after almost being stabbed to death and shot with a pistol by an opposing officer, threw the man down and beat him to death. With this human wave, the lines got mixed and bodies were bouncing into one another.[1161] The battalion said that at 0330 the enemy had "penetrated between" B Company and C Company and by 0400 were overrunning the battalion's CP and rear mortar positions.[1162] The 3d MarDiv noted that when the main thrust at 0400 hit the Marines, the Japanese shouted strange words with one English speaker yelling, "Wake up, American, and die!"[1163]

Right at the battle's apex, Van Stockum arrived at the front. He saw the "enraged" company gunnery sergeant, Albert Hemphill, "pick up a discarded *Samurai* sword" and killed a wounded enemy soldier near him who was trying to rise and fight. He chopped away several times before the man finally lay dead. Screams, shouting, firing and explosions echoed over the ridges. "It was apparent that [the Japanese] objective was to reach the beach in the rear in order to destroy our artillery and the supply dumps." They had penetrated the lines at several points.[1164] Van Stockum borrowed a company radio and reported to headquarters that their line had held, barely. However, the rudimentary radios were not designed to handle the jumble of traffic generated by the attack and it was doubtful Van Stockum's message got through.[1165] (See Gallery 2, Photo 21)

Lieutenant Colonel Roald R. Van Stockum, who was the executive officer of the 1st Battalion, 21st Marines under Lieutenant Colonel Marlowe C. Williams. This was Woody's unit. During the counterattack on Guam on 25 July 1944, also known as the *"Banzai,"* he was in the front lines trying to stop the flood of Japanese soldiers from destroying his battalion. During Iwo Jima, he was stuck out on a troop ship attached to the 3d Marines of the 3d MarDiv. St. Louis Personnel File (In this picture, Van Stockum is now a Colonel).

The attack's main thrust hit B Company head-on, commanded by Captain Donald Beck, which had already been reduced from 200 to 50 men. Beck had suffered injuries on 21 July, but stayed with his men. Against heavy odds, he held the lines and called in mortar and artillery shells to stave off the Japanese who had used their knowledge of the terrain and overpowered most of his company's positions, but he and a few of his men survived the onslaught.[1166] The enemy streamed through the gaps, following a draw to the cliff. The right flank of A Company on the left and the left of C Company on the right pulled back to counter the breakthrough. Tanks parked in the rear killed many "Japs" with their machineguns as the enemy reached and overran them. The enemy's rush upon the tanks resembled a horde of ants. Savagely they swarmed over the vehicles, disregarding the machinegun fire, and frantically pounded, kicked and beat against the turrets in an attempt to get around them and continue their wild rush down the draw to the rear areas of field hospitals, supply depots and staging areas. Most even forgot to use the demolition charges attached to their belts. One Japanese tripped as he ran, igniting his mine and blowing his body into a thousand pieces.[1167] An officer, crazed as he was, slashed at the tank with his sword breaking it and dying under a hail of bullets.[1168] There was much chaos and the lines between defending Marines and attacking Japanese became blended and tangled, looking like snakes wrestling each other. Woody's regiment noted the "situation [was] not clear."[1169] And Bravo Company's captain, Beck, noted this "banzai…did away with my Company."[1170]

The Japanese wave made it to the beach and bypassed the 1st Battalion's CP at the cliff's base. Senior Corpsman for Baker and Charlie Companies, Pharmacist's Mate First Class, Carroll Garnett, was one of the first to recognize "the *Banzai*." In a black comedy of errors, he challenged a misidentified enemy platoon element nearing his position, and in answer received a flying grenade. He dove into a trench, avoiding the blast. This alerted the Marines nearby, giving them invaluable seconds to prepare for the attack. He arose and returned to the CP hundreds of yards to the rear covered by the now engaged Marines. Shocked, he found the men there in a state of "non-alertness" sleeping in their cots and ignorant their "lines had been cut." Cursing and kicking, he tried to get many to rise claiming "five sleepless nights will make the bravest groggy" even during the battle's cacophony. Frantic, he left them and tried to get to Colonel Williams' CP, but his adjutant, thinking Garnett had gone mad, declared "Take it easy Doc, everything is all right!!" Garnett, disgusted, left the adjutant and retired to the CP's rear, attempting to avoid the coming Japanese when a flare went up.[1171]

As he looked over his shoulder, he felt relief thinking he saw Marines, but on peering closer, he recognized them as Japanese when they were "no farther away than 15 paces." He could not believe how fast they moved. Still hoping they might be Marines, he shouted at them and then "heard this distinctive crack and knew a grenade had been struck on the helmet to activate it," a Japanese procedure. American grenades had a pin to pull and Japanese ones had a pressure fuse. Garnett jumped in a trench as it exploded nearby and a grenade splinter hit him in his right buttocks. He then ran to an aid station hundreds of yards behind the moving front, now bleeding "from his ass," and yelled at everyone to get the "hell out of here." In disbelief, Dr. William H. Buchan said, "Take it easy, everything is all right!" Garnett shook his head and thought, "Here we go again!" No one believed him although the sounds of battle came closer. Suddenly, a Nambu machinegun opened up over their heads. Everyone grabbed their weapons and returned fire. Finally, all knew he had spoken the truth. Garnett crawled back from the Japanese as he and the Marines developed a defensive line and fought back against 50-60 Japanese attacking them. As he retreated, an enemy officer charged Garnett, but before he could slash Garnett dead, a sergeant shot and killed him. The Japanese had pushed some to the sea and were spreading out and attacking. Garnett looked for his carbine until he realized it had been strapped around

Senior Corpsman for Baker and Charlie Companies, 1st Battalion, 21st Marines, Pharmacist's Mate First Class Carroll Garnett. He earned the Bronze Star with V for helping warn many men about the *Banzai* attack on 25 July 1944. St. Louis Personnel Records Center

his shoulder the entire time. He now returned fire. Due to this Corpsman, many had warning to protect themselves and they fought back and he later would receive the Bronze Star with V for his heroics.[1172] But not all were so lucky. Far behind the lines, the Japanese slaughtered several unarmed medical staff and wounded in aid stations.[1173] Woody's regiment realized what Garnett had been yelling about noting at 0615, "[CP] surrounded and reinforcements needed."[1174]

The Japanese continued their advance, but the Marines rallied, counterattacked, killed many and drove others back. Woody's regiment reported elements of the 9th Marines "arrived in the nick of time" and were driving the "Japs away from the Command Post toward rear of right flank."[1175] Many Japanese, seeing that their attack had lost momentum and "that they were cut off," blew "themselves up with grenades."[1176] "The… path of the breakthrough in the section could be traced the next day by the litter of dead Japs down the draw along their line of advance."[1177]

One Japanese column attacked the CP of 3rd Battalion, 21st Marines. Cooks, bakers, clerks and mess attendants grabbed their rifles and repelled the attackers. Another group tried to blow up the artillery, but were caught by daylight in front of the division field hospital. "In addition to the corpsmen, ambulatory cases turned out in underwear and pajamas to fight with any weapon they could get hold of; bed patients fired right from their cots in the tents..."[1178] Japanese lost most of their officers in the first assault and could not move up reinforcements because they had none. They reached the artillery positions and supply dumps below *Banzai* Ridge, but it was a Pyrrhic victory as their advance lost momentum and the Marines killed almost all of them.[1179]

Captain Beck kept his men busy on the ridge fighting off the Japanese flanking attacks. Even though surrounded, he rallied his depleted force, defended his CP and kept radio communication with his commander. The reports he sent "enabled his battalion commander to locate the break-through and localize it." During this chaos, Beck's men suffered even more casualties. Even so, they took a heavy "toll from the enemy." One report claimed 200 dead were later found within Beck's CP area.[1180] Later, after the battle, 2nd Battalion commander, 9th Marines, Lieutenant Colonel Robert E. Cushman Jr., said he needed a bulldozer to go over the battlefield to bury 800 dead Japanese near their lines because they had started to reek of decomposition.[1181]

After a full night and tough morning of hard fighting, the Marines retook their positions and 1st Battalion, 21st Marines, noted by 0630, the enemy charge had been broken.[1182] Woody's battalion closed the breach the Japanese had opened.[1183] In the lull of battle, Woody's comrades found Clevenger dead in his foxhole. He fought to the end, taking the initial thrust of the attack: "Death... had stalked with his black shadow before him."[1184] Taking the brunt of the attack, Clevenger had no chance of surviving.

Woody's battalion commander, Lieutenant Colonel Williams, received the Silver Star for his actions that evening. He kept control of his men under this assault and "with mortar and artillery fire falling within his position at the risk of his life, he repeatedly moved along his lines encouraging his men." After the engagement, "forty-three enemy dead and several machine guns were found" within his CP. Garnett had not been exaggerating about how close the enemy had been to his outfit. His battalion had killed hundreds of the attacking Japanese within and behind its lines.[1185]

After the attack waned, the Marines claimed victory, having all but annihilated the attacking Japanese, though it had taken hours of fighting. Later in the day, "many Marines" came to Garnett and thanked him "for the initial warning which gave them enough time to gather their wits before the attack. The adjutant, himself, [even] apologized" for having doubted him.[1186] Garnett and his comrades gave examples of the bravery displayed all along the front during the battle on 25-26 July (especially when they knew they were under attack!). Lieutenant Colonel Cushman Jr. (later, the 25th Commandant of the Marine Corps and Navy Cross Recipient for actions on Guam), commented, "In the large picture, the defeat of the large counterattack on the 26th by many battalions of the 3rd Division who fought valiantly through the bloody night finished the Jap on Guam."[1187] Van Stockum said it more dramatically, "In absorbing this blow, our Marines had broken the enemy's back."[1188]

It is unknown how many enemy participated in the attack. It probably numbered over 5,000 according to accounts: "The wild screams of the charging Japs gave the impression of a wild *Banzai*, but it was far from that type of unorganized suicidal charge. This was a part of an assault in force, planned with care…to drive the Marines from their beachhead."[1189] By noon on 26 July, the Marines had repelled the *Banzai* leaving the Japanese unable to attack, and what was left of their units pulled back. As 3rd Battalion of the 21st Marines near Woody noted: "Nips are in flight to the north."[1190]

As Cushman observed, this *Banzai* was Guam's decisive battle.[1191] The front line of the Marines of the 21st and 9th Regiments that turned back the Japanese covered the main thrust of the Japanese, and although they gave up some territory, they held the front.[1192] In the Guam battles of the 25th and 26th of July 1944, destiny favored Woody.

That same night of 26 July, a few miles away on Orote peninsula, the Japanese launched another attack of 3,500 men. They first got drunk and prepared for their assault. Like their brother soldiers who fought Woody, they were so intoxicated that the Marines facing them described the ruckus they made as akin to a pack of animals on "New Year's Eve in the zoo."[1193] When they attacked between midnight and 0200, many did not have rifles, so they charged with baseball bats, sticks, broken bottles and pitchforks. They yelled, screamed and charged in a dance-like frenzy: "[T]he jolliest of banzais." The surge made it into the Marine lines, although many perished as artillery and mortar fire sent "arms and legs" in the air like "snowflakes." The enemy "screamed in terror until they died."[1194] Observers described this Japanese tactic as "veritable war hysteria."[1195] In three hours, U.S. artillerymen fired 26,000 rounds.[1196] That bombardment ended the other large *Banzai* on Guam, convincing most the end of the Japanese was near. Most likely USMC Major Frank Hough includes this second charge when he notes the *Banzai* during 25-26 July was twice as large as the one on Saipan.[1197] However, more fighting was necessary, although it was largely "mopping up," with small units engaging one another. Many Japanese who had not died in the charges either retreated to caves, bunkers or defensive lines or took their own lives, again refusing to retreat or surrender.[1198] Hough described the charges:

> The [Jap] attack of 26 July was an extraordinary performance, even by the standards of those unpredictable little men. It combined the best—or worst—features of virtually every attack they had ever put on during the entire Pacific war, plus a few new wrinkles apparently dreamed up for the occasion. There were Banzai charges and infiltration, sharp power thrusts and sneaks through the gullies that separated units; there was high courage and hysteria, fanaticism and stupidity; the whole was supported by artillery, mortars, tanks, flares, whiskey and saki. There is evidence that the attack was carefully planned; that the enemy fully expected it to be decisive. The initial phase can justly be termed successful on several counts. Yet the whole petered out in the abject futility as even the Japanese have been able to achieve.[1199]

Hough's description sums up the different elements that comprised this mad dash against the Marine lines that many today lump together as the *Banzai* on Guam. It was a broken wave hitting many sectors. It killed many and the few little victories the Japanese did achieve could not sustain their advantage. Hough's conclusion mirrors Cushman's and Van Stockum's in that after the 26th, the Japanese were clearly defeated in military terms, but still not done fighting.

# Ch. 18: Battle's End for Guam

"No battle plan survives first contact with the enemy's main body."
—Field Marshal Helmuth von Moltke[1200]

THE MORNING AFTER THE *BANZAI*s of 25-26 July, dead and dying Japanese littered Woody's regiment's front. Most of the moribund Japanese, although in extreme pain, still tried to resist. Many Marines were drunk with rage and felt no compassion. After seeing several buddies injured or killed, or witnessing them get blown up or stabbed by wounded Japanese whom Americans were trying to help, many decided the only secure Japanese was a dead one. Emmitt Hays found an officer alive and instead of taking him prisoner, he took out his knife and cut his throat.[1201]

Despite orders from Nimitz and the Joint Chiefs against taking body parts, a few Marines still walked around with them as trophies. But there was a world of difference between the level of decency expected and enforced by the American officers and NCOs and the barbaric behavior encouraged by Japanese officers and NCOs.

Even after experiencing these Japanese human waves of death, in the heat of the battle many Marines did not think much about the destruction surrounding them. Woody did not focus on death when he fought on Guam although he witnessed deaths daily. He felt he had a unique purpose in life and wanted to return to his girlfriend Ruby. During the battles, he did not feel anger or hatred toward the opposition: "How can you feel anger toward someone you don't know?" But he did feel that since they were his enemy and would not surrender, they had to be killed. "They really wanted to kill us. I didn't know why they wanted to kill us. We needed to kill them."[1202] Later Lieutenant General Graves Erskine said, "After we got in the war, I must confess that I didn't feel guilty helping to kill [the Japanese]. It was my duty to do this, and I finally sort of developed a feeling, I guess you'd call it, just like shooting rats."[1203] Erskine stated if any nation threatened his country and killed "his Marines," he would destroy it.[1204] There was no doubt every Marine had to fight the enemy with incredible resolve, but each time Woody killed a Japanese, he felt nothing besides sadness for killing another man. In the end, he killed because if he did not, Woody knew his opponents would kill him. Furthermore, Marines killed because "they were determined to win." When in combat, many things went through Woody's mind, but hesitation was rarely one of them.[1205] For instance, Woody saw up close some of those he killed. On the ridge, his position was about to be attacked when he saw one of the Japanese officers unsheathe his sword and charge him. Woody raised his rifle and dropped the officer. Afterwards, in the lull of battle, Woody took the officer's sword and secured it around his shoulders. He then gathered a personal flag tied around the officer's waist and put it into his pack. After a few days of carrying the sword, it became an irritant and he wrapped it up and sent it to supply. He would never forgot the face or the eyes of the officer he killed. Woody felt that he looked like a man possessed by something evil.[1206]

"Scores of thousands of others fought with fanatic tenacity, and frequently they went berserk in the final battles, allowing themselves to be mowed down in hopeless attacks, blowing themselves up with hand grenades while they were plentifully supplied with ammunition that could have been used against the foe, engaging in bizarre and almost ritualistic dances in the line of fire, charging to their deaths screaming not only the Emperor's name but also outlandish phrases in English like 'Marines, we will kill you!' or 'Blood for the Emperor' or 'Babe Ruth eat shit!' or 'Japanese drink blood like wine!'" In response to these declarations, the Marines' humor showed itself yelling back: "Tojo eat shit" or "Blood for Eleanor [Eleanor Roosevelt, the First Lady]."[1207]

Often, though, Japanese soldiers simply screamed out in a blood-curdling manner "*Banzai.*" Woody described them as madmen recalling they yelled at his front "Diiiiie Marineeeeen."[1208] One American officer said that trying to understand the Imperial Army was like "attempting to describe the other side of the moon."[1209] The enemy Woody faced had shown itself to be a strange combination of a skillful, deadly and unrelenting warrior in the face of defeat, but also one that could be erratic, reckless and insane. This schizophrenic display of warrior-hood perplexed many. Historian John W. Dower explains:

A burnt Japanese soldier lays where he died on Guam after a Marine flamethrower operator got done flaming him to death in July 1944.. Guam Public Library System

> The social controls that had been built up over the centuries in Japan had 'resulted in the Japanese as an individual reacting to the problems of life much as would be expected from a strictly brought up and constantly repressed child. Their behavior after defeat…when all restrictions had clearly gone by the board, was typical; it was strictly comparable to that of a disappointed child, who, losing his temper and with it all control, smashes his toys and kicks his companions…'[1210]

This quote explains some of the Japanese behavior. In the face of defeat, instead of laying down their weapons and raising the white flag, the Japanese became more tenacious. Like cornered animals, instead of surrendering, many chose to die fighting, taking as many Americans with them as possible. They lashed out in imprudent ways and were slaughtered.

Although a large percentage of the Japanese on Guam had been killed from 25-26 July, there was still much fighting needed to hunt down remaining units. So it was not surprising, that soon after taking care of the *Banzai*, Woody's company encountered some stiff resistance.

One of Woody's NCOs, 28-year-old Gunnery Sergeant Albert D. Hemphill, led his men, including Woody, against a Japanese force of 300 hidden in caves on 28 July. These Japanese had held up the Marines and their tanks. Hemphill jumped into action, "snaked 35 yards to the Japs flank," and shot one soldier. When another tried to shoot him hiding under an enemy corpse, he called for a grenade from a nearby Marine. That Leatherneck threw the

Gunnery Sergeant Albert D. Hemphill. He taught Woody how to be a NCO. He was a China Marine and would receive the Purple Heart and Silver Star for his actions on Guam in 1944. He was viewed as one of the toughest NCO's in the entire 3d MarDiv. St. Louis National Personnel Records Center.

grenade to Hemphill, and afterward, was shot "through the forehead." Hemphill took the grenade, pulled the pin and threw it at the sniper, killing him with the explosion. Then he led a group that took out a machinegun position that had pinned them down and, in doing so, killed another three Japanese. Noticing a machinegun nest, Hemphill instructed "bazooka" men to fire on it, which they did, silencing the gun. Although he received shrapnel wounds to his back, he continued fighting. He became a one-man wrecking crew. As this was done, the tanks advanced and engaged the Japanese, blasting the caves. "Some of the Japs…were so stunned they climbed out of the holes" and threw grenades at the tanks, causing little damage. Marines mowed them down. Woody's outfit attacked numerous caves; it took an hour and a half working with the tanks to silence them all. "One tank, only 15 feet from a hole [of a cave], fired three shots directly at it [with its 75mm cannon] before hand grenades stopped popping from its entrance. Two days later, telephones deep in the cave were still ringing. It apparently had been a command post." So Hemphill, Woody and Marine tanks succeeded yet again to take out a tenacious enemy, rendering his cave defenses useless. Hemphill received the Silver Star and Purple Heart for his heroics.[1211] (See Gallery 2, Photo 23)

The Japanese executed hundreds of Chamorros on Guam before the Marines arrived on 21 July 1944. Woody's division would find many of these crime scenes. MARC

Flamethrowers were used as the Marines took the fight to the Japanese although Woody did not use his.[1212] On 26 July, the Marines pushed forward the attack. Close behind the advancing lines, a curious group of unarmed people disembarked from American ships and set up aid stations. They did not bring medical supplies or ammunition. They were far from being armed "with the lethal weapons of war." Instead, they brought "shaving gear, toilet articles, chocolate candy [and] cigarettes." These people who brought such Santa Claus-like gifts were National Red Cross members and they set up tents in the rear. They delivered news to Marines of their "babies born, of mother recovering from illnesses, of sisters being married," and so on. The Red Cross was an excellent purveyor of news going to the front and sending news back home. Also at these tents, they brought magazines bearing news from home and around the world and even played records on phonographs of mostly jazz music. They helped men coming out of the lines rest their nerves, conduct a good shave and feel like a little piece of home was there behind war's nightmare.[1213] After experiencing this for a few hours or a day, the Marines, freshened up, reloaded with ammunition and returned to the front.

Every day as they advanced upon the enemy, the Marines either fought under a relentless tropical sun that baked their steel helmets and sweat-drenched shoulders, or under a torrential rain that soaked everything and lent the air a heavy humidity that made breathing difficult. As the U.S. took over more of the island, civilians promulgated their happiness about their liberation. Living under Japanese occupation had shown them only cruelty. There was absolutely nothing the Japanese had done to improve the lot of Guam's inhabitants. Woody's regiment even found evidence of the atrocities when it came upon 30 beheaded civilians.[1214]

Ironically, the Japanese decision to confine Guam's inhabitants into concentration camps in the southeast had the unintended effect of protecting them from the combat. Nonetheless, many died in the camps or during the forced marches to them.[1215] After several weeks of fighting, the Marines freed the civilians from the camps, and they returned home. Lefty Lee, who by now had returned to his unit, remembered being at one camp after liberating it from the Japanese: "It was horrible how the Japanese had treated these people. The Japs were sadistic killers. Knowing this made it much easier to shoot them."[1216]

Civilians greeted the Marines returning to their supply lines through villages and towns with cheers, hugs and flowers. The "dark days" the citizens had experienced under the Japanese caused them "to welcome [the Americans] with almost pathetic enthusiasm." The Marines had liberated their island "in the most literal sense of the much overworked term."[1217] Palacios Flores exclaimed: "Freedom, freedom, freedom…Free from the brutality and force…free from death…free from starvation and homelessness…free from [suffering]."[1218] Chamorros gave their liberators heartfelt gratitude. Knowing they helped them gain back their freedom boosted the Marines' spirits.

Images of combat patrols on Guam from July-August 1944. National Archives, College Park.

Even in 1994, the praise for their liberators knew no bounds. In that year, Lefty returned for the 50ᵗʰ commemoration of Guam's liberation. Although their plane landed at 11:00pm, hundreds of cheering people lined the terminal hallways and streets outside the airport. As Lefty walked off the plane, an elderly woman grabbed him with tears in her eyes and exclaimed, "You saved my life. I was in a concentration camp and I would've died within a few days had you not rescued us. Thank you." Lefty returned her hug and declared, "If I had to have returned to the plane right then and fly home, the whole trip would have been worth it for that one hug and thank you."[1219] Survivor Edward L.G. Aguon said:

> For three years, we lost our liberty and freedom and lived in a military state. We lived in constant fear of being beaten, tortured, or killed for even the slightest…offense. We had to scrounge for food, or make the best of meager crops to feed our families, while we labored to supply the Japanese…with the best of what we had. We were witness to senseless executions and death of friends and family members, were victims of physical and mental cruelty, were forced out of the homes we built, and saw our church desecrated and used as a social hall….Sixty years later, I still suffer from the physical punishments I endured by the occupiers. No one can truly know the deep pain and constant fear that we had suffered, unless they themselves lived through a brutal occupation.[1220]

Guam's citizens, American nationals, who survived the few thousand deaths the Japanese inflicted on them, had their dignity ripped away from them, lived as slaves and suffered immensely living under the daily threat of death. Aguon described the conditions under the Japanese which consisted of constant oppression, no rights, rape, poor food, fear of death, executions and no hope. So, when the island came under U.S. control again, the citizens practically worshiped the Marines.

On the flip side, the Marines loved the Chamorros. Captain Beck wrote he was glad they were able to free them from "Jap domination." They were "a fine people and all very loyal to the U.S. I have found them to be very kind and interesting."[1221] Although most of the population had been freed after a few weeks of fighting, the Marines still had to search out enemy rogue elements. It was taxing walking several miles through the jungle and along hilly trails in the tropical heat. Daily life was full of sweat, blisters and fatigue, and they suffered sunburns, bug bites, leg rashes and enemy fire in order to secure the island. After Woody and fellow Marines destroyed most of the Japanese on the shore and the ridges, they pursued broken-up units to the north and south.

Tanks were used to take out pillboxes near the rifle range attached to the old Marine barracks. On 29 July, Japanese cut off from their units and defeated on the peninsula started to kill themselves by hurling themselves from cliffs. PFC George F. Eftang said, "I could see the Japanese jumping to their deaths. I actually felt sorry for them. I knew they had families and sweethearts like anyone else."[1222] To the north, the Marines followed a disorganized enemy and when they found caves full of Japanese, their tanks blasted them into dust.[1223] By the end of July, the 3d MarDiv entered the destroyed city of Agaña and reclaimed it for the U.S.[1224]

Marines conducted broad sweeps of the entire island. They sometimes covered two and a half miles a day through jungle, something one commander noted strained the men, since 1,000 yards would have been, in his opinion, satisfactory. The island sweeps physically pushed the Marines, but in conducting them, they had the Japanese on the run.[1225] They found Japanese groups who, lacking ammunition or guns, still attacked using empty rifles, clubs and bamboo sticks. As night fell, they huddled in foxholes and nervously waited in the dark, anticipating *Banzais*. The eerie noises of frogs pierced the darkness. Thinking the Japanese were preparing a sneak attack, Marines fired into the night in the direction of the sounds, probably killing a few amphibians and wasting a lot of ammunition.[1226]

Captain Beck had received replacements and took his company on several of these sweeps. During one of the mopping-up actions, his Protestant Chaplain John P. Lee went along. While on this patrol, Beck's men found a lot of Japanese and "were killing them right and left." Some of his men used "some pretty strong language" which embarrassed Beck being a devout Christian and having the chaplain in tow. Ironically, one of his scouts was a 25-year-old Baptist preacher from North Carolina. "He was in there killing and shouting like a [madman]." "He holds Bible class in the evening twice each week...He's certainly a fine lad."[1227] Some Marines, although steeped in religious indoctrination, had learned to enjoy killing.

One of Beck's replacement lieutenants, Richard Tischler, said the mopping up was tough. They often found the Japanese hiding after they had run away from the Marines and they had to kill them right where they found them usually in a shallow cave. They had to worry most at night because the Japanese would try to infiltrate and drop a grenade on them, but that did not happen frequently. By and large, the Japanese army at this stage was hungry, confused and losing.[1228]

The U.S. Army's 77th Infantry Division, less one regiment held in reserve, arrived ashore on 28 July, relieving the 1st Provisional Marine Brigade by assuming defense of the beachhead (Desmond Doss, the medic who was the hero of the film *Hacksaw Ridge* (2016) and a Medal of Honor recipient, was a member of the 77th). These soldiers patrolled on the right flank of the 3d MarDiv, chasing the enemy to the north, receiving accolades from Marine and army commanders alike.[1229]

A tank battalion accompanied the 77th, which was most welcomed; and along with the 1st Provisional Brigade and 3d MarDiv, they pushed north to clear the Japanese out. Both the M5 Stuart light tanks and M4 Sherman medium tanks supported the ground troops. Unlike the 27th on Saipan, the 77th made an excellent account for itself. The 27th's commander, Major General Ralph C. Smith, had been considered so poor that his superior "Howlin Mad" Smith relieved him.[1230] On Guam, the Dogfaces helped the Marines, effectively using their tanks to provide close support for ground troops and to thrust their way through areas of heavy jungle.

On Guam, both army and Marine tankers were used to fire upon Japanese strongpoints, upon enemy firing from houses, and upon roadblocks set up to halt the Americans' northern advance.[1231] In addition, tanks, including bulldozer tanks and tanks with their turrets reversed, were employed to push through enemy-infested jungle that could not be easily penetrated by infantry alone.[1232]

After the July 28th combat death of Takashina, the Japanese 31st Army's commanding Japanese general for all of the Marianas, Lieutenant General Hideyoshi Obata, who was stuck on the island while making an inspection tour when Saipan's invasion began, took over. During the rainstorm on the 25th and 26th of July, it was Takashina who launched the counterattack mentioned earlier. After the *Banzai*, he withdrew, and in retreat was caught in battle and killed. With the death of their commander and many officers, more Japanese slipped away into the hills and the jungle to continue guerrilla warfare into the following year.

Takashina's superior and successor, Obata, did not have much more success once assuming command. He also ordered counterattacks rather than holding actions, and he organized retreats, both tactics expending his men since Marines pursued him relentlessly. By August 8th, Marines had pushed within a mile and a half of the northern coast and the 1st Provisional Marine Brigade had taken Guam's entire northwestern tip. That night, even Radio Tokyo conceded that nine-tenths of Guam had fallen to American troops.[1233] Taking up his last headquarters, Obata took 55 Chamorro men who had built his bunker complex and executed them to prevent them from escaping and telling the Americans where his CP was.[1234] On 11 August, Obata issued his last order admonishing his troops to fight to the end, and then killed himself. The day before Obata's suicide, Major General Geiger declared the island secure and announced a cessation of hostilities. For his leadership, Geiger received the Distinguished Service Medal. The citation read in part:

> His meticulous attention to detail during the preparatory stages of the campaign, and his fine judgment and inspiring leadership during the assault phase of the operation, cannot be too highly emphasized…Geiger played a vital part in an historic campaign which was outstandingly successful, and his distinguished services are worthy of reward.[1235]

As weeks passed and Japanese numbers dwindled on Guam, Marines still had to exercise caution because the Japanese hid everywhere, conducting guerrilla warfare. The Americans found Japanese lashed to the treetops, sniping at the Marines as they advanced. These snipers could be killed, but often not before shooting a few Marines. While marching forward into a new zone, the Marines frequently spotted dead Japanese hanging limply from tree trunks and branches, like macabre Christmas ornaments, freshly shot by a skillful American marksman.

In review of the battle, a 3d MarDiv's report summed up the Japanese: "[The enemy offered] pointless bravery, inhuman tenacity, cave fighting, infiltration (often indistinguishable from straggling). There was the will to lose hard."[1236] The 3d MarDiv then noted, that in contrast, it exercised "speed, training, initiative, coordination, and the will to win quickly."[1237]

On 11 August, Admiral Nimitz and Commandant Vandegrift steamed into Apra Harbor on the battle-cruiser *USS Indianapolis* and set up the CP from which Nimitz directed the Pacific Fleet for the rest of the war.[1238] After Nimitz took up quarters on Guam, he received a letter from Chamorro elders:

Agana, Guam
August 10, 1944

Admiral Chester W. Nimitz
Commander in Chief of U.S. Pacific Fleet
Pacific Ocean Areas, and
Military Governor of Guam

Dear Sir:

In behalf of the people of Guam, we take this opportunity to express to you and our common nation our heartfelt thanks for the re-capture of Guam by the strong and invincible forces under your command.

The recapture of Guam was opportune. Had it been delayed longer the native inhabitants would have barely withstood the ill-treatments and atrocities received from the Japanese. What kept us up throughout the thirty-two months of Japanese oppression was our determined reliance upon our mother country's power, sense of justice, and national brotherhood.

We rejoice the recapture of Guam and are extremely grateful for the timely relief we are now getting.

In closing we request of you that should you deem it expedient that this our note of appreciation be transmitted to the Honorable President, and to the people of the United States of America.

Gratefully yours,

Guam's Elders sent a thank you letter on 10 August 1944 to Admiral Chester Nimitz for liberating them from the Japanese. They were not being hyperbolic saying Nimitz saved them from obliteration suffering concentration camps and executions. National Archives, College Park

Dear Sir:

[On] behalf of the people of Guam, we take this opportunity to express to you and our common nation our heartfelt thanks for the recapture of Guam by the…forces under your command. The recapture of Guam was opportune. Had it been delayed longer the native inhabitants would have barely withstood the ill-treatments and atrocities received from the Japanese. What kept us up throughout the thirty-two months of Japanese oppression was our determined reliance upon our mother country's power, sense of justice, and national brotherhood. We rejoice the recapture of Guam and are extremely grateful for the timely relief we are now getting. In closing we request…our note of appreciation be transmitted to the Honorable President, and to the people of the United States of America.[1239]

The elders were not hyperbolic. Everywhere Japanese rule came to an end, their brutality only increased and their disregard for any post-war repercussions for their behavior was utterly ignored. Not only were they brutal with foreign citizens, but also with their own as seen on Saipan and Tinian. Everyone on Guam now benefited from the resources America offered and the island knew it was now under a rule of law that honored justice and human rights—in fact, Guam's citizens were soon wearing GI-issued boots and clothes and receiving plenty of food.[1240] Their heartfelt thanks to the U.S. Armed Forces and FDR for returning peace was a gift they would never forget (supposedly even as of the time of this writing, they have the highest percentage of people serving in the U.S. Armed Forces of any other territory or state).[1241] Services of thanksgiving were held throughout the island after having been outlawed during the occupation, and many thanked God for their freedom.

The price tag for Guam was daunting: $900 million dollars ($12.8 billion in 2019 money).[1242] So these campaigns were waged with not only precious lives, but also with financial power. The cash flow for the war came through taxes, war bonds and the U.S. government monetary policies.

Citizens on Guam enjoyed their freedoms after the Japanese were defeated on their island and the U.S. restored its government there. In the top photo, Catholic citizens of Guam, once again, enjoy religious freedom and offer prayers of thanksgiving for their liberation. While the Japanese controlled Guam, the Soldiers of the Sun prevented religious gatherings and public worship. In the bottom photo, young boys flocked to the Boy Scouts and young men hungered to get into the U.S. military. Supposedly, even today, they have the highest percentage of people serving in the U.S. Armed Forces than any other territory or state in the union. National Archives, College Park.

Soon B-29 bombers were flying in force from Guam to Japan, laying waste to her cities, infrastructure and factories.[1243] These battleships of the skies flew at high altitudes, where they had little to fear from fighters and flak. They did, nonetheless, experience casualties, often from mechanical problems or on bombing runs at lower altitudes, where their aim was more accurate but they were more vulnerable to enemy fire (like Fiske Hanley's B-29 that was shot down on 27 March 1945 over Japan—he and only one other crew member exited the plane and survived the attack).[1244] Frustrated with not being able to shoot the majority of them down, Japanese pilots sometimes rammed them.[1245] Using their own planes as weapons was one of the best aerial attack procedures, and sometimes, the only way the Japanese airforce could bring down these bombers, but the Japanese had ever fewer planes and pilots as time passed; they were neither able to produce new aircraft models capable of effectively fighting the technologically advanced American airplanes, nor train the pilots required to man the machines they could produce.[1246] The extent of the foolishness of *Kamikaze* warfare was best exhibited in these suicide flights that cost Japan its desperately needed aircraft and pilots.

After the Americans secured Guam, thousands of Japanese were still on the island. Many stole food from the Americans instead of fighting them. The majority refused to surrender and died of starvation, dysentery or suicide. This part of the invincible IJA of 1931-42 was a defeated, rag-tag group with no supplies. The remains of the IJA on Guam shriveled up on the vine.[1247] Many resorted to eating geckos, frogs, grass, and for some, dead comrades, but despite their efforts, most died horrible deaths. Often, death in war is not glorious, but pathetic and pitiless. A few hid in the jungle, trying to survive just one more day. Woody's regiment noted in October it encountered some Japanese armed and others not, but all in poor shape.[1248] Woody went on patrols from 11 August until 3 November 1944.[1249] As late as 18 January 1945, U.S. elements were engaging, fighting and killing Japanese although the enemy could barely resist.[1250]

In taking Guam, the Marines suffered almost 8,000 casualties with 1,880 dead. For every one Marine who died on Guam, the Marines killed more than 10 Japanese.[1251] The Marines brought 60 of their war dogs with them to Guam; 20 of them were killed or went missing. These dogs not only stood guard to let some get shut-eye without worrying about infiltrators, but also ventured into hideouts to see if any Japanese were waiting in ambush, many paying with their lives.[1252]

Marine Corps' cemetery on Guam. The cost for freedom is never cheap. National Archives, College Park.

U.S. human casualties on Guam numbered half of those on Saipan (8,000 on Guam to Saipan's 17,537), but this was simply due to the size of the Japanese garrison: It was 62% (18,500) of Saipan's 30,000. Although the Marines benefited from several positive factors on Guam, including a heavy thirteen-day pre-landing bombardment, helpful civilian population, and excellent planning and co-ordination among the navy, Marines and army, they actually suffered a higher percentage of KIAs than on Saipan: The kill ratio Japanese to American was 10.2 to 1 on Guam and was 11.4 to 1 on Saipan.[1253] Possibly this kill ratio difference resulted from the "*Banzai*" on Guam being better organized and more a counterattack than a wild, desperate rush at the end of the battle like on Saipan resulting in U.S. servicemen able to kill more enemy per every death they experienced than on Guam. Also, *Banzai* Ridge the Marines assaulted, that was only one mile in from the landing beaches, took the Marines six days to seize fighting uphill the whole time unlike on Saipan where no particular area held up the Marines that long in a difficult attack. These were just a few reasons why it was slightly more difficult to kill enemy soldiers on Guam than on Saipan.

After the island was secured, the Marines on Guam settled into a mundane existence. Patrols lasted a few hours and they always had hot meals, secure tent cities and dry beds to go "home" to. The focus was no longer on Guam, but on the next battle. Woody, Vernon, Beck, Garnett, Hemphill, Lefty and others understood they were headed to a new island, but where, no one knew: "The whole operation in the Pacific and where we were going was secret. And besides, I was a corporal. They didn't tell me anything," Woody said.[1254]

They continued to blow holes with explosives for "heads," burn them and cover them up. They fired their weapons, trained and discussed lessons learned as they searched for the best tactics to vanquish the enemy. Using flamethrowers, they conducted simulated pillbox attacks. A dirty weapon might jam or rust, which got friendly soldiers killed, so they repeatedly broke down their weapons, cleaned, oiled, and then reassembled them. They continued to load on ships, debark on Higgins boats and practice assaulting beaches.[1255]

In addition, Woody, as a demolition man, practiced operating, refueling and repairing flamethrowers. This required knowing how to mix the diesel and gas to the right 50/50 blend, how to re-compress the air tanks using large metal canisters and how to break down the weapon and reassemble it without problems. In addition, as a demolition man, he learned how to operate and repair the bazooka as well as how to make and use pole-charges with simple TNT and dynamite blocks, using both "electric and non-electric blasting devices."[1256] Woody learned how to demolish almost anything. His instructors were strict; the Guam attack "brought out the need for men trained in handling these special weapons" because many were injured or killed from being unfamiliar with the use of these instruments under hostile conditions.[1257] War creates incredible stress, and when dealing with a rifle, making a minor mistake usually does not cost someone his life; however, a small mistake with explosives or flamethrowers meant the death of the operator and all around him. Working with such tools was unforgiving and needed a zero-defect behavior if one wanted to survive and ensure his buddies did too.

After Guam was secured, Woody did not work all the time. On the weekends, they had time off and swam in the bays or explored the island. They wrote letters, careful not to say something that would worry their families. Mail call was a big deal. Receiving letters from loved ones was bittersweet because of the loneliness that came with not being together.

During one day when they had free time and were not writing or reading letters, Woody and one of his boot camp best friends, Ellery B. "Bud" Crabbe, met up with a married couple who had had everything taken by the Japanese. Their home had been demolished and their land destroyed. Feeling pity, Woody and Bud gathered some of their rations and gave them food. The couple was so happy that although they did not have much, they later invited Woody and Bud to their home to have a dinner of rice and beans. They talked of freedom, how good America was, and how thankful they were that men like Woody and Bud had the "courage to come and kill the Nips who were so evil," and give them back their dignity. When the woman said some hateful things about the Japanese, her head hung down and tears dripped from her eyes. Her husband put his arm around her and said, "It will be alright now, Honey." He then slowly gazed back at Woody and Bud and said one of the most "sincere Thank Yous I have ever received in my life," Bud said. Their heartfelt thankfulness gave the Marines' service meaning. In return, the couple did what they could to return these Marines' kindness. What little they had, they gladly shared.[1258] Being with citizens helped with feelings of homesickness. Knowing that they helped better people's lives gave them a sense of satisfaction that helped to compensate for serving far away from home.

While liberating others and helping them taste freedom, Woody thought of his own American dream—namely, marrying Ruby and starting a family. It was one thing that sustained him during moments of melancholy. "I cannot remember being up and down. I was never down. I did miss my home, brothers and girlfriend. I just settled in and did my duty."[1259] He did feel lonely and missed Ruby: "If I knew I was going to be away from you for so long, I would have kissed you many more times."[1260] He would also joke about war, hoping to ease the worry, writing her, "I can run faster than any Jap and I'll be all right as long as I see the Japs first."[1261]

After one mail call, Woody received a letter from Lefty's mother that she had received his package and that Lefty was alive and in a Hawaiian hospital. A few weeks later, Lefty returned to Guam and found Woody at his encampment. He had received news from home that his missing letters had been sent to his loved ones with a note from Woody. Lefty did not know Woody at this time although serving in the same company. He wanted to thank him for sending his letters home, so he entered his tent, asked for Woody, and thanked him for his consideration. Woody was glad Lefty was alive. "When we found a helmet without a body, it usually meant someone had died. I was glad the letters I sent away weren't the last piece of news the family received from their son." Lefty and Woody shook hands and a friendship was born.[1262]

Captain Donald M. Beck received his Purple Heart from 1st Battalion commander, Lieutenant Colonel Marlowe C. Williams, for his wounds suffered on Guam in July 1944 while in charge of Bravo Company. Circa September 1944. He later wrote his wife Ruth about his Purple Heart: "They [the Marines] can have their medals, just so I get home with my dog tags is all that matters to me [when one died, one of the dog tags stayed with the body and the other was sent to the registry—when Beck said he wishes his dog tags to stay together meant he was still alive.]" Author's Collection and Beck Family Archives.

On 6 January 1945, Lieutenant Colonel Marlowe C. Williams, commander of the 1st Battalion, 21st Regiment held a ceremony for his men. He later would say of the battle, especially noting the massive *Banzai* that his battalion mainly destroyed: "We had a job to do no matter what the odds or obstacles encountered, and it was done in the performance of duty."[1263] At the foot of Nimitz Hill, Lefty received his Purple Heart from Lieutenant Colonel Williams. Such ceremonies were held throughout the islands and hundreds received decorations.[1264] After Beck received his Purple Heart, he wrote his wife Ruth, "I hope this is the last one of these that I will be awarded."[1265] He had received shrapnel in his left ankle and right arm on two occasions and felt lucky to be alive.[1266] In another letter, he wrote, "They [the Marines] can have their medals, just so I get home with my dog tags is all that matters to me [when one died, one of the dog tags stayed with the body and the other was sent to the registry]."[1267]

Beck wrote his wife describing his wounds. One might wonder about Ruth's disposition when reading Beck's candid letters—did she appreciate his honesty or have nightmares? To be fair to Beck, the Commandant had sent a telegram to Beck's wife, notifying her that he had been wounded but without providing details. For days the family wondered whether he had lost a leg, his eyesight or something worse. Beck continued writing in his letter to Ruth the following:

> [I] received your letter yesterday and you mention of a telegram from the Commandant in regard to me being wounded. I hope it hasn't caused you undue alarm, because the wounds were not serious. My first wound, which was just above my left ankle, resulted just after hitting the beach…[Shrapnel] tore through my trouser legging and about ¼ inch deep into the flesh. One man in the same boat was killed and five others wounded…I pushed inland about 300 yards and was there only a short while when a shell burst about three yards from me. Shrapnel hit my helmet and pack. Small pieces hit me in the right arm and drew just a very little blood. There is not even a sore there now, but no doubt small fragments are still in the arm. Just a piece went through my legging and shoe on the right foot and didn't tear my sock. That was also a close one. There are other incidences when I had a few

close ones, but I can tell you those later. It happens to a lot of people so why get excited about it.[1268]

Often people read about wounds for the Purple Heart, but rarely do they know what those wounds are. Reading Beck's description of how he received wounds and the circumstances around them, one can see war's "strange arithmetic" with how he was saved by inches. Although it assuaged the family's worry momentarily, the family also knew others were receiving word their sons had given the last full measure, as evidenced by people wearing black ribbons around their arms or putting gold stars in their windows. Just as Beck hoped he never received another Purple Heart, the family hoped they never again received a telegram from the Commandant.

For his survival, Beck relied on a crusty old 6'4" Gunny: Albert Daniel Hemphill. This hard-core Marine inspired his men and had a profound effect on them, especially Woody. "He wanted you to know his name and he made sure you didn't forget it."[1269] He was a China Marine from the Old Corps (one of the few and highly-regarded men who, between 1900-41, were stationed in China to protect American citizens as well as political and business interests).[1270] He had a huge tattoo from his belly button to the bottom of his neck. The story he told was that one night when he was drunk and passed out, the men carrying him back to base decided to stop at a tattoo parlor and give him a surprise. Being generous fellows, they picked a big and colorful tattoo for him. He woke up sore and hung over, but took a liking to his body art. It was a sexy majorette, with baton and all. This salty old Marine often walked around without a shirt, showing off his muscles and how he could flex them to make the majorette dance. He was a Marine's Marine and the men loved and respected him for his toughness, knowledge and humor. His Guam actions displayed that he led by example.[1271] The 3d MarDiv newspaper wrote of him: "His men swore [Hemphill]…was the best combat NCO in the Corps."[1272] He instructed Woody how to operate equipment and be an effective NCO since Woody had become a corporal by October.[1273]

During this time, Woody did not always behave as a Marine should. While they trained, he and his men received two beers a week per man. The officers were allowed hard liquor, but the enlisted men were not permitted to drink, and there was plenty of alcohol on Guam since the Marines captured the IJA liquor depot. One day as Woody and his buddy Halwagner were on guard duty, a truck with what looked like whiskey boxes came into view. It drove slowly enough that Woody and his buddy decided to run behind the truck so the driver could not see, jump on the vehicle's bed and "liberate" a box. Woody vaulted up and grabbed the box while Halwagner remained on lookout. In the Corps, stealing could be punishable with office hours (non-judicial punishment for minor offenses) or with a court-martial for a dischargeable offense. Woody jumped off the truck with his goods, hit the muddy ground and fell over. His comrade helped him up and they ran to a nearby hill, opened the box and prepared to enjoy a few drinks (a serious offense because they were on guard duty). However, to their disappointment, it was hard candy ("Pogie Bait"). They took the contraband back to their tent and shared it with their boys, disappointed they had risked so much for so little.[1274] America sent its youth over to destroy the Empire of Japan, and many were just kids—very deadly armed and trained killers, but at times extremely immature ones.

Guam, like Tinian and Saipan, became a huge base for the U.S. After taking Guam, Seabees built 100 miles of roads in 90 days, expanding the infrastructure from ports to numerous bases throughout the island, as well as fuel storage facilities, training ranges, barracks, port facilities, five hospitals, warehouses, utilities, communications, recreation facilities, and a new Pacific Fleet headquarters. They built runways for B-29s, each 8,500 feet long and 200 feet wide. Guam had the best Mariana deep-water harbor. Before V-J Day, the U.S. placed over 200,000 servicemen there. America was building a highway of death to Tokyo. Seabees worked incessantly, sometimes sleeping in their machines. They fought groups of Japanese using rifles and pistols. Their mottos were "Can do!" and *Construimus Batuimus* ("We Build, We Fight"). One officer explained, they "smelled like goats, lived like dogs and worked like horses."[1275]

Although the battles for Saipan, Tinian and Guam were over, Woody and his brother Marines knew they would have to fight on more islands before the war ended. They got used to the war cycle of preparing, engaging, killing and then recovering. Had they been asked what they would rather be doing, most would have replied they would have preferred dancing to Glenn Miller, drinking beer and being with their "girls." Yet, once put into combat, they fought skillfully, tenaciously and courageously.

Woody and his comrades were proud of what they had accomplished on Guam, but also felt foreboding at preparing for the *next island*. Nonetheless, most wanted to be with their brothers in arms no matter what. They would take great risks to accomplish their mission and protect one another. "These boys would fight to the death for one another. And that motive made them invincible."[1276] Although they prepared for some unknown island and to face an enemy they knew intimately and rightly feared, they were happy to be with their brothers. The War Department noted in 1944, that in contrast to the Japanese who marched off hoping to die for the Emperor, American servicemen went to war "with the intention of causing <u>the enemy</u> to die."[1277]

Losing the Marianas shocked the Japanese. Americans' toughness surprised them. The Japanese indoctrinated their soldiers with how individualistic, luxury-loving, soft, feeble-minded and poorly equipped the Americans were. They claimed "Yanks" were emotionally "unstable."[1278] They viewed Americans as cowardly weaklings who could not match the Japanese' feverish devotion. "They call themselves brave soldiers, yet they have no desire for the glory of their ancestors or posterity, nor for the glory of their family name."[1279] They were told "[a]s soon as they are fired upon, they do not remain in their places, but invariably run into the jungle."[1280] To Japanese surprise, the Marines fought with courage, superior firepower and tactics. Underestimating the enemy is a fatal error and one the Allies had made early in the war. The Japanese had been caricatured as weak-eyed, buck-toothed, backward little men. Their image underwent a radical remaking as they devastated the Pacific Fleet and conquered Allied possessions in 1941-42. The image morphed into feared supermen who were unbeatable.[1281] But overestimating the enemy can be as dangerous as underestimating him. Giving the Japanese too much credit did "incalculable harm in affecting the mental attitude of [American] troops sent out to meet [them] on the battlefields." Many Allied troops felt hesitant to attack, breeding "a sense of initial inferiority" while others, after fighting for a few weeks or months, surrendered when they could have fought on.[1282] After Midway and Guadalcanal, Americans finally could see themselves vanquishing the enemy. At Guadalcanal, Marines learned they could "outthink, outshoot, and outfight the Japanese."[1283] The Japanese defeat in the Marianas brought it home to American leaders and fighting men that the end of the Japanese Empire was now near.

On the other hand, Japanese leaders seemed incapable of acknowledging that the tides of war were reversing. In fact, Imperial Headquarters announced to the public after losing Midway, that instead of losing that battle, they actually had won it and sank many U.S. ships changing the course of the war. To celebrate the so-called victory, "the enthusiastic people of Tokyo staged a flag procession and lantern parade."[1284] Naturally, the Japanese press received all of its information from the military hierarchy, spewing out carefully-crafted lies about the war's status and the character of Americans. When defeats could not be ignored any longer, Japanese leaders invented illogical arguments to mislead their people. For example, after the Gilbert and Marshall Islands' successful campaigns in 1943, Superintendent of the IJA Air Academy Major General Saburo Endo gave a radio address:

> We must continue with unwavering confidence in victory, never thinking we have lost. Only with this frame of mind can war bring…victory. The honorable defeats at Attu, Makin, and Tarawa were but victories. However, the arrogant enemy, relying on…material resources, has not even begun to feel defeatism. Americans have begun a new operation…and have intruded into the sacred Imperial household by stepping on Imperial land with their dirty shoes on. If we have the required air strength the enemy counterattack will be like a summer insect flying into the fire.[1285]

Endo was misleading his listeners that America's material might was giving her a false sense of security. According to Endo, as the Americans overextended their reach in the Pacific, especially coming into the "Japanese sacred realm," they were setting themselves up for failure. Since he ran the Army Air Academy, he felt he and his trained pilots would ultimately lead Japan to victory. Endo's address was one of thousands made by Japan's leaders, and as the war progressed, they would become more fantastic in their lies and promises.

When large numbers of Japanese soldiers died, they were praised for choosing death over surrender in the "Sacred War," creating the perception of a moral victory. An ancient poem *(Umi Yukabai)* was often recited over the radio, praising these deaths: "Across the sea, corpses soaking in the water/ Across the mountains, corpses heaped upon the grass/ We shall die by the side of our lord/ We shall never look back." The Japanese press reported an army of Japanese spirits fighting the Americans, causing mental derangement, suicide, nervous breakdown and morbid fear. Some dead were reported to have taken physical form, continuing the fight.[1286] Their "spiritualism superseded realism."[1287] Defeat and retreat to smaller defense perimeters was described as luring the Americans closer, where they would be annihilated. Many believed the propaganda and those who did not dreaded saying anything for fear of the *Kempeitai* (secret police).[1288] Marines repeatedly took advantage of this underestimation.

Woody knew after Guam, there would be no easy battle. And if he thought Guam was difficult, then he had no idea what was about to hit him on Iwo Jima. Even when he and fellow Marines entertained their worst nightmares, they never doubted the Japanese could and would be defeated; it was just a matter of time. So now the hardened veterans of Saipan, Tinian and Guam prepared for other far-off places like Iwo and Okinawa. The Marines envisioned that they were headed directly to Hirohito and his legions and that they would slaughter every last one of them. By now, Hirohito, "like Hitler," had become "a wartime symbol of a hated enemy, of depravity, of tyranny, and of inhumanity."[1289] Americans knew they needed to kill him and his regime.

# Ch. 19: Events Leading to Iwo Jima

"The weak-willed man makes mistakes.
Willpower is the essence of manhood.
A man's strength is determined by the strength of his will."[1290]
—Lieutenant General Tadamichi Kuribayashi

"War is the Tao of deception. Therefore, when planning an attack,
feign inactivity. When near, appear as if you are far away. When far
away, create the illusion that you are near. If the enemy is efficient,
prepare for him. If he is strong, evade him. If he is angry, agitate him.
If he is arrogant, behave timidly so as to encourage his arrogance. If
he is rested, cause him to exert himself. Advance when he does not
expect you. Attack him when he is unprepared."
—Sun-Tzu, *The Art of War*[1291]

AFTER GUAM, WOODY AND THE men of V Amphibious Corps knew they were headed for another serious battle. He never wanted to be any other place than with Charlie Company. "I wanted to be with my brother Marines, [and] I knew if we did our job, we could win and survive. I would've felt terrible if they said I couldn't go, because my brother Marines were closer to me than my family. Also, we wanted to win the war."[1292]

As Woody prepared for the next "damned island," America and her allies conducted massive global operations. Rome was liberated on 4 June 1944. On 6 June 1944, Allied forces invaded mainland Europe in a meticulously-planned operation, successfully landing 175,000 troops at Normandy, in northern France, in Operation *Overlord*. And *Overlord* was followed in August by Operation *Dragoon*, a second amphibious assault in southern France that pushed north-ward killing 7,000, wounding 20,000 and capturing 130,000 German troops. Unlike the Japanese, many Germans, despite Hitler's fight-to-the-death orders, increasingly surrendered. On the Eastern Front where Germany met Russia's forces, the Red Army battered the *Wehrmacht*, advancing westward with millions of troops. By the time Guam was secure, Allied forces in Normandy had broken out and were pushing the Germans back across France, Belgium and the southern Netherlands into Germany.

By October 1944, General MacArthur's troops had invaded the islands of Morotai and Leyte in the Philippines during the largest naval battle in history in the seas surrounding these islands. The U.S. deployed 34 aircraft carriers, 12 battleships, 24 cruisers, 141 destroyers and destroyer escorts, numerous PT (patrol torpedo) boats and submarines against a Japanese force of four aircraft carriers, nine battleships, 19 cruisers, 34 destroyers and support ships. The U.S. suffered over 3,000 casualties while the Japanese lost 12,500 sailors. Japan lost all four carriers, three battleships, six heavy cruisers, three light cruisers, eight destroyers and six submarines. The U.S. lost a light carrier, two escort carriers and three destroyers. After the battle, the once mighty IJN no longer posed a serious threat.[1293]

By this point in time American pilots had demolished the Japanese air and naval forces to such an extent that they were unable to replace well-trained pilots, planes, or aircraft carriers. Nevertheless, Japan would not abandon the fight in the air; instead, the Empire improvised in a fiendishly clever way with *Kamikazes* in the Divine Wind Special Attack Corps flying planes as human-guided missiles. It was difficult to prevent *Kamikazes* from hitting targets and causing damage and loss of life because the only defense was shooting them out of the sky sufficiently far from the intended goal that they could not reach the target in their descent: "For a steeply diving suicide plane a ship is practically helpless."[1294]

*Kamikazes* began to appear in battle starting in the autumn of 1944 at Leyte Gulf. Almost 2,300 *Kamikazes* would be launched in the following nine months, often encouraged and blessed by Buddhist and Shinto priests exhorting their flocks of suicide bombers to "abandon the cares of this world and adopt a policy of prostration at the feet of a homicidal dictator [Hirohito]."[1295] Admiral William "Bull" Halsey said the *Kamikazes* were the most feared weapon he ever faced.

*Kamikaze* personnel often talked about meeting again in a heavenly bliss at the Yasukuni shrine to regale each other about their exploits in a *Valhalla*-like manner with Valhalla being a hall of heroes killed in battle. No amount of training prepared the Allies for *Kamikazes* and *Banzais*; they were always shocking. Nimitz said, "*Kamikazes* took the Navy by surprise since designed suicide had not been a part of American air doctrine."[1296] "There was a hypnotic fascination to a sight so alien to our Western philosophy," observed Vice Admiral Charles R. Brown, skipper of the carrier *USS Kalinin Bay* (1943-44) and *USS Hornet* (1944-46) and commander of Carrier Strike Group I. He continued, "We watched each plunging *kamikaze* with the detached horror of one witnessing a terrible spectacle rather than as the intended victim. We forgot self for the moment as we groped hopelessly for the thought of that other man up there."[1297] Historian Lee Mandel said, "These suicide aircraft attacks added a new dimension of terror to an already extremely dangerous, stressful environment."[1298] Historian John Toland put it another way: "It was blood-curdling to watch a plane aim relentlessly at your ship, its pilot resolved to blast you and himself to hell."[1299] For Americans, *Kamikazes* expressed an "insane martial spirit."[1300]

And Japanese families rallied around these men and supported them. *Kamikaze* flight instructor Major Hajime Fuji requested from his command to be allowed to fly with his students on their one-way missions. His command refused him, explaining he had a wife and two young daughters to take care of whereas most of the *Kamikazes* he trained were single, and if married, were without children. Fuji's wife Fukuko saw how much this tormented her husband, so one day when he was away on duty, she killed their daughters and herself to release him for the mission he so passionately desired. Liberated from the burden of taking care of his women, his command then granted his wish to fly his plane into an American ship. A few months later, Fuji flew his aircraft, along with another *Kamikaze*, into the *USS Drexler* off the coast of Okinawa sinking the vessel and killing 158 Americans. He had fulfilled his greatest quest for his life.[1301]

The thought of training men to glorify their own deaths was foreign to Americans; but the Japanese never stopped improvising ways to conduct "straight to heaven" attacks. A Japanese officer explained that no one in the IJA thought of death as suffering, but in a divine way, felt martial deaths allowed their spirits to be "further purified."[1302] *Kamikaze* Haruo Araki wrote in his will to his wife Shigeko, "Tomorrow I will dive my plane into an enemy ship. I will cross the river into the other world, taking some Yankees with me. I...will forever protect this nation from [its] enemies."[1303] Haruo believed the gods would give him eternal life, which granted his death meaning. He was convinced his spirit would be "purified" and go to an eternal realm. These beliefs gave Japanese confidence to face death with courage, especially *Kamikazes*. As IJN Captain Rikibei Inoguchi, Chief-of-Staff of the First Air Fleet, explained:

> We Japanese base our lives on obedience to Emperor and Country...[W]e wish for the best place in death, according to *Bushido*. *Kamikaze* originates from these feelings...By this means we can accomplish peace....from this standpoint, the *Kamikaze* deserved the consideration of the whole world.[1304]

Although Inoguchi's request for respect for Japanese spiritual feelings for *Kamikazes* is difficult to accept, their missions were indeed feared and hated. Ships could do little to prepare for suicide attacks other than to learn when they were coming and produce as much antiaircraft fire as possible. As shown throughout the war, neither "radar detection, nor seaplane search, nor American fighter interception, nor the picket line of destroyers and other vessels posted" between the fleets and the attacking planes "were sufficient to keep these planes" from finding targets.[1305]

In November 1944, the U.S. Navy began bombarding Iwo Jima, which was the first piece of Japanese real-estate America would invade. That same month, the first B-29 raids from the Marianas on Japan's pre-war homeland islands began. One month before, Americans started to push their advance for the Philippines' Leyte Island. By January 1945, the U.S. 6th Army invaded Luzon in the Philippines, and carrier forces bombed Japanese-held Indochina (Vietnam, Cambodia and Laos). From 25 October 1944 until the end of January 1945, for the price of 378 *Kamikaze*s sent into battle for the Philippines, Japan sank 22 warships, including two escort carriers and three destroyers, and damaged 110 ships, including 5 battleships, 8 fleet carriers, and 16 light and escort carriers. "Success encouraged repetition" so the Japanese sent *Kamikazes* at Iwo when battle broke out.[1306] Almost 2,000 *Kamikazes* would hit ships off Okinawa a few months later and sink 26 and damage an additional 164.[1307] No one knew how many were earmarked for Iwo, but the navy knew they were coming.

Marines on Guam, however, did not know much about Japan's aerial suicide attacks because, as a general rule, a combatant is aware of little outside his part of the war and what is occurring in his immediate surroundings: this is a timeless observation of human nature during warfare. They practiced their amphibious operations and fighting inland. They diligently trained themselves and the newly-arrived replacements. On 14 October 1944, Woody was promoted to corporal. In November, Woody's regiment received technical training in the improved M2-2 flamethrowers and bazookas and continued to refine their techniques.[1308]

Woody with buddies PFC Ellery B. 'Bud' Crabbe on his right and Philip Cohen on his left on Guam. Cohen was the first Jew Woody had ever met. Fairmont Army Reserve Training Center Museum

During this time, Woody had developed a reputation with some as a hot-head and as "full of himself." Since he had a temper and was a flamethrower operator, they nicknamed him "*Zippo* head." He did not like the name. Bill Schlager, the son of a Marine who knew Woody, said:

> My father [Alexander], coming from Wyoming just didn't click
> with Woody coming from the hills of West Virginia... Don't get
> me wrong, they would defend each other in combat and my fa-
> ther obviously did this for Woody, but in general, my father grew
> to dislike Woody thinking he was too cocky, and those type[s]...
> get you killed. My father knew this since he had been fighting in
> the Pacific since 1942.[1309]

Most military units split up between different cliques, and Woody's and Schlager's disagreements were normal for most young men. Everyone does not get along with everyone in the Marines, but in the Corps, whether friend or not, everyone trained together and learned to support each other when the bullets started to fly.

Woody and Schlager were not alone in their disagreements. For example, during this time, some of the Marines started to think that the B-29 aircrews had cushy jobs and did not encounter much danger like the Leathernecks. One night at the officer's club, a Marine officer started "sounding off" with pretty strong language that the Army Air Force pilots were lazy, sunbathed too much on the beaches and did not really do anything dangerous. Unfortunately, for this officer, Major General Erskine was right next to him listening to his rant. After this officer finished his diatribe, Erskine got on the phone with General LeMay, and the next morning, at 0600, this said Marine officer was on a mission over Tokyo in a B-29. After his one air mission, this grunt-Marine-infantry officer never again said "anything" negative about the B-29 Airedales.[1310]

By November, the 21st regimental commander felt the unit "not ready for combat," noting its training was only 50% completed.[1311] Captain Donald Beck was in charge of the training for the 1st Battalion of 1,000 men and helped the regiment get up to speed. He wrote his wife on 16 November: "I have certainly found it hard to answer my letters here of late. We go from morning until night. Certainly will be glad when these affairs which involve the entire world will be settled."[1312] A few days before, he was philosophical to his wife when writing his son Barry on his 2nd birthday:

> It is unfortunate for little boys like you to begin your life at a time when the whole world is mad with rage and men are engaged in history's greatest conflict. Let us hope these efforts will not be spent in vain but a better world will be made so when you reach manhood it can be lived in peace.[1313]

This letter showed his son how much he loved him. It was also written for the future in case Beck died and never returned. This letter illustrated that Beck desired to make the world a better place. To secure that peace, Beck and his men had a lot of training to do.

Holland Smith ordered Major General Erskine to assume command of the 3d MarDiv from Major General Turnage, and when Erskine did so in October 1944, he found it was "in rather poor condition" and "not functioning as a division" because it was "divided up into three regimental combat teams." While Turnage had been successful at Guam, Erskine thought he had done poorly leading his troops and setting up an organized division.

There was no central 3d MarDiv command controlling all regiments; instead, all three regiments acted independently with 3d MarDiv "support behind" them. Erskine felt the division "should be a very highly integrated organization, and it should be absolutely sensitive to the division headquarters," so Erskine reorganized the command structure. Instead of troops being divided into "four parts" of three combat teams and a fourth comprised of artillery and medical support, Erskine made everything unified under his command. He brought a new staff and relieved Turnage. In preparing for Iwo, Erskine relieved five out of the nine battalion commanders because "they did not cut the mustard the way I thought they should."[1314] Erskine was shaking things up. One of his officers claimed his nickname was "*Flamethrower*" and if you did not "stand up to him, you were a dead duck."[1315] Anyone showing fear of Erskine could be dismissed. Junior officers had to know *everything* about their unit's weapons or there was hell to pay.[1316] He wanted hard, driven men under his command who could tell him the unvarnished truth.

In December, Woody and his men trained to effectively react to a chemical warfare attack, going into chambers to ensure they knew how to use their masks if hit by gas. They continued to practice in their Higgins boats and amtracs for "ship to shore movement." By 31 December, the regiment's commander, Colonel Hartnoll J. Withers, noted it was 80% ready for combat.[1317]

Woody trained for hours using C-2 explosives. He and others worked over open ground, attacked mock-pillboxes and blew them to smithereens. "We used barrels of that stuff learning how to blow things up." Depending on a structure's size, Woody used one to eight blocks (eight were usually in a satchel or on the end of a wood pole). He would put a fuse in a block (timed for ten seconds), pull the fuse lighter with primer cord and then either throw or stick it into the bunker and run for cover. "Pole-charges" were eight feet long on a 2/2 piece of wood. The men practiced repeatedly. If the blocks were hit by bullets or shrapnel, they did not detonate. They only detonated by fuse.[1318]

During training, Woody's commanding general conducted forced marches and personally led them. At 46, the 6'0" and 200 pounds Erskine carried the same gear and frequently led his division in "humping." "I wish he would slow down," Woody said. "He was so tall that for every step he took, I had to take two. He really believed in conditioning...."[1319] Hiking was the best way to "toughen up a soldier" in Erskine's opinion.[1320] "The reason I stayed in the service is because I have a...great love for the Marine Corps and I think...I can train troops better than any other..."[1321] One of Erskine's battalion commanders, Lieutenant Colonel Cushman Jr., supported Erskine's self-assessment, saying:

> [Erskine] was rough. He was not nasty, but he was rough and tough, demanding...[Erskine implemented] changes in training. Training was more realistic in terms of taking risks with live ammunition, I would say, because we set up a series of combined exercises with the artillery and air and lots of live ammunition.[1322]

Erskine put his knowledge about war and preparation into practice with his 3d MarDiv. If his men did well, he rewarded them with more responsibility and trust, but if they did not, he sometimes would first look like he "was going to have apoplexy" and then, he would quickly relieve these men of their responsibilities.[1323]

The days on Guam were not, however, just filled with training. Woody's regiment also had a wonderful base with volleyball and basketball courts and a softball diamond. After training, they had movies and live shows to entertain them. After working hard all day, they enjoyed free-time in the evenings to blow off steam and build camaraderie.[1324] Many of the men would go out to the reefs and collect lobsters and crabs or went fishing, and during the dinner-preparation, they would boil up and/or cook what they caught and have wonderful seafood dishes while drinking some beer.[1325] During the evenings, the regiment played music over the loud PA (public address) system, improving morale.[1326] Daily, the Marines had three solid meals, clean clothes (often freshly washed), and facilities for bathing.[1327] By 31 January, the com-

Colonel Hartnoll J. Withers, commander of the 21st Marines, 3d MarDiv. He approved Captain Donald M. Beck's recommendation for Woody's MOH sending it to his divisional command. Withers would earn two Silver Stars fighting on Guam and Iwo. St. Louis Personnel Records Center (In this photo, Withers was a Captain).

mander Colonel Withers of 21st Regiment confidently wrote, "This regiment is completely ready for amphibious operations against the enemy."[1328]

Many looked up to Withers. He was not an imposing figure, standing 5'7", but his good looks, clear blue eyes and bright blond hair caught people's attention. He had entered the Corps in 1920 and "held every enlisted rank." In 1926, he graduated from the Naval Academy and was appointed a second lieutenant. For years in the Corps, he served as a Legal Aide and Intelligence officer in various billets honing his craft of critical thinking. He fought in Nicaragua and Haiti; and he helped form the first Marine tank outfits and commanded the 3rd Tank Battalion on Bougainville and Guam. On the first day of battle on Guam, he left the steel walls of his tank to direct a bulldozer in its breaking down an embankment, "with complete disregard for his own safety, although continuously under enemy machine gun, mortar and small arms fire," so his tanks and troops could move forward off the beach earning him the Silver Star (signed by Lieutenant General "Howlin Mad" Smith). He was a man with battle experience and knew whether or not his men were ready for combat.[1329]

Sadly, not only did the Marines train for the next fight against the racist, fascist Japanese, but they struggled with their own racism on Guam. Tensions between Caucasians and African-Americans grew from two sources. One, the U.S. practiced segregation and many in its population thought Blacks inferior. Two, the military itself was segregated, with most African-Americans in their own units relegated to non-combat roles due to the misguided belief that they were substandard fighters. These social dynamics created a nasty situation on Christmas night when several armed Black sailors carrying knives and clubs invaded a camp of white Marines. Many were injured in the brawl. A month-long court of inquiry was held, and both Caucasians and African-Americans were convicted and punished.[1330] During this time before segregation was ended, both the military and society saw a slow march to create a more just American society, both outside and *inside* the borders of America and the areas she controlled.

## Lieutenant General Tadamichi Kuribayashi and the Iwo Jima Defense

During the war's early years, the Japanese were uncertain where America would strike; controlling so much of the world's surface, there were many possible points of attack its enemies might try. General Douglas MacArthur had surprised the Japanese repeatedly by attacking in unexpected places, which left large numbers of isolated Japanese stranded, impotent on islands. As Americans fought the war closer to Japan's home islands, the likely points of attack became obvious. Americans looked for harbors for their fleets, airfields for their bombers, and staging areas for the coming invasion of Japan proper. The location of Iwo Jima on air routes to Japan made it a clear target, so Japanese leaders prepared in advance for an effective defense of the island. There would be no surprise attack; only the exact date for invasion was unknown.

In 1944, eight months before the battle of Iwo, Hirohito chose Lieutenant General Tadamichi Kuribayashi to prepare Iwo Jima's defenses 650 miles off Japan's shore and part of Tokyo's Metropolis. He was one of the most skilled generals Japan produced. When called upon by his Emperor to fight against the U.S., Kuribayashi did his best to make Iwo a costly piece of real-estate for Americans to acquire. And Kuribayashi's best was very good indeed turning what U.S. commanders thought would be a short campaign into a several week slug fest. (See Gallery 3, Photo 6)

General Tadamichi Kuribayashi, commander of Iwo Jima, instructing his officers on the island's defense. He would take most of his men underground and in the interior of the island making it very difficult to kill them. In eight months, once he took command of Iwo Jima on 8 June 1944 until the battle began on 19 February 1945, his men reinforced or built eleven miles of tunnels housing thousands of bunkers and pillboxes. Yushukan War Museum

Lieutenant General Kuribayashi on Iwo Jima with his staff review the map of the island as they discuss their defensive strategy. The Asahi Shimbun and Getty Images

Before the battle, Japanese built a fortress on Iwo, the likes of which America had never before seen, and the likes of which few armies had attacked. They did this on a hot and sulfurous island with limited supplies. They stayed filthy since there was insufficient water with which to wash, and rarely had they any downtime. At night, they crawled into their bedrolls in dank tunnels and did not enjoy much in the way of entertainment unlike their American counterparts. Rats and insects infested the tunnels and body lice covered the men.[1331] Yet, they remained committed to their task.

Kuribayashi, commanding officer of Iwo Jima, was an innovative leader. Born into a *Samurai* family on 7 July 1891 in Nagano Prefecture, he attended the IJA Academy finishing second in his class. Unlike most IJA generals, he spent time in North America. He was taller than the average Japanese man of his generation, standing 5'9" and weighing 200 pounds. In 1928, as a captain, IJA posted him in America for an educational tour, a great honor for Kuribayashi and rare privilege for an officer.[1332] For three years, he traveled the U.S. and studied at

Harvard University (like Admiral Isoroku Yamamoto although Yamamoto never finished his class whereas Kuribayashi did). While at Harvard, he took courses in English, American history and U.S. current affairs.[1333] Although he attended this Ivy League University, it seems he did not apply himself because he finished at the bottom of his class in English earning a D+ (it's possible his duties gathering intelligence hurt his studies).[1334] He also attended the U.S. Army War College. He enjoyed Shakespeare and Carl Sandburg's *Lincoln*, watched tackle football and met many Americans. He also acquired a Chevrolet K automobile, learned how to drive and then took a 1,000-mile cross country trip. Because of these experiences, he was convinced: "The United States is the last country in the world Japan should fight. Its industrial potential is huge and fabulous, and the people are energetic and versatile. One must never underestimate the American fighting ability."[1335] He also trained with the U.S. Army at Fort Bliss, Texas, and knew the caliber of the American fighting man. The base commander, Brigadier General George Van Horn Moseley, gave Kuribayashi a signed photograph of himself, writing, "I shall never forget our happy association together in America. Best wishes to you and Japan."[1336] Promoted to major in August 1931, he remained in North America for two and half more years serving as a military *attaché* in Canada at the Japanese Legation until December 1933.[1337]

Besides being a military man and world traveler, he also loved his family and wife dearly (although it appears he enjoyed "Comfort Women" while based in China).[1338] When not at home, he wrote letters to them and even spent time drawing pictures of what he experienced for his son and daughters. He was always gentle with his women in the family and rather harsh with his son, a boy who would grow into an adult to carry on the *Samurai* tradition of his ancestors of the Matsushiro clan. He was a traditional Japanese man raising his men to be warriors and his women to be supportive wives and mothers.[1339]

Many have an imperfect understanding of the Samurai culture (literal meaning of Samurai 侍 is "those who serve"). These men were poets, leaders, politicians, warriors, fathers and husbands. There were different sects, but in general, it was a Spartan mentality and life. Kuribayashi's grandson, Yoshitaka Shindo, explained:

To be *Samurai*, one must be strong and learn how to fight. Yet, to be *Samurai*, one must demonstrate strength for the sake of others, and at the same time, one must be kind to the people who should be protected. Happiness of others must coincide with the happiness of oneself. One must hold the right convictions and the right beliefs so in the view of others, his actions would not bring shame on himself or those he serves or commands.[1340]

With such beliefs, Kuribayashi carried out his duties as an officer rising through the ranks.

Returning to Japan from Canada in 1933, he served in the Main Ordnance, Military Service Department of the War Ministry. Then in August 1936, he took over the 7th Calvary Regiment of 500 men and commanded it until August 1937.[1341] At the time, he obtained the rank of colonel and became Divisional Chief of the "Remount Administrative Section, Military Service Bureau of the War Ministry," which brought him in touch with IJA supply and mobilization capabilities.[1342] His specialty within this ministry was horses. He was tasked with increasing the number, size and strength of horses and became the chief organizer for horse husbandry. Although there were 1.5 million horses in Japan in 1937, he needed to increase this number. He took possession of 7,500 stallions for breeding and coordinated the transfer of thousands of horses to China to support the over one million troops engaged there. For example, when Emperor Hirohito ordered the aggressive takeover of Chinese cities at Beijing and Tientsin in July 1937, he deployed 209,000 troops with 54,000 horses. During this time, most supplies and artillery pieces transported for armies were moved by horses. It was a critical job for militaries during operations (Hitler had sent 625,000 horses in support of his 3.5 million soldiers when they invaded Russia on 21 June 1941). Although no detailed numbers were given, during Kuribayashi's tenure at this post until March 1940, he met with success in every area—he produced stronger and more horses.[1343] Many of these horses helped General Matsui conduct his invasion of China and takeover of Nanking in 1937, an event that held such widespread news attention Kuribayashi could not have been unaware of it.[1344]

Knowing America's capabilities and what the IJA could deploy in case of war, Kuribayashi explained to his superiors before 1941 that U.S. peacetime industry could transform overnight to produce munitions that could overwhelm anything Japan could muster. He told his command about the danger an enemy like America posed, "but they didn't get it."[1345]

Soon after leaving the "Remount Administrative Section," he was assigned as Chief-of-Staff for General Takashi Sakai's 23rd Army stationed in China. During this time, they conducted war games and under secret orders, planned the Hong Kong invasion to coincide with the attack on Pearl Harbor. In preparing the attack on the city, Kuribayashi helped with plans that gave attention to assaulting pillboxes (knowledge he would put to use later at Iwo).[1346] Moreover, he was responsible for disseminating the instructions from an Imperial conference which took place in Tokyo on 5 November 1941 which implored forces under his and Sakai's command to "behave themselves." Since "the eyes of the world would be watching," the Japanese government wanted its troops invading southern China to not repeat "the excesses committed by the Japanese soldiers on the Chinese mainland."[1347] Of course, Kuribayashi knew what his government was asking him to command his troops to do—they were not to practice the *Nanking*-like crimes they had gotten used to doing to the Chinese.[1348] Indeed, Kuribayashi's subordinate, commander of the 38th Division, Major General Sano, did pass on the order to his regiments "to treat any British and other Allied prisoners they might take in Hong Kong with humanity and justice" especially the Indian auxiliaries whom the Japanese felt were forced to fight for their colonial masters.[1349] However, Sakai, Kuribayashi and Sano would fail miserably on passing this directive to troops under their command in such a way that they would obey it as has already been discussed in detail in Chapter 5. As historian Philip Snow observed: "It was one thing to urge moderation on the relatively educated officer caste; quite another to implant it in the line troops, who were for the most part ignorant, xenophobic and brutalized by a training intended to turn them into mindless fighting machines."[1350] Moreover, one could argue it would be difficult to change the behavior of most of these men who had already become accustomed to raping and killing people whenever they went on campaign if for no other reason than that they had no theoretical, moral nor ethical codes that forbade the behavior. Snow continued, "so the...rank and file troops [under Sakai and Kuribayashi]...had learned in the course of [the Second Sino-Japanese war] to perceive the Chinese masses as less than human. And it would seem that in the heat of triumph the habits of a decade were hard to break."[1351] Events would unfortunately prove Snow's observations in spades. And the military success at Hong Kong necessary to inflict those atrocities on the population was largely implemented by Kuribayashi.

One hour after Pearl Harbor was assaulted, the 23rd Army lurched toward The British Crown Colony of Hong Kong on the southern coast of China under the operational code name of *Hara-Saku* (Haller Work) hitting it from the rear and pushing the 12,000 British, Canadian, Indian and Chinese troops toward the sea.[1352] Although Churchill publicly encouraged the garrison to fight to the end, he knew his command could not offer any relief to Hong Kong. The Allied troops' were mortally exposed if a Nationalist army to the north did not attack quickly. It was hoped that they could hang on long enough for the Chinese under Chiang Kai-Shek's general Yu Hanmou, commander of Seventh War Zone (12th Army Group), to hit the IJA 23rd Army from the rear.[1353] In order to launch a proper attack to help the beleaguered defenders of Hong Kong, Hanmou needed until January. It was unclear whether the British could hold out all of one month which the Chinese Nationalist needed in order to help them.[1354]

The 23rd Army was comprised of three divisions, the 18th, 38th and 104th numbering 48,000 men. Kuribayashi issued orders and coordinated several units as Sakai's Chief-of-Staff when the attack was launched on 8 December sending initially 15,000 men across the border. He was in charge of the "Hong Kong capture operation." He played a prominent role with the 38th Division as it penetrated the peninsula.[1355] As the Japanese troops poured into the Kowloon peninsula, Chinese families hide their daughters in basement hovels, in attics or in closets and then locked the other family members in their tiny homes. "Women skulked out of sight of the troops and wore dingy black as a form of protective coloring; to reduce their attractions still further they hunched their backs, daubed their faces with mud and wore sanitary pads irrespective of the time of the month."[1356] Since the city had an estimated one million refugees because of the wars Japan had unleashed throughout China and Manchuria, the citizens knew what awaited them. As a result, General Sakai had his staff (i.e. Kuribayashi) post a "reassurance proclamation" throughout the Kowloon peninsula where Kuribayashi was probably the most senior ranking officer which declared: "We protect Chinese property. The war in Hong Kong is a war against the Whites."[1357] As his troops' actions would soon prove, these were hollow words.

Later, Kuribayashi helped conduct the amphibious assault of Hong Kong Island on 18 December after his forces had pushed the Allied troops off the mainland in full retreat to the isle. The British troops thought they could hold out for months on the Hong Kong Island, but as events revealed, their hopes were unfounded. Preparing for the attack on Hong Kong Island, Kuribayashi issued orders to various military units to coordinate its assault to hit the beaches. As a result, ramped boats gathered on the shores of the mainland and Kuribayashi oversaw the orders to carry at least 10,000 men in an amphibious invasion against the isle traversing a mile of ocean of Vitoria Harbor. In total, they had 3 *Daihatsu* ramped landing craft that were 47 feet long and could carry 70 men, 18 *Shohatsu* landing craft that were 35 feet long and could ferry 35 men and 200 collapsible boats that were 13.6 feet long and could transport 20 men each. They wanted to make sure these assaults were successful unlike a few others from 15-16 December when the Japanese landing parties were caught by searchlights during a night crossing and soundly repulsed. Even with these failures, the landing for 18 December was still thrown together and the planning insufficient. The Japanese admitted later it was a "boar-like blind rush" (*chototsu-mōshin*) "without adequate intelligence about the British or cooperation with Japanese artillery." Nonetheless, the landing proceeded and the Japanese sent their battle hungry troops across the channel in two waves totaling 7,500 men and they made their landfall at the districts of North Point and Shaukeiwan. When the landings occurred, Kuribayashi learned how the British pillboxes' crossfire caused confusion and death, but ultimately the IJA overcame them. Penetrating into the island, the battle later came against the Stanley fortifications by the village. In reducing this defense, Kuribayashi was impressed with how difficult it was to locate weak spots and prevent crossfire. The British utilized crossfire near Stanley Village's narrow entrance by firing a 2-pounder anti-tank gun supported by several machinegun positions and a searchlight beam. They destroyed three Japanese tankettes (Type 94s) and cut down several soldiers. This experience influenced how Kuribayashi designed Iwo and its defenses, having learned from attacking difficult ones himself.[1358] And Kuribayashi's unorthodox and rebel ways were on display at this battle when he defended Colonel Teihichi Doi's aggressive spirit and disobedience to direct orders at the beginning of the battle and visited wounded men in field hospitals (almost an unheard of act by a Japanese senior commander).[1359] Within a week after landing on the island, the Japanese had deployed at least 20,000 men to destroy the British colony.[1360]

By 25 December (known as Black Christmas to the Allied troops), Hong Kong was in Japanese hands thanks in part to Kuribayashi's work.[1361] As the British raised the white flag over Victoria Barracks, the Japanese marched in shouting "*Banzai! Banzai!*" Kuribayashi had helped deliver a devastating "blow to the British in Asia [and helped]…deprive Chiang Kai-shek's [China] of a window to the world."[1362] The British suffered 3,445 casualties and the Japanese 2,118.[1363] Since Hong Kong had become a "symbol of British determination to restrain Japan, as well as a more subtle symbol for British support of Chinese resistance," this victory for Japan had huge psychological as well as military significance.[1364] However, the battle took much longer and produced many more casualties than the Japanese had anticipated. Sakai's planning for the battle, most of which was done by Kuribayashi, "was inadequate at best."[1365] The Japanese benefited by having heavily outnumbered the British forces. Kuribayashi learned from this experience when he had his own command.

Kuribayashi stayed with the 23rd for eighteen months of occupational duties during the oppression of the Chinese in this region. He climbed the ranks to lieutenant general by 1943 and in June, he left China and took charge of the most elite of forces, the Emperor's Tokyo Division (Imperial Guards).[1366] In April 1944, he was appointed to defend Iwo. In May, the army officially selected him for the island's command. On 27 May, he was given a private meeting at 1:45pm with Hirohito, an uncommon "honor for a commoner."[1367] The family remembered when Kuribayashi returned home from this sacred encounter, he was excited. It was described as one of the most amazing events in his life.[1368] This meeting conveyed to Kuribayashi the importance of discouraging the Americans from setting foot on Japan's main islands. The Emperor explained, "Only you among all generals is qualified and capable of holding this post. The entire army and nation will depend on you."[1369] The importance of what Kuribayashi was to do was further evidenced by a meeting with Prime Minister Tojo who admonished him to perform well.[1370]

Soon thereafter, he bid his family goodbye and departed. The day he left, his wife, Yoshi, and he knew he would probably never return home. As a wife of a *Samurai*, she did not show emotion. She would not dishonor what a *Samurai* must do. Yet, their ten-year-old daughter felt the energy. Perhaps she noticed the sword the Emperor had given her father remained on its stand in the home.[1371] Kuribayashi had always taken it on other campaigns, but when he left for Iwo he did not carry it. Shocking many around her, she threw a fit. She did not want her father to leave and clung to his legs. Kuribayashi consoled his daughter. With this unhappy family behind him, Kuribayashi left to take up his command. His wife probably wondered if she would ever see him again.[1372] A few days later, on 8 June 1944, as the American forces were about to invade the Marianas, under orders from Tojo, Kuribayashi officially took charge of the defense of Iwo Jima.[1373]

Kuribayashi knew that defending Iwo was one of the most important responsibilities of the war. Before praying to his ancestors and "begging for their blessing and guidance on his mission," he wrote his brother of his resolve to fight to the end. "I will fight as a son of Kuribayashi, the *Samurai*, and will behave in such a manner as to deserve the name of Kuribayashi. May my ancestors guide me."[1374] These were not hollow words for Kuribayashi. In the Shinto religion, everyone was taught from birth that the *kami*, or spirits of the household clan, the imperial ancestors and legendary heroes intimately influenced one's life. They had to be placated, respected and revered. If not, evil would befall the warrior. In addition to their ancestors, Japanese also worshipped numerous Shinto gods. For Kuribayashi and most of his men, the Iwo Jima defense was a religious experience full of ritual, evidencing reverence for their ancestors, Sun Goddess and god-man ruler, the Emperor. Kuribayashi wrote his wife before battle that his soul would be placed in the Yasukuni shrine and would live on with other warriors since he would soon die.[1375] Kuribayashi's grandson, Yoshitaka Shindo, explored these beliefs:

> What makes Japan dear to the heart? It is the knowledge that the energy, the work, the fruits of the labor of past generations live in the soil, the rocks and the buildings of the land. The cause was indeed to die for the Emperor, but the Emperor was symbolic of those things dear to the Japanese since He and His royal line have been around for 2,600 years. The gods we worship are symbolic of larger feelings about nature...our world and...culture. To be Japanese is to focus on the...nation, past and present, as a nation full of loved ones and those loved ones include *kami*, gods,

spirits—one might say sacred energy or inner voice. Worshiping our ancestors' souls and dying for them and the Emperor make a Japanese Japanese. Not to focus on these spirits and the Emperor would be shame on oneself and family.[1376]

This religious fervor stiffened Kuribayashi's resolve. Responsibility to those warriors past and present put pressure on Kuribayashi to fight to protect his culture, ancestors, family, gods and Emperor. One of his guiding forces to defend Iwo was his conviction that he was protecting his divine country, a country that admonished all to "Honor the Gods and serve loyally their descendants [the Emperors]," creating a martial religion for "a race of warriors."[1377]

Most Japanese on Iwo had similar beliefs to Kuribayashi. U.S. Marines faced an island of thousands who believed they owed a sacred duty to their divine monarch and legions of Japanese dead.[1378] They derided Americans for going into battle without "spiritual incentive" strictly relying "on material superiority."[1379] The majority believed there was an inherent "spiritual power" in Japanese civilization (*seishin-shugi*) that would grant them victory.[1380] Fleet Admiral and Supreme War Counsel member Eisuke Yamamoto believed the gods would ensure victory and claimed in 1935 that in war, Japan would be victorious because of the Shinto spiritual force.[1381] But for all the talk about religious superiority, America's material power worried Kuribayashi.

To meet the U.S. material might, Kuribayashi's innovative way of defending the island departed from orthodox strategies favored by most of his colleagues and caused consternation among subordinates. He generated opposition, and so he confronted it by sacking 18 of his subordinates, "including his own Chief-of-Staff."[1382] He did not tolerate insubordination nor questioning the new strategy although he had received orders to meet the Marines at the shore, an order he himself refused to obey.[1383] He felt those rules on how to defend islands outdated and ineffective. "One may appeal to genius, which is above all rules: which amount to admitting that rules are not only made for idiots, but are idiotic in themselves."[1384] And in order to break with the old rules of defending at the beach, he needed to train his men in new and more effective tactics. At first, he felt many of his soldiers "untrained recruits" led by officers who were "superannuated scarecrows."[1385] He would need to harden them into warriors.

He had conflicts with navy commanders, most notably his highest ranking subordinate, Rear Admiral Tochinosuke Ichimaru. He secured a "truce" with Ichimaru, giving him some material to build a beach defense since Ichimaru believed in the "prevent the landing of the enemy strategy" that proved unsuccessful on every island battle to date. Kuribayashi negotiated "terms whereby half of the munitions and material supplied by the navy would go into building pillboxes on the beach, while the rest would be for the use of the army."[1386] Yet, Kuribayashi had reached an agreement at Imperial Headquarters in August 1944 that once they shifted to a ground operation, units and artillery pieces would revert to his command.[1387] He might have to wait a few hours once battle began to have full control since the navy was responsible for defending the island within the ocean battlespace, but since Kuribayashi wanted to wait to start the active defense until after allowing Marines onto his island, he felt the diversion from his command strategy would not last long. Even with this diversion, Kuribayashi believed he would inflict a higher kill ratio on the Americans than any other commander. This innovative approach created a defensive strategy rare among his fellow generals who relied upon *Banzai*s. Kuribayashi harnessed his soldiers' fanatical fighting spirit to a sound strategy.

He rose at 0330 every day and oversaw the construction of his fortress-island and micromanaged everything with the aim of instructing his men on how to inflict the maximum amount of death on the soon-to-arrive Marines.[1388] He crawled through tunnels, gave advice to improve defenses and directly trained soldiers, "even helping individual soldiers with their shooting."[1389] He constantly walked the island, learned its terrain and offered instruction at unannounced inspections taking interest in all details relating to his soldiers.[1390] The Tokyo radio described him knowing Iwo so thoroughly "that even should he be asked where a certain hole made by the rats is to be found, he would answer quickly without any hesitation."[1391] Not surprisingly, others felt he was too much of a "slave driver" and "despised [his] harsh discipline."[1392]

One may wonder how Kuribayashi and his men felt being on a death mission. Simply, they felt by protecting Iwo, they defended loved ones. Kuribayashi's grandson, Yoshitaka Shindo, explained:

> Human beings draw great strength when they know they're protecting something sacred. It unites them…and gives them a cause

that's worthy to die for....[The] men went to their jobs on Iwo... with joy and determination. It was to keep mainland Japan intact and keep our collective spirit surviving. I don't think they felt forlorn.[1393]

Kuribayashi used this sentiment well and his men labored around the clock to build his fortress.

Before ever reaching the island, Kuribayashi thought about how to defend islands and reasoned by reading reports from lost battles that the existing strategy was not working. As he took command, he received a report on 29 June from Saipan and its lack of success defending the beaches supporting his new strategy.[1394] Studying previous battles, Kuribayashi concluded that beachhead defense and massed *Banzais* were ineffective. As a result, he devised a plan to fight from underground fortifications, tunnels and pillboxes. He would build defensive positions, many of which could survive the battleships' big guns. To execute this defense, he needed a unique type of warrior. First, he forbade *Banzais*. Soldiers would not waste lives this way. Second, to prepare the tunnels and fortifications that would be his defense's backbone, he needed engineers.

In March, the First Company of the 9th Engineers stationed in Manchuria was making preparations to deploy to Truk Island. Using his contacts and skills as an administrator, he was able to divert 300 engineers from going to Truk. They created the foundation of his new strategy. He obtained an additional 700 engineers from various commands and mobilized the 1,233 engineers from the navy already on the island.[1395] As a result, 10% of Kuribayashi's force were engineers who became his new strategy's guiding force—building an underground fortress.

When Kuribayashi took command of Iwo in June of 1944, he felt he could hold the island long enough to provide the navy an opportunity to blindside the U.S. fleet and inflict a terrible blow on the Americans. "When our enemy comes here, we can contain him," he told Major Yoshitaka Horie. "And then our Combined Fleet," he continued, "will come and slap his face. That is to say, our role here is a massive containing action." Horie informed him that much of the fleet was destroyed and they would not receive naval support. Kuribayashi was shocked; Admiral Nimitz's defeat of the navy in the Marianas was kept from commanders in the area. As a result of this IJN loss, Kuribayashi realized there was no hope of resupply, reinforcement or returning to Japan.[1396] He sent his personal effects back home, now fully knowing he would never return.[1397] To maintain morale, he did not tell his staff about this naval defeat.[1398]

This realization of inevitable defeat should not have come as a surprise to him although it appears it did. Before leaving, he met with Tojo who encouraged Kuribayashi to do "something similar to what was done on Attu" in Alaska in May 1943.[1399] At Attu, Imperial forces suffered 6,950 deaths, many in a massive Banzai at the end of battle—only 28 Japanese POWs were taken. It was a major amphibious attack of U.S. Army troops trained by then Major General Smith and his Chief-of-Staff Colonel Erskine.[1400] Using Attu as an example, Tojo indicated to Kuribayashi to go out in a blaze of glory. Even with such a sendoff, Kuribayashi must have harbored hope of returning even if it seemed remote.

With the newly learned knowledge that no IJN ships would coordinate with him in the defense of the island, Kuribayashi had even more of a sense of urgency to prepare his men for battle. During the first two months of his command, Saipan, Tinian and Guam's Japanese forces were defeated. He felt at the rate they were marching across the Pacific, the Americans would attack Iwo in October 1944. Consequently, he demanded that his soldiers work under his engineers around the clock digging trenches and tunnels, and constructing pillboxes. He created a new motto: "Every soldier is an engineer."[1401] All his men's activity was devoted into making the island a fortress. After being on Iwo for a few weeks, Kuribayashi knew that without naval support he could not hold the island. He did not, however, let the lack of naval support deter him from his mission to control the island as long as possible. He wanted the Marines to recognize as they neared Japan's home islands, they would pay a high price for every square mile of Japanese territory. By studying the Mariana campaigns, Tarawa, Roi-Namur, Normandy and other U.S. amphibious invasions, he knew how to inflict heavy casualties on his approaching enemies.[1402]

Soon after arrival, he received a visit from American bombers. On 16 June, he found himself taking cover from a U.S. raid:

> [A] massive bomb landed next to the dugout, setting off a huge explosion. I was convinced that [my] dugout…would be blown to bits, but as luck would have it, I didn't even get a scratch. For

the duration of the ferocious raid, the only thing I could do was wait in the dugout in a state of extreme anxiety, and pray.[1403]

In three raids that month, the Americans destroyed over 100 airplanes and during the attack just described, 40 soldiers died. These events "sapped the morale of the Japanese."[1404] They also made Kuribayashi more frantic to get his fortress built underground to avoid the airpower that would daily hit the island as the Japanese airforces dwindled away into nothing, leaving Kuribayashi's personnel vulnerable to air attack. He was preparing for a battle that would end in a defeat, but he hoped his nation could negotiate peace as a result. He might be able to extract peace from the Americans if they felt the cost for continued warfare too high. He also thought if he could fight tenaciously enough he could delay bombing attacks against Japanese cities, especially Tokyo. He, like many, believed that by defending Iwo, he protected his men's families from bombing raids by constructing a "Japanese Wall" there at Iwo.[1405] He wrote his wife, "if the island I'm on gets captured, there'll be an increase of several hundred enemy planes, and the air raids on the homeland will be many times more savage than now. In the worst case, the enemy may land on the beaches of Chiba and Kanagawa prefectures and penetrate near to Tokyo."[1406] Kuribayashi knew the importance of his command and the force against whom he was defending.

After October, Kuribayashi realized the Americans were not moving in his direction as quickly as anticipated. Instead of coming to him, they diverted west, hitting Peleliu. As a consequence, on 15 November, he issued a new order directing that soldiers spend half the day on building trenches, bunkers and tunnels and spend the other half training. He emphasized two things: First, morale must stay high and commanders must ensure good attitudes. Second, coordination between companies must be practiced. Each unit should know how to work next to the units on their flanks, and Kuribayashi made sure company-level exercises improved these tactics.[1407] By 1 December his fortress was taking on massive proportions so he realized that for his men to know how to use it, he shifted the activity schedule again. Men now dug tunnels, built bunkers and pillboxes 30% of the day and trained for 70%.[1408] In inspecting his fortress, however, he realized he needed to go deeper. On 23 December, he switched the men's focus from building shallow tunnel and trench systems to now deep underground rooms and passageways 10 meters into the earth. U.S. Navy guns would not penetrate fortifications this insulated.[1409] When not digging, they emphasized cover and concealment, marksmanship and anti-tank activities.[1410] Kuribayashi continued to refine the routines to perfect his men's behavior: "Constant practice leads to *brisk, precise,* and *reliable* leadership, reducing natural friction and easing the working of the machine."[1411]

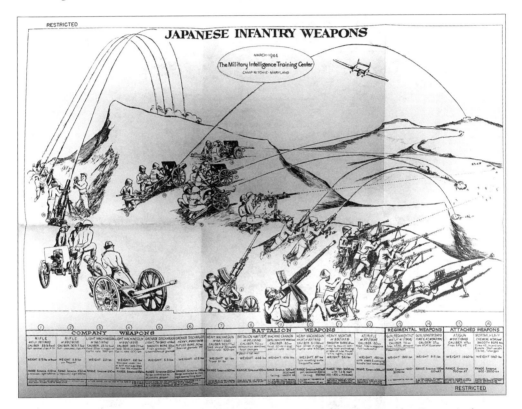

Japanese infantry weapons used on Iwo Jima. San Diego Marine Corps Recruit Depot Archive

He took inspiration from Colonel Kunio Nakagawa and his Peleliu defense by which tunnels, pillboxes and the natural terrain were utilized to inflict casualties. Nakagawa had not fully developed the strategy like Kuribayashi because by the time he realized he should use it, many of the defensive positions had already been built on the beach and the Americans had arrived; nonetheless, Nakagawa's use of interior defenses were superior to beach defenses. For their defense, Hirohito gave the Peleliu garrison *gokashô*—"words of praise from the emperor for their gallant and tenacious fighting—no fewer than ten times."[1412]

Having a defense in depth and letting the Marines land without revealing their positions was key to killing more of them, Kuribayashi reasoned. His defense was like a Ju-Jitsu move of using the momentum of the opponent's attack to bring an enemy into one's body and then flip him onto the ground. When the enemy is close, he cannot hit the Ju-Jitsu practitioner and if he is off balance, he can be twisted to the floor.

An example of one tunnel of the 11 miles of tunnels inside Iwo Jima that General Kuribayashi's engineers built. 9 March 1945. National Archive, College Park

**Photo 1:** Marine Corps Magazine high-lighting a flamethrower taking out the monster-octopus Tojo. The importance of the flamethrower in destroying Japan and the demonization of the enemy are obvious for all to see with this artwork. Author's Collection

**Photo 3:** Anti-Japanese War Poster. The demonization of the enemy is clear here making the enemy out to be a monster, but the image of killing innocent civilians being the *modus operandi* of the IJA was spot on. National Archives, College Park.

**Photo 4:** Anti-Japanese War Poster. It was widely known that the Japanese army raped everywhere it went as depicted in this propaganda. National Archives, College Park.

**Photo 5:** Captain Donald Macaulay Beck as a 2nd Lieutenant in 1941. He was Woody's company commander on Iwo commanding Charlie Company, 1st Battalion, 21st Marines, 3d MarDiv. He earned a Silver Star and Purple Heart on Guam and a Bronze Star with V on Iwo. Author's Collection and Beck Family Archive

Print the complete address in plain block letters in the panel below, and your return address in the space provided. Use typewriter, dark ink, or pencil. Write plainly. Very small writing is not suitable.

No.

To Miss Ruby Dale Meredith
Route #3
Fairmont
West Virginia, U.S.A.

From
Pvt Hershel W. Williams
(Sender's name)
32nd Replacement BN
(Sender's address)
% Fleet Post Office
San Francisco Calif
(Date) 12/24/43

(CENSOR'S STAMP)

My Dearest Ruby,

mistakes in this letter you can take them as kisses, & expect you will get plenty of them. I sure wish you could have them personally. I am shaving now so I would scratch your pretty little face.

is such a swell thing, isn't it? & I have so much love for you, I will always love you Ruby.

so proud to own your love & to be able to give you my love. Ruby you can tell Mom about this letter, I will write her today, but I probably won't be able to write her as much as I have you.

to keep writing, & real often, its so good to hear from you. Your letters mean every thing to me.

I don't get enough of them.

much now to write Ruby, I have about run out of news.

write often as possible, just remember I will always love you, & be thinking of you all the time.

You write often too, and tell every one hello & all my love always
Hershel

Love you Ruby, honest.
P.S. I love you Ruby

Tell Harshel to stop USING CODE + HIS Letters won't Be cut up.
Censor

V...-MAIL

POST OFFICE DEPARTMENT PERMIT NO. 1

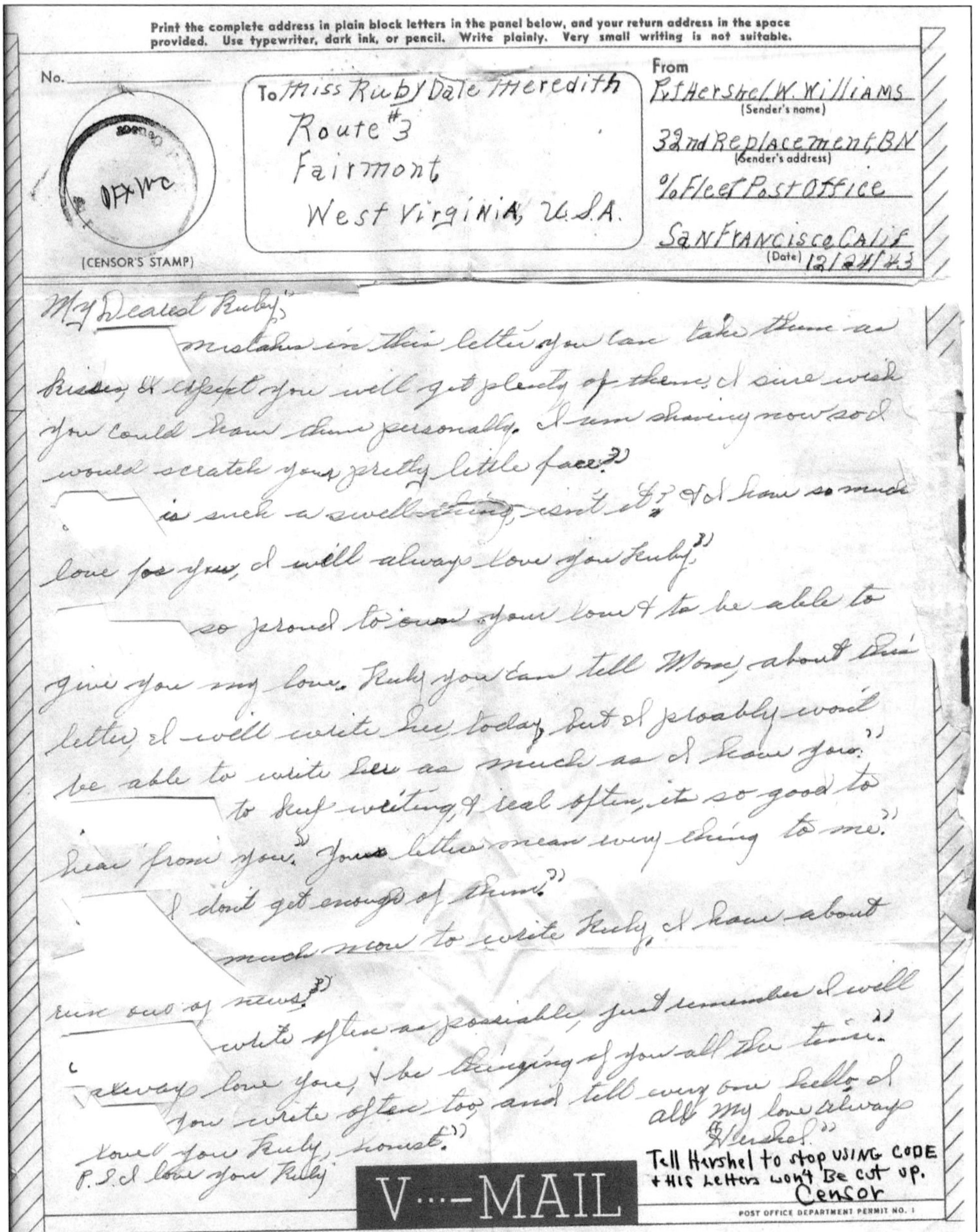

**Photo 2:** For a short time, Woody wrote his girlfriend with a "secret understanding" that he would start each paragraph with the first letter of the name of the island or place he was so she could track him on a map, explaining all this in a letter on 17 November 1943 while stateside, and thus, not yet subjected to censors. Eventually, the censors caught on, because in his letter from 24 December 1943, they scissored out the beginning words of his paragraphs. One told Ruby in a note at the bottom of Woody's letter to her: "Tell Hershel to stop using code and his letters won't be cut up. Censor." Marine Corps Historical Division

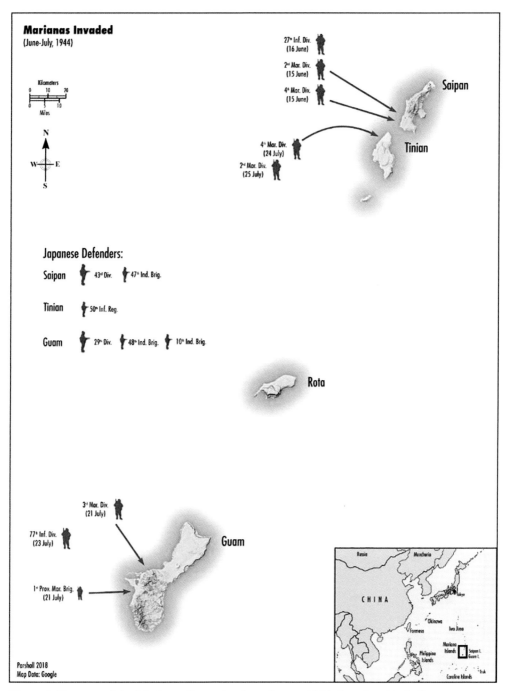

**Photo 6:** The map of the Marianas Campaign June-August 1944. Jonathan Parshall

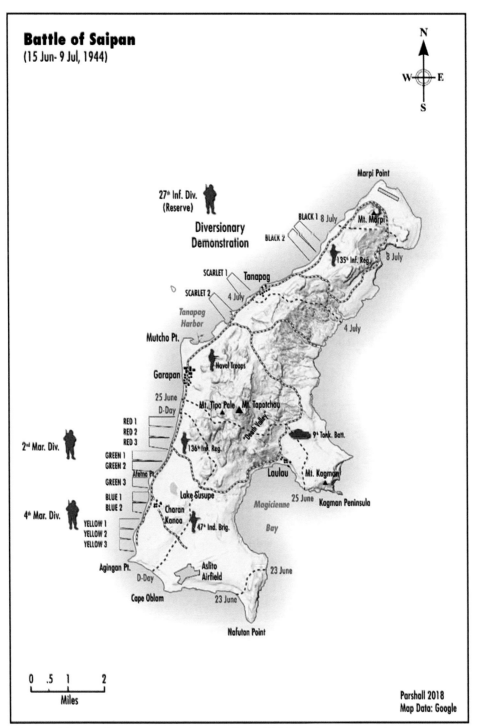

**Battle of Saipan**
(15 Jun- 9 Jul, 1944)

**Photo 7:** The attack on Saipan 15 June-9 July 1944. Marpi Point is noted on this map to the far north of the island where thousands of Japanese citizens killed themselves instead of allowing them and their families to be taken prisoner by the United States. Jonathan Parshall

**Photo 8:** Marpi Point on Saipan where hundreds if not thousands of Japanese citizens and soldiers jumped to their deaths at the end of the battle during July 1944. 20 March 2018. Author's Collection.

**Photo 9:** Marpi Point on Saipan where thousands of Japanese citizens jumped to their deaths instead of surrendering to the Marines. Some mothers threw their infants off these cliffs first before they jumped themselves. These cliffs shot up from the ocean to between 70 and a 100 feet. 19 March 2018. Author's Collection

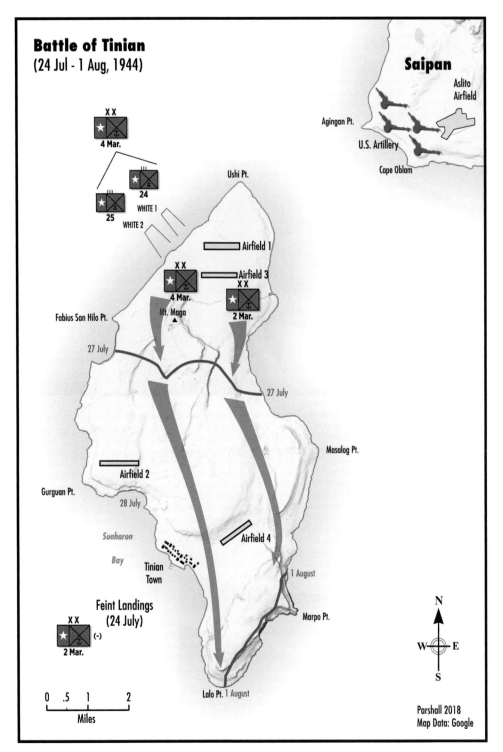

**Photo 10:** Battle of Tinian map with Marpo Point marked where thousands of Japanese citizens killed themselves. Jonathan Parshall.

**Photo 11:** Gunnery Sergeant Keith A. Renstrom. He received the Bronze Star with V for actions on Tinian. He became popular with our tour group often declaring "Death before Dishonor." He was on the Iwo reunion of 2018. 20 March 2018. Author's Collection

**Photo 12:** Here is the author on Green Beach on the Asan Beach on Guam where Woody landed on 21 July 1944. March, 2018.

**Photo 13:** Marpo Point on Tinian where thousands of Japanese citizens jumped to their deaths instead of surrendering themselves to the Marines. Like at Marpi on Saipan, these cliffs were 70-100 feet tall. 20 March 2018. Author's Collection

**Photo 14:** USMC Colonel and historian Charles A. Jones, Woody Williams, and Lefty Lee at Apra Harbor, Guam, March 2015 during the 70th Iwo Jima reunion. Although Lefty served bravely with Woody on Guam, he made up his entire combat experience on Iwo claiming to be one of Woody's MOH witnesses and part of his support team. In reality, he was removed soon after he hit the beach for "cracking up" and suffering shell-shock. Woody had incorrectly used him for years to "prove" what he did on Iwo Jima for his MOH. Author's Collection.

**Photo 15** Author Bryan Rigg, Woody Williams and Justin Rigg at Apra Harbor Guam, March 2015. Woody told me and my son that he could feel his heart bang against his chest as his landing craft made its way to Guam on 21 July 1944 heading toward Asan beach to attack the Japanese. Author's Collection.

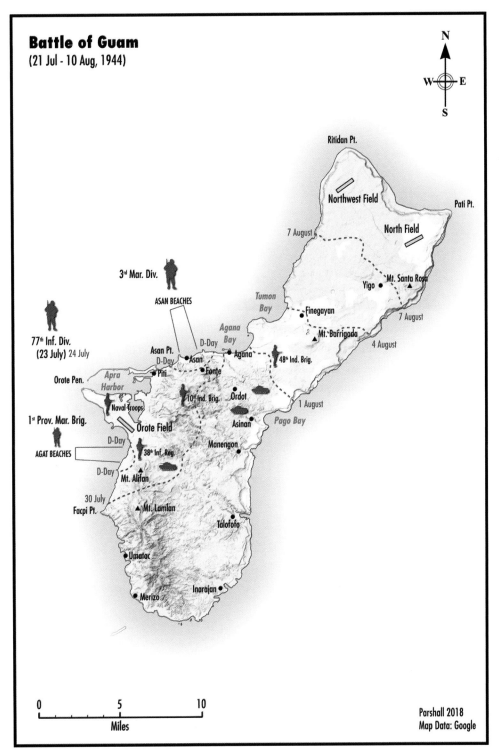

**Photo 16:** The attack on Guam, 21 July-10 August 1944. Jonathan Parshall.

**Photo 17:** Author Bryan Rigg and son Justin with Darol E. "Lefty" Lee. Lefty was wounded on Guam and received the Purple Heart. He made up his entire combat experience with Woody on Iwo although he had just finished telling my son and I an elaborate story about how Woody was killing Japanese and how he, Lefty, led an entire platoon at the attack of Iwo's second airfield. Since he benefitted financially from his lies, he is a classic "Stolen Valor" case. In fact, the World War II museum was going to do a full documentary film with him and Woody, but once his lies were discovered, the trip and film were cancelled. 18 March 2015. Pirates Cove Restaurant, Guam. Author's collection

**Photo 18:** When Woody and Lefty hit Asan beach on 21 July 1944, they were faced with scaling such heights ahead of them to displace the Japanese on Guam. 18 March 2015. Author's Collection

**Photo 19:** A view of Asan beach from Nimitz Hill ("*Banzai* Ridge"). This is the view the Japanese had as night fell on 25 July 1944. March 2015. Author's Collection

**Photo 20:** Author Bryan Rigg with sons Ian and Justin in a Japanese cave on Guam near Nimitz Hill (behind *Banzai* Ridge). This was probably one of the staging areas for the Japanese before they hit Woody's lines on the night of 25 July 1944. 15 March 2018. Author's Collection.

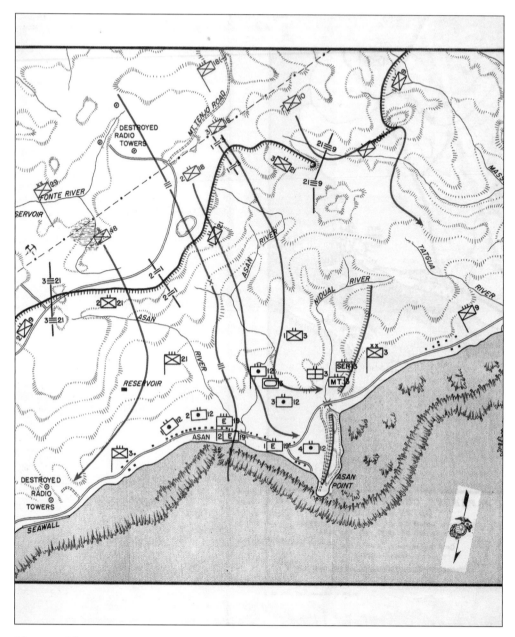

**Photo 21:** The penetration points of the "*Banzai*" attack on Guam 25-26 July 1944. It started at the top of Fonte Ridge and made its way all the way to the ocean, one mile below the heights. USMC

**Photo 22:** Looking up at *Banzai* Ridge from the lines Woody occupied on 25 July 1944. This was where Woody's battalion got hit by 5,000 Japanese during the night of 25-26 July 1944. 19 March 2018. Author's Collection

**Photo 23:** Part of Japanese jaw and shoulder bone and IJA ammunition in a cave near Nimitz Hill. 15 March 2018. Author's Collection.

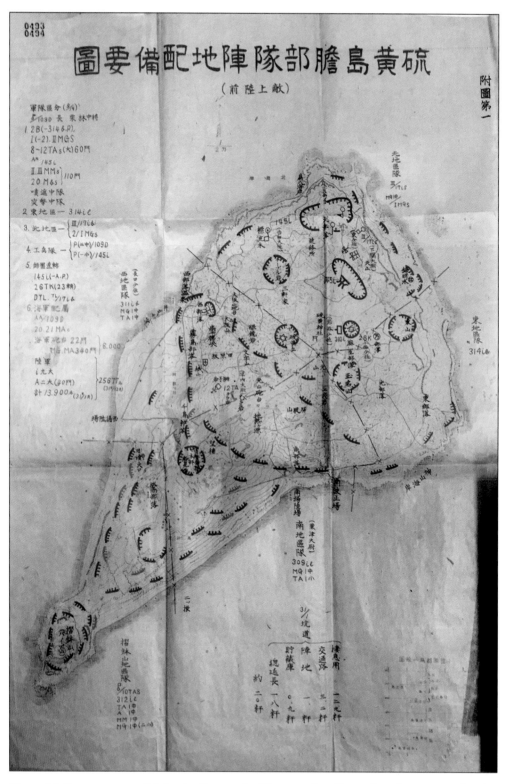

**Photo 24:** Japanese map of defensive lines and pillboxes at Iwo Jima. Circa beginning of 1945.
NIDSMDMA

Distance was a factor in this new defense: Kuribayashi knew he had to get Marines close to him before unleashing his firing positions. He adopted "an all-around defense in depth, utilizing rugged ground in the interior of the island." His tactics were a "*distinct* improvement over the tactics on other small islands."[1413] Kuribayashi "had the correct comprehension of reality [and] refused to be awed by precedent."[1414] (See Gallery 2, Photo 24)

The American delay in attacking gave Kuribayashi eight months to prepare the defenses and he profited from it. This eight square mile island eventually hosted a network of 11 miles of tunnels and thousands of pillboxes and bunkers.[1415] The 3d MarDiv reported, "In the zone of action of one marine division, 800 pillboxes of various types were found in an area approximately 1000 yards square. Most...pillboxes and emplacements were so well camouflaged that pre-invasion photo interpretations revealed only a fraction of the number actually existing."[1416] The island held at least 15,000 cave entrances and pillboxes.[1417]

The Japanese sources give some interesting facts. They verify there was around 11 miles of tunnels (11.25 miles or 18.1 kilometers). A little over eight miles (8.02) were made for living and hiding quarters (12.9 kilometers),1.99 miles for communication networks (3.2 kilometers), .62 miles for firing positions (1 kilometer) and .62 miles for storage areas (1 kilometer). The Japanese report given by Kuribayashi's command noted there were 5,000 areas including pillboxes and caves which his troops could use *to fire* at the Marines, using concealment and defense. Kuribayashi had built this complex within eight months of arriving and he had planned an additional 6.21 miles (10 kilometers) to actually complete the Iwo Jima underground castle had the Marines waited much longer to pay him a visit.[1418]

This network housed 22,000 troops who learned to live troglodytic lives for months. The tunnels boasted large supply depots with trucks, tanks, hospitals and CPs. They had miles of telephone cables, and because they ran underground, it was difficult to sever communications between the Japanese units during battle. Some of the areas traversed almost one hundred feet underground, far below where shells from battleships' big guns could penetrate. "One Brigade headquarters, located near Motoyama, could hold 2,000 troops; it was 75 feet deep and had a dozen entrances." In addition to everything described, millions of cockroaches and ants also resided in the tunnels, crawling over the men as they slept.

Because there was no regular bathing, parasites covered them.[1419] Kuribayashi shared in their deprivation writing: "These caves are airless and humid, and they really are—really truly-awful."[1420] But since the island received regular attacks, it was safer to live underground than above it.

Miles of tunnels were made by breaking earth with pickaxes and removing dirt, sand, and rubble in woven bamboo baskets. In some areas, the geothermal heat was so intense the soles of the troops' *juktabi* "(spilt-toed rubber-soled shoes) melted, and the sulfur gas gave them headaches and made breathing difficult." Sometimes they wore their cumbersome gas masks for protection from sulfur fumes. "Dressed in only loincloths, they would work in five- or ten-minute shifts because of the intense heat" (sometimes measuring 176° Fahrenheit). Kuribayashi turned "his men into supermoles, excavating the hard komhake rock" working in, one could argue, the bowels of Hell. Muscles in men's backs ached and rippled under the workload. They were on half rations with food and water because the island could not support 22,000 troops. Kuribayashi made do with what he had although it meant inflicting privation on his troops, many suffering from malnutrition, dysentery and poisonous sulfuric gas. He wanted no distractions, so he "officially" refused sex slaves. Since there is no record of him opposing his troops using sex slaves in China (he actually "ordered" them for his troops by the thousands), he probably did not allow the "Comfort Women" on Iwo simply because there was not enough food and water for them. He would build a formidable fortress pushing his resources to the limit.[1421] Lieutenant General Smith said as of 1945, Iwo was the most "heavily fortified island in the world."[1422]

To vanquish the highest number of "heathen Americans," Kuribayashi felt he had to put his men through "living Hell."[1423] The ability "to endure privation is one of the soldier's finest qualities: without it an army cannot be filled with genuine military spirit."[1424] Ironically, he had to inflict some of the harshest training in recorded history on his men, and this, unfortunately for the Marines, made them used to deprivation and conditioned them to know how to fight on the island using every advantage the sulfur pit offered. They learned how to mete out overwhelming violence against the Marines who were unprepared for Kuribayashi and his warriors. The months of grueling labor Kuribayashi put his men through hardened them for defensive combat. Correspondent Dick Dashiell observed: "The quality of Jap soldier which fought so bitterly for Iwo Jima was *Nippon's* best, as Marines who were there will verify quickly and grimly."[1425]

However, Dashiell's statement is false. Historian Dan King notes, Iwo was Japan's "Alamo." Only one actual IJA regiment was on the island! The remainder of the units were "slapped together from old reservists and inductees." That Kuribayashi's men fought so well was remarkable because Kuribayashi's troops were a diverse patchwork of all segments of the military and society. Many were not trained warriors, but came from outfits of communication, searchlight, transport, and so on. Some were new recruits without any training while others were hardened China veterans. On average, Kuribayashi got the "dregs of old men and youngsters" since Hirohito was "hoarding" his best units for the invasion of the mainland. Sixty percent of Kuribayashi's force came from the traditions and culture of the army while the others stemmed from the navy numbering 7,500 in the 27th Air Flotilla under Ichimaru's command. As usual, bitter fights erupted between the two services, a common occurrence in any military ("interservice rivalry"). Even so, both soldiers and sailors became physically tough digging tunnels daily, but there was little ammo for rifle practice in order to help them to learn fully their defensive tactics.[1426] Admiral Spruance, Tuner's and Smith's direct boss and commander of the Fifth Fleet, noted that "in view of the character of the defenses and the stubborn resistance encountered, it is fortunate that less seasoned or less resolute [Japanese] troops were not committed."[1427] Knowing this about his men, the "use of cunning" which Kuribayashi strategy's foundation was built on, becomes obvious. As the military philosopher Carl von Clausewitz writes:

> ...the weaker the forces that are at the disposal of the supreme commander, the more appealing the use of cunning becomes. In the state of weakness and insignificance, when prudence, judgement, and ability no longer suffice, cunning may well appear the only hope. The bleaker the situation, with everything concentrating on a single desperate attempt, the more readily cunning is joined to daring. Released from all future considerations, and liberated from thoughts of later retribution, boldness and cunning will be free to augment each other to the point of concentrating a faint glimmer of hope into a single beam of light which may yet kindle a flame.[1428]

In light of this reality, Kuribayashi molded this "randomly cobbled-together" group into a disciplined and focused army.[1429] And they all worked daily to perfect the "cunning" required to give them the advantage against their enemy—the "faint glimmer of hope." Knowing they were going to die, many prayed for the battle to give them relief from the conditions under which they worked. One Japanese survivor said, "There is a saying that 'the way of the warrior is to die' and we so wanted it to be a battle based on that ethos."[1430] Even so, the fortifications and tactics they were learning gave concrete evidence that Kuribayashi was doing everything he could to give his men the advantage in combat. And moreover, these tactics relied heavily on his troops learning how to inflict the maximum amount of harm on the Americans by using methods that would depend on, one might add, Machiavellian sneakiness. The troops' fighting song was:

> In the Lonely mid-Pacific,
> Our sweat a fortress will prepare.
> If the enemy attack us
> Let him come, we will not care.
> Until the hated Anglo-Saxons
> Lie before us in the dust
> Officers and men together
> Work and struggle, strive and trust.[1431]

Kuribayashi shared in their misery. He ate their food and lived in their caves. Often, he would not take extra water or food although he could, because he wanted to show his troops he knew their deprivations.[1432] He shared his rooms with his officers and lived underground. He wrote his wife Yoshi: "To put it in a nutshell, we're living in caves in a barren wasteland…so, depending on how you look at it, our life is hell."[1433] At night, he declined to remove his uniform due to the constant threat of bombings. Men set their watches to his daily routine which started early in the morning before sunrise. An orderly brought him a cup of water to wash the dust from his eyelids.[1434] What was left over would be re-used throughout the day—Kuribayashi tolerated no waste. When he saw an officer "rubbing himself with a towel soaked in water," he said he "deserved to be executed." Then he calmed down and gave him a lecture saying, "on this island, a drop of water is as precious as a drop of blood."[1435]

As the battle neared, Kuribayashi was increasingly cut off from Japan. With each month, the sorties from airfields dwindled as planes failed to return from missions. Ships bringing men and supplies declined as U.S. submarines sank them. "No fewer than 1,500 Japanese troops drowned in the Nanpo Shoto attempting to reach the Bonin Islands" and at least half of Kuribayashi's tanks were sent to the ocean's bottom before reaching his command.[1436] And besides struggling to increase and train his manpower, he continued to grapple with the unstable island he was charged to captain, as the volcanism was active and unstable.

The aptly named Iwo Jima (literally Sulfur Island) was a volcanic island that shot up steam from holes riddling its terrain, and had numerous boiling sulfur springs sprinkled throughout the landscape, making men's lives miserable. It was a waterless, barren land except for scrubby vegetation. Kuribayashi's adjutant Major Yoshitaka Horie said it was as vulnerable as "a pile of eggs" and it would be better for everyone if they sank it to the bottom of the sea.[1437] It looked like a pork chop with the meaty end to the north and the handle pointing down to the south, ending in the extinct volcano center cone of Mt. Suribachi (Japanese for cone-shaped bowl), which rose 556 feet above sea level. It was "an ugly, smelly glob of cold lava squatting in a surly ocean."[1438] Although Horie and Kuribayashi had their misgivings about defending Iwo, they would make the Americans pay for every foot of the god-forsaken island. Iwo would be a maze of death, one of only two places during the Pacific War (the other being Angaur) where the Japanese inflicted more overall casualties on the Americans than Americans inflicted on them.[1439]

Kuribayashi set a goal for his men to kill ten Marines before they died: "Every man will resist until the end, making his position his tomb."[1440] To prepare his soldiers for this unconventional style of fighting, Kuribayashi composed six "Courageous Battle Vows" which were distributed in January 1945. His "stern code of *Bushido*" read as follows:

1. We shall defend this island with all our strength to the end.

2. We shall fling ourselves against the enemy tanks clutching explosives to destroy them.

3. We shall slaughter the enemy, dashing in among them to kill them.

4. Every one of our shots shall be on target and kill the enemy.

5. We shall not die until we have killed ten of the enemy.

6. We shall continue to harass the enemy with guerrilla tactics even if only one of us remains alive.[1441]

These vows were "basically a collection of slogans outlining a soldier's proper state of mind in which to face battle."[1442] Often during "morning assemblies and other occasions," soldiers recited these vows together like religious *mantras*. One man who collected the bones of fallen Japanese after the war claimed Kuribayashi's men "regarded the vows as an article of faith, and, even after their bodies rotted, were still reluctant to let go of them."[1443] Marines found copies of these vows all over the island, in bunkers and pillboxes, in caves and tunnels, on the beaches and on the dead.[1444] A few weeks into the battle, correspondents with Dick Dashiell found these vows and while discussing them, one who had just arrived at the front said it "[l]ooks as if the Japs were living up to their vows." His colleague, a hardened veteran, snorted, "Stick around…It's only the beginning."[1445]

Kuribayashi also composed a set of instructions for soldiers of the "Courage Division":

## Preparations for Battle.

1. Use every moment you have, whether during air raids or during battle, to build strong positions that enable you to smash the enemy at a ratio of ten to one.

2. Build fortifications that enable you to shoot and attack in any direction without pausing even if your comrades should fall.

3. Be resolute and make rapid preparations to store food and water in your position so that your supplies will last even through intense barrages.

## Fighting defensively.

1. Destroy the American devils with heavy fire. Improve your aim and try to hit your target the first time.

2. As we practiced, refrain from reckless charges, but take advantage of the moment when you've smashed the enemy. Watch out for bullets from others of the enemy.

3. When one man dies a hole opens up in your defense. Exploit man-made structures and natural features for your own protection. Take care with camouflage and cover.

4. Destroy enemy tanks with explosives, and several enemy soldiers along with the tank. This is your best chance for meritorious deeds.

5. Do not be alarmed should tanks come toward you with a thunderous rumble. Shoot at them with anti-tank fire and use tanks.

6. Do not be afraid if the enemy penetrates inside your position. Resist stubbornly and shoot them dead.

7. Control is difficult to exercise if you are sparsely dispersed over a wide area. Always tell the officers in charge when you move forward.

8. Even if your commanding officer falls, continue defending your position, by yourself if necessary. Your most important duty is to perform brave deeds.

9. You should be able to defeat the enemy even if you do not have enough food or water. Be brave, O warriors, even if rest and sleep are impossible.

10. The strength of each of you is the cause of our victory. Soldiers of the Courage Division, do not crack at the harshness of the battle and try to hasten your death.

11. We will finally prevail if you make the effort to kill just one man more. Die after killing ten men and yours is a glorious death on the battlefield.

12. Keep on fighting even if you are wounded in the battle. Do not get taken prisoner. At the end, stab the enemy as he stabs you.[1446]

His philosophy was simple: Every soldier must be a focused, killing machine without regard for himself or his comrades' welfare, and obey *all* his orders without question. He made sure his men knew a lesson must be taught to the Marines, and to the world for that matter, that massive casualties would result if an enemy attempted to take any Japanese territory. Most would posit that if a leader motivated his troops by telling them they were all going to die, it would not cause them to fight more effectively. However, Kuribayashi knew his men, and when he explained their goal was to do more than any Japanese soldier had been commanded by killing ten Americans before taking their last breath (on Guam, the soldier was only ordered to kill seven),[1447] they realized how seriously their commanders and nation viewed their mission. They took his new oath "of violence and revenge"[1448] to heart, knowing their deaths were inevitable. In the end, Kuribayashi's instructions for his men clarified that they should never show any mercy for the Marines—*take no prisoners* and *kill them all* was a motto he wanted his men to accept, a motto they would indeed embrace full-heartedly.

Kuribayashi's orders were not mission statements of how to win the battle, but rather orders to inflict such damage on the Americans that they would conclude the cost to defeat and occupy Japan was too high. They were not mission avowals on how to achieve victory, but rather, on how to "not be defeated for as long as possible."[1449] "Kuribayashi clearly understood the historical consequences of culture idealism and harnessed his soldiers to more practical standards."[1450] This shrewd general knew that only in their radical sacrifice could they perhaps 1). buy more time for Japan to build out defensive works in the homeland, 2). wear down the Americans, 3). offer some possibility of securing a semi-favorable cessation of hostilities that might preserve Emperor Hirohito and some of the territorial gains made by the Empire, and 4). save Japan from a dishonorable defeat. Kuribayashi and his command embodied military philosopher Carl von Clausewitz's dictum, "War is merely the continuation of policy by other means (*Der Krieg ist eine bloße Fortsetzung der Politik mit anderen Mitteln*)."[1451] He believed "exacting the maximum bloodshed from the U.S. forces on Iwo… would work in Japan's advantage in negotiating an end to the war."[1452] Once Kuribayashi understood the political conditions he faced, what the current war for Japan hoped to achieve and what it could achieve, it appeared easy for him to chart a certain course. He obviously had great strength of character and strong "lucidity and firmness of mind" to steadily carry out his current plan and to get his men to buy into it fully.[1453] For most armies of the day, especially those in the West, asking them to go on a one-way mission, which would lead to death, would invariably lower "the morale of the troops, corrupt their discipline [and] in short undermine their fighting spirit unless an overwhelming belief in the greatness and infallibility of their commander outweighed all other considerations."[1454] However, Japan's culture afforded Kuribayashi men able to be formed in such a way as to buy into the "enormous exertions and great hardship" and ultimate cruel death all his men faced following his will and orders.[1455]

Guadalcanal Medal of Honor recipient and Brigadier General "Red Mike" Edson, the famed *Raider* leader, warned the public a year earlier on 4 January 1944 that a strategy espoused by Kuribayashi was the norm. Edson cautioned that the "entire Japanese strategy was built on the premise of inflicting casualties, not on achieving battlefield victory." He said, "I think the American people should realize the psychology of the people we are fighting—to make the campaigns as costly as possible because they don't believe we can take it. They are willing to take large losses in the hope that we will be ready to quit before we can lick them."[1456] Although U.S. newspapers picked up Edson's analysis, most Marines did not understand the extent to which the Japanese would take this strategy; and Kuribayashi and his men continued to perfect this *modus operandi* of causing the most harm despite the realization that defeat for them was the result—never a victory. Commanders like Edson knew they needed to defeat the Japanese and show them Americans would not waiver in their resolve to make Japan surrender unconditionally regardless of the cost.

But just as necromancers approach the morning light with horror at the conclusion of their witching hour, Kuribayashi refused to entertain the thought of unconditional surrender although that might happen if they continued to suffer defeats. He knew his Emperor and high command expected him to lose, but they wanted him to do so in a manner that would frame the loss as a Pyrrhic victory for the "heathen" Americans whom he called "devils." Little did he know that Marines pride themselves in being called "Devil Dogs" (*Teufelhunde*) after the nickname Germans gave them during WWI for their crazy warcries.[1457] The people Kuribayashi respected and called friends when in America he now demonized as *kichiku beigun*; namely, "devils," "brutes," and "liars."

To prepare his men, Kuribayashi not only gave orders to condition them that they were going to die, but he also did things to ensure they knew what he expected of them such as blocking their emplacements' exits. After entering pillboxes, many of which were sealed from the outside, the men defending Iwo realized this pillbox was to be their tomb. Once inside, they were *never* leaving.[1458] To this day, many are still in those pillboxes, buried under sand and vegetation (some estimate there are over 11,000 undiscovered bodies on Iwo).[1459] From a military perspective, maybe sealing them inside also camouflaged their dwellings since many structures were buried underground, so this might have been a tactic to prevent Marines from identifying the structures and/or attacking them from a vulnerable rear. There were thousands of cave dwellings and pillboxes organized into a phantasmagoric labyrinth. If many of his soldiers had no way to escape the fortifications, then Kuribayashi probably locked in thousands into these concrete sarcophagi before one Marine put a boot on shore. His men went "contentedly, like a quiet ghost with a clean conscience sitting inside the bars of a snug family vault."[1460] This conditioning of his men traces the extreme form of command of Japanese officers who seemed to be "consumed with one unachieved revengeful desire."[1461]

From the commanders on Saipan and Guam to Kuribayashi on Iwo, a desire burned to fight bravely and sacrifice themselves and their men even when it definitely appeared that defeat was a forgone conclusion. Kuribayashi was "hopelessly holding up hope in the midst of despair."[1462] He was one of the most ruthless commanders of all time, ruling his island with absolute authority, proving indeed that "some dying men are the most tyrannical."[1463] Kuribayashi conducted his battle with skill, knowing that within weeks he would die fighting or commit suicide after almost all his soldiers lay dead. He wrote, "What a pity I have to bring the curtains down on my life in a place like this because of the United States."[1464] His leadership encompassed intelligent design and fantastic fatalism. He was logical and brave in the face of defeat. What has thus far been described of this man, one could argue, was a "result of an excited, or perhaps of a diseased intelligence."[1465]

As he realized he would soon die, Kuribayashi wrote his wife and told her that when the Americans landed on his island, he and his men "must follow the fate of those on Attu and Saipan."[1466] Then he wrote his son Taro: "The life of your father is like a flicker of flame in the wind. It is apparent your father will have the same fate as the commanders of Saipan, Tinian and Guam. There is no possibility of my survival."[1467] To reiterate to his wife that he would indeed never return, on 21 January 1945, just shy of a month before the Marines would land on his beaches, he wrote Yoshi to stop praying for his survival: "I don't care where my grave is located. My ashes will not be returned home and my soul will remain with you and the children. Live as long as possible and please take care of the children."[1468] Although fatalistic by then, Kuribayashi's energy and drive never let up in building out his island defense. His "diseased intelligence" now hungered for the day he would meet the Leathernecks in battle.

## Marines Prepare for Iwo Jima

As the Marines finished up their preparations to attack Iwo, many wrote letters home not knowing when they would be able to do so again. Captain Beck sat down on 8 February and wrote his "Darling Wife": "I believe we can say tonight that the war news is very favorable on all fronts…I just heard over the radio that the Russians are only 7 miles from Berlin."[1469] In January he wrote, "The Army is closing an excellent job in the Philippines."[1470] Beck regularly listened to the radio and read *Time* or *Life* magazines. However, although things were looking good in Europe for the Allies and the Philippines were about to fall to MacArthur, the end for the Pacific War looked distant. Beck wrote, "In just a few days we will have been two years out here. As soon as conditions permit I intend to ask for a leave. It doesn't look like they will ever rotate us officers."[1471] Beck knew they would soon be hitting the beaches of another island still far away from the Empire's heart. In helping his wife Ruth realize he might be away from her and their son Barry longer than they had thought, he wrote:

> What makes it tough is the fact that there are a very small per-
> cent who actually see or fight the enemy and those who do are
> the ones who should be given a break, but they keep the Marines
> and Army fighting men out here as long as they please and the
> Navy, who really never have to fight goes back after 18 months.
> Well that's the way it goes and there is no need of me beating my
> chops about it.[1472]

Beck felt he had put in his time at Bougainville and Guam and yearned for his family.

On 12 February 1945, the 3d MarDiv formed at the docks at Guam and boarded the troopships earmarked for Iwo Jima in Operation *Detachment*. Two days later, Hirohito could have stopped this campaign. Prince Fumimaro Konoe pleaded with him to "surrender immediately." That would have saved a tremendous number of lives and averted great suffering among the Japanese, their captive nations, and the Allies. Hirohito dismissed Konoe, saying, "If we hold out long enough in this war, we may be able to win, but what worries me is whether the nation will be able to endure it until then."[1473] As a result the next battle was still on and Woody's C Company set sail on 16 February on the attack transport *USS President Adams* APA-19.[1474] Some joked with each other, and one character said over the loudspeakers, "Just a pleasure cruise" as they left port.[1475] Battleships, cruisers, destroyers, aircraft carriers and numerous planes squired them into the open sea in pursuit of their next victory. Below deck, the men were packed in like sardines, with bunks stacked five to seven high. Along the way, they daily got a few hours above deck to do calisthenics. Other than that, they stayed below and lay on their bunks and daydreamed or conversed with their buddies while living in their cavernous, hot hole. "I wish they'd had a few fans back then," Woody said, "but they didn't. It was hotter than Hell."[1476]

In their free time, they went topside to try to find shade and look at the ocean and watch the sun playing off the water. Chaplain Rabbi Roland Gittelsohn was impressed at how politely the men treated one another: "[M]en facing the most uncomfortable present and the most frightening future of their lives were wonderfully human toward each other!"[1477] In total, the navy sent 800 ships to Iwo full of such men, from troop ships and supply vessels to battleships and carriers.[1478]

The sun was high in the sky as fleets of airplanes circled above, providing cover. The ships delivered death to Japan in aircraft, naval gun bombardment and amphibious troops as they "voyaged along, through seas so wearily, lonesomely mild, that all space, in repugnance to our vengeful errand, seemed vacating itself of life before our urn-like prow."[1479] Men attended religious services. In fact, several, to "cover their bases" so to speak, attended different religious services, visited the chaplains of all faiths and collected different icons: Crosses from Protestant ministers, crucifixes from Catholic priests and *mezuzas* from rabbis.[1480] One Protestant chaplain read, "Finally, my brethren, be strong in the

Lord, and in the power of His might. Put on the whole armor of God that ye may be able to stand against the wiles of the Devil" (Ephesians 6:11)."[1481] Such words of solace helped calm many, and for the faithful, the thought that God provided protection made many hopeful they would survive. Describing one's enemies as Satan's agents was effective to help religious men justify their killing—since time immemorial, armies have demonized their foes to psychologically acclimate to the thought of killing them. Some chaplains prepared for battle physically, not just spiritually. One chaplain strapped on a Colt .45 pistol. Seeing stunned Marines nearby, the priest said, "Yes, prayer is good, but this thing is quicker."[1482]

The priest was not alone. Rabbi Gittelsohn, with the 5th MarDiv (effectively, second in command of the division's chaplains), trained with a pistol and wore one during Iwo. A common misconception was that the Geneva Convention prohibited medical personnel and chaplains from bearing arms. Americans had little hope the Japanese would honor the Geneva Convention and so almost all medical personnel and chaplains armed themselves "for self-defense."[1483] Many chaplains believed God would protect them, but religious belief could only go so far in building up confidence to meet every situation, especially when fighting the Japanese who did not themselves fear death.

Although it is often said "there are no atheists in foxholes," several Marines had no faith, including Woody at this time. Some became believers and others lost their faith. John Lauriello explained that after what he had seen, he could never again believe in a merciful or just God.[1484] "When I saw someone reading his Bible or praying, I just thought he was wasting his time at this stage in my life," Woody said. He was apathetic about God and divine intervention.[1485]

Aboard an *LST* (Landing Ship, Tanks), Coast Guard correspondent, Vic Hayden, exclaimed he did not believe for one minute that "old crap about there being no atheists in foxholes." He then looked around and told people he was going to set up the Foxhole Atheists Society to protest.[1486] Jerry Yellin, a Jewish American P-51 pilot, said that after all the death he had seen, he could never believe in God and found religious belief a waste of time.[1487]

Men found many of their comrades who were the most religious were among the first to die. Marine veteran Eugene Sledge saw a dead man missing half his face and with one gazing eye he described as staring at him, he imagined the deceased having an imaginary conversation saying, "I prayed like you to survive, but look at me now."[1488]

My uncle, Army 1ˢᵗ Lieutenant, Frank Rigg, training in 1943 before the Angaur Landing 17 September 1944 (center of photo). He served in the 322ⁿᵈ Regiment of the 81ˢᵗ Division. He received a Purple Heart for wounds suffered at Angaur and hated the Japanese often claiming they were a people without morals or ethics. He once said it was a shame "America did not kill more of the fucking bastards." Author's Collection.

So, although many were religious, there were many who felt shocked out of belief by what they saw their fellow men do to others while God, it seemed, would not intervene to make things right. Was He aware of the massive carnage and destruction? Could He hear the desperate prayers over the din of battle? Which country did He favor? Some theologians argue these problems of war and death are a result of "free will" (the power to act at one's own discretion). Yet, ironically, we all have *free will* because we have *no choice*. But even with the prayers and lucky charms, many Christians and Jews had trouble reconciling what they had been taught about a peace-loving God once they witnessed battlefield realities. Was the carnage, suffering and death of good, God-fearing men, part of God's plan? Many prayers went unanswered, causing some to doubt what they learned as children. Many questioned or discarded their beliefs. The author's uncle Frank Rigg, a U.S. Army officer who fought the Japanese on Angaur with the 322nd Regiment of the 81st Division, hated the Japanese and lost his belief in God after what he observed in battle. He claimed only the weak-minded and unobservant need God.[1489] Saipan and Marine veteran, Charlie Toth, rejected his Catholicism and echoed my uncle's belief saying, "God had died a long time ago" and that *He* "must be a very miserable deity to allow even a fraction of the cruelty that goes on in this" war.[1490]

What about the Japanese soldiers' religion? Most Japanese seemed to not lose any faith in their Shinto gods, but this may have been because some Japanese gods practice revenge and hate and are war-gods like *Hachiman Daimyōjin* and *Takemikazuchi-no-kami* or Hirohito, a man-god, who ordered this carnage in the first place.[1491] These gods acted in contrast to the peace-loving Christian God and His son Jesus ("love you enemies" or "whosoever shall smite thee on thy right cheek, turn to him the other also" (Matthew 5:39, 44)),[1492] who to many westerners exemplified divine behavior (although Christian Americans were doing everything but turning the other cheek to get slapped or loving the "Nips"!). In contrast, it seems Japanese religious belief was more in line with reality. Admiral Yamamoto, the embodiment of the warrior class, "loved his Emperor" and believed "his people to be a chosen race, selected by a far-seeing Providence to fulfill an ineluctable destiny" filled with death and destruction ordained by the gods.[1493] He believed they helped Japan win the war against Russia in 1905 and would do so in the future and that men dying in battle would become gods themselves.[1494]

Many Japanese were in a stage of religious development similar to the ancients who worshiped *Ares* (the Greek god of war), *Mars* (the Roman god of war), *Thor* or *Odin* (Norse gods of war and revenge). And Japan's leader, Hirohito, furthered the human tradition of deification as did ancient Egyptian Pharaohs like Tutankhamun or Ramesses II (they were revered as the offspring of a Sun god named *Ra*), and as did Greek leaders like Alexander the Great (son of Ammon-Zeus) or Roman Emperors like Caligula, Nero and Domitian.[1495] In all cases, these men were not only seen as gods, but also were their nations' supreme military commanders. Transcendentalist Ralph Waldo Emerson wrote, "That which dominates our imagination and our thoughts will determine our lives, and character. Therefore, it behooves us to be careful what we worship, for what we are worshipping we are becoming."[1496] Religious commentator Sam Harris could be describing the average Japanese belief in Hirohito and these Shinto gods when he wrote: "The only demons we must fear are those that lurk inside every human mind; ignorance, hatred, greed, and *faith,* which is surely the devil's masterpiece."[1497] The preceding quotation is an oversimplified metaphorical view, but the fact remains the Shinto religion seemed to reinforce the Imperial soldier's will to fight tenaciously, and either die for a merciless Empire or commit suicide rather than surrender.[1498] Shintoism also reinforced the belief that if they died honorably, they could become gods themselves "eternally to protect" their families and clans.[1499] In short, the war was *seisen,* "sacred war."[1500] Harris continued: "As long as it is acceptable for a person to believe that he knows how God wants everyone on earth to live, we will continue to murder one another on account of our myths."[1501] And Japanese myths motivated Japan's soldiers to kill without remorse and die with reckless abandon.

As the Marines approached Iwo, the Japanese went to shrines, offered prayers to their gods and clapped their hands to summon their divinities' attention and ward off evil spirits. Many gazed north and bowed toward the Imperial palace. One soldier's final message to his family read: "As one who has given life in order to serve the Emperor, I was always ready for my corpse to lie in the field of battle. It is my long-cherished desire as a soldier."[1502] Thousands donned their thousand-stitch belts (*Senninbari*) to spiritually "ward off bullets."[1503] These three foot long and six inches wide belts were made by their womenfolk who had brought these *Senninbaris* around their neighborhoods, asking one thousand woman per sash to each put in a stitch in order to create belts that would defend their men from the enemy.

Before the Marines set out from Guam, most Leathernecks did not know where they were going; for operational security and to prevent the Japanese from knowing that Iwo was next, few were told that it was the objective. Only after they were aboard ship did they learn about the island. Looking at the map of Iwo Jima, Woody recalled, "Really, as a farm boy, I can remember thinking two miles was nothing to me [the width of Iwo]. I ran that far back when I was a kid going around the hills of [West Virginia]. Since the island was two by five miles, I really didn't think it was going to be that hard."[1504]

Scenes like this happened everywhere on the transport ships headed to Iwo Jima. While at sea, the men learned of the island they were going to attack and the topography they would be engaged over using these rubber models of the island. National Archives, College Park. Photo Credit Morejohn.

Many Guam veterans thought Iwo would be easy to conquer, with its eight square miles, one third the size of Manhattan, since they had just conquered a 60 square mile island, Guam, in three weeks, and without heavy casualties.[1505] However, unbeknownst to all, it was going to be one of the toughest battles ever. Why it held strategic significance to the U.S. was beyond the ken of most. They reviewed maps and the terrain of Iwo, learning of their mission while sailing into hostile waters. Woody's officers told them that because men were killed trying to take Japanese prisoners on Guam, they should under no circumstances take POWs. The problem was again Japanese treachery, in this case the "feigned surrender." "Kill them all," was the word given to Marines.[1506] Lieutenant General Smith told Nimitz his men had taken the Japanese lesson of no surrender to heart and they "will die to the last man; they never will be taken prisoners."[1507]

The practice certainly was not counter to the wishes of the Japanese in any event; their desire was to die as warriors, not live as prisoners.

When not engaged in learning about Iwo Jima they were about to attack or conducting physical exercises topside, men *en route* to their combat zones inside the transport ships either slept, cleaned weapons or played cards. 3 February 1945. National Archives, College Park, Photo Credit L. Burmeister

Woody and his comrades thought they would not have to practice such draconian measures since they got word they would be held in reserve and not depart the ship. Woody heard the battle would only last five days. His company would be committed only if things got really bad. Woody expected to hang around on the ship for the entire time, like they did during Saipan, when the 3d MarDiv was held in reserve, but never called to battle.[1508] It was stressful because although they were in reserve, they could be sent into battle at a moment's notice. Once anchored in position miles off Iwo's coastline and waiting to perhaps join the battle, the tension and boredom could cause "a lot of friction between the Marines."[1509] For the invasion, the Marines of the 4th and 5th MarDivs would hit the beaches first on 19 February. Woody figured that "[a]fter they secure it, we'll return to our base at Guam for further training."[1510]

By 13 February, Japanese patrol planes had spotted almost 200 ships headed north toward Iwo from Saipan. Kuribayashi alerted his troops to occupy their battle stations. "On Iwo Jima, preparations for the pending battle had been completed, and the defenders were ready."[1511]

By the time of the invasion, planes and naval guns had bombarded the island on and off for 249 days (Fiske Hanley's B-29 had hit it on 24 January 1945).[1512] Fleet Admiral Nimitz observed "no other island received as much preliminary pounding as did Iwo."[1513] Raids by U.S. Army Air Force and navy carrier planes had ended the use of Iwo as a Japanese airfield by 25 December 1944 (one raid on 8 December 1944 dropped 800 tons of explosives on the citadel). Several planes unloading their bombs on the island received "furious gunfire from the fortified peak [Mt. Suribachi]" prompting bomber crews to nickname it "Mount Sonovabitchi." For 72 days leading up to the battle, the U.S. intensified its attacks on Iwo. Aircraft carriers alone sent 2,700 sorties over the island, dropping 5,800 tons of bombs and B-24s from the Marianas flew 2,807 sorties releasing 7,720 tons—aerial reconnaissance photographs showed 5,000 craters in one square mile alone. During these attacks, the Japanese *Ack-Ack* guns were only able to bring down 10 B-24s and four fighters. Kuribayashi hated these raids; all he and his men could do was "hold our breaths, praying to god and Buddha in our bunkers." The U.S. even bombarded Iwo with a few tons of napalm. The island was further "softened-up" (pulverized) for three days by intense naval bombardment prior to the 19 February landing. Battleships, cruisers, destroyers, minesweepers and gunboats probed the defenses, removed obstacles and fired on targets. Some ships were from WWI, like the *USS Texas*, a battleship built at the same time as *RMS Titanic*. She had already seen action in Operation *Torch* (the invasion of North Africa in November 1942) and in Europe with Operation *Overlord* (the invasion of Normandy in June 1944). She was not as fast as the new battleships, but she had big guns and was still powerful. Her battled-hardened gunnery crews were experienced at taking out land targets. By 15 February, the *USS Texas* and other warships had pummeled the island with almost 22,000 shells, with more on the way.[1514] Even today, in her permanent museum berth in La Porte, Texas, one can see a map near her war room marking where in the world she fired her shells. (See Gallery 3, Photos 1-3)

At Iwo, ships like the *USS Texas* shelled the island relentlessly. One intercepted broadcast from Iwo's command back to Imperial Headquarters said, "We are doomed. The enemy is firing on us from both seas."[1515] So many shells and bombs hit the island that if the tons of iron fragments were gathered together, a thick steel blanket would cover the entire island.[1516]

When ships hit positions with 16-inch shells from 1,000 yards away, the explosions must have truly sounded like hell's bells announcing Armageddon.[1517] If one stands outside near one of the battleship guns while firing, the percussion will burst his ear drums and the concussion force will knock him off the ship. When a shell lands near or on target, the destruction is impressive. Japanese veteran, Takahashi Toshiharu, was on the receiving end of this action:

> The guns that were trained on the island all spurted fire at the same time. On the island there was a huge earthquake. There were pillars of fire that looked as if they would touch the sky. Black smoke covered the island, and shrapnel was flying all over the place with a shrieking sound. [Thick trees] were blown out of the ground, roots uppermost. The sound was deafening, as terrible as a couple of hundred thunderclaps coming down at once. Even in a cave thirty meters underground, my body was jerked up off the ground. It was hell on earth.[1518]

A Marine witnessing this display of firepower looked at his buddy and asked, "Do you think there will be any Japanese left for us?"[1519] Observing the bombardment from the sea, *Time* and *Life* correspondent Robert Sherrod wrote he could not help thinking, "nobody can live through this," then added, "but I know better."[1520] Although no previous target to Iwo had "received such a volume of preparatory shelling per square yard of terrain," Kuribayashi had designed his island fortress so well that as a percentage, no other Japanese-held island to date had gone into battle "with so many of its defenses intact," so Sherrod was more right than he could ever imagine.[1521]

To counter this bombardment, Kuribayashi "attempted a meager counterattack" by sending two *Kamikaze* Zero fighters armed with 60-kilogram bombs that had been housed in a concrete structure at Airfield No. 2. As they flew past Suribachi, naval AA guns fired upon them and the two planes, shredded in pieces, fell into the ocean.[1522]

## Navy Frogmen Clear the Way with a Flotilla of LCIs: Events of 17 February

As with other islands, Underwater Demolition Teams (UDTs or Navy Frogmen) cleared the offshore waters on 17 February to remove obstacles that would prevent the landing. Kuribayashi had given strict orders not to fire on the Americans until the Marines were on the beach *en masse*. In a rare violation of his orders, several 5-inch "trigger-happy Nipponese gun crews" hidden deep inside Mt. Suribachi under Rear Admiral Ichimaru along with other units fired at the support vessels for the Frogmen, mistakenly thinking they represented the main landing. They hit the cruiser *USS Pensacola* with numerous rounds, killing 17 and wounding 120 of her sailors. Japanese gunfire further eviscerated the 14 gunboats (*LCIs (G)*) supporting the Frogmen (again, these *LCI* (Landing Craft, Infantry) vessels were 158 feet long, 23 feet wide and weighed around 350 tons carrying a crew of between 25-60 men—on this day, the gunboats had roughly 60 men each since they were there to defend the Frogman operation and not carry troops). To give one a perspective, these ships were between 1,000 to 1,200 yards away from the beach being hit by weapons capable of hitting targets miles away (5-inch guns can hit targets within 13 nautical miles, for example).[1523]

Enemy fire sank landing craft *LCI (G)* 474. *LCI (G)* 473 became disabled from the hits it took and had to be towed out of battle. *LCI (G)* 441 took several shell strikes losing 40% of its crew in combat, but its captain, Lieutenant Junior Grade Forrest W. Bell engaged the enemy and kept fighting later receiving the Navy Cross for his actions. The Japanese killed or wounded 30% of *LCI (G)* 466's crew and *LCI (G)* 449 suffered 60% dead and injured. Most of the ships were on fire or flooding, but their officers and men kept them in the fight by counter-flooding compartments, keeping the ships under steam and returning fire with their cannons, rockets and machineguns (most of these boats had two 40mm guns, four 20mm guns, six .50-calibre machineguns and ten MK7 rocket launchers).[1524]

On *LCI (G)* 449, navy Lieutenant Junior Grade Rufus G. Herring bravely captained his ship near Yellow Beach 1 (where Woody would land a few days later) and directed counterfire against positions on the beach (he and his crew had participated in the battles of Saipan, Guam and Tinian, so they knew each other well and were very familiar with their ship's capabilities). The second round of shells from the Japanese shore batteries that came at him and his men shattered his vessel and knocked out most of his guns hitting his *LCI* with 5-inch shells, large mortars of 37mm and 25mm and machinegun fire. In less than a minute, two large fires spread throughout the ship and Herring's crew had suffered 20 dead and 19 wounded (some dead bodies were on fire). His pilothouse was shattered and one of his crewmen was seen still seated in his chair behind the radio, but now headless. The decks of his ship were covered with spent shells, shrapnel, blood, guts and body parts. Although Herring had *lost* consciousness twice, had almost *lost* his left arm that had been practically severed off at the shoulder by shrapnel (surviving a blast most likely by a 37mm shell blowing him out of

On *LCI (G)* 449, navy Lieutenant Junior Grade Rufus G. Herring, skillfully led his men against Iwo Jima on 17 February 1945 in beach clearing activities. Although severely wounded and losing 60% of his crew in battle, he never "gave up the ship" and continued operations until he had secured his vessels back out to the fleet and made sure his surviving crew members were tended to—he would earn the Medal of Honor for his actions.

the conning tower and onto the lower deck), had *lost* partial use of his right foot due to shrapnel wounds, had *lost* the hearing in his left ear ("perforation of the left drum"), had *lost* copious amounts of blood and had *lost* 60% of crew taking the brunt of the onslaught by the Japanese shore batteries, he never relinquished command and continued to shout orders to put out the fires and carry on the fight, and when unable to fight any longer, to successfully steer his vessel out of the line of fire using his *one good arm*. Out from his "torn flesh and muscle" on his wounded shoulder protruded a large shell fragment, looking like he had grown a dorsal fin. The doctors later wrote just of his arm/shoulder injuries: "The wounds included a large avulsed wound of the deltoid area, a deep 4" laceration along the posterior left axillary border, a 3" laceration of lateral aspect of left arm, another over the 5th metacarpal dorsal aspect left hand and other lacerations and puncture wounds." In layman's terms, "he was fucked up!" Leaning against the ship's rail, Herring took the helm since the helmsman had become a casualty, maintained communication with the engine room and piloted his ship away from the battle to safety saving what remained of his men and *LCI (G) 449* (*he did not give up the ship*). Once he safely had his ship moored next to the 450 foot minelayer *USS Terror* (CM 5), had received two medical officers and two corpsmen and had witnessed his men being taken care of, he *lost* consciousness again. For his heroics, Herring would later receive the Medal of Honor. When the Japanese opened fire on the gunboats, giving away their positions, they exhibited their only instance of breaking fire discipline earmarked for invasion forces except for antiaircraft units that continually tried to take down planes.[1525]

On this day, the U.S. Navy deployed 14 *LCIs* and many of its gunships, as just described, took direct hits and much damage. The flotilla suffered 202 casualties with 47 of them being KIAs or MIAs (25% of the sailors of this tiny fleet became casualties). The U.S. Navy "would lose more ships and men than on D-Day in Europe" assaulting Iwo this day. Even so, every ship did its best to support one another and fight back against the positions inflicting damage on them evidence by the fact that these 14 ships fired a total of 9,000 rounds of 40mm, 12,000 rounds of 20mm and three hundred and fifty rockets against the Japanese on the beaches.[1526] Seeing their "little brothers" taking so much punishment, the fleet started to move its battleships and cruisers in close to the beach.

After the U.S. Navy located those enemy guns and calculated distances and placement, its big ships unleashed a fury of shells. The *USS Nevada*, a survivor of Pearl Harbor, instantly "began counterbattery against the weapons firing at the gunboats."[1527] *Nevada* and many others "fired back with a vengeance and snuffed out the concrete casemates one-by-one."[1528] This was one of the few errors against Kuribayashi's orders his men made. This error, although killing several sailors on the unfortunate gunboats, saved thousands of Marines' lives since the Japanese guns would have taken a "very heavy toll of men and supplies from the outset of the ship-to-shore movement." In fact, "this was unquestionably the most significant role ever played by the bold underwater swimmers and their close covering gunboats in the course of the Pacific War."[1529] Moreover, unlike the almost three dozen amtracs that were sunk assaulting Guam, it was estimated that only five amtracs were sunk assaulting Iwo in the early waves, another benefit of having located and destroyed these Mt. Suribachi guns.[1530] While these sailors and Frogmen did not know it, they had become attractive bait and when the Japanese bit, they allowed the navy ships to make the beach a little easier to assault although there was still more danger awaiting the Marines than any of their commanders had imagined. The Frogmen also collected Intelligence on the beaches and verified for higher command that the Japanese most likely would not flood the beaches with burning oil as was feared.[1531] After Kuribayashi's garrison finished with its attack on 17 February and the Americans withdrew, the Japanese propaganda on the radio stated, "The brave defenders of Iwo Jima have repulsed the Marine Corps landing after the heaviest concentration of fire in the war...The Marines have turned tail and run."[1532] As usual, the Japanese civilians were fed misinformation. Meanwhile, the truth was that Marine leadership wanted more shelling.

## Pre-Invasion Naval Gunfire

One of the controversies of Iwo was the amount of U.S. Navy pre-invasion gunfire. Major General Harry Schmidt, commanding the landing force, asked for a week to ten days of bombardment, but due to the supply of ammo and the pressures of other engagements, naval commander Rear Admiral William H.P. Blandy provided only three days of shelling. Schmidt's boss, Lieutenant General Smith, claimed that had the navy done a longer bombardment, "casualties would have been lower."[1533] Smith wrote, "We had to haggle like horse traders, balancing irreplaceable lives against replaceable ammunition. I was never so depressed in my life."[1534] Blandy, not realizing how well Kuribayashi had prepared the island, decided, "Though weather has not permitted complete expenditure of the...ammunition allowance, and more installations can be found and destroyed with one more day of bombardment, I believe landing can be accomplished tomorrow as scheduled."[1535] Moreover, the navy was aware if they stayed in one place, the danger to their fleet increased dramatically so they limited their time in static positions.

Additional naval bombardment might have decommissioned more pillboxes and fortifications, and according to historian Ronald Spector, "would have shortened the fighting ashore,"[1536] but what is difficult to know is whether a longer bombardment would have been more advantageous than getting troops ashore sooner to attack the gun positions firing on the Marines. The navy had no way of knowing the extent of the underground fortifications, their exact locations nor how deep underground and how well-reinforced they were. In reality, the shelling inflicted limited damage on the defenses, because unlike other earthworks in the Pacific, most Japanese were underground waiting out the bombardment in well-prepared tunnels.[1537]

"The U.S. Navy had studied the effect of bombardment on strong fortifications and the limitations were well understood."[1538] Although a 16" high explosive shell from a battleship had a terminal velocity of 1,500 feet per second and could demolish structures above ground," it had "doubtful penetrating power."[1539] As for planes, historian Samuel Eliot Morison said:

> Aside from a lucky hit, about all that aerial bombardment could do was to blast off camouflage and reveal hidden targets…Kuribayashi had purposely designed his defenses to minimize the effect of bombardment…The defenses were such, by and large, that the only way that they could be taken out was the way that they were taken out, by…infantry and demolitions. Combat engineers of the 5th [MarDiv] destroyed 5,000 cave entrances and pillboxes in their…zone of operations alone.[1540]

The 3rd MarDiv reported that the pre-invasion naval shelling resulted in "only a small percentage of the Japanese installations on Iwo Jima" being neutralized.[1541] Smith summed it up that the shelling "had hardly scratched the fortifications the Japanese had prepared."[1542]

B-29s Superfortress bombers on the North Field Guam, April 1945. These bombers would flatten most major Japanese cities. The large "O" on the tail of these planes denote they were part of the 6th Bombardment Squadron of the 29th Bombardment Group. National Archives, College Park.

Another reason for a shortened period of pre-invasion bombardment was the departure of Task Force (TF) 58 from Iwo to go to Japan's shores to strike her from the sea. Admiral Spruance, the Fifth Fleet commander in overall charge of the Iwo operation, led this attack, and wanted it to be impressive to show that Japan's mainland could be hit from the sea and not just from the air as was being done by the B-29s. For TF 58's raid to be impressive, Spruance took with him several battleships originally allocated for Iwo, thus depriving Marines of shelling before the invasion. The raid was a vain and costly exercise by the navy to prove seapower's worth. Moreover, naval commanders wanted to save ammunition for airfields on Japan's mainland that would be used to hit the approaching U.S. fleet occurring after the landings. This operation would have ships at Iwo return to Japan to neutralize any "air interference that might have developed" from Japan's homeland against American forces at Iwo. These maneuvers supported operations at Iwo by providing a screen against aerial attacks from Japan against the invasion force.[1543] In the end though, even if the navy had pounded Iwo more, it would not have had much effect because the enemy was so entrenched. Moreover, the proposals for more bombing focused on the landing areas and not the middle and north where most casualties would occur due to fighting the toughest defenses that had not been touched by pre-landing bombardment.[1544]

Nine months earlier, when the Japanese lived on the island's topside, bomb attacks did kill more,[1545] but Kuribayashi learned from this experience and took his army underground. While extra days of bombing and shelling might not have made a substantial impact on the defenses, it would have depleted the supplies and deprived the enemy of sleep, which could have affected the Japanese once operations began. Most soldiers in the island survived the preliminary bombing unscathed and were ready to fight as the Marines hit the beaches. Regardless of what more pre-invasion bombardment might have done, the decision was to land the troops with fewer days of bombing than the generals desired.[1546]

Kuribayashi commanded approximately 22,000 mostly Japanese troops— there were 1,600 "Korean laborers who...on the whole fought well."[1547] His troops' morale, health and discipline were excellent. His engineers took advantage of the terrain and "fortified Iwo to near perfection" putting blockhouses, pillboxes and other emplacements where they would be most effective.[1548] Kuribayashi had hundreds of major installations constructed to shelter the big guns and had blockhouses built with 5-foot thick concrete walls, some with 50 feet of earth above them. Mt. Suribachi had a four-story complex and hospital inside it. Kuribayashi made expert use of the terrain in planning the fortifications to stymie the Marines. Besides having these fortifications manned with soldiers with rifles and machineguns, he also had "361 artillery pieces of 75mm or larger caliber, a dozen 70mm mortars, 65 medium (150mm) and light (81mm) mortars, 33 naval guns 80mm or larger, and 94 antiaircraft guns 75mm or larger.

In addition to this formidable array of large caliber guns, the defenses could boast of more than 200 20mm and 25mm antiaircraft guns and 69 37mm and 47mm antitank guns."[1549] A downpour of steel would hit the Marines when battle started. Often they did not know the locations from which it came. Historian Robert Burrell noted Kuribayashi's genius was his use of "indirect fire [firing projectiles that do not rely on a direct line of sight] on the beachheads."[1550] Kuribayashi's men had calculated azimuths and elevations for mortars and artillery shells so they could fire from concealed positions behind ravines, hills, bunkers and trenches to hit the beach without ever having to observe the beach during combat to see where their munitions landed.

To thwart Marine tanks, Kuribayashi installed "scientifically" laid minefields, dug anti-tank ditches, built massive stone walls to block terrain suitable for tanks, and strategically placed anti-tank guns. He partially buried 40 tanks throughout the defensive lines sacrificing their mobility for stationary firepower; by doing so, he made the tanks into pillboxes. The Japanese used mortars proficiently and had huge rocket mortars, including one that was 320mm, weighing almost 700 pounds. Japanese artillery used smokeless powder which made locating guns difficult. Nimitz said, "Iwo Jima appears to have been close to...optimum in size and density of defense."[1551] "Observers who had inspected German fortified areas in both World Wars testified that never had they seen a position so thoroughly defended as Iwo Jima."[1552]

The night before invasion, Chaplain Rabbi Gittelsohn, on the *USS Deuel*, conducted a joint religious service with a Christian chaplain broadcasting over the public address system, reaching thousands. In a confident voice he said, "Men, I'm going to be speaking in these next few minutes quite as much to myself as to you. We're in this thing together. The same fears and doubts and high hopes which fill your hearts tonight are in mine." He spoke about spiritual weapons they had while they fought with their physical ones. He discussed the love their wives, parents and significant others had for them and the pride they all had for the U.S., the Corps and for themselves. "It's you and I who will win this campaign with our tanks and planes...You may fire the shot that will break the enemy's back. You may, by your own example of courage, give strength to fifty other men and they to five hundred more." He concluded by discussing the faith they had in themselves, each other and in God, "a power that makes for righteousness and ensures the triumph of righteousness."[1553] Services like this were being held throughout the fleet, and the men listened in quiet contemplation. It gave many strength. Although in the days leading up to battle, there had been bickering and anti-Semitism among some chaplains, during the Iwo campaign, the religious leaders unified under the threat of mortal harm. Most set aside religious differences and united in a pursuit of a common enemy who intended to kill every last Marine regardless of his beliefs or skin color.

In this religious climate, Lieutenant General Holland Smith prayed to God for guidance. He anticipated "his" Marines would suffer horrendous casualties to capture the island, but he underestimated the extent, thinking only one in four would be killed or wounded.[1554] Weeks before the battle, he wrote the Commandant, Lieutenant General Alexander Vandegrift, that Iwo "wasn't worth the heavy casualties his men would suffer." Such a dire calculation was disconcerting. Smith hoped "to God that something might happen to cancel the operation altogether." Wracked with anxiety, he turned to scripture: "My only source of comfort was in reading the tribulations of leaders described in the Bible. Never before had I the spiritual value, the uplift and solace a man on the eve of a great trial receives from the pages of that book." Although a Methodist, he wore a St. Christopher's medal blessed by the Pope around his neck—apparently, he liked to cover his bases.[1555] Although a self-proclaimed "fatalist," Smith was right to worry because this would be a tough battle and his Marines needed all the support they could receive, both from his fellow Marines and from the heavenly forces he hoped were on their side.[1556]

At Iwo Jima, Smith was the commander of the recently created Fleet Marine Force, Pacific (FMFPAC) which left the immediate control over the operations to subordinates. He only conferred with military officials in the planning and execution of Iwo.

Major General Harry "The Dutchman" Schmidt, commander of V Amphibious Corps (75,144 Marines) was the commander at Iwo and the brains behind the operation.[1557] After the war, he became irritated with how some credited Smith for directing the battle:

> I was the commander of all troops on Iwo Jima at all times...
> Smith never had a command post ashore, never issued a single

order ashore, never spent a single night ashore…Isn't it important from an historical standpoint that I commanded the greatest number of Marines ever…in a single action in [our] entire history?[1558]

The Marines' success on Iwo owes much to Schmidt's intelligence, judgment and experience: "Smith provided a useful role, but Schmidt and his exceptional staff deserve maximum credit for planning and executing the…battle."[1559] For his leadership, Schmidt later received the Distinguished Service Medal.[1560] This confusion about roles stemmed from the fact Smith was commanding general "of all Marines in the Pacific" giving him a superior position to Schmidt in the Corps although he was not directly running the Iwo operation.[1561]

Ironically, this FMFPAC command was given to Smith as a way of kicking him upstairs to deny him future battlefield command after he sacked an army general during the Saipan battle creating a maelstrom in Washington and created the sufficient pressure to motivate Nimitz to remove Smith. Nimitz wrote in one of his Fitness Reports about Smith, "His usefulness has been to some extent impaired by wrangles and disharmony that might have been avoided."[1562] Although he was one of the fathers of amphibious warfare, Smith was relegated to a role with little responsibility for conducting the Iwo battle although much of the doctrine of amphibious warfare used during WWII was created under his direction. Nevertheless, the night before the invasion, during all the services and prayers, men like Smith and Schmidt studied the maps in their command centers, discussed details and then anxiously awaited for the guns to start pounding the shores in the early morning hours signaling the start of battle. As soon as a relatively secure beachhead was cleared, both generals would join their Marines on the fields of combat (with Smith returning at night back to his ship).

# Ch. 20: Iwo Jima Landings Begin

"Victory [at Iwo Jima] was never in doubt. Its cost was."
-Major General Graves B. Erskine[1563]

"Everything in war is very simple, but the simplest thing is difficult."
—Major General Carl von Clausewitz[1564]

D-DAY FOR IWO STARTED ON 19 February 1945 which dawned with ideal weather for an amphibious landing. "A light northerly wind floated fleecy clouds lazily over the island. A calm sea raised no surf on the beaches."[1565] Seeing the armada bearing almost a quarter of a million servicemen and Marines headed to their shore, many Japanese shuddered with awe and fear. Tsuruji Akikusa said the hundreds of ships ringing the island looked like a "huge mountain range had risen up out of the sea."[1566] (See Gallery 3, Photo 11)

Controlling this armada was Vice Admiral Richmond Kelly Turner who commanded Task Force 51, the Joint Expeditionary Force. A 1908 Naval Academy graduate, he served in the navy his whole adult life, including commanding numerous ships; he even spent time in Japan in 1939. He knew the navy and was meticulous with details. He had helped execute many battles leading up to Iwo such as Guadalcanal, Tarawa, Saipan and Tinian and some nicknamed him the "Alligator" due to his "mastery of amphibious operations." The Japanese were defeated in battles in which he commanded, and he now controlled one of the largest amphibious forces in history. Lieutenant General Smith, who worked closely with Turner during many operations, described him: "Turner is aggressive, a mass of energy and a relentless task master. The punctilious exterior hides a terrific determination. He can be plain ornery. He wasn't called 'Terrible Turner' without reason."[1567] Erskine also respected Turner writing, "I…think a hell of a lot of Kelly Turner. I think he was a wonderful admiral in every way. He may have had his faults, but nevertheless Kelly would make a decision: it might not be exactly what you wanted, but he'd make it, and I think that's one of the greatest accomplishments of a commander."[1568]

The mutual confidence and good working relationship between Turner, Smith, Erskine and others was commendable, but it was not always respectful or calm: the men could experience tense moments. For example, during the preparation for Saipan, Turner had a heated discussion with "Blood and Guts" Erskine and threatened to relieve him of command because he wanted to retain control of the amtracs, especially the armored ones. Turner argued that, since amtracs were "boats," the navy should control them. The "Alligator" exclaimed that he was tired of Erskine dictating how to use "naval" resources. Unconvinced, Erskine replied, "to hell with that" because amtracs were a "tactical vehicle" and Erskine planned to have amtracs in the first wave to land on the beach because they had the ability to go inland and use their cannons, machineguns and armor hulls to cover later waves. Thus, amtracs were a Marine asset. With his cold, green-eyes glaring through Turner, Erskine concluded, "You may run a hell of a lot of people in the Navy, but you don't run anybody in the Marine Corps." Eventually, Erskine won this dispute and the Marines kept control of the amtracs. Erskine recalled, "[Turner] got so mad at me one day…that he got up and threw the goddamn [six-inch thick operations] order at me! (Laughs)." "Blood and Guts" Erskine had pledged to himself and his Corps: "[T]hat by God the Navy was not going to take control of those amtracs." Later, after these arguments, however, Turner and Erskine ended "up with having a couple of drinks" and being friends again. Erskine continued to fight with others in the navy and Turner about the amtracs, but he kept control of them because "the uniform" the guy was wearing who drove those amtracs was a "Marine uniform."[1569]

Aerial shots showing ships and landing craft hitting the beach at Iwo Jima on 19 February 1945 and bombs from a B-24 raining down on Iwo before the invasion. National Archives, College Park

Erskine's boss, Lieutenant General Smith entered the fight as well when he learned Turner was trying to control Marine operations ashore. "Howlin-Mad" Smith challenged "Terrible" Turner saying, "I don't try to run your ships and you'd better by a goddam sight lay off my troops."[1570] Whereas Erskine's debate-methods were full of logic and allowed dialogue, Smith's discussion-tactics, true to his character, were curt, brash, confrontational and politically incorrect. Unlike with Erskine, Turner probably did not have a "few drinks" with Smith after this exchange. The historical record shows, though, that both Smith and Erskine won their arguments. Marines and "tactical vehicles" stayed under the Corps' control for the entire war and thereafter. Yet, in general though, for all these confrontations, apparently always right before battle and when the assaults started, Turner worked well with Marines and left them alone to do their jobs (and Erskine and Smith made sure he did!).

On 19 February, Turner, the "devil man" whom the Japanese vowed would never return home alive, arrived at 0600 at the command center and took control of the task force.[1571] Forty-five minutes later, Vice Admiral Turner as Commander, Joint Expeditionary Force, gave the command, "Land the Landing Force." Shortly after daylight, the heaviest pre-H-hour WWII bombardment began (with H-hour being the landing time). Chaplains aboard transports offered up invocations. One had cards printed with Sir Jacob Astley's prayer before the Battle of Edgehill in 1642:

> O Lord! Thou knowest how busy I must be this day;
> If I forget Thee, do not Thou forget me.[1572]

Lieutenant General Smith handed out the prayer on his ship the *USS Eldorado* and on the "eve of their Gethsemane," gave each Marine comfort.[1573]

At 0803, naval gunfire was halted to prevent it from hitting carrier planes. In total, 1,200 planes were deployed that day against the island. After a wave of planes finished its run, the ships resumed their bombardment at 0825. The gunfire pattern was adjusted so planes could strafe the beaches for the last seven minutes before H-hour, when the big naval guns shifted aim 200 yards inland.[1574] "In total, eight thousand more shells landed on the beaches in less than a half hour in the final push to clear the beaches just before the amtracs went ashore."[1575]

Near Woody's troopship, hundreds of Higgins boats circled, timing their landing to coincide with the cessation of the last "thunderous bombardment." Large, fiery columns of dust, rock and metal rose as shells from the naval ships exploded. Berthed five miles off the coast of Iwo, Woody saw the dive bombers flying over the island, delivering their payloads and battleships and cruisers firing their 8-inch and 16-inch shells at enemy targets. One officer said a 16-inch shell weighed as much as a car and then said thousands had hit the island. A "smart-assed" private responded, "Sir, that is an awful lot of Buicks to have coming at you." Once the bombardment stopped, the amtracs launched for the shoreline followed by Higgins boats.[1576] Unlike earlier campaigns, these amtracs had extra armor to protect the troops. At Saipan and Guam, casualties occurred because Japanese fire caused "hull penetration[s]."[1577] The extra armor was just one example of why the Pacific War was one of innovation and improvement—from assault squads to amtracs to M1 rifles to advance bombers and atomic bombs, Americans were developing new weapons due to hard lessons learned from earlier engagements.

An unexploded 16-inch shell in the operation zone of the 1st Battalion, 21st Marines. 9 March 1945. National Archives, College Park. Photo credit Corporal F.E. de Ome.

Earlier, by 0730, control parties had established a line of departure 3,500 yards off the beaches. After the bombing, the landing craft stopped circling above the line of departure and started for the shoreline.[1578] In one Higgins boat, a "sardonic coxswain" wrote inside of the front ramp, "It's too late to worry!" A Leatherneck nearby sang: "Happy D-Day to you, Happy D-Day, dear Marines, Happy D-Day to you!" As the men readied themselves to land on another "goddamn Jap island," a "giant force of B-29s" from Saipan suddenly appeared out of the clouds and dropped hundreds of bombs on Iwo sending earth hundreds of feet into the air in long vertical lines towards to heavens giving one a visual domino effect—one could clearly see the planes' circle arrowhead squadron-symbol on their tail fins.[1579] At 0830, the first assault wave of 68 armored amtracs left the line of departure and landed at H-hour, 0900. Over the next 23 minutes the assault waves landed on schedule and at 0945, artillery and ammunition were loaded in DUKWs (2½ ton amphibious trucks, called "ducks"). With them, arrived craft carrying Sherman tanks. As the landing craft headed for the island, enemy shells hit around the boats splashing seawater thirty feet in the air and shrapnel in all directions. Men kept their heads low and held their weapons close often covered in plastic bags to protect them from getting wet—a dry rifle was always better than a saltwater-soaked one.[1580]

Higgins boat loaded with 4th MarDiv Marines prepare to hit the beach at Iwo Jima on 19 February 1945. National Archives, College Park. Photo Credit Gillespie.

A Higgins boat loaded with Marines heads to Iwo Jima following in the wakes of others. A battleship sits to their left hitting Iwo with shells. Mt. Suribachi looms in the foreground. 19 February 1945. National Archives, College Park. Photo Credit PFC Christian

Marines entering their Higgins boat about to assault Iwo Jima on 19 February 1945. Notice the zinc oxide on their faces to protect from explosions and the sun. National Archives, College Park. Photo Credit Bob Campbell.

Marines assault the beach on Iwo Jima on 19 February 1945. National Archives, College Park.

Many Marines' faces looked odd, covered with zinc oxide (flash paint), an early form of sunscreen and protection from burns "from fires if an amtrac [was] hit."[1581] Cheeks and noses were an eerie white resembling ancient Celtic warriors.

One Marine said in "mock hysteria. 'Someone lied to me. The natives on this island *ain't* friendly.'" A sign greeted the men on the beach reading, "Welcome to Iwo Jima" that Frogmen days before had left.[1582]

As soon as 4th and 5th MarDiv' Marines reached the landing zones, they disembarked and hit the beach at a run, but were slowed by the volcanic cinders. Many had to scale steep, sandy terraces 20-feet high, 20-yards from the shore.[1583] Although there were veterans among them, Iwo was the 5th MarDiv's first battle.[1584] A few amtracs reached the first terrace, but many bogged down in the soil and the Marines struggled to establish an organized landing zone. (See Gallery 3, Photos 4-5)

Aerial views of landing craft ferreting Marines to the shores of Iwo Jima during the assault on 19 February 1945. National Archives, College Park.

For the first time in the Pacific, the navy "employed a rolling naval barrage." Such a barrage was an artillery or naval gunfire barrage that fired ahead of advancing troops to pin down the enemy before friendly troops reached enemy lines; the barrage "rolled" or advanced ahead of the friendly troops then lifted (stopped) when those troops were at the enemy's position. On Iwo, the rolling barrage softened up the areas 200 yards ahead of Marines as they hit the beach. As they advanced, the barrage once again adjusted 200 yards ahead of their lines.[1585]

The bombardment and/or artillery barrages of Iwo Jima by America was so intense that some Japanese were literally blown to bits as seen with this shot. All that was left of this enemy soldier were his legs, nothing else. February 1945. National Archives, College Park. Photo Credit Dreyfuss

The initial waves moved several hundred yards inland while the Japanese held their fire as ordered and did not engage although thousands of eyes watched the Marines. The "tactical conduct of [Kuribayashi's] men can only be described as outstanding."[1586] Dozens of caves were dug into the slopes of Mt. Suribachi and on the northern flank of the beach facing the landing zone that spanned a width of 3,500 yards.[1587] The cave entrances had been positioned at such angles to protect against flamethrowers and explosions. Inside the mountain, there was a vast complex complete with a hospital and steam, water and electricity facilities surrounded by plastered walls. Almost 2,000 Japanese lived inside this one complex. Hardly any died during the bombardments, hidden deep within a human ant mound.[1588]

However, pre-landing bombardment had damaged some Japanese communication lines, making it difficult for them to organize their defense. And the Marines had a secret communications advantage: The Navajo code talkers. The program was Philip Johnston's brainchild, an engineer who grew up the son of a missionary on a Navajo reservation. He realized the Navajo language had a complex grammar structure, was not yet a written language, and was spoken only in the American Southwest, making it ideal as a code for radio communications the Japanese could not crack. It was also much faster, with a three-line message capable of being coded in Navajo, sent, and decoded in 20 seconds in contrast to the 30 minutes necessary for machine translation. Marines recruited Navajos for the program, starting them in boot camp in May 1942 (the movie *Windtalkers* (2002) immortalized these men). The 5th MarDiv had six Navajo code talkers working around the clock transmitting over 800 messages in the first two days, like Sam Holiday, who was on our trip in 2015, did for the 4th MarDiv. Holiday's grandfather had been a medicine-man and taught him to wear the traditional Navajo pouch with "four scared stones and yellow corn pollen" wrapped in a leather binding around his neck to provided protection and calm nerves. On Iwo, Holiday became a medicine-man for the Marines, providing secret ways to ensure life and defeat enemies using his native tongue and cool-head. The presence of these Navajo code talkers and their activities were kept secret to avoid tipping off the enemy who might capture some.[1589] These code talkers so confused the enemy listening in on their traffic that Japanese cryptologists never broke the code. And these brave Native-American Leathernecks proved invaluable to communications and operations.

Possibly one effective but outlawed weapon the U.S. could have used before invasion was poison gas. Gas would have traveled into the trench network killing the Japanese. Because it is heavier than air, it may persist unseen for days in low places and be deadly to any soldier from either side who might walk through it. General of the Army George C. Marshall explored the use of gas on Iwo. After consideration, Fleet Admiral Nimitz felt the U.S. "should not be the first to violate the Geneva Convention," despite the fact that, ironically, an attempt was made in 1925 to outlaw gas via a Geneva Protocol, but neither the U.S. nor Japan signed it. Ultimately gas was not used because Roosevelt's policy prohibited it: Earlier in the war, he issued two declarations to Japan and Germany stating the U.S. would only use gas in retaliation for an enemy who first used it. If gas had been used on Iwo, it would have started a chain reaction and gas would have been used in other battles for the war's remainder. No one knew how long the war would last, but all parties remembered the terror of WWI gas warfare, and most likely all parties to WWII desired to avoid the repetition of utilizing it. Years later, Nimitz lamented it was a shame the U.S. did not use it on Iwo because it would have saved lives. The presidential declaration just mentioned was an executive decision, not a treaty provision, settling the matter by prohibiting the U.S. from using gas first.[1590]

No one knows how many Japanese would have been killed or incapacitated if gas had been used at Iwo. If a total of 91,000 were killed by gas in WWI (half of whom were Russian and poorly equipped, not having masks) where it was employed and where millions fought, the likelihood of gas killing or incapacitating a majority of Iwo's defenders was low if it was not sustained for several days.[1591] Gas would have killed some especially if it was deployed using the element of surprise (i.e. a night attack), but the Japanese were prepared for gas and had trained for it (especially since many had participated in gas attacks against China), while the Americans lacked experience fighting with gas. Woody's battalion noted, "all enemy dead had gas masks and small amounts of decontaminating powder, indicating that gas discipline of the enemy was good."[1592] Thus even if gas had been used, Kuribayashi had prepared his men to weather such an attack. Had gas been deployed, the Marines still would have had to conduct some hard-fought engagements to win against the survivors if they attacked soon thereafter. Considering the inefficiency of gas in killing during WWI and the Japanese preparations, the majority of the Japanese garrison would likely have survived a short gas attack; as a tactical matter, gas is difficult to manage on the battlefield given considerations of weather, wind, terrain, and enemy protective measures. Yet, if the U.S. was willing to resort to using "widely disparaged weapons" like gas, they could have "saturate[d] the entire island" for days which could have had devastating results causing the Japanese to live in their masks underground where the gas would have remained trapped—that could have truly been deadly. Yet, it was not an issue because the US "wanted to preserve a semblance of not being the first to resort" to such despised and "illegal" weapons.[1593]

On the other hand, had the Japanese been able to deploy chemical weapons against Marines, they might have used them as they had against the Chinese.[1594] In fact Woody's boot camp instructors knew Japan had used gas in China and felt the Marines would face it, so they trained for it.[1595] Woody's regiment remained vigilant for chemical warfare.[1596] Japan might have used gas had Kuribayashi been equipped with it, because Japan had tried to use chemical weapons at Saipan. Thousands had died in China succumbing to Japan's use of chemical warfare, especially a virulent strand of bubonic plague. Under Hirohito's orders, Japan used poison gas and released among Chinese civilians fleas and rats infected with the "Black Death."[1597] In March 1939, Hirohito sanctioned one attack that used 15,000 canisters of gas "in the largest chemical attack of the war" against Chinese forces.[1598] Japan was the only nation, outside of Italy, to use such weapons in WWII and the League of Nations "adopted a resolution condemning the Japanese use of" it on 14 May 1938.[1599] The Japanese did not use chemical agents at Saipan because a U.S. submarine sank the ship carrying the men and their supplies before it reached the island.[1600] For the defense of their home islands, Japan had prepared to use gas and bacteriological warfare so it stands to reason that Kuribayashi simply did not use gas at Iwo because he did not have it and/or because he feared it would have caused too much fratricide due to the island's small size.[1601]

One can speculate how Nimitz and Roosevelt would have reacted at Iwo had Americans already been hit by chemical warfare, but the Japanese probably would have used it only if they could have guaranteed surprise and prevented retaliation.[1602] The Japanese were undeterred by legal, ethical and moral ramifications of using such weapons whereas the Americans often struggled with these issues evidenced by the refusal to use gas and debates about using atomic bombs.[1603] As *Atlantic Monthly* correspondent Edgar Jones wrote in July 1945, "The Japanese will do anything to kill their enemy." Therefore, for moral, tactical and legal reasons, the U.S. did not use gas on Iwo, thus forcing Marines to fight over hostile terrain against a murderous garrison that was almost fully intact. Marines, already overburdened with personal equipment and weapons, could watch with a small measure of relief as their gas masks were collected or abandoned later on.

America's twin allies were disease and starvation. Dan King noted the 22,000-man garrison had been depleted somewhat, not by bombings, but by illness: "The number could have been less than a few hundred or a thousand." Even if this were true, 20,000 to 21,000 men was still a larger force than Marine leadership expected.[1604] If left unchecked, dysentery can wreak havoc on troops, especially in confined spaces, and quarters on Iwo were crammed, particularly inside the tunnel system. Sanitation was a problem when squadrons of flies could travel between latrines and mess halls, spreading disease. Tunnel-dwelling soldiers suffered from dysentery, died and were buried in the earthen works besides the latrines to mitigate fetid smells in the cave dwellings. Although disease killed some troops, there were still plenty to fight the Marines. "On the whole, a well-equipped, superbly trained and disciplined garrison awaited the Marines on February 19....The entire island was heavily contested [with] main and secondary lines of fortified positions."[1605]

Arriving on Iwo, Marines faced scaling steep berms of volcanic sand, sometimes 20 feet high and at what seemed like 45 degrees. Climbing the berms was like running up a hill of coffee grounds. There was no traction and boots sank to the shins. If one tried to dig a fox hole, immediately after taking a shovel full of volcanic soil out, it filled in. The Marines in the initial wave at first did not receive any fire. The island looked strangely abandoned. A Marine commented, "There is something screwy."[1606] Most had the "sensation you have when you think someone is looking at you, and you turn around, and you are right."[1607] Thousands of hidden eyes watched the Marines, and the Marines knew it, but could not see where they were. The Japanese were coiled to strike. Marines landed on Iwo's southeastern shore where Kuribayashi predicted.[1608] After Marines massed on the beach, Kuribayashi unleashed his men. Hidden artillery, pillboxes and snipers delivered a storm of metal upon the crowded troops.[1609] Kuribayashi brought his enemy into his Ju-Jitsu hold and threw him down, hard. The violent force and utter surprise of the amount of fire shocked hardened veterans. A Marine captain said, "The honeymoon is over"[1610] while another, radioman John Lauriello said, "It was raining death from the sky." From a nearby Japanese ship-wreck that the Americans had destroyed in an earlier raid, a sniper took out Lauriello's Navajo Code Talker, Paul Kinlacheeney, with a perfectly placed head-shot.[1611] As the Marines hugged the ground, the earth heaved and shook as hundreds of heavy shells came falling down. It was as if Iwo had come to life under the surface, taking in deep breaths.

In war, according to von Clausewitz, *surprise* is extremely difficult to obtain writing, "The principle [of surprise] is highly attractive in theory, but in practice it is often held up by the friction of the whole machine....Its success is often due to favorable circumstances beyond the control of the commander, and it is frequently at the mercy of chance."[1612] Well, with Kuribayashi, the surprises were not left up to "the mercy of chance," and he achieved all of his "shocks" by careful planning and training at a level few commanders have ever reached. Kuribayashi planned for the Marines to arrive *en masse*, to push them back against the sea, and to exterminate as many as possible.[1613] The Marines did exactly what he wanted them to do for his plans to work that first day. His counter to the landings with the "machine of his men" resulted in death with "more than customary bitterness" for the Leathernecks.[1614] Kuribayashi's strategy required many victims, and that first day, exceeded its quota.

This dead Marine took a sniper bullet in the head which can clearly be seen here. He was part of the 4th MarDiv advance off the beach when he was killed on 19 February 1945. National Archives, College Park. Photo Credit Sergeant Bob Cooke.

The island's unforgiving landscape continually pushed Marines back to the ocean with its miniature, black volcanic cinders on steep inclines. Sometimes with a new step a Marine took, the berms brought him back several feet, forcing him to move his legs faster and stand up, which exposed him to the hellfire. The island's terrain gave attacking Marines no cover. Each man carried not only his own weight, but also a weapon, ammunition, and supplies that could weigh between 40-70 pounds.[1615] The extra weight compounded the difficulty of scaling the ever-shifting natural volcanic walls on the beach. A few rookies followed their leader, a veteran named Ernest Lunsford, trying to get off the beach. "We had no idea if this was a bad battle or not. One of the guys yelled, 'Hey, Lunsford, is this a bad battle?' Lunsford shouted back, 'It's a fucking slaughter.' Maybe two minutes later—*Whoom!*—we got hit with a mortar. I ducked and something dropped on my back and rolled off. It felt like a coconut or something. I looked down and saw that it was Lunsford's head. Those were his last words: 'A fucking slaughter.'"[1616] Nearby, 2nd Lieutenant Benjamin Roselle, Jr., had his left foot almost blown off by a mortar barrage. As comrades bandaged his leg, another mortar flew in and landed right on them killing two of his friends and spraying his good leg with shrapnel. As he and one other Marine hugged the ground next to their comrades' lifeless bodies, another round came in and exploded. *Boom!* Shrapnel tore into Roselle's shoulder and blew his comrade's right leg off. Roselle remained in his position as his buddy "silently crawled down the terrace, the stump dragging behind him."[1617] Alone, Roselle started thinking of his parents and of home when a curtain of shells started moving up from the beach to the terrace in which he was hunkered down in. *Kabloom!* Suddenly, Roselle was lifted into the air and slammed back to the ground. He looked at his watch as his arm suddenly received numerous pieces of shrapnel ripping through his wrist and tearing his timepiece off. In the place of where his wristwatch had been, there was a gaping red hole oozing blood. "This is what it feels like to be crucified," he thought to himself.[1618] Kuribayashi had achieved incredible surprise against the Marines and it was confusing them and lowering their morale.[1619] This Japanese commander proved von Clausewitz right: "Only the commander who imposes his will can take the enemy by surprise; and in order to impose his will, he must act correctly."[1620] At this stage of the game, Kuribayashi was acting "correctly" and maintaining the upper hand against the Marines, taking the fight to them instead of the other way around.

By the end of the first day, 548 Marines had been killed, 18 were missing (due to probably being blown to smithereens) and 1,775 were wounded, but with only modest progress.[1621] Lieutenant Cyril P. Zurlinden, a veteran combat correspondent who had seen action at Tarawa, Saipan and Tinian said the ghastliness of Iwo's carnage was far greater than anything he had seen. "Nothing any of us had ever known," he said, "could compare with the utter anguish, frustration, and constant inner battle to maintain some semblance of sanity."[1622] Lieutenant General Smith, who had confronted many Japanese opponents said of Kuribayashi, "I don't know who he is, but the Jap general running the show is one smart bastard."[1623]

As the skies buzzed with aircraft and shells, a fighter plane crashed into the ocean, killing Marines in a landing craft.[1624] Another plane got hit over the ocean and the pilot jumped out of the burning aircraft, but his parachute failed to open; another plane unloaded its payload on its targets, got hammered by antiaircraft fire, flew out into the ocean and just when one thought it had made it to safety, the plane exploded into a ball of fire.[1625] The sky whined and hissed with planes seeking targets.

Well-placed artillery rounds hit whole boatloads of men which disappeared into the deep waters right off the beach. The drop-off into the ocean was steep and the water was a dark, Pacific blue—if a LVT or LCVP got hit, it disappeared below the waves, often taking to a watery grave most of the heavily-loaded Marines aboard.[1626]

Wounded men on stretchers on the beach were shot to pieces or blown up into nothing while waiting evacuation. One Marine, lying next to a dead buddy asked a chaplain to pray for his friend. After the Chaplain finished, the Marine, writhing in pain, asked, "Now, Sir, please say one for me." Marines assigned to unload explosives received enemy fire, causing the explosives to detonate fusing them altogether into a "fireball." Along the beach, bodies floated in the surf.[1627]

Stepping on a landmine when landing on the beach, Sergeant Charles C. Anderson Jr. lost both legs and an arm. He was evacuated to a transport. "In one of the war's tragic ironies," his father was the ship's captain. "I'm feeling pretty good, but I wonder how Mother will take all this," he said in a soft voice as he died in his father's arms.[1628]

The beach became so clogged with the dead that bulldozers and Sherman tanks had no choice but to roll over them in order to get off the beach and engage the enemy.[1629] Tank commander Bill Dobbins said the bone-crunching was one of the hardest things he ever did in combat although he knew they were dead.[1630] The Japanese knocked out many tanks soon after they entered battle. The Japanese used smoke grenades to blind the tanks and surrounding Marines and then attacked them with prepared charges, killing themselves as they ran into the tanks holding the charges like footballs.[1631] In a field full of dead Marines, Sherman driver and Corporal Leighton Willhite almost got taken out by such a crazed Japanese suicide runner. As this enemy neared his tank carrying a mine, its 30mm machinegun opened up and hit the *Kamikaze* sprinter 25-feet away exploding the mine and blowing it and the "Nip" into a rainbow of debris that covered the tank. Later, Willhite and his buddies scrapped off flesh, bone and the skull of their attacker from the tank's outer skin. Willhite was lucky—several other tanks were not able to ascertain an attacker was close until he nailed their tank blowing one section of the steal beast into bent metal, fire and shrapnel wounding or killing some inside and others nearby who were using the tank for cover and as support.[1632] Smith noted, "Our tanks suffered heavily."[1633] For these reasons, and many others, casualties accumulated at such a pace Marines lacked enough stretchers to carry the wounded off the battlefield. They resorted to using blankets and ponchos to drag the dead and wounded back to the beach carving tracks through the soft sand as if they were pulling men through powdered snow. Many Corpsmen, who usually would go into battle to aide those who needed it on the front lines, just remained on the beach due to the heavy stream of wounded coming or being brought back to them. On seeing a Marine, blinded by a blast and having lost both hands, stumble back to where he thought the beach was, a Corpsman got up and ran to him. Others witnessed this poor Marine waving his bloody stumps in the air while tripping over the uneven ground. Zigzagging, the Corpsman braved enemy fire and got this Marine "to safety." One can only image what was said between the men: "Doc, I'm hurt pretty bad, right?" the Marine probably asked. "Hey Buddy, don't worry. We'll get you taken care of now," the Doc most likely answered putting the Leatherneck's arm around his shoulder and getting him away from open ground, and from the heat of battle.[1634]

Robert Sherrod was a veteran correspondent who landed at Iwo on D-Day. As he surveyed the beach the next morning, the sight was awful: "Whether the dead were Japs or Americans, they had died [violently]…Legs and arms lay 50 feet away from anybody…I saw a string of guts 15 feet long" far from the nearest dead.[1635] One Marine, suffering a horrible gut evisceration, "repeatedly, but unsuccessfully, tried to fold his own entrails back into his abdomen. Dazed, desperate, confused, the Marine slipped on his own innards and fell down in the process."[1636] Sherrod said on Iwo, there was "more hell in there than I've seen in the rest of the war put together."[1637] "It reminded one battalion medical officer of a Bellevue [hospital in New York City] dissecting room. Often the only way to distinguish between Japanese and Marine dead was by the legs; Marines wore canvas leggings and Nips khaki puttees. Otherwise identification was…impossible."[1638] For the Marines, even veterans, their pounding hearts filled with fear and awe. Shells continued to tear up earth all around them, whizzing in all directions. The air was full of the sharp cracks of bullets as they rushed forward into the lines of men like an angry nest of wasps, knocking many to the earth with a powerful thud as bone and flesh met the blanket of the earth. The thunder of guns and the rattling of machines made everyone's ears ring. Even the bravest Marine must have been distracted by such a landscape of sensory overload. Many found themselves strangely isolated in this hellish environment "with no recourse save their inner strength to keep them going and their common sense to tell them what to do."[1639] And the plain and simple action they needed to do was to get off the beach.

Examples of Sherman tanks in action on Iwo Jima. 27 Feb. 1945. National Archives, College Park. Photo Credit Corporal Robbins

By the end of D-Day, 30,000 Marines had been ferried to the island.[1640] The Marines had committed six infantry regiments, six artillery battalions and all or part of 12 support battalions. Three Seabee battalions landed on D-Day, transforming the beaches into organized departure points from which the Marines could move from the beach.[1641] Each tank battalion (circa 56 tanks) had a company of flamethrower tanks (circa 14 tanks) which would be quite effective where the terrain permitted their use, but they were spread sparsely across an entire division.[1642] Six hundred aircraft flew support missions, dropping 137 tons of fragmentation bombs and more than 100 napalm bombs.[1643] Marines were well short of their planned objectives for the day, notwithstanding the great amount of pre-invasion bombardment and aerial bombing as well as artillery firing from the beach and the heroics of infantrymen. The 4th MarDiv was bogged down near the taxiway of Airfield No. 1, which they had been tasked to take that day. Assigned objectives were not being seized in the allotted time although its men, like Gunny Keith A. Renstrom, fought "like the devil" to gain ground. "Friction," military philosopher Carl von Clausewitz wrote, causes "[c]ountless minor incidents—the kind you can never really foresee—combine to lower the general level of performance, so that one always falls far short of the intended goal."[1644] "Friction" caused many to fall short of their objectives. Another name for the phenomenon is "the fog of war," which explains why "nothing goes according to the plan" and why in war the simplest task becomes difficult. An old saying is that "the plan goes out the window when the first shot is fired."[1645]

One of the Marines who landed on Iwo Jima's Red Beach Two that first day and experienced this "friction" was Gunnery Sergeant John Basilone, a hero of the Battle for Henderson Field on Guadalcanal. During the battle for this island in 1942, he commanded two sections of machineguns that fought valiantly for two days, repelling Japanese assaults until only Basilone and two other Marines were left standing. He fought, moving through hostile ground to resupply his machine-gunners. After three days without sleep or food, his rifle ammunition exhausted, he fought the enemy with his pistol. Japanese forces in his areas were virtually annihilated—Basilone was credited with killing 38 "Japs." President Roosevelt approved a Medal of Honor for him for his heroic actions.[1646]

The Marines returned Basilone to America on a war bond tour to raise much-needed money for the war effort in 1942, and he could have stayed safely away from combat for the rest of the war. Instead, he refused and at his request was sent back overseas to rejoin the fight as a machinegun section leader with the 5th MarDiv. When enemy fire pinned down his unit on Iwo, Basilone out-flanked a fortified blockhouse, and single-handedly destroyed it with grenades and demolitions. Afterwards, he continued on toward Airfield No. 1, helping a tank trapped in a minefield while huge projectiles landed around Basilone sending immense plums of debris, earth and bodies into the air.

Here is an example of a Japanese Spigot Mortar. These could weigh almost 700 pounds and packed a powerful punch when they landed and exploded. 11 March 1945. National Archives, College Park. Photo Credit Lindsley.

One of the enemy weapons that Basilone probably was experiencing right then had not made it into the U.S. commanders' plans and they did not know how to deal with them; namely, massive mortars. The Japanese employed a 675-pound mortar (320mm) called a spigot mortar, nicknamed the "screaming Jesus" or "screaming MiMi" because of its high-pitched sound. Others described the shells as "giant ash cans as they floated through the sky." These titans of mortars exploded with such force they sent parts of equipment, machines and men 100 feet into the air, flipping the debris in every direction.[1647] At the airfield's edge, near a Japanese defensive line, Basilone may have been killed by one of those mortars; it seemed he was blown up by some projectile although his official casualty record states he died from gunshot wounds. On his left arm, "one could almost see" his tattoo which read "Death Before Dishonor." During a burial detail, Corporal Leighton Willhite had the honor of burying John. Before they placed him in the ground, a Corpsman said, "Hey Marine, do you want to see your hero?" He then unzipped the body-bag Basilone was in and Willhite saw Basilone's peaceful face, "but his whole body cavity was blown out and his entire middle section of vital organs was outside his body. I had to turn away my eyes it was so horrible to look at." This was very different than "gunshot wounds." Basilone was posthumously awarded the Navy Cross (the second highest decoration) by Secretary of the Navy James Forrestal for his heroism on Iwo Jima.[1648] Thus he has the Nation's two highest awards for combat bravery. He also earned a Purple Heart for his death in combat.

Marines encountered at this stage a rain of death of dreadfully destructive spigots, artillery rounds, and mortars of different sizes. Eugene Sledge described this warfare:

> To be under a barrage of prolonged shelling simply magnified all the terrible physical and emotional effects of one shell. To me, artillery was an invention of hell. The onrushing whistle and scream of the big steel package of destruction was the pinnacle of violent fury and the embodiment of pent-up evil. It was the essence of violence and of man's inhumanity to man. I developed a passionate hatred for shells. To be killed by a bullet seemed so clean and surgical. But shells would not only tear and rip the body, they tortured one's mind almost beyond the brink of sanity. After each shell I was wrung out, limp and exhausted.[1649]

One of Erskine's battalion commanders (2nd Battalion, 9th Marines), Lieutenant Colonel Robert E. Cushman Jr., echoed Sledge when he said:

> I hated high explosives. I never had much fear of small arms, but high explosives—its effects are so terrible and I'd seen so much of the effects—it scared the hell out of me when that stuff went off nearby and when it was incoming. On Iwo, they had [those spigots.]…You never knew where the hell it was going to come down. So, I had my own personal fear of that kind of stuff.[1650]

One Marine described these bombardments as, "Big fat shells gliding…down to splatter guts and brains and volcanic ash into a ghastly mulligan stew of death."[1651] In an environment where there were many things to be afraid of, it seems that artillery shells and mortar rounds presented Marines with their most feared, hated and indefensible weapon driving many to the brink of insanity and a few actually over its precipice—there is a reason they call it shell-shock and not bullet-shock, or flamethrower-shock, or tank-shock.

The living crossed a landscape on Iwo Jima of not only sand and natural walls, but also one full of their comrades' body parts and enfilade from all directions.[1652] The Japanese had them in a perfect cross-section of fire, hitting them from the heights of Mt. Suribachi to the south, from the well-concealed pillboxes and trenches in the middle and from the cliffs and high terrain to the north. It was a living nightmare. And then night fell.

"There is something definitely terrifying about the first night on the beach," Lieutenant J.G. Lucas admitted. "No matter what superiority you may boast in men and materiel, on that first night you're the underdog and the enemy is in a position to make you pay through the nose."[1653] That first night, the Marines dug in, expecting a *Banzai*. However, it never came. Kuribayashi performed something more useful and wise than *Banzais*: He "ordered harassing mortar and artillery fire." Sherrod described it as a "nightmare in Hell."[1654] The Japanese also sent specially trained raiders to infiltrate Marines' lines. A landing barge carried 39 raiders who were spotted and killed. One Japanese attack set off a massive explosion in an ammunition dump, destroying critically needed 81mm mortar shells. Enemy artillery hits included the 1st Battalion and 23rd Regiment's CPs which killed the battalion commander, regimental operations officer and others. A fuel dump was blown up at 0400.[1655]

Only 18 hours into the battle, 2,312 Marines had been killed or wounded. During a briefing at the White House on the casualty rate, Roosevelt was stunned. FDR met regularly with his military leaders and had for years received daily briefings from around the world. "It was the first time in the war, through good news and bad," author Jim Bishop explained, "that anyone had seen the President gasp in horror."[1656] "This was worse than anything the Americans had suffered anywhere in World War II… There was no doubt that Marines were in the bloodiest battle since Gettysburg [where 50,000 casualties occurred]."[1657]

The "red harvest of combat" is never easy for a U.S. Commander-in-Chief to swallow.[1658] Iwo was indeed bloody, but it was not yet as bloody as the attack on Oahu on 7 December 1941, which included Pearl Harbor. Almost 600 men died on the first day on Iwo, but 2,403 Americans died in one day during the attack on Oahu. Nonetheless, 7 December was only a day; the battle for Iwo looked like it had no end in sight and would cost more lives than the Oahu attack ever did.

Marines did what they had been trained to do: Move forward, close with the enemy and kill him. This was easier said than done because the Japanese remained hidden and/or well protected in tunnels and fortifications, frustrating Marines who wanted to fight them in open combat. When Lieutenant James P. Mariedas moved forward and jumped into a crater, he was surprised to find a defender left behind by his troops. They started to fight hand-to-hand and Mariedas wrestled the "Jap's bayonet" away, turned it against him, and amidst wild grunting and screaming, stabbed him repeatedly and killed him. "That's one less we'll have to take care of," he yelled back to a comrade.[1659] But there were many more to "take care of," more than what commanding generals had anticipated.

As with Saipan, the intelligence experts had underestimated the Iwo forces. They were off by 45%, pegging the enemy garrison at only 12,000-13,000 soldiers. They extrapolated this figure based upon the amount of rain water the island received and the number of cisterns there; they had no way of knowing Kuribayashi had severely rationed water for his men, who suffered from continued thirst and were allowed one canteen per day of polluted water.[1660] "In military operations enemy opposition often surpasses all expectations," and Kuribayashi did not disappoint.[1661] The incorrect estimate meant Marines would be taking more casualties than expected, with good men like Roselle, Lunsford, Anderson and Basilone suffering wounds or dying everywhere. Moreover, intelligence had little if any knowledge of Kuribayashi's maze of death. This is evidenced by how some U.S. commanders viewed the operation "like a pushover," predicting Marines would secure Iwo in a few days.[1662] Sherman tank driver Leighton R. Willhite remembered his commander telling him the island had only 2,000 enemy and control of the entire island would take at the most a week to accomplish since the "Japs had to be weak and undersupplied."[1663] Major General Schmidt's private estimate was less aggressive, being ten days, but it too was wide of the mark.[1664] Woody's commander, Donald Beck, was told there were 15,000 Japanese on the island

and that it would take the Marines six days to seize.[1665] On Turner's flagship, the *USS Eldorado*, Brigadier General William W. Rogers, as Major General Schmidt's Chief-of-Staff, held a press conference a few days before battle in which he predicted the Japanese would do what they always had done: Defend the beaches and follow up by a nighttime counterattack. He estimated the battle would last five days.[1666] Expectations were simply wrong—the navy and Marines were planning on a massive assault that would roll over a typical Japanese defense. Victory was assured, it was just the cost and time that were not.[1667] What was clear afterwards was that most if not all of the American commanders at Iwo had underestimated in several respects: The length of time and the size of the force necessary to defeat the enemy. And most importantly, they were surprised by the innovations in the tactics and strategy of the general they faced.

The erroneous estimates about Kuribayashi and his garrison are troubling. Certainly, on almost every island the Marines had taken so far, the Japanese had practiced the "doctrine of defense at the water's edge" and most had *Banzai*s. However, that was only half true for Saipan and Peleliu.[1668] Kuribayashi fully removed his forces to the center of and underneath the island, bordering on something novel. He truly knew "the knowledge basic to the art of war is empirical."[1669] So the U.S., in some respects, practiced the faulty strategy of fighting the last war, or in this case, the "last battle." They planned for much that did not happen. For example, on 6 December 1944, Japanese paratroopers attacked an Allied airfield at Leyte and inflicted a lot of damage.[1670] The 4th MarDiv's air officer cited this event in pre-invasion plans, noting it caused considerable "loss of planes and personnel at a critical phase of the operation." He then noted that since the main reason for taking Iwo was to seize the airfields, it was "highly probable" the Japanese might do the same once Americans seized the island. Thus, the Marines needed to develop a SOP (Standard Operating Procedure) to fight an airborne attack, and he laid out a detailed plan.[1671] It is indeed wise to practice the adage "Better to prepare for a situation and not have it then to have a situation and not be prepared." However, it seems that in all the records of Iwo planning for the 4th and 3d MarDivs examined, nothing was mentioned about what should be done if there were far more defenders than expected (a mistake made at other islands) and if all of them fought from underground (like many had at Peleliu). This intelligence failure here cost many lives. In this respect, not using gas, probably the most effective weapon against "tunnel rats", was the wrong decision if saving as many as possible and securing a quick victory were the goals. As a result, the island would have to be taken with assaults that penetrated the surface.

The Americans received enemy fire, but had a difficult time locating their adversary, who pulled back into camouflaged tunnels, holes, caves and pillboxes after firing. Few Japanese were ever seen and no POWs were taken that first day.[1672] On Iwo, Woody found the Japanese were excellent at concealment and camouflage. "They were much better at this than us," Woody explained. "They would hide in a buried drum, pop out, shoot at us and then return to their hideout without us ever seeing him."[1673] The Japanese not only fired from drums, but also from caves, hidden pillboxes and hilltops. Mines were also a constant worry. The engineers wrote, "the enemy made much more extensive use of mines on Iwo Jima than in any other operation in which this [3d MarDiv] has participated."[1674] Engineers could not use their metal detectors because the loose sandy soil retained enough metallic characteristics to render equipment useless, so men had to poke around carefully with sticks and bayonets to locate these mines slowing the attack.[1675] One report noted that also "the prevalence of steel shell and bomb fragments" made the magnetic mine detector (SCR 625) useless—after all, the U.S. had bombed the island with hundreds of tons of ordinance.[1676] Moreover, engineers found on the island a new weapon: Terra cotta mines made of vitrified clay, which eluded metal detectors because of their non-metallic exterior.[1677] So for many reasons, locating mines was tedious and dangerous, and Marines failed to locate and detonate many. From the area in which Woody would eventually land up to the first airfield (a distance of a few hundred yards), engineers would clear 1,000 mines.[1678]

In addition to the terra cotta mines, the minefields on Iwo concealed another nasty surprise: The Japanese buried several armed torpedo warheads within the large minefield between the landing beaches and the southern end of Airfield No. 2. Such warheads, normally designed to sink battleships and aircraft carriers, were used to create large explosions, vaporizing the advancing troops for a radii of several hundred feet and leaving large craters. Several amtracs, weighing 36,400 pounds, were blown into the air and flipped over by several such mines, often killing the entire crew.[1679] Iwo was becoming the worst combat zone in a war filled with horrors.

Just offshore, Woody's division waited in reserve, not knowing how the onshore assault progressed. They thought they would be held in reserve for a few days and then return to Guam. Suddenly on the night of 19 February, an announcement blared over the loudspeakers: "Now hear this! Now hear this!" They were told they were going to battle. Major General Schmidt decided the 3d MarDiv would land the next day.[1680] This changed everything. Reserve Marines aboard the troop ships gathered their gear, cleaned and rechecked their weapons, wrote "final" letters home and prayed. Woody did not pray because he "didn't know God at this time." He believed in himself and his Marines and that gave him strength.[1681] Most surmised if their division was being committed after one day of battle, then things had gone very poorly for the others.

Woody boarded his landing craft at 0500 on the 20th after the traditional last meal before invasion of eating steak and eggs at 0330 (Woody joked and said he was relieved they did not feed them "beans").[1682] Their gunnery sergeant gave them a warning once they boarded their boat, "So listen Bastards, if any of you need to use the head or puke, then you do it off the side of the ship or in the bilge pump in the back. If any of you don't, I will shoot you myself."[1683] It was nerve-wracking, descending the rope ladders in the dark in pitching seas. Woody said, "My scariest moment was getting off the ship by rope ladder. There was no light and we were carrying a backpack that weighed 60 or 70 pounds, and we had to go down that rope ladder 40 or 50 feet. I knew if I fell into the water it would be the end of me. I couldn't swim [very well]."[1684] Woody had a fear of the water since he almost drowned as a child. In fact, if his head went under water, he had the hardest time controlling himself. He could swim on the surface, but he hated having his head underneath. As a result, Woody held on for dear life when he went down the ropes for a boat he could not see because of the dark, but could definitely hear as its tiny hull "smashed" against the "mother ship with resounding thuds" as heavy seas rocked the vessels.[1685]

The landing craft formed up in circles with the line of departure far offshore and waited for the order to hit the beaches, an order that never came. There was no Dramamine to prevent seasickness. Engine fumes were heavy. For hours, they bobbed up and down, awaiting word from the CPs on the beach that the landing area was cleared and ready for the reserve regiments. Troops on the beach had barely advanced at all, and wounded littered the coastline, unable to be evacuated. Debris of destroyed tanks, landing craft, discarded supplies and dead were strewn across the landing zone. Congestion on the shoreline prevented further landings, so Woody's boat continued to rock up and down in the high seas, circling around and around. When someone had to urinate, the group had to rotate inside the Higgins boat for him to get to the bilge pump in the stern's corner. Some got so sick, they sank down and squatted. Sounds of battle continued to echo across the waves and the men waited for the call to hit shore. After circling all day in heavy seas with twelve foot swells and pouring rain, they returned to their ships, unloaded onto the long, single-filed metal ladders and climbed back aboard, being informed they would land the next morning.[1686] A few lost their footing on the ladders and fell into the ocean. Since it was dark, it was difficult to pull them out, and they slipped beneath the waves, pulled down by their gear.[1687] No one in Woody's boat had any problems except for a few bruises and cuts leaving the boat for the ship.[1688] Due to the rough seas, a few ships' davits hoisted boats out of the water. Luckily, in breaking this rule against raising loaded boats by davits, none of the ropes broke due to the weight. Skippers felt they saved men's lives taking this risky action by not losing any more overboard.[1689] Lefty was so tired, he went up to the forecastle, lay down and fell asleep. Woody did likewise; he was deathly seasick and ill from someone regurgitating on him. He found a place on the ship, curled up and went to sleep. "I can't even remember taking off my pack. I don't remember anything until they woke us up the next morning."[1690]

Here is an example of how cluttered the Iwo Jima beach became on the 19th and 20th of February. This was one reason why Woody's Higgins boat could not land on 20 February as originally planned. National Archives, College Park.

Throughout the rest of the night, the continuing thud of artillery rounds and the staccato snaps of small arms fire echoed everywhere. The scene along the horizon was like a macabre Fourth of July firework display. Each flash was either a white or green flare lighting up the landscape to reveal infiltrating Japanese or it was a yellow and white flash of flame of exploding mortars, artillery shells and grenades. It was a panorama of flashes of gold, red and orange illuminating a landscape of body parts, blood and guts.

Correspondent and Associated Press journalist, Sergeant Dick Dashiell, assigned to Woody's 21st regiment, said that during the day, from his ship, they saw Marines "with flamethrowers and satchel charges crawling up the terraces and hills to destroy pillboxes and spider holes…We could see there wasn't a lot of progress being made because of the sand. Also, we were taking a lot of wounded aboard our ship so we knew it wasn't going well."[1691] The wounded told stories of disbelief, shock and anger. The fight was more difficult than anyone had imagined.

# Ch. 21: Woody's Landing on Iwo

"There are only two kinds of Marines on
[Iwo]—the lucky and unlucky."
—3d MarDiv Marine[1692]


"Each day [on Iwo] we learned a new way to die."
—Iwo veteran[1693]


ON THE EARLY MORNING OF 21 February, Woody and his fellow Marines again ate their traditional steak and eggs. "I think we were the only Marines to get two straight meals of steaks and eggs before hitting a beach in the Pacific. Even so, it didn't make us feel *special*."[1694] Woody and his buddies boarded their crafts and by 0930 left for Yellow 1 and Yellow 2 sectors in the landing beach's middle. The seas were rough, and the rope ladders used to bring the men into their crafts slackened as the boats rode over the crest of the waves and then became taut as they dropped to the trough. Gunnery Sergeant Albert Daniel Hemphill and Gunny Durell B. Burkhalter descended the landing net in order to hold it steady in the boat for Woody's platoon. As they climbed down, a large wave "—bigger than any of the others—went swirling by and the boat pitched downward." The rope ladder tightened and sling-shotted the two NCO's into the "angry sea." "There was nothing we could do," claimed Woody. "Our equipment was strapped to our bodies and we could hardly move."[1695] Hemphill was shot over the Higgins boat into the sea while Burkhalter was thrown down between the ship and the craft. Luckily for both, navy personnel fished them out.[1696]

The two gunnery sergeants were both "so shaken by their ordeal that they were unable to make the landing."[1697] They came ashore a few days later and their men did not let them forget their mishap by "busting their chops." After the morning's excitement loading the men on the landing crafts, things settled down and the boats headed for the island. The gunnery sergeants had been lucky—men going down the ropes sometimes lost their footing and skipped between the ship and the craft just as the craft crashed up against the ship's hull, crushing the Marine like a wanton boy would a fly on his arm.[1698] One man had this almost happen, but

Gunnery Sergeant Durell B. Burkhalter. He was one of Woody's superiors. He would die on Iwo taking a head shot on 1 March 1945. St. Louis Personnel Records Center.

luckily his buddy rescued him in the nick of time. When PFC Patrick J. Milligan slipped from the rope ladder with "75 pounds of combat gear," he fell between the "attack transport and the landing craft. Heavy swells tossed him about and banged him against the vessel." As he struggled to grab on to something, the weight of his gear pulled him under. Sergeant Huron J. Lucas saw his friend dying, threw off his pack and jumped into the water. Grabbing a shoulder strap, Lucas hauled Milligan to the side of the ship and wrapped his legs around Milligan's body as he held on to the cargo nets. Marines in a nearby Higgins boat pulled Milligan to safety.[1699] Milligan would survive to fight another day.

Woody's heart pounded in his chest as he looked at his landing craft and lowered himself on the rope ladder. He felt his pack straps dig into his shoulders as he took each descending step into the craft that would take him to the hellish battle. His boondockers were firmly tied, helmet was fastened tightly around his chin, canteens were full of water and pockets were heavy with ammunition.

The navy started landing the 3d MarDiv's 21st Regiment at 1230 and finished by 1600.[1700] The sky had been full of U.S. planes bombing targets and defending against 32 *Kamikazes*. Nonetheless, a few got through and damaged the aircraft carrier *Saratoga*, the escort carrier *Lunga Point* and the anti-torpedo net tender *Keokuk*. One plane attack sank the escort carrier *Bismarck Sea* by the time Woody made landfall. Within a few minutes after taking the *Kamikaze* hit, a tremendous explosion blew out *Bismarck Sea's* stern and she rolled over and sank, taking 218 crewmen to a saltwater grave. Marines around Woody found it disconcerting to see *Kamikazes* dropping from the sky plunging themselves into their targets and transforming ships into giant explosions of smoke and fire.[1701] A plane also hit *LST 477* off Kinami Iwo, which carried equipment and tanks to support Woody's division. Three men were killed, ten men were wounded and tank equipment was damaged.[1702] Maybe the island was safer after all? *LSTs* (Landing Ship, Tanks) were 382-foot-long ships manned by a crew of 98 that could carry 220 troops and numerous vehicles. They could bring supplies and men right to the beach and unload them through their bow doors.

Many *Kamikazes* were dive- and torpedo-bombers with three- or two-man crews. Although these *Kamikaze* planes needed only the pilot, the crews refused to leave their group because they had trained, lived and fought together on the one hand and because they had to die in combat in order to fulfil their religious destiny on the other. They all went to the *Katori Junja* Shinto shrine to pray for their success, "which involved their deaths," then boarded their planes and went on their "straight to heaven" missions.[1703] Pilots told one another they would see each other again at the Yasukuni shrine.[1704] Family members came and watched their sons depart on their final missions. One father prayed holding his Buddhist rosary as his son lifted from the airfield.[1705] Later, some of these suicide planes were rocket-powered. These unique craft, dropped from an airplane, were called "*Baka* bombs" with "*baka*" meaning foolish or stupid in Japanese. The devotion of the young, poorly-trained pilots to their Emperor and country knew no bounds.[1706]

Woody's approach to the island on 21 February was uneventful accept for the sounds of war becoming louder as he got closer to the shore. On the way, thoughts of home and war often swirled in the men's heads.[1707] Many wondered what they would see once they hit the beach. Others worried they would let their buddies down. They understood they had to weed out the enemy in order to win and go home. Most hidden guns that had sprayed the landing craft the first days of battle had been located and silenced by naval gunfire, allowing the craft to arrive largely unopposed. As they moved forward from the landing zone however, they rammed into Kuribayashi's well-designed defense.

They had to flush the enemy from trenches, holes, caves and bunkers so the Marines could confront them. It was laborious since there were thousands of cave entrances. To locate the enemy, they deployed dogs to enter caves and sniff out the Japanese.[1708] One scene was laughable and frustrating at the same time when a fox terrier came up to a group of Marines; the dog was carrying a grenade, "rolling it around" and "flinging it into air." The dog's playfulness sent the group of Marines in all directions. Undeterred, the terrier picked up the grenade and jumped in a foxhole with several Marines in it, forcing the Marines to spring out of the hole and find other cover. Thinking his masters were playing a game with him, the dog retrieved his grenade and "trotted after them, refusing orders to drop it." After much yelling and pointing of fingers, the dog dropped the grenade. As a Marine reached for it to take it away from the dog, he would quickly and playfully snatch it away from his human caretaker and run away toward other Marines, scattering them. The Leathernecks "tempted the dog with food but he would not abandon his new toy." Eventually, the only technique was to ignore the dog and eventually he abandoned his "toy"-grenade.[1709] Someone quickly then grabbed it and many had discussions with one another about future dog training. The men laughed but also shook their heads.

There were other ironic scenes during the landings. Some Marines landing at Iwo had the legend "Rodent Exterminator" stenciled on their helmets' cloth covers.[1710] Marines partook in black humor with a common send-off: "Every Japanese has been told it is his duty to die for the emperor. It is your duty to see that he does so."[1711]

When the 3d MarDiv's men disembarked on Iwo, the scene on the beach was appalling: Sunken boats, destroyed jeeps, landing craft and dead bodies littered the shore. Woody found chaos reigning everywhere, and bodies were stacked like cordwood on the beach. As soon as Lefty set foot on the sand, he saw an arm lying on the ground to his left and a torso to his right; later, he described the place as "a real butcher shop." "The air had a sweet smell of blood in it," Lefty said. "It's hard to explain, but you can smell blood [most likely due to the metallic scent attributed to iron-rich red blood cells]."[1712] One journalist said it was the first time he had seen dirt muddy with blood.[1713] As PFC Charles G.

Fischer's Higgins boat hit the beach nearby Woody and Lefty, a Japanese soldier had somehow stealthily snuck up close enough that as the front ramp lowered and Fischer was about to disembark, a grenade suddenly appeared at the open bow. Seeing the danger, Fischer yelled for two of his comrades to jump away which they did. The explosion missed them, and other Marines nearby must have found the enemy and immediately killed him.[1714] One Marine complained, "[What] the hell did the first ten waves do?" A "smart-ass" neighbor answered, "Die, you dumb bastard [that's what they did for us]."[1715]

The landscape gave testament to what happened to their comrades the past few days. The 21st Regiment had entered the horror of war and were told to walk inland through the battle's flotsam until they reached their front lines. At its zenith, a "Marine had fallen every fifty seconds."[1716] It reminded one of what Dante, Milton or Joyce have described as perdition. James Joyce writes:

> Imagine some foul and putrid corpse that has lain rotting and decomposing in the grave, a jelly-like mass of liquid corruption. Imagine such a corpse a prey to flames, devoured by the fire of burning brimstone… And then imagine this sickening stench, multiplied a millionfold and a millionfold again…[1717]

Joyce did not know it, but he was describing Iwo's battlefield. Woody and fellow Marines had passed through the gates of a foreign world few ever experience of unbridled violence with no quarter asked for and no quarter given. Medal of Honor recipient Douglas Jacobson said, "I was no virgin to battle. But this one wasn't in the books."[1718]

The smell of death made many sick. The fate of many could have been summed up by Herman Melville: "All men live enveloped in whale-lines. All are born with halters round their necks; but it is only when caught in the swift, sudden turn of death, that mortals realize the silent, subtle, ever present perils of life." Many, unknowing at this time, were "dead men on leave."[1719] Often such sights make men question their beliefs about a merciful God. Eugene Sledge wrote:

> I shuddered and choked. A wild desperate feeling of anger, frustration, and pity gripped me. It was an emotion that always would torture my mind when I saw men trapped and was unable to do

anything but watch as they were hit. My own plight forgotten
momentarily, I felt sickened to the depths of my soul. I asked
God, 'Why, why, why?' I turned my face away and wished that
I were imagining it all. I had tasted the bitterest essence of war,
the sight of helpless comrades being slaughtered, and it filled me
with disgust.[1720]

Some veterans calmed themselves, strangely so, with the thought they were "already dead." Feelings of confusion reigned in their minds and the difficult process of trying to understand the chaos created a sensory overload that put most on edge. Woody's commanding general, Graves Erskine, understated as the 21st Marines were committed to battle, that they were "put in a very hot sector and had a pretty rough time."[1721] A few hours after hitting the beach, they received heavy fire from a cave by Airfield No. 1 in the center of the line. It pinned down units and they had a difficult time finding a way to silence the Japanese squads in that cave. In this region, caves and pillboxes intermixed themselves along the terrain of small ravines and gullies, making it difficult to find the enemy hidden in the earth. Demolition man PFC William C. Nolte attached to Able Company, 1st Battalion, strapped on a flamethrower, navigated around the heavy fire and worked his way to the cave's mouth. Firing up his torch, he moved his flame into the cave and held down the trigger for 10-seconds of fire that engulfed the opening and traveled down the cavernous halls. Later, Marines counted "thirty charred Jap bodies" in the cave.[1722] Another flamethrower nearby, a former lumberjack, would yell out "Timber" everytime he flamed an enemy position.[1723]

The first few days, 21st Marines were attached to the 4th MarDiv, commanded by Major General Clifton Cates, and helped this division secure its gains and support its badly mauled units. It remained under Cates until Major General Erskine arrived three days later and officially took it back over on 25 February.[1724]

During the night of the 21st and the day of the 22nd, dark clouds moved over the island, unleashing a rainstorm.[1725] Now the powdered volcanic ash became an abrasive slush. It infested everything; ears, socks, mortar tubes and rifle barrels. Besides fighting the Japanese, the Marines fought the elements. Keeping weapons clean to fire is one of the priorities in battle and this kept the Marines busy. Nights were filled with counterattacks and defense against infiltrating Japanese, and the landscape was full of mortar, artillery fire and flares. Ships' guns and mortars fired illumination rounds to light the battlefield. Marines had barely slept and lived on C-Rations (individual canned, pre-cooked and prepared rations used when fresh food was unavailable), water and adrenaline.[1726] Combat fatigue settled in and morale fell.[1727] Shadows, more numerous than the enemy, kept them alert.

Iwo Jima had three airfields. By 21 February, the third day after the invasion and the day Woody and the 21st Regiment landed, the Marines had still not secured Motoyama Airfield No. 1 (the southernmost airfield closest to the landing beaches), despite countless attempts. By night of the third day, 800 Marines had been killed. North of Airfield No. 1 lay a major line of defense across the island, a complex network of tunnels, pillboxes, and partially buried tanks housing artillery, mortars, light machineguns and rifles.[1728] Around it lay a graveyard of destroyed Zero fighter-planes and Betty bombers housing numerous snipers.[1729] Although the Marines used tanks and tried to overtake the area, they failed. The 24th Marines' 2nd Battalion, where Woody fought, wrote: "Progress has been extremely slow today due to a number of pillboxes that held us up. [Some] were knocked out but a number…remain. Couldn't use tanks due to terrain."[1730] Due to the ash and sandy conditions, the Marines' tanks often bellied out in the silt as their tracks sank into the soft surface. When maneuvering on solid, flat terrain, the steel monsters were lethal, but Iwo's terrain caused difficulty for the Shermans navigating the ashy beaches. The spider web of pillboxes and interlocking fire from defensive positions cut down advancing Marines and armor attacks. The 3d MarDiv noted after battle that 49 Sherman tanks were deployed and 15 damaged "beyond repair."[1731] Consequently, Marines repeatedly had to resort to other means to penetrate the pillboxes.

While Marines tried to secure the first airfield, the Japanese continued to attack. A battalion next to Woody wrote that it feared another counterattack as they already were fighting 800 Japanese on their left flank and dealing with a "counter landing" of 200 enemy to their rear.[1732] On 22 February, on the shore Woody had landed the day before (Yellow Beach 1), beachmaster navy Captain Carl E. "Squeaky" Anderson worked with Marines unloading supplies. That evening, the Japanese attacked with mortars. Chaplain Gittelsohn observed Anderson yelling over a loudspeaker at his men to dig holes and get cover. After the mortars stopped, Anderson got back to work. "Much to his surprise and disgust," Anderson could not get his Marines to leave their holes and to unload supplies. Yelling, cursing and even threatening to report them for insubordination, he could not get the scared men to "emerge from their foxholes." Suddenly an idea came to him. Sneering into his loudspeaker, he said, "And you guys call yourselves Marines!" Immediately, "every last man came out" and resumed his duties.[1733] Out of all the beaches Anderson had commanded "in the Aleutians, the Gilberts, the Marshalls and the Marianas," Iwo was "the toughest beach I have ever seen."[1734] While this unloading was underway, Woody's unit moved into places where the original attack force had been pulverized.

The shelling was so intense and the carnage so dramatic that Lefty Lee could not take it any longer. He "broke down." Many tried to motivate him, but he just shut down. Lefty felt like his "heart was going to push out of my chest." He went to the Corpsman who took him out of the line. He tagged Lefty and sent him to the rear. "I don't remember if I was escorted or walked back alone," said Lefty, "because I was so out of it."[1735] His nervous system had taken all it could and he was now "as helpless as a child."[1736] Woody said, "Some…were so shaken, they couldn't walk back to the first aid stations without help."[1737] The next day, Lefty was sent to ship APA-33 *USS Bayfield* where he was treated for "combat fatigue" and "shell-shock." All Lefty remembered of several days was being sedated. The division registered these men as "non-effectives." Some with such a diagnosis later returned to their units, but Lefty's case was so severe, he would never come back. Not only were men's bodies being ripped apart, but also their minds.[1738] On 5 April 1945, the Marines informed Lee's parents he was "suffering from combat fatigue," but receiving medical treatment.[1739] During WWII "potentially every man had a breaking point when exposed to the rigors of combat," and just a few hours on Iwo was enough for Lefty to reach that point.[1740] Others joined him and by February 26, the Marines logged 657 cases.[1741] More would succumb to such mental breakdowns during the coming weeks, and thousands more would follow in Lefty's footsteps before the island was secured..[1742]

Soon after the killing started, news of this horrific battle appeared in newspapers and radio broadcasts back home. Correspondent Dick Dashiell's father, Professor John Frederick Dashiell, the founder of the University of North Carolina's Department of Psychology, wrote his daughter Dorothy (Dick's sister) about this conflict on 22 February:

Do you know the news, I wonder? This morning the radio announcers were telling that the [3d MarDiv] <u>has</u> joined the Fourth and the Fifth in that awful battle for…Iwo Jima. And that battle is the bloodiest in the long whole history of the Marine Corps since 1775. So I am just awfully anxious about Dick. Better write him at once, write anything long or short, anything that a tired and battle-worn and maybe wounded man would get some comfort from. This is it! This is the great ordeal for which all his training has been shaped. This morning I hung out the big flag that Mother had given Father. I hung it at the side porch, you know, and it has been swinging in the south breeze. But as I hung it, I wasn't thinking much of Washington and his birthday (that serves simply as a pretext for the display). I was thinking of a modern man who has always been just as public-spirited and just as patriotic as Jefferson or Paine or Franklin. For we've had great Americans all through the century and a half of our national history. We've got them today. And, by golly, we've got them in the [3d MarDiv]![1743]

Although the battle had only gone on for a few days, people in the States knew there was something special about it. Tens of thousands of mothers and fathers like Dick's felt "awfully anxious" about their boys going into harm's way. If about 80,000 Marines cycled on and off Iwo, hundreds of thousands of parents, grandparents, siblings and others worried about their fighting men. Professor Dashiell looked at what America was doing at Iwo with her projection of power to defend her freedom and this created a sense that sending off their boys to fight was very American and worth the angst. Professor Dashiell knew that since higher-ups felt the need to deploy his son's division, it must mean the battle was worse than imagined. He was confident since the 3rd MarDiv was made up of men like his son, modern men full of patriotic fervor, that the U.S. would prevail.

In the meantime, 4th MarDiv had been badly mauled and needed reinforcements from the 3rd MarDiv. Destroyed jeeps, broken weapons, dead bodies and discarded gear were strewn along the lines.[1744] Major General Schmidt sent the 21st Regiment, the freshest troops, to relieve the 23rd Marines of the 4th MarDiv.

He placed them in the center of the northeastern front, "in position to crack the main line of resistance at its only vulnerable point...[T]he men of the [21st] spent...days trying to reach and cross the... airfield [No. 2]. The strips were heavily mined and every inch was swept by a crossing fire from a jumble of rocks and ridge lines and from atop the plateau. Antitank projectiles were so thick that tanks were unable to lumber across—only riflemen, assisted by almost point-blank support fire from the tanks' 75-millimeter guns, could turn the trick."[1745] But getting tanks into some positions was impossible.

After Woody and fellow 21st Marines entered the battle near the first air-field, they bogged down in a sector where both Americans and Japanese fought intensely but without any change of the lines. It was akin to two boxers in a cor-ner of the ring exchanging powerful body blows but without any movement of their feet. The regimental report noted the Marines moved "extremely slowly due to numerous pillboxes and emplacements."[1746] Casualties mounted with alarming speed. *Time* and *Life* correspondent Robert Sherrod wrote that the dead, both Marines and enemy, had one thing in common: "They all died with the greatest possible violence. Nowhere in the Pacific War had I seen such badly mangled bodies. Many were cut squarely in half."[1747] They would move forward, but were knocked back. Before Lefty was removed from combat, he observed: "Japs were madder than the devil in this area and they were throwing everything at us."[1748]

After two days' fighting in this area, most officers of Woody's Company, Charlie Company of the 1st Battalion, 21st Marines, were either killed or wounded. After three days of fighting, there were only three of 13 company commanders left in the 21st Regiment.[1749] Even battalion commanders were at risk. Soon after arriving on the island, 1st Battalion commander Lieutenant Colonel Marlowe C. Williams—Woody's original battalion commander—pushed the attack forward. On 22 February, he "continuously exposed himself to enemy fire in the vicinity of the front lines to better control his units and exploit their gains." Although he took a bullet to his torso, he "refused to leave the field for a period of two hours until he had assured himself that his executive officer was thoroughly acquainted with the situation and ready to assume command." For his heroics, Williams received the Bronze Star (fourth highest medal for valor). He was replaced by Major Clay M. Murray who did not last 24 hours. Crouched in a shell crater with Captain Beck, Murray rose up right as a Nambu machinegun sprayed his area, hitting him in the shoulder, neck, face and left hand with seven rounds. He was immediately removed from the island.[1750] The Marines struggled to locate their enemy. "It was now you hear your enemy…but you would never see him," Woody said when describing the difficulty of observing from where bullets hit his buddies and the ground. Locating the Japanese was especially difficult because they fired weapons using smokeless gun powder.[1751] Although enemy positons had been pounded, all that "beating" did not seem to have made a dent. This day, Woody's regiment suffered "many casualties while crossing the eastern part of the Airfield No. 1."[1752]

During the first days, the Marines in this regiment killed several enemy and saw them interspersed among dead Marines. At this stage, retrieving dead Marines was entirely too danger-

Major Clay M. Murray. He took over Woody's 1st Battalion when Lieutenant Colonel Marlowe C. Williams was wounded. Murray would not last 24 hours at the helm of the 1st Battalion before suffering wounds himself being shot seven times from one machinegun burst while trying to spot the enemy over a shell crater. St. Louis Personnel Records Center

ous not only due to ongoing combat in the area but also due to Japanese booby-trapping dead Marines, so they left them unburied along with fallen Japanese. Every morning during those first few days, however, when daylight appeared, they noticed the killed Japanese had been removed during the night. They did not understand how the Japanese retrieved their dead, especially since the units sent up flares throughout the night to prevent sneak attacks. "We were fighting, in some respects, a phantom army," Woody said. "You rarely saw them during the day unless you killed them or blew them out of their holes and trenches, but even so, by morning, the Japanese had removed all their casualties…and left the Marines there to rot. It was an uneasy feeling."[1753] Seeing only one's own dead on the battlefield gave the impression one was losing the battle. The Japanese trained to ensure this, knowing the psychological impact. Whether they mutilated the captured and dead Marines so that living Marines would find them later or removed their own dead to give the impression that only Americans suffered, Japanese burrowed into the minds of Americans and caused them anxiety. The Japanese soldiers' creativity in making war a living horror film, sowing fear, hatred, anxiety and disgust, was unparalleled. After a few weeks, however, the Japanese no longer retrieved their dead as the Marines moved progressively into their lines.

By the morning of 23 February, Woody's unit had not made progress throughout a cold rainy night.[1754] Correspondent Dashiell explained: "[The 21st Regiment] pushes towards the center half of the island. There is little concealment. Suddenly, three Nip machine guns chew noisily. A score of Marines crumple, a look of bewilderment on their faces. Others dive for the dirt, hitting the ground with their chests almost before their feet have left it."[1755] This sketch confirms a grim reality. Locating the enemy usually took a few paying the ultimate price to reveal a firing position.

Marine correspondents were everywhere on the island like Woody's comrade Sergeant Dick Dashiell. They risked their lives to get the stories of ordinary men to the public. Dashiell would write 83 reports when on Iwo Jima while engaged in combat. Although identified as someone else in the archives, it is felt the Marine in this picture with glasses is Dashiell. February 1945. National Archives, College Park. Photo Credit Gene Jones

Dashiell's reports are frequently used and his role on the island was important, since he represented over 100 men who served at the front as journalists. These men carried in one hand a rifle and in the other a Hermes typewriter. Dashiell also had a Colt .45.[1756] He and his men helped document the heroic acts of several men while in the deployed areas and shared in the combat, deprivation and hardships in order to inform the public. Historian Benis Frank wrote that the "Combat Correspondent program could not have worked in any other but an elite organization, such as the Marine Corps, which was magnanimous enough to allow this group of mavericks, individualists and off-beat characters to come in and write about its activities."[1757] Dashiell performed well and after Iwo, General Erskine wrote him a "Letter of Commendation":

> [This award is given to Dashiell] for meritorious service while serving as a…combat correspondent…from 19 February to 16 March 1945. Landing with the first elements of the division [on 21 February], Sergeant Dashiell, despite the fact that he was allowed complete freedom of movement to obtain material, and with complete disregard for his own safety, voluntarily joined those groups involved in the most dangerous fighting in order to gather first-hand material for his news stories. He was constantly exposed to enemy machine gun, mortar, and small arms fire. He voluntarily left positions after they became comparatively secure to observe individuals and units involved in heavier fighting. His courage and professional skill resulted in the production of outstanding news stories, which, widely circulated in the [U.S.], acquainted the American people more fully with the combat phases of the fierce battle of Iwo.[1758] (See Gallery 3, Photo 12)

Much of what has been learned of Woody's unit and experiences came from Dashiell's documentation. To get his stories, he went into the hell of combat: "I was with two other Marines at the top of a crag trying to sight an enemy machine gun position, when a number of enemy riflemen opened up on us and we were forced to slide down the hill on our backs and stomachs to get out of their range of fire."[1759] Since he was a Marine first and a journalist second, putting his life in danger to memorialize his comrades was a given, but the manner in which he did his job and the one thousand pages of articles he wrote were a godsend for historians and for patriots who spout the phrase of "never forgetting" those who sacrificed so much for freedom.

Out of the 366 reports Dashiell penned, 83 were done on Iwo Jima *during* the battle. He wrote these articles in "foxholes, abandoned (naturally) Jap dugouts and pillboxes, and even under the wings of wrecked planes." When completed, he dispatched them to a communications ship or other vessel that "was returning from the island that was available." Once on board, a censor and his command had to sign off on them for public consumption. Once approved, they were flown to Washington and distributed in various newspapers.[1760]

Dashiell was a unique man. He looked like a nerd, wearing glasses and embodying an air of deep thought. He

Frederick "Dick" Dashiell's football picture from the University of North Carolina. He would become a star half-back at North Carolina from 1934-1936. Author's Collection and Dashiell Family Archive

was not impressive physically as he stood 5'10" and weighed 170 pounds. Fellow 3rd MarDiv correspondent Bill Ross called him skinny, scholarly and soft-spoken.[1761] He came from a tradition of education and continued this trend in his family. However, he was not all about books. Behind the unassuming figure lurked impressive athleticism and physicality. He was a star University of North Carolina football halfback from 1934-36 and every year helped his team to winning seasons returning 40 punts for a total of 473 yards. As of 1993, he held the 4th best average for punt returns in UNC's history at 11.8 yards per return. He also held the record for the longest punt at 96 yards. He was called "Slippery" Dick and wore number 79.[1762]

Frederick "Dick" Dashiell catches a pass as the University of North Carolina halfback against the University of Michigan in 1936. He was an excellent punt returner returning 40 punts for a total of 473 yards from 1934-36. Author's Collection and Dashiell Family Archive

Dashiell also came from a strong military heritage. His great-great grand-father James Dashiell fought in the Colonial Wars in the mid-1700's. His great-grandfather John Dashiell served in the Revolutionary War (1776-83). His grandfather Reverend John Thomas Dashiell fought for the Union during the Civil War (1861-65). His other grandfather Milton Emory Jones also served for the Union.[1763] Dashiell's family's Civil War history is presented here to point out something interesting about America. Dashiell had grandfathers who fought against the grandfathers of his commanding officers; namely, Vandegrift, Erskine and Smith. Now, in just two generations, America had built an economic power-house, won the Spanish-American War and WWI and slogged through the Great Depression. Now America was sending these grandsons out shoulder to shoulder to fight against a common foe which posed one of the greatest existential threats to her democracy.

So great was his heritage, while Dashiell fought in the Pacific, so did the destroyer *USS Dashiell* (DD-659), named to honor his great-uncle Robert Brooke Dashiell, an 1881 graduate of the Naval Academy who later became an inventor of "important ordnance mechanisms and an authority on dock construction" and held the office of Assistant Naval Constructor in 1895. The *USS Dashiell* saw action at Tarawa, Saipan, Guam and Okinawa.[1764] With this military pedigree, Dick entered the Corps in August 1943 at 28 years of age and many considered him an "old man." He had tried to enter Candidates' Class (the officer training program), but failed the color blindness test, twice. When drafted, he "opted for the Marine Corps." When he was being processed, a clerk said he would never make it through boot camp because it was "too tough for a

Frederick Knowles "Dick" Dashiell. He was a Marine Corps correspondent attached to Woody's battalion. One of Woody's first interviews about his MOH was given to Dashiell. He helped document many heroic men in Woody's unit. Out of 366 articles he wrote, he apparently only had problems with Woody and how he testified about his MOH actions. St. Louis Personnel Records Center.

guy" his age. Before college, though, he had attended one year at Culver Military Academy in 1931 which gave him a basic understanding of military rituals and discipline. Moreover, he figured his football career enabled him to "handle the physical part, and had experienced the discipline of meeting deadlines and figured I could meet the mental stress which I knew awaited me. Parris Island, of course, was a shocker, but I had not expected a bed of roses." He graduated boot camp and became a correspondent.[1765]

Dashiell soon thereafter joined the 3rd MarDiv to replace the correspondent Sergeant Solomon Israel Blechman (son of Rabbi Nathan Blechman of Mamaroneck, New York), who died during the liberation of Guam on 21 July

1944 doing his best to cover all the units in the front lines to make sure he got the stories correct being hit by shrapnel and machinegun fire in his abdomen. He initially survived the wounds and was taken to the ship *USS DuPage* APA-41, but died soon thereafter and was buried at sea in Apra Harbor.[1766] He would receive the Bronze Star with V for his bravery receiving the rare endorsements for the lowest medal for valor from the likes of Commandant Vandegrift, Fleet Admiral Nimitz and Major General Harry Schmidt (the most highly endorsed Bronze Star this study has documented). It was indeed a "special award" for a correspondent.[1767] His citation read in part:

> Fearlessly making his way forward among the forces who were blasting the enemy's formidable defenses in an advance over the perilously rugged terrain, he repeatedly exposed himself to the devastating Japanese machine-gun, mortar and small-arms fire as he gathered authentic material for his news reports.[1768]

So Dashiell had huge shoes to fill. And on Iwo, he would prove that he could do so performing, like Blechman, some of the most hazardous work any journalist would have to do in order to "get the story right." In many respects, the Corps was lucky to have such talented writers amongst its ranks and this desire to document its heroics by brave men like Blechman and Dashiell is part and parcel of why Marines have become legendary.

When Dashiell wrote his reports on 23 February while his and Woody's 21st Regiment's advance to the North was stalled by defensive positions, the 5th MarDiv's 28th Marines fought to take the key Japanese strongpoint, Mt. Suribachi. This was a moment documented in arguably the most famous photographs of the war, when the American flag was raised by a group of Marines atop Suribachi. Associated Press photographer Joe Rosenthal's photograph of the flag-raising received the Pulitzer Prize.[1769]

The volcanic mountain had to be taken because its high ground dominated the island. From that vantage point, Japanese soldiers with artillery and other weapons could see for miles and fire down on the Marines, including those on the landing beaches. Taking the mountain was daunting, but it had to be done in order to secure the island, including the airfields further north. The enemy well understood the importance of Suribachi and fortified it extensively. "The Japanese were in several hundred emplacements, principally pillboxes, blockhouses, covered guns and grottoes around the base of the mountain, giving way to intricately constructed tiers of caves along the slopes...Tanks, flamethrowers, rockets and demolitions were used, and after three days of hard fighting, the mountain was encircled."[1770] During the beach phase, Marine carrier pilots provided close air support and found many targets. Some requests to bomb targets were carried out in less than 15 minutes and seldom did the Marines have to wait more than 30.[1771] The Marines laid down their brightly painted placards with arrows to indicate where planes needed to drop their ordnance.

The thrilling news that the Marines had seized and now held the high ground on Suribachi was sig-

Marine Corps correspondent Sergeant Solomon Israel Blechman. He died in battle on Guam on 21 July 1944 and earned the Bronze Star with V for his heroics. He was the son of the prominent Rabbi Nathan Blechman of Temple Israel, Mamaroneck, New York. Dashiell would take over Blechman's billet. St. Louis Personnel Records Center

naled by raising the flag atop the mountain. The Stars and Stripes could be seen by Marines and Japanese fighting below, by sailors in the fleet and by airmen flying missions.

Woody heard the ships' fog horns and men cheering echoing through the air that morning as Old Glory was raised over Japanese soil at 1020 on 23 February, the first time an enemy flag had unfurled on Japanese land in four millennia, marking a pivotal event in history.[1772] Woody witnessed men shooting their rifles into the air, and at first he did not know why and felt they had "gone mad." But then he saw the flag, and he too fired in the air "like an idiot." He said, "[T]he flag raising changed the whole attitude on the island. It lifted our spirits and we felt we had a strong momentum going into the next phase." "Once that flag went up, there was a lifting of the spirit that was just, well, you just can't describe it. We were going to win." Woody said it may have helped him build up his courage later to attack pillboxes.[1773] Secretary of the Navy James Forrestal was on his way to the island when the flag went up. He commented to Lieutenant General Smith: "Holland, the raising of that flag on Suribachi means a Marine Corps for the next 500 years."[1774] In a radio broadcast, Forrestal said it was the high point during a difficult week.[1775] Smith wrote the flag raising was a "vision of triumph" and "had an electrifying effect on all our forces…We were in the mood for victory and this glorious spectacle was the spark."[1776]

Attacking Hellcats fighter planes hit Iwo Jima. February 1945. National Archives, College Park.

Later, during an interview with Don Pryor broadcasted live to 30 million Americans and picked up by Columbia, Mutual and Blue networks, Vice Admiral Kelly Turner said on his flagship *USS Eldorado*: "[T]o me, the most impressive incident of the whole campaign so far was the raising of the flag on the top of Suribachi volcano Thursday morning…It affected me very deeply, and I'm sure that General Smith felt the same way about it." Lieutenant General Smith, who was being interviewed with Turner, chimed in:

I certainly did. [The flag raising] was the most outstanding fea-
ture of the operation so far—the assault and capture of that extinct volcano rising 556 feet, with sheer cliffs around the sides.
Its capture was carried out in the face of the most tremendous
difficulties, and I'm sure that only men imbued with the highest
spirit of the offensive, love of country and *esprit de corps* could
have taken [it].[1777]

The famous flag raising on Iwo Jima's Mt. Suribachi, 23 February 1945. This was actually the
second raising of a flag on Mt. Suribachi, one with a bigger flag taken by Joe Rosenthal. The
Associate Press

Turner continued: "I hope the American flag always flies there; that it's never
allowed to come down…And I think that Iwo Jima surely deserves a place—
probably a whole verse—in the Marines' battle song. Don't you?" Not missing
his cue, Smith answered, "Yes, I do." Turner then answered a question whether
this was the "toughest operation in the Pacific war" with, "Yes, in many ways…
The toughest part of the fighting in this campaign, however, is on shore."[1778]

Turner was not being dramatic. Indeed, some of the most brutal fighting in WWII was taking place. U.S. commanders knew this Japanese general was killing more Marines than any IJA commander to date. They were under no more illusions that the battle would be quick and easy. It would take much longer than a week or two. Turner put on a brave face and continued: "We knew…it would be difficult. But it has gone ahead just about as we expected." Smith supported Turner's statement saying, "Yes—it's been difficult in every way on shore, as we knew it would be." He then described the extensive defenses, but then, surprisingly, said, "We had anticipated just such a defense."[1779] Was this posturing for the media? Probably so, knowing they were talking to a journalist and the parents of these men fighting and suffering injuries or dying on Iwo. If it had gone according to plan, as these commanders claimed, then why had the reserve division already been committed to battle?

For men on the island, events were not going "as expected": the fog of war was at work and Marines were dying faster than anyone thought possible. For example, soon after the flag-raising, a Marine entered the CP of the 28th Marines, the regiment responsible for putting the Stars and Stripes on the mountain, and asked about his brother. With wandering eyes, a few told him his brother had been killed. The shocked Leatherneck sat on the ground "and held his head in his hands and cried for a while." After a few minutes, he pulled himself together, wiped his tears away, looked at the men and said, "It sure made us feel good to see that flag go up there." As he walked away and returned to his unit, everyone started to realize that casualties were becoming a real problem on the island. Woody witnessed on average 20 men disappear from his company per day, either from death or from being pulled from battle due to wounds and/or combat fatigue/shell-shock. As Smith and Turner were being interviewed, Woody's company was probably down to 22% of its original strength. In light of the heavy casualties, the flag raising was a focal point for Turner and Smith that gave the U.S. clear evidence America had turned the momentum of this battle. The flag was visible proof that control of the island was changing hands. Turner said, "we're on the offensive, and not the defensive. We plan to use Iwo Jima as you would a gun captured on the battlefield. We'll turn it against the enemy."[1780] Plans were in the works to turn the first airfield into an offensive weapon as a runway to start sending planes to attack Japan's mainland. What must be remembered is that notwithstanding what

Turner and Smith said and the flag-raisings, many additional days of difficult fighting faced Marines before Iwo would be conquered, and Woody was part of that "offensive drive" to accomplish Turner's and Smith's objectives.

Although the top of Mt. Suribachi had been conquered, there were still countless Japanese inside the mountain and further north who had to be hunted and killed or sealed in rocky tombs. Japanese veteran Tsuruji Akikusa points out, "The Marines may have raised their flag, but we still held most of the island."[1781] It took six more days to secure the mountain, but going forward it would be the Marines and naval gunfire officers on the top of the mountain observing the Japanese and directing fire at them.[1782] A key advantage of capturing the mountain was to stop Japanese firing from it onto the three divisions as they moved north.

Immediately after Marines seized Suribachi, they saw the devastating psychological impact on the Japanese. A suicide ritual all over the Pacific was for a defeated Japanese soldier to lie down, with his rifle muzzle in his mouth, and squeeze the trigger with his big toe or to hold a live grenade next to his head or heart and commit *jiketsu*. In one cave, Marines found 150 Japanese who had committed *jiketsu* because they had lost Suribachi.[1783] Before the Marines landed on Iwo, the Japanese flew a flag with the characters of *Namu Hachiman Daibosatsu* on it, meaning "We Believe in the Merciful God of War."[1784] Well, no god of war showed mercy to these Japanese warriors. Along the coast, the Marines fought northward, trying to outflank the Japanese in the center with the Marines killing more Japanese than Marines being lost, but progress was difficult.

Although the Marines were doing well, mistakes occurred elsewhere and it was a chaotic battlefield. Attacking aircraft destroyed a "friendly" tank in the area where Woody fought by dropping a napalm bomb on it.[1785] The 24th Marines' 2nd Battalion next to Woody documented that artillery was falling in the lines of the battalion's E Company, but the Marines could not determine whether it was friendly or not. Moreover, the same company was being attacked by its own tanks. The tank unit officer was informed and apparently his tanks pulled away, but this demonstrates Marines had not only a significant job in killing the Japanese, but also they had a huge responsibility to communicate and keep track of their neighboring units and weapon systems to know what was going on to minimize fratricide, something unavoidable in war, which has long been called organized chaos.[1786]

# Ch. 22: The Pillboxes

Then I heard the voice of the Lord saying, "Whom shall I send?
And who will go for us?" And I said, "Here am I; send me."
—Isaiah 6:8[1787] (See Gallery 3, Photo 13)

ON 23 FEBRUARY, AFTER A heavy artillery barrage softened up the enemy ended at 0835, three battalions including the 1st Battalion, 21st Marines (1/21), attacked on open ground near Airfield No. 1. The enemy had meticulously prepared defenses: "[T]he defending positions were so extensive that as soon as one center of resistance was penetrated, the attacker was forced to face another."[1788]

When their attack stopped, 1/21 had advanced all of 50 yards. The battalion report noted: "Jumped off in [attack]… but were unable to advance due to extremely heavy resistance."[1789] The commanding officer of 1/21 had been wounded. The gap left in the line by the uneven advance of different battalions forced the 26th Marines to relinquish all the hard-fought gains they had made the day before on 22 February. Under merciless Japanese fire, forward observers and battalion commanders called in fire support.

Some tanks tried to enter the battle, but had a difficult time maneuvering because of the ash. In short order, direct hits knocked out five that C Company's commander, Donald Beck, had requested to support his men's attack. "Marines lay grotesquely on what had become their biers" and the cry for Corpsman rang out over the battle's chaos. Woody's platoon leader, Lieutenant Howard F. Chambers, reported to his company commander, Captain Beck, that his unit had been reduced by 50%.[1790] While the men in the 21st Regiment discussed what to do, the Japanese rained down mortar, rocket and artillery fire on Woody's outfit "along the entire front causing many casualties."[1791] Woody's regiment of an original 3,006 had suffered already 427 casualties.[1792] Witnessing all this "violent death" put an icy "chill" through Woody's veins.[1793]

First Battalion, 21st Marines command post at No. 1 Airfield on Iwo. Circa 23 February 1945. National Archives, College Park. Photo Credit Corporal F.E. de Ome.

In all this pandemonium, Woody and his comrades were lucky to have Chambers as their platoon leader. He never asked his men to do anything he would not do himself. Chambers was born 9 September 1919 in Punxsutawney, Pennsylvania. After he graduated high school in 1938, he attended Thiel College where he lettered in football for three years and was the team's captain and known as "Hud." After taking an interception back for a touchdown while playing defensive tackle one game, he became a Big-Man-On-Campus for this and many other athletic accomplishments, "[becoming] one of Thiel's football immortals." He was a stocky, 5'10¼", 190-pound linebacker with blond hair, piercing blue eyes and good looks. He would have played football for four years, but during his senior year, the season was cancelled due to the war. Enjoying the Greek system, he led his fraternity, Lambda Chi Alpha, as president for three years. In his student yearbook, it was proudly noted that it was rare he had held this position since a sophomore writing, he "rivaled FDR for third term fame as prexy of the Lambda Chis." Showing promise as a leader long before the Corps, besides being the college's football captain and his fraternity's president, he also served on the Interfraternity Council for three years and was its president for one year; was vice-president on the Student Council his junior year; was the vice-president for the Student Union also during his junior year; was class president of his class his junior year; and was a Who's Who in American Colleges during his senior year. "Hud" enjoyed staying busy and taking control of groups and situations and proved he knew how to interact with diverse people and lead them. He received a bachelor's of science in economics graduating in May 1943. After graduation, he started his career in the Marines, rising to the top once again, finishing 58 out of 213 graduates in his Officer's Training Course class on 25 August 1943. He served in action on Guam in July 1944 and trained with Woody and his 1st Platoon for six months before Iwo.[1794] He knew his men.

The enemy had the advantage of protection in excellent defilade

Second Lieutenant Howard Frederick Chambers. He was in charge of 1st platoon, Charlie Company, 1st Battalion, 21st Marines. He was the officer over Woody on 23 February 1945 when he earned his MOH. Chambers refused to endorse Woody's MOH for reasons unknown. He would receive the Silver Star and Purple Heart for his actions on Iwo Jima. On the day Woody earned the MOH, he was part of Chambers' platoon that in total broke through 32 emplacements and moved the front 300 yards north closer to Kuribayashi's headquarters and to ultimate victory all under Chambers' leadership. Chambers' Silver Star citation read in part: "Frequently advancing beyond the front lines to supervise tank and demolition personnel, [Chambers] skillfully coordinated the attack of his unit." St. Louis Personnel Records Center.

or fortifications from which they fired at Chambers' pinned-down Marines. Correspondent Dashiell described it as being a network of camouflaged pillboxes housing numerous Nambu machineguns spraying the Marines with withering fire.[1795] Dashiell wrote: "Hellish Jap cross-fire chewed up Marines and pinned others to the ashy ground which Jap mortar shells churned with Leatherneck flesh."[1796] At this juncture, many Marines looked into Death's black eye socket.

Captain Donald Beck, commanding C Company (Woody's company) in 1/21, described the scene as such: "Large pillboxes, constructed from concrete, thickly cut clay blocks and pieces of scrap aircraft, covered over with four to five feet of sand, were located among sandy ridges where tanks could not maneuver and where supporting overhead fire had proven ineffective."[1797] Against this terrain, Beck had to move forward and attack. But it troubled Beck exactly how he was to do so. When discussing this situation, 1st platoon leader, 2nd Lieutenant Chambers told Beck, "Every time my men try to move forward they're hacked to pieces."[1798] And if machineguns were not stopping the Marines, then mortar shells did so: "A mortar shell whangs and explodes, catching two Leathernecks in the legs, and another in the face. The latter is a dead lieutenant."[1799] Flamethrower Joseph Anthony Rybakiewicz, attached to Beck's unit, described another officer's death: "We both were looking around a big rock trying to find where the enemy was. He looked. Then I looked. He looked again and was shot in the chest and immediately killed."[1800] The Marines knew this type of activity was dangerous, but they could not fight an enemy they could not locate. Combat photographer and Dashiell's foxhole buddy, Staff Sergeant Joe Franklin, peeked around some rocks to see where the enemy was, but found none. A sergeant nearby "must have figured everything was clear" so he crept away from his cover and stuck his head right by Franklin's. Wham. He took a bullet that passed along his temple region and "laid both eyes out on his cheeks," but he was still alive with the bullet missing his brain. He rolled over on his back "with those blobs of jelly on his face" yelling in pain, "I can't see."[1801] In war, one must never forget that if you can see the enemy, the enemy can see you.

Even in this murderous environment, the day before, Beck's company had taken out 25 pillboxes "after it had walked into a trap of crisscrossing fire. A less hardy band probably would have been destroyed, but under the leadership of… Beck, the Marines fought frantically and not only escaped the trap, but exterminated more men than they lost."[1802]

During the morning, Woody's Company confronted some Japanese with grenades. "It was sort of like a baseball game," said 2nd Lieutenant Charles (Mike)

Henning, "only we weren't trying to catch what the Japs threw." Henning was nicknamed "Ox" because he stood 5'11" and weighed 215 pounds, all muscle. In several hours of combat, led by Woody and Sergeant Lyman Southwell who were described by "Ox" as "crack shots" with grenades, they took out 50 Japanese. Henning said, "I don't know how many Nips they got, but it was plenty." It was deadly combat though, and Woody's company lost an additional 18 men. Southwell explained, "There was only one thing to do and that was to fight. I guess you know we did."[1803]

Seeing the situation for what it was, Captain Beck gathered his officers to discuss options. Beck had been with his Marines on Bougainville and Guam. He had distinguished himself on Guam under the punishing *Banzai* and earned a Purple Heart and Silver Star—"he was no combat neophyte."[1804] Beck's

Second Lieutenant Charles Wesley Henning was one of Captain Donald Beck's platoon leaders who Woody served under. On Iwo, Henning led an attack and defeated a Japanese position capturing a 47mm cannon, six 25mm dual-purpose weapons and killing many Japanese soldiers. He would receive a Silver Star for his actions although supposedly Beck put him in for a MOH. St. Louis Personnel Records Center

commander, Major Clay M. Murray, described Beck as "the bravest man he had ever known." Another superior, Lieutenant Colonel Ronald Van Stockum, said Beck had the coolest nerves under fire he had ever seen. Whereas some cracked up when a grenade went off near them, Beck could go through the hell of battle and continue to make calm, rational decisions.[1805] According to von Clausewitz, when the clash of arms begins, the "preeminence of the moral forces of war" rules the day often coming in the form of a commander's need "for resolution, self-confidence and *coup d'oeil*," and Beck exhibited all of these powerful characteristics.[1806] With the cool calculation of a scientist, Beck gathered the facts about his situation. The area "was clearly one of the most bitterly defended and best fortified sectors of the entire despicable island."[1807] Chambers and numerous men tried for hours that morning to break the Japanese lines, but failed, suffering casualties. He felt uneasy asking "anybody to do the job [to attack positions in their front]. It was murder."[1808] Yet, Chambers and Beck knew doing nothing was not an option. Gazing, peering, squinting and scowling at the lines they knew they needed to assault, they devised a plan. Iwo veteran Al Ragliano described the dilemma:

> Giving orders which may lead to the death of a comrade is… difficult. Officers in high command did it, to be sure, but it's vastly different to send thousands of men into a battle, far easier in fact on one man in command than to say, 'Steve, you and Hulitzsky move out and stop that damned machine gun.' The personal touch was an unfair, yet necessary cruelty. Could it be this executioner-like responsibility that urged, indeed demanded, my growing up?[1809]

Chambers and Beck knew some of their men would die carrying out their orders, making their "blood and pulses shudder."[1810]

They debated their options while crowded in a shell crater to protect themselves from grazing fire. So many officers had become casualties or were dead that only two other officers of C Company were there. By mistake, First Sergeant William R. "Hose Nose" Elder ordered Woody to attend the meeting since he thought "non-commissioned officers" like Woody were supposed to be there too, especially because he was a demolition and flamethrower expert. The highly-valued new flamethrower M2-2 was called the perfect weapon for Iwo, and they were supplied 24 per battalion (81 per regiment).[1811] Out of the seven flamethrower operators assigned to the Special Weapons Unit attached to Charlie Company's headquarters, Woody's recollection is that the other six had been killed or injured in roughly two days of fighting.[1812] Knowing Woody was trained with the flamethrower, Woody recalls Beck asking him whether he thought he could use it to take out the pillboxes. Woody reports he was heard to say he would try.[1813] Woody later claimed, "I figured, what the hell, I might as well join [my comrades] wherever they were."[1814] People around him must have viewed him as a "dead man walking." (See Gallery 3, Photo 15)

Beck recalled it differently: Writing about this event after it happened, Beck radioed for another demolition sergeant to replace the one who had become a casualty in order to punch through the Japanese line. In fact there were still another four active demolition men attached to or nearby Woody's company, Marquys Kenneth Cookson, William Nolte, Joseph Anthony Rybakiewicz and Nathan Pitner.[1815] Woody, "overhearing [Beck's] conversation, quickly volunteered."[1816]

In yet another version, Woody told correspondent Dashiell in 1945 that his platoon commander Howard F. Chambers had asked him if he could take out the pillboxes and

First Sergeant William Reid "Hose Nose" Elder Jr. Elder was one of Woody's MOH witnesses. He would receive the Purple Heart for his wounds suffered on Iwo Jima on 2 March 1945. St. Louis Personnel Records Center.

Woody responded with a response that he would "see what I can do."[1817] While later memories conflicted, the central fact is that Woody volunteered to attack the pillboxes. "In war everything is uncertain, and calculations have to be made with variable quantities."[1818]

After the war, Beck stated that the only way to destroy this Japanese line of defense was to use flamethrowers and demolitions.[1819] Beck had almost died that day by a sniper's bullet that hit his pack ripping through its lining and the handle on his metal spoon. Had the bullet hit a few inches closer to Beck, he would have been killed or wounded.[1820] "Since danger is the common element in which everything moves in war, courage, the sense of one's own strength, is the principal factor that influences judgement. It is the lens, so to speak, through which impressions pass to the brain."[1821] Beck was in the front lines and always close to the action. His judgement was influenced by knowing firsthand what was going on. He had courage, and he needed it in order to make proper decisions about how to respond to what the enemy was doing to his men. "So courage is not simply a counterweight to danger, to be used for neutralizing its effects: it is a quality on its own."[1822]

Two days prior to this meeting, Charlie Company had attempted to advance without success and lost a lot of men. Woody said, "I wanted to eliminate those pillboxes because they had interlocking fields of fire, meaning if you tried to escape from one, another one would get you. I stayed focused on my job. I needed to concentrate on that if I wanted to live."[1823] One of Napoleon's maxims in battle was never do what the enemy wants you to do. The Japanese wanted Woody's comrades to stay entrenched. Woody and his Marines needed to change that and move forward. His captain estimated 50 pillboxes held up their advance.[1824] Over the next four hours, Woody secured a small victory against one of the toughest defenses the Marines ever faced. Beck wrote the Japanese on Iwo were well led and did not conduct "large scale" *Banzai*s which "attributed to the enemy being a powerful fighting force…until the very end."[1825]

Under Kuribayashi's leadership, Iwo was the second out of almost 80 island battles where the Japanese inflicted more casualties on the Americans than the Americans did on the Japanese.[1826] He "waged a campaign of unprecedented bloodshed and endurance."[1827] Throughout the Pacific theatre, for every American who died in battle, U.S. forces killed ten Japanese, "a more acceptable price for most Americans."[1828] At Iwo, that average was reduced to only three Japanese killed for every Marine killed. This "kill ratio" created controversy in the U.S.[1829]

Woody needed to improve upon this average on 23 February using

Sergeant Lyman D. Southwell. He was in Woody's platoon and was one of his witnesses for the MOH. He would earn a Silver Star for his bravery on Iwo. On the day Woody earned the MOH, Southwell had killed dozens of Japanese with grenades. St. Louis Personnel Records Center.

flamethrowers and demolitions to help destroy a defense that had frustrated whole battalions for days. Destroying the pillboxes would require intense fighting using some of the best weapons the Marines had—liquid fire from flamethrowers followed by explosives. And in this section the "Japs were everywhere," as Lyman D. Southwell, one Marine assigned to protect Woody, declared. "Fire came from sandpiles, ditches, tunnels, pillboxes—everywhere. I never thought I would get out alive."[1830] Southwell suffered a close call a few days later, almost being killed by a bullet through his helmet. He did not realize it until the next morning when he combed his hair and felt a welt on his scalp. Then he analyzed his helmet and almost passed out noticing a hole in his "steel pot."[1831]

After volunteering to attack the pillboxes, Woody went to the supply depot at the company headquarters and obtained a flamethrower and demolition charges ("he grabbed a twelve pound pole charge filled with 15 blocks of TNT").[1832] First Sergeant Elder, in charge of headquarters logistics, watched him gear up and start off on his first attack. He witnessed Woody come back for more flamethrowers over the next few hours.[1833] Knowing what he had to do, Woody was "nervous and scared. But I couldn't let my Marines down."[1834] In the general area, there was a group of Corpsmen and navy doctors working "silently over the torn and bleeding men" who had been evacuated from the lines Woody was marching toward. The stretcher bearers were "bringing back a stream of wounded."[1835] At one of these stations, Dashiell remembered a wounded Marine screaming at a doctor, "Don't take my leg off. Don't take my leg off" as medical personnel gazed at mangled muscle and bone that used to be a thigh and shin making their faces grow twisted as they heard his pleas.[1836]

To protect him, Woody was assigned two men armed with Browning Automatic Rifles (BAR men), two M1 riflemen and one demolition man (sometimes Woody says he picked the four and other times he says Beck assigned them to him). It seems there was also a Marine with a bazooka who, if he could, would try to blow the concrete structures with a clear shot which would enable Woody to move in to deliver the *coup de grâce*.[1837] In interviews Woody gave in 1956 and 1966, he mentioned that a bazooka man fired at the large pillbox he took out.[1838] Without these men, Woody alone had no chance of getting close enough to use his flamethrower against the pillboxes. Any chance of success required a coordinated team approach, executed with a high level of skill in dangerous and stressful conditions. Failure of skill or nerve on the part of any team member could cause the mission to fail. The BAR and riflemen would lay down suppressing fire against the pillboxes' slits and enemy soldiers on the flanks of these structures so Woody could get into position to use his flamethrower on the concrete monsters. After the pillbox was flamed, the demolition man (and sometimes this was Woody too), would use a wooden pole to place a charge in the pillbox. The charge was designed to kill any soldiers inside who were still alive, and render the structure useless. According to the action reports, Woody did both these acts at least twice. And using a pole charge "provides a bit of standoff, and also makes it very difficult/impossible for the defenders to toss the explosives back (as they might if it was just a satchel charge…)."[1839]

Kuribayashi's tunnel system was constructed so that if the soldiers were killed in a pillbox, bunker or cave, other soldiers could often later get inside and use it to kill Americans by traveling through interlocking, underground burrows. The Marines wanted to make sure that the Japanese could not use positions after the original inhabitants had been killed. As Woody notes, his special unit members were trained to either "burn it or blow it," or in this case, do both.[1840] This was an effective tactic to destroy the thousands of pillboxes and caves and Woody was one of hundreds of Marines destroying pillboxes.[1841] To destroy them in his sector, Woody's responsibility was to keep the flamethrowers and explosives of the Special Weapons Unit for Charlie Company prepped for fighting. When these weapons were not being used, Woody and fellow Special Weapons Unit comrades were riflemen and used as infantrymen. But when they were needed for demolitions work, they put down their rifles, strapped on a Colt .45 pistol and picked up their flamethrowers to bring "hellfire" to the battle and incinerate the enemy. While the battle raged, the *New York Times* ran an advertisement for a chemical company supporting flamethrower technology, showing a GI blasting away Japanese defenses with the heading, "Clearing Out a Rat's Nest."[1842]

After Beck and Chambers briefed Woody on his mission, Woody in turn explained to the men in his charge how they should provide cover while he worked his way into position to attack. While the enemy inside a pillbox could shoot at a wide range of targets from the protection of the pillbox, Marines only had a small aperture on each pillbox to target. The pillbox fortification gave the defenders a distinct advantage over the attacking Marines who fought in the open, exposing their entire bodies to enemy fire.

Woody wearing a M2-2 Flamethrower while in the Reserves. Circa 1949. Fairmont Army
Reserve Training Center Museum

Today, Woody has difficulty piecing together what he did in the fast-mov-
ing sequence of events. The evidence is unclear about exactly what he accomplished and precisely how he went about it.[1843] Looking back at his words in the
article he gave in 1966, one could say that he is "upon the brink of remembrance,
without being able, in the end, to remember."[1844]

On 23 February 1945, the beach look like this when Woody attacked the pillboxes just northeast of this photo. National Archives, College Park. Photo Credit Sergeant P. Seheer.

Woody, who had fought through many tough situations, felt more fore-boding when confronted with taking out the pillboxes than he had during other engagements. This stress alone may account for how someone has so few vivid memories of one of the most significant events in his life. One truism about what we now term post-traumatic stress disorder (PTSD), which Woody experienced, is that men who suffer from it often avoid anything that might trigger painful memories.[1845] Although Woody made it through the day without being physi-cally wounded, he still suffered from witnessing comrades being injured or killed and from personally incinerating soldiers. "There is nothing more horrible than the smell of burning human flesh," Woody said.[1846]

Many wonder how one could not recall such heroic acts afterward. The an-swer to this question is complicated, and probably related to brain chemistry as well as a distortion of past events since we tend to remember only the good we do.

First, as the stress hormones are released in the bloodstream, a human's ability to remember diminishes. Stress can even cause long-term damage in the brain, especially when hormones called "glucocorticoids," the most prominent being cortisol, are released. Cortisol impedes the ability of the hippocampus to encode and recall memoires. These stress hormones hinder the brain from encoding memories by diverting glucose away from the brain and delivering it to muscles. Historian Richard Frank explains:

> During that time, extreme physical and emotional stress pummeled participants in an endless series of kill-or-be-killed moments while they witnessed countless killed or maimed. They endured on mere shards of sleep. This combination shredded the brain's ability to retain and to organize memories—not to mention the brain's conscious or unconscious work to suppress terrifying…memories.[1847]

The body has that which ends up being a physiological mercy mechanism that reduces one's worst memories. Woody had been in situations that would make anyone afraid and emerged displaying strength. The assault on the pillboxes took him to a new frontier of bravery but it was highly stressful, making it difficult to recall every detail as it happened. Many had fallen while attempting to knock out fortifications like those blocking Charlie Company's advance. With no escape from the terror of having to go out and confront the enemy strong points would make the most experienced warrior fearful. Dante wrote: "One ought to be afraid of nothing other than things possessed of power to do us harm,"[1848] and the Japanese had plenty of power to harm Woody.

Second, if Woody had not accomplished as much as others reported he had, then he may have gone along with their story and only "remembered" what others said. Also, he may have exaggerated what he did, and later, when called upon to prove it, he may have developed techniques to divert people from suspecting anything might be amiss. To the contrary, this idea departs from his initial interview before the MOH was awarded when he reported his accomplishments *with confidence* to journalists on Guam in 1945.[1849]

Surprisingly, the one person who supposedly saw everything, Woody's platoon leader Lieutenant Chambers, never provided a written report about Woody's actions on 23 February; Captain Beck, in his report, cited Chambers as a source proving Woody had destroyed the pillboxes that Beck believed Woody had incinerated. But Chambers' refusal to endorse Beck's report shows he most likely did not agree with how Beck had remembered their conversation.[1850] Although Chambers was asked by a general several times to submit a report for Woody, Chambers refused to do so. As a result, there are many problems with documenting what Woody did.

Historians know the essence of what the Marines accomplished under harrowing conditions north of the first airfield, but the details of precisely how they did it and who performed the brave acts must be recreated from a variety of evidence by the historian's deductive reasoning. This includes reading statements of witnesses, after-action reports, news stories, contemporaneous investigations by reporters and the military, and old letters and notes written while events were still fresh in people's minds. Historians must examine the terrain where the events occurred along with analyzing fortifications and handling weapons that were used or the same type as those used. As historians reconstruct events from long ago, it is necessary to also look for corroboration from different sources to try to determine whether sources are reliable. If the person who is being studied is alive, historians must examine what he has said and question his motivation and behavior throughout his life. As a result, much of what is written here about these events leading to Woody's MOH is pieced together from evidence in such a way that the resulting account is the most probable description of what happened. The memories, albeit imperfect, bonded with contemporary and investigative sources help create a clearer picture of what happened to Woody, and what he did on one of the most important days in his life and in Marine history.

During the attack on 23 February 1945, events occurred quickly and Woody needed strength, skill and laser-like focus to accomplish his mission while several enemy tried hard to kill him. Some of what happened Woody recalled as transpiring in a dream-like state. After Beck's briefing, the short, 165-pound "Hillbilly" from West Virginia, Woody, lost no time starting his mission. He went to the depot and hoisted the flamethrower's 70-pound tanks onto his back and picked up the 15-pound wand or gun attached to the tanks. All in all, with his pistol, helmet and other gear, Woody entered battle with at least 100 pounds of equipment. If he carried the pole charge with explosives rather than handing it to someone else, then he carried another 12 pounds. This was the load carried by a young man fearful he would not amount to much due to his small size. Once engaged in the attack, he would have laid the pole charge down, burned the structure, then returned with the pole charge to blow it up.[1851]

Marine flamethrower operators or men carrying flamethrowers to the front for operators to use move toward the battle lines. Flamethrower operator Donald Graves is convinced these men in this photo were taking full flamethrowers to actual operators at the front waiting to go into battle with full tanks. In other words, these men shown here would take these flamethrowers to the front and unload them for actual demolition men to man and engage in combat. This allowed flamethrower operators to rest between engagements with the Japanese. National Archives, College Park.

After getting his gear together, a row of full tanks awaited Woody's return as he departed for battle (in earlier drafts for this work, he claimed Vernon Waters was there helping him, but the documents prove he was in a different company at this time).[1852] At that point, he had no idea how many flamethrowers he would have to use to accomplish his mission. He would have a better idea of how many tanks he would need after he took out the first pillbox and surveyed the enemy's lines. That meant working his way through enemy fire in broad daylight to get close enough to the fortification for him to fire his flamethrower at the small slit in the pillbox and neutralize it, allowing his battalion to advance. After dropping his empty weapon, he would have to run back to the supply depot to put on another 70-pound fully tanked M2-2 flamethrower, grab a new pole-charge and then run back to rejoin the battle. And that was if everything went right. Countless things could go wrong. Marine correspondents observed: "Flamethrowers were among the most important and…vulnerable men on Iwo."[1853]

As Woody and his men neared the front, they walked toward the sound of battle on better ground than the volcanic, ash-covered beaches. As they approached the combat line, they saw their brother Marines spread throughout craters created by artillery rounds or prone on their stomachs firing at the enemy. The sounds of men engaged in killing one another echoed everywhere. Carrier planes flying close air support continued to locate targets, rain down their bombs and strafe enemy positions. As many as 1,600 sorties were flown in a day to

PFC Wesley R. Strickland. He was with Woody on 23 February 1945 and showed Woody the first pillbox to destroy. St. Louis Personnel Records Center.

support the Marines' advance.[1854] Luckily for the Marines, the Japanese airforce was unable to send planes against the Americans leaving the U.S. with almost full air supremacy in the battlespace.

For Woody, an unknown number of Japanese fired at his unit from inside the pillboxes, including the one he was about to attack. As Woody neared the line of departure, PFC Wesley Strickland pointed out three pillboxes for him to start "worming" to around 60 yards away.[1855]

In this theatre of death, Woody crawled toward the concrete structure with the "mad clangor of arms"[1856] rattling away. As he did so, his BAR and riflemen started firing toward the pillbox's opening. "But their fire against the concrete and steel emplacements were like bee-bee shot on a cement sidewalk."[1857] He crept toward a point that was slightly higher in elevation from where he could approach the pillbox on a gradual slope. PFC Alexander Schlager followed Woody with his demolition charge strapped to a wooden pole to administer the *coup de grâce* after

Woody's attack.[1858] They moved from crater to crater to get closer to the pillbox. Around them, the sounds of battle echoed and a mortar round exploded near the crater in which they hid, sending a cascade of sand down the sides of the embankment, "loosed by the shock waves of the explosion."[1859] The men trying to protect Woody could do nothing "about the homicidal mortar shells which continued to plumb down."[1860] The environment baffled and intoxicated him. Woody had entered a region of human emotions nobody wanted to explore and few survive.

In this picture, a Marine Flamethrower team is about to go into action on Iwo Jima. The Flamethrower here is Sergeant Wayne B. Helton. National Archives, College Park

As the Marines approached those pillboxes, trenches and spider holes, the gunners inside the first pillbox tried to "depress the muzzle of their machine gun" to get Woody into their crosshairs. They missed him, hitting the sand just in front of his face, sending sharp particles into his cheeks with stinging force. Bullets stitched up the ground. Men behind him unleashed fire at the slit where the machinegun was shooting, forcing the Japanese to take cover "back from the gun." As Woody edged forward, an enemy soldier suddenly materialized, apparently out of a buried oil drum. Their eyes met and both were startled. Woody said, "I could see the surprise in his eyes. We were so close."[1861] Before the enemy could aim his rifle, Woody turned his nozzle and engulfed him in a circle of flame: "*Pluff!* He was gone."[1862]

Corporal Alexander Schlager. He was one of Woody's MOH witnesses. Schlager did not like Woody and thought he took too much credit away from their platoon for things it accomplished during his MOH's actions. He was very upset with Woody's article in *Man's Magazine* from 1966. St. Louis Personnel Records Center

The effect of a flamethrower hit is gruesome. Engulfed in intense heat many bodies lost their moisture and ignited—3,500 degrees does a lot of damage. The flesh quickly incinerates "revealing glistening bones."[1863]

Flamethrower fire was spotted instantly, attracting counterfire since the Japanese targeted men with "special equipment" such as radios, fully automatic weapons, crew-served weapons, or demolitions equipment—they especially hated flamethrowers, whether operated by a Marine or shot from a tank. A mortar round exploded outside the trench Woody used for protection. He was approximately 40 yards from the pillbox, and taking small arms fire. Bullets careened off the tanks on his back "with whining sounds."[1864] Woody kept moving. He capered closer to the structure, "cutting all manner of fantastical steps."[1865] As Woody left one crater, Schlager grabbed his demolition charge and stood out of the crater, exposing himself. He immediately took a bullet to his helmet, knocking him back into the crater from which he and Woody had just emerged. Woody peeked over the lip of the crater, expecting to see a dead Schlager, but fortunately, the bullet had gone through the top of his helmet, missing his head. Even so, the force of the impact had stunned him such that he was unable to move. Woody had to finish the pillbox without him.[1866]

In a 1966 interview, Woody gave another version of this story. Instead of Schlager, it was Southwell who was with him and instead of the first pillbox, it was the third pillbox. Southwell, a sergeant and Drill Instructor, outranked Woody and suggested they both jump out of the crater and "split the target. You go one way and I'll go the other." Woody "had no objections." "Ready?" said

Southwell. "Ready," answered Woody. "Let's go!" Southwell yelled. As the 6'1"
Southwell emerged from the crater, a bullet "smacked into [his] helmet. The impact straightened him up and bowled him backward. Another slug ripped into
the helmet, driving him back even faster." As he tumbled backwards into the
crater, he grabbed Woody's shoulder out of instinct to break his fall, and both
"of them toppled backward and landed with jarring force in the bottom of the
shellhole."[1867]

Marine flamethrower Sergeant Leonard J. Shoemaker on Iwo takes out a Japanese position during
mopping-up operations. 11 March 1945. National Archives, College Park. Photo Credit Sergeant
R. Francis

"Are you hurt?" Woody asked. "I don't know. Am I alive?" Southwell asked
stunned. "You look it," Woody replied. Southwell took off his helmet, viewed
the damage, shook his head and thanked God he was alive.[1868] The men then
re-engaged the enemy.

Regardless of which comrade was stunned, these stories demonstrate how
chaotic and confusing battle can be and how memory of combat can be confused
or contradictory. Woody has mixed up Schlager and Southwell with the same
event. Both cannot be right. Today, he believes this event happened to Schlager
and it was right before he hit the first pillbox.

When Woody got close to the first pillbox, he placed his fingers on the triggers ready to fire. Woody rose up on one knee, hit the forward trigger to light a match with his left hand and then pulled the rear trigger with the other, releasing the compressed air that would push the mixture of fuels out through the nozzle of the wand, igniting as it left his weapon. "The distinctive *Whomp! Whoose!* sounds" ushered from the flamethrower as the "crackle and roar" of the fire left his weapon.[1869] He rolled his flame toward the aperture moving his wand from pointing toward the ground to pushing it up level with his hip.[1870] The "fiery serpent" of flame fell hissing to the ground as Woody moved closer to the position.[1871]

He moved toward the first pillbox pumping the flame toward the front of the structure. Sometimes 50 feet of sand covered the top of the pillbox, so direct hits from above would not do much—only the flame could silence it.[1872] As soon as he took up his position, his protection force diverted their fire from the pillbox to avoid hitting Woody. They refocused firing at other emplacements to the right and left of the assaulted pillbox in order to suppress the enfilading fire from those positions which could kill Woody. Each man knew his role and followed through. What the men did is a testament to Marine training and the ethic of the Corps focusing on teamwork: Although these men had never worked together before, they took care of each other, saw what needed to be done and made sure they did their part. "We were Marines," Woody said, "and thus, we simply told each other what we needed to get done and then we got it done. Although I didn't know the men sent with me… I knew I could depend on them."[1873] In describing what Woody faced, reporter Robert Sherrod does a nice job writing about one blockhouse:

> Its outer walls were reinforced concrete, 40 inches thick. The vent did not open toward the sea, but slantwise toward the upper beaches: the 120mm gun inside could fire on the beaches and some of our ships, but could not be hit except from a particular angle. There was no sign that it had been touched by anything but a flamethrower. Beside it lay the bodies of eight Marines— the apparent cost of taking what was only one of several hundred similar positions, nearly all of which have to be knocked out by men on foot with explosive charges or flamethrowers.[1874]

Scenes like Sherrod described danced inside Woody's head as he began his mission. They had traversed a landscape with the skeletal remains of numerous other pillboxes covered with blood and corpses. The Japanese knew he was coming and saw what he was delivering.

A Marine flamethrower takes out a cave opening under some rocks on Iwo. National Archives, College Park.

When he "got up to within a dozen yards of the first one," he "emptied his flamethrower."[1875] As Woody turned the first pillbox into a burning oven, the acrid smell of burnt flesh and spent ammunition filled the air. Some Marines hated the Japanese so much they described the smell of "roasting flesh" as "the sweetest that they had ever smelled."[1876] Ignited ammunition in the pillbox created a raucous noise from multiple small explosions. According to one report, Woody blew up at least two pillboxes with demolition charges, so in this case, he would have had to remove his tanks, take up a pole charge that he had brought with him or that Schlager had dropped and return to the pillbox he had torched and then stick the C2 explosives on the end of the long wooden pole into the pillbox's opening and blow it sky high.[1877] Woody stuck pole-charges "through narrow openings of emplacements" of the pillboxes he attacked on four occasions. During two of them, two of the charges failed to explode, so he worked his way back to the pillboxes and "squirted flame into [them]" meaning he had to again strap the tanks that still had enough mixture remaining or obtain new tanks and then re-engage the structure.[1878]

Here is an example of a poll-charge on Iwo (Corporal Dale Billings is holding the one here). These were often shoved into pillboxes or caves to blow them shut. 29 February 1945. National Archives, College Park. Photo Credit R.G. Simpson

Woody emptied his tanks on the first bunker which probably held two to four men and then used his pole-charge against it leaving a smoking, crumbled structure. He then most likely hit the ground and crawled back to his lines, all while his protection force covered his escape. As he retreated in a low crawl, sweat poured down his face and sand and gravel crackled under his knees and elbows. Once past his lines and into a secure zone, Woody stood up and ran back to his supply area. He checked his gear and then hoisted a new flamethrower on his back to go out yet again for the second enemy position, accompanied by his security detail. While resupplying, his security men remained in place on the line, awaiting his return.

During this first attack, Southwell said that although he and his comrades provided cover for Woody, they "had had a hell of a day and were low on ammo," so they could not provide as much covering fire as they would have liked. "And everybody had to hug the ground," Southwell continued, "so we didn't have such good fighting."[1879] Woody accomplished much of what he did without the optimal suppressing fire such action required. Suppressing fire is not necessarily to kill or injure the enemy; rather it is intended to keep the enemy ducking for cover to give friendly forces the opportunity to move without getting hit.

As Woody and his Marines neared the second target, the Japanese sprayed the area with machinegun and rifle fire. Around the pillboxes, Japanese popped up from tunnels or oil drums hidden in the earth, and fired or lobbed grenades at surprised Marines. It was a haunted house-like defense, with enemy around the pillboxes appearing suddenly from nowhere. Woody said, "the Japanese would disappear as soon as you thought you had them. They were fighting a totally different way than on Guam. There I could see my enemy, here they were phantoms. At least I knew if I could get close to the pillbox, there were enemy in there to kill because they never stopped firing at us."[1880] Like mist on a mirror, often the Marines got a glimpse of Japanese soldiers before they vanished into the land using caves and tunnels like gophers or prairie dogs. "The defender" here clearly had the "advantage: [T]he ground he" occupied was "better known to him" than it was to the attacking Marines "in the same way as a man can find his way around his own room in the dark more easily than can a stranger. He can find and round up all the component parts of his forces more quickly than can his assailant."[1881] Regardless, the Marines continued to move forward, often stumbling through dark rooms and foggy hallways.

Although the sun was high in the sky a little past noon, the enemy was difficult to see but the effects of his fighting were heard and seen everywhere. Wind blew throughout the island inland off the rough seas, blowing smoke, debris, dirt and putrid smells of death over the island's surface. Marines peeked over their hideouts as they looked for their next target.

Here is a dead Japanese soldier burned by a flamethrower on Iwo Jima. 22 February 1945. National Archives, College Park. Photo Credit W. O. Newcomb

When Woody crawled into position to take out the second pillbox, according to Woody a group conducted a *Banzai*, targeting him. They knew why he was there and were determined to kill him before he took out another emplacement. They probably witnessed what had happened to their comrades in the first pillbox, and wanted to spare their remaining comrades a similar fate. Seeing them charge, Woody turned his flamethrower, spraying fire on them. They were unable to get one bullet in Woody before the flame hit them. Woody remembers, "They went from running full speed at me to almost slow motion. Once that flame lapped them up, they seemed to freeze. Their clothes were on fire and they just fell."[1882] In 1966, Woody claimed there were three men in this charge, but in interviews with the WWII Museum and with this study, he has often claimed there were five to six.[1883] Even if only three screaming, determined Japanese charged him, that charge was sufficient to make anyone's hair stand on end. Woody responded to this counterattack of men who probably had escaped their bunker through sally ports. Dashiell reported Woody having explained that his "flamethrower stopped frantic Japs who ran out of their hiding places and tried to get him with rifles, bayonets, and grenades."[1884]

A Marine flamethrower operator takes out a pillbox at Iwo Jima with two Marine rifleman in support. February 1945. National Archives, College Park. Photo Credit Christian

The very few Japanese Iwo survivors claim they feared napalm and the flamethrowers more than any weapon.[1885] As American researchers discovered, if a flamethrower got close to any structure, the defenders were doomed and they knew it.[1886] One Japanese soldier who witnessed such attacks said his comrades were "chickens being fried."[1887] One U.S. veteran wrote that once the intense flame hit the enemy, their "bodies flared up like pieces of celluloid."[1888] Once killed by flame, removing their bodies from the battlefield proved difficult. "The blackened corpses were gross, and as you tied radio wire around their limbs to remove them from the front lines, their legs and arms would rip away from their bodies because the heat had destroyed the integrity of joints just like pulling a leg off a well baked Turkey."[1889] Woody thus left behind grisly evidence of his kills and the air was foul with the stench of rotting and charbroiled bodies.

In this chaos, as the dead Japanese lay in a haze of blazing fluid, Woody returned his attention to the pillbox. As he focused on this next task, the Japanese on the ground turned into black, bubbling pools of flesh; their skin and subcutaneous fat sizzled, baked by 3,500 degrees of searing heat. Having neutralized this security force, Woody neared the second pillbox and pumped his flame into it until once again the familiar nauseating smell of burning flesh permeated the air. Woody thinks it was at this second pillbox when he first noticed the structure had been sealed *from the outside*. Those men, once in there, were never leaving.[1890]

Destroying the pillboxes was the toughest thing Woody ever had to do in combat and the fear he experienced while completing his mission was unlike any he had ever felt. When asked how he dealt with the fear, he laughed and said he remembered shaking his head at what he was doing. He had escaped death in one form or another only to be delivered into now a worse situation facing, one could argue, the "king" of all pillboxes.

The third fortification he had to take out was far larger than the first and second pillboxes he had just destroyed. It was a sizable concrete structure with several Japanese manning it. Discarding his tanks after destroying the second complex (and possibly also blowing it up with a pole-charge), Woody returned to headquarters' supply to get another flamethrower.[1891]

At this point, Woody remembered feeling like he was in a dream. He knew what he had to do, knew that death surrounded him, but had a weird feeling he would survive and see his girlfriend Ruby again. When asked if he was scared, Woody said of course he was scared—"at first!" "After that you don't care anymore. You don't give a darn whether you get hit or not. Just keep moving... or rather stumbling forward...."[1892] He became fatalistic and indifference replaced his fear. This allowed cool thinking and quick reaction. Holocaust survivor and psychologist Viktor Frankl notes, "Apathy [indifference]...was a necessary mechanism of self-defense."[1893]

Hatred of the enemy was also a motivator. Woody had had a lot of Japanese trying to kill him, had fought off *Banzais*, had witnessed the Japanese desecration of corpses and, most importantly, had seen his buddies wounded or killed. Woody said, "I suppose the only thing that kept us going was the incentive to avenge the loss of our buddies [who] were killed. After a while you'd feel there was only one thing you wanted to do in the world. To get the guy [who] got your buddy."[1894] After being in combat for only a brief amount of time, most Marines learned how to hate and do so with passion.

In the midst of this chaos, Woody strapped on his third backpack of tanks, grabbed his 15 pound wand, turned toward the pillboxes and returned to the storm of battle. In this section of the island, there was no solid ground, pulverized as it had been into gravel and sand. One did not sink into it like on the beach, but one did wobble on it somewhat as he ran. Everywhere the growl of missiles, mortars and shells pierced the air.

At least four dead Marines are in this crater with a fellow flamethrower operator. A Japanese mortar came down on their position killing them all instantly. 27 February 1945. National Archives, College Park. Photo Credit Corporal Robbins.

Nearing the large blockhouse which was probably around 300 to 500 square feet and buried in a sand dune, Woody could not get a good line of fire. Destroyed and badly damaged vegetation or small trees and bushes lay everywhere, obstructing his approach. This enemy blockhouse was Woody's main objective because it was the "one really giving us fits."[1895] Supporting Marines fired away at the slits about eight inches wide and around four feet high along the whole face of the pillbox, but the Japanese inside laid down murderous fire against those in sight, and Woody, with large metal tanks strapped to his body was their obvious focus. He was scared, but he did not ever think about dying although everywhere he went, bullets flew around him as the enemy tried to kill him. Woody said, "You probably didn't think about being actually killed because that would cause you to quit and I wasn't going to quit."[1896] Military philosopher Carl von Clausewitz could have been describing Woody when he wrote:

> Courage in face of personal danger is also of two kinds. It may be indifference to danger, which could be due to the individual's constitution, or to his holding life cheap, or to habit. In any case, it must be regarded as a permanent condition. Alternatively, courage may result from such positive motives as ambition, patriotism, or enthusiasm of any kind. In that case courage is a feeling, an emotion…These two kinds of courage act in different ways. The first is the more dependable; having become second nature, it will never fail. The other will often achieve more. There is more reliability in the first kind, more boldness in the second. The first leaves the mind calmer; the second tends to stimulate, but it can also blind. *The highest kind of courage is a compound of both.*[1897]

These three Marines were killed instantly when their fellow flamethrower comrade took a direct hit on his equipment blowing them all to kingdom-come. 25 February 1945. National Archives, College Park. Photo Credit PFC R. R. Dodds.

It appears Woody indeed had both kinds of courage. Regardless of what was motivating Woody, the Japanese knew they could not let Woody get close to them. He was one of the most feared, hated and targeted Marines for those hundreds of Japanese facing him from inside and around those pillboxes.

Here is an example of a bazooka team like the one that was most likely with Woody on the day he did what he did for his MOH. March 1945. National Archives, College Park.

Watching the Japanese fortification carefully and looking around the area to see what approach he should take, Woody noticed smoke curling out of the top of the structure out of a pipe, most likely coming from the weapons they fired inside or from smoke or debris generated by grenades or bazookas hitting the outside of the structure and being drawn in by the suction of air caused by the explosions.[1898] According to Woody's 1966 interview, he recalled:

> I motioned for the bazookaman to fire a round at the center pill-box. He did and I could see it hit just below and to the left of the slit. Some of the white smoke from the explosion was sucked into the concrete opening and I noticed a wisp of it curling out the center top of the pillbox. This told me that there was an opening of some kind on the top.[1899]

Woody saw several rifles and one Nambu machinegun firing. He remembers rounds striking his tanks and ricocheting off of the metal skin as he used different trenches to approach the structure. "The bullets coming off my tanks sounded like a jackhammer." He moved forward "with an oddly crab-like gait" and the Japanese gunner could not depress his machinegun any lower to hit him. Woody had maneuvered himself out of the field of fire. "Had I gone the other way, he would've zeroed in on me and killed me."[1900] Also, while here, "a number of grenades thrown by the enemy" landed near Woody, but did not reach him.[1901]

Private George B. Schwartz. He was one of Woody's MOH witnesses on 23 February 1945. St. Louis Personnel Records Center

According to some reports, he tried several times to get flames into the pillbox, perhaps using the wall of fire he projected partly as a screen and partly as an offensive weapon. The larger structures the Japanese had built often had burn alleys inside these constructions to divert the flame away from the men and to house defensive chambers to protect them when hit by fire. The Japanese knew the Marines attacked with flamethrowers, so they engineered responses to this weapon and this pillbox probably had such characteristics, making Woody's job even more challenging. According to witness George Schwartz, Woody used two, possibly three flamethrowers since he "had a hell of a time with" this massive bunker.[1902] The Japanese even used their own flamethrowers against the Marines, so they knew their capabilities and destructive power.[1903] In the initial attack, Woody did not have success. He had worked himself into a fury trying to dispose of the threat. Frustrated and determined, Woody decided to perform a maneuver for which he had never trained and that most people would never attempt.

A Marine flamethrower moves into action spewing flames ahead of his advance. 19 February 1945. National Archives, College Park. Photo Credit Farnum

A Marine flamethrower operator from 'E' Company, 9th Marines in Woody's division runs into action. February 1945. National Archives, College Park. Photo Credit Christian

As the Japanese directed massive fire against the lines where Woody fought, Woody zigzagged toward the pillbox. In response, the Japanese threw grenades at him to push him back. He worked his way to the rear of the pillbox with Corporal

Tripp telling him, "Stay put and keep me covered." Tripp "moved his M1 into a better firing position shunting the bamboo-like pole charges to one side."[1904] Then, to the amazement of everyone, Woody climbed up the side of the pillbox onto its roof. Once on top, Woody found the ventilation pipe he had spotted earlier. The aperture was the same circumference as the nozzle on his wand, so he stuck the tip into the pipe and pulled the trigger, unleashing a firestorm into the bunker.[1905] He heard "the flames rushing down the pipe and spilling out with a roar as they reached the greater air volume inside."[1906] The whole structure erupted into flames as tongues of fire shot out from inside the compound. Those not killed by the searing heat died soon thereafter

Corporal Alan Tripp. One of Woody's witnesses and comrades on Iwo Jima. Tripp would receive a "Letter of Commendation" from 3d MarDiv commander Major General Graves Erskine for his bravery. St. Louis Personnel Records Center

due to the inferno sucking up the oxygen, suffocating them. Roasted flesh permeated the air. (See Gallery 3, Photos 7-10)

While Woody stood atop the bunker pumping the structure with the flaming diesel and gas mixture, it seemed like the whole battlefield reached a fever pitch of violence and gunfire, yet somehow no bullet hit him. After he was sure everyone in the bunker was dead, Woody crawled off its roof, and worked his way back to retrieve another flamethrower. As he left the structure, a mortar round landed and exploded near him, but Woody escaped harm again. He was constantly moving so that the Japanese could not home in on him. The mortars landing around were probably intended to bracket Woody so they could kill him, yet they failed to do so. Upon leaving the bunker, Woody ordered a Marine, likely Tripp, with a demolition charge to blow up the structure.[1907]

A Marine flamethrower from 3rd Battalion, 21st Marines in Woody's division attacks a Japanese position on Iwo Jima. National Archives, College Park. Photo Credit Staff Sergeant J. F. Galloway

An example of a concrete reinforced Japanese pillbox on Iwo Jima that represents something that Woody faced, especially with the third and large pillbox. Here there was a dead IJA soldier to the right. February 1945. National Archives, College Park. Photo Credit Staff Sergeant Kauffman

Woody reckoned recently that up to 17 enemy were in the third pillbox as well as in his 1966 interview with Bill Francois, but in 1956 he estimated 21, although other primary sources do not corroborate these figures.[1908] Schlager said he counted eight inside.[1909] If Tripp had placed demolitions inside the bunker, then how could Schlager have counted the dead assuming the blast destroyed the structure somewhat if not totally and probably blew the bodies to pieces? Woody's attack may have silenced at least one Type 92 Heavy Nambu 6.5mm machinegun, having a three-man team and capable of firing 400 to 500 rounds per minute. Nearby, hundreds more enemy soldiers waited for Woody and his brother Marines. If there were so many Japanese in this structure, that meant there were hundreds more in nearby rifle pits, trenches, spider holes, pillboxes and tunnels. The Marines had to kill many more to secure the area.

Demolition Marines in action on Iwo Jima. Often, when Marines found a Japanese position, they would throw C-2 explosives into it and detonate it. National Archives, College Park

Flamethrowers not only delivered incineration, they also poisoned and suffocated enemies. Researchers working on the flamethrower found "studying the toxicology of flame [attacks] in poorly ventilated enclosed spaces like those found in Japanese bunkers" that three events happened "at the moment of the flame attack quite aside from the penetration of the flaming fuel itself: there was a sudden jump in temperature, lethal concentrations of carbon monoxide" and low levels of oxygen. They discovered:

> …that 70 percent carbon monoxide in the blood resulted in unconsciousness and frequently in death and that this accumulation was obtainable in a flame attack within two minutes. Furthermore, only one-tenth of one percent carbon monoxide in the air was sufficient to maintain this lethal blood level, and it was present in bunkers for seven to ten minutes after [a] flame attack.[1910]

Corporal Warren Harding Bornholz. He died on 23 February 1945 possibly defending Woody while he attacked the pillboxes. St. Louis Personnel Records Center.

PFC Charles G. Fischer. He died on 23 February 1945 possibly defending Woody while he attacked the pillboxes. St. Louis Personnel Records Center.

Also, up to fifteen seconds after a flamethrower attack, there was no oxygen in the pillbox under attack which resulted in the defenders falling into "unconsciousness" almost instantaneously.[1911] Woody's flamethrower brought a plethora of fatal attributes to kill those underground. If the flames, lack of oxygen or carbon monoxide did not kill the enemy, then the pole charges used against them afterwards did.

Although Woody escaped unharmed from knocking out the enemy fortifications, his team did not get away unscathed. According to Woody, enemy fire fatally wounded two men covering him. Only much later, after returning to Guam, did Woody learn they had died. They probably were Corporal Warren Bornholz and PFC Charles Fischer.[1912] A sniper shot Fischer in the chest while moving into the attack and Bornholz received head and neck shots during combat.[1913] Had Woody known of their deaths, he had no time to mourn. He had more pillboxes to destroy and infantrymen to protect.[1914] Two other men were quickly assigned to cover him.

To appreciate Woody's heroism, one must consider the environment in which he found himself. While attacking this structure, he heard thousands of rounds of gun fire, much of it directed at him. Every few seconds, artillery or mortar rounds exploded. Airplanes dropped their ordnance on and strafed enemy positions. Ships fired at targets far enough from the Marines that the shells would not hit them too. Antiaircraft guns defended the fleet from attacking *Kamikaze*s, sending columns of fire to the heavens. Wounded screamed in pain calling for corpsmen, friends, God and mothers. The "howls of desperation"[1915] rebounded throughout the landscape. Others passed information by yelling instructions over the din of battle. As Woody explained, "Few can truly comprehend the absolute terror and chaos of war."[1916] As any combat veteran knows, the sound of even a small group firing weapons and taking fire with mortars and artillery in the background is overpowering. The movies and TV never get this right. To give one an idea of the amount of gunfire expended during this battle, during the battle of Peleliu which took place from September to November 1944, the Marines and army expended on average 1,539 rounds of ammunition for each Japanese soldier killed. That figure does not include the tremendous amount of enemy counter-gunfire. If the same ratio of Peleliu is used for Iwo, the Marines fired almost 34 million bullets at the Japanese on the island (during the 36-day Iwo campaign, that meant around 1,300 bullets, both Marine and Japanese, were fired every minute). Around 28,000 tons of infantry cartridges were shipped for the troops at Iwo alone. The noise of gunfire was overwhelming.[1917] These numbers do not take into account the mortars, grenades and other instruments of war exploding as well. Woody's battalion used 8,525 grenades during the campaign.[1918] In such an unforgiving environment with the loudest tumult one can imagine, Woody remained focused for hours while he strategized the best way to silence concrete monsters blocking his company's advance. Unlike situations where a combatant goes into a fervid rage or focused trance, Woody's feats manifested an ability to think clearly under protracted and acute stress.

Returning to the aftermath of destroying the large pillbox, in one of the reports from 1945, it seems many of the Japanese nearby tried to escape—they probably had anchored their defense to this pillbox and once it was gone, they knew they were vulnerable. Woody reported that he killed some fleeing the area.[1919] Scenes such as those surrounding the circumstances in which Woody found himself occurred in other areas. Another MOH recipient and member of the 3d MarDiv., Private Wilson D. Watson, after killing four Japanese with a grenade in a pillbox and pushing the advance forward like Woody, caught the enemy massing on the opposite side of a ridge to counterattack his unit. Commanding the high ground after scaling to the ridge's top, he spotted the enemy and opened up with a BAR and killed 60.[1920] Woody was not alone in pushing forward the attack, and in doing so, causing confusion among the Japanese whose uncoordinated actions allowed the Marines to inflict harm on a fleeing enemy.

As Woody was incinerating pillboxes, many Japanese on "seeing what was in store for them, made a dash to the rear. The Marines had a field day as they killed them one by one by one. Some hopelessly rushed toward the Leathernecks while wielding sabers."[1921] Woody's activity with his flamethrower made some panic, reveal their positions and then attempt to flee, but in so doing, they made themselves vulnerable and the Marines slew them.

By now, Woody's uniform was dripping with sweat, his legs burned from exhaustion, and he became almost blinded from the flames that spewed from his weapon. He was moving and crawling to the next targets on second-nature impulses. Woody had already spent hours with the flamethrower in hazardous duty, but he never allowed himself to think he would not survive. He was lucky, because the life expectancy of most flamethrowers was measured in minutes, not hours or days. The after-action report for Battalion Landing Team 2/21 stated that the casualty rate for flamethrower operators was 92% but that no question existed that the weapon was valuable and necessary in the destruction of emplacements and neutralization of caves.[1922] Yet Woody persevered, still healthy and fighting. He had tempted fate and won thus far.

As Woody returned to headquarters supply for a fourth time, Marines present must have shaken their heads in disbelief seeing him still alive. First Sergeant Elder and Corporal Alan B. Tripp witnessed him going back to the headquarters command post six times for re-supply.[1923] Woody focused on his mission. He barked commands to his detail to protect his backside. He then strapped on his fourth tank, looked at the slope he had descended, viewed the three smoking pillboxes he had just taken out, and then looked for the next target. His security detail tried to keep the Japanese from honing in on him and killing one of the best hopes the company had of minimizing casualties and breaking through to the North to secure the first airfield.

Marine PFC Thomas N. Brown refuels a flamethrower. The terminology "Dry" on the tank Brown was using here was added to tanks with the nitrogen designation. These were 300 series refill-tanks also known as full-sized-end-user-tanks. National Archives, College Park. Photo Credit Corporal R. G. Simpson.

Woody's remaining attacks against pillboxes are difficult to describe. Even the description to this point may be out of sequence. The chaos of battle can present problems for those who were present to recount the events in order. Woody is bothered that he cannot remember the details. In fact, he has no memory of getting new flamethrowers.[1924] A journalist interviewed Woody in 1966 when his memory was fresher but he still struggled to recall details. Journalist Bill Francois described Woody's conundrum:

> Once again the tanks were empty and Williams unbuckled the flamethrower, let it fall away from him so that he could return more quickly to this own lines. He was conscious of an overwhelming fatigue whenever he had to run a short distance in order to reach the protection of a crater or dune. Bullets tried to cut him down. Mortar rounds reached for him, their throaty roars muffled in the depth of the sand. As weariness engulfed him, the terror of those sounds diminished. The passing of time went unnoticed. Another flamethrower. Heavier. Each one heavier than the last. Elbows and knees rubbed raw. Vague, momentary spasms of fear as some new danger threatened. Red fireballs engulfing, penetrating, destroying. Corporal Williams had slipped into a merciful void. After he had destroyed the third pillbox he could recall no chronology of events. Time had ceased for him.[1925]

Now did time "cease for him" because his memory was wiped clean by stress and fatigue or because some or all of the later reported events did not occur? The problem here is probably more that Woody's humanity was stressed beyond the knowledge of others who have never experienced such extreme pressure making one struggle with later describing what really happened or filling in holes one wished to explain. Not only did Woody admit in 1966 he had no memory of the four remaining pillboxes he thinks he took out (although in 1945 he did), but also the other primary sources are also unclear on this point. Yet while Woody originally in 1945 claimed to have destroyed seven pillboxes in total, no single eyewitness submitted a statement he saw *all* seven. Captain Beck later claimed Woody's platoon leader Chambers pointed out seven destroyed pillboxes and

Beck assumed Woody had destroyed them. Woody's MOH citation does not state the number of pillboxes destroyed. Thus, in conclusion, what remains unknown is how the four additional pillboxes were destroyed and whether or not there were indeed seven eliminated by Woody's hand.

Without question, Woody destroyed at least three, maybe four pillboxes. Perhaps the fourth pillbox was described by William J. Naro who was covering Woody. Corporal Naro testified Woody "was subjected to *considerable* small arms fire" while he

Corporal William Naro. He was one of Woody's MOH witnesses. St. Louis Personnel Files

zigzagged through the communication trenches to get behind one structure.[1926] According to Dick Dashiell, this is maybe where Woody "gave the Nambu gunners the business from the rear."[1927]

An example of a Japanese pillbox from the rear that Woody could have faced. 28 February 1945. National Archives, College Park

Marines supporting Woody killed a number of the enemy on their own, especially those "who made a break for the Jap lines from their by-passed trenches." Naro put that number at up to 25 enemy soldiers "bypassed in spider traps and perfectly concealed trenches."[1928] Schlager, who had recovered from nearly having his head blown off, shot dead a lone charging soldier armed with a *Samurai* sword; it took three bullets to stop him.[1929] Lieutenant Chambers was so close to two Japanese that he killed them both with his Colt .45 pistol.[1930]

Dashiell recounted that while another flamethrower worked over a pillbox, two Japanese ran out. Immediately, one was cut in half by supporting fire from surrounding Marines and the other, seeing what just happened to his buddy, killed himself with a grenade.[1931]

Leaving the remains of pillboxes, Woody completed his operation without a single wound, reporting, "Darn if I know why I didn't get bumped off that day."[1932] Many must have stood in awe of what Woody and his team had accomplished, especially climbing on top of one pillbox to flame it: "[H]e was a madman when he did it—a raving maniac through sheer fright" rolling, crawling, dodging, flaming and killing over the battlefield.[1933] But a madman probably would not have survived. Woody also displayed intervals of cool calculation, however emotionally charged he was at times.

Dashiell wrote: "[Woody killed] an uncounted number of Japs."[1934] Woody said to Dashiell he counted 21 dead during one part of the operation. Yet, no one was likely to achieve an accurate count of those killed given the charred and blasted state of the remains. However, more important than the number of enemy killed was that neutralizing the deadly fortifications helped enable thousands in his division who were exposed to enemy fire break through those defenses and move forward. He was part of an overall small operation led by his brave platoon leader, Lieutenant Chambers, who broke through 32 emplacements and moved the front 300 yards north closer to Kuribayashi's headquarters and to ultimate victory. Due to Woody and his supporting crew's deeds under Chambers' leadership, their battalion reported in the afternoon that the "main resistance had been reduced and B[attalion] was able to advance."[1935] "*Tactical* successes are of *paramount importance* in war" because only with them can an army give birth to great strategic victories.[1936] (See Gallery 3, Photo 14)

Until the island was secured, Marines and airmen would continue to die, and the bombing campaign with B-29 Superfortress bombers based in the Marianas flying unescorted to Japan would be more dangerous. Iwo would give the Army Air Forces an island on which to base fighters that could escort the B-29s on missions over Japan, permitting the B-29s to approach their targets at a much lower level where they were more accurate. Major General Curtis LeMay, commander of B-29s in the Pacific, said, "Without Iwo, I couldn't bomb Japan effectively."[1937]

Others in C Company also accomplished acts of valor the day Woody took out the pillboxes, receiving medals for their bravery. For instance, Chambers earned the Silver Star for his role: "Frequently advancing beyond the front lines to supervise tank and demolition personnel, [Chambers] skillfully coordinated the attack of his unit."[1938] Woody's Company Commander, Captain Beck, received the Bronze Star with V for combat on the company level (the V for "valor" indicated he earned the Bronze Star in combat).[1939] His citation read: "His excellent coordination of mortar and artillery fire, in addition to his personal direction of a tank platoon, enabled riflemen to assault and take several emplacements...he led his company forward 300 yards completely eliminating the enemy."[1940]

Beck's, Chambers' and Woody's courageous actions supported by others were "directly instrumental in neutralizing one of the most fanatically defended Japanese strong points encountered by [their] regiment and aided vitally in enabling [their] company to reach its objective" to move forward, secure the first airfield and start attacking the second one.[1941]

Stiff Japanese opposition in this section had held up the Marines for days. Despite the U.S. onslaught of bombers and naval gunfire for at least two months before the battle, artillery, mortar and small arms fire had stopped the Marines dead in their tracks. Lieutenant General Smith, who had released Woody's 21st Regiment for this duty, called the area that Beck, Chambers and Woody helped break "the very heart of the defense belt Kuribayashi had prepared."[1942] As shown on many occasions, American assault troops often charged "with more élan" when supported by flamethrowers and Woody, by his example in destroying the pillboxes, increased his fellow Marines' morale and eagerness to attack that day.[1943] One of Woody's platoon leaders, Lieutenant Charles Wesley (Mike) "The Ox" Henning, said that one of his men, maybe Woody, went absolutely "mad with the flamethrower" that day and helped save thousands.[1944] The reason Woody had a "few" platoon leaders was because he was made available by Charlie Company's headquarters to go out and support each of the three platoons when needed to bring more firepower to a given sector of the front. As Carl von Clausewitz wrote:

Destruction of the enemy's force is only a means to an end, a secondary matter. If a mere demonstration is enough to cause the enemy to abandon his position, the objective has been achieved; but as a rule the hill or bridge is captured only so that even more damage can be inflicted on the enemy.[1945]

Woody's actions allowed the Marines to move forward to inflict "even more damage" on the Japanese.

In breaking this defense, 3d MarDiv commander "Blood and Guts" Major General Erskine sent a report to his units on 25 February:

Your determined relentless assault and skillful use of your weapons in two days advance against highly organized enemy positions and his fanatical defense demonstrated a superior degree of offensive spirit. The Commanding General is proud of you and commends all hands...We can and will go anyplace the Jap can go, dig him out and kill him.[1946]

Erskine's analysis described what Beck, Chambers and Woody had done. These Marines ascertained where the enemy fire originated, placed destructive weapons fire on the areas there and then obliterated that resistance. The "offensive spirit" Beck, Chambers, Woody and others showed had such an effect on their commanding general that he took notice of it in real time as the battle evolved, knowing his men's acts yielded more land and combat success. Progress was slower than Erskine desired, but in the end, when his men met opposition, they prevailed in breaking a major line of resistance. Destroying pillboxes, according to Lieutenant General Smith, was a "Herculean task" and Beck's company helped achieve it.[1947]

Although the Marines had broken through a major defensive line, the Japanese defending that area showed unshakable devotion to duty and sustained bravery for numerous days, holding their posts and fighting until the end. One must consider what went through Japanese minds as they watched Woody destroy one pillbox after the other, knowing he would be coming back for them. They had probably killed and injured many Marines, manning their posts under horrendous conditions for five days. What medals might these Japanese have earned had they survived? British Brigade Major John Masters fought the Japanese in

Burma under the victorious, multi-ethnic army led by British General Bill Slim and wrote that IJA soldiers "are the bravest people I have ever met." He continued, "In our armies, any of them, nearly every Japanese would have had a Congressional Medal [of Honor] or a Victoria Cross. It is the fashion to dismiss their courage as fanaticism, but this only begs the question. They believed in something, and they were willing to die for it, for any smallest detail that would help to achieve it. What else is bravery?"[1948]

In the IJA, medals and accolades were awarded differently than in the U.S. However, one could argue that during the first few days, Kuribayashi awarded the equivalent of four Medals of Honor or *jôbun*, meaning the soldier's feat "was reported to the Emperor," an exceptional honor. Once Kuribayashi documented a man's heroic acts, he wrote up a *Kanjo*. The *Kanjo* was sent to the Emperor where it was read to him. Once this happened, it became a *jôbun* and was reported in the newspapers. To illustrate what a soldier had to do to get this award, on 19 February, 2nd Lieutenant Sadao Nakamura, a "platoon leader," immobilized 20 amtracs and 1 tank dozer, hit three *LSM* ships (Landing Ships, Medium Transports) and killed 30 men with a 47mm cannon. Located at the northern end of the landing beaches, he operated a rapid fire artillery piece *by himself*. Kuribayashi wrote his *Kanjo* on 20 February, sent it to Tokyo, and soon thereafter, it was read before the Emperor officially giving Nakamura a *jôbun*. Kuribyashi also wrote Nakamura personally and said, "he did well."[1949] Nakamura would prove historians Jeter A. Isely and Philip A. Crowl's analysis correct when they wrote: "The Japanese…[enjoyed] a superiority with their highly accurate and mobile 47-millimeter antitank and antiboat weapons" compared to anything the Marines could put in the field.[1950]

The day before Woody stepped off in his MOH attack, his unit was attacked by Captain Masao Hayauchi's 12th Independent Anti-Tank Battalion's cannons. In a few hours, Hayauchi destroyed several tanks with rapid fire artillery pieces, the 47mm cannon yet again. After his cannons had been rendered useless, he led a charge against the remaining Shermans. "Clutching to his chest a charge with the fuse lit, he splayed himself against a tank and blew himself up."[1951] On 24 February, Kuribayashi wrote a *Kanjo* for Hayauchi and it was soon read before the Emperor making it a *jôbun*.[1952] "Citations were very rarely given to individuals" in the IJA showing Kuribayashi's uncustomary care for his troops.[1953]

In defeating such brave men, Woody and his fellow Marines had excellent leadership on the battlefield. Commandant Vandegrift shared with his forces before the battle a report that speaks to the importance of leaders having the confidence of their men and inspiring them to action in the face of a dangerous enemy:

> The most important single thing in battle is to find men who will share responsibility with their commander for accomplishing a job. Issuing orders is the easy part; seeing that they are carried out calls for every ounce of energy and initiative an officer or non-commissioned officer possesses. One of the most difficult things in the world is to get men to move forward.[1954]

Woody "shared responsibility" with Beck and Chambers, all of whom understood the mission and led the Marines moving forward. Beck and Chambers did exactly what Vandegrift's report claimed was necessary for battlefield success. Beck explained: "Pillboxes were often difficult to reduce due to the buried embrasure; however, our flamethrower could always do the job."[1955]

Beck's Marines faced an enemy who showed great skill, tenacity and bravery. Woody fought tough, well-armed and -prepared troops on as well-defended a battlespace as any. At Iwo, Woody, Beck, Chambers, Erskine and thousands of other Marines earned their place on the right side of history gaining a victory for the Allies at one of the most brutal battles of WWII.

# Ch. 23: The Attack on Iwo Continues

The law of war: "Thou Shalt Not Establish a Habit Pattern."[1956]

"There will be security on Iwo Jima
when the last Jap has been killed."
—Sergeant Dick Dashiell[1957]

AFTER WOODY AND HIS MEN finished taking out the pillboxes, Charlie Company, no longer pinned down by deadly fire, started moving northward with other companies to capture the second airfield. After Woody threw off his last empty flamethrower on 23 February, he took up his M1 rifle and grenades, joined his platoon and led a squad of six men. He must have shaken off what he had just done as one would a bad dream and started to scan the horizon for the next target. Everything he had done was part of his mission, and he did not give much thought to what an incredible achievement it was. "[T]o a greater degree than was necessary, taking Iwo Jima was the throwing of human flesh against reinforced concrete."[1958]

After Marines had broken through the defensive line near Airport No.1, they encountered intense opposition on the rising hills on the Motoyama Plateau north of the first airfield and *en route* to their next objective: Airfield No. 2, defended by Colonel Masuo Ikeda's 145th Regiment. This was Kuribayashi's "most powerful" unit; it manned the strongest area of the general's defense and had undergone years of combat experience in China.[1959] While Woody and his team were occupied with the pillboxes near Airfield No. 1, another Marine in his regiment, Corporal John Wlach of Fox Company, 2nd Battalion, 21st Marines, helped secure the left flank above it. He fought the line of defense to the right of Woody. Wlach took over his unit when their leader "shifted to another outfit." Due to casualties, the platoon had gone from 45 to 19 men. Facing what he and his men called *Hand Grenade Hill*, they encountered over 200 grenades flying around them within a period of 15 minutes. As a former "crack amateur and semi-pro baseball player," Wlach, without hesitation, picked up several of the grenades and "pegged" them back at the Japanese before they exploded. Witness PFC Cecil Matheney commented: "Wlach is the guy who's mainly responsible for taking that hill. He was a wild man, but he kept his head. He kept the boys right up on the skirmish line." Wlach exposed himself to fire, stood up and screamed commands. Matheney continued:

> You see, we had gotten in front of these Japs almost before we knew it. They were so well concealed—even their pillboxes—that we'd have never known they were there if they hadn't opened up. But Johnny didn't let us get caught with our breeches down. He snapped us right out of our stunned surprise and had us going just as though we had been there for hours.

Marines move out in the advance to the side of Airfield No. 1 headed north. Circa 23 February 1945. National Archives, College Park. Photo Credit Gene Jones

Seeing some allowing fear to stymie action, he stood up on a mound and yelled, "Come on, you fellows, the only thing that can happen to you is to get killed!"[1960] With that declaration, his men took the hill.

The 3d MarDiv in the center fought their way to the northeastern part of the island and moved "directly into the face of the heaviest enemy defenses."[1961] It attacked the hills designated 362C and 382.[1962] The area between the hills was referred to as the Amphitheater, but the Marines called it the "Meatgrinder." "They fought the Japanese in ankle-deep sand with rifle butts, bayonets, and knives. The survivors opened a gap in the main enemy line." Sounds of men screaming and barking orders were heard between the deafening sounds of mortars and gunfire. Lieutenant General Smith told correspondents, "The fight is the toughest we've run across in a hundred and sixty-eight years." Nearby, Forrestal, now a deeply concerned and worried Secretary of the Navy, concurred.[1963]

Corporal John Wlach of Fox Company, 2nd Battalion, 21st Marines, fought right near Woody on 23 February 1945. Due to his platoon commander having transferred to another outfit, Wlach took over his unit. During a fierce battle with the Japanese, he threw many of the grenades landing in his area back at the attacking Japanese. Seeing some allowing fear to stymie action, he stood up on a mound and yelled, "Come on, you fellows, the only thing that can happen to you is to get killed!" With that declaration, he and his men charged the Japanese and took the enemy position. St. Louis Personel Records.

Woody's squad moved to Hill 283. On 24 February, he was ordered to dig in north of this position in preparation for attacking the second airfield. When he told his men to move out, one of them, Private Thomas E. Scrivens, refused. Woody yelled, "Come on Scrivens, get moving." Scrivens said he could not. Woody became disgusted and lost his ability to be "rational, considerate and compassionate."[1964] Seeing a Marine break down and refuse to keep fight-

Private Thomas E. Scrivens. When Scrivens could not go forward on 24 February 1945 due to shell-shock or fatigue near Hill 283 on Iwo, Woody threatened Scrivens to get on the line or he would shoot him. Scrivens luckily got back on the line since Woody admitted on 20 July 2016 he would have shot Scrivens and killed him had he not done so. St. Louis Personnel Records Center

ing alongside his brothers infuriated Woody. He pulled his rifle to his shoulder and said, "Scrivens, I mean it, if you don't get moving I'll shoot you myself."[1965] Reconsidering, Scrivens realized Woody would not let him drift behind the lines when his squad needed him, so he decided he could advance after all.

Woody believes he was so angry, he might have shot Scrivens had he not taken his place beside his comrades.[1966] No one knows what would have happened if Scrivens had not obeyed. If he believed Scrivens was unable to carry on due to combat fatigue, Woody might have sent Scrivens back for treatment.[1967] John Lauriello saw many such cases since he was a radioman and behind the lines, and said they "had a frantic look in their eyes as if they were being chased by the devil."[1968] If Woody thought Scrivens able to obey, but chose not to, he could have written Scrivens up for disobeying a lawful order which could have meant some minor, non-judicial punishment by his officer or even referral to a court-martial where Scrivens, if found guilty, would have faced a fine, reduction in rank, or even brig (military jail) time. Scrivens was entitled to consideration of the extenuating and mitigating circumstances, including his prior combat record as well as the stress of being in some of the toughest fighting for days with little rest. Soon though, he would be removed from combat on 1 March due to "shell-shock/combat fatigue." Woody kept him in the game for a few more days, but eventually, Scrivens succumbed to the strain.[1969]

If Woody had shot Scrivens, he would have been in a world of trouble. Woody would have faced serious charges at a court-martial. While the Axis powers and the Soviet Union lacked respect for justice, due process, and the rule of law, America took these issues seriously. Japan, on the other hand, did not. The Japanese general commanding forces fighting British General Bill Slim's Allied Army in Burma, issued an order telling his men if they performed well, their names would be sent up the chain of command, but if they performed poorly, their officers were to exercise summary discipline with their swords, i.e. decapitate them. These were not hollow words. When an officer failed to carry out orders in Nanking in 1937, his superior beheaded him.[1970]

Disobedience of a superior's lawful order or deserting one's duty in battle were general court-martial felony offenses in the Articles for the Government of the U.S. Navy (called "Rocks and Shoals") applicable to sailors and Marines.[1971] A properly convened court-martial could impose a sentence of a fine, reduction in rank, imprisonment, or death for one found guilty. Woody, as a squad leader, had no authority to be judge, jury and executioner, and everyone was taught that at boot camp.[1972] Even minor punishment (no more than 20 days confinement) required action by a commissioned officer who was entitled to order a general or special court-martial.[1973] Even a summary court-martial, which could sentence an enlisted defendant to no more than two months in the brig and could order a man discharged, required a trial before three qualified officers, with a court reporter, and gave the defendant an automatic appellate review after the trial.[1974] Woody had been taught that Scrivens, for disobeying an order could be "reduced [in rank]. Two years confinement, dishonorable discharge, and other accessories of sentence…"[1975] Woody may not have remembered the exact consequence of disobedience, but he was taught that it typically entailed dire repercussions. On the other hand, if someone was suffering from combat fatigue or was somehow mentally or physically unable to carry out an order, that was a valid defense and charges should not even be referred to a court-martial, as was the case with Lefty and Scrivens. This was not a system that allowed capricious, summary executions by a low-ranking NCO or condoned drum-head courts, but rather a system that guaranteed the accused a speedy and fair trial. When Woody acted as he did, he was treading on thin ice. As it turned out, both Scrivens and Woody were able to fight and continued to make a contribution and to finish out 24 February with honor.

War placed such stress on the combatants they did things they later regretted when they had time to reflect. In 1957, Scrivens went to Woody to apologize, but asked, "Do you think you really would've shot me?" Woody answered, "Yes, I would've." Woody explains he would react differently as a mature adult, knowing all he does now, but after the horror he had experienced, being stressed out and sleep deprived, a high school dropout with no medical training and no real scientific understanding of the psychological effects of extended periods of combat on the ability to continue to function, he was appalled one of his men would abandon comrades in time of need. Woody remembered seeing some crack up and never resume fighting, but they were not going to be him or men under him.[1976]

Woody saw several men give up and the possibility he might feel similar despair scared him. "I kept repeating to myself, 'I'm not going to die. I'm not going to die.' I would never let myself think that I was not going to make it. I had to get back to that girl in Fairmont who I wanted to marry. That is what kept me sane."[1977] These phrases, repeated in his self-talk, motivated him to move forward and removed the icy chill from creeping into his heart to create doubt and fear.

Holocaust survivor Viktor Frankl recognized this phenomenon as the psychosocial concept of logotherapy which holds that those who find meaning in life live the most productive ones and can survive horrendous events. Those who survived Auschwitz usually did so because they looked forward to seeing a loved one, enacting revenge or writing a book. They had a reason to live. Woody's main reason was to get back to Ruby, and that goal kept him "sane." Marines like Woody had choices unlike Holocaust victims; however, both groups faced death under cruel circumstances on a minute-by-minute basis. Frankl observed that people who face life-threatening situations condition themselves mentally for something else in order to deal with the death often staring them down.

Philosopher Friedrich Nietzsche said, "He who has a *why* to live can bear with almost any *how*."[1978] Why did Woody do what he did? Japan had attacked his country and threatened his people, family, friends and way of life. He did it to protect his buddies, because by killing more Japanese he reduced the danger they presented to him and to those around him. Defeating the enemy did not mean wealth, power and ruling over conquered people. It did not mean the freedom and license to rape defenseless women like many of the IJA soldiers he was killing had done all throughout China. It was the chance to return home to his beloved Ruby and live in freedom. This goal bore fruit in his behavior, with profound results. It gave everything he did meaning.[1979]

And maybe Woody also fought aggressively because he did not want others as well as himself to think he was a coward. Transcendentalist Henry David Thoreau said, "Public opinion is a weak tyrant compared with our own private opinion. What a man thinks of himself, that it is which determines... his fate."[1980] Woody's fate was to love and be loved by Ruby and to get that, he had to fight like a tiger to survive—society and women do not tolerate cowards very well. Woody and fellow Marines became fiercely determined warriors because most latched on to some reason to live (not die like the Japanese). They also wanted to eliminate the enemy forces. Woody said, "We felt no mercy [for the Japanese]."[1981] Radioman John Lauriello said, "The Marines made us into animals. We had to fight these little barbaric animals called Japanese...so our leaders had to make us into big vicious animals, and they indeed succeeded."[1982] As Lieutenant General Smith wrote, the Japanese "were tough. But my Marines were tougher."[1983] He continued:

> To accomplish our task on Iwo...we had to produce men who were tougher than the Japanese, who could beat them at their own game, whose patriotism transcended that of the enemy, who also could reach the heights of *Bushido*, but in the American way, which is based on cool reason and methodical efficiency instead of blind obedience to the dictates of fanaticism.[1984]

Another example to describe this American *Bushido* was exhibited by one of Woody's witnesses, Sergeant Lyman D. Southwell, while Woody was getting Scrivens back on the line to attack the Japanese. Running headlong once again into a line of pillboxes, Southwell and Woody's company faced a dire situation yet again. This time, instead of relying on flamethrowers, Southwell's platoon had a Sherman tank. Moving with the iron monster, Southwell "volunteered" to direct its fire against the enemy positions. Ascertaining that the tank's "telephone was out of order" (tanks actually had a phone on the outside for troops to use to speak to men on the inside), Southwell "boldly exposed himself to the enemy fire to beat upon the ports of the tank and attract attention." In doing this act, he placed himself in full view of the enemy "with complete disregard for his own safety," but had he not, the tank would have attacked in the dark. The "insane" Southwell actually remained in his exposed position and directed the tank to fire on five positions which were all eliminated with his instruction enabling his platoon to continue the advance. For his bravery, Southwell earned the Silver Star (his citation was signed by Lieutenant General Roy S. Geiger).[1985]

When one observes how Woody, Southwell, Basilone, Wlach, Schlager, Chambers, Beck, Williams, Murray, Erskine, Smith and others performed on Iwo, one can see they were grinding the Japanese down, overpowering their defensive positions and beating them. Although the Japanese were extreme, the Marines matched their commitment and surpassed them in combat.

After taking out the pillboxes, Woody still had a lot of fighting to do. He had more psychological challenges still to overcome in addition to war's physical trials. Woody said taking out the pillboxes was the hardest thing he had ever done, but a close second was fighting at night. Every minute during the blackness set nerves on edge. It was like walking through a dark house knowing that behind every door or under every bed was a demon ready to jump out and attack you. Woody explained:

> I never slept very well for that entire month. If I was not fighting
> the Japanese at night, I was dreaming about fighting the Japanese
> at night and that naturally woke me. I was either fighting a night-
> mare or dreaming one. What Japanese did and could do at night
> scared me more than anything else.[1986]

The IJA, like the IJN, trained for night combat. "Night infiltrations and minor counterattacks were constant; day and night men appeared from overrun caves and tunnels, necessitating a continuous mopping up of seized ground."[1987] At night, the U.S. Navy fired star shells illuminating the landscape to prevent a *Banzai* or an enemy unit from sneaking into Marine lines. The starbursts illuminated weird forms along the horizon, sometimes showing the Japanese moving along ravines and hill crests "like little devils running through Hell...All they needed were pitchforks."[1988]

At night, a command post behind the lines loudly played music of Glenn Miller, Benny Goodman, Tommy and Jimmy Dorsey, the Andrews sisters and the Ink Spots. The sounds of these bands "wafted loud and clear among the ravines and cliffs." Hearing this dance music "must have convinced the Japanese the Marines were, indeed, crazy." On the other hand, many on both sides loved American music despite the efforts of Japanese leaders to persuade their people to shun American culture. Gene Jones said,

> I'll never forget how sweet and beautiful that music was…how mar-
> velous and comforting. Out across the shattered landscape, across
> the bodies of dead Marines, we heard swing music, dance music;
> languorous and poignant. We had won so little ground and lost so
> many men and were so tired, but we all listened to the music. It
> meant so much to hear these tender sounds…amid the killing.[1989]

Earlier, when Lefty was in combat, he heard Billie Holiday's *I'll Be Seeing You*
and thought of his girl Marian Hengel which gave him hope.[1990] The Marines
needed hope because by 25 February, 1,600 Marines had died and 4,500 had
been wounded.[1991] By 25 February, Woody's regiment alone had suffered 518
casualties.[1992]

To inspire his men, Woody's divisional commander, Erskine, landed on the
island on 24 February with the 9th Marines and the division's artillery and fought
with his men remaining with them for four days.[1993] He was in the heat of battle:
"Our headquarters…was not over 500 yards, I don't think, from the front line.
We'd frequently get a spray of machine gun fire."[1994] While in his CP, Erskine's
signal officer was talking on the phone when suddenly he grabbed his carbine
and shot two attacking "Japs" right before they entered the area.[1995] This was
common among Marine generals, to be close to the troops. On Guadalcanal,
for example, four IJA soldiers reached General Vandegrift's "command tent and
killed a…sergeant with a sword" before Marines nearby shot them down.[1996]

After taking Airfield No. 1, Woody and the 21st Marines moved to Airfield
No. 2, methodically taking every square foot of island along the way by killing enemy soldiers who emerged from caves, tunnels, bunkers and buried oil
barrels.[1997] Dashiell described that walking by one cave, when he, "like a fool,"
stopped and looked in. "I didn't see anything so I kept walking. A minute later
a squad that was behind me got ambushed from the same cave. The only reason they didn't get me was they didn't want to give their position away for one
man."[1998] The small island was crowded with Japanese: It was 5,120 acres or one
third the size of Manhattan.[1999] Kuribayashi's 22,000 soldiers could have put four
men on every acre if spread evenly. America would land over 80,000 Marines,

sailors and soldiers plus many thousand tons of equipment and supplies on Iwo, keeping the main attack force on the island around 45,000 to 50,000 as the dead and casualties were rotated out and replacements rotated in.[2000] At the start of the battle, around 11,500 Marine replacements were earmarked for the entire Pacific theatre and some feared they would all have to be used at Iwo.[2001]

Sergeant Frederick K. "Dick" Dashiell on Iwo 23 February 1945. He is the one with glasses and a carbine rifle. The Marine to his right is unknown. In this photo, they were to the west of the First Airfield by a wrecked Japanese plane. Author's Collection and Dashiell Family Archive

Ships with at least 150,000 sailors surrounded the island, providing Marines with essential weapons, ammunition, fire support, water, food, fuel and medical wards.[2002] The U.S. stacked the cards well in its favor by sending a massive force to capture the island. The general rule of thumb for an attacking force to have success is that it must have a three-to-one superiority; the U.S. pushed this number to around ten-to-one. As Napoleon said, "God is on the side of the big battalions."[2003] Carl von Clausewitz echoed Napoleon in preaching that the "first rule of strategy" was to always "put the largest possible army into the field."[2004] However, even with the odds stacked in their favor, the Marines were still attacking a defensive position, and defense "is nothing but the stronger form of combat."[2005] The Marines needed superior numbers or they would not win.

The Japanese tried to mount airstrikes of their own, but with limited success. One bomber was able to fly over the island and drop bombs on the night of 24 February. As the plane flew its bombing run, a crewmember described Iwo in the distance resembling a "bed of glowing, writhing hot coals." Radioman Yamada said, "It was what Hell must look like." "Here and there, long flames of fire stretched out and disappeared," as if fiery tongues were sticking themselves in and out of volcanic surface mouths; however, it was not lava creating the flashes, but the constant use of machineguns, grenades, artillery and mortars. The Japanese on the ground marked their battle lines for their pilots with flares and the bombs fishtailed down close to where Woody was. Kuribayashi thanked his command and asked for more.[2006] These bombs did not hit any Marines, but the "great amount of flak" fired at this plane came falling back down and caused one casualty at the 2nd Battalion, 24th Marines' CP—a freakish case of friendly fire.[2007]

At night, Woody never left his foxhole unless he needed to relieve himself. Field sanitation broke down under the constant fighting. Under normal conditions, field armies placed their bathroom facilities away from the troops in the rear. However, during this type of combat, infantrymen on both sides could not leave their area of protection or foxhole under any conditions without making themselves targets. James "Big Moon" Mulligan of the 28th Marines got out of his hole to have a bowel movement and while he was "taking his shit, they shot him through his ass and belly and he just fell over and bled out."[2008] Consequently, most did their "business" in their hole and then threw it over the lip in the open. As a result, tens of thousands of men, both Japanese and Marines, defecated and urinated wherever positioned. During the nights, a Marine might use an empty grenade canister or ration can and throw it outside his hole. Maybe during the day, if he had the opportunity to do so, he might scoop some dirt over it, but it was usually the last thing on his mind. The smell of raw sewage added to the stench of rotting corpses making a revolting, odorous cocktail. Most worried little about keeping their waste hidden and in one place. This helped make the island's landscape "inconceivably vile."[2009]

During the first week's chaos, the entire battlefield was strewn with dead bodies. Fissures of sulfurous steam spewed from cracks in the ground and mixed with the aforementioned stench of human waste and decomposing cadavers. During the day, the heat turned the corpses dark and they swelled with gas. Black and florescent green-blue, and hairy, blowflies feasted on a smorgasbord of cadavers and laid their eggs in any openings they found along the carcasses, filling the bodies full of maggots and the "island became infested" with the insects. One female fly can lay up to 300 eggs in one laying that will hatch in a single day, but hundreds of these bugs would actually land on a corpse to deploy their larva so the dead bodies had tens of thousands of ova within minutes of dying since flies can smell death miles away (and get to their targets quickly to do their dive-bombing runs since they can beat their wings 100-times per second to fly at 5 mph). And the ova would quickly hatch maggots that immediately began feasting on the necrotic tissue. Since 156 grams (.34 pounds) of meat can produce 231 flies, and since the muscular system accounts for 50% of a person's body mass, a 180-pound dead man can theoretically produce 61,481 flies within a day (that means, as few as 100 KIAs could yield over six million blowflies within 24 hours).

The flies became so bloated on Iwo that if swatted, they squashed into a pool of blood, the very blood taken from dead bodies. Flies landed in latrines where food was prepared and on people's faces. They flew from Japanese zones to Marine areas as uninvited guests. Dysentery can be a great killer. The unhygienic situation with the flies became so bad, the island was sprayed during the battle on the 23rd with DDT insecticide. Two torpedo bombers flew from a carrier and sprayed the chemical from shoreline to shoreline.[2010] Some Japanese mistook the DDT for poison gas and must have put on their masks.[2011] Behind the front lines, the number of dead Marines carried back overwhelmed divisional troops assigned to bury them and they did their best to spray the rotting bodies with disinfectant.[2012] Woody said, "It was hard to see all those young boys, lifeless, taken back to be buried."[2013] DDT was often sprayed over the island; on 28 February, it was noted the entire island was sprayed again.[2014]

Rabbi Gittelsohn found it difficult to bury so many who were often unidentifiable. While at the cemetery one day, he had to bury a hand, a piece of a foot, 12 inches of somebody's torso, bloated stumps and burnt bodies. Yet, he, along with other chaplains, dutifully gave the body parts a Protestant prayer, the Catholic Last Rites and the Jewish *Shema*[2015] to cover all bases for the unidentified person. They sent souls on their journeys as they tended to the ghastly business of helping the wounded and burying the dead. Tragically, sometimes the men who had just plowed the burial plots for the dead would be caught, unexpectedly, by enemy shells, and "were buried in the grave they had just dug."[2016]

During the battle, the red harvest of war had to be dealt with and burials happened daily on Iwo. 3 March 1945. National Archives, College Park. Photo Credit Staff Sergeant M. A. Cornelius

Due to unburied decaying bodies, the island reeked of rotting flesh. Sometimes the smell was so powerful, it made Marines vomit suddenly.[2017] Some dead lay where they had fallen, and many lay close to those pinned down and unable to move for days, much less able to bury the dead. Many corpses had been burned by napalm or flamethrowers, as evidenced by their blackened and shriveled bodies as if someone had left them too long on a grill during a cannibalistic barbecue. "The smell of burnt flesh," recalls Woody, "never left my nostrils. It stayed with me the whole time on the island."[2018] The odor seeped into everything, from one's hair, to clothes and boots. "It is difficult to put words to the smell of decomposing humans. It is dense and cloying, sweet but not flower-sweet. Halfway between rotting fruit and rotting meat."[2019] The Marines fought in a place "so degrading I believed we had been flung into hell's own cesspool."[2020]

Chaplain E. Gage Hotaling performs the last rites for a Marine on Iwo. March 1945. Hotaling found it especially difficult to bury so many men with young families writing: "It is the saddest things I know that war should do this to people, to take the heads of families and send them off to die, leaving thousands of little children to grow up without any daddies." National Archives, College Park. Photo Credit PFC Bud Lindsley

During their 70th Iwo Jima reunion, veterans like Sergeant William "Bill" Pasewark Sr., explained how hard it was for them to see the dead, especially after several days. The bloated, putrid, decaying bodies still haunted their dreams all those decades later. Often, when they tried to move the dead to the rear lines, the skin would roll off of legs and arms when grabbed. Forensics call this gloving.[2021] Sometimes the bodies were so degraded, they fell apart. Body bearer Robert Hensely of the 3rd Service Battalion, 3d MarDiv went out with his comrade to pick up a dead Marine. They found the fallen man face down in the sand. Hensely set the stretcher next to the body, grabbed the shoulders and rolled him over. His buddy grabbed the legs and as they lifted the corpse up, it had been so badly damaged in the midsection that it split in two. Hensely's comrade lost control of himself, threw down the legs, and ran screaming like a madman. "I was left alone with the dead man, now in two parts. I just placed the legs on top of his torso and then did my best to drag my fellow Marine to his final resting place."[2022] Woody stated, "Often, I found myself feeling guilty thinking I didn't like seeing these dead men and wish they would just go away because they bothered me and took me away from focusing on killing the enemy."[2023]

Three dead Japanese soldiers lay where they died defending their pillbox. National Archives, College Park. Photo Credit D. Fox

Here is an example of a dead Marine who had probably been buried in the hot, volcanic soil for a few days. One clearly can see the violent death he met and now he was mummified by the island's heat. National Archives, College Park. Photo Credit Sergeant Mulstay

A Japanese officer commits suicide by holding a grenade to his face. Corporal Walter Killen kneels by the dead enemy. The white dots on the dead IJA officer's face, or what is left of it, are maggots and the dark dots all over his body and uniform are flies. Notice he is missing his right hand due to him probably having held the grenade that killed him with this hand against his face. 26 February 1945. In the background, one can barely see an example of a well-camouflaged, and well-hidden, Japanese bunker. National Archives, College Park. Photo Credit Schwartz.

The advance on Iwo was absolutely vile and full of dead bodies as evidence in this picture as a bloated corpse of an IJA soldier was right behind the advancing Marine, Corporal Marlin Hoge. 28 February 1945. National Archives, College Park. Photo Credit Corporal Schwartz

Liquids oozed out of every orifice of the bodies when the remains were moved to graves. This landscape of corpses had an "atmosphere which had no affinity with the air of heaven…a pestilent and mystic vapor, dull, sluggish, faintly discernible, and leaden-hued."[2024] A modern reader rarely understands death and decomposition, especially so on that hot island that often had daytime sunshine heating up the black sand and whose volcanic geothermal temperatures exceeded 160° Fahrenheit.[2025]

Exertion in that climate can lead to heat exhaustion or stroke. Woody explained, "The first few nights, I got sick in my foxhole because I was suffering as a heat casualty. I was dehydrated and constantly feeling the warmth radiate up through my body at the bottom of my foxhole. It got so hot, I had to put cardboard or wood underneath me to not feel uncomfortable."[2026] Now in this hellish landscape of heat and sun, add thousands of dead bodies decaying at different rates, and even Dante would avert his gaze and cover his nose.

As soon as life ceases, the body begins to decay, with death working its way from inside out. Human cells use enzymes to break down molecules into compounds they can mobilize for their biological functions. In living tissue, the cells keep these enzymes in check, "preventing them from breaking down the cells' own walls."[2027] Upon death, these enzymes "operate unchecked" and eat through the cell structure, "allowing the liquid to leak out."[2028] Soon the skin starts to come off bodies, peeling in strips across the frame. "The brain liquefies very quickly. It just pours out the ears and bubbles out of the mouth."[2029] Flies quickly lay their eggs in every point of entry they can find as already mentioned: Wounds, eyes, mouths, noses, ears, and genitalia. Soon the body, now turning blue, squirms with maggots that feast on dead tissue. They gravitate toward fat and devour it. And now, the body's cells rupture because they're eating themselves and pour their fluid into the body's bacteria colonies. Bacteria then multiply creating gas, a waste product of bacterial metabolism, which the body can no longer expel as flatulence, so the bodies bloat. The dead lack workable stomach muscles and sphincters, and thus cannot expel gas, so it inflates the corpses into rotting balloons. The bellies, mouths and genitalia get large on the dead; testicles can swell to the size of softballs. At this stage, the eyes are gone—"X's. In real life as it is in cartoons."[2030] When close to a corpse at this time, one can hear the maggots feasting because it sounds like "Rice Krispies." Eventually, the body will explode with a "rending, ripping noise." Then, the bodies "collapse and sink in upon themselves and eventually seep out onto the ground."[2031]

Veteran Eugene Sledge wrote, "It was gruesome to see the stages of decay proceed from just killed, to bloated, to maggot-infested rotting, to partially exposed bones—like some biological clock marking the inexorable passage of time."[2032] Six days into the battle, the island was covered with the sight and smell of thousands of dead decomposing to "ashes to ashes and dust to dust" (Genesis 3:19).[2033] Sledge wrote about the maggot-filled battlefield, saying, "[I]t is too preposterous to think that men could actually live and fight for days and nights on end under such terrible conditions and not be driven insane. But I saw much of it…and to me the war was insanity."[2034] Woody and his comrades constantly moved through an unburied cemetery to kill the enemy, always having to crawl across the bloody by-product of war to do so.

North of Airfield No. 1, Woody was involved in the push to take Airfield No. 2. Looking over the wide expanse of the airfield, the Japanese had full view of the next attack and the Marines feared that for many this would be their last day. It was a time almost of madness as they contemplated what they had to do and felt death sneaking up on them. In fact it was worse than being mad, "because one was aware of it."[2035]

As the sun rose on the 24th, Woody looked over the ground they had to cover to take the second airfield. Before they stepped off, an artillery barrage hit the other side erupting the earth into geysers of rock and dust. Air became acrid with the smoke of the guns and explosions. A bitter taste set itself on the tongue as the powder from cannons and spent shells settled on the battlefield. There was an eerie unapproachable silence when it stopped. The order to charge was given, and without hesitation, the Marines attacked, following a rolling barrage. It was one of the most amazing assaults the Marines had made, as hundreds raced across the airfield's midsection to close with the enemy. Men fell everywhere. One shell removed a Marine's head from his shoulders as he ran and "his headless body" continued for several steps "before collapsing."[2036]

The men never wavered in their attack.[2037] "It was one of the most resolute charges since Pickett's at Gettysburg," but with a better result. When the Marines reached the enemy side of Airfield No. 2, the Leathernecks flung themselves at concrete enclosures, armed with grenades, bazookas and satchel charges. When volcanic ash jammed or clogged weapons, they surged at the enemy, pummeling them with "rifle butts—even entrenching tools."[2038]

After taking Airfield No. 2, the Marines continued over the ridge to the north leaving the sands and entering "a wild, barren stretch of rocky ridges, cut into crags, chasms, and gulleys." It reminded Sergeant Alvin Josephy of "the Bad Lands of the American West—or, as someone said, like hell with the fire out." As soon as Woody entered this area, Japanese poured out of caves and hidden bunkers, throwing many Marines back. The Marines regrouped and charged at the Japanese, retaking the area above Airfield No. 2 in a few hours of fighting.[2039] In typical Marine fashion, gallows humor sprang up everywhere. Signs went up by landmarks that read:

**Suribachi Heights Realty Company**
**Ocean View**
**Cool Breezes**
**Free Fireworks Nightly!**

**Or**
**Ichimoto's Inn**
**Under New Management**
**Soon Available to Personnel**
**Of the U.S. Army (We Hope)**[2040]—(after islands were secured by Marines, the army garrisoned them).

One commentator said, "Battle humor is grim and often not funny in the telling." In one sector, Marines laughed hysterically when a Leatherneck jabbed his bayonet at a Japanese's ass as the soldier ran from a pillbox. Before the Marine could stab him, one of his buddies shot the soldier dead. It was as if they were playing a game, "Who can kill him first" and everyone around chuckled. Another time, Marines surrounded a pillbox and in the ensuing stalemate, the Marines started making catcalls "to the Japs inside." Suddenly, an angry Japanese appeared with a machinegun, climbed on the structure's top and was putting it in place to fire when erupting fire cut him down. A Marine called out, "That guy couldn't take a ribbing." Laughter echoed throughout the area.[2041] At another battle, after obliterating a Japanese unit, Marines captured its carrier pigeon. The

Jarheads released the bird with a "typical Marine taunt: 'We are returning your pigeon. Sorry we cannot return your demolition engineers."[2042]

Seabees got involved with the humor and as they buried the dead, one "comedian" placed five dead Japanese in a mass grave with the headstone: "Here Lie Five Nips 1945: [Names] 1. Hir No Iwo 2. Si No Iwo 3. Spik No Iwo 4. Du No Iwo 5. No Mo Iwo. By Jima." The burial detail obviously was made of callous men, "but given the times and context who can blame them."[2043]

A mass grave for five Japanese soldiers. The Marines and Seabees had become quite cynical as evidenced by this tombstone, but after what they had seen on the island, who could blame them. National Archives, College Park. Tech Sergeant Kress.

Gallows humor notwithstanding, the Marines still had a lot of fighting to do, and it was no laughing matter. Every day, Woody and his squad would close with an enemy trench or pillbox and they would destroy it. When they found a cave entrance, Woody blew it closed with C-2 explosives.

Here is an example of a flamethrower tank (*aka* "Zippo" tank) engaged in combat on Iwo. Such tanks could send a stream of flame out over 100 yards. Notice there are five Marines trailing this tank, three on the left side of the road and two on the other side. The training Woody and others had at Camp Jacques Farm, that taught many Leathernecks how to work with tanks, was paying dividends on Iwo. National Archives, College Park. Photo Credit Kaufman

By the time Woody and his team destroyed the pillboxes on the 23rd, the Seabees had used metal Marston Mat (pierced steel plank) roadway to get tanks across the volcanic ash on the beach and had built enough makeshift roads to get the Shermans into battle. There were almost 150 tanks plus armored bull-dozers in combat at that point. As the dozers started to cancel out the Japanese terrain advantage, they became targets themselves, but in doing so, the procedure helped nearby Marines target the enemy's firing positions and destroy them. Some thought that without these tanks "the assault would have failed."[2044] Nearly all units were equipped with tanks with small flamethrowers shooting from the machinegun ports, but the newer model, fitted with a flamethrower launcher tube in place of the 75mm gun, was especially effective. Major General Erskine noted: "The flamethrower tank…was found to be most effective against emplace-ments and caves."[2045] Unfortunately for the 21st Marines, most went to other divisions. But, by 24 February, Woody's regiment was largely supported by new tanks on loan from other divisions that delivered larger blasts of flame than Woody's shoulder pack. The *Zippos*, named for the popular lighter of the day, could shoot flames "approximately" 100-150 yards. *Zippo* tanks were often M4 Sherman tanks outfitted with a flamethrower cannon carrying 150-290 gallons of fuel.[2046] A Japanese Iwo veteran testified: "The flamethrower tanks were awful. I heard our troops screaming. The sounds combined to form a buzzing sound like radio static."[2047] Major General Harry Schmidt said, "Flamethrower tanks were indispensable in the Iwo…operation."[2048] A tank officer noted that the 3d MarDiv fighting on islands did not use "*panzer* attacks," referring to the German general Heinz Guderian's edict and his massive waves of armor, but rather, tanks supported the troops in a "close-in, inch-by-inch advance."[2049] Woody agreed saying that having the tanks as mobile pillboxes with cannons provided support in taking out Japanese positions in the middle of the island.[2050] Marines would fire tracer rounds at the pillboxes they wanted taken out and the tank gun crews then hammered or flamed away at those areas marked for destruction.[2051]

Now, busting bunkers became easier than it had been a few days before. According to the regimental report, a dozen tanks supported the Marines at the fringes of Airfield No. 2, although the Japanese knocked a lot of them out with mines and anti-tank cannons.[2052] They continued to reach the battlefield from the beach using Marston matting, allowing them to negotiate the pumice.[2053] With the tanks' help, Marines took Airfield No. 2, but it was still ringed to the north with lines of Japanese soldiers. The night of the 24th, Marines experienced several counterattacks, and although they repulsed them, they suffered "considerable casualties."[2054] Even so, as the Marines moved northward, they left piles of dead Japanese. They were sprayed with DDT and were buried where they fell. Often, Korean POWs were used to dig the graves for them.[2055] A few Japanese POWs were taken too. When one was taken prisoner, he pleaded with a Marine: "No chop head off? No chop head off?" Unlike what the Japanese had done everywhere they occupied, slaughtering POWs and defenseless citizens, the Marines "didn't chop his head off, or do anything else to him."[2056]

The Marines sweated and labored by day, but at night, due to the cold air, wind and rain, they shivered in their holes. One shaking Marine hated the cold nights so much he wished they were back on Guadalcanal.[2057] Woody remembered intense fighting on the northward march. When a Japanese officer wielding a *Samurai* sword attacked Leo Jez, Leo caught the attacker's hand and part of the sword, snatched the weapon from the enemy and then used it to chop off the officer's head. Later, Jez arrived at the aid station with his thumb almost severed away from his palm, a large gash in his hand and a bloody sword, now his. Another Marine lay in wait on top of a bunker for a Japanese gunner to leave. Once the enemy felt safe, he crept out of the shelter. Like a cat, the Marine jumped off the bunker and onto his enemy's back, ramming the knife into his neck. Such hand-to-hand combat became commonplace as the two opposing forces comingled in close quarters.[2058] Even with such primitive combat, Marines still brought incredible firepower to the battlefield. In the first week alone, they launched over 32,000 mortar shells at the Japanese.[2059]

They took small steps, but they churned forward. On 28 February, around the Second Airfield, they encountered resistance after starting their advance at 0815. Within 100 yards, Woody's battalion was "stopped and pinned down by extremely heavy fire from all types of Japanese Inf[antry] W[eapons]…from caves, pillboxes, trenches and dugouts. No further advance was made prior to 1800 although constant aggressive action was taken to effect a penetration of these positions."[2060] To keep the lines of communications open, Woody's 21st Regiment's radiomen "laid approximately 40 miles of wire" during the first six days of combat "under sniper and mortar fire" to keep units in touch. Bent over and carrying up to half-mile reels of wire, they worked tirelessly to maintain communications.[2061] Woody and his men measured progress in feet and the number of Japanese killed.

Woody's friend in Bravo Company, Vernon Waters, encountered stiff resistance on this day of 28 February, and now, it was Waters' turn to show his bravery. Bunkers held them up once again. There were two "mutually supporting pillboxes… ahead of the unit" and Waters volunteered to take them out. With only a squad using small arms fire to support him, Waters worked his way across the expanse with satchel charges. He was a huge target standing 6'4¼", but in a remarkable feat of dexterity, he traversed 30 yards to reduce one structure by throwing explosives into it while receiving fire from the other pillbox, and then, undeterred, turned his attention to the remaining structure and blew it up as well. After taking out the second pillbox, Waters confronted a small *Banzai* of three armed Japanese emerging from a cave. Waters "coolly eliminated [them] with a hand grenade." "His heroic action undoubtedly saved the lives of many… and immeasurably contributed to the advance of his unit." For his actions, Waters received the Silver Star (signed by Lieutenant General Roy S. Geiger).[2062]

SAMPLE CITATION FOR AWARD OF SILVER STAR MEDAL, CASE OF CORPORAL VERNON "J.H" WATERS, (854581), UNITED STATES MARINE CORPS RESERVE.

For conspicuous gallantry and intrepidity in action against the enemy while attached to a Marine rifle company on IWO JIMA, VOLCANO ISLANDS on 28 February, 1945. While serving as the demolitions sergeant of his company, Corporal WATERS unhesitatingly volunteered to go forward to destroy two heavily fortified mutually supporting pillboxes which had held up the advance of his company with deadly automatic weapons' fire, and which were inflicting heavy casualties. In the face of this fire, and with complete disregard for his own personal safety, he worked his way across thirty yards of open terrain with two demolitions charges. Unmindful of heavy fire aimed directly at him, he crawled to one emplacement, placed his charge and destroyed it. Under similar hazardous conditions, he also destroyed the remaining emplacement. At this time he was attacked by three heavily armed enemies from a nearby cave, but he coolly eliminated them with a hand grenade. Corporal WATERS heroic actions, brilliantly executed in the face of these tremendous dangers were an inspiration to his comrades and immeasurably contributed to the general success of his unit. His conduct throughout was in keeping with the highest traditions of the United States Naval Service.

ENCLOSURE (A)

Corporal Vernon Waters' Silver Star Citation (it would be endorsed and signed by Lieutenant General Roy S. Geiger later). Waters' citation here is concise and detailed, quite different from Woody's final MOH citation. Waters destroyed two pillboxes with satchel charges and killed three charging Japanese with a grenade. St. Louis Personnel Records Center

```
1740-55-80              HEADQUARTERS,
0296/360          FLEET MARINE FORCE, PACIFIC,
               c/o FLEET POST OFFICE, SAN FRANCISCO.        RECEIVED
                                                           25 JUL 1945
Serial:
       56298                                            JUL 24 1945

                          3d Endorsement on
                          CO, 21stMars, 3d
                          MARDIV, 1tr 1740-5
                          SBG-rdp dtd 7May45.

From:             The Commanding General.
To:               The Secretary of the Navy.

Via:              (1) The Commander in Chief, U. S. Pacific Fleet
                      and Pacific Ocean Areas.
                  (2) Commandant of the Marine Corps.

Subject:          Award of the Silver Star Medal, case of Corporal
                      Vernon "J" "H" WATERS, (854581), U. S. Marine
                      Corps Reserve, (Posthumous) - recommendation
                      for.

      1.          Forwarded, recommending the award of the Silver
Star Medal, posthumously, to Corporal WATERS as a suitable deco-
ration for the act cited.


                                        ROY S. GEIGER.


Copy to:
   CG, 3dMARDIV
```

Corporal Vernon Waters' Silver Star 3rd Endorsement signed by Lieutenant General Roy S. Geiger, 7 May 1945. St. Louis Personnel Records Center

The same day, one of Woody's and Waters' platoon commanders, Lieutenant Charles Wesley (Mike) "The Ox" Henning, led his men in an action that earned him a medal too. Henning, seeing his company held up by an entrenched force "defending a strategic hill," moved his men into assault positions and pressed ahead. Henning "courageously led his men in an aggressive attack in the face of fierce enemy mortar and machine-gun fire." He pushed 1,000 yards forward in an effort to seize the hill, but was repulsed twice. Units on both sides were stuck themselves. Henning reorganized his men, led a third attack and defeated the Japanese while capturing a 47mm cannon and six 25mm dual-purpose weapons. He received the Silver Star (signed by Lieutenant General Roy S. Geiger).[2063]

None other than Woody's, Waters' and Henning's regimental commander, Colonel Hartnoll J. Withers, led all of them from the front on this day. Lieutenant General "Howlin Mad" Smith took note of Wither's incredible leadership in subjecting himself "to the most intense fire in order to move out with his advanced elements." Seeing one's commander leading from the front and "practicing what he preaches," inspired men like Waters and Henning to perform noteworthy deeds. "By heroically leading his regiment in person, he contributed in great measure to the high morale of his men." He

Corporal Vernon Waters, Woody's friend and fellow flamethrower. He received the Silver Star for single-handily destroying two pillboxes and killing many Japanese on 28 February 1945. It is my opinion that he probably killed at least 11 Japanese on this day. He died on Iwo Jima on 3 March 1945. St. Louis Personnel Records Center

helped the numerous attacks by his brave Marines push forward their advantage and they dislodged the enemy "from his main line of resistance" shoving him back 1400 yards. Withers was awarded his second Silver Star for these heroics on 28 February (signed by Lieutenant General "Howlin Mad" Smith).[2064]

By 1 March 1945, Lieutenant General Smith "extended congratulations and admiration to all officers and men of the 3d MarDiv for exemplary conduct and slashing attack against the enemy positions on Iwo Jima." Erskine added he was "proud of the honor of being able to accept [this praise from Smith] on behalf of" his gallant officers and men.[2065]

Erskine's 3d MarDiv, in a little over a week since deployment, played one of the most important roles on Iwo. It spearheaded the middle thrust up the island "confronting the hostile main [enemy] battle position," taking over the main airfields and destroying key fortifications.[2066] It was the "most heavily fortified part of the island."[2067] Although it may have been given this vital role because the 5th and 4th MarDivs had suffered so many casualties the first days of battle when the 3d MarDiv sat safely off shore, and thus had rested men, it now had the responsibility of taking the island's airfields that stretched across Iwo's middle.[2068] Since Kuribayashi had placed most of his defenses in the interior, Erskine faced a large burden for victory, a burden he skillfully bore.[2069]

Major General Graves B. Erskine (with his hand pointing to the sky) with Lieutenant General Holland M. "Howlin Mad" Smith (middle) and Major General Harry Schmidt (commander of all ground forces on Iwo), 26 February 1945. National Archives, College Park.

In comparison to earlier campaigns, Erskine's thrust through the island's middle proved successful. In contrast, back in 1944, Army Major General Ralph C. Smith failed to lead his division aggressively when it led the charge up Saipan's middle, so Major General Holland Smith relieved him of duty. The hesitation exposed both Marine divisions' flanks. In juxtaposition to Smith, Erskine drove his men forward keeping the integrity of both the 4th and 5th MarDivs' lines intact ensuring a coordinated attack. General Smith thanked Erskine for leading the division that broke "the back bone of the Jap resistance on the island."[2070]

Lieutenant General Holland M. "Howlin Mad" Smith, commander of the Fleet Marine Force (FMFPAC) and Major General Graves B. Erskine, commander of 3d MarDiv. Smith complimented Erskine for breaking "the back bone of the Jap resistance on the island." National Archives, College Park. Photo Credit Tech Sergeant J.A. Mundell

Erskine's *modus operandi* explored the enemy's defenses and as soon as he "sensed a weak spot," he brought all the fire power he had to hit the enemy at the spot of least resistance. When he gained momentum, "he exploited the situation by committing reserves at the flanks and through the gaps." His tactics during this "critical phase of the operation" took full advantage of his pivotal position for the campaign's success. By seizing the initiative and moving forward past the two airfields and beyond "the shambles that was Motoyama Village," Erskine's division "cut its way through the main line of resistance into the guts of Iwo Jima."[2071] Erskine soon met heavy resistance to the north in a terrain of rocky hills, cliffs, ravines, trenches and tank traps covered by mortar, rifle, grenade launcher, and machinegun fire and buried mines.[2072] Erskine maintained momentum up "a narrow front and envelop[ed] the bypassed enemy positions from the rear rather than [utilizing] frontal assault[s] up the entire zone of action. These ideas were a distinct departure from assault doctrine used in most previous Marine campaigns."[2073] Woody personally benefitted from that excellent leadership. In one major action to the north, as flamethrowers and engineers from Woody's battalion worked over a cave, around 150 Japanese in a nearby cavern, seeing what was about to happen to them, "ran from the frying pan into the fire." "It was a brilliant maneuver," correspondent Dashiell wrote, "except that two companies of Marines were waiting for them with such tools as rifles, BARs and machine-guns."[2074]

Reports about how well the Marine divisions progressed made it back to Washington. In FDR's last address to Congress on 1 March 1945, he said, "The Japs now know what it means to hear that the United States Marines have landed." When he said this, Congress erupted in applause. He then claimed Iwo was "well in hand."[2075] Everyone took note of what happened on Iwo Jima and Americans took pride in their men.

Woody's regimental executive officer, Lieutenant Colonel Eustace Rogers Smoak noted at this juncture that the "enemy strength and material continues to dwindle, but there is no indication that his defense has collapsed and he apparently maintains control of his remaining forces and materiel." Woody and his regiment fought on and suffered some casualties, but inflicted 701 casualties on the enemy from 28 February to 1 March.[2076]

A few days before Lieutenant General Smith sent his praise to the 3d MarDiv and Erskine and after FDR praised the Marines, Woody's company attacked around Motayama, a small town, now largely destroyed, north of Airfield No. 2.[2077] The battalion noted after it "jumped off" in the attack at 0800, it was "held up by very heavy" machinegun, artillery and mortar fire at the "southern edge of Motoyama Air Field No. 3."[2078] Woody was engaged in blowing caves and tunnels shut and finding Japanese and shooting them. Marines continued to be wounded and die, but Woody's guardian angel stayed nearby. For example, one of Woody's Gunnery Sergeants, Durrell Burkhalter took a bullet to the head killing him instantly on 1 March. Also, on this day, near Woody, Gunnery Sergeant Keith A. Renstrom's luck once again ran out like on Tinian as a grenade blast peppered him with shrapnel over his face, chest and left arm and he was again removed from combat and sent to a hospital on Saipan (he would receive his second Bronze Star with V for actions this day leading a platoon once again after its leader had become a casualty).[2079]

In addition to Burkhalter and Renstrom, three of Woody's witnesses had become casualties by then or the next day: Corporal Schlager on 1 March from being shot in the buttocks, Lieutenant Chambers on 1 March from severe shrapnel wounds and First Sergeant Elder on 2 March 1945 from taking shrapnel from a grenade to his left leg. Chambers' wounds were so severe, he was left for dead on the battlefield. He had been taking cover behind a tank when a shell exploded spraying his entire body with fragments and shattering his lower extremities. When body bearers took him to the grave, he lifted a hand showing he was alive and then the litter bearers jumped into action to get him aid. It appears he was hit with a Spigot mortar that killed men both in front of and behind him. His catalogue of severe injuries included his left leg broken in two places, his right shoulder broken and his index finger on his right hand snapped. Shrapnel had sprayed his legs, arms, right knee and around his eyes with hundreds of pieces of metal. His left eye "contained foreign bodies imbedded in the temporal side of the conjunctive" and later "X-ray films revealed numerous small metallic fragments in the left facial area." His left eardrum was "perforated in its antero-inferior aspect" and "draining purulent fluid" despite "penicillin therapy." He suffered partial loss of hearing.[2080] From what he was later told, Lieutenant Henning was the only platoon leader who stayed "combat effective" during the whole campaign who landed with the 21st Marines on 21 February even though he too was wounded on 12 March.[2081] Similar to Chambers, Schlager was also mistaken for being dead. While sitting under a rock eating C-rations, he stood up and was "shot in the ass." This made him hit his head against the rock, knocking him out. The body bearers came, picked him up thinking him dead and when they dumped him out he woke up and returned to the living. "He was always thankful he was not buried alive."[2082]

So many officers and NCO's had been wounded or killed in C Company by that time, Corporal Nathan G. Pitner Jr. was the acting platoon leader for five days since his lieutenant (most likely Chambers) and senior non-commissioned officers "had been knocked out of action." He took his responsibility seriously. He rescued several who had become WIAs and later "destroyed a Jap field piece with a satchel charge. He crept up on the weapon, which was hidden in a defilade, from the top of a ridge which had shielded it from the Marines for days. He dropped the charge on the gun and its operators, and seconds later," all that remained were scraps "of Japanese steel and flesh."[2083] Pitner received his second Purple Heart for wounds suffered on 3 March 1945 (his first one came from Guam), but he refused to leave the battle since it was "only" a bullet to his left arm. Later on 8 March, he would receive a Bronze Star for removing five wounded to "positions of…safety" when his platoon came under fire: "Pitner's heroic action…[saved] the lives of these men."[2084]

During this time, as Lefty healed in a psych ward on Saipan, he read negative reports about Iwo and that Marines had taken too many casualties. He knew it was bloody, "but there simply was no other way to take that rock away from the Japanese besides the way we did it." During his recovery, he often worried about his buddies.[2085]

Although there were a few on Iwo like Major General Cates, 4th MarDiv commander, who did not tolerate combat fatigue cases and called them "yellow [i.e. cowards]," Lefty benefited from excellent medical care that few armies in the world provided.[2086] The Corps showed Lefty and others like him kindness when they mentally broke down. In the IJA, if a soldier "cracked," he was shot or had his head lopped off.

Another new medical advantage the U.S. had that saved many lives was "transfusions of whole blood. Critically wounded men had it in their veins within a week after it was given by donors on the West Coast and flown six thousand miles to the battle in special ice-packed containers." Blood was more crucial in saving lives than plasma because it countered hemorrhaging. Medical men on Iwo used a total of 12,600 pints.[2087] In contrast, if a Japanese lost much blood, he died.[2088]

On 3 March, one of the first *large* planes set down on Airfield No. 1, a dirt strip which Woody and his compatriots had secured a little over a week beforehand (small observation "grasshopper" planes had been using it since 26 February). It was a navy C-47 hospital plane from the Marianas carrying mail and numerous boxes of medical supplies. To the surprise of the Marines offloading the materials from the aircraft, out jumped a woman reporter from Reuters, Barbara Finch. As explosions and bullets danced around them, a Marine yelled at her, "How the hell did you get here?" First, men pushed her into a tent and then under a parked jeep nearby. Finally when the fighting cleared up somewhat, she "was hustled back aboard the plane" which "then lumbered down the runway to return to Saipan."[2089]

While Ms. Finch left the island, Woody fought near the area called the "Meat Grinder." Japanese Iwo veteran Satoru Omagari understood why the Marines nicknamed this section as such: "Men didn't just die on Iwo Jima, they were ripped apart, torn to shreds and scattered. I saw torsos with no limbs, dismembered legs, arms and hands, and internal organs splashed onto rocks."[2090]

If the original island defense force had 22,000 servicemen, then by 4 March, some estimates put the Japanese dead at around 8,000. In contrast, the Marines and sailors had by now suffered 16,000 casualties of which 3,000 had died and 13,000 had suffered wounds.[2091] At this point, the Japanese were running out of food and water and resorting to scavenging the Marines' trash.[2092] One battalion noted that by 28 February, the Japanese were sending parties at night to get water—they were running out not only from use, but also because the Marines targeted their cisterns and destroyed them.[2093] In contrast to the dwindling forces and supplies of Japan, the Americans continued to land fresh troops and provisions and kept tens of thousands of men on the front line well supplied, relentlessly bringing the battle to the Japanese. One engineer battalion produced 339,375 gallons of water in 17 days from desalination machines (Iwo veteran and Water Purification Engineer Ed Graham on our trip in 2015 was busy in such an outfit bringing fresh water to his fellow Marines on a daily basis).[2094]

When it became clear the moment of defeat approached, the Japanese destroyed everything that bore the sacred sixteen-petal imperial mum crest called the *kiku no gomonshō*. Religiously, it was not allowed to fall into enemy hands. Everything within command posts and bunkers bearing this symbol had to "be defaced or destroyed." During the war's chaos, the Japanese burned or destroyed their money with this crest as well as documents, codebooks, letters and even the crest stamped on rifles.[2095] It's interesting to note that the American flag and regimental colors were similarly destroyed before the surviving U.S. troops surrendered to the Japanese in the Philippines in 1942. When a symbol rises to represent a god or country, it cannot be abandoned to a heathen, as both sides regarded their enemy.

As the Marines fought, many reassured each other that if they died, they would see each other in heaven. If possible, last rites were given to a Catholic to ensure he entered heaven having received forgiveness for his sins. Marines were great warriors, but few considered themselves saints. Without forgiveness, many Catholics feared their sins might condemn them to Hell. Having seen Iwo, that was not a chance they wanted to take. Likewise, often Japanese told one another they would wait for each other at the *Sunzu* river, the river that Buddhists believe separates the living from the dead, like the river Styx in Greek mythology.[2096] The spiritual rewards Japanese believed they would inherit by dying for such beliefs "created a remarkable combat ideology."[2097] Neuroscientist Sam Harris wrote, "Without death, the influence of faith-based religion would be unthinkable. Clearly, the fact of death is intolerable to us, and faith is little more than the shadow cast by our hope for a better life beyond the grave."[2098] If a Japanese soldier leaned more toward Zen Buddhism than Shintoism, then his spiritual leaders told him that if he were to die, "corporeal annihilation really means a rebirth of [the] soul, not in heaven, indeed, but here among ourselves."[2099] So, if one were to die, he had the chance to either become a god with Shinto beliefs or be re-born and reincarnated with Buddhist ones. A Japanese youth had many options to ease his anxiety. He really would never die even if his physical body was annihilated on the battlefield.

On 4 March, a distressed B-29 illustrated the value of Iwo. *Dinah Might* had finished bombing Tokyo, and on its return flight to Tinian, had complications with a fuel valve. Luckily, it landed safely for repairs at Airfield No. 1, which Woody and fellow Marines had secured on 23 February. All 11 crew members became strong "evangelists for the Marine Corps." Before the fighting died down on Iwo on 26 March, more than 40 B-29s made successful forced landings on the island, potentially saving 440 crewmembers. Thousands more utilized the island for landings during the war. Admiral Spruance, Fifth Fleet commander, watched the landing of *Dinah Might* from his flagship *USS Indianapolis'* quarter-deck, the same ship that later carried parts for the atomic bombs to be flown by B-29s from the Marianas to Hiroshima and Nagasaki. He felt the B-29 landing justified his "urgent request to occupy the island."[2100] War correspondent Robert Sherrod stated: "To the Marines, Iwo looked like the ugliest place on earth, but B-29 pilots who made emergency landings months later called it the most beautiful. One pilot... said, 'Whenever I land on the island, I thank God and the men who fought for it.'"[2101]

As on Guam, the Seabees built a large airbase on Iwo Jima, including a 5,500 foot long airstrip, several miles of roads, shops, defense works, hangars, repair centers, armories, and supply depots. They often worked only a few hundred yards from the deadly combat, transforming the island into an unsinkable aircraft carrier. Iwo was 600 miles closer to Japan than the Marianas, allowing the fighter planes to escort bombers. Secretary of the Navy Forrestal said in 1945, "The Seabees have carried the war in the Pacific on their backs" and that they transformed Iwo into an airbase was a testament to that work ethic.[2102] P-51 pilot Jerry Yellin, who had only months before conducted strafing missions against Kuribayashi's garrison, now could land on his Iwo island on a transformed airstrip that was state of the art. "However, once I opened my canopy, I realized I wasn't in Kansas anymore. The smell was horrible. You could smell death everywhere and nearby our landing strip, there were bulldozers pushing piles of dead Japanese into a mass grave."[2103] Dumbfounded by what he was witnessing, Yellin just gazed out from his cockpit until a Marine yelled at him, "Lieutenant, get out of the plane!...We're in a war zone. We've gotta get you to a foxhole!"[2104] As Yellin left his plane, he described how the Japanese cadavers were full of maggots and how tons of flies hovered over the bodies like a dark, roving cloud. Yellin was there to help support the Marines' march north by taking out positions marked by the Leathernecks. Yellin said, "Those were the shortest missions I ever flew. As soon as I got airborne, I located my target and dropped my bombs."[2105]

On the day *Dinah Might* landed on the island, Kuribayashi radioed the Army Vice Chief-of-Staff a report, apologizing for his inability to hold off the Marines. He feared the U.S. would soon use Iwo to mount an invasion of Japan:

> Our forces are making every effort to annihilate the enemy. But we have already lost most guns and tanks and two-thirds of officers...We may have some difficulties in future engagements... Now I, Kuribayashi, believe that the enemy will invade Japan

proper from this island…I am very sorry because I can imagine the scenes of disaster in our Empire…Although my own death approaches, I calmly pray to God for a good future for my motherland…I would like now to apologize to my senior and fellow officers for not being strong enough to stop the enemy invasion… My soul will always assault the dastardly enemy and defend the lands of the Empire forever.[2106]

Kuribayashi had become fatalistic and was confronting the seemingly inevitable with his command. He feared what his countrymen on the mainland would soon face, especially after personally seeing how powerful and relentless the U.S. military wave of attack was. In a rare but expected act of humility, he apologized to his commanders for not stopping the enemy although he and his command, especially Tojo, knew that such a fate awaited him and his men even before his boots ever crushed Iwo earth in October 1944. And now Kuribayashi realized that instead of being a major dam stopping the flow of American forces, he was actually just a downed tree limb forcefully pushed aside. His hope for an *Ermattungsstrategie*, a strategy of attrition, although somewhat effective in killing more Marines than the Corps thought possible, still did not achieve his desire to deter the U.S. from marching forward toward Japan.[2107] America would continue to build out massive amphibious forces to attack Japan. Yet, interestingly, Kuribayashi felt heaven could be pleaded with to change the course of events. Furthermore, he felt his own death would not bring eternal rest. Instead, once dead, there would be a spiritual realm where he would continue to battle the Americans.[2108] The afterlife was the only place he could hope to defeat the Americans because in the physical realm, the gods failed to deliver the support Kuribayashi needed.

While Kuribayashi radioed this report, Woody's regiment continued to encounter his elaborate trench, tunnel and cave systems. A Japanese would fire at them from one hole, disappear down a tunnel and then reappear "in another firing position," as regimental commander Colonel Withers noted. Underground, it was difficult to target the enemy which had freedom of movement in the part of the battlespace the Marines did not control.[2109] Veterans have described how blood-curdling it was being in a trench or foxhole at night hearing the Japanese move underneath them as they moved into secret positions for the next day.[2110]

On 4 March, another witness of Woody's actions became a casualty, taking shrapnel to his right hand: Private George B. Schwartz.[2111] This casualty may have taken place in fighting off a counterattack when Kuribayashi launched 200 men at a gap he discovered between Woody's regiment and the 9th Regiment. The Marines repulsed this incursion inflicting 166 casualties on the "Japs."[2112]

By 5 March, Woody's 278-man company had been dramatically reduced and Woody can only remember around 17 were left from his core group.[2113] Colonel Withers noted Woody's battalion was at 18% strength.[2114] The 21st Regiment reported this day was used "to resupply, improve defensive positions and conduct mopping up operations."[2115] The "toll had been so gruesomely severe that Major General Schmidt" declared March 5th a "day of rest, regrouping and replacement" for the divisions. Holding their line that night, the 21st Regiment received their replacements, bringing their units' numbers back to where the Marines could offer effective combat.[2116] The force of the attack over the preceding two weeks had gradually diminished units and the Marines needed to regroup. A and C Companies, where Woody was, withdrew from the lines and traveled to the rear at Motoyama Village. The report noted only "remnants" were pulled back.[2117] A correspondent with Dashiell said, "All those guys who lived through the *Banzai* attack on Guam are either dead or wounded. We've got only three company commanders left in the whole regiment."[2118] Woody explained the day of 5 March was "anything but resting."[2119] They may not have been fighting, but they had to prepare for night infiltrations and strategize for the next attack, which was a daily occurrence. Any operations with a company that had gone from 278 to 60 seemed reckless. To function properly, Charlie Company needed more replacements. Woody's commanders made "every effort…to rehabilitate the reorganized troops on the line. Plenty of rations, water and ammunition supplied to all hands. Blankets were sent up to supplement those already on the line."[2120]

Before exploring what Woody did when his company received its replacements, one needs to consider what Woody is telling us when he says his company had dwindled to around 17 combat veterans. It has proven difficult to get the exact muster roll data on the men rotating out (dead and wounded) and rotating in (replacements and those healed), but there is an after-action report of a battalion stationed next to Woody's outfit that helps explain his situation. When his company went down to around 18% of its original strength, it barely could hold the line, much less surge forward.

A breakdown of dead and wounded for 2<sup>nd</sup> Battalion, 24<sup>th</sup> Marines adjacent to Woody's group represents what Woody's battalion experienced (and these numbers below do not even match up on a daily basis illustrating the chaos):

**19 February** 2 Dead, 14 Wounded. No Replacements. Total Effectives 878. Original Number: 894

**20 Feb.** 11 Dead, 48 Wounded, 7 Sick. No Replacements. Total Effectives 802.

**21 Feb.** 17 Dead, 55 Wounded. 1 Returned from Wounds/Sickness. No Replacements. Total Effectives 732.

**22 Feb.** 1 Dead, 19 Wounded, 7 Sick. 1 Returned from Wounds/Sickness. No Replacements. Total Effectives 706.

**23 Feb.** 3 Dead, 15 Wounded, 6 Sick. No Returns. No Replacements. Total Effectives 682.

**24 Feb.** 19 Dead, 23 Wounded, 24 Sick. 2 Returned from Wounds/Sickness. 7 Replacements. Total Effectives 615.

**25 Feb.** 0 Dead, 3 Wounded, 13 Sick. 4 Returned from Wounds/Sickness. No Returns. No Replacements. Total Effectives 601.

**26 Feb.** 0 Dead, 1 Wounded, 3 Sick. 4 Returned from Wounds/Sickness. 60 Replacements. Total Effectives 661.

**27 Feb.** 0 Dead, 0 Wounded, 3 Sick. 7 Returned from Wounds/Sickness. 177 Replacements. Total Effectives 782.

**28 Feb.** 7 Dead, 15 Wounded, 7 Sick. 9 Returned from Wounds/Sickness. 20 Replacements. Total Effectives 772.

**1 March** 13 Dead, 1 MIA, 35 Wounded, 13 Sick. No Returns. No Replacements. Total Effectives 666.

**2 March** 24 Dead, 43 Wounded, 9 Sick. 6 Returned from Wounds/Sickness. No Replacements. Total Effectives 644.

**3 March** 9 Dead, 32 Wounded, 11 Sick. 11 Returned from Wounds/Sickness. No Returns. No Replacements. Total Effectives 603.

**4 March** 14 Dead, 61 Wounded, 0 Sick. 7 Returned from Wounds/Sickness. 3 Replacements. Total Effectives 533.

**5 March** 9 Dead, 22 Wounded, 1 Sick. No Returns. 9 Replacements. Total Effectives 510.

The battalion started with 894. With replacements (276), a total of 1,170 had served in the battalion by 6 March. From its landing until March 6, it suffered 129 dead (11%) and 386 wounded (33%). Total casualties in 15 days numbered 515(44%). There were 104 sick including those with illnesses like dysentery and shell-shock/combat fatigue. In total in 15 days, the battalion lost 619 (53%) due to death, wounds and sickness. There was still three weeks of fighting left (they would suffer 850 casualties (73%) out of 1,170 original members and replacements before leaving the island!). Only 45 out of the 490 sick and wounded returned to the lines. However, this unit had difficulty finding qualified leaders and replacements, giving rise to the battalion's claim that morale was falling.[2121] Commanding general Clifton Cates described the land the 24th fought over: "That right flank was a bitch if there ever was one" chewing up his men.[2122] Now Woody's regiment had the following update by 7 March 1945:

**Killed:** 14 Officers and 184 Enlisted Men

**Wounded:** 38 Officers and 750 Enlisted Men

**MIA:** 26 Men

**Non-Effectives:** 447

**Returned:** 1 Officer and 88 Enlisted Men

From March 5-6, the regiment received 1 Officer and 148 Enlisted Men

From March 6-7, it received 24 Officers and 655 Enlisted Men[2123]

Woody's battalion went through a lot of turnover and problems plugging the holes in their units as their members continued to become KIAs, MIAs and WIAs. Woody's regiment suffered losses of 1,459 men, almost half its strength. Between returning Marines and replacements, the regiment regained 917 men but it was well short of its need to get back to 3,006 men.[2124]

Captain Donald Beck (center), Gunnery Sergeant Albert Hemphill (to Beck's right and with a Gunnery Sergeant's insignia on his left arm) and unknown Marine at Iwo Jima. Circa March 1945. National Archives, College Park. Photographer Corporal F. E. De Ome

Woody's C Company received replacements, but none had seen combat. Many did not know much about their weapons, and they did not know anything about fighting. Since Woody's company was to go into battle on the 6th, he and fellow veterans faced a dilemma: Either get some much-needed sleep before the attack in the morning and go into battle the next day with men who did not know what they were doing, or stay up, and train the "greenhorns" as well as they could. They opted to stay up and instruct the men on weapons handling and tactics.[2125] In this chaos, Woody's captain, Donald Beck, had given his NCOs guidance on what to do and made sure his men were cared for. They were tired, scared, under strength and about to go into battle again. Beck took a few moments in this reprieve to write his family: "My darling wife and son. Today I am able to write a short note to you. We are in combat on Iwo…and I am alright. We are having a hard time, but should soon get their goat. This is no paradise, and I like everyone else will be glad to get out of here."[2126] Beck knew the Marines were gaining the advantage, but he was struggling to keep everything together and yearned to get away. This letter was the last he wrote on Iwo.

Units everywhere experienced the same dilemma and "morale was plummeting."[2127] A neighboring battalion reported by 2 March it had lost so many NCOs and officers that "leadership [was] now an acute problem."[2128] The next day, it was noted: "Men very tired and listless, lack leaders."[2129] When replacements showed up, often the hardened veterans looked upon them as "boots." One machine-gunner said half to himself and half to a buddy, "Those guys, they won't be 'boots' in the morning." "Yeah, I know what you mean," said a scruffy sergeant with cold memories flashing in his eyes.[2130] But Woody and Beck realized the men looked to them, as veterans, to keep things together. They put on a brave face, went to each group, told them what was expected, and ensured they knew how to handle their weapons. It was painstaking and frantic work, but the 17 veterans Woody had in his group circulated among the replacements (around 140) to explain weapons, tactics and the enemy to scared, naïve newcomers.[2131] Most had not seen their 20th birthdays and Woody was only a year older. Many died in their first hours of battle because they had not developed the skills necessary to survive. PFC Donald "Don" Graves of D Company, 2nd Battalion, 28th Marines said:

> We had some replacements…from the 3d MarDiv. One guy came
> to my crater where two buddies and I were…I had binoculars and
> was searching for a sniper who was hitting our position. The new
> replacement asked 'Hey, can I look for the Nip?' to which I re-

sponded, 'Hell no, you just got here. You'd get killed.' Then one of my buddies said, 'Hell Graves, let the guy look for the sniper. That's why he's here so you can get some rest.' So, like an idiot I gave him some instructions and the binoculars and sat down...A few minutes later, a shot rings out and the new guy's head slings back and he flips over. One of the Nip snipers got him...in the forehead sending the bullet out the back of his head... His body fell at my feet and his upside down helmet skipped next to me. As I looked into the webbing..., I saw a picture of a...young woman with an infant...—the dead Marine's wife and child...I lost it. I stood up...like an idiot and started yelling 'Fuck the Marine Corps! Fuck Iwo Jima! Fuck God!' My buddy, seeing that I'd lost it, body slammed me to the ground...so I wouldn't get shot...I started laughing like a madman for few minutes. Then I stopped, picked up my rifle, returned to my post and me and my buddies never discussed it. Anyone in battle long enough will know he suffered from combat fatigue at some time or other. You cannot help but feel bouts of insanity when you are living in a madhouse.[2132]

Such scenes involving replacements happened everywhere, and Woody was about to enter combat with 140 "virgin warriors." In fact, Kuribayashi wrote up two snipers in a *Kanjo* on 3 March for shooting many Marines. Engaging the enemy constantly from 27 February to 3 March, Corporals Kunimori Chaen and Hirke Kaenko killed 20 men each, one of whom may have been the new replacement Graves talked about since they were in his area of operation. Their *Kanjo* was read before the Emperor making it a *jōbun*. By then, Kuribayashi had awarded his men four of the Japanese version of the Medal of Honor.[2133]

As an illustration of how snipers like Chaen and Kaenko worked, the following story is telling. When Private James Myers in Company F, 2nd Battalion, 21st Regiment, 3d MarDiv noticed one of his buddies "lying wounded among the rocks and crevices" in the island's northern part where his division had pushed through, he naturally wanted to help. "Jap sharpshooters made the air whistle with their bullets." One of these snipers, working with others, had probably intentionally wounded this man knowing his buddies would come for him. Myers took the bait. He would not leave a comrade to suffer. "He crawled. He creeped. He snaked along on his stomach. He heard the snipers' slugs digging into the earth sounding like someone beating a pillow with a fist. He heard them strike rocks he hid behind, then zing away. He kept going—for half a hundred yards." His friend yelled for help and moved with pain on the ground. Finally, Myers reached his comrade. As he started to drag his buddy back, making himself even a bigger and slower target than before, the sniper put a bullet in his head killing him. Soon thereafter, a bullet ended his buddy's life. Both lay there, intertwined as if two brothers were holding each other in a loving embrace. Myers made sure his friend knew he would never be left alone, and if he must die, he would not die alone. The snipers marked up two more kills that day.[2134]

Meanwhile, Woody worked with his replacements to prevent events such as had happened to both Graves and Myers. He hoped the brief training he gave would help them survive their coming baptism of fire. Major General Erskine noted that replacements were poorly trained; combat efficiency dropped and casualties increased when they arrived at their units.[2135] These men were indeed aggressive, but did not know how to operate within the fire team framework and "lacked knowledge of the proper use of cover and concealment."[2136] One battalion near Woody reported: "Replacements inexperienced and require a lot of attention. Few experienced men left on the front lines. Troops are tired and lethargical because of being constantly in attack and from the ceaseless enemy mortar and artillery fire being received."[2137] Woody was probably correct in noting 17 were all that was left of the "experienced men," but that the rest required babysitting and training. As Lieutenant General Smith's operations officer wrote after Iwo: "A technique or the indoctrination and integration of replacements prior to and during combat operations remains one of our *most urgent problems.*"[2138] One man who helped with getting replacements up to speed was Corporal Billie Griggs serving in Woody's 21st Regiment. Major General Erskine pointed him out after battle for special mention writing: "Serving under heavy fire, Corporal Griggs displayed outstanding leadership in converting new replacements into an efficient combat team. His leadership was a valuable asset to his company..."[2139] However, Griggs was rare and many replacements needed men like Griggs to help them acclimate to combat and unfortunately, those men were either lacking in units or dead and wounded.

A strange piece of history arose at that time. If replacements were needed, especially trained ones, why did the majority of 3d MarDiv's 3rd Regiment remain out on the ocean in reserve? If trained Marines were needed, which the men in the 3rd Marines were, why were greenhorns deployed instead of them? Lieutenant General Smith kept them in reserve because this regiment held the last units he had left and he knew he would need later at the rate Kuribayashi was depleting his forces.[2140] He felt a good commander probably should not commit all his reserves when there were questions whether the enemy force had more men or capabilities than anticipated, and so far Smith and his generals had been surprised too much. Indeed the reserve was there to "counter unforeseen threats."[2141]

However, a commander should always fight the battle in front of him, not the one planned for in the future and the unwillingness or the refusal

Corporal Billie Griggs. He participated in the Iwo Jima reunion in 2015. He did a tremendous job on Iwo helping replacements understand the realities of combat while on the island. Major General Erskine pointed him out after battle for special mention writing: "Serving under heavy fire, Corporal Griggs displayed outstanding leadership in converting new replacements into an efficient combat team." St. Louis Personnel Records Center.

to deploy this unit caused unneeded casualties as Graves' story illustrates. More importantly, a few weeks into the battle, the Marine command structure should have known what they were dealing with and adjusted their dispositions to the enemy's actions. As von Clausewitz wrote, "all forces intended and available for a strategic purpose should be applied *simultaneously*; their employment will be the more effective the more everything can be concentrated in a single action at a single moment."[2142] The Marines missed an opportunity to maximize their "economy of force."[2143]

Another reason why the 3rd Marines was not fully deployed (one battalion from it was indeed put into the lines) was the possibility that there was not room on the island to put more into action. Vice Admiral Turner agreed with Smith not to allow the regiment to land stating "there were enough Marines ashore."[2144] This may have been a concern at the outset while the beachhead was the totality of the landing area; however, Smith's command found a way to land two 3d MarDiv regiments (21st and 9th) there so it stands to reason he could have put the 3rd Regiment into action with half the island in Marines' hands. Historian Charles Neimeyer claims that Erskine pleaded with Smith to do so, but Smith refused his request.[2145] Regardless, the 3rd Regiment languished while Woody and thousands of Marines had to make do with inexperienced replacements—and those replacements suffered higher casualty rates due to that inexperience. The V Amphibious Corps operations officer said Turner's and Smith's decision not to release the regiment "was wrong and increased casualties."[2146] Likewise, Erskine protested saying it was necessary to replace "tired and worn out" troops with trained ones from the 3rd Marines, but Smith and Turner countered: "You keep quiet, we've made the decision."[2147]

This strange replacement policy stemmed from "an organizational innovation employed for the first time in the Marianas and subsequently on Iwo Jima." V Amphibious Corps had attached to the divisions at Iwo six replacement drafts of initially 7,188 men and officers. All were recent arrivals from training. The original plan was to feed these replacements into the fighting units piecemeal allowing the "new guys" to slowly integrate with the hardened old-timers. Before they ever hit a combat unit, they were to be first used in shore parties to acclimate them to campaigning. Unfortunately, this plan did not work due to the enormous casualties Marines had already suffered. Instead of this gradual integration, these men were shoved into the thick of battle and often, "it appeared that progress was being hindered rather than helped by the presence of the new men."[2148] V Amphibious Corps' Chief-of-Staff, Brigadier General William Rogers, disingenuously noted, those were the only replacements the Marines had and they had to make do.[2149] In fact on 5 March, a few battalions of the 3rd were actually shipped back to Guam![2150]

Sleepy, physically exhausted and worried about his men, Woody looked up from his position to analyze the Japanese positions they were about to attack on

6 March. Before the attack at 0900 for which Woody's unit was earmarked, an artillery barrage pummeled the positions he and fellow Marines were to assault. Unfortunately, the Japanese were so well dug in, the artillery had little effect.[2151] This was remarkable since 11 artillery battalions had fired "2,500 rounds of 155mm howitzer ammunition and 20,000 rounds of 75mm and 105mm shells." In addition to this awe-inspiring iron curtain of explosives, navy ships offshore fired on their targeted positions sending 50 rounds of 14-inch shells and 400 rounds of 8-inch ammunition into the Japanese lines. As usual, numerous planes flew close air support missions, many of them now flying from Airfield No. 1 to the south on Iwo.[2152] The earth shook and plumes of dust and debris rose hundreds of feet into the air. After the shelling, Woody and the men of the 21st moved forward, with the goal of taking Hill 362C, one of the highest points in the north. They lobbed grenades over every rock, into every cave and around every corner of every ravine they encountered. Woody often carried six grenades and used them liberally.[2153]

During this attack on 6 March, Woody believed his friend Vernon Waters' luck ran out. According to Woody, while closing with the enemy, firing away and trying to push past a defensive position, a Japanese grenade from a Type-89 launcher (often named a "knee-fired 50mm mortar") hit Waters in the head and killed him. He was around 30 yards in front of Woody when the grenade exploded (other times Woody has said he was only six feet away from Waters). Seeing Vernon fall over limp, Woody ran to his friend. He shook him and tried to bring his life back, but there was nothing he could do. Vernon, the strapping 6'4¼" boy from Froid, Montana, had died instantaneously.[2154] Rage and sadness welled up inside of Woody and he felt a burning desire to kill the enemy. He grabbed Waters' shoulders and yelled inside his head or aloud, he doesn't quite remember, "No, No, No!"[2155]

Woody recalled that he and this buddy had made a promise to one another. They each carried a ring—Woody's was a five-and-dime ruby ring from his girlfriend that she bought at a Murphy's 10-cent store, and Waters' was a ring he received from his father upon High School graduation with a lake pictured on it. They pledged to one another while on Guam that if either one of them died, the survivor would take the ring back to the family of the deceased. Although it was strictly forbidden to take anything off a fallen Marine, Woody spit on Waters' hand, rubbed the saliva on his buddy's index finger and removed the ring.[2156] He placed it in his pocket and continued to lead his green Marines in combat. Soon body bearers arrived, tagged Vernon, placed him on a stretcher and carried his corpse away.[2157] Woody refused to watch. All Woody could do was move on. The battalion noted that during this day of fighting, "Progress was slow throughout the day with heavy small arms fire being received and also fire from a high caliber flat trajectory weapon."[2158]

It has been difficult to verify Woody's story about Waters. The records and a few other veterans bring out discrepancies. From archival sources, Waters died on 3 March, not 6 March.[2159] Woody gets his "friend's" death out of date by four days and misplaces the action that did so—is it reasonable to give Woody a hall-pass due to the chaos of combat or is he now making up information? Moreover, the archival report wrote Waters died from a gunshot wound and not a grenade blast to the skull.[2160] Sometimes, the grave registrars assumed a gunshot wound when it was shrapnel or some other trauma, but one would think a blown up head would be clear for all to see—and a blown up head would normally not result from a bullet, but a grenade.[2161] But there was no mention of an exploded skull. Also, in the report, Waters' fingerprints were unable to be taken right before burial due to decomposition, meaning Vernon had been left out in the open for six days before being buried on 9 March.[2162] So the body bearers picking him up probably did not happen as Woody reported. And if Waters was left in the open for days, it was most likely because the unit did not move forward enough to allow the stretcher bearers to get to Waters soon after death since he was obviously bloated and decayed by the time of burial. Also, Iwo veteran Don Graves thinks running to a dead friend in battle out of the ordinary:

> No one when actively engaged in battle would've run up to a buddy and then taken off a ring. Think about this. If you're suddenly attacked and one of your buddies got shot, you would hit the ground, find out where the fire was coming from and then engage the enemy. You don't want any other guys to buy it especially yourself. If any of us started messing around with a dead buddy during battle, we would have got our butt kicked by our NCO. You might try and get your buddy if he was wounded into a safe area, but if someone was dead, you would just leave him and

fight the enemy especially on an active battlefield. I mean, if you are taking grenades into your position, the last thing you should be thinking about is taking a ring off a buddy's finger because of some child-like promise you made...[2163]

Moreover, since Waters was in B Company and Woody in C Company, Graves highly doubts they were even in the same area much less in the same skirmish line although Woody believed Waters was in his company at the time.[2164] "That sounds strange to me," Graves continued. "When you were the demolition and flamethrower guy for your company you supported your own boys, not the boys in another company."[2165] Iwo veteran John Lauriello also doubts the mixing of units would happen, but observed:

> The battlefield was so chaotic anything could happen. Rarely did I go over to different units, but there was indeed some mixing of companies. Remember, a lot of people were getting killed and all the buddies you were fighting with the week before were usually gone the next week. However, we still tried to keep our chain of command solid and rarely did the units get that mixed up where demolition guys from a different unit were on the line with us.[2166]

Waters' B Company had been removed from the 1st Battalion on 4 March and attached to 2nd Battalion "to cover the gap between the 9th and 21st Regiments when lines were consolidated" so even if Waters was still alive at this time, he was nowhere near where Woody was.[2167]

Another troubling testimony concerning Woody's narrative comes from fellow flamethrower Joseph Anthony Rybakiewicz, who claimed to be Vernon Waters' best friend and was there the day Waters died. In fact, Rybakiewicz's testimony mirrors Waters' death certificate. In Rybakiewicz's story, "Big Waters" as they called him got shot in the neck and hit the earth from the bullet wound. Rybakiewicz ran to Waters, got him into a secure area and tried to get the bleeding stopped. Waters was alive and struggling to breathe. Blood was spurting out of his neck so Rybakiewicz put his arm around Waters' huge shoulders and neck and took his free fist and tried to press it with all his strength against the open jugular to stop the bleeding, but it was pointless. As "Rybak" sat there with his friend, Waters died in his arms. He does not remember Woody being there or him taking off Waters' ring right after he died.[2168] So Woody's version of Waters' death is dramatic but not plausible. Rybakiewicz's grandson, Zack Rybak writes: "[I am convinced] that this one fact—who was really there—matters, in the same way that I used to believe naming a thing could matter, could be cathartic, could heal."[2169] (See Gallery 3, Photo 16)

PFC John Lauriello. He was a radioman on Iwo who had trained with the Navajo Code-Talkers. He landed on Iwo the first day of the invasion and said, "It was raining death from the sky." He participated in the Iwo reunion in 2015. St. Louis Personnel Records Center

During this same day's fighting, a piece of shrapnel hit Woody on the inside of his left leg. He placed his hands on his wound and yelled "Corpsman!" One went to his aid. The Corpsman felt around the wound and pushed against a lump of metal. He then took out forceps and removed the shrapnel. He asked Woody, "Do you want this?" as he held a piece of metal. Woody responded, "I sure do!" Woody took the piece of steel still hot to the touch. The Corpsman then poured sulfa powder on the wound, wrapped Woody's leg with a bandage and tagged Woody with information that an aid station behind the lines would use to treat him further. Woody then put the piece of shrapnel into his pocket.[2170]

"Now get back to the first aid station and have that looked at," the Corpsman said.

"No, I can't," Woody replied.[2171] He further explained, "The new men in my section got their first training with many of our weapons the night before. I couldn't possibly leave those guys." He told the Corpsman, "I'm not going."[2172]

"You know the rules," the Corpsman said. "When I tag you, you have to go back."

Corporal Joseph Anthony Rybakiewicz on Guam before Iwo
1945. He was a fellow flamethrower with Woody and Vernon
Waters' good friend. He received the Bronze Star with V and
the Purple Heart on Iwo Jima. According to his son Gene,
Rybakiewicz had issues with how and why Woody received
the MOH. Author's Collection and Rybak Family Archive

Woody reached for the tag, ripped it off and said, "I don't have a tag on me."[2173]

He was not happy with Woody. He bellowed, "You crazy son-of-a-bitch, get back to the aid station." Another victim yelled for the Corpsman. The Corpsman shook his head, turned away from Woody and left to tend to another. Woody looked down at his bleeding leg and felt he could continue fighting although there had been a piece of metal in his flesh a few minutes before. He knew his men needed him to face the enemy. He took up his rifle and resumed firing.[2174] He recounted to a correspondent a few weeks later, "Then a mortar shell fragment '*bounced*' off my leg while I was in a foxhole one night. Didn't hurt much and I stayed in action [author's italics]."[2175] Nonetheless, for days, Woody walked with a "big, long limp."[2176] Here is also a discrepancy. Woody has claimed recently the shrapnel entered his leg and a Corpsman pulled it out with tongs, but in the 1945 interview about it, the shrapnel never entered his leg and just "bounced off" his inner thigh, bruising him. Either way, he was injured by a flying piece of metal, but we have two stories, both from Woody, about what that metal did. He did receive a Purple Heart, so it had to have been reported as a "wound" (sometimes the Marines were actually generous with Purple Hearts giving them to men who did not actually "bleed" like possibly Woody or like his comrade Scrivens who received one for a "blast concussion").[2177] Regardless of whether or not a piece of metal penetrated his flesh or only bruised it, it was remarkable Woody never *really* got injured. By March, Woody was one of 82,000 Marines who had fought on Iwo.[2178] Many showed courage in the face of what the Japanese threw at them and it was a battle that took incredible concentration and physical stamina. Although Kuribayashi had developed a strategy that was blunting several of the Marine stabs gabbed at him, several thrusts nonetheless were hitting vital areas of his defense, wearing him down and slowly bleeding him white.

For instance, later on 6 March and then throughout 7 March, in a light rain, Woody and his men moved close to enemy lines around 0320 before dawn and caught the Japanese sleeping. Their goal was Hill 362C. Shortly before dawn, the illumination starbursts ceased along the entire front and smoke was laid down to hide the Marines' movement. Units from the 3d MarDiv moved forward at 0500 in a rare attack in the dark. As a result, they gained several hundred yards of tough hill terrain without a shot being fired. In other areas, many of the unsuspecting Japanese were killed in their sleep because they did not expect the Marines to have penetrated so far north.[2179] Other Japanese, still in their sleep stupor, "stumbled out of their caves and were killed by Marine flamethrowers."[2180] In the ensuing chaos, Woody and fellow Marines actually conquered the wrong Hill (331) thinking it was Hill 362C, with their objective still 250 yards away. In the vicious battle afterwards with the element of surprise gone, several companies were almost annihilated in an area full of crags and ravines. The regimental report notes the men "made little or no gain."[2181] One man near Woody, also a demolition man, did his best to try to move forward through this chaos. Working in the 2nd Battalion, Private Gerald H. Young, attacked four pillboxes with satchel charges, snaking his way around their lines of fire to get to their openings and then throw twenty-pound sacks of explosives into their embrasures. After he did this to the fourth pillbox and returned to his lines, a mortar shell fragment hit and killed him.[2182]

Eventually, the Marines in Woody's sector took Hill 362C that day. By that time, Woody knew only a handful of men around him. Just a few remained with whom he had trained in San Diego, served on Guadalcanal, fought at Guam and landed on Iwo. Most had been killed, wounded or suffered combat fatigue, like Waters, Marlowe Williams, Renstrom, Clay Murray, Bornholz, Fischer, Scrivens and Lefty. He was alone in a sea of Marines he did not know and felt orphaned. "But that's the strength of the Corps. We were all Marines and as such, we knew we could depend on one another for our lives even though I had never seen my comrades to my left or right before yesterday. We knew we were brothers and we were working together like one collective organism."[2183]

Woody did not strap on another flamethrower for the rest of the campaign after his exploits on the 23 February.[2184] Studying this battle, it is confusing that Woody never used a flamethrower again when his company was in the middle of the island and still facing numerous caves and pillboxes. Additionally, if Woody's assertion was correct that he was the last flamethrower in his company, one would think he would have been utilized all the time to burn out emplacements since that was one of the Marines' primary weapons. USMC veteran Lieutenant Colonel Eugene Rybak, the son of flamethrower Joseph Rybakiewicz, finds it "really…strange" that Woody never did this again.[2185] This is indeed peculiar. If Woody is to be believed that he was the last flamethrower in the unit after the third day of being in battle, then he would have been used repeat-

Corporal Joseph Anthony Rybakiewicz. He earned a Purple Heart and the Bronze Star with V on Iwo as a flamethrower operator. St. Louis Personnel Records Center

edly with the flamethrower since his battalion used flamethrowers 24 days out of the 34 it was on the island (they were in heavy use).[2186] Captain Beck often deployed flamethrowers to break through the numerous defenses his company destroyed—if this was the case, why did Woody not use this weapon on more days than just one?[2187] Maybe he became so scared after what he did against the pillboxes on 23 February that he let others, apparently less trained or even replacements, try their hand at taking out pillboxes with liquid fire. That Woody never engaged the enemy again with a flamethrower when he was the only one left, according to him, in the company with the knowledge of how to operate, repair, fuel and fire a flamethrower does not make sense. Perhaps Woody did not strap on another portable flamethrower due to the fact that he had another type of flamethrower helping him, *Zippo* or *Satan* tanks.

As Woody and the 3d MarDiv advanced through the island's middle hitting lines of defense, flamethrower tanks often destroyed Japanese positions. When Woody and his Marines could not kill the Japanese with rifles, machineguns, grenades or satchel charges, they called up the *Zippos* and they:

> were extremely valuable in the reduction of fortified positions and in the destruction of enemy groups concealed by rubble or brush. The large flamethrower was particularly valuable against positions or personnel in defilade and in caves and ravines because of its long range and the large volume of flame.[2188]

V Amphibious Corps commander, Major General Schmidt, wrote that these weapons "were indispensable in the Iwo Jima operation."[2189] Woody watched in awe as these tanks spewed hundreds of gallons of flame into the tunnels, pillboxes and caves. It was simultaneously a hydra of death and a mobile pillbox. Sometimes, in taking out one blockhouse a few of these tanks would use 800 to 1,000 gallons.[2190] And these tanks did their hazardous duty with much risk. As Corporal Leighton R. Willhite's tank moved into enemy territory near Hill 362A, one of his neighboring Shermans "became disabled" and came under attack "by the enemy in a cave entrance about six feet away." The tank had actually fallen about eight feet into an underground hidden bunker crushing several of the enemy and releasing other angry Japanese survivors who swarmed around the Sherman. Willhite's 30mm machine-

9 1 6 4 8 3

Corporal Leighton R. Willhite. He received the Bronze Star with V on Iwo. He participated in the Iwo Jima reunion in 2015. Before battle, his commander told him the island had only 2,000 enemy and control of the entire island would take at the most a week since the "Japs had to be weak and undersupplied." "Boy, was he wrong," Willhite told the author Bryan Rigg on 10 May 2017. St. Louis Personnel Records Center.

gun opened up and killed all the "Japs" on the right side of the tank, but the left side remained vulnerable to attack and blocked from their view. Since Willhite and his crew were unable, due to the terrain features, of offering any support to eliminate the enemy position on the left side of the tank, he, a comrade and their lieutenant, Blake "Dusty" Leonard, left their tank and rendered aid to the stricken tank and its crew as they evacuated their stricken vehicle. The short, 5'5½" tank driver Willhite, grabbed his .45 Colt pistol and fired at the enemy cave opening. He helped kill three attacking Japanese who emerged from the darkness of the hole. His "daring aggressiveness" covered his comrades and both tanks' contingent of men, shockingly, suffered no casualties. Willhite would receive the Bronze Star with V for his actions and daring. Lt. Leonard, who killed around 8 Japanese with his Tommy-gun would receive the Silver Star.[2191] Although such actions prove the attack on Iwo in general was sluggish, the Marines were nonetheless, slowly but surely, eradicating the Japanese lines of defense.

Hirohito said of the fighting at this stage (disingenuously one might add): "I am fully satisfied that naval units have taken charge of defense and are cooperating very well with the army. Even after the enemy landed, they fought ferociously against much greater forces and contributed to the entire operation."[2192] While fighting around the middle of the island, Woody did indeed find that the enemy was ferocious.

Soon after this engagement around Hill 362C, Woody's company was pulled out of combat to rotate back near the landing beaches to clean up, get some food and rest.[2193] His hair was matted with sweat, dirt and rifle oil, and his whiskers had several days of growth. He must have smelled awful, having fought on ground that was everyone's graveyard, trash heap, firing range, hospital and toilet.[2194] While there, Woody had his leg tended to, got some chow, took a shower, shaved and received a new pair of dungarees. Between the hard wear and tear that uniforms got on the battlefield in all kinds of weather and the rot that broke down the damp and dirty clothing fibers, uniforms of infantrymen soon became "unserviceable" and had to be replaced. After resting, he and his men resupplied themselves with ammunition, cleaned and oiled weapons and went back onto the lines of combat.

A destroyer fires its 5-inch guns on Japanese positions ahead of the Marine lines of advance as the Leathernecks move north on Iwo Jima 2 March 1945. National Archives, College Park.

Woody's company most likely again attacked in force on 8 March. Before the advance, a "ten-minute artillery barrage followed by a rolling barrage lifting [several] yards every seven minutes for 200 yards" occurred.[2195] Erskine "believed in this rolling barrage type" of WWI combat "and close follow-up of assault troops right behind the screens of fire."[2196] From the ocean, destroyers hammered the enemy. The 3d MarDiv history noted: "It is unbelievable that any force could withstand such a pounding day after day. But the Jap did!"[2197] Determined to go out in a blaze of glory, naval commander Captain Samaji Inouye ignored Kuribayashi's orders and led a 1,000 man *Banzai*. They attacked near Woody, hitting elements of the 2nd Battalion, 23rd Marines. The Leathernecks killed 784 Imperial Naval Landing Sailors. The Marine casualties included 90 killed and 257 wounded. One company used 20 cases of grenades fighting the *Banzai* (500 MK-2 grenades). The almost 800 Japanese dead from this engagement was the "largest single-day enemy death toll recorded on Iwo."[2198]

Woody's regiment continued to use flamethrowers and demolitions against caves and pillboxes, destroying numerous positions daily.[2199] On 8 March, PFC Warren Welch in neighboring B Company of the 1st Battalion "worked his way up a side of a sandy ridge and demolished" an emplacement with dynamite. Near it was another pillbox

PFC Warren Welch. He served in Woody's battalion and would receive the Navy Cross for his actions on Iwo. He would die on the island. St. Louis Personnel Records Center.

that Welch closed with and threw grenades into making the occupants catch "a fatal occupational disease [of lead poisoning]." After silencing these "blockhouses," Welch located a "Jap mortar position" several yards ahead of him. He signaled its placement to his own mortar platoon and then directed "automatic rifle fire at the enemy crew pinning it down until" the mortars wiped the Japanese out. The next day, a sniper's bullet found its mark and killed him.[2200] Welch received the Navy Cross posthumously.[2201] On this day, Corporal Joseph Rybakiewicz silenced one cave with his flamethrower, then retrieved a wounded man and brought him to safety. When he tried to bring out another downed Marine, he was wounded himself. For his actions, he received the Bronze Star with V.[2202]

By 9 March, elements of the 21st Regiment had fought from the landing beaches at the southern end of the island to their objective on the northeastern coast, but Kuribayashi held a square mile of the island to the west and he would defend every square inch to the death.[2203] Announcing their arrival on the north coast, one Marine filled a canteen with ocean water and sent it back to Major General Erskine with the CYA ("cover your ass") note, "For inspection, not consumption."[2204] CBS radio reported a patrol in Erskine's division had reached the northern shore and "washed the Japanese stench from their hands" cutting the Japanese defense in half. Exploits of the 3d MarDiv were known to everyone, both on Iwo and back in the States.[2205] Erskine's men pushed through some of the toughest defenses in WWII. Splitting Kuribayashi's force "was like an earth-worm cut in half: both ends were still alive and able to move, but they had lost their natural functions."[2206] It was now Erskine practicing a Ju-Jitsu move on Kuribayashi, taking his legs out from under him, and gearing up to deliver the death blow. And to deliver this blow, Captain Beck continued to deploy flamethrowers. One of them, Rybakiewicz, would burn out "seven pillboxes and caves" while on the island. On 8 March, one of the enemy emplacements encountered was set up to ambush the Marines and it did a good job doing so. "We were in a gulley," Rybakiewicz explained, "and the emplacement was slightly above us. Before we could withdraw in the narrow passage, four of our men were killed and five wounded. I gave first aid to two of them and somehow managed to carry them back to the dressing station. A hand grenade exploded and fragments scraped across my left knuckles, but I didn't think it was serious." Then "Rybak" asked a nearby BAR man to cover him. Now agitated, he moved up with his flamethrower and gave the pillbox a "long squirt and burned it out good. The next day [9 March], a doctor looked" at his bloody knuckles and evacuated him.[2207] After receiving first aid and with his hand wrapped up, he returned to the front. "That was the scariest moment for me, when I realized I had to go back to the fight after getting my hand cared for," Rybakiewicz said.[2208]

On the night of 9 March, 346 B-29 Superfortress bombers left the Marianas and dropped 1,665 tons of firebombs on Tokyo, creating a firestorm that destroyed a sizeable portion of the city (16 square miles) and killed between 80,000 and 100,000 people. B-29 Flight Engineer Fiske Hanley, flying at 4,000 feet, said, "It looked like Hell and we could smell the burning flesh even at that altitude." Commanding general Lemay knew the "slaughter of civilians would be unprecedented," but he wanted to destroy Japan's war-making capabilities and prevent an invasion that he knew would take the lives of hundreds of thousands of American boys. The heat was so intense that metal melted, canals boiled and "buildings and human beings burst spontaneously into flames." Temperatures reached 1,800 degrees, causing "babies to explode on mothers' backs, and cars on streets were 'consumed like crumpled paper.'…People's heads exploded in the heat, the liquid brains in their burst skulls bubbling an eerie fluorescence." In early bombing raids, there were strikes at large industrial targets that hampered, but did not stop production of war materiel. Bombing was inaccurate at first, especially when done from high altitude to avoid antiaircraft fire and fighter planes. Powerful jet stream winds often blew bombs dropped from over 20,000 feet off course. Much of Japanese industry was dispersed, with multitudes of tiny shops and much backyard manufacturing. Only when the firebombing campaign was employed were U.S. planes able to reach the "home production" of Japan's war industry. Japan had intentionally decentralized as much as 90 percent of its war production into subcontractor workshops in civilian districts, making what was left of the war industry largely immune to conventional precision bombing with high explosives. But with the fire-bombings, America was destroying the capacity for Japan to wage war while also taking her revenge for Pearl Harbor, the Bataan Death March, and countless atrocities committed upon POWs and civilians.[2209]

The Japanese suffered the agony it had delivered after having killed tens of thousands in bombing and incendiary aerial raids on China, especially in campaigns in 1932 and from 1937-40 against Chapei, Tientsin, Shanghai, Nanking, Lanchow, Guangzhou and Chongqing (Chungking), just to name a few. At Chongqing alone, Japan conducted 200 air raids from 1937-40 killing 12,000, mostly non-combatants. The difference between its raids and those of the Americans, according to the Tokyo International Military Tribunal, was that Japan conducted its raids to deliberately kill civilians.[2210] The League of Nations Chinese Delegation, represented by Wellington Koo, wrote the *Société des Nations'* General Secretary in 1937 documenting the Japanese terror bombing operations against civilians which violated international law. Koo pleaded with the *Société des Nations* to use its resources to condemn and end such illegal and inhuman war practices.[2211] While U.S. bombing raids against Japan killed thousands of civilians, the strategy was always to cripple the war industry and end the war: *It was never designed to kill civilians* unlike Japan's goal. Nonetheless, by 1945, the majority of Americans believed that Nazi Germany and Imperial Japan "deserved every bomb that fell on their countries."[2212]

Freedom-loving nations brought an end to Japanese aggression from the skies by bombing campaigns against Japan to render their armed forces useless. And the Allies did this with massive firepower. The bombers hitting Japan were much larger and deadlier than anything the Japanese had used on their enemies and many throughout the world took pleasure in knowing the Japanese felt the terror their sons had for years inflicted on others. Within a few months, the bombing campaign against Japan reduced its "major cities to ashes," including Nagoya, Osaka and Kobe; however, Tokyo suffered the worst by far.[2213] The cheering crowds that once lined the streets of these cities clapping and celebrating the destruction of Chinese cities like Nanking or the attack on Pearl Harbor and the military successes thereafter no longer were gleeful. Thousands of dead lined Japan's city streets and neighborhoods and the "stink of death permeated the air."[2214]

Brigadier General Thomas Powers, who led the Tokyo raid, said it caused "more casualties than… any other military action in the history of the world."[2215] Hearing about the destructive Tokyo raid, Kuribayashi's hope to protect his family from air raids had literally disappeared. Colleagues witnessed the commander "slumped, head in hands…He looked like an old man."[2216] He had prophesied: "When I imagine what Tokyo would look like if it were bombed—I see a burned out desert with dead bodies lying everywhere—I'm desperate to stop them carrying out the air raids."[2217] The intelligence chief of the 23rd Army who had served under Kuribayashi, after witnessing the "horrors" Kuribayashi's men had committed in Hong Kong, prophetically quoted the shameful fate of the Mongols that now mirrored that of the Japanese: "[They] conquered a vast empire but failed to give profound thought to their administration…Not only did they leave no trace, but they brought about the ruin of the very land from which they had sprung."[2218] Kuribayashi's vision of American airpower destroying Tokyo had become reality after he failed his mission to defend the capital, and the land he had "sprung" from now lay in ruins. Although his family was safe for the time being, "he had no way of knowing that."[2219] Nevertheless, for all the destruction these raids wrought, they had no direct "influence on the policies of [its] government, whose military die-hards had not yet been convinced they should surrender."[2220] For example, a few weeks later in April 1945 the government issued a *Decree of the Homeland Decisive Battle* that stated: "Every soldier should fight to the last moment believing [in]…final victory."[2221] Japanese rulers readied their citizens to behave as their soldiers did on Iwo: Fight the "foreign devils" and die rather than surrender. Although Kuribayashi was disheartened, he would never surrender. It would "take even greater shocks than B-29 fire raids to end the war,"[2222] and hundreds of thousands of Woody's fellow Marines and soldiers would continue the fight in one more island battle on Okinawa, which garrisoned five times the number of troops than on Iwo.

In Iwo's north, the terrain differed from the south, requiring modification of the tactical engagement. The Japanese hid behind rocks and hills, often emerging from their caves to fire at the Marines. In response, the Leathernecks aggressively captured the area and killed most Japanese they found because most continued to refuse to surrender.

However, one combatant did surrender to Woody after crawling out of a spider hole. He was unarmed and dressed in shredded clothes. He had not eaten for a while and was haggard and dirty. Woody placed a Marine on guard duty while he went for an interpreter. Upon returning, he found the enemy curled up in a ball, dead, in a bloody puddle spread across the ground. The guard said, "The Japanese begged for my bayonet so he could commit suicide. I gave it to him and the crazy bastard kneeled down and then rammed the knife into his midsection until he bled out and died." Woody was upset. They had lost an opportunity to interrogate the POW and perhaps gain some Intelligence. He would have punished the Marine, but had patrols to conduct, and so, went on about his duty.[2223]

Woody's battalion tried to take prisoners contrary to earlier "orders" not to, yet it was difficult to know if a Japanese wanted to surrender. For instance, in front of Woody's battalion lines, a Japanese appeared 40 yards away ramrod straight holding his hands high. The battalion's Intel Section wanted the prisoner alive. So Corporal Gedeon LaCroix was given the task to lead six men to apprehend him. "I couldn't tell whether the Jap was bluffing," said LaCroix as he moved forward cautiously. "He just stood still, like a statue. When I got within seven yards of him, I saw that he held a conical hand mine. Just as I threw myself on the deck, he turned around and jumped into a shell crater. I hollered for the fellows behind me to watch out." Suddenly, machinegun fire erupted from a hidden bunker and a camouflaged pillbox "throwing cross-fire at us." LaCroix crawled back to his lines as his men covered him. Knowing what awaited them and where to attack, LaCroix and his men re-organized and "wiped out the pillboxes and the Nip in the shell crater."[2224] They did not try to take any more prisoners during the second time around.

Often, the Japanese were never given a chance to surrender. When a tank drove around a boulder, the crew saw a "Jap sitting down with no clothes on" crying and apparently blind from a recent attack. The tank's machine-gunner fired upon him. The burst took off the "Jap's left arm at the elbow." Blood started leaping out of his body "in great spurts and in rhythm with his heart beat." The machine-gunner fired again hitting the enemy in the chest slamming the man against the ground "with a great invisible force."[2225]

A few Japanese did successfully surrender to Woody's battalion which enabled them to find hideouts. When Marines took these prisoners to their old tunnel entrances, they were encouraged to talk "fellow Japanese into surrendering." Although this tactic was used liberally, it was met with "little success."[2226] With no response, the Marines sealed the caves with explosives. When demolition personnel in Woody's regiment came upon a cave with a dozen Japanese, one Marine who spoke Japanese asked them to surrender. They replied, "If somebody will come in, we'll come out." Hearty laughter came from those Marines who heard this and the demolition squad "wondered whom the Japs thought they were fooling" and "blew up the tunnel."[2227] Many POWs Woody's unit captured were not Japanese but Korean. They were not eager to die for the Emperor or confident they would become a god since they were not Shintoists.[2228]

In the few cases when a Japanese did surrender, the Marines had to be careful. There was a case in Woody's lines when the enemy tried to use a "suicide surrenderer" in an attempt to draw in Marines to place them in a kill zone, but his act was noticed and the surrendering man was shot dead.[2229] Even after becoming a prisoner, there were times that it was a ruse. After one wounded Japanese was taken to a hospital, he escaped, commandeered a weapon and shot several patients before Marines nearby killed him.[2230] Most continued to fight to the death or commit *seppuku* or *jiketsu*. There were some who gave up, but very few. Some of those who had not made the decision to kill themselves in the face of surrender were forced by their superiors to do so. One prisoner who emerged from a cave that had been blown up by Seabees told his captors the explosion caused so much confusion it allowed him to leave the cave, preventing "his officers from killing him."[2231] There were many reasons why it was difficult to take prisoners, not primarily due to every Japanese being a fanatic; it seems that when a Japanese was not fanatical, he had a difficult time escaping the majority of those around him who wanted him to die in their death cult.

In front of Woody's regiment, a Japanese officer committed *jiketsu* (literally self-determination, but in this context, suicide other than disembowelment). Here, he held a grenade against his heart blowing a huge hole into his chest. Few Japanese ever surrendered on Iwo Jima. February 1945. NACPM. Photo Credit Christian

Sometimes, the Marines would not blow the cave or bunker shut or flame it with a portable or tank flamethrower, but rather, would pour a mixture of gasoline and seawater pumped from the ocean into the underground enclosures, light it and then roast "the soldiers within to death."[2232] Japanese Iwo veteran Tsuruji Akikusa's cave suffered such an attack and he was the only one to make it outside before the Marines blew it up after setting the gasoline/seawater afire. As he left the entrance, he fell down unconscious. Black smoke traveled high in the sky carrying the smell of burnt fuel and flesh. When Akikusa awoke, he was an American POW.[2233]

Meanwhile, the Marines in Woody's unit continued to suffer wounds. Two more of his witnesses had become casualties: Corporal William Naro on 10 March suffered a "concussion blast" and Sergeant Lyman Southwell on 11 March took shrapnel throughout his chest (receiving his second Purple Heart).[2234] Naro was taking an injured buddy out of the battle zone when an explosion took him out.[2235] Naro's initial luck turned—days before, three Japanese had "prowled around his one-man foxhole" during the night, but apparently did not see him and left.[2236] At this stage, of the nine or more men who had witnessed some of Woody's acts on 23 February, two had died and five had suffered wounds.

Another who ran security for Woody during 23 February, Alan Tripp, also almost died on 11 March. Coming under heavy fire from "mortar and small arms fire," Tripp "voluntarily exposed himself to maintain communications between the company command post and the platoon command posts." Although "severely dazed" by a shell blast, he "refused evacuation," repaired the severed lines and kept communications open. In war, men die when there is no communication between higher command and front-line units, and they usually die fast. Tripp knew this and helped his Marines know what was going on. His acts were so extraordinary his commanding general, Erskine wrote a "Letter of Commendation" for Tripp, noting his "outstanding devotion to duty contributed immeasurably to the success of his unit."[2237] On 12 March, Woody's other Lieutenant, Charles Henning, took shrapnel to the right knee, but he stayed in the fight.[2238] The longer one fought on Iwo, the higher the chance was he would suffer wounds or die and Woody and the few remnants of his initial group were strongly tempting fate by now being on the island for three weeks. Many must have felt their time to live was running out.

On 13 March, although the battle raged on, several Marines hunted for souvenirs. Men walked around with sabers and Japanese rifles. Woody's regiment, the 21st Marines, expressed its irritation at this behavior and sent a message through the chain of command that this must stop.[2239] A few days later, even unauthorized personnel who were not supposed to be at the front looked for souvenirs among the dead—it became an obsession with some.[2240] Iwo Jima was a brutal battle fought by kids full of mischief and curiosity. Scavenging trophies was a dangerous activity since some of the gear was "booby trapped."[2241] Searching for items to take home, one might encounter alive and armed Japanese, as happened to five sailors on Saipan during Christmas 1944. Armed Japanese ambushed the men and all five died trying to find trinkets.[2242] This problem got so bad on Iwo, authorities assigned military police to prevent such souvenir collecting.[2243]

As Marines continued to prevail in battle and destroy underground bunkers, Japanese continued *jiketsu* rather than surrender. Most notable was the suicide of Major General Sadasue Senda. By 14 March, his 2nd Mixed Brigade, initially 4,600 strong, had been reduced to dozens. On that day, while Marines dedicated their cemeteries on Iwo, Senda gathered together 50 men, almost the entirety of those remaining in his brigade, and said he was going to commit *jiketsu* rather than "surrender to the heathen Marines." He declared, "Let us all meet again at Yasukuni Shrine." He distributed grenades to his men, who then placed them next to their stomachs and blew themselves inside out. During this death ritual, Senda took out his pistol and blew out his brains.[2244] "Many [religious faithful like Senda] were eager to sacrifice happiness, compassion and justice in this world, for a fantasy of a world to come."[2245] The sadistic, brutal and intolerably arrogant Senda "honored" his gods and ancestors by ending his own life and those of his men rather than live as POWs under their enemies. Woody's regiment was in the sector that defeated Senda so it helped kill a merciless fanatic and war criminal. Woody maintains that knowing he played a role in taking out such a leader helps him value what he did.[2246]

On other fronts, Japan's contempt for the welfare of conquered civilians in its Greater East Asia Co-Prosperity Sphere was once again on display in Manila, the Philippine capital. It fell to the U.S. Army as Iwo was falling to the Marines. Sailors and soldiers under Rear Admiral Sanji Iwabuchi ignored Lieutenant General Tomoyuki Yamashita's orders to withdraw, and instead made Manila a battleground. Subjected to incessant pounding and facing certain death or capture, the beleaguered Japanese took out their anger and frustration on civilians, committing acts later known as the Manila Massacre. Violent mutilations, rapes, and massacres of the populace accompanied the battle. One woman, while fighting off her attacker got blindsided by his comrade who sliced off her head. Her body fell to the ground and the soldier who had just decapitated her pulled down his pants, mounted the dead body and sexually assaulted the still warm, lifeless corpse. Some tried to have necrophilia with another murdered woman, but were unable to do so due to rigor mortis preventing them from spreading her legs. "Hospital patients were strapped to their beds and set afire; babies' eyeballs were gouged out and smeared on walls like jelly… others were ordered to dig their own graves and then shot down." Army private Cam Dowell remembered his shock at the main bank downtown when he found the vaults downstairs full of the decaying corpses of civilians the Japanese had raped and murdered and placed there to horrify the Americans. At least 100,000 Manilians lost their lives. Historian James M. Scott described Manila as receiving a "tsunami of barbarity" and the war crimes committed there as "one of the worst human ca-tastrophes" of WWII. Iwabuchi and his officers committed *seppuku* on 26 February. He was posthumously promoted to Vice Admiral having just "presided over one of the most barbaric massacres" of the war.[2247] MacArthur received a report about this slaughter by one of his staff which declared the responsibility of the Manila crimes "rests with the Japanese High Command and the government of Japan, represented by the Emperor, while the people of Japan itself cannot ultimately escape the aw-ful weight of moral participation and moral guilt."[2248] Journalist Henry Keys of the *London Daily Express*, in discussing Manila, wrote, "At last the Japanese have matched the Rape of Nanking."[2249] Whether in Manila or on Iwo, the Japanese were dying by the thousands at that time and inflicting death every way they could.

While battling northward up through the middle of Iwo, over ravines and cliffs, many Marines fought using only grenades. Woody used them generously.[2250] Baseball had helped many throw grenades with accuracy.[2251] Countless stories document how Marines showed better precision with them than Japanese. The battle was being fought with bullets and handheld bombs, and the Americans, with their culture of throwing things like footballs, basketballs and baseballs ex-celled in this type of warfare. In one sector to the north, a bunch of Japanese and Marines engaged in a veritable "grenade-tossing contest…The Japs tossed theirs too far. They went over the heads of the Marines and exploded harmlessly behind them. The Marines threw theirs with better aim. One hit a man wired with a picric acid charge. He blew up, a human bomb, killing the companions nearest him."[2252] If not for Babe Ruth, Lou Gehrig and Joe DiMaggio being the idols of some of these young Americans, these Marines may not have been any more accurate than their counterparts. A Japanese report admitted the Americans were "skillful with Grenade Throwers."[2253] Although the Japanese loved baseball and played it throughout Junior High and High School and had professional teams, they seemed not to be as good at throwing grenades as the Americans since young boys in the U.S., at this time, often played it all the time on playgrounds, at parks and during recess at school, activity not as often seen in Japan's society. Moreover, while training for Iwo on Guam, the 3d MarDiv's companies had fielded numerous baseball teams and there was an entire league on the island, keeping the Marines' throwing arms in good shape. In contrast, Kuribayashi's division definitely did not have a baseball league on Iwo Jima prior to the battle.

By mid-March, food and water were scarce on Iwo for the Japanese, so they resorted to brutal tactics to conserve supplies. Many of the caves remained in Japanese hands, and from these a few were "tossed out once every two or three days to reduce the population. This was called *kirikomi-tai* (death by a squad of sword-wielding soldiers). The rule was that those who left the shelter were never allowed back in. Before leaving, they were given two grenades, a canteen of water and a pack of dry bread. The grenades were to commit suicide."[2254] These were the men Woody faced: Roving, rejected small groups abandoned by their comrades to die in battle or by suicide. Such treatment illustrates the difference between Marines and IJA soldiers. Even today, Marines laud the concept of "a band of brothers" as in Shakespeare's phrase form *Henry V*: "We few, we happy few, we band of brothers; For he today that sheds his blood with me shall be my brother."[2255] The "brothers" concept was unimaginable in Japanese culture while Marines who volunteered for dangerous missions, jumped on grenades to save their buddies and rendered aid to the wounded under fire were a testament to how Marines "take care of one another no matter what."[2256]

As war dragged on, some Japanese realized they were not invincible and the gods would not save them. Contrary to their government's propaganda, the American fighting men were brave, tough and relentless. A few IJA servicemen started to change their views, a hard task for religious fanatics. Lieutenant Satoru Omagari was such a man. He ventured out of his cave in the north carrying dynamite for his final mission. He waited for nightfall, crept out until he found several dead Japanese near a logical route for the Americans, and then lay down with the cadavers so he could kill more Marines. He took out his bayonet, sliced open a dead comrade, gathered his innards and smeared the fallen comrade's intestines over his own body. He then feigned death and awaited his prey. "The dead were no longer seen as human beings, but as objects. Even the dead were called to fight," Omagari said. He thought, "[W]ho will be using my guts tomorrow?" He continued:

> I didn't feel fear anymore, but I could sense the dead. Their wide-open eyes became 1,000 sharp arrows, piercing my skin, my flesh and my bones. I tried hard to stay calm, clenching my teeth, fighting against nausea caused by this inhuman, brutal ordeal… Lying among the bodies, I waited for the enemy tanks. I floated

in and out of consciousness and didn't know if I was alive or dead anymore. Suddenly, maggots crawling around my neck and face brought me to myself. I became one with the bodies that had their guts taken out. I could be one of them tomorrow. 'This is war,' I told myself, cursing.[2257]

Another corpse lying nearby had his mouth open, filled with what looked like rice, "but was a pile of fly larvae." Large bottle-blue flies buzzed around his head like mad hornets. Throughout the night, and as morning came, the Marines never arrived yet Omagari waited, sweating in the sun and experiencing the horrifying rot of his dead countrymen with whom he had lain down. Thoughts penetrated his mind about duty, country and death. Was this what he had been "educated and trained for?" His elders had ingrained in him that it was glorious, honorable, spiritual and expected to die for the Emperor.

Omagari revered the tale of the 47 *rōnin*, believing they exemplified qualities of the Japanese fighting man. The legend of the 47 *rōnin*, also known as the *Akō* vendetta or the *Genroku Akō* incident, is a tale describing a band of *rōnin* (leaderless *Samurai*) who avenged their feudal lord's death. Plotting for a year, they finally kill the man who murdered their master. Then, they all commit *seppuku*. Some believe this is the best-known example of the *Samurai* code of *bushidō*, the warrior code. As Omagari lay among the corpses, planning the death of Marines, he believed the legend of the 47 *rōnin* was emblematic of the devotion, sacrifice and honor people should follow. That was what he had been taught, but by degree, he began to question it.

When darkness fell, he broke the strings of now-hardened intestines that bound him to the bodies, and crawled back to his cave. He tried to clean himself up, but the smell of death "clung tenaciously." The next day, he again half-buried himself among the dead awaiting the Marines, who again did not come. While there, he agonized about the meaning of life and Japanese culture. When darkness came again, he changed his mind about Japanese beliefs and swore he would never again try to be a human bomb. The battle had changed him and he started to see the weakness of his religious creeds.[2258] Omagari was rare in questioning his fanatic upbringing. He was raised with a faith that despised the "mind and free individual" and preached "submission and resignation, and that regard[ed] life as a poor and transient thing" preventing him and comrades from "self-criticism."[2259] Luckily for Omagari, he reflected upon Japan, the Emperor, his service and started to question it all.

A Japanese soldier is playing possum and his hand is close to a grenade. He wants an American to come near. The Marine in the picture was wise to the trick and was trying to talk the "Jap" out of suicide. 20 February 1945. National Archives, College Park. Photo Credit Ferneyhough

Omagari saw the inhumanity among his countrymen as they fought each other over water and refused to tend to the wounded, thereby making their environment devolve into a world of "every man for himself." The chain of command broke down and many started to strike out alone instead of taking orders.[2260] Some killed others for water. Omagari lost his faith in humanity as his comrades killed one another during a losing battle.[2261] As often happens, cultural norms, religious creeds and moral codes broke down under extreme physical and mental stress. The murderous martial discipline inculcated in these men and the barbaric religious creeds of superiority, military domination and invincibility withered away under the strain of military defeat. Omagari ruminated on these thoughts as he waited to waylay Marines.

It is unclear what would have happened had the Marines stumbled on the pile of corpses where Omagari lay. Often, when Marines found heaps of seemingly dead enemy, they fired into the bodies and burned them with a flamethrower to ensure there were no suicide sleepers. They had learned the hard way to be careful.[2262] For instance, a group of five Japanese had played possum in Woody's area and suddenly "sprung to life," shooting two officers before they were killed.[2263] Almost all procedures that are in the Marine manual about fighting techniques are written in blood. One rule they had learned to follow was, "If it's possible a dead Japanese soldier is really alive, act as if he is." Woody's 3d MarDiv concluded: "The Japanese probably have used more deception in the present war than has ever been practiced in any other campaign in history."[2264] Woody's boot camp instructors had exclaimed: "If it don't stink, stick it."[2265]

Omagari's reflection during the battle was not unique, however. Battle, where death is never far away, makes one reflective. On Iwo, a few Japanese came to realize the futility of their beliefs about the cult of death and their complete, unquestioning devotion to the Emperor, so they started to change like Omagari. Evidently rejecting his indoctrination, Omagari surrendered, something almost unheard-of for officers. "I became a hostage, but I felt nothing; no shame, sadness or any other emotion. I was an empty, worn-out husk."[2266] The war's horror made him realize that, to use a phrase from Melville, "a man's religion is one thing, and the practical world quite another."[2267] Omagari realized he made himself a *musekimono* (an outcast) by disgracing his family, military and nation by surrendering.[2268] Despite the outcome of the war and the repudiation of the Empire's tenets, remnants of his Japanese code can still be found among old soldiers today. At the 70th commemoration of the battle at Iwo Jima, Japanese POW and Iwo veteran Tsuruji Akikusa had to sit with the Americans instead of the Japanese at the ceremony because he had surrendered and, in their minds, disgraced his nation.

# Ch. 24: Battle on Iwo Winds Down

"I drew this gallant head of war
And cull'd these fiery spirits from the world,
To outlook conquest and to win renown
Even in the jaws of danger and of death."
　　　　　　　　—Shakespeare, *King John*, Act V, Scene 2

As FIGHTING CONTINUED INTO MARCH, the days rolled into one another. Digging in on the north shore, Woody said he did not know if it was March or May. Time had no significance and men just tried to survive the next hour. They killed and did not think much about it after having eradicated dozens, hundreds or more.[2269] Marine veteran E. B. Sledge said, "Time had no meaning, life had no meaning. The fierce struggle for survival…eroded the veneer of civilization and made savages of us all. We existed in an environment totally incomprehensible to men behind the lines—service troops and civilians."[2270]

On 14 March, a group of Marines on Iwo, including Smith, commanding general of the FMFPAC, participated in another flag-raising on an incinerated bunker using an 80-foot flag pole. Smith wrote, "Tears filled my eyes when I stood at attention and saluted the flag. The ceremony marked the capture of Iwo Jima and the end of the most terrific battle in the history of the Marine Corps."[2271] Although Major General Schmidt was commanding officer on the ground, Smith was higher up the chain of command and it was appropriate he attend. A colonel read Nimitz's proclamation, who had overall charge of the central Pacific:

…forces under my command have occupied this and other of the Volcano Islands. All powers of government of the Japanese Empire in these islands so occupied are hereby suspended. All powers of government are vested in me as military governor and will be exercised by subordinate commanders under my direction...[2272]

Many still on patrol on the island wished the Japanese would heed Nimitz's proclamation. Iwo had become U.S. property. Control of the island was securely in American hands, but there was still a lot of fighting to be done. A wise guy exclaimed, "Who does the admiral think he's kidding? We're still getting killed."[2273] The Japanese would fight to the last man's last breath. As Woody held,

> The Japanese never let up. One would think that after taking over so much of the island and seeing all the materiel might we brought, that they would see the writing on the wall and surrender. But no sir, they seemed sometimes to fight even harder, if that's possible, as we cornered them to the north.[2274]

Photos of dead Japanese taken by Sergeant Dick Dashiell on Iwo. Often Marines had to be careful with IJA dead bodies because the Japanese would booby-trap them. Author's Collection and Dashiell Family Archive

Admiring their bravery in the face of defeat, a Marine, exhausted and shaking his head while looking at a dead combatant slumped over his rifle, asked Dick Dashiell, "What other guys would fight like these birds with this place surrounded by 800 ships?"[2275]

That evening Kuribayashi lowered the Japanese flag and burned it so Americans could not capture it. He radioed Tokyo: "The battle is approaching its end. Since the enemy's landing, the Gods would weep at the bravery of the officers and men under my command... I regret very much [though] that I have allowed the enemy to occupy a piece of Japanese territory...I sincerely hope my soul will spearhead a future attack. Praying to God for the final victory and safety of our motherland." He soon issued his final order: "Everyone will fight to the death. No man will be concerned about his life. I will always be at the head of our troops."[2276] Later that day, Kuribayashi listened to a Tokyo broadcast in which schoolchildren sang the "Song of Iwo Defense." The show ended with children from Kuribayashi's hometown praying to their Shinto gods for victory. From the shelter of his cave, he issued a message to Tokyo thanking the people of Japan for the presentation.[2277]

That same day, 307 B-29s from bases in the Marianas bombed Kobe, killing 2,669 people and injuring 11,289 others. A quarter of a million people were rendered homeless and 66,000 houses destroyed. Bombs fell on Japan weekly.[2278] According to several B-29 crews, they called this treatment of the Japanese as "being Lemayed" in honor of their commander General Curtis LeMay.[2279] Veteran John Lauriello said, "After what they did to us at Pearl Harbor and what they would've done to us if they could, then all I got to say about the bombing we gave them is 'Payback is a bitch, Japs.'"[2280]

Meanwhile, Woody's unit moved northward and continued to kill Japanese infiltrators and seal caves with explosives. On 15 March, the battalion reported Japanese suddenly appeared from openings in the ground, threw grenades and disappeared. The problem with fighting these types of attacks was the enemy was appearing from caves both in front of and behind Woody's lines, such that the Marines had to maintain a lookout both ahead and behind. It was difficult and tense fighting, but eventually the report noted that 12 Japanese were killed.[2281]

During this time, Vernon Waters platoon commander, 2nd Lieutenant Richard Tischler, saved his B Company, and most likely Woody's C Company, from an ambush. While moving forward over difficult ground to the north, Tischler noticed a suspicious position ahead of his men along a path that was the logical route for his Marines. For Tischler to have noticed this meant he had a keen eye and was leading his Marines from the front. Also, after being on the island for four weeks, he had learned how the Japanese used concealment and camouflage. He stopped his men, placed them in defensive positions and radioed back to Colonel Withers at the regimental CP. Coordinates were probably given and then mortars most likely soon landed on the suspicious position blowing it up into thousands of pieces. After the attack, Tischler and his men inspected the position and noticed several dead Japanese. Had this position not been ascertained, the enemy had them in a perfect kill zone and Tischler would have lost several of his Leathernecks. Later, Colonel Hartnoll J. Withers wrote a "Letter of Commendation," saying Tischler "Is hereby cited for the submission of vital intelligence information of the enemy. This information enabled the regimental commander to expedite operations, and accomplish our mission with the least possible loss of time and life."[2282] Brave platoon leaders like Tischler, and like Chambers shown earlier, helped the Marines slowly but surely take more and more of the island. (See Gallery 3, Photo 17)

Daily the Marines inched closer to the northern terminus of the island until on 16 March, elements of the 21st Regiment finally occupied Kitano Point.[2283] This was a 3d MarDiv badge of honor since, apparently Erskine was irritated with the 5th MarDiv's lack of success at Kitano Point and pushed his men to secure it, which he did independently of the 5th using Colonel Wither's 21st Marines to do so.[2284] During the previous week, Woody's company attacked areas that artillery barrages had first softened up; killed the remaining Japanese; and dug in. They would wait for another barrage, advance, kill more Japanese, halt for the night, settle into ravines, shell craters or trenches and wait for the next day. The Japanese naval commander on Iwo, Rear Admiral Toshinosuke Ichimaru, wrote, "The Americans only advance after making a desert out of everything before them. Their infantry advance at a speed of about ten meters an hour. They fight with a mentality as though exterminating insects."[2285]

After artillery bombardments, Marines routinely closed with the enemy and fought in gullies, on hilltops and throughout the hundreds of bunkers the Japanese had built. The artillery batteries "expended 450,000 shells" during the campaign and that figure did not include the mortars, grenades and rockets used as well.[2286] The northern terrain on the high plateau was rocky, so Woody and his comrades could not dig foxholes to protect himself. Through many fissures in the rock, hot mist and sulfur foam rose up. It was like another planet and the land hissed with steam. Over such harsh terrain, the Marines had one mission: Take the fight to the enemy. They rarely held their defensive positions. As "Howlin Mad" Smith said, "We're not accustomed to occupying defensive positions. It's destructive to morale." Smith's quotes and his pugnacious behavior earned him the title "Patton of the Pacific."[2287] On 16 March, Woody's battalion noted it had destroyed "24 caves, 2 pillboxes, and 36 of the enemy."[2288] It was just a typical day for a Marine battalion in its advance on Iwo. They improved upon those numbers on 17 March, killing 90 enemy soldiers and closing 25 caves—the Marines found many of these caves using dogs.[2289] Since Woody was a demolition man, some of these caves probably were "closed" by him.

Being a volcanic island, Iwo was boiling under the surface. Hot pools of water that collected on the surface allowed men to bathe behind the lines. Marines in Woody's outfit placed their cans (C-rations) of string beans, tomatoes, vegetable hash, ham, eggs, corned beef and other rations between the rocks where steam rose from the island's belly and within a few minutes, they had a hot meal, heated to over 100 degrees.[2290] In contrast to the Marines, the Japanese suffered from thirst and hunger. To get food, they were reduced to scavenging Marine trash dumps at night.

Humorous scenes occurred everywhere despite the pervasive tragedy. The Marines made light of horrible situations and witnessed bizarre acts of heroism and desperation daily. The Corps' philosophy is that everyone is a rifleman first, so when a cook suddenly found himself in combat in Woody's regiment, he pulled out his rifle and started shooting several Japanese caught in a ravine. Running out of ammunition, in frustration, the cook picked up rocks and started pelting them. "I reckon I hit about a dozen on the side of the head before more ammunition was brought up."[2291] Sometimes bullets would hit the Marines from hidden areas and they dodged and ran for cover. One Marine in Woody's regiment said, "I missed winning the Purple Heart, but if they'd seen me running to get out of the way of some of those bullets, they'd have given me the Distinguished Flying Cross—and that's damned hard to win in the infantry."[2292]

Although Marines had been ordered to "take no prisoners," and indeed many of the enemy were killed who probably could have been taken prisoner, there was still considerable effort to get the Japanese to surrender when they were discovered in caves. In one instance, the Marines pleaded with around 15 Japanese to exit. They responded with a request to give them 15 minutes to discuss it.[2293] The men could hear the enemy speaking loudly with one another. Hearing them debate with one another, one Marine said, "When the war is over we must abolish the [Japanese] language: it's an ugly one."[2294] Eventually the enemy stopped speaking and instead of surrendering, they blew themselves up with grenades. Sometimes, the Marines set up loudspeakers and had native speakers plead with those inside to come out. In one case, they used a captured officer as their spokesman. Marines met with little success using these methods and explosive experts like Woody had to blow the caves shut because the Japanese, if left alive, would do all within their power to kill Americans.[2295] Also, when the Japanese refused to come out, Marines used flamethrowers on them. Woody's fellow flamethrower Joseph Rybakiewicz said: "Those caves, especially in the last days, when the Japanese were running out of ammunition—you never knew how many people were hiding in a cave. I never knew how many people I was killing with the flamethrower...I still wonder how many people died from my flamethrower."[2296]

In a few cases, *Nisei* (second generation Japanese Americans) interpreters assigned to the U.S. Army tried to help the Marines save some before they committed suicide or before Marines ran out of patience and sealed them in their caves for eternity. Bravely—insanely—they entered the caves, calling out to their "blood brothers" to lay down arms. Although the report noted that surprisingly, none of the *Nisei* died this way, they also met with little success and many of the Japanese blew themselves up after allowing their *Nisei* cousins to leave. When questioned about their bravery, one *Nisei* sarcastically said that he felt safer with the enemy in the caves than with the Leathernecks on the surface since he had "been shot at by more Marines than the enemy" due to misidentification.[2297] Besides using *Nisei*, the 3d MarDiv used POWs to enter the caves and persuade the enemy to surrender when possible.[2298]

Erskine even tried his own hand at getting Japanese to surrender, and on 16 March he had a message written up and sent to Colonel Masuo Ikeda, commanding officer of the 145th Regiment directly to his front. He used two Japanese POWs, Ueno Yasuo and Yoshio Yameda, to deliver the message, written in both English and Japanese. One can only imagine how the two Japanese POWs looked at each other when Erskine gave them this mission. The message read in part:

> Our forces now have complete control and freedom of movement on the island…except in the small area now held by the valiant Japanese troops just south of Kitano Point. The fearlessness and indomitable fighting spirit which has been displayed by the Japanese troops on Iwo Jima warrants the admiration of all fighting men. You have handled your troops in a superb manner but we have no desire to completely annihilate brave troops who have been forced into a hopeless position. Accordingly, I suggest that you cease resistance…and march, with your command, through my lines to a place of safety where you and all your officers and men will be humanely treated in accordance with the… (Geneva Convention).[2299]

Erskine then sent this message off with the POWs with a two-way radio. In a 24-hour strange course of events, the POWs went to several caves and tried to convince the occupants to surrender and get the message to Ikeda. In some respects, it was a suicide mission Erskine sent them on although he had provided them another note telling those confronted that these POWs had "been captured while unconscious and unarmed and still professed a desire to die for their country if necessary."[2300] Remarkably, going from one cave to the next delivering the message, even to one cave with a full house of troops, they were not killed although dressed like the enemy and pleading for their comrades to violate their codes. Failing at their mission, the two POWs returned to the Marine lines, but encountered 5th MarDiv units instead of Erskine's. Over the radio, the 5th MarDiv informed Erskine's CP they had two crazy prisoners, dressed in Marine uniforms and with a U.S. two-way radio claiming to be 3d MarDiv members, in good standing and under *his* orders. One can imagine the thoughts going through the 5th MarDiv's CP after hearing this story and asking, "What the hell is going on?" Erskine's command verified they were indeed his and thus ended the Marines' "strongest effort to persuade Japanese officers to behave like the officers of a civilized military power."[2301] The Japanese ignored Erskine's request across the board.

That Erskine took the time to reason with an enemy commander to try to get him to stop fighting so Erskine would not be forced to annihilate him and his men was a testament to American ethical values.

Although he searched out Ikeda, Erskine had no way of knowing that the colonel had already died a few days before in battle and that Elmer Bechtold of Company D, 2nd Battalion, 28th Marines, 5th MarDiv had found the colonel after most likely having killed him; pillaged his personal effects; and by 16 March was the proud owner of the Japanese officer's "postcard, an envelope, money, and calling cards."[2302] Ikeda's worldly possessions, although not amounting to much, were Bechtold's war booty. Erskine's request was addressed to a dead man with a shattered command (by March 14, his once 2,700 strong regiment was down to only six men).[2303]

On 16 March, Erskine once again addressed his men and said:

Although considerable mopping up remains, a big job is done and you have captured Iwo Jima, a lane deep in the heart of the empire's defenses and I salute the gallant men of [3d MarDiv] and its supporting units who have carried this attack to the enemy relentlessly for 26 consecutive days. A well done to all units of all services participating in the seizure of this island.[2304]

A day later, he further stated:

You have fought a large and well-equipped Jap force across unbelievable, difficult and heavily fortified terrain. Victory has been the result of indomitable courage, skill and determination on the part of every officer and man. It has been a great Honor to be associated with such an outstanding organization and we look forward to similar associations on the future road to Tokyo.[2305]

Erskine knew this battle was conditioning his men for another invasion, one which would be a huge undertaking. Even with all the horror of Iwo, Erskine, not knowing the atomic bombs would obviate the need for a homeland invasion, tried to keep his men focused on the goal—the defeat of Japan. He felt proud of all that his men had accomplished knowing they were not even supposed to be in this battle in the first place. They not only achieved victory on this difficult battlefield, but also earmarked the advance by taking up the main thrust through the island's center. Although Erskine's statements implied the battle was over, it was not. In conducting "mopping up" operations to which Erskine alluded to on March 16 (above), Leathernecks still had to be alert. On many occasions, the Japanese were found to be in unfit fighting shape due to substance abuse—just like their comrades on Guam with sake. In the few weeks leading to the end of Iwo's defense, the Japanese often engaged the Marines while drunk on sake. Knowing imminent defeat awaited them, many killed themselves with grenades while others got wildly drunk and conducted uncoordinated attacks or ventured on one-man suicide missions to commit "suicide by Marine."[2306]

As I walked in many of the caves in 2015, my feet often crunched the broken glass of sake bottles. According to Don Graves of the 28th Marines, 5th MarDiv:

> The Japs were so liquored up…that it took several bullets to bring them down. One Nip came at us and I unloaded my entire clip of my M1 in him and he still kept running at us. His heart must have stopped by then, but the adrenaline pushed him forward and guys in a foxhole near me finally brought him down. When the Japanese got drunk on sake…they behaved as if we could not shoot them, but when we did, they continued to fight until they bled out. These people were…nuts.[2307]

Graves remembered seeing two Japanese walking out of their cave in their G-string underwear with sake bottles in their hands, laughing; Dashiell reported on a small *Banzai* where the enemy soldiers howled hysterically as they charged. Were they drunk or just crazed in the knowledge they were conducting a suicide attack?[2308]

Graves found other evidence that showed the Japanese had women with them, although many historians doubt this. According to Graves, as they neared the island's northern end, he found women among the dead Japanese.[2309] At least six other Iwo veterans have verified they also found women on Iwo.[2310] In one of the first histories on Iwo, written in 1945 by several Marine correspondents and Iwo veterans, two women were found in the wreckage of Motoyama after artillery and ship "guns had smashed it."[2311] George Bernstein of the 4th MarDiv witnessed women jumping off cliffs on the northern shore just like at Marpi Point on Saipan and Marpo Point at Tinian.[2312] If they were indeed killing themselves, they were not necessarily "Comfort Women" and maybe some, if not all, of them were "Japanese prostitutes" Kuribayashi had shown he preferred while in charge of the Hong Kong region back in 1942.[2313]

If there were women on the island, where did they come from? In June 1944, Kuribayashi is known to have moved civilians off the island, but if the accounts are accurate, some women stayed or were "Comfort Women" and had been sneaked in or brought in unofficially. Perhaps some were Geisha girls kept on the island by "their" officers. In fact, Yamamoto made no secret about taking his favorite Geisha girl, Chiyoko Kawai, on his flagships although he was married.[2314] Therefore, if there were no women on Iwo, this would be the first Japanese garrison that did not have prostitutes, "Comfort Women" or geisha girls. In some respects, it would be odd if there were no women on the island, knowing the IJA's *modus operandi*.

On 16 March, Kuribayashi sent his final message to Imperial Headquarters:

> So sad to fall (in battle), our ammunition is exhausted, we are unable to fulfill the heavy duty for the Nation. I will pick up my sword, though my body lay decaying in the field, I shall reincarnate seven times to seek revenge. My earnest thoughts will go to the Empire long after this island is overgrown with ugly vines.[2315]

The word *chiru* ("fall") Kuribayashi used in this message "evokes an image of the cherry trees at the Yasukuni War Shrine in Tokyo." This is a place where "fallen cherry blossoms," or fallen warriors' spirits, gather in the afterlife.[2316] Kuribayashi wrote about being reincarnated seven times to continue to fight, a religious concept of the afterlife common to Japanese.[2317] In Judeo-Christian dogma, individuals on the brink of death may take a summation of their lives, will often ask for forgiveness and will generally affirm the Almighty in the hope they will enter into His heavenly realm. In contrast, Kuribayashi yearned to avenge his country for seven different lifetimes, a significant number in Buddhism leading to a state of Nirvana. The word for revenge in Japanese is *katakiuchi* which means "attack enemy." These were Kuribayashi's last thoughts—never view your enemy as worthy of anything less than death even in the face of your own demise.[2318] The day that Kuribayashi made these declarations, 16 March, the government decided to promote Kuribayashi to full general.[2319]

On 21 March, Imperial General Headquarters in Japan announced over the radio the Iwo Jima defenders had all died glorious *Samurai* deaths on 17 March. As usual, the Japanese government continued to lie to its people; the *Samurai* troops were starving, thirsty and killing each other. Having run out of ammunition, they were incapable of offering resistance. To feed themselves, they stole and scavenged food from the enemy. They reverted to a primitive state in an attempt to survive for another day. The glory, if there was any, was found in *Banzai*s and *jiketsu*s because surrender was not glorious in their belief systems. To those who rejected those beliefs, however, the end of the Japanese resistance on Iwo was anything but glorious—it was, as Omagari noted, "shameful" and "disgusting."[2320] Kuribayashi reported on 21 March to the Chichi Jima garrison, "My officers and men are still fighting. The enemy front line is 200 to 300 meters from us and they are attacking by tank…They advised us to surrender by loudspeaker, but we only laughed at this childish trick."[2321] He then wrote the Emperor, "We have not eaten or drunk for five days. But our fighting spirit is still high."[2322] On 21 March, Prime Minister Kuniaki Koiso issued a radio broadcast about the effort their countrymen made on Iwo. He praised Kuribayashi and his men for their "heroic resistance," embodying the "Japanese spirit."[2323] Though they were suffering physically and spiritually, surrender was not a viable option for the vast majority of those remaining alive.

On 21 March, the 147th Army Infantry Regiment of 2,952 soldiers landed on the southwest Purple Beach near Suribachi and were put under the command of Major General Erskine. It was there to help with mopping up operations and to relieve some of the exhausted Marines. The regiment's assignment to relieve the 2nd and 3rd Battalions of the 21st Marines by 23 March meant that elements of Woody's 1st battalion would stay and fight alongside the "dog faces" remaining under Erskine's control until 4 April.[2324]

From 20 to 26 March, Woody's company patrolled the north, securing the areas they had won.[2325] Erskine ordered them on 19 March to "maintain contact with the… enemy encountered until he is annihilated."[2326] Marines now fought Japanese soldiers dressed in Marine uniforms, so attired partly because they wanted to practice subterfuge, and partly because their own tattered uniforms had rotted off their bodies.[2327] As the end neared, the Japanese continued their suicide. Marines started to hear explosions throughout the night from caves as the Japanese blew themselves up.[2328] So many rotting bodies covered the island producing millions of flies that once again it was sprayed with DDT with "crop-dusters," or better said, "corpse-dusters."[2329]

As the battle wound down, the Marines increased the hunt for souvenirs. The ritual of trophy taking from enemy dead is as old as human existence. Once again, higher command ordered this practice stopped. After the battle, Marines wanted trophies for their hard-fought victory, and thus, many disregarded the prohibition.[2330] Captain Beck described this:

> It must have been interesting for everyone to see the collection of sou-venirs from the two different theaters. I could have collected truck loads if I would have had the time and means to carry them. When we overran defensive areas there is no limit to the bounty at hand, but when one is responsible for so many…you can't neglect them and gather Jap gear. I never did allow the men to do it unless we were pa-trolling then I would always let the men who shot the Japs collect what was on the dead Japs. I have quite a collection on hand but have never gotten around to sending it [back home].[2331]

Even as the souvenir hunting was going on, there was a lot of "mopping-up" or spo-radic fighting. Marine demolition experts like Woody continued to discover caves and blow them shut. If groups of Japanese offered resistance, they killed them. Marines often found caves previously sealed were "re-opened by the Japanese" a few days later. As a result, Marines resealed them using more explosives than nec-essary to ensure that no escape was possible. They often heard the faint groans of the dying and muffled explosions of grenades from deep inside the earth as men committed *jiketsu* when it dawned on them the cave was a tomb from which they would never escape.[2332] One such cave caused a horrific night for nearby Marines as the entombed Japanese "hollered" and "squealed" under their feet until dawn. Eventually, the cacophony dwindled to occasional "moans," and by morning, only an eerie silence.[2333] For decades, Japanese had been indoctrinated with the belief their fighting spirit could overcome any enemy's material superiority which seemed to have been proven during the Russo-Japanese War of 1904-05 and during the first year of WWII when the Japanese defeated superior forces.[2334] However, on Iwo that belief was shattered and many died probably with broken hearts having realized the lessons their leaders had given them were fallacious.

The Japanese were out of provisions. They had no functional supply chain. They were alone and starving. Marines knew they were without water, and a man could only, theoretically, go three days without water or die. If Marines had not killed them, the Japanese would have died from lack of hydration.[2335]

Behind the lines near Mt. Suribachi, the Seabees and Marines built a functional base. Barracks, tents and supply depots were everywhere and hundreds of planes used the runway daily. A hot mineral shower, "large enough for a full platoon," was constructed to improve hygiene and utilize the naturally heated, spa-like mineral water.[2336] The infrastructure was so well done, that on

First Lieutenant Harry L. Martin. He countered the last, desperate *Banzai* attack on Iwo rallying his men (some of whom were "Negroes") and helped defeat the Japanese earning him the MOH. He would die in this action on 26 March 1945. St. Louis Personnel Records Center

22-23 March, an outdoor theatre showed the movie *Two Tickets to London* (1943) starring Michèle Morgan and Alan Curtis. There were 550 men from Woody's regiment in attendance, having been pulled off the lines for a few days of rest. From 23-24 March, the theatre showed *Hey Rookie* (1944) starring Ann Miller, Joe Besser and Larry Parks to 850 men of the 21st Regiment.[2337] In a surreal display of military magic, in just a few weeks, the land that had been occupied by the Japanese military machine just a few weeks before had now become a film center. American culture had arrived to Iwo Jima.

On 26 March, while Woody and his men covered ground along the shoreline in the north, a hidden underground barracks behind their lines and just north of Airfield No. 2 housed almost 300 Japanese plotting a massive attack. They had used their underground tunnel system to travel from the north to infiltrate the rear echelons. From their position behind the lines, they had the element of surprise and were poised to hit the Americans.

In the early morning around 0200, a Japanese commander (some think Kuribayashi), cunningly picked an area west of the second airfield in the island's middle where there were many non-combat U.S. troops in the rear. Sleeping comfortably in their tents were pilots, crewmen, supply troops, shore parties, anti-aircraft gunners and Seabees. Emerging from underground, the Japanese sliced through tents, killing many who were not armed for or accustomed to this type of warfare. Most pilots carried only pistols, and many died in their cots not knowing what hit them. Unlike concurrent *Banzais* in areas that were not behind enemy lines, these Japanese were silent and sober when they struck. As they attacked, they screamed, slashed, fired and threw grenades to kill as many Americans as they could. The spearhead of this attack also hit the 5th Pioneer Battalion which had many "Colored" Marines. However, Marines have always been trained as combat troops throughout history epitomized by the 29th Commandant of the Marine Corps Al Gray's declaration: "*Every Marine is, first and foremost, a rifleman. All other conditions are secondary.*" As a result, even these "Negro" Marines knew fire team tactics and proper weapon handling. Thus, 1st Lieutenant Harry L. Martin was able to organize a skirmish line largely manned by these African-American Marines. Martin, a 34-year-old White reserve officer, counterattacked along with the Black Marines beside him, overrunning a machinegun position and killing four Japanese with his pistol while "yelling abuse at them." Seeing the Japanese using a ridge to hide themselves in order to mass another charge, Martin "let out a yell to 'Follow me; I can hold the bastards for a while.' Then I heard him yell, 'Come on out you little yellow bastards or I'm coming in and get you,' and before any of us below could move, he had run down towards the rear of the ridge." [2338] Several Leathernecks followed Martin to join his charge attacking and killing the stunned Japanese many of whom were wielding swords.[2339] Although he had been hit in the head and buttocks with shrapnel, he refused to leave the battle.[2340] A grenade finally killed Martin while he led his men from the front. He received the Medal of Honor posthumously. His Black compatriots fought tenaciously and the Corps shore party commander "was highly gratified with the performance of these colored troops… while in direct action against the enemy for the first time. Proper security prevented their being taken unawares, and they conducted themselves with marked coolness and courage."[2341] As one sees, occasionally in war, racism disappears and merit trumps all, especially in the moments of life and death.

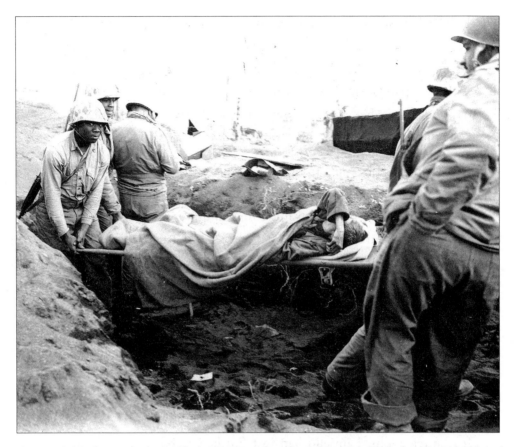

On Iwo Jima, there were many Black Marines serving, mostly in supply outfits. Towards the end of the battle, however, 1st Lieutenant Harry L. Martin, organized a group of "Negro" Marines to counter a *Banzai* and these brave African-Americans fought well. The Corps shore party commander said he "was highly gratified with the performance of these colored troops." As one sees in this photograph, Black Marines were armed on Iwo, proving once again that every Marine is a Rifleman. In this picture, they are carrying back a Japanese POW who was in need of medical care. 23 February 1945. National Archives, College Park. Photo Credit Sgt. Don Fox

This was the last "major" battle on Iwo. In this *Banzai*, almost all the Japanese died, the American Air Corps suffered 44 dead and 88 wounded, and the Marines suffered 9 dead and 31 wounded.[2342] Kuribayashi's biographer Kumiko Kakehashi, has argued that this last all-out attack was particularly skillful because it was led by the general.[2343] According to a Japanese survivor, Kuribayashi led this attack, and addressed his men beforehand:

Even if I should perish before you in the battle, the glorious exploits that you have carried out will never be forgotten. Japan may now be losing this battle, but the people of Japan are burning at your loyalty and your patriotism; they are praising your glorious deeds; and the day will come when they offer silent prayers for your ghosts. Be easy in your minds and sacrifice yourself for your country.[2344]

Kuribayashi inspired his men to fight heroically although there was no prospect for victory or survival. As he acknowledged they had lost the battle, he took pride they had fought well. His goal to inflict maximum destruction on the Marines had succeeded beyond anything his predecessors had accomplished.

This last *Banzai* caught the Americans off-guard. At the beginning of Iwo, the Americans looked for a *Banzai* that never happened. Letting their guard down, believing the Japanese had no more offensive capabilities, they got way-laid. This "counterattack" did indeed cause much havoc and death, but not at the ratio that Kuribayashi had hoped for since he had asked his men to actually kill 100 Marines each in this attack before they died. In that light, it was another failure. Still the charge afforded him and his men slightly more than the opportunity to give their nation a "worthwhile death."[2345]

In the rear, John Lauriello had just finished a shower as the attack commenced and he learned how surprised everyone in that area was when the Japanese struck. He sarcastically remarked, "What the hell were they doing with their guard down? Did they think they were staying at the Bellevue-Stratford in Philadelphia?"[2346] Jerry Yellin, a Jewish American P-51 pilot, observed the aftermath saying, "It was a horrible sight...Our dead had been removed, but piles of Japanese dead lay everywhere. If I'd been with this squadron, I would've been killed. It was a bloody nightmare and the bodies lay in a grotesque, tangled mess."[2347] Compared to other battles, apart from this small *Banzai* and the one earlier by Rear Admiral Iwabuchi, Erskine stated, "large-scale uncontrolled so-called 'Banzai' attacks" were "noticeably absent [at Iwo]."[2348] This was possibly the reason the Marines were not expecting such an attack in the middle of their new base at the battle's end. As this final skirmish raged, Woody and his men progressed westward along the northern shore looking for hidden Japanese. He knew nothing of this Japanese last-ditch effort and learned later "a bunch of poor airmen bought it sleeping in their tents."[2349]

On 26 March, the Corps sent a letter to Woody's mother informing her he had suffered wounds. Although he continued fighting for weeks after his injury, his mother was told he had received wounds "against the enemy on 6 March at Iwo." For all Mrs. Williams knew, Woody had lost a leg, an eye or an arm. The letter's writer could have told her his wounds were not life-threatening and he was still in the fight, but nothing of the sort was revealed. The officer, Major F. Belton, continued, "Your anxiety is realized, and you may be sure that any additional information received will be forwarded to you at the earliest possible moment. Meanwhile, you have been furnished all the facts available."[2350] Belton was probably tasked with reporting that Woody had been a WIA (wounded-in-action) in a form letter not knowing details. Woody wrote, "During [the war], she received 2 telegrams telling her that [two of] her sons had been wounded, but she never got any details as to how bad or chance of recovery so she had no idea of the severity. This not knowing took a great toll on her...health."[2351]

Eventually, Woody and his men reached the northern beaches. Touching the ocean, he felt safe, but he received Intelligence from a captured major that 300 enemy were holed up in a cave on a hillside above the beach. As a demolition expert, Woody was ordered to blow it closed.[2352] The Marines did not want another surprise *Banzai* like that of 26 March. Woody grabbed his C-2 explosives (an eight-pound satchel), inserted a charge with a 10-second fuse, pulled the fuse lighter with his prime cord, threw it into the truck-sized cave and ran for cover. Nothing happened except the sound of a light thud, but no explosion. He repeated the action with the same result. He realized his blasting caps were damp and malfunctioning, so he went back to supply, got dry ones, returned to the cave, and threw in a third explosive. This time, it exploded, igniting the other two within as well. The whole section of the hill collapsed.[2353] "You couldn't tell there had been an opening in that hillside" afterward, Woody said.[2354] The regimental report noted one of the duties of the men was to "close caves" and Woody did his part.[2355] It is unknown whether this catacomb was the one where Kuribayashi had been—his body has never been found. Woody may have ended the battle for the general the way Kuribayashi had directed, with the final defensive position becoming his burial chamber.[2356]

According to another report, on 27 March Kuribayashi moved to another cave on the northern shore with his staff officer, Colonel Kaneji Nakane.

Kuribayashi faced north in the direction of the Imperial Palace where Hirohito reigned supreme, knelt and solemnly bowed three times to the man-god. He then pulled his sword, thrust it in his stomach and bowed his head. Nakane raised his own sword, taking on the role as Kuribayashi's "Second," and sliced off the general's head in ritualistic *seppuku*, *Samurai*-fashion. The colonel then buried his commander's body in an unmarked grave.[2357] Other stories about Kuribayashi's demise include shellfire blowing the general to bits or a sniper shooting him.[2358]

In view of Kuribayashi's unorthodox ways, especially his logic about war and resources that transcended those of other commanders, he probably reasoned it would be unwise to kill himself when he could still fight. Moreover, killing himself would violate at least two of his Battle Vows, namely, (1) "We shall defend this island with all our strength to the end" and (5) "We shall not die until we have killed ten of the enemy."[2359] If he had the ability to fight and kill a few more Americans, he would live and die by his vows. Everything Kuribayashi did as a commander had sound logic behind it; breaking with tradition was something he did frequently. IJA officers usually went to Europe for overseas study and training, but Kuribayashi went to the U.S. and Canada. IJA officers usually met the Americans at the beach, but Kuribayashi conducted a defense from the interior. IJA officers usually got on poorly with IJN personnel, but it seems Kuribayashi got on better with his naval officers than most IJA commanders.[2360] IJA officers, when in full command, usually had "Comfort Women" at their bases, but Kuribayashi probably denied himself and his men these "pleasures" on Iwo. Lastly, the majority of IJA commanders killed themselves at the conclusion of battles they lost performing ritual *seppuku*, and it is this study's belief that Kuribayashi, in contrast to his peers, did not do this; killing yourself when you could still kill the enemy was wasteful. In his 16 March farewell telegram to General Headquarters, he hinted at what he most likely would do in the end and it was not *seppuku*. He wrote: "Our ammunition is gone and our water dried up. Now is the time for us *all* to make *our* final counterattack and fight gallantly, conscious of the Emperor's favor, not begrudging *our* efforts though they turn *our* bones to powder and pulverize *our* bodies [author's italics]."[2361] He did not write that his men would attack; instead, he used the plural "us" and "our," thus including himself in the counterattack plan. There were no words or hint of ritual suicide. On 17 March, Kuribayashi called his remaining men together after all had burned their "insignia of rank" and addressed them while grasping his sword's pommel:

> Even if you have to eat grass, bite the earth, or throw yourselves on the ground…you will fight and, in so doing, find a way out of this fatal situation. With things as they are, each one of you must kill one hundred—there is nothing else for it. I believe in your devotion. Please do *as I do* [author's italics].[2362]

He asked his men to attack just as he was going to attack, giving themselves a "worthwhile death."[2363] He donned a common soldier's uniform, picked up a rifle or a sword and led his surviving men in one more attack.[2364] Kuribayashi biographer Kakehashi and renowned historian Michael Burleigh claim with conviction that he led and died in this charge and perhaps this theory was not verified because Americans threw Kuribayashi in his common uniform into a pit for burial or burning, either of which would have kept him anonymous.[2365] Another source supports this final act of Kuribayashi, and it comes from his radioman Sergeant Shezuo Oda although it gives a slightly different ending than dying in the all-out charge. Oda supports that Kuribayashi did lead the charge, but somehow, he was able to disengage after the attack petered out and his staff member Colonel Tadashi Takaishi and his Chief-of-Staff Kanji Nakane shot the general before they killed themselves by shooting into their own hearts.[2366] One way or the other, in the end, how he met his end remains a mystery.

Regardless of how he died, the Marines were just glad he was dead. According to Shinto belief, if Kuribayashi's body was not properly cremated, his spirit (*Tama*) has not been freed and is in a perpetual state of suffering, yearning to be released from the Shinto/Buddhist purgatory that awaits those improperly "disposed of." Such spirits are believed to haunt living relatives, pestering them to find their remains and liberate their souls to escape the "unholy form of limbo."[2367] Even though most Japanese soldiers left a fingernail and a lock of hair behind at home to be later cremated in case their bodies were lost to the chaos and violence of war, families *always* desired to know that their men were properly sent off to the next life.[2368] If not, then even with burning the lone fingernail and bundle of head hair, the warrior's spirit still was apt to make his relatives' lives miserable according to their beliefs.

General Kuribayashi's widow, Yoshi Kuribayashi, gets notice of her husband's posthumous promotion to full general after the fall of Iwo Jima. She is handed the official promotion certificate personally from the Japanese Army Minister Hajime Sugiyama on 6 April 1945 in Tokyo, Japan. The Asahi Shimbun and Getty Images

As mentioned earlier, recognizing Kuribayashi's heroic defense, the IJA officially promoted him to full general on 17 March. It appears this promotion was given several days after this day in March, but the official document was backdated to correspond with when Imperial Headquarters thought he died. The Emperor also awarded Kuribayashi new status decorations with court rank, Senior Fourth and Junior Third Rank medals.[2369]

Meanwhile, back on the home islands, even women trained to defend against the American invasion, some with spears. American submarine torpedoes and mines (both navally and aerially deployed) sank merchant ships, and B-29s bombed much of Japan's industry, cutting production of military essentials and consumer goods.[2370] Rationing grew stricter, and almost everyone went hungry. People were expected to work harder and eat less. Reports of civilians committing suicide in large numbers on Saipan and of garrisons being wiped out while fighting to the last man like on Iwo Jima were treated with sorrow, yet many still glorified these acts as proof of Japan's unique character. Censorship grew tighter, with new explanations as to why the series of stunning American victories which the government had concealed or lied about, were a trick to lure Americans closer so they would be destroyed. The Emperor was considered infallible, so his policies were correct and his past statements had to be rationalized. The feared *Kempeitai* secret police grew more aggressive, seeking out "subversive elements" who questioned government propaganda or the war effort, or sympathized with foreigners.[2371]

Japanese women with spears in Japan in 1945 train to meet an invasion by the U.S. as the nation mobilized all its citizens for war. United States Marine Corps Museum

As Woody and his men reached the ocean to the north, they settled down to clean their weapons, get some chow and rest. Every few minutes, a few of them would gaze up at the P-51s that flew round-the-clock patrols around the island searching for enemy ships and planes. Two fighter Groups had at least eight fighter planes airborne circling Iwo at any given time. From P-51 pilot Jerry Yellin's perspective, the island, at this time, looked like it was fully in Leatherneck's hands.[2372] By capturing the island, the Marines had accomplished their mission, surviving one of the bloodiest battles in history. Major General Smith said, "Iwo Jima was the most savage and the most costly in the history of the Marine Corps. Indeed, it has few parallels in military annals."[2373] Historian Robert Burrell wrote: "Far from glorious, combat on Iwo Jima was perhaps the most brutal, tragic, and deadly in American history."[2374]

An example of a religious service held on Iwo (this one was a Protestant gathering with Chaplain Valbracht). Every day, one could observe Catholic, Protestant or Jewish services being held everywhere throughout the island. 5 March 1945. National Archives, College Park

Services were held throughout the island and eulogies delivered. As a testament to the U.S. pluralistic society, 5th MarDiv head chaplain Warren Cuthriell, a Baptist minister, chose Chaplain Roland B. Gittelsohn, a 34-year-old Reform rabbi, to deliver the sermon at a combined, non-denominational ceremony dedicating the division's cemetery. Gittelsohn gave a eulogy titled "The Purest Democracy" on 21 March 1945:

Here before us lie the bodies of comrades and friends. Men who until yesterday or last week laughed with us...Men who fought with us and feared with us... Here lie officers and men, Negroes and whites, rich men and poor—together. Here are Protestants, Catholics, and Jews—together. Here no man prefers another because of his faith or despises him because of his color. Here there are no quotas of how many from each group are admitted or allowed. Among these men there is no discrimination. No prejudices. No hatred. Theirs is the highest and purest democracy. Whosoever of us lifts his hand in hate against a brother, or who thinks himself superior to those who happen to be in the minority, makes of this ceremony and the bloody sacrifice it commemorates, an empty, hollow mockery. To this then, as our solemn sacred duty, do we the living now dedicate ourselves: To the right of Protestants, Catholics, and Jews, of White men and Negroes alike, to enjoy the democracy for which all of them have here paid the price...We here solemnly swear this shall not be in vain. Out of this and from the suffering and sorrow of those who mourn this, will come, we promise, the birth of a new freedom for the sons of men everywhere. AMEN.[2375]

He had ministered to Marines of all faiths in the battle, witnessing firsthand the carnage. Gittelsohn's sermon attracted wide attention and was read by radio announcers and covered in newspapers.[2376] Such wise religious voices for an aspirational U.S. were not universal. Gittelsohn's eulogy expressed the ideal that America is a land where all can participate in democracy and government, but in order to do so, every American must be willing to put on a uniform and defend that freedom. His eulogy challenged people to realize we are all part of one race, the human race. He affirmed that neither ethnicity nor religion should divide us from the common kinship we share as humans. (See Gallery 3, Photo 18)

Gittelsohn lived during a time when hatred of racial, religious or political minorities had been violently advocated and legally required by the governments of Germany, Italy, Japan and Russia, and had been warmly embraced by bigoted elements of society in other countries including even the United States. Some, even people of education, swallowed whole the pseudoscientific, doctrinaire religious, or extremist political justifications for hatred. Many of these prejudices

had been around for centuries, infect-
ing generations. Exposure of the stark
results of hatred promoted by fascist
regimes and the devastation produced
by them increasingly caused repul-
sion of such ideas among Americans
in general and religious leaders in
particular, as well as among other
civilized peoples around the world.
Gittelsohn's dedication came near the
war's end in which hundreds of thousands of Americans risked their lives
to prevent the triumph of fascism. Yet,
even so, many Americans still harbored prejudices. As the 5th MarDiv
cemetery was about to be dedicated,
Catholic and some Protestant chaplains of the division tried to prevent
"the Jew" Gittelsohn from giving the
eulogy because he was a minority, an
"unbeliever," going to Hell, a Christ-
killer, and a "Negro lover" (he often

Chaplain Roland B. Gittelsohn. He gave one
of the most famous speeches of WWII "The
Purest Democracy" while dedicating the 5th
MarDiv's cemetery on Iwo on 21 March 1945.
St. Louis Personnel Records Center

held racially mixed discussion groups). Catholic chaplains, hobbled by church
policy, objected to any joint ceremony with Protestants or Jews. The Vatican II
encyclical (1962-65) *Nostra Actate*, which absolved modern Jews of any fault in
Christ's death and recognized Judaism as a sister religion to be accorded tolerance to cleanse the Church of the scourge of anti-Semitism, was not adopted
until years later. Pope Pius XI was so focused on the "threat" posed by communism, the competition with Protestant denominations and the limits on church's
temporal power in democracies, that he failed to recognize the danger posed by
fascism until shortly before he died in 1939. He never banned Hitler's book *Mein
Kampf*, never excommunicated Hitler—a Catholic—and signed the first treaty
with the Nazi state in July 1933.[2377] Pope Pius the XII, who rose to the papacy in
1939, chose not to denounce Hitler, Mussolini and Japanese Fascism and bring

the full moral authority of the Roman Catholic Church against the racist Axis powers. Rather, he supported and admired the Catholic-educated Hitler. The grave irony was that the Catholic chaplains working on Iwo to support the U.S.'s fight against Fascist Japan also took oaths of allegiance to a Pope who enabled Hitler and refused to denounce the Nazi dictator. Knowing this, it is not at all surprising that some Catholic chaplains were intolerant of other chaplains, especially Jewish ones.[2378]

As a battlefield chaplain, navy Lieutenant Gittelsohn ministered not only to the 1,500 Jewish Marines on Iwo Jima, such as Woody's comrade Warren Bornholz, Beck's platoon leader Richard Tischler, and MOH recipient Tony Stein, but to Marines of all faiths.[2379] An interesting Marine Corps-Jewish factoid should be mentioned here: Many of these Jewish Marines and their gentile comrades on Iwo had traveled to the beaches of this island in Higgins boats that had been perfected by another Jewish Marine, Victor H. Krulak (by 1945 he had converted to Christianity but had been raised as a Jew). While he was serving in the Marine Corps at this time, the Nazis had murdered his paternal grandparents.[2380] So American Jews had a proud tradition in the Corps and Gittelsohn took care of many of them on Iwo. Knowing these facts and the truth that Gittelsohn had performed heroic acts on Iwo, 5th MarDiv's Chaplain Warren Cuthriell overrode the complaints and insisted Gittelsohn give the dedication declaring the right of a Jewish chaplain to preach a sermon was "one of the things for which we are fighting the war." To avoid dissension, Gittelsohn requested three services—Protestant, Catholic and Jewish—which was agreed to. Three Protestant chaplains were so incensed by their colleagues' prejudice that they boycotted a separate Christian service to attend Gittelsohn's gathering. These chaplains had the rabbi's sermon printed up and distributed, resulting in it becoming widely known. Ironically, the Catholic priests' efforts in denying Gittelsohn's eulogy, rather than marginalizing him, made his sermon famous. He not only praised those who fought and died for the cause of freedom, but he also called on his Christian colleagues and all Americans to reject religious prejudice, racism and discrimination.[2381]

Having the rabbi give this eulogy was a watershed moment in American history. After conquering the territory of racist, militaristic, bigoted, anti-democratic Imperial Japan—a nation allied with Nazi Germany—it was fitting that a

Jew supported by Protestant clergymen led the way to promote freedom, brotherhood and tolerance by showing what type of society should rule in the new world—one where racism and religious bigotry have no claim.

Rabbi Roland Gittelsohn giving his "Purest Democracy" speech at the 5th MarDiv cemetery on Iwo, 21 March 1945. Catholic priests had tried to prevent him from giving this speech because he was Jewish and an "unbeliever" showing their religious racism. National Archives, College Park

By the end of March 1945, the U.S. Army had relieved almost all Marine units on Iwo. The 147th Regiment slowly relieved Woody's regiment as it took over garrison duty. The Marines had conquered the island and it was the army's job to take over the "mopping up" and then occupy it.[2382] Much effort was still made to encourage the remaining Japanese to surrender, but meeting, once again, with little success. As a result, Woody's regiment sent out the following report asking those who broadcasted over loudspeakers to refrain from using the word "surrender" because that seemed to elicit a suicide response. Instead, the report stressed they should try to convince the Japanese there was no dishonor in laying down their arms, "since their Emperor wants them to live for their country instead of dying uselessly."[2383] Even with this new tactic, it seemed the Marines continued to fail gathering most of the beaten IJA soldiers they had tried to detain. For example, on 30 March, the regiment discovered a cave with 12 Japanese in it. After pleading with them to lay down arms, they refused so, the Marines blew it shut.[2384] One rare prisoner, a major, said the Marines had their work cut out for them. According to him, there still were 1,000 Japanese at large on Iwo with a lot of fight left in them.[2385] On 30 March, a cave was captured and the Americans were "astonished to discover" that the dwelling housed "two cows, a vegetable garden, chickens, medical supplies, and a stockpile of ammunition for a 75mm howitzer." Evidently, not all Japanese were running low on supplies after 41 days of combat.[2386] On 1 April, a group of 25 Japanese commandeered a barge and attempted a raid infiltrating behind Marine lines. A plane spotted them, sank the barge and killed the Japanese.[2387] This last desperate amphibious assault mirrored what was apparent for all. The island of Iwo by now had received so many staggering hits from the air, sea and land that it was barely able to keep standing. Lieutenant General Smith said Iwo Jima was a knockout blow.[2388]

By 4 April, Erskine gave control of the 147th to the army, and Iwo was mostly in its hands to continue the operation. Erskine wrote, "The 147th Infantry regiment displayed in their debarkation, movement into positions and execution of assigned missions a fine spirit of cooperation and a commendable eagerness for combat [and was] an inspiration to all hands." The 147th regiment still had to go out on patrols, but the Marines had done most of the heavy lifting. The regiment conducted over 6,000 patrols and fought 2,500 Japanese by 30 June. They would eventually kill 1,602 and capture 867. The 147th would suffer 144 WIAs, 15 KIAs and lost dozens of cases to combat fatigue and sickness.[2389] Compared to the Marines, their "kill ratio" was much stronger and their casualties dramatically less.

Here are two examples of bombers, one a PB4Y2 and the other a B-29, which had to use Iwo Jima as a place for emergency landings. National Archives, College Park.

# Ch. 25: Justification for Iwo Jima

"We have to 'continue the need for history.'"
—Woody Williams[2390]

SOME HISTORIANS ESTIMATE TAKING THE island may have saved more lives than it took since, they argue, 25,000 airmen may have used the island as a safe haven after experiencing mechanical problems or combat damage on the flights to and from their Mariana bases *en route* to Japan. However, historian Robert Burrell argues the operation did not come close to saving that many. Ronald Spector and Gerhard Weinberg say there were 2,400 "emergency" landings (if an average crew had 11 men, then one gets roughly 25,000 crewmembers out of these landings), but dispute that those landings saved all the crews. Richard Frank says there were actually 3,092 landings due to battle damage or fuel shortage, although he further writes there is no way of knowing whether they "saved" the men because they could have made it on to the Marianas or could have successfully ditched and been recovered.[2391] By this reckoning, that these landings actually saved so many is an "illusion."[2392] Iwo indeed provided some planes a place to land when they needed a place to repair, refuel or rearm, but most of their crewmembers were not "saved" by the island changing hands.

The Twentieth Army Air Force lost 218 B-29s out of 2,148 in combat. At its zenith, XXI Bomber Command consisted of 1,000 B-29s, so Iwo could not have "saved" 2,400 planes. In fact there were only 2,242 B-29s at war's end. The overstatement that the takeover of the island saved tens of thousands was used to justify the losses incurred while taking Iwo. Admiral Charles Adair, a "senior amphibious operations planner" of the 7th Fleet, said the invasion did not save anywhere near 25,000 airmen. Marine correspondents and historians have written that within three months of securing Iwo, more than 850 B-29s made emergency landings there—up to 9,350 crew members may have been saved, which was more than died taking the island if these numbers are accurate, but definitely much less than 25,000 cited in earlier works.[2393] General of the Army Henry "Hap" Arnold wrote in the summer of 1945 that the island supported three landing fields for three "Groups" of fighter escort P-51s (circa 480 planes)[2394] and had provided safe landings for 1,299 "crippled planes or planes out of gas." If these landings saved their crews, then 14,289 men could have possibly been spared.[2395] This is a far cry from 25,000 and an impossibility to prove these landings saved those men. Iwo did save lives, but it is controversial as to the actual count.

The *sole* reason given by General of the Army Arnold and Fleet Admiral Nimitz for taking Iwo was to provide a fighter base to escort bombers. The range of fighters was not as long as that of bombers, and specifically did not extend from the Marianas to Japan. Having fighter planes on Iwo allowed fighters to accompany B-29s on their bombing runs, protecting the crews and allowing them to bomb from lower altitudes, although the Japanese airforce threat was basically finished by this time.[2396]

There were other, less compelling, reasons for taking the island. Because of its position on the route the B-29 bombers used, American commanders concluded that conquering the island prevented the Japanese on it from using their radar to detect the bombers and warn the mainland to scramble its fighters and alert antiaircraft batteries. Yet there were other islands nearby like Chichi Jima, and the Japanese there could also warn the mainland once they saw the same aircraft. On the other hand, even the logic of this argument breaks down when one thinks about the inefficient manner of such a warning system. True, the enemy could see planes in the air over their Bonin Islands, but they had no way to know where these bombers would attack; and alerting the entire nation that planes were coming was not practical. Considering the Japanese homeland extends from north to south 1,869 miles, it would be like seeing planes in the sky over Hawaii, and then facing the decision of which West Coast city or state to warn planes were coming—Oregon? California? Portland? San Diego?

Some have argued taking the island prevented Japan from using it as an "island aircraft carrier" from which to send planes to attack bombers on their way to and returning from Japan. In addition, taking the island, they argued, would prevent Japanese planes from using it as a staging area to attack Saipan which they had done in October 1944, leaving the base there with "burned and blasted B-29s."[2397] Notably, by February 1945, the Iwo airfields were not functioning and Kuribayashi had few airplanes left. Iwo had been neutered as an airbase by months of bombing attacks and naval shelling. Chichi Jima and Haha Jima nearby had airfields, but by February they also had been rendered useless by bombing, and the higher-ups did not conclude they had to be conquered to decommission them. The U.S. Army Air Force's history concluded: "Japanese raids against B-29 bases, though troublesome, were not important enough alone to have justified the cost of capturing Iwo Jima."[2398]

Taking Iwo could possibly have saved billions of dollars' worth of hardware instead of losing it to the ocean's depths when planes ditched (in 1945, one B-29 cost $640,000 which is equivalent to $9,033,096.63 in 2019).[2399] There were indeed numerous B-29s that used the island instead of the ocean to land. With over 117 planes having needed Iwo to land, one could in the quartermaster's office calculate Iwo had saved $1 billion worth of hardware, not to mention the millions expended training the planes' personnel. So, an economic argument could justify taking the island if not for the fact that most citizens do not accept trading riches for lives. In the final analysis, there was one reason for taking the island that many had no clue about, but was indeed an important one, for having the island under American control allowed the U.S. to have a Plan B for the atomic bombs.

In case Nimitz had any doubts about the battle, a Washington emissary, navy Commander Frederick L. Ashworth, hand delivered a top-secret letter from Major General Leslie Groves of the Manhattan Project just before Iwo that emphasized the importance the island held for a "super bomb." Nimitz pleaded with Ashworth to use atomic bombs on Iwo and Okinawa for which he was planning attacks. Ashworth told Nimitz they were on a strict schedule, but securing Iwo was imperative to deploying this weapon. Nimitz was disappointed, but assured Ashworth he would take the island before August. Had Colonel Paul Tibbets, the *Enola Gay* pilot who dropped the first atomic bomb, had mechanical problems flying from Tinian, he would have landed at Iwo and transferred his crew and bomb to another B-29 aptly named *Top Secret*. A loading pit and an emergency crew had been readied to effect this Plan B.[2400]

Had the U.S. been forced to conquer Japan with more conventional means, Iwo Jima would have played a major role in bringing down Hirohito's Empire. There was no way of knowing whether deploying atomic bombs would convince the Japanese to surrender, so having the island for the continuation of a conventional war was part of the strategic groundwork required to attack Japan if necessary.[2401] As the truism states, it is better to be prepared for a situation and not have it than to have the situation and be unprepared. The victory at Iwo allowed for the preparation for what would have been the largest amphibious invasion in history. Knowing this, it is easier to accept that Iwo was taken at a horrible cost "for an objective that never fulfilled the intended purposes."[2402]

So there were many benefits, realized and unrealized, exaggerated and real, for taking the island, and in the end, it was one of the best stepping stones necessary to get closer to Japan to defeat it by bringing weapons, men and ships closer to its heartland, thus saving lives by ending the war sooner. In conclusion, this analysis of justifying the horrible carnage on Iwo is complicated. "There is no way for the calculus of war to achieve closure, nor for the calculus of life."[2403] In the end, Iwo's conquest helped the war strategically and provided a forward base for the U.S. in case it had to execute the invasion of Japan.

# Ch. 26: After Iwo Jima

"[Iwo] is a place where the spirits of the dead sleep."
—Yoshitaka Shindo, Kuribayashi's grandson[2404]

"Every Marine…Rifleman will be forever linked
to the warriors who fought on Iwo Jima."[2405]
—31ˢᵗ Commandant of the Marine Corps
Charles C. Krulak (1995-1999)

Woody feels he left the island on 1 April 1945, but his file indicates it was on 26 March that he boarded a Higgins boat and returned to the fleet aboard the *USS Zaurak* (Colonel Hartnoll Withers, Captain Beck, 1ˢᵗ Lieutenant Henning, Sergeant Dick Dashiell, Corporal Alan B. Tripp, Corporal and fellow Flamethrower operator Joseph Anthony Rybakiewicz joined Woody on this same ship). Two days later, he and his brother Marines sailed for Guam, arriving there on 1 April, and disembarked the next day.[2406] As Woody and his fellow Marines passed the 5th MarDiv cemetery on Iwo Jima, they read the admonition:

When you go home
Tell them for us and say
For your tomorrow
We gave our today

Rows of white crosses and Stars of David stood as testimony of the battle's cost. Most were young, vibrantly alive boys six weeks before, and thereafter they lay buried in Japanese soil where they gave their lives. One Marine in Woody's regiment said, "Those white crosses in the cemetery we passed on Iwo coming down to board ship are damned tough to take."[2407] Many felt numb. The tension fell away and the reality of what had happened dawned on them. They "began to remember what had passed rather as a frightful dream from which [they] had been happily awakened, than as events which had taken place in sober and naked reality."[2408] And the events that had transpired in "sober and naked reality" made up one of the most brutal battles in history, and many must have known they had just experienced some of the most trying hardships humans have ever experienced.

Behind the advancing lines, dead Marines waiting for burial on 23 February were sprayed with DDT to cut down on flies and disease. This was the day Woody earned his MOH and he was lucky to still be alive. After enlarging this picture to count all the dead shown in this photograph, the author concludes that there are 83 fallen Marines waiting for burial in this picture alone. Maybe Corporal Warren Bornholz and PFC Charles Fischer, both who possibly protected Woody during his MOH actions, were among the dead here since they died the day this picture was taken. NationaArchives, College Park. Photo Credit A. Kelly Jr.

Some of the 3rd MarDiv walked between the rows to find their buddies. Two stopped in front of one grave. "There's the captain," one of them murmured. Ambulances drove new bodies to the cemetery's freshly dug rows. Behind the crosses, a bulldozer worked on a new trench for the fallen, scooping out large portions of earth with its claw-like device known as a ripper. Scores of dead lay nearby on a bank covered by ponchos with their leggings and boots sticking out. Some sprayed the corpses with DDT from shoulder held tanks. Occasionally, ocean wind ruffled the flap of a poncho revealing ashen, pale and bloated faces. One man's rigor mortis had stiffened his left arm high in the air as if to protest this final act of taking him to the underworld. Chaplains had their men bedeck each body with a flag as it was placed in its final resting place and then the chaplain administered a last prayer. Across the road where trucks brought back the dead, a large white sign in the ground with big red letters read "Mines." To the north, the sounds of machineguns and rifle fire echoed across the landscape.[2409] Sammy Bernstein, a Jewish Marine who had attended Gittelsohn's famous speech *Purest Democracy*, later put together a poem after visiting the graves:

Oh, I just saw a sight to see,
A sight that will always live in me,
And there they were, row on row,
The graves of boys who gave their all.

Here a cross and there a star,
Try to see it, 'cause here they are,
A Catholic, Protestant, and a Jew,
All American boys we once knew.

And though you read, "so many thousands dead"
You know not what you really read
'Cause only those who see their graves
Will ever know and be amazed.

So to the ones who must receive
A Notice that they've been bereaved,
The boys they died for four great rights,[2410]
We alive, for all time, must keep them bright.

And when it's over, God make it soon,
Let's not forget ere we're doomed,
That war is hell and pray we must
To keep the peace they gave to us.[2411]

Bernstein's poem shows he had taken to heart Gittelsohn's earlier message and that he understood that these men's sacrifices, regardless of their religions, helped preserve the democracy that gave Americans their freedoms. And those freedoms Bernstein believed they all fought for and currently had were the four "great rights" FDR had proclaimed; namely, the freedom *of speech*, the freedom *to worship*, the freedom *from want* and the freedom *from fear*. In a reading for a film in 2015, Bernstein choked up with emotion as he read the words that took him back to walking those rows of "boys who gave their all," hoping to convey to Americans today that freedom is never free and we all owe a debt we can never repay to those young men who fought, killed and died on Iwo to defeat the Empire of the Sun.

As Woody walked by the cemeteries, he thought of his buddies like Waters. Most who had landed with him on 21 February were either dead or wounded. The majority of the men with whom he walked back to the Higgins boat had not trained with him and did not know him well if at all. Out of the original 3,006 who had landed with Woody with the 21st Marines, few remained.[2412] As Carl von Clausewitz rightly noted, as though describing Woody, "No other human activity [i.e. combat] is so continuously or universally bound up with chance. And through the element of chance, guesswork and luck come to play a great part in war… In the whole range of human activities, war most closely resembles a game of cards." [2413] Woody must have felt this luck in his bones. The after-action report listed 305 KIA, 1,332 WIA, 1 MIA, and 391 "Non-Effectives" (shell-shock/combat fatigue and sickness) for a total of 2,029 casualties the regiment suffered.[2414] According to correspondent Dashiell, he was one of only five original Marines of his company out of 250 to walk off the island.[2415] Dashiell was not exaggerating here. Deterioration was precipitous in all Iwo units until the very end. Between 60% to 70% of the original infantry personnel were out of action by the struggle's end. "Had no battle replacements been committed the percentages would have been higher. Some rifle companies and platoons, not counting the replacements, set their casualty figures at almost 100 per cent."[2416] Most survivors had no comrades left whom they knew. Gittelsohn's words aptly described the situation: "Some of us have buried our closest friends here. We saw these men killed before our very eyes. Any one of us might have died in their places. Indeed, some of us are alive and breathing at this very moment only because men who lie here beneath us had the courage and strength to give their lives for ours."[2417] Reflecting, Woody said seeing "row upon row of white crosses indicating what was to be their final resting place for eternity...I believe many are standing guard on the streets of heaven like all Marines who sacrifice their lives for the protection of others."[2418]

At the 3d MarDiv's cemetery, which was just off Airfield No. 1 where Woody earned his *Medal*, Major General Erskine also gave a dedication. With tears in his eyes, he said:

> There is nothing I can say which is wholly adequate to the occasion... Only the accumulated praise of time will pay proper tribute to our valiant dead. Long after those who lament their immediate loss are themselves dead, these men will be mourned by the nation. They are the nation's loss...Victory was never in doubt. Its cost was. The enemy could have displaced every cubic inch of volcanic ash on this fortress with concrete pillboxes and blockhouses, which he nearly did, and still victory would not have been in doubt...Let the world count our crosses...Then when they understand the significance of the fighting for Iwo...let them wonder at how few there are...Let us do away with names, with ranks and rates and unit designations, here. Do away with terms, *regular, reserve, veteran, boot, old timer, replacement*. They are categorizing words which belong only in the adjutant's dull vocabulary. Here lie only...Marines.[2419]

Erskine fought in some of the bloodiest WWI battles, including Chateau-Thierry, Belleau Wood and St. Mihiel as a platoon leader in 2nd Battalion, 6th Marines, earning the Silver Star and Purple Heart. After having his right leg riddled with bullets and lying on the ground, he shot one attacking German and made "two others captive." He also served at the battles of Attu, Kiska, Kwajalein, Saipan, Tinian and now Iwo in WWII, so he knew whereof he spoke when giving this tribute.[2420]

According to General Anthony Zinni, Commander-in-Chief of U.S. Central Command (CENTCOM) from 1997 to 2000, Erskine was one of the Corps' most brilliant general officers during WWII. Historian Robert Burrell agrees with this study's assessment as well as Zinni's observation writing that Erskine was "one of the most experienced and intelligent Marine officers of the Pacific."[2421] Erskine expressed the brotherhood all Marines share. In the Corps, one has to earn the title of Marine. And, once a Marine always....It is an identity and not a job. Erskine affirmed this identity, calling all who died Marines regardless of rank. Erskine's vision for the Marines' future was clear by declaring his belief that whatever the Japanese confronted the Marines with, they would overcome. Not knowing about the atomic bombs or their potential effect on the Emperor, he believed they would soon battle the Japanese on their four main islands and it would be tough. Iwo proved to Erskine the Marines would prevail despite the Japanese fanatic resistance.

For his leadership on Iwo, Erskine was awarded the Distinguished Service Medal endorsed by President Truman and Secretary of the Navy Forrestal. This medal is given to those who have exhibited extraordinary service with great responsibility. In all, 27 Medals of Honor were awarded to brave men on Iwo, but only six merited a DSM. His citation read in part:

> For exceptionally meritorious service... in a duty of great responsibility as Commanding General of the [3d MarDiv], prior to and during the seizure of enemy Japanese-held Iwo...Inculcating in the officers and men of his regiments his own indomitable spirit of determination...Erskine welded his organization into a formidable fighting command... he deployed his units in support of the assault divisions according to plan, quickly assumed control of a difficult sector of the line and waged fierce battle against the fanatic Japanese garrisons. A bold tactician, he maintained his division well in the forefront of the assault, pushing the relentless advance inch by inch through the enemy's intricate network of defenses to make tortuous but steady progress over the fire-swept terrain, blasting strongly fortified gun positions and fighting off repeated counterattacks as he moved his battalions inexorably forward. Ultimately breaking through to the north coast...

he succeeded in splitting the defending Japanese forces into two disorganized and vulnerable groups. Constantly rallying his tried, depleted units…Erskine by his undaunted valor, tenacious perseverance and resolute fortitude in the face of overwhelming odds, inspired his stouthearted Marines in heroic effort…. His dynamic leadership and decisive conduct throughout were important factors in the successful conclusion of the…Campaign.[2422]

It was rare for generals in WWII to lead their men from the front, but Erskine was in the fighting lines during combat from 24-28 February. He knew the best way to lead was to get to the front to "see what is happening," and he proved this was not just empty bravado. He had been shot up by Germans attacking their "machine gun nests"[2423] at Thiaucourt in 1918 and now, had faced one of the most skillful Japanese generals of the war and helped beat him. His DSM citation notes he knew firsthand what Woody and others had to do in order to secure victory, and he was right there pushing his men forward. Leadership and energy trickle down the ranks and most in a unit reflect their leader. On an island where Marines proved they were an elite force, Erskine showed that his 3d MarDiv performed above the average. Indeed this status might have not only occurred due to his leadership, but also due to the fact they were given the toughest task of hitting the Japanese where their defenses were strongest. The 3d MarDiv Marines' unique status was forged in the cauldron of fighting against the best on the island and learning how to defeat them. The cost was high—most who landed with the 3d MarDiv on 21 February were no longer in the ranks by 1 April when its cemetery was full. In pushing his men, Erskine's forces suffered heavy casualties, and he and others had solemn moments reflecting on the battle. When Erskine spoke at the cemetery where thousands of his boys lay, he had no idea he would be so praised as in his DSM citation. Erskine inspired his men and his leadership garnered the praise of Truman and Forrestal. Erskine's Marines and those of the other two divisions, according to Secretary of the Navy Forrestal, had just taken a critical step "in the sequence of doom for Japan."[2424] U.S. Army Air Force general Curtis LeMay said, "Having Iwo Jima, Guadalcanal, Guam, Saipan and other islands in the chain as airbases, really won the war for us…without them, it would've taken another year or longer to win, maybe 100,000 more American lives lost, until we dropped the atomic bombs."[2425] And due to Erskine's brilliant planning of the amphibious attacks on Saipan and Tinian and his combat leadership on Iwo, he helped generals like LeMay accomplish their goals of bringing the war to a quick end through air power.

After walking past the cemetery, Woody boarded the ship that would ferry him back to Guam. He did not give much thought to what he had survived and was happy to be out of the fight, alive and clean after taking another shower on board. He could not believe he was among the living, and enjoyed relaxing on his bunk. Chaplain Gittelsohn wrote, "How a single one of us left Iwo alive is a miracle." When Gittelsohn returned onboard ship, he broke down in uncontrollable sobs.[2426]

Their departure traffic from Iwo in late March was further evidence of the horrific casualties the Marines suffered. When the 5th MarDiv embarked from Hawaii to attack Iwo, it required 22 transport ships to carry its men into battle. When it left Iwo Jima, it needed only eight.[2427] Many of the almost 20,000 wounded had been taken to Guam or Hawaii, or still were healing on hospital ships, or, like Lefty or Chambers, were already back home in the United States. Most of the dead had been buried in Iwo's black, volcanic ash where they had taken their last breaths. For the living, it was sobering to reboard ships where most did not know each other and the comrades to whom they had "bidden farewell" just seven weeks before were now memories.

The island Woody left was dramatically different from the one at which he had arrived. It had been transformed from a barren landscape with warrens of underground tunnels full of Japanese soldiers to a thriving base with hundreds of aircraft and command stations, built as a huge staging ground for the invasion of Japan. In a few weeks, there would be over 7,600 Seabees working on the island, transforming it into one of the most impressive U.S. bases in the Pacific.[2428] The total financial cost of equipping the island with three functional airfields, roads, Quonset huts, tank farms, harbor construction and docks, water development and signal communications came to a total of $29,450,000 (in 2019 dollars $415,663,587.08).[2429]

During the journey back to Guam, many lay in the ships' sick bays, suffering from wounds; several perished during the return journey. Every day, ceremonies were held on ships for those who had died the night before. Covered in

American flags, they were consigned to the depths of the ocean, buried at sea. Iwo casualties were still mounting as these sobering scenes unfolded.[2430] Sadly, some deaths on board were self-inflicted. One man who had had his penis and testicles blown off by a grenade crept topside at night and threw himself into the sea.[2431] He must have felt foreboding feeling that no woman would ever love him, that he would never be able to sire a child and that everyone, especially his fellow man, would *always* look upon him with pity and sadness. He would not embrace such a future. Body counts from battles continue to mount long after the rifles and grenades were laid down.

Although Iwo had been declared secure by the time Woody left, Marines, Seabees and now soldiers still engaged in combat throughout April and May, killing an additional 1,600 Japanese; not surprisingly, few surrendered. Many U.S. soldiers mocked the Marines' claim that the island was "secured" on 16 March and the battle all but finished on 26 March when they continued fighting. One joked, "I wish the Marines had told the Japanese the island was secure when they turned it over to us for garrison duty." Often, when Seabees knocked down a hill, or dug into the ground, several Japanese suddenly emerged, like ants from a disturbed mound, and the Americans were faced with "unsurrendering" Japanese warriors. Two Japanese held out and remained undiscovered until 1949.[2432] One who had surrendered was so distraught that, while atop Mt. Suribachi four years later after the battle on a return trip to Iwo, he threw himself off the cliff and killed himself.[2433]

# Ch. 27: Receiving the Medal of Honor

"Our society craves heroes."
—Edward Bernays, premier U.S. public-relations
counselor during WWII[2434]

"The Marine Corps trains all Marines emphasizing…that the most
important person to you is the Marine on your right and left…"
—Woody[2435]

"When I get to heaven, one of my first
questions to God will be: 'Why me?'"
—Woody[2436]

AFTER RETURNING TO GUAM, 3D MarDiv's men had a well-deserved and sorely needed opportunity to relax, eat and heal for several weeks. Woody's regiment noted that since it had had "a large personnel turnover a complete training program will be necessary to prepare it for combat."[2437] Then on 12 April 1945, President Franklin D. Roosevelt died. Many were saddened by FDR's death because they saw him as a father figure and heroic leader who had brought America out of the Great Depression and to victory in the war. Mussolini had long since been overthrown by his own countrymen and his execution by firing squad would come before month's end. Most of Europe had been liberated. If Roosevelt had lived a few more weeks, he would have witnessed a defeated Hitler commit suicide followed by Germany's unconditional surrender. Most of the Philippines had been liberated. The Japanese army in Burma was being routed. Devastating B-29 bombing runs from the Marianas pounded Japan. First elected in 1932, FDR was the only president many could remember. However, in the middle of a war, there was little time to mourn for lives lost, even for this one. Vice President Harry S. Truman, the Missouri senator and WWI army officer, was sworn in as president and the fight against Japan continued.[2438]

During this time of repose on Guam, some "strange things began to happen." First Sergeant Elder from the company HQ questioned Woody about the "pillbox attacks." After Elder finished getting the answers he wanted, Woody asked him why, and "the topkick shrugged his shoulders" and simply replied, "For the division's history."[2439] Then, soon thereafter, a PFC from Salem, West Virginia sought out Woody and explained that as a clerk, he had written up an officer's recommendation for a big medal for Woody. As a fellow West Virginian, he felt excited for Woody, but Woody did not understand what this PFC was saying.[2440]

> If he mentioned that the medal I was getting was indeed *the Medal of Honor*, I wouldn't have really known what he was talking about. I thanked him for telling me what he did and for his congratulations. I shook his hand and he left…I then went back to spending time with my buddies, drinking my two beers a week…and blowing heads in the coral rock to maintain sanitation. I had no idea the significance of why I, out of all the other brave men…was being highlighted for special thanks. I just started to focus on the next campaign and what I would have to do then since now, the men in my platoon looked to me as a seasoned veteran. I would now be teaching others how to fight.[2441]

Interestingly, no one who witnessed what Woody did on 23 February thought it special. They were not speaking about it in the Quonset hut. Woody in particular did not think he had done anything remarkable although it seems he was not reticent about describing it to First Sergeant Elder, correspondent Sergeant Dick Dashiell and 21st Regimental journalist Private Paul B. Hoolihan. That this clerk sought out Woody indicates he was in possession of interesting information buzzing around the officers' duty station. Since his job was to write up reports, he had a stronger understanding of what someone had to do to merit a high award and was excited for Woody.

Many letters from servicemen were written during this time. Traditionally, Marines—especially officers—wrote to the families of the fallen to inform them what happened to their sons; families were obsessed with knowing how their sons died, and more importantly, they wanted assurance their sons had not suffered. For example, the mother of Charles G. Fischer was "extremely" relieved when one of Fischer's comrades from California personally visited her in Somers, Montana after the war and verified he died "instantly" from a head shot and did not experience any pain dying (she had already received letters hinting this, but it was better hearing it verbatim from a fellow comrade).[2442] And while Woody's recommendations for his Medal of Honor were sent to higher authorities, Lieutenant Richard Tischler from B Company wrote a letter to Vernon Waters' mother:

> It is with the deepest and heartfelt sympathy on behalf of your son's comrades that I take this means of expressing their and my own personal condolences. There are few words that can be said to comfort you in your bereavement…Your son gave his all for the cause that is to insure the future security and posterity of our nation and the world as a whole. His sacrifice and your's [sic] will not have been in vain… under extreme enemy fire during several actions, he acquitted himself most gallantly…the memory of him will serve as an inspiration to those of us who remain to carry on.[2443]

Tischler showed compassion and profound insights into what Waters' death meant for the Corps, nation and world. He was not being overly dramatic when he wrote Waters' actions would have a lasting effect on the health of society and the global community by bringing down Hirohito's regime. After receiving this news, Mrs. Waters informed the family of her son's death and let his girlfriend, Marceline Sorenson, know her love had died.[2444]

Similarly, Captain Beck turned his attention to the family of a fallen Marine, Charles Fischer once again, who gave his life possibly protecting Woody. Beck wrote Mrs. Fischer: "Your son was a most cheerful person at all times. He was well liked by all…Charles was mortally wounded by a sniper on Feb. 23 as we moved out in the attack. His thoughts were only of you and his loved ones."[2445] Unfortunately, officers sent thousands of letters like these. The thoughtful letters represented a sad tallying of the cost of battle. These letters followed those

Second Lieutenant and platoon leader Richard Tischler in B Company, 1st Battalion, 21st Marines. Tischler commanded Vernon Waters' platoon the day Waters died and most likely was the one who wrote Waters up for a Silver Star after the battle. St. Louis Personnel Records Center.

sent by the Commandant, General Alexander A. Vandegrift, offering his condolences by saying that Fischer "nobly gave his life in the performance of his duty [and may this thought] comfort you in this sad hour."[2446]

In all the files looked at in this study, the only family members who responded to these condolence letters were Rabbi Blechman and his wife Esther, the parents of Sergeant and 3d MarDiv correspondent, Solomon Israel Blechman (Dashiell's predecessor). After receiving a heartfelt letter upon the death of their son from Major General DeWitt Peck, Assistant Commandant to the Marine Corps, Solomon's parents responded:

Accept our gratitude for what you have been to us in our great sorrow. We know that God will reward our faith in His own way. May the All Compassionate establish His world in your days upon the everlasting foundation of His Law. May He bring about His Kingdom of righteousness and love for which our Solomon and other heroes and saints made the supreme sacrifice.[2447]

Rabbi Blechman not only sacrificed a son for the preservation of the nation, but he would later serve the nation himself (probably to honor the memory of his son) by taking care of veterans as a Chaplain at Veterans' Hospital No. 81, now known as James J. Peters VA Medical Center, in the Bronx, New York.[2448]

As the casualties continued to rise for the Marines in their march to Japan from Gaum to Iwo Jima to Okinawa, there were going to be more "heroes and saints." Many hoped that such a prayer as that of Rabbi Blechman would come true and that the Almighty would finally "bring about His Kingdom of righteousness and love."

After weeks of recovery, the Marines prepared for the next battle or *Blitz* (short for *Blitzkrieg*) as many termed it.[2449] They trained for the invasion of Kyushu, Japan (the southernmost island of the four making up Japan). The Kyushu invasion was called Operation *Olympic*, and it along with Operation *Coronet* encompassed the overall invasion of Japan called Operation *Downfall*. It should now read as follows: Based on the fanaticism and preparedness of the Japanese, intelligence studies concluded that this invasion would have resulted in the "biggest blood bath in the history of modern warfare."[2450] (See Gallery 4, Photo 2)

In preparation for the invasion, Woody and his men learned street fighting as they would see a different landscape for battle than they had previously experienced. The Kyushu attack was scheduled for 1 November and approved by Truman on 18 June 1945 unless Japan surrendered. When Japan received the Allied Potsdam Declaration on 26 July 1945 stating that it faced utter destruction unless it unconditionally surrendered, Japan ignored it. As a result, invasion preparations continued. Over 3,000 ships were readied for this invasion. The Marine divisions that had fought on Iwo had been dramatically reduced through death and injury and needed to be "rebuilt with [mostly] teenage replacements before they could fight again."[2451] At the age of 21, Woody was considered an "old man" and had started to earn the rank for that sentiment as an acting temporary Sergeant on 9 July 1945 (this meant he had duties of that rank and was called Sergeant, but was not paid for that grade).[2452]

Woody was part of the 21st Regiment's flamethrower school on Guam under Captain N. A. Weathers. The school intended to teach every Marine how to handle a M2-2 flamethrower since it had proven so effective on Iwo. Woody was mentioned as one of the excellent instructors and he and others discussed with the trainees the importance of flaming bunkers, caves and pillboxes "so that the 'blowers,' the demolition boys, can get up or in there to place their explosive charges."[2453] The information here came from the *21st Regimental Newsletter*, which derived much of this information from Woody. It discussed the proper distance, fuel mixture and nomenclature of the machine. Woody was imparting the lessons he had learned to a new freshman class.

To better familiarize Marines with urban fighting, engineers built "false fronted buildings, like it was a street," and they learned how to fight in that environment. Navigating this urban *milieu* in which they would fight, Woody doubted he would see home again:

> I wasn't a particularly religious person at the time, but I was optimistic I would always live. The only time I was worried was after Iwo when we trained to take over Japan's cities. Their whole population had been taught that to die for the Emperor was glorious and since I would've been involved with the first wave, I wouldn't probably be here today. However, I had my orders and a corporal doesn't question what his superiors tell him to do so I would've invaded Japan had I been ordered and would've killed or flamed as many as I could before some fanatic would've ended up killing me.[2454]

Many shared Woody's fatalism, especially after Iwo. Japan would be a conflagration and the knowledge of what they were training to do overwhelmed them. Lieutenant Tischler echoed Woody's feelings when writing his parents:

> Well, Iwo is now history and I guess it really did make history, because we can now say that we have just begun to win this war. However, it will still take plenty of these expensive winnings before it is all over. Iwo was very rough; but I have no illusions about the next campaign, which will probably be even rougher. Old Lady Luck stayed right with me all the way. I had so many close calls that it wasn't even funny anymore. I saw men get violently killed, or wounded, on all sides of me. It was a happy day when we left that rock and I hope that I never have to see another one like it…The Nips were plenty tough and hardly any of them surrendered. Both of my tent-mates were casualties: Joe Curtis was killed and Paul Rudell had his left arm blown off. I was near them when they were hit…Capt. [Melvin Robert] Voorhees came down with shell-shock and sickness…Beck came through all the way with me…I'm going to try to forget what took place at Iwo.[2455]

Both Woody and Tischler felt lucky they had survived Iwo and knew the future looked bleak. Tischler was not hyperbolic when he said the future battles with the Japanese would be expensive in lives. During this time, Tischler trained his new platoon and felt he would soon see the enemy again, something he anticipated with foreboding. Although he told his parents he was going to try to forget Iwo, it dominated his thoughts.

After 30 April, the Marines received the welcome news that Hitler had killed himself and then, on 8 May, that Germany had surrendered. That was celebrated as Victory in Europe or VE Day. Everyone in the Pacific theatre was glad to hear the news that the Third Reich had finally been demolished and, for the most part, were pleased his brothers, cousins, friends and fellow Americans who had fought the Nazis had survived and could take a break. However, while VE Day was important, war still raged in the Pacific. The conflict for those in the Pacific looked as if it would continue for years; a common saying was "Golden Gate in '48." The men in Europe now turned their attention to whether they would fight in the Pacific. Many felt they had to travel several miles along war's path until they could sleep soundly.[2456]

The Marines were exhausted; many, like Woody, Tischler and Beck, had been serving for years and they longed to return home. Three "characters," "tired of Pacific duty," put a song together set to the score of *Embraceable You* and called it *Replace Me*:

> Replace me, the years have passed to make two,
> Replace me, I can't go home without you,
> Two years is too long to spend overseas,
> Time has passed and finds me beneath the coconut trees
> Believe me, you know I've earned my month's leave,
> Believe me, my girl is yearning for me.
> Don't be a one-way guy, but pack your bag and come overseas.....
> My time is up – hurry here, please.[2457]

But despite this sentiment in this "Replace Me" song, Americans knew these trained men were not coming home anytime soon. Facing a radicalized people willing to die to the last man would be difficult, and Saipan, Tinian, Guam, Iwo Jima and Okinawa showed the U.S. forces what they faced in conquering a Japanese citizenry steeped in religious extremism: a "fanatically hostile population."[2458] The United State would need many more millions than they currently had in uniform to conquer Japan.

One U.S. intelligence officer wrote on 21 July 1945 that Japan's "entire population…is a proper Military Target… THERE ARE NO CIVILIANS IN JAPAN."[2459] That was not entirely accurate; Japan had tens of millions of civilians, mostly unarmed, who were distinct from the legitimate targets of soldiers, sailors, airmen and government officials. Although heavily indoctrinated from an early age and fed a diet of propaganda during the war, doubts and unease grew as conditions on the home front worsened and reports filtered in about how badly the Imperial forces were faring. Yet many still viewed dying for their Emperor as a pathway to reward in the hereafter, "a supreme form of spiritual cleansing."[2460] As stated in earlier chapters, every Japanese adult and possibly even older children had to be viewed as a potential enemy, including those who armed themselves with devices that could be used as weapons, but that did not mean the U.S. needed to treat every Japanese as a combatant. Americans demonstrated this compassion with civilians on Saipan, Tinian and Okinawa. The reason why the intelligence officer wrote that there were no civilians was that many trained to fight and were intermixed with combat units. Such a melting pot of uniformed combatants comingled with armed citizens would have created an unimaginable number of unnecessary deaths.[2461]

In invading Japan, the Allies would have encountered more *Kamikaze* planes than any U.S. military force to date, as many as 10,000. Father of the *Kamikazes*, Vice Admiral Takijiro Ōnishi, believed if the nation was willing "to sacrifice twenty million Japanese lives in a special attack [*Kamikaze*] effort, victory will be ours!"[2462] One Japanese Zen Buddhist priest, Dr. Reihō Masunaga, wrote of these *Kamikazes*: "The source of the spirit of the Special Attack Forces lies in the denial of the individual self and the rebirth of the soul, which takes upon itself the burden of history. From ancient times Zen has described this conversion of mind as the achievement of complete enlightenment."[2463] So *Kamikaze* pilots were enlightened Buddhas who would gain the gift of reincarnation and "complete enlightenment" at the moment they rammed their planes into U.S. Navy ships. Japanese religion had truly embraced extreme militarism and radical, self-destructive principles. Despite being aware that Japan would lose the war, Zen Master Sōgaku Daiun Harada wrote in late 1944 that it "is necessary for all one hundred million subjects [of the Emperor] to be prepared to die with honor…If you see the enemy you must kill him; you must destroy the false and establish the true—these are the cardinal points of Zen."[2464] With defeat and destruction of their nation a certainty, Japanese leaders, both military and religious, encouraged their citizens to abandon the "cares of this world and adopt a policy of prostration at the feet of a homicidal dictator [Hirohito]."[2465] Japanese leadership "willfully consigned" their countrymen to death. "It was a recipe for extinction."[2466] Prime Minister Kantarō Suzuki said in June 1945:

> If our hundred million people fight with the resolve to sacrifice their lives, I believe it is not at all impossible to attain the great goal of preserving the essence of Japan…None of our fighting men can understand how it is that Germany, with such a large army left, was not able to hold out until the end. In quantities of arms and supplies, we may not compare favorably with the enemy, but our determination as we stand on the firing line is peculiar to us alone. With this formidable strength we must fight to the end, the entire population uniting as one body.[2467]

*Kamikaze* hits a U.S. Destroyer, 25 October 1944. National Archives, College Park.

*USS Bunker Hill* received *Kamikaze* hits at Okinawa on 11 May 1945. The ship suffered 390 dead and 264 wounded from these attacks. National Archives, College Park.

The frightening fact emerging from this monologue promoting action that could result in the eradication of Japan's culture and nation asking all to fight until no one was left standing was that the Japanese citizenry were actually willing to follow such mandates. Most were conditioned to fight to the bitter end, probably the only people in the world who would do so under such circumstances. Suzuki brought out an interesting point about how his citizens viewed the Nazis. If the average Japanese did not understand how passionate Nazis could lay down their arms in the face of defeat, that indeed shines a light on how fanatical the Japanese were. And that fanaticism would have carried them into the next phase of war if not for Hirohito surrendering to avoid annihilation. Suzuki's interpretation of German acquiescence was that although most German cities had been bombed flat, Germany was split in two and largely occupied and the *Luftwaffe* was neutered, Suzuki and many of his cohorts expected the Nazis to continue fighting. To his point, nobody will ever know how much longer the Germans would have fought had their charismatic leader not committed suicide, but without a functioning army, there was little left with which to fight—a fact that was clearly overlooked by Suzuki. By his calculus, whereas the Germans no longer had the stomach to do so, Suzuki would have expected every Japanese citizen to fight until his own house was conquered and his own family was dispatched to the afterlife. It was indeed, as historian Richard Frank wrote, "a recipe for extinction."[2468] Nonetheless, the leadership of Japan continued with their final plans "for the suicidal defense of the Homeland—Operation Decision (*Ketsu-Go*)."[2469]

## History of Woody's Medal of Honor

While Woody trained to confront this fanatic population, Captain Beck, commander of Woody's C Company, remembered his conversation on 23 February with 2nd Lieutenant Chambers. Based on this conversation, Beck believed he should recommend Woody for a medal for his actions on 23 February. In the heat of battle, Chambers "pointed out" to Beck seven destroyed pillboxes (Chambers never claimed Woody took them out). Later, Beck wrote he believed Woody destroyed these seven but does not explain who told him this since he had not observed Woody's acts. Beck felt Chambers believed Woody worked alone because Beck mentioned only Chambers and no one else in connection with Woody's deeds.

After Chambers pointed out the pillboxes, Beck had someone count the dead "in or just outside" those structures. He was told the number was 21 and assumed Woody had done this too. The only person claiming to have counted 21 dead Japanese and accounted for seven pillboxes in the documents was Woody. He made these claims to 1). Sergeant Dashiell in May 1945, to 2). Private Paul B. Hoolihan with the 21st *Regimental Newsletter* in May 1945, to 3). George Lawless at the *Charleston Gazette* in May 1956, to 4). Bill Francois with *Man's Magazine* in 1966 and to 5). Peter Collier in 2005 for his book when Woody apparently gave one-on-one interviews with these men.[2470] In April 1945 when Beck started his write-up after 23 February and after a lot of combat, he felt Woody deserved the *highest* medal for valor the U.S. can bestow. Consequently, he initiated the process to award Woody the Medal of Honor by writing a report of what he thought happened and then sent it up his chain of command.

Recommendation for Award of Congressional Medal of Honor; case
of Corporal Hershel Woodrow WILLIAMS, 854310, USMCR.

STATEMENT OF MAJOR DONALD M. BECK, U.S. MARINE CORPS RESERVE:

On 23 February, 1945, the first platoon of Company "C",
1st Battalion, 21st Marines, was assigned the mission of reducing a strong enemy fortified sector which had held up infantry units on three previous days. Large pillboxes, constructed from concrete, thickly cut clay blocks and pieces
of scrap aircraft, covered over with four to five feet of
sand, were located among sandy ridges where tanks could not
maneuver and where supporting overhead fire had proven ineffective.
Upon receiving word that the demolition Sergeant attached to
the first platoon of Company "C" had become a casualty, the demolition Sergeant of the support platoon was ordered to replace him. Corporal Williams, who was attached to company headquarters as demolition Sergeant, overhearing the conversation,
quickly volunteered. After he had reported to the 1st platoon,
I observed him repeatedly return to the company command post
for filled flame throwers and demolition charges. When the objective was taken, seven (7) strong emplacements were pointed
out to me by the platoon leader of the first platoon, (Second
Lieutenant Howard F. Chambers, USMCR, who was later wounded and
evacuated), and twenty-one (21) enemy dead were located in or
just outside the pillboxes. All emplacements, as much as terrain
would allow, were mutually supporting, and connected by deep
communication trenches, and were located among short, thick vegetation, thus affording excellent camouflage around all emplacements. By close of the day, Williams had labored steadily for
four hours before he had accomplished the destruction of these
seven emplacements. At no time during this action was Williams
covered by more than the fire of four riflemen.

*Donald M Beck*
Donald M. Beck,
Major, USMCR,
Commanding Company "C".

Captain Donald Beck's recommendation for Woody to get the Medal of Honor. St. Louis
Personnel Records Center

To support his recommendation, Beck gathered six eye-witness accounts of
Woody's deeds. Considering several units on Iwo lost most of their original personnel, it is remarkable that enough witnesses survived to testify. Many units lost
50% of their contingents within a few days after landing on Iwo. One battalion
of the 4th MarDiv suffered 50% casualties the first day of battle! In A Company
of Woody's battalion, only three original men remained on duty of its original
200-plus contingent. Fourteen officers and 377 men had been killed or wounded

out of the battalion including the replacements. Woody's company experienced a similar fate. This attrition made finding witnesses to medal actions difficult.[2471] Woody learned that two in his unit, Warren Bornholz and Charles Fischer, who possibly protected him, died near him on 23 February, so not all of his witnesses survived.[2472]

The witness statements were one paragraph attestations of what they had observed Woody do. The only thing Beck had seen Woody do that day was return "to the company command post for filled flamethrowers and demolition charges."[2473] The reports gathered did not include a statement from Chambers. He had been injured on 1 March and evacuated stateside arriving at Fort Eustis, Virginia by 10 May 1945.[2474] So, the statements Beck gathered provide but fragments of the story he thought Chambers had given him. Here is a breakdown of the witnesses' observations. All these men served in Charlie Company with Woody.

> **Elder and Tripp:** In supporting Beck's observations, First Sergeant Elder and Corporal Tripp echoed him observing Woody return six times to the CP for flamethrowers and demolitions.
>
> **Southwell:** Sergeant Southwell wrote Woody destroyed three pillboxes and killed eight.
>
> **Naro:** Corporal Naro wrote Woody neutralized one pillbox from the rear.
>
> **Schlager:** Corporal Schlager wrote he witnessed Woody climb on a pillbox and flame the enemy from the inside out through a vent killing eight Japanese.
>
> **Schwartz:** PFC Schwartz wrote he saw Woody use a flamethrower and three demolition charges to reduce two pillboxes.

Although many MOH citations included actions over several days, why Beck did not get a statement from Lieutenant Henning about Woody helping Southwell earlier in the day, killing 50 Japanese using grenades, is unfortunate.[2475] Usually, when recommending Medals of Honor, several acts were included to improve the chances of approval.[2476]

None of the supporting cast witnessed seven pillboxes destroyed or counted 21 dead Japanese mentioned in Beck's report. Why Chambers highlighted seven pillboxes and why Beck believed Woody destroyed them remains unknown. In Dashiell's report, Woody told him this exact number: "Williams accounted for seven pillboxes."[2477] Woody apparently gave the same account to Private Hoolihan, a journalist working for the *21ˢᵗ Regimental Newsletter* who, after interviewing Woody, wrote: "Outstanding among the boys with the 'know-how' [with flame-throwers] is Corp. Hershey [sic] Williams, the stubby, pleasant-faced fellow who wiped out seven Jap pillboxes at Iwo with his flamethrowing savvy."[2478] Today, however, Woody states he cannot recall hardly any of his actions that earned him *the Medal.*[2479] In 1966, his recall was very strong though. For example, Woody described Tripp as being his "right-hand man" taking out the big pillbox (quoting Tripp as saying "I'll be with you"!), but Tripp's 1945 statement does not mention this pillbox (or any pillbox for that matter).[2480] This 1966 article for which Woody supplied information, was written by Drake University professor of journalism and WWII paratrooper and veteran, Bill Francois, and thus, one would think he would have tried to have interviewed Woody extensively and those he could find who were also there; however, it looks like all the information came from Woody, Dashiell's article and history books.[2481] Comparing Tripp's 1945 statement and Woody's recollection some twenty-one years later of what Tripp did, it is hard not to find Tripp's 1945 statement much more credible, i.e. Tripp was not there with Woody, although in 1966, Woody places Tripp right next to him during the action against the big pillbox. Reversing this story, in 2017, after reading Hallas' book *Uncommon Valor on Iwo Jima* where Tripp is mentioned, Woody affirmed Tripp wasn't there that day which is different than what he said in 1966, but in line with Tripp's statement in 1945: "Unless I have completely lost my memory…Tripp was never a pole man, never crawled with me and so far as I know never had anything to do with the burning out of the Pillboxes."[2482] When studying Woody, often one gets totally different stories about what happened causing confusion about what he really did getting events and people continually confused.

The Iwo battlefield was indeed chaotic which could cause some confusion after the fact, but after interviewing hundreds of *Wehrmacht* and Marine Corps WWII veterans, Woody appears to be one of the most confused veterans I have ever encountered. Nonetheless, to describe some of the chaos surrounding Woody that day, Dashiell reported Woody's platoon killed between 20-25 "Japs." Chambers shot two dead with his pistol "neatly between the eyes" and Schlager dropped a sword-wielding one: "Schlager put a slug in one Nip 20 feet away. The Jap lurched, but kept coming. Schlager hit him again in the chest. The Jap wavered, almost collapsed, then started towards Schlager again. Schlager's third bullet did the trick."[2483] In the middle of such action, Woody and none of the Marines with him that day were calm, detached observers.

Presumably, the best positioned witnesses to Woody's action would have been his protection team members. But two of them were apparently killed in the process, Bornholz and Fischer. Another two were there the entire time; leaving two who may have died later—nobody knows. Furthermore, taking out the large bunker was the highlight of the entire action and was the only "pillbox" action that would make it into Woody's MOH citation. That only one person witnessed it seems strange. Southwell and Strickland may have been there, but Southwell does not describe the massive pillbox and Strickland did not give a statement possibly due to wounds suffered on 12 March. Southwell wrote that Woody destroyed three pillboxes and he counted eight enemy dead inside those structures. Perhaps one of the three pillboxes was indeed the massive one, but he did not delineate the difference.[2484] Second Lieutenant Chambers organized all the attacks of Woody and others that day in first platoon, but he would never support Woody's MOH package although he must have observed some if not all of Woody's actions. At the time of Beck writing Woody up for the MOH, no one knew where Chambers actually was so he had not been officially asked to provide an affidavit for Woody. Apparently, from the evidence gathered, no one witnessed all the pillboxes Woody destroyed or the dead enemy resulting from his actions that Beck had reported him doing and that Woody had been bragging about performing.

And none of the witness statements discussed Woody flaming a *Banzai* of three to six enemy. This incident materializes in the endorsement of the Medal of Honor recommendation by Colonel Hartnoll Withers, who commanded the 21st Marines, and his endorsement does not give the names of those providing this info. Beck's son, Barry, also a Marine officer and trained lawyer observed:

This is a curious appearance of an event in a document train that has no source. Who witnessed this? How did this event make its way into the documents? Why didn't the Awards Board not ask for proof for this, what one could say, was a major part of the action that day? This is quite confusing.[2485]

Killing the Japanese during the charge is one of Woody's few memories when giving talks and interviews, although he cites different numbers of how many attacked.[2486] When Dashiell interviewed him in 1945, he stated he killed some who attacked him after leaving bunkers; this could be the source of Withers' story since he was not a witness.[2487] None of Woody's comrades who submitted statements supporting the MOH recommendation stated they saw him incinerating a wave of attacking Japanese. If at least four riflemen were covering him, they must have fired into that wave of enemy fighters and helped drop them along with Woody's flames. One would think the witness statements would have mentioned the charge since it is in the official MOH citation, but the statements do not reference it. Waters' Silver Star citation makes it clear he stopped a three-man *Banzai*: "[Waters] was attacked by three heavily armed enemies from a nearby cave, but he coolly eliminated them with a…grenade."[2488] Nothing like this appears in the witness statements for Woody's *Medal* recommendation.

Since Woody never acted alone on the battlefield, all his actions must have been observed by men who survived that day and the war, like Chambers, Schlager and Southwell, just to name three, who all helped him survive. This was especially the case for a flamethrower operator. By

PFC Donald G. Graves. He served in the 5th MarDiv on Iwo and was a flamethrower operator. He finds many of Woody's claims highly dubious. St. Louis Personnel Records Center.

way of illustration, on 26 February 1945, near Hill 362A, Don Graves attacked one cave with his flamethrower and "roasted" the enemy inside. As he finished, his rifleman, Robert "Bob" Mueller, yelled, "Watch out to your left Graves!" Three "Nips" charged from a neighboring cave. As Graves turned to engage, he tripped and fell. "Luckily," Graves said, "Mueller fired over my body and into the wave of men and killed them with his BAR. If he'd not been there, I would've been a goner." Graves had another rifleman there who must have fired at the wave as well. "No flamethrower attacked without support. If Woody really killed six attacking Japanese, that would've been observed and the whole platoon would have known about it within hours—remember, Marines are…worse with telling stories and rumors than old ladies at a retirement center…It's indeed strange that *no one* directly involved saw this and put it into his witness statement."[2489] (See Gallery 3, Photo 19)

Combining the eye-witness statements, one could ascertain that Woody destroyed at least three, maybe six pillboxes, but not seven pillboxes, and killed at least eight Japanese, maybe 16, but not 21. What is documented is an *incredible feat* and, from a layman's perspective, definitely *worthy* of a medal. But this history addresses candidly the tension between what can be documented with first-person reports versus what was *initially* used to write up Woody's endorsement using second- and third-hand stories. In a 1966 interview, Woody told Francois he only remembered taking out three pillboxes but allowed the journalist to write he took out seven; however, in interviews for this book, Woody does not remember in detail destroying any of the pillboxes except the massive one although he often mentions the seven in passing.[2490] In 2005 and 2012, he told author Peter Collier and *Charleston Gazette* reporter Sandy Wells that he "knocked out seven pillboxes" like he did with Dashiell and Hoolihan in May 1945 although after the official Navy Board of Awards and Decorations recommendation in September 1945, Woody's final MOH citation was changed and this number deleted because there was no supporting evidence for it.[2491] Also, the men who gave statements for Woody's MOH could have been reporting the same pillboxes being taken out. Realistically, the eight dead counted could have been counted twice since that exact number was given twice: eight.

SAMPLE CITATION FOR AWARD OF THE CONGRESSIONAL MEDAL OF HONOR, CASE OF CORPORAL HERSHEL W. WILLIAMS, (854310), UNITED STATES MARINE CORPS RESERVE.

"For gallantry and intrepidity at the risk of his life above and beyond the call of duty and without detriment to the mission of his command in combat with the enemy on IWO JIMA, VOLCANO ISLANDS, on 23 February, 1945, while serving as a demolitions sergeant attached to a Marine rifle company. On this date Corporal WILLIAMS' company was stopped by heavy enemy machine gun fire originating from a series of exceptionally strong pillboxes on high ground commanding MOTOYAMA AIRFIELD No. 1, which for three days had prevented an advance in that area. When an artillery barrage had failed to reduce the enemy positions and since tanks could not operate over the steep sandy ridges, Corporal WILLIAMS volunteered to go forward alone to attempt the reduction of the formidable emplacements. Supported only by the fire of four riflemen stationed just in front of his own lines, WILLIAMS proceeded to attack the pillboxes in succession. Finding a series of shallow connecting trenches which led to the rear entrances of the enemy emplacements, he crawled forward with flame thrower and demolition charges to burn out and blast each in turn despite intense machine gun fire directed at him from these mutually supporting pillboxes. Working continuously for nearly four hours, he continued with his mission, crawling back to his own lines six times to obtain additional demolition charges and refilled flame throwers. At the end of this period, he had destroyed seven enemy pillboxes killing twenty-one Japanese soldiers, thus clearing the way for the advance of his company. His courageous and gallant act contributed immeasurably to the success of his unit and served to inspire his comrades during the remainder of the attack. His outstanding heroism was in keeping with the highest traditions of the United States Naval Service."

ENCLOSURE "AA"

This is the original MOH citation for Woody's actions with details of Woody destroying seven pillboxes and killing 21 Japanese. These "facts" would be removed in the final citation due to lack of evidence. St. Louis Personnel Records

854310
DGP-298-ebg

Cincpac
App. SofN 9-22-45
Serial 36222
Signed    OCT 4  1945

      The President of the United States takes pleasure in presenting the MEDAL OF HONOR to

      CORPORAL HERSHEL W. WILLIAMS, USMCR.,

for service as set forth in the following

    CITATION:

      "For conspicuous gallantry and intrepidity at the risk of his life above and beyond the call of duty as Demolition Sergeant serving with the First Battalion, Twenty-First Marines, Third Marine Division, in action against enemy Japanese forces on Iwo Jima, Volcano Islands, 23 February 1945.  Quick to volunteer his services when our tanks were maneuvering vainly to open a lane for the infantry through the network of reinforced concrete pillboxes, buried mines and black, volcanic sands, Corporal Williams daringly went forward alone to attempt the reduction of devastating machine-gun fire from the unyielding positions.  Covered only by four riflemen, he fought desperately for four hours under terrific enemy small-arms fire and repeatedly returned to his own lines to prepare demolition charges and obtain serviced flamethrowers, struggling back, frequently to the rear of hostile emplacements, to wipe out one position after another.  On one occasion he daringly mounted a pillbox to insert the nozzle of his flame thrower through the air vent, kill the occupants and silence the gun; on another he grimly charged enemy riflemen who attempted to stop him with bayonets and destroyed them with a burst of flame from his weapon.  His unyielding determination and extraordinary heroism in the face of ruthless enemy resistance were directly instrumental in neutralizing one of the most fanatically defended Japanese strong points encountered by his regiment and aided vitally in enabling his company to reach its objective.  Corporal Williams' aggressive fighting spirit and valiant devotion to duty throughout this fiercely contested action sustain and enhance the highest traditions of the United States Naval Service."

R--Fairmont, West Virginia.
B--Fairmont, West Virginia.

                      /S/ HARRY S. TRUMAN

This is the final version of Woody's MOH citation with all the previous listings of actual numbers of destroyed pillboxes and killed Japanese deleted. This is the version President Truman signed. St. Louis Personnel Records Center.

Iwo veteran and flamethrower Don Graves finds two things troubling with Woody's story. First, in his regiment, they trained that once one finished a pill-box with a flamethrower he discharged it, took cover and radioed back to the CP for another to be brought up. "You never left the line." Graves said. "Think about it. You don't want to be using those tanks while exhausted…Using flamethrower tanks for four hours would make the most fit, muscular guy absolutely spent… you wanted to conserve your energy for the next action and there were other guys in the unit who would bring you new ones." "And last," Graves says, "No one on Iwo stayed around after they killed a lot of Japs and counted how many… they killed. That…could get you killed. How and why were these men 'count-ing' the dead during a major battle?"[2492] To counter Graves' comment, however, units were suffering on Iwo, and if one wanted fresh tanks, then one might have had to get them himself. If Woody did return to the CP to get tanks he *actually used six times*, that would have been beyond remarkable. Whether he used those flamethrowers himself and on seven pillboxes remains unknown although Beck felt he did and that made all the difference.

While all Medal of Honor witness statements cannot be expected to be iden-tical often describing very different types of actions while in combat, the lack of similarity raises questions. In contrast to Woody's case, Iwo MOH recipient Corpsman John H. Willis' five witnesses all attest to the same act: that Willis left a first aid station to render aid to his platoon and while helping a wounded comrade, he threw out eight grenades from the crater while administrating first aid. When the ninth came in, he picked it up and tried to throw it away, but it exploded, killing him and his comrade. No discrepancies existed in the witness statements and Lieutenant General Geiger (FMFPAC Commanding General), Fleet Admiral Chester Nimitz (Commander-in-Chief, Pacific Fleet and Pacific Ocean Areas), and Fleet Admiral Ernest King (Chief of Naval Operations) all endorsed Willis' "package" without question.[2493] Woody's statements do not describe the same act, are confusing and did not receive endorsements from Geiger or Nimitz.

Regardless of the statements' incongruity in Woody's case and the absence of a statement from Woody's platoon leader, Lieutenant Chambers, Beck sub-mitted the recommendation for a Medal of Honor. Woody was lucky his officer was not only willing, but also alive to do so. A Marine cannot receive a valor award if an officer does not submit a recommendation or if the witnesses are dead or missing. Beck seems to have been the only original company commander to survive out of a regiment that suffered 70% casualties.[2494] One 21st Marines' battalion reported all its "company commanders and company executive officers were casualties after three days of fighting."[2495] Graves said hardly anyone in his Dog Company, 2nd Battalion, 28th Marines, 5th MarDiv received medals because "all the officers were killed or wounded and no one wrote any of us up. [NCOs] ran the show and witnessed…the acts for valor but they weren't allowed to write anyone up for medals because the brass didn't accept their reports."[2496]

When Beck returned to Guam, he wrote his wife on 4 April, "I must write citations for all those deserving and believe me there are many of them and it will take a long time to get them out."[2497] Woody was on his list. Beck was good at recommending men for medals and did this also after Guam, submitting many, even three for Navy Crosses.[2498] Of the thousands who served in the 3d MarDiv at Iwo, only three received the MOH, and Woody would be one of them, due, in part, to how Beck constructed his recommendation.

After Beck wrote up his recommendation and gathered supporting statements, he submitted his "package" for Woody to his chain of command. He submitted it by 14 April because on that day, he left Guam to return home to his wife and son whom he had not seen in years.[2499] After thirty days leave, Beck reported to Camp Pendleton and did not return to Guam. Had he become aware of Geiger's, Nimitz's and Chambers' refusals to endorse Woody's case, he would not have been in an ideal position after 14 April to amend his report and/or gather more evidence. Or if he did learn of the controversy, maybe he neglected or refused to get involved with what would turn out to become a "pissing" war between several high level commands over a medal.

Beck's recommendation moved up the chain of command and, on 27 April, the 21st Regiment's commander, Colonel Hartnoll J. Withers, endorsed Beck's recommendation and wrote the formal recommendation for Woody. In his report, Withers misrepresented some details from Beck's report, stating Woody killed 21 *in* the pillboxes. He also added some details, without citing evidence, claiming Woody took out a *Banzai* mentioned previously.[2500] Withers submitted the formal recommendation on 27 April 1945 for action by the next level of command, the 3d MarDiv.[2501]

The approval process started with an examination by Major General Erskine's Divisional Board of Awards headed by senior member, Colonel Howard N. "Red" Kenyon, who had also fought at Iwo as a RCT leader (9th Marines). He received a Navy Cross for leading an attack on 25 February north of Airfield No. 2 that "proved to have been the decisive factor in the defeat of the enemy" there. After Kenyon gathered and analyzed the sources, he and his staff of between two and four officers, recommended approval or disapproval of MOH recommendations. If approved, Kenyon presented the recommendation to Erskine for endorsement.[2502]

While Woody's witness statements were examined, Dashiell, the correspondent attached to Woody's regiment, learned Woody had been recommended for a MOH so he interviewed Woody and some of his witnesses, which led to further discrepancies. First, Schwartz was interviewed and instead of saying Woody destroyed two pillboxes with three demolition charges and a flamethrower, he said he destroyed one pillbox with two to three flamethrowers. When interviewed about how many Japanese he killed, loaded with adrenaline-fueled testosterone, Woody claimed, "I counted 21 of 'em that tried to get away and I killed 'em all with flames."[2503] Beck's recommendation claimed some of the dead were inside the pillboxes, but what is troubling is apparently Woody was the one who counted the dead for Beck. He also gave this exact number in 1956 to George Lawless at the *Charleston Gazette*, but changed the story to saying the 21 kills were inside the big pillbox. And in 1966, in the interview with Bill Francois for *Man's Magazine*, he changed the number slightly to 17 for IJA dead in this pillbox.[2504] Yet ever since being interviewed for this study, he does not have any knowledge about KIA numbers. Dashiell did not interview Beck because Beck was on leave.[2505]

*I wrote this after return to Guam when I learned Williams had been recommended for Medal of Honor. An almost unbelievable story.*

*See # 208*

dashiell
man with flames-iwo jima

dashiell #267

*Later — Presented to him by President Truman at White House in July 1945.*

(Note:Williams has been recommended for the
Medal of Honor for the action covered in this
story.)

By Sergeant Dick Dashiell of 5 Westwood Drive,
Chapel Hill, N.C., a Marine Corps Combat Correspondent formerly
of The Associated Press, Charlotte, N.C.

Iwo Jima -- (Delayed) -- Withering Jap machine gun
fire from Nambus concealed in a network of camouflaged pillboxes
and bunkers halted, on D-day plus four, Charlie company of the 21st
Marines in the searing Third Marine Division advance towards
the central Iwo Jima airfield.

Hellish cross-fire chewed up Marines and pinned
others to the ashy ground which Jap mortar shells churned with
Leatherneck flesh. Nip riflemen popped out of by-passed spider
traps to splatter lead into backs of well and wounded. The strongpoint was clearly one of the most bitterly defended and well planned on the entire *dirty little* ~~despicable~~ island.

"The Japs were everywhere," said Sergeant
Lyman D. Southwell, a squad leader from 1883 Race Street, Denver,
Colo. "Fire came from sandpiles, ditches, tunnels, pillboxes --
everywhere. Never, I thought, would I get out alive."

The big thing was the bunkers and pillboxes
situated on high ground commanding the battlefield-plateau and
Motoyama Airfield No. 1. Somebody had to clean them out, one
way or another. Some foot trooper. Artillery couldn't fire into
the Marine-clogged area, and the lattice-work of graveled hummocks
and ditches made it next to impossible for tanks to maneuver.
(more)

*after the war. He was peacefully painting a dairy farm. ... I met him at a 3rd Mar Div reunion*

Marine Correspondent Sergeant Dick Dashiell's unedited article where Woody claimed he killed 21
Japanese ["killed 'em all with flames"] and "accounted for 7 pillboxes." National Archives, College Park

-2-                                                          267-2

dashiell #267

Second Lieutenant Howard Chambers, platoon leader
from 121 Gilpin Street, Punxsutawney, Pa., didn't want to order
anybody to do the  job. It was murder. His platoon already had
been ripped badly from the time it jumped off in the attack at
8 o'clock that morning, six hours earlier. Five men, for instance,
were killed in the first minute of the  assault.

What was needed was a volunteer. And that was
what he got when  Corporal Hershel (cq) Williams (854310), a stocky,
blue-eyed demolitions man from Route 4, Fairmont, W. Va., said he
would "see what I can do".

First thing the 21-year-old former Sharon, Pa.,
steel worker did was to strap on a flame thrower and grab a 12-
pound pole charge filled with 15 blocks of TNT. Then he crouched
and wormed his way towards the first of three Jap pillboxes
which Private First Class Wesley Strickland of Route 4, Holly
Springs, Miss., had spotted. They were from 50 to 60 yards away.

Covering Williams --"Willie" to his buddies --
with Browning automatic and Garand rifles were Southwell, Corp-
orals William Naro of 6072 Annunciation Street, New  Orleans, La.;
Allan Tripp of 2035 17th East, Salt Lake City, Utah; and Alex
Schlager of 2509  East B Street, Torrington, Wyo.; and Private
First Class George Schwartz of 212  Jefferson Heights, Jefferson
Parish, La.

But their fire against the concrete and steel emplacements were like bee-bee shot on a made cement sidewalk. And
there was nothing they could do about the homicidal mortar shells
which continued to plump down.
                    (more)

267-3

dashiell #267

-3-

"Willie moved right up to those pillboxes," South-
well said. "He got up to within a dozen yards of the first one, gave
it a squirt, came back for another flame thrower, ran out again,
burned out the second, returned, took another flame thrower, and
knocked out the  third. He had cover, but we couldn't give him
a lot because we had had a hell of a day and were low on ammo.
And everybody had to hug the ground, so we didn't have such good
sighting."

Four times "Willie" threw the pole charges through
narrow openings of emplacements. Two faulty ones failed to
explode, placing "Willie" in a predicament. The Japs inside knew
he was around when the pole charges dropped in even if they hadn't
seen him before. Notwithstanding, "Willie" worked his way back
and gave them the squirts.

Twice, he weaved and bobbed over the sandy fis-
sures until he was behind the foremost Jap defense lines and gave
the Nambu gunners the business from the rear. He crawled on top
of one pillbox inserted the nozzle through the air vent and killed
the occupants.

"Willie had a hell of a time with one bunker,"
Schwartz recalled. "He used two or three flame throwers on it
before he finished the job.

"Willie" also threw the blaze at frantic Japs who
ran out of their hiding places and tried to get him with rifles,
bayonets and grenades.

"I counted 21 of 'em that tried to get away," he
(more)

-4-

267=4

dashiell #267

said, "but I killed 'em all withk flames."

Naro told of 20 or 25 Nips, bypassed in spider traps and perfectly concealed trenches. Seeing what was in store for them, they made a dash for their lines. The Marines had a field day as they exterminated them one by one. Some, hopelessly, rushed toward the Leathernecks while wielding sabres.

Schlager put a slug in one Nip 20 feet away. The Jap lurched, but kept coming. Schlager hit him again in the chest. The Jap wavered, almost collapsed, then started towards Schlager again. Schlager's third bullet did the trick.

Two of the enemy jumped up near Chambers. The Lieutenant quickly shot both neatly between the eyes with two rounds from his .45.

After four sweating, perilous hours, Williams account- ed for seven pillboxes and uncounted numbers of Japs with the make changes TNT and the liquid fire of six flame throwers.

"Darned if I know why I didn't get bumped off that day, "Williams mused. "Didn't even get a scratch. Then a mortar shell fragment bounced off my leg while I was in a foxhole one night. Didn't hurt much and I stayed in action."

Following Williams' little task, Charlie company dug in for the night in comparative safety. The next morning it moved out through the area Williams had cleared and helped to spearhead the 21st Marines' drive to the second airstrip.

Williams, whose mother, Mrs.Lurenna Williams, lives at the Fairmont address, attended Fairmont High School and worked with the Civilian Conservation Corps. He joined the Marines in Charleston, W.Va., in May, 1943, attended "boot" camp in San Diego,

(more)

dashiell $267

Cal., and shipped overseas in December, 1943. He was promoted to
private first class the next month.

He fought in the Guam campaign as a demol-
itions man and was promoted to his present rank last October.
He is now acting demolitions sergeant.

-USMC-

Did anyone notice the incongruities? Out of Dashiell's 366 reports, the one he wrote on Woody was the *only one* where he made a suggestive note that something was awry, scribbling: "An almost unbelievable story."[2506] Was he in doubt of or was he impressed with Woody? Dashiell's son Fred believes his father felt something was amiss with Woody's claims.[2507] (See Gallery 3, Photo 20)

Yet, in addition to Dashiell quoting Woody having killed 21, Dashiell also quoted Woody having killed numerous men with demolitions and flames.[2508] So, he could have killed more than 21 according to his interview here. And if Woody let people believe that he had killed 21 at the third, massive pillbox and then went on to take out another six emplacements, then he had to have known that others would have logically concluded that the KIAs he inflicted on the enemy would have numbered into the multiple dozens when taking out the other six pillboxes.[2509] Iwo veteran John Lauriello doubts anyone counted bodies on the battlefield *during* the fighting: "The battlefield was so chaotic, it would've been impossible to have counted all the people one killed and when you're in the thick of it, that would've been the last thing on someone's mind."[2510] The record is unclear about Dashiell's feelings about Woody's testimony, but he submitted the article and it was printed in newspapers in mid-May 1945. Concurrently, Woody's *21st Regimental Newsletter* also published information about Woody taking out seven pillboxes, so Woody was busy giving interviews.[2511]

Returning to Woody's kills, Beck's recommendation states 21 dead were counted in or around the pillboxes. Colonel Withers wrote 21 dead were tallied in pillboxes. And when interviewed in May 1945, Woody stated he killed 21 Japanese "fleeing" and counted them. In the 1956 interview, Woody changed his story saying he killed 21 Japanese in the large pillbox.[2512] And then in 1966, Woody altered his story yet again and claimed 17 killed for this third pillbox.[2513] As mentioned earlier, again, if Woody let people believe he killed 21 (or even 17) in one pillbox then he must have known readers would assume he killed more than 21 in destroying other pillboxes. Historian Colonel Jon T. Hoffman (USMCR), in describing Chesty Puller, could have been describing Woody: "Given the Marine penchant for sea stories—small threads of fact woven into tall tales—it is no surprise that myth soon mixed with reality."[2514] Veteran Graves explains:

> During the first week on Iwo, we never saw the Japanese above the surface. We definitely didn't see them in groups. If we did catch a glimpse of one, then it was only for a moment as he ran from one hideout to another. Moreover, we didn't count the dead. And anyone who says they stuck their head in a pillbox after they took it out to count how many…they killed was suicidal. All those caves and pillboxes were connected by tunnels. You may have killed some…in a pillbox…but you never knew how soon that very same pillbox would be reoccupied. You might've peeked into a pillbox and been met with a barrel of a rifle and then it would've been *Sayōnara*. And besides, if every pillbox Woody took out was blown up according to Woody…, how the Hell would you have been able to count the dead in the pillbox when you couldn't see what and who was there since it was blown to millions of pieces? Doesn't make sense.[2515]

Although in 1945 Woody was clear he killed 21, he has repeatedly said since then, except in the 1956 *Charleston Gazette* and the 1966 *Man's Magazine*, that he has no clue how many he killed: "There was so much chaos and potential death around each corner that I wasn't going to just walk around a battlefield counting dead bodies. I had other things to do," Woody said in July 2016, which differs dramatically from what he was saying as a 21-year-old kid after surviving Iwo in April/May 1945, and later as a 32-year-old young adult in May 1956 and as a 42-year-old middle-age man in January 1966.[2516]

Withers and Beck did not address these conflicting stories among the various reports in 1945 and Withers sent Woody's MOH recommendation to Erskine for review. The sample citation at this stage mentioned seven pillboxes and 21 dead. Without addressing these discrepancies noted above, the 3d MarDiv Board of Awards and Erskine recommended approval of the MOH and forwarded the recommendation via the chain of command to Secretary of the Navy James Forrestal on 19 May 1945.[2517] Then on 16 July 1945, Major General Harry Schmidt of the V Amphibious Corps, senior to the 3d MarDiv, forwarded his recommendation for approval to Forrestal as well. On 24 July, Lieutenant General Roy S. Geiger, Commanding Marine FMFPAC, forwarded his recommendation of approval provided a statement was obtained from Woody's platoon leader, 2nd Lieutenant Howard F. Chambers, that substantiated the proposed citation and Beck's recommendation. Geiger and/or his staff noticed the problem of not having a report from this witness. Consequently, he wrote Chambers' account was "necessary in order to establish incontestable justification in this case."[2518] Geiger followed the regulations requiring that the case for "heroism or gallantry" above the call of duty be "supported by incontestable proof in the form of affidavits from at least two individuals who were eye-witnesses of the deed."[2519] In this case, no eye-witness to date had seen Woody do what Beck and Withers had reported second- and third-hand: destroyed seven pillboxes, killed 21 Japanese and obliterated a *Banzai* charge. The guidelines for a MOH stated "that second-hand evidence would not be accepted in Medal of Honor cases" meaning Beck's and Wither's letters could not be used as evidence.[2520] Interestingly, on the same day of 24 July 1945, Geiger approved Woody's friend, Vernon Waters, for the Silver Star for destroying two pillboxes and killing three Japanese conducting a *Banzai*.[2521]

```
1740-55-50              HEADQUARTERS,
0296/110          FLEET MARINE FORCE, PACIFIC,
               c/o FLEET POST OFFICE, SAN FRANCISCO.        RECEIVED
                                                          Navy Department
Serial:                                              SEP  Secy's Office

                         6th Endorsement on
               832       CO, 21stMars, 3d
                         MARDIV ltr 1740-5
                         over HJW-rmg, dtd
                         27Apr45.                   SO9  19  35

From:          The Commanding General.
To  :          The Secretary of the Navy.

Subject:       Award of the Congressional Medal of Honor, case
                 of Corporal Hershel W. WILLIAMS, (854310),
                 U.S. Marine Corps Reserve - recommendation for.

Reference:     (a) CINCPAC dispatch #140304, dated 14Sep45.

Enclosure:     (A) Copy of CG, FMF, PAC. Serial #68975, to
                 2dLt Howard F. CHAMBERS, (029195), U.S.
                 Marine Corps Reserve, via MARPAC, dated
                 27Aug45.

    1.             Forwarded, in compliance with reference (a),
recommending the award of the Medal of Honor to Corporal WILLIAMS
as a suitable decoration for the acts cited.  The case is incom-
plete, however, in that a statement of an additional witness is
believed to be necessary.  Attempt is being made to procure this,
as indicated by enclosure (A).

    2.             By copy of this endorsement the Commanding General,
Department of the Pacific, is directed to expedite the procure-
ment and forwarding of the statement of Second Lieutenant Howard
F. CHAMBERS, (029195), U.S. Marine Corps Reserve, as requested in
enclosure (A).  In compliance with reference (a), the statement
should be forwarded direct to the Secretary of the Navy, with
information copies to the Commander in Chief, U.S. Pacific Fleet
and Pacific Ocean Areas, and the Commanding General, Fleet Marine
Force, Pacific.

                                          ROY S. GEIGER.

Copy to:
CINCPAC     MARPAC (w/copy of reference (a))   CG, 3dMARDIV
```

Lieutenant General Roy S. Geiger's letter to Secretary of the Navy Forrestal saying Woody's MOH is incomplete and needs more evidence. He specifically focuses on getting evidence from Lieutenant Howard Chambers, Woody's platoon leader. He sends this letter via Admiral Nimitz's office. 27 April 1945. St. Louis Personnel Records Center

```
1740-55-50                  HEADQUARTERS,
0296/360              FLEET MARINE FORCE, PACIFIC,
                  c/o FLEET POST OFFICE, SAN FRANCISCO.

    Serial:                                          JUL 24 1945

            5C337     4th Endorsement on
                      CO, 21stMars, 3d
                      MARDIV ltr 1740-5          SO9  19   35
                      over HJW-rmg dtd
                      27Apr45.


    From:          The Commanding General.
    To:            The Secretary of the Navy.

    Via:           (1) The Commander in Chief, U. S. Pacific Fleet
                       and Pacific Ocean Areas.
                   (2) Commandant of the Marine Corps.

    Subject:       Award of the Congressional Medal of Honor, case
                       of Corporal Hershel W. WILLIAMS, (854310),
                       U. S. Marine Corps Reserve - recommendation
                       for.

         1.        Forwarded, recommending approval, provided a
    statement from the principal witnesses to the events recounted
    therein by Second Lieutenant Howard F. CHAMBERS, U. S. Marine
    Corps Reserve, commander of the platoon to which Corporal
    WILLIAMS belonged, is obtained and substantiates the citation.
    Second Lieutenant CHAMBERS was evacuated from United States
    Naval Hospital #10 on 9 April 1945 to the Department of the
    Pacific on a medical status.  His present whereabouts is un-
    known.  A statement from him is believed to be necessary in
    order to establish incontestible justification in this case,
    and it is recommended that the Commandant procure such state-
    ment before forwarding the case to the Secretary of the Navy.
```

ROY S. GEIGER.

```
    Copy to:
      CG, 3dMARDIV
```

Lieutenant General Roy S. Geiger's second letter to Secretary of the Navy Forrestal saying Woody's MOH is incomplete and needs more evidence. He particularly asks for a statement from 2nd Lieutenant Howard Chambers. 27 April 1945. St. Louis Personnel Records Center

```
1740-55-50
0296/360

                         HEADQUARTERS,
                    FLEET MARINE FORCE, PACIFIC,
               c/o FLEET POST OFFICE, SAN FRANCISCO.

Serial:  68975                                Aug 27 1945


From:          The Commanding General.
To:            Second Lieutenant Howard F. CHAMBERS, (029195),
               U. S. Marine Corps Reserve.

Via:           The Commanding General, Department of the Pa-
               cific.

Subject:       Statement to support the award of the Congres-
               sional Medal of Honor, case of Corporal
               Hersehel W. WILLIAMS, (854310), U. S. Marine
               Corps Reserve - request for.

Enclosures:    (A) Copy of Sample Citation.
               (B) Copy of statement of Major Donald M. BECK,
                   U. S. Marine Corps Reserve.

        1.        It is the intention of this Headquarters to ap-
prove an award of the Congressional Medal of Honor to the sub-
ject named man, provided that the narrative contained in en-
closure (B) can be substantiated by a sworn statement from you
as an eye witness.

        2.        You are directed to forward to this Headquarters
immediately a statement as to what you actually observed in con-
nection with the incident related in enclosure (A).




                                   A. F. HOWARD,
                                   Deputy Commander.
- - - - - - - - - - - - - - - - - - - - - - - - - - - - - - -




                                              C O P Y
```

Lieutenant General Roy S. Geiger's writes Woody's platoon leader, 2nd Lt. Howard Chambers, directly asking for him to attest to the facts he witnessed Woody do and to the truth of Captain Beck's endorsement. Chambers would ignore Geiger's request. 27 April 1945. This is a copy of the letter Geiger sent to Chambers that is in Chambers' Personnel File. St. Louis Personnel Records Center

Cinepac File

F15/EF37(1)

Serial 36222

UNITED STATES PACIFIC FLEET
AND PACIFIC OCEAN AREAS
HEADQUARTERS OF THE COMMANDER IN CHIEF

19 AUG 1945

5th Endorsement on
CO,21stMars,3rd
MarDiv, ltr. dated
27 April 1945.

SO9 19 35

From:        Commander in Chief, U. S. Pacific Fleet.
To:          Commanding General, Fleet Marine Force, Pacific.

Subject:     Award of the Congressional Medal of Honor,
             case of Corporal Hershel W. Williams,
             (854310), U. S. Marine Corps Reserve -
             recommendation for.

1.           Returned.

2.           This correspondence has been referred to the
Pacific Fleet Board of Awards for review.  It is the recommendation of that Board that the Commanding General, Fleet Marine
Force, Pacific, obtain the statement of substantiation mentioned
in the 4th endorsement prior to submission of this case to
higher authorities; the recommendation is returned herewith for
such action.

J.F. NEWMAN,
By direction

Admiral Chester Nimitz and his board agreed with Lieutenant General Roy S. Geiger that the evidence from 2nd Lieutenant Howard Chambers was necessary to complete Woody's case and did not endorse the medal for him. 19 August 1945. St. Louis Personnel Records Center.

Yet, at the time, no one knew where Chambers was. Meanwhile, Fleet Admiral Nimitz's office—with Nimitz being in the chain of command above Geiger—and his own Award's Board of between three to five officers wrote Geiger's office on 19 August agreeing that Chambers' report was needed to legitimize *the Medal*: "It is the recommendation of [the Pacific Fleet Board of Awards] that the Commanding General [Geiger], FMF, Pacific, obtain the statement of [Chambers]…prior to submission of this case to higher authorities; the recommendation is returned herewith for such action."[2522] By 27 August 1945, Geiger had located Chambers and he wrote him, requesting that he review Beck's statement and the first draft of the MOH citation for Woody mentioning the seven pillboxes and 21 dead Japanese. Geiger noted he intended to approve Woody's case "provided that the narrative contained in [Beck's letter] can be substantiated by a sworn statement from you as an eye witness." Then it asked for him to forward to Geiger's office "immediately a statement as to what you actually observed in connection with the incident related to [the draft of the citation]."[2523]

```
IN REPLYING
REFER TO NO.

1740-55-50
0296/360
                              UNITED STATES MARINE CORPS
                                    HEADQUARTERS,
                              FLEET MARINE FORCE, PACIFIC
                          c/o FLEET POST OFFICE, SAN FRANCISCO.

        68975                                          AUG 2 7 1945


        From:           The Commanding General.
        To:             Second Lieutenant Howard F. CHAMBERS, (029195),
                        U. S. Marine Corps Reserve.

        Via:            The Commanding General, Department of the Pa-
                        cific.

        Subject:        Statement to support the award of the Congres-
                        sional Medal of Honor, case of Corporal
                        Hershel W. WILLIAMS, (854310), U. S. Marine
                        Corps Reserve - request for.

        Enclosures:     (A) Copy of Sample Citation.
                        (B) Copy of statement of Major Donald M. BECK,
                        U. S. Marine Corps Reserve.

            1.          It is the intention of this Headquarters to ap-
        prove an award of the Congressional Medal of Honor to the sub-
        ject named man, provided that the narrative contained in en-
        closure (B) can be substantiated by a sworn statement from you
        as an eye witness.

            2.          You are directed to forward to this Headquarters
        immediately a statement as to what you actually observed in con-
        nection with the incident related in enclosure (A).


                                              A. F. HOWARD,
                                              Deputy Commander.
```

Lieutenant General Roy S. Geiger's letter to 2nd Lieutenant Howard Chamber requesting him to give evidence in support of Woody's MOH case. Chambers refused to respond to Geiger. 27 August 1945. This is the actual letter Chambers received from Geiger's office requesting information about Woody's MOH package. Author's Collection and the Chambers Family Archive

Surprisingly, Chambers ignored Geiger's request. Thus, Geiger did not forward the recommendation for approval. As late as 19 September 1945, due to Chambers' report having not been obtained, Woody's recommendation package was still in limbo.[2524] Although Forrestal was informed of the problem and although Geiger had asked Chambers to submit a report again on the 4th, by late September, Chambers had not done so and gave no reason why he would not give what is called a Summary of Action Report.[2525]

```
029195
1740-55-50              First Endorsement           4 September 1945.
11.mbf-O  HEADQUARTERS, DEPARTMENT OF THE PACIFIC, MARINE CORPS,
             100 Harrison Street, San Francisco 6, California.

From:      The Commanding General.
To  :      First Lieutenant Howard F. Chambers, (029195),
               U. S. Marine Corps Reserve.
Via :      The Commanding Officer, Marine Barracks, Naval Mine
               Depot, Yorktown, Virginia.

Subject:   Statement to support the award of the Congressional Medal
               of Honor, case of Corporal Hershel W. Williams,
               (854310), U. S. Marine Corps Reserve, request for.

     1.    Forwarded for compliance with the directive contained in
     paragraph two of basic letter.


                                          ALBERT E. BENSON,
                                          By direction.
```

This is Lieutenant General Roy S. Geiger's second attempt to get the supporting evidence from 2nd Lieutenant Howard Chambers about Woody's MOH action. Geiger sent his request via the commanding officer at the Naval Mine Depot at Yorktown, Virginia. Chambers would also ignore this request. 4 September 1945. Author's Collection and Chambers Family Archive

In the meantime, Chambers' home town newspapers reported he had first been taken to the Philippines after suffering his wounds for "splendid medical attention" and had written his parents that he was fine. Then by 12 March, he arrived at Saipan's 148th General Hospital for treatment where he underwent surgery on his knee to remove shrapnel. Weeks later, he sailed from Saipan and arrived at Aiea Heights, Hawaii where he stayed a few weeks. On 18 April, he boarded the *USS Matsonia* and sailed from Honolulu, Hawaii for the mainland. He arrived at San Francisco on 23 April and called his parents. Over the next days, he made his way to Fort Eustis Naval Hospital, Virginia. Then in June, he had leave and spent time with his wife and parents in Punxsutawney, Pennsylvania.[2526] He had no problem speaking with his parents, traveling around the country, writing letters, making calls, showing up to medical facilities in San Francisco, Philadelphia and Fort Eustis, but there is no explanation why he was remiss in answering Geiger's two requests. He had a full cast on his left leg and walked with a cane, but he could have talked and wrote a letter.[2527] If his wife and parents could locate him, speak to him on the phone, welcome him for visits, receive letters from him and know his whereabouts, one would think one of the key players in defeating Japan—the Commanding General, FMF, Lieutenant General Geiger—would be able to locate Chambers and get a simple "yes" or "no" response. Lieutenant General "Howlin Mad" Smith's office had tracked Chambers down and had his Purple Heart awarded that June.[2528] Then on 8 August 1945, Chambers received his Silver Star citation signed by Geiger while at Fort Eustis.[2529] Other MOHs candidates' files received letters from witnesses from hospitals stateside in support of their recommendations, including Jack Lucas, Wilson Watson and Tony Stein, to complete cases without issues.[2530]

Chambers' family archive shows Chambers received Geiger's requests at this time and these documents were substantially all he saved from the war.[2531] But Geiger's request for Chambers to "immediately" give a statement about the evidence was ignored.[2532] Compared to others from whom Geiger asked for more evidence concerning a MOH case, Chambers was the *only* one who did not respond. Was this evidence of intended neglect or PTSD? Unfortunately, we don't know. Retired Marine and combat veteran Major Paul Stubbs observed: "There had to be something wrong with this whole process if Woody's own boss wouldn't endorse him--the very guy who Beck relied on to even start this process."[2533] Marine Corps Commandant, General Charles C. Krulak, echoed Stubbs declaring, "If Woody's platoon leader refused to support his MOH package, then Woody must have not merited the award. That is the strangest thing I have ever heard of when discussing a Medal of Honor case."[2534] Chambers' son-in-law, retired Army Colonel, Robert Olson wrote: "Something had to bug...[Chambers] about this. For one thing, he saved no news clippings of his island hopping, but he saved [the first draft of Woody's MOH] citation."[2535] Apparently on hearing the award was going through without Chambers' report, Geiger wrote Forrestal on 19 September that Chambers' report was needed, that he would do all he could to procure it, and to have it sent directly from Chambers to the navy. "The case is incomplete," Geiger wrote, and as soon as he got the report, he would have it sent to Nimitz and Forrestal.[2536] (See Gallery 3, Photo 21)

After Nimitz's review, the next one was by the Navy Department Board of Awards and Decorations, whose senior member was Rear Admiral Robert Ward Hayler, recipient of three Navy Crosses. On 20 September 1945, Hayler recommended to the Secretary of the Navy that Woody's award be approved; he also wrote that his board did not believe the missing testimony sought by Geiger necessary "to complete this case."[2537] His staff and board were comprised of nine officers (a Marine colonel, six navy captains, one commander and one lieutenant commander) who reviewed the case and gave their opinions.[2538] Hayler's action seems surprising: The Board of Awards and Decorations disagreed with a lieutenant general field commander and the Pacific Fleet commander concerning proof for a MOH. Equally startling was the absence of an endorsement by the Commandant of the Marine Corps; General Vandegrift never endorsed Woody's recommendation one way or another which USMC Commandant Charles C. Krulak finds "very strange."[2539] Geiger had encouraged Vandegrift not to submit Woody's case to Truman until Chambers' recommendation was obtained (most MOHs reviewed for this study had Vandegrift's endorsement).[2540] While Marines like recognizing heroes, Geiger was not going to do so if he felt more evidence was needed. Erskine, who had recommended approval of the MOH for Woody, was treating final approval by the president as a *fait accompli*: Erskine had cut orders to move Woody to Washington D.C. in preparation to award him *the Medal* on 18 September 1945.[2541] Apparently no one took issue with the absence of the Commandant's endorsement, and no one other than Geiger and Nimitz and their boards saw the absence of Chambers' statement as problematic.

```
ADDRESS REPLY TO

    SecNav                      NAVY DEPARTMENT
AND REFER TO                    WASHINGTON 25, D. C.
    QB4-EMP
    20 September 1945


    End 7 on CO 21stMars 3rdMar
    Div Ltr. 1740-5 JHW-rmg of
    27 April 1945.

    From:   Navy Department Board of Decorations and Medals.
    To:     The Secretary of the Navy.
    Via:    Commander in Chief, United States Fleet.

    Subj:   Award of the Congressional Medal of Honor in the Case of
            Corporal Hershel W. Williams, USMCR - Recommendation for.

    1.  The Board has considered the recommendation for the award of the
    Medal of Honor to Corporal Hershel W. Williams, U. S. Marine Corps
    Reserve, and is of the opinion that the additional testimony requested
    by the Commanding General, Fleet Marine Force, Pacific, in his sixth
    endorsement is not necessary to complete this case.  The Board recom-
    mends the award of the Medal of Honor to Corporal Hershel W. Williams,
    U. S. Marine Corps Reserve, in recognition of his outstanding heroism
    in action on Iwo Jima on 23 February 1945.


                                    R. W. HAYLER.


    CC:  CominCh
         Bd of Dec & Medals
         USMC
```

Rear Admiral Hayler and his board overrode Lieutenant General Roy S. Geiger's grave concerns about the lack of evidence for Woody's MOH deeds. 20 September 1945. St. Louis Personnel Records Center.

In the end, Secretary of the Navy Forrestal granted Hayler *carte blanche* to recommend whether awards should be approved or disapproved within the Navy Department (and the Marine Corps is under the Navy). Forrestal wrote to Hayler and conferred the following to him:

> The Board shall be responsible for considering and making recommendations to the Secretary of the Navy concerning the following subjects:...(b) Legislative matters, General Orders, and Executive Orders concerning decorations, awards, and campaign medals including any changes therein...(e) Definition of policies for guidance of officers to whom authority to make certain awards is delegated.[2542]

As President Roosevelt had declared, the Secretary of the Navy was the only person who "could issue direct orders to the different chiefs" in the Department of the Navy, and with this mandate above, Forrestal had given Halyer free reign over medal awarding giving him authority over superior officers when it came to the final decision on valorous awards.[2543] So even if Geiger, Nimitz and their awards' boards were not in agreement with Hayler's recommendation, they could do nothing about it after Hayler recommended Woody be awarded a Medal of Honor exercising the powers Forrestal had given him.

## Biography: Rear Admiral R.W. Hayler

Hayler was born on 7 June 1891 in Sandusky, Ohio and attended the Naval Academy from 1910 to 1914. During WWI, he served on the battleship *USS Oklahoma* at Scapa Flow, Scotland. After the war, he studied ordnance engineering at the Massachusetts Institute of Technology (MIT) and rose through the ranks. From June 1939 to June 1942, he served as Inspector of Ordnance in charge of the Naval Torpedo Station. His Navy Commendation Medal citation from 1945 claimed his duties allowed this department, once war started in 1941, to become a:

> highly successful and efficient wartime production and ranging station for torpedoes…This splendid record of achievement in the manufacture and ranging of torpedoes is a monument to the brilliant administration of…Hayler, who contributed vitally to the success of the entire underwater ordnance program by his successful inauguration of this station's multiple and varied activities.[2544]

However, when one analyzes the U.S. Navy's torpedo problems after the outbreak of war, a debacle that cost the lives of thousands of sailors, then one finds his recognition was without merit. For the first year of the war, U.S. submariners and ships went off to combat with defective torpedoes. As a result, enemy vessels were notified of the submarines' presence and were able to counterattack them. Other times, these defective torpedoes turned around and sank the crafts from which they had been fired like with *USS Tullibee* and *USS Tang*. Due to these defects, defects that the Inspector of Ordnance in charge of the Torpedo

Station, Captain Hayler, was in part responsible for, the submarine service earned "the highest casualty rate of any branch of the military service."[2545] Many torpedoes, developed under Hayler, were dysfunctional.[2546] Several torpedo parts that were in use under Hayler's tenure had been developed before his time, but the defective engineering was not ascertained during his duties in Newport, something that should have been identified and fixed since he was supposed to make sure all the "parts worked together."[2547] Luckily, the navy fixed many of these problems by 1943 and U.S. submarines then decimated the Japanese navy and merchant marine fleets.[2548] Hayler received his award in 1945 for his activities with torpedo development when many knew the torpedoes from 1939-42 were faulty. This seems to be the only black mark on Hayler's otherwise stellar career; however, for these purposes, it exposes that medals were occasionally awarded but not earned.

Rear Admiral Robert Ward Hayler. He was in charge of the Navy Department Board of Awards and Decorations. He ignored Lieutenant General Roy S. Geiger and Admiral Chester Nimitz's requests for more evidence in Woody's MOH case and approved the *Medal* anyway for Woody. St. Louis Personnel Records Center

Six months after war broke out, Hayler left the torpedo problems to others to solve to go to sea to fight the enemy. Captain Hayler took command of the 600-foot cruiser *USS Honolulu* and its 870 crew from 18 June 1942 to 7 March 1944. During this command, he participated in the August 1942 bombarding of Kiska, Alaska.[2549] Later, he took part in the battle of Tassafaronga at Guadalcanal on the night of 30 November 1942 where his "fine seamanship" and "gallant leadership under fire" turned back a Japanese force that had damaged several U.S. ships. He received the Navy Cross for his actions. Later, he supported the landings at New Georgia on 5-6 July 1943 and engaged the enemy at the Battle of Kula Gulf, displaying superior leadership, for which he received a second

Navy Cross. Then on 12-13 July 1943, he combatted forces off Kolombangara Island in the British Solomon Islands, leading a "cruiser line of battle into action with an opposing Japanese force of six enemy vessels." He directed "accurate fire upon the...enemy formation and was in large measure responsible for the... destruction of at least four and probably all of these Japanese ships." In this battle, Japanese destroyers launched four torpedoes at Hayler. He evaded three, but the fourth crumpled his bow. "With complete calmness and fortitude... Hayler directed the control of this damage and brought his ship safely into port." For these actions, Hayler received a third Navy Cross.[2550] Inexplicably, a Silver Star was also awarded to Hayler, but in that citation, only the battle action was mentioned, not the saving of his vessel. The wording for the battle action in both the Silver Star and Navy Cross citations is the same. He was awarded two medals for the same action with Admiral Halsey endorsing his Silver Star.[2551] In February 1947, the Awards and Decorations Board noticed this duplication and another review of his Silver Star was held, but the board kept the *status quo*.[2552] Yet only a few months later, "by direction" of Captain P.W. Steinhagen, Hayler's office wrote the Commanding General of FMFPAC for Aircraft about another case that "since only one decoration or medal may be awarded for the same act or acts, it is necessary to set aside and cancel previous awards."[2553] However, this criterion was not applied to Hayler—was that due to his position? And why did Hayler later not act on this apparent violation of regulations which made him a beneficiary of a double award for one action? One could argue this is possibly a second black mark on Hayler's career using a double standard when he himself was considered—and the sad thing is that he did not need to do so. Three Navy crosses were enough without adding a Silver Star that was, in the end, against regulations. However, Hayler was not alone. Major Gregory "Pappy" Boyington was also awarded two awards for actions cited in both citations for the Navy Cross and Medal of Honor. In making this exception, one overseeing officer of the process claimed: "Yet, of all our air heroes, Major Boyington has been one of the least recognized."[2554] So, according to this staffer, Boyington was "owed" these awards because he deserved the recognition since he had been ignored too long. So Hayler was not an isolated case having received special consideration due to his position—others also got "double" medals for their heroics, although Hayler and Boyington were indeed rare.

However, there was another case that actually got *three* medals for the same actions that were noted in his MOH citation. This happened to none other than the Commandant of the Marine Corps, General Alexander Archer Vandegrift. He received a Medal of Honor, Navy Cross and Distinguished Service Medal for all the actions conducted on Guadalcanal from 7 August to 9 December 1942. A staff member noticed that this was against regulations, but then noted that "Nimitz did not recommend that the Navy Cross be recalled."[2555] And Vandegrift, like MacArthur, got the award for actions that lay outside the parameters of why one got a Medal of Honor; namely, for doing something way beyond the call of duty and putting one's life in horrendous danger. So, maybe one could say getting multiple awards for the same actions was very rare for most members of the service, especially enlisted men, but if someone was at the top of the food chain, then regulations were not always followed especially when political considerations were at stake and a "good-old-boys network" in play. These men in power, who did indeed perform heroic acts and often brilliant leadership, seemed to also display a "childish" need to decorate each other like English royalty do with one another or as one Marine derisively described: "Like Mexican generals."[2556] The only problem with Hayler in these cases when looking at Boyington and Vandegrift was that he knew better and had the power to change things according to the regulations he had sworn to enforce when taking over the duties of the Navy's Awards and Decorations office in 1945. Of course Boyington and Vandegrift could have also self-declared that they got too many awards for the same actions violating the regulation that double awards were not allowed, but they did nothing. Maybe Vandegrift did so because he hungered for his third-star which he indeed received after Guadalcanal in July 1943 after barely being a two-star general for a year! Having such medals definitely did not hurt his review with his promotional board. And Halyer, if he knew about Boyington and Vandegrift, did not complain about them or his double awarding once he had power to actually do something about them (also complaining about just them would have highlighted his own hypocrisy which of course he would not have wanted). This has been a troubling revelation of this research, but it is also all too human—people like to be praised and they like more accolades instead of less for what they have done. And in looking at numerous officer and enlisted men's files, this study has found that double awarding only happened to officers, never to enlisted men. Knowing this fact and how generous people had been with Hayler—and he had to have known how generous people in his department had been with Boyington and Vandegrift too—one then wonders if he would also be generous with Woody once his case came across his desk after Geiger offered him opposition. Time would only tell.

Returning back to Halyer's service record before he took over the Awards and Decoration office, on 7 March 1944, Hayler was promoted to Rear Admiral and given command of Cruiser Division Twelve of four ships taking part in the assaults on Saipan, Tinian, Guam and Palau from June through August 1944.[2557] He also repulsed air attacks and rescued downed pilots during the Mariana Turkey Shoot in June 1944.[2558] Halsey wrote, "Hayler...performed his duty as Commander Cruiser Division TWELVE in an outstanding manner. He was alert, thorough and efficient. He continued the exceptional training and performance of this crack cruiser division."[2559] With his fleet, he helped take over Ulithi Atoll in August and September 1944 and provided fire support for the Angaur invasion where my uncle, 1st Lieutenant Frank Rigg, fought. Both Rear Admirals Kingman and Blandy wrote excellent reports on Hayler.[2560] He participated in Leyte's occupation in the Philippines on 19 October until 29 October. His superior, Vice Admiral Jesse B. Oldendorf, wrote,

> Hayler...understands the Japanese enemy well, and...has always succeeded admirably against him...[H]e was...successful in assisting in preparing the landing beaches for the landing of the SIXTH Army, which landed with minimum casualties. This is an accomplishment worthy of the highest praise.[2561]

A few days later, Hayler took part in the Battle of Surigao Strait in Leyte Gulf on 25 October 1944. Throughout the Leyte campaign, Hayler weathered numerous *Kamikazes*.[2562] On 11 December, he was transferred to Washington D.C. to the Bureau of Naval Personnel and took as "additional duty" the Board of Awards and Decorations on 9 June 1945.[2563] When Woody's file landed on his desk, Hayler had many medals and a distinguished career. As head of this board, Hayler was the man responsible for anointing America's future heroes.

## Hayler and Woody's MOH

When a potential Medal of Honor recipient's file crossed Hayler's desk, he had the power to award or deny him *the Medal*. Hayler was the gatekeeper for most of the Iwo Jima and Okinawan Marine and navy Medals of Honor (27 at Iwo Jima and 14 at Okinawa). Vice Admiral Frank Jack Fletcher wrote of Hayler in his role conducting his duties at the board as "a flag officer of unusual intelligence and judgement who combines diligence with common sense."[2564] Admiral Arthur J. Hepburn stated:

> The four citations for service in combat sufficiently attest to [Hayler's] outstanding qualities of decision, judgement and the ease with which he assumes and discharges responsibilities of the highest order. His work on the General Board, of an entirely differently nature, reflects these same qualities.[2565]

He was the highest authority in the land for navy, Marine and Coast Guard servicemen in the summer of 1945 when it came to awarding medals, and he took his responsibilities seriously.

Yet in the case of Woody's MOH recommendation, the absence of a recommendation from the Marine Corps Commandant is puzzling; the Commandant is required to make a recommendation before action by the Secretary of the Navy and the president. What seemed to be occurring was that some authority in Washington D.C. wanted an answer about Woody's case quickly so the president would be prepared for the MOH ceremony taking place soon. On 5 September 1945, Secretary of the Navy Forrestal put pressure on Commandant Vandegrift, writing:

> I should like to accumulate the cases of living Congressional Medal of Honor winners who have not yet received their medals, including [Gregory "Pappy"] Boyington, and have the President present eight or nine medals to navy and Marine personnel toward the end of this month. Will you please ask your people to work on this basis?[2566]

On 7 September 1945, Forrestal informed Vandegrift that Truman had set aside the time to present MOHs to those who "have been awarded the Medal but have not received it. He has been told specifically that Major Boyington will be in the group. The President requests that as many members of the families of the men be present as possible. He would like to have the Marine Band play and would like an Honor Guard from the Navy."[2567] Feeling this pressure apparently coming directly from Truman, Secretary of the Navy Forrestal had his staff member, retired Lieutenant Commander Richard Wagner, make a direct request to Admiral Ernest King's Aide and Flag Secretary, Captain Neil K. Dietrich, that Woody's case along with three others be expedited because, "if approved, the President desires to have these men at the presentation on 5 October, and a considerable delay will be experienced in preparing the Citations for his signature and in having the medals engraved [if a further postponement is incurred]."[2568] Under this political pressure, Forrestal was overly aggressive writing the likes of the Commandant of the Marine Corps, General Vandegrift, who still had not endorsed Woody's package, and Admiral King, the absolute very top of the United States Navy's chain-of-command, to push through Medal of Honor packages that had yet to gain the full chain-of-command endorsements and/or evidence necessary to complete the review process. Apparently, Forrestal's campaign worked because after he wrote Fleet Admiral King's Aide and Flag Secretary Dietrich on 20 September, King signed the final endorsement on 21 September for Woody completing his case for the *Medal*. Thus Woody's MOH may have been the result of a desire by the White House to have more MOH's awarded, or at the very least to have those who were expected to be awarded, done without delay, causing errors to have been overlooked or worse, purposely ignored. According to documents obtained from the Harry S. Truman Library, the Secretary of the Navy sent by memorandum dated 2 October 1945 the nine names of Medal of Honor candidates to the president recommending approval; Woody was among the nine. The president approved Woody's award on 3 October. As USMC Commandant, General Krulak, observed, "At the end of the day, POTUS [President of The United States] wanted a bunch of medals to award and he got what he wanted."[2569]

ADDRESS REPLY TO
SecNav

AND REFER TO
QB4-ILK
20 September 1945.

NAVY DEPARTMENT
WASHINGTON 25, D. C.

FOR VICTORY
BUY
UNITED
STATES
WAR
BONDS
AND
STAMPS

MEMORANDUM FOR CAPTAIN NEIL K. DIETRICH, USN, ROOM 3047, NAVY DEPARTMENT:

It is requested that the attached four cases for the award of the Medal of Honor be expedited as much as is practicable, in view of the fact that, if approved, the President desires to have these men at the presentation on 5 October, and a considerable delay will be experienced in preparing the Citations for his signature and in having the medals engraved.

RICHARD WAGNER,
Lieutenant Commander, USN (Ret.)

On 20 September 1945, retired Lieutenant Commander Richard Wagner in Secretary of the Navy Forrestal's office, wrote the Aide and Flag Secretary of Admiral Ernest King, Captain Neil K. Dietrich, to approve Woody's case as quickly as possible. Although Woody's Medal of Honor package had not received the endorsements of Lieutenant General Geiger and Fleet Admiral Nimitz, and their boards, and Commandant Vandegrift, Forrestal was feeling the pressure from President Truman to get him more alive Medal of Honor recipients for the big ceremony being planned to take place at the Rose Garden at the White House on 5 October 1945. St. Louis Personnel Records

So how do we interpret Forrestal's and Hayler's actions in disregarding Geiger's and Nimitz's concerns, and, most importantly, Chambers' refusal to support his Marine, and recommending Woody's MOH? We don't know for sure, but one curious interpretation is that the recommendation had reached Forrestal and Hayler and there was urgency to decide on the award before a major ceremony involving the president. This was a big event in a president's life, with reporters, high level brass, radio announcers and dignitaries present. This was an event full of pomp and ceremony starring the president, with a strong supporting cast of MOH recipients. Chambers' son-in-law Colonel Olson (USA) wrote:

> Besides the flag raising, [the Marines] needed heroes, they needed to demonstrate to the American people they could get the job done. I'm sure there was pressure to produce results meriting the medal. The

Corps found its man and I have to believe that Williams actually did merit the award. But the Corps maybe had no problem with the embellishment, and Howard did. He was a good leader and I simply refuse to believe he would deny a [Marine] what was earned.[2570]

Chambers did not want to deny Williams the award, but did not endorse statements that, by his silence, appeared to have caused him discomfort. Historian Roger Cirillo observes: "[Chambers'] silence is thunderous. He obviously had problems with this whole process."[2571] Another historian and former USMC Judge Advocate General (JAG) officer, Colonel Charles A. Jones, said it more dramatically: "Geiger smelled a rat when he asked TWICE for the Chambers' statement. Geiger was no dummy, after all, he made CG FMFPAC as a lieutenant general…[H]ave you ever seen a higher [command] endorsement asking for, and not getting, further useful information for a MOH case?…bizarre."[2572] Historian Colonel Jon T. Hoffman writes about going forward with Woody's MOH without Chambers' report explaining:

> [Commanders] were eager to finalize the senior awards for Iwo…if nothing else to get them done prior to the next big operation (if the invasion of Japan was still pending) or prior to everyone leaving the Corps (if the war was over by then). Then too, Burrell's point about the Corps (and everyone else) wanting to find a justification for Iwo, also may have played a role—in this case, getting the public to focus on the heroism, and less on the casualties and whether the result was worth the cost.[2573]

Hoffman echoed Colonel Olson's assessment of the politics behind the award (as apparently seen with General Vandegrift and Admiral King being pressured by the White House (via Forrestal) to approve more MOHs than were possibly deserved). But in going forward, the men in authority did have to modify Woody's record, and the writers of Woody's official citation (possibly Hayler's board) found the numbers problematic and *deleted* the references to seven destroyed pillboxes and 21 Japanese killed from the final draft because the evidence was *unsubstantiated*. Of the 17 Iwo Jima MOH cases reviewed, this was the most dramatic change in downgrading the evidence for any citation. In the end, *no one* could fully explain what Woody did. In contrast, Woody's friend Vernon

Waters' citation notes his actions clearly; namely, he destroyed two pillboxes with satchel charges and killed three charging soldiers by throwing a grenade at them. This report was concise and detailed without ambiguity. Everyone from Geiger, Nimitz and Vandegrift to Forrestal never doubted anything Waters did. More importantly, these officials all endorsed his recommendation.

Many performed acts like Waters and Woody. PFC Johnny Nosarzewski, in Company K, 3rd Battalion, 21st Regiment, 3d MarDiv, was called a "flame-thrower extraordinaire" by Dick Dashiell. "None in the company was more rugged or gutty or nervy than" Nosarzewski. He first flamed a 75mm cannon emplacement and its crew firing out of a high crevice to their front. He then crept to a mortar pit and took it out: "The Nips had one less death-dealer." Next, he incinerated a pillbox killing some of the inhabitants and in the process ran out of fuel. Throwing off his empty tanks, he took out his pistol and shot two surviving Japanese trying to escape and "felled them all, fatally." His file, like Waters, was clear and detailed about what he did. He received the Navy Cross (Signed by Lieutenant General Roy S. Geiger).[2574]

Luckily for Woody, the record shows the absence of a statement or an unconditional endorsement from Geiger were overlooked; his case was not reviewed again. In addition to political reasons, perhaps Woody's case being pushed through resulted from the fact that some high level officers wanted enlisted men recognized for their heroism, something not being done according to some generals. Lieutenant General "Howlin Mad" Smith wrote all three Iwo divisions on 15 July 1945 requesting more medals be awarded to enlisted men to improve morale since he felt too many officers were getting recognized. Smith had put his finger on a potential sore point for morale and his administration may have also pushed Woody's file along.[2575] The 1953 edition of *Navy and Marine Corps Awards Manual* echoes Smith's view: "A judicious use of decorations provides incentive to greater effort and builds morale; an injudicious use will destroy basic value. Promptness with which decorations are presented is an important morale factor."[2576] Napoleon said it another way: "A soldier will fight long and hard for a bit of colored ribbon."[2577] Smith wanted to improve his community and build the *esprit de corps*, and awarding medals and decorations was one way to do so.

In pursuit of awarding enlisted men and asking officers to quit writing awards for each other, maybe Beck took Smith's request to heart. At a reunion,

**Photo 1:** Some ships that bombarded Iwo Jima were from WWI, like the *USS Texas*, a battleship built at the same time as *RMS Titanic*. She had already seen action in Operation *Torch* (invasion of North Africa in November 1942) and in Europe with Operation *Overlord* (invasion of Normandy in June 1944). Her battled-hardened gunnery crews were experienced at taking out land targets. By 15 February, the *USS Texas* and other warships had pummeled Iwo Jima with almost 22,000 shells, with more to come. 23 August 2019, La Porte, TX. Author's Collection

**Photo 2:** Author Bryan Mark Rigg on the forward deck of *USS Texas* next to four of the ships' 14-inch guns. 23 August 2019, La Porte, TX. Author's Collection.

**Photo 3:** Three 14-inch shells stacked on the deck of the *USS Texas*. These shells weighed 1500 pounds each and the U.S. Navy hit Iwo Jima with thousands of them during the campaign against the island. 23 August 2019, La Porte, TX. Author's Collection.

**Photo 4:** Landing beach on Iwo Jima taken 21 March 2015 from the top of Mt. Suribachi. Author's Collection

**Photo 5:** My son Justin Rigg tries to climb up a steep berm as fast as he can on Iwo Jima during our visit to the island on 21 March 2015. One can see the track he has left in front of himself as the ashy incline continued to carry him backwards. Marines (standing in the background) part of the military contingent for the joint ceremony with the Japanese to happen later that day had some free time on the beach and included my 12-year-old son in experiencing the trials of trying to get off the beach as Marines struggled with back on 19 February 1945. Here, Justin is carrying a 40-pound backpack of gear and it took him a few minutes to get to the top of this 30-foot-plus berm. One of the Marines yelled at Justin as he complained about how hard it was: "Hey kid, at least no one is shooting at you." 21 March 2015. Author's Collection.

**Photo 6:** Lieutenant General Tadamichi Kuribayashi, commander at Iwo. Unfortunately, he also was responsible for horrible war crimes in China, especially at Hong Kong where he was Chief-of-Staff of the 23rd Army. Tens of thousands of murders and rapes happened within only a few months of taking over the city in December 1941 with his knowledge and possibly full approval. USMC Photo

**Photo 7:** Painting of Woody taking out the third, and very large pillbox. According to one eye-witness, he climbed up on this massive pillbox in the middle of the battle, found the ventilation pipe opening on the roof and then inserted his flamethrower nozzle into it and flamed the Japanese from the inside out on 23 February 1945. Fairmont Army Reserve Training Center Museum

**Photo 8:** Ventilation pipe on a massive bunker on the WWI battlefield Hartmannsweilerkopf, France. This probably resembled what Woody would have found on the large, third pillbox he took out on Iwo on 23 February 1945. 26 December 2016. Author's Collection.

**Photo 9:** This might be the big and third pillbox Woody took out on 23 February 1945 (side defensive opening). I relayed the coordinates of where Woody feels the large pillbox was to Lieutenant Colonel Gaddy who was also on Iwo Jima in March 2018 with the Marine Corps' military contingent. He later located the largest pillbox he could find from Kuribayashi's line of defense where I think Woody fought on 23 February 1945, took pictures of it, and then sent them to me. Photo Credit Lieutenant Colonel Jason Gaddy USMC

**Photo 10:** Possible big and third pillbox Woody took out on 23 February 1945 (front view). The pillbox's apertures the Japanese used to fire from back in 1945 are hidden now by shrub growth coming out of the openings seen in this picture along the lower front of the structure. Photo Credit Lieutenant Colonel Jason Gaddy USMC

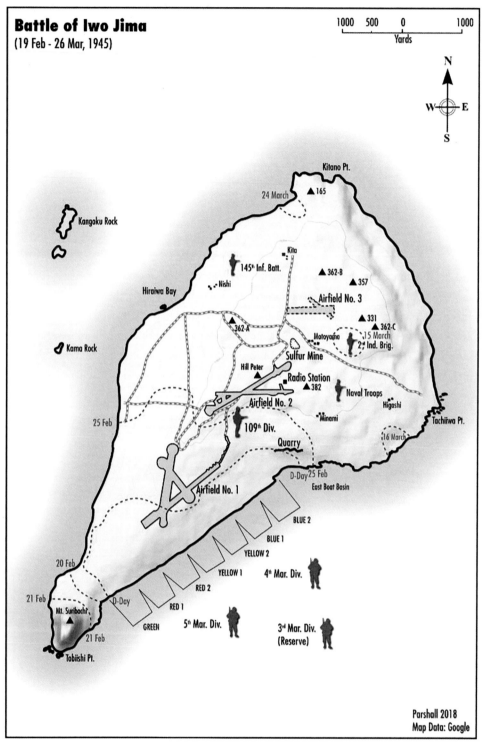

**Battle of Iwo Jima**
(19 Feb - 26 Mar, 1945)

1000  500  0  1000
Yards

N
W E
S

Kitano Pt.
24 March
▲ 165

Kangoku Rock

Kita
145ᵗʰ Inf. Batt.
▲ 362-B
▲ 357
Nishi

Hiraiwa Bay
Airfield No. 3
▲ 331
▲ 362-C
▲ 362-A
Motoyama
15 March
Kama Rock
Sulfur Mine
2ⁿᵈ Ind. Brig.
Hill Peter
Radio Station
▲ 382
Naval Troops
Higashi
Airfield No. 2
Minami
109ᵗʰ Div.
Tachiiwa Pt.
25 Feb
Quarry
16 March
D-Day 25 Feb
East Boat Basin
Airfield No. 1
BLUE 2
BLUE 1
YELLOW 2
YELLOW 1
4ᵗʰ Mar. Div.
20 Feb
RED 2
21 Feb
D-Day
RED 1
Mt. Suribachi
GREEN
5ᵗʰ Mar. Div.
3ʳᵈ Mar. Div.
(Reserve)
21 Feb
Tobiishi Pt.

Parshall 2018
Map Data: Google

**Photo 11:** The attack on Iwo Jima 19 February-26 March 1945. Jonathan Parshall

**HEADQUARTERS**
**3D MARINE DIVISION F.M.F.**
**IN THE FIELD**

*THE COMMANDING GENERAL, 3D MARINE DIVISION,*
*FLEET MARINE FORCE, TAKES PLEASURE IN COMMENDING*

SERGEANT FREDERICK K. DASHIELL,
UNITED STATES MARINE CORPS RESERVE,

*FOR MERITORIOUS SERVICE AS SET FORTH IN THE FOLLOWING*
**CITATION:**

"For meritorious service while serving as a Marine Corps combat correspondent with a Marine division on IWO JIMA, VOLCANO ISLANDS, from 19 February to 16 March, 1945. Landing with the first elements of the division, Sergeant DASHIELL, despite the fact that he was allowed complete freedom of movement to obtain material, and with complete disregard for his own safety, voluntarily joined those groups involved in the most dangerous fighting in order to gather first-hand material for his news stories. He was constantly exposed to enemy machine gun, mortar, and small arms fire. He voluntarily left positions after they became comparatively secure to observe individuals and units involved in heavier fighting. His courage and professional skill resulted in the production of outstanding news stories, which, widely circulated in the United States, acquainted the American people more fully with the combat phases of the fierce battle of IWO JIMA. His courageous and unselfish conduct was in keeping with the highest traditions of the United States Naval Service."

G. B. ERSKINE,
Major General,
U. S. Marine Corps,
Commanding.

**Photo 12:** Major General Graves Erskine wrote a commendation letter for Sergeant Frederick K. "Dick" Dashiell for his meritorious service on Iwo Jima documenting the experiences of many Marines in Woody's unit including Woody himself. Out of the 366 reports Dashiell wrote up, he made only notes doubting Woody's personal claims about his acts and surprisingly, never question any other Marine he interviewed. Author's Collection and Dashiell Family's archives.

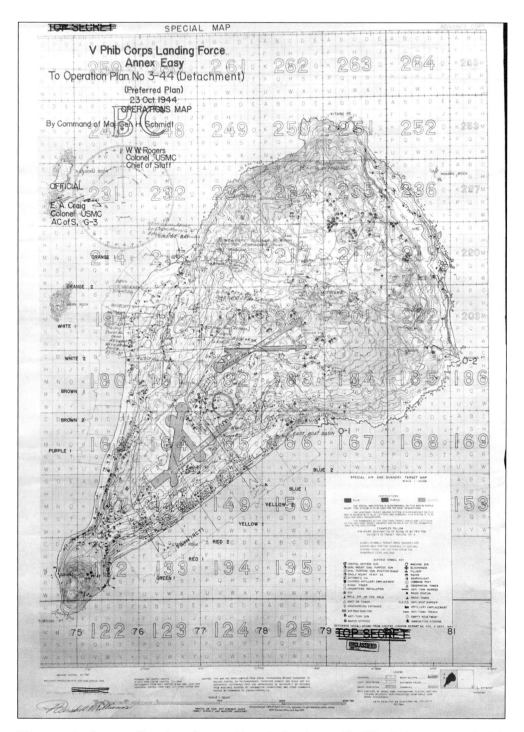

**Photo 13:** Battle map of Iwo Jima from 1944 marked and signed by Woody (signature in the left hand corner of the map) of where he thinks he took out the pillboxes (see the circle in pencil in the middle of the map). All the dots on the map represent bunkers and pillboxes that had been identified by aerial reconnaissance. There would be over 5,000 on the island. Author's Collection.

**UNITED STATES MARINE CORPS**

**HEADQUARTERS**
**FLEET MARINE FORCE, PACIFIC**
**C/O FLEET POST OFFICE, SAN FRANCISCO**

In the name of the President of the United States, the Commanding General, Fleet Marine Force, Pacific, takes pleasure in awarding the SILVER STAR MEDAL to

SECOND LIEUTENANT HOWARD F. CHAMBERS,
UNITED STATES MARINE CORPS RESERVE

for service as set forth in the following
CITATION:

"For conspicuous gallantry and intrepidity in action against the enemy while attached to a Marine infantry battalion on IWO JIMA, VOLCANO ISLANDS, on 23 February, 1945. As a rifle platoon leader, Second Lieutenant CHAMBERS was assigned an attack mission which covered a sector of rough, sandy ridges, among which were concealed several enemy emplacements. In addition to fire being received from these positions, the enemy had a high velocity field piece and a heavy machine gun emplaced along high ground to the rear. Second Lieutenant CHAMBERS, after a careful reconnaissance of the sector, deployed his unit, then only half of its original strength, in a series of maneuvers which resulted in the destruction of the mutually supporting emplacements. By day's end his unit had destroyed thirty-two of these emplacements and had pushed on an additional three hundred yards. The task had been accomplished with a minimum of casualties because of Second Lieutenant CHAMBERS' skillful leadership, which often took him forward of the front lines, supervising tank and demolitions personnel. His heroic actions were in keeping with the highest traditions of the United States Naval Service."

ROY S. GEIGER,
Lieutenant General,
U. S. Marine Corps.

Temporary Citation

**Photo 14:** Silver Star Citation for 2ⁿᵈ Lieutenant Howard Chambers signed by Lieutenant General Roy S. Geiger. He led his depleted platoon against a tough defensive line and took out 32 emplacements and killed probably over a 100 Japanese. His citation read in part: "Chambers' skillful leadership, which often took him forward to the front lines, supervising tank and demolitions personnel…were [heroic and] in keeping with the highest traditions of the United States Naval Service." Author's Collection, Chambers Family Archives and National Archives, College Park

**Photo 15:** Author Bryan Rigg with a M2-2 Flamethrower like the one Woody used on Iwo Jima. Dallas Executive Airport, Commemorative Air Force Show, 30 October 2016. Author's Collection

**Photo 16:** Author with Corie (left), Gene (center) and Pattie Rybak (right), the daughter, son and daughter-in-law of Corporal Joseph Anthony Rybakiewicz. 19 May 2019 Dallas, TX. The reason why their name is now Rybak is that after the war, their father, under pressure from his wife and their mother, shortened Rybakiewicz to Rybak to make is simpler for people to understand and/or for civil servants to write when filling out documents. Author's Collection.

> HEADQUARTERS TWENTY-FIRST MARINES
> 3D MARINE DIVISION
>
> SEMPER FIDELIS
>
> 1ST. LT. R. TISCHLER "B" CO. 1ST BN. 21ST MARINES, WHILE IN COMBAT
> AGAINST JAPANESE FORCES AT IWO JIMA, VOLCANO ISLANDS IS HEREBY CITED FOR
> THE SUBMISSION OF VITAL INTELLIGENCE INFORMATION OF THE ENEMY. THIS INFORMATION ENABLED
> THE REGIMENTAL COMMANDER TO EXPEDITE OPERATIONS, AND ACCOMPLISH OUR MISSION WITH
> THE LEAST POSSIBLE LOSS OF TIME AND LIFE.
>
> 21 FEBRUARY – 16 MARCH, 1945
>
> H.J. WITHERS
> COL., U.S.M.C.
> COMMANDING

**Photo 17:** After the Iwo Jima battle, 21st Regimental Commander Colonel Hartnoll J. Withers personally wrote 1st Lieutenant Richard Tischler up for a commendation for providing his unit with invaluable intelligence during combat. Author's Collection and Tischler Family Archive

**Photo 18:** Justin Rigg at the 5th MarDiv monument at Mt. Suribachi on Iwo. Author Bryan Rigg and Justin read Gittelsohn's famous speech "The Purest Democracy" here on Mt. Suribachi at the 5th MarDiv memorial. After reading it, they put their copy of Gittelsohn's speech on one of the foundational steps of the structure. 21 March 2015. Author's Collection.

**Photo 19:** Author Bryan Rigg with PFC Donald Graves, flamethrower operator with the 5th MarDiv. Graves does not believe half of what Woody has said about himself and finds his story full of problems. This photo was taken on 7 May 2017 at the 72nd VE Day Anniversary event by Daughters of World War II at George W. Bush Presidential Library, Dallas, TX. It was organized by Laura Leppert, a daughter of an Iwo veteran. Author's Collection

**Photo 20:** Author Bryan Rigg and Fred Dashiell, son of Marine Corps Correspondent Dick Dashiell, 12 June 2018 Los Angeles. Fred feels his father had many problems with Woody's testimony that he gave him in 1945. Author's Collection.

**Photo 21:** Colonel Robert Olson and Barbara Olson, son-in-law and daughter of Lieutenant Howard Chambers, with the author Bryan Rigg. 7 June 2018, Virginia Beach, VA. Colonel Olson feels Chambers had many problems with Woody's medal process and did not approve of it. Author's Collection

one of his platoon leaders, Lieutenant Henning, took Beck aside and asked him why he did not put him in for a MOH for his actions on 28 February 1945 that *only* earned him a Silver Star. Henning had led his men against a defense that netted Beck's company a 47mm cannon and six 25mm dual-purpose machineguns and killed a few dozen enemy. Beck countered that he had put Henning in for a MOH, but had received pressure to remove him from the list because there were too many officers getting awards, so he downgraded his medal. Henning was disappointed, but he returned from the reunion with a better understanding of why he did not receive a MOH and felt content that at least his commander had recommended him for

PFC Johnny Nosarzewski. He was another flamethrower in Woody's regiment. He would receive the Navy Cross for his actions and was called a "flamethrower extraordinaire" by none other than Sergeant Dick Dashiell. St. Louis Personnel Records Center.

it.[2578] Smith's encouragement to 3d MarDiv officers about awards was affected by politics. Maybe Smith's focus on enlisted men influenced Woody receiving *the Medal* and dissuaded Beck from pushing harder for Henning. Marines were better at obtaining MOHs for enlisted men than the navy: for WWII actions, 37 of the 57 MOH awards for the navy went to officers whereas 52 of the 82 for the Marine Corps went to enlisted men.[2579]

Soon after the Navy Department of Awards and Decorations recommended that Woody receive a MOH, on 21 September 1945, the favorable "eighth endorsement" came from Commander-in-Chief, U.S. Fleet and Chief-of-Naval Operations Fleet Admiral Ernest J. King to Secretary of the Navy Forrestal as already mentioned. After Hayler sent Woody's MOH recommendation to King, it was all but final. In the MOHs cases reviewed, every recommendation reaching King received a favorable recommendation.[2580] Forrestal then approved these medals and recommended the president approve the MOH recommendations.

The president or his designee is the Medal of Honor approval authority; contrary to popular belief, the only role Congress had in the MOH process was to create the award's eligibility criteria by statute. The proper name is "Medal of Honor," not "Congressional Medal of Honor" which often is used; by adding the "Congressional," people think Congress is involved in its award, and although Congress can apply pressure to award MOHs, it is uninvolved in the awards process of the classic combat Medal of Honor except on rare occasions.[2581] In the cases for Iwo adjudicated in 1945, King's was the last endorsement required for a MOH before it went to Forrestal and on to the president which, once a case got that far, the president's signature was a formality, but it was indeed the final act in "christening" a MOH recipient.

As a result, once King gave his endorsement for Woody's MOH on 21 September 1945, the men in Woody's chain of command were given explicit instructions to get the Marine to Washington quickly. On 22 September 1945, Woody's approval for the MOH was clear for all to see and he was to report to Washington by 3 October.[2582] Now, what did Geiger and Nimitz and their boards think about all this happening without their approvals? Well the record is silent on this point, but these men probably did not file a complaint or raise a fuss about Woody's MOH not having their endorsements because they did not want to upset their chain of command or embarrass the President who now had given his seal of approval. By their silence, they let the issue go.

While this bureaucratic paper-pushing transpired, Woody and his Marines trained for the invasion of Japan. At that time, "if you were to ask me about the Medal of Honor," Woody claims, "I wouldn't have had the foggiest idea what it was. The medal had no meaning to me at this time. I was just happy to be alive and with my friends."[2583] However, when Dashiell interviewed him in May 1945, Woody claimed "he was being considered for [the] Medal."[2584]

A few weeks before Woody's medal was approved, on 15 August, Woody, Dashiell, Naro, Schlager, Tripp, Tischler, Schwartz, Southwell, Henning, Withers and others were called to formation to listen to an announcement. Over the loudspeaker, they were informed that Hirohito and his nation had surrendered. This became known as Victory over Japan Day or VJ Day. The news reported a new weapon had helped this occur, but none of them knew about the atomic bombs. FDR's Manhattan Project had been kept such a closely-guarded secret even Vice President Harry Truman did not know about it until he was sworn in after Roosevelt's death. After Woody's formation was dismissed, many ran back to their tents, grabbed their rifles and started shooting in the air.[2585] There would be no bloody Operation *Downfall*. The war was over. They were alive, had defeated Japan and were going home. "The abrupt surrender of Japan came more or less as a surprise," General of the Army Henry A. "Hap" Arnold, commander of the U.S. Army Air Forces, commented.[2586]

Soon after the final surrender was signed on 2 September aboard the *USS Missouri*, Fleet Admiral Nimitz sent a message to his fleets setting a different tone than what Japan had done during its time of victory. Nimitz, "displaying "his magnanimous character," wrote:

> However, the use of insulting epithets in connection with the Japanese as a race or as individuals does not now become the officers of the United States Navy. Officers in the Pacific Fleet will take steps to require of all personnel under their command a high standard of conduct in this matter. Neither familiarity nor abuse and vituperation should be permitted.[2587]

General of the Army MacArthur had already set the tone for what Nimitz said when while on the battleship *USS Missouri* on 2 September, he declared to all those assembled:

> We are gathered here, representatives of the major warring powers, to conclude a solemn agreement whereby peace may be restored. The issues, involving divergent ideals and ideologies, have been determined on the battlefields of the world and hence are not for our discussion or our debate. Nor is it for us here to meet, representing as we do a majority of the peoples of the earth, in a spirit of distrust, malice or hatred. But rather it is for us, both victors and vanquished, to rise to that higher dignity which alone benefits the sacred purposes we are about to serve, committing all our people unreservedly to faithful compliance with the understanding they are here formally to assume. It is my earnest hope—indeed the hope of all mankind—that from

this solemn occasion a better world shall emerge out of the blood and carnage of the past, a world founded upon faith and understanding, a world dedicated to the dignity of man and the fulfillment of his most cherished wish for freedom, tolerance and justice.[2588]

MacArthur's declaration on behalf of peace and compassion was surprising to the Japanese, who viewed their enemies with fear and expected severe punishment for a nation that had lowered itself to the depths of disgrace by surrendering. Instead of MacArthur's gloating in victory and enjoying the shame he knew the Japanese must have felt or declaring extreme reparations, he threw away those petty desires and held out a large olive branch of friendship and understanding, very different from the actions of most Japanese leaders, especially Sakai and Kuribayashi after they sacked Hong Kong in 1941. And unlike Sakai, Kuribayashi and a large host of their fellow generals like Tojo, Matsui and Yamashita, MacArthur did not unleash American soldiers to rape women and kill innocent citizens. For that matter, even had MacArthur issued an *illegal* order to do so, the U.S. Army would no doubt have countered such a declaration. Regardless, the American culture would have prevented its warriors from carrying out Nanking-like acts against women and children. By showing kindness to the vanquished, MacArthur indeed heaped "coals of fire upon" their heads (Proverbs 25: 21-22).[2589] He established the goal of "freedom, tolerance and justice" for the world that democratic and freedom-loving people should always embrace and support.

Over the next few weeks, Woody and his comrades relaxed in their tents, played cards, listened to music, drank beer and looked forward to returning home. They still had their duties, one of which for Woody was to burn heads ("latrines in army speech") that were full and ensure new ones were blown into the coral ground to serve the tent city. But they would not invade Japan, and that made them happy whether or not they had an unsavory duty.

Around the middle of September, Woody's First Sergeant "Hose Nose" Elder told him to get his khakis ready because he was going to see the 3d MarDiv's commanding general, Major General Erskine, the next day. The first sergeant had no details about the summons, and Woody was confused. He pulled out his khaki uniform which he had worn for inspections, pressed it and made it presentable.

The next morning, he put on his overseas cap and met a jeep at his tent for the drive to the general's headquarters far from where Woody was bivouacked. Before he entered the general's office, a colonel, probably Erskine's Chief-of-Staff, Colonel Robert "Bobby" Hogaboom, briefed him on how to behave, hold his cover (*aka* hat) and address the general. "I was scared to death to meet the general," Woody said. "More scared than when I had met with the Japanese." After telling Woody to stand at ease, Erskine congratulated him, informing him he was to return to the States under sealed orders. "I can't tell you why, but it's something that will make you proud the rest of your life," Erskine said.[2590] Woody then left and the colonel outside handed him a sealed envelope with his orders and told him that if he opened it before delivering them to his next duty station in D.C., it would be a court-martial offense. "He put the fear of God into me. I wouldn't have opened those orders if someone had put a gun to my head." Woody replied, "Yes, Sir," took his orders and left.[2591] Although some bureaucratic paper shuffling was ongoing for approval of Woody's *Medal*, everyone assumed only formalities were needed; the time had come to get him back to the States to receive his MOH. By 19 September, Woody was officially a "full-blooded" sergeant with full title and responsibilities.[2592] Curiously, he was being moved toward Washington to receive his *Medal* despite not yet having received a favorable recommendation by Geiger at FMFPAC, Nimitz at the Pacific Command or the Navy Department Board of Awards and Decorations.

Woody returned to his tent, packed up his things, said goodbye to his buddies and left for the airport. He remained there the entire day, awaiting a seat on one of the planes. When evening arrived, he radioed back to First Sergeant Elder that he had not left and had not eaten anything all day. Elder picked him up and returned him to his tent for the night.

The next morning, 20 September, he once again packed up everything and said goodbye, this time for the last time. The driver who drove him to the airport now had the colonel from Erskine's HQ with him (probably Colonel Hogaboom again). Woody got the proper documents and later that evening, he boarded a C-47 headed to Hawaii via Johnston Atoll.

Woody had difficulty finding a transport plane because POWs released from Japanese camps had priority to fly home for medical treatment. Sadly, one of the POWs died in Japan before his plane left, opening a seat for Woody, the first aircraft he had ever been on. He took his seat in the cabin with the former POWs.[2593]

"These men had the most hollow cheeks you had ever seen. They were skeletons [Fiske Hanley, on our 2015 trip, weighed 96 pounds when he was liberated][2594]…However, they were the happiest people I'd ever been around in my life," Woody said. Many should have naturally been between 150-180 pounds, but most weighed around 80 to 100 pounds. They looked horrible.[2595] The POW next to him said the Japanese hardly fed him anything and he had to perform hard labor in a coal mine. He told Woody something he has never forgotten: "You never know what freedom is until you have lost it." (Sometimes Woody gives this same story when giving talks, but instead of a WWII POW, he says it was a Vietnam *POW who suffered six years under the Viet Cong*).[2596] During the journey, they made one short stop at Johnston Atoll to refuel and then headed to Pearl Harbor. Due to passing the International Date Line, it was still 20 September when he arrived in the U.S. territory of Hawaii.

While there, Woody waited on base for several days for a flight to the mainland. Going from round the clock training and combat to doing nothing was surreal. Due to a mistake, his orders were opened and he was informed he was receiving the Medal of Honor. He claimed he had no clue what this medal was, why he had received it or the import of it. He had his orders to go to Washington by 3 October and that was all that mattered; as a good Marine, Sergeant Woody Williams went where he was ordered. Eventually, he boarded a larger plane to the mainland on 22 September 1945. The POWs were headed to Letterman Hospital in San Francisco and Woody was headed to the White House. Woody landed in San Francisco on 23 September and started to walk to the train station, but the shore police picked him up for not being in the proper uniform. After he explained himself, they took him to the Marine base in San Francisco and outfitted him in a dressier green uniform. He then boarded a train on 23 September for Washington D.C. Somewhere between Chicago and Washington, he had an unusual meeting.

On this route was a man who enjoyed dressing up like a Confederate Colonel and entertaining passengers with stories and jokes. He took a seat beside Woody, asking him about his story. "We talked for a while and I told him how I wish I could see my family before going to Washington. As it was, I was going to arrive on the 29th and that was several days before I had to report. I explained I hadn't been home in two and half years." The "Colonel" excused himself and a few minutes later, returned with the conductor.

The conductor explained they would not be going through West Virginia, but that he could slow down at Connellsville, Pennsylvania and allow him to jump off at the platform. This was close to Morgantown and then from there, Woody could find a bus or hitchhike to Fairmont and get home. "I can slow the train down to about three miles an hour and allow you to jump off. Can you do this?" Woody said, "Absolutely." A few days later, Woody grabbed his duffle bag and around four o'clock in the morning, jumped off at Connellsville.

Woody found a night watchman on duty and asked if there was a restaurant open. He replied no, but that there was a USO in town open for servicemen. Woody went to the office and got a cup of coffee and a sandwich. The lady on duty heard his story and told him if he wanted, she could drive him to Morgantown, West Virginia when she got off work at 0600 that morning. Grateful, Woody waited until morning in the sleepy little town of Connellsville, eager to see his girlfriend and family. That morning, the woman dropped him off in Morgantown and Woody boarded the 7:00am bus headed toward Fairmont on State Route 73.

While aboard the bus, Woody told the driver his story and asked if he could stop near his girlfriend's house along the highway at Meadowdale. The driver agreed, and when he neared her town, he pulled over and let Woody out. Woody walked the familiar roads he had so often walked a few years earlier. His heart started to beat against his chest with such force he not only felt it inside his ribcage, but also could hear it. Horrible thoughts of combat flooded back through his mind, but he finally was coming back to the person who he felt had helped him survive. Thoughts of kissing, holding and seeing her filled his heart. Sometimes his gait turned into a speed walk and he could not wait to see her home on the hill as he left the valley highway behind. Approaching the house, a silly grin spread across his face. He had made it home, safe and sound. He knocked on the door. Ruby opened it. Her eyes opened wide, she covered her mouth with her hands and joyfully yelled "Woody!" She then "jumped into my arms and kissed me." Everyone was shocked and elated to see Woody. It was 28 September 1945. He had returned to the woman who was foremost on his mind during the war, and who was one of the main reasons he had survived. He "was back in the arms of my Angel."[2597] The next day, he borrowed his soon-to-be father-in-law's car and drove to Quiet Dell to see his mother. The task of running the dairy farm without her boys was too much, and she had sold it. She moved nearby where she found work to support herself. Her ability to write was limited so she did not send many letters, but Woody had written her frequently. She had received two telegrams notifying her two sons had suffered battle wounds. Due to the censors, the boys could not write about what had happened, so she was never sure where her sons were, what they were doing or what the state of their health was. This lack of information took a toll on her well-being. Naturally, when she saw Woody safely home, she was excited beyond belief.[2598]

The Marine Corps Commandant's office requested Woody send a list of attendees for his medal ceremony. In a telegram from 30 September 1945, the Williams family notified General Vandegrift's office that Woody's mother, Lurenna, and "fiancée," Ruby, would attend the White House ceremony.[2599] Regardless of what Woody believed then about only calling Ruby his "girlfriend" because "fiancée" was not a term he felt comfortable using, his "womenfolk" had other ideas. In a 2014 interview, he thought he might be going to D.C. for a Purple Heart, but in light of all the Purple Hearts he had witnessed awarded and the activity surrounding his transit to D.C., the Purple Heart was one of the most unlikely medals to be awarded by Truman (sometimes Woody's statements defy logic).[2600]

On 2 October, Woody boarded a train for D.C. with his mother and fiancée, and the next day, reported to the Commandant's office. He was assigned a sergeant who instructed him on etiquette for the medal ceremony. Soon thereafter, a major appeared to escort Woody and 10 other Marines slated to receive the MOH. On the 4th, the major helped the Marines get their uniforms in order, briefing them on the protocol for meeting their Commander-in-Chief, the President. Woody recalls, "There have only been two times I've been literally nearly scared to death. One time when I almost crashed a truck going over a hundred miles per hour downhill, and meeting the president. I was shaking in my shoes."[2601]

The next day, with uniforms pressed, and brass and shoes polished, his group of 11 Marines, two Corpsmen and one naval officer proceeded to the White House. Woody said:

Going to the White House to meet…Truman, I was with other Iwo Jima Marines, PFC Jacklyn H. Lucas, Sergeant William G. Harrell, PFC Douglas T. Jacobson and Captain Joseph J. McCarthy. We were all in the same hotel with our families and getting ready. I had no idea what was going on. On 5 October, we formed up near two Marine Corps platoons and two Navy platoons, and were seated outside of the White House…Photographers and the press were everywhere.[2602]

After a man's MOH citation was read aloud, he approached Truman, who placed the medal around the man's neck. Some were seriously wounded; Harrell, for example, had lost both hands due to grenade blasts. First, one grenade tore off his left hand. With his remaining hand, he killed two attacking Japanese. After killing the last Japanese, he tried to throw a second grenade away with his right hand, but he was unable to get rid of it quickly enough and it exploded. Before he lost his hands, he had killed five Japanese and remained at his post, protecting his comrades. Woody said, "I couldn't believe this man and his bravery—he had no hands, but he was just happy to be alive and off that island."[2603] When he received his medal from Truman, Harrell proudly walked up, wearing two metal hooks where his hands had been. Woody was in the company of extraordinary men that day. They went alphabetically, so Woody was one of the last ones to be honored. Woody's turn finally came up and Vice Admiral Louis E. Denfeld read his citation as follows. It was the first time Woody had heard it:

For conspicuous gallantry and intrepidity at the risk of his life above and beyond the call of duty as demolition sergeant serving with the 21st Marines, 3d Marine Division, in action against enemy Japanese forces on Iwo Jima, Volcano Islands, on 23 February 1945. Quick to volunteer his services when our tanks were maneuvering vainly to open a lane for the infantry through the network of reinforced concrete pillboxes, buried mines, and black volcanic sands, Cpl. Williams daringly went forward alone to attempt the reduction of devastating machine-gun fire from the unyielding positions. Covered only by 4 riflemen, he fought desperately for 4 hours under terrific enemy small-arms fire and repeatedly returned to his own lines to prepare demolition charges and obtain serviced flamethrowers, struggling back, frequently to the rear of hostile emplacements, to wipe out 1 position after another. On 1 occasion, he daringly mounted a pillbox to insert the nozzle of his flamethrower through the air vent, killing the occupants and silencing the gun; on another he grimly charged enemy riflemen who attempted to stop him with bayonets and destroyed them with a burst of flame from his weapon. His unyielding determination and extraordinary heroism in the face of ruthless enemy resistance were directly instrumental in neutralizing one of the most fanatically defended Japanese strong points encountered by his regiment and aided vitally in enabling his company to reach its objective. Cpl. Williams' aggressive fighting spirit and valiant devotion to duty throughout this fiercely contested action sustain and enhance the highest traditions of the U.S. Naval Service.[2604]

Then they "asked me to come up to get the medal put on me by the President. I had to walk on my toes I was so nervous." As Truman put the nation's highest award around Woody's neck, the man once possibly turned down by a recruiter because he was not tall enough, now stood at attention looking every bit the outstanding Marine NCO he was. Truman told Woody, "I would rather have this medal than be President."[2605] Truman had been a farm boy who enlisted in the Missouri National Guard in 1911, fought as an artillery officer during WWI serving as a captain in the 129th Artillery and served as a regimental commander in the Reserves after the war obtaining the rank of colonel.[2606] He could relate to these combat veterans he was now decorating. After he told Woody that he wanted *the Medal* too for himself, Woody simply replied, "Thank you, Sir" and sat back down. The brazen and confident 17-year-old PFC Lucas, upon hearing the same words of congratulations, replied, "I will trade you, Mr. President."[2607]

After he made the Medal of Honor presentations, Truman said:

This is one of the pleasant duties of the President…These are the young men who represent us in our fighting forces. They said we were soft, that we would not fight, that we could not win. We are not a warlike nation. We do not go to war for gain or for territory; we go to war for principles, and we produce young men like these. I think I told every one of them that I would rather have that medal…than to be

President…We fought a good fight. We've won two great victories. We're facing another fight, and we must win the victory in that. That is a fight for a peaceful world, a fight so we won't have to maim the flower of our young men, and bury them. Now let us go forward and win that fight, as we have won these other two victories, and this war will not have been in vain.[2608]

President Harry Truman puts the MOH around Woody Williams' shoulders outside the White House on 5 October 1945. When one looks right over Woody's shoulders at the men behind him sitting down, the first person one can see looking on is President Truman's Chairman of the Joint Chiefs-of-Staff, Fleet Admiral William D. Leahy. National Archives, College Park. Photo Credit USMC.

Later that day, Woody had an audience with another MOH recipient, USMC Commandant and General Alexander A. Vandegrift, who had earned *the Medal* for service on Guadalcanal as discussed earlier. His connection to Iwo was his son, Lieutenant Colonel Alexander Vandegrift Jr., who was injured in both legs while there. He was also tied to the island because he believed its seizure contributed to his promotion in April to full general, as the first four-star general in Marine Corps history.[2609] The Commandant met privately with each Marine who had just received *the Medal*. Woody later learned the general had prepared different remarks for each honoree. Reporting to him, Woody commented, was like "reporting to God." While Woody was in his office, the Commandant, with no hand gestures and with a stern look, remarked, "That *Medal* doesn't belong to you. It belongs to all those

Marines who'll never come home and paid the ultimate price. It's an honor to have it. Don't do anything to tarnish that *Medal*."[2610] Woody agreed saying, "I'm just a caretaker of that *Medal*." When he wears *the Medal*, he thinks of the Marines, possibly Charles Fischer and Warren Bornholz, who died protecting him.[2611]

Out of the 27 Medals of Honor from Iwo Jima, Woody's seems to stand out somewhat because his action was part of a larger action by his platoon leader, Lieutenant Chambers, in destroying almost three dozen emplacements. These actions ultimately broke the elaborate and unyielding defense encountered by Woody's and Chambers' regiment north of the first airfield, allowing the Marines to surge forward. Therefore, Woody's actions that earned *the Medal* seem to have had

The 18th Commandant of the United States Marine Corps from 1944 to 1947, General Alexander Archer Vandegrift. He became the first four-star general in the history of the Marine Corps and was a recipient of the Medal of Honor for his leadership on Guadalcanal in 1942. National Archives, College Park

larger operational significance than many Iwo *Medals* because his actions helped his officers push forward to fulfill their mission of breaking the defense ahead of the 21st Marines' lines. This observation takes nothing away from the heroism displayed by other MOH recipients. A number received posthumous *Medals* for sacrificing their lives by jumping on grenades to save their buddies like Don Ruhl. Seven earned *the Medal* on Iwo this way; seventeen-year-old Jack Lucas was the only MOH recipient to survive such a heroic act.[2612] Others fought through pain, exhaustion and hardship defending their comrades.[2613] Woody exposed himself to great danger to accomplish his mission and helped save many, pushing the battle forward and opening the front enabling his regiment to use its tanks.

A few of the 27 Medal of Honor recipients had equal or greater effect on the battle than Woody. Like Woody, Sergeant Darrell Cole and Corporal Tony

Stein destroyed several pillboxes and killed many Japanese, which allowed pinned down Marines to move off the beach.[2614] Sadly, soon after they performed their heroic acts, they were killed.

In contrast to Cole, Stein, and Woody's MOH actions, one Marine's actions may have had a greater impact than theirs: PFC Douglas Jacobson of 3rd Battalion, 23rd Marines, 4th MarDiv, single-handedly took out a 20mm gun crew, two machine-gun positions, a tank, two pillboxes and a blockhouse. Jacobson killed 75 Japanese around Hill 382 as the Marines pushed through the island's middle. He simply went "berserk." He did for the middle section of the battle what Woody did for the southern section and what Cole and Stein did on the beaches—he got the

PFC Douglas Jacobson who received the MOH for his actions on Iwo Jima. A panel of "top newspapermen and combat veterans" named Jacobson the most "outstanding living" enlisted WWII Marine and one whose deeds most resembled the exploits of the most famous hero of WWI, MOH recipient Sergeant Alvin York. National Archives, College Park

Marines moving by taking out strong defensive positions.[2615] A panel of "top newspapermen and combat veterans" named Jacobson the most "outstanding living" enlisted WWII Marine and one whose deeds most resembled the exploits of the most famous hero of WWI, MOH recipient Sergeant Alvin York.[2616]

MOH awardees' parade in Washington D.C. 5 October 1945. From right to left in the car looking at this picture (Private Wilson D. Watson, Corpsman Robert Bush, PFC Jacklyn Lucas, Private Franklin Sigler and Sergeant Woody Williams). Woody is sitting in the back on the left side of the car. Marine Corps Historical Division.

After receiving the MOH, Woody and the other *Medal* recipients participated in a parade in the streets of D.C. and then returned to their hotel and celebrated. Many enjoyed a few drinks, especially the teenager Lucas (he had turned 17 on Iwo). When asked if Woody stayed with his fiancée that night, he said absolutely not. "My mother would've jerked me into 19 kinks if I'd tried that," Woody laughs shaking his head.[2617]

The next morning, Woody returned home for ten days of leave. Back home, he was treated to a "rousing homecoming" by the residents of Fairmont and Marion Counties. There was a welcome parade with the state guard mustered and the entire complement of the Legionnaires of Fairmont Post 17 "marching in review." West Virginia Senator C. Howard Hardesty addressed those in attendance, and there were several parties held in Woody's honor, from the police ball to the Chamber of Commerce dinner to special meetings at the Lions Club. Back on Guam, his buddies named the firing range for him.[2618] He became a celebrity overnight. His best friend from boot camp, Ellery B. "Bud" Crabbe, said he and others who knew Woody were excited and the news spread through their company that "one of theirs got *the Medal*." Bud continued, "I was so proud of him for receiving the Medal of Honor. I didn't think many who got *the Medal* survived, but Woody [did]."[2619] Woody was a man many wanted to be around. Moreover, he was now someone not only the Marine Corps was honored to call its own, but also someone the nation took great pride in having as a hero.

As his buddies on Guam gave toasts to Woody and regaled each other with his exploits, they settled down and waited to go home. Woody's regimental diary noted there were *no* "plans and orders" from their superior. Their unit had to make *no*

Woody Williams proudly wearing his MOH after the ceremony at the White House 5 October 1945. National Archives, College Park. Photo Credit USMC.

"principal movements" and there were *no* combat operations in their futures. The Marines were done with the war and "military training has been reduced to a minimum." To illustrate how young these boys were, the only thing the regiment was doing differently at that time was to enact an "educational program…which offers courses on pre-elementary, grammar school, high school and college levels." The command was going to get these boys educated and ready to integrate back into civilian life.[2620] Erskine seemed to be the first Marine general officer to take a survey "within the Division" to see who could teach school:

I felt very strongly that any man who joined the Marine Corps—in particular if he served in my command—should go home and be better qualified civilian than the people who'd stayed home. And I was determined to give them an opportunity to do this. So I thought one way to do it was to start schools—all kinds of schools—the first idea being to get these young fellows' minds back on studying…And then to do everything we could to give him credit for any studies that he did while he was out in the Pacific—that is academic credit.[2621]

Erskine did his best to arm his men with tools they needed to be successful in life, in the world of ideas and in business, after being full-time killing machines. In 1969, Erskine recalled he had "over 6,000 men going to school" daily.[2622] In fact, Erskine's memory seemed to be off since his newspaper reported in October 1945 that actually 8,365 had enrolled in 181 optional classes that had been established. The course on how to be an auto mechanic attracted "the largest enrollment" of Woody's 21st Regiment with 118 in the class.[2623] Such a caring attitude and such an intelligent decision were testaments to Erskine's leadership, and his focus helped many continue their education through the GI Bill or by putting themselves through school. Dick Dashiell became an instructor of English. He was surprised by how little schooling many had, writing, "Christ, some of them hadn't gone past the fourth grade; some even didn't know you began a sentence with a capital letter. And here they had been, putting their lives on the line for a country that left them illiterate."[2624] Erskine was doing his best to change that, using talented teachers like Dashiell.

In addition to their educational pursuits, the division organized a jazz band, an orchestra, a theatre and a movie hall, that showcased weekly performances. Practically every company had a baseball team and the men competed and kept records for the "league," noting in October that Woody's C Company had made the first triple play since the liberation of the island the previous year in November. At the end of the season, they even had an All-Star game between the 3d MarDiv and the Navy Seabees (the Marines won). Also, a boxing federation with the various classes of lightweight to heavyweight sprang up with every unit putting up their best ringmen, and numerous fights were held throughout the week. The 3d MarDiv's newspaper bragged that in the heavyweight class, their "colored" boxer, Riggs, was the toughest and held the title. And large swim parties were organized at Tumon Bay, splitting up the days between the various regiments and service battalions—it became a pseudo-water park with organized bus transportation. It was like the division transformed itself into a full-time college with academic courses, performing arts, leisure activities and sport teams. On 6 October, the division's newspaper recognized Woody, writing, "Cpl. Hershel Williams…flew to Washington late last month. 'Tis rumored he will *receive a certain medal there soon* [author's italics]."[2625] (See Gallery 4, Photo 1)

Guam baseball. Every company had a baseball team and a competitive league with a full schedule for all teams while based on Guam. Sergeant Dick Dashiell took this photo circa May 1945. Author's Collection and the Dashiell Family Archive

While Woody's comrades conducted their studies and extracurricular activities on Guam, he and Ruby were married on 17 October at her home in a ceremony ministered by her uncle, Reverend Thomas Grafton "Tom" Meredith, who "plighted their troth and made their vows."[2626] With Woody on leave, but on active duty, there was neither the time nor the money for the couple to take an extended honeymoon. After the wedding, they spent a few nights at the Hotel Hill in Oak Hill, West Virginia, followed by Woody returning to duty in Washington D.C. Back on base, a sergeant major tried to persuade Woody to remain on active duty. The fact a sergeant major took an interest in persuading Woody to make a career in the Marines is one more indication of Woody's high status. Woody considered the offer, and responded he would serve only if he could have duty that stationed him at the same base for three years. He wanted to be with Ruby and start a family. The sergeant major could not guarantee such duty, so Woody was released from active duty on 6 November 1945 at the Naval Training Center Bainbridge, Maryland. He had spent a total of one year, nine months and 20 days deployed with the Marine Corps.

After leaving the Corps, he took a few weeks just to relax and enjoy his new bride. It was the first time in a long time he could sleep in and not be bound by military routine. When Thanksgiving rolled around, Woody and his new family got ready for the tradition in the West Virginian hills that many families observed—slaughtering a hog. For years, his family had actually killed two to feed everyone and they ate everything from the skin, to the organs to the tail which, in Woody's opinion, "was the best part." He considered it a delicacy. Since Woody was a combat-hardened Marine and an expert with a rifle, his new family gave him the .22 caliber rifle to shoot the hog. So, Woody took the rifle and family members pushed the hog out of the pen, down a shoot from the barn for Woody to kill. Woody took aim at the hog between the eyes, fired and waited for the pig to fall over dead. Instead, the hog squealed, turned tail, and went back into the barn. The family members made fun of the "tough Marine" with "bad aim," but Woody swore to them that he had hit it right between the eyes thinking to himself that he had hit targets from a hundred yards out in the service—it was impossible to miss a hog from a few feet away. The family members once again corralled the hog down the shoot and the pig came back towards Woody and the tough Marine once again, raised the rifle, took special care this time to aim right for the place between the eyes and fired. "EEEEEEE," squealed the pig and it ran back into the barn. The family really started ribbing the Marine with the Medal of Honor who couldn't even shoot a defenseless pig. And besides that, at the rate they were going, they were going to go hungry too. The next time they brought the pig out, Woody took his rifle and basically fired the shot right behind the ear making sure he missed all the bones of the head. With that shot, the pig rolled over and died. He noticed two bullets holes in the head, but was puzzled why they obviously had done nothing to the animal. Later, as they gutted and slaughtered the hog, they found that it had a double skull mutation. As soon as those bullets penetrated the first skull, the second one had stopped the bullets. Woody was thankful for the excuse. Later that day, they all sat down to a feast and Woody started to blend into his new family with Ruby by his side. He was happier than he had ever been in his life.[2627] Woody saw a future of some promise with a wonderful woman on his shoulder and having ended his WWII career with honor. It was still early to say whether or not the MOH would help him in civilian life. For many, it would serve as a huge cross they had to bear, under which many would fall while bearing its weight. Hopefully, something like that would not happen to Woody.

The U.S. Marine Corps War Memorial at Arlington National Cemetery, VA. This sculpture is a copy of the Rosenthal photo of the famous flag raising done on Iwo Jima from 23 February 1945. In this photo, author Bryan Rigg is with son Ian in the front of the memorial on a very sunny day. 24 July 2017. Author's Collection.

# Ch. 28: Observations About Military Awards and Woody's MOH

"Pride goeth before destruction, a haughty spirit before a fall."
—Proverbs 16:18[2628]

"When the legend becomes fact, print the legend."
—*The Man Who Shot Liberty Valance*

"The Bible says, 'The truth will set you free,' [Woody] said.
'That's because when it's the truth, it just is. You don't
have to try to remember what you said.'"[2629]

ONE-FOURTH OF MARINES WHO DIED during WWII did so on Iwo. More medals for valor were awarded during this engagement than in any other U.S. battle: In addition to 27 Medals of Honor (almost 30% of all those from WWII), there were 186 Navy Crosses (17% of all those from WWII), 6 Distinguished Service Medals and 892 Silver Stars (20% of all those from WWII).[2630] Only 82 Marines received MOHs during the Second World War.[2631] Woody claims that for every MOH given, there should be "dramatically more" awarded to others who did not have the witnesses to document their deeds. MOH recipient Wilson Watson echoed Woody, remarking, "[T]here were probably a hundred Marines there that day [the day he earned his MOH] that deserve this medal more than me, but the right people did not see them."[2632]

In this conflict, 22 Medals of Honor were awarded to Marines and 5 to navy personnel. Four were Navy Medical Corpsmen embedded with combat units (accounting for 15% of *the Medals* although only numbering 3% of the force). To Woody and other Marines, Corpsmen, called "Docs," who risked their lives to treat the wounded were considered "really Marines."[2633]

Most actions deserving *the Medal* happen instantaneously. A person might jump on a grenade to shield his comrades or throw one away to save a wounded man in a crater. Actions that earned Woody a medal occurred over four hours. There was plenty of time for numerous people to see different aspects of his deeds. Acts that go on for hours have more opportunity to be witnessed whereas spur of the moment acts of heroism often go unnoticed.

In general, the 22 Medals of Honor given to Marines at Iwo reflect the brave deeds numerous men performed to take the island. Commander-in-Chief of the Pacific, Fleet Admiral Chester W. Nimitz, said of Marines at Iwo Jima that "uncommon valor was a common virtue."[2634] After Iwo, Secretary of the Navy James Forrestal, the civilian cabinet official in charge of the navy and Marines, said, "I can never again see a U.S. Marine without feeling… reverence."[2635]

Beginning this study, I had a high appreciation for MOH recipients and for what *the Medal* represents. Although I still do, I have been surprised by the arbitrariness and subjectivity of the process of receiving *the Medal*. Using Woody's accomplishments as a benchmark, many should have received *the Medal*, but they did not because no one put them in for it and/or witnessed what they did. Using Woody's achievements as criteria, hundreds should have received *the Medal* from Iwo. One must remember that the Japanese had 5,000 emplacements—pillboxes, bunkers and reinforced cave entrances. In that environment and with 22,000 Japanese on the island, killing many enemy combatants and destroying fortifications was common. Many have received a lesser medal than perhaps they deserved. Should Chambers have received *the Medal* rather than the Silver Star for taking out 32 enemy positions and killing untold number of enemy with his depleted unit? Using this logic, should Woody have received the Silver Star instead of *the Medal* since he was one piece of Chambers' chessboard to break through the defense? Maybe Chambers did not receive the MOH because he did not do something extraordinary personally. Skillfully leading his unit in heroic acts was not considered adequate, although the navy "often violated this requirement during WWII."[2636] Jon T. Hoffman offers an insightful explanation differentiating Chambers and Woody:

> I can probably easily explain why Woody got a MOH for something that was part of the action that led to Chambers getting a Silver Star. Woody personally exposed himself multiple times to personally take out enemy fortifications. Chambers was likely very exposed with his men, but wasn't personally carrying a flamethrower or tossing a satchel charge into an embrasure. He gets credit for leading his unit in a difficult and successful assault, but doesn't necessarily rate the same award as men who served under him and performed very particular feats of heroism. I know, the opposite of how Vandegrift and Edson got high medals, but their rank was part of the calculus. As I said, the whole system is very subjective, so it's not even possible to rationalize why one person gets a medal and another doesn't.[2637]

Should both Chambers and Woody have received *the Medal*? For instance, in a two-and-a-half-hour firefight, Army Private Cleto "Chico" Rodriquez, a Mexican kid from Texas, with Army Private John Reese, a Cherokee Indian from Oklahoma, killed 82 Japanese, captured a 20mm cannon and disrupted a skilled enemy defense at Paco Railroad Station in the Philippines on 7 February 1945. Reese would ultimately die in the action, taking a bullet to the forehead. For their heroic combat and bravery, they *both* received *the Medal*.[2638] However, military leadership was not as generous with the Chambers/Woody team. The reality is that the Medal of Honor is not, and cannot be, based on "points," such as how many fortifications one destroys or how many of the enemy he kills. Certain actions, however, seem to ensure a MOH will be awarded (if properly reported): For example, covering a grenade to protect one's comrades almost always resulted in a MOH when recommended. And it did not necessarily have to be covering a grenade—any action that resulted in someone's death while he was protecting or saving his comrades, again when properly documented, earned one a medal, and often the MOH. By way of illustration, Gunnery Sergeant Fred Stockham in WWI, during a gas attack while fighting the Germans in 1918, took off his gas mask and put it on a wounded Marine nearby and "heedless of his own safety," continued to help his men nearby "before collapsing." He died a few days later in a field hospital and received the Medal of Honor posthumously.[2639]

Other Allied and Axis militaries awarded similar medals for such heroics. When the British fought the 23rd Army where Kuribayashi served at Hong Kong in 1941, Canadian Company Sergeant Major John Robert Osborn, felled a few enemy with his bayonet and then threw several enemy grenades back at the Japanese away from his fellow soldiers. When he noticed one that could not be grabbed and thrown, he jumped on it to save comrades nearby. He received the Victoria Cross, the British MOH equivalent (182 of these medals were awarded to 181 recipients in WWII and 85 were done posthumously).[2640]

*Wehrmacht* Lieutenant Colonel Walter Hollaender led his *Sturmregiment* (Storm or Assault Regiment) 195 south of Orel near Dmitrowski on 5-6 July 1943 six kilometers deep into a heavily defended Russian position during the largest tank battle of WWII at Kursk known as Operation *Zitadelle* (Citadel) involving 8,000 tanks and 2.5 million men (25 times bigger than Iwo Jima). Leading his men in the front lines, Hollaender and his regiment broke the enemy's resistance in his sector (one of the few successes Germans had in this northern zone of the conflict where Hollaender was deployed). His regiment destroyed 21 Soviet tanks, numerous enemy positions and killed hundreds of *Ivans*. General Walter Model wrote that through Hollaender's actions, he secured the 9th Army's left flank and cut off the lines of supplies and replacements for the Soviets in his area of operations. The battle was basically a draw and

*Wehrmacht* Colonel Walter H. Hollaender (Awarded the *Ritterkreuz* of the Iron Cross, the German equivalent of the U.S. Medal of Honor). Hollaender led his *Sturmregiment* (Storm or Assault Regiment) 195 in the front lines at the Battle of Kursk on 5-6 July 1943 destroying dozen of enemy tanks and killing hundreds of Russians. Author's Collection.

the Germans would later have to withdraw from many of the areas they had seized. Nonetheless, Hollaender received the *Ritterkreuz* (Knight's Cross of the Iron Cross (*Ritterkreuz des Eisernen Kreuzes*), the German equivalent of the MOH, on 20 July 1943 for his bravery and leadership (7,000 were awarded during the Third Reich).[2641] If Beck or Chambers, or even Woody's Regimental commander Withers, had been in the *Wehrmacht*, under the criteria that helped Hollaender receive his award, they all could have probably received the *Ritterkreuz* for their leadership and bravery on Iwo. So as one has read, the Medal of Honor is not unique to our military and comes from a long tradition in many militaries that want to award their brave warriors whether they be a *jōbun,* Victoria Cross, *Ritterkreuz* or Medal of Honor.

Returning to the MOH, by the time a candidate's recommendation reached high levels of command such as FMFPAC or the Pacific Fleet, scrutiny of the recommendation was likely to be minimal if the candidate died in performing the deeds for which he was nominated or if he died later in battle after performing those deeds. The process, however, may take time as in the case of Darrell Cole, who earned a posthumous MOH for actions on Iwo; his honorific recommendation was initially downgraded to a Navy Cross and then upgraded back to an MOH.[2642]

Heroic acts combined with death in battle seem to help in upgrading an award too. One example is Tony Stein, who was initially recommended for a Navy Cross. But his killing of 20 Japanese, destroying several pillboxes, helping the wounded to safety and then later dying leading men in battle caught the attention of a general officer who felt he warranted a higher medal.[2643]

Stein, who earned a posthumous MOH, was unquestionably heroic. The day the Marines landed on Iwo was hell on earth. Marines could not get off the beach nor see the enemy. Locating positions Stein wanted to attack, he stood up in plain view receiving fire. He accomplished assaults by lugging a Browning .30 caliber machinegun that "had a tremendous rate of fire" of 500 rounds per minute. He was a demolition specialist and had "improvised this weapon himself."[2644] With help from others, Stein had salvaged a Browning machinegun from a crashed aircraft and modified it by "fitting the gun with a M1 Garand buttstock, a BAR bipod, a BAR rear sight, and a fabricated trigger."[2645] When he fired it, men around could hear it above the roar of battle. They called it the *Stinger.*[2646] Carrying a *Stinger,* like the backpack flamethrower, required courage. "He was fearless," a comrade of Stein's said. "He didn't know the meaning of the word fear."[2647] The day Stein earned the MOH, he ran back to the beach to reload his *Stinger* eight times with 100 rounds each time either carrying back a wounded man or helping one to an Aid Station. To improve agility and to enhance visibility, he shed his boots and helmet. He helped get Marines moving off the beach and was a one-man wrecking crew. A week later, he was killed leading a patrol.[2648]

During the review of the command's recommendation that Stein receive the Navy Cross posthumously, Major General James L. Underhill, a member of FMFPAC's Awards Board, believed a better description of Stein's actions might upgrade the award. On 17 May 1945, he wrote Stein's battalion commander, Colonel J. B. Butterfield, for more info.[2649] Eventually, a more detailed report was submitted and sent up the chain

Corporal Tony Stein. He received a MOH posthumously for his actions on Iwo. One comrade said of him, "He didn't know the meaning of the word fear." He was the son of Jewish Austrian immigrants. The author finds it interesting that had his family not immigrated to the United States when they did, they all probably would have perished in one of Hitler's extermination camps. St. Louis Personnel Records Center.

of command and Lieutenant General Smith, Fleet Admiral Nimitz, General Vandegrift, Rear Admiral Hayler, and Fleet Admiral King recommended approval of a MOH for Stein.[2650] Due to upgrading of the recommendation, Stein's MOH process had the most endorsements the study has uncovered—15 compared to Woody's eight. While the medal process was under Smith's command at FMFPAC, Smith was generous. In fact, Nimitz gave Smith in August 1944 free rein in giving all awards except the Medal of Honor on his own authority "in the name of the President" and Smith exercised this privilege often. When it came to Medals of Honor, Smith tried to get his Marines as many as he could.[2651] After

Geiger assumed command of FMFPAC in July 1945, he implemented a Board of Awards that employed a more stringent process of reviewing and documenting MOH recommendations compared to Smith's *laissez faire modus operandi*.

The opposite of what happened to Stein could also occur: A MOH recommendation could get downgraded. A few days after Woody's MOH exploits on Iwo, Captain Francis Fagan, also of the 3d MarDiv, fought above the second Airfield from 25-6 February. Fagan and his men "annihilated" 150 Japanese and took out 30 fortifications: "He led his men with absolute disregard for his own safety in a hand to hand assault of the [enemy] position."[2652] He personally "participated in destroying three pillboxes as he was going from platoon to platoon coordinating the attack and inspiring his men."[2653] Although shot through the shoulder, he refused to leave the battlefield. Colonel Howard N. Kenyon, head of Erskine's Awards Board, witnessed Fagan's leadership. When Fagan was on a stretcher and in pain from a gut-shot that eventually killed him, he still reported the situation to Kenyon, to whom he had delegated command and related his objective was nearly completed. "His performance as a wounded man was definitely beyond the call of duty and contributed directly and immediately to the success of the U.S. Forces."[2654] Thereafter, he was evacuated to the *USS Rinsdale* for medical treatment. Two days later, he succumbed to his wounds and was buried at sea.[2655] According to Air Force Officer and trauma surgeon, Dr. Brian Williams, Fagan probably suffered from multiple organ dysfunction syndrome and died while in a coma. In other words, after suffering his gut shots, defecation most likely leaked out into his body cavity creating harmful bacteria colonies (deadly microorganisms) that overwhelmed his system and shut down his organs due to sepsis. Within 24 to 36 hours, he probably fell into septic shock and then soon thereafter, after his immune system was overwhelmed, his body shut down and he died. Thousands of men perished this way during WWII and according to Dr. Williams, "that conflict ushered in a new approach to how surgeons treated patients with abdominal gunshot wounds—learning how to treat the numerous cases during the Second World War has probably saved hundreds of thousands of lives in the post-war era."[2656] So, on many levels, one can say, Fagan did not die in vain.

Fagan was a strong-willed leader who inspired his men. One of his Marines, T. Fred Harvey, said of him: "He was the strongest and toughest Marine I ever knew. We feared and loved him."[2657] He also demonstrated leadership on Guam on 28 July 1944 when he prevented his men from retreating against a 150-man Japanese attack:

> Fagan…ran [150] yards along the front line through heavy enemy fire to the right flank, where, under intense fire and with screaming enemy only fifteen yards away, he personally and with utter disregard for his…safety, stopped the withdrawal by rallying his men. His action alone prevented a break-through and repulsed the counterattack.[2658]

And as one also just read, he also showed such bravery on Iwo. As a result, Kenyon prepared a MOH recommendation for Fagan and submitted it to Major General Erskine for approval on 19 May 1945.[2659] That same day, Erskine recommended approval and sent his endorsement to Forrestal.[2660] V Amphibious Corps commander, Major General Harry Schmidt, however, recommended Fagan for the Navy Cross. Schmidt gave no reason for the downgrade.[2661] Geiger agreed with Schmidt, recommending Fagan for the Navy Cross.[2662] Later, Nimitz's office agreed with Geiger.[2663] Vandegrift followed suit a few weeks later like Nimitz's office.[2664] Rear Admiral Hayler at the Navy Department of Awards and Decorations gathered the command endorsements concluding:

Captain Francis Fagan. Fagan served in the 3d MarDiv and received two Navy Crosses, one on Guam and the other on Iwo. He would die of his wounds suffered on Iwo and many believe his actions there should have earned him a MOH. T. Fred Harvey, who was on the reunion trip to Iwo in 2015, served under Fagan and said he was the toughest Marine he ever knew. St. Louis Personnel Records Center

> After a studied evaluation of the recommended posthumous award of the Medal of Honor to…Fagan…and a compari-

son with accounts of similar acts of heroism submitted to the Board for consideration, the Board concurs with the endorsements from [Schmidt, Geiger, and Nimitz] and recommends that the proposed [MOH] award…be disapproved and that, in lieu thereof…[Fagan] be posthumously awarded [a second Navy Cross].[2665]

Fagan accomplished one of the most punishing days on the enemy a line officer had on Iwo, but it was not enough for a MOH. Let's not forget that Woody's platoon leader Chambers took more enemy emplacements than Fagan in one day and received a Silver Star. Because many were involved with the mission, the chain of command did not feel awarding *the Medal* to *one* man of many was appropriate. In short, the leader of a collective effort did not necessarily warrant a MOH. Nonetheless, one would think Fagan deserved a MOH especially when one knows Colonel Mike Edson at Guadalcanal received *the Medal* for leading his battalion in a fashion similar to Fagan.[2666] After the war, several men did not like how Fagan was treated and tried to get the Navy Cross upgraded. Historian James Hallas supports this conclusion: "[Fagan] unquestionably should have received a posthumous MOH for Iwo."[2667] But after a second review in 1947, the Navy Department's Board of Awards and Decorations concluded: "The Board has reviewed [Fagan's] case…and recommends…no change."[2668]

One MOH recommendation was upgraded for a Marine who survived Iwo showing that death, although often helping in upgrading an award, was not always a requirement when receiving *the Medal*. Moreover, in this case, numerous acts over days were documented tipping the scale in this Marine's favor. Wilson Watson's case was similar to Tony Stein's. Initially, Watson was recommended for a Navy Cross, but the recommendation was upgraded to a MOH by 3d MarDiv's Awards Board whose senior member was Colonel Kenyon. Erskine agreed with the Board and sent the recommendation to Geiger, and as he did with Woody's case, Geiger wanted more evidence and the two witness statements to be more specific. Initially, the witness statements were written in an "epic" narrative format and this created doubt, according to Geiger, as to whether the men had witnessed Watson's acts firsthand.[2669] Later, they were re-written using first-person language which clarified the witnesses had observed Watson's heroics. The reports discuss Watson's actions over three days. On 25 February, he covered his squad by exposing himself to enemy fire and attacked some pillboxes. On 26 February, he assaulted another pillbox suppressing its fire with his rifle while charging it. He tossed in a grenade and killed many. He then ran to the back of the bunker, waited for retreating Japanese to leave and killed two who emerged. On 27 February, he had one of the biggest killing days of any Marine on Iwo. Rushing a hilltop, he caught dozens of enemy soldiers massing on the other side. Carrying a bandolier full of loaded magazines, he unleashed his fire killing over 60 men.

Private Wilson D. Watson. He served in the 3d MarDiv on Iwo and received a MOH for his heroics on the island from 26-27 February 1945. St. Louis Personnel Records Center.

On 2 March, he led his squad against fortified positions when "he was hit by shrapnel and small arms fire and evacuated."[2670] Although Geiger had obtained new witness statements, they had not arrived in time for Hayler, and again, he declared Geiger's "additional testimony" unnecessary.[2671] Yet, unlike in Woody's case, Geiger's office eventually obtained the evidence requested for Watson.

The path to a Medal of Honor is occasionally not straight either up or lateral. Sometimes it is up, down and then up again. Returning to Sergeant Darrell S. Cole, he was initially recommended for a MOH, but Rear Admiral Hayler recommended it be downgraded to a Navy Cross. In this case, both Geiger and Nimitz (unlike with Woody) firmly endorsed Cole's MOH recommendation, but in the end, Hayler and his Board exerted great influence.[2672] Cole had destroyed three pillboxes and after eliminating the third (after two tries), he died as a result of wounds sustained during the action. A year later in 1946, after his MOH had been downgraded, his file was reviewed again and upgraded back to a MOH which was awarded posthumously to his widow Margaret on 17 April 1947.[2673] There were similar back and forth for other MOH recipients and one of these ordeals happened to Jack Lucas.

Lucas covered two grenades and survived (luckily one of them did not go off). He was recommended for a MOH, but V Amphibious Corps commander Major General Harry Schmidt recommended Lucas receive a Navy Cross. Hayler and the Board ignored Schmidt's recommendation although Geiger and Nimitz wanted more evidence for Lucas before they recommended approval of his MOH recommendation. And as with Woody, Hayler and his board ignored Geiger's and Nimitz's requests for more evidence and recommended that Lucas' MOH recommendation be approved.[2674] One would think jumping on TWO grenades was such a selfless act that with sufficient witnesses, which Lucas had, there would be no problem. Perhaps Schmidt, Geiger and Nimitz were influ-

enced by Lucas' advance disciplinary record (he was listed as a deserter) and by the immaturity and unpredictability he had shown as a teenager (he was 17). Nonetheless, his bravery was recognized with a MOH.

Other men covering grenades did not receive MOHs. PFC George Barlow of G Company, 2nd Battalion, 24th Marines, 4th MarDiv jumped on a grenade for several comrades and died in the process and did not receive any medals. PFC Harry L. Jackson of K Company, 3rd Battalion, 21st Marines jumped on a grenade to save two comrades, Harvey Ellis and Troy Gardner, and survived but with horrible injuries and received the Silver Star (signed by Lieutenant General "Howlin Mad" Smith). According to his citation, he saved three comrades.

Sergeant Darrell S. Cole. He earned the MOH on Iwo and would die of wounds he sustained in the action on 19 February 1945. The U.S. Navy guided-missile destroyer *USS Cole* is named after him (made famous when Islamic terrorists attacked it in 2000 at Yemen's Aden harbor). St. Louis Personnel Records Center

Jackson survived because he first placed his helmet over the grenade and then his body over the helmet. The explosion shattered his helmet and lifted Jackson into the air. He landed back to the earth in extreme pain. Jackson stood up and started hopping on his good leg and holding his right arm. "His right knee-cap had been blown off" and his arm was shredded. Buddies rendered aid to Jackson while others nearby located the soldier who had thrown the grenade; he was in a crater 25 feet in front of the lines, and they riddled him with bullets. Jackson lost his right leg at a field hospital and later, his right arm was amputated in a rear hospital "while the medical officers fought to save his right eye." The "guillotine amputations" took away his right arm "just below the elbow" and his "left leg through [the] mid-thigh." There was no saving his left eye and doctors would hollow out his orb and pack it full of gauze unable to save his eyeball—they "enucleated" it. His vision in his right eye was also reduced.[2675] Private Robert C. Filip, also of the 21st Regiment, was in a crater aiding a wounded friend when a grenade landed amongst them. He could have easily leapt to safety, but in doing so his buddy would have died. Filip moved as quickly as he could to remove the grenade near his comrade and tried to throw it away. He was too slow and the grenade exploded taking off parts of his hand (all five fingers would later be amputated) and sending fragments into his face and legs. His buddy remained safe from the discharge and survived. Filip received the Navy Cross (signed by Lieutenant General "Howlin Mad" Smith).[2676]

Private Robert C. Filip. He received the Navy Cross on Iwo Jima by protecting a wounded comrade and losing five fingers while trying to throw away a live grenade. In doing these heroic acts, he saved his comrade's life. St. Louis Personnel Records Center.

It may be difficult to understand the reasoning that Marines covering grenades were treated differently. Both Barlow and Ruhl jumped on a grenade and died saving fellow Marines, but only Ruhl received the MOH;

perhaps Ruhl's previous actions, also mentioned in his citation, of attacking and killing escaping Japanese from a bunker and rescuing a fallen comrade from the battlefield led to his MOH. Geiger recommended approval for Ruhl without reservation. Ruhl's witness statements all match up with his captain's initial endorsement.[2677] Barlow, however, received nothing despite the fact that his selfless act mirrored those of three other MOH recipients: William R. Caddy, James Dennis La Belle and William Walsh all covered grenades and saved the lives of their buddies with no further action mentioned in their citations, all of which were for posthumous Medals of Honor.[2678] Barlow's comrade John A. Synder gives telling commentary while pleading with the archives in 1994 to help Barlow get a medal: "He was killed [jumping on the grenade] but received no award for heroism

PFC George Barlow served in G Company, 2nd Battalion, 24th Marines, 4th MarDiv. He jumped on a grenade and saved some of his buddies but did not receive any medals. His comrade John A. Synder gave telling commentary while pleading with the archives in 1994 to help Barlow get a medal: "He was killed [jumping on the grenade] but received no award for heroism since I was the only battle survivor to tell the story" (no officer put him up for an award and there was a lack of witnesses for Barlow). He has never received any recognition for his bravery. St. Louis Personnel Records Center.

Gunnery Sergeant William Gary Walsh. He received the MOH posthumously for jumping on a grenade to save his fellow Marines on Iwo Jima on 27 February 1945. St. Louis Personnel Records Center.

since I was the only battle survivor to tell the story" (no officer put him up for an award and there was a lack of witnesses for Barlow).[2679] Jack Lucas jumped on one grenade and then pulled another under his body to protect his comrades (as mentioned before, luckily only one went off). Although horribly injured, he kept both his arms and legs and survived. He saved three Marines and received the MOH. Jackson jumped on one grenade and lost an arm, a leg and an eye saving probably three Marines and he received the Silver Star. Filip tossed a live grenade away from a wounded comrade and lost his hand when he failed to throw it away soon enough. He saved his buddy's life. He received the Navy Cross. All four dealt with live grenades; they saved Marines' lives; they put their lives in danger when they could have sought cover themselves; they suffered horrible wounds or died; but all four received different awards or no awards at all. It is conceivable that the problem lay with the authorities who were often "over hesitant to reward bravery that doesn't result in death," leading to the outcomes in the cases of Jackson and Filip, but not the cases with Barlow and Lucas.[2680] During WWII, "ambiguous regulations produced different outcomes in various commands."[2681]

PFC James D. LaBelle. He would earn the MOH on Iwo for jumping on a grenade to save his comrades' lives on 8 March 1945. St. Louis Personnel Records.

PFC Donald J. Ruhl. He earned a MOH on Iwo by jumping on a grenade to save the lives of his fellow Marines on 21 February 1945. St. Louis Personnel Records Center.

As a result, although jumping on a grenade increased one's chances of receiving the MOH, doing so was not a guarantee of a MOH. The U.S. military does not have criteria for earning its medals requiring that one has to commit specific acts to receive a certain medal.

The military does have at least one number giving status: shooting down five planes makes a pilot an *Ace*. But that achievement does not ensure a medal. Legendary Chesty Puller received five Navy Crosses and many believe he never received the MOH because of politics. Puller did rub many the wrong way and some felt he "butchered" his troops in poorly executed frontal assaults and was dimwitted.[2682] In fact, if one reads about his exploits, it is clear that some actions were equally if not more heroic than those of a lot of MOH recipients. Puller performed similar deeds to Edson who did receive a MOH and Puller's officers recommended him for a Medal of Honor. When some of Puller's other superiors, specifically Lieutenant General Vandegrift, learned of this recommendation, they were "emphatically negative" and prevented him from receiving the MOH.[2683] After the war, in 1952, USMC Colonel and MOH recipient Pappy Boyington solicited President Truman to get Puller a Medal of Honor citing the reason he had not got one was due to politics and not due to a lack of merit. Boyington even claimed his MOH would not have been awarded to him had people not assumed the Japanese had killed him: "I was never awarded a single decoration until after I was presumed dead and lost in action."[2684] He also took the opportunity to "document" that he should have received more medals than he did, but was denied them due to personalities working against him writing: "During this time, according to the standards, I should have received thirty-two Air Medals, fifteen Distinguished Flying Crosses, three Navy Crosses and two Purple Hearts and a Silver Star from the Army Commanding General at Guadalcanal."[2685] Now Boyington maybe warranted a few of these medals, but how in the world he thought he deserved more Navy Crosses and a Silver Star was purely subjective on his part and without evidence. Nonetheless, Truman pushed Boyington's request for Puller with the Marine Corps. However, Puller still never got a Medal of Honor with the Commandant answering Truman: "The records of this Headquarters show that Brigadier General Puller has never been recommended for the award of the Medal of Honor for his services in the Marine Corps."[2686] With that statement, the Corps hoped to put to rest a review of Puller's file for awarding him a MOH.

But the Commandant had further explaining to do. Truman had apparently also enquired with the Corps as to why Boyington did not merit the list of awards he claimed he should have received. The Commandant then took two pages to explain many errors in Boyington's letter with pointing out one of the most glaring ones: "It will be noted that the recommendation for the Medal of Honor was submitted over two months prior to Colonel Boyington's capture by the Japanese."[2687] So the highest medal for valor was being "processed" for Boyington months before he was "presumed dead." While both Puller and Boyington probably should have received more awards, not all acts of a military man will ever be fully documented. Although this "medal" process seems unfair and arbitrary, it does not mean one can dismiss the practice of giving medals. Puller, Boyington, Lucas, Ruhl, Jackson, Filip and Woody were all brave and performed heroic acts; and the medals, in general, reflect that behavior regardless whether their actions merited a lesser or higher award and/or more medals.

## Analyzing Woody's MOH

Some events reported about Woody may have been exaggerated or duplicated. Notably, reports by Darol "Lefty" Lee about Woody's exploits that are showcased in the National Museum of World War II and in the National Museum of the Pacific War are false. Lefty obviously fabricated this story about what he saw Woody do since he was on a ship off-shore of Iwo suffering from shell-shock/combat fatigue on 23 February and not with Woody. At one time, even Woody believed Lefty's tales about his heroism and allowed Lefty to speak publicly about Woody's MOH acts from 23 February everywhere they went, making Lefty's lies Woody's truth.[2688] Woody recommended Lefty be used for this book and had Lefty's misinformation not been discovered, it would have been presented here. Perhaps most illuminating is the absence of a witness statement from Lefty attesting to what Woody did on 23 February (although Lefty fabricated a document and passed it around).[2689] Why could Woody not have ascertained Lefty was not there by listening to his canards? When Lefty's falsehoods were brought to light, people expressed surprise asking, "How could that be? Woody believed in him."[2690] Could Chambers, Schwartz or Beck have been given misinformation from others like Lefty at the time statements were put together? These sources make it difficult to trust the different statements and to ferret out what truly happened.

Although some of what was reported about Woody did occur, an accurate sequence of events was never given. If Woody is unable to remember most of what happened, then necessarily that lack of memory would include the details. John Bradley let people believe he was in the famous flag-raising picture on Mt. Suribachi, including powerful figures in Hollywood and Washington D.C. With this fame, many found his refusal to give interviews strange, but he had falsely represented that he had been in the photograph by remaining silent rather than setting the record straight. Corpsman Bradley earned the Navy Cross for saving men's lives, but after people started telling stories about him and his fame grew, he hid behind "no comment;" and Woody might be behaving similarly.[2691] Woody did admit some things reported about him might not be true. He acknowledges that memory sometimes does not intersect with truth often saying about what happened to him, "They say I did so and so" or "I was told." Woody has said someone claimed he told Beck, "I'll try" when ordered to take out the pillboxes or that he killed 17 Japanese at the big pillbox. But when asked who told him, he cannot name informants.[2692] As he said in 1956, "I can't remember many exact details. Everything is kind of hazy."[2693]

Beck might have assumed the seven destroyed pillboxes Chambers identified were destroyed by Woody when they were not. Whatever Beck's convictions were, they were enough to get the ball rolling to award Woody the MOH. MOH standards are high among the services, but the Corps misidentified who the Iwo flag raisers were *three* times, one of the most famous battlefield pictures in history, and it could have made a mistake about Woody. Although it had witnesses, pictures and film documenting this flag raising, it still made slipups.[2694] Lieutenant General Geiger was worried about such mistakes with Woody's MOH, but Hayler and his board did not agree and recommended to Forrestal, after receiving pressure from his office to make a favorable decision, that Woody's recommendation be approved.

What type of mistakes could Geiger have suspected? Maybe Woody only used seven flamethrowers on two or three pillboxes and another flamethrower operator from a neighboring unit reduced the remaining five. Beck was getting another demolition sergeant to attack the pillboxes when Woody, overhearing this conversation, volunteered his services. What happened with that other flamethrower? Beck would have deployed all the flamethrowers he could to destroy the 32 emplacements his first platoon under Chambers destroyed that day and all 50 facing his company.[2695] What about his company's second and third platoons? What did they do? Could Woody have been bringing some of his flamethrowers to another operator? Veteran Don Graves stated: "Many of those pictures in the archives showing flamethrowers moving to the front are carrying those machines to the front to be used by others, not by themselves. Think about it, if you're carrying them hundreds of yards to the front and then using them at the front, you'd be

PFC Marquys Kenneth Cookson. He was a fellow demolition and flamethrower Marine in Woody's company on Iwo Jima. He served the entire time on the island and received a Silver Star for his bravery. St. Louis Personnel Records Center.

exhausted."[2696] Was the other demolition man possibly PFC Marquys Kenneth Cookson, who like Woody, was attached to C Company and also volunteered his services? In destroying a large pillbox on 9 March, Cookson likewise "crawled on his hands and knees and wriggled on his stomach to the top of the pillbox, then swung a satchel charge filled with 20 pounds of TNT into the aperture." Seconds later, a deafening explosion disintegrated the pillbox killing an "estimated 35 enemy soldiers." Cookson then obtained another satchel charge and belly crawled toward a cave full of enemy. Comrades in Charlie Company fired at the entrance suppressing enemy fire. He then did a twist on the ground as he performed a side discus throw of the satchel charge into the cave opening and then "scooted for safety." These explosives collapsed the cave with a loud detonation. Cookson had sealed the cave and, as a result, received the Silver Star for his actions this day signed by Lieutenant General "Howlin Mad" Smith.[2697] How many he killed with this second attack is not known, but it added to his prior

35 KIAs. Furthermore, since he did this on 9 March, he was likely active on 23 February as well.

In other words, in order for Chambers' men to have taken out the 32 pillboxes, there must have been numerous people with skills like those Woody possessed who were busy on the same front. After interviewing several of the key American leaders from the Pacific War, historians Philip A. Crowl and Jeter A. Isely from Princeton University wrote about a Marine demolition squad which Woody was a part of, explaining:

> [It] contained three teams of four men each, plus the squad leader, but each of these teams had a separate function. The 'pin-up' team, with the general mission of bringing a large amount of fire to bear on a particular target such as a pillbox or a cave, was heavily armed with the bazooka, two automatic rifles, and the Garand. While the Japanese were pinned up by the bazooka team, assisted by rifle assault squads, one or both of the other demolition teams would move in for the kill. One was equipped with two sections of Bangalore torpedo and at least four heavy charges of explosives, while the other carried two Ronson flamethrowers which in turn were covered by two protective riflemen.[2698]

So as one can read here, these teams which Woody was a part of worked closely together to bring numerous different weapon systems to the battlefield when taking out entrenched enemy. Apparently, *no one person* ever took out a pillbox without significant help. In addition, when pillboxes were attacked in force, *two flamethrowers* were deployed. We know Beck actually had access to several flamethrowers on 23 February, all of which contradicts Woody's claim that he was the last flamethrower in his company at that moment since all the others had been killed or wounded.[2699]

There were many chess pieces on 23 February and Woody was just one and there is confusion about what he did.[2700] The problem could be memory, youth, misidentification, trauma, embellishment or a combination of many of these factors that explain why Woody and his witnesses offered conflicting accounts which, in the end, troubled Geiger and other high ranking officers.

When Alexander Schlager read *The Man's Magazine* 1966 article written by Bill Francois, he told his son Bill, "Woody didn't do everything he says he did in that article. The platoon did a lot of what he's now trying to claim for himself [echoing Crowl and Isely in the above quote about demolition squads]. He has a big head and is full of himself."[2701] Schlager's younger brother Lawrence explained, "My brother was more nuanced with me than with my nephew [Bill], but he basically said after reading [Francois'] article: 'You know, many…did not get the medals they deserved and others got medals they shouldn't have gotten.'"[2702] Since the article is primarily about Woody, this is revealing commentary. When Schlager's son Bill was asked for more details, he said before his father died, his father wanted to set the record straight. Schlager claimed his commanding officer, Lieutenant Chambers, was an amazing leader, and that Woody took credit for taking out pillboxes his platoon destroyed with demolition charges, grenades, machineguns and bazookas. This bothered Schlager for years but he did not go public with this information because he did not want to tarnish *the Medal*.[2703]

Two flamethrower operators move into action to take out a Japanese position on Iwo Jima. National Archives, College Park.

Fellow flamethrower Joseph Anthony Rybakiewicz was also disturbed by Woody's receipt of the MOH. When his son Eugene asked him about Woody's award, he dryly responded, "I knew Hershel [Woody]," but did not offer any details or words of praise. His son said it was an awkward conversation; it was obvious his father was ambivalent about Woody and his *Medal*. "Clearly my Dad didn't want to explore the story with me. Basically he said Woody was in his unit and that was all. He was never one to speak ill of someone, so his silence spoke volumes about what his true feelings might be."[2704] During our trip to Iwo Jima, Saipan, Tinian and Guam in 2018, a Marine Corps Master Sergeant and former Drill Instructor, Keith A. Renstrom, after hearing Woody speak and being around him for a few days echoed Rybakiewicz and Schlager simply saying: "There is something not right with his story and he is making things up."[2705] (See Gallery 4, Photo 3)

In numerous interviews throughout the years, Woody has told different versions of his story. For instance, although numerous accounts mention that Woody carried and used demolition charges, Woody has no memory of using them on 23 February and has tried to edit them out of this book. Even the final MOH citation says he "prepared" them, but he obviously disagrees with it.[2706]

Today he claims there was never a bazooka in his unit when attacking the pillboxes,[2707] but in 1956 and 1966 articles, he claims there was one and gave elaborate, and different details of its use. In 1956, he said:

> Someone scored a hit through one of the portholes with a bazooka shell and the smoke curled out of the vent pipe which they had camouflaged up the hill. I had my men throw all the fire they could at the box and I managed to get up the hill to the pipe...(There were 21 Japs in the pillbox; all burned to death...)[2708]

But in 1966, he reported the bazooka as written earlier in the book:

> I motioned for the bazookaman to fire a round at the center pillbox. He did and I could see it hit just below and to the left of the slit. Some of the white smoke from the explosion was sucked into the concrete opening and I noticed a wisp of it curling out the center top of the pillbox. This told me that there was an opening of some kind on top, so I decided that our only chance was to work around and attack the center one from the rear.[2709]

In Woody's interview in 1956, he reported someone operated the bazooka without his direction, but in 1966, he gave the order to the man to fire (and Woody had actually requested this bazooka man support him before the mission ever took place while speaking with Beck in the shell crater). In 1956, Woody reported the bazooka round went through the pillbox's aperture, but in 1966, he reported it hit right below the slit and to the left. In the 1956 article, Woody marched right up on top of the pillbox (which he has claimed in interviews for this book), but in his report from 1966, he mounted the pillbox from the rear. And in notes and letters to me from 2016-17, Woody repeatedly claimed there was never a bazooka during his engagement with the pillboxes. Which report and which details should be believed? It was indeed part of Marine Corps training and doctrine to always have a bazooka man as part of demolition assault teams when taking out "pillboxes and other fortified positions,"[2710] so why would Woody want to deny this fact later in life after affirming it for the first twenty years after the war? Was it due to lapse of memory and old age or a tendency to funnel all action to his experience? Was it also due to the fact that in Woody's earlier testimony, he had mounted a pillbox that had taken a direct hit from a bazooka in its main chamber, and was thus possibly less dangerous to destroy, and now later, in 1966, he wanted it to appear more dangerous, so now the bazooka round hit the outside of the bunker leaving most of the Japanese inside alive and still able to fight?

In Dashiell's 1945 interview, Woody claimed he killed 21 Japanese and took out seven pillboxes, but in the articles, speeches and interviews seen for this study, Woody has no memory of how many he killed (except in the 1956 *Charleston Gazette* and 1966 *Man's Magazine* articles) or how many pillboxes he took out although he often mentions the seven parenthetically and references other people having told him he killed 17 in the big pillbox.[2711]

Further analysis is needed about the events on 23 February. In Beck's report, Chambers "pointed out" seven pillboxes. Beck ascertained second-hand that 21 dead "Japs" were in or around those pillboxes. Beck never claimed who saw Woody take out the pillboxes or documented who counted the dead. Since several Marines attacked those pillboxes, many probably helped kill those 21 combatants. Beck's claim of Woody killing 21 Japanese matches Woody's later assertion of the same to Dashiell.[2712] To our knowledge, Beck never had access to

Dashiell's story. Woody never had access to Beck's report that cited Chambers' testimony until recently. Beck used the number 21, but did not claim to have counted that number of dead "located in or just outside the pillboxes."[2713] Woody gave this figure to Dashiell in 1945 and to Lawless at the *Charleston Gazette* in 1956 although after Hayler and the Navy Department's Awards and Decorations Board endorsed Woody's MOH recommendation, they or someone in the chain of command for this decoration noticed the evidence did not prove this figure and he or they removed it from the citation.[2714] Until recently, Woody did not know someone in authority with knowledge of his case had removed the "fact" of 21 killed from his record.[2715] How did Woody count them during the Iwo chaos? When asked about this today, Woody alleges he would not have counted those he killed.[2716] After 1956, Woody has no memory of the 21, but from 1945 to 1956, he said he killed 21. Moreover, the only documents claiming firsthand knowledge of seven pillboxes was Woody's statement to Dashiell that he "accounted for seven pillboxes" and his reflections reported to 21st Regimental journalist Hoolihan, both done in 1945.[2717] Chambers' depleted platoon of between 18-20 men, down from 42, had just taken out 32 emplacements, but did he actually sum up who took out which pillboxes after the battle?[2718] Did his men keep track? Chambers shot two Japanese dead with his pistol during this action. What if a pillbox went up in smoke right when he killed them? If so, he could not have witnessed who took out the pillbox at that moment. Reading Chambers' Silver Star citation (signed by Geiger), one knows he was busy taking these emplacements and killing dozens with his men that day.[2719] The "counting" must have happened after they reconsolidated their lines and tallied the pillboxes left in their wake of destruction, but how could Chambers have known who took out which one? During this counting, only Woody counted seven for himself. *None* of his witnesses counted seven for him.

## Reviewing Several Case Studies of Valorous Decorations

In contrast to Woody, MOH recipient Sergeant Ross Franklin Gray's witness statements contained firsthand observations with thorough documentation that Gray, in the 4th MarDiv, took out six pillboxes with satchel charges, and destroyed a machinegun and a small field piece. Later, 25 dead Japanese were counted in the first two pillboxes. Some of the places he blew up only had one entrance, so the dead were not counted, but one could guess he killed over 50 Japanese from reading the reports. None of the endorsements from Major General Clifton Cates, Major General Harry Schmidt, Lieutenant General "Howlin Mad" Smith, Rear Admiral Blandy, Fleet Admiral Nimitz, and General Vandegrift to Rear Admiral Hayler had any problems with Gray's case, a case that had ample evidence with lengthy statements to support the initial recommendation documenting Gray's acts—they were consistent and clear.[2720] Something similar was stated for Woody at the beginning of his process, but was never proven like it was with Gray. Gray's smooth process of receiving a MOH illuminates several problems with the evidence supporting Woody's *Medal*. After reading multiple MOH citations though, it is clear the process is flawed. Still, at least we have an "awards process" to recognize bravery and often the medals are deserving. It is simply imperfect.

Sometimes, it takes years to receive an award. Navy Corpsman Francis J. Pierce was one of the lucky ones, since he only had to wait three years to receive his medal. Of all the Iwo MOHs this study explored, Pierce was one of the bravest men on not only Iwo, but also during WWII. His commander, navy Lieutenant James D. Carter, wrote, "I have never seen before nor do I ever hope to see again such a display of courage and devotion to duty."[2721] Pierce defied death to rescue countless lives, kill Japanese and point out enemy positions to attacking forces.

Reading the 12 detailed witness statements collected for his MOH recommendation, one stands in awe with what this man did: "He deliberately exposed himself in open terrain to enemy fire in order to distract the enemies' attention from the casualties and litter bearers. They describe his actions as 'suicidal.'"[2722] In the middle of the battlefield, Pierce suddenly "appeared in the distance down the road carrying a wounded man on his back. Because of his burden he was unable to run. Enemy bullets passed on all sides of him. Finally he fell to his knees after having traversed about 200 yards, and we went to his aid." After Pierce turned over this wounded man to Carter and the stretcher bearers, he "told me that there was one patient left and that he had to get him out before the snipers killed him. I told him it would be suicide to attempt to rescue the man and ordered Pierce to return to the aid station. All this time we were being fired upon….Without warning, Pierce suddenly jumped up and dashed toward the remaining casualty… about 15 minutes later he came struggling down the road carrying the last wounded man on his back" with bullets turning up earth around him.[2723]

Later, Pierce "volunteered to lead [a] patrol to the snipers as he knew their exact location" and Marines eliminated the threat. "On numerous occasions he volunteered to rescue men under intense enemy fire when company commanders ventured the suggestion that rescue was not to be attempted until things quieted down...He was completely fearless." [2724] Unlike Beck, Carter observed his man first-hand and wrote up a three-page first-person narrative. Men serving with Pierce wrote:

> As we were carrying three…wounded out Pierce stood up to draw the Jap fire his way, and at the same time he was firing back to cover up. Pierce spotted and shot one Jap [who] was not over 20 to 30 yards from us. It wasn't long before we ran out of ammunition, and then [Pierce] started giving first aid to the other two wounded, while the Japs were still firing upon him. Having treated the men properly, Pierce proceeded to carry the two men out of fire and into safety… It is my belief…that men as brave and courageous as Pierce are rare..."[2725]

This statement, coming from Private Paul J. Neuman who served the entire 36 days on Iwo, spoke volumes for Pierce. Marine Captain Walter Ridlon described Pierce as an "invaluable aid to my command" who always knew where different units were located. Although seriously wounded, Pierce refused to remove himself. "He volunteered to lead more men with demolitions back to the area and neutralize the position [they had just attacked]. Because of the seriousness of his wounds this request had to be refused."[2726] PFC Francis S. Brown wrote:

> [Pierce] showed utmost courage, and bravery [throughout] the campaign...I have seen [him] go into places pinned down by machine-gun fire and mortar and administer first aid to casualties then carry… many of them to safety on his back when stretcher bearers were [unavailable]. He was an inspiration…to all of us on the front lines because we knew that we had a Corpsman on the job [who] would never let us down....I have seen this Corpsman be on the go from sunrise to sunset tending and evacuating wounded.[2727]

Another Private, Stanley H. Nuttall, wrote:

> I was with Pierce for 22 days on Iwo…and I have seen things done by him that I would never do…Twice I have witnessed Pierce carry a casualty in his arms where the terrain was too rough for more than one man to carry anyone. The first time I witnessed the incident a mortar shell exploded about 20 feet from him with the concussion almost knocked him down but he still continued without fear.[2728]

Unlike Woody's witnesses, Pierce had countless testimonies giving detailed corroborative first-hand evidence for his deeds. Pierce never bragged or was the source of information for his acts.[2729]

Pierce was seemingly everywhere tending to wounded Marines and under all circumstances. Initially, his MOH was downgraded to a Navy Cross (Geiger again wanted more evidence—probably to link all his actions over several days).[2730] Pierce was also given a Silver Star, breaking up his actions into two acts, but at least keeping the acts separate. Eventually, the Awards and Decorations Board reviewed Pierce's case, put his awards together and upgraded them into a single decoration: a MOH. On 26 June 1948, three years after Woody received his MOH, Pierce stood before Truman and received his *Medal*.[2731]

At the time Pierce's Navy Cross case was reconsidered and upgraded, another Navy Cross recipient was also being reconsidered: Captain Walter J. Ridlon from the 4th MarDiv (Pierce's commander). On 21 February on Iwo, when his platoon leaders and NCO's were wounded trying to take a position, Ridlon moved forward and attacked two bunkers alone. While doing so, he ordered a flamethrower to flame one bunker while he held the defenders of both bunkers at bay with grenades and rifle fire. While the flamethrower sprayed fire into the bunker, he was put out of action, but the enemy position was destroyed. Ridlon continued the attack on the next bunker and was "painfully wounded." Even so, he continued fighting and neutralized the second bunker. He then commanded his company to join him as he pushed forward and overtook the enemy area his men had just attacked.[2732] In review for his upgrade, although the Secretary of the Navy Forrestal believed Ridlon's case for a "Medal of Honor appears to be the proper award," the Pacific Fleet Admiral Louis E. Denfeld and his board decided to keep it a Navy Cross.[2733] So unlike Pierce, Ridlon did not receive the upgrade and his Navy Cross remained in force.

WWII cases are continually being reviewed. During Bill Clinton's presidency (1993-01), many Japanese-Americans who fought in Europe in WWII were awarded MOHs; *the Medals* had been denied them during or after the war due to bigotry. Many of these MOHs were presented due to political reasons according to historian James Corum and not because their military actions warranted them.[2734]

Perhaps one of the most infamous politically motivated Medals of Honor was awarded to General Douglas MacArthur in 1942 after Roosevelt ordered him to leave the Philippines.[2735] General Marshall, the Army Chief-of-Staff, admitted to the Secretary of War, "there is no specific act of...MacArthur's to justify the award of the [MOH] under a literal interpretation of the statutes;" nonetheless, MacArthur still received *the Medal* for "services that he has rendered." Marshall felt he could help MacArthur's public image that had taken some criticism for losing the Philippines in 1942, by recognizing MacArthur's brave leadership fighting superior forces at Bataan and Corregidor.[2736] Moreover, in giving this award to MacArthur, Marshall showed he was "more concerned about the prospect for Japanese propaganda to claim MacArthur abandoned his men."[2737] When Marshall tried to get a MOH for MacArthur's successor in the Philippines, Major General Jonathan Mayhew Wainwright IV, MacArthur, hypocritically one might add, blocked and prevented Wainwright from getting *the Medal*. MacArthur was angry at Wainwright for surrendering the Philippines although he fought and led his troops bravely until the end. MacArthur argued that Wainwright's actions did "not warrant this great distinction and it would be an injustice to others who had done far more." After the war, Wainwright received the MOH from President Truman after several other officers pushed his case forward.[2738] All the fighting and political maneuvering people did to get their MOHs, especially when looking at MacArthur's, made General Dwight D. Eisenhower, Supreme Commander for the European Theatre of War, turn away in disgust. After the successful landings in North Africa, Eisenhower was put in for a Medal of Honor. Eisenhower declined *the Medal* claiming he did so "because he knew of a man who had received one for sitting in a hole in the ground [i.e. MacArthur]."[2739]

Another medal was given for actions never done or not meriting the award when Admiral Nimitz pinned the Navy Cross on submarine commander Lieutenant Commander John A. Scott for attacking the carriers *Taiyo* and *Chuyo*. Although Nimitz and Scott knew his attacks "had failed" due to faulty torpedoes, they went along with the ceremony. They knew the assaults failed from the secret service info-gathering network called ULTRA, but the officers kept up pretenses to preserve the code breakers' secrecy and keep morale high for the sailors that their mission had been "successful." Scott's citation has misinformation, but "his tactics were worthy of commendation, even if the torpedoes were not."[2740]

Other times, a higher award was not given possibly due to a commander's jealousy for those under their command having a higher medal than their superior officer. Returning to Gunnery Sergeant Keith A. Renstrom's actions on Tinian on 24-25 July 1944, commanders in the field initially put him in for a Navy Cross. When looking at what he did, one could conclude his recommendation could have maybe been upgraded for even a Medal of Honor had it indeed started as a Navy Cross endorsement. In short, leading from the front Renstrom coordinated actions of his platoon and others nearby that destroyed six Japanese tanks and killed at least 100 IJA soldiers. His platoon did not suffer one casualty (2nd Platoon, F Company, 2nd Battalion, 25th Marines, 4th MarDiv). In his file, the Navy Cross recommendation apparently got downgraded to a Silver Star and then the Silver Star was disapproved on 2 April 1945. Eventually, Renstrom received some recognition and got a Bronze Star with V for his actions on Tinian *over five years later* on 2 September 1949. Apparently some officer in his chain of command noticed the problem and eventually got him some consideration. In the meantime, he had already received a Bronze Star with V for actions on Iwo from 4th MarDiv commander Major General Cates. Later, Renstrom learned his battalion commander, Colonel Lewis C. Hudson, had blocked his medal for his Tinian actions because, according to Renstrom and others, Hudson had no valorous awards himself at the time (somewhat similar to what MacArthur did to Wainwright mentioned above). At a 4th MarDiv reunion in Memphis in 1975, Renstrom confronted Hudson about his behavior. Hudson admitted he had blocked Renstrom's Navy Cross and then confessed he should not have done so and was too hard on many of his Marines. Renstrom, perplexed, said, "Colonel, the more medals that your battalion take makes you look a lot better as a leader." Hudson dropped his head and again admitted he had been too strict when awarding medals. Ironically, Hudson eventually received a Navy Cross for actions on Iwo which mirrored Renstrom's accomplishments on Tinian: "[I]n the face of intense hostile fire, Colonel Hudson continuously exposed himself in the forward areas...to encourage and direct his subordinates in the attack. By his personal example of fearlessness, he inspired his men to move forward in the attack..." Hudson was willing to receive an award for reasons he was unwilling to give one.[2741] And as often been the case, when enlisted men did the same thing as their officers, they usually, when reported properly, got a higher award. If this observation from this study is true, then Renstrom should have been considered for the Navy Cross and possibly even the MOH.

And sometimes, MOHs were not awarded because the chain of command felt the person already had too many medals. USMC First Sergeant Dan Daly, famously known for one time motivating his Marines to attack a German line by yelling, "Come on, you sons-of-bitches, do you want to live forever?," was up for a third MOH for actions against the Germans in June 1918. His command thought awarding Daly a third MOH was excessive, so it downgraded his MOH recommendation to a Navy Cross. Although he extinguished an ammunition dump fire, visited his front units under "violent bombardment," neutralized a machinegun emplacement using grenades and a pistol and rescued wounded "under fire," his MOH recommendation was downgraded not because his actions were unworthy of a MOH, but because an army officer in his command felt it excessive to award a man *three* Medals of Honor.[2742] So although Daly's bravery warranted consideration for another MOH, people above him could not tolerate so much recognition for just one man (and that U.S. Army officers were over this Leatherneck also probably played a factor in him not receiving *the Medal* again).

In contrast to Daly's stellar record, if character or psychological stability played a role in receiving awards, some most likely would not or should not have received the MOH. By way of illustration, Jack Lucas might have been lucky he jumped on grenades and earned the MOH because he was in trouble. His career to that point was embarrassing, but how much could one expect from a teenager? First, he lied about his age and entered the Corps in 1942 at the age of 14 by falsifying documents from his birthdate to his mother's affidavit. Then he tried to quit and leave the service in 1944.[2743] Even his officer wrote: "his mental development is so immature as to make him of little or no value to the Marine Corps for another year or two."[2744] Even with this damning critique, Vandegrift relayed down his chain of command that Lucas should finish out his enlistment and not be allowed to skirt his duty. The Corps decided to retain Lucas although several wanted him discharged.[2745] But Lucas' troubles continued. By September 1944, he went AWOL, stole beer and assaulted another serviceman "with intent to do bodily harm, willfully, maliciously and without justifiable cause."[2746] He spent a month in the brig for his crimes.[2747] On 10 January 1945, he deserted his post and became a stowaway on the *USS Deuel* carrying 5th MarDiv Marines earmarked for Iwo.[2748] There was a fleet-wide "Wanted Poster" for Lucas with a reward of $50 as if it were the Wild West.[2749] Lucas explained:

> I was dissatisfied with my duties as a helper to the Company Police Sergeant. My only duties…were to sweep out the recreation hall and police up the recreation hall…My cousin, Samuel Lucas, a member of the 5th MarDiv, was in port and I came aboard to visit with him and found out that this was a good organization so I decided to stow away until we were at sea, and then give myself up in the hopes that I would be allowed to become a member of this combat team and in some way avenge the death of my [step] brother, who was killed in action aboard a destroyer in the Pacific…I am an able bodied marine, joined the Marine Corps for combat duty, and not that of a "Gold Brick."[2750]

Remarkably, 5th MarDiv commander, Major General Keller Rockey, requested that "Howlin Mad" Smith allow him to take Lucas, noting Lucas would be disciplined. Lucas was therefore subsequently part of C Company, 1st Battalion, 26th Marines.[2751] Smith wrote Commandant Vandegrift saying Lucas' move into the 5th MarDiv was approved and the reward for Lucas' apprehension for desertion should be dismissed. However, by the time Smith's letter to Vandegrift had arrived, Lucas had already done his MOH heroics.[2752] Obviously, the MOH was given to men, not for their character, but for a brave act or acts they had performed. It is a testament to the honor of *the Medal*'s intent to reward extraordinary heroism without regard to character.

Sometimes these men were not unruly like Lucas, but mentally sick. If the mental disorder was known, this makes one wonder why they received the award or, if unknown, what would have happened had his superiors known beforehand. Weeks after Truman tied the MOH around PFC Franklin E. Sigler's neck, Sigler was admitted to St. Albans Naval Hospital in New York. Then he was moved to Bethesda Naval Hospital, Maryland and then St. Elizabeth's in Washington, D.C. The institutions diagnosed him as a paranoid schizophrenic unsuitable for duty. He mumbled to himself, heard voices in his head and laughed spontaneously. Perhaps he had these symptoms before the war or perhaps combat precipitated them.[2753] Also, he scored 87 on the IQ test.[2754] Months before while in combat, Sigler did the following: He took command of a squad on 12 March 1945 when its leader became a casualty and led a "bold charge" "fearlessly" against a gun installation

PFC Jacklyn Harrel Lucas. He earned the MOH by jumping on two grenades to protect his comrades on 20 February 1945 (luckily only one exploded). He shockingly survived the action keeping all his body parts. Only one other Medal of Honor Marine has survived jumping on a grenade documented in this study, and that was of Corporal Kyle Carpenter who did this to save a comrade during the Battle of Marjah on 21 Nov. 2010. St. Louis Personnel Records Center.

"which had held up…his company" for days. When he reached the position, he attacked the emplacement with grenades and annihilated the enemy crew. When enemy soldiers fired from concealed tunnels and caves, he carried out "a furious one-man assault" against these positions. Wounded, he refused evacuation and helped direct machinegun fire and bazooka barrages on the "cave entrances." Seeing three men become wounded, he raced to their aid and pulled them to safety although injured himself. He then took up his position and continued to fight until a superior officer, seeing his injuries, "ordered [him] to retire for medical treatment."[2755] So, as one can see, the MOH was given to brave men, but not necessarily always good men, or ones who always behaved in the right way or even men who were mentally stable or intelligent (Woody had an IQ of 107 and a below average GCT score of 112 and MOH recipient and New York Giants professional football player Jack Lummus never finished at Baylor University and when he left after four years, his GPA was a 1.6 (C- average)—maybe he just was more interested in football and girls than school).[2756] One did not have to be good, mentally stable or smart to get the MOH; he just had to be *insanely brave.*

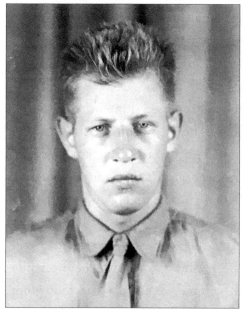

First Lieutenant Jack Lummus. He was a former NFL football player (NY Giants) and received the MOH for actions on Iwo. He died of his wounds having lost his legs to a mine blast on 8 March 1945. He told a doctor trying to save him, "The Giants lost a damn good End today." St. Louis Personnel Records Center

PFC Franklin Earl Sigler. He earned the MOH on Iwo on 14 March 1945. He was at the ceremony with Woody at the White House on 5 October 1945 and one week thereafter, he had a psychotic break and was hospitalized for weeks. St. Louis Personnel Records Center.

Sometimes, cases existed when the person probably would have received a MOH but for the fact he was a civilian, although he had received a medal for *valor* already for the same act (in this case, a Navy Cross in *absentia*). During Wake Island, Dr. Lawton Ely Shank, a U.S. Army reserve officer, but "not serving as an officer…at the time he served his country and fellow men…was in charge of the civilian hospital maintained by the Pacific Naval Air Base."[2757] He knew the island, having served there with Pan American Airways before the war. On 8 December 1941, when the Japanese attacked the airstrip, Shank, with total disregard for his safety, rushed to the scene of battle and saved many by carrying them to the hospital's underground magazine. The next day, when the Japanese returned and bombed the island, he again ran toward the action to save lives and rescue the wounded.[2758]

On 23 December, as the Japanese conquered the island, Shank went out of his underground shelter and confronted them, explaining he and his patients were unarmed and wounded. Master Sergeant Walter J. Kennedy, who was in Shank's care, said, "due to his courageous act…we were undoubtedly saved from being shot or bayoneted."[2759] Knowing how the Japanese treated British troops in a hospital when Hong Kong fell, Kennedy was not being hyperbolic. Although Shank had an opportunity to leave the island on 12 January 1942 with 250 others, he refused to abandon the wounded.[2760] Repeatedly, he stood up to his captors demanding better treatment (a bold action since the two doctors in Hong Kong at St. Stephen's College who tried this were immediately killed by Kuribayashi's men).[2761] Shank often had success obtaining better care than otherwise would have happened due to his *chutzpah*, but he "took much punishment" for doing so. He conducted operations under difficult situations, but saved lives.[2762] On 24 February 1942, when U.S. Forces bombed Wake Island under the Japanese, he helped move 50 patients to a safe dugout and then returned to his facility to save surgical equipment so he could continue to care for his charges (an hour after the attack, he had to use those instruments during emergency surgery).[2763] Eventually, his luck ran out and the "Japanese sadistic beasts" collected him and 97 other POWs on 7 October 1943, blindfolded them, marched them to the beach, tied them to stakes and machine-gunned them. No one survived.[2764]

At first, since Shank served as a reserve Army 1st Lieutenant, his MOH recommendation was put up through U.S. Army chain of command, but when the bureaucrats discovered he had "performed under Navy jurisdiction and not under… MacArthur's command," a recommendation by USMC Captain John Hamas was returned with instructions to send it to the Navy Department.[2765] Hamas and others made sure the U.S. Navy became aware of Shank's heroics and it was petitioned to award him a MOH. In addition to Hamas' recommendation, Marine Master Technical Sergeant Jesse L. Stewart, whose life was saved by Shank, wrote: "due to his gallant actions above and beyond the call of duty, and to the circumstances surrounding his execution, he should be posthumously awarded the...[MOH]."[2766] He also received endorsements from Colonel James P. S. Devereux, senior Marine Officer and Captain Winfield S. Cunningham, Senior Naval Officer at Wake Island.[2767] Although he seemed to meet MOH requirements, Rear Admiral Hayler wrote, "the Medal of Honor, as provided for by the act of August 7, 1942, is awarded only to any person who distinguishes himself 'while in the Naval service of the [U.S.]'. Since...Shank was a civilian he is not eligible for this award."[2768] Although this statue was written nine months after Shank's heroics, this discrepancy was overlooked (remember, Lindbergh as a civilian had received the MOH a few years beforehand for his flight across the Atlantic). As with most cases, once a judgement from on high came from Hayler, a person's medal fate was sealed. Shank, a civilian, reserve army officer and hero remained a Navy Cross recipient. He was the only civilian to ever receive this medal during World War II.[2769]

In some cases a MOH is awarded by upgrading a lower award years after the fact because the merits of the action were finally proven. After more evidence was reviewed, Lieutenant Garlin Murl Conner was posthumously awarded a MOH in March 2018 for "stopping more than 150 German troops, destroying...[six] tanks and 'disintegrating the powerful enemy assault force and preventing heavy loss of life in his own outfit.'"[2770] He called in artillery on his position for three hours having shells sometimes hit 25 yards from his hideout. Conner died in 1998 and never knew about his award.

And although many men have received MOH by receiving upgrades like Conner, few MOH recipients ever get downgraded, especially during the past hundred years. For example, recent MOH Marine recipient Dakota Meyer may not have done all that was reported to receive his medal in 2009, but will not be further reviewed because once awarded in modern times, it is not revoked. Indeed, in 1917, a special "Review Board" analyzed 2,625 Medal of Honor recipients, mostly from the Civil War, revoking 911 of them because they failed to meet the statutes. For example, many MOHs during the Civil War were simply given for re-enlistments and not heroic deeds.[2771] Consequently, since WWI, a more thorough process has been implemented that often proves the *Medal* was earned by documenting bravery in battle. However, the process is not perfect and sometimes outside forces can enter into the calculation to approve a medal for gallantry.

For instance, there are recent examples in which receiving valorous awards was political, even moreso than MacArthur's case. By way of illustration, in 2003 when PFC Jessica Lynch was rescued from the Iraqis, she was awarded a Bronze Star with Valor with both the U.S. Army and press excited to have a female combat veteran to praise. However, although she was injured when her Humvee crashed and suffered as an Iraqi POW at Saddam Hospital in Nasiriya being gang raped, her actions did not warrant a Bronze Star with V.[2772] She merited the POW Medal and the Purple Heart, but she, by her own admission, did not merit a Bronze Star with V since she never engaged in combat. The errors started when media falsely claimed her convoy was attacked and she defended herself and others by firing her M-16; however, her medal was never revoked.[2773]

Similar to the Lynch case was the case of the NFL football player and Army Ranger Pat Tillman, who posthumously received a Silver Star for actions he never took in 2004. Officials in the army submitted this medal recommendation to honor him and cover up how he really died—through friendly fire. His medal was never revoked due to consideration of not adding to the family's trauma since they had discovered how the U.S. Army had actually behaved.[2774]

And politics and medals can be used not only to create a hero like with Lynch or cover up a horrible event like with Tillman, but also they can simply be used to save a man's career. Politically, John F. "Jack" Kennedy was probably saved by his father, Ambassador Joseph P. Kennedy Sr., after he lost his PT-109 boat in combat when a Japanese destroyer smashed it sinking it within minutes. In fact, out of the hundreds of PT-boats built, Kennedy's was the *only one "ever rammed in the entire war."*[2775] There were also some reports that he was not a good skipper with his superior officer, Thomas G. Warfield, writing Kennedy was not "a particularly good boat commander."[2776] This assessment seemed to be also supported by the fact Jack lost his PT-Boat, which was much faster than a destroyer, possibly due to him not following proper procedures (not having his engines already fired up) and/or being situationally unaware (not hearing or seeing the 367-foot-long destroyer *Amagiri* bearing down on him).[2777] The optics looked bad. This would be like a person sitting in a Ferrari on the salt flats of California getting hit by a School Bus after watching it for 20-30 seconds coming towards him. Kennedy's father, former U.S. Ambassador to the United Kingdom (1938-1940) and 1st Chair of the U.S. Maritime Commission (1937-

1938), knew he needed to get ahead of the story to protect his son.[2778] Also, the U.S. Navy, just like the U.S. Army with Tillman, maybe wanted also to get ahead of the story that their famous poster-child had had serious problems under fire instead of performing magnificent deeds. Despite Jack's courageous and noble reasons for serving, he still would have dilemmas to face. For instance, he suffered from several medical complications with his intestines and back requiring his father "to pull some strings on his behalf" so he could enter the Navy. And eventually, Jack obtained his request for sea duty and for combat when he could have safely stayed out of the war.[2779] These facts were all noble truths to have on a resume, especially of a privileged and wealthy Harvard man. However, Jack's crowning achievement while an officer was unmitigated failure, not glory.

What JFK became known for was that he lost his ship and two of his men apparently "neglecting" his duty. After the sinking, he did tow a wounded sailor to a nearby island and rescued his life, but in the end, the disaster started with his poor decision as the captain of the ship. Due to these potential negative reports getting out, father Kennedy informed the press about PT-109's ordeal and spun the story positively, making it one of bravery, survival and sacrifice in favor of his son. Soon, Jack "became headline news: KENNEDY'S SON IS HERO IN PACIFIC AS DESTROYER SPLITS HIS PT BOAT, the *New York Times* disclosed."[2780] The *Boston Globe* also chimed in declaring, "KENNEDY'S SON SAVES 10 IN THE PACIFIC."[2781] The Secretary of the Navy, James Forrestal, soon thereafter made sure Kennedy was awarded the Navy and Marine Corps Medal writing, "For extremely heroic conduct as Commanding Officer of Motor Torpedo Boat 109 following the collision and sinking of that vessel in the Pacific War area on August 1-2, 1943. Unmindful of personal danger…Kennedy unhesitatingly braved the difficulties and hazards of darkness to direct rescue operations…His outstanding courage, endurance and leadership contributed to the saving of several lives."[2782] Yet, Jack could have also just as easily received a dishonorable discharge claiming he lost his ship resulting in the death of two crewmen. In many respects, he failed in his duty.

Although praised as a "hero," Jack's self-esteem took a hit and he felt humiliated. A lesser naval officer without the political clout he enjoyed could have suffered tremendously from those who were not intimidated by and/or wanted to help a son from a celebrated family. To Jack's credit, he approached what happened to him with wry humor and humility. One time when asked how he became a hero, he replied "It was easy. They cut my PT boat in half."[2783] Another time when asked about his heroism, he would turn serious and claim, "The real heroes are not the men who returned, but those who stay out there, like plenty of them do, two of my men included."[2784] He was able to regain some of his pride though by returning to combat duty several months later where he, this time, was able to inflict "some damage on the enemy." He got some payback, but, in the end, he would have never survived the disaster of PT-109 had he not have come from the political and prominent family that he did. As historian Steven M. Gillon declares, "Without Jack's father's intervention with the press and contacts with the government and navy, Jack would have never recovered from this mistake of his and possibly would have been dishonorably discharged from the service and possibly never been able to run for President because of it."[2785] So as one can see with Lynch, Tillman and Kennedy, political considerations often make up the mathematics of certain awards and the medals and citations often do not tell the full story of what really transpired.

## Exploring Further Problems with Woody's MOH

Back to Woody's MOH process, compared to others, its vetting process failed to establish "incontestable proof because the command amalgamated various facts in ways that cast doubt on the veracity of the recommendation package which was known but never corrected because of the time pressure."[2786] Although seven Marines (including Beck) gave statements about some of Woody's actions, they each gave a small and incomplete piece of what Beck wrote in his endorsement. It seems Beck wrote up what he thought Woody did and then gathered witness statements to support his claims, not a proper way to conduct discovery. One must go where the evidence leads one and not try to prove something by collecting evidence to support the hypothesis. When put together, there is a picture of the flow of events, but the statements from the enlisted Marines do not support Beck's statement. Besides Beck talking with the witnesses, Dashiell interviewed a few of them too (Naro, Schlager, Schwartz, Southwell, and Tripp), but Dashiell failed to find and interview the key witness: Chambers. Since he was in a hospital in Fort Eustis, Virginia, Chambers was only 150 miles from Marine Corps headquarters in D.C. so someone could have been dispatched to conduct an interview, but that was not done. And since Beck was stateside, Dashiell also did not have the benefit of speaking with him.

The men Dashiell interviewed do not document that Woody destroyed seven pillboxes and killed 21 Japanese. But these numbers were part of Beck's statement that did not list its sources other than to say (1) Chambers pointed out seven destroyed pillboxes to Beck and (2) 21 dead were located inside or outside the pillboxes. While Chambers pointed out the pillboxes, the source of 21 dead remains unknown. And traveling to the CP six times to carry flamethrowers hundreds of yards each way and then engage in combat for hours would be exhausting. "If I just used up a…flamethrower, I wouldn't leave the line to get another. That would be stupid. I'd call back on the field telephone and one would be brought up to me. If you spent hours, like Woody said, running back and forth carrying a 100 pounds of gear, you'd be so worn out an enemy would be able to locate and shoot you dead. I think there's something wrong with his story," claims flamethrower veteran Don Graves.[2787] Chambers could have verified this, but never did.

The absence of Chambers' statement bothered Geiger, who focused on Woody's case finding it lacked evidence. Thus, Woody's MOH did not receive the sanction of the Marine Corps, since Geiger never unconditionally recommended approval due to Chambers' apparent refusal to give a statement. Chambers' son-in-law, Colonel Bob Olson said, "Clearly Chambers had problems with what Woody was reported to have done and most likely did not approve of Woody's behavior in some way."[2788] Woody's case did not benefit from an "unbroken command endorsement" process. Geiger not once, but twice, requested evidence from Chambers. Not only Geiger, but Nimitz and Forrestal supported getting Chambers' statement.

And here is another problem with this story—no matter the justification in his own mind for not giving a statement, Chambers' decision not to answer Lieutenant General Geiger was absolutely the wrong one. As Barry Beck, Donald Beck's son, said, Chambers ultimately had two options when receiving Geiger's requests and "silence was not one of them." Also as a former combat veteran and Marine officer, Barry Beck further states Chambers "ignoring General Geiger's request is inexcusable, unworthy of a Marine officer."[2789] In an otherwise stellar career, this was indeed a mistake on Chambers' part. He could have responded and verified everything Captain Beck had written for Woody and/or written that Beck misunderstood him and that Woody did not do all that was reported. If the latter, then Woody's medal might have been downgraded to a Navy Cross or Silver Star, but we would not be having this controversy today had Chambers told the truth of what he witnessed.

Other Marines could also have given statements for Woody's MOH recommendation; their statements could have helped clarify what Woody did but they were not asked to make such reports. Woody was brave, but whether he did all for which he has been given credit is clearly questionable. Although witnesses discuss heroic acts, they only saw the battle from certain viewpoints.

Captain Beck started the recommendation process for Woody's MOH, but he never observed Woody in combat since Beck was at the company CP during Woody's MOH actions. Ultimately, to know what Woody had done, Beck relied on his interactions with Chambers and witness statements. Beck depended on a count of enemy dead and destroyed pillboxes that only Woody has been documented as providing *verbatim*. Such testimony is called self-reporting and unreliable. Criteria for eyewitness statements for the Marines read as follows: "… two notarized eyewitness statements [are required]…and *neither statement* may be from *the individual* being recommended for the award [author's italics]."[2790] Although Woody submitted no statement supporting his MOH recommendation, it seems the source for Beck's facts about Woody were in part documented as coming from *statements made by* Woody. Maybe this confusion could have been clarified had Woody reached out to the witnesses after the war, but he did not. Fellow Iwo veteran, George Bernstein, finds this strange:

> Hell, if I'd received the Bronze Star from someone, I would've done all I could to have thanked them. I find it strange Woody never reached out to the men who were responsible for making him a celebrity. Also, if I didn't remember what I'd done to get such a…[MOH], I would've surely gone to those…who did remember to let me know what the Hell I'd done. I was on…Jima for the entire time and I remember most of the…events that happened…and the men who protected me.[2791]

B-29 radio operator David Fisher, who flew 22 missions over Japan, echoes Bernstein: "If I'd received the [MOH], I would sure have thanked the men personally who helped me get it. That is strange Woody never did that…when he could have. Now, of course, they're all dead."[2792]

Why did Woody receive *the Medal*? He had an officer in Beck who liked him and considered his actions extraordinary. Beck felt, although the documentation was incomplete, that sufficient facts had been gathered to tip the scale in Woody's favor for a MOH. And Beck knew how to write recommendations. Charles Neimeyer, retired Director and Chief of Marine Corps History at Marine Corps University, wrote: "The strength of the write-up and numbers of witnesses seem…key for a MOH nomination to make it…successful."[2793] In all likelihood, if Beck had not survived the battle and if he had not spent the time writing a witness statement, the public would never know Woody. Everything else he did in life never merited a mention in a history book although his service was honorable. We know about Woody's life not because of what Woody did, but because Beck felt Woody's actions MOH worthy.

This book's recognition of Woody has brought to light the bravery of other courageous men like Beck, Fagan, Chambers, Hemphill, Henning, Cookson, Dashiell, Rybakiewicz, Tischler, Waters, Renstrom and many others. Until this book, they were unknown except to their families.

Studying Woody's MOH illustrates that more Medals of Honor should have been awarded for Iwo. A study of it also shows the process can be subjective and flawed because Woody's award is full of questions. Historian Dwight Mears wrote: "[World War II] produced some anomalous awards" and Woody is one of them.[2794] Mears continues: "Woody…might well have qualified [for a MOH] but the evidentiary process in place apparently wasn't satisfied, so the outcome is unclear."[2795] In the end, how the process started and where Beck received the information of 21 enemy dead and that Woody destroyed seven pillboxes remains unknown. When Woody boasted he had done that, could he have taken credit from others? Different pilots claimed credit for shooting down Yamamoto (the claims led to a 9th Circuit court of appeals case), and recently, several Navy SEALs have claimed credit for killing Osama bin Laden.[2796] When Woody told people about his exploits, he did not do so in order to get *the Medal*. He was a kid excited about what he had done. The unintended consequence of the command learning about his stories motivated Beck to ensure Woody obtained recognition. But the facts reported by Woody and the actual facts given by the witnesses are different. Woody performed some amazing, death-defying acts, but were his acts worthy of a MOH in comparison to others like Waters, Henning, Cookson, Rybakiewicz, Chambers and Fagan especially if Woody did far less than what Beck thought? If Woody's recitation of facts led Beck to recommend him for a MOH, is such self-reporting fair? If Woody knows he did this, should he report this fact to the military now?[2797]

This case shows the difficulty historians face in determining what happened. Woody's story spun out of control and what he said about himself was excluded from his final citation. The quote used at the chapter's beginning best describes the end of Woody's review process when people involved fell into the trap of: "When the legend becomes fact, print the legend." We as a society "are often more comfortable with treasured legends than with hard facts."[2798] Colonel Warren Wiedhahn (USMC) of the Iwo Jima American Association (IJAA), when confronted with these facts behind Woody's MOH, proclaimed, "Bury it. Don't let anyone know about it."[2799] Although it is understandable why he encouraged me to do so because he does not want to see his Corps and board IJAA (Woody is an honorary board member) experience any embarrassment, one cannot ignore facts. As Medal of Honor recipient, Navy Seal and Navy Lieutenant, Commander Michael E. Thornton, said, "What Woody has done is disturbing and the facts surrounding his medal is very troubling. But facts are facts and they need to be known."[2800]

Facts can show a person not to be as grandiose as desired. Legends, however, often present a hero without flaws and one whose deeds are unblemished. People want to know only the best about a person. In letting the story take on the proportions it has and not correcting the facts, allowing them to get bigger (in early drafts, Woody allowed "facts" to be written that he killed 50 and destroyed more than seven pillboxes), one sees Woody probably wants the legend to grow.[2801] Recently, he has supported a petition to President Trump to hold a state funeral for the last WWII MOH veteran to die. Since Woody is in the best shape out of the two remaining MOH recipients, he shows that he probably hopes that that State Funeral to be held in the future will be his funeral since he probably will be the last man standing. On the one hand, it would be great to honor an enlisted man instead of a general or a president. On the other hand, this is clear evidence that Woody seems to want his memory to strongly continue even when he is dead.[2802]

In documenting Woody's life, one could not ignore what the sources were saying. Examination was not done lightly. As scholars, we must weigh evidence against interest. One would gladly report Woody destroyed seven pillboxes and killed 21, but the evidence does not support this. In the end, what was originally submitted for Woody for the MOH was not proven.

In contrast, there was a Marine who did similar acts, but did not receive recognition. A flamethrower in Woody's battalion who had also been in combat on 21 February, Joseph Rybakiewicz, destroyed seven pillboxes or caves and rescued a wounded comrade, but no officer saw or was informed of everything he did. Chambers, having been wounded, was no longer in C Company, and most C and B Company officers had been killed or wounded.[2803] In taking out the emplacements, Rybakiewicz probably killed dozens.[2804] At one cave, two buddies were cut down by hostile fire. Disregarding his safety, he rushed the cave with a flamethrower and burned it out. He then rescued one of the fallen. He returned to try and rescue the second, but shrapnel cut him down.[2805] For these actions, he received a Bronze Star with V for a portion of his activities since the citation only mentioned one of the caves he destroyed and the rescuing actions. What would have happened if Rybakiewicz had been put in for all his actions, actions Dashiell verified?

Corporal Joseph Anthony Rybakiewicz receives his Purple Heart for wounds suffered on Iwo when back on Guam. Circa May 1945. A "Full-Bird" Colonel presented this award to Rybakiewicz in this photo. Author's Collection and the Rybak Family Archive

Contrasting this case and Woody's, there are many discrepancies and questions. Besides focusing on why there were problems with Woody's medal process and his interviews throughout the years, one might ask the larger question why the nation needs heroes. Edward L. Bernays, the premier U.S. public-relations counselor during WWII, declared: "Our society craves heroes."[2806] Since procedure was not followed, one could argue that Woody's case was pushed through in order to create the heroes this nation not only craved in 1945, but also wants today. Woody has stated he did not desire *the Medal*, but once it came to him, he has enjoyed the benefits of having it. Our heroes, whether they be Tillman, Lynch, Kennedy, Waters, Rybakiewicz, Dashiell, Woody or Chambers, put on the uniform and risked their lives for their country. They did not do it for fame, but for their fellow Americans. The U.S. recognized their bravery by awarding medals. Although the award process is flawed, it is a meaningful testament that the U.S. continues to honor those who put their lives into harm's way to defend America's democracy, constitution and people. Medals show the nation's appreciation for the willingness to defend these sacred elements.

Woody was one of the few to receive the MOH and thereby be the spokesman for the sacrifice of WWII Marines. When a president invites extraordinary Americans to the White House to receive medals, it was one of the few things a president does that is apolitical. It brings everyone together when a president drapes a MOH around someone's neck to display a grateful nation's appreciation. When that medal goes around one's neck, the awardee's identity is forever changed. That person must now answer questions most servicemen never have to deal with; namely, what does *the Medal* mean? How many did you kill to get it? What did you do for *the Medal*? Why did you get it and not others? *The Medal* gives power to command an audience and receive respect from superiors: for example, the tradition is that everyone in the military salutes a Medal of Honor recipient, so general officers salute a corporal who has the MOH. "Often MOH winners pile untruths upon their story until the original heroics get buried in even more grandiose events and conquests," said historian Roger Cirillo.[2807] Navy Cross and MOH recipient Pappy Boyington said, "'I'm a psychopathic liar,' probably intending to say" pathological. In reviewing Pappy's memoir, historians have found he was being truthful about being a liar.[2808] Although heroic and deserving of a MOH (receiving high praise and an un-broken line of endorsements from the likes of Admirals Halsey,

Nimitz and King and USMC Commandant Vandegrift),[2809] he doctored his autobiography, mixing truth with lies. Many of these facts and myths became immortalized in the popular TV series *Baa Baa Black Sheep* (1976-78).

The honor and unquestioned loyalty MOH recipients receive goes against the knowledge they were common men who did something extraordinary. Feeling they are expected to be more than what they are, many give false tales like Boyington or allow their stories to grow bigger than they actually were. Possibly, Woody may have done this, especially in his interviews with journalists. Many Iwo veterans acted beyond the call of duty: the reality remains that Woody's actions were not unique. What was unique is that officers in his command *were still alive* after Iwo and thought he *deserved* a MOH. In the end, however flawed the process was, we must put faith in their decisions and accept that Woody merited *the Medal* in their eyes. One does wonder, though, what would have happened had Hayler and his board known the information in this book. Would they have continued with the award or returned to Geiger's and Nimitz's concerns and allowed them to ultimately corner Chambers and get the one piece of evidence they required to make their final decision on Woody's case. For now, no one knows the answers to these questions. When the 31st USMC Commandant, General Charles C. Krulak, was asked about Woody's Medal of Honor process, he declared that there were so many problems with it that it should never have gotten as far as it did up the chain of command. He explains, "It should have been a Navy Cross (at best). No real eye-witness statements...short blurbs that really don't paint the picture. No supporting signature of Platoon Commander [and] no support from [the] highest levels [of the Corps]. Woody obviously took heroic action but without all the right documentation, *it should have not been a MOH.* I fault the media, Woody... and the President (author's italics)."[2810]

# Ch. 29: Life After the War

"There is…a time to kill and a time to heal."
—Ecclesiastes 3:1-3[2811]

"Though he should conquer a thousand men in the
battlefield a thousand times, yet he, indeed, who would
conquer himself is the noblest victor."
—Buddha[2812]

WOODY RETURNED HOME AND MOVED into a house his in-laws owned. It had no running water, an outhouse, an ice block refrigerator and not much furniture, but he was with his wife and happy. American industries that used to manufacture cars, appliances, clothing and consumer goods had during the war produced planes, tanks, ships, weapons, and other materiel. It took time to revert to non-war production to meet the demand for consumer goods. Woody worked as a laborer for a few months and then joined the Veterans Administration (currently the Department of Veterans Affairs) in January 1946. The VA searched out MOH recipients for employment because of their prestige and not having to do lengthy security clearances. When they offered Woody a salary of almost $3,000 a year, Woody jumped at it. "That was more money than I ever had seen and more money than I could've ever earned on the farm. I was a rich man," Woody says. At first his title was Contact Representative, but later it changed to Veterans Service Officer. He helped veterans learn about their benefits. He did not choose this work, but he enjoyed it. He also focused on his education and soon completed his GED. Soon thereafter, West Virginia honored those who went off to war before completing high school, so Woody received a high school diploma from the state in addition to his GED. During this time, he continued working for the VA: "In fact, I wouldn't call it work because I loved it so much because every day I found ways to help others." For 33 years he helped veterans until retiring in 1979.[2813]

Before starting his VA career, he wanted to fulfill his promise to his friend, Vernon Waters, a man closer to him "than a brother."[2814] According to Woody, he carried Vernon's ring throughout the remainder of the war, mindful of his pledge to return it to his family. He and Ruby drove to Vernon's home in Montana. A friend who had an automobile dealership loaned Woody a red 1942 Dodge convertible for the journey. Along the way, Woody told Ruby how sad he was and how Vernon died. Their conversation mixed between sorrows of war and the happiness they felt having a future together.[2815]

At Vernon's home, there was a Gold Star in the window showing the family gave a son to the nation and was in mourning. He and Ruby stepped out of the car and approached the front door. The Waters family noticed the car and opened the door before he even knocked. They all were giants compared to Woody and Ruby (5'3"). Vernon's mother (5'7") and his three brothers were between 6'4" and 6'6". His brother, George, had fought on Saipan with the 2nd MarDiv and was there with his wife Kathryn. She explained, "We were so happy to see Woody and hear what happened to our beloved Vernon. We all sat down at the kitchen table and we gathered round Woody and let him talk. He was quiet, calm and methodical with how he talked about his gentle giant of a friend. We cried together and we were glad he came. It meant the world to us."[2816] According to Woody, he began the meeting giving Vernon's ring to the mother (in some interviews Woody claims he gave it to his father which was impossible since he was already dead).[2817] Woody said, they "responded as if I had brought them all the gold at Fort Knox"[2818] or "the greatest diamond that ever existed."[2819] Vernon's mother Verna had been writing the Marines during the summer of 1945 to find her son's Parker Fountain Pen, wrist watch, black wallet and "*Montana agget ring* [sic.]"[2820] The Marines never found these items to return to her (probably either stolen by some Japanese since he was left on the field of battle for so long or taken by a fellow Marine).[2821] The only problem with Woody's story of returning the ring, which Woody has only given in recent interviews (this story is not in his lengthy 1945, 1956 and 1966 interviews cited throughout this book), is that the Waters family has no oral tradition that Woody gave the ring back to them and the family today still does not know where the ring is.[2822] "I don't remember Woody giving us a ring," Kathryn said.[2823] Regardless whether he gave the ring back or even had it to begin with, Woody sat with them for hours recounting his time with Vernon. Upon departing, he felt satisfied that he had honored his promise to his friend.[2824] He never contacted the family again.

After parades and welcome-home parties there was readjustment to civilian life. Returning home was difficult for many who had not seen their loved ones for years. Many had a young child whom they had never met. Veterans had been scarred by experiences they wanted to forget and were difficult to explain to one who was not there. In 1946, the divorce rates shot up.

As soon as Woody returned home, Woody adapted to the fact that he was famous. Everyone wanted to get close to him, and he was routinely invited to grand openings, cocktail parties and politicians' campaigns. People invited him to give talks everywhere. Many wished he would run for office, but he felt if he did so, he would "have to sell my soul and I wasn't going to become a politician."[2825] He was given perks such as receiving a small stipend every month for the MOH (currently around $1200 per month tax-free), an open invitation to every presidential inauguration (which he has attended since Kennedy's) and the ability to receive immediate acceptance for any son who wanted to go to any of the academies ("I unfortunately only had daughters," Woody says [West Point did not go co-ed until 1976 and the Naval Academy did not do so until 1979]).[2826] Right after the war, Commandant Vandegrift wrote the Judge Advocate General that MOH recipients should not have to wait until the age of 65 to receive $10 per month, but should receive $30 per month and a one time gift of $2,500.[2827] From the records, such financial rewards were indeed given and have continued to be increased as Woody's monthly stipend today displays. And he had authorization "to ride as a passenger on Armed Service aircraft when space is available on a regularly scheduled flight within the continental United States." None other than the Commandant and Iwo veteran, General C. B. Cates, wrote Woody this notification in 1948.[2828]

With all these perks, Woody understood *the Medal* afforded him special consideration everywhere. He received respect most could only dream of and few obtain. He took incredible pride in having *the Medal*. Woody was so confident about his new status, he felt emboldened to write the Commandant, General Vandegrift, a personal letter to send him a new, and improved "lapel button to wear on civilian clothes." In this June 1946 letter, Woody explained his lapel pin had fallen apart:

> Along with many other…Medal of Honor winners, I am very proud of my medal and desire very much to wear something that will indicate that we were awarded the [MOH]. I am positive if the…Government could make a button that was more substantial than the one that is given for wear for civilian life, it would please every Medal of Honor winner…[2829]

In less than a year, this shy West Virginia "Hillbilly" went from being terrified when meeting General Erskine, Commandant Vandegrift and President Truman to writing Vandegrift directly about a lapel pin. This was the second letter Woody wrote the Commandant after not receiving an answer in February. Woody wrote Vandegrift, "As I am the winner of a [MOH], I am wondering if you could tell me where I could obtain a lapel button representing the same, as the one I was given at the time of receiving my medal has come apart and I am unable to wear it."[2830] The *Medal* gave him a new confidence. And interestingly, Vandegrift's office sent Woody a thank you note and then in July 1946, Vandegrift wrote Forrestal, apparently reacting to Woody's second letter: "1. Forwarded, recommending favorable consideration for the adoption of a more substantial [MOH] lapel button. 2. It appears that the men who win the highest honor bestowed by our Government should be given a lapel button which will be practicable and long lasting."[2831]

Across the ocean, many Japanese families who sent their men to Iwo knew they would never see them again. They had forever disappeared and all that remained of their memory was that they fought bravely and died horribly. One family that struggled with not knowing a soldier's fate was Kuribayashi's wife and three children. Although Mrs. Yoshi Kuribayashi knew her husband would never return, it was still difficult for her and her children to absorb. When Kuribayashi's grandson, Yoshitaka Shindo, was asked how his family knew he was never coming back, he said,

In all the deployments and battles my grandfather had participated in, he always took the Emperor's sword given to him upon graduating the military academy. Once he received his assignment for Iwo, he left the sword…home. He never told his wife the place he had been assigned, but my grandmother realized…her husband would never come back by seeing his important sword he just left at home. However, she never mentioned it and my grandfather just left and my grandmother only saw him off. What did he aim to tell her by leaving his sword from the Emperor at home? [Simply], he would die for his country.[2832]

One could also argue that leaving the sword was passing the baton to his son to carry on the fight as a *Samurai*. Moreover, this gesture also might have shown he wanted his family to be strong and remember that where that sword remained, there was hope. The war was over and the country in ruins. The public, instead of praising families like Kuribayashi's, blamed people like him for "getting themselves into hot water."[2833] As a result, there was no discussion about Kuribayashi or his achievements for a long time. There was no more talk of military pedigree and honor. However, the sword continued to be displayed in the home. Kuribayashi's wife also displayed a picture of her husband in uniform near the sword. Shindo said,

> When I was a little child, my mother always sang the same song her mother sang to her while bathing me. It was a song called *The Fall at Hometown (Sato no aki)*. This song is about a mother and a child praying for the father to come home safely from the southern island. I didn't realize when I was a child, but now, after my mother has passed away, I know that a young girl was singing the song in remembrance of her father to express her dream which never came true.[2834]

Unlike Americans, Kuribayashi, a war hero who should have been awarded for his heroic defense, had no honors from a grateful nation. He disappeared and his family tried to find food and rebuild their lives. "My grandmother never complained [or] raised her voice to express her anger. However, I had never seen her smiling from the bottom of her heart."[2835] Later in the interview, Shindo did claim, "However, my grandmother did express pleasure when the U.S. returned

Iwo Jima to Japan. The night before, my grandmother had a vision from the general [Kuribayashi]. He came to her and told her he had returned home."[2836] From this spiritual encounter, one sees this event played a huge role in the lives of Shindo's family members. Now a famous politician, the Honorable Shindo has resurrected his grandfather's legacy using his image and heroic defense at Iwo in posters and slogans to promote his political party's policies of defense especially with the territorial disputes with China. He has turned his family's sadness into something positive for him and his party, the Liberal Democratic Party of Japan.

But as Kuribayashi's family struggled with reality in 1945-46, Woody also had a difficult time. It was not all a bed of roses for him either. The events leading to his receiving *the Medal* haunted him. He turned to drinking and smoking three packs a day to dull the pain, but these offered little comfort. Woody struggled with the deaths he had caused and the misery, injuries and death he witnessed. He did not understand why he had survived when others did not. Why did some testify on his behalf for the MOH? Was he really worthy? For years, he continued to have the rank, cloying odor of burnt flesh in his nose. The pains of combat never left him and the faces of his buddies haunted him. Although he knew Japan had started the war and made itself America's enemy, he felt bad he had killed people. He realized they wanted to kill him, but taking life was never easy.[2837]

Certain sounds reminded him of war. In 1947, when he was filling the car with gas, one of the rear tires blew an inner tube and exploded with an earsplitting bang. "It sounded like a loud gun," Woody said. He dove under the car, ripping up his clothes and skinning his elbows and knees. His wife watched in bewilderment at Woody's speed as he disappeared under the car. The split second, sub-conscious reactions that were so vital in combat were out of place in civilian life. But the instincts forged in combat that ensured survival were not easily discarded. Even when he was 77-years-old visiting the WWII museum in New Orleans, Woody responded instinctively to unexpected noises. At the museum, a video showed the Hiroshima atomic bomb explosion accompanied by loud sound effects. Not anticipating the detonation would echo raucously through the hall, the sounds sent Woody diving for cover under a table. His family was baffled, helped him up from the floor and laughed, but Woody's military memory enabled him, even as an old man, to react with catlike reflexes. Such skills determined survival in combat, and Woody had them.[2838]

As tough as it was for Woody to deal with the memories of the war, it was worse for some others. His two brothers who served in Europe wrestled with memories of their experiences fighting the Nazis. His closest brother, Gerald, "who was one of the gentlest and kindness guys you would ever know," struggled mightily with his time in combat. In fact, he "cracked up," and had to be removed from the front lines during the Battle of the Bulge during the winter of 1944-1945 and never recovered mentally. At the beginning of the battle, he was injured, sent to a field hospital, treated and then returned to the front lines. While on a troop train returning to the lines, a German fighter pilot strafed the cars. The men jumped from the train, taking cover in the woods, but the plane continued strafing them, following the men into the trees. This was where Gerald lost it, suffered "psychoneurosis" and collapsed. He was sent back to England and eventually home. He would die in his 40s after having struggled for years with depression and alcoholism. "He simply gave up and died," Woody said.[2839]

Woody's older brother, June (Lloyd Jr.) who, according to Woody, had a mean streak, was in logistics and did not experience much combat, although the war had put a lot of stress on him too. When he returned home, he, like his brother Gerald, also drank too much. Moreover, he hated the fact that Woody was a war-hero when he was just simply a veteran. He would make fun of Woody, yelling in public, "There is my HEROOOOO brother. Look at my HEROOOOO." Woody ignored him. One day, however, June got drunk and entered a bar where Woody was shooting pool and began "making fun" of him. Woody, feeling he had endured years of abuse from his brother, had heard enough. He raised his billiard stick and struck his brother across the head, knocking him to the floor (hitting each other with clubs, sticks and other weapons seems to have been a common occurrence for the boys). That short-temper of Woody's raised its head once again and he reacted quickly and with violence. According to Woody, that was the last time June picked on him. Woody claims his bully-brother crawled back into his shell and never achieved much of anything in his life. "He often would call me up from jail and I would have to bail him out. He was addicted to alcohol and disorderly." When asked why he bothered to help a brother who treated him so badly, Woody said, "well, the Bible teaches me to love my enemies."[2840] June would die in his early 50s of cirrhosis of the liver and other medical problems. Woody sums up June's life and epitaph in a sad way, saying he was "a people abuser."[2841]

While his brothers struggled to deal with their wartime experiences, Woody continued to suffer too after the war. Ruby often heard him struggle with nightmares. She saw his panic attacks. As a result of this distress, she became irritated when the Marine Corps Quartermaster General wrote them in 1947 that Woody needed to pay back $54.57 in overpayment. It requested immediate action and was cold and bureaucratic. Since Woody was on a business trip, Ruby answered the Quartermaster. She quoted Woody's MOH citation. Then she gave her own thoughts:

> It was for you and your children—for me and my children and for all the other [Americans]…that Woody…went beyond the call of duty and risked his life. He, along with thousands…saved this government and our very way of living. How could we possibly repay him…much less overpay him for giving his youth and offering his life to save us[?] When our [men] were overseas fighting for us, we read every day of the things we must all do for them to show our gratitude when they returned. Must we destroy their faith in our promises in what they fought for and let them learn that we have forgotten our debt to them? Yes, now we shall show them that we measure what they did for us in dollar & cent values. For helping to save all the things we hold dear Corporal Williams was overpaid $54.57. In your letter of Nov. 25, 1947, you asked that [he] advise you by return mail as to his intentions relative to refunding the $54.57. Due to the fact that my husband is working out of town and that you wish an immediate reply, I am…answering the letter myself. We shall refund this overpayment as soon as it is possible…We will not be able to send it in a lump sum but we will start making payments on it in January. Sincerely, Mrs. Ruby Dale Williams." [2842]

Ruby was upset. She and her husband had started a family and were trying to earn a living for themselves. She saw a typical government letter as an insensitive assault on her husband's service. The least the Corps could do was to "forgive" this overpayment that was its mistake and not her husband's. She hoped the officer would waive this payment.

In military fashion, the Corps ignored Ruby's complaints and Captain Robert L. Williams wrote back: "You are informed that this office will agree to frequent and regular payments toward the liquidation of the overpayment… your husband [received] during his…service…Your cooperation in liquidating this indebtedness, will be appreciated." [2843] To be fair, Woody was only entitled to what his rank and service deserved, nothing more and nothing less. Corners cannot be cut in the military even when doing so might show unwarranted favoritism. Woody had to pay back unearned funds, and he did, although he continued to suffer from the war. He had difficulties with his emotions especially since as a man of his generation, he was taught to show no weakness. His father had declared, "Boys don't cry. Man up and don't do that. Women do that." Woody explained: "We may have cried, but we didn't do it openly."[2844] Raised not to deal with emotional pain, Woody suffered in private.

His suffering ceased a little over a decade later and he almost made a total recovery. Easter Sunday 1962, things changed when his wife convinced him to attend church. He did not want to go, but he went for her. The minister talked about God's healing power and salvation and described how they had nailed Christ to the cross because of their sins.[2845] The minister slammed his fist into his other open hand while looking at Woody. He explained Christ sacrificed "his life just for us!" That hit home for Woody because of the men who had sacrificed their lives for him. "They didn't have to give their lives, but they did, protecting me," Woody said. In the middle of the service, Woody got up, went to the pulpit and asked the minister to pray with him and ask "Jesus into his heart." Woody felt an incredible peace wash over him. He felt the love and forgiveness, as he explains, from Jesus, "the Son of God…the only one who could save me from my nightmares." After that transformative experience Woody claims he never drank alcohol or smoked again. He never had nightmares about cremating the enemy. He ceased feeling animosity towards the Japanese. It was a pivotal moment. He practices his faith in many ways; he became the Medal of Honor Society's official chaplain and a lay minister. He gives sermons and spreads the message of "God's love to this day." It helped him forgive the Japanese, his brother June and most importantly, himself for the things he did in his life, especially killing other human beings.[2846]

He feels the only way people can be saved, healed from their pain and sin is through Jesus. He gave a sermon at Bethesda Church in 2010:

> Funny how about everyone wants to go to Heaven…provided they
> do not have to believe what Jesus says…you cannot get to the Father
> except by me…Romans 10:13. Anyone who calls on the name of the
> Lord will be saved…Acts 4:12…Jesus is the only one who can save
> people. His name is the only power in the world that has been given
> to save people.[2847]

Woody believes if someone does not believe in Jesus, like all those Japanese he killed, they go to Hell. As for all the Jewish Marines with whom he fought who died like Tony Stein, if they "did not accept Jesus," Woody believes that they "are in Hell."[2848] Woody's conversion took him from hardly any belief to a full evangelical system full of hellfire and damnation (a true exclusionary theology). He feels it healed his soul and gave him a second life, so he is passionate about what he feels are requirements for salvation although some would now call what he believes religious racism ("either believe the way I do or God will punish you *forever*").

Throughout the 1960s and beyond, America struggled with the Vietnam War, the urgent demand for civil rights and women's liberation. After several years in the Marine Reserves, Woody was promoted to Chief Warrant Officer 4, a skilled leader in a special field.[2849] He tried to become a lieutenant, but was rejected because his aptitude scores were too low (GCT Score of 112), he only had an 8th grade education, he was too old and he had not displayed the necessary leadership skills:

> There is no positive indication that his abilities consist of anything in
> the way of officer-like qualities or leadership ability than that which
> may be occasioned because of his…civilian connections. The actions
> for which he was awarded the [MOH] do not, in themselves, indicate
> leadership ability, although heroic.[2850]

The minimum for an officer was to have a GCT score of 120 and a college degree, both of which Woody did not have.[2851] So Woody rose through the Warrant Officer ranks and served honorably. His main occupation was working explaining veteran benefits in the VA. In 1967, he volunteered to go to Vietnam in that VA capacity. Although he did not think America should have been in Vietnam, he wanted to educate veterans on their entitlements. The VA would award him "the only" Administrator's Commendation certificate of the regional office in West Virginia and a $500 stipend.

That is Woody's nature; he helps people. He helped dedicate a bridge to be named after the distinguished veteran Bob Vandenlinde, a Korean War Silver Star recipient. Woody helped Vandenlinde get his Silver Star since the paperwork had not been filed properly decades prior. Woody made sure the Department of Highways put up signage for the Veterans Administration Office in Barboursville, West Virginia. When someone needed help with the VA, he made sure they got it. When he noticed other MOH recipients in West Virginia were not getting what he deemed proper recognition, he made sure DVA Medical Centers and State Veterans Nursing facilities put up walls of honor describing them. He has even been instrumental in helping transfer WWI MOH recipient, U.S. Army First Sergeant, Chester H. West, from an overgrown, forgotten cemetery in Mason County State Forest, West Virginia to the Donel C. Kinnard Memorial State Veterans Cemetery in May 2018 ensuring West received a police escort and a proper headstone with the MOH emblem and a formal military funeral. Since West killed several Germans and destroyed two "*Kraut*" machinegun nests, he is a kindred spirit of Woody's on many levels.[2852]

Because of his accomplishments, Woody has earned other distinctions. The West Virginia Legislature included him in its Hall of Fame, naming him a Distinguished West Virginian in 1980 and 2013. He is on the Huntington Civic Center's "Wall of Fame." His hometown Quiet Dell in Marion County displays a sign outside town limits as being the original home of MOH recipient Hershel "Woody" Williams. The military recognized him with the christening of T-ESB 4 (Expeditionary Sea Base Ship 4), an 800-foot man-of-war vessel commissioned *USNS Hershel Woody Williams*. Also noteworthy is the $32 million Hershel "Woody" Williams Armed Forces Reserve Center in Fairmont, West Virginia. The VFW (Veterans of Foreign Wars) Post 7048 in Fairmont, the main bridge in Barboursville and the Veterans Affairs Hospital in Huntington are named after him. The United States Post Office honored Woody with a stamp in his likeness in 2013. His legend continues to grow when during a rally in Huntington on 3 August 2017, President Trump praised Woody in public and at Trump's 4th of July 2019 parade in Washington, D.C., Woody was right by the president during the celebration in a seat of honor.[2853] He has crafted a keen sense of what to say when given the microphone highlighting those who died for his safety and those who never returned, claiming that is why he wears *the Medal*.[2854] (See Gallery 4, Photos 4-6)

Recently, Woody has focused on helping set up Gold Star Families Monuments run by his *Hershel Woody Williams Medal of Honor Foundation* which helps community leaders throughout America establish memorials in communities. His goal is to put a Gold Star Families Monument in each state, and, as of the date of publication, he has erected dozens throughout the U.S. These black granite monuments commemorate the sacrifices of families who have lost a relative to war. Rarely do such families get recognized. Woody's goal is "to show the story of the loved one's homeland, having a family, going to war, losing their life and the whole family grieving the loss."[2855] To give an example of the cost to family members, one of Woody's comrades, Douglas Dorman, died during the *Banzai* on Guam on 26 July 1944. When Dorman's mother found out her son had died, she suffered a heart attack and was hospitalized for two weeks.[2856] Most civilians fail to realize the cost to everyone associated with servicemen and women killed-in-action.

Many often ask how the Gold Star came into being. Especially during WWII, if a family had someone in the service, it displayed a flag with a blue star in the window or on the door to let others know a loved one was serving. If that family member died in battle, the family replaced the blue star with a gold star. The home would be called one of a "Gold Star Mother," recognizing the mothers who sacrificed a son for the nation's freedom. Often, when Woody erects these monuments, he hears from many they never knew what the Gold Star meant. Pointing to a map of the Gold Star monuments going up in parks, cemeteries, military bases and other locations throughout the nation, he says, "hundreds and thousands of children will now walk by these memorials and ask what the Gold Star means, and many…will find…that people like them, walking on the same ground and going to the same schools as they are, one day in the past died for them to be free." Woody describes this project as both honoring those who died and those who are grieving that death.[2857] In doing these projects, he remembers his childhood friend Leonard Brown. He died from antiaircraft fire while perched in a B-24 Liberator bomber nose turret in the U.S. Army Air Force during a mission to bomb the Mako Japanese naval base. The family did not know this and felt he had disappeared in action. They hoped and waited seven years after the war for news until the government said he had "been lost."[2858] Since Woody regarded the Browns as his second family, Woody thinks of them often when erecting monuments. Woody helped rename the indoor training facility at the base named after him in Fairmont, West Virginia *The Leonard Brown Memorial Drill Hall* with a plaque and picture. When this was done, Woody also erected a Gold Star Families monument at the base's entrance.[2859] In a meeting with other MOH recipients during President Donald Trump's inauguration, former Marine Commandant and now Chairman of the Joint Chiefs-of-Staff under Presidents Barack Obama and Trump, General Joseph Dunford, singled out Woody for his service with the Gold Star project, a rare honor among a distinguished crowd.[2860] (See Gallery 4, Photo 7)

Most parks, veterans' cemeteries and military bases are open to having these Gold Star memorials, and Woody invites the community to participate during the opening ceremonies, often honoring families who have sacrificed their children for America. He often helps the organizations with a starter check of $5,000 to alleviate the expense, usually around $40,000 total, to complete the black granite monument that measures six-feet tall, 20-feet long and a foot thick. Unfortunately, Woody does not always get the reception he desires and certainly deserves. At the National Museum of the Marine Corps Museum, he spoke with the president of the Marine Corps Heritage Foundation, Lieutenant General Robert R. Blackman Jr. to get his monument in the park surrounding the museum. According to Woody, General Blackman was not receptive. So Woody did what he had done with everyone else. "General, to get things started here, I have a check here for $5,000 to help start things. It is for the families." "Are you trying to bribe me?" the general responded according to Woody. "I am shocked and offended you would ask me such a thing General," Woody responded. Then Woody tore up the check in front of the General and walked out.[2861] Woody's temper often gets the best of him. He spoke of the general in negative terms although when the general was questioned about this event, he was offended at how Woody remembered the event and explained that what Woody was asking of the museum did not fit into its ground plans or budget ("Woody's $5,000 probably represented about two and half a percent of the cost"). Moreover, the image Woody wanted on the granite stone was of a soldier and not a Marine which confused Blackman since he ran a Marine Corps organization.[2862] According to the general, he was trying to have a candid discussion with Woody about what was possible, but the Iwo veteran apparently did not seem open to it. Nonetheless, most often, wherever Woody goes, such as the Fairmont Army Reserve Training Center, the Eastern Kentucky Veteran Cemetery, the Barboursville Public Park, or the McKinney, Texas Veterans Memorial Park, he is welcomed and his requests are most often in line with the organizations' budgets. These projects and events do a lot to strengthen communities. When he erected the first monument in Texas at Cedar Park on 23 September 2017, a few hundred attended including the First Lady of Texas, Mrs. Abbot. At the ceremony, there were 38 mothers who had lost children in the nation's service. (See Gallery 4, Photo 8)

In Woody's pursuit of service, he never forgets when talking about *the Medal* to acknowledge others' sacrifices. Before his speech in Indianapolis in 2013, he would not speak unless a man in attendance who had protected him on Iwo would put the MOH around his neck. Lefty Lee rose up from the auditorium, walked to the front and tied *the Medal* around Woody's neck, receiving warm applause. Sadly, Lee was still pretending to be one of Woody's security men at this juncture.[2863] Nevertheless, Woody believed he had been there protecting him and he wanted to thank him. As for Woody's commander, Captain Beck, Woody wrote Beck's sons in 2016 this note: "Because of your Dad, my life took on a new meaning. Because of your Dad, I received *the Medal*. Because of your Dad, I will have a ship with my name on it. I'll always be grateful. Semper Fi… Woody."[2864] Woody says that one should never forget those who have supported him throughout life because, no one "is here alone. We all need one another."[2865]

*The Medal* for Woody also represents America's greatness. Woody believes that when he was born, he was given the greatest gift a person could receive: Freedom. This freedom "is a gem given to me when I was born…that is impossible to buy. There's not enough money in the world that can buy it. So to me, *the Medal* stands for sacrifice. The sacrifice Americans have always been willing to give since the founding of this great country. It also stands above all else for Service. If we don't give that, we will lose that precious freedom millions of Americans have died for throughout the centuries to allow us today to have the ability to live free."[2866]

# Conclusion

"Truth is the daughter of time."
—Gordan W. Prange[2867]

"On Iwo Jima, we came to respect each other's strengths.
However, we also came to the conclusion that we should
never fight each other again. We must keep our relationship
between the U.S. and Japan strong."
—Yoshitaka Shindo, General Kuribayashi's grandson[2868]

"No one who has not experienced it can realize how difficult it is to
track the shadow of truth through the fog of war." Wilfred J. "Jasper"
—Holmes, U.S. Naval Intelligence Officer[2869]

"If we listen carefully, we can hear the soft voices of those
fallen warriors [of Iwo Jima] telling us 'The deed is everything.
The glory nothing. Remember our deeds. Remember our deeds.'"
—31st Commandant, General Charles C. Krulak[2870]

WOODY'S, CHAMBERS', BECK'S AND ERSKINE'S stories teach us much about war, society and life. They epitomize the sacrifice many made to rescue the world from the twin thunderclouds of Imperial Japan and Nazi Germany that had enveloped much of the globe by 1941. It took the entire U.S. nation, military and civilians, to strike back against Hirohito, Hitler and their Axis minions to win the war. Without the support of American warriors and supplies, Great Britain, Canada, Australia, the U.S.S.R. and China might have been defeated by Germany and Japan. No Asian or Pacific island country was capable of defeating Japan without American help. China under Chiang Kai-shek offered sustained, stubborn resistance, killing at least 500,000 Japanese troops, but could not defeat Japan alone. The remaining free combatants in Asia were only in China, whose most prosperous regions were occupied by Japan years before Pearl Harbor, and lightly-populated Australia and New Zealand, whose supply line to the U.S. was in danger of being cut until America captured Guadalcanal. Many of Australia's soldiers fought in North Africa and Malaya, but without the U.S., it had little hope of resisting Japan possibly suffering occupation.

It took the Japanese six months to secure much of the Pacific during 1941-42. American isolationists naïvely placed great reliance on the protection the Atlantic and Pacific Oceans provided; however, without the United Kingdom's navy, the Atlantic could have turned into a snake pit of Nazi submarines. Absent the Pearl Harbor attack, the U.S. would have let Japan rule most of the western Pacific, content to control the eastern part. Looking back now, it's clear that Japan's hubris cost the country the opportunity to rule most, if not all, of the Orient had it not attacked the U.S. at Pearl. Brazil joined the Allies sending forces to Italy, but powerful Nazi sympathizers in Argentina might have made a deal with the Axis Powers if they had conquered Britain and prepared to go after the U.S. Without Great Britain in the war, America would not have had the British Isles as an unsinkable base from which to launch bombing sorties, assemble resources and invade Nazi-occupied Europe. By 1941, Great Britain was going broke and experienced shortages of war materiel and food. American aid, including the Destroyers for Bases trade with the U.K. and the Lend-Lease Act pushed through by FDR over opposition from the isolationists, was vital for Britain's and America's survival. Nonetheless, pacifists called Roosevelt a "Dictator" who was committing "Acts of War."[2871] By spring of 1941, a Gallup Poll showed 55% of Americans favored Lend-Lease without qualification, but most were disinterested in going to war.[2872] In fact, until attacked, "at no point did the [U.S.] choose to go to war against the Axis powers and Japan."[2873]

To their everlasting credit, Democrats like Roosevelt and interventionist

Republicans like Wendell Wilkie, William Allen White, Henry Stimson and Frank Knox put their country first and refused to play politics in pushing for more action against fascist regimes. Had America remained neutral any longer, the world would have been a scarier place. Who could say with confidence that freedom and democracy would have triumphed? Japan's attack on America's Pacific Fleet, followed by Germany's and Italy's declarations of war days later, pushed America to join forces with Great Britain, the Commonwealth countries, China and the U.S.S.R. Japan's sneak attack dispelled the myth that neutrality safeguarded the U.S. from war and America's fighting men would soon prove they were strong and courageous, contrary to the weak image of freedom-loving people the totalitarian aggressors spread.

Shortly after Pearl Harbor was attacked, America and Great Britain agreed on making victory in Europe a priority. The lion's share of America's men, weapons and supplies would go there.[2874] However, Admiral William D. Leahy (the president's Chief-of-Staff or Chairman of the Joint Chiefs-of-Staff) and Admiral Ernest King (Chief-of-Naval Operations and Commander-in-Chief, U.S. Fleet) made sure that considerable resources were devoted to the Pacific. By December 1943, Allied resources were more evenly distributed between Europe and Asia. The Allies realized much effort was required to take back occupied lands from Japan.[2875] The capture of Saipan and Tinian, and the liberation of Guam enabled the U.S. Army Air Forces to conduct bombing raids on Japan's home islands. Japan, for a while, had control of the skies in the western Pacific after devastating Pearl Harbor and crushing MacArthur's airpower in the Philippines and, thus, won major victories because of it. America later achieved competiveness and then air superiority over the Japanese, which proved vital to its destruction of Imperial ships, planes and factories. Surpassing the Japanese both in aircraft technology and precision in flight was also essential for amphibious assaults. This achievement was responsible for the success of America's two-pronged island-hopping campaign and its capacity for bombing Japanese cities. Eventually, the Allies rolled back all of Japan's gains. Pearl Harbor did the exact opposite of what Japanese leaders hoped for. Historian Samuel Eliot Morison wrote:

> The surprise attack on Pearl Harbor… was a strategic imbecility. One
> can search military history in vain for an operation more fatal to the

aggressor. On the tactical level, the… attack was wrongly concentrated on ships rather than permanent installations and oil tanks. On the strategic level it was idiotic. On the high political level it was disastrous [it united Americans like nothing else could].[2876]

Yale University political scientist Paul Kennedy said the attack "bordered on the incredible—and the absurd."[2877] The attack "defies the logic of economic and industrial thinkers on both sides of the Pacific that Japan with coal and steel production only one-thirteenth that of the United States, should nonetheless have launched the Greater East Asian War. The results were predictable."[2878] Indeed, everything the Japanese had witnessed the U.S. do internationally up to December 1941 gave them many "reasons to be optimistic" that America might "sue for peace rather than pay the price for victory," but history has shown the Japanese miscalculated the potential returns for the risk they took making the U.S. their arch-enemy.[2879] Simultaneously, other fascist regimes acted foolishly declaring war on the U.S. at a time when Mussolini was trying to secure the Mediterranean and Hitler was still fighting Britain and was engaged in a massive war with Russia which, in December 1941, was not going well for him. Luckily for the world, the fascist leaders had picked a fight with the wrong country in the U.S.

Japan had devoted tremendous resources for years before the war to create large, modern and well-equipped military forces. But when attacked, America mobilized rapidly, transforming its production from consumer goods to weapons. At the time of the Pearl Harbor attack, the U.S. already produced "twelve times the steel, five times the amount of ships, one hundred and five times the number of automobiles, and five and a half times the amount of electricity than Japan did."[2880] The U.S. out-innovated and out-produced Japan in almost every category of weapon, from aircraft carriers, to tanks, from submarines, to rifles. Here are a few examples: From 1941-44, the U.S. produced 261,826 airplanes while Japan produced only 58,822, often of inferior quality. In 1944 alone, Japanese production of rifle and machinegun ammunition was 6.5% of the U.S. and tank production was just 4.7%. The U.S. output of tanks reached 20,357 in 1944 whereas Japan was around 1,000. And their tanks, like with their planes, were second-rate. From 1939-45, the U.S. produced 2,382,311 trucks while Japan could only muster 165,945.[2881] Japan marched off "to war against the West with no means to defeat the U.S., much less an allied coalition."[2882] Historian Edward Drea stated that the Japanese armed forces "produced a military strategy that the nation could not afford."[2883]

So, why did Japan do it? In its long history, Japan had never been conquered; to many Japanese, that was unimaginable. The Japanese hoped the Soviets and British would soon fall in defeat to Germany and that the U.S. would fail to rise to the occasion and reorganize its entire nation to project power across Asia and the Pacific. If these things happened, Japan felt it could conquer China, seize American territory along with European colonies and control Asia. With these hopes, Japan proved that it overplayed its hand. Great Britain and its Commonwealth of Nations remained in the fight against Hitler and Hirohito and made good on Churchill's pledge to "never surrender." Russia, a Japanese neighbor and hated enemy, pounded the Nazi armies on the march toward Berlin. The U.S., while at first devoting most of its effort to defeating the Nazis, eventually took the fight to the Japanese everywhere. And China continued to stubbornly fight Japan.

By 1945, three years after Pearl Harbor, America had rebuilt and expanded its Pacific Fleet; mobilized, trained and armed its army infantry and Marines, and created a modern airforce. By spring 1945, Americans stood at the enemy's gate, able to not only sink battleships like the Japanese did at Pearl Harbor, but also effectively destroy the Japanese navy, cripple its economy and kill hundreds of thousands of its people. No longer the dominant predator, the Japanese military and Hirohito knew they were now the hunted and Americans were coming for them. Still, Japanese leaders thought that if they could not defeat America, they could fight a tenacious defensive action that would break America's will to fight, enabling Japan to win a favorable peace.

However, the U.S. would never entertain a truce like the one the Japanese desired. And the fighting ability of America's soldiers, sailors, airmen and Marines proved more than a match for Japan's servicemen in order to force Hirohito to surrender unconditionally. In analyzing the quality of the fighting men of Iwo Jima, Lieutenant General "Howlin Mad" Smith, commander of FMFPAC said, "The teamwork of all services and the grit, devotional skill of the Marines who refused to be stopped reduced those defenses, won the island, and gave convincing proof to the Japanese that the best they had to offer was not good enough."[2884]

Besides its tenacious warriors and superior volume of materiel, another reason America prevailed over Japan was the coordination of the different services in contrast to the Empire's navy and army:

> The paralysis in Tokyo was so great that not only did the Navy and Army separately exercise veto power over policy and even the makeup of governments, but were constantly at loggerheads with each other. The high commands in Tokyo could not control hotheads in the field who made their own military decisions and had demonstrated that they were prepared to kill superiors who showed insufficient zeal or reverence for the 'Imperial Will.'[2885]

In fact, while Marine General Clifton Cates was stationed in China from 1936-39, he witnessed a Japanese army unit engage in a firefight with a navy unit. "The Jap army and the navy hated each other as much as anyone could."[2886] Yamamoto and Tojo bickered with each other like little "children" over the roles each service should play and how many planes should be built.[2887] Yamamoto felt Tojo and his ilk were a "pack of 'damned fools'" and that their entire conduct during the China War was a "costly drain of manpower and military resources."[2888] The army and navy were "pitted against each other in an eternal brawl over funding, political influence, and the allocation of resources."[2889]

America benefited from a unified chain of command under civilian control of the President, who was Commander-in-Chief. The civilian Secretaries of War and the Navy worked in close cooperation. The heads of the navy, army and army air force met regularly to plan the war and issued orders to their subordinates. Any issue they could not resolve could be decided by President Roosevelt. Effective cooperation between the navy, Marines, army and army air force in planning and execution was exemplified at many battles like Guam and Okinawa. Although there were disagreements between Nimitz and MacArthur, and sometimes lack of respect and coordination between the services, like the Marines and army displayed on Guadalcanal, Saipan and Peleliu, still, compared to Japan, America's forces worked well together.[2890]

Marines worked closely with sailors, especially Seabees. On Iwo, the Marines served with 7,000 of them, and every Marine appreciated them: Lieutenant General Smith said they "never let us down."[2891] Seabees served on the front lines, proving themselves fast and effective in the construction and repair of air fields, roads, and fortifications; conveyed men and equipment over the reefs; and provided water desalination systems. Sailors transported Marines and soldiers over long distances to the battlefield, with coxswains steering the variety of specialized landing craft carrying Marines and soldiers along with their weapons, ammunition, food, water, communications equipment, artillery, tanks and supplies under dangerous conditions, through enemy obstacles and fire and onto the beaches. They helped evacuate the wounded to hospital ships where navy doctors, dentists and nurses cared for them. Corpsmen moved with Marines on the battlefield and provided first aid and evacuation of casualties to medical stations. Gunnery teams went ashore with Marines to direct naval gunfire on targets, as well as to assist fighters and bombers flying close air support. Marine and navy pilots and the army air force helped destroy enemy aircraft and airfields that could be used to bomb and strafe Marines and their fleet. Navy chaplains appealed to God with servicemen and offered what comfort they could along with last rites and burials. In contrast to the IJA and IJN internecine battles, U.S. Marine, army and navy leaders worked well together and respected each other enough to achieve victories.

Japan also lacked flexible command—if an officer died in combat his men were often at a loss of who should assume leadership.[2892] As Princeton University historians Philip A. Crowl and Jeter A. Isely wrote, the Japanese soldier was "unimaginative in battle."[2893] They also employed outdated tactics. Although the Japanese military exhibited skill, determination and boldness of leadership, they glorified death, preferring *Banzais* to less glorious but more effective methods. One historian estimates that out of the 1,140,000 Japanese army dead, 200,000 of them died in inefficient charges. The highest value to which a Japanese subject could aspire was "death in the nation's service, followed by apotheosis as a national deity."[2894] It has been said that "Religion's surest foundation is the contempt for life."[2895] Japanese disregard for life gave them *Kamikazes* and glorified the desire of hundreds of thousands of their men *to die in battle*. It also caused them to leave their men on ignored isles until they perished, often from starvation, rather than relocating them to more secure and better defended islands.[2896] Of the 2.3 million military deaths Japan suffered, 1.4 million were due to "starvation, malnutrition and…diseases."[2897] Military strategist Bernard Brodie wrote:

"The islands and other territories seized by Japan following her Pearl Harbor attack in December 1941 turned out to be liabilities in the latter stages of the Second World War, when her garrisons were isolated and immobilized."[2898] Japan had "outrun its supplies, especially of liquid fuel" and it failed to continue its massive wave of momentum after the summer of 1942. And soon thereafter, its logistics capabilities dwindled as enemy attacks from the sea and air killed its supply chain throughout its Empire.[2899] Historian Yuki Tanaka derisively calls his countrymen's war from 1941-45: "The Japanese war of starvation."[2900]

And not surprisingly, the Japanese policies in the lands they occupied only created hatred toward them as they subjugated and slaughtered civilians everywhere they ruled. Had they spent half the time building up the communities and being kind to the locals as they did pulling down their pants and raping women and tying up victims and slicing off their heads (pursuing *randori*),[2901] they might have received more support and supplies from their East Asian Co-Prosperity Sphere. Instead of winning hearts and minds, they committed innumerable atrocities and incurred the hatred from the populations they controlled.

In contrast, the U.S. military had several attributes allowing it to overcome its enemies. Marines, along the chain of command, were taught to be leaders and to assume command of their units if their superior was wounded or killed. Basic platoons (roughly 44 men) were broken into three squads (10-13 men each) and each squad was split into four fire-teams (3-4 men each). It had a platoon leader (lieutenant), a platoon sergeant, squad leaders, a runner, a guide and a Corpsman. These "maneuver units" gave "the squad leader greater tactical flexibility to employ the increased combat power at his disposal."[2902] Each group knew how to operate within the larger whole. It was like multiple football teams went into battle with the same playbook and played all the positions so everyone, regardless of what position they took, knew the plays. When taking heavy casualties, some Corporals who ordinarily led squads found themselves acting as platoon sergeants. During the Pacific War, "the Marine rifleman had developed into such a finely honed instrument of destruction that squad leaders, platoon commanders, and company commanders wielded the majority of combat power" and during training, they often role-played others' positions to ensure the command structure never collapsed while under fire like was seen on Tinian and Iwo Jima when Gunny Renstrom took over the leadership of entire platoons after their lieutenants and sergeants became casualties.[2903] Chambers' platoon proved the value of this flexibility when his depleted group of *only* 20 men took out 32 emplacements on 23 February 1945. It is a certainty that these 20 had never operated as a single group before, but when called to do so, their training took over and they burned, exploded and killed their way through one of Kuribayashi's toughest defenses as a coordinated team and helped Woody earn the MOH.[2904] "Charging forward into sheer death and destruction, the fighting spirit of the amphibious assault warrior seemingly matched the fatalistic fanaticism of his Japanese foe."[2905] Marines did not glorify death, but they willingly risked their lives to aid comrades and accomplish their mission.

In addition, U.S. military personnel, unlike those of Imperial Japanese, created friendship and bonds of trust in the communities they took over as seen with Woody's and Beck's friendly interactions with the natives of Guadalcanal and Guam and those of their comrades on Saipan and Tinian like Reed and Renstrom displayed. They did not exploit the areas economically or rape women under their control, even in occupied Japan when the Japanese expected them to do so—due to Japan's propaganda, the U.S. shocked them by not immediately doing to the Japanese what Japan had done to others throughout its Empire. In the end, the U.S. conferred on the Japanese a peaceful government with self-determination and full representation, giving the country's women the right to vote. When General of the Army MacArthur enacted this policy, his Military Secretary, Brigadier General Bonner Fellers, said, "The Japanese men won't like it." MacArthur quickly retorted, "I don't care. I want to discredit the military. Women don't want war."[2906] And since 1945, Japan has not waged war and has had relatively peaceful dealings with its neighbors thanks largely to the U.S. A Co-Prosperity Sphere was created wherever the U.S. conquered in WWII, not only in name, but also, and most importantly, in deed.

Woody, Beck, Chambers, Dashiell, Renstrom, Hemphill, Erskine and other Marines, sailors, coastguardsmen, soldiers and airmen helped bring about the victory that ensured this new world, both with their training and with their behavior. The victory these men helped seize at Iwo has played large in American cultural consciousness. Many books and movies have highlighted it. From John Wayne's famous role in the *Sands of Iwo Jima* (1949) to Clint Eastwood's productions of *Flags of Our Fathers* (2006) and *Letters from Iwo Jima* (2006), the battle has cultural appeal. The second flag-raising on Mt. Suribachi is probably the most iconic war photograph in history, and is the inspiration for one of the best-known Washington D.C. monuments. It has been commemorated in countless ways, from stamps to campus monuments; from art shown in the Smithsonian to the architectural form of the Marine Corps Museum in Quantico, Virginia. Naval vessels have been named after it like the *USS Iwo Jima LHD-7*, and the men who fought there, including the navy guided-missile destroyer *USS Cole* (made famous when Islamic terrorists attacked it in 2000 at Yemen's Aden harbor, killing 17 of her crew) and the *USNS Hershel Woody Williams*.

This battle has become one of the most symbolic events in American history exemplifying heroism, bravery and success. Iwo veteran Billie Griggs said, "We got the job done. I cheated death so many times, I'm surprised I'm still here. However, if my country would ask me to do Iwo Jima again, I'd do it. I wouldn't want to, but I'd still do it."[2907] The mentality of these men moving forward on Iwo against incredible danger exemplified the "can-do" American spirit. Rabbi Gittelsohn reflected:

> The wonder is not that we suffered such grievous losses, but that we succeeded in taking the island at all!...No one will ever again be able to use the words 'American' and 'impossible' in the same sentence to me...Along with a humble respect for the average American's courage, I carried back with me from Iwo an admittedly egotistical pride in the fact that for [an American], nothing is impossible.[2908]

American victory on Iwo came at a terrible cost, though. Almost 7,000 Marines and navy sailors lost their lives over the 36-day battle. To put the battle in context for its brutality, more men died in this battle than all the deaths the U.S. has suffered in Afghanistan and Iraq in almost two decades of war since 2001. Additionally, over 19,000 were wounded there.[2909] Many carried their wounds for the rest of their lives. Men suffered gut shots, stab wounds, shrapnel lacerations, and traumatic brain injuries. Some lost limbs or eyesight. Like every war, many had psychological wounds, now identified as post-traumatic stress disorder (PTSD), originally called battle/combat fatigue, shell-shock or psychoneurosis.[2910]

As one has read, many around Woody during the campaigns became casualties of war like Lefty, Scrivens, Bornholz, Fischer and Waters. Many paid the ultimate sacrifice for freedom. That America was able to bring such a nation as Japan to its knees while also devoting its resources to defeating Hitler and Mussolini was a testament to its strength. The cost was great with 400,000 dead Americans plus the millions who sustained injuries.[2911]

On a personal note, my uncle, Army 1st Lieutenant Frank H. Rigg, fought on Angaur in the Palau islands in 1944. He was injured and left with a deformed right hand. He never got over his combat, hated the Japanese and, like Woody, was driven to drink after the war. My Aunt Doris often found him alone in the garage, staring at the wall suffering flashbacks. She remembered him screaming in his sleep and jumping out of bed. Woody's platoon leader, Howard Chambers, struggled like my uncle. "When he returned from the war," his daughter Barbara Olson said, "He was a shattered man. At first, he drank way too much. Every time he tried to talk to us about the war, he'd break down in tears."[2912] Chambers also went from a fit 190 pound 5'10" hard-as-a-rock Marine in 1945 to a limping, full-of-pain, partially-deaf and overweight man of 260 pounds by 1955.[2913] He struggled physically and psychologically, although, eventually, like Woody, he pulled it together and lived a productive life, taking care of his kids, playing tennis regularly and being a successful businessman.[2914]

Woody suffered too, but Woody's religious conversion and his wife's love are credited with saving him. Woody's postwar experiences remind us of the psychological cost a nation's warriors and their families bear during and after war, and the importance of taking care of our veterans.

Frank Rigg's, Chambers' and Woody's stories detail how the Pacific War was won. America faced in the Japanese enemy, a belligerent, racist, religiously fanatical and criminally brutal adversary. Nowhere in its annals of warfare had America met with a more determined foe, armed with state-of-the-art weapons, and backed by millions of civilians. Overwhelming violence had to be brought against Japan for it to surrender, its Emperor to renounce his divine status, its military aggression to cease and its government to embrace liberal democracy. Defeating the Japanese on Guam, Iwo and throughout the Pacific, they had to be shot, shelled, sealed in or blown up in their caves, tunnels and fortifications or cremated with flamethrowers. On Iwo alone, 10,000 gallons of flamethrower mixture were used daily for both troop-carried models and *Zippo* tanks (360,000 gallons total for the 36-day campaign).[2915] The 3d MarDiv utilized 172 flamethrowers during the 36-day campaign; Woody alone used 6.[2916] Eric Hammel wrote, "without flamethrower assaultmen willing to take enormous gambles, the advance on Iwo would have bogged down at the beach."[2917] Japanese bone collectors who yearly explore the island to find the remains of soldiers to cremate and dispose of find two types of bones: "White bones" of men "killed by bombs, explosives and bullets," or those "who got progressively weaker and died" and "black bones" of men "burned to death by flamethrowers."[2918]

Woody's battalion wrote its assessment of explosives: "The following weapons were, in order of listing, found most effective against fortified installations. (a) Shape Charges...[and] (b) Composition C.2 (in 18# Knapsack as 'Satchel Charge')."[2919] Woody, as a demolition man, was responsible for these options that blew up enemy positions often after they were incinerated. The report then moved to flamethrowers:

> [P]ortable flamethrowers used by this B[attalion] were found to be... extremely effective against caves, bunkers, and pillboxes encountered during this operation. Normal employment consisted of moving the flamethrower operator into position from a flank, covering him then meanwhile with rifle and BAR fire. Upon arrival within range of target, the operator would shoot his flame across the cave or emplacement entrance or embrasure, and advancing quickly under cover of his own fire, place himself in position to direct his fire <u>directly</u> into the cave opening or embrasure. When so directed, flamethrower fire was...effective in silencing the cave or emplacement to...allow the placing of a demolitions charge.[2920]

Woody's battalion commander, Lieutenant Colonel Marlowe C. Williams, the report's author, best explained what men like Woody and support teams did. Williams' description of this weapon as "extremely effective" conveys its importance as one of the most lethal instruments against fortified positions. Out of the 34 days Woody's battalion was engaged on Iwo, it used flamethrower operators for 24. The after-action report noted, "excellent use of the flamethrower was made in this operation and functioning was very good."[2921] Historian and U.S Army Brigadier General John Mountcastle wrote, "In Pacific ground combat, the flamethrower was frequently regarded as the most valuable weapon on the field."[2922] He further stated, "The Marines were the best at using the flamethrower during WWII."[2923] Men like Woody, Rybakiewicz and Nosarzewski, using them the way they did, helped add to this reputation. And Woody ensured what he learned he taught to others in real time.

Woody's responsibility to instruct and lead during combat displayed the Marines' resourcefulness in teaching tactics and weapons *during battle*. Knowing three companies (A, B, and C) made up Woody's 1st Battalion and that they suffered dreadful losses, one understands why few of the original flamethrower/demolition men were left by campaign's end. The 21st Regimental report noted flamethrower operators suffered 92% casualties in the 2nd Battalion alone.[2924] If companies in Woody's battalion suffered similar attrition, roughly only three flamethrower/demolition men serviced the battalion instead of the original 21 after weeks of battle. Because these weapons were utilized throughout the campaign, Woody not only used the most effective weapons against defense systems (explosives and flamethrowers), but also would have been responsible for teaching these weapons to new arrivals although he also had flamethrower operator/demolition men Joseph Anthony Rybakiewicz and Marquays Kenneth Cookson there to help too. Both survived 34 days on Iwo like Woody. In fact, Rybakiewicz reported that five out of 15 flamethrower operators in his unit, including him and Woody, survived, so there were possibly two others there besides Cookson, Rybakiewicz and Woody.[2925] Replacements quickly learned how to use explosives and flamethrowers against enemy positions, requiring difficult on-the-job-training. As Woody noted, replacements rushed to the front were unprepared to use most of their weapons, arriving directly from boot camp. Marine leadership at Iwo prepared for the worst-case scenario placing the 3d MarDiv in reserve to be used in case of trouble although it expected weak resistance and believed the division would not be utilized. Instead 3d MarDiv regiments were deployed by the third day of battle to everybody's surprise and replacements were used by the thousands to help replenish the large losses from the 3d, 4th and 5th MarDivs.[2926]

Since flamethrowers were crucial and so few could operate them, Woody's battalion commander, Lieutenant Colonel Williams, wrote on 6 April 1945 that *all* Marines should learn to use them.[2927] By May 1945, Woody's *21st Regimental Newsletter* wrote, "3d Division Marines are indoctrinated at an early hour into the niceties of spreading this flaming death. The Marine Corps wants every combat man to know the rudiments of flamethrower operation—just in case."[2928] Even with deadly attrition of flamethrowers, the 2nd Battalion's commanding officer of Woody's regiment noted: "[T]here is no question that the [weapon] is a valuable and necessary one in the reduction of [emplacements] and the neutralization of caves."[2929] Major General Erskine echoed his battalion commanders, remarking the flamethrower was an "excellent weapon against emplacements," and "very effective" for closing caves. However, due to high casualties, "it was difficult to keep flamethrowers manned with personnel of any experience at all."[2930] Thus, toward battle's end, Woody's know-how helped ensure success at his front. Woody knew every time someone touched that weapon, his chances of dying increased dramatically, but the enemy's did even more. The flamethrower was one of the most effective tools to counter the tactic of fighting from underground. The importance of this weapon was also evidenced by the fact that the U.S. produced 14,000 M1-A1 first generation and 25,000 M2-2 second generation flamethrowers during WWII.[2931] At the war's beginning, the Germans and Japanese had superior flamethrowers, but by war's end, Americans had developed flamethrower technology that surpassed that of their enemies.[2932]

The effectiveness of this weapon justified the loss of the majority of the troops using it. These reports demonstrate what a crucial role Woody played during his time on Iwo. He was one of the last survivors of his original cohort with this unique specialty, and one of the few in the battalion who had been trained with flamethrowers and explosives. This placed him in a critical position educating others how to use these weapons in actual combat based on real battlefield experience. These lessons learned on Iwo would be disseminated in the next campaign, once again on Japanese soil: Okinawa. This engagement was fought on a bigger island (466 square miles) south of the home islands. This last Pacific battle took place from 1 April to 22 June 1945 just as Woody left Iwo. The Japanese also had adopted the fortified defense in depth approach that they had used on Iwo, Angaur and somewhat on Peleliu. The largest amphibious invasion of the Pacific War was mounted there on Okinawa numbering 1,457 Allied ships, transporting 541,866 troops.[2933] The offensive techniques used on Iwo were used again on Okinawa, only moreso. Of the 107,539 Japanese who died on this island, 27,769 were sealed in caves or burned alive by flamethrowers often followed by C2 explosives.[2934] Demolition men and flamethrowers continued to lead American warriors on the front. On Iwo alone, an official history wrote, the only way to get most Japanese was with "the flames and the grenades from men who were willing to crawl close enough to throw them." That took courage, and those who undertook such perilous duty helped America win the war.[2935]

Woody's, Chambers', Beck's and Erskine's experiences on Iwo illustrate how tactics had changed for both Japanese soldiers and U.S. Marines after Guadalcanal. The Japanese no longer fought an aggressive, mobile offensive supported by naval gunfire, fighters and bombers with extensive reinforcements. They abandoned their tactics of opposing beach landings and quit using *Banzais*. Instead, Kuribayashi proved the advantage of developing defense in depth and using men sparingly, deploying them in expertly-designed pillboxes, blockhouses, caves, tunnels, and buried tanks, and using nighttime infiltration tactics. Lieutenant General Smith wrote of Kuribayashi's defense: "Iwo Jima was notable in that organized resistance did not collapse after the first few days, but continued to the end."[2936] Later, Smith called Kuribayashi Hirohito's "most redoubtable" commander and the best general "in the whole Japanese Army."[2937] Kuribayashi knew that if the attack passed through his defense, he would fail to continue to inflict casualties on the Marines, so obviously part of his training was to teach his men how to stage organized retreats to make sure the integrity of his defensive lines remained strong and that there were many of them to fall back to when the need arose.

Kuribayashi's main strategic goal was to inflict massive casualties against the Americans. His battle plan was sound, but his "object mad."[2938] It was sound in that he was prepared to inflict maximum damage on the enemy, but mad in that by doing so, there was no hope America would settle for less than total victory and unconditional surrender. He made the Americans feel the way he wanted and forced them to rethink their strategy, but not necessarily in the manner he had hoped. Instead of causing the U.S. to shrink from war, it motivated America to intensify its bombing of Japan and drop the atomic bombs on two of its major cities in an effort to prevent more Iwos. Neither Kuribayashi nor Hirohito realized that making victory by conventional means too costly made deployment of super-weapons against Japan a certainty.[2939]

After Iwo, Japan prepared its citizens to psychologically embrace *gyokusai*—the conviction to die heroically in battle rather than surrender. The government produced a slogan *ichioku gyokusai*, meaning "the shattering of the hundred million like a beautiful jewel." Leaders taught their citizens to embrace their own extermination, feeling it would provide spiritual purification.[2940] Fighter ace Saburō Sakai wrote, "There was no doubt in anyone's mind that the end was near, that soon the fighting would be transferred to our soil. There was no possibility of surrender. We would fight to the last man."[2941] His wife Hatsuyo carried a knife she would have used on herself instead of falling into American hands.[2942] At the highest levels of the government, leaders believed "Japan must fight to the finish and choose extinction before surrender."[2943] This mentality permeated the Japanese strategic vision (Japan must be defended at all costs) and emotional bearing (Japan is worthy to die for even in the face of defeat).

Iwo Jima, along with other battles like Guadalcanal, Tarawa, Saipan, Guam and Peleliu, highlighted this nationalist fanaticism and how dreadful the prospect of a land war with Japan was. Lieutenant General Smith explained: "In two wars I have fought Germans and Japanese. Throw a hand grenade into a German pillbox and they come out with hands reaching for the sky…shouting '*Kamerad*!' Throw a hand grenade into a Japanese pillbox and they throw it right back at you."[2944] Lieutenant General Erskine echoed Smith: "The Germans had sense enough to get the hell out or surrender when it was hopeless. The Japanese didn't."[2945] By way of illustration, often when GI's presented flamethrowers near a *Wehrmacht* bunker and pumped a few serpent streams of fire in the air, the Nazis came out with hands high. This study has not documented a case where the Japanese surrendered in the face of being incinerated.[2946] The 4th MarDiv general, Clifton Cates, describing the discrepancies between these Axis troops, said, "thank God the Japs didn't have the German know-how and thank God the Germans didn't have the Jap's tenacity and [lack of] fear to die. Because if you put them both together, it would be bad. The Japanese had the courage. I must admit, it was dumb courage and the Germans had the know-how and the training. But [the Germans] would give up."[2947] The IJA, in general, had a fatality rate of almost 100% in battle.[2948] Persuading the Japanese to hoist the white flag was difficult because when closer to defeat, the Japanese increased their ferocity, as illustrated by *Banzai*s and *Kamikaze*s and their exhortation for citizens to die to the last person in case of invasion. The Japanese, who had rapidly modernized their industry and military, lived in a grotesquely warped, repressive, totalitarian society whose growth "was stunted in every way; in cultural evolution and in mental and emotional development, both as individuals and as a group."[2949] It was difficult to change the conviction of their ruler not to surrender. So Iwo among other battles made American leaders aware of what they faced when invading Japan.

If America had to send over 800 ships[2950] and over 230,000[2951] men to invade and conquer eight square miles of territory defended by almost 22,000[2952] fanatical troops like those at Iwo, how much more would it have had to expend conquering Japan's 146,000 square miles with 71 million citizens (4 million servicemen and an additional 18 million citizen soldiers (men between the ages of 15-60 and women between ages 17-40))? Children were trained "to carry backpacks of explosives and to throw themselves under the treads of tanks [as] 'Sherman carpets.'" The government renamed schools "National Schools" and assigned teachers "the crucial task" of educating the "children of the Emperor" to sacrifice themselves for Japan. In fact, children 12 years and older were trained to fight the invasion. Even married women exercised with bamboo spears to combat invaders.[2953] Land war on Japan's mainland would have seen not only some of the bloodiest battles of WWII, but also some of the bloodiest in history.

Secretary of War Henry Stimson's staff estimated conquering Japan would cost 1.7 to 4 million casualties, including up to 800,000 dead (double what the U.S. had already experienced). It would cost the Japanese between 5 to 10 million dead (half of the 381,550 Japanese in the Philippines, and, at least a third, if not one-half, of the 300,000 Okinawan and the 40,000 Saipan/Tinian Japanese civilians died during these battles alone).[2954] In preparation for invading Japan, America produced 500,000 Purple Hearts. After all the bloody wars since WWII, there are still 120,000 of them left in stock after thousands were awarded to personnel from campaigns in Korea, Vietnam, Iraq, Afghanistan and Syria.[2955]

These medals were not needed because the bombing campaign launched from the Marianas culminated in the detonation of two atomic bombs that ended the war. These massive attacks convinced Hirohito to command his subjects to lay down arms.[2956] The weapon's destructiveness had the necessary psychological impact to make Hirohito alter his thinking. He said on 7 August, "we should lose no time in ending the war so as not to have another tragedy like this."[2957] After the first bomb's deployment, the might of the Soviet Union bore down on

Japanese forces in Manchuria with 1.5 million troops killing thousands after Stalin *officially* declared war on Japan on 8 August. This made Hirohito want to move more quickly on the surrender agreement. Then with the explosion of the second bomb on 9 August, he knew he needed to act. While he waited to address his nation, thousands of his 800,000 citizens in Manchuria started to kill themselves fearing surrender and the enemy, like those on Saipan, Tinian and Okinawa had done. Once again, they killed themselves with grenades and cyanide handed out to them by none other than IJA soldiers in spiritual harmony. IJA units numbering 665,000 in Manchuria started to disappear by the thousands under the red wave of Soviet superior armor, planes and infantry. The IJA offered mad, useless resistance while still in loyal service to their nation and Emperor.[2958] Sadly, just like the Japanese had done to their conquered people, Japanese civilians and prisoners were subjected by the Russians to "rape, pillage and murder."[2959]

These events prompted Hirohito to sue for peace immediately, saying "I have given serious thought to the situation prevailing at home and abroad and have concluded that continuing the war can only mean destruction for the nation and prolongation of bloodshed and cruelty in the world...the time has come when we must bear the unbearable."[2960] Even with an imminent invasion of Japan by two major powers and with no chance of a "negotiated peace" through Russian diplomats, Hirohito waited days to address his nation due to fear of retribution from military diehards who wanted to continue fighting.[2961] Even after the U.S. Navy's submarines had tightened their coil around Japan, sinking most supply ships coming and going from Asia, after the introduction of the B-29 bombers in destructive bomb raids and after atomic bombs devastated two cities, IJA reactionaries attempted to capture Hirohito, destroy his recorded surrender message and kill his advisors to prevent capitulation. They stormed the Imperial Palace, shooting and beheading those opposing them. They searched for the as yet unbroadcasted phonograph recording of Hirohito's surrender announcement, but did not locate it. The coup failed on the day of the broadcast. The conspirators, seeing they had failed, committed suicide in front of the palace.[2962] After this threat was eliminated, on 15 August, Hirohito announced over the radio with his "high-pitched voice" "to a weeping nation" Japan's surrender. It was the first time his people had heard him speak:[2963]

Japanese POWs on Guam on hearing the surrender announcement of Hirohito bow their heads in disbelief. Many of the men's faces here also show pain and anguish at what Hirohito was telling his nation: i.e. Japan gave up because it could not take the punishment any longer its enemies was dishing out on it. MARC

We declared war on America and Britain out of Our…desire to assure Japan's self-preservation and the stabilization of East Asia, it being far from Our thought either to infringe upon the sovereignty of other nations or to embark upon territorial aggrandizement…Despite the best that has been done by everyone—the gallant fighting of military… forces, the diligence and assiduity of Our servants of the State and the devoted service of Our one hundred million people,[2964] the war situation has developed not…to Japan's advantage, while the general trends of the world have all turned against her interest. Moreover, the enemy has begun to employ a…cruel bomb, the power of which to do damage is indeed incalculable, taking the toll of many innocent lives. Should We continue to fight it would not only result in the… obliteration of the Japanese nation, but also it would lead to the total extinction of human civilization. Such being the case how are We to save the millions of Our subjects or to atone Ourselves before the hollowed spirits of Our Imperial Ancestors? This is the reason why We have ordered the acceptance of the Joint Declaration of the Powers…We are…aware of the inmost feelings of all ye, Our subjects [that we ask you to surrender!]…We have resolved to pave the way for…peace for all…generations to come.[2965]

As he had repeatedly done throughout WWII, Hirohito packed his rescript with lies. Not only had he shown himself during the war to be a Holocaust-minded, warmongering dictator, in his radio address he again showed he was a liar. His brazen claim that Japan's actions were for the good of Asia and that Japan had never wanted to "infringe" upon other nations' rights was clearly false. Historian Saburō Ienaga said, "It's a complete lie to say that Japan made aggressive moves in Asia to liberate Asians from the control of Europe and America."[2966] Hirohito's hyperbolic assessment that if Japan did not lay down arms, it would result in the demise of its culture and civilization as a whole took his speech from "understatement to exaggeration."[2967] Without surrender, Japanese culture would have taken more blows from Allied forces, but civilization in the form of Western democracy would have continued without Japan. The fate of civilization was not in question, but the survival of Japan was. Nevertheless, he said something that psychologically was important here by asking his subjects to think of their ancestors. By doing so, he, as a god, could re-shape people's views implying the gods wanted this peace because they would not otherwise have allowed such a bomb to hit Japan. This part of his talk was a wise tactic since he was issuing a new version of the religious code of "never surrender."

Hirohito's risky and bold move worked. The bombs turned his thinking around, and he was the only one who could persuade his people to lay down arms. Weeks later, on 2 September 1945, a Japanese delegation signed the surrender on the *USS Missouri* in Tokyo Bay. Since his subjects believed he was a god, only such authority could end the 2,600-year span from Japan's origins of never surrendering to a foreign power. They accepted the decision because "His Majesty's orders come before anything else."[2968] Knowing Nagasaki and Hiroshima's destruction changed the mind of a Japanese god to surrender and prevent deaths that would have become staggering, millions of weary GIs, Marines and their families thanked *their* God for the atomic bombs. Had Japan not surrendered, the U.S. had prepared more atomic bombs in case the first two failed.[2969] As President Truman said after Hiroshima, "If [the Japanese] do not now accept our terms, they may expect a rain of ruin from the air, the like of which has never been seen on this earth."[2970] Truman was not full of bravado here. Two more bombs were being readied on Tinian to be dropped on Japan on 13 and 16 August.[2971] America wanted to avoid facing massive Japanese armies full of soldiers and civilians who would fight to the very last death. Luckily for the Japanese citizenry, and America, it took only two atomic bombs and not four or eight to do the trick. "Political objects can greatly alter during the course of the war and may finally change entirely *since they are influenced by events and their probable consequences*" which the atomic bombs indeed did. These bombs brought about the exhaustion of Japan's *"physical and moral resistance."*[2972]

Picture of Hiroshima's destruction after the atomic bomb attack from 6 August 1945. The aircraft commander on the *Enola Gay*, Captain Robert A. Lewis, witnessed the "massive blinding flash of the explosion," and then shouted to his crewmembers, "My God, look at that son-of-a-bitch go!" At that time, roughly 80,000 souls were immediately killed and many more would follow in the next days. National Archives, College Park

Picture of the blast plum after the second atomic bomb attack on 9 August 1945 on Nagasaki. When a similar plutonium bomb had been tested in the U.S. a month before "generating a fireball with a temperature four times" that of the Sun's center, its inventor J. Robert Oppenheimer quoted the apocalyptic *Bhagavadgita* saying, "the radiance of a thousand suns…I am become as death, the destroyer of worlds." National Archives, College Park

The blast plum of the atomic bomb attack on Hiroshima on 6 August 1945. President Truman declared after he heard this bomb was successful, that this accomplishment "was the greatest thing in history." National Archives, College Park

Historian Richard Frank said, "In face of [the] evidence, it is fantasy, not history, to believe that the end of the war was at hand before the use of the atomic bomb."[2973] Veteran Paul Fussell wrote: "[F]or all the fake manliness of our facades [of attacking Japan], we cried with relief and joy [when the bombs were dropped]. We were going to live. We were going to grow up to adulthood after all."[2974] Now men started saying to one another, "Home Alive in '45."[2975] Although many Americans understood neither the science of atomic bombs nor the destruction they inflicted, they knew the impact they had; namely, they ended the war that most felt was years away from concluding. Servicemen knew that the longer the fighting lasted, the more likely they were to be wounded or killed. Fussell wrote that the bombs brought life to many by sparing them from deadly combat and they were relieved. General LeMay said: "It did not take a genius to realize, that after what the military and civilians did on Saipan, and later Okinawa [suicide and useless resistance], that not dropping the two atomic bombs would have been an exercise in fucking stupidity."[2976] Churchill echoed LeMay telling the Commons on 16 August, "[A]fter the amount of money, time, expertise and effort that had gone into building the Bomb, it would have been unacceptable for soldiers to have died in their hundreds of thousands to salve politicians' consciences about using it. It was approved overwhelmingly by the public, and especially by the armed forces, at the time."[2977] Churchill also rightly described the atomic bomb as the "Second Coming in wrath."[2978]

As mentioned before, the atomic bombings were approved in response to learning and believing in the absolute resolve the Japanese had with regard to martyrdom. As President Truman said, "Nobody is more disturbed over the use of Atomic bombs than I…The only language [the Japanese] seem to understand is the one that we have been using to bombard them. When you have to deal with a beast you have to treat him as a beast."[2979] If the atomic bombs had not worked, Japanese leadership would have sacrificed millions in futile suicide actions. When Truman learned of the successful atomic bomb test, he decided to use it against Japan. He wrote in his diary that this was necessary because the Japanese "were savages, ruthless, merciless, and fanatic."[2980] After the bombs had been dropped, Truman said:

> We have used [atomic bombs] against those who attacked us without warning at Pearl Harbor, against those who have starved and beaten and executed American prisoners of war, against those who have abandoned all pretense of obeying international laws of warfare. We have used it in order to shorten the agony of war, in order to save the lives of thousands of young Americans.[2981]

One can say concerning the bombs' use, LeMay, Churchill and Truman were kindred spirits.

The Japanese convinced Americans of their suicidal zeal as warriors and in response, the U.S. gave them a strong message. It was a message of *Vernichtungsstrategie*, the strategy of annihilation.[2982] Until the atomic bombs were dropped, it really did appear that every Japanese leader and citizen was willing to die defending their homeland. After Iwo, the 3d MarDiv noted, "The use of various types of suicide tactics, including those involving aircraft [*Kamikaze*s], piloted rocket bombs, small boats [*Shin-yō*], suicidal Navy Frogmen [*Fukuryu Tokko Taiinzo*] manned-torpedoes [*Kaiten*], has been developed greatly and can be expected to increase as the Allies approach closer to or invade the Japanese home island."[2983] In other words, Japanese fanaticism only increased as the Allies got closer to Japan's home islands and Emperor. After the bombs were dropped, Hirohito realized the destruction the bombs could render. He saw that the war's continuation would bring about Japan's annihilation. By dropping the bombs, the U.S. took the war to another extreme and forced its enemy to do its "will."[2984]

The dropping of the bombs and Japan's surrender had another effect on Japanese culture. On 1 January 1946, Hirohito, in his "Humanity Declaration" rescript and his last act as a supposed deity, renounced his "claims to heavenly descent," breaking a putatively divine line of his ancestors spanning from 660 B.C.E. Japan's defeat brought down *its supposed god*. The infallible and divine Hirohito became a mere mortal like his subjects.[2985] Yale University historian James B. Crowley wrote, "The nature of this defeat [of Japan] was overwhelming, numbering among its victims the efficacy of a religious myth—a belief in the divine qualities of the Imperial institution."[2986] This shows the war's dramatic impact on Japan's psyche and its beliefs. The psychosocial change was immense. The war changed how an entire nation thought about its gods, origins and future. Americans and many around the world thanked Roosevelt and American military leaders for developing the bomb ahead of the Nazis and Japanese. And they were thankful Truman was willing to use it. General LeMay, the man responsible for carpet and fire-bombing Japan, was more blunt about the reasons for using the bombs:

> If Truman had been an idiot humanitarian, and not wanted to drop those damned bombs, I would have probably resigned. The clarity of

such an action was unmistakable. Only a foolish asshole would not use any and every weapon at his disposal to win a goddamned war.[2987]

Moreover, the atomic bomb attacks that ended the war saved Japan from Communism. After invading Manchuria, Stalin would have continued to Japan's mainland. A delayed end of the war could have resulted in a North and South Japan after the end of hostilities (like East and West Germany in Europe and North and South Korea).[2988] Moreover, the bombs saved a few million people from death in battles for Japan and others from continued warring (and raping) on the Asian continent and on islands Japan still occupied. Historian Richard Frank wrote:

> What is clear beyond dispute is that the minimum plausible range for deaths of Asian noncombatants each month in 1945 was over 100,000 and more probably reached or even exceeded 250,000. Any moral assessment of how the Pacific war did or could have ended must consider the fate of these Asian noncombatants and the POWs.[2989]

Many are thankful for the bombs since they ended Japanese slaughter and saved the lives of millions: American, Japanese and a host of Asians. Historian Frank continued, "Any soldier or Marine infantryman slated for *Olympic* who believed the atomic bomb saved him from death or wounds had solid grounds for this belief."[2990] Woody agrees: "Without those bombs, I wouldn't be here today."[2991]

And these bombs destroyed one of the most horrible regimes mankind has produced. As Chief of Counsel for the International Military Tribunal for the Far East, Joseph B. Keenan, wrote in 1946:

> A very few throughout the world, including these accused [Japanese Class-A War Criminals like Tojo, Matsui, Sakai], decided to take the law into their own hands and to force their individual will upon mankind. They declared war upon civilization. They made the rules and defined the issues. They were determined to destroy democracy and its essential basis—freedom and respect of human personality; they were determined that the system of government of and by and for the people should be eradicated and what they called the 'New Order' established instead. And to this end they joined hands with the Hitlerite group;

they did formally, by way of treaty, and were proud of their confederacy. Together they planned, prepared and initiated aggressive wars against the great democracies [throughout the world]. They willingly dealt with human beings as chattels and pawns. That it meant murder and the murder and the subjugation and enslavement of millions was of no moment to them. That it encompassed a plan or design for the murder in all parts of the world of children and aged, that it envisaged the entire obliteration of whole communities, was to them a matter of complete indifference. That it should cause the premature end of the very flower of the youth of the world—their own included—was entirely beside the point. Treaties, agreements and assurances were treated as mere words—bits of paper—in their minds, and constituted no deterring influence on their efforts. Their purpose was that force should be unloosed upon the world. They thought in terms of force and domination and entirely obscured the end of justice. In this enterprise millions could die; the resources of nations could be destroyed. All of this was of no import in their mad scheme for domination and control of Eastern Asia, and as they advanced, ultimately the entire world. This was the purport of their conspiracy.[2992]

This indictment supported the justification for all living free men and women of the Allied nations to be grateful for the atomic bombs. And for all the nameless and named victims of Japan's Holocaust, from Nanking to Canton, to Shanghai to Hong Kong, to Hankow to Manila (names on a par with Dachau and Auschwitz), they too, if they could still have spoken, would have thanked those who developed and dropped the atomic bombs. With the atomic bombs, their cries were heard across the ocean throughout Japan's society.

And having engineered such fantastic and destructive weapons like the uranium and plutonium bombs and then implemented them showed how far the U.S. had come in three and a half years since December 1941 in the evolution of weapons. And the atomic bombs were just one weapon out of thousands America created and used.

The Marines evolved from using water cooled machineguns, M1903 Springfield bolt action rifles, M1 flamethrowers and decent airplanes like the

Grumman F4 Wildcat to using M2-2 flamethrowers, bazookas, satchel charges, air-cooled machine guns, shorter-barreled carbines, M1 semiautomatic rifles, *Zippo* tanks, Higgins boats and far superior airplanes to those of the Japanese, like the Vought F4U Corsair.[2993] The manner in which Marines used these weapons in a combined arms manner in 1945 was much more efficient than at Guadalcanal in 1942. The Japanese had lost air superiority, their great advantage in torpedoes, superiority in night warfare at sea, and ability to mount a credible naval threat to the U.S. Navy or Army and Marine forces on the battleground and in the skies. Japanese soldiers could no longer look for resupply, reinforcement or evacuation by sea and many were bypassed and left to die of starvation and disease. During war, as Carl von Clausewitz writes, a commander must "make sure that all forces are involved...that no part of the whole force is idle," because those elements not busy with defeating the enemy "are being wasted, which is even worse than using them inappropriately."[2994]

By 1944, Americans had killed the majority of Japan's best-trained pilots. Although Japan produced some improved aircraft models, they still fell short of U.S. quality. The Zero was now no match for new American aircraft and tactics. Fighter pilot Saburō Sakai said, "Japan couldn't produce new planes in any numbers since the country didn't have the industrial strength. More than ten upgrades of the Zero weren't enough."[2995] As a result, the Japanese introduced the *Kamikaze* human-guided flying bomb, which was difficult to defend against but still could not prevent America's rolling military tide. Using the Higgins boats to assault beaches from Normandy to Iwo allowed the U.S. to hit the enemy from the sea, and hit him fast with tens of thousands of armed men. Lieutenant General Smith stated the Higgins boat "contributed more to our common victory than any other" piece of equipment.[2996] General of the Army Eisenhower echoed Smith saying the boat "won the war for us."[2997] So the evolution of war took many forms for both nations.

Woody's training in 1943 and 1944 with the newly-developed M2-2 flamethrower showed one example of how the Marines and other service branches were continually adapting their methods and, upgrading their weapons and organization on the fly in order to combat the enemy's tactics. Militaries in the U.S. and in Japan both knew about flamethrowers before the war, but the way the U.S. designed, manufactured, tested and used them made the difference. The

Japanese were adopting elaborate, static defenses and the flamethrower was the weapon to counter them.

Besides developing state-of-the-art weapons, the U.S. also developed engineering units that were the best any military had ever used in history. The navy employed roughly 325,000 men in Naval Construction Battalions ("Seabees").[2998] "The war in the Pacific, then, was an engineer's war."[2999] They knew over 60 skilled trades. In the Pacific War, where 80% of Seabees served, they built 111 major airstrips, 441 piers, tanks for 100 million gallons of gas and hospitals for 70,000 patients. They figured out how to get landing craft, amtracs and tanks over coral reefs and other obstacles. In many battles, like Saipan, the Seabees landed in the second wave. They also fought in combat earning over 2,000 Purple Hearts. They lost approximately 200 men in war and more to accidents. Yale University political scientist Paul Kennedy rates their role on Guam, Saipan and Tinian in Operation *Forager* as possibly their greatest during the war.[3000]

Years of isolationism left the U.S. unprepared to fight a world war at the time of Pearl Harbor. America, with great natural resources, skilled workers, universities and industrial might beyond the enemy's reach, evolved its equipment and tactics faster than the Japanese which gave America a decisive edge in combat inflicting more casualties and damage on the Japanese than they did on Americans. Moreover, the U.S. was excellent at motivating, equipping and training its men on the new tactics and equipment used on the battlefield. Helping each Marine learn intimately how to use his weapons fostered confidence. Refugee scientists made great contributions, as with the Manhattan Project, to develop the atom bomb. There were two destructive arsenals which could bring massive destruction to Japan: chemical weapons and atomic bombs. The atomic bombs were deployed against Japan to good effect, but why did the U.S. not use chemical warfare? Poison gas might have been efficient in killing the Japanese on the Mariana Islands and Iwo, and although Nimitz and FDR contemplated its use, the government ultimately rejected it as a violation of treaty obligations.

In February 1945, Commander Dick Ashworth delivered a letter to Nimitz from his boss, Fleet Admiral Ernest King, ordering him to support the Manhattan Project. But there were only two atomic bombs being readied, Germany fought until May, and FDR had ordered the atomic bomb would first be used against Germany, the priority target, if it did not surrender.[3001] Even if he did know of the atom bomb project, Nimitz could not be certain in February that the new super-weapon would work, be ready by August or be replicated enough to drop more than one on Japan. Scientists and policy makers were concerned that in actual use, the atomic bombs might malfunction. Finally, there was the matter of safely getting the weapon close enough to deliver it. It was the cruiser *USS Indianapolis* that delivered parts of the *Little Boy* atomic bomb to Tinian on 26 July 1945, which contained about half of the uranium-235 that existed in the world. Four days after the successful delivery, Japanese submarine I-58 sank the *USS Indianapolis*, resulting in the death of 879 men, one of the greatest losses of life in the U.S. Navy's history; only the *USS Arizona* was worse, with 1,177 dead.[3002] So, it was difficult to get the atomic bombs into position to even be used, much less truly tested in battle.

But at this stage, gas was a known quantity and deploying it could have destroyed Japan's ability to wage war just as the atomic bomb would. However, America struggled with using both weapons, and ultimately decided against using gas.

We know Japan would have used chemical warfare against the Americans if shipments of lethal agents sent for the Saipan battle had not been sunk. Once chemical weapons were introduced, they would have been used in other battles around the world. Would American chemical weapon usage in 1945 have ended the war faster? In WWI, poison gas caused injuries or death to an estimated one million soldiers and hundreds of thousands of civilians and increased the misery for those who wore gas masks in the trenches. Major General Clifton Cates, 4th MarDiv commander on Iwo, had his entire company wiped out by a German gas attack in WWI.[3003] However, this form of warfare was not the major killer or the decisive weapon of the war. For one thing, its use was dependent on the weather, which could delay attacks making it ineffective or blowing it back on its deployers. Chemical weapons were heavier than air and could linger for days in invisible clouds, causing injuries or death to friendly forces later moving through the area.

Hirohito, knowing his men were fighting for a losing cause when he could have surrendered, was willing to let hundreds of thousands of soldiers and civilians die in the Battle of Okinawa, in the Battle of Luzon, in China, in Burma, on bypassed Pacific islands and in the destructive firebombing campaigns. Even after Mussolini was overthrown and Hitler committed suicide and Germany waved the white flag in May 1945, Hirohito ordered his military to continue fighting.

Even when over a million Russian troops massed on the Manchurian border stretching their leashes to attack in the summer of 1945, Hirohito seemed to ignore Stalin's message. The Soviets could focus all their energies on Japan now that they had defeated the Nazis. So had gas been used in February 1945, Hirohito most likely would still not have stopped the war.

In war, the two primary goals are to destroy the enemy's will and ability to fight. Gas may have been effective at Iwo if the winds co-operated and the Japanese were not prepared for it. Also, dropping gas canisters on cities would have helped weaken Japan besides the firebombings. Such attacks might have sent a powerful message to Tokyo. Although firebombing was destructive to people and property in flammable Japanese cities, Hirohito did not surrender, so one could make the same argument against gas. However, one could also make an argument that using gas would have saved American lives until atomic bombs were used. Nimitz later concluded it would have saved Marines' lives on Iwo.[3004] Lieutenant General Smith said people often confronted him on why he did not use gas, citing a Washington newspaper pleading with the military in March 1945 to give the Marines a break and "gas the Japs." Smith replied:

> Certainly, gas shells smothering the island, or gas introduced into caves and tunnels would have simplified our task, but naturally the use of this prohibited weapon was not within the power of a field commander [although ships in the fleet had it]. The decision was on a higher level. It was in the hands of the Allied Powers, who alone could authorize its use in a war which would have announced even more frightful proportions had gas been allowed.[3005]

The U.S. was changing its mind about gas, though, and it might have used it during the invasion of Japan to prevent more Iwos.[3006] In March 1945, General of the Army Marshall suggested using gas against the Japanese at Okinawa to prevent similar casualties to those seen on Iwo.[3007] Had gas preceded the atomic bombs, it might have also helped push Hirohito to where the Allies needed him just like the Soviet Manchurian invasion helped push him closer to accepting the Potsdam Declaration. As one sees, the implications for the use of chemical weapons went far beyond one battle or theater, entering into a realm of ethics, morality, politics and psychology.

Regardless of what weapons were used and the reasons for or for not doing so, the U.S. demonstrated those approved for war were not only superior to the enemy's, but also utilized by men who knew them intimately. Whether Americans were trained to drop atomic bombs, man aircraft carriers, call in artillery, deploy naval gunfire, conduct airstrikes, fire M-1 rifles or spew liquid flame at pillboxes, the U.S. showed it had developed a superior method of training and deployment of its weapons compared to its enemies. Woody's training enabled him to control his fear. He said, "The flamethrower was just another weapon of war. We trained to the point that one could do what had to be done without having to pause and think about it… Naturally, there is planning before the action, then you follow the plan."[3008] The months on Guadalcanal and Guam when Woody trained with his team helped them to operate as a unit, do things instinctively, and rely on second nature to combat the enemy.

As part of the excellent training, the Corps instilled in Woody the right attitude; namely, perseverance. "Giving up was not an option, because the whole purpose was to win," Woody said. "If a person goes into combat with the thought of giving up, they will not be able to do the job at their best."[3009] So along with the training came also the goal—defeat the enemy. This focus on winning for your nation and protecting comrades made men do extraordinary things and respond to situations with the proper behavior. In the Corps, they call it situational awareness—the ability to adapt to your surroundings in real time. Woody observes, "It is not *what* happens in life, it is *what we do* with what happens in life."[3010] Captain Beck asked Woody, Southwell, Tripp, Naro, Schwartz, Fischer, Bornholz, Schlager and their platoon leader Chambers to take out several pillboxes, while being targeted by Japanese firing thousands of rounds of high-caliber machinegun and rifle bullets, throwing hundreds of grenades, launching barrages of mortar rounds, calling in artillery strikes and charging with bayonets and swords at any moment. The fact that Chambers and his men went repeatedly to face numerous enemy capable of and intent on inflicting catastrophic injuries or death upon them had to be a daunting prospect. However, they dug deeply and relied on their training. Woody also remembered Ruby and his buddies who depended on him. He was not going to let himself, fellow Marines or his girlfriend down. Woody relied upon his squad, and more importantly, upon himself.

He picked up the flamethrower and went toward danger to eliminate the enemy threat. Woody was a young man from a West Virginia dairy farm who served in the Marines when drafted. He was short in stature, but strong in heart. As Admiral William "Bull" Halsey said of his, and thousands of others' service, "There aren't any great men. There are just great challenges that ordinary men… are forced by circumstances to meet."[3011]

Battles are made of millions of small decisions, and usually the best-trained, best-led and best-equipped forces win. Training Marines received in boot camp and continuing throughout their service assured the highest level of performance. Woody and Beck performed bravely at Guam, and when time came to put their training and experience to work at Iwo Jima, they performed well. In a campaign where thousands of Marines exhibited the skill, devotion and bravery to make their country proud, Woody's, Chambers' and Beck's actions stood out as being the best of the best.

Yet, some events Woody has allowed the public to believe may not have happened. Woody's inability to remember, the different stories he has given throughout the years and the uncoordinated fashion of the MOH witness statements raise questions. In his friend Waters' Silver Star report, he took out two pillboxes and killed three Japanese. If four men were in each pillbox, then Waters could have killed 11 men. Was that the number required for a Silver Star? If he had killed 21, would he have received a MOH? We know one of Waters' and Woody's flamethrower comrades, PFC Merrill L. Christopher, during the night of 3 March 1945, volunteered to take out a pillbox in the dark. With no covering fire, he "hosed the Jap pillbox" and killed three. This netted Christopher a Bronze Star with V.[3012] So had Christopher or Waters taken out more pillboxes, would that have tipped the scales for them to have received a higher award? Were events under- or over-reported for them? And conversely, if Woody only took out three pillboxes instead of seven like Beck thought, would his medal recommendation have been different from the outset? From the evidence, this appears to be the case, and Commandant Krulak, as mentioned before, strongly believes Woody should have "*only*" received a Silver Star or, "at best," a Navy Cross.[3013] This study concurs with Krulak's assessment.

Combat's chaos indeed makes it difficult to remember events. In the air, pilots often thought they shot down more planes than they did, and after-action reports indicate that U.S. squadrons eliminated more enemy aircraft than were flying. Japanese reported their attack planes and *Kamikazes* sank more ships than they did in actuality.[3014] These facts illustrate the complexity of finding the truth from history and one sees that this study also struggles with human memory, recorded facts, and how medals are awarded. As a result, some questions remain about Woody's postwar behavior. If Woody is bothered by why he cannot remember what he did and since he has known for years the names of some who put in statements for his MOH, then why did he *never* make an effort to get in touch with them to ask them what they remembered? He immediately reached out to the Waters family after the war, but not to the Fischer or Bornholz families to thank them for their sons' sacrifice or the seven MOH witnesses who gave testimony for him. He never reached out to Beck, who endorsed him. For a man who has shown gratitude for families who sacrificed a child for this nation with his Gold Star Families project, this behavior is confounding.

Woody's MOH witness statements are short and often comprise a few lines. Until this study, Woody thought there were only four instead of seven who gave statements.[3015] So what does the evidence tell us that really happened? Woody performed acts of courage, but were these exploits, especially if they were not described in detail by the witnesses and cannot be remembered by the participant, actually worthy of *the Medal*? As already mentioned, Commandant Krulak and this study do not think so.[3016] But, in the end, the Navy Board of Awards and Decorations felt Woody deserved *the Medal* and that made the difference. This study shows how arbitrary it is for one to receive military medals and that it is not by any stretch of the imagination an objective and pure process. After reviewing a majority of the Iwo MOH cases, Woody's case is the most vague and the only one for which the initial citation draft dramatically differs from the final one because of disputed evidence. Dramatically, it appears to be the only Iwo Marine MOH that never received the Marine Corps and Pacific Fleet's endorsements. Ironically, Woody is now a huge hero for the Marine Corps as a MOH recipient although *the Corps of 1945 did not endorse his actions for a MOH.*

Moreover, if the men who gave witness statements for Woody were instrumental in getting him *the Medal*, one would think they would have reached out to Woody after the war. Of course, many wanted to return to their lives and forget the war. Alan Tripp was an intelligent and quiet man who never discussed what he did in combat.[3017] Howard Chambers would often not finish stories about his men because he choked up with emotions.[3018] Waters' officer Richard Tischler struggled to utter the words necessary when talking about his Marines dying and holding some of them as they took their last breaths.[3019] Regardless of the discrepancies, enough men pushed Woody's case forward and gave him one of the greatest awards the U.S. nation can bestow on a serviceman.

Woody's *Medal* process is a fascinating case study of how flawed medal awarding is in the military. Often, historians, like the officers reviewing Woody's *Medal*, are given 100 pieces to a puzzle that requires 500 to complete. One puts those 100 pieces together, does his best to interpret the partial images those pieces give and then analyzes what the holes in the story, or in this case the picture, should be in relation to what is seen. A historian will never get the complete picture; however, a historian most often knows, when looking at the pieces he has, that the picture has a definite theme. And often, he knows what probably happened by looking at the evidence around the missing pieces. And whereas the officers reviewing Woody's MOH only did so over a few weeks, and sometimes only probably a few days, this study has gathered more pieces of the puzzle than they after having looked at the evidence for *five* years giving this study a much clearer, and problematic image than what the men had in their offices back in 1945.

Besides revealing the challenges a historian faces with documenting the past, Woody's story also shows the difficult decisions NCO's like Woody had to make. Woody had to coordinate his squad running protection for him the day he took out the pillboxes. Good decisions and leadership by NCOs were vital to success in battle. Woody's actions combined to make one piece of the success seen on Iwo. No one before the battle thought Woody would be able to do as much as he did, not even himself. The day after he performed his MOH deeds, Woody was faced with a tough situation when a squad member, Thomas Scrivens, was on the verge of quitting.

His rash decision to point his weapon at a Marine to force him back in line, which he had no authority to do, may have convinced Scrivens not to quit, but it also could have instigated Scrivens reporting Woody to their commanding officer resulting in Woody facing disciplinary action. Scrivens, had he refused, might have faced discipline himself for disobedience unless he was found to be medically unfit. Had Woody shot Scrivens, he would have been court-martialed instead of awarded the MOH. Fortunately, this difficult situation did not end in tragedy as Scrivens continued fighting. Knowing Woody's short temper, Scrivens was lucky he got back on the line. Obviously, Woody wasn't nicknamed *Zippo* head without reason.

Was Woody right to threaten Scrivens when after difficult, dangerous and exhausting combat, he was on the verge of losing his nerve? Viktor Frankl cautions, "No man should judge unless he asks himself in absolute honesty whether in a similar situation he might not have done the same."[3020] There was an established legal procedure in the Corps for dealing with one accused of disobeying a lawful order, desertion or another serious offense. To many, it would seem to have been the absolutely wrong decision for Woody to have patted Scrivens on the head, told him it would be alright, and permitted him to abandon the fight. Both would have suffered for such behavior, and no military can tolerate such softness and disobedience in the face of overwhelming violence. However, it certainly was not for Woody, a 21-year-old high school dropout and low-ranking NCO with no medical or legal expertise, to shoot a Marine, even if he was sure the Marine's conduct was criminal. Neither his company first sergeant nor his company commander had the right to shoot Scrivens. Not even Commandant Vandegrift or Fleet Admiral Nimitz had the right to shoot a Marine who had violated regulations. Scrivens was an American warrior and as such had rights; it was Woody's responsibility to protect those rights. Woody sums up this situation:

> Combat creates desperate situations and removes all normalcy. The mind does not weigh the consequences, just the circumstances at the moment. If every Marine stopped to weigh the killing of a person, another human being who had not given a cause for killing, a war could not be successfully accomplished.[3021]

The ability to kill without thought in the moment is a necessity for a warrior if he wants to survive. The unwillingness to kill and support one's unit causes nations to lose wars. What Woody did against the Japanese and his willingness to threaten Scrivens shows Marines had been conditioned most to concentrate all behavior to the task at hand: Annihilating the enemy and making sure everyone was similarly focused.

Scrivens's fear was understandable after what his division had been through, but so was Woody's concern about keeping his squad in the fight. Many reached a point where they, like Lefty Lee, could no longer function. It was up to the authorities to determine if Scrivens was ill or had committed an offense and, if he had, what punishment within the limits of Rocks and Shoals would be imposed on him. Interestingly, Scrivens went to Woody after the war and apologized for his behavior. The only other comrade of Woody's to reach out to him was also another PTSD case who was removed from Iwo for shell-shock, Lefty. The two close to Woody's combat experiences who were taken off the island for psychological reasons were two who searched him out after the war. Maybe they both wanted to learn if Woody knew they had been removed for shell-shock or perhaps they just wanted to be around a famous person.

It is a sad fact that many veterans, like Lefty, feel a need to embellish their stories, or to associate their actions with a famous person to give them more luster. For decades, Lefty told his family, fellow Marines, archivists at national museums, journalists, historians and even Woody that he was there on the island supporting Woody and protecting him. Lefty also fabricated stories of fighting at the second airfield, killing ten Japanese soldiers and being blown "thirty feet in the air" by a mortar. Lefty was so embarrassed about being taken away for a mental rather than physical injury that he could not bring himself to stand on the unvarnished truth. When he finally admitted to his lies after months of argument and fighting, he begged, "Please, please don't tell my children."[3022] In 1945, he felt society and his girlfriend would not have respected him if they knew he was removed for combat fatigue instead of being blown up by a Spigot mortar as he has been claiming (maintaining he deserved a second Purple Heart). At boot camp, they were taught that desertion, which Lefty felt he did, was unforgiveable:

> [A deserter's] family, if he has one or if he can find a girl who will be willing to have him, will be continually ashamed of him. When his kids want to know what kind of a hero he was, or when the boys start talking about all they did, he will have nothing to say—because he'll know what they think of him and there will be nothing he can do about it; he can't even re-enlist…to redeem himself…[3023]

These words must have had a powerful impact on recruits and even more so for those who felt they let down their buddies. Lefty's tale started by adding a little bit to his real story and he soon enjoyed the reaction he got so much he added a little more and a little more in the retellings. The Marines indeed told his parents he had been removed for combat fatigue, so maybe his family was also in on the lie and told him not to reveal what happened. In his August 1945 medical survey, the examiner wrote that Lefty's hands shook "with palmar hyperhidrosis," and that he experienced anxiety, had nightmares and suffered "from a psychoneurotic disorder which renders him unfit for Marine services," something that had started on Guam and concluded on Iwo.[3024]

When Lefty was interviewed, he had fabricated an entire episode of his story and presented it to this study, the National World War II Museum, congressmen and senators, Woody and Headquarters of the Marines with falsified documents. For decades, he had convinced Woody he was killing Japanese while Woody incinerated their pillboxes. Lefty praised Woody in public and private for years and was one of the first people Woody wanted interviewed to support his story. Woody even built a case to get Lefty more recognition for what he supposedly had done, giving supporting testimony to a Congressman and a Senator from Minnesota and to the Marine Corps Awards Board in May 2017 for a medal for valor as well as a Purple Heart—the petition had to be withdrawn once sources in the National Archives proved Lefty had been medevacked soon after he hit the beach on Iwo.

Lefty's actions highlight that mental injuries were not well understood by the general public and, for some, did not seem genuine and serious. As mentioned before, after many intense conversations, Lefty admitted his lies to the author and Woody. Soon thereafter, Woody forgave Lefty and told him he would still be his friend. People come to Woody hungering for understanding or redemption putting Woody in a constant state of telling his Iwo stories. Woody shares his story as a way to honor his Marines and country, and that is how he has dealt with his mental scars and trauma.

Although Woody was disappointed in Lefty and Scrivens, he acknowledges the sacrifice many have had to make mentally to defend this country:

We need to show some compassion here. Many…probably suffered greatly with PTSD. We didn't call it that back in my day—we called it psychoneurosis [combat fatigue and shell-shock]. This can affect people in troubling ways and maybe they did the best they could, but the PTSD just got the best of them. War is never easy on a person and sometimes they don't deal with that stress in productive and honest ways. However, instead of condemning the person, we should try to understand and help him.[3025]

This sensitivity and understanding shows that Woody understands how veterans can suffer. Moreover, he feels that he experienced PTSD too, so he can relate to the stress of returning to "normal life" after experiencing pain, suffering and loss.[3026]

Woody's history reveals how difficult it is to live with one's experiences in combat. No matter how necessary the mission or how willing the participants, each person has his own limits beyond which he may not safely go without serious and sometimes permanent psychological consequences. Woody struggled when he returned home. Eventually, he found ways to cope with PTSD. Woody found religion, focused on his family and was often reminded, with justification, that what he did was necessary to win the war, save his buddies and survive. This is often a theme of "existentialism: To live is to suffer, to survive is to find meaning in that suffering."[3027]

Woody and his comrades had that reason for suffering clear in their minds because the Marines drilled it into them—they had to defend their country and one another. Many veterans developed relationships under life and death situations that became sacred. This helped them after the war although this also gave some survivor's guilt. Woody said, "I still don't know why I was picked to survive when so many others died before me. Why me? I just don't know why." But then Woody would catch himself, "The only way I feel good about what I had to do and why I survived is to now serve others, help others and be a good citizen."[3028] However, it took years to come to this conclusion. The pain and resignation ebb and flow between nightmares and quiet contemplation. Woody's experience taught him that it is vital for veterans to find meaning in their service, and psychological support for their deeds. These thoughts are prominent in their consciousness as observed at the reunion in which Woody took part in 2015. Veterans recalled events that took place on that island and wondered whether it was worth the sacrifice. As Herman Melville wrote in *Moby Dick*:

Would to God these blessed calms would last. But the mingled, mingling threads of life are woven by warp and woof: calms crossed by storms, a storm for every calm. There is no steady unretracing progress in this life; we do not advance through fixed gradations, and at the last one pause:— through infancy's unconscious spell, boyhood's thought-less faith, adolescence' doubt (the common doom), then scepticism, then disbelief, resting at last in manhood's pondering repose of If. But once gone through, we trace the round again; and are infants, boys, and men, and Ifs eternally. Where lies the final harbor, whence we unmoor no more? In what rapt ether sails the world, of which the weariest will never weary? Where is the foundling's father hidden? Our souls are like those orphans whose unwedded mothers die in bearing them: the secret of our paternity lies in their grave, and we must there to learn it.[3029]

As these men approach the twilight of their lives, they ponder what it means to survive the hell of battle and to look after the survivors after they "beat their swords into plowshares and their spears into pruning hooks" (Isaiah 2:4).[3030] The "what ifs" haunt many veterans, especially Iwo survivors. Luckily for Woody and others, they won at Iwo and returned as heroes to a grateful nation. Although this realization helped many feel better, they faced a long road of healing and experienced survivor's guilt. Yet they knew that had they not fought, the world would be a darker place for having allowed thugs like Hitler, Hirohito, Mussolini, Tojo, Matsui and Kuribayashi to run amok.

American WWII veterans take comfort in the righteousness of their cause. Family and friends depended upon their protection to live as free people. Defeat would have eradicated freedom and justice, perverted Western Civilization, caused the deaths of millions of civilians and the cruel, degrading subjugation of millions of others. While Americans were confronting the enemy on foreign shores, Hitler's bureaucrats drew up lists of people to be rounded up in the U.S. when the Nazi invasion and occupation commenced. America's continued freedom depended upon a favorable outcome to the war.

There is a huge difference in how the U.S. conducted war compared to its WWII enemies. The U.S. exhibited a higher moral standard than its enemies in war and after their defeat. Almost everywhere the Japanese conquered, they ignored international law, treated POWs brutally and instituted a tyrannically, oppressive rule, one of deprivation, humiliation, beatings, rape and arbitrary slaughter.

In contrast, when the U.S. and its allies fought they respected international law, were welcomed by grateful, liberated people and when they conquered Europe and Japan, were neither cruel nor oppressive to the aggressor nations who had subjugated two continents and conducted two of the most evil slaughters known to man—the Nazi genocide and Japan's Holocaust. Uncle Sam not only helped his long-suffering allies, he also helped restore freedom and democracy to the Axis countries and helped these mortal enemies emerge from the devastation of war with the Government Aid and Relief in Occupied Areas, Marshall Plan, World Bank, International Monetary Fund, General Agreement on Tariffs and Trade (GATT) and other institutions.

America made mistakes such as allowing colonial territories in Asia to remain under their imperialist masters as India under the British or returning Vietnam to the French, but all in all, it was nothing like what the Nazis or Japanese had done. The North Atlantic Treaty Organization (NATO) was established to deter future aggression and WWIII. America also dispersed aid in Asia and the Pacific. Postwar American rule of conquered Japan resulted in a liberal, democratic government, the renunciation of war, and an improved status for women. The U.S. gave Japan some $2.44 billion in aid ($35 billion in 2019 dollars). It has become a reflection of the strong democratic society of the U.S. This is a testament to the American love of freedom and compassion.

America was a democratic country working toward a more just society at the time. It was one of the most tolerant and lawfully-minded in the world among the super-powers although it did have serious problems with racial discrimination. Compared to countries like Nazi Germany, Imperial Japan or the Soviet Union and compared to the British and French colonial systems, the U.S. pursuit of a more just society was evident.

Discrimination took many forms during WWII, but in the end, compared to the Germans, Japanese and Italians, America was further along on the racial equality and political scales although it did have further to go and still needs improving. One could argue that the brave feats of all minorities in America caused many to reconsider their prejudices and laid the foundation for their struggle for equal rights after the war. WWII brought over 16 million American men and women into uniform and forced many to work together whether they were Protestant, Catholic, Jewish or Mormon; Black, Asian, Hispanic, Native

American or Caucasian. They were from big cities and small towns. They were West Virginia "Hillbillies" like Woody or Montana farmers like Vernon Waters. They were South Dakota Danish farmers like Beck and white collar Heinz-57 college boys from Pennsylvania like Chambers. Woody never knew an African-American or Jew until he joined the Marines. Hardly anyone in Woody's outfit had ever met a "Hillbilly." And thousands of women joined the Marine Corps, like Woody's ex-girlfriend Eleanor G. Pyles, and helped with administrative work—"They called us BAMS (Broad-Ass Marines)," said WWII Marine Corps veteran, 2nd Lieutenant, Gertrude "Trudy" McKitterick. One of the most famous female WWII Marines was Staff Sergeant Bernice Frankel (Bea Arthur or Dorothy Zbornak of the famous TV series *Golden Girls*).[3031] This was the first time many men saw women in uniform. The melting pot of American society, especially in boot camps, broke down many of the racial and sexist stereotypes because in working in such close quarters, one saw the common bond all shared as humans. If you fight alongside another human being, it becomes more difficult to demonize that person merely because of his skin pigmentation, political differences or religious creeds. Yes, Black military personnel returned to a largely segregated North and a Jim Crow South. Even Navajo Code Talker, Sam Holiday, had problems with racism as a Native-American when he returned home, saying, "I fought next to a Black man, White man, Mexican and came back to segregation."[3032] However, soon after World War II, probably largely due in part to the different "races" mixing so much in the ranks, President Truman would integrate the military in 1948 and the civil rights movement would gain momentum. (See Gallery 4, Photo 9)

Woody has also shown after the war his willingness to reach out to the Japanese with an olive branch. After climbing Mt. Suribachi in 2015 before the Reunion of Honor ceremony, as part of the healing and reconciliation, Woody embraced the hand of Japanese Iwo veteran Senior Seaman Tsuruji Akikusa. Later, he said, "I made a new friend today." For years, one of the most active and premier veteran's organizations that unites the Japanese and American veterans from this campaign has been facilitated and maintained by the Iwo Jima Association American (IJAA) under the leadership of the late Lieutenant General Lawrence Snowden (USMC), and Woody had finally agreed to return to Iwo with it to continue the healing. For decades, it has brought veterans together in the fellowship of peace and as allies. Snowden was an Iwo veteran himself and was wounded during the battle. Interestingly, one of the biggest supporters of this tradition in Japan is General Kuribayashi's grandson, Yoshita Shindo, a prominent member of the Liberal Democratic Party and House of Representatives in Japan. Born into a family of one of the Marines' most hated enemies, he now is one of their strongest allies. He was a friend of Snowden's and they even showed up together during a joint session of Congress when Prime Minister Abe addressed the government on 29 April 2015 and pointed out that it is a miracle that these two men now stood together in friendship:

> Ladies and Gentlemen, in the gallery today is Lt. General...Snowden. Seventy years ago in February, he landed on...Iwo Jima as a captain in command of a company. In recent years, General Snowden has often participated in the memorial services held jointly by Japan and the U.S. on Iwo...He said, and I quote, 'We didn't and don't go to Iwo Jima to celebrate victory, but for the solemn purpose to pay tribute to honor those who lost their lives on both sides.' Next to General Snowden sits Diet Member Yoshitaka Shindo, who is a former member of my Cabinet. His grandfather, General...Kuribayashi, whose valor we remember even today, was the commander of the Japanese garrison during the Battle of Iwo Jima. What should we call this, if not a miracle of history? Enemies who had fought each other so fiercely have become friends...To General Snowden, I say that I pay tribute to your efforts for reconciliation. Thank you very much.[3033] (See Gallery 4, Photo 10)

Although Abe failed to mention how horrible Kuribayashi's earlier actions were in Hong Kong and that besides his valor, his atrocities should be remembered and his victims not forgotten, the fact that Abe made overtures of peace at the seat of the U.S. government is indeed a step toward healing the wounds of WWII. He also honored a true American hero in Snowden, a Purple Heart recipient from Iwo Jima and a veteran of the battles of Roi-Namur, Kwajalein, Saipan and Tinian. Abe did indeed admit that Japanese "actions brought suffering to the peoples in Asian countries. We must not avert our eyes from that," which was remarkable knowing that he rarely offers such contrition. Yet, he then almost nullified everything he had just said by mentioning his grandfather and former Prime Minister and Class-A

War Criminal Nobusuke Kishi, praising him but failing to mention he was also "Hirohito's devil," a mass murderer and personally responsible for that "suffering" of Asian peoples.[3034] So with this knowledge, Snowden's friendship with the Japanese, especially with Shindo and Abe despite their grandfathers' criminal pasts, is unique and good, but full of unresolved issues that are not openly discussed; namely, what are Japan's responsibilities to the world about documenting and admitting its war crimes? In the end, the fact that these men came together to honor one another and the dead of Iwo Jima shows the significance this battle continues to have in the landscape of social memory and political action.

During the IJAA Symposium in Washington D.C. in February 2018, Shindo made a special trip there to pay his respects to the veterans and reiterate the bonds of friendship this battle now holds for modern-day Japanese and Americans. Veterans and their offspring at the symposium event in 2018 embraced Shindo as one of their own and he in turn welcomed their familial open arms. He was given an IJAA award and it hangs prominently in his Tokyo office. If such groups who once swore to annihilate one another can join together in friendship, then one takes hope that other enmities currently in the world can also be bridged. This is the only reunion from two warring nations from WWII. Woody honors what Shindo is doing and was part of the reunion Shindo and IJAA helped organize in 2015 in the pursuit of "peace and reconciliation." This was a huge step for Woody because for years, he had sworn never to go back to the island to open mental wounds and to be subjected to "Japanese telling me what to do and where to go." Describing this change of attitude, Woody said, "I think it is a powerful display of humanity and forgiveness that we and the Japanese have come together on the bloody battlefield of Iwo Jima to strengthen our friendship. And I want to keep that friendship strong because I don't ever want them as an enemy again."[3035]

Shindo agrees with Woody. He says, it is crucial that Americans and Japanese stay strong. When asked if he thinks it would have been better for Japan had it won the war, he said,

> At the time, many nations were colonizing the world. England had the whole of India. The Dutch had the whole of Indonesia. The French had all of Vietnam, Laos and Cambodia. Why were we highlighted as demons when we did the same thing in China? Yes, many bad things

happened under the rule of colonizers. So, I don't agree with those bad actions. The totalitarian regime in Japan at that time wasn't good. There was no freedom of speech. Had we won, that wouldn't have been good for global health. I think what we tried to do, imperfect though it was, helped many nations become independent afterwards. That could be looked at as good. We were human beings trying to do something worthwhile although misguided in its implementation. So many lessons can be drawn from this time with how we do things in the future and how things have been done.[3036]

And one of those main lessons is that Japan and the U.S. should never go to war against each other again. As long as both nations have a strong rule of law, government checks and balances and competing political parties, then there is much hope for both nations in the future.

The battle also helped save the Marine Corps, an instrument to preserve democracy, from being shut down. Politically and psychologically, the "glorification of Iwo had an enormous positive impact on the Marines in the postwar era."[3037] There was a danger of the Corps being disbanded; many army hardliners argued it was a redundant service maintaining that the U.S. did not need two ground armies. President Truman supported the integration of the Corps within the army.[3038] Even in 1949, Secretary of Defense Louis A. Johnson wrote, "There's no reason for having a... Marine Corps. General [of the Army Omar] Bradley tells me that amphibious operations are a thing of the past. We'll never have any more amphibious operations. That does away with the Marine Corps [the next year, the Inchon landing in Korea turned this argument on its head]."[3039] However, since Iwo and other Pacific battles were well-known and the value of amphibious troops was proven, these facts probably saved the Corps from extinction. By 1950, Truman had to relent, signing a National Security Act amendment that gave the Marine commandant a permanent "position on the Joint Chiefs-of-Staff." "The sense of sacrifice and bravery Americans associate with the flag raising and the conquest of the island" proved the value of this force.[3040] As historian Allan Millett put it, "the capture of Iwo Jima in 1945 gave the Corps a heroic public image and doctrinal confidence" that has created a lasting and formable force for America's Armed Forces.[3041] Millett continues writing, "Today...the Marine Corps enjoys the special privilege of having its force structure written into law, the Douglas-Mansfield Act of 1952 (Public Law 416, 82d Congress). No other service has the same legitimation."[3042] And there is no doubt that the victory at Iwo helped push the government to issue this law to preserve the Corps.

In exploring these deep historical issues, one must also realize that Woody's story within these pages is about much more than lessons learned from WWII albeit those lessons are important to study. It is also more than a story of how the war was won and its impact on American youth of that time and Americans today. It is a personal story of triumph about what one man can do with the opportunities afforded him. Woody was a simple teenager who had a limited education when he entered the Marines. He returned home as a NCO who had served with such distinction that Truman hung the Medal of Honor around his neck. This brought him into a realm for which he had not been born and had never been groomed—society's elite. People who had accomplished great things in their own fields looked up to Woody. He took this status and used it to help others although he enjoys the recognition of doing so—becoming one of the main voices for doing good and helping families of fallen warriors. Like a charismatic preacher, he brings the good news to others, but likes the praise that he is indeed God's messenger. And in his quiet moments, when he reflects on his past and maybe thinks about the possibility that if he indeed inflated facts that helped him receive the MOH, maybe then his activities as an elderly man are stemming more from a guilty conscience, rather than pure motives. Regardless of the motivations, Woody has done a lot of good with his charitable foundation.

He served in the Marine Corps on active duty from 1943-45 and then in the reserves from 1948-49 and then again from 1954-69. He also did a short stint in the U.S. Army National Guard in 1948 while living in New Mexico, but he hated their culture. "They were just not like the Corps," he claimed, so he quit them and re-joined the Corps. He served in a reserve unit and taught the next generation everything he knew about combat.

He also took care of veterans for 33 years until 1979 when he retired from the VA. He helped men get medical help, financial assistance and educational grants. After retirement, he built up a horse farm called *Royal Winds Farms*. He won five blue ribbons at the Kentucky State Fair during the 1980's in various classes. He is a simple man who has done extraordinary things.

Unlike many who are thrust into the limelight and become famous overnight, Woody has done a decent job of not letting fame go too much to his head. He does enjoy being praised and does not like being questioned, but by and large, he has done far more good than harm using his fame. Many MOH recipients struggled with their celebrity and combat, something Woody has indeed done also, but compared to others, he has been able to conquer many of his demons and live a productive life largely free from depression, addiction and anxiety. Renowned Marine Corps historian Allan Millett said Woody has done a better job than most dealing with the challenge of "living a long life as a war hero of modest means and education…An MOH creates serious problems for a kid previously unknown outside his hometown."[3043]

But, in the end, the MOH was awarded to men not for their character and ability to carry it with dignity, but for a moment in their lives when they went beyond human bounds of self-preservation and did things for their fellow man that made others stand in awe. If character was the basis for the award, then Pappy Boyington or Jack Lucas would have *never* received it.[3044]

Besides being a story of war, medals and psychology, Woody's narrative highlights the love of his dear Ruby during difficult times. He and his beloved Ruby were married for 63 years until she passed away in 2007. One cannot overstate the power and gift she was to him. (See Gallery 4, Photo 12)

He was in the living room when Ruby died in their bedroom one evening. Woody wondered why she did not answer him after having retired to bed only a few minutes earlier. When he walked into the bedroom, he recognized she was gone. Her head hung limp; she appeared cold, still and without life. He had not been near a dead body in a long time, but he had seen death so often in the Pacific, he knew when someone had left this mortal coil. The emotions of pain and loss welled up inside of him, but he remained calm and checked her vital signs just to make sure. He sat down gently, took her hand and cried. She had saved his life during the war by giving him a reason to live. She had saved him after the war by supporting him as he battled with PTSD, and endured his drinking, smoking and struggles with pain. She introduced him to God by her gentle example and she coaxed him to accompany her to the Pea Ridge Methodist Church in Barboursville that fateful day in 1962 that healed his soul and changed him into a calmer and more peaceful man. She had made him stronger. In that moment of death and farewell, he knew the essence of life— "And this maiden she lived with no other thought/ Than to love and be loved by me."[3045] When someone has true love and loyalty throughout life, it is a rare gift and Woody had it. As Viktor Frankl noted:

> The truth—that love is the ultimate and the highest goal to which man can aspire. Then I grasped the meaning of the greatest secret that human poetry and human thought and belief have to impart. *The salvation of man is through love and in love.* I understood how a man who has nothing left in this world still may know bliss, be it only for a brief moment, in the contemplation of his beloved.[3046]

Woody's letters during the war show what he most thought about was Ruby and returning safely to her. After all the pain he found in war, he knew it would be worthwhile if only he could hold her, kiss her, love her. She was his reason for living. As he lay there by her, he said his goodbyes. He told her he would see her again and that now it was once more his time, to fight in this world alone while she awaited him. For a second time she would be watching out of a window for her beloved to meet her while he stayed in the world to carry on his work. It was comforting for him to believe that later, when he goes through those heavenly gates, Ruby will be standing there with open arms for him just as she did in 1945.

Her death made him contemplate his own mortality. What would people say about him when he died? He hoped they would say, "Woody loved people and the United States of America." He kissed her goodbye and then called the ambulance. Once again, he watched as the body bearers came to take a hero from the field of life.

Woody knew love during the war and this more than anything else sustained him. We cannot always send our youth to war knowing they have a wonderful love to come home to, but this love was one reason why Woody was so successful in battle. He had fantastic training, excellent equipment and tough Marines by his side, but in the end, the love of his devoted Ruby was crucial.

One can argue that if one has a person in his life whom he values more than himself, then one is far richer than anything he can have in his bank account. Ned Hallowell, a noted Harvard psychiatrist, claims that most who are strong and healthy can point to one person in their lives who believed in them; for Woody, that was

Ruby.[3047] As Frankl said, feeling unconditional love from someone is ultimately the "salvation of man." He continued: "Love goes very far beyond the physical person of the beloved. It finds its deepest meaning in his spiritual being, his inner self. Whether or not he is actually present, whether or not he is still alive at all, ceases somehow to be of importance."[3048] Woody was a better Marine because of Ruby (one could also argue Beck was a better officer also due to his wife Ruth's love). Many fought bravely without the silent support of strong and beautiful women waiting at home for them, but in Woody's and Beck's cases, having that love gave them hope and drive. Yes, war is random and death that comes with it picks its victims indiscriminately. As one Marine said, "There are only two kinds of Marines on [Iwo Jima]—the lucky and unlucky."[3049] Lieutenant General Smith was not being hyperbolic when he wrote the Pacific campaign was "two years of corroding, soul-destroying war."[3050]

On Iwo, a constant stream of men suffered "combat fatigue" and returned to the rear echelons; hardened veterans like Lefty, suddenly shut down after years of heroic service. "[T]hey needed rest. Just rest."[3051] As few as 2,700 and as many as 8,000 Marines suffered mental breakdowns on Iwo and Woody witnessed some letting fear get "the upper hand" until they became "absolutely useless."[3052] He and his comrades called these guys "cracked-up," and often, "they had to be helped back to the rear echelons after they were tagged by a Corpsman because they suffered such a bad nervous breakdown."[3053] Woody's commander, Donald Beck, said of most of his men that their courage was "unbelievable…They seem to have no regard at all for their lives." But then he mentioned the ones who suffered mentally saying, "However some were too nervous to leave on the line."[3054] Many started mumbling to themselves or sat staring into the distance at nothing, deaf and numb; they seemed "to be brought to a species of second childhood, generally simpering in their expression."[3055] Maybe what set Woody above such chaos was the love he had for Ruby. What kept him going when others ran out of steam was the constant thought: "I have to get back to my girlfriend" giving his deeds the "right action and right behavior" he needed.[3056] Frankl said, "It is a peculiarity of man that he can only live by looking to the future—*sub specie aeternitatis.*"[3057] Ruby was Woody's future. In many respects, she still is. As he wrote her in April 1944: "If I couldn't write you, I would go completely crazy… I know you love me and that you are waiting for me and that is plenty for me until we can be together forever."[3058] This statement is as true in 2019 as it was in 1944.

Woody embraced the ideals of honor, courage and commitment, and, empowered by love, achieved greatness in war and goodness in peacetime. He has made a difference. He is the type of citizen a nation relies on to protect its sovereignty in times of war (like he did serving in the Marines) and preserve its legacy in times of peace like he is doing with the Gold Star Families Monument project. He has benefited having *the Medal* by ensuring fame and respect, and at the end of his life, has even tried to benefit financially from it, but all in all, he has shown he has lived a life well lived by most human standards of decent conduct.

However, as one has read throughout this book, this study has often been critical of Woody and how he has tried to control his story and reported his tales. Everyone in the legal and historical community whom I know and who have learned that Woody has tried to use a lawsuit to stop this work have expressed disappointment that he has done so. Barry Beck, son of Donald Beck and Marine combat veteran and trained lawyer, said, "That's one of the stupidest things Woody could've ever done."[3059] The lawsuit Woody waged shows his fear of the revelation of many of the truths in this book. Perhaps Woody got nervous once I found out that one of his favorite side-kicks, Lefty Lee, had fabricated everything he witnessed Woody do on Iwo since he was never there. This had to be embarrassing since Woody had used him often to support his stories he told in public. Once Woody learned what type of researcher I was finding out and revealing Lefty's lies he had been able to maintain and hide for seventy years fooling even his family, the WWII Museum in New Orleans and government officials, he may have gotten nervous. Or perhaps he got nervous when he realized I was not going to write a "puff-piece" about his life and, instead, would do serious research surrounding his biography. Only Woody can tell us why he has behaved as he has. His behavior has elicited responses from the Beck, Henning, Rybak, Tischler, Waters, Dashiell and Schlager families denouncing how he has behaved with the Schlager family stating that what Woody has done has "tarnished" the *Medal*. In fact, William Schlager wrote that his father, Alexander, who was responsible for writing Woody a recommendation for the MOH would now "be deeply disappointed in you."[3060]

Moreover, out of the thousands I have interviewed the past 25 years, Woody seems to be the most difficult interview partner when I challenged him on his narrative—rarely have I found a person so difficult. When I interrupted him and stopped him during his story-telling, instead of being gracious and listening to my line of reasoning or allowing me to ask questions (I was after all trying to tell his story accurately), he often would get irritated asking, "Why are you interrupting me?", "Let me finish my story" or "I'm getting to that. Let me tell it how I want." In other words, red flags slowly started to raise themselves, as they always do, when a subject gets angry or irritable when his story is disputed or analyzed. He even threatened to pull his support early if I used the term *fiancée* to describe Ruby because that was not a term "we used back then!" although it was a term his mother and soon to be wife used when they wrote the White House in September 1945 about their attendance at Woody's MOH ceremony.[3061] After all, he had asked Ruby to marry him and she accepted, so they were actually engaged. Woody usually gets angry if he does not get his way—maybe great attributes in combat, but not for people who are trying to write an objective history of events and using terminology that describes situations properly.

## Exploring and Analyzing the Problems with Woody's Narrative

Many have asked me why I think Woody has created so many discrepancies about his narrative throughout his life. It appears every time Woody's testimony changed, it was almost always in his favor for building out the case that he was the "underdog" overcoming incredible odds. He wants people to buy into this theme that he has developed throughout each stage of his life. It is a character arc most love and what movies are made of—the underestimated and picked on guy becoming the hero of the story who defeats the evil villains and gets the girl. This formulae is as timeless as recorded history shown by such stories of Homer's *Odyssey* to William Shakespeare's *Henry V* to Tom Clancy's *The Hunt for Red October* to Stephen King's *Shawshank Redemption*. Notably, there are many elements to this type of story which Woody comes by naturally: He is below average height; he came from a rural West Virginia town offering few opportunities; he was probably picked on a lot; his father died when he was young during the Great Depression; he dropped out of high school and was uneducated; he survived one of the largest *Banzais* of the war; he cheated death knocking out fortifications with a flamethrower on Iwo; he endured 34-days of combat on the Sulphur Island. There are many things that are true to the "underestimated hero overcoming incredible odds story" making his embellishments or fictitious additions that much more perplexing. I am not a psychologist, but here is my plebian take on the matter.

Woody grew up with a Napoleon complex and later, when he was given a moment where everyone viewed him as a giant (surviving Iwo and performing heroic acts), he took the opportunities to slowly grow his legend. It started off small. When he returned to Guam from Iwo in March 1945, his exploits probably went from being *taking out a few pillboxes* to *seven* and from killing *several* Japanese to *at least 21 in one pillbox and God only knows about the probably over 50 in the other six*! In building out this theme of being an underdog, he added to the story throughout the years receiving admiration from his listeners.

## 1). Problems with His Birth and Dead Siblings

Woody was born last in the line of purportedly eleven births his parents had and he was the smallest of the litter at *three pounds*. Against incredible odds, he survived those early months when most babies died. In fact, the number at first was that his mother lost less than six, sometimes four and other times five, but then it grew to six out of her brood died, but Woody survived yet again! And as it has already been noted, only three dead siblings have been found in his hometown cemetery: Samuel (died in 1912 at 5 months old), Brady (died in 1913 at 8 ½ months old) and Lila (died in 1915 at 1 ½ months old). Although he said many if not all were killed in the Spanish flu epidemic from 1918-19, none have been found dying during that time. Losing three babies for Mama Williams was traumatic enough and devastating for the family so why Woody continued to add to it seems bizarre unless he continued to add to the number for dramatic effect. (See Gallery 4, Photo 11)

## 2. Height and Enlistment (Volunteer or Draftee?)

Then we get to him trying to "volunteer" for the Marines soon after war broke out, but being rejected yet again since he was an underdog—according to Woody, he was too short at 5'6" when the minimum was 5'8". But he stayed at it until his recruiter told him they lowered their standards in spring 1943 and he could volunteer. This all flies in the face of the facts that he was within height requirements and that he was actually drafted (he was especially within height requirements if the height he has reported all his life was then believed: either 5'6" or 5'7"). MOH recipient Jack Lucas entered the Corps in August 1942 at the same height as Woody at 5'5¼" and was accepted without any waiver or problems due to his measurements.[3062] Don Graves was even shorter than Woody at 5'5", had no issues and entered the Corps in August 1942 too.[3063] Finally, one of the innovators and proponents for the Higgins boat, later Lieutenant General Victor H. Krulak, was 5'4" (he was a 113 pounds and 5'2" when he entered the Naval Academy at 16 in 1930). Since Krulak was at the top of his Naval Academy class, then Chief of the Bureau of Navigation, Vice Admiral William D. Leahy, allowed him to join the fleet.[3064] With all these examples and with the facts about height requirements, Woody is being less than truthful about how he remembered the issues surrounding his enlistment.

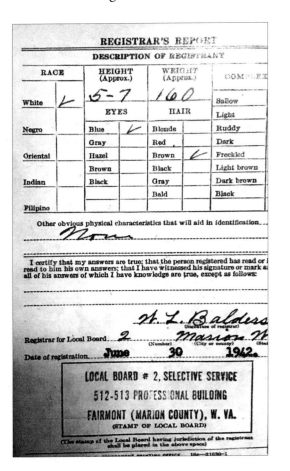

Here are the two pages that made up Woody's draft card. This card was filled out by Woody. It is clear he was trying to be taller than he was by lying about his height writing 5'7". However, in Woody's Service Book during his induction to the Corps on 26 May 1943 when accurate measurements were done at the in-processing station, it was clear that Woody was 5'5¼." Ancestry.com & St. Louis Personnel Files

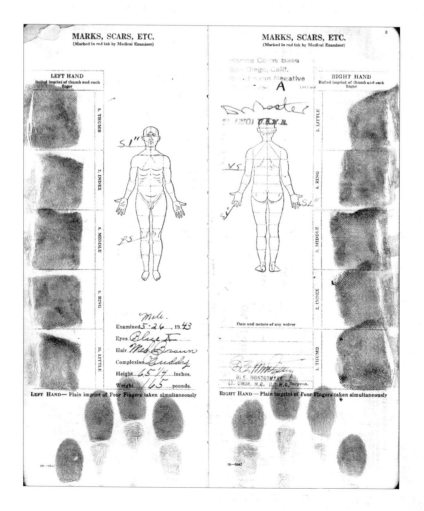

## 3). Boot Camp Accomplishments

Woody likes to tell the story that when he started boot camp, he never thought he would measure up, but by the end he was in the top ten of his class, something not shown in the Marine Corps records, but, nonetheless, added to the mystique.

## 4). Mysterious Comrade on Guam

When we get to Guam, he talks about at the last minute before the *Banzai* hit his lines, he was taken away from the extreme front and his foxhole/lookout position was given to Clevenger (he has made it a point numerous times of spelling his comrade's name exactly as shown here) who ended up dying as soon as the action started. Clevenger or Cleavenger or Klevenger or Klievenger etc. (I tried numerous spellings to try to confirm the man in the archives), according to the National Archives and Marine Corps, never existed. In the list of war dead from Guam, there is not a name that is remotely similar to this one and the author has looked through them all (especially looking at all the C Company, 1st Battalion, 21st Marines, 3d MarDiv dead for July 1944). But, according to Woody, on that bloody night of July 25-26, he again cheated death to continue the noble fight. When people hear this story, they breathe a sigh of relief that Woody once again escapes the dark shadow of death.

## 5). Accomplishments on Woody's MOH Day

In Woody's account of his most memorable combat on Iwo, he tells us that the advance of his regiment had ground to a halt in front of interlocking enemy pillboxes. He was his unit's last hope to break through the line of fortified positions with all the other flamethrowers having been killed or wounded. *No one else is* there to help and Beck and all the others have to depend on him, the runt from West Virginia who was too small to enter the Corps was now the man everyone relied on! Yet, we find out from the records that there were at least three if not five other flamethrower/demolition men along Captain Beck's front that day. Nonetheless, Woody gives the impression that if he does not take out the pillboxes with his flamethrower (the only guy left with the know-how to operate it in the company according to him), his unit will be stuck and continue to suffer casualties in their exposed condition. As a result, he rose up and attacked in the face of murderous fire and helped win the day for the Marines.

Back on Guam after March 1945, when people asked about his exploits, Woody offered a perfect narrative of taking out seven pillboxes and killing at least 21 Japanese and counting them personally: "[I] accounted for seven pillboxes"[3065] and "I counted 21 of 'em that tried to get away and I killed 'em all with flames."[3066] Beck sent this story up his chain of command to get Woody the MOH, but some in the highest levels of the Marine Corps led by Lieutenant General Geiger and the U.S. Navy led by Fleet Admiral Nimitz and their boards found the necessary documentation for it lacking and Rear Admiral Hayler and/ or his staff felt it necessary to revise the citation right before it was sanctioned, something Woody did not know about for decades if ever.

From 1945 until the major article by Francois in 1966, Woody has an elaborate memory of events and dialogue and gives "facts," some of which were never confirmed by the MOH process (something he had no knowledge of). After the Francois article in 1966 in *Man's Magazine*, probably due to some of his comrades calling him up and being upset (we know Schlager was unhappy about the article and stewed about it for years)[3067] or getting negative feedback from others who were there, Woody no longer discussed the events of 23 February in the detail he had given to journalists like Dashiell, Hoolihan, Lawless and Francois from 1945-1966. Thereafter, his memory seems more cloudy and he talks about the events he can prove (the two that made the cut in his citation) or he relies on the testimonies of others who remain unnamed and unprovable saying "people said I took out seven" or "they told me I killed 17." At the beginning of this study on 19 July 2016 at Woody's home in West Virginia, he ridiculed journalists throughout the decades as always getting his story wrong and that we, he and I, were now going to write a work that was "*FACTUAL*" disregarding information in most of those articles much closer in time to the events that, in the end, had actually relied on him for the facts![3068] What has never been proven, but definitely explored, is the possibility that Woody knows that his MOH process was later started from all the inflated "facts" he was bragging about when back on Guam—of course, he was not doing so to get a medal, but the unintended consequence of that type of bravado and possible exaggeration led to him possibly getting *the Medal*.

Think about the previously mentioned problems. From 1945 until 1966, Woody has perfect recollection that he killed 17 to 21 Japanese and destroyed seven pillboxes and regaled journalists with his exploits, often with swagger (changing the story somewhat along the way, but not the basic gist that he did something amazing using startling statistics to support his claims). How could one suddenly not remember such facts after 1966? If these facts were true and Woody recalled them to numerous people, especially 3d MarDiv comrades like Hoolihan and Dashiell, they were facts that he should have never forgotten. Statements like "[I] accounted for seven pillboxes"[3069] and "I counted 21 of 'em that tried to get away and I killed 'em all with flames"[3070] given in 1945 or "I killed 21 in the massive pillbox" in 1956[3071] are very different from "I don't know how many I killed" or "I can't remember how many pillboxes" or "people tell me that I took out seven" or "killed 21 or 17" or "I would not have counted those I killed" uttered after 1966. Flaming 21 Japanese and counting every one of their dead, charred and smoking bodies in 1945 is different from "I don't remember" of today. Why Woody suddenly after 1966 never mentions these figures again in his first-person story telling denotes that he did not suddenly "forget" these facts, but that he was "suddenly" editing them out of his narrative because someone or something had brought it to light that these "facts" were untrue. Just as Hayler and his team ascertained in 1945 that these facts were unsubstantiated when they changed his citation *deleting* these figures, Woody was also scrubbing them from his narrative soon after 1966 because he knew he could not back them up with evidence because some event had changed how he either could or would continue to tell his story.

## 6). The Appearing and Disappearing Bazooka Man

Throughout the years, some of the stories Woody continued to "remember," told and could "prove" continued to increase in their dramatic effect in his favor. This was especially the case with the disappearing bazooka tales when one explores the timeline of Woody's firsthand renditions of this action. In 1956, Woody says a bazooka man without his instruction fired at the massive pillbox and the round went through the aperture and exploded inside the inner chamber.[3072] Such an explosion in the enclosed space must have killed many from its detonation or the resulting concussion inside those thick concrete walls. Woody declares at this time that the explosion *inside* created smoke and debris that then exited the top of the bunker through a vent, giving him an idea of how to destroy the pillbox. Consequently, Woody mounted it from the front or side and flamed it from the inside out through the vent on top of the structure rendering it useless for the time being and killing anyone who might have escaped the earlier explosion.

But in 1966, his interview has a bazooka man now taking his orders to fire at the structure. This bazooka man had actually been requested to go on the mission by Woody himself when discussing the operations plan with Captain Beck—Woody places himself now more in the center of the action actually making critical decisions and giving orders: "I like to have four men who can provide some protective fire with rifles and BAR's…One of them should be armed with a bazooka. I'll also need the demolitions man."[3073] Beck, if anything, never missed a detail on the battlefield and he would not have depended on a corporal, like Woody, to remind him of small unit tactical engagement procedures for demolition squads. He would have already made sure a bazooka man was able to go with Woody, especially after Chambers had briefed him on the line of defense he had to attack. Woody claiming he asked for a bazooka man, and other key personnel, from Beck would be similar to a quarterback in discussion with his coach before a big play requesting that a center come into the game to snap the ball. Once Woody affirmed to Beck that he thought he could take on the pillboxes, of course Beck would rally his men together and tell them to support Woody and a bazooka man would have been part of that team. There were many things in Woody's interview from 1966 that do not make sense logically and/or match up with the facts.

During the action described in 1966 by Woody, this unnamed bazooka Marine then fired a round that now does not enter the pillbox like from the story of 1956, but hits below the aperture and "below and to the left" of the open slit (remarkable detail). From this explosion, Woody saw smoke exit the vent and then made his way behind the pillbox, mounted and then flamed it.[3074] He even claims in 1966, that another bazooka round "exploded harmlessly off the concave front" of a machinegun nest (once again extraordinary detail) after the attack on the large pillbox.[3075] In 1966, Woody described the landscape full of bazooka rounds, but throughout 2016 and 2017, he repeatedly said that there was never a bazooka man there at all.[3076] Since 1966, none of his interviews discovered describe a bazooka man although most flamethrower assault teams, by mandate, had a bazooka man with them.[3077] In all recent recappings of this event, he gives a tale that he did not have any "heavy equipment" in support of his attack and he was on his own. Each level of re-telling continued to increase the danger Woody faced making one, indirectly, feel Woody was overcoming incredible odds. And even if the 1956 version was the real story (which most historians usually would claim was closer to the truth since it was one of the first "tellings" of the story), Woody was still subjected to incredible dangers. Corporal Schlager, who did not like Woody but was nonetheless one of his MOH witnesses, claimed that in attacking this third pillbox, Woody was "subjected to a number of grenades thrown by the enemy."[3078] So even if the men inside the pillbox were dead once he attacked the bunker from the front and top, he still was operating in a hostile environment. Nonetheless, we once again see a story increasing in its magnitude in order to present a guy overcoming incredible odds eventually taking on the pillbox without a bazooka man because he disappears from the story altogether and Woody just focuses on himself taking on the pillbox alone and against a full contingent of enemy soldiers unhurt by any explosions.

## 7). Was Climbing on Top of a Pillbox on Iwo Jima Unique?

And when Woody has described this attack on the third, large pillbox, he often gives the impression that he was unique in climbing on top of the pillbox braving enemy fire. He presents his act as an extreme form of courage. Now to be fair, what he did was indeed heroic, no doubt, but it seems it was done by others too, possibly even two other flamethrowers that day on his front.[3079] So although this event did make its way into one affidavit supporting his MOH package, it seems that others were doing this same tactic too, something Woody has not admitted to, making one think he desires others to believe what he did extremely rare. From the sources, people were climbing on top of pillboxes all the time to attack them, like Woody's comrade PFC Marquys Kenneth Cookson, also in Charlie Company, who crawled and "wriggled on his stomach to the top" of also a large pillbox, and then hammer-threw a satchel filled with 20-pounds of TNT into the aperture from above. Seconds after he rolled off the top, the whole pillbox erupted into an explosion killing an "estimated 35 enemy soldiers."[3080] That also was very brave and a better way to bring down the odds of getting killed (crawling verses standing narrows the space in which bullets can hit you!).

Another man who climbed on a pillbox/cave network and destroyed it through a vent on Iwo was Gunnery Sergeant Keith A. Renstrom. Like Woody, he observed a vent sticking out from a massive network for many caves and pillboxes on 22 February. So, Renstrom took four men to the supply depot on the beach, found a 55-gallon drum of gasoline, and then using two bars, carried the drum back to their battle line on a makeshift stretcher. Then during either very late at night or in the early hours of the morning (using the cover of darkness), he and his Marines climbed on top of the large pillbox/cave structure and poured the fuel down the vent. Then the Gunny probably got some shut-eye, waiting several hours to ensure the fumes had spread out through the enemy tunnels and hovels. He knew that vaporized gasoline is 3-4 times heavier than air and would travel quickly throughout the lower regions of the Japanese network. Once he felt the deadly gas had spread throughout his target area, he went into action right before the sun rose on 23 February.[3081]

Moving cautiously and carefully, he crawled back on the structure, keeping a low profile, and threw a phosphorous grenade down the vent. He then rolled off quickly, and ran for cover. Within a few seconds, the whole side of the hill and the land around it shuttered and out of three openings in the earth, one known to Renstrom and his men and two not, long streams of fire shot over a hundred feet in the air. While Woody's flamethrower could create 3,500° Fahrenheit of heat, gasoline fumes can "become a 'fireball' with a temperature of 15,000 degrees F." When one knows that the sun's surface is 10,000° F, one understands how much power Renstrom had just unleashed.[3082] Renstrom just brought one of the most destructive forces to the island and instantaneously "disintegrated" any IJA soldiers within hundreds of feet of the underground blast zone.[3083] There would be no Japanese corpses to bury in Renstrom's area of operations this day. Pebbles and small rocks bounced off the surface of the land as the earth heaved with expansive, hot air and the secondary explosions of all the Japanese ordinance. The blast was so powerful, ships radioed Renstrom's battalion's CP wondering what had just happened.[3084] Remember, Woody was able to put a maximum of 4.5 gallons of fuel into his pillbox, and it was fuel already aflame. Renstrom just poured 55 gallons into one section, allowed for it to vaporize and spread out hundreds of feet, then set it on fire, unleashing an inferno—a mega-flamethrower. One gallon of vaporized gasoline "has the explosive energy of 83 sticks of dynamite."[3085] Renstrom basically put thousands of sticks of dynamite into this one area alone. So Woody was not alone climbing on structures and taking them out. Marines everywhere were destroying pillboxes and caves by crawling all over their tops and finding their openings and then attacking.

## 8). Was Standing on Top of a Pillbox Smart?

Moreover, although most people think this event of Woody taking out the third pillbox by mounting it quite heroic, often being read aloud at public events he speaks at, one needs to analyze this act more closely and ask whether or not it was smart. In discussing this with Lieutenant Colonel Eugene Rybak, a Marine veteran and the son of flamethrower operator Joseph Anthony Rybakiewicz, he finds this a strange thing to do on Iwo: "Why in the world would you expose yourself on top of a pillbox in clear view of the enemy when you could theoretically flank the pillbox and slowly move around the corner to pump fire into its aperture and keep your silhouette below the horizon away from the enemy? This act, although sounding very brave, sounds very foolish."[3086] Flamethrower operator and Iwo veteran, Donald Graves, echoed Rybak saying, "What Woody did was idiotic. Yes, we did climb on some pillboxes to attack them through their vents, but we did so with grenades and by crawling low, not standing up. Woody's action on that pillbox was a dumb move."[3087] As one can see, Woody's brave act can be construed as also an unwise tactic. As we say in the Marine Corps, *"It is easy to be hard, but it is hard to be smart."* One can say what Woody did was indeed courageous, but maybe not the smartest tactic to employ. Since we know Woody's temper sometimes can get the best of him, maybe his rage took him to act as such and he was lucky he was not killed standing upright in the middle of a battlefield on top of a huge pillbox with 70-pounds of flamethrower gear. Woody's guardian angel was indeed hard at work on the island keeping him safe. And looking at both Cookson's and Renstrom's techniques compared to Woody's, one sees they were just as brave; however, one also observes, their actions were smarter and, most likely more effective in killing Japanese. Their guardian angels, apparently, did not have to work as hard as Woody's since they applied more intelligence and stealth in their attacks.

## 9). Where did Woody's *Banzai* Story on Iwo Jima Come From and How Many Japanese Attacked Woody Using this Tactic?

Next, there are problems with Woody's *Banzai* story. First, there are no first-hand testimonies from *anyone* witnessing Woody taking out a wave of six charging Japanese with his flamethrower. In reports and files looked at from Beck, Strickland, Tripp, Schwartz, Southwell, Henning, Chambers, Dashiell, Naro, Elder and Schlager concerning Woody's actions this day, *no one ever* mentioned a *Banzai* wave eliminated by Woody's hand. As mentioned before, had he accomplished this, one would think someone would have remembered such a hair-raising event. Throughout time, the only person mentioning anything about flaming the Japanese is Woody which he did with Dashiell in 1945 (claiming he killed 21: "I killed 'em all with flames") and with Francois in 1966 (he killed three).[3088] The story made its way into his MOH citation without any evidence to support it in the document trail. Woody's 21st Regimental commander, Colonel Hartnoll J. Withers, was responsible for putting this in his endorsement which led to it being put in the citation since it is the only source in the chain of documents submitted for the MOH, but where did he get his information? We know Dashiell learned about Woody being considered for *the Medal* while Beck was doing his write-up and collecting documents in late March and early April (Beck left Guam on 14 April and never returned to the unit). We know that Dashiell's articles had to be approved by his command before they ever were published in newspapers (probably going across Wither's desk). In the heavily marked up and edited version of Dashiell's article in the National Archives, it is written prominently at the beginning of the write-up: "Note: Williams has been *recommended* for the <u>Medal of Honor</u> for the action covered in this story [author's italics]."[3089] We know men at the highest levels knew about Dashiell's reporting and noted it in real time with even the commander of the division, Major General Graves B. Erskine, writing a "Letter of Commendation" for Dashiell in June 1945.[3090] Withers' MOH endorsement for Woody was sent up the chain of command on

27 April most likely after Dashiell had interviewed Woody and written his article. Dashiell's article got approval at about the same time as Withers wrote his report. Soon thereafter, Dashiell's article started to appear in newspapers in May 1945, four months *before* final approval for Woody's MOH happened.[3091]

Did Withers get his information from reading Dashiell's article that was relying on Woody for information? Withers was indeed nuanced about how many *Banzai* attackers there were with stating "enemy riflemen" without specifying a number.[3092] But the only place where Withers could have found information about Woody taking out a *Banzai* attack documented in this study was either from Woody's own mouth or Dashiell's report which got the information in the first place *from Woody*.

And in Francois' 1966 article, Woody goes from saying he "flamed" 21 in 1945, to just three. Why the change? Maybe he got enough negative feedback in 1945 for saying he "flamed" 21 that he revised his numbers and lowered them to a more believable sum of only three. But then, throughout the years, the number starts to once again march higher. In interviews for this study and for the World War II Museum, the *Banzai* wave grows into being five or six IJA soldiers.[3093] Once again, one sees the danger Woody overcomes to increase as the years have marched on and with the re-tellings to audiences. It seems he modified his story to what he thinks he "can get away with," and then tips the scales always to increasing danger for himself.

## 10 ). Why Chambers Possibly Did Not Support Woody

And what about Woody's platoon leader, 2nd Lieutenant Howard Chambers? Woody never mentions the man who was primarily responsible for directing all the action on the battlespace where Woody did his deeds to get *the Medal*. Several who served with Chambers mention what a great leader he was, but Woody utters nothing about his direct boss especially nothing about him during the day when Chambers and his men helped take out 32 emplacements and kill numerous Japanese. Did Woody know Chambers was asked about his *Medal* and did not give an endorsement for it? After Marines trained with one another for six months leading up to Iwo Jima, they all knew their platoon leaders well—it would be rare not to include these leaders in their sea tales. Yet, Woody apparently has never mentioned Chambers in his interviews and Chambers never endorsed Woody for the *Medal* for many of the actions he must have witnessed Woody do although Fleet Marine Force, Pacific Commander, Lieutenant General Roy Geiger wrote him *twice* to submit evidence to prove Woody's heroism. Chambers' silence to Geiger may indicate that Chambers had serious reservations about Woody's MOH. We know Woody's MOH troubled Chambers his *entire* life because the documents about Woody were basically the only ones he *saved* from his entire time in the Pacific. One would think if a lowly lieutenant failed to provide the evidence requested by General Geiger, he would not have kept the document trail to brood over it throughout the years unless the situation bothered Chambers to such a degree that he wanted to contemplate about either the injustice, the inaccuracies, and/or the problems with the whole process.

Moreover, maybe another reason why Chambers also refused to support Woody was quite possibly due to having learned that Woody had been willing to murder one of his squad members, Thomas E. Scrivens, because he refused to move out into a skirmish line. Maybe some of the other Marines nearby were so horrified that Woody was going to kill a Marine under his charge, that they reported Woody to Chambers for this assault against one of their own. Any Marine officer worth his salt would never train his men to threaten death to any member who was not willing or unable to follow orders—if this were the case, we would have chaos within the ranks allowing young kids and men to exercise murderous authority over those deemed unworthy of their command. Moreover, unless more evidence shows other facts unknown about this event, if Chambers knew about what Woody has been claiming he almost did, then he should and probably would have been very disappointed that Woody had behaved as he did. When Woody "bragged" about this event back in July 2016, he did so in a manner that would make one think he was trying to prove his manhood or how tough he was, something that Chambers would not have agreed with at all and felt unbecoming of a Marine.[3094]

The only person alive who can give us some clarification about why Chambers might have had problems with Woody getting the MOH is Woody himself because every other witness searched for who could have given some local color on this process during this study was found to be already dead. As historian Ian W. Toll wrote, "History, like nature, abhors a vacuum"[3095] and there is a large one when one tries to document the events that led to Woody's MOH, but hopefully this book has gotten one closer to understanding what can be proven to have happened versus what probably has been creatively added after the battle.

## 11). Why Did Beck Not Talk About Woody?

And besides the absence of any written or oral statement given by Chambers in support of Woody or Woody's lack of any statements about Chambers *all* being strange, one also finds it somewhat incommodious that Beck *never* spoke of Woody after he put in his package for the MOH by 14 April 1945. Did Beck also later receive information about the disputed facts Woody had been giving journalists and fellow Marines that explains his silence? Did Beck learn that Rear Admiral Hayler or his board had to change his initial draft of Woody's MOH citation and remove the two main facts of seven pillboxes and 21 dead Japanese soldiers killed out of it because they both were unconfirmed? Beck was responsible for making one Marine out of 669,100 in WWII a war hero and famous (and remember, only 82 Marines got the MOH for actions they performed during the war).[3096] Woody was one of two we know of who Beck endorsed for a Medal of Honor out of the three battles he fought in: Bougainville, Guam and Iwo Jima (Lieutenant Henning was the other MOH endorsement). But Woody was *the only* MOH endorsement Beck put in that was successful.

Medal of Honor recipient Desmond Doss was in regular contact with his officer and even appeared on the famous TV show *This is Your Life* and in historical documentaries with his commander. They were close friends.[3097] In the hundreds of documents, letters and archival sources reviewed for Beck, he *never* mentions Woody and *never* contacted him after he left Guam in 1945. And in the hundreds of pages of letters, articles and unpublished memoir material Woody let me look at, there is not one letter, email or note of a phone call he exchanged with Beck. From the sources, they were *not friends*. One would think Beck would have been proud of the achievement of getting one of his own a MOH and would have often spoken about it. Was this due to him later finding out that he had acted on information that appears to have come from Woody or finding out why his platoon commander, Chambers, had actually refused to endorse his recommendation for Woody? We do not know, but here is another missing piece of the puzzle, and yet again, another relationship surrounding Woody that is difficult to explain.

## 12). Woody's Stories About His "Friend" Vernon Waters

There are other troubling relationships surrounding Woody's story. He recounts the tale of his close friendship with Vernon Waters and that he was there when he was killed. In reality, the evidence shows he was not there getting the date wrong, the unit out of place (Vernon was in a different company), the cause of death wrong (his head was not blown up by a Type 89 50mm grenade, but rather, Waters was either shot near the bottom of the skull or in the neck), the movement of the lines incorrect (they moved backwards or sideways and not forwards), the handling of Waters remains wrong (body bearers were not observed by Woody taking Vernon away from the scene since he was left in the open for days thereafter), and, most importantly, the tale of returning Waters' ring to his family left unsubstantiated. Woody's version is a beautiful story and one that possibly started as a way to impress or entertain his new bride who accompanied him on the journey to Montana and/or later listeners he thought would enjoy such a yarn, but all the evidence shows Woody was incorrect on almost everything about Waters except that he was tall, well-liked and killed. However, the story that the tallest guy became friends with the shortest in the unit; that they had this sacred pact to return the fallen guy's ring and tell the stories of death to their families in case the Grim Reaper visited one of them; and that when, in reality, the Reaper did take Vernon in front of Woody followed by Woody adhering to their pledge and getting his ring to return to the family is indeed a Hollywood tale. Woody's tall-tale is one full of power and is the stuff of legend, but there is no support for that version in the document trail uncovered to date. Vernon's nephew George Waters concluded, "It makes me *very angry* that Woody has used my uncle's death to better his image using false stories to do so. He has dishonored the memory of his friend."[3098]

Sadly, Woody gets other facts about Water's wrong. Before Waters' death,

**Photo 1:** Woody and Ruby's Wedding Day, 17 October 1945. Marine Corps Historical Division

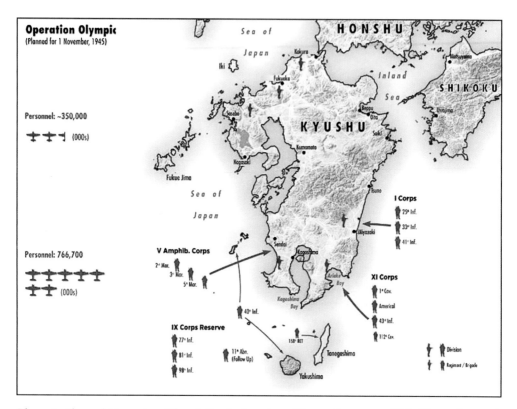

**Photo 2:** Planned Operation *Olympic* for the invasion of Japan. Secretary of War Henry Stimson's staff estimated conquering the main islands of Japan would cost 1.7 to 4 million casualties, including up to 800,000 dead. It was planned to commence on 1 November 1945. Jonathan Parshall

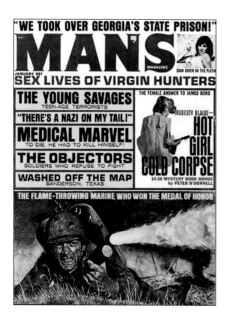

**Photo 3:** The *Man's Magazine* of January 1966 that contains the article written by Bill Francois about Woody's MOH exploits. Most of the information was given to Francois by Woody and the article has incredible detail about Woody's actions from 23 February 1945 given to Francois by none other than Woody. Author's Collection

**Photo 4:** The *USNS Woody Williams* Christening, San Diego, 21 October 2017. Author's Collection.

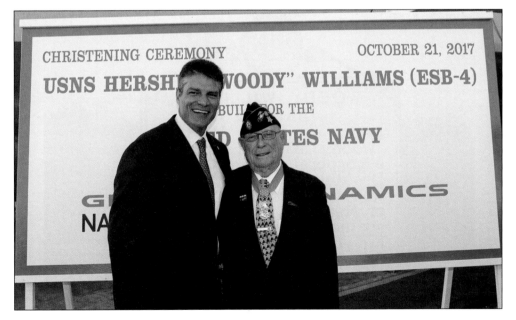

**Photo 5:** Author Bryan Rigg and Woody Williams at his ship christening on 21 October 2017 at San Diego. Author's Collection.

**Photo 7:** Author Bryan Rigg and Woody Williams at Cedar Park, TX during the dedication of one of his foundation's Gold Star Memorial Monuments. There were 38 mothers who had lost children in the nation's service in attendance. 23 September 2017. Author's Collection.

**Photo 6:** Woody Williams' Bridge, Cabell County, Barboursville, West Virginia. 20 July 2016. Author's Collection.

**Photo 8:** Before one of Woody's speeches in Indianapolis in 2013, he would not speak, he said, unless a man in attendance would put the MOH around his neck who protected him on Iwo Jima. Darol E. "Lefty" Lee rose up from the auditorium, walked to the front and tied *the Medal* around Woody's neck, receiving warm applause. Sadly, Lee was pretending to be one of Woody's security men still at this juncture fooling Woody and hundreds in the audience about his role during the day when Woody did the acts that earned him the Medal of Honor. Author's Collection.

**Photo 9:** Woody and Japanese Iwo veteran Tsuruji Akikusa on top of Mt. Suribachi. Laura Leppert, founder of Daughters of World War II, and daughter of Iwo veteran George Broderick of the 5ᵗʰ MarDiv is in the background. Woody described his time back on the island as such: "My emotions are like a tornado. They go up, and then something happens, and they go down. So they've been roller-coastering all day." Dave Shively

**Photo 10:** General Kuribayashi's grandson Yoshita Shindo, a prominent member of the Liberal Democratic Party and House of Representatives, takes the hand of USMC Lieutenant General Lawrence Snowden after Prime Minister Abe, during his address to a joint session of the U.S. Congress, claimed it was a miracle that the grandson of Iwo Jima garrison is now standing next to a combat U.S. Marine veteran from Iwo and prominent Marine Corps general in friendship. Lieutenant General Snowden was the current leader of the Iwo Jima American Association that strives to build bridges of understanding and reconciliation with the Japanese. It is the only organization after World War II that brings veterans from both sides of the conflict together. 29 April 2015. Associated Press

**Photo 11:** Woody's siblings' grave stone at E.T. Vincent Cemetery, Quiet Dell, West Virginia (Note there are only three buried here although Woody often claims there were six siblings who died). Author's Collection

**Photo 12:** Woody and wife Ruby circa. 2004. Marine Corps Historical Division

**Photo 13:** Politician Yoshitaka Shindo uses images of Kuribayashi in campaign posters saying he defends Japan like his grandfather. Knowing his grandfather committed crimes during WWII, one would think Shindo would be wise not to do this. Author's Collection

during the actions on 23 February for the pillboxes, when Woody put on all his gear and/or returned to the CP for more flamethrowers, he claimed Vernon Waters was there to help him with his equipment when we now know Waters was in Baker Company, not Charlie, and under a different lieutenant, Tischler, not Chambers. Maybe Beck had requested Waters to help Woody since he knew the men of Baker Company well, having commanded them during the battle of Guam and trained them until he took command of Charlie Company one month before Iwo, but the documents are unclear on this point and it is highly unlikely Beck would have "taken" another captain's flamethrower away to support Woody when that commander also needed all *his* men deployed at his sector of the front.[3099]

And many facts about the Waters' family by Woody are bizarrely incorrect; namely, he incorrectly claimed all the Waters men, including Vernon, had a strange finger mutation where the index and middle fingers were fused and he stated that Vernon's father was at the meeting although the father had been dead for several years.[3100]

## 13). Woody's Women

Concerning his romantic exploits, Woody builds up the perfect love story of wanting to get back to his fiancée Ruby during his wartime deployment, but we find out he had another girl in the wings. He continued to write his ex-girlfriend Eleanor and had her spying on Ruby to make sure she "was behaving." Eleanor Pyles was either divorced or still married and had a small child (Marion County has no records of her divorcing her first husband so she probably had been married for at least two years at the time Woody started dating her). She had to leave the Marine Corps after only serving one year since she had secretly gotten married *again* and was pregnant.[3101] After the war, she lied and claimed to have received the *Good Conduct Medal* in a periodical published in her state of West Virginia when in reality, her records disprove she ever got this medal (another Stolen Valor case like Lefty's).[3102] In fact, she had to leave the Corps because she had not disclosed her marriage and had gotten pregnant (her marks as a Marine were not that great either)—she was discharged "*under* honorable conditions," which means her service was satisfactory, but did "not deserve the highest level of discharge for performance and conduct" due to "minor misconduct." The Corps explicitly noted in her discharge report that she would never be allowed to return: "WILL NOT BE REENLISTED."[3103] Knowing all these facts about Pyles, one understands there were many reasons why Woody would want to keep his love affair with Eleanor hidden and his apparent "double-timing" with her and Ruby secret.

## So Why are There So Many Different Stories Coming From Woody About His Life?

These uncomfortable truths all go to show that the image Woody wants to create is one of a person overcoming incredible odds to become a hero and a man with few, to no flaws. This is all too human, but it goes to show that throughout the years, Woody created a more detailed story, very interesting and compelling, but at best not entirely based on what can be proved and at worst, faked. Everything he has talked about indeed is based on a foundation of truth, but has grown trees full of tall-tales that seem to continue to grow. Every time he got up in front of an audience, he changed his narrative to match the emotional response he wanted (*My Mom lost three babies* [the audience of women cover their mouths in horror]. *My Mom lost four babies* [women shake their heads and some cry]. *My Mom lost SIX babies and I barely survived weighing three pounds* [gasps of "Oh my God" ring out from the crowd]. *All the flamethrowers were dead or wounded and I was the last one Beck could depend on* [Marines nearby puff up their chests with admiring eyes]. *I was the last hope for us all!!!*) *I destroyed three, now wait, four, no, actually seven pillboxes* [replacements on Guam in the flamethrowing school in May 1945 look at Woody with venerating gazes]. Almost every discrepancy I found in his story was *always in his favor*, adding to the legend. Notably, even with all these issues, he still is a tremendous hero who fought bravely and effectively including at an important time in a major battle—he did not need to embellish. If all his heroic acts on Guam and Iwo were properly documented and put together, he probably would have merited a few MOHs as we have seen with other MOH recipients' cases and with just regular Marines doing their jobs as explored in this book like with Rybakiewicz, Chambers, Cookson, Wlach, Ridlon, Renstrom,

Waters, Graves and Fagan just to name a few. Nonetheless, embellishers do not like looking in the mirror and they do not like their real stories to be told even if they are still heroic because in their eyes, the real story is simply not good enough. In looking at what probably happened versus what was reported by Woody, we see a huge gulf.

As one can rightly assume, telling the truth and not compromising the narrative's integrity as a historian often gets in the way of friendships. As a result, I am no longer in contact with Woody. Often in recording the past, one encounters those who do not understand the craft of history and the quest for truth. Many are supportive at first, but if the work departs from their beliefs about themselves, their politics, their religion or in Woody's case, their narrative, then often people will part ways and occasionally do so ungracefully. As Plato said, it is better to deal with the ugly truth than with beautiful lies, but many have trouble with this motto.

Simon Wiesenthal told me in 1996 not to write *Hitler's Jewish Soldiers*. I said the truth must be told and I continued writing and finally published my research in 2002. It was one of the most complete works on *Mischlinge* (partial-Jews) in the *Wehrmacht* and won the William E. Colby Award in 2003 for the best military history book of that year (documenting men like "half-Jew" Colonel Walter Hollaender, the *Ritterkreuz* recipient discussed earlier). I was threatened by a powerful ultra-orthodox religious group Chabad not to publish my books *Rescued from the Reich* (2004) and *The Rabbi Saved by Hitler's Soldiers* (2016), but I did not give into their threats and have published works about one of the most amazing WWII rescues and a critical analysis of their leader Rebbe Schneersohn, a religious fanatic. And ironically, since I did not want to give into Woody's desire to keep the book a certain way (hagiography) or reveal unflattering facts about his medal process, he has tried to stop the work. I did not give into the threats and published a book I hope will give fresh insight into the Pacific War, the Japanese Imperial military, the Marine Corps and humanity during times of crisis. And although I have had misunderstandings with Woody, this book also shows he was indeed an American hero and fought bravely.

A job of a historian is to tell the truth, and in doing so, try to push humanity into a better understanding of why we do the things we do and how we can continue to make a better and more just society. If I had been hired to paint a portrait, of course as a paid artist, I would have painted a beautiful piece and made Woody look better than he was in life and leave off several of the blemishes or wrinkles of behavior and personality that I have observed. However, I am a historical doctor, and not a painter of portraits for this book. When I looked at the CT scan of Woody and his story, my duty as a historian was to interpret everything I saw and reveal my findings. If I and other historians do not do so, then the truth of history would die in poor diagnostic care. Historian Jon Hoffman supports this conclusion that a critical view of our heroes is a must when writing history. He explains:

> There are those who believe such a reexamination of a military icon does a disservice to the institution because it removes some of the luster from the burnished image held up as an example to be emulated. There is a definite and vital benefit in recalling the shining moments when an organization and its members were at their best. But it is equally important to highlight those occasions when things went wrong, for failure, more often than not, is the breeding ground for lessons that can improve future performance.[3104]

The study of Woody illuminates the timeless struggle of humanity. All of the themes are present throughout Woody's 96 years of what one might say, to borrow from Joseph Campbell's work, is a *hero's journey*. It is a story of a simple and flawed man accomplishing heroic deeds. Most specifically, his months in battle encapsulate the enduring issues freedom-loving people have faced through the ages when forced into conflict over religious, political and sociological problems. Freedom is never free and in learning about Woody, one learns what it takes to enjoy self-determination and security within one's own nation. And to "enjoy that self-determination and security," democracy's sons and daughters must be willing to fight and die for freedom and defeat fascist warmongers whenever they raise their ugly heads of exclusivity, racism and genocide.

## What Does All This History Just Explored Mean?

Leaving the micro-level with Woody and returning to the macro-level of this study, one sees this history has explored socio-religious issues that led to mass murder, racism and intolerance so we may do a better job in the future of recognizing these dangers sooner and destroy them. In particular, this book has examined Shintoists' and Zen Buddhists' beliefs under their High Priest and god, Hirohito, which led the Japanese to embrace fascist beliefs that fueled their rape and conquest in the hope of helping prevent such things from happening again. Young American boys' lives should not be interrupted like Woody's or destroyed like Waters' to prevent the plagues of murderous regimes like Hitler's and Hirohito's. Their lives should instead focus on making this world a better place through science, industry, education and invention, not on how to kill people. Yes, we should always have a portion of our population trained as efficient and willing warriors, but the vast majority of each generation should focus on more productive and useful activities like curing cancer, Parkinson's disease or Alzheimer's. In recent times, it truly does just take a small percentage of America's population to fight its wars. The victory our men and women secured in WWII probably has ensured that reality to date. Let us hope that we never again have to mobilize our entire population to focus on war and killing millions to preserve democracy. However, we must always be willing to do so if that democracy is to survive the ages.

So apart from honoring men like Waters, Beck, Chambers, Woody, Erskine and other American heroes and to show how evil Kuribayashi and his minions were in furthering Hirohito's lust for Empire, this book has another mission: To challenge readers to think about their beliefs, education, society and actions and whether they are making the world better. Since I began this research in 2015, genocide continues to be a recurring event. Violence against women is widespread especially in territory controlled by fundamentalist Muslims; they would have made many Imperial Japanese commanders proud. America is still at war against various groups of viciously intolerant Islamic radicals in a global war on terror, one of the most protracted and complicated conflicts in world history. Many problems that faced humanity from 1931-45 still remain. This book explicitly explores how a nation founded on freedom of religion responded to religious fanaticism, how we defeated Imperial Japan, how our military operated within the moral rules of engagement and how we lent our humanitarian support after the swords were pounded into plowshares. When looking at WWII, we get high marks in all these areas: We destroyed Hirohito's religious fanatics; we saved millions of additional Asians from being added to the list of Japan's

Holocaust; we adhered to the laws of war; and we re-built Japan's country in a massive, postwar humanitarian effort and *actually gave* the territory back to them so that they could rule themselves yet again! We fought the good fight, but we still have more to do.

In recent years, the world seems to have become more dangerous and unstable. Maybe this has been the case since history began, but most of the world throughout time did not have nuclear weapons spread out over the earth as is true now. I am hoping this work will get more people, at least in a small way, interested in the issues threatening us and ready to take action against the criminals who are making this world worse. Moreover, I hope this book will help people fight against the bad ideas that motivate the wicked. Although this book explores several heroic men and notable WWII battles, it also explores deeper and more enduring issues of humanity such as: How do we create a good society?

Reflecting on the U.S.' Imperial Japanese enemy in light of religious fundamentalism, I realize that fundamentalist groups, regardless of who they are, tend to suppress and disregard information when it conflicts with their dogmas and often kill those people who bring such information to light: People who claimed Hirohito was not a god and evil in 1940 Japan got their heads cut off, and people in radical Islamic lands who claim Mohammed is not Allah's messenger and a murderer get slaughtered as well. Getting rid of these types of people who commit such crimes against free thought and expression is one way to "create a good society." Noticing this, I have come to greatly admire America's founding fathers' wisdom that recognized the danger of combining religion with civil government, as the fusion is harmful to both religious freedom and good governance.

America's colonies were overwhelmingly Protestant, from a variety of denominations. "Typically the proponents of all these forms of Protestantism saw a high regard for natural science, reason, common sense, self-evident rights, and ideals of liberty as fully compatible with their Protestant heritage" although they looked at each other with skepticism. The founding fathers' task was to build a cohesive, voluntary civilization out of competing subgroups. America is fortunate that Thomas Jefferson, Benjamin Franklin, George Washington and James Madison, to name a few, were the architects of the religious and political freedoms that prevented government entanglement with religious beliefs as expressed in the First Amendment. They recognized that the separation of church and state is a cornerstone of democracy, laying a foundation to prevent government from restricting people's rights to follow their spiritual convictions and/or political beliefs. The freedom and tolerance in America, where so many had come to escape Old World religious and ethnic persecution as well as monarchist oppression, were a nurturing environment for religious minorities.[3105] President George Washington pledged to the Jews of Newport, Rhode Island: "To bigotry no sanction; to persecution no assistance."[3106] To prevent bigotry and persecution, Washington knew no religious institution should be permitted to hold power in government. Moreover, one could interpret Washington as also saying that fanaticism has no part in governance. This doctrine is a foundational tenet of American civics. American beliefs about how government should rule were bequeathed to the post-WWII Japanese and German governments, and demonstrate that such American mandates create democratic, non-Fascistic and anti-theocratic regimes.

This history is much more than a narration of Woody, Kuribayashi, the Guam and Iwo Jima battles or the atomic bombs. It explores how history is documented, how ideas shape action and how knowledge is preserved. To explore such themes, authors rely on the freedom of speech and of the press. Ironically, one of the biggest fights this author had in writing this work was with Woody himself who tried to suppress this work through threats and legal action, demanding the right to approve what I write and seeking financial compensation. He failed to realize history should not be censored and this book is not *solely about him.* Fortunately for history and this work, the U.S.' *First Amendment* won out and the publication of this book has not been suppressed. Freedom of speech triumphs again. Barry Beck, the son of Donald Beck, the officer primarily responsible for starting the MOH process to get Woody the MOH, said: "My father would've been terribly disappointed in Woody with how he has behaved with the writing of this book and how he tried to control his narrative."[3107]

I was saddened by the obstruction I encountered especially since Woody is indeed an American hero and patriot who, ironically, fought to defend our constitution and its rights. My experience with him makes me realize that when a historian gathers the truth from history, he should never make a Faustian arrangement with any of his subjects nor give them the final word on the research. If this became the norm, we would have limited ourselves to censored history, not truthful history. The sad conflicts between Woody and me probably stem more from his desire to have a book exclusively focused on him and my desire to use his story to tell the history of that period and explore historical truths and other heroes besides him. Moreover, he also probably did not like his flaws shown to the world, especially since they fly in the face of the narrative he has presented throughout his post-war life. I am a historian and I write the truth, uncompromisingly. Those qualities that ensured Woody's survival in war—stubbornness and tenacity—made him a poor partner when documenting the full history within this book. The book he wanted was narrow in scope and avoided rational self-criticism, whereas this book explores larger issues of Marine Corps history, battle accounts, amphibious warfare, war crimes, medal awarding, politics, government, religion and most importantly, how we, as a people, improve our world.

And in making this world a better place, one sees that we as a society need to be ever vigilant against dangerous religions and political movements (which often mirror religious organizations). As ex-evangelical Dan Barker writes, "[I]f we consider ourselves to be moral (or ethical), we ought not to refrain from denouncing religious teachings and practices that cause harm."[3108] Albert Camus, in his novel *The Plague*, described not only many religions, but also dangerous ideologies, as having much intolerance buried away in their foundational creeds and manuscripts. Camus wrote using the disease as a symbol of hateful ideas:

> [T]he plague bacillus never dies or disappears for good; that it can lie dormant for years and years in furniture and linen-chests; that it bides its time in bedrooms, cellars, trunks, and bookshelves; and that perhaps the day would come when, for the bane and the enlightening of men, it would rouse up its rats again and send them forth to die in a happy city.[3109]

In other words, we should guard against malevolent ideas, both political and religious, that find their breeding ground in holy books and hate groups, and shun and marginalize those elements before they infect society and emerge with terrible violence in times of social strain as seen with Imperial Japan and Nazi Germany. "We must keep *ourselves* in check, remaining ever vigilant against the darker side of our natures by which we are constantly threatened."[3110] More importantly, we need to remember that what a person believes is not private because beliefs are a "fount of action *in potentia*."[3111] Some beliefs, like "Jews are bad" or "infidels should die" or "non-believers go to Hell" or "Gentiles are inferior" or "homosexuals are deviant and sinners" or "the gods say Japanese should rule everyone" or "Chinks are bad" are beliefs that can give birth to dangerous movements as history attests.

Imperial Japan continued a human tradition of religious and ethnic prejudice expressed in mass slaughter which is not unique. Consider the Biblical story of 12,000 Israelites killing probably over 300,000 Midianites and raping 32,000 "virgins" in a divinely ordered orgy of ethnic cleansing (see Numbers 31)[3112] for enticing Israelites to be unfaithful to the Lord, or the Christians killing Jews and Muslims during the Crusades for not being Christian.[3113] Consider the one million Christian Armenians slaughtered by the Ottoman Muslim Turks from 1915-17.[3114] Consider the Nazis or Maoists who massacred millions in the name of ideology, which was just a replacement for religion. Hitler or Mao Tse-Tung were stand-ins for the Messiah or leader chosen by Providence and both were creating a "utopia/heaven" on earth—for Hitler, the thousand-year *Reich* and for Mao, the new China reborn through Marxism.[3115] Historian Jonathan Steinberg warns that when followers of men like Hitler, Mao or Hirohito make gods out of them, they will suffer dire consequences.[3116] By way of illustration, Mao is possibly the greatest slaughterer of all times, exceeding Hirohito, Hitler and Stalin combined, with a likely total of 70 million dead largely because his reign as "god" lasted for five decades.[3117] In these examples, we have people motivated by their misguided convictions to act with complete disregard for individual freedom and well-being and do harm to specific religious, ethnic and political groups.

Thus, political or religious fanaticism destroys everything it touches. But these movements can only take hold if the organization that gives birth to this fanaticism remains unchallenged. Ideas, leaders and movements should always be questioned—in fact, every society or government should preserve this ability to challenge, through the press or through oppositional parties for example, so that no one person or group obtains unquestioned authority. Fanatics like Hitler, Hirohito, Stalin and Mao remained in power for so long because they could not be questioned. This absolute power allowed them to turn into monsters. Jonathan Steinberg writes:

> Nobody is always right. No arguments are 'unanswerable.'... Unreason
> knows no limits. It cannot measure profit against loss or assess means

and ends. It rejects liberty of the mind and threatens the person of the thinker. It cannot tolerate free speech, blasphemous books, satire and irreverence. It mobilizes the turbulent energies, emotions and wishes inside each of us and hurls them against the limits of the human condition. In doing so it destroys itself and lays waste its surroundings.[3118]

To explore such truths as Steinberg has just mentioned, this book highlights the results of extreme forms of religious and political rule by men like Hirohito and his "Zen-obedient zombies."[3119] Shintoist-Zen Buddhist-Imperial Japan prevented its adherents from seeing their own atrocities and biases. Blaise Pascal said, "Men never do evil so completely and cheerfully as when they do it from religious conviction."[3120] Quite simply, fanatic beliefs, whether religious or political, prevent their adherents from seeing injustices around them so they may correct them.

In today's world, fighting genocide and religious totalitarianism should be two focuses of enlightened and democratic societies. In our battle against the Axis fanatics, America became what Roosevelt called "the arsenal of democracy,"[3121] beginning in the period while we were still neutral, and eventually played a crucial role in bringing Hitler's and Hirohito's regimes to their knees. However, Germany and Japan could have been defeated earlier if the U.S. had acted expeditiously against them instead of waiting for them to arm, militarize, threaten, and attack.

Long before his rise to power, Hitler revealed in *Mein Kampf* a murderous intolerance, and he continued preaching of his hate in the diatribes he made to anyone who would listen. Starting as early as 1927, Hirohito's regime committed atrocities in its rampage across Asia. The lack of any major international condemnation only emboldened both of these obsessive leaders. Although the U.S. played one of the most important roles in WWII to defeat Hitler and Hirohito, it was disastrously slow to pick up the sword, and did so not on moral grounds, but for the sake of self-preservation.

Other democracies also made mistakes at this time. British Prime Minister Neville Chamberlain sold out Europe with his policy of appeasement when he signed the Munich Agreement in September 1938. The Allies allowed Hitler to violate numerous border and treaty obligations from 1935-38 without any serious repercussions. Several Allied nations knew about the Rape of Nanking and of Japanese atrocities in cities like Beijing, Canton, Shanghai, and Tientsin throughout 1937-38, yet they were unwilling to risk war to help prevent China from falling under Japan's cruel shadow. As WWII MOH recipient and later Vice Admiral John D. Bulkeley, who had personally witnessed Japanese atrocities, wrote, "Official Washington and the American people could stick their heads in the sand and ignore the holocaust that the empire of Japan was inflicting on the Chinese people."[3122] Similarly, the U.S. and her Allies seemed to continually turn a blind eye to the Nazi's escalating violence against Jews as seen in the Night of Broken Glass (*Kristallnacht*: 9-10 November 1938), the *MS St. Louis* refugees (May 1939), the ghetto system in Poland (1939-40) and then the full slaughter of Jews once Russia was invaded (21 June 1941). These failures of action by the Allies in response to these fascist men's deeds told Hitler and Hirohito that they held the political and military high ground. There was a tragically high price in blood and treasure that was paid for years of neutrality in the face of the evil displayed in events around the world. This dangerous inaction in the face of injustice remains a lesson to enlightened and democratic nations of today that when they see an evil regime or dictator, they would be wise to destroy it or him *immediately*. And more importantly, that they maintain militaries that can and will be able to destroy such immorality.

We need to do our best to punish those who persecute and exterminate others for their thoughts, ethnicity, and religions and when possible, defeat these fanatical groups as we did with the Nazi Germans and Imperialistic Japanese, and as we are currently doing with Muslim fanatics in the Taliban, Al-Qaida or ISIS.

Hopefully, this book will also show we need to act rather than wait for a God to save us. History has shown repeatedly that humans cannot count on God taking action in time to prevent us from slaughtering each other, quite often in *His* name. In speaking about the Holocaust, Sam Harris writes, "If having half of your people [the Jews] systematically delivered to the furnace does not count as evidence against the notion that an all-powerful God is looking out for your interests, it seems reasonable to assume that nothing could."[3123] One could also say, if your god in Hirohito actually led your nation to suffer defeat ending with two atomic bombs, maybe the Shinto gods you depended on don't care about you. We only have ourselves to rely upon to eradicate these evils of war and persecution around us and cannot depend on any god out there regardless of what religion created him, her or them. Kem Stone writes about God commenting on Albert Camus' *The Plague*:

> The idea that even if God exists he would rather we act as though he doesn't is a very powerful idea... Whether we believe that God does not exist or that he exists is unknowable, but we should take matters into our own hands and improve the plight of the world rather than rely on his infinite grace to take care of everything for us.[3124]

In conclusion, besides reading about remarkable and disturbing events of WWII, I hope the reader thinks about the ideas and motivations that set the stage for the events between 1927 and 1945. In addition to being interested in the varied subject matters explored, I hope you are challenged to think about how you live your life and whether you would have the courage to stand up to evil. History shows us we need more citizens like the men documented in this book who acted with a high sense of moral integrity to save lives and preserve democracy. Michael Berenbaum wrote about a prisoner in the Sachsenhausen concentration camp who would tell the new arrivals to that Nazi hell about "the darkness that awaited them. He told them what was to be—honestly, directly, and without adornment." He would end with the admonition: "I have told you this story not to weaken you. But to strengthen you. Now it is up to you!"[3125] In the end, I hope this book will make you a more open-minded person, a more aggressive champion of justice, and a defender of the helpless. We must constantly remind ourselves, as Socrates taught, that "the unexamined life is not worth living" and it is in the pursuit of truth that we achieve enlightenment rather than in the attainment of it.[3126] Because once one thinks he has the truth, then he stops to grow and learn. And those who think they know everything are the most ruthless and dangerous men amongst us.

# Afterword

Japan's atrocities and murder of millions during WWII
are the worst of any nation during that time period. They are
the worst not only for their scope, but also because this needless
violence was done by what many considered a cultured nation.
They are the worst because Japan was too wise to have conducted
such slaughter, but not wise enough to follow its conscience.
—Japanese Fighter Ace Saburō Sakai[3127]

Moral imagination requires us to take responsibility for
past wrongdoings and, at the same time, stimulates us
to project our thoughts toward a more human
future through the creative examination of our past…
investigating the war crimes and atrocities committed
by the Japanese is, therefore, to master the past.
—Japanese historian Yuki Tanaka, *Hidden Horrors:
Japanese War Crimes in World War II*[3128]

As I WAS WORKING ON this book, an interesting disagreement between General Kuribayashi's grandson, the Honorable Yoshitaka Shindo, and me arose. After supporting me for almost a year, now on reflection, he wanted to be out of my book and pull his interview. When I had an hour and a half interview with him in his Tokyo office in April 2018 with an official translator, he was very candid and gave me lengthy answers to many of the pre-screened questions provided to his government. I was strongly encouraged by his honesty about the time period. I was especially impressed that he told me to write the truth if I discovered war crimes his grandfather committed in Hong Kong, which I indeed have found. He also claimed that Imperial Japan was wrong to do many of the aggressive acts it did from 1927-45.

The reason I was impressed is that many Japanese, and Japan's politicians for that matter, refuse to admit Japan did anything wrong during WWII, and they censor textbooks and curtail reparations as already documented in this book. This is especially the case with politicians in his current ruling party, The Liberal Democratic Party (which is really ultra-conservative, not liberal). Prime Minister Shinzo Abe, Shindo's colleague and friend, has tried to whitewash Japan's history and even attempted to rescind apologies to other

Fighter ace Sub-Lieutenant Saburō Sakai signed photograph. Officially he had 28 aerial kills although his autobiography claims he had 64. He was one of the few famous Japanese military personalities from the war who admitted Japan slaughtered millions and should feel shame for doing so. Author's Collection and Colin Heaton's Personal Archive

nations for Japan's "Comfort Women" and slaughter of millions during that time. He denies "Comfort Women" even existed! He is viewed by many in the world as a right-wing nationalist, a revisionist historian and, one might add, a "Japanese Holocaust denier."[3129] He seems to be a new Japanese nationalist who wants to produce "an evangelical edition of history with the inconvenient facts omitted."[3130] Knowing his grandfather, Nobusuke Kishi, was a Class-A war criminal and helped carry out the brutal Japanese rule in Manchuria as one of the economic ministers there, maybe we should not be surprised by Abe's behavior since he respects his grandfather. Abe has never come out to condemn his grandfather, who believed in the *Yamato* race theory (describing the Chinese as dogs), supported the opium trade, conducted sexual abuse against women and facilitated the death of hundreds of thousands of people in the puppet state of *Manchukuo*.[3131] So after my interview, knowing Shindo has aspirations of running for Prime Minister in Japan, I felt I had a friend and colleague in the pursuit of truth and in the quest of preserving some of the virtues democratic nations fight for; namely, for the "freedom of speech," "the freedom of the press," historical truth and the protection of victims of crimes against humanity. I had hoped that Shindo might change his party for the betterment of Japan.

After I met with him, with the help of my interpreter Ms. Uo of SIMUL, I put together a transcription of our interview and the information Shindo gave me. I sent it off to him and his government contact in the Ministry of Foreign Affairs, Iba Takamasa, and then waited for six months to get a reply. When I received the reply, he gave corrections on half of the notes I gave him and then he told me to take the other half out entirely—he did not want to be quoted saying what he had said. What I assume happened in the meantime was that once his remarks were translated and reviewed by him and his advisors, they felt they were too controversial for his conservative, and one might add, radical support base. Moreover, he may have even received pressure from some of his colleagues to break ties with an American *gaijin* who was going to start once again the controversial discussion of Japan's wartime past. This interview and the criminal information about his grandfather's wartime past may be politically devastating for Shindo since he has used the image of his grandfather Kuribayashi in political posters defending Japanese islands from China, Korea and Russia or trying to get back islands these three nations seized after WWII. Historians Philip A. Crowl and Jeter A. Isely prophetically declared the danger of what Shindo is doing when they wrote in 1951, "[Kuribayashi] was a man to be feared alive, but is probably even more dangerous to America dead, since he is capable of becoming a hero of a resurgent nationalism in Japan."[3132] Shindo's unguarded comments spoken truthfully for this research might jeopardize his position within his own party, a party that continually has difficulty apologizing properly to the world for Japan's crimes against humanity and for *its Holocaust*, and which, for all intents and purposes, is creating that misguided nationalism Crowl and Isely warned their readers about. Shindo even made a disguised threat to the Iwo Jima American Association (IJAA) that he might shut down Iwo Jima to America if they did not do something about my book and my role on the board (he claimed that the airport might be under repair and unable to receive the Americans!).[3133]

In reality, I should not be surprised by Shindo's behavior since he is a member of the revisionist lobby *Nippon Kaigi*, which aims to erase Japan's crimes from all history books, promote nationalistic education, support official visits to the Yasukuni Shrine and foster a nationalist interpretation of Shinto for all citizens. It wants the nation to return to having a standing army and one of its influential leaders, Hideaki Kase, unbelievably, wants it to return to a monarchy and Imperial state.[3134] Moreover, it stands for xenophobic, misogynistic, anti-modern, anti-liberal and anti-democratic principles—one could call it a fascist's organization which National Review journalist Josh Gelernter indeed claims.[3135] It is fascist for all the reasons just listed and more, since it supports the myth that Imperial Japan tried to "liberate" East Asia from Western colonial powers, that the Tokyo War Trials were illegitimate (a clear violation of Article 11 of the San Francisco Peace Treaty) and that the IJA massacre at Nanking in 1937 is fabricated. Moreover, it fights against feminism, LGBT rights and the 1999 Gender Equality Law.[3136] *Nippon Kaigi* would make Tojo, Hirohito, Matsui, and the butchers of Hong Kong Lieutenant General Sakai and General Kuribayashi proud. Shindo would be wise to listen to historian Liu Yizhend when he writes: "Historical records can be used to contribute to national prosperity and world peace or to spell disaster by covering up or even glorifying the evil."[3137] It looks like Shindo, unfortunately, is doing the latter. He might be wise to harken to the admonition by Chief of Counsel of the International Military Tribunal for the Far East, Joseph B. Keenan, when he said of people like Shindo's grandfather: "If there is no justification for punishment of individuals who have already brought civilization to the brink of disaster, then justice itself is a mockery."[3138] (See Gallery 4, Photo 13)

Ironically, such revisionists like Shindo and Abe, who support organizations and pursue activities that downplay, excuse and deny WWII atrocities, perform a "pivotal role in publicizing" Japan's mass murder during the Pacific War "beyond national boundaries."[3139] These revisionists' intense resistance to admit Japan's crimes, and the history and memory of massacres like Nanking, according to historian Takashi Yoshida, make the events now headline news: "[these atrocities] might have remained a domestic issue rather than becoming an international symbol of Japan's wartime aggression" had Japanese leaders just been honest about their past.[3140] So although Shindo and Abe think their behavior will make the "issues" go away, they are actually doing the opposite. In short, according to historian Zhang Sheng and President of the Society for Research on the History of the Nanjing Massacre by Japanese Invaders, people like Shindo and Abe have "distorted East Asia's history and played havoc with its present."[3141]

If Japan and its leaders like Abe and Shindo refuse to acknowledge their country's past and instead support jingoistic organizations like *Nippon Kaigi*, they dishonor not only the victims of twentieth-century Japanese atrocities in Korea and China (just to name a few), but also all those men like Woody, Beck, Chambers, Fagan, Stein, Erskine, Renstrom, Waters and millions of others who fought for freedom against totalitarian regimes like Hirohito's Japan. Moreover, they dishonor the deaths of their own countrymen who, although they died for an evil regime, at least could have taken hope that their deaths indirectly helped create a better government for their children and grandchildren. Although the government of Japan today is far better than that of Imperial Japan, it continues to fail to recognize the full shape of its country's criminal history. Until this denial ends, leaders like Shindo will dishonor the positive attributes of his grandfather's leadership seen on Iwo Jima and will implicitly condone its catastrophic behavior as seen in Hong Kong. Shindo's not speaking about Japan's horrific past enacts a "violence of silence." Shindo, and other leaders like Abe, seem to be perverting what the *Nippon Times* asked its readers to embrace in 1945:

> If we use this pain and this humiliation [of our defeat and surrendering] as a spur to self-reflection and reform, and if we make this self-reflection and reform the motive force for a great constructive effort, there is nothing to stop us from building, out of the ashes of our defeat, a magnificent new Japan free from the dross of the old which is now gone, a new Japan which will vindicate our pride by winning the respect of the world.[3142]

Sadly, Abe and Shindo are not getting the "respect of the world," but instead distrust for what the historian Michael Weiner calls their "mytho-history."[3143] It is hoped, however unlikely this may be, that this book might help people like Abe and Shindo accept their ancestors' wrongs so that such misdeeds will never again be done by Japan. Their choice will expose to their people, and to the world, the direction in which they want to take their nation.

# Acknowledgements

"We have 'fought the good fight,' and now let us 'keep the faith'
and make sure that we will not send our children out
to fatten the cemeteries of another Iwo Jima."
—Chaplain D.G. Creech[3144]

A good book is never written alone. As historian James E. Scott wrote, "Nonfiction books are like historical scavenger hunts," and along the way, there are many who help a historian gather his stories, words and analysis.[3145] First, I would like to thank my Yale University Professors Paul Kennedy and James Crowley for supporting my study of the Pacific War while I was an undergraduate. Kennedy was generous with his time, conducting a small seminar with me to review the history of the war with Japan. Also, I thank famous author Richard Selzer, who did three independent studies with me at Yale. He was a wonderful mentor and taught me to explore language and find good stories.

I thank Japanese theologian Dr. Masao Uenuma, President of Japanese Bible and Theology Ministry, for checking the theological issues raised in this book and the reasons behind the Japanese religious fervor of WWII. His clarification of Japanese religious beliefs was most helpful. Also, many thanks to historian Mark Driscoll for looking over my chapter on Japan's Holocaust and its quest for "Pan-Asianism" and for offering insightful feedback.

My past two books have benefited greatly from the review and feedback of Marine Corps General Anthony Zinni, author of *Before the First Shots are Fired*. Besides being the former Commander-in-Chief of U.S. Central Command, he has conducted several staff tours of Iwo Jima and became a resident expert of the battle. His support for this book has been greatly appreciated, especially since he proof-read it *twice* and provided a forward. I am also incredibly grateful to the 31st Commandant of the Marine Corps (1995-1999), General Charles C. Krulak, for his careful reading of the manuscript and for his forward. Moreover, I am thankful for the information he gave me about amphibious warfare, especially about the Higgins boat and how his father, Lieutenant General Victor H. Krulak, helped develop it. On a side note, we both attended Phillips Exeter Academy and were members of the same dorm (Peabody), facts we both are very proud to have on our resumes. Exeter's motto is *Non Sibi*, not-for-self, and General Krulak's support, feedback and time he gave for this book shows he lives by that creed. Also of interest is the fact that his godfather was none other than the father of amphibious warfare, Lieutenant General Holland M. "Howlin Mad" Smith, so it is appropriate for many reasons to have had General Krulak proof-read, endorse and write a forward for this work.

I am also thankful for the support from another distinguished Marine, the 29th Commandant General Al Gray (1987-1991). When I was transitioning out of the Marine Corps in 2001, my superior officer, Colonel John Allen and General Gray helped me with my new activities at American Military University. In conversations with General Gray, I told him how disappointed I was to not be able to fulfill my dream as serving as a pilot in the Marine Corps due to my injury. I had been at The Basic School (TBS) for almost two years in the holding company (Mike Company) trying to get healed up enough to return to active duty—I think I have the record here of being in this particular company the longest. However, during this time at TBS, Colonel Allen made sure I used the skills I had as a PhD from Cambridge and I took care of the library, gave lectures and developed professional reading guides for the Commandant's Reading List working in the "War Fighting Lab" under Major Brian Gudmundsson. Hearing how disappointed I was that I was not able to finish my time in the Corps, General Gray just simply told me: "Whether a librarian or pilot Rigg, you have accomplished something few people do and many people envy and that is you have earned the title Marine. Never forget that." I have never forgot those words and I am honored beyond words that General Gray, now 91-years of age, took the time to give me insightful feedback on my work and his endorsement. As possibly the only "Librarian" in the history of the Marine Corps, I hope I have used my skills well to write a work here that justifies all the time I spent in the TBS library and at the Al Gray Research Center at the Marine Corps University (I probably also have the record of being in that place more than any other USMC 2nd Lieutenant while serving at that rank).

Many thanks go to Don Farrell, a historian of the Guam battle. While touring the battlefield of Guam in 2015, he was our tour guide, providing useful insights into the conflict. He has been gracious, helping me find sources and providing excellent proofreading of my manuscript on the section on Guam. His book *The Pictorial History of Guam Liberation 1944* is a good and thorough exploration of that battle. In addition to Farrell, I would like to thank Ken Bingham, who wrote *Black Hell*, a comprehensive work on the Seabees on Iwo, for his feedback.

My dear friend Frank Hytken has once again gone beyond the call of duty in proofreading and fact-checking this book. A former Marine Corps Captain, JAG officer of the 3d MarDiv (FMFPAC) in the 1970's, Military Judge and an amateur historian, he has spent countless hours looking over my work and reading books about the conflict to ensure my facts are correct. He has done more for this book than most dissertation advisors would do for their students, having gone through the work *four times*.

For her professional editing, I am indebted to Angela Langlotz, J.D. She served as an articles editor on the Journal of Small and Emerging Business Law at Northwestern School of Law at Lewis and Clark College and is herself a published author. She edited this manuscript *twice* and gave me excellent legal advice about issues relating to intellectual property.

Many thanks go to the renowned historian Richard Frank, author of *Downfall*, for his support, friendship, proofreading and counsel. While on Iwo Jima in 2015 and 2018 for the commemoration of the 70th and 73rd anniversaries of that battle, we discussed many of the facts in this book and the possibility of writing Woody's biography. His support has proved extremely helpful; he proof-read the manuscript *twice*. It is a true honor to have had the support of such a mind.

I thank Charles Neimeyer, retired Director and Chief of Marine Corps History at Marine Corps University, for his careful reading and insightful comments and constructive criticism of the work having read it *three times*. Special thanks go to the Lieutenant Colonel (USMC) and distinguished author of *The*

*Ghosts of Iwo Jima*, Robert Burrell for his careful reading of the manuscript *twice*; he saved me from many errors and helped me add humor when possible.

I thank Dan King, one of the foremost experts on the IJA, for his detailed and careful reading of this book. His book, *The Tomb Called Iwo Jima*, helped construct my understanding of how and why Iwo was defended the way it was. Moreover, he helped me understand the IJA soldier and what drove him militarily, religiously and ideologically. I also thank renowned historian and USMC Colonel Allan Millett for his detailed reading of the work and for his feedback.

Many thanks go to retired navy Captain Lee Mandel for his edits and feedback. His work *Unlikely Warrior* on Rabbi Roland E. Gittelsohn, the first Marine Corps Jewish Chaplain, was valuable in helping me understand Gittelsohn and his role on Iwo.

I appreciated the Japanese Deputy Consul General in Houston, Ryuji Iwasaki, for spending time with me and with the Honorary Japanese Consul John Stich discussing Japan and its wartime past. Iwasaki provided some documents about his government's policies toward its history and convinced me that Japan still has a long way to go in being honest with its Holocaust, rape and slaughter of so many of the people under its control from 1927-45.

I thank Iwo Jima veterans Jerry Yellin, Hanse Hamilton, John Lauriello, Don Graves, Ellery B. "Bud" Crabbe, Leighton Willhite, and Darol E. "Lefty" Lee for their countless hours of interviews and proofreading. I especially thank retired Brigadier General Ronald R. Van Stockum, Woody's battalion Executive Officer at the battle of Guam. His insights and feedback on the battle of 25-26 July 1944 proved invaluable in understanding what happened that night. At 103, he is remarkably astute and clear-headed. I thank Barry Beck, the son of Donald Beck, Woody's company commander. Besides his proof-reading of the manuscript *three times*, Barry gave me countless documents that have enriched this book.

Once again, I'd like to thank my mentor and friend Michael Berenbaum. He is a renowned Holocaust historian and has helped me with all my books since 2001, especially in proof-reading and reviewing this book. His support, insights and feedback throughout the years means more to me than he can possibly know.

Since I was a student at Cambridge University, historian and fellow Marine Colin Heaton has consistently shared information with me from his vast collection of oral histories. He has published numerous books on WWII and has an incredible knowledge of the personalities from the war. All my books have been enriched by his support and sources.

I thank Kimberly Bauer and Chris McDougal at the National Museum of the Pacific War, home of the Fleet Admiral Nimitz Museum in Fredericksburg, Texas, for their help and assistance with the book. Especially, I would like to thank McDougal, who is the Associate Archivist and Librarian at the museum for his time in helping me answer questions about numbers and facts about the Pacific War and learn more about the events surrounding Woody's experiences.

I thank the staff at the National Park Service center of War in the Pacific at Guam, especially the Park Ranger Kina Doreen Lewis. They provided useful contacts and feedback for my manuscripts. Park Ranger Lewis not only put me in touch with Monique C. Storie, Director of MARC (Micronesian Area Research Center) at the University of Guam and the Library Technician Lourdes T. Nededog, both of whom helped me gather several primary sources, but she also went to this center with me on her off day to look through the files with me. I also thank the lead of admissions Jonathan Barnhart and Director Dominica Tolentino of Guam Museum, for their support and help with finding primary sources and documents, and especially for putting me in touch with Colasita Gumabon, who heads up the archives for the Museum.

Many thanks go to Ms. Jinghua Shi who helped organize my research trip to Shanghai and Nanking in May 2019. While on this trip, I had a wonderful translator, my daughter Sophia Rigg, who helped me read documents in the archives and museums and navigate the train, bus and taxi system of China (not an easy task). After spending a year in China in 2018 with School Year Abroad (SYA), she had remarkably become fluent in Chinese and could help me with my book (a good return on investment for Dad). While at the Shanghai Library Historical archives, we were greatly helped with finding sources by Ms. Lian Yu-Cheng and Director Chu Man-Ching. While in Nanking, we were taken around the city and shown historical sites by Professor Jianfend Ju and his lovely wife Lin Xian. All these individuals contributed remarkably to my understanding more about Japan's Holocaust and Japan's inhumane actions in China from 1927-45.

My friend Robert Citino, a renowned *Wehrmacht* historian currently at the WWII Museum in New Orleans, has been a great supporter and helpful critic throughout the years. His reading of the manuscript pointed out several areas where I needed to do more research and revise my analysis. Like Citino, historian Edward Westermann has offered constructive criticism and careful readings of my last two books and his support is greatly appreciated. Also Dwight Mears, author of *The Medal of Honor*, provided excellent insights in understanding MOH history and provided welcomed feedback on the book.

I am deeply thankful to accomplished Marine Corps historian, Colonel Jon T. Hoffman, who gave me a tough and detailed report on my work in 2017 and helped me reshape this work dramatically. He helped me find mistakes and fix many errors. For his careful reading and thorough review, I thank him from the bottom of my heart. And then in 2018, while undergoing cancer treatment, he *still* read a *second* and much larger draft and provided once again indispensable feedback and support. I cannot tell him how much I value his mentorship and friendship.

James Ginther and Fred Allison at the Marine Corps Historical Division in Quantico, Virginia proved helpful in directing me to primary sources about numerous units Woody served in, as well as oral histories from Woody's commanders in addition to those from Woody himself. Not only did both prove helpful in finding original source material, but also they have encouraged me and listened to me explore ideas and analysis for countless hours on the phone and in person.

The helpful staff at the National Archives in College Park, Maryland has once again been invaluable to my research. In the documents section, I thank Tim Nenninger and Nathaniel Patch for helping me locate sources. Aaron Arthur of the photographs archive not only helped me locate rarely-seen photographs, but he also showed me how to use the scanning machines, something often difficult to navigate in an archive. Also Archival Room Supervisor Eric Kilgore at the National Archives Personnel Records in St. Louis helped me tremendously in locating several never before seen personal records of men in this book. His supportive staff of James Herbert, Dean Gall, Alexa Kitchen, Michelle Johannes and Sarah Rigdon were also most helpful and kind to me during my stays at their center. I thank Ellen Guillemette and Joanie Schwarz in the MCRD archive for their help in finding new information for me about WWII Marine Corps boot camp as well as Jan Warßischek and Daniel Schuler at the Bundesarchiv/Militärarchiv in Freiburg, Germany for their help in finding material on the IJA and Hirohito.

I give special thanks to Tatsushi Saitō and Colonel Yukio Yasunaga at the National Institute for Defense Studies, Ministry of Defense of Japan. Mr. Saito works in the Military History Division and Colonel Yasunaga is a Senior Fellow for the Center for Military History. These two men worked hours pulling sources to answer my many questions. Then, they sat for two days in the reading room at their center in Tokyo going over the names, facts and interpretation of the sources with me. They have corrected many mistruths in the historiography that have made their way into the literature over the last several decades. With their help, this work will present one of the most complete studies of Kuribayashi in the English language.

In understanding General Kuribayashi, I had the opportunity to spend time with his grandson, the Honorable Yoshitaka Shindo of the Liberal Democratic Party in Japan. Shindo is a busy man and holds a prominent position in Prime Minister Abe's government. His spending time with me to answer personal and historical questions about his family was a privilege. Unfortunately, like Woody to some extent, he has tried to control what has been written here and used strong arm tactics to try and prevent some of what I discovered, but I responded to him that I do not write censored history, but history. He tends to only want to focus on his grandfather's excellent military record and not on Kuribayashi's crimes—this is probably all too human but it is not historically honest. Special thanks go to Shindo's colleague in the U.S., Political Counsellor at the Embassy of Japan, Iba Takamasa. He supported me in arranging the meeting with Shindo and providing invaluable contacts with professional interpreters at SIMUL International and researchers at the military archives in Tokyo. Because he helped me with my research and due to the fact that my research did not always reveal Kuribayashi in a positive light, Shindo then made Takamasa's life difficult: at the writing of this book, Takamasa may have to leave the Ministry of Foreign Affairs due to Shindo's displeasure.

Moreover, in understanding Kuribayashi's actions in China, I thank Professor Chi Man Kwong at Hong Kong Baptist University. He went out of his way to send me essays and documentation to help me better understand the battle of Hong Kong of 1941 and Kuribayashi's role in the IJA takeover of it and the crimes committed by him and his troops.

Many friends and lay readers have provided invaluable feedback and corrections. I am thankful to Dr. David Alkek (USAF), Shannon Christine (an excellent editor who read the book *four times*), Scott Drescher (his detailed edits were incredibly helpful), Colonel Jeff Gault (USA), Donald Graves (USMC), Mike Hillyard (USMC), Sara Horton, Earl Kirkpatrick (USMC), Dr. Horst Knapp, Jack M. Kuykendall, Kimberly Larkin, Gary Lawson (USA), Robert Leonard (USMC), Laura Leppert, Jacob Levitt, Alex Y. Lim (from the Philippines, a retired IT head and BPO executive, who colorized the famous photos from Iwo shown on the cover), Maria Elizabeth Lynn, Catherine Nolan, James Oelke Farley (his battle-field tours and support mean the world to me), Jonathan Parshall (many thanks for his incredible maps), John Renstrom, Mike Renstrom, David Rigg (USMC), Lieutenant Colonel Eugene Rybak (USMC), Brian Solecki (USMC), Major Paul Stubbs (USMC), Amy Goings-Thornton, Steve Tischler, Colin Traver, Masaaki Tsuda, Captain Maurice Uenuma (USMC), historian Geoff Warwo, Phillips Exeter Academy English teacher David R. Weber (his detailed and thorough edits made the English here stronger) and Rabbi Edgar Weinsberg. I especially thank Laura Leppert, founder of *Daughters of World War II*, for her support on getting me on the IJAA board and for helping me find primary sources.

My special thanks to Colonel Charles A. Jones U.S. Marine Corps Reserve (Retired), who is a self-taught expert on the Pacific War, particularly 7 December 1941 and Iwo Jima. He wrote two books about WWII: *Pearl Harbor's Hidden Heroes* about the Medal of Honor recipients in the Hawaiian Islands during WWII and *Iwo Jima's Battlefield Promise*, a comprehensive review of the Battle of Iwo Jima. His corrections, insights, and suggestions (particularly about the MOH) made this book possible.

Lastly, I thank Woody and his family for their support from 2015-18. Woody has refused for years to write a book or have someone write a book about his life: Taking a critical view of one's life is never easy to do. His story is interesting and helps tell the stories of many other veterans; and his support helped add local cultural color to many events. As mentioned before, his life highlights many others' stories, so this book now honors people like Beck, Chambers, Erskine, Waters, Hemphill, Rybakiewicz, Tischler, Renstrom, Bornholz and Fischer as well as a host of others. In Woody's desire to "keep the focus on others," we have accomplished that with these pages although when the shift went to others, he sometimes did not like it. It is never easy to write a book about a living person and working with him on it was difficult. But Woody for a long time proved to be a good helper, reading manuscripts, going over ideas and talking for hours in person and on the phone. All his quotes and information about him as the source were reviewed by Woody and this final version you have read here was given to Woody for him to give his feedback: He refused to do so. As time marched on, we have not seen eye-to-eye on how this book has evolved. He wanted a strict biography without any criticism or historical analysis and I quickly learned, that although Woody's life is worthwhile to study, it is really useful in telling the larger story of the war and about the countless men who never returned home. As a result of this focus and other issues, we parted ways. I thank him for giving me that opportunity to highlight others and to better understand a history that needs telling of what America sacrificed, which is the larger focus here than why someone was awarded a MOH.

# Bibliography

## Books:

The 3d Marine Division Bulletin, Vol. II, 6 Oct. 1945, Number 32.

Alexander, Joseph H., "*Closing In: Marines in the Seizure of Iwo Jima*," WWII Commemorative Series, Marine Corps Historical Center, Washington D.C., 1994.

Allen, J. Michael and Allen, James B., *World History From 1500*, NY, 1993.

Arendt, Hannah, *Eichmann in Jerusalem.* New York, 1984.

Aurthur, Robert A., and Cohlmia, Kenneth, *The Third Marine Division*, Washington D.C., 1948.

Axelrod, Alan, *Miracle at Belleau Wood: The Birth of the Modern U.S. Marine Corps*, Lyons Press, 2007.

Barker, Dan, *Godless: How an Evangelical Preacher Became One of America's Leading Atheists*, Berkeley, 2008.

Barry, John M., *The Great Influenza: The Story of the Deadliest Pandemic in History*, NY, 2004.

Ben-Sasson, H. H. (ed.), *A History of the Jewish People*, Cambridge, 1976.

Berenbaum, Michael, *The World Must Know: The History of the Holocaust as told in the United States Holocaust Memorial Museum*, Baltimore, 2007.

Berry, Henry, *Semper Fi, Mac*, NY, 1982.

Bingham, Kenneth E., *Black Hell: The Story of the 133rd Navy Seabees on Iwo Jima February 19, 1945*, CreateSpace Publishing, 2011.

Bix, Herbert P., *Hirohito and the Making of Modern Japan*, NY, 2000.

Black, Conrad, *Franklin Delano Roosevelt: Champion of Freedom*, NY, 2003.

Blum, John Morton, *V Was for Victory: Politics and American Culture During World War II*, NY, 1976.

Bradley, James and Powers, Ron, *Flags of Our Fathers*, NY, 2000.

Bradley, James, *Flyboys: A True Story of Courage*, NY, 2003.

Bright, Richard Carl, *Pan & Purpose in the Pacific: True Reports of War*, Trafford Publishing, 2014.

Breuer, William B., *Sea Wolf: A Biography of John D. Bulkeley, USN*, Novato, CA, 1989.

Bryan, J. and William F. Halsey, *Admiral Halsey's Story*, NY, 1947.

*The Teaching of Buddha*, Buddhist Promoting Foundation, Tokyo, Japan, 1992.

Burkett, B. G., and Whitley, Glenna, *Stolen Valor: How the Vietnam Generation was Robbed of its Heroes and its History*, Dallas, 1998.

Burleigh, Michael, *Moral Combat: Good and Evil in World War II*, NY, 2012.

Burns, Edward, *Western Civilization: Their History and Their Culture*, NY, 1954.

Burrell, Robert S., *The Ghosts of Iwo Jima*, Bryan, TX, 2006.

-- Review of *Flamethrower*, 26 Feb. 2018.

Cameron, Meribeth E., Mahoney, Thomas H.D., and McReynolds, George E., *China, Japan and the Powers: A History of the Modern Far East*, NY, 1960.

Camp, Dick, *Leatherneck Legends: Conversations with the Marine Corps' Old Breed*, Minneapolis, MN, 2006.

Campbell, Joseph, *The Power of Myth with Bill Moyers*, NY, 1991.

Cantor, Norman L., *After We Die: The Life and Times of the Human Cadaver*, Univ. of Georgetown Press, 2010.

Carroll, John Mark, *A Concise History of Hong Kong*, 2007.

Ch'êng-ên, Wu, *Monkey*, trans. Arthur Waley, NY, 1943.

Chang, Iris, *The Rape of Nanking: The Forgotten Holocaust of World War II*, NY, 1997.

Chang, Jung and Halliday, Jon, *Mao: The Unknown Story*, London, 2006.

Churchill, Winston S., *The Grand Alliance*, NY, 1950.

--*Memories of the Second World War: An Abridgement of the Six Volumes of The Second World War*, Boston, 1987.

Coffey, Thomas M. *Iron Eagle : The Turbulent Life of General Curtis LeMay*, NY,1987.

Cohen, R., *Soldiers and Slaves: American POWs Trapped by the Nazis' Final Gamble*, NY, 2005.

Collier, Peter, *Medal of Honor: Portraits of Valor Beyond the Call of Duty*, NY, 2005.

Coogan, Michael D., (ed.), *Eastern Religions: Origins, Beliefs, Practices, Holy Texts, Sacred Places*, Oxford, 2005.

Cook, Haruko Taya and Cook, Theodore F., *Japan at War: An Oral History*, NY, 1992.

Cooper, Matthew, *The German Army, 1933-1945: Its Political and Military Failure*, NY, 1978.

Cornwell, John, *Hitler's Pope: The Secret History of Pius XII*, NY, 1999.

Coram, Robert, *Brute: The Life of Victor Krulak, U.S. Marine*, NY, 2010.

Craig, Gordon A., *Germany, 1866–1945*, NY, 1978.

Crowley, James B., *Japan's Quest or Autonomy: National Security and Foreign Policy 1930-1938*, Princeton, 1966.

Crowl, Philip A. and Isely, Jeter A., *The U.S. Marines and Amphibious War Its Theory and Its Practice in the Pacific*, Princeton, 1951.

Dallek, Robert, *An Unfinished Life: John F. Kennedy 1917-1963*, NY, 2003.

---*Franklin D. Roosevelt: A Political Life*, NY, 2017.

Dallin, Alexander, *German Rule in Russia 1941-1945: A Study of Occupation Policies*, NY, 1957.

Dante Alighieri, *The Divine Comedy of Dante Alighieri: Inferno*, trans. Allen Mandelbaum, NY, 1980.

Darby, Jean, *Douglas MacArthur*, NY, 1989.

Davis, Burke, *Marine: The Life of Chesty Puller*, NY, 1991.

Dawkins, Richard, *The God Delusion*, NY, 2006.

Dimont, Max I., *Jews, God and History*, NY, 1964.

*Documents on the Tokyo International Military Tribunal Charter, Indictment and Judgements*, Boister, Neil and Cryer, Robert (eds.), Oxford, 2008.

Dower, John W., *War Without Mercy: Race & Power in the Pacific War*, NY, 1986.

Divine, Robert A., Breen, T.H., Fredrickson, George M., and Williams, R. Hal, *America: Past and Present*, London, 1987.

Drea, Edward J., *Japan's Imperial Army: Its Rise and Fall, 1853-1945*, Kansas, 2009.

Driscoll, Mark, *Absolute Erotic, Absolute Grotesque: The Living, Dead, and Undead in Japan's Imperialism, 1895-1945*, Chapel Hill, 2010.

Dunn, Susan, *A Blueprint for War: FDR and the Hundred Days that Mobilized America*, New Haven, 2018.

Edgerton, Robert B., *Warriors of the Rising Sun*, NY, 1997.

*Families in the Face of Survival: World War II Japanese Occupation of Guam 1941-1944*, Guam War Survivors Memorial Foundation, Guam, 2015.

Farrell, Don A., *Liberation-1944: The Pictorial History of Guam*, Tinian, 1984.

—*Saipan: A Brief History*, Tinian, 2016.

Ferguson, Ted, *Desperate Siege: The Battle of Hong Kong*, Scarborough, Ontario, 1980.

Flores, Judy S. (ed.), *Growing Up in the War Years: Elders' Stories of Childhood Memories*, University of Guam, 2002.

Foote, Shelby, *The Civil War: Fort Sumter to Perryville*, NY, 1958.

Frank, Richard B., *Downfall: The End of the Imperial Japanese Empire*, New York, 1999.

—Review of *Flamethrower*, 25 May 2017.

—*Tower of Skulls*: A History of the Asia-Pacific War July 1937-May 1942, NY, 2020.

Frankl, Victor, *Man's Search for Meaning: An Introduction to Logotherapy*, NY, 1963.

Gandt, Robert, *The Twilight Warriors: The Deadliest Naval Battle of World War II and the Men Who Fought It*, NY, 2010.

Gannon, Robert, *Hellions of the Deep: The Development of American Torpedoes in World War II*, Penn State Univ. Press, 2009.

Garand, George W. & Stobridge, Truman R., *Western Pacific Operations: History of U.S. Marine Corps Operations in World War II*, Quantico, 1971.

Gaster, Theodor H., *Myth, Legend, and Custom in the Old Testament*, NY, 1969.

Giangreco, Dennis M., *Hell to Pay: Operation Downfall and the Invasion of Japan, 1945–1947*, Annapolis, 2009.

Gilbert, Martin, *The Second World War: A Complete History*, NY, 1989.

Gill, Anton, *An Honourable Defeat: A History of German Resistance to Hitler 1933-45*, NY, 1994.

*The Goebbels Diaries, 1942–43*, ed. and trans. by Louis P. Lochner, New York, 1948.

Goldberg, Harold, *D-Day in the Pacific: The Battle of Saipan*, Indiana Univ. Press, 2007.

Gow, Ian, *Military Intervention in Pre-war Japanese Politics: Admiral Katō Kanji and the 'Washington System'*, London, 2004.

Graves, Kersey, *The World's Sixteen Crucified Saviors: Christianity Before Christ*, NY, 1875.

Groves, Leslie, *Now it Can be Told: The Story of the Manhattan Project*, NY, 1962.

*Guam War Claims Review Commission: Report on the Implementation of the Guam Meritorious Claims Act of 1945*, Guam, 2004.

Haas, Kurt & Haas, Adelaide, *Understanding Sexuality*, New Paltz, NY, 1993.

Hallas, James H., *Saipan: The Battle that Doomed Japan in World War II*, Lanham, MD, 2019.

---*Uncommon Valor on Iwo Jima: The Stories of the Medal of Honor Recipients in the Marine Corps' Bloodiest Battle of World War II*, Mechanicsburg, PA, 2016.

Hammel, Eric, *Iwo Jima*, Minneapolis, MN, 2006.

Hanayama, Shinsho, *The Way of Deliverance: Three Years with the Condemned Japanese War Criminals*, NY, 1950.

Hane, Mikiso, *Modern Japan: A Historical Survey*, Oxford, 2001.

Hanley II, Fiske, *Accused War Criminal: AN American Kempei Tai Survivor*, Dallas, TX, 2020.

Hardacre, Helen, *Shinto and the State, 1886-1988*, Princeton Univ. Press, 1991.

Harris, Meirion and Harris, Susie, *Soldiers of the Sun: The Rise and Fall of the Imperial Japanese Army*, NY, 1991.

Harris, Sam, *The End of Faith: Religion, Terror, and the Future of Reason*, NY, 2004.

Harris, Sheldon H., *Factories of Death: Japanese Biological Warfare, 1932-1945, and the American Cover-up*, NY, 2002.

Harvey, T. Fred, *Hell Yes, I'd Do it All Again*, Alpine, Texas, 2011.

Haynes, Fred & Warren, James A., *The Lions of Iwo Jima: The Story of Combat Team 28 and the Bloodiest Battle in Marine Corps History*, NY, 2008.

Henri, Raymond, *Iwo Jima: Springboard to Final Victory*, NY, 1945.

Henri, Jim G., Beech, W. Keyes, Dempsey, David K., Josephy, Alvin M., and Dunn, Tom, *The U.S. Marines on Iwo Jima*, NY, 1945.

Herman, Arthur, *Freedom's Forge: How American Business Produced Victory in World War II*, NY, 2013.

Herman, Jan K., *Battle Station Sick Bay: Navy Medicine in World War II*, Annapolis MD, 1997.

Hilberg, Raul, *Destruction of the European Jews*, NY, 1961.

Hirsh, Michael, *The Liberators: America's Witnesses to the Holocaust*, NY, 2010.

Hitchens, Christopher, *god is not Great: How Religion Poisons Everything*, NY, 2007.

--*Hitch-22: A Memoir*, NY, 2010.

--*Mortality*, NY, 2012.

Hitchens, Christopher, (ed.), *The Portable Atheist: Essential Readings for the Nonbeliever*, NY, 2007.

Hitler, Adolf, *Mein Kampf*, Boston, 1971.

Hoffman, Carl W., *The Seizure of Tinian*, Historical Division Headquarters U.S. Marine Corps,

Quantico, 1951.

Hoffman, Jon T., *Chesty: The Story of Lieutenant General Lewis B. Puller*, NY, 2001.

—*Once A Legend: "Red Mike Edson of the Marine Raiders*, Novato, 1994.

Holiday, Samuel, and McPherson, Robert S., *Under the Eagle: Samuel Holiday Navajo Code Talker*, Univ. Press of Oklahoma, 2013.

Hotaling, Kerry, *Go Forward into the Storm: An Iwo Jima Journal*, Christopher Matthews, 2016.

Hough, Frank O., *The Island War: The United States Marine Corps in the Pacific*, NY, 1947.

Iriye, Akira, *The Origins of the Second World War: In Asia and The Pacific*, NY, 1987.

*Japan at War*, Time Life Books, NY, 1980.

Johnson, Paul, *A History of Christianity*, NY, 1995.

—*A History of the American People*, NY, 1997.

Josephy, Jr., Alvin M., *The Long and the Short and the Tall: The Story of a Marine Combat Unit in the Pacific*, NY, 1946.

Kagan, Robert, *Dangerous Nation: America's Place in the World from Its Earliest Days to the Dawn of the Twentieth Century*, NY, 2006.

Kakehashi, Kumiko, *So Sad to Fall in Battle*, NY, 2007.

Keegan, John, *The Second World War*, NY, 1990.

Kennedy, Paul, *Engineers of Victory*, NY, 2013.

—*The Rise and Fall of the Great Powers: Economic Change and Military Conflict from 1500 to 2000*, NY, 1987.

Kershaw, Ian, *Hitler, 1936–1945: Nemesis*, NY, 2000.

—*The Hitler Myth: Image and Reality in the Third Reich*, Oxford, 1990.

Kertzer, David, *The Pope and Mussolini: The Secret History of Pius XI and the Rise of Fascism in Europe*, Oxford, 2014.

King, Dan, *A Tomb Called Iwo Jima: Firsthand Accounts From Japanese Survivors*, North Charleston, SC, 2014.

Kirchmann, Hans, *Hirohito: Japans letzter Kaiser Der Tenno*, München, 1989.

Kozak, Warren, *Lemay: The Life and Wars of General Curtis Lemay*, Washington D.C., 2009.

Krakauer, Jon, *Where Men Win Glory: The Odyssey of Pat Tillman*, NY, 2009.

Krulak, Victor H., *First to Fight: An Inside View of the U.S. Marine Corps*, Annapolis, 1999.

—*Japanese Assault Boats, Shanghai, 1937*, CMP Productions, 2017.

Kugel, James L., *How to Read the Bible: A Guide to Scripture Then and Now*, NY, 2007.

Kuhn, Dieter, *Der Zweite Weltkrieg in China*, Berlin, 1999.

Kwong, Chi Man and Tsoi, Yiu Lun, *Eastern Fortress: A Military History of Hong Kong, 1840-1970*, Hong Kong, 2014.

LaFeber, Walter, *The American Age: United States Foreign Policy at Home and Abroad Since 1750*, NY, 1989.

Lai, Benjamin, *Hong Kong 1941-45; First Strike in the Pacific War*, NY, 2014.

Lauren, Paul Gordon, *Power and Prejudice*, London, 1988.

*Law Reports of Trials of War Criminals*. Vol. III. The United Nations War Crimes Commission, 1948, Trial of General Takashi Sakai.

Levin, Harry (Edited), *The Portable James Joyce*, "A Portrait of the Artist as a Young Man," NY, 1976.

Linton, Suzannah (ed.), *Hong Kong's War Crimes Trials*, Oxford, 2013.

Manchester, William, *Goodbye Darkness*, NY, 1979.

Mandel, Lee, *Sterling Hayden's Wars*, Jackson MS, 2018.

—*Unlikely Warrior: A Pacifist Rabbi's Journey from the Pulpit to Iwo Jima*, Gretna, 2015.

Maslowski, Peter and Millett, Allan R., *For the Common Defense: A Military History of the United States of America*, NY, 1984.

Mears, Dwight, *The Medal of Honor: The Evolution of America's Highest Military Decoration*, Kansas, 2018.

Melson, Robert, *Revolution and Genocide: On the Origins of the Armenian Genocide and the Holocaust*, University of Chicago, 1996.

Melville, Herman, *Moby Dick*, Oxford, 2008.

Miller, Donald L., *The Story of World War II*, NY, 2001.

Millett, Allan R., and Maslowski, Peter, *For the Common Defense: A Military History of the United States of America*, NY, 1984.

Millett, Allan R., *Semper Fidelis: The History of the United States Marine Corps*, NY, 1991.

Modder, Ralph, *The Singapore Chinese Massacre: 18 February to 4 March 1942 Were 5,000 or 50,000 Civilians Executed by the Japanese Army?*, Singapore, 2004.

Morison, Samuel Eliot, *History of the United States Naval Operations in World War II, Vol. VIII*, NY, 1975.

Morison, Samuel Eliot, and Henry Steele Commager. *The Growth of the American Republic*. Vol.2, Oxford, 1958.

Moore, Aaron William, *Writing War: Soldiers Record the Japanese Empire*, Cambridge, 2013.

Moore, J.D., "Iwo Jima Eyewitness," Pass in Review, Feb. 1989.

Mosley, Leonard, *Hirohito: Emperor of Japan*, NJ, 1966.

Moskin, J. Robert, *The U.S. Marine Corps Story*, NY, 1992.

Mountcastle, John W., *Flame On: U.S. Incendiary Weapons, 1918-1945*, Mechanicsburg, PA, 2016.

Muggenthaler, Karl August, *German Raiders of World War II: The First Complete History of Germany's Mysterious Naval Marauders*, Suffolk, 1978.

Mullener, Elizabeth, *War Stories: Remembering World War II*, NY, 2002.

Murkoff, Heidi and Mazel, Sharon, *What to Expect When You're Expecting*, NY, 2016.

Murphy, R. Taggart, *Japan and the Shackles of the Past*, NY, 2014.

Murray, Kevin Charles, *Sgt. A. F. "Kelly" Murray U.S.M.C.: A Hoosier Hibernian In The Great Pacific War*, Bloomington, IN, 2011.

Nalty, Bernhard C., Shaw, Henry I. Jr. and Turnbladh, Edwin T., *Central Pacific Drive: History of* the U.S. marine Corps Operations in World War II, Vol. III, Quantico, 1966.

"Navajo Code Talker Samuel Holiday Dies in Ivins at Age 94," St. George News, 12 June 2018.

Ness, Leland, *Jane's World War II Tanks and Fighting Vehicles: The Complete Guide*, London, 2002.

Newcomb, Richard F., *Iwo Jima*, NY, 1965.

Newpower, Anthony, *Iron Men and Tin Fish: The Race to Build a Better Torpedo during World War II*, Santa Barbara, CA, 2006.

O'Brien, Cyril J., *Liberation: Marines in the Recapture of Guam*, World War II Commemorative Series, Marine Corps Historical Center, Washington D.C., 1994.

O'Brien, Phillips Payson, *The Second Most Powerful Man in the World: The Life of Admiral William D. Leahy, Roosevelt's Chief of Staff*, NY, 2019.

O'Donnell, Patrick K., *Into the Rising Sun: World War II's Pacific Veterans Reveal the Heart of Combat*, NY, 2002.

Oliver, Robert T., *A History of the Korean People in Modern Times: 1800 to the Present*, Newark, 1993.

Olson, Lyne, *Those Angry Days: Roosevelt, Lindbergh, and America's Fight Over World War II, 1939-1941*, NY 2013.

Owen, Frank, *The Fall of Singapore*, NY, 2001.

Parshall, Jonathan, and Tully, Anthony, *Shattered Sword: The Untold Story of the Battle of Midway*, Nebraska, 2005.

Paterson Thomas A. (ed.), *Major Problems in American Foreign Policy: Volume II: Since 1914*, Third Edition, Lexington, MA, 1989.

Patterson, James T., *American in the Twentieth Century: A History—Third Edition*, Orlando, 1976.

Picker, Henry, *Hitlers Tischgespräche im Führerhauptquartier, 1941–42*, ed. Percy Ernst Schramm, Stuttgart, 1976.

Pike, Francis, *Hirohito's War: The Pacific War 1941-1945*, NY, 2015.

Plato, *The Trial and Death of Socrates: Apology*, Cambridge, 1975.

Poe, Edgar Allan, *Poetry and Tales*, The Library of America, NY 1984.

Prange, Gordan W., *At Dawn We Slept: The Untold Story of Pearl Harbor*, NY, 1981.

Redlich, Fritz, *Hitler: Diagnosis of a Destructive Prophet*, Oxford, 1998.

Rhodes, Richard, *The Making of the Atomic Bomb*, NY, 1986.

Rice, Edward, *Ten Religions of the East*, NY, 1978.

Rigg, Bryan Mark, *Hitler's Jewish Soldiers: The Untold Story of Nazi Racial Laws and Men of Jewish Descent in the German Military*, Kansas, 2002.

—*Lives of Hitler's Jewish Soldiers: Untold Tales of Men of Jewish Descent Who Fought for the Third Reich*, Kansas, 2009.

—*The Rabbi Saved by Hitler's Soldiers: Rebbe Joseph Isaac Schneersohn and His Astonishing Rescue*, Kansas, 2016.

Roach, Mary, *Stiff: The Curious Lives of Human Cadavers*, NY, 2003.

Roberts, Andrew, *The Storm of War: A New History of World War II*, New York, 2012.

---*Churchill: Walking with Destiny*, New York, 2019.

Robertson, Jennifer E., *Politics and Pitfalls of Japan Ethnography: Reflexivity, Responsibility, and Anthropological Ethics*, NY, 2009.

*Rogers, Robert F.,* Destiny's Landfall: A History of Guam, Univ. of Hawaii Press, *1995.*

Ross, Bill D., *Iwo Jima: Legacy of Valor*, NY, 1985.

Russell, Edward Frederick Langley (known as Lord Russell of Liverpool), *The Knights of Bushido: A History of Japanese Crimes During World War II*, NY, 2016.

Sakai, Saburo with Caidin, Martin and Saito, Fred, *Samurai!: The Unforgettable Saga of Japan's Greatest Fighter Pilot—The Legendary Angel of Death*, NY, 1975.

Salmood, John A., *The Civilian Conservation Corps CCC 1933-1942: A New Deal Case Study*, NY, 1967.

Samuels, Richard J., *Machiavelli's Children: Leaders and Their Legacies in Italy and Japan*, Cornell Univ. Press, 2005.

Sasser, Charles, *Two Fronts, One War: Dramatic Eyewitness Accounts of Major Events in the European and Pacific Theaters of Operations on Land, Sea and Air in WWII*, London, 2014.

Scott, James M., *Rampage: MacArthur, Yamashita, and the Battle of Manila*, NY, 2018.

Sherrod, Robert, *On to Westward: The Battles of Saipan and Iwo Jima*, Mt. Pleasant (SC), 1991.

*Singing the Living Tradition*, Universalist-Unitarian Song Book, Meditations.

Skya, Walter A., *Japan's Holy War: The Ideology of Radical Shinto Ultranationalism*, Duke Univ. Press, 2009.

Sledge, E. B., *With the Old Breed: At Peleliu and Okinawa*, NY, 1981.

Smith, Larry, *Iwo Jima: World War II Veterans Remember the Greatest Battle of the Pacific*, NY, 2008.

Smith, Holland M., *Coral and Brass*, Arcadia Press, 2017.

Smith, Seven B., *Spinoza, Liberalism, and the Question of Jewish Identity*, New Haven, 1997.

Snow, Philip, *The Fall of Hong Kong: Britain, China and the Japanese Occupation*, New Haven, 2003.

Snyder, Louis L., *Encyclopedia of the Third Reich*, NY, 1989.

Snyder, Timothy, *Bloodlands: Europe Between Hitler and Stalin*, NY, 2012.

*Soldiers Guide to the Japanese Army*, Military Intel Service, War Dept., Washington D.C., 15 Nov. 1944.

Spector, Ronald H., *Eagle Against the Sun: The American War Against Japan*, NY, 1985.

Speer's Albert, *Inside the Third Reich*, New York, 1970.

Spence, Jonathan, *The Search for Modern China*, NY, 1990.

Stalecker, Gene E., *Rolling Thunder*, Mechanicsburg, PA, 2008.

Steinberg, Jonathan, *All or Nothing: The Axis and the Holocaust 1941-1943*, NY, 1991.

Steinberg, Rafael, *Island Fighting*, Alexandria Virginia (Time Life Books), 1978.

Stenger, Victor J., *God The Failed Hypothesis: How Science Shows that God Does Not Exist*, Amherst, 2007.

Stoltzfus, Nathan, *Resistance of the Heart: Intermarriage and the Rosenstrasse Protest in Nazi Germany*, NY, 1996.

Sulzberger, C.L., *The American Heritage Picture History of World War II*, NY, 1966.

Tanaka, Yuki, *Hidden Horrors: Japanese War Crimes in World War II*, Lanham, MD, 2018.

Taylor, Jon E., *Freedom to Serve: Truman, Civil Rights, and Executive Order 9981*, NY, 2013.

*Third Marine Division's Two Score and Ten History*, authored and produced by Third Marine Division Association, Inc., Paducah, KT, 1992.

*Third Marine Division's Two Score and Thirteen History*, authored and produced by Third Marine Division Association, Inc., Paducah, KT, 2002.

Thompson, Paul W., Doud, Harold, and Scofield, John, *How the Jap Army Fights*, NY, 1942.

Toland, John, *The Rising Sun: The Decline and Fall of the Japanese Empire 1936-1945, Vol. I-II*, NY, 1970.

Toll, Ian W., *The Conquering Tide: War in the Pacific Islands, 1942-1944*, NY, 2015.

--*Pacific Crucible: War at Sea in the Pacific, 1941-1942*, NY, 2012.

Totani, Yuma, *The Tokyo War Crimes Trial: The Pursuit of Justice in the Wake of World War II* (Cambridge, MA, 2008).

Trevor-Roper, R., *The Last Days of Hitler*, NY, 1947.

Tuchman, Barbara, *Stilwell and the American Experience in China 1911-45*, NY, 1971.

U.S. Marine Corps Publication, "Expeditionary Operations," MCDP 3, 16 April 1998.

—"How to Conduct Training," MCRP 3-0B, 25 Nov. 1996.

—"Intelligence," MCDP 2, 7 June 1997.

—"Planning," MCDP 5, 21 July 1997.

US Army Infantry Human Research Unit, "Japanese Studies on Manchuria, Vol. XI, Part 3, Book A, "Small Wars and Border Problems: The Changkufeng Incident."

Vachon, Duane, "A Hoosier Hero—Dr. Lawton E. Shank, WWII, Civilian (1907-1943)," Hawaii Reporter, 14 April 2013.

Van Ness, C. P., *Exploding the Japanese Superman Myth*, Washington D.C., 1942.

VandeLinde, Bob L., *Respect: Forgotten Heroes*, Victoria, Canada, 2008.

Victoria, Brian Daizen, Zen at War, NY, 2006.

---*Zen War Stories*, NY, 2004.

Von Clausewitz, Carol, *On War*, (eds.) Howard, Michael & Paret, Peter, Princeton, NJ, 1976.

Von Lang, Jochen, *The Secretary: Martin Bornmann*, NY, 1979.

Wakabayashi, Bob Tadashi (edited), *The Nanking Atrocity, 1937-38: Complicating the Picture*, New York, 2017.

Wallace, David and Williams, Peter, *Unit 731: Japan's Secret Biological Warfare in World War II*, NY, 1989.

Wallance, Gregory, *America's Soul in the Balance: The Holocaust, FDR's State Department, and the Moral Disgrace of an American Aristocracy*, Austin, 2012.

Watson, Caroline N., *The Medal of Honor: Our American Heroes*, Columbia S.C., 2010.

Weinberg, Gerhard, *A World at Arms: A Global History of World War II*, NY, 1999.

Weiner, Michael (ed.), *Race, Ethnicity and Migration in Modern Japan*, NY, 2004.

Werrell, Kenneth P., *Blankets of Fire: U.S. Bombers over Japan During World War II*, Washington D.C., 1998.

Wheeler, Keith, *The Road to Tokyo*, Alexandria Virginia (Time Life Books), 1979.

Williams, Kathleen Broome, *The Measure of a Man: My Father, the Marine Corps and Saipan*, Naval Institute Press, 2013.

Yagami, Kazuo, *Konoe Fumimaro and the Failure of Peace in Japan, 1937-1941: A Critical Appraisal of the Three-time Prime Minister*, Jefferson NC, 2006.

Yahil, Leni, *The Holocaust*, Tel Aviv, 1987.

Yahara, Hiromichi, *The Battle for Okinawa: A Japanese Officer's Eyewitness Account of the Last Great Campaign of World War II*, NY, 1995.

Yellin, Jerry (also with Don Brown), *The Last Fighter Pilot: The True Story of the Final Combat Mission of World War II*, Washington D.C., 2017.

Yenne, Bill, *Hap Arnold: The General Who Invented the U.S. Air Force,* Washington D.C., 2013.

*--Panic on the Pacific: How America Prepared for a West Coast Invasion*, Washington D.C., 2016.

Yoshida, Takashi, *The Making of the 'Rape of Nanking': History and Memory in Japan, China, and the United States*, Oxford, 2006.

Yoshimi, Yoshiaki and O'Brien, Suzanne, *Comfort Women*, NY, 2002.

Zabel, Mortan Dauwen (Ed.), *The Portable Conrad*, "Heart of Darkness," NY, 1976.

Zeiler, Thomas W., *Unconditional Defeat: Japan, America, and the End of World War II*, Wilmington, DE, 2004.

Zich, Arthur, *The Rising Sun*, Alexandria, Virginia (Time Life Books), 1977.

## Articles/Journals/Poems:

Adams, Bob, "WWII Hero Praises GIs in Vietnam," Huntington Advertiser, 28 July 1967.

Anderson, Patrick, "MOH Winner Visits Winona Veteran," Winona Daily News, 7 July 2010.

"Another Attempt to Deny Japan's History," NY Times, 2 Jan. 2013.

Asbury, Kyla, "Well-known World War II vet Woody Williams sues author, publisher over book," West Virginia Record, 5 June 2019.

---"Author files motion to dismiss lawsuit filed by World War II veteran," West Virginia Record, 18 Oct. 2019.

Bartlett, Duncan, "Japan Looks Back on 17th-Century Persecutions," BBC News, 24 Nov. 2008.

Bechtel, Isabelle, "The Making of a Marine," Semper Fi Magazine, Spring 2019.

Berger, Michael, "Print the Legend," Notes to the Editor, NY Times, 30 Jan. 2000.

Bernstein, Adam, "Arthur Jackson, Medal of Honor Recipient for WWII 'one-man assault at Peleliu, dies," Washington Post, 17 June 2017.

"Bill to Aid Britain Strongly Backed," NY Times, 9 Feb. 1941.

Black, Richard, "Those Who Deny Auschwitz Would be Ready to Remake it," Jewish News, 27 Jan. 2015.

Bomphrey, Richard J., Walker, Simon M. and Taylor, Graham K., "The Typical Flight Performance of Blowflies: Measuring the Normal Performance Envelope of *Calliphora vicina* Using a Novel Corner-Cube Arena, US National Library of Medicine National Institutes of Health, 18 Nov. 2009.

Bradsher, Keith, "Thousands March in Anti-Japan Protest in Hong Kong," NYT, 18 April 2005.

Brooke, James, "Okinawa Suicides and Japan's Army: Burying the Truth?", NYT, 20 June 2005.

Brophy, Leo P., Miles, Wyndham D. and Cochrane, Rexmond C., *The Chemical Warfare Service: From Laboratory to Field*, Washington D.C., 1988.

Burke, Matthew M., "Marines Old and New Return to Iwo Jima," Stars and Stripes, 22 March 2015.

Burrell, Robert S. *Crucibles: Selected Readings in U.S. Marine Corps History,* Bel Air, 2004.

--"Did We Have to Fight Here?, WWII Magazine, Special Collector's Addition for the 60th Commemoration of Iwo Jima.

Byrd, Jason H., "Featured Creatures: Hairy Maggot Blow Fly," Univ. of Florida, Jan. 1998.

Carney, Matt, "Iwo Jima: US, Japanese Veterans Recall Horror of Pivotal World War II Battle, 70 Years On," ABC News, 28 March 2015.

The Carryall, Iwo Jima, February 19, 1945—October 15, 1945, 133rd NCB Navy Seabee Souvenir Newsletter, Reprint by Ken Bingham, Vol. III, No. 8, Gulffort MS, 2011.

Casto, James E., "Huntington VA Medical Center Renamed to Honor Woody Williams," Charleston Gazette, 11 Oct. 2018.

Cho, Kap-jae: "Spit on My Grave—The Life of Park Chung-hee," Chosun Ilbo, Seoul, Korea, article no. 104-116, 1998.

Clayton, James, D., "Strategies in the Pacific," Makers of Modern Strategy from Machiavelli to the Nuclear Age, ed. Peter Paret, Princeton NJ, 1986.

Copeland, Larry, "Life Expectancy in the USA Hits a Record High," USA Today, 9 Oct. 2014.

Crovitz, L. Gordon, "Defending Satire to the Death," WS Journal, 12 Jan. 2015.

Crozier, William and Schild, Steve, "Uncommon Valor: 3 Winona Marines at Iwo Jima, 1945, Winona Post, 27 March 2015.

Dalton, Donald, "Interview with Woody Williams," Story for JMC 530.

Daly, Kyle, "The Legacy of Holland M. Smith," Leatherneck Magazine, Oct. 2018.

Dashiell, Dick, Marine Corps Combat Correspondent Dispatch, 17 May 1945. (Sometimes this source is cited from the Fairmont Times where it was also published).

Davis, James Martin, "Operation Olympic: An Invasion not Found in History Books," Omaha World Herald, November 1987.

Dolan, Michael, "Conversation: Forged in Flame," WWII Magazine, 15 March 2016).

Dower, John W., "Lessons from Iwo Jima," Perspectives on History, Sept. 2007.

Durdin, F. Tillman, "All Captives Slain: Civilians Also Killed as Japanese Spread Terror in Nanking," NY Times, 18 Dec. 1937.

Editorial Board, "A Lesson From the Holocaust: Never Stop Telling the Terrible Stories," Chicago Tribune, 12 April 2018.

Francois, Bill, "Flame-Throwing Marine Who Won the Medal of Honor," The Man's Magazine, Vol. 14, #1, Jan. 1966.

Frank, Richard B., "Flags, Memories and Iwo Jima," WWII Museum blog, http://www.nww2m.com/2016/06/flags-memories-and-iwo-jima/

Geggel, Laura, "Jesus Wasn't the Only Man to Be Crucified: Here's the History Behind This Brutal Practice," Live Science, 19 April 2019.

Gelernter, Josh, "Japan Reverts to Fascism," National Review, 16 July 2016.

Ghosh, Palash, "Germany to Pay Out $1 Billion in Reparations For Care of Aging Holocaust Survivors," International Business Times, 29 May 2013.

Glum, Julia, "San Francisco Statue Honoring 'Comfort Women' Sex Slaves from World War II Infuriates Japan," Newsweek, 30 Oct. 2017.

Greiner, Jack, "Strictly Legal: War hero at war with book author," The Enquirer (Cincinnati.com), 5 Nov. 2019.

Guthrie, Sean M., "Leadership Competency: Intellectual deficiencies in the officer corps," Marine Corps Gazette, Vol. 100, Issue 6, June 2016.

Harris, Sheldon H., Chapter 16, "Japanese Biomedical Experimentation During the World War II Era," LaGuardia Community College Files.

Harl, Van, "He is Out in the Garage: Medal of Honor Winner," 16 Oct. 2014.

Harwood, Richard, "A Close Encounter: The Marine Landing on Tinian," Marines in World War II Commemorative Series, 1994.

Heller, Charles E., "The U.S. Army, the Civilian Conservation Corps, and Leadership for World War II, 1933-1942," *Armed Forces & Society* (2010) Vol. 36 No. 3.

Heneroty, Kate, "Japanese Court Rules Newspaper Didn't Fabricate 1937 Chinese Killing Game," Jurist Legal News and Research, Univ. of Pittsburgh School of Law, 23 Aug. 2005.

Hogg, Chris, "Victory for Japan's War Critics," BBC, 23 Aug. 2005.

Hoolihan, Paul B., "Flame Throwers Class at 21st Teaches the Art of Roasting Japs," 21st Regimental Newsletter, May 1945.

"How the Blow-fly Flies and Why Engineers are All Excited About It," Haaretz, 26 March 2014.

"Japan Changes Name of Island of Iwo Jima," Associated Press, 20 June 2007.

"Japan Halts UNESCO Funding Following Nanjing Massacre Row," The Guardian, 14 Oct. 2016.

"Japan PM Abe Demands Apology for South Korean Comments on Emperor Akihito," The Straits Times, 12 Feb. 2019.

"Japanese Minister Yoshitaka Shindo Visits Yasukuni Shrine Provoking China's Ire," South China Morning Post, 1 Jan. 2014.

Jitsuhara, Takashi, "Guarantee of the Right to Freedom of Speech in Japan—A Comparison with Doctrines in Germany," Contemporary Issues in Human Rights Law, Oct. 2017, 169-91.

Jones, Kevin L., "'Comfort Women' Statue Strains 60-Year San Francisco-Osaka Alliance," Art Wire, 28 Sept. 2017.

Journal of Nanjing Massacre Studies (JNMS), Vol. 1, No. 1, Nanking, China, 2019.

Karmath, Melissa, "Humble Farmer Now Legendary Marine," Feb. 2015.

Kato, Norihiro, "Tea Party Politics in Japan," NY Times, 12 Sept. 2014.

Klineberg, Otto, "Racialism in Nazi Germany," in *The Third Reich,* ed. Maurice Baumont, John H. E. Fried and Edmond Vermeil, NY, 1955.

Knapton, Sarah, "Small Man Syndrome Really Does Exist, US Government Researchers Conclude," The Telegraph, 25 Aug. 2015.

Lawless, George, "of Such Stuff are Heroes," Charleston Gazette, 20 May 1956.

Linton, Suzannah, "Rediscovering the War Crimes Trials in Hong Kong, 1946-48," Melbourne Journal of International Law, Vol. 13, No. 2, 2012.

Mandel, Lee, "Combat Fatigue in the Navy and Marine Corps During World War II," Navy Medicine, July-Aug., 2001.

Matthews, Dylan, "Six Times Victims Have Received Reparations—Including Four in the US," Vox, 23 May 2014.

McCurry, Justin, "Shinzo Abe, an Outspoken Nationalist, Takes Reins at Japan's LDP, Risking Tensions with China, South Korea," Global Post, 28 Sept. 2012.

McNeill, David, "Even the Dead Were Being Forced to Fight," The Japan Times, 13 Aug. 2006.

--"Reluctant Warrior," World War II Magazine, Special Collector's Addition for the 60th Commemoration of Iwo Jima.

Metzger, LtGen Louis (USMC), "Guam 1944" Marine Corps Gazette, Col. 78, Issue 7, July 1994.

Michaels, Jim, "Voices: An Enduring Hero of Marines' Most Iconic Battle," USA Today, 12 Nov. 2015.

Miller, Donald L., "Deathtrap Island; Where a Stone Age Battle Set the Stage for the Atomic Era," World War II Magazine, Special Collector's Addition for the 60th Commemoration of Iwo Jima.

Nash Sr., Douglas E., "Army Boots on Volcanic Sands: The 147th Infantry Regiment at Iwo Jima," Army History, Fall 2017.

"Nationalist 'Japan Conference' Building Its Clout: Ten Days after the Meeting, Abe Officially Addressed the Issue of Revising the Pacifist Constitution," Korea JoongAng Daily, 3 May 2013.

Neimeyer, Charles P., "Eyewitness to Iwo," Marine Corps Gazette, Vol. 92, Issue 8, Aug. 2008.

—"Paladins at War: The Battle for Saipan, June 1944," Marine Corps Gazette, Vol. 78, Issue 7, July 1994.

Onishi, Norimitsu, "Abe Rejects Japan's Files on War Sex," NYTimes, 22 July 2018.

Pearson, Richard, "Vice Admiral R.W. Hayler, Navy Hero, Dies," Washington Post, 22 Nov. 1980.

Picard, Joseph, "Iwo Jima Not Pretty as Suribachi Picture: Clark Veteran Returning 50 years Later," News Tribune (NJ), 26 Feb. 1995.

Rayner, Gordon, "Sir Winston Churchill May Have Had 'Short Man Syndrome,' Suggest Boris Johnson," London Telegraph, 10 Oct 2014.

*Researching Japanese War Crimes Records: Introductory Essays* by Edward Drea, Greg Bradsher, Robert Hanyok, James Lide, Michael Petersen and Daqing Yang, Washington D.C., 2006.

Reinwald, Mary H., "Seventy Years Later—Was a Mistake Made?: Honoring all who fought on Iwo Jima," Marine Corps Gazette, Aug. 2016.

Rizzo, Johnna, "How the Romans Used Crucifixion—Including Jesus's—As a Political Weapon," Newsweek, 4 April 2015.

Roberts, James C., "Letter from Iwo Jima," American Veterans Center, April 2015.

Roland, Charles G., "Massacre and Rape in Hong Kong: Two Case Studies Involving Medical Personnel and Patients," Journal of Contemporary History, Vol. 32 (I), 43-61, 1997.

Rybak, Zack, "Herefrom," Narrative Magazine, June 2019.

Samuels, Richard J., "Kishi and Corruption: An Anatomy of the 955 System," Working Paper No. 83, Dec. 2001, Japan Policy Research Institute.

Sanger, David, "New Tokyo Minister Calls 'Rape of Nanking' a Fabrication," NYT, 5 May 1994.

Scarborough, Rowan, "Marines Returning to Iwo Jima on 70th Anniversary of Famed World War II Battle," The Washington Times, 18 Feb. 2015.

Senthilingam, Meera, "People Lie to Seem More Honest, Study Finds," CNN, 30 Jan. 2020.

Sharp, Tim, "How Hot is the Sun?," Space.com, 19 Oct. 2017.

Sieg, Linda, "Historians Battle Over Okinawa WW2 Mass Suicides," Reuters, 6 April 2007.

Smith, Charlotte Ferrell, "Ona Medal of Honor Recipient Hershel 'Woody' Williams Honored on Stamp," Charleston Gazette, 10 Nov. 2013.

Speelman, Patricia Ann, "Medal of Honor Recipient Shares Tales, Shakes Hands," 11 May 2013.

Spitzer, Kirk, "Apology Question Hounds Obama's Planed Visit to Hiroshima," USA Today, 21 May 2016.

Steelhammer, Rick, "WV Supreme Court Oks Moving Medal of Honor Recipient's Remains," Charleston Gazette-Mail, 7 June 2017.

Tally, Steve, "Farmers Should Use Extra Caution with Gasoline," Purdue News Service, 4 June 1999.

UNESCO Nomination Form, International Memory of the World Register, Documents of Nanjing Massacre, 2014-50.

"UNESCO Strikes Political Nerve with Nanking Massacre Documents," The Japan Times, 19 Oct. 2015.

"US House Passes Resolution to Name Huntington VA for 'Woody' Williams," Charleston Gazette, 21 May 2018.

Van Stockum, Ronald R., "The Battle of the Philippine Sea: 'The Great Marianas Turkey Shoot' [and] Planning and Training on Guadalcanal," 2017.

--"The Battle for Guam (1944) (Concluded)," 2017.

--"Japanese Counterattack Plan 25-26 July 1944," June 2017.

Veasey, John, "Williams Foundation Leader in Establishing Permanent Gold Star Family Memorial Monuments," Times West Virginian, 4 Dec. 2015.

Vernuccio Jr., Frank V., Flynn, Daniel J., Tyrrell Jr., R. Emmett, "Stolen Valor: The Fake History a Real Historian That Fooled Presidents and Publishers," The American Spectator, 25 May 2017.

Volokh, Eugene, "The Medal of Honor Recipient vs. The Historian, and the Right of Publicity," The Volokh Conspiracy (found at reason.com) 25 Oct. 2019.

Wells, Sandy, "Medal of Honor Recipient Devotes Life to Veterans," Charleston Gazette, Sunday Gazette-Mail, 27 May 2012.

Williams, Hershel "Woody", "Medal of Honor Recipient Hershel 'Woody' Williams Visits Gettysburg: Take Out the Enemy: Reflections on the Mears Party's Valor," American Battlefield Trust, 25 March 2019.

Winona Daily News, "Purple Heart Presentation," 6 Jan. 1945.

Yoshida, Reiji, "Japan Withholds UNESCO Funding After Nanjing Massacre Row," Reuters, 14 Oct. 2016.

Zdon, Al, "A Marine in the Pacific," Minnesota Legionnaire, May 2015.


## Internet Sources

"Ship Force Levels 1917-present". History.navy.mil.

http://www.hwwmohfoundation.org/gold-star-monument.html

http://www.nationalww2museum.org/learn/education/for-students/ww2-history/ww2-by-the-numbers/world-wide-deaths.html?referrer=https://www.google.com/

http://www.nww2m.com/2016/06/flags-memories-and-iwo-jima/

https://www1.cbn.com/video/HOL58v2/easter-holds-special-meaning-for-ww2-hero

https://www.youtube.com/watch?v=DeJ2JHmQsFY

https://archive.org/details/1945RadioNews/1945-02-25-CAN-Secretary-Of-The-Navy-James-Forrestal-On-The-Battle-Of-Iwo-Jima.mp3

https://archive.org/details/1945RadioNews/1945-02-xx-NBC-Battle-for-Iwo-Jima-Bud-Foster.mp3

https://archive.org/details/1945RadioNews/1945-02-19-CAN-Sgt-Mawson-On-Iwo-Jima-Landings.mp3

https://archive.org/details/1945RadioNews/1945-02-19-CAN-Live-Coverage-Of-US-Marines-Landing-On-Iwo-Jima.mp3

https://archive.org/details/1945RadioNews/1945-03-11-CBS-World-News-Today.mp3

https://en.wikipedia.org/wiki/DUKW#cite_note-6

https://en.wikipedia.org/wiki/Anti-Japanese_sentiment

http://ww2-weapons.com/us-navy-in-late-1941/

https://en.wikipedia.org/wiki/American_women_in_World_War_II

http://www.nationalww2museum.org/learn/education/for-students/ww2-history/ww2-by-the-numbers/us-military.html?referrer=https://www.google.com/

https://www.usmcu.edu/historydivision/casualty-card-database (USMCU Casualty Web Data Base)

https://www.youtube.com/watch?v=2QvDk316BKo

https://archive.org/details/gov.archives.arc.36072

http://www.winonapost.com/Article/ArticleID/43439/Uncommon-valor-3-Winona-Marines-at-Iwo-Jima-1945

http://www.mnlegion.org/may2015MinnesotaLegionnaire_Layout-1.pdf

http://www.naval-history.net/WW2UScasaaDB-USMCbyNameF.htm

https://findagrave.com/cgi-bin/fg.cgi?page=gr&GSln=Bornholz&GSbyrel=all&GSdyrel=all&GSob=n&GRid=2619943&df=all&

https://findagrave.com/cgi-bin/fg.cgi?page=gr&GSln=FIscher&GSfn=Charles&GSbyrel=all&GSdy=1945&GSdyrel=in&GSob=n&GRid=3774625&df=all&

https://www.exeter.edu/about-us/academy-mission

https://www.poetryfoundation.org/poems/46473/if---.

https://www.nationalww2museum.org/war/articles/history-through-viewfinder-23

https://www.revolvy.com/page/Yoshitaka-Shind%C5%8D

https://www.freedomforuminstitute.org/

www.daughtersofww2.org.

http://www.wva-ccc-legacy.org/index.php

https://www.youtube.com/watch?v=O9TqQWLtBJk

https://www.youtube.com/watch?v=eh-06Hy45Qc

https://australianmuseum.net.au/learn/science/decomposition-fly-life-cycles/

https://valorguardians.com/blog/?p=73960

https://www.cob.org/services/safety/education/pages/gasoline.aspx

http://coolcosmos.ipac.caltech.edu/ask/7-How-hot-is-the-Sun-

## Award's Board, Washington, D.C.

The Congressional Medal of Honor, Awarded to Hershel Woodrow Williams, By the President of the U.S. Harry S. Truman, The White House, 5 Oct. 1945

## Bundesarchiv-Militärarchiv (BA-MA), Freiburg, Germany

Bibliothek, *Japan's Eintritt in den Krieg*, Hrsg. v. der Kaiserlich Japanischen Botschaft, Berlin, 1942.

BMRS, File Walter Hollaender

NS 19/3134

RH 2/ 1848

RH 67/52

RH 67/53

RM 11/74

RM 11/77

RM 11/79

RM 11/81

RM 12 II/250

## Films/Documentaries

*A Savage Christmas: The Fall of Hong Kong 1941*, Canadian Documentary, 12 Jan. 1992.

*Battle of Saipan: Full Battle of Saipan Documentary*, History Channel, 2014.

*The Battle for the Marianas*, Military Heritage Institute, 2011.

*X-Day: The Invasion of Japan*, History Channel, 20 Aug. 2017.

## University of Georgia School of Law (UGSL)

J. Alton Hosch Papers, Tokyo War Crimes Trials

## Harry S. Truman Presidential Library & Museum

Public Papers Harry S. Truman 1945-1953, 160, "Remarks at the Presentation of the Congressional Medal of Honor to Fourteen Members of the Navy and Marine Corps."

## Harvard School Student Archives

File Osami Nagano

File Tadamichi Kuribayashi

File Isoroku Yamamoto

## IJAA 2018 Symposium Washington D.C.

Lecture Iwo Jima veteran Ira Rigger, 17 Feb. 2018

## Marine Corps Historical Division, Quantico, VA, (MCHDQV/MCHDHQ)

Biographical File Donald Beck

Beck, Mark, "Biography of Donald Beck," Dec. 2002

General Information, File Woody Williams

Headquarters, 1st Battalion, 21st Marines, 3d MarDiv, FMF, In Field, Report Lt. Col. Marlowe C. Williams, 6 April 1945

Headquarters, 3d MarDiv, FMF, In the Field, Report Lt. Col. H. J. Turton (D-3), 19 Aug. 1944

Headquarters, 2nd Battalion, 21st Marines, 3d MarDiv, FMF, In Field, 12 April 1945

Interviews Woody Williams

Letter Woody to Donald Cordell, 20 April 2014.

Letter Woody to Ruby Meredith, 18 Nov. 1943.

Ibid., 19 Nov. 1943.

Ibid., 3 Dec. 1943.

Ibid., 1 Dec. 1943.

Ibid., 12 April 1944.

Report by Woody, "Reflections About Life and Service Above Self," Oct. 2016.

Un-authored article from the 1945, "Sgt. Hershel Woody Williams."

Williams, Woody (Unpublished memoir), "The Destiny of a Farm Boy: A Farm Boy Who Became a Marine Honored by His Country," cir. 1988.

Williams, Woody, "Looking Back On Earlier Times," Undated.

Williams, Woody, "My Experience in the Civilian Conservation Corps," 28 July 2013.

## Oral History Division: Marine Corps Historical Division.

Bert Banks, Ion Bethel, George Burledge, Clifton Cates, Robert E. Cushman Jr., Graves Erskine, Frederick D. Hunt, August Larson, Holland Smith, Henry Schmidt, Samuel G. Taxis, Van Ryzin, Woody Williams

## Marine Corps Recruit Depot Archive, San Diego (MCRDASD)

2$^{nd}$ Blt., 24$^{th}$ Marines, 4$^{th}$ MarDiv, Operation Journal

2$^{nd}$ Blt., 24$^{th}$ Marines, 4$^{th}$ MarDiv, Battle Reports, Iwo, Narrative

4$^{th}$ Tank Battalion Report, 4$^{th}$ MarDiv., Iwo.

4$^{th}$ Engineer Battalion Report, 4$^{th}$ MarDiv., Iwo.

C-2 Special Study, Enemy Situation, VAC, 6 Jan. 1945

HQ 8$^{th}$ Marines, 2$^{nd}$ Marine Brigade, July-Aug. 1942, Japanese Tactics, Warfare and Weapons.

MCRD Training Book, File 1-2, 1944, By Sgt. Lawrence Henry Pepin

*Soldiers Guide to the Japanese Army*, Military Intel Service, War Dept., Washington D.C., 15 Nov. 1944

## Micronesian Area Research Center (MARC), Univ. of Guam

*Guam War Reparations Commission* (GWRC)

Olivia L.G. Cruz Abell, Box 1, Rec. 2090, Maria T. Abrenilla, Box 1, Rec. 40, Fausto Acfalle, Box 1, Rec. 120, Francisco Acfalle, Box 1, Rec. 2230, Frutuoso S. Aflague, Box 1, Consuelo Aguon, Box 1, 1980, Candelaria Aguon, Box 1, Rec. 1910, Dolores F. Aguon, Box 1, Rec. 380, Edward L.G. Aguon, Box 1, Felix Aguon, Box 1, Felix Cepeda Aguon, Box 1, Ignacio T. Leon Guerrero, Box 20, Rec. 39680, Diana Leon Guerrero, Box 20, 39502, Francisco Q. Leon Guerrero, Box 20, 39620, Jesus Leon Guerrero, Box 20

GC 64 SC 324737

GC 65—80 G 241236, 279846, 279849,301770, 301776, 301784, 324737, 332346, 333809, 356560,490319, 490364

GC 94

## National Archives, College Park, Maryland, (NACPM)

RG 24, Bureau of Naval Personnel

RG 80, Box 5 (Navy Department Board of Decorations and Medals)

RG 111-SC Box 675

DOD Dir 5200.9, Enclosure D, 21ˢᵗ Marines, Action Report (This report is also found in RG 127 series, but it is an old copy in Woody Williams archive).

Microfilm Room, Roll 1499, #86, #101, #113

RG 127 Container 14, Frederick Dashiell, Chapel Hill, N.C. Folders 1-3.

RG 127 Boxes 14-16 (3d MarDiv)

RG 127 Box 18 (3d MarDiv)

RG 127 Box 20 (3d MarDiv)

RG 127 Box 21 (4ᵗʰ MarDiv)

RG 127 Box 22 (4ᵗʰ MarDiv)

RG 127 Boxes 50-1 (Guam)

RG 127 Box 51 (20ᵗʰ, 21ˢᵗ & 22ⁿᵈ Marine Rgts.)

RG 127, Box 83 (Iwo Jima)

RG 127 Box 96 (Iwo Jima)

RG 127, Box 96, VAC, (Report Inspection Forward Areas)

RG 127 Box 90 (Ground Combat) MSI-1168, Vol. 24

RG 127 Box 90 (Ground Combat) MSI-1174 Vol. 24

RG 127, Office of the Commandant, Boxes 821-25

RG 127 Box 318 (Saipan-Tinian)

RG 127 Box 329 (Saipan-Tinian)

RG 165, Boxes 2151-53 (Japan)

RG 226, Box 368 (Records of Office of Strategic Services)

RW 127 Box 319 (Iwo Jima: Photo Archive), Flag Raising 14 March 1945, Gen. Smith with Gen. Erskine.

SC-299291

## National Archives, St. Louis Personnel Files, (NASLPF)

Warren Bornholz, Gregory "Pappy" Boyington, Durell B. Burkhalter, William Caddy, Howard Chambers, Merrill Christopher, Darrell Cole, Marquys Kenneth Cookson, Frederick Knowles Dashiell, William Elder, Graves Erskine, Francis Fagan, Robert C. Filip, Charles Fischer, Beatrice Frankel (Bernice Arthur), Carroll M. Garnett, Roy Geiger, Ross Franklin Gray, Donald Graves, Harry Lee Jackson, Howard Kenyon, Gedeon LaCroix, Darol E. Lee, Jacklyn Lucas, Jack Lummus, Albert Hemphill, Charles W. Henning, Harry Lee Jackson, James La Belle, Harry Martin, Clay Murray, William Naro, William Nolte, Johnny Nosarzewski, Lawrence Henry Pepin, Francis Pierce, Nathan Pitner, Eleanor Genevieve Pyles, Keith Arnold Renstrom, Donald Ruhl, Joseph Rybakiewicz, Alexander Schlager, George Schwartz, Thomas Scrivens, Franklin Sigler, Holland M. Smith, Lyman D. Southwell, Tony Stein, Wesley Strickland, Richard Stanley Tischler, Alan Tripp, Alexander Archer Vandegrift Sr., William Walsh, John Warner, Vernon Waters, Wilson Watson, Warren Welch, Leighton Roy Willhite, Jack Williams, Hershel Woody Williams, Marlowe C. Williams, John Willis, Hartnoll J. Withers, John Wlach

## National Institute for Defense Studies, Ministry of Defense Military Archives (NIDSMDMA), Tokyo, Japan

Akira Fukuda, JDF Civilian Instructor, "Regarding Army Engineering at Iwo Jima."

Collection Iwo Jima Operation, #1

Collections for Lt. General Kuribayashi, Bio info.

Collections of History Calvary Rgts

*Kai-Ko-Sha Kiji*, October 1938 Vol. 769, Kuribayashi, "Building New Remount Administrative Plan," 83-92.

—, March 2001, "Fund for the Production to Increase Strength of the Control by the Military."

*Kambu Gakko Kigi*, (JGSDF Staff College), Col. Fugiwara, "Commemorating General Kuribayashi," August 1966, Tokyo

Records of Showa Emperor, (ed.) *Kunaichō*, Tokyo, Sept. 2016.

*Senshi-Sosho*, Operation Central Pacific, Operation #2, Peleliu, Angaur, Iwo Jima, Vol. 13, 1971

—, "Operations Hong Kong & Chosa," War History Series, Vol. 47, 1971.

## Info from the display of the National Museum of the Pacific War, Home of the Admiral Nimitz Museum, Fredericksburg, TX (NMPWANM), 13 July 2017.

National Park Service, War in the Pacific Online web page resource.

## Punxsutawney Area Historical & Genealogical Society (PAHGS)

Articles about Howard Frederick Chambers and Family

## Renstrom Family Archives

Keith A. Renstrom Letters

Interview Keith Renstrom by John Renstrom, 20-23 June 2016.

## The Second Historical Archives of China, Nanjing, China, Memory of the World: Documents of Nanjing Massacre, Vol 1-20 (TSHACNCDNM)

## Shanghai Library Historical Archives, China (SLHA)

Far Eastern Illustrated News, No. 1-61

The Young Companion, 1940, 150-155

The Young Companion, 1941-1945, 167-172.

## World War II Museum, New Orleans, Oral History Division

Darol L. "Lefty" Lee

Keith A. Renstrom

Hershel Woodrow "Woody" Williams

## Interviews by Bryan Mark Rigg

Barry Beck, 4 Sept. 2017, 10 March 2018, 5 April 2018, 25 June 2019

Sherri Bell, 30 March 2018, 26 June 2019

George Bernstein, 22 March 2018, 23 March 2018, 26 March 2018

B. K. "Jug" Burkett, 20 Nov. 2018

Ellery "Bud" Crabbe, 23 July 2017

Brent Casey, 30 Nov. 2017

Roger Cirillo, 12 May 2018

Jerry Crage 26 June 2017

Mary Dalbey-Rigg, Summer 1996

Fred Dashiell, 11 June 2018

James Dowell, 13 May 2018

Edward J. Drea, 17 Dec. 2017

Jean S. Elliot, 29 March 2018

Gary Fischer, 11 Jan. 2020

David Fisher, 26 March 2018

Amelia Gehron, 30 March 2018

Don Graves, 7 May 2017, 8 May 2017, 23 May 2017, 14 June 2017, 3 April 2018, 4 April 2018, 2 May 2018, 7 May 2018, 4 July 2018, 11 Jan. 2020

James Oelke Farley, 23 March 2015, 8 Nov. 2017, 17 March 2018, 25 March 2018

Richard Frank, 20 July 2017, 22 March 2018, 23 March 2018

Billy Griggs, 20 March 2015

Fiske Hanley, 20 March 2015

T. Fred Harvey, 2 Sept. 2018, 4 March 2019

Bonnie Haynes, 27 Feb. 2019

Mant Hawkins, 16 Feb. 2019

Colin Heaton, 17 Dec. 2017, 8 June 2018, 16 Feb. 2019

Robert (Bob) Hensely, 18 Feb. 2018

Shayne Jarosz, 12 May 2017, 25 March 2018

Charles C. Krulak, 6 Dec. 2019

John Lauriello, 19 Nov. 2016, 10 May 2017, 6 June 2017, 14 June 2017

Lefty Lee, 16 March 2015, 24 Nov. 2015, 19 July 2016, 22 July 2016, 8 Aug. 2016, 15 Aug. 2016, 21 Oct. 2016, 17 March 2017, 4 May 2017, 8 Aug. 2017, 7 May 2018

Master Johnny Kwong Ming Lee, 10 April 2019

Mike Lee, 16 June 2017

Laura Leppert, 13 May 2017, 24 May 2019

Sigmund L. Liberman, 10 June 2013

James Livingston, 8 June 2018

Tommy Lofton, 30 April 2018

Ann R. Mandel, 17 June 2019

Lee Mandel, 17 June 2019

Dwight Mears, 10 May 2018

Martin K.A. Morgan, 10 Dec. 2017

John Mountcastle, 14 Jan. 2018

David Naro, 28 April 2018

Kelly Naro, 28 April 2018

William J. Naro Jr., 28 April 2018

Charles Neimeyer, 15 Dec. 2017

William "Bill" Pasewark, 12 Jan. 2020

Keith A. Renstrom, 22 March 2018

John Renstrom, 17 Nov. 2019, 16 Jan. 2020, 22 Jan. 2020, 23 Jan. 2020

Mike Renstrom, 17 Nov. 2019

Ken Roseman, 23 Feb. 2015

Corrie Rybak, 19 May 2019.

Eugene Rybak, 28 May 2018, 29 June 2018, 15 May 2019, 11 Jan. 2020

Tatsushi Saito, 9-10 April 2018

Bill Schlager, 27 April 2018, 17 May 2018

Lawrence Schlager, 27 April 2018

Williams Schlager, 27 April 2018

Jinghua Shi, 22 May 2019

Yoshitaka Shindo, 9 April 2018

Dennis Showalter, 18 May 2019

Jim Skinner, 17 March 2015

Lt. Colonel, Paul Stubbs, 4 April 2018

Michael E. Thornton, 14 Jan. 2020

Bryce Tripp, 3 April 2018

Erin Susannah Tripp, 29 March 2018

George Waters, 29 May 2018, 21 Oct. 2019

Kathryn Waters, 24 April 2017

Gerhard Weinberg, 2 Sept. 2005

Edgar Weinsberg, 6 Dec. 2019

Glenna Whitley, 7 Jan. 2020

Leighton Willhite, 10 May 2017, 30 Nov. 2019

Brian Williams, 2 June 2019

Woody Williams, 17 March 2015, 24 March 2015, 24 April 2015, 6 June 2015, 8 June 2015, 19 July 2016, 20 July 2016, 21 July 2016, 22 July 2016, 23 July 2016, 21 Oct. 2016, 12 Feb. 2017, 13 Feb. 2017, 19 June 2017, 13 July 2017, 20 July 2017, 29 July 2017, 15 Aug. 2017, 28 Aug. 2017, 3 Sept. 2017, 5 Dec. 2017, 14 Dec. 2017, 21 Dec. 2017, 29 Dec. 2017, 11 Jan. 2018, 18 March 2018

Yukio Yasunaga, 10 April 2018.

Jerry Yellin, 22 March 2015, 18 Aug. 2016, 2 June 2017, 6 July 2017

## Interviews By Scott Farber

Generations Broadcast Center, Project SFMedia Consultants, Inc., WWII Interviews

Jim Reed, 31 Jan. 2013

## Interviews By Colin Heaton

General Curtis LeMay, June 1986

Captain David McCampbell, Aug. 1986

Saburo Sakai, 1990

Alex Vraciu, July 1987

## United Nations Archives

Security Microfilm Programme 1998, UNWCC, PAG-3, Reel no. 61, Summary Translation of the Proceedings of the Military Tribunal, Nanking, on the Trial of Takashi Sakai

## Disclosure

Regarding the Marine Corps Emblems on the cover: Neither the United States Marine Corps nor any other component of the Department of Defense has approved, endorsed, or authorized this book.

# Appendix #1

## (Legal document of Rigg's motion to dismiss Woody's frivolous lawsuit)

THIS LEGAL DOCUMENT IS GIVEN here because I believe this gives an excellent description of the legal battle I experienced with Woody and gives an excellent legal history of why I was right and Woody wrong with what he tried to do—i.e. "violate" the U.S. Constitution's First Amendment protecting the freedom of speech and the press.

UNITED STATES DISTRICT COURT
FOR THE SOUTHERN DISTRICT OF WEST VIRGINIA
HUNTINGTON DIVISION

**Hersehlf Woodrow "Woody" Williams, Plantiff**
**V.**
**Bryan Mark Rigg and John Doe Publihsing Company, Defendants.**
**Civil Action No. 3:19-CV-00423**
**bryan rigg's motion to dismiss original verified complaint**

Bryan Rigg, defendant, filed this Motion to Dismiss Original Verified Complaint and Motion for More Definite Statement and, in support thereof, shows the Court as follows:

# I.

# INTRODUCTION

Plaintiff's suit is meritless. Mr. Rigg, an accomplished historian, has spent the last four years working on his newest book which covers the Pacific campaign in World War II. Plaintiff, a historical figure, is featured in the book and has been interviewed by Mr. Rigg numerous times.[1] While Plaintiff initially welcomed the attention, after several years of work on the book he began to demand monies. As that topic was being discussed and written contracts being exchanged, Plaintiff became aware that Mr. Rigg's historical research was revealing unflattering aspects of his life and service. When Mr. Rigg refused – as a historian – to ignore the documented facts, Plaintiff filed suit in an effort to prevent this book from being published and hide the truth. Plaintiff's claim, however, fails as a matter of law as any alleged contract (or promissory estoppel claim) is – besides being clearly refuted by the various draft written contracts and correspondence exchanged between the parties – barred by the statute of frauds. Similarly, Plaintiff's claim for harm to his right of publicity is inapplicable to historical biographies such as presented here. Finally, Plaintiff's claim for conversion is not pled to give Mr. Rigg notice of the claim. The Court, therefore, should dismissed the Original Complaint in its entirety.

# II.

# BACKGROUND FACTS

Mr. Rigg is an award-winning historian who has received considerable acclaim for his books on World War II. His hallmark is his thorough research, incredible detail, and insistence on primary proof of the facts he presents. His books all have hundreds of footnotes that cite to the documentary evidence that he develops while researching and writing about his subjects. As a former United

---

[1]     This is nothing unusual, of course, as Plaintiff regularly gives interviews and speeches about his past as even the most rudimentary Google search demonstrates.

States Marine who obtained history degrees at Yale and Cambridge, Mr. Rigg developed a keen academic interest in World War II and has published four books detailing his new discoveries in the area.

### A. Mr. Rigg Begins New Book and Meets Plaintiff.

In 2015, Mr. Rigg had decided to write a new book about the Pacific campaign of World War II. As part of his research, he began interviewing numerous veterans and gathering their written materials – both American and Japanese – about their experiences in the war and he traveled to numerous archives throughout the world in China, Japan, Guam, Germany and the United States to gather primary source materials. As part of his work, he traveled to Guam and Iwo Jima with several Americans who had fought on the islands. While there, he met Plaintiff in 2015.

Plaintiff is currently a 96-year old living legend. Plaintiff was drafted into the Marines in 1943 and was sent to the Pacific theater where he saw action at the Battle of Guam. In February of 1945, he fought at the Battle of Iwo Jima as a corporal and earned renown for his use of a flamethrower to destroy enemy pillboxes in battle. For his heroism, he was awarded the Medal of Honor, the highest U.S. military decoration for valor, which was personally pinned on him by President Harry Truman at the White House on October 5, 1945.

Mr. Rigg was pleased to meet Plaintiff and interviewed him numerous times for the book he was writing. Plaintiff was gracious, willing to talk, and shared stories from his life. While most of those stories had already been reported – not surprisingly, Plaintiff has been interviewed for hundreds (if not thousands) of articles about his life and exploits – Mr. Rigg began to think that Plaintiff's life could provide an effective vehicle to tell readers the stories of all the servicemen who fought in the campaigns mentioned above

Thus Mr. Rigg quickly decided that the book he was writing about the Pacific campaign would tell the stories of many brave Marines and Sailors and would spend considerable time talking about two men on opposite sides of the Battle of Iwo Jima: Plaintiff and Japanese General Tadamichi Kuribayashi.

## B.  Mr. Rigg Begins the Long Process of
## Research and Writing His Book.

Critically, as Plaintiff knew and was repeatedly told, Mr. Rigg is a historian. Mr. Rigg is not a biographer and he does not write accounts of people's lives as they tell them. Facts are researched, the truth is discovered and supported with evidence, and facts – both good and bad – must be told.

Plaintiff placed no limitations on Mr. Rigg and asked for nothing. While researching the book, Mr. Rigg interviewed fifty-six subjects, including Plaintiff. Mr. Rigg also traveled around the world locating primary source material on the war. Mr. Rigg interviewed Plaintiff numerous times and, each time, conducted further research or interviews as needed to flesh out the stories or find support for the facts mentioned by Plaintiff. For example, Plaintiff was very adamant that Mr. Rigg interview Corporal Darol E. "Lefty" Lee who Plaintiff claimed was a witness to all his heroic acts on Iwo Jima. Indeed, prior to Mr. Rigg, Cpl. Lee had been interviewed numerous times about Plaintiff and has constantly shared stories of Plaintiff's heroism that he personally witnessed. However, during his research, Mr. Rigg found out that Cpl. Lee had lied about his experiences on Iwo Jima and never witnessed anything Plaintiff did while fighting on the island (as Cpl. Lee was aboard the USS Bayfield being treated for "combat fatigue/shell-shock" on the day in question). In fact, Mr. Rigg's discovery helped the World War II Museum stop a historical documentary it was about to do about Plaintiff and Cpl. Lee once they saw that Cpl. Lee was a "Stolen Valor" case who lied about his combat experiences. This is just one example showing why Mr. Rigg insists on first-hand evidence and has earned a reputation for finding information and separating fact from fiction.

When Mr. Rigg began to write the book, he shared various early drafts with Plaintiff to obtain his comments and thoughts. Plaintiff would occasionally suggest edits or clarifications that Mr. Rigg would accept or not as he saw fit.

Although historical accounts are rarely commercial successes, Mr. Rigg began to think that this book would present a compelling story. As he shared drafts of the book with potential publishers and editors, his thoughts were confirmed. As such, he began to raise with Plaintiff the idea of doing publicity for the book together.

Suddenly, something changed with Plaintiff and those close to him.

## C. Plaintiff Requests a Contract and Money.

In February of 2017, approximately two years (and twelve interviews of Plaintiff) after they first met, Mr. Rigg attended a meeting of Plaintiff's family. The book was on draft nine, which had already incorporated the interviews and comments to date. During that meeting, Mr. Rigg suggested that, should the book be successful, he would like to donate some of the proceeds to the Plaintiff's charity, the Herschel "Woody" Williams Medal of Honor Foundation ("HWWMOH Foundation"). This was the first time money was ever discussed between Plaintiff and Mr. Rigg. (That meeting was recorded by Plaintiff and any listening of that recording will confirm both that fact and what happened next.)

The mention of money and the possible success of the book changed the entire tenor of the conversations. Suddenly, those around Plaintiff – especially his heirs – shifted the entire conversation to money and the sharing of proceeds. After the meeting concluded, Plaintiff and Mr. Rigg discussed the issue further as they drove around. Mr. Rigg noted that it would be extremely unorthodox for a historian to share book proceeds with a subject in the book, as such an arrangement might lead to accusations that the book was not historically accurate. Nonetheless, due to the extreme respect that Mr. Rigg had developed for Plaintiff, he indicated that he was willing to discuss an agreement whereby royalties from the book could be shared with the HWWMOH Foundation and used to honor Gold Star Families after expenses were recuperated.

In that regard, Mr. Rigg had his attorney draft up an agreement which he sent to one of Plaintiff's grandsons. That grandson, Mr. Casey, then sent back an agreement that was utterly unacceptable as it suddenly assumed that Mr. Rigg was acting as a personal writer of a biography rather than a historian.

Drafts of proposed contracts went back and forth for more than a year with no agreement ever being reached. Plaintiff's representatives kept adding additional terms that they wanted, many of which Mr. Rigg could not accept.

By January of 2018, the book was on draft nineteen and was still evolving. Nonetheless, no contract could be agreed to as Plaintiff's grandsons kept changing the proposed agreement. Plaintiff's main attorney, Dale Egan, confided in Mr. Rigg that he was as frustrated with the grandchildren as Mr. Rigg was.

On March 7, 2018, negotiations finally broke down and all offers were withdrawn. By mid-May of 2018, Plaintiff and his grandchildren began to threaten legal action. One of Plaintiff's grandsons demanded that Mr. Rigg sign a contract or he would make sure the book was never published and would tell everyone that Mr. Rigg was a liar. When rebuffed by Mr. Rigg, Plaintiff contacted Mr. Rigg's then publisher and stated he would sue them if any book was ever published without compensation to Plaintiff. That publisher, unwilling to face suit, withdrew from its contract with Mr. Rigg.

### D. Plaintiff's Children Threaten to Sue Over Money; Plaintiff Threatens to Sue Over the Truth.

At the same time, some members of Plaintiff's family began to express concern over certain facts that were being discovered about Plaintiff.

The more research Mr. Rigg did for the book, the more discrepancies appeared. As to Plaintiff, while much of this could be ascribed to a 90+ year old individual trying to recall facts from more than seventy years ago, the simple fact is Plaintiff – while undoubtedly a heroic and legendary figure – has uttered many statements over the years that simply were proving to be false. For example, Plaintiff had always insisted that he volunteered for service as soon as he could (and that his height kept him from volunteering earlier). But the government's records clearly show both that the height requirement was not an impediment to his service and that, rather than volunteering, Plaintiff was drafted. Moreover, there are problems with Plaintiff's story of how one of his friends, Vernon Waters, died; how many children his mother had and how many died; how he behaved with his girlfriends; how he remembered his platoon commander, Howard Chambers; how he reported his actions on Iwo Jima; how he has changed his story since a *Man's Magazine* article in 1966; and how he remembered actions on Guam. Of more concern, however, was what the historical records said about Plaintiff's Medal of Honor.

To be clear, all of the research performed by Mr. Rigg shows that Plaintiff was a true hero on Iwo Jima. However, there are numerous discrepancies and unusual circumstances that surround Plaintiff being awarded the Medal of Honor. (Purely as one example, Plaintiff was the only Iwo Jima Medal of Honor recipient who never received an endorsement for the award from either the Fleet Marine Force, Pacific Commander, Lt. General Roy S. Geiger, and his board; Marine Corps Commandant Alexander A. Vandegrift; or the Pacific Fleet Commander, Fleet Admiral Chester

Nimitz, and his board, all of whom were concerned about the lack of evidentiary support for Plaintiff's actions.) Indeed, many of the "facts" that were put into the draft of his Medal of Honor citation had to be dramatically altered at the last moment, right before President Truman signed off on the award, due to unsubstantiated facts having to be removed, many of which seem to have been entered into the historical record due to Plaintiff's self-reporting of the event. Moreover, had these original "facts" not been put into the record, there was a good chance that Plaintiff's captain would never have recommended him for the Medal of Honor.

As a historian, Mr. Rigg felt the need to accurately report the facts as he discovered them. Plaintiff's representatives, however, were only interested in articles that were fully flattering to him. Once made aware of the discrepancies in the book, they tried to shut it down.

Plaintiff had his lawyer write to Mr. Rigg in early 2018 and demand that Mr. Rigg cease writing anything about Plaintiff. Mr. Rigg refused but, wanting to make sure of the accuracy of the book, he asked Plaintiff to let him know of any inaccuracies that he saw contained in the latest draft of the book (draft twenty-three) so that they could be corrected. Plaintiff did not respond.

Plaintiff also asked for the return of all materials he had loaned to Mr. Rigg. While it took some time to gather them all, Mr. Rigg eventually sent them to Plaintiff.

Mr. Rigg's book has undergone numerous revisions since discussions with Plaintiff ended. Although the book still has a lot of information about Plaintiff, it is now much more than a book about just one Marine. Much of the book is now focused on many other heroic Marines and American commanders. Moreover, the book also focuses on General Kuribayashi and Japanese atrocities that were committed under his command and under the command of his fellow officers. Significant controversy arose over some of the facts discovered about General Kuribayashi, and numerous groups have asked Mr. Rigg to not report those facts for fear of angering the Japanese government. Mr. Rigg has not given in to such threats since they violate the First Amendment and the pursuit to remember the victims of fascist regimes of World War II.

Mr. Rigg had assumed that the only controversy surrounding his book was the reported facts about General Kuribayashi until he received the Original Complaint here that asks for an injunction to prevent the publication of Mr. Rigg's book.

As noted below, other than the facts that all of the documents – and the recordings of their meetings – exchanged between the parties show that the facts outlined in the Complaint are false, the Complaint should be dismissed as it fails as a matter of law.

# III.

# ARGUMENT AND AUTHORITIES

Plaintiff has filed suit alleging causes of action for breach of contract, promissory estoppel, interference with right of publicity, and conversion and is seeking a prior restraint injunction to prevent Mr. Rigg from publishing his book. For the reasons stated below, the claims fail as a matter of law.

## A. Plaintiff's Causes of Action for Breach of Contract and Promissory Estoppel are Not Valid as Pled.

Plaintiff's causes of action for breach of contract and promissory estoppel are barred pursuant to the statute of frauds. The applicable statute of frauds holds that "[n]o action shall be brought . . . upon any agreement that is not to be performed within a year . . . unless the offer, promise, contract, agreement, representation, assurance, or ratification, or some memorandum or note thereof, be in writing and signed by the party to be charged thereby or his agent." W. Va. Code § 55-1-1(f). The issue of whether an alleged agreement is unenforceable under the statute of frauds is a question of law for the Court. *Commonwealth Film Processing, Inc. v. Courtaulds U.S., Inc.*, 717 F. Supp. 1157, 1159 (W.D. Va. 1989). Here, the alleged oral agreement is barred.

The purpose of the statute of frauds is critical in cases like this. As the Supreme Court of Appeals of West Virginia explained:

The purpose of the statute of frauds was and is to make difficult the establishment of perjured and fraudulent claims. The method of making perjured claims difficult, under this type of statute, is to refuse to admit oral testimony as to the existence of terms of certain classes of contracts. The statute reflects a judgment that parole evidence of certain types of agreements is so inherently suspect that it should not even be presented to a jury, an institution otherwise generally considered capable of distinguishing fact from invention. On these matters, the statute implies, even a jury cannot be trusted to recognize truth.

*Thompson v. Stuckey*, 300 S.E.2d 295, 298 (W. Va. 1983) (internal quotes and citations omitted). Indeed, wanting to stress the importance of the statute of frauds, West Virginia mandates that for any situation where a party claims the existence of an oral contract that was not performed in a year – as here – the plaintiff must present "clear and convincing evidence" of such oral contract before it can be presented to a jury. *Id.*; *accord Yanero v. Thompson*, 342 S.E.2d 224, 226 (W. Va. 1986). Here, the statute of frauds prevents Plaintiff's cause of action.

First, there is no dispute that there is no signed contract between the parties. Indeed, the parties circulated draft contracts back and forth for more than a year – and they do not match the terms that Plaintiff's Complaint alleges were agreed to verbally some time before – but no agreement could be reached.

Second, the contract was not "to be performed within a year." Mr. Rigg recognizes that the standard for reviewing such claims is not whether the contract was performed in a year, but rather whether the contract could be performed within a year, even if only by some improbable event. *See Thompson*, 300 S.E.2d at 297. However, even under this liberal test, the alleged oral contract fails.

According to the Complaint, the alleged oral agreement had five terms: (1) Plaintiff would collaborate with Mr. Rigg; (2) Mr. Rigg would use that information plus whatever else he found on his own research, to write a book; (3) the information would be factual; (4) both parties would have input into, and authority over, the content of the book; and (5) the parties would share equally in the profits of the book. Complaint ¶ 43. While, as a fact, items 1-4 could never be completed in a year (as historical treatises take years to research, write, and edit), Mr. Rigg concedes that under the "remotely possible" standard, the Court must imagine a historical treatise that could be somehow researched, written, and printed within a year. However, even with that suspension of belief, the contract would still be barred under the statute of frauds due to the final term.

The key term of Plaintiff's claim is that he is entitled to share in the profits of the book in perpetuity. Thus, Plaintiff concedes that he is to be paid for decades as every book sale brings him his share of the royalties. It is for this reason that courts around the country have held that oral agreements to share profits for the sales of books, records, and movies fall within the statute of frauds because they necessarily cannot be performed within a year. *See, e.g., Grossberg v. Double H. Licensing Corp.*, 86 A.D. 2d 565, 566 (N.Y. App. 1982) (pursuant to an alleged oral agreement to share profits for a record, "defendants' liability endured so long as a single record of the [] performance was sold anywhere in the world. In these circumstances the agreement could not be performed within one year and the statute is applicable"). Under Plaintiff's alleged oral agreement, he would get his share of the book royalties every time a book was sold. As courts have repeatedly held, such contracts are barred by the statute of frauds.[2]

That same thinking applies to other alleged contracts for the sharing of profits that are of such indefinite duration. *See, e.g., United Beer Distrib. Co., Inc. v. Hiram Walker (N.Y.) Inc.*, 163 A.D.2d 79, 80 (NY App. 1990) ("Since the agreement called for performance for an indefinite duration and could only be terminated within one year by its breach during that period, it is not one which by its terms could be performed within one year."); *Computech Int'l, Inc. v. Compaq Computer Corp.*, 2002 U.S. Dist. LEXIS 20307, at *3 (S.D.N.Y. Oct. 24, 2002) ("[C]ontracts of indefinite duration are deemed to be incapable of being performed within a year, and thus fall within the ambit of the Statute of Frauds"); *Madison Oslin, Inc. v. Interstate Res., Inc.*, No. MJG-12-3041, 2015 U.S. Dist. LEXIS 37587, at *16-17 (D. Md. March 25, 2015) (holding that oral joint venture could not be performed within a year as it involved profit payments that, by their nature, were long term).

Moreover, because the statute of frauds bars Plaintiff's contract claim, his claim for promissory estoppel fails as well. As the Supreme Court of Appeals of West Virginia has held"

---

2    Further, although an agreement capable of indefinite continuance may not fall within the Statute of Frauds where it can be terminated by either party at any time, *Am. Credit Servs., Inc. v. Jay Robinson Chrysler/Plymouth, Inc.*, 206 A.D.2d 918 (N.Y. Supp. 1994), the types of agreements like those in this case in which one party's continued liability is "wholly dependent upon the act of [a] third party," *i.e.*, those purchasing the records, are held to be unenforceable under the Statute of Frauds, *Levine v. Zadro Prods., Inc.*, 2003 U.S. Dist. LEXIS 9637, at *4 (S.D.N.Y. 2003) (internal quotes and citation omitted).

Where, as here, the alleged "promise" is identical in substance to the terms of the alleged oral agreement and the performance requested is the performance of the terms of the oral agreement, a litigant cannot escape the application of the Statute of Frauds to that oral agreement through the mere expediency of asserting "promissory estoppel."

*Long v. Long*, No. 11-0865, 2012 W. Va. LEXIS 445, at *3 (W. Va. June 8, 2012); (citing *Paper Corp. of the U.S. v. Schoeller Technical Papers. Inc.*, 724 F. Supp. 110, 118 (S.D.N.Y. 1989)).

Because the alleged oral agreement involved the payment of book royalties in perpetuity, Plaintiff's causes of action for breach of contract and promissory estoppel are barred by the statute of frauds and must be dismissed.

### B. Plaintiff has No Valid Claim for "Interference with Right of Publicity."

Plaintiff has also asserted a cause of action for "Interference with Right of Publicity." This cause of action in inapplicable here.

The "right of publicity" is a right that is recognized in many states. The right is designed to protect the commercial value of the names and likenesses of famous figures. *Haelan Labs., Inc. v. Topps Chewing Gum, Inc.*, 202 F.2d 866, 868 (2nd Cir. 1953). Specifically, the right is an "injury to the pocketbook" of a celebrity who contends that he should have been allowed to profit on the sale of his likeness or performance. *Allison v. Vintage Sports Plaques*, 136 F.3d 1443, 1446 (11th Cir. 1998). As the Supreme Court has noted, the right of publicity is designed not to provide privacy, but to ensure that an entertainer can still profit. *Zacchini v. Scripps-Howard Broadcastibng Co.*, 433 U.S. 562, 573 (1977). Thus, while a television station may freely report on a performance by an entertainer, it may not show the entre performance to which the entertainer normally charges admission to watch. *Id.* This right is not relevant here.

First, it is critical to note that West Virginia has not formally recognized such a cause of action. The right of publicity is a state law claim. *C.B.C. Dist. & Marketing, Inc. v. Major League Baseball Advanced Media, L.P.*, 505 F.3d 818, 822 (8th Cir. 2007) (citing *Zacchini*, 433 U.S. at 566). The only mention of the right in West Virginia case law is a footnote in *Crump v. Beckley Newspapers, Inc.*,

320 S.E.2d 70, 85 n. 6 (W. Va. 1984). That being said, at least one court in this district has already determined that such a right would be cognizable in the state. *See Curran v. Amazon.com, Inc.*, No. 2:07-0354, 2008 U.S. Dist. LEXIS 12479, at *12-13 (S.D.W. Va. Feb. 19, 2008).

However, even if such a right did exist in West Virginia, it would not apply here. The right of publicity cannot stop "the free dissemination of thoughts, ideas, newsworthy events, and matters of public interest" guaranteed by the First Amendment to the United States Constitution. *Time, Inc. v. Hill*, 385 U.S. 374, 382 (1967) (internal citations omitted) (cited by *Curran*, 2008 U.S. Dist. LEXIS 12479, at *25). This district specifically held that any right of publicity in West Virginia does not extend to writings about public figures or matters of legitimate public interest. *Curran*, 2008 U.S. Dist. LEXIS 12479, at *25-26 (citing to *Crump*, 320 S.E.2d at 85; *Rosemont Enterprises, Inc. v. Random House, Inc.*, 294 N.Y.S.2d 122, 129 (1968) ("Just as a public figure's 'right of privacy' must yield to the public interest so too must the 'right of publicity' bow where such conflicts with the free dissemination of thoughts, ideas, newsworthy events, and matters of public interest.")). As one commentator wrote:

> Courts long ago recognized that a celebrity's right of publicity does not preclude others from incorporating a person's name, features or biography in a literary work, motion picture, news or entertainment story. Only the use of an individual's identity in advertising infringes on the persona.

George M. Armstrong, Jr., *The Reification of Celebrity: Persona as Property*, 51 La. L. Rev. 443, 467 (1991) (citing *Rogers v. Grimaldi*, 695 F. Supp. 112, 121 (S.D.N.Y.1988), *aff'd*, 875 F.2d 994 (2d Cir.1989)).

There can be no doubt that the First Amendment bars Plaintiff's claim here. Mr. Rigg is a historian writing a book about history. Plaintiff is a public figure. This district has already held that the public interest exception of the First Amendment applies to books, such as this. *Curran*, 2008 U.S. Dist. LEXIS 12479, at *28 (citing *Dallesandro v. Henry Holt & Co.*, 4 A.D.2d 470, 471 (N.Y. Supp. 1957)). Indeed, allowing the "right of publicity" to prevent historians from researching and writing biographies of historical figures would be catastrophic. That is why courts routinely hold that any right of publicity does not extend to unconsented works of an individual's life story, such as an unauthorized biography. *E.g.*, *Matthews v. Wozencraft*, 15 F.3d 432, 436 (5th Cir. 1994) (no right to privacy related to publication of biographical novel); *Rosa & Raymand Parks Inst. for Self Dev. v. Target Corp.*, 90 F. Supp. 3d 1256, 1263 (M.D. Ala. 2015) (no right to publicity related to sale of biographical books and movies of Rosa Parks); *Seale v. Gramercy Pictures*, 949 F. Supp. 331, 337 (E.D. Pa. 1996) (no right of publicity for founding member of Black Panthers to prevent movie and book about his accomplishments); *see also* RESTATEMENT (THIRD) OF UNFAIR COMPETITION § 47 cmt. (1995). ("[T]he right of publicity is not infringed by the dissemination of an unauthorized print or broadcast biography").

Because the law clearly allows Mr. Rigg to include information about Plaintiff in a historical novel about the Pacific campaign in World War II, the cause of action for right of publicity must be dismissed.

### C. Plaintiff's Conversion Cause of Action is not Properly Pled.

Plaintiff has also asserted a cause of action for conversion. This claim, however, is not properly pled. Pursuant to FED. R. CIV. P. 8(a), a party's complaint must state the grounds for their cause of action and put the defendant on notice so that he can defend himself. If such is not done, a party can move for a more definite statement under FED. R. CIV. P. 12(e).

Here, Mr. Rigg has no earthly idea what he is alleged to have converted. To the best of his knowledge, Mr. Rigg has no property of Plaintiff. In the Compliant, Plaintiff says he is the "owner of the items of the personal property identified and described herein." Complaint ¶ 65. The Complaint further states that those items of personal property include "handwritten journals, notes, photographs, and artifacts." *Id.* No further description is provided.

Mr. Rigg has no idea what Plaintiff is referring to. To the extent he has ever had any of Plaintiff's property, he has returned it (including gifts he was given by Plaintiff). If Plaintiff believes Mr. Rigg has any of Plaintiff's property, Mr. Rigg would happily return it. But he cannot do so when Plaintiff does not identify the property allegedly converted. The Court, therefore, should order Plaintiff to identify the items that he contends are in the possession of Mr. Rigg.

Moreover, Plaintiff has pled for damages that may not be recovered in his conversion claim. In the complaint, Plaintiff states that his damages include "expenses incurred in pursuing the property." *Id.* ¶ 70. These damages – such as for attorney's fees – are not recoverable in a conversion claim and must be dismissed. *See Belknap v. Baltimore & O. R.R.*, 91 S.E. 656, 661 (W. Va. 1917) (noting that the only damages that may be recovered for conversion is any loss of value and loss of use).

### D. Plaintiff's Claim for Injunctive Relief is Not an Independent Claim.

Finally, Plaintiff has asked the Court for injunctive relief and asks the Court to issue a "prior restraint" injunction to prevent the publication of his book. Obviously, an injunction is a remedy and not an independent cause of action. Because, as noted above, Plaintiff has not properly pled a valid cause of action against Mr. Rigg, its request for an injunction must fail.

Because the request for injunctive relief is not an independent cause of action, Mr. Rigg will not fully respond to the particular allegations and request for relief at this time. Should Plaintiff actually seek injunctive relief at some point, however, Mr. Rigg will note that prior restraint injunctions have been repeatedly held to violate the First Amendment. As the Supreme Court has noted, prior restraints may be issued only in rare and extraordinary circumstances, such as when necessary to prevent the publication of troop movements during time of war, to prevent the publication of obscene material, and to prevent the overthrow of the government. *Near v. Minnesota*, 283 U.S. 697, 716 (1931). Nothing of that nature is present here.

For the foregoing reason, the Court should dismiss the Complaint in its entirety.

# Endnotes

1     https://www.exeter.edu/about-us/academy-mission

2     https://www.nationalww2museum.org/war/articles/history-through-viewfinder-23

3     Jon Krakauer, *Where Men win Glory: The Odyssey of Pat Tillman*, NY, 2009, xxiii.

4     Carol von Clausewitz, *On War*, (eds.) Howard, Michael & Paret, Peter, Princeton, NJ, 1976, 170.

5     Two board members did defend my research and pushed against those who sided with Shindo over me. They were Laura Leppert, IJAA member and founder of *Daughters of World War II* (her father fought on Iwo) and Bonnie Haynes, IJAA member and widow of famous USMC Major General and Iwo veteran Fred Haynes. Mrs. Leppert even scolded the current head of the IJAA, Lt. General Norman Smith telling him he should support me because that would be akin to supporting the First Amendment of our Constitution, but instead of listening to her, he "hushed her." Interview Laura Leppert, 24 May 2019; Conversation between Rigg and Iba Takamasa, 18 Nov. 2018; Conversation between Rigg and Shayne Jarosz, 26 Feb. 2019; Conversation between Rigg and Jarosz, 21 May 2019.

6     Telephone conversation with Gen. Lt. Norm Smith, 16 Jan. 2019; Letter 5 April 1968 Takeo Miki of the MOFA to U.S. Ambassador U. Alexis Johnson (National Archives Diplomatic Correspondence concerning Treaties with the U.S. and Japan); Letter Shindo to IJAA, DD Dec. 2018.

7     Conversation with VP of Military History Tours & IJAA Staff Member, Lt. Col. Raul "Art" Sifuentes, 28 Feb. 2019; Conversation with Gen. Lt. Norm Smith, IJAA President and CEO, 16 Jan. 2019; Conversation with Shayne Jarosz, VP, Educational Programs Military History Tours & Former IJAA Board Member, 22 Feb. 2019; Conversation with Bonnie Haynes, IJAA Board Member, 27 Feb. 2019; "Nationalist 'Japan Conference' Building Its Clout: Ten Days after the Meeting, Abe Officially Addressed the Issue of Revising the Pacifist Constitution," Korea JoongAng Daily, 3 May 2013; "Japanese Minister Yoshitaka Shindo Visits Yasukuni Shrine Provoking China's Ire," South China Morning Post, 1 Jan. 2014; https://www.revolvy.com/page/Yoshitaka-Shind%C5%8D; Yuki Tanaka, *Hidden Horrors: Japanese War Crimes in World War II*, Lanham, MD, 2018, xxvii. Although Tanaka has done an excellent job with this work on Japan's WWII crimes, his analysis of comparing rapes done by German and American military personnel during and right after WWII with what Japan did from 1931-45 needs to be furthered analyzed. According to historian Manfred Messerschmidt, the *Wehrmacht* was quite strict about rape and executed thousands of soldiers for this crime. Although there were indeed rapes conducted by *Wehrmacht* personnel, it was not institutionalized and supported by the armed forces like with Japan. Moreover, the rapes that did occur after the war by American personnel in occupied Japan was again not nearly as severe or numerous as that of what Japan did during its oppression of other countries. And last, Tanaka's assessment about the atomic bombs being dropped solely for political reasons *vis a vis* the Soviet Union is historically incorrect and revisionist. See Tanaka, 112, 259.

8     Conversation with Bonnie Haynes, IJAA Board Member, 27 Feb. 2019. Right before I met with Kuribayashi's grandson, Yoshitaka Shindo of the Liberal Party on 9 April 2018, he had met with a delegation with members from the Ministry of Defense, Health, Labor and Welfare organizing

the retrieval of the remains still on the island. He claimed that his grandfather commanded 21,900 troops. However, only "10,000 have been returned to Japan." What Shindo means here, is that only 10,000 have been found and their bones cremated releasing their souls to return to Japan in Japanese cultural thought. As he said during the interview, "The Battle is still not over until all the souls are found and returned to Japan." Interview Shindo, 9 April 2018.

9    J. Tanner Watkins, Dinsmore & Shohl, LLP, to Rigg, Hershel Woodrow "Woody" Williams Book Agreement: Cease and Desist, 3 May 2018; J. Tanner Watkins, Dinsmore & Shohl, LLP, to Rigg, Hershel Woodrow "Woody" Williams Book Agreement: Cease and Desist, 18 June 2018.

10    U.S. District Court for the Southern District of WV, Hershel Woodrow "Woody" Williams verse Bryan Mark Rigg, Case 3:19-cv-00423, Doc. 1, Filed 31 May 2019, 1-20.

11    Geoffrey Harper, Winston & Strawn, LLP to J. Tanner Watkins, Dinsmore & Shohl, LLP, 9 July 2018. See also United States District Court for the Southern District of West Virginia Huntington Division, Hershel Woodrow "Woody" Williams v. Bryan Mark Rigg and John Doe Publishing Company, Civil Action No. 3:19-CV-00423, "Bryan Rigg's Motion to Dismiss Original Verified Complaint, Doc. #9, 18 Oct. 2019 (See Appendix 1). Legal scholar and lawyer Mark Weitz wrote that Woody's "publicity claim is just not supportable" and the "state claims regarding contract are likewise weak, if not totally unsupportable." Email Mike Briggs to Rigg, 25 Dec. 2019.

12    https://www.freedomforuminstitute.org/

13    Book Agreement proposal approved by Woody for Rigg, 6 March 2018, p. 1, Sec. 7. Book Content and Publication. Draft of legal agreement drawn up by Dale Egan, VP Hershel Woody Williams Medal of Honor Foundation (HWWMHF).

14    Book Agreement proposal approved by Woody for Rigg, 6 March 2018, p. 1, Sec. 10. Royalties. Draft of legal agreement drawn up by Dale Egan, VP of HWWMHF. It was the original understanding some book proceeds would be given, after author's expenses incurred were recouped, to Woody's charitable foundation, *never* to him personally since he has said he never wanted to financially benefit from his MOH; Interview Brent Casey, 30 Nov. 2017. However, the possibility of sharing royalties were discussed, but in the end, the terms Woody wanted were unacceptable. When I told Woody and his grandsons, Bryan and Brent Casey, I was not going to engage anymore with such negotiations and revoked all previous business talks as "null and void" and *all* proposals were "withdrawn" and I was going to just be a historian (Email Rigg to Woody and Brent Casey, 7 March 2018), Brent told me: "We will tell everyone you're a liar and destroy you." Telephone conversation with Brent Casey, 28 March 2018; Email with Dale Egan, 28 March 2018. Brent made good on his promise and called up Bruno Nechamkin, a son of a flamethrower operator in the Pacific George Nechamkin and who was working on a film about Woody's life declaring that I "was a liar and cheat." Telephone conversation Brent with Bruno Nechamkin, 18 March 2018 (Interview with Nechamkin, 22 Dec. 2019). Also, Brent has written to son Bill Schlager of one of Woody's MOH witnesses Alexander Schlager, claiming that I "dupe people" and make them believe lies (Email Brent to Bill Schlager, 26 Nov. 2019). Also, due to these problems, I resigned from Woody's Gold Star Memorial Foundation as a board member on 21 March 2018 where I had served for three years. Medal of Honor recipient and Navy Seal Michael E. Thornton finds Brent Casey's behavior troubling and said he also had several problems with him abusing his position in Medal of Honor circles and with his own organization. He simply called Brent "a snake and untrustworthy." Interview Thornton, 14 Jan. 2020. In discussing Brent Casey with MOH recipient Major General Livingston, Livingston told historian Colin Heaton: "Brent Casey is a loud-mouth, fat-ass jerk. He is a royal pain the ass." Interview Colin Heaton with Major General Livingston, 8 Feb. 2020.

15    Telephone conversation with Alex Novak, 18 June 2018.

16    Kyla Asbury, "Well-known World War II vet Woody Williams sues author, publisher over book," West Virginia Record, 5 June 2019; Kyla Asbury, "Author files motion to dismiss lawsuit filed by World War II veteran," West Virginia Record, 18 Oct. 2019; Eugene Volokh, "The Medal of Honor

Recipient vs. The Historian, and the Right of Publicity," The Volokh Conspiracy (found at reason.com) 25 Oct. 2019; Jack Greiner, "Strictly Legal: War hero at war with book author," The Enquirer (Cincinnati.com), 5 Nov. 2019.

17    Telephone conversation with Judith Schnell, 12 Dec. 2019; Email Rigg to Schnell, 12 Dec. 2019.

18    Telephone conversation with Joe Skaggs, 10 Jan. 2020; Telephone conversation with Heather Carter,10 Jan. 2020.

19    Telephone conversation with Geoffrey Harper, 23 Nov. 2019; Telephone conversation with Michael Berenbaum, 12 Dec. 2019.

20    Michael Weiner (ed.), *Race, Ethnicity and Migration in Modern Japan*, NY, 2004, Weiner, "Introduction," 3; Fred Varcoe, "Is Japan Becoming an Enemy of Press Freedom?" Number 1 Shimbun, 11 Sept. 2017; Takashi Jitsuhara, "Guarantee of the Right to Freedom of Speech in Japan—A Comparison with Doctrines in Germany," Contemporary Issues in Human Rights Law, Oct. 2017, 169-91; Tanaka, xv. To give a personal example of Japan censoring people's freedom of speech, the following story is revealing. When my book *Rescued from the Reich* was published in Japan in 2005, I wanted to dedicate it to the millions of Asians the Japanese slaughtered during WWII. My Japanese editors at *Namiki Shobo* (並木書房) and translator Yoshito Takigawa responded that what they did during WWII was no different than what we had done in Vietnam or The Gulf War. I responded that the U.S. did not unleash whole armies to brutally subjugate, murder and rape like the Japanese and that our criminals, like Lt. William Calley Jr. at the My Lai massacre in Vietnam, were brought to justice however imperfectly. Moreover, My Lai is an exception to the rule, whereas atrocities were the norm in the Japanese army from 1927-45. Because of this dispute with the publisher, my book was almost not published and my Japanese editors, not to mention John Donatich of Yale Univ. Press, failed to include my dedication and did not support me.

21    Philip A. Crowl and Jetek A. Isely, *The U.S. Marines and Amphibious War Its Theory and Its Practice in the Pacific*, Princeton, 1951, vi.

22    Tanaka, 1.

23    The Founding Era Collection, Thomas Jefferson Papers, Docs. Jan. 1819- 4 July 1836, Jefferson to Henry Lee, 15 May 1826.

24    Andrew Roberts, *Churchill: Walking with Destiny*, NY, 2019, 617.

25    Don Brown and Captain Jerry Yellin, *The Last Fighter Pilot: The True Story of the Final Combat Mission of World War II*, Washington D.C., 2017.

26    See his autobiography Fiske Hanley, *Accused War Criminal: An American Kempei Tai Survivor*, Dallas, TX, 2020.

27    See his autobiography T. Fred Harvey, *Hell Yes, I'd Do it All Again*, Alpine, Texas, 2011.

28    See his autobiography, Samuel Holiday and Robert S. McPherson, *Under the Eagle: Samuel Holiday Navajo Code Talker*, Univ. Press of Oklahoma, 2013.

29    John Lauriello lost his friend Paul Kinlacheeney. Lauriello named his son, Paul, after him. Interview Lauriello, 20 Nov. 2016.

30    David Anderson, son of David E. Anderson, 25th Marines, 4th MarDiv who died 14 March 1945 and Kathy Dunn Painton, daughter of George Addison Dunn, Fox Co., 2nd Btl., 28th Marines, 5th MarDiv who died 19 Feb. 1945.

31    www.daughtersofww2.org. Broderick served in Co. A in the 1st Btl., 26th Marines, 5th MarDiv. He was wounded on the island on 4 March and evacuated.

32    Herman Melville, *Moby Dick*, Oxford, 2008, 508.

33    Holland M. Smith, *Coral and Brass*, Arcadia Press, 2017, 14. Richard Frank says Smith's memoir must be used with caution because many claims "are highly suspect." Review *Flamethrower* Richard Frank, 8 Aug. 2018.

34    Mortan Dauwen Zabel (ed.), *The Portable Conrad,* "Heart of Darkness," NY, 1976, 492.

35    "Japan Changes Name of Island of Iwo Jima," AP, 20 June 2007; Interview Oelke Farley, 23 March 2015.

36    Interview James R. "Jim" Skinner, 24 March 2015. Woody agrees with Skinner. U.S. Marines Historical Div. (MCHDQV), Oral History, Woody Williams 13800A; Rowan Scarborough, "Marines Returning to Iwo Jima on 70th Anniversary of Famed World War II Battle," The Washington Times, 18 Feb. 2015.

37    Kumiko Kakehashi, *So Sad to Fall in Battle*, NY, 2007, 115. Although Kakehashi's book contains valuable information about Kuribayashi, it needs to be used with caution because she never cites her sources and makes countless historical mistakes. Moreover, she wrote a book "that hero-worships Kuribayashi." Email King to Rigg, 14 June 2017.

38    Gordan W. Prange, *At Dawn We Slept: The Untold Story of Pearl Harbor*, NY, 1981, 242. This quote is ironic in that Kimmel, according to Washington, was caught "flat-footed" with the attacked on 7 Dec. 1941. He was in charge of naval forces at Pearl Harbor and was relieved of command ten days later for his poor leadership. His leadership failures are still debated with some arguing he showed poor judgement while others claiming he acted as best he could with the information at hand. In the end, Kimmel lost this battle to the Japanese and could have done more to have prevented the attack. Interestingly, Woody's 3d MarDiv made a dig at Kimmel in an Intel report from 27 Aug. 1943 when D-2 Intel officer Lt. Col. H.J. Turton wrote: "Pearl Harbor certainly awoke us to the importance of intelligence and the necessity of paying a little attention to it after we get it. It was an outstanding example of what results when the commander failed to realize his intelligence responsibilities." NACPM, 127 Box 14 (3d MarDiv), HQ, FMF, Intel. Sec., Subj.: Combat Intel., Turton 27 Aug. 1943, 2.

39    Publius Flavius Vegetius Renatus, *De Rei Militari*, Book III.

40    Susan Dunn, *A Blueprint for War: FDR and the Hundred Days that Mobilized America*, New Haven, 2018, 9-10; Peter Collier, *Medal of Honor: Portraits of Valor Beyond the Call of Duty*, NY, 2005, 281; Walter LaFeber, *The American Age: United States Foreign Policy at Home and Abroad Since 1750*, NY, 1989, 363-6.

41    Robert Dallek, *Franklin D. Roosevelt: A Political Life*, NY, 2017, 291; James T. Patterson, *American in the Twentieth Century: A History—Third Edition*, Orlando, 1976, 243, 247-8; Robert B. Edgerton, *Warriors of the Rising Sun*, NY, 1997, 246; Akira Iriye, *The Origins of the Second World War: In Asia and The Pacific*, NY, 1987, 48-9; Lee Mandel, *Unlikely Warrior: A Pacifist Rabbi's Journey from the Pulpit to Iwo Jima*, Gretna, 2015, 108; Leonard Mosely, *Hirohito: Emperor of Japan*, NJ, 1966, 171-3; Herbert P. Bix, *Hirohito and the Making of Modern Japan*, NY, 2000, 340; Mandel, 109; Leonard Mosley *Hirohito: Emperor of Japan*, NJ, 1966, 173-4; Conrad Black, *Franklin Delano Roosevelt: Champion of Freedom*, NY, 2003, 427-8; BA-MA, RM 11/79, Bericht Marineattachés Deut. Botschaft Tokyo v. 20.12.1937. Zwischenfälle "Panay" u. "Ladybird", 1; John Toland, *The Rising Sun: The Decline and Fall of the Japanese Empire 1936-45*, Vol I, 60-1; Richard Frank, *Tower of Skulls: A History of the Asia-Pacific War* July 1937-May 1942, NY, 2020, 101-2. Barbara Tuchman, *Stilwell and the American Experience in China 1911-45*, NY, 1971, 179; The Second Historical Archives of China, Nanjing, China, *Kangri Zhanzheng Zhengmian Zhanchang*, Vol. 1., 10; Toll, *Pacific Crucible*, 114; LaFeber, 370; Phillips Payson O'Brien, *The Second Most Powerful Man in the World: The Life of Admiral William D. Leahy, Roosevelt's Chief of Staff*, NY, 2010, 113. Historian Bob Tadashi Wakabayashi claims the attack on *Panay* may have been justified since it "furtively escorted Chinese troops and war materiel to safety under U.S." colors. Bob Tadashi Wakabayashi (ed.), *The Nanking Atrocity, 1937-38*: Complicating the Picture," NY, 2017, Wakabayashi, "The Messiness of Historical Reality," 18. Although Wakabayashi has put together a good book of insightful articles by various authors, one must use the articles he wrote with caution. Here are just a few problems with his work. 1. He claims the rapes Japan conducted during WWII was not unique, which from what I have studied about the WWII British, U.S. and German troops is absolutely false (xxxviii). One can claim the Soviet troops committed horrible rapes too, but they did not conduct the rapes

to the extent of the Japanese especially once on garrison duty. 2. He further states the Japanese did not conduct *Lebensraum* campaigns, but when one looks at the takeover of Manchuria where Japan brought in 800,000 settlers, then his argument breaks down (xl). 3. As a Jew, I take offense at his comparison of Chang's use of faulty documents to people using the horribly anti-Semitic *Protocols of the Elders of Zion* which have been used to persecute and kill Jews throughout the last 115 years since it came out under Tsarist Russia after its defeat in 1905 by the Japanese (xlii). Wakabayashi's insensitivity here bespeaks of a lack of historical basic knowledge and understanding of the history of anti-Semitism. 4. Although the U.S. did make major mistakes in its conduct of the Vietnam War, to place our presidents at the time at the level of Tojo, Matsui, Sakai or Yamashita in claiming they possibly should have been brought to a war trial speaks of an absolutely grotesque attempt to equate our leaders on the level with such unethical and barbaric men just listed which is so ahistorical one has a hard time believing Wakabayashi could write such nonsense (127). Our leaders did not unleash armies of wholesale rape and mass slaughter like the Imperial Japanese government under Hirohito did during WWII murdering millions. 5. Wakabayashi claims the U.S. killed half of the Okinawans during the island campaign there in 1945 (259). If one studies this battle, one learns that many of the islanders killed themselves, were killed by their own IJA personnel and died during the battle being caught in the cross fire of both Japanese and American forces. To say *America killed half the population* is horribly inaccurate and shows that this man has problems often of trying to prove too much always in favor, to some degree, of minimizing what Japan did during WWII. Although one would not necessarily call him a revisionist historian, his arguments seem to lean toward equivocating Japan's actions often with America's which is wrong.

42    Tuchman, 179, 198.

43    MCHDQV, Oral History, Robert E. Cushman Jr., 118.

44    Edgerton, 246; Mosley, 171-3; Bix, 340; BA-MA, RM 11/79, Bericht Marineattachés Deut. Botschaft Tokyo v. 20.12.1937. "Panay" und "Ladybird", 1; Haruko Taya Cook & Theodore F. Cook, *Japan at War: An Oral History*, NY, 1992, 52; Edward J. Drea, *Japan's Imperial Army: Its Rise and Fall, 1853-1945*, Kansas, 2009, 201; Toland, Vol. I, 60-1. The *Ladybird* attack was not the first act of Japanese aggression against the British. On 27 Aug. 1937, Japanese planes attacked a car bedecked with a Union Jack driving between Nanking and Shanghai. Inside was British Ambassador Sir Hugh Knatchbull-Hugessen who was "gravely wounded in the back." Mosley, 171.

45    Brian Daizen Victoria, *Zen War Stories*, NY, 2004, 12; Wakabayashi (ed.), *The Nanking Atrocity, 1937-38*, Akira, "The Nanking Atrocity: An Interpretive Overview," 37.

46    BA-MA, RH 67/53, Deut. Botschaft Militär- und Luftattaché Wash. D.C., Bericht 28/40, 4 Oct. 1940, 1 & Bericht No. 11/40, 8 April 1940, 4; Patterson, 272.

47    Tuchman, 179.

48    MCHDQV, Oral History, Robert E. Cushman Jr., 117. Even though the Marines were not fighting the Japanese, the Marines stationed there knew war was coming with Japan. MCHDQV, Oral History, Victor H. Krulak, 36-7.

49    Review *Flamethrower* Richard Frank, 25 May 2017.

50    Patterson, 255; BA-MA, RH 67/53, Deut. Botschaft Militär- und Luftattaché Wash. D.C., Bericht No. 11/40, 8 April 1940, 4.

51    Peter Maslowski, and Allan R. Millett, *For the Common Defense: A Military History of the United States of America*, NY, 1984, 401; Charles Sasser, *Two Fronts, One War: Dramatic Eyewitness Accounts of Major Events in the European and Pacific Theaters of Operations on Land, Sea and Air in WWII*, London, 2014, 3; Frank, "Review *Flamethrower*," 25 May 2017 & 8 Aug. 2018; Gerhard Weinberg, *A World at Arms: A Global History of World War II*, NY, 1999, 260. While Japan did not ratify the Geneva Convention, it had ratified the Hague Convention which outlawed going to war without prior notice to the party attacked. That treaty obligation did no more to shape Japan's conduct than it did to limit attacks by Germany.

52      Martin Gilbert, *The Second World War*, NY, 1989, 273-4; BA-MA, RM 12 II/250, Kriegstagebuch (KTB), T. 38, Marineattachés u. M(Ltr) Grossetappe Japan-China, Sept. 1941, 17, 113-4; Toll, *Pacific Crucible*, 59.

53      BA-MA, RM 12 II/250, KTB, T. 38, Marineattachés u. M(Ltr) Grossetappe Japan-China, Sept. 1941, 17, 233.

54      Frank O. Hough, *The Island War: The United States Marine Corps in the Pacific*, NY, 1947, 27.

55      Kazuo Yagami, *Konoe Fumimaro and the Failure of Peace in Japan, 1937-41: A Critical Appraisal of the Three-Time Prime Minister*, Jefferson NC, 2006, 97; Prange, 9, 736; Spector, 79-82; Toll, *Pacific Crucible*, 117. Yamamoto's Naval Chief-of-Staff, and commanding officer of the Pearl Harbor attack, Admiral Nagano, also claimed he could run wild for six months after a surprised attack. Johnson, *A History of the American People*, 777.

56      BA-MA, RM 12 II/250, KTB, T. 38, Marineattachés u. M(Ltr) Grossetappe Japan-China, Sept. 1941, 17, 114; Toll, Pacific Crucible, 68; Edgerton, 257.

57      Dwight Mears, *The Medal of Honor: The Evolution of America's Highest Military Decoration*, Kansas, 2018, 13-14; NASLPR, Woody Williams, Woody to Vandegrift, 14 Feb. 1946 & 20 June 1946; U.S. Dept. of Veterans Affairs, Office of Public Affairs, "America's Wars," Fact Sheet, May 2017.

58      Mears, 15-6.

59      James Bradley and Ron Powers, *Flags of Our Fathers*, NY, 2000, 217. Frank says Bradley's book needs to be used with caution since it is unreliable for historical facts. Frank, "Review *Flamethrower*," 25 May 2017

60      (Transcriptions of all interviews with Woody are in the MCHDQV) Interview Woody, 19 July 2016; Interview Bill Schlager, 17 May 2018, 27 April 2018; Francois, 43.

61      MCHDQV, HQ, 3d MarDiv, FMF, Field, Report, Lt. Col. H. J. Turton (D-3), 19 Aug. 1944, 8; Harold Goldberg, *D-Day in the Pacific: The Battle of Saipan*, Indiana Univ. Press, 2007, 167–94; J. Robert Moskin, *The U.S. Marine Corps Story*, NY, 1992, 329; Ronald H. Spector, *Eagle Against the Sun: The American War Against Japan*, NY, 1985, 316, 320; Robert A. Aurthur, and Kenneth Cohlmia, *The Third Marine Division*, Wash. D.C., 1948, 152, 154; Drea, 240; Cyril J. O'Brien, *Liberation: Marines in the Recapture of Guam*, WWII Commemorative Series, Marine Corps Historical Center, Wash. D.C., 1994, 27-8.

62      NACPM, RG 127 (3d MarDiv), G- Periodic Report, No. 21-22; Interview Woody, 19 July 2016; Donald Dalton, "Interview with Woody Williams," Story for JMC 530, 9. In a 1966 interview Woody gave, he said 17 remained out of an original 265 in his company. However, today he believes the original number of his Company was 278. Bill Francois, "Flame-Throwing Marine Who Won the Medal of Honor," The Man's Magazine, Vol. 14, #1, Jan. 1966, 94. In Larry Smith's book, the number 289 is given (Smith, *Iwo Jima*, 66). Moreover, in a note given by Woody on 25 June 2017, Woody remembers the number being given to him of only 17 left in his company by this date. It has been difficult to find an exact figure still fighting in the company at that time because there are no muster rolls for companies on a daily basis in the National Archives (Email Nenninger to Rigg, 19 Dec. 2017); MCHDHQ, Unattributed article from 1945 "Sgt. Hershel Woody Williams," 3.

63      Interview Woody, 21 July 2016.

64      Larry Copeland, "Life Expectancy in the USA Hits a Record High," USA Today, 9 Oct. 2014; Interview Woody, 12 Feb. 2017.

65      Interview Woody, 19 July 2016.

66      Interview Woody, 23 July 2016, 14 Feb. 2017, 23 March 2018.

67      Crowl & Isely, 501; Toland, Vol. II, 82; Moskin, 373; Smith, *Coral and Brass*, 18.

68      *Minchas Chinuch*, Com. 37. The Talmudic phrase actually comes in the form of commentary on the Bible verse Leviticus 19:16: "Do not go about spreading slander among your people. Do not do anything that endangers your neighbor's life" (NIV), p. 125 or "Thou shalt not go up and down as

a talebearer among they people: neither shalt thou stand against the blood of thy neighbour: I am the Lord" (KJV), p. 194. Various translations have varied word formations, but the above quote in the text is the one the author and Rabbi Edgar Weinsberg believe is the most accurate rendition of the Hebrew. Interview Edgar Weinsberg, 6 Dec. 2019.

69 Paul Kennedy, *The Rise and Fall of the Great Powers: Economic Change and Military Conflict from 1500 to 2000*, NY, 1987, 298; Patterson, 255; LaFeber, 383.

70 Toland, Vol. I, 185; Philip Snow, *The Fall of Hong Kong: Britain, China and the Japanese Occupation*, New Haven, 2003, 54; Toll, *Pacific Crucible*, 247.

71 Bryan, J., and Halsey, William F., *Admiral Halsey's Story*, NY, 1947, 75–6; Toll, Pacific Crucible, 41.

72 Toland, Vol. I, 279.

73 Ibid., Vol. I, 285-6. See also Aaron William Moore, *Writing War: Soldiers Record the Japanese Empire*, Cambridge, 2013, 169.

74 Toll, Pacific Crucible, 36; LaFeber, 385.

75 Robert A. Divine, T.H. Breen, George Fredrickson and Hal Williams, *America: Past and Present*, NY, 1987, 783; LaFeber, 369-70; O'Brien, *The Second Most Powerful Man in the World*, 78, 96, 108, 119.

76 Mandel, 116.

77 Divine, Breen, Fredrickson and Williams, 787.

78 Hitler was not required to declare war against the U.S. The Tripartite Pact required him to support his allies only if they were attacked (Toland, Vol. I, 305). His alliance with Japan was recent: He had only decided to side with Japan against China in 1938. Beforehand, he had troops in China training the Nationalists, also allied with the U.S. There were many strange geopolitical relationships at this time. The world was already shocked by the 1939 Nazi-Soviet Non-Aggression Pact causing confusion in the U.S. and Japan, both of which felt Stalin was Hitler's mortal enemy (which was the case, but was not fully realized until Hitler invaded Russia on 21 June 1941). Patterson, 254; Iriye, viii, 25; Bryan Mark Rigg, *Hitler's Jewish Soldiers: The Untold Story of Nazi Rail Laws and Men of Jewish Descent in the German Military*, Kansas, 2002, 278; Weinberg, 35. The Japanese were excited about Hitler's decision to come into the war. Japanese Ambassador Hiroshi Oshima wrote on 20 Dec. 1941 to the German government that "shoulder to shoulder", they would achieve victory and "build a new world." BA-MA, Bibliothek, *Japan's Eintritt in den Krieg*, Hrsg. v. Kaiserlich Japanischen Botschaft, Berlin, 1942.

79 Toland, Vol. I, 306.

80 Ibid.

81 Divine, Breen, Fredrickson and Williams, 778.

82 NACPM, 127 Box 14 (3d MarDiv), HQ, FMF, Intel. Section, Subj: Combat Intel., Lt. Col. Turton 27 Aug. 1943, 2.

83 Divine, Breen, Fredrickson and Williams, 790-1.

84 Toland, Vol. 1, 282.

85 Ibid.

86 NMPWANM, Fredericksburg, TX, 13 July 2017.

87 O'Brien, *The Second Most Powerful Man in the World*, 170.

88 http://ww2-weapons.com/us-navy-in-late-1941/ is the source for 1941 ship levels of 352.; https://www.history.navy.mil/research/histories/ship-histories/us-ship-force-levels.html gives the ship level of 6,768 for the U.S. as of Aug. 1945; The WWII section at the Smithsonian in Washington D.C. on the American People lists the fact America produced 8,800 warships during the war. Maslowski, and Millett, 408; Paul Johnson, *A History of the American People*, NY, 1997, 780.

89 Ian W. Toll, *The Conquering Tide: War in the Pacific Islands, 1942-1944*, NY, 2015, 91-3; O'Brien, *The Second Most Powerful Man in the World*, 181.

90 http://ww2-weapons.com/us-navy-in-late-1941/; "Ship Force Levels 1917-present". History.navy.

mil; Maslowski, and Millett, 408. WWII Museum web site says there were 458,365 at the end of 1940. http://www.nationalww2museum.org/learn/education/for-students/ww2-history/ww2-by-the-numbers/us-military.html?referrer=https://www.google.com/; Thomas W. Zeiler, *Unconditional Defeat: Japan, America, and the End of World War II*, Wilmington, DE, 2004, 5. Historian Paul Johnson has the number pegged at slightly higher than 16.3 throughout the war writing that there were 11,260,000 soldiers, 4,183,466 sailors, 669,100 Marines and 241,093 coastguardsmen amounting to a total of 16,353,659. Johnson, *A History of the American People*, 780.

91    Weinberg, 338; Cook & Cook, 261; Ian W. Toll, *Pacific Crucible: War at Sea in the Pacific, 1941-1942*, NY, 2012, 211-8, 371; LaFeber, 397. Concerning the Battle of the Coral Sea, some historians view it as a strategic victory for the U.S. since it stopped the Japanese landings at Port Moresby.

92    Edgerton, 281; Bill D. Ross, *Iwo Jima: Legacy of Valor*, NY, 1985, 10; Drea, 228.

93    D. Clayton James, "Strategies in the Pacific," 718; Toland, Vol. II, 555.

94    Doolittle's raid "shocked Japanese authorities." Drea, 227; John Morton Blum, *V Was for Victory: Politics and American Culture During World War II*, NY, 1976, 62. Gen. Hap Arnold wrote of Doolittle: "The selection of Doolittle to lead this nearly suicidal mission was a natural one. He was fearless, technically brilliant, a leader who not only could be counted upon to do a task himself if it were humanly possible, but could impart his spirit to others." Bill Yenne, *Hap Arnold: The General Who Invented the U.S. Air Force,* Washington D.C., 2013, 97. This raid only killed 50 civilians and caused little damage to Tokyo, but it had a huge psychological effect on the Japanese leaders. Yamamoto was so shocked by the raid, he "locked himself in his room, pale and shaken," embarrassed to his core. Edgerton, 274. Since the Doolittle raid used Chinese airfields, the Japanese ordered an offensive around Chekiang. They sent 100,000 soldiers to destroy the area. When it was done, "at least 250,000 Chinese civilians lay dead." Edgerton, 275; Scott, 510. Chiang Kai-Shek cabled FDR: "The Japanese troops slaughtered every man, woman, and child in those areas." Toll, *Pacific Crucible,* 300. The raid caused Yamamoto to put aside his plans to invade Australia, Samoa, Fiji and Hawaii and focused on Midway Island to prevent future attacks like Doolittle's. Edgerton, 275, 281; Cook & Cook, 55; Bix, 450. Some Doolittle airmen captured were executed because they had disturbed the *Hakko Ichiu* or Imperial Spirit. NACPM, 127 Box 14 (3d MarDiv), HQ, FMF, Intel Bulletin #4-44, Capt. Whipple, <u>Disturbance of Hakko Ichiu</u>, 5; Weinberg, 332; LaFeber, 393.

95    Blum, 49.

96    Frank, "Review *Flamethrower*," 25 May 2017; Patterson, 261; Interview Yellin, 2 June 2017.

97    Ross, 11; Weinberg, 339.

98    This quote comes from Elie Wiesel's essay "Why I Write" in his book *From the Kingdom of Memory*.

99    Johnson, *A History of the American People*, 461; Application Type: Regular Membership in Capt. James Neal's Ch. of WV of ancestors of men who fought in the Rev. War: Application documenting Woody's descent from George Jacob Helsley, Jr.: NSSAR #179686, svc; DAR Natl# 582926, Anc: Jacob Helsley, svc.

100    MCHDQV, Oral History, Woody Williams 13801A; Interview Woody 21 July 2016; Discussion Richard Frank and Woody Williams, Guam 23 March 2018; Patricia Ann Speelman, "Medal of Honor Recipient Shares Tales, Shakes Hands," 11 May 2013.

101    MCHDQV, Oral History, Woody Williams 13801A; Interview Woody 21 July 2016; Discussion Richard Frank and Woody Williams, Guam 23 March 2018; Speelman, "Medal of Honor Recipient Shares Tales, Shakes Hands," 11 May 2013. Interview David Alkek, 20 Aug. 2018.

102    MCHDQV, Oral History, Woody Williams 13801A; Interview Woody 21 July 2016; Discussion Richard Frank and Woody Williams, Guam 23 March 2018; Speelman, "Medal of Honor Recipient Shares Tales, Shakes Hands," 11 May 2013. In Speelman's article, Woody said a chicken incubator was used to keep him alive. In drafts for this book, he has edited that out. Interview Ann R. Mandel, 17 June 2019. See also Heidi Murkoff and Sharon Mazel, *What to Expect When You're Expecting*, NY, 2016, 308.

103    Interview Ann R. Mandel, 17 June 2019. Navy Medical doctor, Captain Lee Mandel, concurs with everything nurse Ann Mandel has just stated. Interview Lee Mandel, 17 June 2019.

104    Divine, Breen, Fredrickson and Williams, 750.

105    Ibid., 750.

106    O'Brien, *The Second Most Powerful Man in the World*, 71.

107    Divine, Breen, Fredrickson and Williams, 756-74.

108    Ibid., 756.

109    Hallas, *Iwo Jima*, 122; Interview Woody, 19 July 2016; Larry Smith, *Iwo Jima: World War II Veterans Remember the Greatest Battle of the Pacific*, NY, 2008, 59; MCHDHQ, Un-authored article from the 1945, "Sgt. Hershel Woody Williams," 1. Woody often has claimed the flu epidemic killed six siblings. However, when exploring their graves at E.T. Vincent Cemetery, several died before the epidemic (Samuel B. Williams, Born 1-13-1912 Died 6-7-1912, Brady L. Williams, Born 4-29-1913 Died 12-13-1913 and Lila Z. Williams, Born 2-8-1915 Died 3-27-1915). Archivist Janice in Marion County Court records says Brady was actually 1 year and six months old and died of colitis and Lila (in the birth records, it is spelled Lela) was born on the 9th and not the 8th of February and died on the 26th of March and not the 27th of pneumonia. For his claims about them dying in the epidemic see Sandy Wells, "Medal of Honor Recipient Devotes Life to Veterans," Charleston Gazette, Sunday Gazette-Mail, 27 May 2012. He also said this during the interview from 19 July 2016. In his unpublished biography "The Destiny of a Farm Boy," 1, Woody wrote only five died. This document is in MCHDHQ. And Woody sometimes claims the flu epidemic did not kill all six, but killed only four. Caroline N. Watson, *The Medal of Honor: Our American Heroes*, Columbia S.C., 2010, 201: Barry, 452; Interview Woody, 19 July 2016; Smith, *Iwo Jima*, 59-60; Wells, "Medal of Honor Recipient Devotes Life to Veterans."; Dolan, "Conversation: Forged in Flame."

110    John M. Barry, *The Great Influenza: The Story of the Deadliest Pandemic in History*, NY, 2004, 452.

111    Interview Woody, 19 July 2016; Bob L. VandeLinde, *Respect: Forgotten Heroes*, Victoria, Canada, 2008, 367.

112    Interview Woody, 19 July 2016. Sometimes Woody says there were 35 cows. See Isabelle Bechtel, "The Making of a Marine," Semper Fi Magazine, Spring 2019, 38; Hallas, *Iwo Jima*, 122.

113    Bechtel, "The Making of a Marine," 38. In Woody's unpublished biography, "Destiny…Farm Boy," 1, Woody writes he was 10 when his father died. A copy of it is at MCHDHQ. Hallas, *Iwo Jima*, 122.

114    Interview Woody, 23 July 2016; MCHDHQ, Woody Williams, "Destiny…Farm Boy," 1; Interview Lefty, 20 July 2016. All documents in the Marine Corps Historical Division about Woody are also placed in the Pritzker Military Library.

115    KJV (King James Version), p. 120 for Exodus 20:13 and p. 299 for Deuteronomy 5:17. The version of the KJV used in this book is *The Holy Bible: Old and New Testaments in the King James Version*, Thomas Nelson Publishers, Nashville, 1976.

116    Interview Woody, 19 July 2016; MCRDSDA, MCRD Training Book, File 1, Discipline.

117    William Manchester, *Goodbye Darkness*, NY, 1979, 4. Manchester must be used with caution, according to Jon T. Hoffman, since his autobiography is a mixture of "memoir and fiction." Report *Flamethrower* Hoffman, 19 Dec. 2017; Interview, James Hallas, 1 May 2018; Frank V. Vernuccio Jr., Daniel J. Flynn, R. Emmett Tyrrell, Jr., "Stolen Valor: The Fake History a Real Historian That Fooled Presidents and Publishers," American Spectator, 25 May 2017. Manchester fabricated much of his military career and falsified sources. Churchill biographer Andrew Roberts said, "He made up whole quotes in writing his biography of Churchill proving he should not be trusted ever as a source." Lecture, Arlington and Lee Park Pavilion, Dallas, TX 18 Feb. 2019.

118    Interview Woody, 19 July 2016.

119    Divine, Breen, Fredrickson and Williams, 778; Weinberg, 24.

120    Interview Woody, 19 July 2016. Genesis 1:27 on p. 25 reads "So God created man in his own

image, in the image of God he created him; male and female he created them." (NIV—New International Version). The version of the NIV used throughout this book is: "The Student Bible: New International Version, Notes by Philip Yancey and Tim Stafford, Zondervan Bible Publishers, Grand Rapids, Michigan, 1987."

121     Interview Woody, 19 July 2016; MCHDHQ, Woody, "Destiny...Farm Boy," 3.

122     Ibid.; Smith, *Iwo Jima*, 63.

123     Interview Woody, 19 July 2016; MCHDHQ, Woody, "Destiny...Farm Boy," 1, 4.

124     Interview Woody, 23 July 2016.

125     Watson, 202; Interview Woody, 23 July 2016.

126     Interview Woody, 21 July 2016, 29 July 2017.

127     Interview Lefty, 8 Aug. 2016. Woody disputes he ever sang this song, but it's believed that Woody is disputing this fact because it goes against his narrative of being a shy young man without vices. Lefty heard him sing this song many times while they were stationed on Guam.

128     Interview Woody, 21 July 2016.

129     Interview Woody, 21 July 2016; Marine Corps Historical Div., Quantico, VA (MCHDQV), Gen. Info, File Woody Williams, Interview 1982, 13, 17; Interview Ellery "Bud" Crabbe, 23 July 2017.

130     MCHDQV, Gen. Info, File Woody Williams, Interview 1982, 13, 17.

131     Interview Woody, 21 July 2016; Hallas, *Iwo Jima*, 121-2.

132     MCHDQV, Gen. Info, File Woody, Interview 1982, 13, 17; MCHDHQ, Woody, "Destiny...Farm Boy," 5.

133     The Molotov-Ribbentrop Pact, or the Nazi-Soviet Nonaggression Pact was signed on 23 Aug. 1939. Beforehand, Germany and the U.S.S.R were ideological enemies. Afterwards, the world was shocked that these countries signed a peace pact and it changed geopolitical relations. When Hitler invaded Russia with operation *Barbarossa* on 21 June 1941, the pact was discarded and the two countries returned to being enemies.

134     Jean Darby, *Douglas MacArthur*, NY, 1989, 47; Divine, Breen, Fredrickson and Williams, 760; http://www.wva-ccc-legacy.org/index.php; Discussion Richard Frank and Woody Williams, Guam 23 March 2018.

135     George Lawless, "of Such Stuff are Heroes," Charleston Gazette, 20 May 1956, 68.

136     Hallas, *Iwo Jima*, 122.

137     MCHDHQ, Woody, "Destiny...Farm Boy," 2, 5; MCHDHQ, Woody, "My Experience in the CCC," 28 July 2013.

138     Ibid.

139     Ibid.

140     Interview Woody, 21 July 2016.

141     John A. Salmood, *The Civilian Conservation Corps CCC 1933-42: A New Deal Case Study*, NY, 1967; Interview Woody, 19 July 2016; MCHDHQ, Woody, "Destiny...Farm Boy," 2, 5-6; Smith, *Iwo Jima*, 60-61; MCHDHQ, Woody Williams, "My Experience in the Civilian Conservation Corps," 28 July 2013.

142     Divine, Breen, Fredrickson and Williams, 760. Although there were other editors for this book, the principal one, Robert Divine, is only cited in the text here.

143     Charles E. Heller, "The U.S. Army, the Civilian Conservation Corps, and Leadership for World War II, 1933-42," *Armed Forces & Society* (2010) Vol. 36, No. 3, 439-53.

144     Ibid.

145     Bradley & Powers, *Flags of Our Fathers*, 53.

146     Smith, *Coral and Brass*, 25.

147     Toland, Vol. II, 599.

148     Edgerton, 255; Patterson, 262; Crowl & Isely, 74.

149     Patterson, 262.

150     Prange, 738.

151     Iriye, 183.

152     Weinberg, 263.

153     Frank, "Review *Flamethrower*," 25 May 2017. The USMC recruited 6,000 men weekly in the aftermath of Pearl Harbor. Hoffman, *Once a Legend,* 155.

154     Brian Daizen Victoria, *Zen at War*, NY, 2006, 86-7; Christopher Hitchens, *god is not Great: How Religion Poisons Everything*, NY, 2007, 203.

155     Cook & Cook, 21; See also Moore, 6.

156     Ibid., 25.

157     Tanaka, xvi; Drea, 168; Cook & Cook, 23; Burleigh, 15-6; Lord Russell of Liverpool, 7; Edgerton, 311; LaFeber, 337-8. Hitler did the same thing with Poland in 1939. He created a false attack and then countered with an invasion. Bryan Mark Rigg, *The Rabbi Saved by Hitler's Soldiers: Rebbe Joseph Isaac Schneersohn and His Astonishing Rescue*, Kansas, 2016, 11-8. Mukden today is Shenyang. *Manchukuo* was the name the Japanese gave the occupied land and their puppet government in Manchuria (1931-45) which was 600,000 square miles (roughly the size of Texas, California and Montana together). Japan took over this area that had a population of 43,233,954. In 1933, Japan sent 800,000 colonists there. NACPM, Microfilm Room, Roll 1499, #101, 19557, U.S. Intelligence Report from Reference Division, 7 July 1942, Regarding: The Economy of Manchuria, Population Oct. 1940; Haruko Taya Cook and Theodore F. Cook, *Japan at War: An Oral History*, New York, 1992, 33-4, 125, 156; Information from the display of the National Museum of the Pacific War, Home of Admiral Nimitz Museum (NMPWANM), Fredericksburg, TX, 13 July 2017.

158     MCHDQV, Oral History, Erskine from 1969-70, 149.

159     Univ. of Georgia School of Law (UGSL), J. Alton Hosch Papers, Tokyo War Crimes Trials, Japanese Aggression Against China, Part B, Ch. V, 644-7; Tuchman, 141; Meirion Harries and Susie Harries, *Soldiers of the Sun: The Rise and Fall of the Imperial Japanese Army*, NY, 1991, vii, 230, 246.

160     Tuchman, 155; Harries & Harries, 179, 243, 480.

161     MCHDQV, Oral History, Erskine from 1969-70, 445. See also Bix, 257.

162     Ross, 7; Sasser, 3; Cook & Cook, 211, 339; Weinberg, 258; Johnson, *A History of the American People*, 772.

163     Cook & Cook, 52.

164     Ross, 7; Sasser, 3.

165     Winston Churchill, *Memories of the Second World War: An Abridgement of the Six Volumes of the Second World War*, Boston, 1987, 505-6.

166     Interview Woody, 19 July 2016; Smith, *Iwo Jima*, 62; MCHDHQ, Woody "Destiny…Farm Boy," 6, 10. In Woody's essay "My Experience in the Civilian Conservation Corps," 28 July 2013, he gave a different story. On p. 3, he wrote that shortly after he turned 18 in Oct. 1941, he tried to enlist in the Marines but was rejected because of his height. In this story, he tried right away to enlist. However, in his older essay "Destiny of a Farm Boy," he said he left the CCC in March 1942 and then later that fall, he tried to enter the Corps and was rejected.

167     NASLPR, Woody Williams, USMC Report of Separation; Hallas, *Iwo Jima*, 123. For years, Woody claimed he was not legal age to get a DL, but according to WV state documents, one could obtain a driver's license at 16 so he was able to get a license.

168     Interview Woody, 21 July 2016; MCHDHQ, Woody, "Destiny…Farm Boy," 6.

169     MCHDHQ, Woody, "Reflections About Life and Service Above Self," Oct. 2016, 1.

170     MCHDHQ, Letter Woody to Ruby, 12 April 1944. Woody misspelled his girlfriend's name as Eleanore when it was Eleanor. National Census Records, 1940, Eleanor G. Pyles, born 26 June 1923, Marion County, WV. In 1940, it was noted she was married at the age of 16. Interview Beth Orman Shuff, 24 June 2019. There are no records of her ever getting divorced from the marriage she

had when 16. Conversation with Janice in the Marion County Court House Archives, 24 June 2019. The son's name was Jeff and Beth Orman Shuff has a letter that possibly shows that Woody might have been the father although there has been no evidence found which supports this assumption.

171 MCHDHQ, Woody, "Destiny…Farm Boy," 10; Interview Woody, 21 July 2016; Collier, 272; Bechtel, "The Making of a Marine,", 38; Francois, 43; Hallas, *Iwo Jima*, 123.

172 Interview Woody, 19 July 2016; Smith, *Iwo Jima*, 62; MCHDHQ, Woody, "Destiny…Farm Boy," 6.

173 Kathleen Broome Williams, *The Measure of a Man: My Father, the Marine Corps and Saipan*, Naval Institute Press, 2013, 16; Mandel, 224. According to Article 2-121, ¶ (6) of the 1940 Marine Corps Manual, Recruiters were required to examine potential recruits as follows: "Recruiters will test his vision, hearing, and color sense; measure and weigh him; examine his teeth, hands and feet. If he wears a shoe smaller than 5-D or larger than 12-F, or wears a hat smaller than 6-1/2 or larger than 7-3/4, or is otherwise unable to wear issue clothing, or does not come within the heights of 66 and 74 inches [5'6" to 6'2"], he will be rejected." Since Woody claims repeatedly he was 5'6" when he tried to volunteer and the only paperwork found before his service, his draft documents, actually list him as 5'7", then he was well within the height requirement to have been accepted. And since there are no documents yet found that prove Woody ever "volunteered" for the Marines before he was drafted, it is highly unlikely he was ever subjected to the manual's requirements listed in Article 2-121, ¶ (6). Looking at these criteria, it looks like they were not followed that closely since Woody's supposedly best friend, Vernon Waters, was 6'4¼" tall (76 ¼ inches), well outside the height parameters given here. Likewise, Don Graves (65 inches) and Jack Lucas (65 ¼ inches) both were also outside the parameters, but still joined in summer of 1942 without any issues. NASLPR, Don Graves & Jacklyn Lucas. And in the case of Rabbi Gittelsohn, who was 5'5¾", the Navy, which the Marine Corps is under, granted him a waiver for his "physical defect" of being too short. He needed to be 5'6". The regulation cited in Gittlesohn's case was Public Law 816, 77th Congress, Sec. Session, approved 18 Dec. 1942. NASLPR, Gittelsohn, Report of Physical Examination, 14 June 1943 & Director, Naval Officer Procurement, Randall Jacobs to Gittelsohn 18 May 1943; O'Brien, *The Second Most Powerful Man in the World*, 86. So if Woody's height had actually been bumped down from how he reported it during 1942 to his real height of 5'5¼", then he could have applied for a waiver to get in—no documents concerning these matters have been found to date. And when looking at Graves and Lucas' files, it seems like as long as someone was at least 65 inches, it was not a problem. One must not forget, America was at war and needed men.

174 Review *Flamethrower* Hoffman, 28 Aug. 2018.

175 <u>Ibid</u>.

176 Local Board #2, Selective Service, Fairmont, WV, 1-2, Hershel Woodrow Williams (accessed ancestry.com: https://www.fold3.com/image/609326633-4).

177 NASLPR, Jacklyn Lucas, Marks, Scars, Etc., 6 Aug. 1942.

178 NASLPR, Tony Stein, Marks, Scars, Etc., 22 Sept. 1942.

179 NASLPR, Donald Graves, Marks, Scars, Etc., 17 Aug. 1942.

180 Interview Graves, 5 July 2019; NASLPR, Robert C. Filip, Service Book, Marks Scars, Etc. Section.

181 Hallas, *Iwo Jima*, 123.

182 Review *Flamethrower* Hoffman, 28 Aug. 2018.

183 NASLPR, Woody Williams, Service Record, 1.

184 Email Hoffman to Rigg, 5 Sept. 2018. Roosevelt ended volunteering in December 1942 "and placed all manpower in the" 18 to 36 age group "under Selective Service. Allan R. Millett, *Semper Fidelis: The History of the United States Marine Corps*, NY, 1991, 374; Hallas, *Iwo Jima*, 123.

185 Interview Woody, 22 July 2016; MCHDHQ, Woody, "Destiny…Farm Boy," 3, 10.

186 LaFeber, 393; Bill Yenne, *Panic on the Pacific: How America Prepared for the West Coast Invasion*,

Wash. D.C., 2016; Tuchman, 230-1.

187    Toll, *Pacific Crucible*, 23-4.

188    <u>Ibid.</u>, 43.

189    Weinberg, 330, 1019; Toll, *Pacific Crucible,* 265-6.

190    Toland, Vol. I, 297.

191    USMC Publication, "How to Conduct Training," MCRP 3-0B, 25 Nov. 1996, 7-1.

192    Patterson, 267.

193    http://www.usmcpress.com/heritage/usmc_quotations.htm

194    Hershel "Woody" Williams, "Medal of Honor Recipient Hershel 'Woody' Williams Visits Gettysburg: Take Out the Enemy: Reflections on the Mears Party's Valor," American Battlefield Trust, 25 March 2019.

195    Marines are called "Leathernecks" because they used to wear leather around their necks to protect against sword attacks.

196    NASLPR, Woody Williams, Service Record book, Marks, Scars, etc.; Discussion Richard Frank and Woody, Guam 23 March 2018; Interview Woody, 21 July 2016; Francois, 43.

197    https://www.nhlbi.nih.gov/health/educational/healthdisp/pdf/tipsheets/Are-You-at-a-Healthy-Weight.pdf; Interview Woody, 20 Nov. 2017.

198    Hallas, *Iwo Jima*, 123-4.

199    Millett, 348.

200    Manchester, 119.

201    Manchester, 119; NASLPR, Woody Williams, USMC Recruiting/Induction Station, US Court House, Charleston WV, Lt. Commander B.E. Montgomery, 27 May 1943, Orders.

202    Interview Woody, 19 July 2016; Smith, *Iwo Jima*, 62; VandeLinde, 366.

203    MCRDSDA, MCRD Training Book, File 1, <u>Good Luck Marines: Current History</u>, 1.

204    <u>Ibid.</u>

205    E. B. Sledge, *With the Old Breed: At Peleliu and Okinawa*, NY, 1981, 8.

206    Berry, 110; Notes from Woody, 12 Feb. 2017. Woody benefited from a new breed of DI at this time. By 1942, both Parris Island and San Diego had implemented a strict screening process by which the most skilled and best suited noncommissioned officers were selected to train the next generations of Marines. These new DI schools, that every DI Woody encountered had produced, created stronger, more educated and tougher Marines than before. Krulak, 165.

207    Victor H. Krulak, *First to Fight: An Inside View of the U.S. Marine Corps*, Annapolis, 1999, 161.

208    Interview Woody, 19 July 2016; Smith, *Iwo Jima,* 62; VandeLinde, 366.

209    Manchester, 120.

210    Krulak, 161.

211    Millett, 360; Manchester, 121; Sledge, 163.

212    Berry, 35; Paul Fussell, intro to *With the Old Breed*, xv; Notes from Woody, 12 Feb. 2017; Blum, 62-3.

213    The reason why it was called "head" is because sailors used to have to go to the ship's front, the "head of the ship," in order to relieve themselves.

214    Sledge, 10.

215    Interview Woody, 3 Sept. 2017.

216    <u>Ibid.</u>, 21 July 2016.

217    Krulak, 160.

218    Interview Woody, 22 July 2016.

219    Krulak, 155; USMC Publication, "Leading Marines," FMFM 1-0, 3 Jan. 1995, 7. See also Millett, xvii.

220    Interview Woody, 19 July 2016; MCHDHQ, Woody, "Destiny…Farm Boy," 10; Smith, *Iwo Jima,*

61.

221    NASLPR, Woody Williams, Classified Service-Record Book, Professional and Conduct Record. Woody maintains it was thirteen weeks. Discussion Richard Frank with Woody, Guam 23 March 2018; Interview Woody, 21 July 2016. Woody may feel it was 13 weeks because he is counting the first weeks of "in-processing" as part of the boot camp which his personnel file does not reflect. Boot camp for a short time during WWII even got truncated to four weeks, but that was quickly changed and a minimum of eight weeks was implemented. Krulak, 167.

222    MCRDSDA, MCRD Training Book, File 1, Gen. Info, 2.

223    Ibid., Chow, 2.

224    Bradley & Powers, Flags of Our Fathers, 50

225    MCRDSDA, MCRD Training Book, File 1, Miscellaneous.

226    Interview Woody, 3 Sept. 2017; Krulak, 160.

227    Hitchens, god is not Great, 202-3.

228    Toland, Vol. I, 457, 475; Crowl & Isely, 138.

229    MCHDQV, Dashiell File, #50, Coughlin-Leyte.

230    Ibid., #303, Deadly Errand of Mercy.

231    Dower, 64; Jim G. Henri, W. Keyes Beech, David K. Dempsey, Alvin M. Josephy, and Tom Dunn, The U.S. Marines on Iwo Jima, NY, 1945, 114; Crowl & Isely, 185.

232    Richard B. Frank, Downfall: The End of the Imperial Japanese Empire, NY, 28; Sledge, 33-4; MCHDQV, Oral History, Vandegrift, 6-7.

233    Tanaka, 19, 79, 255; James M. Scott, Rampage: MacArthur, Yamashita, and the Battle of Manila, NY, 2018, 99; Roberts, The Storm of War, 99.

234    Frank, "Review of Flamethrower," 25 May 2017; Drea, 158.

235    Edgerton, 263, 269-70; Burleigh, 562; Drea, 223-4; Lord Russell of Liverpool, 242-4, 261; Andrew Roberts, The Storm of War: A New History of World War II, NY, 2012, 275-7; "Bataan Death March," Encyclopedia Britannica, https://www.britannica.com/event/Bataan-Death-March accessed May 28, 2017; Ralph Modder, The Singapore Chinese Massacre: 18 February to 4 March 1942 Were 5,000 or 50,000 Civilians Executed by the Japanese Army?, Singapore, 2004, vii. Col. Tsuji was responsible for many crimes done in Singapore. Toland, Vol. I, 367.

236    Interview Jim Skinner, 24 March 2015; Interview Lee, 23 Nov. 2015; Cook & Cook, 5; NACPM, Atrocities/China 208-AA-132N; Tanaka, 69; Scott, 43.

237    Sledge, 41.

238    NMPWANM, 13 July 2017.

239    MCHDQV, Oral History, Bert Banks, 4-5.

240    Toland, Vol. I, 372.

241    R. Cohen, Soldiers and Slaves: American POWs Trapped by the Nazis' Final Gamble, NY, 2005; Lord Russell of Liverpool, 57; Interview Morgan, 10 Dec. 2017.

242    Bradley & Powers, Flags of Our Fathers, 138; Dower, 47-8; Chang, 173; Timothy Snyder, Bloodlands: Europe Between Hitler and Stalin, NY, 2012, 416; NACPM, RG 127 Box 96 (Iwo) A45-2, 21st Marine Rgt. Journal, 16 March 1945, R-3 Log, Sent 1200.

243    Roberts, The Storm of War, 566.

244    Scott, 3-4.

245    Bradley & Powers, Flags of Our Fathers, 64; Rafael Steinberg, Island Fighting, Alexandria VA (Time Life Books), 1978, 18; Manchester, 166.

246    NACPM, 127 Box 14 (3d MarDiv), HQ, FMF, Intel Bulletin #4-44, Capt. Whipple, Atrocities, 7; See also Moore, 202.

247    NACPM, 208 AA 132R-6; Cook & Cook, 380; Scott, 79.

248    NACPM, 127 Box 14 (3d MarDiv), HQ, FMF, Intel Bulletin #4-44, Capt. Whipple, American Soldier, 7-8.

249 NASLPR, Alexander Archer Vandegrift Sr., Nimitz to Secretary of the Navy, Subj: Recommendation for the award of the Distinguished Service Medal to Maj. Gen. Vandegrift, 11 Dec. 1942 & Nimitz to Secretary of the Navy, Subj: Congressional Medal of Honor, award of, case Maj. Gen. Alexander Vandegrift, 16 Dec. 1942.

250 MCRDSDA, MCRD Training Book, File 1, Good Luck Marines: Current History, 1.

251 Ibid., Introduction, 1-2. See also Millett, 355.

252 Ibid.

253 Ibid.

254 Ibid., Little Friendly Advice, 5.

255 Ibid., Vocational Features of Service in the Marine Corps.

256 Ibid., Introduction, 3 & Personal Hygiene, 1.

257 MCRDSDA, MCRD Training Book, File 1, Personal Hygiene, 1-2; NASLPR, Lawrence H. Pepin, Transcript Medical Record Taken From Health Record of Pepin, 7 July 1943 & Com. Gen. to Com. Officer, Marine Corps Base, San Diego, CA, Subj. Award Purple Heart, Sgt. Pepin, 9 Sept. 1943 & USMC Enlisted Man's Qualification Card, No. 27 & File 267306 Duty Stations & Service Book, Expeditions, Tulagi and Guadalcanal, 14. When Woody showed up to MCRD in May 1943, Pepin soon followed him and was stationed there by 9 June 1943.

258 MCRDSDA, MCRD Training Book, File 1, Personal Hygiene, 1-2.

259 Ibid.

260 Ibid.

261 Ibid. Interestingly, Pepin, who wrote most of this educational guide, was a closet homosexual. He would later be dishonorable discharged when caught having sex with Sailor Wendell D. Willey. NASLPR, Pepin, USMC Report Separation & File 267306 Report Duty Stations, Special note: "[Discharged due to] scandalous conduct tending to the destruction of good morals" & Service Book, Station or Vessel, Dishonorable Discharge, 6, Offenses, 10 & MCRD Area Gen. Court-Martial Order 19-46, Lawrence H. Pepin, 28 Jan. 1946 & US Naval Disciplinary Barracks to Forrestal (Judge Advocate Gen.), 15 Nov. 1946 & USN Disciplinary Barracks, Present Offense and Sentence, 14 Feb. 1946 & Prisoner's Request, Pepin, Case No. 144836 & Statement of Willey about Pepin (267306).

262 Interview Woody, 22 Jan. 2018.

263 Discussion Lee Mandel, 10 Dec. 2017.

264 Hoffman, Chesty, 124-5. This was a problem with many militaries. The "VD infection rate" for British sailors stationed in Hong Kong in 1938 was 25%. Chi Man Kwong and Yiu Lun Tsoi, Eastern Fortress: A Military History of Hong Kong, 1840-1970, Hong Kong, 2014, 81.

265 Letter Dick Dashiell to Vivian Dashiell, 7 Sept. 1945, 3.

266 MCRDSDA, MCRD Training Book, File 1, Washing, 3 & Huts, 4.

267 Ibid., Chow, 2 & Little Friendly Advice., 4.

268 Ibid., Lecture on Marine Corps; Hoffman, Once a Legend, 108.

269 MCRDSDA, MCRD Training Book, File 1, Lecture on the Marine Corps.

270 Millett, xviii.

271 As of 1983, the attrition rate of boot camp classes was 12%, half of them being due to physical injury. A similar attrition rate probably was also going on in 1943. Krulak, 173.

272 Ross, 194.

273 MCRDSDA, MCRD Training Book, File 1, Church, 3.

274 Interview Woody 21 July 2016.

275 Henry Berry, Semper Fi, Mac, New York, 1982, 110; Millett, 360-1.

276 MCRDSDA, MCRD Training Book, File 1, Wearing Apparel.

277 Sledge, 12.

278    MCRDSDA, MCRD Training Book, File 1, <u>U.S. Rifle Text</u>, 2.

279    <u>Ibid.</u>, <u>Marine Rifle Creed</u>.

280    Interview Woody, 29 Dec. 2017.

281    MCRDSDA, MCRD Training Book, File 1, <u>U.S. Rifle Text</u>, 1; See also MCHDQV, Oral History, Erskine, 45; Robert Coram, *Brute: The Life of Victor Krulak, U.S. Marine*, NY, 2010, 12.

282    MCRDSDA, MCRD Training Book, File 1, 2.

283    Krulak, 160.

284    <u>Ibid.</u>

285    Interview Woody, 19 July 2016.

286    MCRDSDA, MCRD Training Book, File 1, <u>Lecture on Marine Corps.</u>

287    NASLPR, Woody Williams, Conduct Record. From May until Oct., his record shows he was average (grade of "Good"). However, from Jan. 1944 until his discharge in 1945, he had better marks (grade of "Excellent"). For example, at completion of boot camp, on a scale from 1-5, Woody scored 3.5 for Military Efficiency, Neatness and Military Bearing and Intelligence and two 5s for Obedience and Sobriety. Woody claims he was in the top ten of his class, but his personnel file does not document this and one would think, with such scores, he was not in the top. His "GPA" using the Marine Corps grading system was 4.1 which using our grading system is a B which in most classes would not put one in the top ten of around 63 men many of whom had High School educations.

288    Interview Woody, 20 July 2016; MCHDHQ, Woody, "Destiny…Farm Boy," 10.

289    MCRDSDA, MCRD Training Book, File 1, <u>Discipline</u>.

290    Hough, 14.

291    MCRDSDA, MCRD Training Book, File 1, <u>Discipline</u>.

292    Discussion given at MCRD, BayView Restaurant, 20 Oct. 2017.

293    L. Gordon Crovitz, "Defending Satire to the Death," WS Journal, 12 Jan. 2015; Hitchens, *god is not Great*, 264.

294    Editorial Board, "A Lesson From the Holocaust: Never Stop Telling the Terrible Stories," Chicago Tribune, 12 April 2018.

295    Tuchman, 178. IJA soldiers had actually, after the battle, taken their films to Shanghai shops to have them developed. Shop keepers made sure copies of the horrific pictures made their way to western correspondents. Tuchman, 178. Nanking is often spelled *Nanjing* in many documents today.

296    Dante Alighieri, *The Divine Comedy of Dante Alighieri: Inferno*, trans. Allen Mandelbaum, NY, 1980, Canto IV 31, 7-8.

297    Dante, *Inferno*, Canto V 45, 90.

298    Dante, *Inferno*, Canto III 23, 56-57.

299    Michael Burleigh, *Moral Combat: Good and Evil in World War II*, NY, 2012, 12; Mosley, 95; Bix, 171.

300    Edgerton, 271; Drea, 137; Richard B. Frank, *Downfall*, 88.

301    Quote comes from Drea, vii. See also *Researching Japanese War Crimes Records: Introductory Essays*, Edward Drea, Greg Bradsher, Robert Hanyok, James Lide, Michael Petersen and Daqing Yang, Wash. D.C., 2006; *Documents on the Tokyo International Military Tribunal Charter, Indictment and Judgements,* Neil Boister & Robert Cryer (eds.), Oxford, 2008; The Second Historical Archives of China, Nanjing, China, Memory of the World: Documents of Nanjing Massacre (TSHACNCDNM), Vol. 10, 295, Statement Joseph B. Keenan, Chief of Counsel, International Military Tribunal for the Far East, 4 June 1945, 33; Wakabayashi (ed.), *The Nanking Atrocity, 1937-38*, Akira, "The Nanking Atrocity: An Interpretive Overview," 29; Harries & Harries, 142. Often historians start chronicling Japan's aggressive acts with 1931 in Manchuria forgetting it had already deployed troops and committed crimes in China starting in 1927.

302    Drea, viii; 29.

303    Victoria, *Zen at War*, xiv.

304    Tanaka, 237.

305    Mosley, 131.

306    Bix, 21. Hirohito took the throne on 25 Dec. 1925. After his father's death, he took possession of "the three sacred regalia, a sword, jeweled necklace and mirror, signifying courage, benevolence and wisdom." A few days later, he took the name *Shōwa* which means "illustrious peace." Unfortunately, most of his rule would be the antithesis of what all of these ceremonies and words stood for. Burleigh, 12; Mosley, 95; Bix, 171.

307    Bix, 39.

308    Bix, 119-121. Bix's biography is superior to Mosley's. So, Mosley's work needs to be used with caution. Historian Drea says Mosely's work is seriously flawed. Interview Drea, 17 Dec. 2017.

309    Victoria, *Zen at War*, 88.

310    The quotes come from Bix, xvi, 8, 12, 387. See also Burleigh, 12; Drea, 52, 164; Cook & Cook, 442.

311    Quote comes from Bix, 8. See also Toland, Vol. I, 29. Historians Meirion and Susie Harries wrote: "[A]s part of the...the constitution, it had been agreed that the Emperor possessed two separate military powers: *gunrei*, the power of command...and *gunsei*, the power of military administration." Harries & Harries, 37.

312    Edward Frederick Langley Russell (Lord Russell of Liverpool), *The Knights of Bushido: A History of Japanese Crimes During World War II*, NY, 2016, xv; *Documents on the Tokyo International Military Tribunal Charter*, xlvi.

313    Mosley, 165.

314    Quote comes from Drea, viii. See also Lord Russell of Liverpool, 38. Drea, "Introduction," in *Researching Japanese War Crimes Records*, 15; Victoria, *Zen at War*, 230; Tanaka, xii.

315    Quote comes from James B. Crowley, *Japan's Quest or Autonomy: National Security and Foreign Policy 1930-38*, Princeton, 1966, xiii. See also Drea, 1; Harries & Harries, 5, 10-11.

316    R. Taggart Murphy, *Japan and the Shackles of the Past*, NY, 2014, 73.

317    Murphy, 53-4; Mark Driscoll, *Absolute Erotic, Absolute Grotesque: The Living, Dead, and Undead in Japan's Imperialism, 1895-45*, Chapel Hill, 2010, x; Jonathan Spence, *The Search for Modern China*, NY, 1990, 155-9; Moore, 5; Harries & Harries, 8.

318    Mikiso Hane, *Modern Japan: A Historical Survey*, Oxford, 2001, 77; Harries & Harries, 8, 18.

319    Paul Kennedy, *Engineers of Victory*, NY, 2013, 284; Drea, 7, 15; Tanaka, 237, 242.

320    Helen Hardacre, *Shinto and the State, 1886-1988*, Princeton, 1991, 123-31.

321    Drea, 38.

322    Murphy, 74; Eugen Weber's *Peasants into Frenchmen: The Modernization of Rural France, 1870-1914*.

323    Crowley, xiii; 176; Drea, 29; Harries & Harries, 19.

324    Drea, 7-8, 10-34; Harries & Harries, 13-17, 30-1; Moore, 22.

325    Many ideas explored here come from Richard Dawkins and Christopher Hitchens. Richard Dawkins, *The God Delusion*, NY, 2006, 37-9; Christopher Hitchens (ed.), The Portable Atheist: Essential Readings for the Nonbeliever, NY, 2007, Hitchens, "Introduction," xvii-xviii & Introduction to Omar Khayyam by Hitchens, 7. See also Harries & Harries, 4; Toland, Vol. I, 30.

326    Seven B. Smith, *Spinoza, Liberalism, and the Question of Jewish Identity*, New Haven, 1997, 30.

327    Harries & Harries, 19. These missionaries promulgated the "Three Great Principles," which included 1). "love of country," 2.) "reverence for the Emperor," and 3). "obedience to the will of the Court."

328    Drea, 31; Hardacre, 122; Hane, 33, 66; Crowley, 86; Burleigh, 12; Cook & Cook, 217; NACPM, RG 165, Box 2153, War Dept., 6910 (Japan), Memo on Japanese Vulnerability, <u>Tradition (a)</u>; Edward Rice, *Ten Religions of the East*, NY, 1978, 116; Michael D. Coogan (ed.), *Eastern Religions: Origins, Beliefs, Practices, Holy Texts, Sacred Places*, Oxford, 2005, 446; Burleigh, 12; Mosley, 32, 102, 131-3; Bix, xvi, 2, 7, 16.

329     Murphy, 73; Coogan, 420-1, 440, 464; Cook & Cook, 232, 309, 328; Drea, 154; Mosley, 17-8, 99; Bix, xv; Smith, *Spinoza*, 31.

330     NACPM, Microfilm Room, Roll 1499, #86, To: The A.C. of S., G-2, War Dept., Wash. D.C. (26 June 1942), Subj.: 'The Imperial Forces Will Surely Win,' Adm. Kato, 1933; Ian Gow, *Military Intervention in Pre-war Japanese Politics: Admiral Katō Kanji and the 'Washington System'*, London, 2004, 1.

331     NACPM, Microfilm Room, Roll 1499, #86, To: The A.C. of S., G-2, War Dept., Wash. D.C. (26 June 1942), Subj: 'The Imperial Forces Will Surely Win,' Admiral Kato, 1933.

332     Murphy, 75; Coogan, 420-2, 428-9; Crowley, 86; Rice, 122.

333     Coogan, 423; Victoria, *Zen at War*,11.

334     Victoria, *Zen at War*, 12.

335     Cook & Cook, 238-9.

336     Smith, *Spinoza*, 95.

337     Smith, *Spinoza*, 3. Yale Univ. Professor Steven Smith writes that James Madison in Federalist No. 10 concluded that by defending religious pluralism and creating a multiplicity of sects would lead to more questioning of what was right and prevent violence as a method used to convert others to one way since there would be many "ways of believing." Japan unfortunately did not embrace such notions.

338     Smith, *Spinoza*, 20.

339     Victoria, *Zen at War*, 192-3, 223.

340     Ibid., xiv.

341     Ibid., 78, 80.

342     Ibid., 82.

343     Ibid., 91, 104-5.

344     Ibid., 25, 91, 152, 154, 156, 167.

345     Ibid., 88.

346     Tanaka, 243.

347     Bix, 8-10; Robert Kagan, *Dangerous Nation: America's Place in the World from Its Earliest Days to the Dawn of the Twentieth Century*, NY, 2006, 294.

348     Quote comes from Wakabayashi (ed.), *The Nanking Atrocity, 1937-38*, Fogel, "The Nanking Atrocity and Chinese Historical Memory," 282. See also Harries & Harries, 60; Drea, 86.

349     Drea, 109; Harries & Harries, 89, 91.

350     Drea, 109; Mosley, 14; Bix, 9; Kagan, 294-5; Churchill, 580.

351     O'Brien, *The Second Most Powerful Man in the World*, 33-4.

352     Ibid., 33-4, 40, 77, 104-7, 113-4, 118-9, 170, 192, 202, 206-7, 214-5, 217, 226, 237, 268, 281, 284-5.

353     Isely and Crowl, 25-7 (loc. 696-748); Krulak, 76-8; Smith, *Coral and Brass*, 37.

354     Walter A. Skya, *Japan's Holy War: The Ideology of Radical Shinto Ultranationalism*, Duke Univ., 2009, 190.

355     Burleigh, 12; Mosley, 5-6, 94; Bix, xv, 21-2, 88.

356     Drea, 128, 140, 147, 150-1, 153, 175.

357     Kennedy, *Engineers of Victory*, 284; Frank, "Review *Flamethrower*," 25 May 2017.

358     Divine, Breen, Fredrickson and Williams, 782; Burleigh, 18; Bix, 261; TSHACNCDNM, Vol. 10, 295, Statement Keenan, Chief of Counsel, International Military Tribunal Far East, 4 June 1945, 33; Harries & Harries, 141, 150; Toll, *Pacific Crucible*, 89; LaFeber, 339. The Japanese made Manchuria one of the most industrialized areas of Asia outside Japan. In fact, Japan invested more capital into Manchukuo than "Britain devoted to India" in 200 years of imperial rule. Frank, *Tower of Skulls*, 12.

359    Drea, 194; Bix, 357; O'Brien, *The Second Most Powerful Man in the World*, 106.

360    Quote comes from Wakabayashi (ed.), *The Nanking Atrocity, 1937-38*, Akira, "The Nanking Atrocity: An Interpretive Overview," 30. See also Bix, 359-60; Tanaka, xvi; Frank, *Tower of Skulls*, 1, 9.

361    Hitler at first saw his union with China as a way to fight communism, but as his alliance shifted to Japan, especially with the anti-Comintern Pact of 1936, he slowly but surely pulled out his advisors and focused on building good relations with Japan ending his support of China finally by July 1938. Wakabayashi (ed.), *The Nanking Atrocity, 1937-38*, Wakabayashi, "The Messiness of Historical Reality," 17; Tuchman, 141; Frank, *Tower of Skulls*, 22.

362    Tuchman, 102, 131-2, 141, 144, 152, 156, 167; Wakabayashi (ed.), *The Nanking Atrocity, 1937-38*, Akira, "The Nanking Atrocity: An Interpretive Overview," 34; Harries & Harries, 204; Zeiler, 24; Moore, 6, 38-9, 46; LaFeber, 336, 393; Frank, *Tower of Skulls*, 1, 3, 19.

363    Lord Russell of Liverpool, 40.

364    Mosley, 170. The Second Sino-Japanese War not only disturbed the world powers for its crimes, but it also made them upset for hitting them in their pocketbooks since it disrupted their international trade. Mosley 170.

365    BA-MA, RH 2/1848, OKH, Anliegend wird eine Ausarbeitung Erfahrungen und Netrachtungen aus dem japanisch-chinesischen Feldzug 1937-38 übersandt, 15.3.1938. IJA was called "Imperial" to connect the army to the Emperor. It has also been used as an apologist term, which is not the intent when used in this book. In other words, some Japanese today say it was the Imperial warlords who committed crimes during the Pacific War and not Japan to whitewash its crimes. Drea, 161; Review of *Flamethrower* by Robert Burrell, 26 Feb. 2018.

366    Cook & Cook, 2.

367    John W. Dower, *War Without Mercy: Race & Power in the Pacific War*, NY, 1986, 126.

368    Crowley, 86; Bradley & Powers, *Flags of Our Fathers*, 37; Review *Flamethrower*, Burrell, 26 Feb. 2018.

369    Suzannah Linton (ed.), *Hong Kong War Crimes Trials*, Oxford, 2013, Daqun, "Forward," vi & Linton, "Major Murray Ormsby: War Crimes Judge and Prosecutor 1919-2012," 226; Paul W. Thompson, Harold Doud, and John Scofield, *How the Jap Army Fights*, NY, 1942, 13; Crowley, 86; Kennedy, *Rise and Fall of Great Powers*, 301; Bradley & Powers, *Flags of Our Fathers*, 40; Bix, 55.

370    Quote comes from Saburo Sakai, *Samurai!: The Unforgettable Saga of Japan's Greatest Fighter Pilot— The Legendary Angel of Death*, New York, 1975, 3, 8. See also Harries & Harries, 7.

371    Bradley & Powers, *Flags of Our Fathers*, 137.

372    Frank, *Downfall*, 28.

373    Lord Russell of Liverpool, 4.

374    Quote comes from Meribeth E. Cameron, Thomas H.D. Mahoney, and George E. McReynolds, *China, Japan and the Powers: A History of the Modern Far East*, NY, 1960, 537-8. See also Burleigh, 12; Drea, 134; Interview Morgan, 10 Dec. 2017.

375    Quote comes from Dower, 207, 217, 269. See also Hane, 33; Coogan, 419; Lord Russell of Liverpool, 13; Modder, viii.

376    Crowley, 203; Mosley, 126-9, 131-2; Lord Russell of Liverpool, 286-7.

377    Victoria, *Zen at War*, 117-27.

378    Ibid., 117-9, 121.

379    Adolf Hitler, *Mein Kampf*, Boston, 1971, 65; Raul Hilberg, *Destruction of the European Jews*, NY, 1961; BA-B, NS 19/3134, Bl. 1-2; Werner Maser, *Adolf Hitler. Legende Mythos Wirklichkeit*, München, 1971, 282; Henry Picker, *Hitlers Tischgespräche im Führerhauptquartier, 1941–42*, ed. Percy Ernst Schramm, Stuttgart, 1976, 45; Jochen von Lang, *The Secretary: Martin Bornmann*, NY, 1979, 156; Max I. Dimont, *Jews, God and History*, NY, 1964, 331-32; Bryan Mark Rigg, *Lives of Hitler's Jewish Soldiers: Untold Tales of Men of Jewish Descent Who Fought for the Third Reich*, Kansas,

2009, 238; Rigg, *Hitler's Jewish Soldiers*, 185; Kertzer, 50-1, 58, 106, 109, 280, 404; Dawkins, 274; Hitchens, *god is not Great*, 4, 240; John Cornwell, *Hitler's Pope: The Secret History of Pius XII*, NY, 1999, 7, 137, 199-215.

380    Victoria, *Zen at War*, 116, 125.

381    Toll, *Pacific Crucible*, 115.

382    Weiner (ed.), *Race, Ethnicity and Migration in Modern Japan*, Weiner, "Race, Nation and Empire," 8; Toll, *Pacific Crucible*, 115.

383    Sarah Knapton, "Small Man Syndrome Really Does Exist, US Government Researchers Conclude," The Telegraph, 25 Aug. 2015.

384    Gordon Rayner, "Sir Winston Churchill May Have Had 'Short Man Syndrome,' Suggest Boris Johnson," London Telegraph, 10 Oct 2014; O'Brien, *The Second Most Powerful Man in the World*, 258.

385    Weiner (ed.), *Race, Ethnicity and Migration in Modern Japan*, Weiner, "Race, Nation and Empire," 8.

386    Weiner (ed.), *Race, Ethnicity and Migration in Modern Japan*, Louise Young, "Rethinking Race for Manchukuo: Self and Other in the Colonial Context," 280; Harries & Harries, 95; LaFeber, 427; O'Brien, *The Second Most Powerful Man in the World*, 33.

387    Victoria, *Zen at War*, xi, 3, 10, 13, 30, 81, 134; Hane, 7-9, 14; TSHACNCDNM, Vol. 2, 214, 221-2 from H. J. Timperley, *What War Means: The Japanese Terror in China*, London, 1938, 164. Timperley was a paid propagandist for the Nationalists' International Propaganda Division and may have falsified numbers. However, the crimes in general he described did occur. Wakabayashi (ed.), *The Nanking Atrocity, 1937-38*, Askew, "Part of the Numbers Issue: Demography and Civilian Victims," 97; Harries & Harries, 4, 7, 44; Toll, *Pacific Crucible*, 64; Email Drea to Rigg, 17 Jan. 2020.

388    Victoria, *Zen at War*, 3; Hane, 7-9.

389    Victoria, *Zen at War*, 15.

390    Victoria, *Zen at War*, 64. See also Moore, 213.

391    <u>Ibid.</u>, 137. Interestingly, across the ocean into Asia, Chinese Buddhist priest were praying for the Nationalists' success against Japan. In 1940, large gatherings of 4,000 Lama priests at the Ta-Er Temple gathered and prayed for victory against Japan. In fact the *Living Buddha* of the Golden Tile Temple, one of the leaders of Kokonor, was "active in the campaigns in support of the war [against Japan]." Shanghai Library Historical Archives (SLHA), China, The Young Companion, 1940, 150-5, "4000 Lama Priests Pray for Chinese Victory." Which prayers was the Buddha listening to during WWII? Ones espoused by Zen Buddhists or ones by Lama Buddhists? Such religious behavior taking on political overtones starts to make a mockery of both belief systems. One could also make the same argument about the U.S. and Nazi Germany. Both had thousands of Christian chaplains in the ranks. Did the Christian God listen more to the U.S. chaplains than the Nazi ones?

392    Toland, Vol. I, 375; Tanaka, 80.

393    Sakai, 3, 8. See also Modder, 21; Tanaka, 243; *Hong Kong's War Crimes Trials*, Daqun, "Forward," vi & Linton, "Major Murray Ormsby: War Crimes Judge and Prosecutor 1919-2012," 226.

394    Tanaka, 239; Frank, *Tower of Skulls*, 57.

395    Dan King, *A Tomb Called Iwo Jima: Firsthand Accounts From Japanese Survivors*, Charleston, SC, 2014, 68, 70; Burleigh, 20; Cook & Cook, 318; Drea, 33, 68, 134-5, 161; Bix, 51-2; Takashi Yoshida, *The Making of the 'Rape of Nanking': History and Memory in Japan, China, and the United States*, Oxford, 2006, 18; Tanaka, 239; Toll, *Pacific Crucible*, 96; Moore, 189, 194.

396    Mosley, 99. Supposedly teachers who made similar mistakes of stumbling "over the words" of an Imperial Rescript as the officer mentioned in the text did also committed suicide. Harries & Harries, 41.

397    Quote is from Edgerton, 310. See also James Bradley, *Flyboys: A True Story of Courage*, NY, 2003, 16-7; Coogan, 465; Dower, 46. *Gaizin* or *Gaijin* can mean simply foreigner, and today, that is how many Japanese view the term. However, it can also be used as meaning non-human and during fascist Japan, that is what it meant when used by a Japanese citizen. See also Frank, *Tower of Skulls*, 58.

398    Cook & Cook, 153, 155, 325.

399    Dower, 46.

400    Takashi Yoshida, 40.

401    Bradley, *Flyboys*, 16-7; Edgerton, 310; Coogan, 465; Winston S. Churchill, *The Grand Alliance*, NY, 1950, 579; Duncan Bartlett, "Japan Looks Back on 17th-Century Persecutions," BBC News, 24 Nov. 2008; NACPM, Trial of Col. Ryoichi Tazuka (typist in document misspelled Tazuka's name Tozuga), SC 249223, 27 June 1946 & SC 203673-S; Lord Russell of Liverpool, 260-1; Victoria, *Zen at War*, xiv.

402    Quote from Modder, 28. See also Toll, *Pacific Crucible*, 115.

403    Hitler, 290–1; Toll, *Pacific Crucible*, 115.

404    Albert Speer's *Inside the Third Reich*, NY, 1970,145.

405    *Hitler's Tischgespräche im Führerhauptquartier,* Einführung von Picker, 310, 398 n. 388; *The Goebbels Diaries, 1942–43,* ed. and trans. by Louis P. Lochner, NY, 1948, 51, 60, 77, 79, 86, 91, 138; Raul Hilberg, *Destruction of the European Jews*, NY, 1961, 45; Paul Gordon Lauren, *Power and Prejudice*, London, 1988, 124; Nathan Stoltzfus, *Resistance of the Heart: Intermarriage and the Rosenstrasse Protest in Nazi Germany*, NY, 1996, 42; Leni Yahil, *The Holocaust*, Tel Aviv, 1987, 71; . Louis L. Snyder, *Encyclopedia of the Third Reich*, NY, 1989, 170; Gordon A. Craig, *Germany, 1866– 1945*, NY, 1978, 696; R. Trevor-Roper, *The Last Days of Hitler*, NY, 1947, 21–2; Otto Klineberg, "Racialism in Nazi Germany," in *The Third Reich,* ed. Maurice Baumont, John H. E. Fried and Edmond Vermeil, NY, 1955, 859; Fritz Redlich, *Hitler: Diagnosis of a Destructive Prophet*, Oxford, 1998, 149; Ian Kershaw, *Hitler, 1936–1945: Nemesis*, NY, 2000, 504; Rigg, *Hitler's Jewish Soldiers*, 184; Burleigh, 13.

406    UGSL, J. Alton Hosch Papers, Tokyo War Crimes Trials, The Military Domination of Japan and Preparation of War (Vol. 2), 519. U.S. Secretary of State Cordell Hull noted the Tripartite Pact created a partnership "enabling Hitler to take charge of one-half of the world and Japan the other half." Toland, Vol. I, 175.

407    Some take umbrage with this study's use of the term Holocaust to describe Japan's mass murder. However, this study builds upon Iris Chang's declaration that the Rape of Nanking is WWII's "Forgotten Holocaust" and Merriam-Webster dictionary when it defines Holocaust as "mass slaughter." Also, it builds on what *The Memorial Hall of the Victims in Nanjing Massacre by Japanese Invaders* calls "a Human Holocaust." As a result, using the term Holocaust correctly defines what Japan did during WWII. Moreover, many Jewish scholars dislike the term Holocaust because the Jews were not a sacrificial burnt offering which is the original meaning of the word although it universally is understood today as meaning mass murder. Many Jewish scholars, especially in Israel, prefer the term *Shoah* which in Hebrew means "catastrophe," but is uniquely describing the Nazi extermination of Jews.

408    Bradley, *Flyboys*, 36.

409    Cook & Cook, 66, 334; Mosley, 68; Bradley, *Flyboys*, 36; Lord Russell of Liverpool, 1.

410    Mosley, 133; Cook & Cook, 78; Bix, 365; Hans Kirchmann, *Hirohito: Japans letzter Kaiser Der Tenno*, München, 1989, 65.

411    Smith, *Spinoza*, 45.

412    Victoria, *Zen at War*, 89.

413    Lord Russell of Liverpool, xv.

414    TSHACNCDNM, Vol. 10, 295, Statement Keenan, Chief of Counsel, International Military Tribunal Far East, 4 June 1945, 33; Tuchman, 118.

415    Crowley, 182; Weinberg, 79; Burleigh, 17; Driscoll, x, xiv; Cook & Cook, 23, 33-4, 56, 125, 156; NACPM, Microfilm Room, Roll 1499, #101, 19557, U.S. Intel Report Ref. Div., 7 July 1942, Regarding: Economy of Manchuria, Population Oct. 1940; Drea, 168; Burleigh, 15-6; Lord Russell of Liverpool, 7; Edgerton, 311; Info from the National Museum of the Pacific War's display, Home

of Adm. Nimitz Museum (NMPWANM, 13 July 2017). In this book, info collected from this museum will be cited. Although it is unorthodox, Michael Berenbaum encouraged the use of such data since from his experience at the Holocaust Museum in D.C., most information displayed in good museums is put together using committees of experts.

416    Crowley, 186, 350.
417    Divine, Breen, Fredrickson and Williams, 789; Drea, 213; Bix, 380.
418    Bix, 383.
419    Drea, 212, 217-8; LaFeber, 379. In response to Japan's completion of taking over Indochina in 1941, FDR froze Japanese assets in the U.S. and embargoed oil exports to Japan.
420    Frank Owen, *The Fall of Singapore*, NY, 2001, 65; Toland, Vol. I, 298-303.
421    Divine, Breen, Fredrickson and Williams, 778.
422    Quote comes from Bix, 12. One of the reasons for Japan taking Attu and Kiska in Alaska's Aleutian island chain was to prevent air attacks on the mainland coming from that direction. This reason is often overlooked and many claim it was just a diversion from the main attack on Midway. Bix, 12. It is interesting to note, Thailand was already free from colonial rule and independent so the Japanese were not "liberating" it from the "white man." Weinberg, 322.
423    Tanaka, 17.
424    Quote from Steinberg, *Island Fighting*, 8, 18-21. See also Sasser, 3; Cook & Cook, 55; Weinberg, 3; Toll, *Pacific Crucible*, 254.
425    Manchester, 77-8.
426    Crowley, xvii.
427    Dower, 8; Manchester, 78; Crowley, 190.
428    Interview Shindo, 9 April 2018. Special thanks to the translation company SIMUL and translator expert Ms. Mariko Uo who helped screen and formulate my questions for Shindo and then reviewed his answers for accuracy after the interview's conclusion.
429    Crowley, xiii; Bix, 176; Toland, Vol. II, 566; Frank, *Tower of Skulls*, 13.
430    Don A. Farrell, *Liberation-1944: The Pictorial History of Guam*, Tinian, 1984, 10.
431    Bix, 1-3. His younger brother, Prince Mikasa, said at the war's end to his subordinate Sadao Magami: "Even if Japan were to win this war, do you think the one billion people of Asia would follow Japan after the things we did in China [i.e.] genocide?" Cook & Cook, 456.
432    Cook & Cook, 69; Bix, 1.
433    Robert Burrell, Review *Flamethrower*, 26 Feb. 2018.
434    BA-MA, RM 11/81, Marine-Kriegsberichter-Halbkompanie Kriegsberichter Werner Jörg Lüddecke, "Die glorreiche Ankunft. Aus den Erfahrungen einer Blockadereise nach Japan," 5 June 1942, 298.
435    Ibid., 12 II/250, KTB, T. 38, Marineattachés u. M(Ltr) Grossetappe Japan-China, Sept. 1941, 17, 223.
436    Bix, xx; Prange, 40. U.S. commanders had failed to learn from the Battle of Taranto 11-2 Nov. 1940 when the British surprised the Italian navy at port and took out several warships. Prange, 40. Japan learned from it and studied it for their Pearl Harbor attack. Churchill, 585.
437    Weinberg, xiv; Harries & Harries, 271; Toll, *Pacific Crucible*, 123.
438    Edgerton, 312.
439    Mosley, 194; Toland, Vol. I, 76; Tuchman, 199.
440    Toland, Vol. I, 101.
441    Bix, 401, 408; Patterson, 256-7; Thomas A. Paterson (ed.), *Problems in American Foreign Policy: Vol. II: Since 1914*, Third Ed., Lexington, MA, 1989; Robert Dallek, "Roosevelt's Leadership, Public Opinion, and Playing for Time in Asia," 215; Edgerton, 251, 312; Jon T. Hoffman, *Chesty: The Story of Lieutenant General Lewis B. Puller*, NY, 2001, 127; Cook & Cook, 23, 44-6; NMPWANM, 13 July 2017; Toland, Vol. I, 96, 98; Toll, *Pacific Crucible*, 122; LaFeber, 379, 382; O'Brien, *The Second Most Powerful Man in the World*, 155; Frank, *Tower of Skulls*, 193-94.
442    Clausewitz, *On War*, Commentary by Bernard Brodie, "A Guide to Reading of On War," 690.

443    Tuchman, 264; SLHA, The Young Companion, 1941-45, 167-72, "A Review of 4 Years of War in Figures". The Young Companion here says the Nationalists had 5 million troops, but since Tuchman is a trained scholar and the YC was a propaganda magazine, her figure is used.

444    Quote from SLHA, The Young Companion, 1940, 150-5, "Chin's Another Million". See also Tuchman, 264. By 1942, Kai-Shek had around 3 million men under arms.

445    Toll, *Pacific Crucible*, 250' O'Brien, *The Second Most Powerful Man in the World*, 255.

446    Wakabayashi (ed.), *The Nanking Atrocity, 1937-38*, Akira, "The Nanking Atrocity: An Interpretive Overview," 32.

447    Cook & Cook, 71-2.

448    Bix, 436-7, 441.

449    Ibid., 443.

450    Drea, 206; Cook & Cook, 380; Victoria, *Zen at War*, 230; Tanaka, 111.

451    UGSL, J. Alton Hosch Papers, Tokyo War Crimes Trials, Conventional War Crimes (Atrocities), Part B, Ch. VIII, 1011-19; Iris Chang, *The Rape of Nanking: The Forgotten Holocaust of World War II*, NY, 1997, 4, 6, 87, 91, 103, 155, 211; Bradley & Powers, *Flags of Our Fathers*, 65; Burleigh, 19-20; Arthur Zich, *The Rising Sun*, Alexandria, VA (Time Life Books), 1977, 23; Spence, 448; Drea, 197; Edgerton, 14, 246-8; Hane, 297-8; Iriye, 48; Cook & Cook, 39, 206; Conversation with Frank at Iwo, 21 March 2015; Drea, 236-7; Bradley & Powers, *Flags of Our Fathers*, 65; Bix, 334-5; Dieter Kuhn, *Der Zweite Weltkrieg in China*, Berlin, 1999, 91; Hane, 297-8; Lord Russell, 41-2, 294; Kirchmann, 78; Weinberg, 322; NMPWANM, 13 July 2017; Burleigh, 18, 20; Daqing Yang, "Diary of a Japanese Army Medical Doctor, 1937," in *Researching Japanese War Crimes Records*, x; Toland, Vol. I, 63; Tanaka, 5; Journal of Nanjing Massacre Studies (JNMS), Vol. 1, No. 1, Nanking, China, 2019, Sun Zhaiwei, "How Many Chinese Military Personnel Were Among the Nanjing Massacre Victims," 16-7; Moore, 118-24, 144-5. A special note on Iris Chang: Although she has done a good job of bringing public awareness to the Rape of Nanking, her work needs to be used with caution. She was not a trained historian and makes historical errors. Wakabayashi (ed.), *The Nanking Atrocity, 1937-38*, Wakabayashi, "Iris Chang Reassessed," xxi-lii & Akira, "The Nanking Atrocity: An Interpretive Overview," 34; Moore, 268.

452    JNMS, Vol. 1, No. 1, Zhaiwei, "How Many Chinese Military Personnel Were Among the Nanjing Massacre Victims," 16.

453    Spence 448; Wakabayashi (ed.), *The Nanking Atrocity, 1937-38*, Wakabayashi, "The Messiness of Historical Reality," 18 & Akira, "The Nanking Atrocity: An Interpretive Overview," 37, 43; Harries & Harries, 221. Tang was joined by many of his fellow officers in abandoning their troops causing much confusion. Moore, 117. Also see Frank, *Tower of Skulls,* 40, 46, 56.

454    Quote from Tuchman, 194. See also LaFeber, 393; O'Brien, *The Second Most Powerful Man in the World*, 301.

455    Victoria, *Zen War Stories*, 12. See also JNMS, Vol. 1, No. 1, Zhaiwei, "How Many Chinese Military Personnel Were Among the Nanjing Massacre Victims," 19-20.

456    JNMS, Vol. 1, No. 1, Zhaiwei, "How Many Chinese Military Personnel Were Among the Nanjing Massacre Victims," 16, 25; Wakabayashi (ed.), *The Nanking Atrocity, 1937-38*, Akira, "The Nanking Atrocity: An Interpretive Overview," 38-9, 43-5 & Tokushi, "Massacres Outside Nanking City," 58 & Kenji, "Massacres Near Mufushan," 77-9, 81-2. See also Frank, *Tower of Skulls*, 51-2, 56.

457    JNMS, Vol. 1, No. 1, Zhaiwei, "How Many Chinese Military Personnel Were Among the Nanjing Massacre Victims," 18; Tanaka, 18-19; *Hong Kong's War Crimes Trials*, Zahar, "Trial Procedure at the British Military Courts, Hong Kong, 1946-1948," 15.

458    JNMS, Vol. 1, No. 1, Zhaiwei, "How Many Chinese Military Personnel Were Among the Nanjing Massacre Victims," 18; Wakabayashi (ed.), *The Nanking Atrocity, 1937-38*, Akira, "The Nanking Atrocity: An Interpretive Overview," 40.

459    Tanaka, 88. Matsui was also in control of the Shanghai Expeditionary Army (SEA) or Shanghai

Expeditionary Force (SEF) which merged together with the 10th Army to make the Central China Area Army (CCAA) under Matsui's overall command. Wakabayashi (ed.), *The Nanking Atrocity, 1937-38*, Akira, "The Nanking Atrocity: An Interpretive Overview," 31-2.

460    Wu Ch'êng-ên, *Monkey*, trans. Arthur Waley, NY, 1943, 28.

461    Victoria, *Zen War Stories*, 13, 234 fn.5.

462    Spence, 448.

463    Bix , xvii-xviii; NMPWANM, 13 July 2017. The Japanese Ministry of Foreign Affairs calls it the "Nanjing Massacre" and admits its army killed "a large number of noncombatants," and looted their belongings. However, it denies the number of 300,000 dead and denies it committed rape. "History Issues Q&A by the Ministry of Foreign Affairs of Japan" given to author by Deputy Consul Gen. in Houston Ryuji Iwasaki, 10 Dec. 2017; Bundesarchiv-Militärarchiv (BA-MA), RH 2/1848, OKH, Ausarbeitung Erfahrungen u. Netrachtungen aus dem japanisch-chinesischen Feldzug 1937/1938 übersandt, 15.3.1938, S. 31; Interview Drea, 17 Dec. 2017; Wakabayashi (ed.), *The Nanking Atrocity, 1937-38*, Wakabayashi, "The Messiness of Historical Reality," 3-4 & Akira, "The Nanking Atrocity: An Interpretive Overview," 33-4, 50; Tuchman, 178; Harries & Harries, 229, 241.

464    TSHACNCDNM, Vol. 2, 214, 221-2 from H. J. Timperley, *What War Means*, 91.

465    TSHACNCDNM, Vol. 2, 214, 221-2 from H. J. Timperley, *What War Means*, 94-5; Wakabayashi (ed.), *The Nanking Atrocity, 1937-38*, Akira, "The Nanking Atrocity: An Interpretive Overview," 40; Harries & Harries, 228. In many cities, the Chinese "huddled around foreign missionaries in their churches," but it seems without the success like was scene at Nanking. Moore, 82.

466    TSHACNCDNM, Vol. 10, 290, Statement Keenan, Chief Counsel, International Military Tribunal Far East, 4 June 1945, 28.

467    Tanaka, 249-50; Wakabayashi (ed.), *The Nanking Atrocity, 1937-38*, Akira, "The Nanking Atrocity: An Interpretive Overview," 32, 36 & Tokushi, "Massacres Outside Nanking City," 58, 64; Harries & Harries, 481.

468    Wakabayashi (ed.), *The Nanking Atrocity, 1937-38*, Tokushi, "Massacres Outside Nanking City," 64; Frank, *Tower of Skulls*, 42, 47, 56, 57..

469    Bradley, *Flyboys*, 59; Chang, 59. There has been postwar controversy about the competition and historian Richard Frank claimed he has seen some evidence that it may not be true. However, a Japanese court reviewed the evidence and concluded it had occurred. Chris Hogg, "Victory for Japan's War Critics," BBC, 23 Aug. 2005; Kate Heneroty, "Japanese Court Rules Newspaper Didn't Fabricate 1937 Chinese Killing Game," Jurist Legal News and Research, Univ. of Pittsburgh School of Law, 23 Aug. 2005. Both Noda and Mukai were found guilty of war crimes at the Nanking War Crime Trials in 1948 and executed. Historian Bob Tadashi Wakabayashi makes a strong argument though that this competition did not happen. See Wakabayashi (ed.), *The Nanking Atrocity, 1937-38*, Wakabayashi, "The Nanking 100-Man Killing Context Debate, 1971-75," 115-46. It is this study's conclusion that both men probably committed crimes against civilians since they were part of the Nanking military campaign. Yale University professor Jonathan Spence actually believes the numbers and the sources for the full competition. Mukai said he played along with the article and lied about killing many to get a "good wife" which his wife admitted to working! The fact these men felt that by bragging about killing people was something to be proud of probably denotes that only after the war, did they claim they made it up to try to save their lives. Noda supposedly did claim after the war that he slaughtered innocent POWs and even told such stories to school children.

470    Conrad, "Heart of Darkness," 586.

471    TSHACNCDNM, Vol. 2, 214, 221-2 from H. J. Timperley, *What War Means*, 92-3.

472    Crowley, 357, 396; Cook & Cook, 23.

473    Kirchmann, 78; Bix, 339.

474    JNMS, Vol. 1, No. 1, Meng Guoxiang, "How the Japanese Military Invaders Destroyed and Exploited Buddhism in Nanjing," 97-125; Toll, *Pacific Crucible*, 99. Buddhist monks had been serving in the IJA since its inception in the 1860s. Harries & Harries, 12.

475    JNMS, Vol. 1, No. 1, Zhaiwei, "How Many Chinese Military Personnel Were Among the Nanjing
       Massacre Victims," 18-9 & Zhu Tianle & Zhu Chengshan, "The Legally Affirmed Chain of
       Evidence for the Nanjing Massacre, 71.

476    *The Memorial Hall of the Victims in Nanjing Massacre by Japanese Invaders*, Doc. Display, 11 May
       2019.

477    Quote from Chang, 91. See also Edgerton, 246; Lord Russell, 45; Bix, 336; Shinsho Hanayama, *The
       Way of Deliverance: Three Years with the Condemned Japanese War Criminals*, NY, 1950, 185–6.
       Slicing open a young woman's vagina in order to rape her seems to have been common among IJA
       troops. See Scott, 263, 471.

478    TSHACNCDNM, Vol. 2, 214, 221-2 from H. J. Timperley, *What War Means*, 182.

479    Ibid., 91.

480    Ibid.

481    TSHACNCDNM, Vol. 2, 214, 221-2 from H. J. Timperley, *What War Means*, 174; Wakabayashi
       (ed.), *The Nanking Atrocity, 1937-38*, Akira, "The Nanking Atrocity: An Interpretive Overview," 49;
       Frank, *Tower of Skulls*, 49.

482    TSHACNCDNM, Vol. 2, 214, 221-2 from H. J. Timperley, *What War Means*, 178, 181.

483    Ibid., 182-3.

484    TSHACNCDNM, Vol. 2, 214, 221-2 from H. J. Timperley, *What War Means*, 92-3, 181. See also
       Wakabayashi (ed.), *The Nanking Atrocity, 1937-38*, Akira, "The Nanking Atrocity: An Interpretive
       Overview," 32, 36; Tuchman, 178; Harries & Harries, 227; Moore, 83; Frank, *Tower of Skulls*, 59-61.

485    TSHACNCDNM, Vol. 2, 163-4.

486    UGSL, J. Alton Hosch Papers, Tokyo War Crimes Trials, Conventional War Crimes (Atrocities),
       Part B, Ch. VIII, 10013; Chang, 4, 6, 59, 86, 87-9 91, 94-5, 103, 155; Interview Master Johnny
       Kwong Ming Lee, 10 April 2019; Scott, 184-5, 260; JNMS, Vol. 1, No. 1, Tianle & Chengshan,
       "The Legally Affirmed Chain of Evidence for the Nanjing Massacre, 79; TSHACNCDNM, Vol. 2,
       214 from H. J. Timperley, *What War Means*, Chapter II, 61; The Memorial Hall of the Victims in
       Nanjing Massacre by Japanese Invaders, White Stone Display Wall, Testimony Chang Zhiqiang, 12
       May 2019; Tuchman, 178; Frank, *Tower of Skulls*, 43-4, 49.

487    TSHACNCDNM, Vol. 3, No. 10, 18-23, Rabe to Japanese Embassy, 18 Dec. 1937.

488    Chang, 4, 6, 59, 86, 87-91, 94-5, 103, 155; Toland, Vol. I, 62; Frank, *Tower of Skulls*, 48.

489    Chang, 59.

490    Chang, 4, 6, 59, 86, 87-9 91, 94-5, 103, 155; Bradley, *Flyboys*, 58-9, 62, 111-2, 152, 200-1, 225,
       228-232, 245, 316; Edgerton, 16, 246-7; Burleigh, 561; Cook & Cook, 155, 273-4, 462; Lord
       Russel, 41-2, 57; NACPM staff, "Japanese War Crimes Records National Archives: Research Starting
       Points," in *Researching Japanese War Crimes Records*, 102-3; Bradley, *Flyboys*, 200; Greg Bradsher,
       "The Exploitation of Captured and Seized Japanese Records Relating to War Crimes, 1942-1945,"
       in *Researching Japanese War Crimes Records*, 152); MCHDQV, Dashiell File, #1, Davis-Filipinos;
       Mark Felton, *Slaughter at Sea: The Story of Japan's Naval War Crimes*, South Yorkshire, 2007; UGSL,
       J. Alton Hosch Papers, Tokyo War Crimes Trials, Conventional War Crimes (Atrocities), Part B,
       Ch. VIII, 1033, 1067; Tanaka, 129, 139; JNMS, Vol. 1, No. 1, Zhang Sheng, "The Seven Why-
       Questions in the Writing of History," 10 & Zhang Lianhong, "A Hero of Nanjing: Austrian Mechanic
       Rupert R. Hatz During the Nanjing Massacre," 27; TSHACNCDNM, Vol. 2, 214, 221-2 from H.
       J. Timperley, *What War Means*, 36, 42-3, 62, 65, 184-5, 189-91 & Vol. 3, 28-49, Cases of Disorder
       by Japanese Soldiers in the Safety Zone; Moore, 84-107; Frank, *Tower of Skulls*, 48-50.

491    Tuchman, 168; Moore, 84-5; Frank, *Tower of Skulls*, 45.

492    Tuchman, 171.

493    Chang, 4, 6, 59, 86, 87-9 91, 94-5, 103, 155; Bradley, *Flyboys*, 58-9, 62, 111-2, 152, 200-1, 225,
       228-232, 245, 316; Edgerton, 16, 246-7; Burleigh, 561; Cook & Cook, 155, 273-4, 462; Lord
       Russel, 41-2, 57; NACPM staff, "Japanese War Crimes Records National Archives: Research Starting

Points," in *Researching Japanese War Crimes Records*, 102-3; Bradley, *Flyboys*, 200; Greg Bradsher, "The Exploitation of Captured and Seized Japanese Records Relating to War Crimes, 1942-1945," in *Researching Japanese War Crimes Records*, 152); MCHDQV, Dashiell File, #1, Davis-Filipinos; Mark Felton, *Slaughter at Sea: The Story of Japan's Naval War Crimes*, South Yorkshire, 2007; UGSL, J. Alton Hosch Papers, Tokyo War Crimes Trials, Conventional War Crimes (Atrocities), Part B, Ch. VIII, 1033, 1067; Tanaka, 129, 139; JNMS, Vol. 1, No. 1, Zhang Sheng, "The Seven Why-Questions in the Writing of History," 10 & Zhang Lianhong, "A Hero of Nanjing: Austrian Mechanic Rupert R. Hatz During the Nanjing Massacre," 27; Dante, *Inferno*, Canto XXXIV, 115; TSHACNCDNM, Vol. 2, 214, 221-2 from H. J. Timperley, *What War Means*, 36, 42-3, 62, 65, 184-5, 189-91 & Vol. 3, 28-49, Cases of Disorder by Japanese Soldiers in the Safety Zone; Moore, 210.

494     Wu Ch'êng-ên, 284.

495     *Silence of the Lambs* is a 1991 horror film about a serial killer who tortures and eats his victims.

496     Conrad, "Heart of Darkness," 495. Conrad is describing Belgium King Leopold II's brutal Congo rule from the 1890's until 1908.

497     Cook & Cook, 40-4; Victoria, *Zen at War*, 76-7; Toll, *Pacific Crucible*, 112.

498     Bix, 15. See also Bix, 336-8.

499     Weinberg, 894; Dower, 47, 295-7; Bix, 4; Interview Gerhard Weinberg, 2 Sept. 2005; Edgerton, 250, 272, 284; Frank, 162-3, 325, 329; Hane, 361-2; Karl August Muggenthaler, *German Raiders of World War II: The First Complete History of Germany's Mysterious Naval Marauders*, Suffolk, 1978, 270; Cook & Cook, 74, 99; Drea, 236; Robert Hanyok, "Wartime COMINT Records National Archives about Japanese War Crimes in the Asia and Pacific Theaters, 1978-1997," in *Researching Japanese War Crimes Records*, 142; Spence, 496. Frank, *Downfall*, 162-3, 325, 329; Frank writes that recent scholarship puts the number at 18 million civilian deaths under the Japanese alone. Here is the breakdown of death by the Japanese from 1931-45 by country using historian Robert Newman's data: China (10,000,000); Java (Dutch West Indies Indonesia) (3,000,000); Outer Islands (Dutch Indies) (1,000,000); Philippines (120,000); India (180,000); Bengal Famine (1,500,000); Korea (70,000); Burma-Siam railroad (82,500); Indonesian, Europeans (30,000); Malaya (100,000); Vietnam (1,000,000)**; Australia (30,000); New Zealand (10,000); U.S. (100,000). Altogether, Frank cites 17,222,500 killed by the Japanese during WWII. Frank, *Downfall*, 163. Frank also notes these numbers would be higher if one also included the deaths that happened after 1945 when the effects of the Japanese medical experiments and famine on the populations they controlled were tabulated. Frank, *Downfall*,162, 325, 329. **A special note on Vietnam, according to the National Archives, 1,000,000 Vietnamese starved alone from 1944-45 due to Japanese policies, so this nation probably lost more than what is cited in Frank's work. Robert Hanyok, "Wartime COMINT Records in the National Archives about Japanese War Crimes in the Asia and Pacific Theaters, 1978-97," in *Researching Japanese War Crimes Records*, 142. Historian Yuki Tanaka puts the number even higher to 2 million Vietnamese. Tanaka, 250.

500     NMPWANM, 13 July 2017; The breakdown follows of those citizens killed by the Japanese: U.S. 1,542 , Malaya 100,000, Indonesia (Europeans) 30,000, Korea 378,000-533,000 (includes forced laborers and IJA soldiers), Philippines 120,000, Indo China (Vietnam) 500,000, India 26,000, Dutch East Indies 4 million, Burma 250,000, China 10-17 million and Australia 700. Bix, 4; Tanaka, xix, 8-9, 251.

501     Dinah Shelton, ed. *Encyclopedia of Genocide and Crimes of Humanity*, NY, 2005, 171; Dower, 295-7; Chang, 8; Spence, 464; Drea, 245; NMPWANM, 13 July 2017; Kuhn, 30; Interview Morgan, 10 Dec. 2017; Interview Heaton, 17 Dec. 2017; Burleigh, 562. Chang has the number between 10-19 million. Considering how the Japanese behaved, this study finds 10 million low. The general consensus is that Japanese killed many more in China than can be documented because they murdered everywhere they went for fifteen years unlike the Nazis who only murdered intensively

for four years from 1941-4. This study believes the Japanese murdered over 20 million. Renowned Yale Univ. historian Jonathan Spence claims in one region in Communist Mao-held lands, due to Japanese "counterattacks" and their "immense cruelty," the population dropped from 44 million to 24 million, with many if not most dying. In fact, the Japanese often destroyed "whole villages" in their march of destruction. Spence, 464. Drea writes that 10 million Chinese soldiers died fighting the Japanese and "civilian casualties certainly surpassed that number." Drea, 245. Dieter Kuhn writes that between 1937-45, the Japanese murdered over 19 million. Kuhn, 30. Marty Morgan and Colin Heaton say the Japanese probably murdered over 40 million. Interview Morgan, 10 Dec. 2017; Interview Heaton, 17 Dec. 2017.

502    NMPWANM, 13 July 2017.

503    Frank, *Downfall*,162; Edgerton, 14.

504    Driscoll, 5.

505    Interview Morgan, 10 Dec. 2017; Weiner (ed.), *Race, Ethnicity and Migration in Modern Japan*, Weiner, "Introduction," 3. Bob Tadashi Wakabayashi quotes sources that 35 million died during the Second Sino-Japanese War, but does not delineate how many were due to atrocities and how many were due to battle. Wakabayashi (ed.), *The Nanking Atrocity, 1937-38*, Wakabayashi, "The Messiness of Historical Reality," 9; Dower, 43; Spence, 469; Frank, *Downfall*,109,160; Edgerton, 249-50; Drea, 214; Weinberg, 2, 20-1, 28; Patterson, 254-5; Zeiler, 3.

506    Drea, 214; Bix, 365.

507    Quote from Hane, 361. See also Cook & Cook, 2.

508    Lord Russell of Liverpool, 52.

509    Spence, 448.

510    Statement Prime Minister Junichiro Koizumi, 15 Aug. 2005.

511    Interview, Japanese Deputy Consul Gen. Ryuji Iwasaki, 12 Dec. 2017; Ministry of Foreign Affairs of Japan's take on "What is the view of the Government of Japan on the incident known as the "Nanjing Massacre"?" It answers some of this question with "…there are numerous theories as to the actual number of victims, and the Government of Japan believes it is difficult to determine which the correct number is." In Prime Minister Shinzo Abe's statement on WWII from 14 Aug. 2017, he never mentioned his country murdered millions. Statement PM Abe on 70th commemoration of WWII's end.

512    Note by Colin Heaton, 28 May 2017.

513    Cook & Cook, 10; Interview Drea, 17 Dec. 2017; Weiner (ed.), *Race, Ethnicity and Migration in Modern Japan*, Weiner, "Introduction," 2. Historians Meirion and Susie Harries write about Japan's Holocaust: "The exhaustive and cathartic examination of the Holocaust by Germans in recent years has no Japanese equivalent. There has been no easy explanation of Japanese atrocities, no Nazi party to act as the scapegoat for collective war guilt. Because the Japanese have not come to terms with their own past, neither have others." Harries & Harries, ix.

514    Honda Katsuichi, *The Nanjing Massacre: A Japanese Journalist Confronts Japan's National Shame*, London, 1999, 287 (Nanjing and Nanking are two spellings for the same city); Bradley, *Flyboys*, 61; Edgerton, 16; Cook & Cook, 116, 446-7; Drea, vii; Interview Morgan, 10 Dec. 2017. See also *Hong Kong's War Crimes Trials*, Daqun, "Foreword," viii.

515    Cook & Cook, 441; Edward J. Drea, "Introduction," in *Researching Japanese War Crimes Records*, 6; Interview Drea, 10 Dec. 2017; Tanaka, xv; Wakabayashi (ed.), *The Nanking Atrocity, 1937-38*, Akira, "The Nanking Atrocity: An Interpretive Overview," 39, 51.

516    Chang, xii, (Foreword William C. Kirby); Bradley & Powers, *Flags of Our Fathers*, 62; Drea, "Introduction," in *Researching Japanese War Crimes Records*, 5.

517    Edgerton, 249. David Sanger, "New Tokyo Minister Calls 'Rape of Nanking' Fabrication," NYT, 5 May 1994.

518   Yashukan War Museum in Tokyo, Japan, 9 April 2018

519   Tanaka, xvi.

520   Richard Black, "Those Who Deny Auschwitz Would be Ready to Remake it," Jewish News, 27 Jan. 2015.

521   Meeting at AJC's Dallas office with Consul-General of Japan Houston, Tetsuro Amano, 30 Jan. 2018.

522   Chang, 12, 222. As of 2012, Germany has given $90 billion in reparations. Dylan Matthews, "Six Times Victims Have Received Reparations—Including Four in the US," Vox, 23 May 2014; Palash Ghosh, "Germany to Pay Out $1 Billion in Reparations For Care of Aging Holocaust Survivors," International Business Times, 29 May 2013; Source Minister of Foreign Affairs of Japan given to author by Ryuji Iwasaki, 12 Dec. 2017. Lord Russell, 41. Although, Japan's reparations are miserly, there has been a lot of economic activity that supposedly Japan has used as a form of reparation in China that needs further exploration. See Wakabayashi (ed.), *The Nanking Atrocity, 1937-38*, Wakabayashi, "The Messiness of Historical Reality," 11 & Wakabayashi, "The Nanking 100-Man Killing Context Debate, 1971-75," 116 & Yamamoto, "A Tale of Two Atrocities: Critical Appraisal of American Historiography," 294; Alexander Dallin, *German Rule in Russia 1941-1945: A Study of Occupation Policies*, NY, 1957, 427.

523   Wakabayashi (ed.), *The Nanking Atrocity, 1937-38*, Wakabayashi, "The Messiness of Historical Reality," 10.

524   Moreover, most Japanese war criminals showed no remorse for what they had done. *Hong Kong's War Crimes Trials*, Linton, "Major Murray Ormsby: War Crimes Judge and Prosecutor 1919-2012," 226, 236; Harries & Harries, 478.

525   Bradley & Powers, *Flags of Our Fathers*, 65-6; Richard Rhodes, *The Making of the Atomic Bomb*, NY, 1986, 734-43; John Keegan, *The Second World War*, NY, 1990, 584; Cook & Cook, 383; Scott, 510.

526   Weiner (ed.), *Race, Ethnicity and Migration in Modern Japan*, Weiner, "Introduction," 2-3; Tanaka, xv, xvii, xxii.

527   Kevin L. Jones, "'Comfort Women' Statue Strains 60-Year San Francisco-Osaka Alliance," Art Wire, 28 Sept. 2017; Julia Glum, "San Francisco Statue Honoring 'Comfort Women' Sex Slaves from WWII Infuriates Japan," Newsweek, 30 Oct. 2017; Heather Knight, "Japanese Mayor Cuts Ties Between SF and Osaka Over Comfort Women Statue," San Francisco Chronicle, 3 Oct. 2018.

528   Interview Yukio Yasunaga, 10 April 2018.

529   Yoshiaki Yoshimi and Suzanne O'Brien, *Comfort Women*, NY, 2002; Lord Russell, 43, 243, 257, 266, 294; Jones, "Comfort Women," Art Wire, 28 Sept. 2017; Julia Glum, "San Francisco Statue Honoring 'Comfort Women' Sex Slaves from WWII Infuriates Japan," Newsweek, 30 Oct. 2017; Tanaka, 102; Scott, 263-4, 280; TSHACNCDNM, Vol. 2, 240 from H. J. Timperley, *What War Means*, 64-5, 188-9, 192; Moore, 134, 148-153; Frank, *Tower of Skulls*, 48.

530   Tanaka, 102.

531   Ibid., 104.

532   UNESCO Nomination Form, International Memory of the World Register, Documents of Nanjing Massacre, 2014-50; "Japan Halts UNESCO Funding Following Nanjing Massacre Row," The Guardian, 14 Oct. 2016; Reiji Yoshida, "UNESCO Strikes Political Nerve with Nanking Massacre Documents," The Japan Times, 19 Oct. 2015; "Japan Withholds UNESCO Funding After Nanjing Massacre Row," Reuters, 14 Oct. 2016.

533   Wakabayashi (ed.), *The Nanking Atrocity, 1937-38*, Wakabayashi, "The Messiness of Historical Reality," 11.

534   Cook & Cook, 15.

535   Interview Walter Frank, 18 April 1997; Kirchmann, 180.

536   Interview Woody, 23 July 2016.

537    Bix, 6.

538    Cook & Cook, 16; Lord Russell of Liverpool, xiii.

539    Interview Ryuji Iwasaki, 12 Dec. 2017.

540    *Documents on the Tokyo International Military Tribunal Charter*, xlvi.

541    Interview LeMay by Colin Heaton June 1986.

542    Cook & Cook, 6-7.

543    Bix, 5, 16-7.

544    Kirk Spitzer, "Apology Question Hounds Obama's Planed Visit to Hiroshima," USA Today, 21 May 2016.

545    Tanaka, xvii.

546    Ibid., xi.

547    Spitzer, "Apology Question Hounds Obama's Planed Visit to Hiroshima."

548    Ibid.

549    Burleigh, 562.

550    TSHACNCDNM, Vol. 10, 268-272, Statement Keenan, Chief Counsel, International Military Tribunal Far East, 4 June 1945, 7-10; *Hong Kong's War Crimes Trials*, Linton, "Introduction," 2-3.

551    NIDSMDMA, Collections Kuribayashi, Bio info; NIDSMDMA, *Senshi-Sosho*, Central Pacific, Op. #2, Peleliu, Angaur, Iwo, Vol. 13, 278-9. The Marine history called *The U.S. Marines on Iwo Jima* incorrectly claims Kuribayashi fought at Lake Nomanhan against the Soviets in 1939, but the Tokyo NIDS archives do not support this claim. Henri, et al., 6.

552    Mosley, 115.

553    UGSL, Hosch Papers, Tokyo War Crimes Trials, Conventional War Crimes (Atrocities), Part B, Ch. VIII, 1008.

554    Tanaka, 4.

555    Snow, 81; Keith Bradsher, "Thousands March in Anti-Japan Protest in Hong Kong," NYT, 18 April 2005; Suzannah Linton, "Rediscovering the War Crimes Trials in Hong Kong, 1946-48," Melbourne Journal of International Law, Vol. 13, No. 2, 2012, 22-3; Ted Ferguson, *Desperate Siege: The Battle of Hong Kong*, Scarborough, Ontario, 1980, 54, 76, 94, 211-2, 217-8; UGSL, Hosch Papers, Tokyo War Crimes Trials, Conventional War Crimes (Atrocities), Part B, Ch. VIII, 1033, 1038; Drea, 224; *Hong Kong's War Crimes Trials*, Daqun, "Foreword," vi; Frank, *Tower of Skulls*, 315.

556    Dower, 42; Fred Haynes & James A. Warren, *The Lions of Iwo Jima: The Story of Combat Team 28 and the Bloodiest Battle in Marine Corps History*, NY, 2008, 147.

557    Interview Mary Dalbey-Rigg, Summer 1996. The interview with the nurse was conducted at Yale Univ. School of Nursing circa 1957.

558    Ferguson, 216.

559    Wakabayashi (ed.), *The Nanking Atrocity, 1937-38*, Tokushi, "Massacres Outside Nanking City," 58.

560    Ferguson, 211; Toland, Vol. I, 316; Snow, 80; Tanaka, 91-2; Hong *Kong's War Crimes Trials*, Daqun, "Foreword," v; Haynes & Warren, 147; Frank, *Tower of Skulls*, 314.

561    Snow, 80.

562    Ibid.

563    Ferguson, 212.

564    Ferguson, 216; Interview Dalbey-Rigg, Summer 1996; Tanaka, 97. Often Japanese soldiers were observed to break out in laughter as they committed horrible crimes. After seizing ten "unresisting" British Royal Army Medical Corps officers and Privates and eight wounded Canadians, the Japanese stabbed them with their bayonets and cut off their heads "amid shouts of laughter." Snow, 79-80. In numerous sources, the Japanese were found to laugh when tormenting and killing their victims. See Scott, 80, 261, 382.

565    Ferguson, 217. Japanese often enjoyed raping wives in front of their husbands. See Scott, 210.

566    Kurt Haas & Adelaide Haas, *Understanding Sexuality*, New Paltz, NY, 1993, 564; Harries & Harries,

479.

567     Haas & Haas, 564-5.

568     Ibid., 565.

569     Haas & Haas, 565; Historian Richard Frank wrote, "physical violence against women was rampant in Japanese society." Frank, *Tower of Skulls*, 57.

570     Haas & Haas, 565.

571     Ibid.

572     Ibid.

573     Email Chi Man Kwong, Hong to Rigg, 30 April & 5 May 2018; Ferguson, 101, 218; Kwong and Tsoi, 181, 184; Snow, 80; *Hong Kong's War Crimes Trials*, Daqun, "Foreword," vi.

574     Snow, 82; *Hong Kong's War Crimes Trials*, Daqun, "Foreword," v.

575     *A Savage Christmas: The Fall of Hong Kong 1941*, Canadian Documentary, 12 Jan. 1992.

576     Dower, 44; Lord Russell, 63.

577     Ferguson, 80, 144-7, 158, 166, 170, 176, 199, 210-1; Kwong and Tsoi, 218; Benjamin Lai, *Hong Kong 1941-45; First Strike in the Pacific War*, NY, 2014, 56, 79, 89; UGSL, Hosch Papers, Tokyo War Crimes Trials, Conventional War Crimes (Atrocities), Part B, Ch. VIII, 1033, 1038; Drea, 224; Snow 79-80; *Hong Kong's War Crimes Trials*, Daqun, "Foreword," v & Linton, "War Crimes," 96.

578     Ferguson, 211.

579     *Hong Kong's War Crimes Trials*, Daqun, "Foreword," v.

580     Bradsher, "Thousands March in Anti-Japan Protest in Hong Kong"; Ferguson, 25; Snow, 79; *Hong Kong's War Crimes Trials*, Daqun, "Foreword," vi & Linton, "War Crimes," 107-9, 113-4.

581     Snow, 86.

582     Ibid., 105

583     Ibid., 86.

584     *Hong Kong's War Crimes Trials*, Daqun, "Foreword," vi.

585     Snow, 87.

586     Snow, 94-5.

587     Kakehashi's biography *So Sad to Fall in Battle*, cited throughout this book. Burrell, *Ghosts of Iwo Jima*, 43; Email Tatsushi Saito to Rigg, 18 March 2019; Toland, Vol. II, 797; Crowl & Isely, 468.

588     Kwong and Tsoi, 177-8. Kwong writes that Kuribayashi was removed from "active staff duty" when he confronted Sakai over the treatment of Col. Doi, but this is wrong. He just moved from the front and returned to headquarters of the 23rd Army to the north. It was unclear whether this was punishment or just a normal placement for Kuribayashi for his duties. Interview Tatsushi Saito, NIDS, 9 April 2018; Email Chi Man Kwong to Rigg, 10 March 2019 & 18 March 2019; Email Tatsushi Saito to Rigg, 17 March 2019.

589     *Hong Kong's War Crimes Trials*, Zahar, "Trial Procedure at the British Military Courts, Hong Kong, 1946-48," 15 & Linton, "War Crimes," 133.

590     Charles G. Roland, "Massacre and Rape in Hong Kong: Two Case Studies Involving Medical Personnel and Patients," Journal of Contemporary History, Vol. 32 (I), 43-61, 1997. However, historian Chi Man Kwong defends Kuribayashi somewhat saying he was on the Kowloon side and most of the atrocities took place "on the island side," but nonetheless, atrocities also took place in Kowloon as well. Email Chi Man Kwong to Rigg, 5 May 2018; *Hong Kong's War Crimes Trials*, Linton, "War Crimes," 133.

591     Linton, "Rediscovering the War Crimes Trials in Hong Kong, 1946-48," 56-8; Carroll, John Mark. (2007). A concise history of Hong Kong, 123; Alexander, 6; Edgerton, 264-5; Burleigh, 331; Kakehashi, xi, 3-4; *Law Reports of Trials of War Criminals*. Vol. III. The UN War Crimes Commission, 1948, Trial Gen. Takashi Sakai; NACPM, RG 165, Box 2151, War Dept., 6900-5 (Japan), memorandum Consul Gen. Robert S. Ward: Japanese attack on and Capture of...Hong Kong, 17; NACPM, Microfilm Room, Roll 1499, #144, 23838, Current Intel Sec. A-2, Interview

Charles Schaefer, Dist. Manager, Hong Kong—PAM Airlines, 28 Aug. 1942; UN Archives (UNA), Security Microfilm Programme 1998, UNWCC, PAG-3, Reel no. 61, Summary Translation of the Proceedings of the Military Tribunal, Nanking, Trial of Takashi Sakai, 3; *Hong Kong's War Crimes Trials*, Honorable Justice Liu Daqun, "Foreword," vi & Linton, "War Crimes," 96.

592    Linton, 58.

593    UNA, Security Microfilm Programme 1998, UNWCC, PAG-3, Reel no. 61, Summary Translation of the Proceedings of the Military Tribunal, Nanking, Trial Takashi Sakai, 1.

594    Ibid., 3.

595    Snow, 81.

596    Ibid., 82.

597    Linton (ed.,) Linton, "Introduction," 1.

598    Interview Dennis Showalter, 18 May 2019.

599    Snow, 80. It seems there was only one case where the Japanese command tried to re-instate some semblance of justice. Three days after the fall of Hong Kong, nine IJA troops who had raped British nurses in Happy Valley were executed. However, this study has seen no evidence that IJA personnel were ever punished for harming and killing Chinese civilians in Hong Kong. Snow, 81.

600    Hong Kong's War Crimes Trials Collection, Case No. W0235/1030, Maj. Gen. Ryossaburo Tanaka & Lt. Gen. Takeo Ito, Case No. W0235/1107; Lai, 32, 89; Snow, 80; *Hong Kong's War Crimes Trials*, Linton, "War Crimes," 99.

601    Snow, 80.

602    *Hong Kong's War Crimes Trials*, Totani, "The Prisoner of War Camp Trials," 75, 78-9.

603    Ibid., 78.

604    Ibid., Daqun, "Foreword," vi & Linton, "Major Murray Ormsby," 231-2, 240-1.

605    Interview Mant Hawkins, 16 Feb. 2019.

606    Scott, 489.

607    Ibid., 501.

608    Ibid., 502.

609    Sheldon H. Harris, *Factories of Death: Japanese Biological Warfare, 1932-45, and the American Cover-up*, NY, 2002, 86-7, 127, 229; Sheldon H. Harris, Ch. 16, "Japanese Biomedical Experimentation During the World War II Era," 477, 481, LaGuardia Community College Files; Email Saito to Rigg, 4 April 2019; *People's Daily Online*, New Proof of Japan's WWII Invasion Found in Guangzhou, 17 April 2005.

610    Email Saito to Rigg, 4 April 2019; (以下、栗林が参謀長の時の第23軍の話です。1941.9-1943.6)
第23軍司令部は、この間、"広東"にありました。
まず、香港攻略以後、栗林のいる間は大きな作戦は実施していません。
そして、香港の占領地行政は、陸軍大臣直轄の香港占領地総督部が行います。
香港の占領地行政には第23軍は、ほとんど関与しません。
香港攻略戦の後、第23軍が何をていたかという史料は少なく、詳しくはわかりません。
しかし、ここには、いくつかの参謀長名の電報と、レポートがあります。
この史料から推測すると、占領地域（広東、香港地域など）の治安維持、飛行場などの施設
整備、経済対策（金融、通貨、物資配給・統制）などである。
その他、当然ながら、軍隊の教育訓練、維持管理、兵站基地としての機能もあったと思われる。
上記の参考となった史料名、以下2点です。]
①南支那方面第23軍関係電報綴（中央―作戦指導重要電報―42）
②波集団司令部「波集団経済封鎖情報月報」（支那―大東亜戦争南支―3）See also Moore, 219.

611    Ibid.

612    Burrell, *Ghosts of Iwo Jima*, 41.

613    Snow, 102.

614    Tanaka, 105.

615    Ibid., 106.

616    Snow, 102; Tanaka, 110.

617    Tanaka, 107.

618    Ibid., 110.

619    Snow, 131.

620    Ibid., 132.

621    Ibid., 133-4.

622    Crowl & Isely, 468; Henri, et al., 6.

623    Snow, 159. Renaming many landmarks and cities after Japanese names was common throughout
       Japan's conquered Empire. See Scott, 62.

624    Snow, 160.

625    Ibid., 163-4.

626    Ibid., 106, 114, 118.

627    NIDSMDMA, Senshi-Sosho, Central Pacific, Op. #2, Peleliu, Angaur, Iwo, Vol. 13, 391; MCHDQV,
       Dashiell File, #196, Iwo Items; Larry Smith, Drill Sergeants, 29; Henri, et al., 13; Smith, Iwo Jima,
       xxi; Wheeler, 41; Richard F. Newcomb, Iwo Jima, NY, 1965, 35; Edgerton, 226; Robert T. Oliver, A
       History of the Korean People in Modern Times: 1800 to the Present, Newark, 1993, 110–24; Cook &
       Cook, 48; Drea, 232. Thanks to Dr. Kwan-sa You for translating this info from Kap-jae Cho: "Spit on
       My Grave—The Life of Park Chung-hee," Chosun Ilbo, Seoul, Korea, No. 104-16, 1998; Snow, 95.

628    Interview Shindo, 9 April 2018.

629    NIDSMDMA, Senshi-Sosho, Central Pacific, Op. #2, Peleliu, Angaur, Iwo, Vol. 13, 280-83.

630    Ibid, 340-41; Ross, 344; Henri, et al., 186; Newcomb, 188; Bix, 348, 351; Cook & Cook, 83-4;
       Drea, 204-5; Weinberg, 6; Email Saito to Rigg, 12 June 2019; US Army Infantry Human Research
       Unit, "Japanese Studies on Manchuria, Vol. XI, Part 3, Book A, "Small Wars and Border Problems:
       The Changkufeng Incident," 27, 88. The Battle of Lake Khasan is known as the Chokoho Incident
       in Japanese.

631    Wheeler, 41; Burrell, Ghosts of Iwo Jima, 42; Garand & Stobridge, 454-5; Harries & Harries, 339.

632    Frank, "Review Flamethrower," 25 May 2017; Toland, Vol. II, 823.

633    Kennedy, The Rise and Fall of Great Powers, 350; NACPM, 127 Box 14 (3d MarDiv), HQ, FMF,
       Intel Bulletin #4-44, Capt. Whipple, Disposition Japanese Ground Strength, 1; Bix, 396.

634    Wakabayashi (ed.), The Nanking Atrocity, 1937-38, Akira, "The Nanking Atrocity: An Interpretive
       Overview," 40.

635    Ibid., 48.

636    UGSL, Hosch Papers, Tokyo War Crimes Trials, Conventional War Crimes (Atrocities), Part B, Ch.
       VIII, 1023; Modder, 23.

637    Carroll, 123; Bradsher, "Thousands March in Anti-Japan Protest in Hong Kong"; Documents on the
       Tokyo International Military Tribunal Charter, 541; UGSL, Hosch Papers, Tokyo War Crimes Trials,
       Conventional War Crimes (Atrocities), Part B, Ch. VIII, 1023-4.

638    Hough, 20; Toll, Pacific Crucible, 248, 269.

639    Hong Kong's War Crimes Trials, Linton, "War Crimes," 98.

640    Zeiler, 3.

641    Documents on the Tokyo International Military Tribunal Charter, 538; Interview Yellin, 2 June
       2017. See newsreel "Why We Fight: The Battle for China (1944)." https://www.youtube.com/
       watch?v=2QvDk316BKo; https://archive.org/details/gov.archives.arc.36072; Takashi Yoshida, 23,
       28, 38-9; Toland, Vol. I, 289-90; F. Tillman Durdin, "All Captives Slain: Civilians Also Killed as
       Japanese Spread Terror in Nanking," NY Times, 18 Dec. 1937.

642    Patterson, 247; Mosley, 171.

643    Interview Lee, 4 April 2017; Notes by Woody on manuscript 25 June 2017; NASLPR, John Warner.

644    MCHDQV, Oral History, Cates, 52-3. MOH recipient and China Marine Mitchell Paige witnessed the same crimes in Shanghai in 1937 as Cates. Larry Smith, *Beyond Glory*, 10.

645    NACPM, RG 226, Box 368 (Records Off. US Strategic Services), Offices of Jap Forces, 8 July 1944, 3. When one asks the question why did America not do more to stop Japan's "ethnocide" in China since it knew about it, an answer that can be given is simply, the U.S. did not care. Historian Yuki Tanaka wrote: "It is believed that the racism against Asians prevalent among Westerners at that time was a major contributing factor to their indifference toward Japanese atrocities" (Tanaka, 254).

646    Toll, *Pacific Crucible*, 37.

647    Ibid., 46.

648    Edgerton, 314.

649    Weiner (ed.), *Race, Ethnicity and Migration in Modern Japan*, Weiner, "Introduction," 2.

650    Bradley, *Flyboys*, 316; "Japan PM Shinzo Abe Visits Yasukuni WW2 Shrine," BBC News, 26 Dec. 2013; Coogan, 439, 511; Burleigh, 561; Yuma Totani, *The Tokyo War Crimes Trial: The Pursuit of Justice in the Wake of World War II* (Cambridge, 2008), 135; Cook & Cook, 151, 405; "Japanese Minister Yoshitaka Shindo Visits Yasukuni Shrine Provoking China's Ire," South China Morning Post, 1 Jan. 2014; Keegan, 590.

651    Wakabayashi (ed.), *The Nanking Atrocity, 1937-38*, Wakabayashi, "The Messiness of Historical Reality," 13.

652    Hitchens, *god is not Great*, 203. See also *Hong Kong's War Crimes Trials*, Daqun, "Foreword," viii.

653    Chang, 12; Cook & Cook, 448.

654    Ibid.

655    Cook & Cook, 453; Drea, 50.

656    Email Dan King to Rigg, 22 Feb. 2018.

657    Interview Shindo, 9 April 2018.

658    "Japanese Minister Shindo Visits Yasukuni Shrine Provoking China's Ire," South China Morning Post, 1 Jan. 2014.

659    *Hong Kong's War Crimes Trials*, Daqun, "Foreword," viii.

660    Interview Woody, 23 July 2016.

661    Dower, 130; Bix, 10.

662    Victor Frankl, *Man's Search for Meaning: An Introduction to Logotherapy*, NY, 1963, 155; Cook & Cook, 14-5.

663    Edgerton, 241-2; NACPM, 127 Box 14 (3d MarDiv), HQ Div, FMF, Intel Bulletin #4-44, Capt. D. Whipple, Intel Technique on Attu, 3; Drea, 120; Bix, 251. Shirō Azuma said before he left for war, his mother gave him a dagger and said: "If, through some misfortune, you are captured by the Chinese troops, use this to cut your stomach open and die. I have three boys already, so if I just lose you, it won't bother me." Azuma then wrote, that instead of being horrified, that his mother's words made him happy and that he was impressed with what a wonderful mother she was. Moore, 77.

664    Edgerton, 310; Toland, Vol. I, 375.

665    NACPM, 208-AA-132T-23 & 25 (on the typewritten backs of the two photographs cited by the U.S. Authority, see 43150-FA and 43152-FA). On some of the photographs, the reference number is 208-AA-132S-25.

666    Ibid.; Smith, *Coral and Brass*, 104; Moore, 170.

667    Cook & Cook, 271-2, 278; *Soldiers Guide to the Japanese Army*, Military Intel. Service, War Dept., Wash. D.C., 15 Nov. 1944, 12.

668    Frank, *Downfall*, 29; Drea, 17-8, 45, 119-20, 172-3; Interview Drea, 17 Dec. 2017.

669    Kennedy, *Rise and Fall of Great Powers*, 298.

670   Dower, 248-9; Cook & Cook, 17.

671   John Toland, *The Rising Sun: The Decline and Fall of the Japanese Empire 1936-1945, Vol. 2*, NY, 1970, 798, 808; King, 96.

672   Henri, et al., 8; Newcomb, 41; Hane, 10; The word *Kamikaze* comes from historical storms that saved Japan from invasions. In 1274 and 1282, typhoons destroyed powerful enemy Mongolian fleets of Kublai Kahn bent on invading and conquering Japan. Japanese referred to it as the Divine Wind—i.e. *Kamikaze*. See Coogan, 436.

673   Dante, *Inferno*, Canto III 21, 9.

674   Maslowski, and Millett, 404; Frank, "Review *Flamethrower*," 25 May 2017.

675   Maslowski, and Millett, 316.

676   Ibid., 318.

677   Harries & Harries, 210; Drea, 191; Mosley puts the attacking army's size at 160,000 which was too low. Mosley, 171; Daqing Yang, "Diary of a Japanese Army Medical Doctor, 1937," in *Researching Japanese War Crimes Records*, ix; Toland, Vol. I, 63; JNMS, Vol. 1, No. 1, Sheng, "The Seven Why-Questions in the Writing of History," 14; Wakabayashi (ed.), *The Nanking Atrocity, 1937-38*, Akira, "The Nanking Atrocity: An Interpretive Overview," 32.

678   Interview Yellin, 6 July 2017; Dallek, *Franklin D. Roosevelt*, 444; NMPWANM, 13 July 2017.

679   Drea, 259.

680   Ibid., 262.

681   Sam Harris, *The End of Faith: Religion, Terror, and the Future of Reason*, NY, 2004, 134.

682   Scott, 453.

683   Dante, *Inferno*, XV 137, 78.

684   Ibid., XV 135, 67-69.

685   TSHACNCDNM, Vol. 10, 263, Statement Keenan, Chief of Counsel, International Military Tribunal Far East, 4 June 1946, Encl. 4.

686   *The Teaching of Buddha*, Buddhist Promoting Foundation, Tokyo, 1992, 130.

687   Interview Woody, 22 Jan. 2018.

688   NASLPR, Woody Williams, Service File, 31 & Notice Woody completed Inf. Rifleman School, 5 Nov. 1943.

689   Interview Woody, 20 July 2016; Interview Woody, 17 March 2017; MCHDHQ, Woody, "Destiny…Farm Boy," 10-11; Smith, *Iwo Jima*, 62; NASLPR, Woody Williams, Medical History 12-211115, 2.

690   Kevin Charles Murray, *Sgt. A. F. "Kelly" Murray U.S.M.C.: A Hoosier Hibernian In The Great Pacific War*, Bloomington, IN, 2011, 34-6; Interview 24 June 2016.

691   Interview Woody, 20 July 2016; MCHDHQ, Woody, "Destiny…Farm Boy," 11.

692   Berry, 122; Hoffman, *Once a Legend*, 151.

693   Interview Woody, 20 July 2016; O'Brien, *Liberation*, 6; James H. Hallas, *Uncommon Valor on Iwo Jima: The Stories of the Medal of Honor Recipients in the Marine Corps' Bloodiest Battle of World War II*, Mechanicsburg, PA, 2016, 77.

694   Toll, *Conquering Tide*, 232, 235; Conrad, 931.

695   Edgerton, 286.

696   MCHDQV, Oral History, Robert E. Cushman Jr., 121-22. Cushman also called it a "terrible volcanic upheaval" that "rattled you in your foxhole like a pea in a pod." MCHDQV, Oral History, Robert E. Cushman Jr., 154.

697   *Third Marine Division's Two Score and Ten History*, authored and produced by Third Marine Div. Assoc., Inc., Paducah, KT, 1992, 8, 19. This work will be cited hereafter as *Two Score and Ten*.

698   Morison, 187.

699   NACPM, RG 127, Box 90, 21st Marines, Reinforced, 8 July 42-3 Nov. 43, Personnel.

700   Ibid., Box 84 (Iwo), Actual & Authorized Strength, 3d MarDiv, G-1 Periodic Report, 18-19 Feb. 1945.

701     *Two Score and Ten*, 8, 19.

702     Letter Dashiell to Vivian Dashiell, 6 Sept. 1945, 3.

703     MCHDHQ, Letter Beck to Ruth Beck, 18 Aug. 1943.

704     Interview Woody, 21 July 2016; Frank, "Review *Flamethrower*," 25 May 2017.

705     MCHDQV, Dick Dashiell File, #280, Lawrence C. Skiba, Silver Star Action.

706     Interview Woody, 19 July 2016. Newcomb, 209.

707     MCHDQV, 2nd Btl., 21st Marines, 3d MarDiv, FMF, In Field, G.A. Percy, 12 April 1945, 15; NACPM, RG 127, Box 83, (Iwo), AB-1, Com. Gen. 3d MarDiv Erskine to Commandant, Report, 3 June 1945, 36 & Box 90, 21st Marines, Reinforced, 8 July 42-3 Nov. 43, Personnel.

708     Sledge, 118; Letter Woody to Rigg, 16 Nov. 2017.

709     MCHDHQ, Woody Williams, "Looking Back On Earlier Times," Undated, 2; Interview Woody, 19 July 2016.

710     Michael Dolan, "Conversation: Forged in Flame," WWII Magazine, 15 March 2016.

711     Mountcastle, 8-18; Brophy, Miles and Cochrane, 139.

712     Brophy, Miles and Cochrane, 146; Edgerton, 262; NACPM, Photograph Archives, RG 111-SC 334278; Mountcastle, 81; NACPM, 127 Box 14 (3d MarDiv), HQ, FMF, Intel Sec., Subj.: Sec. I. Translation Japanese Documents. (a) Flamethrower Model 93 Small, Lt. Col. Coleman, 6 March 1943, 1; Cook & Cook, 270; Ferguson, 208; Crowl & Isely, 140.

713     Brophy, Miles and Cochrane, 139; Mountcastle, 50, 82.

714     Interview Woody, 21 July 2016; Brophy, Miles and Cochrane, 140; Dolan, "Conversation: Forged in Flame."

715     Henri, et al., 204.

716     Paul Hoolihan, "Flame Throwers Class at 21st Teaches the Art of Roasting Japs," 21st Regt. Newsletter, May 1945.

717     NACPM, RG 127 Box 21, (4th MarDiv), (1540) (1 of 2) Folder 5, HQ, 4th Tank Bn., Subj.: Air Compressors-Flame Throwers, 16-17 April 1945, Lt. Col. R. K. Schmidt; NACPM, RG 127 Box 21 (4th MarDiv), (1540) (1 of 2) Folder 5, HQ, 24th Marines, 4th MarDiv, FMF, Com. officer Jordan to Com. Gen., 4thMarDiv 19 March 1945, 2.

718     Interview Woody, 19 July 2016; MCHDHQ, Woody Williams, "Looking Back On Earlier Times," Undated, 2; NACPM, RG 127 Box 51 (20th, 21st & 22nd Rgts), Co. "C", 1st Btl., 20th Marines (Engrs), 4th MarDiv, FMF, Pendleton, Oceanside CA, 2nd Lt. Robert B. Reynolds to Com. officer 20th Marines (Engrs)., 3 Jan. 1944, 1 & Box 329 (Saipan-Tinian), A14-1, 4th MarDiv Operations Report 15 June to 9 July, HQ 24th Marines, 4th MarDiv, FMF, Brig. Gen. F.A. Hart, 28 Aug. 1944, (5) Fuel Flamethrowers, (a); Dolan, "Conversation: Forged in Flame."

719     MCHDQV, HQ, 2d Btl., 21st Marines, 3d MarDiv, FMF, In Field, G.A. Percy, 12 April 1945, 20.

720     Interview Woody, 19 July 2016; MCHDHQ, Woody Williams, "Looking Back On Earlier Times," Undated, 2. Woody claims the fuel mixture lasted 72 seconds, but if the tanks only held 4.5 gallons and the weapon fired half a gallon a second, then there was no way the tank had 72 seconds of fuel once the triggered was pulled. Interview Graves, 4 July 2018.

721     MCHDQV, Oral History, Cates, 17.

722     Sledge, 36.

723     Interview Woody, 21 July 2016; Smith, *Iwo Jima*, 68.

724     Dolan, "Conversation: Forged in Flame."

725     Manchester, 377.

726     Mountcastle, 33; Email Mountcastle to Rigg, 13 Feb. 2018; Mountcastle, 34.

727     Mountcastle, 50.

728     Ibid., 52.

729     Ibid., 82.

730     Ibid., 88.

731   Mountcastle, 89; Email Mountcastle to Rigg, 13 Feb. 2018.

732   Mountcastle, 89.

733   NASLPR, File Woody Williams, Duration of National Emergency, 854310 & Professional Conduct Record; MCHDHQ, Letter Woody to Ruby Meredith, 16 March 1944.

734   MCHDHQ, Letter Woody to Ruby, 12 April 1944.

735   Ibid.

736   Interview Woody, 5 July 2017; Letter Woody to Rigg, 14 July 2017. Woody became so upset that I used this word for Ruby, he threatened to stop working with me in 2017.

737   MCHDHQ, Letter Woody to Ruby, 12 April 1944.

738   NASLPR, Eleanor G. Pyles, Enlistment Photo & Professional and Conduct Record of Pyles, 4; World War II Young American Patriots, 1941-1945, WV, Marion County, 442.

739   Interview Woody, 19 June 2017.

740   Ibid.

741   MCHDHQ, Letter Beck to Ruth Beck, 3 Dec. 1944.

742   Ibid., 18 Jan. 1945 (Commentary Mark Beck).

743   Ibid., 18 Dec. 1944.

744   Ibid., 18 Jan. 1945.

745   Ibid., 7 Jan. 1945.

746   Ibid., 18 Jan. 1945.

747   MCHDHQ, Undated Letter, Woody to Ruby, cir. Summer 1944.

748   MCHDHQ, Letter Woody to Ruby, 17 Nov. 1943.

749   Ibid., 24 Dec. 1943.

750   Manchester, 262.

751   Toland, Vol. II, 647; Hoffman, Once a Legend, 263.

752   Mandel, 250.

753   Interview Woody, 23 July 2016.

754   NACPM, 127 Box 14 (3d MarDiv), HQ, FMF, In Field, Restricted, Training of Btl. and Rgt. Intel Personnel, Col. Robert Blake, Chief-of-Staff of Maj. Gen. Turnage, 25 Sept. 1945, 7-8.

755   Ibid.

756   Ibid.; Dick Camp, Leatherneck Legends: Conversations with the Marine Corps' Old Breed, St. Paul, 2006, 124-5.

757   Cook & Cook, 25.

758    KJV (King James Version), p. 120 for Exodus 20: 13 and p. 1404 for Matthew 5:44 .

759   Johnson, The History of the American People, 75-80, 270-274; Hoffman, Once a Legend, 231; LaFeber, 426.

760   Mosley, 154.

761   Marine Corps Recruit Depot, San Diego, Archive (MCRDSDA), MCRD Training Book, File 1, Wearing Apparel.; LaFeber, 427.

762   NACPM, 127 Box 14 (3d MarDiv), HQ, FMF, Camp Elliott, Subj.: Jap. Military Org., 29 Oct. 1942, 3.

763   Thompson, Doud & Scofield, 20; Soldiers Guide to the Japanese Army, Mil. Intel Serv., War Dept., D.C., 15 Nov. 1944, 5.

764   MCHDHQ, Letter Beck to Ruth Beck, 13 June 1943. Copies of this file can also be found in the Pritzker Military Library.

765   Ibid., 29 March 1944.

766   MCHDHQ, Mark Beck, "Bio of Donald Beck," Dec. 2002, 5-6.

767   Edgerton, 286; Interview Lee, 23 Nov. 2015.

768   Interview Lee, 23 Nov. 2015 & 17 March 2017; Email Frank to Rigg, 14 Aug, 2017; Lee Mandel, "Combat Fatigue in the Navy and Marine Corps During WWII," Navy Medicine, July-Aug., 2001;

NACPM, RG 127, Box 83 (Iwo), Aa-26.1 [1 of 2], Iwo, HQ VAC Landing Force in Field, 3 March 1945, Report of casualties evacuated to ships (Report #11), Name D.E. Lee, Serial Number 813957, Ship APA-33 *USS Bayfield*; NACPM, RG 24, Bureau of Naval Personnel, Causality Branch, *USS Bayfield*; https://www.usmcu.edu/historydivision/casualty-card-database (hereafter known as USMCU Casualty Web Data Base).

769    Interview Lee, 23 Nov. 2015.

770    Toland, Vol. II, 552.

771    Bradley & Powers, *Flags of Our Fathers*, 87.

772    Interview Lee, 14 July 2016. See also Moore, 213.

773    Interview Woody, 19 July 2016; Edgerton, 287; *Soldiers Guide to the Japanese Army*, Mil. Intel Service, War Dept., D.C., 15 Nov. 1944, 12; Moore, 210.

774    Interview Gene Rybak, 15 Mary 2019; Zack Rybak, "Herefrom," Narrative Magazine, June 2019.

775    Interview Lee, 23 June 2016.

776    Dower, 64-5; Newcomb, 240; Blum, 84; Burrell, *Ghosts of Iwo Jima*, 238, fn. 37.

777    Jan K. Herman, *Battle Station Sick Bay: Navy Medicine in World War II*, Annapolis MD, 1997, 113.

778    Toll, *The Conquering Tide*, 196; NACPM, RG 127 Box 96 (Iwo) A45-2, 21st Marine Rgt. Journal Iwo, 12 March 1945, R-3 Log, Time 1220; NACPM, RG 127 Box 96 (Iwo) A45-2, 21st Marine Rgt. Journal Iwo, 20 March 1945, R-3 Log, Sent 1900.

779    Newcomb, 35.

780    Interview Skinner, 17 March 2015.

781    Keith A. Renstrom, *Keith A. Renstrom Recounts Battle in Tinian and Awards*, 30 Aug. 2017; Interview Keith Renstrom by John Renstrom, 20-23 June 2016.

782    Interview Sigmund L. Liberman, 10 June 2013. The Nazis murdered around 20,000 people at Mittlebau-Dora.

783    Michael Hirsh, *The Liberators: America's Witnesses to the Holocaust*, 2010, 196.

784    Van Stockum, "Battle of the Philippine Sea," 2017, 2; Cook & Cook, 262, 267; Aurthur and Cohlmia, 137-9; MCHDQV, Oral History, Robert E. Cushman Jr., 158-9.

785    Drea, 238.

786    *Johnson, The American People, 409.*

787    Steinberg, *Island Fighting*, 166.

788    *Rogers, 116; Interview James Oelke Farley, 17 March 2018.*

789    *Farrell, Saipan, 36.*

790    Crowl & Isely, 200; Toll, *The Conquering Tide*, 436.

791    O'Brien, *The Second Most Powerful Man in the World*, 284-5.

792    Robert Burrell, *Ghosts of Iwo Jima*, Bryan TX, 2006, 26; Newcomb, 20; O'Brien, *The Second Most Powerful Man in the World*, 298.

793    Burrell, *Ghosts of Iwo Jima*, 25.

794    Morison, 157; Farrell, *Saipan*, 41-2; Millett, xviii.

795    When MacArthur invaded Leyte Gulf, he designated it A-Day to delineate it from D-Day which the public associated with Normandy. Toland, Vol. II, 675.

796    Farrell, *Liberation-1944*, 13.

797    Krulak, 30.

798    Camp, 118; Hoffman, *Once a Legend*, 279.

799    Camp, 121.

800    Farrell, *Liberation-1944*, 13-4; Steinberg, *Island Fighting*, 166; Toll, *The Conquering Tide*, 511; Hoffman, *Once a Legend*, 279; Frank, *Downfall*, 306.

801    Interview Woody, 23 July 2016.

802    This is a reference for someone who has seen combat, referring to Hannibal's attack on Rome over

the Pyrenes in 218 B.C.E when he led his troops with a host of elephants. When the Roman soldiers engaged him in combat, they were confronted with fighting elephants and hence came the question soldiers ask one another ever since: "Did you see the Elephant," meaning, did you see combat?

803    MCHDHQ, Letter Woody to Ruby Meredith, 17 Nov. 1943.

804    MCHDHQ, Letter Beck to Ruth Beck, 1 Sept. 1943.

805    Ibid., 12 Jan. 1944.

806    Interview T. Fred Harvey, 20 March 2015.

807    Edgerton, 308-9.

808    Samuel Eliot Morison, *History of the United States Naval Operations in World War II, Vol. VIII*, NY, 1975, 149; *Robert F. Rogers,* Destiny's Landfall: A History of Guam, Univ. of Hawaii Press, *1995, 116;* Jon T. Hoffman, *Once A Legend: Red Mike Edson of the Marine Raiders*, Novato, 1994, *38-9.*

809    Toll, *Conquering Tide*, 459.

810    O'Brien, *The Second Most Powerful Man in the World*, 16.

811    Morison, 377; *Families in the Face of Survival*, 126. Gregorio S.N. Aguon, Nicholas S.N. Fergurgur, Francisco Reyes Mafnas, Vicente Gegue Meno, Jose Sanches Quinata and Francisco Umpingco Rivera died on the *USS Arizona*; Andreas F. Mafnas died on the *USS Nevada*; Ignacio C. Farfan and Jesus F. Garcia died on the *USS Oklahoma*; and Jose S.N. Flores, Jesus M. Mata and Enrique C. Mendiola died on the *USS West Virginia*.

812    Morison, 151.

813    National Park Service, War in the Pacific: Outbreak of the War accessed 10/22/16 at https://www.nps.gov/parkhistory/online_books/npswapa/extContent/wapa/guides/outbreak/sec6.htm    and    /sec.3.htm.

814    *Families in the Face of Survival*, 126; Hough, 20.

815    *Families in the Face of Survival*, 126; Camp, 120.

816    Hough, 21; Thomas Wilds, "The Japanese Seizure of Guam", *Marine Corps Gazette*, July, 1955.

817    Moskin, 236; Farrell, *Liberation-1944*, 9.

818    Farrell, *Liberation-1944*, 9; Edgerton, 294.

819    Morison, 373; Hough, 284.

820    Farrell, *Liberation-1944*, 27; *Guam War Claims Review Commission*, 5.

821    MCHDQV, Oral History, George Burledge, 9-11; Toland, Vol. II, 748-9.

822    Hough, 282; MARC, Guam War Reparations Commission (GWRC), Olivia L.G. Cruz Abell, Box 1, Rec. 2090 & Francisco Q. Leon Guerrero, Box 20, 39620 & Ignacio T. Leon Guerrero, Box 20, Rec. 39680; *Families in the Face of Survival*, 127; *Guam War Claims Review Commission*, 5; National Park Service, A Guide to the War in the Pacific.

823    Wakabayashi (ed.), *The Nanking Atrocity, 1937-38*, Akira, "The Nanking Atrocity: An Interpretive Overview," 48 & Tokushi, "Massacres Outside Nanking City," 66.

824    Hough, 282; MARC, Guam War Reparations Commission (GWRC), Olivia L.G. Cruz Abell, Box 1, Rec. 2090 & Francisco Q. Leon Guerrero, Box 20, 39620 & Ignacio T. Leon Guerrero, Box 20, Rec. 39680; *Families in the Face of Survival*, 127; *Guam War Claims Review Commission*, 5; National Park Service, A Guide to the War in the Pacific.

825    MARC, GWRC, Francisco Acfalle, Box 1, Rec. 2230 & Ignacio T. Leon Guerrero, Box 20, Rec. 39680.

826    Wakabayashi (ed.), *The Nanking Atrocity, 1937-38*, Akira, "The Nanking Atrocity: An Interpretive Overview," 49.

827    National Park Service, A Guide to the War in the Pacific.

828    MARC, GWRC, Frutuoso S. Aflague, Box 1 & Ignacio T. Leon Guerrero, Box 20, Rec. 39680; National Park Service, A Guide to the War in the Pacific; *Guam War Claims Review Commission*, 6.

829    Hough, 284; MARC, GWRC, Edward L.G. Aguon, Box 1 & Felix Aguon, Box 1; Flores, Camacho 2, Palacios 6; National Park Service, *A Guide to the War, in the Pacific*.

830    MARC, GWRC, Fausto Acfalle, Box 1, Rec. 120.

831    *Guam War Claims Review Commission*, 5-6; MARC, Guam War Reparations, Jesus Leon Guerrero, Box 20 & Francisco Q. Leon Guerrero, Box 20, 39620 & Diana Leon Guerrero, Box 20, 39502.

832    *Guam War Claims Review Commission*, 5; Email Farrell to Rigg, 22 Aug. 2016; Farrell, *Liberation-1944*, 3, 27-53; Interview James Farley Oelke, 25 March 2018; MARC, GWRC, Olivia L.G. Cruz Abell, Box 1, Rec. 2090 & Maria T. Abrenilla, Box 1, Rec. 40 & Felix Cepeda Aguon, Box 1 & Dolores F. Aguon, Box 1, Rec. 380; *Families in the Face of Survival*, 151.

833    Farrell, *Liberation-1944*, 27.

834    Hough, 284.

835    Farrell, *Liberation-1944*, 12; Hough, 284.

836    Farrell, *Liberation-1944*, 7, 27-53.

837    Morison, 373; Hough, 284; MARC, GWRC, Olivia L.G. Cruz Abell, Box 1, Rec. 2090; MARC, GWRC, Edward L.G. Aguon, Box 1. The concentration camps were Atate, Malojloj, Mata, Manenggon, Asinan, Maimai and Tai.

838    Moskin, 335.

839    Hane, 362.

840    Moskin, 312; Steinberg, *Island Fighting*, 166; Spector, 301; Farrell, *Liberation-1944*, 15.

841    U.S. Marine Corps Publication, "Expeditionary Operations," MCDP 3, 16 April 1998, 95.

842    Ibid., 95-6.

843    Morison, 344. These were called "fleet trains" and America created an incredible network of supply ships to get supplies to its troops and forward operational vessels. O'Brien, *The Second Most Powerful Man in the World*, 117.

844    Ibid.

845    Toll, *Conquering Tide*, 457.

846    Farrell, *Liberation-1944*, 15.

847    Farrell, *Saipan*, 45; Toland, Vol. II, 615; James H. Hallas, *Saipan: The Battle that Doomed Japan in World War II*, Lanham, MD, 2019, 99.

848    Lyne Olson, *Those Angry Days: Roosevelt, Lindbergh, and America's Fight Over World War II, 1939-1941*, NY, 2013, 28-9.

849    BA-MA, RH 67/53, Deutsche Botschaft Militär- und Luftattaché D.C., Bericht 28/40, Beurteilung eines vollen Warenembargos der Vereinigten Staaten gegenüber Japan, 4 Oct. 1940, 5.

850    By 1941, there were 500,000 personnel in the U.S. Armed Forces according to the WWII Museum in New Orleans.

851    Rigg, *Hitler's Jewish Soldiers*, 127; NACPM, RG 165, Box 2152, War Dept., 6906 (Japan), Gen. Distribution Japanese Forces, 1 Nov. 1941; Tanaka, 250.

852    Lord Russell of Liverpool, xv.

853    Kennedy, *The Rise and Fall of Great Powers*, 343.

854    Prange, 10-1, 736-7.

855    Paul Kennedy, *The Rise and Fall of the Great Powers*, NY, 1987, 331-2; Maslowski, and Millett, 389; Edgerton, 257; Patterson, 262. Even with reservations, once he learned of the success at Pearl Harbor, Yamamoto got caught up in the enthusiasm and rejoiced with *sake* and *surume* (dried squid) giving numerous toasts. Toland, Vol. I, 279.

856    Toland, Vol. I, 190.

857    Ross, 20; Newcomb, 8; McNeill, "Reluctant Warrior," 39; Edgerton, 256-7; Kakehashi, 112; Bradley & Powers, *Flags of Our Fathers*, 148.

858    Paul Johnson, *A History of the American People*, NY, 1997, 780.

859    John Myers Myers, *The Alamo*, London, 1948, 39.

860    Gregory Wallance, *American's Soul in the Balance: The Holocaust, FDR's State Department, and the Moral Disgrace of an American Aristocracy*, Austin, 2012, 13.

861    Frank, "Review *Flamethrower*," 25 May 2017.

862    NMPWANM, 13 July 2017.

863    Wallance, 202.

864    Anthony Newpower's *Iron Men and Tin Fish: The Race to Build a Better Torpedo During World War II*, Santa Barbara, CA, 2006; Zeiler, 5; O'Brien, *The Second Most Powerful Man in the World*, 169. .

865    Paul Kennedy, *Engineers of Victory*, NY, 2013, 294.

866    <u>Ibid.</u>

867    Toll, *Conquering Tide*, 523; Cook & Cook, 33-4, 125.

868    Cook & Cook, 325.

869    Victoria, *Zen at War*, 132.

870    Toll, *Conquering Tide*, 532.

871    Steinberg, *Island Fighting*, 74; Edgerton, 287; *Japan at War*, 125; Hane, 341. Frank, "Review *Flamethrower*," 25 May 2017. On learning of Yamamoto's death, Admiral Nimitz described his enemy as "a tenacious and shrewed foe who did his job, sometimes too well to suit me." Ross, 11.

872    Moskin, 316; Steinberg, *Island Fighting*, 166; Frank, *Downfall*, 216.

873    Spector, 314; Hough, 3; Crowl & Isely, 62; Millett, 325-6, 373.

874    Burrell, *Ghosts of Iwo Jima*, 51. Crowl & Isely, *The U.S. Marines and Amphibious War*.

875    Crowl & Isely, v, 3; Smith, *Coral and Brass*, 22-3; Krulak, 82-3.

876    NASLPR, Holland M. Smith, Monthly Chronological Data 1917-1918. Even though Smith did not like many U.S. Army officer, he did respect General Harbord that he was one of America's "great soldiers." Smith, *Coral and Brass*, 32.

877    NASLPR, Holland M. Smith, Synopsis of Military History of Lieutenant Colonel Holland M. Smith, USMC, 13 Feb. 1933, 2.

878    <u>Ibid.</u>, Monthly Chronological Data, March 1919, p. 1-2.

879    Kyle Daly, "The Legacy of Holland M. Smith," Leatherneck Magazine, Oct. 2018, 14-20; Smith, *Coral and Brass*, 32.

880    Kennedy, *Engineers of Victory*, 219, 308; Crowl & Jeter, *The Marines and Amphibious War*; Paterson (ed.), *Major Problems in American Foreign Policy*," John Braeman, "American Military Power and Security," 154; Coram, 50, 73; MCHDQV, Oral History, Victor H. Krulak, 41.

881    Hough, 4.

882    <u>Ibid.</u>, 3-4.

883    Hoffman, *Once a Legend*, 134-6; Crowl & Isely, 66.

884    Coram, 69-71; Richard Goldstein, "Victor H. Krulak, Marine Behind U.S. Landing Craft, Dies at 95," NYT, 4 Jan. 2009; Millett, 340. Krulak was the assistant intelligence officer attached to the 4th Marines in China. In 1937, he personally observed the Japanese outside Shanghai at Woosung make amphibious landings with ramped boats like the Higgins boats. His observations and reports helped push General Smith more to develop and support the Higgins boats of WWII. Hoffman, *Once a Legend*, 118; Krulak, 90; MCHDQV, Oral History, Victor H. Krulak, 22, 67.

885    Email Charles C. Krulak to Rigg, 5 Dec. 2019. Ironically, Krulak got his ideas about how to help Higgins develop the amphibious boat by analyzing how the Japanese utilized their own outside of Shanghai. General Charles C. Krulak said that turning this technology against them, his father helped the Corps use "A Japanese system to defeat…the Japanese." Email Charles C. Krulak to Rigg, 14 Dec. 2019.

886    Email Charles C. Krulak to Rigg, 14 Dec. 2019; Victor H. Krulak, *Japanese Assault Boats, Shanghai, 1937*, CMP Productions, 2017; Krulak, 90-4; MCHDQV, Oral History, Victor H. Krulak, 23-5, 45, 54-56.

887    Arthur Herman, *Freedom's Forge: How American Business Produced Victory in World War II*, NY, 2013, 204-6; Smith, *Coral and Brass*, 44-7.

888    Smith, *Coral and Brass*, 45; Burrell, *Ghosts of Iwo Jima*, 19.

889    Kennedy, *Engineers of Victory*, 311

890    Smith, *Coral and Brass*, 13.

891    Johnson, *The History of the American People*, 780; Smith, 13, 52; Millett, 361. Lieutenant General Holland M. Smith cites the number of 599,693 Marines serving during the Second World War in his book mentioned here. Since Smith's memoir is not known for its historical accuracy and does not cite its sources, this number is not used. Johnson cites the total number of Marines during WWII being 669,100 and cites a reputable source for his figures in L.P. Adams, *Wartime Manpower Mobilization*, NY, 1951. See Johnson, *The History of the American People*, 1039 fn. 150. As a result, Johnson's number will be used.

892    Hough, 5; Crowl & Isely, 202.

893    Crowl & Isely, 4.

894    Krulak, *First to Fight*, 36. J. F. C. Fuller, a noted British military historian, "described the amphibious assault as perfected and practiced by the U.S. Marines as 'in all probability…the most far reaching tactical innovation of the war.'" Krulak, 87. See Krulak 100-9 and MCHDQV, Oral History, Victor H. Krulak, 49-54 about the amtracs or "alligators."

895    MCHDQV, Oral History, Erskine, 203-4; Hoffman, *Once a Legend,* 136; MCHDQV, Oral History, Victor H. Krulak, 61; Ross, 39.

896    Smith, *Coral and Brass*, 140; *Two Score and Ten*, 5.

897    National Archives, St. Louis Personnel Files, (NASLPR), Graves Erskine; MCHDQV, Oral History, Erskine from 1969-1970, 2, 15, 51, 56, 67. Ross incorrectly says Erskine had blue eyes and was under 6'0". When one studies Erskine's personal file, one learns he had green eyes and stood at 72 inches. See Ross, 38.

898    Hoffman, *Once a Legend,* 306-8.

899    Hough, 12.

900    Crowl & Isely, 6.

901    Crowley, 35.

902    Toll, *The Conquering Tide*, 459; NACPM, RG 127 Box 318 (Saipan-Tinian) HQ Northern Troops and Landing Force, Marianas Phase I (Saipan) 1. Civ. Affairs Report 2. Liaison Officers Report 2. Public Relation Report, 9 Nov. 1944, HQ VAC, Corps Civil Affairs Officer to Com. Gen., 13 Aug. 1944, Donald T. Winder, 3.

903    Toll, *The Conquering Tide*, 459; Smith, *Coral and Brass*, 88; NACPM, RG 127 Box 328 (Saipan-Tinian), 4th MarDiv Ops. Report—Saipan, Annex B, Intel, E. Enemy, 23; NASLPR, Gen. Erskine, Record Graves Blanchard Erskine, #0268, Card #11 or 12, Note Jan. 1945; Toland, Vol. II, 611; Hallas, *Saipan*, 55.

904    Hallas, *Saipan*, 87.

905    Crowl & Isely, 330.

906    Hallas, *Saipan*, 79.

907    Toland, Vol. II, 615-6.

908    Hallas, *Saipan,* 86.

909    NACPM, RG 127 Box 318 (Saipan-Tinian) HQ Northern Troops and Landing Force, Marianas Phase I (Saipan) 1. Civ. Affairs Report 2. Liaison Officers Report 2. Report, 9 Nov. 1944, HQ VAC, Corps Civil Affairs Off. to Com. Gen., 13 Aug. 1944, Donald T. Winder, 3; Toll, *Conquering Tide*, 508. Some citizens leaving Saipan in March 1944 found their ship under attack and sunk by U.S. submarines. Possibly up to 1,700 Japanese citizens died on 6 March when the *USS Nautilus* sank the *Amerika Maru*. Hallas, *Saipan*, 25-6.

910    Toll, *Conquering Tide*, 508.

911    Moskin, 328; Interview Woody, 20 July 2016; Email Farrell to Rigg, 23 Aug. 2016.

912     Toland, Vol. II, 613. William Manchester mentions this scene in his book, but in typical fashion, he changes the information and misquotes Toland. Manchester, 253.

913     Moskin, 314; C. J. Sulzberger, The American Heritage Picture History of WWII (American Heritage Publishing Co. 1966) 536; Spector, 302, 304; Sledge, 48; Hoffman, *Once a Legend*, 284-5; Toland, Vol. II, 615.

914     Bix, 364; Edgerton, 284; Wallace and Williams, 23, 81; Justin McCurry, "Japan Unearths Site Linked to Human Experiments," The Guardian, 21 Feb. 2011; Bradley, *Flyboys*, 113; Bradley & Powers, *Flags of Our Fathers*, 59-60; Frank, *Downfall*, 324-5; Toland Vol. II, 612-3; Crowl & Isely, 336; Hallas, *Saipan*, 60-1; *Documents on the Tokyo International Military Tribunal Charter*, 539-40.

915     MCHDQV, Oral History, Erskine, 162-3.

916     Hiromichi Yahara, *The Battle for Okinawa: A Japanese Officer's Eyewitness Account of the Last Great Campaign of World War II*, NY, 1995, 4.

917     Ibid.

918     Charles P. Neimeyer, "Paladins at War: The Battle for Saipan, June 1944," Marine Corps Gazette, Vol. 78, Issue, 7 July 1994.

919     Moskin, 316.

920     Cook & Cook, 262; Scott, 49; Crowl & Isely, 183. See Kenneth P. Werrell's *Blankets of Fire* for a description of aerial dropping of mines.

921     Lecture Richard Frank on Saipan, 21 March 2018; Hallas, *Saipan*, 25-6, 28-9, 30-1.

922     Morison, vol. VIII, 216.

923     Toll, *Conquering Tide*, 470.

924     Ibid., 451.

925     Ibid., 469.

926     Morison, vol. VIII, 232.

927     Ibid., 242.

928     Smith, *Coral and Brass*, 90.

929     Hallas, *Saipan*, 193.

930     Frank, *Downfall*, 89.

931     Morison, vol. VIII, 257

932     Frank, 89.

933     Maslowski, and Millett, 444; Shores, 205; Spector, 310; Drea, 239; Smith, *Coral and Brass*, 91; Van Stockum, "Battle of the Philippine Sea," 2017, 9; Hough, 235; Toland, Vol. II, 631.

934     Maslowski, and Millett, 444; Spector, 310; Keegan, 307.

935     Morison, vol. VIII, 233.

936     Toll, *Conquering Tide*, 452.

937     Morison, vol. VIII, 276.

938     Ibid., 321

939     Manchester, 267.

940     Ibid.

941     NACPM, 127 Box 14 (3d MarDiv), HQ, FMF, Intel Sec., Subj.: Combat Intel., Lt. Col. Turton 27 Aug. 1943, 5.

942     Toland, Vol. II, 635.

943     Edgerton, 284; Cook & Cook, 288, 329; Smith, *Coral and Brass*, 106.

944     Hough, 236.

945     Smith, *Coral and Brass*, 99.

946     MCHDQV, HQ, 3d MarDiv, FMF, In Field, Report Turton (D-3), 19 Aug. 1944, 8; Harold Goldberg, *D-Day in the Pacific: The Battle of Saipan*, Indiana Univ., 2007, 167–94; J. Robert Moskin, *U.S. Marine Corps Story*, NY, 1992, 329; Spector, 316-20; Aurthur and Cohlmia, 152,

154; Edgerton, 289; Toll, 506; Hough, 244-5; O'Brien, *Liberation*, 27-8; Toland, Vol. II, 643; Millett, 414.

947    Hough, 245; Cook & Cook, 357.

948    Toll, *Conquering Tide,* 505.

949    Toland, Vol. II, 642.

950    Ibid., 643.

951    Hough, 245.

952    NACPM, RG 127 Box 329 (Saipan-Tinian), RCT 23 Report Forager Phase I Saipan, Enclosure (A)., 10, 12.

953    RG 127 Box 318 (Saipan-Tinian) HQ Northern Troops Landing Force, Marianas Phase I (Saipan) 1. Civ. Affairs Report 2. Liaison Off. Report 2. Public Rel. Report, 9 Nov. 1944, HQ VAC, Corps Civ. Affairs Off. to Com. Gen., 13 Aug. 1944, Donald T. Winder, 3; Hoffman, *Once a Legend,* 220.

954    Manchester, 270-1; Toland, Vol. II, 640.

955    NACPM, 127 Box 14 (3d MarDiv), HQ, FMF, Intel Bulletin #4-44, Capt. Whipple, American Soldier, 3, 6-7.

956    Burrell, *Ghosts of Iwo Jima*, 47.

957    NACPM, RG 127 Box 329 (Saipan-Tinian), RCT 23 Report Forager Phase I Saipan, Enclosure (A), D. Propaganda Sec. 2; Cook & Cook, 359-65; Hallas, *Saipan*, 429.

958    NACPM, RG 127 Box 329 (Saipan-Tinian), A14-1, 4th MarDiv Report 15 June to 9 July, HQ 24th Marines, 4th MarDiv, FMF, Brig. Gen., F.A. Hart, 28 Aug. 1944 & RCT 23 Report Forager Phase I Saipan, Enclosure (A)., 7.

959    Moskin, 329; Dower, 45; Toland, Vol. II, 640; Steinberg, *Island Fighting*, 170-1; Frank, *Downfall*, 29; Melville, 19; Hane, 345; Cook & Cook, 289-92; Smith, *Coral and Brass*, 107; Hough, 246; Toland, Vol. II, 647-8.

960    NACPM, RG 127 Box 329 (Saipan-Tinian), RCT 23 Report Forager Phase I Saipan, Enc. (A)., 9-10.

961    Spector, 317; Frank, *Downfall*, 29; Toland, Vol. II, 648.

962    Interview Keith Renstrom by John Renstrom, 20-23 June 2016.

963    Spector, 317-8.

964    Cook & Cook, 289, 342; Toland, Vol. II, 648.

965    NACPM, RG 127 Box 329 (Saipan-Tinian), A14-1, 4th MarDiv Report 15 June to 9 July, HQ 24th Marines, 4th MarDiv, FMF, Subj.: Report Saipan Op., Com. Off. to Com. Gen., 4th MarDiv, Brig. Gen., F.A. Hart, 28 Aug. 1944.

966    Interview Keith Renstrom, 22 March 2018; Interview Keith Renstrom by John Renstrom, 20-23 June 2016.

967    Toll, *Conquering Tide*, 535; Cook & Cook, 339.

968    Drea, 240.

969    Frank, *Downfall*, 29-30; Drea, 240; Keegan, 307. Military Heritage Institute 2011 film *The Battle for the Marianas.*

970    Interview Woody, 24 July 2016.

971    NACPM, RG 127 Box 318 (Saipan-Tinian) HQ Northern Troops Landing Force, Marianas Phase I (Saipan) 1. Civ. Affairs Report 2. Liaison Offs. Report 2. Public Rel. Report, 9 Nov. 1944, HQ VAC, Corps Civ. Affairs Off. to Com. Gen., 13 Aug. 1944, Donald Winder, 3; Interview Frank, 21 July 2017; Keegan, 307; Interview Frank, 20 July 2017; Toland, Vol. II, 650.

972    Cook & Cook, 291-2.

973    *Hong Kong's War Crimes Trials*, Linton, "War Crimes," 104.

974    NACPM, RG 127 Box 329 (Saipan-Tinian), HQ, 25th Marines, 4th MarDiv, FMF, San Francisco, CA, Enc. A, Report Saipan, 6 & 7 July 1944 D Plus 21 & 22 day & Box 328 (Saipan-Tinian),4th MarDiv Report Saipan 15 June to 9 July 1944, Subseq. Ops. (10 July to 16 July), 37.

975    NACPM, RG 127 Box 329 (Saipan-Tinian), RCT 23 Report Forager Phase I Saipan, Encl. (A)., 9 & Enc. (A)., II. (g); Hallas, *Uncommon Valor on Iwo Jima*, 332.

976    NACPM, RG 127 Box 329 (Saipan-Tinian), RCT 23 Report Forager Phase I Saipan, Encl. (A)., II. (f.); Drea, 240.

977    Dower, 144.

978    Sledge, 31; MCRDSDA, MCRD Training Book, File 1, <u>Miscellaneous</u>.

979    Drea, 173.

980    Generations Broadcast Center, Project SFMedia Consultants, Inc., WWII Interviews, Jim Reed, 31 Jan. 2013.

981    <u>Ibid.</u>

982    Interview Keith A. Renstrom, 22 March 2018; Interview Keith Renstrom by John Renstrom, 20-23 June 2016; Email John Renstrom to Rigg, 9 Jan. 2020 (11:53 CT); Email John Renstrom to Rigg, 9 Jan. 2020 (12:57 CT).

983    <u>Ibid.</u>

984    Ibid. In 2016, during the interview about this event, Renstrom said with tears streaming from his eyes, "I often wonder whatever happened to her."

985    Linda Sieg, "Historians Battle Over Okinawa WW2 Mass Suicides," Reuters, 6 April 2007.

986    Frank, "Review *Flamethrower*," 25 May 2017.

987    Toland, Vol. II, 641.

988    Smith, *Coral and Brass*, 107; Toland, Vol. II, 642.

989    Moskin, 316, 328; Toland, Vol. II, 642; O'Brien, *Liberation*, 5; Farrell, *Liberation-1944*, 19; Steinberg, *Island Fighting*, 170; Hane, 345.

990    Moskin, 335.

991    Frank, *Downfall*, 29; Hough, 247.

992    Smith, *Coral and Brass*, 107; Hough, 244; MCHDQV, Oral History, Erskine, 165; Mosley, 68.

993    Farrell, *Liberation-1944*, 25; USMC Publication, "Expeditionary Operations, MCDP 3, 16 April 1998, 98; Toland, Vol. II, 658; Crowl & Isely, 310.

994    Toll, *Conquering Tide*, 536.

995    Lecture Richard Frank, at Saipan, 21 March 2018.

996    Frank, *Downfall*, 5.

997    USMC, "Expeditionary Operations, MCDP 3, 16 April 1998, 98.

998    Interview LeMay by Colin Heaton in June 1986.

999    <u>Ibid.</u> LeMay was correct. After Saipan's fall, the Imperial Japanese government's Lord Keeper of the Privy Seal, Marquis Koichi Kido, wrote air attacks were going to increase against the mainland. Crowl & Isely, 310.

1000   Edgar Allan Poe, *Poetry and Tales*, Library of America, NY 1984, "The Cash of Amontillado," 849. See also Zeiler, 41.

1001   Crowl & Isely, 391.

1002   NACPM, RG 127 Box 318 (Saipan-Tinian) HQ Northern Troops Landing Force, Marianas Phase I (Saipan) 1. Civ. Affairs Report 2. Liaison Offs. Report 2. Public Rel. Report, 9 Nov. 1944, HQ VAC, Corps Civ. Affairs Off. to Com. Gen., 13 Aug. 1944, Donald Winder, 1.

1003   <u>Ibid.</u>, 6.

1004   NACPM, RG 127 Box 318 (Saipan-Tinian) HQ Northern Troops Landing Force, Marianas Phase I (Saipan) 1. Civ. Affairs Report 2. Liaison Offs. Report 2. Public Rel. Report, 9 Nov. 1944, HQ VAC, Corps Civ. Affairs Off. to Com. Gen., 13 Aug. 1944, Donald Winder, 3; Drea, 240.

1005   RG 127 Box 318 (Saipan-Tinian) HQ Northern Troops Landing Force, Marianas Phase I (Saipan) 1. Civ. Affairs Report 2. Liaison Offs. Report 2. Public Rel. Report, 9 Nov. 1944, HQ VAC, Corps Civ. Affairs Off. to Com. Gen., 13 Aug. 1944, Donald Winder, 4.

1006    Letter Renstrom to Mother and Dad, 17 July 1944, Renstrom Family Archive.

1007    Smith, *Coral and Brass*, 108. Historian Allan R. Millett said, "Tinian represented a high point in amphibious warfare." Millett, 416.

1008     Morison, Vol. VIII, 353.

1009    MCHDQV, Oral History, Erskine, 331-2; Carl W. Hoffman, *The Seizure of Tinian*, Historical Division Headquarters U.S. Marine Corps, Quantico, 1951, 151, 161.

1010    Moskin, 334; Spector, 319; USMC Publication, "Intel," MCDP 3, 16 April 1998, 85; Hough, 248; Hoffman, *Once a Legend,* 298; Crowl & Isely, 356, 363; Millett, 416; Richard Harwood, "A Close Encounter: The Marine Landing on Tinian," Final Days, Marines in WWII Commemorative Services, Washington D.C., 1994, 6, 9.

1011    NASLPR, Keith Arnold Renstrom, Gold Star in lieu of a second Bronze Star Medal for actions on Tinian; Interview John Renstrom, 17 Nov. 2019; Keith A. Renstrom, *Keith A. Renstrom Recounts Battle in Tinian and Awards*, 30 Aug. 2017; Interview Keith Renstrom by John Renstrom, 20-23 June 2016; Interview John Renstrom, 16 Jan. 2020; Interview John Renstrom, 22-23 Jan. 2020; https://www.ww2online.org/view/keith-renstrom; Harwood, 17. The sergeant was John David Pruitt who got injured. He lost his arm on the beach. See USMCU Casualty Web Data Base.

1012    Ibid.

1013    Berry, 7, 196-216.

1014    NASLPR, Keith Arnold Renstrom, Gold Star in lieu of a second Bronze Star Medal for actions on Tinian; Keith A. Renstrom, *Keith A. Renstrom Recounts Battle in Tinian and Awards*, 30 Aug. 2017; Interview Keith Renstrom by John Renstrom, 20-23 June 2016; Interview John Renstrom, 16 Jan. 2020; Interview John Renstrom, 22-23 Jan. 2020; https://www.ww2online.org/view/keith-renstrom; Harwood, 17. In some reports, Renstrom says there were 10 soldiers in the initial scouting party that night and that it started at around 2300 on the 24th of January. Records have not been located to ascertain the exact enemy squad number and the exact time of first contact.

1015    Ibid.

1016    Ibid.; Quote comes from Hoffman, *The Seizure of Tinian*, 65.

1017    Keith A. Renstrom, *Keith A. Renstrom Recounts Battle in Tinian and Awards*, 30 Aug. 2017; Interview Keith Renstrom by John Renstrom, 20-23 June 2016; Interview John Renstrom, 16 Jan. 2020; Interview John Renstrom, 22-23 Jan. 2020; https://www.ww2online.org/view/keith-renstrom.

1018    Keith A. Renstrom, *Keith A. Renstrom Recounts Battle in Tinian and Awards*, 30 Aug. 2017; Interview Keith Renstrom by John Renstrom, 20-23 June 2016; Interview John Renstrom, 16 Jan. 2020; Interview John Renstrom, 22-23 Jan. 2020; https://www.ww2online.org/view/keith-renstrom; Hoffman, *The Seizure of Tinian*, 67.

1019    Keith A. Renstrom, *Keith A. Renstrom Recounts Battle in Tinian and Awards*, 30 Aug. 2017; Interview Keith Renstrom by John Renstrom, 20-23 June 2016; Interview John Renstrom, 16 Jan. 2020; Interview John Renstrom, 22-23 Jan. 2020; https://www.ww2online.org/view/keith-renstrom.

1020    Ibid.

1021    Hoffman, *The Seizure of Tinian*, 64.

1022    Keith A. Renstrom, *Keith A. Renstrom Recounts Battle in Tinian and Awards*, 30 Aug. 2017; Interview Keith Renstrom by John Renstrom, 20-23 June 2016; Interview John Renstrom, 16 Jan. 2020; Interview John Renstrom, 22-23 Jan. 2020; https://www.ww2online.org/view/keith-renstrom; Harwood, 17; Hoffman, *The Seizure of Tinian*, 65.

1023    Harwood, 17; Hoffman, *The Seizure of Tinian*, 65.

1024    NASLPR, Keith Arnold Renstrom, Gold Star in lieu of a second Bronze Star Medal for actions on Tinian; Keith A. Renstrom, *Keith A. Renstrom Recounts Battle in Tinian and Awards*, 30 Aug. 2017; Interview Keith Renstrom by John Renstrom, 20-23 June 2016; Interview John Renstrom, 16 Jan. 2020; Interview John Renstrom, 22 Jan. 2020; https://www.ww2online.org/view/keith-renstrom; Harwood, 17.

1025    Ibid.

1026    Hoffman, *The Seizure of Tinian*, 66; Interview Don Graves, 24 Jan. 2020.

1027    Keith A. Renstrom, *Keith A. Renstrom Recounts Battle in Tinian and Awards*, 30 Aug. 2017; Interview Keith Renstrom by John Renstrom, 20-23 June 2016; Interview John Renstrom, 16 Jan. 2020; Interview John Renstrom, 22 Jan. 2020; https://www.ww2online.org/view/keith-renstrom; Harwood, 17.

1028    Hoffman, *The Seizure of Tinian*, 66.

1029    NASLPR, Keith Arnold Renstrom, Gold Star in lieu of a second Bronze Star Medal for actions on Tinian; Keith A. Renstrom, *Keith A. Renstrom Recounts Battle in Tinian and Awards*, 30 Aug. 2017; Interview Keith Renstrom by John Renstrom, 20-23 June 2016; Interview John Renstrom, 16 Jan. 2020; Interview John Renstrom, 22 Jan. 2020; https://www.ww2online.org/view/keith-renstrom; Harwood, 17. In the WWII museum interview, Resntrom explained why they killed the injured Japanese: "Well you barely have time to care for your own wounded let along the enemy's wounded. If you did leave them there alive then your leaving yourself open to the possibility of them being able to snipe someone else later." As historian Colonel Carl Hoffman wrote of the battlefield where Renstrom was at: "Another tank incurred minor damage when a Japanese rose fro the dead around him [holding] a magnetic mine. A hail of Marine fire dropped him almost where he had lain before." Hoffman, *The Seizure of Tinian*, 66.

1030    USMC Museum, Quantico VA, display Marines during WWII; Hough, 258; Crowl & Isely, 357.

1031    Hough, 259.

1032    Crowl & Isely, 371.

1033    USMC Publication, "Intel," 84-6.

1034    Steinberg, *Island Fighting*, 194; Smith, *Coral and Brass*, 112-3; Hough, 259.

1035    Harwood, 29.

1036    Harwood, 29; Hough, 259.

1037    Steinberg, *Island Fighting*, 194; Smith, *Coral and Brass*, 112-3; Hough, 259.

1038    Interview Woody, 19 July 2016; Interview 21 March 2018.

1039    Steinberg, *Island Fighting*, 194.

1040    Prange, 591.

1041    Sasser, 25, 32-3; Info from NMPWANM, 13 July 2017; Smith, *Coral and Brass*, 62.

1042    Morison, 373; Hough, 262; Interview James Oelke Farley, 8 Nov. 2017; O'Brien, *Liberation*, 7; MCHDHQ, Woody Williams, unpublished article, "Leaving Guadalcanal June 1944," 1.

1043    *Guam War Claims Review Commission*, 6.

1044    NACPM, RG 127 Box 50 (Guam), HQ 21st Marines, 3d MarDiv, FMF, In Field, Com. Off. 21st Marines to Com. Gen. 3d MarDiv, Report, Forager Operation, Butler, 16 Aug. 1944, 10.

1045    Larry Smith, *Drill Sergeants*, 29; Henri, et al., 13; Smith, xxi; Wheeler, 41; Newcomb, 35; Edgerton, 226; Oliver, 110–24; Cook & Cook, 48, 73, 173, 192-3; Drea, 213-4, 232, 285; "Spit on My Grave," Chosun Ilbo, Seoul, Korea, article no. 104-16, 1998; Weinberg, xiii; NACPM, RG 226, Box 368 (Off. US Strategic Servs.), Y-125, Enemy Troop Movements, 20 June 1944 & Enemy Conscription 10 June 1944; Dower, 285; Interview Graves, 2 May 2018; Victoria, *Zen at War*, 156; Tanaka, 42-3, 77.

1046    Morison, 373.

1047    Ibid., 374.

1048    O'Brien, *Liberation*, 8.

1049    Morison, 376-7.

1050    Ibid., 379; Crowl & Isely, 381.

1051    Crowl & Isely, 373, 384.

1052    Email Farrell to Rigg, 23 Aug. 2016; Farrell, *Liberation-1944*, 16; Smith, *Coral and Brass*, 20.

1053    Letter Dick Dashiell to Vivian Dashiell, 7 Sept. 1945, 3.

1054    O'Brien, *Liberation*, 5-6; Farrell, *Liberation-1944,* 16; NASLPR, Woody Williams; LtGen Louis Metzger (USMC), "Guam 1944" Marine Corps Gazette, Col. 78, Issue 7.

1055    Metzger, "Guam 1944."

1056    Farrell, *Liberation-1944,* 19, 63. Phillips Payson O'Brien gives slightly different numbers. He writes that for Operation *Forager*, the U.S. Navy deployed 15 aircraft carriers, seven battleships, 20 cruisers and 67 destroyers. He also notes this is a dramatically more powerful force than what was at Normandy at the same time. *Overlord* only had three older battleships, three cruisers and 34 destroyers. This proves, in O'Brien's opinion, that the U.S. was actually focusing more on the Pacific than Europe at this time. O'Brien, *The Second Most Powerful Man in the World*, 285.

1057    Moskin, 335; O'Brien, *Liberation*, 2.

1058    Farrell, *Liberation-1944,* 20.

1059    Toll, *Conquering Tide*, 512.

1060    Hough, 270.

1061    Berry, 205; Sledge, 182.

1062    Morison, 382.

1063    USMC History, Vol. III, 458.

1064    O'Brien, *Liberation*, 1.

1065    Morison, 197.

1066    Ibid., 381.

1067    O'Brien, *Liberation*, 8.

1068    Harries & Harries, 25.

1069    MARC, GWRC Candelaria Aguon, Box 1, Rec. 1910; Manchester, 284.

1070    MCHDHQ, Letter Don Beck to Ruth Beck, 8 July 1944.

1071    Toland, Vol. II, 809.

1072    Frankl, 14.

1073    Interview Woody, 23 July 2016; VandeLinde, 368.

1074    Interview Woody, 23 July 2016.

1075    Farrell, *Liberation-1944,* 59, 102; Moskin, 336; O'Brien, *Liberation*, 1; Hough, 241, 243.

1076    Woody called his craft at Guam a "pointed-nose ship," maybe an amtrac. Interview Woody, 22 July 2016.

1077    Farrell, *Liberation-1944*59, 102; Moskin, 336; O'Brien, 1; Hough, 241, 243; Crowl & Isely, 325; Hallas, *Saipan*, 91.

1078    Moskin, 336; O'Brien, *Liberation*, 1; Farrell, *Liberation-1944,* 59.

1079    Aurthur and Cholmia, 146; Woody, unpublished article, "Leaving Guam June 1944," 1; NACPM, RG 127 Box 50 (Guam), Narrative Battle of Guam, HQ 3d MarDiv, FMF, In Field, 19 Aug. 1944, 1.

1080    Ronald R. Van Stockum, "Battle of the Philippine Sea," 2017, 5.

1081    Farrell, *Liberation-1944,* 59.

1082    Interview Woody, 21 Oct. 2016. This phrase came from Captain Evans F. Carlson, famous from Carlson's Raiders from Guadalcanal. He had been stationed in China and had taken this *Gung-ho* actually from Mao's Communists. Tuchman, 175. The name actually comes from the CCP Industrial Cooperative (*Gonghe*), but Carlson told his Marines that the term "called for self-discipline and implicit belief in the doctrine of helping the other fellow." Moore, 155.

1083    Zdon, "A Marine in the Pacific," 8. The horrible hydrographic intelligence was well discussed in the 25th Commandant, General Robert E. Cushman Jr.'s, oral testimony when he declared, "so, we landed on Guam and we were so poorly prepared in terms of maps and things, that it was just a disgrace to the United States, I figured. We had held that island since 1898, and there wasn't any map of the reefs, there wasn't any map worth a doggone of the island whatsoever—just a green blob, you know. So, we got our intelligence by dragooning some of the Navy Chamorro stewards working in the galleys and so on amongst the ships that were in the assault force and questioned them about the reefs and so on, which of course, you can imagine, [was poor] [sic.]." MCHDQV, Oral History, Robert E. Cushman Jr., 160-1.

1084 Moskin, 336; Farrell, *Liberation-1944,* 63, 71, 83.

1085 Morison, 383.

1086 Hoffman, *Once a Legend,* 302-3.

1087 NACPM, RG 127 Box 50 (Guam), Narrative Guam Battle, HQ 3d MarDiv, FMF, In Field, 19 Aug. 1944, 1.

1088 Morison, 386; O'Brien, *The Second Most Powerful Man in the World,* 252.

1089 Clausewitz, *On War,* 353. As historian Allan R. Millett wrote about Guam, "The key to the campaign, then, became control of hills dominating the beaches." Millett, 417.

1090 Toll, *Conquering Tide,* 512.

1091 Morison, 387.

1092 Interview Woody, 21 July 2016; NACPM, RG 127, Box 83, (Iwo), AB-1, Com. Gen. 3d MarDiv Erskine to Commandant, Report, 3 June 1945, 48.

1093 Interview Woody, 21 July 2016; VandeLinde, 367.

1094 O'Brien, *Liberation,* 8.

1095 Ibid..

1096 Interview Woody, 20 July 2016.

1097 O'Brien, *Liberation,* 28; Farrell, *Liberation-1944,* 113.

1098 Farrell, *Liberation-1944,* 25.

1099 Ibid., 63.

1100 Interview Woody, 21 July 2016.

1101 O'Brien, *Liberation,* 28, 31; Farrell, *Liberation-1944,* 75; MCHDQV, HQ, 3d MarDiv, FMF, In Field, Report Lt. Col. Turton (D-3), 19 Aug. 1944, 4; MCHDQV, Dashiell File, #50, Coughlin-Leyte.

1102 Kennedy, *Engineers of Victory,* 227.

1103 Woody Williams, unpublished article, "Leaving Guam in June 1944," 1.

1104 Interview Woody, 19 July 2016.

1105 Frankl, 104.

1106 Moskin, 336.

1107 Aurthur and Cholmia, 147-50.

1108 Interview Woody, 24 July 2016; MCHDHQ, Woody, "Destiny...Farm Boy," 11.

1109 Dolan, "Conversation: Forged in Flame."

1110 Woody claims Waters and he were in Boot Camp together, but their personnel files indicate they attended different ones. Also, the picture of Vernon's Boot Camp platoon does not have Woody in it. In later interviews with George Waters, Woody said Waters became friends on Guadalcanal changing the story of when they first met. Interview George Waters, 24 Aug. 2018; Hallas, *Iwo Jima,* 135.

1111 MCHDHQ, Woody, "Destiny...Farm Boy," 11; Smith, *Iwo Jima,* 64; Woody, unpublished article, "Leaving Guam in June 1944," 1; Interview Woody, 21 Oct. 2016; NASLPR, Vernon Waters, Death Certificate; Shelby Foote, *The Civil War: Fort Sumter to Perryville,* NY, 1958, 17.

1112 MCHDQV, HQ, 3d MarDiv, FMF, In Field, Report Lt. Col. Turton (D-3), 19 Aug. 1944, 6.

1113 NACPM, RG 127 Box 51 (20th, 21st & 22nd Marine Rgts), 1st Btl., 21st Rgt. Log, 21 July 1944.

1114 Interview Lee, 24 June 2016; Zdon, "A Marine in the Pacific," 8-9.

1115 Ibid.

1116 Ibid.

1117 Interview Woody, 12 Feb. 2017; Woody's sermon, River Cities Church, 29 March 2003.

1118    Lefty has maintained he quickly returned to his unit, but according to naval medical officer, Lee Mandel, if a he suffered a sucking chest wound, he would not have returned to combat for days. Interview Mandel, 12 Aug. 2017.

1119    Donald Beck's citation for Leo Simon; USMCU Casualty Web Data Base.

1120    NACPM, RG 127 Box 60 (Guam) -1 21st MarRegt Reports 23 July-3 Nov. 1944, 1800 July 21 to 1800 July 22, Report 2, 1.

1121    NACPM, RG 127 Box 60 (Guam) -1 21st MarRegt Reports 23 July-3 Nov. 1944, 1800 22 July to 1800 23 July, Report 3, Log. 2215-0300, 1 & 1800 23 July to 1800 24 July, Report 4; NACPM, RG 127 Box 51 (20th, 21st & 22nd Marine Rgts), 1st Blt., 21st Rgt. Log, 22 July 1944, 2330 & 23 July 1944.

1122    NASLPR, Carroll M. Garnett, Garnett to Lt. Col. Marlowe C. Williams, 15 Oct. 1950.

1123    NASLPR, Garnett, Garnett to Lt. Col. Williams, 15 Oct. 1950 & Bronze Star Citation.

1124    NACPM, RG 127 Box 60 (Guam) -1 21st MarRegt Reports 23 July-3 Nov. 1944, 1800 22 July to 1800 23 July, Report 3, Log. 2215-0300, 1 & 1800 23 July to 1800 24 July, Report 4; NACPM, RG 127 Box 51 (20th, 21st & 22nd Marine Rgts), 1st Blt., 21st Rgt. Log, 22 July 1944, 2330 & 23 July 1944.

1125    NACPM, RG 127 Box 60 (Guam) -1 21st MarRegt Reports 23 July-3 Nov. 1944, 1800 22 July to 1800 23 July, Report 3, Log. 2215-0300, 1 & 1800 23 July to 1800 24 July, Report 4; NACPM, RG 127 Box 51 (20th, 21st & 22nd Marine Rgts), 1st Blt., 21st Rgt. Log, 22 July 1944, 2330 & 23 July 1944.

1126    NACPM, RG 127 Box 60 (Guam) -1 21st MarRegt Reports 23 July-3 Nov. 1944, 1800 24 July to 1800 25 July, Report 5, 2.

1127    Crowl & Isely, 376.

1128    Ibid., 385.

1129    Krakauer, 205.

1130    Interview Woody, 19 July 2016.

1131    Ibid., 12 Feb. 2016.

1132    Van Stockum, "Japanese Counterattack Plan 25-26 July 1944," June 2017, 2.

1133    Ibid.

1134    NACPM, RG 127 Box 60 (Guam) -1 21st MarRegt Reports 23 July-3 Nov. 1944.

1135    Conrad, "Heart of Darkness," 543.

1136    NACPM, RG 127 Box 50 (Guam), HQ 3d MarDiv, FMF, In Field, Appendix No. 3 to Anne B, Intel, Pass Word System, Lt. Col. Turton, 13 May 1944.

1137    NACPM, RG 127 Container 14, Dashiell, #223.

1138    Interview Woody, 21 July 2016; Interview Woody, 21 Oct. 2016.

1139    Commentary by Gen. Van Stockum on Flamethrower, 23 June 2017.

1140    NACPM, RG 127 Box 51 (20th, 21st & 22nd Rgts), 1st Btl.,21st Rgt Log, 25 July 1944.

1141    MCHDQV, HQ, 3d MarDiv, FMF, In Field, Report Lt. Col. Turton (D-3), 19 Aug. 1944, 7.

1142    MCHDQV, Dashiell File, #82, Wagoner; Interview Lee, 21 July 2016.

1143    Aurthur and Cholmia, 154; NACPM, RG 127 Box 50 (Guam), HQ 21st Marines, 3d MarDiv, FMF, In Field, Com. officer 21st Marines to Com. Gen. 3d MarDiv, Report, Forager, A.H. Butler, 16 Aug. 1944, 5, 14; Hough, 274.

1144    Farrell, Liberation-1944, 21-2; Spector, 320; Hough, 260.

1145    O'Brien, Liberation, 9, 22-5; Aurthur and Cholmia, 151; Woody Williams, unpublished article, "Leaving Guam in June 1944," 1; Interview Woody, 21 July 2016; Farrell, Liberation-1944, 102-3.

1146    Farrell, Liberation-1944, 98.

1147    MCHDQV, HQ, 3d MarDiv, FMF, In Field, Report Lt. Col. Turton (D-3), 19 Aug. 1944, 8.

1148    Van Stockum, "Japanese Counterattack Plan," 3.

1149    NACPM, RG 127 Box 51 (20th, 21st & 22nd Rgts), 3rd Btl., 21st Rgt. Log, 25 July 1944 (3rd notation of date).

1150    Donald Beck's Bronze Star citation for PFC Mack Drake; NACPM, RG 127 Container 14, Dashiell, #143.

1151    Aurthur and Cholmia, 154.

1152    Farrell, *Liberation-1944,* 103. For a similar act like Basilone's that happened on Guam, see MCHDQV, Dashiell File, #34, Mulcahy-Award.

1153    MCHDHQ, Un-authored article from the 1945, "Sgt. Hershel Woody Williams, 2; VandeLinde, 367-8.

1154    NACPM, RG 127 Box 60 (Guam) A31-1 3rd Btl. 21st Regt Rpts 21 July-1 Nov., 2400 25 July to 2400 26 July, 1.

1155    Van Stockum, "Japanese Counterattack Plan," 3.

1156    Interview Woody, 22 July 2016; MCHDQV, HQ, 3d MarDiv, FMF, In Field, Report Lt. Col. Turton (D-3), 19 Aug. 1944, 8.

1157    Poe, "The Tell-Tale Heart," 555.

1158    Van Stockum, "Japanese Counterattack Plan," 4; MCHDHQ, Letter Beck to Ruth Beck, 23 April 1943.

1159    Ibid., 4-5.

1160    O'Brien, *Liberation*, 24; Patrick K. O'Donnell, *Into the Rising Sun: World War II's Pacific Veterans Reveal the Heart of Combat*, NY, 2002, 139.

1161    O'Donnell, 138.

1162    NACPM, RG 127 Box 51 (20th, 21st & 22nd Rgts), 1st Btl., 21st Rgt. Log, 26 July 1944; NACPM, RG 127 Box 50 (Guam), HQ 21st Marines, 3d MarDiv, FMF, In Field, Com.off. 21st Marines to Com. Gen. 3d MarDiv, Report, *Forager*, Butler, 16 Aug. 1944, 4-5.

1163    NACPM, RG 127 Box 50 (Guam), HQ 21st Marines, 3d MarDiv, FMF, In Field, Com. Off. 21st Marines to Com. Gen. 3d MarDiv, Report, *Forager*, Butler, 16 Aug. 1944, 4.

1164    Van Stockum, "Japanese Counterattack Plan," 4-5; Email Van Stockum to Rigg, 18 June 2017.

1165    Commentary Van Stockum on *Flamethrower*, 23 June 2017.

1166    Van Stockum, "Japanese Counterattack Plan," 6-7; NACPM, RG 127 Box 50 (Guam), HQ 21st Marines, 3d MarDiv, FMF, In Field, Com. Off. 21st Marines to Com. Gen. 3d MarDiv, Report, *Forager*, Butler, 16 Aug. 1944, 4.

1167    Aurthur and Cholmia, 153; Sulzberger, 541.

1168    MCHDQV, August Larson, 86.

1169    NACPM, RG 127 Box 60 (Guam) A31-1 3rdBtl. 21st Regt. Reports 21 July-1 Nov., 2400 25 July-2400 26 July, 1.

1170    MCHDHQ, Letter Beck to Ruth Beck, 18 Jan. 1945.

1171    Ibid., Letter Carroll Garnett to Beck, 24 July 1990.

1172    Ibid.; NASLPR, Garnett, Report of Casualty, Case of Status, Wound-Frag Rt. Buttocks, Date of Casualty, 26 July 1944 & Bronze Star Citation & Sample Bronze Star citation written by Dr. Buchan.

1173    Interview Woody, 21 July 2016.

1174    NACPM, RG 127 Box 60 (Guam) A31-1 3rdBtl. 21st Regt. Reports 21 July-1 Nov., 2400 25 July-2400 26 July, 1.

1175    Ibid.

1176    NACPM, RG 127 Box 50 (Guam), HQ 21st Marines, 3d MarDiv, FMF, In Field, Com. Off. 21st Marines to Com.Gen. 3d MarDiv, Report, *Forager*, Butler, 16 Aug. 1944, 5.

1177    Aurthur and Cholmia, 153; Sulzberger, 541.

1178    Morison, 387-8.

1179    Van Stockum, "Japanese Counterattack Plan," 6-7; Aurthur and Cholmia, 153; 541; Sulzberger, 541.

1180    NASLPR, Beck, Silver Star Citation.

1181    MCHDQV, Oral History, Robert E. Cushman Jr., 167.

1182    NACPM, RG 127 Box 51 (20th, 21st & 22nd Marine Rgts), 1st Btl., 21st Rgt. Log, 26 July 1944.

1183    Hough, 274.

1184    Hallas, *Iwo Jima*, 124; Interview Woody, 22 July 2016; Poe, "The Tell-Tale Heart," 557. Woody is convinced the Marine who died in that foxhole was Clevenger. However, in the casualty cards, files and documents at the National Archives in College Park, MD and in St. Louis, MO and on the memorial wall at Guam's Asan Beach overlook, there is no mention of a Clevenger dying on Guam. Email James Herbert to Rigg, 31 May 2018.

1185    MCHDQV, Dashiell File, #159, Williams-Silver Star; NASLPR, Marlowe Williams, Silver Star Citation.

1186    NASLPR, Garnett, Garnett to Lt. Col. Marlowe C. Williams, 15 Oct. 1950.

1187    O'Brien, *Liberation*, 24, 26.

1188    Van Stockum, "The Battle for Guam," 1.

1189    Aurthur and Cholmia, 152, 154; Steinberg, *Island Fighting*, 174; Spector, 320; MCHDQV, HQ, 3d MarDiv, FMF, In Field, Report Lt. Col. Turton (D-3), 19 Aug. 1944, 8; Hough, 274.

1190    NACPM, RG 127 Box 51 (20th, 21st & 22nd Rgts), 3rd Btl., 21st Rgt. Log, 27 July 1944.

1191    Morison, vol. XIII, 388.

1192    O'Brien, *Liberation*, 25.

1193    Hough, 279; Farrell, *Liberation-1944*, 112-3; Steinberg, *Island Fighting*, 173.

1194    Hough, 279; Farrell, *Liberation-1944*, 112-3; Steinberg, *Island Fighting*, 173; MCHDQV, Oral History, August Larson, 84-5.

1195    Thompson, Doud & Scofield, 16.

1196    O'Brien, *Liberation*, 28.

1197    Hough, 274.

1198    Ibid., 274.

1199    Ibid., 271-2.

1200    U.S. Marine Corps Publication, "Planning," MCDP 5, 21 July 1997, 61. The phrase has been slightly modified from the translation used in the cited publication since the author knows German: "*Kein Operationsplan reicht mit einiger Sicherheit über das erste Zusammentreffen mit der feindlichen Hauptmacht hinaus.*"

1201    O'Donnell, 135, 137.

1202    Interview Woody, 20 July 2016.

1203    MCHDQV, Oral History, Erskine, 166.

1204    Ibid.

1205    Interview Woody, 21 July 2016; Letter Woody to Rigg, 16 Nov. 2017.

1206    Interview Woody, 21 July 2016.

1207    Dower, 144, 341 fn. 33; Edgerton, 285. Frank, "Review *Flamethrower*," 25 May 2017; Toland, Vol. I, 503; Larry Smith, *Beyond Glory: Medal of Honor Heroes in Their Own Words*, NY, 2003, 5, 12.

1208    Interview Woody, 19 July 2016.

1209    Frank, *Downfall*, 28. Quote comes from Thompson, Doud, and Scofield's *How the Jap Army Fights*, 7.

1210    Dower, 143-4.

1211    NACPM, RG 127 Container 14, Dashiell, #145, 1-3; NASLPR, Albert Hemphill, Description Medal or Badge; *Third Marine Division's Two Score and Thirteen History*, authored and produced Third Marine Division Association, Inc., Paducah, KT, 2002, 104.

1212    NACPM, RG 127 Box 50 (Guam), HQ 21st Marines, 3d MarDiv, FMF, In Field, Com. officer 21st Marines to Com. Gen. 3d MarDiv, Report, *Forager*, Butler, 16 Aug. 1944, 12.

1213 MCHDQV, Dashiell File, #127, Red Cross.

1214 O'Brien, *Liberation*, 41; Farrell, *Liberation-1944,* 27-53; MARC, Guam War Reparations, Consuelo Aguon, Box 1, 1980.

1215 Farrell, *Liberation-1944,* 3, 34-6; Prange, 12; MARC, GWRC, Olivia L.G. Cruz Abell, Box 1, Rec. 2090; Flores, Palacios 6.

1216 Interview Lee, 4 May 2017.

1217 Hough, 263; Farrell, *Liberation-1944,* 18.

1218 MARC, GWRC, Flores, Palacios, 7.

1219 Interview Lee, 4 May 2017.

1220 MARC, GWRC, Edward L.G. Aguon, Box 1.

1221 MCHDHQ, Letter Beck to Ruth Beck, 12 Nov. 1944.

1222 O'Brien, *Liberation*, 30.

1223 NACPM, RG 127 Box 51 (20th, 21st & 22nd Rgts), 3rd Btl., 21st Rgt. Log, 28 July 1944 (1st notation of date) & 29 July 1944 (2nd notation of date).

1224 Aurthur and Chlomia, 159.

1225 NACPM, RG 127 Box 16 (3d MarDiv), (2295) 14 Nov. 42-27 Nov. 1945 (1 of 2), HQ, 3rd Marines, 3d MarDiv, FMF, In Field, 26 Aug. 1944, Major J.A. Scott, 1.

1226 Interview Woody, 23 July 2016.

1227 MCHDHQ, Letter Beck to Ruth Beck, 5 Dec. 1944.

1228 Interview Ben and Steven Tischler with Richard Tischler, 2008, NYC.

1229 Crowl & Isely, loc. 8302; Millett, 418.

1230 MCHDQV, Oral History, Erskine, 321-3; Hough, 239; Hoffman, *Once a Legend,* 291, 293; Sledge, 202; Kakehashi, 98; Neimeyer, "Paladins at War"; NMPWANM, 13 July 2017; Smith, *Coral and Brass*, 72, 92-7; Neimeyer, "Paladins at War"; Millett, 413-4.

1231 Gene E. Stalecker, *Rolling Thunder*, Mechanicsburg, PA, 2008, 206.

1232 Stalecker, 203.

1233 Stalecker, 207; Newcomb, 6, 18; Kakehashi, 49.

1234 Interview James Oelke Farley, 25 March 2018.

1235 NASLPR, Roy Stanley Geiger, Smith to Forrestal, 27 Sept. 1944 & King to Forrestal, 27 Oct. 1944.

1236 NACPM, RG 127 Box 50 (Guam), Narrative Battle of Guam, HQ 3d MarDiv, FMF, In Field, 19 Aug. 1944, 1; MCHDQV, HQ, 3d MarDiv, FMF, In Field, Report Turton (D-3), 19 Aug. 1944, 3.

1237 Ibid., 2.

1238 O'Brien, *Liberation*, 42-3; Frank 3-19.

1239 NACPM, 127 GW-1386, Chamorro Elders to Nimitz, 10 Aug. 1944.

1240 NACPM, 127 GW-1381.

1241 Interview James Oelke Farley, 25 March 2018; Email from Park Ranger Kina Doreen Lewis of the National Park Service center of War in the Pacific at Guam, 9 May 2019.

1242 Aurthur and Cholmia, 140.

1243 O'Brien, *Liberation*, 42-3; Frank 3-19.

1244 Hanley II, xiii, 97-101.

1245 Frank, *Downfall*, 54; Toland, Vol. II, 779-80.

1246 Frank, *Downfall*, 54, 59; Cook & Cook, 455.

1247 O'Brien, *Liberation*, 43.

1248 NACPM, RG 127 Box 60 (Guam) A 30-2, 21st MarRegt Reports 27 Oct.-1 Nov. 1944, Report 2, 26-7 Oct. and A31-2, 21st MarRegt Journal 16 Sept.-2 Nov. 1944, Report 21 Sept. 1944.

1249 NASLPR, Woody Williams.

1250 NACPM, RG 127 Box 51 (Guam), HQ, 3d MarDiv, FMF, In Field, Enemy Contact, Lt. Col. Turton, 1801 K, 18 Jan 45 to 1800 K, 19 Jan. 45. For years, Japanese soldiers who did not know the war was over were discovered in Asia and the Pacific. The last Japanese soldier on Guam who surrendered was Shōichi Yokoi in January 1972. Cook & Cook, say he emerged in 1973, but it seems the true date was 1972. Cook & Cook, 404.

1251 Aurthur and Cholmia, 162; MCHDQV, HQ, 3d MarDiv, FMF, In Field, Report Lt. Col. Turton (D-3), 19 Aug. 1944, 10; NACPM, RG 127 Box 16 (3d MarDiv), (2295) 14 Nov. 42-27 Nov. 1945 (1 of 2), HQ, Third Marines, 3d MarDiv, FMF, In Field, 26 Aug. 1944, Major J.A. Scott, 1.

1252 Morison, 376; MCHDQV, HQ, 2nd Btl., 21st Marines, 3d MarDiv, FMF, In Field, G.A. Percy, 12 April 1945, 16; MCHDQV, HQ, 1st Btl., 21st Marines, 3d MarDiv, FMF, In Field, 6 April 1945, Report Lt. Col. Williams, 15; NACPM, RG 127 Box 16 (3d MarDiv), (2295) 14 Nov. 42-27 Nov. 1945 (1 of 2), HQ, Second Marine Raider Rgt. (Provisional), In Field, Com. Off. to Com. Gen., First Amphibious Corps, 15 Nov. 1943, 1.

1253 Morison, 373.

1254 Interview Woody, 20 July 2016; Interview Lee, 4 May 2017.

1255 MCHDQV, HQ 1st Btl., 21st Marines, 3d MarDiv, FMF, In Field, 6 April 1945, Report Lt. Williams, 13.

1256 NACPM, RG 127 Box 16 (3d MarDiv), (2295) 14 Nov. 42-27 Nov. 1945 (1 of 2), HQ, 3rd Marines, 3d MarDiv, FMF, In Field, 26 Aug. 1944, Major Scott, 2.

1257 Ibid.

1258 Interview Ellery B. "Bud" Crabbe, 23 July 2017.

1259 Interview Woody, 21 July 2016; Zdon, "A Marine in the Pacific," 10.

1260 MCHDHQ, Letter Woody to Ruby Meredith, 12 April 1944.

1261 MCHDHQ, Un-authored article from the 1945, "Sgt. Hershel Woody Williams," 2.

1262 Interview Woody, 12 Feb. 2017; Woody's sermon, River Cities Church ,29 March 2003.

1263 NASLPR, Carroll M. Garnett, Marlowe C. Williams to Dr. William H. Buchan, 27 Oct 1950.

1264 Winona Daily News, "Purple Heart Presentation," 6 Jan. 1945.

1265 MCHDHQ, Letter Beck to Ruth Beck, 16 Dec. 1944.

1266 Ibid., 28 Aug. 1944.

1267 Ibid., 3 Dec. 1945.

1268 Ibid., 28 Aug. 1944.

1269 Francois, 44.

1270 Robert S. Burrell (ed.), Crucibles: Selected Readings in U.S. Marine Corps History, Bel Air, 2004, Heather P. Marshall, "China Marines and the Crucible of the Warrior Mythos, 1900-41," 83-96; Hoffman, Once A Legend, 111-24.

1271 Interview Woody, 21 July 2016.

1272 The 3d Marine Div. Bulletin, Vol. II, 6 Oct. 1945, Number 32, Dick Dashiell, "Regimental Ramblings."

1273 NASLPR, File Woody Williams, Duration of National Emergency, 854310; Interview Woody, 21 July 2016.

1274 Interview Woody, 12 Feb. 2017. MCRDSDA, MCRD Training Book, File 1, Wearing Apparel.

1275 Steinberg, Island Fighting, 194.

1276 Bradley & Powers, Flags of Our Fathers, 147.

1277 NACPM, RG 165, Box 2153, War Dept., 6910 (Japan), Memo Japanese Vulnerability, Tradition (b); Crowley, 86.

1278 NACPM, 127 Box 14 (3d MarDiv), HQ, FMF, Intel. Bulletin #4-44, Capt. D. Whipple, American Soldier, 7.

1279 Ibid.

1280 Ibid. See also Toland, Vol. I, 455-6.

1281 Hough, 21; Weinberg, 322.

1282   Lt. Col. (USMC) C. P. Van Ness, *Exploding the Japanese Superman Myth*, Wash. D.C., 1942, 1.

1283   Van Ness, 2. See also Millett, 369.

1284   Toland, Vol. I, 428, 431-3.

1285   NACPM, 127 Box 14 (3d MarDiv), HQ, FMF, Intel. Bulletin #4-44, Capt. D. Whipple, Marshalls, 4.

1286   Toll, *Conquering Tide*, 444-5; Cook & Cook, 174-5, 259; D. Clayton James, "Strategies in the Pacific," 720; Toland, Vol. I, 287.

1287   Cook & Cook, 175.

1288   Toll, *Conquering Tide*, 523-42.

1289   Paterson (ed.), *Major Problems in American Foreign Policy*," Bernstein, "Atomic Bomb and Diplomacy," 321.

1290   David McNeill, "Reluctant Warrior," WWII Magazine, Special Collector's Addition for the 60th Commemoration of Iwo Jima, 43.

1291   Toll, *Pacific Crucible*, 5.

1292   Interview Woody, 20 July 2016.

1293   Hane, 350.

1294   Frank, *Downfall*, 181.

1295   Hitchens, *god is not Great*, 203. See also Drea, 248; Hane, 353; Mandel, "Combat Fatigue," 24; NACPM, RG 127 Container 14, R.N. Davis, Report #688.

1296   Maslowski, and Millett, 445; Rice, 122-3; Cook & Cook, 353; Robert Gandt, *The Twilight Warriors: The Deadliest Naval Battle of World War II and the Men Who Fought It*, NY, 2010, 42.

1297   Toland, Vol. II, 883.

1298   Mandel, "Combat Fatigue," 24.

1299   Toland, Vol. II, 883.

1300   Blum, 46.

1301   Shahan Russell, "The Tragic Tale of Hajime Fuji: A Kamikaze Fighter Who Crashed Into & Sunk the USS Drexler," War History Online, 16 Sept. 2016.

1302   Dower, 232; Maslowski, and Millett, 445.

1303   Cook & Cook, 327.

1304   Crowl & Isely, 536.

1305   Ibid., 558.

1306   Cook & Cook, 265.

1307   Crowl & Isely, 539, 558.

1308   NACPM, RG 127 Box 90 (Records Ground Combat) MSI-1174 Vol. 24, Jan 44-April 1945, HQ, 21st Marines, 3d MarDiv, FMF, In Field, Report Col. Butler, 6 Nov. 1944, 1.

1309   Interview Bill Schlager, 27 April 2018.

1310   MCHDQV, Oral History, Robert E. Cushman Jr., 166.

1311   NACPM, RG 127 Box 90 (Records Ground Combat) MSI-1174 Vol. 24, Jan 44-April 1945, HQ, 21st Marines, 3d MarDiv, FMF, In Field, Report Col. Butler, 6 Nov. 1944, 1 & Report Col. Butler, 30 Nov. 1944, 1.

1312   MCHDHQ, Letter Beck to Ruth Beck, 16 Nov. 1944.

1313   Ibid., 22 Oct. 1944.

1314   MCHDQV, Oral History, Erskine, 354-6, 371; Hoffman, *Once a Legend,* 303; MCHDQV, Oral History, Robert E. Cushman Jr., 166.

1315   MCHDQV, Oral History, Van Ryzin, 51.

1316   Ibid., 54.

1317   NACPM, RG 127 Box 90 (Records Ground Combat) MSI-1174 Vol. 24, Jan 44-April 1945, HQ, 21st Marines, 3d MarDiv, FMF, In Field, Report Col. H.J. Withers, 31 Dec. 1944, 1-2.

1318   Interview Woody, 23 July 2016: MCHDQV, Oral History, Woody Williams 13801A.

1319   Interview Woody, 19 & 23 July 2016.

1320   MCHDQV, Oral History, Erskine, 59-61.

1321   Interview Woody, 5 Dec. 2017; NASLPR, Gen. Graves B. Erskine, Record of Erskine, Physical Disability Appeal Board, Navy Dept., Wash. D.C., Lt. Gen. Erskine, 26 June 1953, 3.

1322   MCHDQV, Oral History, Robert E. Cushman Jr., 168-170. Cushman commanded 2nd Battalion, 9th Marines.

1323   MCHDQV, Oral History, Robert E. Cushman Jr., 170.

1324   NACPM, RG 127 Box 90 (Records Ground Combat) MSI-1174 Vol. 24, Jan 44-April 1945, HQ, 21st Marines, 3d MarDiv, FMF, In Field, Report Col. A.H. Butler, 6 Nov. 1944, 2. MCHDQV, Oral History, Robert E. Cushman Jr., 121-22.

1325   MCHDQV, Oral History, Robert E. Cushman Jr., 165.

1326   NACPM, RG 127 Box 90 (Records Ground Combat) MSI-1174 Vol. 24, Jan 44-April 1945, HQ, 21st Marines, 3d MarDiv, FMF, In Field, Report Col. A.H. Butler, 6 Nov. 1944, 2.

Ibid., Report Col. H.J. Withers, 31 Dec. 1944, 3.

1327   Ibid., 1-3.

1328   Ibid., 3.

1329   MCHDQV, Dashiell File, Biographies, Withers; NASLPR, Hartnoll J. Withers, Record of Withers, Hartnoll J., Sheet 5-6 & Card #8.

1330   Jon E. Taylor, *Freedom to Serve: Truman, Civil Rights, and Executive Order 9981*, NY, 2013, 35-6; Millett, 375. Over 15,000 Blacks served in the Marine Corps during WWII. Millett, 375.

1331   NACPM, RG 127, Box 83, (Iwo), AB-1, Com. Gen. 3d MarDiv Erskine to Commandant, 3 June 1945, 79.

1332   Interview Shindo, 9 April 2018; Interview Tatsushi Saito, 10 April 2018 (Military History Div., National Institute for Defense Studies, Ministry of Def.); Interview Col. Yukio Yasunaga, 10 April 2018 (Sen. Fellow, Military History Div., National Institute for Def. Studies, Ministry of Def.); MCRDSDA, C-2 Special Study, Enemy Situation, VAC, 6 Jan. '45, Unit Commanders, 11.

1333   Many sources have Kuribayashi listed as a Deputy Military *attaché* in Washington, but in the Japanese Defense Military archives, there is no evidence he held such a position. Burleigh, 331; Kakehashi, 37, 112. Yamamoto also attended Harvard (although he dropped out after a month in 1920 (Harvard Univ. Student Archives File Yamamoto)). Prange, 111; Sledge, 304; Cook & Cook, 90-1. Several historians incorrectly list Yamamoto as a "Harvard-educated admiral," which is dead wrong like historian Bob Tadashi Wakabayashi and Ian W. Toll. See Wakabayashi (ed.), *The Nanking Atrocity, 1937-38*, Wakabayashi, "The Messiness of Historical Reality," 17; Toll, *Pacific Crucible*, 69.

1334   Harvard Univ. Student Archives, Tadamichi Kuribayashi, Summer 1928; Email King to Rigg, 2 Aug. 2017.

1335   Ross, 20; Newcomb, 8; McNeill, "Reluctant Warrior," 39; Edgerton, 256-7; Kakehashi, 112.

1336   Keith Wheeler, *The Road to Tokyo*, Alexandria, VA (Time Life Books), 1979, 40; McNeill, "Reluctant Warrior," 39; Kakehashi, 113-4.

1337   NIDSMDMA, Collections Kuribayashi, Bio info; NIDSMDMA, *Senshi-Sosho*, Central Pacific, Op. #2, Peleliu, Angaur, Iwo, Vol. 13, 278-9; Burrell, *Ghosts of Iwo Jima*, 40; Alexander, 6; Henri, et al., 6; MCRDSDA, C-2 Special Study, Enemy Situation, VAC, 6 Jan. '45, Unit Commanders, 11; Frank, *Downfall*, 282; Edgerton, 264-5; Kakehashi, 87; Cook & Cook, 83-4; Drea, 204-5; Weinberg, 6.

1338   Snow, 102; Tanaka, 110.

1339   Interview Shindo, 9 April 2018.

1340   Ibid.; Notes Shindo 3 Oct. 2018. Many of Shindo's thoughts mirror what Zen Buddhists believed back during the Imperial age. See Victoria, *Zen at War*, 58-9; Harries & Harries, 4-5.

1341   NIDSMDMA, Collections Kuribayashi, Bio info; NIDSMDMA, History Calvary Rgts, 508, 513, 515; Email Tatsushi Saito to Rigg, 23 April 2018.

1342 NIDSMDMA, Kuribayashi, Bio info; MCRDSDA, C-2 Special Study, Enemy Situation, VAC, 6 Jan. 1945, <u>Unit Commanders</u>, 11.

1343 NIDSMDMA, *Kai-Ko-Sha Kiji*, Oct. 1938 Vol. 769, Kuribayashi, "Building New Remount Administrative Plan," 83-92; NIDSMDMA, *Kai-Ko-Sha Kiji*, March 2001, "Fund for the Production to Increase Strength of the Control by the Military"; Wakabayashi (ed.), *The Nanking Atrocity, 1937-38*, Akira, "The Nanking Atrocity: An Interpretive Overview," 30; Rigg, *Lives of Hitler's Jewish Soldiers*, 160; Matthew Cooper, *The German Army, 1933-1945: Its Political and Military Failure*, NY, 1978, 279. A IJA infantry division, on average, fielded between 6,000 to 7,000 horses, so Kuribayashi had alot of responsibility to help the army remain strong in "mobility and logistics." Frank, *Tower of Skulls*, 29

1344 TSHACNCDNM, Vol. 2, 214, 221-2 from H. J. Timperley, *What War Means*, 36, 75.

1345 Kakehashi, 36.

1346  Kwong and Tsoi, 168; Lai, 32, 62.

1347 Snow, 78. Such directives were given to many commands, but they were not obeyed. Harries & Harries, 480-1.

1348 Snow, 78.

1349 <u>Ibid.</u>

1350 Ibid.

1351 Snow, 82.

1352 Email Tatsushi Saito to Rigg, 23 April 2018; NIDSMDMA, *Senshi-Sosho*, "Operations Hong Kong & Chosa," War History Series, Vol. 47, 1971, 53, 180-1, 323; Kwong and Tsoi, 171; Ferguson, 32, 48; Snow, 53. Most sources say Hong Kong's attack took place after Pearl Harbor, although Lai writes it took place nine hours before the attack on the U.S. Lai, 8. There are discrepancies about when Hong Kong's attack actually happened. Historian Benjamin Lai writes it happened nine hours before. Ferguson writes it happened six hours after Pearl Harbor. However, the Tokyo War Crimes Trials notes it happened about six hours after the attack on Pearl Harbor. UGSL, Hosch Papers, Tokyo War Crimes Trials, The Pacific War, Part B, Ch. VII, 984. Kwong and Tsoi write that Hong Kong's attack transpired one hour after Pearl Harbor.

1353 Roberts, *Churchill*, 698; Snow, 63.

1354 Snow, 74-5. In July 1941, Chiang Kai-shek had offered, generously one might add, to provide the British 200,000 Nationalist troops for the defense of Hong Kong. This British, in hindsight, unwisely rejected this offer. Frank, *Tower of Skulls*, 311.

1355 NIDSMDMA, *Senshi-Sosho*, "Hong Kong & Chosa," War History Series, Vol. 47, 1971, 53, 180-1, 323; Interview Saito, 9-10 April 2018; Kwong and Tsoi, 192, 194, 223; Email Kwong to Rigg, 6 May 2018; Lai, 13, 32; 118-20, 137, 140; Snow, 53, 64-5; Email Tatsushi Saito to Rigg, 18 March 2019.

1356 Snow, 82-3; Interview Master Johnny Lee, 9 April 2019; Scott, 66, 266. There was indeed one case documented when a Japanese man found 15 year old Priscilla Garcia in Manila on her period, he refused to rape her so maybe there was some taboo amongst Japanese not to rape someone menstruating. Scott, 262.

1357 Snow, 78-9, 82-3; Ferguson, 25, 76, 94, 205, 218.

1358 NIDSMDMA, *Senshi-Sosho*, "Hong Kong & Chosa," War History Series, Vol. 47, 1971, 53, 180-1, 323; Interview Saito, 9-10 April 2018; Kwong and Tsoi, 192, 194, 223; Email Kwong to Rigg, 6 May 2018; Lai, 13, 32; 118-20, 137, 140; Snow, 53, 64-5.

1359 Kwong and Tsoi, 177-8; Lai, 13; Snow, 54; Ferguson, 70-2.

1360 Snow, 65.

1361 NIDSMDMA, *Senshi-Sosho*, "Hong Kong & Chosa," War History Series, Vol. 47, 1971, 53, 180-1, 323; Tanaka, 92.

1362    Kwong and Tsoi, 161.

1363    Ibid., 222.

1364    Ibid., 115.

1365    Ibid., 223.

1366    NIDSMDMA, Kuribayashi, Bio info; Burrell, *Ghosts of Iwo Jima*, 41; Garand & Stobridge, 451; MCRDSDA, C-2 Special Study, Enemy Situation, VAC, 6 Jan. '45, Unit Commanders, 11.

1367    Ross, 20; Newcomb, 4-5, 7; NIDSMDMA, Records Showa Emperor, (ed.) *Kunaichō*, Tokyo, Sept. 2016, 359; Interview Shindo, 9 April 2018; NIDSMDMA, Collections Kuribayashi, Bio info.

1368    Interview Shindo, 9 April 2018.

1369    Burrell, *Ghosts of Iwo Jima*, 41.

1370    Toland, Vol. II, 797.

1371    Kakehashi, 110.

1372    Interview Shindo, 9 April 2018.

1373    NIDSMDMA, Kuribayashi, Bio info; Interview Shindo, 9 April 2018; Bradley & Powers, *Flags of Our Fathers*, 148.

1374    Ross, 20.

1375    Kakehashi, 94.

1376    Interview Shindo, 9 April 2018.

1377    Cameron, Mahoney and McReynolds, 133.

1378    Cameron, Mahoney and McReynolds, 133, 537; Bradley & Powers, *Flags of Our Fathers*, 140; Interview Shindo, 9 April 2018; Henri, et al., 13; Hough, 328); Rice, 120 & Coogan, 421, 435-7.

1379    Toland, Vol. II, 808.

1380    Crowley, 89-90; Drea, 132-3, 147-8, 151, 157; Bix, 533-4.

1381    BA-MA, RM 11/69, Gottesreich Japan v. Adm. Yamamoto Mitglied des Obersten Kriegsrates, 3.

1382    Alexander, 6; Newcomb, 19; McNeill, "Reluctant Warrior," 40; Kakehashi, 62; Garand & Stobridge, 458.

1383    John W. Dower, "Lessons from Iwo Jima," Perspectives on History, Sept. 2007.

1384    Clausewitz, *On War*, 184.

1385    Burrell, *Ghosts of Iwo Jima*, 43.

1386    Kakehashi, 55; Burrell, *Ghosts of Iwo Jima*, 43-5.

1387    NIDSMDMA, *Senshi-Sosho*, Central Pacific, Op. #2, Peleliu, Angaur, Iwo, Vol. 13, 308.

1388    McNeill, "Reluctant Warrior," 40.

1389    Kakehashi, 70.

1390    NIDSMDMA, *Kambu Gakko Kigi*, (JGSDF Staff College), Col. Fugiwara, "Commemorating Gen. Kuribayashi," Aug. 1966, Tokyo, 63.

1391    Crowl & Isely, 468.

1392    Burrell, *Ghosts of Iwo Jima*, 43.

1393    Interview Shindo, 9 April 2018.

1394    NIDSMDMA, *Senshi-Sosho*, Central Pacific, Op. #2, Peleliu, Angaur, Iwo, Vol. 13, 280-83.

1395    NIDSMDMA, Akira Fukuda, JDF Instructor, "Regarding Army Engineering at Iwo Jima," 3-5. It was strange these engineers were earmarked for Truk in March 1944 since after February, the Japanese knew Truk was no longer a viable base after the Gilbert Islands fell. Moreover, after the April 1944 U.S. raids there, Truk's base was rendered useless so it was wise Kuribayashi "stole" these engineers from being sent to a losing battle. Zeiler, 38.

1396    Toland, Vol. II, 796-8; Wheeler, 41; Burrell, *Ghosts of Iwo Jima*, 41.

1397    Kakehashi, 10.

1398    Burrell, *Ghosts of Iwo Jima*, 45.

1399    Kakehashi, 18; Hough, 27.

1400 NACPM, 127 Box 14 (3d MarDiv), HQ, FMF, Intel. Bulletin #4-44, Capt. D. Whipple, Intel Technique on Attu, 2-3; NACPM, 127 Box 14 (3d MarDiv), HQ, FMF, Camp Elliott, Subj.: Japanese Mil. Org., 29 Oct. 1942, By Com. of Maj. Gen. Barrett, Col. A.H. Noble, Chief-of-Staff, (9.) Deception practiced by Japanese (6); Drea, 231; Smith, *Coral and Brass*, 60; NASI PR, Erskine, Commandant Lt. Gen. T. Holcomb to Sen. David Walsh, 3 Nov. 1943.

1401 NIDSMDMA, *Senshi-Sosho*, Central Pacific, Op. #2, Peleliu, Angaur, Iwo, Vol. 13, 320-1.

1402 Alexander, 4; Bradley & Powers, *Flags of Our Fathers*, 141; King, 76; Hough, 257; Kakehashi, 65.

1403 Kakehashi, 8.

1404 Ibid., 8.

1405 Ibid., 179.

1406 Ibid., 43-4.

1407 NIDSMDMA, *Senshi-Sosho*, Central Pacific, Op. #2, Peleliu, Angaur, Iwo, Vol. 13, 329-30.

1408 Ibid., 335.

1409 Ibid.

1410 Ibid., 329.

1411 Clausewitz, *On War*, 153.

1412 Kakehashi, 65; Hough, 257.

1413 NACPM, 127 Box 14 (3d MarDiv), Intel. Bulletin: Japanese Tactics and Strategy, 29 Oct 1942-29 May 1945 Folder 7, Col. Robert E. Hogaboom, 29 May 1945, 1, 7, Section (1) (a) and (3) (b) (2); Kakehashi, 60, 64.

1414 Kakehashi, 65.

1415 Interview Shindo, 9 April 2018; Interview King, 21 July 2016; Henri, et al., 92; Wheeler, 42-3; Garand & Stobridge, 456; NACPM, 127 Box 14 (3d MarDiv), Intel. Bulletin: Recent Trends Japanese Defenses, 29 Oct 1942-29 May 1945 Folder 7, Col. R. E. Hogaboom, 4 May 1945, 1, Sec. (1.) Emplacements; Smith, *Coral and Brass*, 142; Kakehashi, 123; McNeill, "Reluctant Warrior," 40; Interview Oelke Farley, 8 Nov. 2017.

1416 NACPM, 127 Box 14 (3d MarDiv), Intel. Bulletin: Recent Trends in Japanese Defenses, 29 Oct 1942-29 May 1945 Folder 7, Col. R. E. Hogaboom, 4 May 1945, 1, Sec. (2.) Emplacements.; Hough, 345; Interview Oelke Farley, 8 Nov. 2017; Crowl & Isely, 485.

1417 Burrell, *Ghosts of Iwo Jima*, 47; NIDSMDMA, *Senshi-Sosho*, Cen. Pacific, Op. #2, Peleliu, Angaur, Iwo, Vol. 13, 348.

1418 NIDSMDMA, *Senshi-Sosho*, Cen. Pacific, Op. #2, Peleliu, Angaur, Iwo Jima, Vol. 13, 348.

1419 Toland, Vol. II, 801-2; McNeill, "Reluctant Warrior," 40; Kakehashi, 7, 17, 156.

1420 Kakehashi, 8.

1421 King, 31, 59-60; Burrell, *Ghosts of Iwo Jima*, 39; McNeill, "Reluctant Warrior," 40; Burleigh, 330; Kakehashi, 34, 66, 125; Harries & Harries, 245; Snow, 102; Tanaka, 110. Dan King doubts women were on Iwo. Email King to Rigg, 9 May 2017. However, veteran Don Graves says he saw women on Iwo. Interview Graves, 7 May 2017. Shayne Jarosz, Ex. Dir. of the IJAA, says Iwo veterans George Alden and Jack Lazarus claimed the same thing and that they found these women "dressed in uniforms wearing thousand stitch belts." Email Jarosz to Don Farrell, 11 May 2017; Interview Shayne Jarosz, 12 May 2017. In *Black Hell*, Bingham writes veteran Earl Stephenson saw women in the tunnels. Bingham, 159. Historian Jeffrey Ethell interviewed a Japanese Iwo veteran who said some officers had brought their Korean mistresses/comfort women with them to Iwo. Conversation with Heaton, 27 May 2017. Historian James Oelke Farley believes the stories of women being on the island. He said it was common for IJA officers to bring their prostitutes and comfort women with them in combat zones. Interview Farley, 8 Nov. 2017; IJAA symposium lecture by Iwo veteran Ira Rigger, 17 Feb. 2018. Rigger was a Seabee on the island and personally found a woman in the tunnels. Iwo veteran George Bernstein also claimed he found women on the island. Interview

George Bernstein, 22 March 2018. On other battlefronts, Japanese females were also found to have fought against the Marines. See Moore, 209.

1422   Henri, et al., vii.

1423   McNeill, "Reluctant Warrior," 40.

1424   Clausewitz, *On War*, 331.

1425   NACPM, RG 127 Container 14, Dashiell, Folders 1-2, #211.

1426   Email Dan King to Rigg, 20 Feb. 2018.

1427   Crowl & Isely, 316, 432.

1428   Clausewitz, *On War*, 203.

1429   Kakehashi, 11; Burrell, *Ghosts of Iwo Jima*, 42.

1430   Kakehashi, 42.

1431   Burrell, *Ghosts of Iwo Jima*, 39.

1432   Kakehashi, 29-30.

1433   Ibid., 10.

1434   Burrell, *Ghosts of Iwo Jima*, 38.

1435   Kakehashi, 28.

1436   Burrell, *Ghosts of Iwo Jima*, 48; Toland, Vol. II, 799.

1437   Burrell, *Ghosts of Iwo Jima*, 4; Wheeler, 40. See also Henri, et al., 97.

1438   Manchester, 337; Toland, Vol. II, 795.

1439   Blair & DeCioccio, 255.

1440   Toland, Vol. II, 800-1; Henri, et al., 8; Bingham, 7.

1441   Ross, 151; Kakehashi, 39.

1442   Kakehashi, 39.

1443   Ibid., 40.

1444   NIDSMDMA, *Senshi-Sosho*, Cen. Pacific, Op. #2, Peleliu, Angaur, Iwo, Vol. 13, 351; Ross, 150; Kakehashi, 40.

1445   Alvin M. Josephy, Jr., *The Long and the Short and the Tall: The Story of a Marine Combat Unit in the Pacific*, NY, 1946, 170.

1446   Kakehashi, 159-60; NIDSMDMA, *Senshi-Sosho*, Cen. Pacific, Op. #2, Peleliu, Angaur, Iwo, Vol. 13, 350-1.

1447   Farrell, *Liberation-1944,* 102.

1448   Melville, 159.

1449   Kakehashi, 41.

1450   Burrell, *Ghosts of Iwo Jima*, 44.

1451   Clausewitz, *On War*, 69,87 & Brodie, "A Guide to Reading of On War," 645.

1452   Kakehashi, 45.

1453   Clausewitz, *On War*, 178.

1454   Ibid., 180.

1455   Ibid.

1456   Hoffman, *Once a Legend*, 262.

1457   Moskin, 113-4.

1458   Interview Woody, 23 July 2016.

1459   Interview Shindo, 9 April 2018; Notes by Shindo, 3 Oct. 2018.

1460   Melville, 205.

1461   Ibid., 203.

1462   Ibid., 203.

1463   Ibid., 427.

1464   Kakehashi, 90.

1465    Poe, "Murders in the Rue Morgue," 403.

1466    Toland, Vol. II, 802.

1467    *Ibid.*, 803.

1468    Ibid.

1469    MCHDHQ, Letter Donald Beck to Ruth Beck, 8 Feb. 1945.

1470    *Ibid.*, 20 Jan. 1945.

1471    Ibid., 13 Feb. 1945.

1472    Ibid., 20 Jan. 1945.

1473    Bix, xxi.

1474    MCHDHQ, Woody's archive, "No Word From Jim Hamilton Following Battle of Iwo Jima," 1 (original sources are from the personal records of the Hamilton family, from the *Times News* and the *Buhl Herald* from 1945 and 1946); National Archives, DOD Dir 5200.9, Enclosure D, 21$^{st}$ Marines, Action Report, 6 April 1945, 1; Miller, "Deathtrap Island," 12; Burrell, "Did We Have to Fight Here?," 62; Blum, 113; MCHDQV, HQ, 1$^{st}$ Btl., 21$^{st}$ Marines, 3$^{rd}$ MarDiv, FMF, In Field, 6 April 1945, Report Lt. Col. Williams, 1.

1475    Francois, 43.

1476    Interview Woody, 20 July 2016; Hallas, *Uncommon Valor on Iwo Jima*, 304.

1477    Mandel, 260.

1478    Henri, et al., 21. Bradley & Powers, *Flags of Our Fathers*, 144.

1479    Melville, 210.

1480    Mandel, 16, 263-4.

1481    Newcomb, 54. The chaplain here is quoting the King James Bible verbatim. See KJV (King James Version), p. 1720 for Ephesians 6:11.

1482    *Ibid.*, 58.

1483    Mandel, 239, 240, 290.

1484    Interview Lauriello, 19 Nov. 2016.

1485    Interview Woody, 21 July 2016.

1486    Newcomb, 74.

1487    Interview Yellin, 22 March 2015.

1488    Sledge, 270.

1489    Discussion with Frank Rigg's sister, Mary Dalbey née Rigg, Aug. 2001; James H. Hallas, *The Devil's Anvil: The Assault on Peleliu*, London, 1994, 254; NACPM, 127 Box 14 (3d MarDiv), Intel. Bulletin: Recent Trends Japanese Defenses, 29 Oct. 1942-29 May 1945 Folder 7, Col. R. E. Hogaboom, 4 May 1945, 2, Sec. (3.) Use of Caves and Dugouts (c).

1490    Hallas, *Saipan*, 477.

1491    Hane, 12; Edgerton, 253; King, 149; McNeill, "Reluctant Warrior," 40.

1492    KJV (King James Version).

1493    Prange, 12; Toland, Vol. I, 256.

1494    BA-MA, RM 11/69, Gottesreich Japan von Adm. Yamamoto Mitglied des Obersten Kriegsrates, 1-2.

1495    Paul Johnson, *A History of Christianity*, NY, 1995, 6; George Hart, *A Dictionary of Egyptian Gods and Goddesses*, NY, 1986, 179-82.

1496    *Singing the Living Tradition*, Univ.-Unitarian Song Book, Meditations, 563.

1497    Harris, 226.

1498    Hane, 12.

1499    Edgerton, 323.

1500    Cook & Cook, 26; Toland, Vol. I, 376; Moore, 64.

1501    Harris, 134.

1502    Kakehashi, 146.

1503    Cook & Cook, 5, 124; Newcomb, 41.

1504 MCHDHQ, Dalton, "Interview Woody Williams," 4.

1505 Toland, Vol. II, 795. Woody, unpublished article, "Leaving Guam in June 1944," 2.

1506 Interview Woody, 19 July 2016; NACPM, 127 Box 14 (3d MarDiv), HQ, FMF, Intel. Bulletin #4-44, Capt. D. Whipple, Prisoners of War., 2 & Camp Elliott, Subj.: Japanese Mil. Org., 29 Oct. 1942, Command Maj. Gen. Barrett, Col. A.H. Noble, Chief of Staff, (9.) Deception… by Japanese (9) & (11).

1507 Smith, *Coral and Brass*, 91.

1508 Interview Woody, 19 July 2016; Interview Woody, 12 Feb. 2017; Zdon, "A Marine in the Pacific," 8; Smith, *Iwo Jima,* 65; O'Brien, *Liberation*, 5; Interview Woody, 12 Feb. 2017.

1509 MCHDHQ, Woody, unpublished article, "Leaving Guam in June 1944," 1.

1510 Francois, 43.

1511 Garand & Stobridge, 461.

1512 King, 33; Hanley II, 379.

1513 Bingham, 15; Mandel 257.

1514 Crowl & Isely, 470; Bradley & Powers, *Flags of Our Fathers*, 144-5; Toland, Vol. II, 804-5; King, 79; Henri, et al., 15, 19; Wheeler, 32; Eric Hammel, *Iwo Jima*, Minneapolis, MN, 2006, 41 (Jon T. Hoffman says one needs to use Hammel with caution since he makes many mistakes and is not a historian. Report *Flamethrower*, Hoffman, 19 Dec. 2017); Newcomb, 28; Miller, "Deathtrap Island," 10; McNeill, "Reluctant Warrior," 40; Smith, *Coral and Brass*, 129; Burrell, *Ghosts of Iwo Jima*, 52, 105. Bradley and Henri, Beech, Dempsey, Josephy and Dunn write the bombing of Iwo persisted for 72 days and Miller writes it was 74 days, but Dan King says when you take into account aerial as well as naval bombardment, the island had been hit for 249 days (King, 33).

1515 Henri, et al., 28.

1516 Kakehashi, 119; The Carryall, Iwo Jima, Feb. 19, 1945—Oct. 15, 1945, 133rd NCB Navy Seabee Souvenir Newsletter, Reprint Ken Bingham, Vol. III, No. 8, Gulffort MS, 2011, 14.

1517 Henri, et al., 23.

1518 Kakehashi, 101-2.

1519 Ibid., 104.

1520 Wheeler, 44; Blum, 54; Toland, Vol. II, 807.

1521 Crowl & Isely, 465.

1522 Burrell *Ghosts of Iwo Jima*, 60.

1523 King, 113; Newcomb, 66-7; Spector, 498; Burrell, *Ghosts of Iwo Jima*, 59; Crowl & Isely, 464, 468; Hallas, *Iwo Jima*, 4-5; NASLPR, Rufus G. Herring, MOH 1st Endorsement, Commander LCI (G) Flotilla Three to Commander Amphibious Forces, U.S. Pacific Fleet, 14 April 1945 & Commander LCI (G) Group Eight to Commander Amphibious Forces, U.S. Pacific Fleet, 24 March 1945. Ross believes Kuribayashi actually ordered this attack on the LCIs on 17 February 1945 because he believed "the invasion had begun." See Ross, 48-51. However, this seems unlikely because he had studied the USMC's tactics and would or should have known that amtracs and Higgins boats announced when an invasion was truly underway.

1524 Hallas, *Iwo Jima*, 4-5; NASLP, Rufus Geddie Herring, MOH Citation & MOH 1st Endorsement, Commander LCI (G) Flotilla Three to Commander Amphibious Forces, U.S. Pacific Fleet (USPF), 14 April 1945 & Commander LCI (G) Group Eight to Commander Amphibious Forces, USPF, 24 March 1945.

1525 King, 113; Newcomb, 66-7; Spector, 498; Burrell, *Ghosts of Iwo Jima*, 59; Crowl & Isely, 464, 468; Hallas, *Iwo Jima*, 4-5, 9-10; NASLPR, Herring, MOH Citation & MOH 1st Endorsement, Commander LCI (G) Flotilla Three to Commander Amphibious Forces, USPF, 14 April 1945 & Commander LCI (G) Group Eight to Commander Amphibious Forces, USPF, 24 March 1945 & Report of Medical Survey, U.S. Naval Hospital, Bethesda, MA, Herring, 26 Jan. 1946 & Report of Medical Survey, U.S. Naval Hospital, Quantico, VA, Herring, 14 May 1945; Ross, 49.

1526    NASLPR, Herring, Commander LCI (G) Group Eight to Commander Amphibious Forces, USPF, 24 March 1945. Quote came from Ross, 48.

1527    Crowl & Isely, 469.

1528    King, 113.

1529    Crowl & Isely, 469.

1530    Ibid., 478.

1531    Ibid., 469. See also Henri, et al., 5.

1532    King, 114; Spector 498.

1533    Bradley & Powers, *Flags of Our Fathers*, 144-5; Newcomb, 26-7; Smith, *Coral and Brass*, 17; Burrell, *Ghosts of Iwo Jima*, 54.

1534    Smith, *Coral and Brass*, 130.

1535    Bradley & Powers, *Flags of Our Fathers*, 145.

1536    Spector, 499.

1537    Alexander, 31.

1538    Toll, *Conquering Tide*, 365.

1539    Harry W. Hill, CCOH Naval History Project, No. 685, Vol. 3, 305.

1540    Morison, 73.

1541    NACPM, 127 Box 14 (3d MarDiv), Intel. Bulletin: Recent Trends Japanese Defenses, 29 Oct 1942-29 May 1945 Folder 7, Col. R. E. Hogaboom, 4 May 1945, 1, Sec. (1.) General.

1542    Smith, *Coral and Brass*, 126; Weinberg, 868.

1543    Smith, *Coral and Brass*, 130; Weinberg, 867.

1544    Frank, Review *Flamethrower*, 8 Aug. 2018.

1545    Sakai, 272.

1546    Spector, 499. A few weeks into the battle, the Japanese had difficulty with their water supply. Yet, according to Burrell, before the battle the Japanese had "approximately four months' of rations and forage." So they had plenty of food to weather months of bombardment. Burrell, *Ghosts of Iwo Jima*, 48.

1547    NIDSMDMA, *Senshi-Sosho*, Central Pacific, Op. #2, Peleliu, Angaur, Iwo, Vol. 13, 391; Crowl & Isely, 483 (loc. 10526); Newcomb, 12, 35; McNeill, "Reluctant Warrior," 40. Crowl & Isely write that 700 Koreans were on the island, but NIDSMDMA write 1,600 were garrisoned there.

1548    Crowl & Isely, 485 (loc. 10571); Burrell, *Ghosts of Iwo Jima*, 47.

1549    Garand & Stobridge, 454.

1550    Burrell, Review *Flamethrower*, 26 Feb. 2018.

1551    Crowl & Isely, 483-5 (loc. 10526).

1552    Ibid., 485-6 (loc. 10580).

1553    Mandel, 262-3.

1554    Moskin, 359; Henri, et al., 1, 11, 52, 166.

1555    Ross, 47; Toland, Vol. II, 806; Newcomb, 24, 30, 75; Burrell, *Ghosts of Iwo Jima*, 8; Smith, *Coral and Brass*, 134.

1556    Smith, *Coral and Brass*, 31.

1557    Frank, *Downfall*, 60. Pronounced "Fifth Amphibious Corps."

1558    Alexander, 3-4.

1559    Alexander, 3-4; Newcomb, 22.

1560    NACPM, RG 127, Box 96, VAC, DSM citation Schmidt signed Forrestal.

1561    Henri, et al., 11.

1562    NASLPR, Holland M. Smith, Report on Fitness of Officers of the United States Marine Corps, Reporting officer Admiral C. W. Nimitz, 19 Sept. 1944. See also Millett, 413-5.

1563    USMC Museum, Quantico VA, display on WWII Marines.

1564    Clausewitz, *On War*, 119.

1565    Morison, 33.

1566 King, 111; Newcomb, 23.

1567 Smith, *Coral and Brass*, 63; Burrell, *Ghosts of Iwo Jima*, 51.

1568 MCHDQV, Oral History, Erskine, 183-6; Gandt, 51.

1569 MCHDQV, Oral History, Erskine, 184-7. Interesting, members in the Bureau of Ships debated the semantics of how to identify amtracs and whether one should call it a boat or vehicle. Ultimately, since amtracs moved through the water "by a track rather than by a propeller, it was called a vehicle." Krulak, 104.

1570 Millett, 392.

1571 Morison, 65.

1572 Smith, *Coral and Brass*, 134; Morison, 32.

1573 Smith, *Coral and Brass*, 134.

1574 Morison, 32, 35; Raymond Henri, 15, 20.

1575 Mandel, 265.

1576 Francois, 43; Interview Woody, 21 July 2016; Dalton, "Interview with Woody Williams," 5; Henri, et al., 22; VandeLinde, 365, 368.

1577 NACPM, RG 127 Box 21 (4ᵗʰ MarDiv), (1540) (1 of 2) Folder 5, HQ, 4ᵗʰ MarDiv, FMF, Com. Gen. to Com. Gen., FMF, Pacific, 21 April 1945, 3.

1578 Francois, 43; Interview Woody, 21 July 2016; Dalton, "Interview with Woody Williams," 5; Henri, et al., 22; VandeLinde, 365, 368; Bingham, 58.

1579 Raymond Henri, 27; Toland, Vol. II, 808; Henri, et al., 24-5; Ross, 62.

1580 Morison, 37-8; Newcomb, 95; NACPM, RG 127 Box 21 (4ᵗʰ MarDiv), (1540) (1 of 2) Folder 5, HQ, 4ᵗʰ MarDiv, FMF, Com. Gen. to Com. Gen., FMF, Pacific, 21 April 1945, 4.

1581 Review *Flamethrower* Jon Hoffman, 28 Aug. 2018.

1582 Mandel, 261; Ross, 62.

1583 Toland, Vol. II, 808. Drea mistakenly writes all three divisions hit Iwo on 19 Feb. Drea, 246.

1584 NACPM, RG 127, Box 96 (Iwo Jima), A46-2, 3ʳᵈ Btl., 21ˢᵗ Marines Rgt., Don Pryor interview with Adm. Turner & Gen. Smith, 28 Feb. 1945, 2.

1585 Smith, *Coral and Brass*, 137.

1586 Crowl & Isely, 468.

1587 MCRDSDA, C-2 Special Study, Enemy Situation, VAC, 10 Jan. 1945, Supplement No. 1 to CINCPAC-CINCPOA Bulletin No. 122-44.

1588 Toland, Vol. II, 800-1, 811; Henri, et al., 26.

1589 *Two Score and Ten*, 56; "NAVAL HISTORICAL CENTER, Navajo Code Talkers: World War II Fact Sheet". *History.navy.mil.* 1992-09-17;*NACPM,* RG 127 Box 21, 4ᵗʰ MarDiv (1530), Folder 3, HQ, 4ᵗʰ MarDiv, FMF, San Francisco, CA, E.A. Pollock (By Direction) Com. Gen. to Com. Gen., VAC, 10 July 1945; "Navajo Code Talker Samuel Holiday Dies in Ivins at Age 94," St. George News, 12 June 2018.

1590 Toland, Vol. II, 819; Frank, 143. Rhodes, 594; Newcomb, 204-5; Weinberg, 867; Lord Russell of Liverpool, 54; Lee Mandel, *Sterling Hayden's Wars*, Jackson, 2018, 99; Frank, *Downfall*, 143; Edgerton, 314; Drea, 203; Bix, 349.

1591 Wallace and Williams, 9.

1592 MCHDQV, HQ, 1ˢᵗ Btl., 21ˢᵗ Marines, 3d MarDiv, FMF, In Field, 6 April 1945, Report Lt. Col. Williams, 22; Kakehashi, 68.

1593 Hoffman, Review *Flamethrower*, 2 Sept. 2018.

1594 Drea, 203; Cook & Cook, 44-5; Frank, "Review *Flamethrower*," 25 May 2017; NACPM staff, "Japanese War Crimes Records at the National Archives," in *Researching Japanese War Crimes Records*, 89-91; NACPM, RG 165, Box 2153, War Dept., 6910 (Japan), Military Attaché, 22 Oct. 1942, 3. (6) Use of Gas; MCRDSDA, HQ 8ᵗʰ Marines, 2ⁿᵈ Marine Brigade, July-Aug. 1942, Japanese Tactics, Warfare and Weapons, 15 Aug. 1942, 3.

1595 MCRDSDA, MCRD Training Book, File 1, <u>Chemical Warfare</u>; MCHDQV, HQ, 1ˢᵗ Blt., 21ˢᵗ Marines, 3d MarDiv, FMF, In Field, 6 April 1945, Report Lt. Col. Williams, 22. Cook & Cook, 44-5, 162-5, 199-202; Drea, 203; Bix, 361-4.

1596 NACPM, RG 127 Box 96 (Iwo) A45-2, 21ˢᵗ Regt. Journal Iwo, 25 Feb. 1945, R-3 Log, Time 2345 to CO-9.

1597 Cook & Cook, 44-5, 162-5, 199-202; Drea, 203; Bix, 361-4; Takashi Yoshida, 29-30; Moore, 127-9, 227. Bubonic plague, or *Yersinia pestis*, happened often throughout the Middle-Ages in Europe. It got the name, "Black Death," due to a symptom people displayed when suffering with this bacteria; namely, lymph nodes turn black and swell once someone was infected with this disease.

1598 Bix, 362.

1599 <u>Ibid.</u>, 739 fn. 8. See also Frank, *Tower of Skulls*, 79.

1600 Wallace and Williams, 23, 81, 247; Cook & Cook, 146-51, 159; Justin McCurry, "Japan Unearths Site Linked to Human Experiments," The Guardian, 21 Feb. 2011; Bradley, *Flyboys*, 113, 293; Bradley & Powers, *Flags of Our Fathers*, 59-60; Frank, 324-5; Spence, 496; Drea, 261; NACPM staff, "Japanese War Crimes Records at the National Archives: Research Starting Points," *Researching Japanese War Crimes Records*, 97; Edgerton, 284; Tanaka, 153-4. Frank, *Downfall*, 324-5.

1601 Davis, "Operation Olympic," 21; Bix, 362.

1602 NACPM, RG 127 Box 23 (4ᵗʰ MarDiv), (1975) for 1944, Folder 2, HQ, 20ᵗʰ Marines (eng.), 4ᵗʰ MarDiv, FMF, San Francisco, 6 Aug. 1944, (c) Intel, point 3.

1603 Brophy, Miles and Cochrane, 74; Rhodes' *Making of the Atomic Bomb*; O'Brien, *The Second Most Powerful Man in the World*, 298.

1604 Discussion with King, 15 July 2016; King, 90; Newcomb, 34; NACPM, RG 127, Box 83, (Iwo), AB-1, Com. Gen. 3d MarDiv Erskine to Commandant, Report, 3 June 1945, 79.

1605 Crowl & Isely, 487, (loc 10604).

1606 Henri, et al., 28; Newcomb, 97.

1607 Manchester, 314.

1608 Kakehashi, 104.

1609 Henri, et al., 26, 30.

1610 <u>Ibid.</u>, 30.

1611 Interview Lauriello, 19 Nov. 2016. In the USMC casualty reports, his name is spelled Kinlacheeny.

1612 Clausewitz, *On War*, Brodie, "A Guide to Reading of On War," 660.

1613 Kakehashi, 53.

1614 Poe, "The Pit and the Pendulum," 494.

1615 Interview Woody, 21 July 2016; Spector, 499.

1616 Bradley & Powers, *Flags of Our Fathers*, 158-9; USMCU Casualty Web Data Base.

1617 Toland, Vol. II, 810; Henri, et al., 34-5.

1618 <u>Ibid.</u>, 810-1.

1619 Clausewitz, *On War*, 198.

1620 <u>Ibid.</u>, 200.

1621 Moskin, 363-4; Alexander, 19.

1622 Alexander, 19; Henri, et al., 34.

1623 Ross, 80.

1624 Muellener, 219.

1625 NACPM, RG 127 Container 14, Dashiell, Folders 1-2, #184, 2.

1626 Toland, Vol. II, 809; Francois, 43; VandeLinde, 369.

1627 Bradley & Powers, *Flags of Our Fathers*, 163, 165: Henri, et al., 49.

1628 Ross, 69; Newcomb, 225; USMCU Casualty Web Data Base. When the Navy chaplain called on Mrs. Anderson to tell her of her son's death, she said, "A force stronger than ours has taken charge and our beloved son is with us on Earth no more." She then put on her outfit and went to her volunteering station at the Naval Hospital in Bethesda. Newcomb, 225. Ross places Anderson's death at the campaign's beginning while Newcomb places it at the end.

1629    Ross, 124.

1630    Bill Dobbins served in the US Navy assigned to Company B, 2nd Medical Btl., 2nd MarDiv. and he served with the 3d MarDiv's 3rd Tank Btl. See also Ross, 124.

1631    NACPM, RG 127, Box 83, (Iwo), AB-1, Com Gen. 3d MarDiv Erskine to Commandant, Report, 3 June 1945, 34.

1632    Interview Leighton Willhite, 30 Nov. 2019; Henri, et al., 100.

1633    Smith, *Coral and Brass*, 141.

1634    MCRDSDA, 2nd Blt., 24th Marines, 4th MarDiv, Battle Reports, Iwo, Narrative, 5; Henri, et al., 38-40.

1635    Morison, 47.

1636    Burrell, *Ghosts of Iwo Jima*, 63.

1637    Miller, "Deathtrap Island," 13.

1638    Manchester, 340.

1639    Clausewitz, *On War*, Introductory Essay by Michael Howard, "The Influence of Clausewitz," 35.

1640    Toland, Vol. II, 804; Crowl & Isely, 482; Moskin, 363; Newcomb, 115; Miller, "Deathtrap Island," 13; Bradley in *Flags of Our Fathers*, 143.

1641    Morison, 70.

1642    Hammel, 109.

1643    Hammel, 93; Burrell, *Ghosts of Iwo Jima*, 219, fn 85; MCRDSDA, 2nd Blt., 24th Marines, 4th MarDiv, Operation Journal, 1 March 45, Times 0935 & 1015. Some napalm bombs had difficulties detonating. See Crowl & Isely, 510.

1644    U.S. Marine Corps Publication, "Planning," MCDP 5, 21 July 1997, 22; Clausewitz, *On War*, Brodie, "A Guide to Reading of On War," 648-9.

1645    U.S. Marine Corps Publication, "Planning," MCDP 5, 21 July 1997, 61.

1646    "Medal of Honor recipients". *World War II (A – F)*. United States Army Center of Military History. *2009-06-08. Archived from* the original *on June 16, 2008. Retrieved 2009-06-08.* The Story of Gunnery Sergeant John Basilone Part 3". *John Basilone Parade Website. Retrieved October 5, 2005.*

1647    Henri, et al., 32; Wheeler, 46; Hallas, *Uncommon Valor on Iwo Jima*, 46; Interview Woody, 21 Oct. 2016; Newcomb, 94; Miller, "Deathtrap Island," 13; Smith, *Coral and Brass*, 120.

1648    Henri, et al., 32; Wheeler, 46; Hallas, 46; Interview Woody, 21 Oct. 2016; Newcomb, 94; Miller, "Deathtrap Island," 13. According to Manchester, the large mortars were also referred to as "screaming meemies, box-car Charlies, and flying seabags." Manchester, 382. Gen. Smith called them "buzzbombs." Smith, *Coral and Brass*, 120; Interview Leighton Willhite, 30 Nov. 2019.

1649    Sledge, 74.

1650    MCHDQV, Oral History, Robert E. Cushman Jr., 177.

1651    Hallas, *Uncommon Valor on Iwo Jima*, 210.

1652    Toland, Vol. II, 812.

1653    Morison, 200.

1654    Toland, Vol. II, 811-2; Henri, et al., 8.

1655    Hammel, 95.

1656    Ross, 81.

1657    <u>Ibid.</u>, 104.

1658    Manchester, 242.

1659    Henri, et al., 47.

1660    Alexander, 5; Bingham, 7; Kakehashi, 66; Burrell, *Ghosts of Iwo Jima*, 29; Smith, *Coral and Brass*, 129; MCRDSDA, C-2 Special Study, Enemy Situation, VAC, 6 Jan. 1945, 1. Enemy Order of Battle. A.) Estimate Enemy Strength—Iwo, 3. The Marines had made the mistake also at Eniwetok of underestimating enemy strength in Feb. 1944 thinking 800 defended the Atoll when in reality, 3,500 fought there. Zeiler, 40.

1661 Nalty, Shaw and Turnbladh, 586.

1662 Manchester, 337.

1663 Interview Leighton Willhite, 10 May 2017.

1664 Newcomb, 65; Miller, "Deathtrap Island," 12; Sledge, 48; Burrell, *Ghosts of Iwo Jima*, 32.

1665 Beck's lecture on Iwo Jima, NROTC Midshipmen, Purdue Univ., 1952, 3-4.

1666 Kakehashi, 101; MCRDSDA, C-2 Special Study, Enemy Situation, VAC, 6 Jan. 1945, 17-8.

1667 Kakehashi, 106.

1668 Ibid., 49-50. See also Toland, Vol. II, 665.

1669 Clausewitz, *On War*, 170.

1670 Donald Miller, *The Story of World War II*, NY, 2001, 422; Toland, Vol. II, 741-2.

1671 NACPM, RG 127 Box 22 (4ᵗʰ MarDiv), (1875) (2 of 2), HQ, 4ᵗʰ MarDiv, FMF, San Fran., CA, Lt. William R. Wendt, Div. Air Officer to Com. Gen. (Cates), 1 Jan. 1945; NACPM, 127 Box 14 (3d MarDiv), Intel Sec. 9ᵗʰ Marine 3d MarDiv, FMF, In Field, Intel Bulletin #2-44, Capt. Robert Campbell, 2; Drea, 225.

1672 Moskin, 363-4.

1673 Interview Woody, 22 July 2016; Hoffman, *Chesty*, 269; Crowl & Isely, 469.

1674 NACPM, RG 127 Box 83 A 13-8 3d MarDiv AR, 19 Feb. 25 March 3ʳᵈ Eng. Btl., 5.

1675 Alexander, 23.

1676 NACPM, RG 127 Box 21, (4ᵗʰ MarDiv), (1540) (1 of 2) Folder 5, HQ, 4ᵗʰ Eng. Btl., 4ᵗʰ MarDiv, FMF, In Field, Com. officer N. K. Brown to Com. Gen., 4ᵗʰ MarDiv., 16 March 1945, 2.

1677 NACPM, RG 127 Box 83 A 13-8 3d MarDiv AR, 19 Feb. 25 March 3ʳᵈ Eng. Btl., 5; Crowl & Isely, 485.

1678 Mandel, 282.

1679 NACPM, 127 Box 14 (3d MarDiv), Intel. Bulletin: Recent Trends Japanese Defenses, 29 Oct 1942-29 May 1945 Folder 7, Col. R. E. Hogaboom, 4 May 1945, 3-4, sec. (4) Minefields (b); Hallas, *Uncommon Valor on Iwo Jima*, 145; Henri, et al., 43.

1680 Nalty, Shaw and Turnbladh, 573; Francois, 43.

1681 Interview Woody, 20 July 2016; VandeLinde, 377; MCHDHQ, Notes on manuscript by Woody Williams, Oct. 2016.

1682 Francois, 43.

1683 Interview Woody, 20 July 2016.

1684 Smith, *Iwo Jima*, 65.

1685 MCHDQV, Oral History, Woody Williams 13801B; Henri, et al., 78.

1686 Interview Lee, 24 June 2016; Zdon, "A Marine in the Pacific," 10; Interview Woody, 20 July 2016; MCHDHQ, Woody, unpublished article, "Leaving Guam in June 1944," 2; Henri, et al., 78-9; VandeLinde, 368; Wells, "Medal of Honor Recipient Devotes Life to Veterans."

1687 Interview Woody, 20 July 2016; Francois, 44.

1688 Francois, 44.

1689 Ibid.

1690 Interview Lee, 24 June 2016; Zdon, "A Marine in the Pacific," 10; Interview Woody, 20 July 2016; Woody, unpublished article, "Leaving Guam in June 1944," 2; Henri, et al., 78-9; VandeLinde, 368; Wells, "Medal of Honor Recipient Devotes Life to Veterans."

1691 J.D. Moore, "Iwo Jima Eyewitness," Pass in Review, Feb. 1989, 9.

1692 NACPM, RG 127 Container 14, Dashiell, Folders 1-2, #201.

1693 Charles Neimeyer, "Eyewitness to Iwo," Marine Corps Gazette, Vol. 92, Issue 8, Aug. 2008.

1694 Talk given by Woody at MCRD in the museum, 19 Oct. 2017.

1695 Interview Woody, 23 July 2016.

1696 MCHDQV, Oral History, Woody Williams 13801B; Hammel, 75.

1697   Interview Woody, 23 July 2016.

1698   Interview Lauriello, 19 Nov. 2016; Shakespeare's *King Lear*, Act 4, Scene 1.

1699   MCHDQV, Dashiell File, #307, Life-Saver.

1700   National Archives, DOD Dir 5200.9, Enclosure D, 21ˢᵗ Marines, Report, 6 April 1945 & 10 April 1945, 5.

1701   Moskin, 364; Toland, Vol. II, 813; Newcomb, 136-7; Burrell, *Ghosts of Iwo Jima*, 68. Ross, 115-118. Most studies, like in Ross' book, estimate that there were 50 Kamikazes attacking ships off of Iwo on 21 February 1945. However, historian Robert Burrell cites a credible, and, possibly, a definitive study that proves there were "only" 32 Kamikazes in the air this day. See Burrell, 223 fn 150.

1702   NACPM, RG 127 Box 83 A 13-7, 3d MarDiv. Tank Btl. Report, 2 April 1945, 1. Ross, 118

1703   King, 123-4.

1704   Cook & Cook, 324.

1705   Ibid., 219.

1706   Frank, 182; Keegan, 568-72; Frank, Review *Flamethrower*, 25 May 2017; Bix, 485; Hane, 353; Edgerton 297-9.

1707   https://archive.org/details/1945RadioNews/1945-02-19-CAN-Live-Coverage-Of-US-Marines-Landing-On-Iwo-Jima.mp3

1708   Toland, Vol. II, 812. MCHDQV, HQ, 1ˢᵗ Btl., 21ˢᵗ Marines, 3d MarDiv, FMF, In Field, 6 April 1945, Report Lt. Col. Marlowe C. Williams, 15; NACPM, RG 127, Box 83, (Iwo), AB-1, Com. Gen. 3d MarDiv Erskine to Commandant, Report, 3 June 1945, 38; Morison, 73.

1709   Toland, Vol. II, 812.

1710   Dower, 92; Edgerton, 283.

1711   Dower, 53; Newcomb, 104; Henri, et al., 64.

1712   MCHDHQ, Dolan, "Conversation: Forged in Flame"; Interview Lee, 24 June 2016; Zdon, "A Marine in the Pacific," 10; Scott, 293; Moore, 137.

1713   https://archive.org/details/1945RadioNews/1945-02-xx-NBC-Battle-for-Iwo-Jima-Bud-Foster.mp3

1714   Interview Gary Fischer, 11 Jan. 2020.

1715   Hallas, *Uncommon Valor on Iwo Jima*, 143.

1716   Ross, 173.

1717   Harry Levin (ed.), *The Portable James Joyce,* "A Portrait of the Artist as a Young Man," NY, 1976, 375.

1718   Hallas, *Uncommon Valor on Iwo Jima*, 140.

1719   Melville, 254; Christopher Hitchens, *Hitch-22: A Memoir*, NY, 2010, 3; Martin Gilbert, *The Second World War: A Complete History*, NY, 1989, 10.

1720   Sledge, 60, 242.

1721   MCHDQV, Oral History, Erskine, 356-7.

1722   MCHDQV, Dashiell File, #304, Cave Burner; NASLPR, William Nolte, Bronze Star Citation; NASLPR, William Naro, Naval Speedletter, 1 May 1945.

1723   Henri, et al., 70.

1724   MCHDQV, Oral History, Cates, 93; Crowl & Isely, 491.

1725   Toland, Vol. II, 814.

1726   Moskin, 366; Henri, et al., 75.

1727   Hammel, 145; NACPM, RG 127 Box 83 (Iwo), Aa-26.1 Folders 1-2, HQ VAC Land Force in the Field.

1728   Toland, Vol. II, 801; King, 108; Wheeler, 65; MCRDSDA, 2ⁿᵈ Blt., 24ᵗʰ Rgt., 4ᵗʰ MarDiv, Op. Journal, 1 March 45; Kakehashi, 54; Crowl & Isely, 485.

1729   Henri, et al., 49; Interview Woody, 21 July 2016.

1730    MCRDSDA, 2nd Blt., 24th Marines, 4th MarDiv, Op. Journal, 21 Feb. 45.

1731    NACPM, RG 127 Box 83 A 13-7, 3d MarDiv. Tank Btl., Report, 2 April 1945, 1, Section F.

1732    MCRDSDA, 2nd Blt., 24th Marines, 4th MarDiv, Op. Journal, 22 Feb. 45.

1733    Mandel, 284.

1734    Crowl & Isely, 517.

1735    Interview Lee, 8 Aug. 2017.

1736    Dick Dashiell, "Dashiell Describes Dangerous Fighting of Marines on Iwo," The Asheville Times, March 1945.

1737    Interview Woody, 13 July 2017.

1738    NACPM, RG 127, Box 83 (Iwo Jima), Aa-26.1 [1 of 2], HQ VAC Landing Force in Field, 3 March 1945, Report casualties evacuated to ships (Report #11), D.E. Lee, Serial Number 813957, Ship APA-33 USS Bayfield; NACPM, RG 127, Muster Rolls, March 1945, 1st Btl., 21st Marines, 3d MarDiv; Email Nenninger to Rigg, 14 June 2017; Mandel, "Combat Fatigue in the Navy and Marine Corps During World War II."; NACPM, RG 127 Box 15 (3d MarDiv), G-1 Periodic Report No. 21-22; Interview Lee, 8 Aug. 2017.

1739    NASLPR, Darol E. Lee, Capt. L. B. Brooks to Mr. and Mrs. Lee, 5 April 1945.

1740    Mandel, "Combat Fatigue," 25.

1741    Hallas, Uncommon Valor on Iwo Jima, 139.

1742    Burrell, Ghosts of Iwo Jima, 87. Probably many of these 8,000 were just blast concussions and fatigue, but most likely around 2,700 of these were true cases of men who mentally broke down. See Miller, "Deathtrap," 9.

1743    Letter Prof. J. F. Dashiell to Dorothy Dashiell, 22 Feb. 1945.

1744    Henri, et al., 79.

1745    Crowl & Isely, 491 (loc. 10691). See also Henri, et al., 81.

1746    National Archives, DOD Dir 5200.9, Enclosure D, 21st Marines, Report, 6 April 1945, 2.

1747    Alexander, 28.

1748    Interview Lee, 24 June 2016; Hallas, Iwo Jima, 126.

1749    Henri, et al., 91; NACPM, RG 127 Box 90 (Ground Combat), MSI-1168, Vol. 24, 21st Marines, War Diary, 8 July 42-3 Nov. 1943, figures opposite 1.

1750    NASLPR, Marlowe C. Williams, Bronze Star Citation; Newcomb, 152; Henri, et al., 85-8; NACPM, RG 127 Container 14, Dashiell, Folders 1-2, #171, 25 Feb. 1945; NASLPR, Clay M. Murray; Barry Beck, Speech at "All Forces Dinner," Wichita Falls, TX, 16 Feb. 2013, 2. Beck's testimony counters the testimony given in Henri, et al., where the Marine Corps correspondents say Murray was just shot once through the jaw while on the radio. Since Beck was with Murray, his testimony is used. See Henri, et al., 88.

1751    Aurthur and Chlomia, 246-7; Crowl & Isely, 469.

1752    National Archives, DOD Dir 5200.9, Enclosure D, 21st Marines, Report, 10 April 1945, 5; Zdon, "A Marine in the Pacific," 10.

1753    Interview Woody, 12 Feb. 2017; Sledge, 247-8l; Hough, 231.

1754    Crowl & Isely, loc. 10640.

1755    MCHDQV, Dashiell File, #164, Iwo Jima.

1756    Henri, et al., v; Letter Dashiell to Fred, Tom and Mary Dashiell, 4 Jan. 1985. According to Marine correspondent, Bill Ross, three journalists and photographers were killed and 13 wounded while on Iwo. Ross, xv.

1757    Benis M. Frank, "Marine Corps Combat Correspondents, Photographers and Artists," Dashiell Archive, 1985. Admiral Turner had even said before Iwo: "It is the express desire of the Navy Department that a more aggressive policy be pursued with regard to press, magazine, radio, and photographic coverage of the military activities in the Pacific Ocea areas [be pursued]." Ross, 45.

1758    NASLPR, Frederick Knowles Dashiell, Letter of Commendation by Erskine for Dashiell, 17 June 1945.

1759   Ibid., Personnel Claim, 15 March 1946.

1760   Letter Dashiell to Fred Dashiell, Tom Dashiell and Mary Dashiell Gregory, 4 Jan. 1985. Frank, *Downfall*, 182.

1761   Ross, 152; Letter Bill Ross to Dick Dashiell, 9 Mach 1985.

1762   Smith Barrier, *On Carolina's Gridiron 1888-1936: A History of Football at the University of North Carolina*, Durham, N.C., 1937; Univ. North Carolina Football Brochure, 1990; Interview Fred Dashiell, 13 June 2018.

1763   Dick Dashiell, *Dashiell Family Records*, 13, 78, 108, 122; M.A. Farr, "Rev. John W. Dashiell, 85th Annual Session, Indiana Conference; Notes Edith Dashiell about Dashiell family's experience in Civil War; National Archives, Gen. Services Administration, SC-471396, Milton E. Jones.

1764   *American Naval Fighting Ships*, Vol. II, 1963.

1765   Frank, "Marine Corps Combat Correspondents, Photographers and Artists," Dashiell Archive, 1985 (parenthetical phrase by Dick Dashiell); Letter Dashiell to Ray Truelove, Editor Caltrop, 26 April 1986.

1766   Frank, "Marine Corps Combat Correspondents, Photographers and Artists," Dashiell Archive, 1985 (parenthetical phrase by Dick Dashiell); Letter Dashiell to Ray Truelove, Editor Caltrop, 26 April 1986; NASLPR, Solomon Israel Blechman, Bronze Star Citation & Service Record Book, Professional and Conduct Record of Blechman, Solomon I., 4 & Commanding Officer, Casualty Division to Bazell, S.N., Lt. Comdr. Chc USNR, 19 Oct. 1945 & Notifying Report of death, Solomon I. Israel, 3 Aug. 1944 & Sample citation Bronze Star Medal (Posthumous) SSG Solomon I. Blechman (this report says mortar fire killed him) & Official Bronze Star Medal citation (in this report it mentions Blechman was "killed by a burst of enemy gunfire"). There are two different reports claiming Blechman either was hit by mortar fire or was shot by machine gun fire. As a result, both are mentioned in the text.

1767   NASLPR, Solomon Israel Blechman, 5th Endorsement, Commandant Vandegrift to Forrestal, Subj: Recommendation for the award for the Bronze Star Medal in the case of SSG Blechman, USMCR, deceased, 10 Jan. 1945 & 4th Endorsement, Admiral Nimitz to Forrestal, Subj: Award of the Bronze Star Medal, recommendation for—case of SSG Blechman, (396234), USMCR, 13 Dec. 1944 & 3rd Endorsement, Maj. Gen. Schmidt to Forrestal, Subj: Award of Bronze Star Medal, recommendation for, case of SSG Blechman, 396234, USMCR, 21 Nov. 1944 & Lt. Col. Martin Fenton to Forrestal, Subj: Special Award, recommendation for, case of SSG Blechman, 8 Sept. 1944. One might notice that in the text, Blechman is referred to as Sgt. when in all these endorsements, made after his death, refer to him as a SSG. This is due to the fact that he was promoted to Staff Sergeant posthumously.

1768   NASLPR, Solomon Israel Blechman, Bronze Star Citation.

1769   Burrell, *Ghosts of Iwo Jima*, xvi.

1770   Crowl & Isely, loc. 10631.

1771   Ibid., loc. 11129.

1772   Bradley & Powers, *Flags of Our Fathers*, 205; Toland, Vol. II, 815; McNeill, "Reluctant Warrior," 39.

1773   Interview Woody, 20 July 2016; Smith, *Iwo Jima*, 66; MCHDHQ, Gen. Info, File Woody Williams, 14.

1774   Smith, *Coral and Brass*, 137-8; Moskin, 364; Alexander, 27; Toland, Vol. II, 815; Burrell, *Ghosts of Iwo Jima,* 131; Bradley & Powers, *Flags of Our Fathers*, 207; Wheeler, 50.

1775   https://archive.org/details/1945RadioNews/1945-02-25-CAN-Secretary-Of-The-Navy-James-Forrestal-On-The-Battle-Of-Iwo-Jima.mp3

1776   Smith, *Coral and Brass*, 137.

1777   NACPM, RG 127 Box 96 (Iwo) A45-2, 21st Rgt. Journal Iwo, 17 March 1945, R-3 Log, Time 2256, CG statement.

1778   NACPM, RG 127, Box 96 (Iwo), A46-2, 3rd Btl., 21st Rgt., Unit Reps, Don Pryor interview Adm. Turner and Gen. Smith, 28 Feb. 1945, 1-2.

1779    Ibid.; Smith, *Coral and Brass*, 63.

1780    Henri, et al., 73; NACPM, RG 127, Box 96 (Iwo), A46-2, 3rd Btl., 21st Rgt., Unit Reps, Don Pryor interview Adm. Turner and Gen. Smith, 28 Feb. 1945, 3.

1781    King, 124.

1782    Crowl & Isely, loc 10645.

1783    Bradley & Powers, *Flags of Our Fathers*, 206-7.

1784    Sakai, 253; King, 33.

1785    MCRDSDA, 4th Tank Btl. Report, 4th MarDiv, Iwo, Op. Report, 18 April 1945, 9 & 2nd Blt., 24th Marines, 4th MarDiv, Op. Journal, 23 Feb. 45, Time 1100.

1786    Ibid., 2nd Blt., 24th Marines, 4th MarDiv, Op. Journal, 23 Feb. 45, Time 1100.

1787    NIV—New International Version, 609.

1788    Crowl & Isely, 487 (loc 10619); MCHDQV, HQ, 1st Blt., 21st Marines, 3d MarDiv., FMF, In Field, 6 April 1945, Report Lt. Col. Williams, 3.

1789    MCHDQV, HQ, 1st Blt., 21st Marines, 3d MarDiv, FMF, In Field, 6 April 1945, Report Lt. Col. Williams, 3.

1790    Francois, 44.

1791    National Archives, DOD Dir 5200.9, Enclosure D, 21st Marines, Report, 10 April 1945, 5.

1792    NACPM, RG 127, Box 84 (Iwo), 3d MarDiv G-1 Report No. 12. Casualties.

1793    Francois, 44.

1794    PAHGS, Howard F. Chambers' Obituary, #7, Spirit 29 May 1943; NASLPR, Howard Chambers, Chambers Standing of the 35th Reserve Officers' Course, 25 Aug. 1943; Email John Hauser, head archivist for Thiel College, to Rigg, 10 Jan. 2020.

1795    NACPM, RG 127 Container 14, Dashiell, Folders 1-2, #267, 1.

1796    Ibid.; Raymond Henri, *Iwo Jima: Springboard to Final Victory*, NY, 1945, 60; NASLPR, Dashiell.

1797    NASLPR, Woody Williams, Recommendation for Award of MOH, Case of Cpl. Hershel Woodrow Williams by Donald Beck, Enclosure "A", 157.

1798    Francois, 44.

1799    MCHDQV, Dashiell File, #164, Iwo Jima.

1800    Interview Eugene Rybak, 28 May 2018; Interview Rybak, 29 June 2018.

1801    Excerpts of a letter written by S/Sgt. Joe Franklin, a Combat Photographers with 21st Maines, who was wounded and evacuated during Iwo, 1.

1802    MCHDQV, Dashiell, #208, Baseball Game-Iwo Jima.

1803    Ibid.

1804    MCHDQV, Bio File Donald Beck, 1 May 1968, 1; Hallas, *Uncommon Valor on Iwo Jima*, 127; MCHDHQ, Letter Beck to Ruth Beck, 16 Dec. 1944; Email Barry Beck to Rigg, 19 June 2017; Letter Charles Henning to Parents, 23 Jan. 1945.

1805    Interview Barry Beck, 4 Sept. 2017.

1806    Clausewitz, *On War*, Howard, "The Influence of Clausewitz," 34.

1807    NACPM, RG 127 Container 14, Dashiell, Folders 1-2, #267, 1.

1808    Ibid., 2.

1809    Essay by Al Ragliano done for college in 1947. He served in the 3rd Btl., 25th Rgt., 4th MarDiv.

1810    Dante, *Inferno*, Canto 17, 90.

1811    In 1942, the 1st MarDiv had a total of 27 flamethrowers. By April 1944, a Rgt. had 27 flamethrowers (81 per division). By 1945, a Btl. had 24 flamethrowers. In order to service 27 flamethrowers, each Rgt. had 405 fillings, 42 cylinders (commercial N2), 14 cylinders (commercial H2) and 2025 gallons of fuel (41 50-gallon drums). NACPM, RG 127 Box 21, 4th MarDiv (1540) (2 of 2), Folder 6, HQ, 1st Btl., 20th Marines (Eng.), 4thMarDiv., FMF, San Fran, 29 April 1944, R. G. Ruby to Ass. Chief-of-Staff D-4, 1; Burrell (ed.), Burrell, "The Prototype U.S. Marine," 192; Crowl & Isely, 515 (loc. 11236).

1812   MCHDHQ, Unattributed article from 1945 "Sgt. Hershel Woody Williams," 3; Hallas, *Iwo Jima*, 127. Woody claimed in 1945 that one had died and five were wounded. Interview Woody 21 July 2016; Interview Lefty Lee, 16 March 2015.

1813   Interview Woody, 21 July 2016; NACPM, RG 127 Container 14, Dashiell, Folders 1-2, #267, 2; Lawless, "of Such Stuff are Heroes," 68.

1814   Hallas, *Iwo Jima*, 128.

1815   NASLPR, Marquys Cookson, Silver Star Citation: "Demolition Man HQ and Serv. Co., 21st Marines" & Enlisted Man's Qualification Card, No. 27 Classification in MOS 607 Mortar Crewman, 729 Pioneer, No. 29 Record of Current Service H&S Co., 21st Marines & Service Book, Prof. and Conduct Record Cookson, 4, Pay Account Record, 26; NASLPR, William Nolte, Report Separation, Principle Military Duty, Flamethrower (781) & Service Book, Prof. Conduct Record of Nolte, Duty Station HqCo, 1st Btl., 21st Marines & Enlisted Man's Qualification Card, No. 27 Flamethrower Op. (781); MCHDQV, Dashiell File, #289, Pitner; NASLPR, Nathan Pitner, Service Book; NASLPR, Joseph Rybakiewicz, Report Separation, Principle Mil. Duty, Chemical (870) & Enlist man Qualification Card, No. 27, Chemical NCO (870), No. 28, Sat.Com—Demolition & Certificate Completion Chemical Warfare School, Schools Rgt., Training Command, FMF, Camp Lejeune, NC, Rybakiewicz. On 28 Nov. 1994, Rybakiewicz applied to the VA for disability for PTSD and wrote five out of 15 in his flamethrower unit survived Iwo, one of them being Woody. Review *Flamethrower*, Eugene Rybak, 28 June 2018. Although Rybakiewicz was in Baker Company, not Charlie, it does seem these demolition/flamethrower men were sometimes transferred between companies depending on the need at the front, although according to Donald Graves, this was indeed rare. So when he writes the VA that there were 15 in his unit, knowing that on average each company had 7 flamethrowers attached to it, he probably was lumping in all flamethrowers for both Baker and Charlie companies. Since Captain Beck knew the men so well in both companies, there might have been more coordination between the two units than normal although the sources are silent on this point. From the reports, Cookson, Pitner and Nolte were all in Charlie Company *with Woody*.

1816   NASLPR, Woody Williams, Recommendation for Award of MOH; case of Cpl. Hershel Woodrow Williams by Beck, Enclosure "A", 157.

1817   NACPM, RG 127 Container 14, Dashiell, Folders 1-2, #267, 2.

1818   Clausewitz, *On War*, 136.

1819   Beck's lecture of Iwo to NROTC Midshipmen, Purdue Univ., 1952, 4.

1820   Barry Beck, Speech at "All Forces Dinner," Wichita Falls, TX, 16 Feb. 2013, 4.

1821   Clausewitz, *On War*, 137.

1822   Ibid., 138.

1823   Interview Woody, 20 July 2016; Smith, *Iwo Jima*, 67; MCHDQV, Gen. Info, File Woody Williams, Interview 1982, 16. Correspondent Dashiell reported in 1945 it was actually Woody's platoon leader, Lt. Chambers, who had requested a volunteer to take out the pillboxes. Woody disputes this saying it was Beck. Possibly Chambers had put the request to Beck and then Beck asked Woody. Dashiell, "Combat Correspondent Dispatch," 17 May 1945, 1.

1824   Beck's lecture of Iwo to NROTC Midshipmen, Purdue Univ., 1952, 5.

1825   Ibid., 6.

1826   According to Kimberly Bauer (Visitor Services) and Chris McDougal (Associate Archivist and Librarian) at the National Museum of the Pacific War, there were 22 major island campaigns with 78 battles on different islands. One campaign might be the campaign in the Marianas which included the islands of Saipan, Tinian and Guam. Gen. Smith notes that the Marines made one hundred landings during WWII. Smith, *Coral and Brass*, 15.

1827   Kakehashi, x.

1828   Maslowski, and Millett, 443; Paul Fussell's introduction to Sledge's *With the Old Breed*, xiv.

1829    Toland, Vol. II, 818-9.

1830    Hallas, *Uncommon Valor on Iwo Jima*, 127; Dick Dashiell, Marine Corps Correspondent Dispatch, The Fairmont Times, 17 May 1945.

1831    Henri, et al., 203-4. Woody disputes this, thinking it was only Schlager who got hit in the helmet. Notes Woody Williams, Oct. 2016. However, in one of Dashiell's reports, he notes Southwell did take a bullet through his helmet, grazing the top of his head so maybe both Schlager and Southwell escaped close calls. NACPM, RG 127 Container 14, Dashiell, #206.

1832    NACPM, RG 127 Container 14, Dashiell, #267, 2. Woody claims Vernon Waters was there to help him, but the records show Waters was in Company B, so he probably was not there. Interview Woody, 21 July 2016.

1833    Congressional Medal of Honor, Awarded to Hershel Woodrow Williams, By President of U.S. Harry S. Truman, The White House, 5 Oct. 1945 (known hereafter as Woody Williams Congressional MOH Report 5 Oct. 1945), 7.

1834    Interview Woody, 20 July 2016.

1835    Henri, et al., 172.

1836    Moore, "Iwo Jima Eyewitness," Pass in Review, Feb. 1989, 10.

1837    Dalton, "Interview Woody Williams," 7; Francois, 93; Hallas, *Uncommon Valor on Iwo Jima*, 130; VandeLinde, 369; Lawless, "of Such Stuff are Heroes," 68-9; Notes Woody Williams, October 2016; NACPM, RG 127, Box 83, (Iwo), AB-1, Com. Gen. 3d MarDiv Erskine to Commandant, Report, 3 June 1945, 36; Interview Woody, 19 July 2016;https://www.youtube.com/watch?v=O9TqQWLtBJk; https://www.youtube.com/watch?v=eh-06Hy45Qc.

1838    Francois, 93; Lawless, "of Such Stuff are Heroes," 68.

1839    Hoffman, Review *Flamethrower*, 28 Aug. 2018.

1840    Interview Woody, 19 & 21 July 2016. See also Henri, et al., 65.

1841    Crowl & Isely, 297.

1842    Dower, 92.

1843    Interview Woody, 21 July 2016.

1844    Poe, "The Murders in the Rue Morgue," 421.

1845    Interview Woody, 21 July 2016; Zdon, "A Marine in the Pacific," 10; Interview Lee, 24 June 2016.

1846    NASLPR, Woody Williams, Beck's MOH Recommendation for Woody, Encl. "A", 157; Smith, *Iwo Jima*, 69; Interview Woody, 21 July 2016.

1847    Richard Frank, "Flags, Memories and Iwo Jima," WWII Museum blog.

1848    Dante, *Inferno*, II 17, 88-9.

1849    NACPM, RG 127 Container 14, Dashiell, #267; Paul Hoolihan, "Flame Throwers Class at 21st Teaches the Art of Roasting Japs," 21st Regt. Newsletter, May 1945.

1850    NASLPR, Howard F. Chambers, Bio record, sheet 3; Interview Sherri Bell, 30 March 2018; Interview Amelia Gehron, 30 March 2018.

1851    VandeLinde, 369. Many reports say the tanks weighed together not 70, but 72 pounds. For evidence he also took a pole charge with him, see NACPM, RG 127 Container 14, Dashiell, #267, 2. There are several details this day Woody cannot remember. For example, he knows he took six flamethrowers into battle, but he does not remember going back for them. Interview Woody, 28 Aug. 2017. For years, Woody has mentioned Waters was at the CP helping him with the supplies, but this study has documented Waters was in Co. B, not C, so he would not have been there.

1852    Interview Woody, 19 July 2016; NASLPR, Vernon Waters; Francois, 44. Woody even claims B Company was on his flank on 23 February, and not with C Company in the rear in support of C Company.

1853    Henri, et al., (caption on picture of flamethrower between pp. 72-3).

1854    Kakehashi, 166-7.

1855   NACPM, RG 127 Container 14, Dashiell, #267, 2.

1856   Poe, "Eleonora," 473.

1857   NACPM, RG 127 Container 14, Dashiell, #267, 2.

1858   Interview Woody, 19 July 2016.

1859   Francois, 44; VandeLinde, 370.

1860   NACPM, RG 127 Container 14, Dashiell, #267, 2.

1861   Francois, 93.

1862   Francois, 93; Hallas, *Uncommon Valor on Iwo Jima*, 129-30.

1863   King, 146.

1864   Francois, 93.

1865   Poe, "The Devil in the Belfry," 303.

1866   Interview Woody, 21 July 2016.

1867   Francois, 94; NASLPR, Lyman D. Southwell, Service Book, Sec. Marks, Scars, etc., p.2.

1868   Francois, 94. In Southwell's military file, he received this head wound one day later, on 24 Feb. NASLPR, Southwell, HQ, 3d MarDiv, Com. Gen. Erskine to Southwell, Subj: Gold Star in lieu of Second Purple Heart Medal, for wounds incurred 24 Feb. 1945, Gunshot, Scalp. Also, in Francois' article, Woody said Southwell had "coal black hair." However, in Southwell's personnel file, it says he had brown hair. NASLPR, Southwell, Service Book, Marks, Scars Etc., p.2.

1869   Mountcastle, xi.

1870   Interview Woody, 21 July 2016; MCHDHQ, Woody, "Destiny…Farm Boy," 7.

1871   Mountcastle, 8.

1872   Toland, Vol. II, 801.

1873   Interview Woody, 21 July 2016.

1874   Burrell, *Ghosts of Iwo Jima*, 48.

1875   Hallas, *Uncommon Valor on Iwo Jima*, 133.

1876   Mountcastle, xii.

1877   Woody Williams Congressional MOH Report 5 Oct. 1945, 12. Another report claims that four times, Woody "threw pole charges through narrow openings of emplacements." VandeLinde, 372. Dashiell wrote that when he went after the first pillbox, Woody also had "grabbed a 12-pound pole charge filled with 15 blocks of TNT." Dashiell, "Combat Correspondent Dispatch," 17 May 1945, 2. Woody's witness George Schwartz testified "I personally observed him as he used his flame thrower and used three demolition charges in reducing two of the pillboxes." NASLPR, Woody Williams, PFC George Schwartz's Statement, Enc. "G", 151.

1878   Dashiell, "Combat Correspondent Dispatch, 17 May 1945, 2.

1879   Interview Lee, 11 May 2017; Dashiell, "Combat Correspondent Dispatch, 17 May 1945, 2.

1880   Interview Woody, 22 July 2016.

1881   Clausewitz, *On War*, 274.

1882   Interview Woody, 22 July 2016; Dalton, "Interview with Woody Williams," 9; Dashiell, "Combat Correspondent Dispatch," 17 May 1945, 3. Woody maintains there were only five to six Japanese who tried to kill him. However, back in 1945, he reported he counted 21 who had tried to get him with "rifles, bayonets and grenades." NACPM, RG 127 Container 14, Dashiell, #267, 2-3.

1883   Francois, 94; Interview Woody, 21 July 2016; WWII Museum Oral History Project, 3 Feb. 2011; Dalton, "Interview with Woody Williams," 9.

1884   NACPM, RG 127 Container 14, Dashiell, #267, 3.

1885   King, 161;Mountcastle, 83.

1886   Brophy, Miles and Cochrane, 166.

1887   Wheeler, 69.

1888   VandeLinde, 371.

1889    Interview Lauriello, 6 June 2017.

1890    Interview Woody, 20 July 2016; Kakehashi, 81; Francois, 93.

1891    Interview Woody, 21 July 2016; MCHDHQ, Woody, "Destiny…Farm Boy," 11; Smith, *Iwo Jima*, 69. Woody claims Waters was there keeping the supplies ready for him. However, Waters was attached to Co. B and not there.

1892    MCHDHQ, Unauthored article from 1945 "Sgt. Hershel Woody Williams," 2.

1893    Frankl, 43; Sledge, 128.

1894    MCHDHQ, Unauthored article from 1945 "Sgt. Hershel Woody Williams," 2.

1895    Francois, 44.

1896    Interview Woody, 22 July 2016.

1897    Clausewitz, *On War*, 101.

1898    Both Francois and Hallas mention the smoke was generated from a bazooka hit when some of the smoke was sucked inside the pillbox and then exited the vent pipe. This info came from a Woody quote. Woody does not now think this happened. He does not remember any bazooka being there and the witnesses do not mention this weapon being used. "A bazooka could have been used," Woody said, "but I just do not remember it being there." Interview Woody, 21 July 2016; VandeLinde, 369; Spector, 499; NACPM, RG 127 Box 21, 4th MarDiv (1540) (1 of 2) Folder 5, HQ, 24th Marines, 4th MarDiv, FMF, Com. officer Jordan to Com. Gen., 19 March 1945, 1.

1899    Francois, 93.

1900    Mountcastle, 9; Interview Woody, 21 July 2016.

1901    Woody Williams Congressional MOH Report 5 Oct. 1945, 11.

1902    NACPM, RG 127 Container 14, Dashiell, #267, 3.

1903    NACPM, RG 127, Box 83, (Iwo), AB-1, Com. Gen. 3d MarDiv Erskine to Commandant, 3 June 1945, 69.

1904    Francois, 93. This information about Tripp apparently was given to Francois by Woody because when one reads the Tripp statement about Woody's actions, Tripp just says he was at the CP witnessing Woody come back for flamethrowers six times. Tripp never mentioned the third, large pillbox. There are two reasons for this. First, if Tripp was there, then he apparently felt someone else took out the pillbox. Last, if he was not there, then he gave an accurate affidavit for Woody saying he only witnessed him from an area he was assigned that day, the CP and not the front. See NASLPR, Woody Williams, Statement Cpl. Alan Tripp, 153.

1905    Interview Woody, 19 July 2016; Dashiell, "Combat Correspondent Dispatch, 17 May 1945, 2; NASLPR, Woody Williams, Statement Cpl. Schlager. Many structures of Iwo had "air shafts 30 or 40 feet deep," and sometimes there built with sharp turns and angles as "protection against flamethrowers." Henri, et al., 10.

1906    Francois, 93.

1907    Ibid. Woody wrote in May 2017 in his notes on the draft of the book, "I did not have a pole charge man to help me. Schlager was my pole charge man and a bullet hit his helmet before we hit the first pillbox." However, Schlager had returned to the battle and was active according to Dashiell's article.

1908    Interview Woody, 21 July 2016; Interview Woody, 23 July 2016; NACPM, RG 127 Container 14, Dashiell, #267, 3-4; https://www.youtube.com/watch?v=fGts5WeLEgk; Lawless, "of Such Stuff are Heroes," 68; Francois, 93.

1909    WWII Museum Oral History Project, Woody Williams, 3 Feb. 2011; https://www.youtube.com/watch?v=DeJ2JHmQsFY; Interview Woody, 20 July 2016; Francois, 93; Dalton, "Interview with Woody Williams," 8; Smith, *Iwo Jima,* 68; NASLPR, Woody Williams, Statement Schlager, 152.

1910    Brophy, Miles and Cochrane, 165-6.

1911    Ibid.

1912    Hallas, *Uncommon Valor on Iwo Jima*, 134; Interview Woody, 22 July 2016; Smith, *Iwo Jima*, 69; VandeLinde, 375. To see Bornholz and Fischer's KIA dates, see http://www.naval-history.net/WW2UScasaaDB-USMCbyNameF.htm; https://findagrave.com/cgi-bin/fg.cgi?page=gr&GSln=Bornholz&GSbyrel=all&GSdyrel=all&GSob=n&GRid=2619943&df=all& ; https://findagrave.com/cgi-bin/fg.cgi?page=gr&GSln=FIscher&GSfn=Charles&GSbyrel=all&GSdy=1945&GSdyrel=in&GSob=n&GRid=3774625&df=all&; Email Capt. Corydon S. Cusack, Protocol Officer, MCRD San Diego to Patrick O'Leary, UPS Veterans Affairs Manager, 10 Aug. 2017; USMCU Casualty Web Data Base. Bornholz and Fischer were in the same company as Woody and seem to be the only two who died this day from Woody's outfit leading one to believe they were the two who died protecting him.

1913    NASLPR, Charles Fischer & Warren Bornholz. According to Fischer's nephew, Gary Fischer, Charles was killed while supporting a flamethrower in action. Interview Gary Fischer, 11 Jan. 2020.

1914    Interview Woody, 22 July 2016.

1915    Dante, *Inferno*, Canto 19, 115.

1916    Interview Woody, 19 July 2016.

1917    MCHDHQ, Historical fact sheet in Woody's archive about Iwo entitled "The Battle for Iwo Jima," 2; MCHDQV, HQ, 1st Btl., 21st Marines, 3d MarDiv, FMF, In Field, 6 April 1945, Report Lt. Col. Williams, 20.

1918    Ibid.

1919    NACPM, RG 127 Container 14, Dashiell, #267, 2-3; NACPM, RG 127, Box 83, (Iwo), AB-1, Com. Gen. 3d MarDiv. Erskine to Commandant, Report, 3 June 1945, 66-8.

1920    Hallas, *Uncommon Valor on Iwo Jima*, 157, 161.

1921    NACPM, RG 127 Container 14, Dashiell, #267, page Text under the flap of paper on p. 3.

1922    MCHDQV, HQ, 2nd Btl., 21st Marines, 3d MarDiv, FMF, In Field, G.A. Percy, 12 April 1945, 15.

1923    Woody Williams Congressional MOH Report 5 Oct. 1945, 7, 10. In 1945, Woody reported that he only returned four times. MCHDHQ, Unattributed article from 1945 "Sgt. Hershel Woody Williams," 4.

1924    Interview Woody, 24 July 2016.

1925    Francois, 94.

1926    Woody Williams Congressional MOH Report 5 Oct. 1945, 9.

1927    NACPM, RG 127 Container 14, Dashiell, #267, 2.

1928    Ibid., 4. In Dashiell's original report, he said Woody's group took out between 20 to 35. However, in revised versions, he wrote 20-25. MCHDQV, Dashiell File, #267, Man with Flames-Iwo Jima.

1929    Woody disputes this claim saying Schlager was out of commissioned once he had been knocked out by his helmet taking a bullet to it. Interview Woody, 23 Oct. 2017; MCHDQV, Dashiell File, #267, Man with Flames-Iwo Jima.

1930    Dick Dashiell, Marine Corps Correspondent Dispatch, Fairmont Times, 17 May 1945; VandeLinde, 372.

1931    NACPM, RG 127 Container 14, Dashiell, #184, 4.

1932    Dashiell, Marine Corps Correspondent Dispatch, Fairmont Times, 17 May 1945.

1933    Poe, "A Descent into the Maelström," 444.

1934    Dashiell, "Combat Correspondent Dispatch, 17 May 1945, 1; NACPM, RG 127 Container 14, Dashiell, #267, Cover Page.

1935    MCHDQV, HQ, 1st Btl., 21st Marines, 3d MarDiv, FMF, In Field, 6 April 1945, Report Lt. Col. Williams, 3.

1936    Clausewitz, *On War*, 228.

1937    Burrell, *Ghosts of Iwo Jima*, 10.

1938    Hallas, *Uncommon Valor on Iwo Jima*, 127; NASLPR, Howard F. Chambers.

1939    MCHDQV, Biographical File Donald Beck, 1 May 1968, 1.

1940 NASLPR, Beck, Bronze Star Citation.

1941 Woody Williams Congressional MOH Report 5 Oct. 1945, 2.

1942 Smith, *Coral and Brass*, 140.

1943 Mountcastle, 88.

1944 Interview James Henning, 21 April 2018.

1945 Clausewitz, *On War*, 96.

1946 NACPM, RG 127 Box 96 (Iwo) A45-2, 21st Rgt. Journal Iwo, 25 Feb. 1945, R-3 Log, Time 1809, Time Received 1811, CO-9; Discussion with Tim Nenninger, 2 June 2017.

1947 Smith, *Coral and Brass*, 137.

1948 Frank, *Downfall*, 28; Drea, 238.

1949 NIDSMDMA, Senshi-Sosho, Cen. Pacific, Op. #2, Peleliu, Angaur, Iwo, Vol. 13, 368; Kakehashi, 161.

1950 Isely & Crowl, 515.

1951 NIDSMDMA, *Senshi-Sosho*, Cen. Pacific, Op. #2, Peleliu, Angaur, Iwo, Vol. 13, 381-2; Newcomb, 105, 170.

1952 Ibid.

1953 Harries & Harries, 324.

1954 NACPM, RG 127 Box 16 (3d MarDiv), (2295) 14 Nov. 42-27 Nov. 1945 (1 of 2), HQ USMC, Commandant to list of divisions and regiments, 2 Jan. 1945. Report ETOUSA, HQ USA Ground Forces, 21 Oct. 1944, 3.

1955 Beck's lecture, Iwo Jima to NROTC Midshipmen, Purdue Univ., 1952, 8.

1956 Prange, 73

1957 MCHDQV, Dashiell File, #712, No Rest for Weary.

1958 Crowl & Isely, 475; Moskin, 373.

1959 Email Saito to Rigg, 12 June 2019; 歩兵第145連隊は、1938年、鹿児島で編成された。編成後は、主に中国戦線で戦っていたが、詳細は史料がないので不明である。1943.7には、サイパンの米軍を撃滅する作戦が計画され、その実行部隊として同連隊が鹿児島で待機していた。しかし、これが中止となり、硫黄島に転用された。歩兵第145連隊は、主は現役兵だったので、硫黄島の部隊の中では精強であった。

1960 MCHDQV, Dashiell File, #268, Johnny Wlach; *Two Score and Ten*, 149; NASLPR, Letter of Commendation Gen. Erskine for Cpl. John Wlach, 16 June 1945.

1961 Crowl & Isely, doc 10654.

1962 There were three hills on Iwo that were 362 feet causing confusion. To avoid confusion, the troops labeled them 362A, 362B and 362C. In this study, often the documents don't clarify which Hill 362 Woody's unit attacked, but the only one in 3d MarDiv's section was 362C. Nalty, Shaw and Turnbladh, 581-2 fn. 18, 583.

1963 Ross, 162; Wheeler, 50-1.

1964 Interview Woody, 21 July 2016.

1965 Ibid.

1966 Ibid., 23 July 2016.

1967 See file NACPM, RG 127, Box 83 (Iwo), Aa-26.1 [1 of 2].

1968 Interview Lauriello, 14 June 2017.

1969 USMCU Casualty Web Data Base; NASLPR, Thomas Scrivens. The USMCU database has Scrivens listed as suffering Shell-Shock while his personnel file has it listed as a blast concussion.

1970 NMPWANM, 13 July 2017; Article 4, Articles for Gov. of the US Navy, 1930, Dept. of the Navy - Bureau of Navigation, (US Gov. Printing Office, DC 1932). https://www.history.navy.mil/research/library/online-reading-room/title-list-alphabetically/a/articles-government-united-states-navy-1930.html accessed June 2, 2017.

1971    Article 4, Articles for Gov. of the US Navy, 1930, Dept. of the Navy - Bureau of Navigation, (US Gov. Printing Office, DC 1932). https://www.history.navy.mil/research/library/online-reading-room/title-list-alphabetically/a/articles-government-united-states-navy-1930.html accessed June 2, 2017.

1972    MCRDSDA, MCRD Training Book, File 1, <u>Crime Doesn't Pay</u>, 1.

1973    Article 64, Articles for the Gov. of the U.S. Navy, 1930.

1974    Articles 27 and 30, Articles for the Gov. of the U.S. Navy, 1930.

1975    MCRDSDA, MCRD Training Book, File 1, <u>Crime Doesn't Pay</u>, 1.

1976    Interview Woody, 23 July 2016.

1977    Wells, "Medal of Honor Recipient Devotes Life to Veterans."

1978    Frankl, xi, 121, 164.

1979    <u>Ibid.</u>, 164.

1980    Manchester, 130.

1981    Bob Adams, "WWII Hero Praises GIs in Vietnam," Huntington Advertiser, 28 July 1967.

1982    Interview Lauriello, 19 Nov. 2016.

1983    Smith, *Coral and Brass*, 15.

1984    <u>Ibid.</u>, 126.

1985    NASLPR, Lyman D. Southwell, HQ, FMFPAC, Silver Star Citation, Signed by Lt. Gen. Roy S. Geiger.

1986    Interview Woody, 23 July 2016.

1987    Crowl & Isely, loc. 10824; Hough, 341.

1988    Henri, et al., 180.

1989    Ross, 206.

1990    Interview Lee, 8 Aug. 2016; Newcomb, 143.

1991    Newcomb, 159.

1992    NACPM, RG 127, Box 84 (Iwo Jima), 3d MarDiv, G-1 Periodic Report No. 14.

1993    NASLPR, Erskine, Record Graves Erskine, #0268, Card #11 or 12, Note Feb. 1945; Crowl & Isely, loc. 10681.

1994    MCHDQV, Oral History, Erskine, 46.

1995    <u>Ibid.</u>, 361-2.

1996    Crowl & Isely, 144.

1997    National Archives, DOD Dir 5200.9, Enclosure D, 21ˢᵗ Marines, Report, 10 April 1945, 1.

1998    Moore, "Iwo Jima Eyewitness," Pass in Review, Feb. 1989, 10.

1999    Miller, "Deathtrap Island," 9.

2000    Ross, xiii, 115. Bradley, *Flyboys*, 213; Elizabeth Mullener, *War Stories: Remembering World War II*, NY, 2002, 220; Moskin, 359; Henri, et al., 1, 11, 52, 57, 166; Newcomb, 23; Hammel, 51; Alexander, "Closing In", 3; Burrell, *Ghosts of Iwo Jima*, 3; Bingham, xiii; Wheeler, 196; Miller, "Deathtrap Island," 10; Smith, *Coral and Brass*, 132; Hough, 335.

2001    Hoffman, *Once a Legend,* 313; Miller, "Deathtrap Island," 10; Burrell, *Ghosts of Iwo Jima*, 58.

2002    Ross, 115.

2003    Frank, *Downfall*, 135; Sledge, 46; Crowl & Isely, 205.

2004    Clausewitz, *On War*, 194-5 & Brodie, "A Guide to Reading of On War," 660.

2005    Clausewitz, *On War*, 484.

2006    King, 126-8.

2007    MCRDSDA, 2ⁿᵈ Blt., 24ᵗʰ Marines, 4ᵗʰ MarDiv, Op. Journal, 23 Feb. 45 & Reports, Iwo, Narrative, 8.

2008    Interview Graves, 2 May 2018.

2009    Sledge, 143.

2010    Ross, 206; Mullener, 223; Maslowski, and Millett, 462; King, 114; Norman L. Cantor, *After We Die: The Life and Times of the Human Cadaver*, Univ. of Georgetown Press, 2010, 80-3; https://

australianmuseum.net.au/learn/science/decomposition-fly-life-cycles/; "How the Blow-fly Flies and Why Engineers are All Excited About It," Haaretz, 26 March 2014; Richard J. Bomphrey, Simon M. Walker and Graham K. Taylor, "The Typical Flight Performance of Blowflies: Measuring the Normal Performance Envelope of *Calliphora vicina* Using a Novel Corner-Cube Arena, US National Library of Medicine National Institutes of Health, 18 Nov. 2009; Jason H. Byrd, "Featured Creatures: Hairy Maggot Blow Fly," University of Florida, Jan. 1998; Roach, 65. The synthetic organic pesticide DDT is an abbreviation for *dichlorodiphenyltrichloroethane*. See also Cecie Starr and Ralph Taggart (eds.), *Biology: The Unity and Diversity of Life*, Belmont, California, 1992, 842-3.

2011   King, 114: Newcomb, 178.

2012   Hammel, 163; NACPM, 127 GW 312.

2013   Interview Woody, 22 July 2016.

2014   NACPM, RG 127, Box 83 (Iwo), HqServBn, CorHqTrps, VAC, 28 Feb 1945 to 1 March 1945.

2015   The Biblical declaration that there is only one God. *Shema Yisrael* means "Hear, [O] Israel" and are the first two words of a verse in the Jewish Bible called the *Torah* (Pentateuch or first five books of the Bible), and is the title (sometimes shortened to *Shema*) of a Jewish prayer that serves as a centerpiece of what it means to be Jewish. It comes from Deuteronomy 6:4 which reads: "Hear, O Israel, the Lord is God, the Lord is one" (NIV—New International Version, p. 176 & note *a*).

2016   Quote comes from Henri, et al., 57; Mandel, 278.

2017   Hoffman, *Once a Legend,* 256.

2018   Interview Woody, 21 July 2016.

2019   Mary Roach, *Stiff: The Curious Lives of Human Cadavers*, NY, 2003, 70.

2020   Sledge, 253.

2021   Roach, 65; Lecture given by William "Bill" Pasewark Sr. at the Outrigger Hotel in Guam, 20 March 2015; Interview Bill Pasewark, 12 Jan. 2020.

2022   Interview Robert (Bob) Hensely, 18 Feb. 2018.

2023   Interview Woody, 21 July 2016.

2024   Poe, "The Fall of the House of Usher," 319.

2025   Newcomb, 36; Kakehashi, 65.

2026   Interview Woody, 19 July 2016; Hallas, *Uncommon Valor on Iwo Jima*, 196.

2027   Roach, 65-8.

2028   Ibid.

2029   Ibid.

2030   Ibid.

2031   Roach, 65-8.

2032   Sledge, 143.

2033   The phrase "ashes to ashes, dust to dust" comes from the traditional English Burial Service. It originates from the Biblical text, Genesis 3:19 which reads "In the sweat of thy face shalt thou eat bread, till thou return unto the ground; for out of it wast thou taken: for dust thou art, and unto dust shalt thou return" (KJV—King James Version, p. 5).

2034   Sledge, 260.

2035   Conrad, "The Secret Sharer," 670.

2036   Hallas, *Uncommon Valor on Iwo Jima*, 178.

2037   National Archives, DOD Dir 5200.9, Encl. D, 21st Marines, Action Report, 10 April 1945, 1.

2038   Toland, Vol. II, 816; Bradley & Powers, *Flags of Our Fathers*, 190.

2039   Toland, Vol. II, 817.

2040   Ibid.,Vol. II, 818.

2041   Henri, et al., 67; MCHDQV, Dashiell, #160, Josephy.

2042   Toland, Vol. II, 879.

2043  Bingham, 201.

2044  Crowl & Isely, loc. 11255.

2045  NACPM, RG 127, Box 83, (Iwo), AB-1, Com. Gen. 3d MarDiv to Commandant, Report, 3 June 1945, 36, 102.

2046  NACPM, RG 127 Box 21, 4th MarDiv (1520), Folder 2, HQ, VAC, CG VAC released by Col. V.F. Sham (USA), 21 Feb. 1944, 1; John W. Mountcastle, *Flame On: U.S. Incendiary Weapons, 1918-1945*, Mechanicsburg, PA, 2016, 90; Capt. Beck's lecture of Iwo to NROTC Midshipmen, Purdue Univ., 1952, 9; Leo P. Brophy, Miles and Cochrane, Wyndham D. Miles and Rexmond C. Cochrane, *The Chemical Warfare Service: From Laboratory to Field*, Washington D.C., 1988, 155; Hoffman, *Chesty*, 266; Crowl & Isely, 516; Henri, et al., 42-3.

2047  King, 131; Wheeler, 121; Interview Willhite, 10 May 2017; Sasser, 144; Beck's lecture, Iwo to NROTC Midshipmen, Purdue Univ., 1952, 9; NACPM, RG 127 Box 21, 4th MarDiv (1540) (1 of 2) Folder 5, HQ, FMF, Com. Gen. to Commanding Gen., FMF, Pacific, 21 April 1945, 1; NACPM, RG 127 Box 21, 4th MarDiv. (1520), Folder 2, HQ, VAC, CG VAC released by Col. V.F. Sham (USA), 21 Feb. 1944, 1.

2048  NACPM, RG 127 Box 21 (4th MarDiv) (1540) (Folder 6) (2of 2), Com. Gen. VAC to Com. Gen. FMFPAC, 7 April 1945.

2049  Ibid., Box 83 A 13-7, 3d MarDiv. Tank Btl. Report, 2 April 1945, 1, Sec. F.

2050  Interview Woody, 23 July 2016.

2051  MCRDSDA, 2nd Blt., 24th Marines, 4th MarDiv, Op. Journal, 22 Feb. 45.

2052  National Archives, DOD Dir 5200.9, Encl. D, 21st Rgt., Report, 10 April 1945, 5; Moskin, 366; King, 120.

2053  Crowl & Isely, 526.

2054  National Archives, DOD Dir 5200.9, Encl. D, 21st Rgt., Report, 10 April 1945, 6.

2055  NACPM, (Iwo) 127 GW 313; NACPM, RG 127, Box 83, (Iwo), AB-1, Com. Gen. 3d MarDiv to Commandant, Report, 3 June 1945, 78; NACPM, RG 127 Box 50 (Guam), Col. Ray A. Robinson, CoS by command Maj. Gen. Turnage, 13 May 1944, Intel, 7.

2056  MCHDQV, Dashiell File, #258, DeGroff.

2057  Henri, et al., 65. See also MCHDQV, Oral History, Robert E. Cushman Jr., 170-1.

2058  Ibid., 66.

2059  Manchester, 341.

2060  MCHDQV, HQ, 1st Btl., 21st Rgt., 3d MarDiv, FMF, In Field, 6 April 1945, Report Lt. Col. Williams, 4.

2061  MCHDQV, Dashiell File, #190, Wiremen-Iwo Jima.

2062  NASLPR, Vernon Waters, S.B. Griffith II to SECNAV, 7 May 1945.

2063  http://www.hqmc.marines.mil/Portals/61/Docs/FOIA/WWII%20SS%20Citations/USMC%20WWII%20SS%20(Letter%20H)%20(HAAG%20-%20HERZOG).pdf?ver=2013-03-26-190343-197; NASLPR, Charles Wesley Henning, Silver Star citation.

2064  MCHDQV, Dashiell File, Biographies, Col. Withers; NASLPR, Hartnoll J. Withers, Record of Withers, Hartnoll J., Sheet Card #8 of 9.

2065  NACPM, RG 127 Box 96 (Iwo) A45-2, 21st Rgt. Journal Iwo, 1 March 1945, R-3 Log, Sent 1535, Received 1558, CG to 9th and 21st Rgt.

2066  NACPM, RG 127, Box 83, AB-1, Com. Gen. 3d MarDiv to Commandant, Report, 3 June 1945, 7.

2067  Nalty, Shaw and Turnbladh, 571.

2068  Smith, *Coral and Brass*, 140.

2069  In discussions with Gen. Zinni, he mentions Erskine was the most brilliant general the Corps produced in WWII. Kakehashi, 52; NACPM, 127 GW 319 (Iwo: Photo Archive), Flag Raising 14 March 1945, Gen. Smith with Erskine.

2070    NACPM, 127 GW 319 (Iwo: Photo Archive), Flag Raising 14 March 1945, Smith with Erskine.

2071    Nalty, Shaw and Turnbladh, 579-80; Crowl & Isely, 494.

2072    Ibid., 583-4.

2073    Burrell, *Ghosts of Iwo Jima*, 71.

2074    MCHDQV, Dashiell, #224, 150 Nips—Iwo Jima.

2075    https://archive.org/details/1945RadioNews/1945-03-01-FDR-Last-Address-Before-Congress.mp3

2076    NACPM, RG 127 Box 96 (Iwo) A45-2, 21st Rgt. Journal Iwo, MarDiv Records, Report No. 5, 1 March 1945.

2077    National Archives, DOD Dir 5200.9, Enclosure D, 21st Rgt.,Report, 10 April 1945, 1.

2078    MCHDQV, HQ, 1st Btl., 21st Rgt., 3d MarDiv, FMF, In Field, 6 April 1945, Report Lt. Col. Williams, 4.

2079    NASLPR, Durrell Burkhalter, Certificate of Death; NASLPR, Keith Renstrom; Interview Keith Renstrom by John Renstrom, 20-23 June 2016.

2080    Interview Sherri Bell, 30 March 2018; PAHGS, Punxsy News 14 March 1945, Punxsy News, 15 May 1945; NASLPR, Howard Chambers, Temporary Pers. Classification Record, Off. and Enlisted, Chambers, 3 Jan. 1946 & Medical History, 6 Dec. 1945, 1 & 24 Aug. 1945, 1-2.

2081    Interview James Henning, 21 April 2018. One of Erskine's battalion commanders, Lieutenant Colonel Robert E. Cushman Jr. (2nd Blt. 9th Marines), said he went through an entire two sets of platoon leaders—lieutenants while on Iwo Jima. MCHDQV, Oral History, Robert E. Cushman Jr., 172.

2082    Interview Bill Schlager, 27 April 2018.

2083    MCHDQV, Dashiell File, #289, Pitner.

2084    NASLPR, Nathan G. Pitner, Bronze Star & Purple Heart Citations, 28 July 1944 & Naval Speedletter, 724 3046, Casualty #606, 15 May 1945.

2085    Mandel, 286; Toland Vol. II, 818; Interview Lefty Lee, 24 June 2016.

2086    Burrell, *Ghosts of Iwo Jima*, 87.

2087    Ross, 182; Wheeler, 51; Newcomb, 174; Raymond Henri, 74-5. MCHDHQ, Historical fact sheet, Woody's archive, "Battle for Iwo Jima," 3; MSHDQV, Dashiell File, #182, Albumen-Iwo Jima; Crowl & Isely, 528-9.

2088    James, "Strategies in the Pacific," 717-8.

2089    Toland, Vol. II, 820. Toland says this plane was the first one, but the National Archives photo division proves the first plane to land was piloted by 1st Lt. Harvey Olson and was a Grasshopper observation plane (127-GW-308). These planes were heavily used by the ground troops for intelligence. Henri, et al., 102.

2090    King, 131.

2091    Bradley & Powers, *Flags of Our Fathers*, 234; Alexander, 1.

2092    King, 132, 154. MCRD Training Book, File 1

2093    MCRDSDA, 2nd Blt., 24th Marines, 4th MarDiv, Op. Journal, 28 Feb. 45, Time 1410.

2094    Bix, 483. MCRDSDA, 4th Engineer Btl., 4th MarDiv, Iwo, Op. Report, 27.

2095    King, 137.

2096    Ibid., 139.

2097    Crowley, 86.

2098    Harris, 39.

2099    Victoria, *Zen at War*, 26.

2100    Toland, Vol. II, 820; Wheeler, 55; Raymond Henri, 19.

2101    Morison, 75.

2102    Steinberg, *Island Fighting,* 194; Wheeler, 51; Burrell, *Ghosts of Iwo Jima*, 100.

2103    Interview Jerry Yellin, 22 March 2015. See also Brown & Yellin, 15-6.

2104    Brown & Yellin, 16.

2105    Interview Jerry Yellin, 22 March 2015; Brown & Yellin, 69.

2106  Toland, Vol. II, 821.

2107  Clausewitz, *On War*, Howard, "The Influence of Clausewitz," 35.

2108  Toland, Vol. II, 821.

2109  NACPM, RG 127 Box 96 (Iwo Jima) A45-2, 21ˢᵗ Rgt. Journal, Report 7, 4 March 1945, Col. Withers, 1.

2110  Toland, Vol. II, 812.

2111  USMCU Casualty Web Data Base; NASLPR, George Schwartz.

2112  Nalty, Shaw and Turnbladh, 585.

2113  MCHDHQ, Dalton, "Interview Woody Williams," 9; Interview Woody 22 July 2016. In the 1966 Francois interview, Woody claims 17 Marines were left again, but he says his company was a different strength of 265. Francois, 94. In the Congressional MOH Foundation Biographical Essay, Woody yet again gives a different number of 279 as the original number.

2114  NACPM, RG 127 Box 96 (Iwo) A45-2, 21ˢᵗ Rgt. Journal, Report 8, 5 March 1945, Col. Withers, 2.

2115  National Archives, DOD Dir 5200.9, Encl. D, 21ˢᵗ Rgt., Report, 10 April 1945, 1; MCHDHQ, Dalton, "Interview Woody Williams," 9; Interview Woody 22 July 2016.

2116  Crowl & Isely, 497; Moskin, 370. Moskin writes their units were brought back to the original numbers, but the 1ˢᵗ Btl., 21ˢᵗ Marine report contradicts this. If the original company numbers were almost 300 and the report claims they were brought up to around 160, then Woody's company was not getting its full replacement. MCHDQV, HQ, 1ˢᵗ Btl., 21ˢᵗ Rgt., 3d MarDiv, FMF, In Field, 6 April 1945, Report Lt. Col. Williams, 5.

2117  MCHDQV, HQ, 1ˢᵗ Blt., 21ˢᵗ Rgt., 3d MarDiv, FMF, In Field, 6 April 1945, Report Lt. Col. Williams, 5; Nalty, Shaw and Turnbladh, 585.

2118  Josephy, 170.

2119  Interview Woody 19 July 2016.

2120  MCRDSDA, 2ⁿᵈ Blt., 24ᵗʰ Rgt., 4ᵗʰ MarDiv, Reports, Iwo, Narrative, 20-1.

2121  Ibid., 1-21, 36-7. When one looks at the casualty reports from 19 Feb., it reads that the Btl. had 2 KIAs, 14 WIA's and 878 effective troops for a total of 894. However, on p. 37, it gives the original number for the Btl. of 850. It also gives the replacement number of 205. However, in tallying the replacement number from 19 Feb. to 5 March 1945, the Btl. received 276 replacements. Looking at the numbers in detail, this study feels the total number of the original contingent for this battalion plus replacements was 1,170 and not the 1,064 listed on p. 37.

2122  MCHDQV, Oral History, Cates, 94.

2123  NACPM, RG 127 (3d MarDiv), G-1 Periodic Report, No. 21-22.

2124  NACPM, RG 127, Box 84 (Iwo), Actual & Authorized Strength, 3d MarDiv, G-1 Report, 18-19 Feb. 1945.

2125  Interview Woody, 19 July 2016. In Jan. 1945, Vandegrift issued the Observer's Report for the ETOUSA (Europe Theatre of Operations) to Marine divisions. NACPM, RG 127 Box 16 (3d MarDiv), (2295) 14 Nov. 42-27 Nov. 1945 (1 of 2), HQ USMC, Commandant to Distribution list of all MarDivs and Rgts., 2 Jan. 1945. Report ETOUSA, HQ Army Ground Forces, 21 Oct. 1944, 1.

2126  Barry Beck, Speech at "All Forces Dinner," Wichita Falls, TX, 16 Feb. 2013, 3; MCHDHQ, Letter Beck to Ruth Beck, 5 March 1945.

2127  Moskin, 370.

2128  MCRDSDA, 2ⁿᵈ Blt.,24ᵗʰ Marines, 4ᵗʰ MarDiv, Reports, Iwo, Narrative, 17.

2129  Ibid., 18. See also Nalty, Shaw and Turnbladh, 586.

2130  Dashiell, "Clean Iwo Replacements Soon Soiled by Battle," March 1945.

2131  The Btl. report notes that A and C companies were brought up to 160 men each. If true and Woody's company the day before numbered 17, then they received roughly 140 replacements. MCHDQV, HQ, 1ˢᵗ Btl., 21ˢᵗ Rgt., 3D MarDiv, FMF, In Field, 6 April 1945, Report Lt. Col. Williams, 5; Interview Woody, 19 July 2016; Hallas, *Uncommon Valor on Iwo Jima*, 293.

2132   Interview Graves, 14 June 2016.

2133   NIDSMDMA, *Senshi-Sosho*, Central Pacific, Op. #2, Peleliu, Angaur, Iwo, Vol. 13, 393.

2134   MCHDQV, Dashiell File, #269, Myers; See also Haynes & Warren, 147.

2135   NACPM, RG 127, Box 83, (Iwo), AB-1, Com. Gen. 3d MarDiv to Commandant, Report, 3 June 1945, 102.

2136   Nalty, Shaw and Turnbladh, 592.

2137   MCRDSDA, 2nd Blt., 24th Marines, 4th MarDiv, Battle Reports, Iwo, Narrative, 21.

2138   Crowl & Isely, 457-8.

2139   NASLPR, Billie Griggs, Special Letter of Commendation from Maj. Gen. Erskine about Cpl. Griggs.

2140   Smith, *Coral and Brass*, 140. See also Henri, et al., 95.

2141   Clausewitz, *On War*, 210.

2142   Ibid., 209.

2143   Ibid., 213.

2144   Moskin, 369.

2145   Interview Charles Neimeyer, 15 Dec. 2017; Moskin, 369.

2146   Moskin, 369; Burrell, *Ghosts of Iwo Jima*, 71.

2147   MCHDQV, Oral History, Erskine, 357.

2148   Nalty, Shaw and Turnbladh, 592-3.

2149   Ibid., 593.

2150   Crowl & Isely, 528.

2151   Interview Woody, 19 July 2016; Nalty, Shaw and Turnbladh, 587; Moskin, 370; MCRDSDA, 2nd Blt., 24th Rgt., 4th MarDiv, Battle Reports, Iwo, Narrative, 22.

2152   Nalty, Shaw and Turnbladh, 587; Crowl & Isely, 506-7. According to several Marine correspondents who put together a history in 1945, the VAC artillery actually put 45,000 rounds on the Japanese positions before this attack on 6 March. The historians cited here are thought to be more accurate, and thus they are used for the information in the text, but the actual archival artillery files have not been analyzed for this study. Henri, et al., 111.

2153   Interview Woody, 21 July 2016; MCHDQV, HQ, 1st Btl, 21st Rgt., 3d MarDiv, FMF, In Field, 6 April 1945, Report Lt. Col. Williams, 5.

2154   Interview Woody, 21 July 2016; Hallas, *Iwo Jima*, 135; Smith, *Iwo Jima*, 64, 69-70; MCHDHQ, Woody, "Destiny...Farm Boy," 11; VandeLinde, 379; USMCU Casualty Web Data Base; Interview Waters, 28 Aug. 2017; NASLPR, Vernon Waters, Death Certificate & Deceased Unidentified take fingerprints Both hands document & Professional and Conduct, 7.

2155   Interview Woody, 23 July 2016.

2156   Interview Woody, 21 July 2016; Wells, "Medal of Honor Recipient Devotes Life to Veterans."; Hallas, *Uncommon Valor on Iwo Jima*, 135. Woody remembers the ring was not worn on his middle finger because Vernon's middle two fingers were grown together. However, according to Vernon's nephew, George Waters, Vernon did not have this abnormality. Woody even claimed his brothers had this deformity which was not the case.

2157   Interview Woody, 20 July 2016; VandeLinde, 378-9.

2158   MCHDQV, HQ, 1st Btl., 21st Rgt., 3d MarDiv, FMF, In Field, 6 April 1945, Report Lt. Col. Williams, 5-6.

2159   NASLPR, Vernon Waters, Casualties Statistics, KIA, 3-3-1945, 30 & Death Certificate Waters, 3-3-1945, 28.

2160   NASLPR, Vernon Waters, Combat Casualties, 7-7-45, sheet 903 1032, 31. The report reads: "Wound, Gunshot, head." This report was verified by Gene Rybak as being what his father Joseph witnessed. Interview Eugene Rybak, 28 May 2018.

2161   Interview Tommy Lofton, 30 April 2018

2162    NASLPR, Vernon Waters, "If Deceased Unidentified, Take Fingerprints of Both Hands," 59; NASLPR, Vernon Waters, Professional/Conduct, 7.

2163    Interview Graves, 7 May 2018.

2164    NASLPR, Vernon Waters, "Professional/Conduct Record of Vernon Waters," 42; Interview Woody 20 July 2016.

2165    Interview Graves, 7 May 2018.

2166    Interview Lauriello, 7 May 2018.

2167    Aurthur and Chlomia, 239.

2168    Interview, Gene Rybak, 28 May 2018; Rybak, "Herefrom."; Interview Corrie Rybak, 19 May 2019.

2169    Rybak, "Herefrom." USMC Major General James Livingston said that it appears Woody "lied" about his comrade Waters' death, and if "this is the case," Livingston said, "then that is disgraceful and unbecoming of a Marine." Interview James Livingston, 8 June 2018.

2170    Interview Woody, 20 July 2016; Melissa Karmath, "Humble Farmer Now Legendary Marine," Feb. 2015; Wells, "MOH Recipient Devotes Life to Veterans"; Hallas, *Iwo Jima*, 134-5.

2171    Interview Woody, 20 July 2016.

2172    Francois, 94.

2173    Interview Woody, 21 Oct. 2016.

2174    Interview Woody, 20 July 2016; Smith, *Iwo Jima*, 69.

2175    Dashiell, "Combat Correspondent Dispatch, 17 May 1945, 3.

2176    Melissa Karmath, "Humble Farmer Now Legendary Marine," Feb. 2015.

2177    NASLPR, Scrivens, Purple Heart Citation, 9 Nov. 1945.

2178    Miller, "Deathtrap Island," 10.

2179    Crowl & Isely, loc.10833; MCHDQV, Oral History, Robert E. Cushman Jr., 172.

2180    Nalty, Shaw and Turnbladh, 589.

2181    National Archives, DOD Dir 5200.9, Enclosure D, 21st Marines, Report, 10 April 1945, 1; Nalty, Shaw and Turnbladh, 588-9.

2182    MCHDQV, Dashiell File, #301, Fourth and Last.

2183    Interview Woody, 21 July 2016. See Henri, et al., 205.

2184    Ibid.

2185    Rybak, Review *Flamethrower* by, 28 June 2018.

2186    MCHDQV, HQ, 1st Btl. 21st Rgt., 3d MarDiv, FMF, In Field, 6 April 1945, Report Lt. Col. Williams, 21.

2187    Beck's lecture, Iwo Jima to NROTC Midshipmen, Purdue Univ., 1952, 8.

2188    NACPM, RG 127 Box 21, 4th MarDiv (1540) (1 of 2) Folder 5, HQ, 4th MarDiv., FMF, Com. Gen. to Com. General, FMFPAC, 21 April 1945, 1.

2189    NACPM, RG 127 Box 21, 4th MarDiv (1540) Folder 7, HQ, VAC, C/o FPO, San Fran, CA, Com. Gen. to Com. Gen., FMFPAC, 7 April 1945.

2190    MCRDSDA, 4th Tank Btl. Report, 4th MarDiv, Iwo, Report, 18 April 1945, 8.

2191    NASLPR, Leighton R. Willhite, Bronze Star Citation signed by Gen. Rockey, 5th MarDiv Commander & Service Record Book, Marks and Scars, Etc.; Interview Leighton Willhite, 30 Nov. 2019. Although Willhite's citation says he shot three, Willhite in 2019 only remembers shooting two.

2192    Bix, 483.

2193    Interview Woody, 19 July 2016; MCHDQV, HQ, 1st Bt., 21st Rgt., 3d MarDiv, FMF, In Field, 6 April 1945, Report Lt. Col. Williams, 6-7.

2194    NACPM, RG 127, Box 96, VAC, Forward Areas, 1750-50, Inspection Board to Schmidt, 22 March 1945, 1-3.

2195    Essay in Woody's archive, "No Word From Jim Hamilton Following Battle of Iwo Jima," 1 (original sources are from the Hamilton family's personal records, from the *Times News* and the *Buhl Herald* from 1945 & 1946); Wheeler, 56; Hallas, *Uncommon Valor on Iwo Jima*, 318; Burrell, *Ghosts of Iwo Jima*, 74.

2196  MCHDQV, Oral History, Van Ryzin, 80.

2197  Aurthur and Chlomia, 240.

2198  Essay in Woody's archive, "No Word From Jim Hamilton Following Battle of Iwo Jima," 1; Wheeler, 56; Hallas, *Uncommon Valor on Iwo Jima*, 318; Burrell, *Ghosts of Iwo Jima*, 74.

2199  NACPM, RG 127 Box 96 (Iwo), A45-2, 21st Rgt. Journal Iwo, 9 March 1945, R-3 Log, Sent 0700 from CG.

2200  MCHDQV, Dashiell File, #302, PFC. Warren Welch.

2201  NASLPR, Warren Welch, Navy Cross Citation.

2202  NASLPR, Joseph Rybakiewicz, Bronze Star Citation. In file Rybakiewicz, on his USMC Enlisted Man's Qualification Card, it was stated he received the Silver Star on 22 June 1945. However, a citation has not been located.

2203  Moskin, 370; National Archives, DOD Dir5200.9, Enclosure D, 21st Marines, Report, 10 April 1945, 1.

2204  Smith, *Coral and Brass*, 143; Wheeler, 56.

2205  https://archive.org/details/1945RadioNews/1945-03-11-CBS-World-News-Today.mp3.

2206  Clausewitz, *On War*, 292.

2207  MCHDQV, Dashiell File, #291, Report Cpl. Joseph Rybakiewicz; NASLPR, Joseph Rybakiewicz, Purple Heart & Bronze Star Citations.

2208  Interview Eugene Rybak, 28 May 2018.

2209  Maslowski & Millett, 457-8; Moskin, 372; Dower, 40-1; Bradley, *Flyboys*, 273; Bradley & Powers, *Flags of Our Fathers*, 240; Chang, 6; Toland, Vol. II, 671, 834-8; Thomas M. Coffey, Iron Eagle: The Turbulent Life of General Curtis LeMay, NY,1987; Frank, *Downfall*, 3-19, 44, 109, 160, 258, 294; Rhodes, 734; Hane, 348; Edgerton, 300; Kennedy, *Rise and Fall of Great Nations*, 356; Cook & Cook, 340, 345-7, 350; Weinberg, 870; Mountcastle, 102; Frank, 3-19, 160; Gandt, 101.

2210  *Documents on the Tokyo International Military Tribunal Charter*, 36; Bix, xvii, 364; Tanaka, xvii, 5-6; SLHA, Young Companion, 1941-1945, 167-72, "A Review of 4 Years of War in Figures" & 1940, 150-5, "Air Raids in Lanchow"; Tuchman, 136, 166; Interview Fiske Hanley, 20 March 2015; LaFeber, 426; O'Brien, *The Second Most Powerful Man in the World*, 107. ; Frank, *Tower of Skulls*, 133-6.

2211  BA-MA, RM 11/79, Mitteilung d. chinesischen Delegation an den Generalsekretär (Société des Nations), 21 Sept. 1937, 26 & RM 11/74, Marineattaché Tokyo an OKM, Auszug 31 Okt. 1938, 26.

2212  Toland, Vol. II, 839.

2213  Cook & Cook, 340-1, 500.

2214  Toland, Vol. II, 837-8.

2215  Edgerton, 317.

2216  McNeill, "Reluctant Warrior," 42.

2217  John W. Dower, "Lessons from Iwo Jima," Perspectives on History, Sept. 2007; Kakehashi, 93.

2218  Snow, 96.

2219  Kakehashi, 192.

2220  Maslowski, and Millett, 458.

2221  NMPWANM, 13 July 2017.

2222  Maslowski, and Millett, 458.

2223  Interview Woody, 23 July 2016; Interview Woody, 24 July 2016.

2224  MCHDQV, Dashiell File, #286, Fake Surrender; NASLPR, William Naro, Naval Speedletter, 1 May 1945.

2225  Burrell, *Ghosts of Iwo Jima*, 74.

2226  MCHDQV, HQ, 1st Btl., 21st Rgt., 3d MarDiv, FMF, In Field, 6 April 1945, Report Lt. Col. Williams, 12.

2227  NACPM, RG 127 Container 14, Dashiell, #229.

2228  NACPM, RG 127 Box 96 (Iwo) A45-2, 21st Marines Journal Iwo, Report 19, 16 March 1945 & Unit Report 20, 17 March 1945; MCHDQV, Oral History, Ion Bethel, 92; Cook & Cook, 193-8; Hough, 333; Crowl & Isely, 487. It seems most IJA POWs taken at Kwajalein in Feb. 1944 were also Korean and not Japanese. Zeiler, 37.

2229  MCHDQV, HQ, 1st Btl., 21st Rgt., 3d MarDiv, FMF, In Field, 6 April 1945, Report Lt. Col. Williams, 13.

2230  Mandel, 289.

2231  Bingham, 76.

2232  Kakehashi, 125; Toland, Vol. II, 910.

2233  Matt Carney, "Iwo Jima: US, Japanese Veterans Recall Horror of Pivotal World War II Battle, 70 Years On," ABC News, 28 March 2015.

2234  USMCU Casualty Web Data Base; NASLPR, William Naro, Naval Speedletter, 1 May 1945.

2235  Interview David and Kelly Naro, 28 April 2018.

2236  MCHDQV, Dashiell, #206, Close Calls—Iwo Jima.

2237  NASLPR, Alan Tripp, HQ 3d MarDiv, FMF, Com. Gen, 3d MarDiv, Award Letter of Commendation to Cpl. Tripp, USMC, 16 June 1945.

2238  NASLPR, Charles W. Henning, 1st Lt. M. G. Craig to Mr. & Mrs. Harley P Henning, 21 June 1945.

2239  NACPM, RG 127 Box 96 (Iwo Jima) A45-2, 21st Rgt. Journal Iwo, 13 March 1945, R-3 Log, Sent 1220; See also MCHDQV, Dashiell File, #53, Racey-Leyte.

2240  NACPM, RG 127 Box 96 (Iwo) A45-2, 21st Rgt. Journal Iwo, 16 March 1945, R-3 Log, Sent 1900.

2241  MCRDSDA, 2nd Blt., 24th Marines, 4th MarDiv, Journal, 4 March 45, Time 0835.

2242  MCHDQV, Oral History, Samuel G. Taxis, 173; See also Haynes & Warren, 147.

2243  Crowl & Isely, 520.

2244  Ross, 344; Toland, Vol. II, 822-30; Keegan, *Second World War*, 566; Rhodes, 594; Frank, *Downfall*, 61; Raymond Henri, 91-3.

2245  Harris, 223.

2246  Interview Woody, 20 July 2016.

2247  Dower, 44-5; Frank, *Downfall*, 26; Maslowski, and Millett, 461; Lord Russell of Liverpool, 252-5; Edgerton, 17, 292-3; Connaughton, R., Pimlott, J., and Anderson, D., 1995, The Battle for Manila, London: Bloomsbury Publishing; Gilbert, *Second World War*, 642. Greg Bradsher, "The Exploitation of Captured and Seized Japanese Records Relating to War Crimes, 1942-1945," *Researching Japanese War Crimes Records*, 152; Interview James Dowell, 13 May 2018; NACPM, Trial of Yamashita, SC 220798, 13 Nov. 1945; NACPM, 208 AA 132T 8 Feb 1945 & SC-2030320-S; Scott, 251-2, 278-9, 282, 338, 418, 422.

2248  Greg Bradsher, "The Exploitation of Captured and Seized Japanese Records Relating to War Crimes, 1942-1945," *Researching Japanese War Crimes Records*, 154-5; NACPM, Microfilm Room, Roll 1499, #101, 22607, Subj: Conditions Philippine Islands Japanese Occupation, 23 Aug. 1942: Atrocities; Bix, 445; Toland, Vol. II, 840.

2249  Scott, 386.

2250  Interview Woody, 21 July 2016.

2251  Toland, Vol. II, 726.

2252  Henri, et al., 70, 215-6.

2253  NACPM, 127 Box 14 (3d MarDiv), HQ 3d MarDiv, FMF, Intel Bulletin #4-44, Capt. Whipple, "The American Soldiers", 7.

2254  David McNeill, "Even the Dead Were Being Forced to Fight," Japan Times, 13 Aug. 2006.

2255  *Henry V*, Act IV, Scene iii.

2256  Interview Woody, 21 July 2016.

2257   King, 147-8, 152-6; Toland, Vol. II, 824-5; McNeill, "Even the Dead Were Being Forced to Fight."

2258   King, 147-8, 152-6; Toland, Vol. II, 824-5; McNeill, "Even the Dead Were Being Forced to Fight"; Wheeler, 56; Mosley, 31; *Japan at War*, 139.

2259   Hitchens, *god is not Great*, 204.

2260   McNeill, "Even the Dead Were Being Forced to Fight"; Wheeler, 56.

2261   King, 156.

2262   McNeill, "Even the Dead Were Being Forced to Fight."

2263   NACPM, RG 127 Container 14, Dashiell, #300; NACPM, 127 Box 14 (3d MarDiv), HQ, FMF, Camp Elliott, Subj: Japanese Military Org., 29 Oct. 1942, By Command of Maj. Gen. Barrett, Col. Noble, CoS, (9.) Deception as practiced by Japanese (7).

2264   NACPM, 127 Box 14 (3d MarDiv), HQ, FMF, Camp Elliott, Subj.: Japanese Military Org., 29 Oct. 1942, By Command of Maj. Gen. Barrett, Col. Noble, CoS, (9.) Deception as practiced by Japanese; Hallas, *Uncommon Valor on Iwo Jima*, 294.

2265   MCRDSDA, MCRD Training Book, File 1, Jungle Warfare.

2266   McNeill, "Even the Dead Were Being Forced to Fight."

2267   Melville, 66.

2268   Toland, Vol. II, 909.

2269   Interview Woody, 21 July 2016;Henri, et al., 212-3.

2270   Sledge, 63, 121; Dower, 63.

2271   Smith, *Coral and Brass*, 144

2272   Toland, Vol. II, 826-7; King, 150; Wheeler, 57.

2273   Mandel, 300.

2274   Interview Woody, 21 July 2016.

2275   NACPM, RG 127 Container 14, Dashiell, #167, 3.

2276   Toland, Vol. II, 826-7; King, 150; Wheeler, 57; Kakehashi, ix.

2277   Newcomb, 237.

2278   Ibid., 238-9.

2279   Heaton, Review *Flamethrower*, 22 May 2017.

2280   Interview Lauriello, 17 May 2017.

2281   MCHDQV, HQ, 1st Btl., 21st Rgt., 3d MarDiv, FMF, In Field, 6 April 1945, Report Lt. Col. Williams, 9.

2282   Headquarters Twenty-First Marines, 3D Marine Division, 1st Lt. R. Tischler "B" Co. 1st BN, Tischler Letter of Commendation, Col. H.J. Withers, 21 February-16 March 1945. Tischler Family Archive.

2283   Moskin, 372; National Archives, DOD Dir5200.9, Enclosure D, 21st Rgt., Report, 10 April 1945, 1.

2284   Burrell, *Ghosts of Iwo Jima*, 76.

2285   Kakehashi, 156.

2286   Smith, *Coral and Brass*, 144.

2287   MCHDHQ, Historical fact sheet in Woody's archive about Iwo entitled "The Battle for Iwo Jima," 3.

2288   MCHDQV, HQ, 1st Btl., 21st Rgt., 3d MarDiv, FMF, In Field, 6 April 1945, Report Lt. Col. Williams, 9.

2289   Ibid., 9-10.

2290   NACPM, RG 127 Container 14, Dashiell, #216; Smith, *Coral and Brass*, 141; Hough, 331.

2291   NACPM, RG 127 Container 14, Dashiell, #86.

2292   MCHDQV, Dashiell File, #243, DFC-Iwo Jima.

2293   NACPM, RG 127 Box 96 (Iwo) A45-2, 21st Marine Rgt. Journal Iwo, 16 March 1945, R-3 Log, Time 1200; NACPM, RG 127 Container 14, Dashiell, (Report in mid 200's).

2294   Excerpts of letter by SSG Joe Franklin, a 21st Maines Photographer, who was wounded and evacuated during operations on Iwo, 2. Dashiell's archives.

2295   NACPM, RG 127 Box 96 (Iwo) A45-2, 21st Rgt. Journal, 16 March 1945, R-3 Log, Time 1200; NACPM, RG 127 Container 14, Dashiell, (Report in the mid 200's).

2296 Joseph Picard, "Iwo Jima Not Pretty as Suribachi Picture: Clark Veteran Returning 50 years Later," *News Tribune* (NJ), 26 Feb. 1995.

2297 NACPM, RG 127 Container 14, Dashiell, #263, 1-3.

2298 NACPM, RG 127, Box 83, (Iwo), AB-1, Com. Gen. 3d MarDiv to Commandant, Report, 3 June 1945, 28; Samuel Eliot Morison and Henry Steele Commager, *The Growth of the American Republic*. Vol. 2, Oxford, 1958, 788; Paul Gordon Lauren, *Power and Prejudice*. London, 1988, 132–3; NMPWANM, 13 July 2017; Mears, 197-8.

2299 Hough, 356.

2300 Ibid.

2301 Ibid.

2302 Arizona Republic, "Marine Veteran Bothered by Souvenirs of Iwo Jima," The Baltimore Sun, 15 March 1995.

2303 Newcomb, 8, 230.

2304 NACPM, RG 127 Box 96 (Iwo) A45-2, 21ˢᵗ Marines Journal, 16 March 1945, R-3 Log, Time 2248, CG statement.

2305 Ibid., 17 March 1945, R-3 Log, Time Sent 2256, CG statement.

2306 MCRDSDA, 2ⁿᵈ Blt., 24ᵗʰ Rgt., 4ᵗʰ MarDiv, Battle Reports, Iwo, Narrative, 35.

2307 Interview Graves, 7 May 2017. Leighton Willhite also verifies Graves' testimony. He was a driver in the Sherman tank *Lorraine* attached to the 5ᵗʰ MarDiv. Interview Willhite, 10 May 2017.

2308 *Documents on the Tokyo International Military Tribunal Charter*, 37; Interview Graves, 7 May 2017; NACPM, RG 127 Container 14, Dashiell, #204.

2309 Interview Graves, 7 May 2017.

2310 Interview Graves, 7 May 2017; Interview Bernstein, 22 March 2018; Interview Farley, 8 Nov. 2017; IJAA symposium lecture by veteran Ira Rigger, 17 Feb. 2018.

2311 Henri, et al., 109.

2312 Interview Bernstein, 22 March 2018; Interview Bernstein, 23 March 2018.

2313 Snow, 102.

2314 Cook & Cook, 83; John Dower, "Lessons from Iwo Jima," Perspectives on History, Sept. 2007; Bingham, 1; Crowl & Isely, 484; Toll, *Pacific Crucible*, 75, 272.

2315 King, 149; McNeill, "Reluctant Warrior," 40.

2316 Ibid.

2317 Victoria, *Zen at War*, 102; Moore, 75; Gandt, 87.

2318 Prange, Gordan W., Goldstein, Donald M., and Dillon, Katherine V., *God's Samurai: Lead Pilot at Pearl Harbor*, D.C., 2004, 199-204; Kakehashi, xix.

2319 Newcomb, 233.

2320 King, 156-7.

2321 Newcomb, 234.

2322 Ibid.

2323 Kakehashi, xviii.

2324 Douglas Nash Sr., "Army Boots on Volcanic Sands: The 147ᵗʰ Inf. Rgt. at Iwo," Army History, Fall 2017, 9-10.

2325 MCHDQV, HQ, 1ˢᵗ Btl., 21ˢᵗ Rgt., 3d MarDiv, FMF, In Field, 6 April 1945, Report Col. Williams, 9-11.

2326 NACPM, RG 127 Box 96 (Iwo) A45-2, 21ˢᵗ Rgt. Journal, 19 March 1945, R-3 Log, Time 0926, CG statement.

2327 Ibid., 20 March 1945, R-3 Log, Sent 2123; MCRDSDA, 2ⁿᵈ Blt., 24ᵗʰ Marines, 4ᵗʰ MarDiv, Op. Journal, 3 March 45, Times 1510 & 1516.

2328 Ibid., 21 March 1945, R-3 Log.

2329 Ibid., 22 March 1945, R-3 Log, Sent 1400.

2330   Ibid., 23 March 1945, R-3 Log, Sent 1229.

2331   MCHDHQ, Letter Beck to Ruth Beck, 5 Dec. 1944.

2332   NACPM, RG 127 Box 96 (Iwo) A45-2, 21st Rgt. Journal, 24 March 1945, R-3 Log, Sent 1645.

2333   NACPM, RG 127 Container 14, Dashiell, #200-1.

2334   See Drea, 147-57, 206.

2335   NACPM, RG 127 Box 96 (Iwo) A45-2, 21st Rgt. Journal, 23 March 1945, R-3 Log, sent 0715 & 27 March 1945, R-3 Log, Sent 1510.

2336   Crowl & Isely, 527.

2337   NACPM, RG 127 Box 96 (Iwo) A45-2, 21st Rgt. Journal, Report No. 27, 24 March 1945, Col. Withers, 3 and Report No. 28, 24 March 1945, Col. Withers.

2338   NASLPR, Harry Linn Martin, Testimony of 1st Lt. Norris L. Bowen, Jr., 10 April 1945, 1-2.

2339   Ibid., 2.

2340   Ibid., 1.

2341   Crowl & Isely, 500.

2342   Ross, 335-7; Newcomb, 241; Kakehashi, xvi; Hallas, *Uncommon Valor on Iwo Jima*, 343. To read how the airmen responded to this attack, see Brown & Yelling, 91-100.

2343   Kakehashi, 192-6; Burrell, *Ghosts of Iwo Jima*, 80; Hallas, *Uncommon Valor on Iwo Jima*, 344.

2344   Kakehashi, 194.

2345   Ibid.

2346   Interview Lauriello, 10 May 2017.

2347   Interview Yellin, 18 Aug. 2016.

2348   NACPM, RG 127, Box 83, (Iwo), AB-1, Com. Gen. 3d MarDiv to Commandant, Report, 3 June 1945, 31.

2349   Interview Woody, 20 July 2017.

2350   NASLPR, Woody Williams, Major F. Belton to Lurenna Williams, 26 March 1945.

2351   MCHDHQ, Woody, "Destiny…Farm Boy," second page inserted between pp. 1-2.

2352   Interview Woody, 19 July 2016; Francois, 94; Smith, *Iwo Jima*, 70; VandeLinde, 372-3.

2353   Interview Woody, 19 July 2016.

2354   Dolan, "Conversation: Forged in Flame."

2355   National Archives, DOD Dir 5200.9, Encl. D, 21st Marines, Report, 10 April 1945, 8.

2356   Moskin, 372-3.

2357   Toland, Vol. II, 830; Wheeler, 57.

2358   McNeill, "Reluctant Warrior," 43; Burleigh, 331-2.

2359   Ross, 151; Kakehashi, 39.

2360   Kakehashi, 170.

2361   Ibid., 186.

2362   Ibid., 189.

2363   Ibid., 194.

2364   Kakehashi, 117; Frank, Review *Flamethrower*, 8 Aug. 2018.

2365   Burleigh, 331-2; Smith, *Coral and Brass*, 144; Hough, 358.

2366   NIDSMDMA, Collection Iwo Operation, #1, 298-311; NIDSMDMA, *Senshi-Sosho*, Cen. Pacific, Op. #2, Peleliu, Angaur, Iwo, Vol. 13, 340.

2367   King, 145; Coogan, 501.

2368   Toland, Vol. II, 884-5; Gandt, 25, 85.

2369   NIDSMDMA, Bio info Kuribayashi.

2370   Frank, "Review *Flamethrower*," 25 May 2017; Warren Kozak, *Lemay: The Life and Wars of General Curtis Lemay*, D.C., 2009, 239.

2371   Toll, *Conquering Tide*, 537; Cook & Cook, 208-9; Toland, Vol. II, 783.

2372   Brown & Yelling, 80-1; Interview Yellin, 2 June 2017.

2373    Moskin, 373; Newcomb, 240.

2374    Burrell, *Ghosts of Iwo Jima*, 5.

2375    Mandel, 217, 239, 307-10; Discussion with Mandel, 15 May 2017.

2376    Mandel, 310, 315, 317-8; David Stout, "Roland Gittelsohn, 85, Rabbi and a Marine Chaplain on Iwo Jima" NY Times, Dec. 15, 1995; Ross, 323. Interview Ken Roseman, 23 Feb. 2015.

2377    Hitchens, *god is not Great*, 4, 240; Harris, 103-5; Cornwell, 7; Victor J. Stenger, *God The Failed Hypothesis: How Science Shows that God Does Not Exist*, Amherst, 2007, 249; Anton Gill, *An Honourable Defeat: A History of German Resistance to Hitler 1933-1945*, NY, 1994, 74; David Kertzer, *The Pope and Mussolini: The Secret History of Pius XI and the Rise of Fascism in Europe*, Oxford, 2014, 12, 50-1, 58, 106, 109, 404. Listen to Thomas Childers' lecture "A History of Hitler's Empire, 2nd Edition," with Teaching Company, 2001; Gill, 65.

2378    Cornwell, 199-215; Kertzer, 373, 376, 386-7; Johnson, *A History of Christianity*, 487-3; Hitchens, *god is not Great*, 4, 239-40.

2379    National Jewish Welfare Board, Bureau War Records, Master Card System, Death, Warren H. Bornholz. Interestingly, Tony Stein's parents emigrated from Austria, so had they not left Europe, he and they would have most likely perished in the Holocaust.

2380    Email Charles C. Krulak to Rigg, 6 Dec. 2019; Email Charles C. Krulak, 15 Dec. 2019.

2381    Mandel, 11, 254, 303-22; Ross, 323. Interview Roseman, 23 Feb. 2015.

2382    NACPM, RG 127 Box 96 (Iwo) A45-2, 21st Rgt. Journal, 27 March 1945, R-3 Log, Sent 1510; Nash, 12.

2383    NACPM, RG 127 Box 96 (Iwo) A45-2, 21st Rgt. Journal, 29 March 1945, R-3 Log, Sent 1307.

2384    Ibid., 30 March 1945, R-3 Log, Sent 1910.

2385    Ibid., 31 March 1945, R-3 Log, Sent 1620.

2386    Nash, 14. Another time, Marines found a goat and her kid in a cave and a rooster on the battlefield. The Japanese seem to have had a lot of animals on the island for food. Henri, et al., 71, 81.

2387    NACPM, RG 127 Box 96 (Iwo) A45-2, 21st Rgt. Journal, 1 April 1945, R-3 Log, Sent 1945.

2388    Smith, *Coral and Brass*, 13.

2389    NACPM, RG 127 Box 96 (Iwo) A45-2, 21st Rgt. Journal, 27 March 1945, R-3 Log, Sent 1510; Nash, 12, 14.

2390    Van Harl, "He is Out in the Garage: Medal of Honor Winner," 16 Oct. 2014 (http://www.chuck-hawks.com/woody_williams.htm). This claim is different than what he declared to Dashiell that he was going to receive *the Medal*.

2391    Spector, 502; Frank, *Downfall*, 61, 375; Newcomb, 243; Weinberg, 869; Newcomb, 20. Donald L. Miller says there were 2,251 emergency landings. Miller, "Deathtrap Island," 9; Burrell, "Did We Have to Fight Here?," 63; Burrell, *Ghosts of Iwo Jima*, 6; Hallas, *Uncommon Valor on Iwo Jima*, 207; Millett, 432.

2392    Burrell, *Ghosts of Iwo Jima*, 5.

2393    Henri, et al., 232; Hough, 334; Burrell, "Did We Have to Fight Here?," 66; Burrell, *Ghosts of Iwo Jima*, xvii, 231, fn. 82.

2394    An average squadron had 12-20 planes (I will use 16 as an average for the number in the text to calculate how many planes would have been in three Groups). However, Jerry Yellin said his P-51 squadron had 37 planes. If the squadrons were this big on Iwo Jima, then instead of 360 P-51s, there were then 1,110. Brown & 49; Email Mant Hawkins to Rigg, 27 Dec. 2019.

2395    Yenne, *Hap Arnold*, 240.

2396    Burrell, *Ghosts of Iwo Jima*, 29-31, 35-6; Interview Morgan, 17 Sept. 2018.

2397    Miller, "Deathtrap Island," 12.

2398    Burrell, *Ghosts of Iwo Jima*, 98; Crowl & Isely, 471. P-51 pilot Jerry Yelling conducted a few missions against Chichi Jima to render the airfield there useless.

2399 Moskin, 373; Mullener, 221; Bradley & Powers, *Flags of Our Fathers*, 247; Wheeler, 196; Newcomb, 28; Hammel, 243.

2400 Miller, "Deathrrap Island," 12, 15.

2401 Frank, "Review of *Flamethrower*," 25 May 2017.

2402 Burrell, *Ghosts of Iwo Jima*, 5.

2403 Hammel, 243.

2404 McNeill, "Reluctant Warrior," 43.

2405 Commandant Charles C. Krulak's Marine Corps 223rd Birthday address to the Marine Corps from Iwo Jima, 10 Nov. 1998.

2406 Interview Woody, 23 July 2016; Smith, *Iwo Jima*, 71; NASLPR, Woody Williams.

2407 NACPM, RG 127 Container 14, Dashiell, #249.

2408 Poe, "Arthur Gordon Pym," 1114.

2409 Dashiell, "Marines on Iwo Jima Visit Graves," March 1945; NACPM, (Iwo), 127-RW-312.

2410 Many U.S. servicemen in WWII carried a pamphlet in their pockets that listed what they were fighting for which named, among other freedoms: the freedom of *religion, the press, the right to vote and free speech*. Also FDR often mentioned the four freedoms dear to Americans; namely, freedom of speech, freedom to worship, freedom from want and freedom from fear.

2411 From Lee Mandel's personal archive in Virginia Beach, VA. It is also in the historical documentary *In the Shadow of Suribachi: Sammy's Story*. A copy of it can also be found under Sammy Bernstein's name in the Marine Corps Historical Division Archive.

2412 NACPM, RG 127, Box 84 (Iwo), Actual/Authorized Strength, 3d MarDiv, G-1 Report, 18-19 Feb. 1945.

2413 Clausewitz, *On War*, 85-6.

2414 NACPM, RG 127, Box 51, 3d MarDiv Casualties Iwo Jima, 21st Rgt. & Box 83 (Iwo), AB-1, 3d MarDiv, AR & Enclosure A, IV., A. Admin. 1. Casualties suffered by Div.: 21st Marines 1,567.

2415 Moore, "Iwo Jima Eyewitness," Pass in Review, Feb. 1989, 10.

2416 Crowl & Isely, 528.

2417 Henri, et al., 233.

2418 MCHDHQ, Signed document by Woody answering a student's question: "What memory sticks out the most...about...Iwo Jima?" Undated.

2419 Ross, 323-4; Raymond Henri, 92; Aurthur and Chlomia, 252.

2420 Newcomb, 25; NASLPR, Erskine, Record of Graves Blanchard Erskine, #0268 & HQ, 2nd Btl., 6th Rgt., USMC, AEF, 19 Sept., Report Major Ernest C. Williams, Commander of 2nd Btl.

2421 Burrell, *Ghosts of Iwo Jima*, 70-1.

2422 NASLPR, Graves B. Erskine, Record of Erskine, #0268, Card #11 or 12, Note July 1945.

2423 Ibid., #0268, Sheet 3 of 4.

2424 Henri, et al., vii.

2425 Interview Gen. LeMay by Colin Heaton in June 1986.

2426 Mandel, 276.

2427 Ibid., 312.

2428 Morison, 70.

2429 The Carryall, Iwo Jima, Feb. 19, 1945—Oct. 15, 1945, 133rd NCB Navy Seabee Souvenir Newsletter, Reprint Ken Bingham, Vol. III, No. 8, Gulffort MS, 2011, 13.

2430 Interview Lauriello, 19 Nov. 2016.

2431 Interview Laura Leppert, 13 May 2017. Leppert heard this story from her father and Iwo veteran George Broderick who served in the 1st Btl., 26th Marines, 5th MarDiv.

2432 Moskin, 373; Toland, Vol. II, 911; Bingham, 72-6, 79, 135; http://www.wanpela.com/holdouts/profiles/kufuku.html.

2433   Burrell, *Ghosts of Iwo Jima*, 84-5; Toland, Vol. II, 911-2.

2434   Blum, 53.

2435   Hershel Woodrow "Woody" Williams, "Dedication of Highway for Rodney A. Breedlove."

2436   Hallas, *Uncommon Valor on Iwo Jima*, 138; Wells, "Medal of Honor Recipient Devotes Life to Veterans."

2437   NACPM, RG 127, Box 90, 21$^{st}$ Marines, HQ, 6 April 1945.

2438   Blum, 4-5; Sledge, 201; Toland, Vol. II, 869.

2439   Francois, 94.

2440   Interview Woody Williams, 13 Feb. 2017. See also Hallas, *Uncommon Valor on Iwo Jima*, 136.

2441   Interview Woody Williams, 13 Feb. 2017. See also Hallas, *Uncommon Valor on Iwo Jima*, 136.

2442   Interview Gary Fischer, 11 Jan. 2020.

2443   NASLPR, Vernon Waters, 1$^{st}$ Lt. Richard Tischler to Mrs. Verna Laura Waters, 20 April 1945.

2444   Letter George Waters to Rigg, 10 May 2018.

2445   NASLPR, Charles Fischer, Major Beck to A. W. Christianson, April 1945.

2446   NASLPR, Vernon Waters, Vandegrift to Verna Laura Waters, 24 March 1945; NASLPR, Charles Fischer, Vandegrift to Mrs. Fischer, 15 March 1945.

2447   NASLPR, Solomon Israel Blechman, Maj. Gen. Peck to Mr. and Mrs. Blechman, 21 Aug. 1944 & Note to Maj. Gen. Peck from Rabbi Nathan Blechman and Esther Blechman.

2448   NASLPR, Solomon Israel Blechman, Rabbi Nathan Blechman, Chaplain, Veterans' Hospital, No. 81 to HQ USMC, April 1949.

2449   Sledge, 171.

2450   Davis, "Operation Olympic," 45.

2451   Millet, 462; Davis, "Operation Olympic," 18; Frank, *Downfall*, 232-4; Hoffman, *Once a Legend,* 315.

2452   NASLPR, Woody Williams, Duration of National Emergency, 854310.

2453   Hoolihan, "Flame Thrower's Class at 21$^{st}$ Teaches Art of Roasting Japs." VAC commander, Harry Schmidt, wrote after Iwo "in view of the probable future operations against other heavily fortified positions, more intensive training of individual personnel in the use of flamethrowers, demolitions, and siege tactics is indicated, as well as continued emphasis on small unit leadership training." Crowl & Isely, 457.

2454   Smith, *Iwo Jima*, 71; Interview Woody, 22 July 2016; Melissa Karmath, "Humble Farmer Now Legendary Marine," Feb. 2015; Dalton, "Interview with Woody," 11; Sledge, 312.

2455   Letter Tischler to Parents, 2 April 1945.

2456   Sledge, 223.

2457   NACPM, RG 127 Container 14, Dashiell, #363.

2458   Frank, *Downfall*, 190.

2459   Ibid., 189.

2460   Dower, 216.

2461   Interview Frank, 20 July 2017.

2462   Frank, *Downfall*, 311; Rhodes, 744; Edgerton, 298. The number of 7,500 *Kamikaze* aircraft defending the mainland comes from the *Yushukan* War Museum in Tokyo in its displayed labelled: "Available Military Strength for Defense of the Homeland." Richard Frank says the number was more like 10,000. Frank, Review *Flamethrower*, 8 Aug. 2018. Toland agrees with Frank. Toland, Vol. II, 934, 1025.

2463   Victoria, *Zen at War*, 138-9.

2464   Ibid., 138.

2465   Hitchens, *god is not Great*, 203.

2466   Frank, *Downfall*, 190.

2467   Toland, Vol. II, 929-30.

2468   Frank, *Downfall*, 190.

2469   Toland, Vol. II, 934.

2470   NACPM, RG 127 Container 14, Dashiell, #267, 4; Lawless, "of Such Stuff are Heroes"; Hoolihan, "Flame Thrower's Class at 21ˢᵗ Teaches Art of Roasting Japs."

2471   MCHDQV, Dashiell File, #357, Able Company, 21ˢᵗ Marines, 15; Wheeler, 56; Crowl & Isely, 528. The battalion in the 4ᵗʰ MarDiv was the 3ʳᵈ Battalion in the 25ᵗʰ Marines.

2472   Interview Woody, 21 July 2016.

2473   Woody Williams Congressional MOH Report 5 Oct. 1945, 6.

2474   NASLPR, Howard Chambers, Medical Survey, 24 Aug. 1945, 1.

2475   MCHDQV, Dashiell, #208, Baseball Game-Iwo Jima. MOH cases of Watson and Pierce that will be explored in Chapter 28 will show that when multiple actions were collected in a MOH candidate's package, it could tip the case in his favor to receive a MOH.

2476   NASLPR, Wilson Watson; NASLR, Francis Pierce.

2477   NACPM, RG 127 Container 14, Dashiell, #267, 4.

2478   Hoolihan, "Flame Thrower's Class at 21ˢᵗ Teaches Art of Roasting Japs."

2479   Interview Woody 19 July 2016; Interview Woody, 21 July 2016; Interview Woody 23 July 2016; Interview Woody, 21 July 2017.

2480   Francois, 44, 93; NASLPR, Woody Williams, Statement Cpl. Alan Tripp, 153.

2481   In reading Francois' article carefully, he basically only quotes Woody for statements, facts he must have used from some history book/books and resident information about Woody's witnesses that has only been seen coming from Dashiell's article about Woody's actions. These facts lead the author to conclude that the three sources Francois used were the interviews with Woody Francois conducted (which Woody admitted giving Francois), secondary sources about the history of the Pacific War and/or Iwo Jima and Dashiell's article.

2482   Letter Woody to Rigg, 15 July 2017.

2483   NACPM, RG 127 Container 14, Dashiell, #267, 4.

2484   NACPM, RG 127 Container 14, Dashiell, #267, 3; NASLPR, Woody Williams, Statement Sgt. Lyman Southwell; USMCU Casualty Web Data Base.

2485   Interview Barry Beck, 5 April 2018. See also Ross, 159. Here Woody had given a report that he flamed five IJA soldiers. In 2015, Woody signed this page in the book that is in the author's collection and did not dispute the facts written on this page.

2486   Interview Woody 21 July 2016; *Semper Fi: The Magazine of the Marine Corps League,* May/June 2015, "Back to Iwo Jima," 16-17. He sometimes says three, but other times he says four, five or six.

2487   WWII Museum Oral History Project, 3 Feb. 2011; https://www.youtube.com/watch?v=DeJ2JHmQsFY; Interview 21 July 2016; NACPM, RG 127 Container 14, Dashiell, #267, 2-3.

2488   NASLPR, Vernon Waters, Silver Star Citation, 24. One must remember though, when receiving a Silver Star, it indeed required less evidence than a MOH. Even so, it still had to go through a review process to get awarded with witnesses attesting to the acts.

2489   Interview, Don Graves, 4 July 2018.

2490   Francois, 94; Interview Woody 21 July 2016; Interview Woody 22 July 2016; https://www.youtube.com/watch?v=fGts5WeLEgk; Patricia Ann Speelman, "MOH Recipient Shares Tales, Shakes Hands," 11 May 2013; Wells, "MOH Recipient Devotes Life to Veterans"; Collier, attached DVD to the book; Interview Woody, 17 March 2015. Earlier in the book, this study claimed Woody probably took out three to four pillboxes which is what probably happened. It is claimed here that possibly one could push that number to six if everyone of the witnesses gave statements unique to them and not overlapping with the experiences of the other Marines who gave affidavits, which, is highly unlikely. As a result, this study believes Woody probably took out three, possibly four pillboxes as mentioned before.

2491  NASLPR, Woody Williams, MOH citation Second Draft for Woody Williams. Sandy Wells' article from the *Charleston Gazette* from 2012 is not included in the list of sources on p. 577 where Woody talks about the pillbox numbers and KIAs he had because it is unclear how Wells actually received these numbers. When Woody showed the author this article, Woody claimed he did not know where Wells got his information although, once again, like with many journalists, it appears Wells got the information from Woody. Interview Woody, 19 July 2016.

2492  Interview Graves, 4 April 2018.

2493  NASLPR, John Willis, MOH Endorsement Train.

2494  NACPM, RG 127 Box 51, Casualties 3d MarDiv Iwo, 21st Rgt.; MCHDQV, Dashiell File, #169, Costly Battle-Iwo.

2495  MCHDQV, Dashiell File, #199, High Price-Iwo Jima.

2496  Interview Graves, 2 May 2018.

2497  MCHDHQ, Letter Beck to Ruth Beck, 4 April 1945.

2498  Ibid., 10 Sept. 1944.

2499  MCHDHQ, Beck, "Biography of Donald Beck," Dec. 2002, 16.

2500  NASLPR, Woody Williams, Withers to Forrestal, 27 April 1945, 159-60.

2501  Ibid.

2502  Ibid, Howard N. Kenyon, Record of Kenyon, Howard N. Marine Corps, Card #9-10; MCHDQV, Oral History, Robert E. Cushman Jr., 145. Interesting, later 25th Commandant and General Robert E. Cushman Jr. wrote that Kenyon, when he was in combat, was "unsure of himself" and always asking Cushman for advice. For such leadership, it is interesting that Kenyon got a Navy Cross for his actions on Iwo Jima. It seems that this medal may have been politically motivated since his citation does not really state any particular actions done by the colonel that can be documented as he personally doing himself. However, he did want to get into the action and he was there on Iwo Jima, so one has a difficult time saying that he possibly did not deserve this Navy Cross. However, sometimes, Navy Crosses were given to men who just were in a combat arena and behaved bravely, like what happened to Fleet Admiral William D. Leahy in WWI, when he was given a Navy Cross "because he had captained a ship in a war zone." O'Brien, *The Second Most Powerful Man in the World*, 49.

2503  NACPM, RG 127 Container 14, Dashiell, #267, 2-3.

2504  Lawless, "of Such Stuff are Heroes," 68.

2505  MCHDHQ, Beck, "Biography of Donald Beck," Dec. 2002, 16.

2506  MCHDQV, Dashiell File, #267, Man with Flames—Iwo. In discussion with Dashiell's son, Fred, he strongly feels his father had problems with Woody's testimony. Interview Fred Dashiell, 11 June 2018.

2507  Interview Fred Dashiell, 11 June 1945.

2508  Dashiell, Combat Dispatch, 17 May 1945, 1; NACPM, RG 127 Container 14, Dashiell, #267, Cover Page.

2509  During the interview conducted with Woody on 19 July 2016, I asked: "If you took out 21 in only on pillbox Woody, you must have taken out probably an additional 30 with the remaining seven pillboxes, right?" Woody responded, "Yes, that is correct. Maybe even more." Historian James H. Hallas also came to this conclusion. Interview James H. Hallas, 20 June 2016; Hallas, *Iwo Jima*, 133.

2510  Interview Lauriello, 7 May 2018.

2511  Hoolihan, "Flame Thrower's Class at 21st Teaches Art of Roasting Japs."

2512  Lawless, "of Such Stuff are Heroes," 68.

2513  Francois, 93.

2514  Hoffman, *Chesty*, x.

2515  Interview Graves, 4 April 2018. The Japanese were very good about re-using their pillboxes and bunkers thought to have been knocked out by the Marines by using "underground tunnels." See Henri, et al., 107.

2516   Interview Woody, 21 July 2016; George Lawless, "of Such Stuff are Heroes," Charleston Gazette, 20 May 1956; Bill Francois, "Flame-Throwing Marine Who Won the Medal of Honor," The Man's Magazine, Vol. 14, #1, Jan. 1966.

2517   Woody Williams Congressional MOH Report 5 Oct. 1945, 3-4, 14.

2518   NASLPR, Woody Williams, Geiger to Forrestal, 24 July 1945, 164; Woody Williams Congressional MOH Report 5 Oct. 1945, 14-6.

2519   Mears, 102.

2520   Mears, 102; NASLPR, Jacklyn Lucas, Geiger to Forrestal, 14 Sept. 1945 & Hayler to Forrestal, 20 Sept. 1945 & Nimitz to Forrestal, 6 Sept. 1945 & Schmidt to Forrestal, 21 July 1945.

2521   NASLPR, Vernon Waters, Geiger to Forrestal, 24 July 1945, Approving Silver Star Medal for Vernon Waters.

2522   NASLPR, Woody Williams, Nimitz to Geiger, 19 Aug. 1945, 165; Mears, 101.

2523   NASLPR, Woody Williams, Geiger to Forrestal, 24 July 1945, 164; NASLPR, Woody Williams, Geiger to Chambers, 27 Aug. 1945, 166; Chambers Family Archive, Geiger to Chambers, 27 Aug. 1945.

2524   NASLPR, Woody Williams, Geiger to Forrestal, 19 Sept. 1945, 149.

2525   Chambers Family Archive, Geiger to Chambers, 4 Sept. 1945; NASLPR, Chambers, Thompson to Bureau of Personnel, HQ Marine Corps, D.C., 20 Oct. 1945.

2526   NASLPR, Howard Chambers, 148th General Hospital, APO 244, San Fran. CA, 12 March 1945 & Travel Update, Naval Barracks, U.S. Naval Dry-docks, 23 April 1945; PAHGS, Punxsy News 1 May 1945, Punxsy News 15 May 1945, Punxsy News 1 June 1945, Punxsy News, 8 June 1945, Spirit, June 1945.

2527   PAHGS, Punxsy News 1 June 1945, Spirit June 1945.

2528   NASLPR, Howard Chambers, Bio History, Jun. 1945.

2529   Ibid., Chambers to Commandant, 5 June 1946.

2530   NASLPR, Jacklyn H. Lucas, Extract Letter of 1st Lt. Lay L. Bailey from U.S. Naval Convalescent Hospital, Harriman, NY & Statement PFC Allan C. Crowson, Casual Co. Number One, Marine Detachment, U.S. Naval Hospital, Oakland, CA, 28 Sept. 1945; NASLPR, Tony Stein, Underhill to Com. Officer, 1st Btl., 28th Marines, 17 May 1945 & Com. Officer to Com. Gen., FMF, 25 May 1945; NASLPR, Wilson Watson, Geiger to Forrestal, 16 Sept. 1945 & Statement Lt. Thurman H. Nolan & MOH Citation.

2531   Chambers Family Archive, Geiger to Chambers, 27 Aug. 1945.

2532   Ibid.

2533   Interview Paul Stubbs, 4 April 2018.

2534   Telephone conversation with General Charles C. Krulak, 6 Dec. 2019.

2535   Email Bob Olson to Rigg, 4 April 2018.

2536   NASLPR, Woody Williams, Geiger to Forrestal, 19 Sept. 1945, 149.

2537   NASLPR, Woody Williams, Hayler to Forrestal, 20 Sept. 1945, 150; Richard Pearson, "Vice Admiral R.W. Hayler, Navy Hero, Dies," Washington Post, 22 Nov. 1980; NASLPR, R. W. Hayler.

2538   NASLPR, R. W. Hayler, Forrestal to Hayler, Subj. Navy Dept. Board Decorations and Medals, 17 Aug. 1945, 1.

2539   Telephone conversation with General Charles C. Krulak, 6 Dec. 2019.

2540   NASLPR, Woody Williams, HQ, FMFPAC, 4th Endorsement, Geiger to Forrestal, 27 April 1945.

2541   Ibid., 3d MarDiv, FMF, Transfer Order Williams 854310, Hershel, by command Maj. Gen. Erskine, 18 Sept. 1945.

2542   NASLPR, Hayler, Forrestal to Hayler, Subj. Navy Dept. Board Decorations and Medals, 17 Aug. 1945, 1-2.

2543   O'Brien, The Second Most Powerful Man in the World, 91.

2544   NASLPR, Hayler, Capt. Robert Ward Hayler, U.S. Navy, Service of, 27 March 1944, 1-3 & Chief Bureau of Ordnance Vice Admiral G. F. Hussey, Jr. to Forrestal (Board Decorations/Medals), 29 Nov. 1945.

2545    Newpower, xi-xii; Interview Anthony Newpower, 11 July 2018; Hallas, *Saipan,* 29; Toll, *Pacific Crucible,* 237.

2546    Newpower, 59-68: Interview Newpower, 11 July 2018; Toll, *Pacific Crucible,* 256, 258.

2547    Interview Anthony Newpower, 11 July 2018. Newpower explains Hayler had many he was working with so he was not solely responsible for the failure, but since he was in charge of putting torpedo parts together and ensuring they worked, he indeed made "some critical mistakes." Robert Gannon, *Hellions of the Deep: The Development of American Torpedoes in World War II,* Penn State Univ. Press, 2009, 43, 76-82; Zeiler, 16-7.

2548    Toland, Vol. II, 599-600, 610.

2549    NASLPR, Hayler, Commander Task Group Eight Point Six to Commander Cruisers, Pacific Fleet, Subj.: Capt. Hayler, 28 Oct. 1942.

2550    Ibid., Capt. Robert Ward Hayler, U.S. Navy, Service of, 27 March 1944, 1-3, & Citations—Night Action off Kolombangara 12-13 July 1943 by Adm. W. L. Ainsworth, & Second Gold Star in lieu of Third Navy Cross to Capt. Hayler Citation.

2551    Ibid., Silver Star Medal to Capt. Hayler.

2552    Ibid., Sen. Member, Board Review for Decorations and Medals, Adm. F. J. Horne to John Sullivan, 4 Feb. 1947.

2553    NACPM, RG 80, Box 5 (Navy Dept. Board Decorations/Medals), Steinhagen Navy Dept. Board Decorations and Medals to Com. Gen., Aircraft, FMF, Pacific, 1 Oct. 1947.

2554    NASLPR, Gregory "Pappy" Boyington, Memorandum for Marine Corps Commandant, Subj: Recommendation for award of the Navy Cross in case of Maj. Boyington, USMCR, in absentia, 27 April 1944.

2555    NASLPR, Alexander Archer Vandegrift Sr., Nimitz to Secretary of the Navy, Subj: Recommendation for the award of the Distinguished Service Medal to Maj. Gen. Vandegrift, 11 Dec. 1942 & Nimitz to Secretary of the Navy, Subj: Congressional Medal of Honor, award of, case Maj. Gen. Vandegrift, 16 Dec. 1942 & HQ U.S. Marine Corps, Memorandum for Navy Cross, 5 Jan. 1943 & Commandant to Maj. Gen. Vandegrift, Subj: Award of Navy Cross and Citation, 11 Jan. 1943 & Navy Cross Citation & MOH Citation.

2556    Interview Don Graves, 20 Nov. 2019.

2557    NASLPR, R. W. Hayler, Vice Adm. Turner to Chief of Naval Personnel, Fitness Report Hayler, 27 Aug. 1944 & Officer Bio Sheet File No. 8611, 16 Aug. 1949.

2558    Ibid., Com. Carrier Division Four Adm. J.W. Reeves, Jr. to Chief of Naval Pers., Fitness Report Rear Adm. Robert W. Hayler, 6 July 1944.

2559    Ibid., Com. Third Fleet Halsey to Chief of Naval Pers., Fitness Report about Hayler, 23 Sep. 1944.

2560    Ibid., Com. Amphibious Group One Adm. W.H.P. Blandy to Com. Cruisers, U.S. Pacific Fleet, Subj: Report on the Performance of Duty of Rear Adm. R.W. Hayler, 21 Dec. 1944 & Com. Battleship Div. Two Adm. H.F. Kingman to Com. Cruisers, U.S. Pacific Fleet, Subj: Fitness Report Robert Ward Hayler, 29 Sep. 1944.

2561    Ibid., Com. Battleship Sq. One Vice Adm. J. B. Oldendorf to Chief Naval Personnel, Fitness Report Rear Adm. Hayler, 27 Dec. 1944.

2562    Ibid., Capt. Hayler, U.S. Navy, Service of, 27 March 1944, 1-3, & Com. Cruiser Div. Four Adm. J.B. Oldendorf to Chief of Naval Personnel, Subj: Performance of Duty Rear Adm. Hayler, 13 Aug. 1944 & Report Fitness of Officers, R. W. Hayler, 27 Feb. 1945.

2563    Ibid., Report of Compliance with Orders, 13 June 1945 & Records of Officers, U.S Navy, No. 20.

2564    Ibid., Hayler' Fitness Report, 1 March 1947.

2565    Ibid., 2 Feb. 1946.

2566    NASLPR, Off. of Commandant, Box 821, Folder MOH 1740-55-50, Folder (1 Jan. 1944-31 Dec. 1947), Forrestal to Vandegrift, 5 Sept. 1945.

2567    NACPM, RG 127, Off. of Commandant, Box 821, Folder MOH 1740-55-50, Folder (1 Jan. 1944-31 Dec. 1947), Forrestal to Vandegrift, 7 Sept. 1945.

2568  NASLPR, Woody Williams, Wagner's memorandum for Capt. Neil K. Dietrich, 20 Sept. 1945, 100.

2569  Email General Charles C. Krulak to Rigg, 24 Dec. 2019.

2570  Email Bob Olson to Rigg, 4 April 2018.

2571  Interview Roger Cirillo, 12 May 2018.

2572  Email Jones to Rigg, 26 June 2018.

2573  Email Hoffman to Rigg, 28 March 2018.

2574  MCHDQV, Dashiell File, #228, Nosarzewski-Iwo Jima; NASLPR, John Nosarzewski, Navy Cross Citation; https://valor.militarytimes.com/hero/8222.

2575  NACPM, RG 127 Box 20 (3d MarDiv), (Correspondence) Gen. Smith to distribution list, subj.: Increase awards of enlisted men, 15 July 1945

2576  *Navy and Marine Corps Awards Manual*, 1953, Part I.—Per. Decors., Sec. 1. Gen. Instructions (1) Purpose, 1.

2577  The French Emperor created the modern development of awarding men for valor by issuing decorations for acts both in battle and in civilian life that "saw the adoption of medals for bravery in virtually every European nation" at that time. Mears, 10.

2578  Interview James Henning, 21 April 2018.

2579  Mears, 98.

2580  NASLPR, Woody Williams, King to Forrestal 21 Sept. 1945; NASLPR, Jacklyn Lucas, King to Forrestal, 21 Sept. 1945; NASLPR, John Willis, King to Forrestal, 14 Sept. 1945.

2581  Charles A. Jones, Review *Flamethrower*, 29 May 2018.

2582  Woody Williams Congressional MOH Report 5 Oct. 1945, 21-5; NASLPR, Woody Williams, King to Forrestal 21 Sept. 1945 & Memorandum Naval Aide to the President; NASLPR, John Willis.

2583  Interview Woody, 23 July 2016; MCRDSDA, MCRD Training Book, File 1, <u>Financial Rewards of the Service.</u>

2584  MCHDQV, Dashiell, #208, Baseball Game-Iwo; MCHDHQ, Van Harl, "He is Out in the Garage: MOH Winner," 16 Oct. 2014.

2585  Interview Woody, 22 July 2016; Sledge, 312; O'Brien, *The Second Most Powerful Man in the World*, 340.

2586  Paterson (ed.), *Major Problems in American Foreign Policy*," Bernstein, Barton J., "The Atomic Bomb and Diplomacy," 319.

2587  Frank, *Downfall*, 330.

2588  Toland, Vol. II, 1076.

2589  KJV—King James Version, 996.

2590  Interview Woody, 22 July 2016; Francois, 94; Smith, *Iwo Jima*, 71; VandeLinde, 373.

2591  MCHDQV, Oral History, Woody Williams 13801B.

2592  NASLPR, Woody Williams, Professional and Conduct Record.

2593  Interview Woody, 22 July 2016; Smith, *Iwo Jima,* 71.

2594  Hanley II, 328.

2595  Smith, *Iwo Jima,* 71.

2596  Interview Woody, 23 July 2016; Wells, "Medal of Honor Recipient Devotes Life to Veterans." However, in a sermon Woody gave at River Cities Church on 29 March 2003, he said that POW who was in a North Vietnamese camp for six years told him this saying: "You can never really know freedom…until you have lost it."

2597  Interview Woody, 23 July 2016.

2598  Interview Woody, 23 July 2016.

2599  NASLPR, Woody Williams, Telegram Lurenna Williams to Vandegrift, HQ USMC, 30 Sept. 1945, 93.

2600  Van Harl, "He is Out in the Garage: Medal of Honor Winner," 16 Oct. 2014. Woody has apparently also used the term fiancée in 1966 when he gave an interview to Francois, but when I attempted to use it, he strangely threatened to stop working with me if I did not write girlfriend instead of fiancée. See Francois, 94.

2601 Interview Woody, 23 July 2016.

2602 Ibid., 22 July 2016.

2603 Newcomb, 191-2; Interview Woody, 12 Feb. 2017.

2604 Woody Williams Congressional MOH Report 5 Oct. 1945, 2.

2605 Interview Woody, 22 July 2016.

2606 O'Brien, *The Second Most Powerful Man in the World*, 332.

2607 MCHDHQ, Unattributed article from 1945 "Sgt. Hershel Woody Williams," 4.

2608 Harry S. Truman Presidential Library & Museum, Public Papers Harry S. Truman 1945-53, 160, "Remarks at the Presentation of the Congressional Medal of Honor to Fourteen Members of the Navy and Marine Corps."

2609 Burrell, *Ghosts of Iwo Jima*, 152.

2610 Interview Woody, 21 July 2016; VandeLinde, 376; Newcomb, 159; Hallas, *Iwo Jima*, 137.

2611 Dalton, "Interview with Woody Williams," 11; Smith, *Iwo Jima*, 69; Bechtel, "The Making of a Marine,", 38.

2612 Toland, Vol. II, 826; Hallas, *Uncommon Valor on Iwo Jima*, 299; Newcomb, 127, 131.

2613 Bradley & Powers, *Flags of Our Fathers*, 184; Moskin, 337; King, 119; Henri, et al., 67-8; O'Brien, *Liberation*, 9.

2614 Hallas, *Uncommon Valor on Iwo Jima*, 20-39; Henri, et al., 42.

2615 Moskin, 369; Ross, 223-4; Wheeler, 51 Hallas, 149.

2616 Hallas, *Uncommon Valor on Iwo Jima*, 150.

2617 Interview Woody, 19 July 2016.

2618 MCHDHQ, Unattributed article from 1945 "Sgt. Hershel Woody Williams," 4.

2619 Interview Ellery B. "Bud" Crabbe, 23 July 2017.

2620 NACPM, RG 127, Box 90, 21st Marines, War Diary, 1-31 Oct. 1945.

2621 MCHDQV, Oral History, Erskine, 381-4.

2622 Ibid., 384.

2623 3d MarDiv Bulletin, Vol. II, 6 Oct. 1945, Number 32, Dick Dashiell, "Regimental Ramblings."

2624 Letter Dashiell to Ray Truelove, Editor Caltrop, 26 April 1986.

2625 3d MarDiv Bulletin, Vol. II, 6 Oct. 1945, Number 32, Dick Dashiell, "Regimental Ramblings."

2626 MCHDHQ, Unattributed article from 1945 "Sgt. Hershel Woody Williams," 4.

2627 Interview Woody, 11 Jan. 2018.

2628 NIV—New International Version, 574.

2629 Speelman, "Medal of Honor Recipient Shares Tales, Shakes Hands," 11 May 2013. Woody is quoting Jesus here from John 8:31-2: "To the Jews who had believed him Jesus said, 'If you hold to my teaching, you are really my disciples. Then you will know the truth, and the truth will set you free.'" NIV-New International Version, 937.

2630 Burrell, *Ghosts of Iwo Jima*, xv, 151, 205 fn. 5; L. Miller, "Deathtrap Island; Where a Stone Age Battle Set the Stage for the Atomic Era," WWII Magazine, Special Collector's Addition for the 60th Commemoration of Iwo, 2006, 9; Mullener, 225; Smith, *Coral and Brass*, 18; Hough, 358-9; NACPM, RG 127 Box 20 (3d MarDiv), Erskine to all Units, Subj: Increase awards to enlisted, 8 Aug. 1945.

2631 Mears, 98.

2632 Hallas, *Uncommon Valor on Iwo Jima*, 165.

2633 Burrell, *Ghosts of Iwo Jima*, 75; Bradley & Powers, *Flags of Our Fathers*, 10; Interview Woody, 23 July 2016; Berry, 110; Notes from Woody Williams, 12 Feb. 2017; Hallas, *Uncommon Valor on Iwo Jima*, 173; Henri, et al., 37.

2634 Crowl & Isely, 501; Toland, Vol. II, 82; Moskin, 373; Smith, *Coral and Brass*, 18.

2635 Newcomb, 74; Mandel, 330-1.

2636 Dwight Mears, Review *Flamethrower*, 15 Oct. 2018.

2637 Email Hoffman to Rigg, 28 March 2018.

2638 Scott, 240-2.

2639 Alan Axelrod, *Miracle at Belleau Wood: The Birth of the Modern U.S. Marine Corps*, Lyons Press, 2007, 217.

2640 Kwong and Tsoi, 203-4; Lai, 63; Ferguson, 168-9; Gilbert, *The Second World War*, 280.

2641 BA-MA, BMRS, File Walter Hollaender; Rigg, *Hitler's Jewish Soldiers*, 42-3. Hollaender was actually a "half-Jew "or *Mischling 1. Grades* in the *Wehrmacht* and had received special clemency from Adolf Hitler to remain in the army (Hitler actually declared him of "German Blood," i.e. he received the Führer's *Deutschblütigkeitserklärung*).

2642 NASLPR, Donald Ruhl; NASLPR, James Dennis La Belle; NASLPR, Darrell Cole; NASLPR, William Caddy; NASLPR, Harry Linn Martin.

2643 NASLPR, Tony Stein, MOH Citation & J.B. Butterfield to Smith, Commander FMF, 25 May 1945, 1-2.

2644 NASLPR, Tony Stein, J.B. Butterfield to Smith, Com. FMF, 25 May 1945, 1; Hallas, 32.

2645 Hallas, *Uncommon Valor on Iwo Jima*, 30-1.

2646 Ibid.

2647 Ibid., 31.

2648 NASLPR, Tony Stein, J.B. Butterfield to Smith, Com. FMF, 25 May 1945, 1-2 & Statement Sgt. Kent F. Stegner, 23 April 1945.

2649 Ibid., J. L. Underhill to Com. Officer, First Btl., 28th Marines, 5th MarDiv, 17 May 1945.

2650 Ibid., King to Forrestal, 9 Oct. 1945 & Hayler to Forrestal, 27 Sept. 1945 & Vandegrift to Forrestal, 21 Sept. 1945 & Nimitz to Forrestal, 8 Sept. 1945 & Smith to Forrestal, 8 June 1945.

2651 NASLPR, Holland M. Smith, Nimitz to Smith, Subj: Authority to make Medal Awards, 16 Aug. 1944.

2652 NACPM, RG 127 Box 96 (Iwo) A45-2, 21st Marine Rgt. Journal, 26 Feb. 1945, R-3 Log, Time Sent 0800; NASLPR, Francis Fagan, Sample MOH Citation.

2653 NASLPR, Francis Fagan, Statement of 1st Lt. James Papnis.

2654 Ibid., Kenyon to Forrestal, 30 April 1945.

2655 Ibid., Navy Cross Citation for Iwo & Airmail Brief Notifying Report of Death, 28 March 1945 & Subj: Combat Casualties, Report of. 2400 25 Feb45, 2400 26Feb45, 802 1127.

2656 Interview Brian Williams, 2 June 2019.

2657 Interview T. Fred. Harvey, 4 March 2019.

2658 NASLPR, Francis Fagan, Navy Cross Citation for Guam.

2659 Ibid., Kenyon to Erskine, 19 May 1945.

2660 Ibid., Erskine to Forrestal, 19 May 1945.

2661 Ibid., Schmidt to Forrestal, 16 July 1945.

2662 Ibid., Geiger to Forrestal, 31 July 1945.

2663 Ibid., Hoover to Forrestal, 7 Aug. 1945.

2664 Ibid., Vandegrift to Forrestal, 28 Aug. 1945.

2665 Ibid., Hayler to Forrestal, 6 Sept. 1945.

2666 Hoffman, *Once a Legend,* 208; Mears, 99.

2667 Email Hallas to Rigg, 29 March 2018.

2668 NASLPR, Francis Fagan, Collins to Vandegrift, 16 Jan. 1947.

2669 NASLPR, Wilson Watson, Geiger to Forrestal, 16 Sept. 1945.

2670 Ibid., Statement of Lt. Thurman H. Nolan & MOH Citation.

2671 Ibid., Hayler to Forrestal, 20 Sept. 1945.

2672 NASLPR, Darrell Cole, Geiger to Forrestal, 17 July 1945.

2673 Ibid., Hayler to Forrestal, 6 Sept. 1945, Horne to Forrestal, 23 May 1946, Blissard to Forrestal, 12 April 1945, 1-2.

2674    NASLPR, Jacklyn Lucas, Schmidt to Forrestal, 21 July 1945, Geiger to Forrestal, 14 Sept. 1945, Hayler to Forrestal, 20 Sept. 1945 & Nimitz to Forrestal, 6 Sept. 1945.

2675    NASLPR, George Barlow, Letter John A. "Jack" Synder to NASLPR, 8 Feb. 1994; MCHDQV, Dashiell File, #293, Smothers Grenade; Email Hallas to Rigg, 29 March 2018. Email Hallas to Rigg, 4 April 2018; http://www.hqmc.marines.mil/Portals/61/Docs/FOIA/WWII%20SS%20Citations/ USMC%20WWII%20SS%20(Letter%20J)%20(121%20pgs).pdf?ver=2013-01-25-133556-970; NASLPR, Harry Lee Jackson, USNH, Mere Island, CA, 25 April 1945 & US Naval Hospital, Philadelphia, PA, 30 Oct. 1945& Silver Star Citation.

2676    MCHDQV, Dashiell File, #305, Hand Blown Off; NASLPR, Robert C. Filip, Navy Cross Citation & Report of Board of Medical Survey on Robert Filip, U.S. Naval Hospital, Philadelphia, PA, 6 November 1945.
     http://www.homeofheroes.com/members/02_NX/citations/03_wwii-nc/nc_06wwii_usmcE.html

2677    NASLPR, Donald Ruhl, MOH Citation, Geiger to Forrestal, 10 May 1946, Capt. Dave E. Severance to Forrestal, 2 Jan. 1946 & Statement of 1st Lt. John K. Wells, Statement Sgt. Everett M. Lavelle & Statement 2nd Lt. John A. Daskalakis. See also Henri, et al., 67-8.

2678    NASLPR, James Dennis La Belle, MOH Citation; NASLPR, William R. Caddy, MOH Citation; NASLPR, William Walsh, MOH Citation.

2679    NASLPR, George Barlow, Letter John A. "Jack" Synder to NASLPR, 8 Feb. 1994.

2680    Mears, 1.

2681    Ibid., 108.

2682    Email Hoffman to Rigg, 28 March 2018; Hoffman, Chesty, 21, 48, 55-7, 88, 121, 145, 147, 244, 291, 294, 537-8.

2683    Hoffman, Chesty, 195.

2684    NASLPR, Gregory "Pappy" Boyington, Boyington to Truman, 28 Feb. 1952.

2685    Ibid.

2686    NASLPR, Boyington, Memorandum From: Marine Corps Commandant, To: Naval Aide to the President, Subj: Awards; cases of BRIGEN Lewis B. Puller, USMC and Col. Boyington, USMC (Ret'd), information concerning Boyington's 28 Feb. 1952 letter to Truman. According to historian Jon T. Hoffman, Gen. Puller had been recommended for the MOH, but it had just been denied by higher ups. See Hoffman, Chesty, 195. So either the staff that put this letter together for Truman had an incomplete file on Puller or they hid the fact that Puller had been considered but denied so that Truman would not order a further investigation.

2687    NASLPR, Boyington, Memorandum From: Marine Corps Commandant, To: Naval Aide to the President, Subj: Awards; cases of BRIGEN Lewis B. Puller, USMC and Col. Boyington, USMC (Ret'd), information concerning Boyington's 28 Feb. 1952 letter to Truman.

2688    Interview Woody 19 July 2016; Interview Lefty Lee, 24 June 2016; Interview Woody 17 March 2015; Interview Lefty Lee, 16 March 2015. Woody even sent me a book by amateur historian, Richard Carl Bright, who was on the trip to Iwo Jima in 2015 with us highlighting sections about Lefty and his association with Woody Williams. Bright writes: "Covered by only four riflemen (one which may have been Platoon Sergeant Darol "Lefty" Lee), he fought for four hours under terrific enemy small-arms fire…" Besides telling the lies that Lefty was there with Woody, this book also regurgitates Lefty's lies that he was a Platoon Sergeant, which he never was. Richard Carl Bright, Pan & Purpose in the Pacific: True Reports of War, Trafford Publishing, 2014, 494, 497.

2689    See "This Ain't Hell: But You Can See it From Here." The article I helped Jonn Lilyea write is entitled "Darol 'Lefty' Lee: Phony Iwo Jima Hero," 14 Aug. 2017 (https://valorguardians.com/ blog/?p=73960). The author gave Lilyea a copy of Lefty's fabricated document "proving" he was one of Woody's witnesses and Lilyea scanned a copy of it and put it on this web page.

2690    Interview Barry Beck, 10 Aug. 2017; Interview Gerry Crage, 17 April 2018.

2691 Frank, "Flags, Memories and Iwo Jima," WWII Museum blog; Daniella Diaz, "Marine Corps Says it Misidentified Man in Iwo Jima Photo," CNN News Article, 23 June 2016; Mary H. Reinwald, "Seventy Years Later—Was a Mistake Made?: Honoring all who fought on Iwo Jima," Marine Corps Gazette, Aug. 2016, 27-31.

2692 Lawless, "of Such Stuff are Heroes," 68. He claimed in 1945 that he said "I'll see what I can do." MCHDHQ, Unattributed article from 1945 "Sgt. Hershel Woody Williams," 3; Interview Woody, 21 July 2016; Francois, 93.

2693 Lawless, "of Such Stuff are Heroes," 68.

2694 Until 2016, people believed the man in the picture's middle was Bradley, but it was Harald Schultz. For the first two years of the picture, the Corps believed the person to the far right was Hank Hansen when it was Harlon Block. And only in 2019, did the Corps discover that Corporal Harold "Pie" Keller, not PFC Rene Gagnon, was the man behind the central figure (Schultz) in the middle of the picture.

2695 Beck's lecture of Iwo to NROTC Midshipmen, Purdue Univ., 1952, 5.

2696 Interview Don Graves, 5 April 2018.

2697 MCHDQV, Dashiell File, #288, Cookson; NASLPR, Marquys Kenneth Cookson, Silver Star Citation. Cookson would be removed from the island on 27 March for Combat Fatigue. http://www.hqmc.marines.mil/Portals/61/Docs/FOIA/WWII%20SS%20Citations/USMC%20 SSM%20WW2%20-%20C%20(283pgs)ocr.pdf?ver=2017-05-10-105212-627

2698 Crowl & Isely, 456-7. See also Henri, et al., 84. The Marine Corps correspondents also write of flamethrower and bazooka teams as if they always went into battle together.

2699 MCHDHQ, Unattributed article from 1945 "Sgt. Hershel Woody Williams," 3. Woody claimed in 1945 that one had died and five were wounded. Interview Woody 21 July 2016; Hershel "Woody" Williams, "Medal of Honor Recipient Hershel 'Woody' Williams Visits Gettysburg: Take Out the Enemy: Reflections on the Mears Party's Valor," American Battlefield Trust, 25 March 2019.

2700 NACPM, RG 127 Container 14, Dashiell, #267, 3; NASLPR, Woody Williams, Statement PFC Schwartz, Enclosure "G", 151; Francois, 93. NASLPR, Woody Williams , Statement Cpl. Tripp, 153. In discussing Hallas book on Iwo, Woody says Tripp was never there with him at the pillbox (4th paragraph, p. 129 and 3rd and 4th paragraphs, p. 130)). Letter Woody to Rigg, 14 July 2017. However, Tripp was right there according to the correspondent who covered Woody's story Dick Dashiell (NACPM, RG 127 Container 14, Dashiell, #267, 2) and according to Tripp's affidavit in Woody's Personnel file (NASLPR, Woody Williams , Statement Cpl. Tripp, 153).

2701 Interview Bill Schlager, 27 April 2018. MOH recipient Maj. Gen. Livingston who knew many Iwo re-cipients and was himself a Marine general has doubted many things Woody has said for years. In fact, he has even spoken with some of Woody's comrades at reunions years ago and they echo what Schlager claimed about Woody that he "is full of himself." Interview Colin Heaton, 8 June 2018. Also, Schlager might have been upset with Woody since Woody claimed Tripp was at the third, large pillbox with him and not Schlager. The records show Tripp was never in combat with Woody this day. However, the records do show that Schlager was there at the third pillbox. Did Woody delete Schlager from the story because he disliked him? That could have happened knowing Woody's vindictive nature.

2702 Interview Lawrence Schlager, 27 April 2018.

2703 Interview Bill Schlager, 27 April 2018.

2704 Interview Eugene Rybak, 28 May 2018; Rybak, Review Flamethrower, 28 June 2018.

2705 Interview Keith Renstrom, 22 March 2018; Interview John Renstrom, 17 Nov. 2018.

2706 Interview Woody, 23 July 2016; Interview Woody, 20 Oct. 2017.

2707 Letter Williams to Rigg, 15 July 2017; Notes on book May 2017 referencing Hallas' book p. 159, Woody writes, "There was not any bazooka involved."

2708 In discussing Hallas' book on Iwo, Woody says it is false that a bazooka fired at the pillbox (2nd paragraph p. 130). Letter Woody to Rigg, 14 July 2017. However, he said there was indeed one in 1956 and 1966 (Lawless, "of Such Stuff are Heroes," 68-9; Francois, 93).

2709 Francois, 93.

2710 Crowl & Isely, 267. See also Henri, et al., 104. Also, sometimes the flamethrower would actually also use a bazooka together with his attacks. See Henri, et al., 106.

2711 Interview Woody, 21 July 2016; Interview Woody, 23 July 2016; NACPM, RG 127 Container 14, Dashiell, #267, 3-4; https://www.youtube.com/watch?v=fGts5WeLEgk. In Francois article, Woody claimed 17 were killed in the big pillbox. Francois, 93, Collier, DVD interview attached to book.

2712 NACPM, RG 127 Container 14, Dashiell, #267, 2-3.

2713 NASLPR, Woody Williams, Recommendation for Award of Congressional MOH & Case of Cpl. Hershel Woodrow Williams by Beck, En. "A", 157.

2714 Lawless, "of Such Stuff are Heroes," 68, NASLPR, Draft of first MOH & MOH Citation.

2715 Around 2007 or 2008, Woody obtained the official record of his Award process. Interview Woody, 19 July 2016.

2716 Interview Woody, 21 July 2016.

2717 NACPM, RG 127 Container 14, Dashiell, #267, 4; Hoolihan, "Flame Thrower's Class at 21st Teaches Art of Roasting Japs."

2718 PAHGS, Punxsy News 1 June 1945, Spirit June 1945.

2719 NASLPR, Howard F. Chambers, Silver Start Citation, Lt. Gen. Roy S. Geiger, 24 July 1945.

2720 NASLPR, Ross Ferdinand Gray, George J. Burns to Forrestal, 16 April 1945, 1-4 & Statement Cpl. Riley K. Stone & Statement Cpl. Thomas W. Lynch Jr. & Statement Cpl. John J. Wallace.

2721 NASLPR, Francis Pierce, Statement James D. Carter, 10 May 1945, 1-3.

2722 Ibid., 2.

2723 Ibid.

2724 Ibid., 3.

2725 Ibid., Statement PVT. Paul J. Neumann.

2726 Ibid., Statement Capt. Walter J. Ridlon, Jr., 12 May 1945.

2727 Ibid., Statement PFC Francis S. Brown, 10 May 1945.

2728 Ibid., Statement Private Stanely H. Nuttall, 17 May 1945.

2729 Hallas, *Uncommon Valor on Iwo Jima*, 325-9.

2730 NASLPR, Francis Pierce, Com. Off. HQ, 2nd Btl., 24th Marines, 4th MarDiv, FMF, to Forrestal, 12 May 1945, 1-3; Hallas, *Uncommon Valor on Iwo Jima*, 327-8.

2731 Hallas, *Uncommon Valor on Iwo Jima*, 328.

2732 https://valor.militarytimes.com/hero/7900; NACPM, RG 127, Commandant's Office, Box 821, Folder MOH 1740-55-50, Folder (1 Jan. 1944-31 Dec. 1947), Commandant to Sec. of Navy, Subj: Rec. for MOH in cases of Pharmacist's Mate First Class Francis J. Pierce, USN. & Capt. Walter J. Ridlon, Jr., USMCR.

2733 NACPM, RG 127, Commandant's Office, Box 821, Folder MOH 1740-55-50, Folder (1 Jan. 1944-31 Dec. 1947), Commandant to Secretary of the Navy, Subj: Rec. for award of MOH in the cases of Pharmacist's Mate First Class Pierce, USN. & Capt. Ridlon.

2734 Samuel Eliot Morison and Henry Steele Commager, *The Growth of the American Republic*. Vol. 2, Oxford, 1958, 788; Paul Gordon Lauren, *Power and Prejudice*. London, 1988, 132–3; Interview Dwight Mears, 10 May 2018. See NMPWANM, 13 July 2017.

2735 Interview Dwight Mears, 10 May 2018; Mears, 144, 203.

2736 Mears, 143-4; Frank, Review *Flamethrower*, 8 Aug. 2018; Scott, 11-2.

2737 Frank, Review *Flamethrower*, 8 Aug. 2018; Scott, 12.

2738 Toland, Vol. I, 395.

2739 Scott, 12.

2740 Newpower, 4-5, 150.

2741 NASLPR, Keith A. Renstrom; Keith A. Renstrom, *Keith A. Renstrom Recounts Battle in Tinian and Awards*, 30 Aug. 2017; Interview Mike Renstrom, 19 Nov. 2019; Interview John Renstrom, 19 Nov.

2019; https://valor.militarytimes.com/hero/8051; Interview Keith Renstrom by John Renstrom, 20-23 June 2016.

2742 NACPM, RG 127, Off. of Commandant, Box 821, Folder MOH 1740-55-50, Ostermann to Nicholson, 29 Oct. 1941; Axelrod, 15, 144-5; David Zabecki, "Path to Glory: Medal of Honor Recipients Smedley Butler and Dan Daly," HistoryNet, 26 Dec. 2007; Krulak, 175.

2743 NASLPR, Jacklyn Lucas, Lucas to HQ Co., 6th Base Depot, VAC, 6 Feb. 1944.

2744 Ibid., Com. Off. H.F. Chambers to Vandegrift, Subj: recommend discharge of Lucas, 6 Feb. 1944.

2745 Ibid., Vandegrift to Com. General, VAC, 9 March 1944.

2746 Ibid., To the Director, Pers. Dept. USMC, HQ, D.C., Summary Court Memorandum, 15 Sept. 1944 & To the Director, Pers. Dept. USMC, HQ, D.C., Summary Court Memorandum, 1 June 1944.

2747 Ibid.,To the Director, Pers. Dept. USMC, HQ, D.C., Summary Court Memorandum, 1 June 1944.

2748 Ibid., Com. Off. of Troops, USS Deuel, APA 160 to Com. Gen., Dept of Pacific, Subj: Report of Straggler, case PFC Jacklyn H. Lucas, 8 Feb. 1945.

2749 Ibid., Reward, Gen. Supply Co., 6th Base Depot, Supply Service, FMF, Deserter Lucas, 10 Feb. 1945.

2750 Ibid., Statement PFC Jacklyn H. Lucas (403248).

2751 Ibid., Rockey to Smith, 14 Feb. 1945.

2752 Ibid., Smith to Vandegrift, 25 Feb. 1945.

2753 NASLPR, Franklin Sigler, Report Board of Medical Survey, 7 March 1946 & Report Medical Survey, 16 Nov. 1945 & 19 Nov. 1945, 1-3.

2754 NASLPR, Franklin Sigler, Report of Board of Medical Survey, 19 Nov. 1945, 2; https://carra-lucia-books.co.uk/2017/03/17/what-does-an-iq-of-less-than-105-mean-for-your-child/

2755 NASLPR, Franklin Sigler, MOH citation & Statement of PFC Rudolf T. Mueller.

2756 NASLPR, Woody Williams, Office Memorandum, To: Dir. of Pers., From: Dir., Marine Corps Reserve, 31 Dec. 1954. J.C. Burger. The GCT (Gen. Classification Test) "was designed to have a mean score of 100 with a standard deviation of 20." The GCT scores "were found to be highly correlated to success at OCS." The "average score for college graduates was a 130." Knowing this, the score definitely would have convinced many in authority to view Woody's likelihood of succeeding at OCS as remote. Sean M Guthrie, "Leadership Competency: Intellectual deficiencies in the officer corps," Marine Corps Gazette, Vol. 100, Issue 6, June 2016. To see how the IQ was arrived at, see web site conversions at http://colloquysociety.org/col83eqv.htm. This site shows one that in order to arrive at an IQ number, divide the GCT score by 1.05 which then gives an approximation of an IQ score which makes Woody receive a score of 106.67. NASLPR, Jack Lummus, Baylor Univ. Transcript, Honorable Dismissal, 15 June 1942 & Letter NY Giants President John V. Mara to Commandant Marine Corps, 22 June 1942.

2757 NACPM, RG 127, Commandant's Office, Box 821, Folder MOH 1740-55-50, Folder (1 Jan. 1944-31 Dec. 1947), Capt. John Hamas (USMC) to Supreme Commander Allied Powers, Gen. HQ, 5 Nov. 1946, 2.

2758 Ibid., Capt. John Hamas (USMC) to Supreme Commander Allied Powers, Gen. HQ, 5 Nov. 1946, 1-2 & Statement by MTechSgt. Jesse Leon Stewart, 8 Oct. 1945, 1-3.

2759 Ibid., Statement Master Sgt. Walter J. Kennedy, 22 Oct. 1946.

2760 Ibid., Statement MTechSgt. Jesse Leon Stewart, 8 Oct. 1945, 1.

2761 NACPM, RG 127, Commandant's Office, Box 821, Folder MOH 1740-55-50, Folder (1 Jan. 1944-31 Dec. 1947), Capt. Hamas (USMC) to Supreme Commander Allied Powers, Gen. HQ, 5 Nov. 1946, 2; Ferguson, 208-9.

2762 NACPM, RG 127, Commandant's Office, Box 821, Folder MOH 1740-55-50, Folder (1 Jan. 1944-31 Dec. 1947), Capt. Hamas (USMC) to Supreme Commander Allied Powers, Gen. HQ, 5 Nov. 1946, 2.

2763 Ibid.

2764 Ibid., Capt. Hamas (USMC) to Supreme Commander Allied Powers, Gen. HQ, 5 Nov. 1946, 2.

2765  Ibid., AG to Capt. John Hamas, 26 Nov. 1946.

2766  Ibid., Statement by MTechSgt. Jesse Leon Stewart, 8 Oct. 1945, 2-3.

2767  Ibid., Commandant to Forrestal, 24 Feb. 1947 & Recommendations Devereux and Cunningham for Shank, Navy Cross, 3 July 1946.

2768  Ibid., Hayler to Forrestal, 26 Nov. 1946.

2769  Duane Vachon, "A Hoosier Hero—Dr. Lawton E. Shank, WWII, Civilian (1907-43)," Hawaii Reporter, 14 April 2013.

2770  Charlsy Panzino, "MOH Approved for World War II Hero," Army Times, 30 March 2018.

2771  Jonathan S. Landay, "Marines Promoted Inflated Story for MOH Recipient," McClatchy Newspapers, 14 Dec. 2011; Interview Sgt. Saint Awards Board, 21 April 2018; J. Mark Powell, "When the Army Took Back the Congressional Medal of Honor," American History, 12 Sept. 2015.

2772  Krakauer, 177-8; Interview Capt. Lee Mandel, 15 April 2018.

2773  John Kampfner, "The Truth About Jessica," The Guardian, 15 May 2003; Krakauer, 178-83.

2774  Krakauer, 297-8, 322; Interview Dwight Mears, 10 May 2018; Mears, 125-6.

2775  Robert Dallek, *An Unfinished Life: John F. Kennedy 1917-1963*, New York, 2003, 95.

2776  Dallek, *An Unfinished Life*, 95.

2777  Ibid.

2778  Telephone conversation with Steven Gillon, 15 Dec. 2019.

2779  Dallek, *An Unfinished Life*, 82.

2780  Ibid., 98.

2781  Ibid.

2782  https://www.jfklibrary.org/learn/about-jfk/media-galleries/world-war-ii

2783  Dallek, *An Unfinished Life*, 98.

2784  Ibid.

2785  Telephone conversation with Steven Gillon, 15 Dec. 2019.

2786  Dwight Mears, Review *Flamethrower*, 15 Oct. 2018.

2787  Interview Don Graves, 3 April 2018.

2788  Interview Bob Olson, 4 April 2018.

2789  Email Barry Beck to Bryan Rigg, 28 Jan. 2019.

2790  Email James Ginther to Rigg, 9 April 2018.

2791  Interview George Bernstein, 25 March 2018.

2792  Interview David Fisher, 25 March 2018.

2793  Email Neimeyer to Rigg, 28 March 2018.

2794  Mears, 108.

2795  Dwight Mears, Review *Flamethrower*, 15 Oct. 2018.

2796  Lee Ferran, "So Who Actually Killed Osama Bin Laden?", ABC News, 6 Nov. 2014;"WWII Research Team Marks Death of Pearl Harbor Mastermind," CBS News, 16 April 2018. Other interesting military stories the reader may want to explore that were made up out of thin air were the sea tales Captain Mitsui Fuchida and Ensign George H. Gay gave during and after the war. Fuchida did indeed lead the first wave of air attacks at Pearl Harbor, but after the war, he made up the myth that there were debates about launching a third attack wave at Pearl Harbor to destroy the fuel depot and that he was the one arguing with Vice Admiral Nagumo for it—a totally made up story. He also made up stories of the Battle of Midway and how the decks were when the American dive bombers hit the ships. Historian Jon Parshall basically says Fuchida "is full of crap." Email Jon Parshall to Rigg, 28 Dec. 2019; Jonathan Parshall and Anthony Tully, *Shattered Sword: The Untold Story of the Battle of Midway*, Nebraska, 2005, 437-42; Edgerton, 260-1. Ensign Gay was a torpedo pilot that the Japanese shot down during the battle of Midway—the sole survivor of his squadron. He claimed he saw the follow on attacks by the dive bombers, but recent research shows he did not see all that he claimed to have seen. He, like Fuchida, was just trying to make their stories bigger and better than what they were to add luster to their own fame, something that Woody apparently has also done as shown in this book. Email Jon Parshall to Rigg, 28 Dec. 2019.

2797   If Woody knows his self-reporting possibly ended up getting him the *Medal* since we know he learned about the process sometime in the 1990's and had to have seen how his initial story of seven pillboxes and 21 dead Japanese was removed from the citation because there was no support for it, should he have notified the USMC? Since the evidence we have shows that only Woody gave these "facts," he had to have known in the 1990's when he received the documentation on his MOH process that he most likely was the source for this information that initially started his MOH package. If he now knows this and does not report it, under civil-law, there is theoretically a cause for action according to many lawyers consulted. In Texas, this is called Fraud by Non-Disclosure and in law in general, Fraud by Omission (as it is so named in the state of West Virginia). Since MOH recipients receive tax-free income for their entire lives for earning the MOH, there is a lot of money at stake if someone knows he received the *Medal* by self-reporting his own heroics which would violate the statutes and could even be a federal crime according to lawyers John "Jack" Kuykendall, Debra S., Tom Hancock and Richard Wallace. Interview John M. "Jack" Kuykendall, J.D., 2 Jan. 2020; Interview Debra S., J.D., 2 Jan. 2020; Interview Roger Rippy, J.D., 2 Jan. 2020; Interview Tom Hancock, J.D., 10 Jan. 2020; Interview Richard Wallace, J.D., 9 Jan. 2020; https://www.jdsupra.com/legalnews/texas-supreme-court-holds-that-a-fraud-46351/. Fraud by Non-Disclosure reads as follows: "Fraud by non-disclosure…occurs when a party has a duty to disclose certain information and fails to disclose it. To establish fraud by non-disclosure, the plaintiff must show: (1) the defendant deliberately failed to disclose material facts; (2) the defendant had a duty to disclose such facts to the plaintiff; (3) the plaintiff was ignorant of the facts and did not have an equal opportunity to discover them; (4) the defendant intended the plaintiff to act or refrain from acting based on the nondisclosure; and (5) the plaintiff relied on the non-disclosure, which resulted in injury…" When talking with lawyers Jack Kuykendall, Debra Sims and Roger Rippy, all lawyers practicing law in Texas, Woody could be guilty of many of these charges. 1.) Woody could have failed to reveal facts that the story he originally gave Dashiell and others was not true and that unsubstantiated facts entered his record. 2.) Since Woody has received hundreds of thousands of dollars in tax-free money throughout his life because of the MOH, he has a duty to the U.S. government if this money was received illegally to report the fact that maybe his embellishments resulted in him receiving the MOH. 3.) As we see, due to political pressure from Truman, the government (i.e. Geiger and Nimitz's request) did not have the time to discover what really happened—not Woody's fault, but something that Woody had to have known once he got the official record in the mid-1990s when he received the reports. 4). Once Woody possibly knew that information he gave was not substantiated and that this information was the reason why he received the MOH, he probably should have reported such information to the "plaintiff" (i.e. the U.S. government) so they could know if it should continue on with payment to Woody (i.e. government pension benefits reserved only for MOH recipients). 5). Since Woody possibly did not disclose the misinformation he knew that came from him, the government continued to give him money when it did not know that possibly Woody's embellishments had begun the process in the first place. If the military and government had known that possibly Woody's embellishments started the whole MOH to begin with, they most likely would have never given him the MOH even with the political pressure that has been documented as having happened. Woody told the author on 19 July 2016 that he received the "The Congressional Medal of Honor, Awarded to Hershel Woodrow Williams, By the President of the U.S. Harry S. Truman, The White House, 5 Oct. 1945" documenting his case sometime in 1995 or 1996. Interview Woody, 19 July 2016. When Glenna Whitley, co-author of the *Stolen Valor* book that inspired the legislation with that name, was informed of all the problems with Woody's MOH and that he possibly caused his own medal with his reporting events that were not documented, she said that he was "clearly a Stolen Valor case. No question." Interview Glenna Whitley, 7 Jan. 2020. Echoing his co-author, the researcher of *Stolen Valor*, B. K. "Jug" Burkett said, "Woody clearly has an obligation to report the truth. It is a complicated Stolen Valor case, but nonetheless, it appears he has benefited greatly from an award he should not have received especially since he apparently received it for information he gave that never was proven." Interview Jug Burkett, 20 Nov. 2018.

2798    Michael Berger, "Print the Legend," Notes to the Editor, NY Times, 30 Jan. 2000.

2799    Interview Warren Wiedhahn, 25 April 2018.

2800    Interview Michael E. Thornton, 14 Jan. 2020.

2801    In draft 21 of this work, which Woody read and approved, it was written at least three times that he killed at least 50 IJA soldiers: (1.) "…according to this study's estimate, Woody probably burned alive upwards of 50 men in taking out these fortifications" (p. 365), (2.) "In addition to Woody killing possibly 50 Japanese soldiers…" (p. 386) and (3.) "As mentioned before, he took out at least 50 Japanese during his assault on the pillboxes" (p. 387). Woody never countered these claims.

2802    https://worldwar2salute.org/State-Funeral-Blog/5025029; Maj. Gen. Livingston was even asked by the government about such a funeral for Woody and he endorsed it. However, after reading this book and learning the real story about Woody, Livingston has now second thoughts about the support he has given for Woody.

2803    MCHDQV, Dashiell File, #291, Report Cpl. Joseph Rybakiewicz; NASLPR, Rybakiewicz, Purple Heart Citation.

2804    Interview Eugene Rybak, 28 May 2018.

2805    MCHDQV, Dashiell File, #291, Report Cpl. Joseph Rybakiewicz; NASLPR, Rybakiewicz, Purple Heart Citation.

2806    Blum, 53.

2807    Interview Roger Cirillo, 12 May 2018.

2808    Bill Stone, Review of Bruce Gamble's *The Black Sheep: The Definitive Account of Marine Fighting Squadron 214 in World War II*, 5 July 1998, http://books.stonebooks.com/reviews/980705.shtml

2809    NASLPR, Boyington, Endorsements for the MOH for Boyington from Halsey, Nimitz, King and Vandegrift.

2810    Email General Charles C. Krulak to Rigg, 23 Dec. 2019. General Krulak further wrote about Woody's MOH, "In today's world, it would have not passed the USMC Awards Board as submitted. It would have required more documentation and signatures up the [entire] line. Woody's [MOH process] was at a different time with different pressures." Email General Charles C. Krulak to Rigg, 24 Dec. 2019.

2811    NIV—New International Version, 591-2.

2812    *The Teaching of Buddha*, Buddhist Promoting Foundation, Tokyo, Japan, 1992, DHAMMAPADA (103).

2813    Interview Woody, 21 Dec. 2017; MCHDQV, Oral History, Woody Williams 13800A.

2814    MCHDHQ, Woody, "Destiny…Farm Boy," 7; Hallas, *Iwo Jima*, 135, 137.

2815    MCHDHQ, Woody, "Destiny…Farm Boy," 7; Hallas, *Iwo Jima*, 135, 137; Interview Woody, 21 July 2016.

2816    Interview Kathryn Waters, 24 April 2017.

2817    U.S. Department of Veterans Affairs Interview with Woody Williams, "Woody Williams Remembers the Battle of Iwo Jima, Part 6", Feb. 2019; Interview George Waters, 29 May 2018; Hallas, *Iwo Jima*, 137.

2818    Interview Woody, 21 July 2016; U.S. Department of Veterans Affairs Interview with Woody Williams, "Woody Williams Remembers the Battle of Iwo Jima, Part 6", Feb. 2019.

2819    Hallas, *Iwo Jima*, 137.

2820    NASLPR, Vernon Waters, Verna L. Waters to the Marine Corps, 11 July 1945 (rubber stamped 25 July 1945).

2821 NASLPR, Vernon Waters, 1ˢᵗ Lt. M. G. Craig to Verna L. Waters, 3 Aug. 1945.

2822 Interview George Waters, 29 May 2018.

2823 Interview Kathryn Waters, 24 April 2017.

2824 Interview Woody, 22 July, 2016; VandeLinde, 379-80; NASLPR, Vernon Waters, 1ˢᵗ Lt. M. G. Craig to Verna L. Waters, 3 Aug. 1945 & Verna L. Waters, 25 July 1945 to the Marine Corps. Woody has claimed that Waters' father was there, but the father had died years beforehand. Interview George Waters, 29 May 2018.

2825 Interview Woody, 23 July 2016.

2826 MCHDQV, Oral History, Woody Williams 13800A.

2827 NACPM, RG 127, Commandant's Office, Box 821, Folder MOH 1740-55-50, Folder (1 Jan. 1944-31 Dec. 1947), Vandegrift to Judge Advocate Gen., 8 Jan. 1946.

2828 NASLPR, Woody Williams, Cates to Williams, 2 March 1948, 123.

2829 Ibid., Williams to Vandegrift, 20 June 1946, 89. In a 24 July 2016 interview with Woody, he ridiculed me for using the term "winners" when describing MOH recipients. However, he himself used this term soon after he earned the medal in his correspondence. Saying this, one should probably not use the term "winner" because the recipient was not competing for it necessarily although prominent historian Allan R. Millett uses it in his definitive work on the Marine Corps. See Millett, xiv.

2830 Ibid., 14 Feb. 1946, 87.

2831 NASLPR, Woody Williams, Col. A.E. O'Neil, 29 July 1946, 88; NACPM, RG 127, Commandant's Office, Box 821, Folder MOH 1740-55-50, Folder (1 Jan. 1944-31 Dec. 1947), Vandegrift to Forrestal, 29 July 1946.

2832 Interview Shindo, 9 April 2018; Notes by Shindo, 3 Oct. 2018.

2833 Ibid.

2834 Ibid.

2835 Ibid.

2836 Ibid.

2837 Interview Woody, 21 July 2016, 12 Feb. 2017.

2838 Interview Woody, 12 Feb. 2017.

2839 Interview Woody, 23 July 2016; Woody, "The Destiny of a Farm Boy," 3.

2840 Interview Woody, 21 July 2016 & 23 July 2016.

2841 Interview Woody, 23 July 2016; Woody, "The Destiny of a Farm Boy," page inserted 3 and p. 4.

2842 NASLPR, Woody Williams, Ruby Williams to Quartermaster Gen. USMC, 30 Nov. 1947.

2843 Ibid., Capt. Robert L. Williams to Ruby Williams, 9 Jan. 1948.

2844 Interview Woody, 19 July 2016.

2845 The minister was being overly dramatic here. The people in his church did not nail Christ to the cross although he was claiming their "future sins" made them guilty for why Christ had to come to earth in the first place, i.e. sacrifice himself for their sins in order give them forgiveness and to provide them with eternal life. Technically though, the Romans drove the nails into Jesus for him being a law breaker and calling for the downfall of the Roman Empire. Yes, the Jewish court at the time turned him over to the Romans as being a law-breaker and blasphemer, but the Romans did the actual execution and were very good about stamping out Messiah-figures like Jesus crucifying many during Jesus's lifetime. Johnna Rizzo, "How the Romans Used Crucifixion—Including Jesus's—As a Political Weapon," Newsweek, 4 April 2015; Laura Geggel, "Jesus Wasn't the Only Man to Be Crucified: Here's the History Behind This Brutal Practice," Live Science, 19 April 2019; Kersey Graves, The World's Sixteen Crucified Saviors: Christianity Before Christ, NY, 1875.

2846 Interview Woody, 19 July 2016; Smith, Iwo Jima, 72; https://www1.cbn.com/video/HOL58v2/easter-holds-special-meaning-for-ww2-hero, 29 March 2018; Woody Williams, "Hershel Woody Williams Christian Experience," Undated Essay; Hallas, Uncommon Valor on Iwo Jima, 138; Hallas, Iwo Jima, 138. In Hallas' book, he writes this spiritual awakening happened in 1961, but in Woody's interviews for this book and writings, it was in 1962.

2847   MCHDHQ, Sermon by Woody Williams to Bethesda Church, 11 April 2010.

2848   Interview Woody, 22 July 2016. On this day, Woody even was so brazen as to tell me that due to my own Jewish background and ethical humanistic convictions, *I was going to hell*. Woody later gave me Ed Moore's religious book *Prayer Force One* and told me he would pray for me so I might be saved. His grandson, Brent Casey, told me the same thing and explained that is why Woody gave me *Prayer Force One*.

2849   He served with the 98th and 25th Spl. Inf. Co's. in Clarksburg, WV and 83rd Rifle Co. in Huntington, WV. NASLPR, Woody Williams.

2850   NASLPR, Woody Williams, Office Memorandum, To: Dir. of Pers., From: Dir., Marine Corps Reserve, 31 Dec. 1954. J.C. Burger.

2851   Notes on Flamethrower by Allan Millett, 25 Sept. 2019.

2852   Interview Woody, 14 Dec. 2017; Rick Steelhammer, "WV Supreme Court Oks Moving MOH Recipient's Remains," Charleston Gazette-Mail, 7 June 2017; Caity Coyne, "WWI MOH Recipient West Laid to Rest in Dunbar," Charleston Gazette-Mail, 12 May 2018.

2853   https://www.youtube.com/watch?v=Rn9WxFO5Ymo&feature=youtu.be   (ABC   news   event); Charlotte Ferrell Smith, "Óna Medal of Honor Recipient Hershel 'Woody' Williams Honored on Stamp," Charleston Gazette, 10 Nov. 2013; "US House Passes Resolution to Name Huntington VA for 'Woody' Williams," Charleston Gazette, 21 May 2018; James E. Casto, "Huntington VA Medical Center Renamed to Honor Woody Williams," Charleston Gazette, 11 Oct. 2018.

2854   Bechtel, "The Making of a Marine," 38-9. Woody's speech, Texas at Cedar Park on 23 Sept. 2017.

2855   MCHDHQ, Report by Woody, "Reflections About Life and Service Above Self," Oct. 2016, 2.

2856   Letter Mrs. Russell J. Dorman to Donald Beck, 1 Nov. 1944.

2857   MCHDHQ, Report Woody, "Reflections About Life and Service Above Self," Oct. 2016, 2-3; Interview Woody, 23 July 2016.

2858   John Veasey, "Williams Foundation Leader in Establishing Permanent Gold Star Family Memorial Monuments," Times West Virginian, 4 Dec. 2015; MCHDHQ, Report Woody, "Reflections About Life and Service Above Self," Oct. 2016, 1; Interview Woody, 23 July 2016.

2859   Veasey, "Williams Foundation Leader," Times West Virginian, 4 Dec. 2015; Interview Woody, 23 July 2016.

2860   Interview Brent Casey, 10 Jan. 2017.

2861   Interview Woody, 23 July 2016.

2862   Email Blackman to Rigg, 27 April 2017.

2863   Zdon, "A Marine in the Pacific," 8.

2864   Letter Woody to the Beck Family, 25 Jan. 2016.

2865   Zdon, "A Marine in the Pacific," 8.

2866   Interview Woody, 23 July 2016.

2867   Prange, 724.

2868   Interview Yoshitaka Shindo, 9 April 2018.

2869   Toll, *Pacific Crucible*, 387.

2870   Commandant Charles C. Krulak's Marine Corps 223rd Birthday address to the Marine Corps from Iwo Jima, 10 Nov. 1998.

2871   Divine, Breen, Fredrickson and Williams, 787.

2872   "Bill to Aid Britain Strongly Backed," NY Times, 9 Feb. 1941; Wallance, 198-200.

2873   Johnson, *A History of the American People*, 769.

2874   Frank, *Downfall*, 23; Divine, Breen, Fredrickson and Williams, 789; *Two Score and Ten*, 19; Patterson, 265; Prange, 287; Drea, 226; Bix, 385; Toland, Vol. I, 169, 186, 321, 333; Toland, Vol. II, 548-9; Crowl & Isely, 84.

2875   D. Clayton James, "Strategies in the Pacific," Makers of Modern Strategy from Machiavelli to the Nuclear Age, ed. Peter Paret, Princeton NJ, 1986, 720; Toland, Vol. II, 571; O'Brien, *The Second Most Powerful Man in the World*, 205, 214-5, 217, 226, 237.

2876  Hane, 331-2; Weinberg, 261-2; Paterson (ed.), *Major Problems in American Foreign Policy*," Dallek, "Roosevelt's Leadership, Public Opinion, and Playing for Time in Asia," 222.

2877  Kennedy, *The Rise and Fall of Great Powers*, 303.

2878  Cook & Cook, 171.

2879  Hoffman, Review *Flamethrower*, 28 Aug. 2018.

2880  Bix, 749 fn. 1.

2881  Cook & Cook, 171; Leland Ness, *Jane's WWII Tanks and Fighting Vehicles: The Complete Guide*, London, 2002, 13; Drea, 232; Toland, Vol. I, 121; Zeiler, 5.

2882  Drea, 221.

2883  Drea, 257; BA-MA, RM 12 II/250, KTB, T. 38, Marineattachés M(Ltr) Grossetappe Japan-China, Sept. 1941, 17.

2884  Henri, et al., vii; Burrell (ed.), John J. Reber, "Pete Ellis: Amphibious Warfare Prophet," 175-90 & Robert Burrell, "The Prototype U.S. Marine: Evolution of the Amphibious Assault Warrior, 1942-1945," 191; Hoffman, *Once a Legend*, 47.

2885  Murphy, 86; Weinberg, 323, 917; Drea, 60, 100-3; Churchill, 581; Burleigh, 13; Kennedy, *The Rise and Fall of Great Powers*, 301; Frank, "Review *Flamethrower*," 25 May 2017; Kakehashi, 110-1

2886  MCHDQV, Oral History, Cates, 51.

2887  Victoria, *Zen at War*, 163-4.

2888  Toll, *Pacific Crucible*, 113.

2889  Ibid., 120.

2890  Blair and DeCioccio write about the Army, Marine and Navy conflicts in *Victory at Peleliu* and Burrell does as well in *Ghosts of Iwo Jima*. Blair & DeCioccio, 26-9; Burrell, 5, 11-34; Millett, 391-2.

2891  Bingham, xiii; Newcomb, 144; Smith, *Coral and Brass*, 5.

2892  Farrell and King talk about this in their books *Liberation-1944* (p. 119) and *A Tomb Called Iwo Jima*.

2893  Crowl & Isely, 140.

2894  Hane, 362; Hardacre, 133; Coogan, 439; Cook & Cook, 4, 307.

2895  Christopher Hitchens, *Mortality*, NY, 2012, 90.

2896  Tanaka, 37, 141, 144; Scott, 49; Zeiler, 41.

2897  Tanaka, 8, 250.

2898  Clausewitz, *On War*, Brodie, "A Guide to Reading of On War," 657.

2899  Ibid., 699. Moore, 233; Tanaka, 8, 250.

2900  Tanaka, 251.

2901  Wakabayashi (ed.), *The Nanking Atrocity, 1937-38*, Wakabayashi, "Iris Chang Reassessed," xxi.

2902  Burrell (ed.), Burrell, "The Prototype U.S. Marine," 191-2; Hoffman, *Once a Legend*, 272; Crowl & Isely, 341.

2903  Burrell (ed.), Burrell, "The Prototype U.S. Marine," 202; Interview Keith Renstrom by John Renstrom, 20-23 June 2016.

2904  Henri, et al., 92: Dashiell, "Combat Correspondent Dispatch," 17 May 1945, 1.

2905  Burrell (ed.), Burrell, "The Prototype U.S. Marine," 202.

2906  Toland, Vol. II, 1070.

2907  Interview Billy Griggs, 20 March 2015.

2908  Mandel, 286.

2909  Moskin, 373; Mullener, 220; Keegan, 566; Hammel 243; Hane, 351; Drea, 236; Wheeler, 57; Robert S. Burrell, "Did We Have to Fight Here?," World War II Magazine, Special Collector's Addition for the 60th Commemoration of Iwo Jima, 62; Miller, "Deathtrap Island," 9; Edgerton, 296; Weinberg, 869.

2910  NACPM, RG 127 Box 15 (3d MarDiv), 3rdMarDiv, G-1 Periodic Report No. 21-2.

2911  The WWII Museum in New Orleans says there were 416,800 military deaths during WWII. http://www.nationalww2museum.org/learn/education/for-students/ww2-history/ww2-by-the-numbers/world-wide-deaths.html?referrer=https://www.google.com/

2912  Interview Barbara Olson, 30 March 2018.

2913  NASLPR, Howard Chambers, Report Medical Examination, 21 Feb. 1955.

2914  Interview Sherri Bell, 11 Jan. 2020.

2915  Wheeler, 57.

2916  NACPM, RG 127, Box 83, (Iwo), AB-1, Com. Gen. 3dMarDiv Erskine to Commandant, Report, 3 June 1945, 65.

2917  Hammel, 174.

2918  Kakehashi, 125.

2919  MCHDQV, HQ, 1ˢᵗ Btl. 21ˢᵗ Rgt., 3d MarDiv, FMF, In Field, 6 April 1945, Report Lt. Col. Williams, 14.

2920  Ibid. For the flame's range, see Brophy, Miles and Cochrane, 140-1, 145.

2921  Ibid., 21.

2922  Mountcastle, 111.

2923  Interview John Mountcastle, 14 Jan. 2018.

2924  MCHDQV, HQ, 2ⁿᵈ Btl., 21ˢᵗ Marines, 3d MarDiv, FMF, In Field, G.A. Percy, 12 April 1945, 15.

2925  Joseph Rybakiewicz's petition to the VA for benefits due to PTSD. Rybak, Review *Flamethrower*, 28 June 2018.

2926  Interview Woody, 23 July 2016; Mandel, 297.

2927  MCHDQV, HQ, 1ˢᵗ Btl, 21ˢᵗ Rgt., 3d MarDiv, FMF, In Field, 6 April 1945, Report Lt. Col. Williams, 27; Interview Graves, 23 May 2017.

2928  Hoolihan, "Flame Thrower's Class at 21ˢᵗ Teaches Art of Roasting Japs."

2929  MCHDQV, HQ, 2ⁿᵈ Btl., 21ˢᵗRgt., 3d MarDiv, FMF, In Field, G.A. Percy, 12 April 1945, 15.

2930  NACPM, RG 127, Box 83, AB-1, Comm. Gen. 3d MarDiv Erskine to Commandant, Report, 3 June 1945, 36.

2931  Brophy, Miles and Cochrane, 143, 368.

2932  Ibid., 166.

2933  Crowl & Isely, 535; Sledge, 178, 190; Edgerton, 297.

2934  Harries & Harries, 443; Sledge, 312; Hane, 352-3.

2935  Henri, et al., 19.

2936  Smith, *Coral and Brass*, 135; Crowl & Isely, 485.

2937  MCHDQV, Oral History, Smith, 2; Smith, *Coral and Brass*, 135; Kakehashi, 98.

2938  McNeill, "Reluctant Warrior," 39; Melville, 166.

2939  Kakehashi, 48; NMPWANM, 13 July 2017; Drea, 246.

2940  Dower, 232-33; Burrell, *Ghosts of Iwo Jima*, 46; Bradley, *Flyboys*, 5, 293; Davis, "Operation Olympic," 20-1; Miller, "Deathtrap Island," 10; Edgerton, 300; Cook & Cook, 14-5, 263-4, 364; Drea, 231.

2941  Sakai, 309.

2942  Ibid., 308-9.

2943  Frank, *Downfall*, 95, 107.

2944  Smith, *Coral and Brass*, 18; Drea, 46, 107.

2945  MCHDQV, Oral History, Erskine, 46.

2946  Mountcastle, 59, 62-4.

2947  MCHDQV, Oral History, Cates, 24.

2948  Frank, *Downfall*, 29; Drea, 17-8, 45, 119-20, 172-3; Interview Edward Drea, 17 Dec. 2017.

2949  Dower, 144.

2950  Newcomb, 23; Moskin, 359; Henri, et al., 21; Bradley & Powers, *Flags of Our Fathers*, 144; Miller, "Deathtrap Island," 10; Alexander, 6.

2951  Alexander, 3; Ross, 115; Miller, "Deathtrap Island," 10; Burrell, *Ghosts of Iwo Jima*, 58.

2952  Interview King, 19 July 2016; Henri, et al., 13; George W. Garand & Truman R. Stobridge, *Western Pacific Operations: History of U.S. Marine Corps Operations in World War II*, Quantico, 1971,

448; The Carryall, Iwo Jima, Feb. 19, 1945—Oct. 15, 1945, 133rd NCB Navy Seabee Souvenir Newsletter, Reprint Ken Bingham, Vol. III, No. 8, Gulfport MS, 2011, 9; NIDSMDMA, *Senshi-Sosho*, Cen. Pacific, Op. #2, Peleliu, Angaur, Iwo Jima, Vol. 13, 348; Cameron, Mahoney and McReynolds, 133, 537; Bradley & Powers, *Flags of Our Fathers*, 140; Interview Shindo, 9 April 2018; Hough, 328.

2953   Dower, 232-3; Bradley, *Flyboys*, 5, 293; Cook & Cook, 172, 184, 324-5, 375; James Martin Davis, "Operation Olympic: An Invasion not Found in History Books," Omaha World Herald, Nov. 1987, 20-1; Miller, "Deathtrap Island," 10; Edgerton, 300; Bix, 480, 482; Frank, *Downfall*, 188; Frank, Review *Flamethrower*, 8 Aug. 2018; Toland, Vol. II, 934-5; Victoria, *Zen at War*, 138.

2954   Frank, *Downfall*, 188, 243, 257, 340, 351-2; Davis, "Operation Olympic," 18; Hane, 353; Cook & Cook, 341-2, 367, 459-60; Drea, 247-8, 250; Sledge, 312; Bix, 485; Sieg, "Historians Battle Over Okinawa WW2 Mass Suicides"; James Brooke, "Okinawa Suicides and Japan's Army: Burying the Truth?", NYT, 20 June '05; Toland, Vol. II, 897-8; NACPM, RG 127 Box 318 (Saipan-Tinian) HQ Northern Troops Landing Force, Marianas Phase I (Saipan) 1. Civ. Affairs Report 2. Liaison Offs. Report 2. Public Rel. Report, 9 Nov. 1944, HQ VAC, Corps Civ. Affairs Off. to Com. Gen., 13 Aug. 1944, Donald Winder, 3; Harwood, 29; Hough, 259. Ironically, Fleet Admiral and Chairman of the Joint Chiefs-of-Staff William D. Leahy opposed the invasion. He also opposed using the atomic bombs, but his argument against the invasion and the resulting high casualties that would happen putting troops on the ground on mainland Japan, that it convinced President Truman of the necessity of actually using the atomic bombs. O'Brien, *The Second Most Powerful Man in the World*, 347-8.

2955   Dennis M. Giangreco, *Hell to Pay: Operation Downfall and the Invasion of Japan, 1945–1947*, Annapolis, 2009.

2956   Frank, *Downfall*, 19, 242.

2957   Ibid., 272.

2958   Cook & Cook, 56, 403, 407-9; Kuhn, Epilog; Patterson, 296; Bix, 508; Toland, Vol. II, 998; Moore, 242.

2959   Toland, Vol. II, 1068; Tanaka, 113.

2960   Frank, *Downfall*, 295-6.

2961   Toland, Vol. II, 998-9.

2962   Maslowski, and Millett, 466; Frank, *Downfall*, 316-9, 330; Rhodes, 745; Edgerton, 302; Burleigh, 13, 551; Kennedy, *The Rise and Fall of Great Nations*, 356; Toland, Vol. II, 1045, 1052; Scott, 49.

2963   Cook & Cook, 403; Bix, 81; Toland, Vol. II, 1054.

2964   Hirohito included Koreans and Taiwanese as "his people" in addition to 70 million Japanese to arrive at this number. Interview Col. Yukio Yasunaga, 9 April 2018.

2965   Cook & Cook, 401.

2966   Ibid., 445.

2967   Yenne, *Hap Arnold*, 255.

2968   Sakai, 310; Frank, *Downfall*, 186-7; Edgerton, 303.

2969   Miller, "Deathtrap Island," 9.

2970   Toland, Vol. II, 979.

2971   Toland, Vol. II, 1015.

2972   Clausewitz, *On War*, 92-3.

2973   Frank, *Downfall*, 239.

2974   Rhodes, 736.

2975   Toland, Vol. II, 647.

2976   Interview Gen. LeMay by Colin Heaton in June 1986.

2977   Roberts, *Churchill*, 890.

2978   Toland, Vol. II, 952.

2979   Paterson (ed.), *Major Problems in American Foreign Policy*," Bernstein, "Atomic Bomb and Diplomacy," 324.

2980   Dower, 142; Frank, *Downfall*, 269, 302-3, 327; Rhodes, 744; Paterson (ed.), *Major Problems in American Foreign Policy*," Bernstein, Barton J., "Atomic Bomb and Diplomacy," 319. An additional seven atomic bombs would have been ready by 31 Oct. Frank, *Downfall*, 312.

2981   Bix, 502. See also O'Brien, *The Second Most Powerful Man in the World*, 356-6.

2982   Clausewitz, *On War*, Howard, "The Influence of Clausewitz," 35.

2983   NACPM, 127 Box 14 (3d MarDiv), Intel Bulletin: Japanese Tactics and Strategy, 29 Oct 1942-29 May 1945 Folder 7, Col. Hogaboom, 29 May 1945, 1, Sec. (1) (f).

2984   Clausewitz, *On War*, 75.

2985   Rice, 116; Coogan, 421, 464-5; Cook & Cook, 15, 405; Mosley, 16.

2986   Crowley, xiii-xiv.

2987   Interview LeMay by Heaton in June 1986.

2988   Davis, "Operation Olympic," 45; Frank, *Downfall*, 90, 217, 324.

2989   Frank, *Downfall*, 163, 359.

2990   Ibid., 195.

2991   Interview Woody, 19 July 2016.

2992   TSHACNCDNM, Vol. 10, 264, Statement Keenan, Chief Counsel, International Military Tribunal Far East, 4 June 1945, 2.

2993   Crowl & Isely, 171; Millett, 379.

2994   Clausewitz, *On War*, Brodie, "A Guide to Reading of On War," 665.

2995   Cook & Cook, 138, 142.

2996   Smith, *Coral and Brass*, 19, 40, 54; MCHDQV, Oral History, Erskine, 209.

2997   Coram, 71.

2998   Farrell, *Liberation-1944,* 17; Steinberg, *Island Fighting*, 194. Kennedy, *Engineers of Victory*, 329-32.

2999   Spector, 299.

3000   Kennedy, *Engineers of Victory*, 329-32.

3001   Leslie Groves, Now it Can be Told: The Story of the Manhattan Project, NY, 1962, 184. Some Japanese argue the atomic bombs were only dropped on them due to racism. However, since the Americans wanted to drop these bombs on the Germans first disproves this argument. Moreover, the Allies killed twice as many civilians in Germany as they did in Japan in bombing campaigns showing the Allies were motivated by winning the war and killing as many fascists as it took to secure the peace. The breakdown according to historian Morgan was that around 750,000 German citizens died in bombing raids whereas Japanese lost around 375,000 in the Allies raids on Japan. Interview Morgan, 10 Dec. 2017.

3002   Rhodes, 710.

3003   MCHDQV, Oral History, Cates, 11.

3004   Toland, Vol. II, 819.

3005   Smith, *Coral and Brass*, 144-5.

3006   Brophy, Miles and Cochrane, 411.

3007   Burrell, *Ghosts of Iwo Jima*, 119.

3008   MCHDHQ, Letter Woody to Donald Cordell, 20 April 2014.

3009   Ibid.

3010   Ibid.

3011   Bradley & Powers, *Flags of Our Fathers*, 529.

3012   MCHDQV, Dashiell File, #339, Christopher; NASLPR, Merrill Christopher, Bronze Star Citation & USMC Enlisted Man's Qualification Card, No. 28 Remarks & D. Routh to Mrs. Christopher, 19 April 1945 & Service Book, Offenses, 10 & Summary Court Memorandum, Merrill Christopher, AWOL, 21 April 1945.

3013    Email General Charles C. Krulak to Rigg, 23 Dec. 2019. General Krulak further wrote about Woody's MOH, "In today's world, it would have not passed the USMC Awards Board as submitted. It would have required more documentation and signatures up the [entire] line. Woody's [MOH process] was at a different time with different pressures." Email General Charles C. Krulak to Rigg, 24 Dec. 2019.

3014    For such wartime exaggerations, see Toland, Vol. I, 405-6, 415, 460; Toland, Vol. II, 707; Toll, *Pacific Crucible,* 331-2, 377; Gandt, 80.

3015    MCHDHQ, Statement signed by Woody answering the question "Looking back on your entire experience during the war, what are you most proud of?" In answer, Woody wrote, "That my Commanding Officer on Iwo Jima thought me worthy to be recommended for our Nation's Highest award for Valor, The MOH and that 4 of my fellow Marines by their testimony verifying the details of the action on Feb. 23, 1945, thought me worthy, and gave credence to the recommendation assuring consideration of the nomination." Undated document, but most likely done in 2003.

3016    Email General Charles C. Krulak to Rigg, 23 Dec. 2019.

3017    Interview Erin Tripp, 29 March 2018.

3018    Interview Barbara Olson, 5 April 2018.

3019    Interview Steve Tischler, 3 May 2018.

3020    Frankl, 75.

3021    Letter Woody to Rigg, 24 Nov. 2017.

3022    Interview Lefty, 1 Aug. 2017. The author helped the late Jonn Lilyea write up the full report of Lefty's lies with supporting documentation on the web site "This Ain't Hell: But You Can See it From Here." The article I helped Lilyea write is entitled "Darol 'Lefty' Lee: Phony Iwo Jima Hero," 14 Aug. 2017 (https://valorguardians.com/blog/?p=73960). In the forgery Lefty Lee presented to this study (he doctored Dashiell's article from 1945), Lefty added to the document: "Lee killed 10 with his B.A.R (Browning Automatic Rifle)." Lefty Lee's Dashiell Forgery, page 3 (Author's Collection).

3023    MCRDSDA, MCRD Training Book, File 1, <u>What 'Dishonorable Discharge' For Desertion in Time of War Means to the Man Who Gets One.</u>

3024    NASLPR, Darol E. Lee, Capt. L. B. Brooks to Mr. and Mrs. Lee, 5 April 1945, Medical Survey 2 Aug. 1945.

3025    Interview Woody, 19 July 2016; Sledge, 257.

3026    Interview Woody, 19 July 2016.

3027    Frankl, x-xi.

3028    Interview Woody, 23 July 2016; Hallas, *Uncommon Valor on Iwo Jima*, 137.

3029    Melville, xiii-xiv, 437.

3030    NIV—New International Version, 606.

3031    NASLPR, Beatrice Arthur née Frankel. Over 20,000 women served in the Corps during WWII. Millett, 374.

3032    "Navajo Code Talker Samuel Holiday Dies in Ivins at Age 94," St. George News, 12 June 2018.

3033    Bright, 515-6. A few grammatical errors in Abe's speech have been corrected by the author.

3034    Bright, 517; Richard J. Samuels, *Machiavelli's Children: Leaders and Their Legacies in Italy and Japan*, Cornell Univ. Press, 2005, 441.

3035    Interview Woody, 24 March 2015; MCHDQV, Oral History, Woody Williams 13800.

3036    Interview Shindo, 9 April 2018. This is one of the quotes Shindo wanted remove from this book. One reason for this might be that his sentiments expressed above contradict the mandate of the revisionist group he is a member of called *Nippon Kaigi* which denies Japanese WWII atrocities. These candid remarks might hurt Shindo politically when his real, and one might add moral, opinions as so expressed. And if this be the case, is he playing at being a racists with *Nippon Kaigi* in order to strengthen his political career or was he trying to give the impression that he was not a racist with such quotes in order to hide his true feelings?

3037    Burrell, "Did We Have to Fight Here?," 62, 66; Burrell, *Ghosts of Iwo Jima*, 157-88.

3038    Burrell, "Did We Have to Fight Here?," 62, 66.

3039    U.S. Marine Corps Publication, "Expeditionary Operations," MCDP 3, 16 April 1998, 61; Krulak, 71.

3040    Burrell, "Did We Have to Fight Here?," 62, 66; Burrell, *Ghosts of Iwo Jima*, xvii.

3041    Millett, xviii.

3042    Ibid., xv. See also Krulak, 58.

3043    Email Millett to Rigg, 7 Jan. 2018.

3044    NASLPR, Jacklyn Lucas; NASLPR, Boyington. Boyington had to resign from the USMC because of his excessive drinking and inability to take care of his debts in 1941. Had he not done so well in the Flying Tigers and there not been an urgent need for fighter pilots in 1942, he would have not made it back into the USMC. Moreover, his lying about his military record during his time in the Corps and thereafter is a permanent stain upon his legacy.

3045    Poe, "Annabel Lee,"102. Phrase taken from Poe's famous poem *Annabel Lee*. The pastor Woody prayed with was Rev. Robert Chenoweth. MCHDHQ, Woody, "Hershel Woody Williams Christian Experience," Undated Essay.

3046    Frankl, 58-9.

3047    Conversation with author at the Shelton School, Dallas TX, fall of 2004.

3048    Frankl, 60.

3049    NACPM, RG 127 Container 14, Dashiell, #201.

3050    Smith, *Coral and Brass*, 11.

3051    Henri, et al., 86-7; Hoffman, *Chesty*, 282, 279.

3052    Wells, "Medal of Honor Recipient Devotes Life to Veterans."; Historical fact sheet in Woody's archive about Iwo entitled "The Battle for Iwo Jima," p. 1; Miller, "Deathtrap Island,"9; Burrell, *Ghosts of Iwo Jima*, 87; ^ "Soviets declare war on Japan; invade Manchuria." History.com. A&E Television Networks, n.d. Web. 6 July 2014; Frank, *Downfall*, 71, 127; Sledge, 312.

3053    Interview Woody, 13 July 2016.

3054    MCHDHQ, Letter Beck to Ruth Beck, 12 Jan. 1944.

3055    Poe, "Arthur Gordon Pym," 1092.

3056    Frankl, 122.

3057    Ibid., 115.

3058    MCHDHQ, Letter Woody to Ruby Meredith, 12 April 1944.

3059    Interview Barry Beck, 25 June 2019.

3060    Email William E. Schlager to Woody Williams, 26 Nov. 2019; Steven Tischler to Woody, 27 Nov. 2019; Letter George Waters to Woody, 19 Nov. 2019; James Henning to Woody, 1 Dec. 2019.

3061    NASLPR, Woody Williams, Telegram Lurenna Williams to Vandegrift, HQ USMC, 30 Sept. 1945, 93.

3062    NASLPR, Jacklyn Lucas, Marks, Scars, Etc., 6 Aug. 1942.

3063    NASLPR, Donald Graves, Marks, Scars, Etc., 17 Aug. 1942.

3064    MCHDQV, Oral History, Victor H. Krulak, 2, 5; O'Brien, *The Second Most Powerful Man in the World*, 86-7; Email General Charles C. Krulak to Rigg, 1 Jan. 2020.

3065    NACPM, RG 127 Container 14, Dashiell, #267, 4.

3066    NACPM, RG 127 Container 14, Dashiell, #267, 2-3.

3067    Interview Bill Schlager, 30 June 2019.

3068    Interview Woody, 19 July 2016.

3069    NACPM, RG 127 Container 14, Dashiell, #267, 4.

3070    Ibid., 2-3.

3071    Lawless, "of Such Stuff are Heroes," 68, NASLPR, Draft of first MOH & MOH Citation.

3072    In discussing Hallas' book on Iwo, Woody says it is false that a bazooka fired at the pillbox (2nd paragraph p. 130). Letter Woody to Rigg, 14 July 2017. However, he said there was indeed one in 1956 and 1966 (Lawless, "of Such Stuff are Heroes," 68-9; Francois, 93).

3073    Francois, 44.

3074    Ibid., 93.

3075    Ibid., 94.

3076    Interview Woody, 19 July 2016, 21 July 2016 and 8 March 2017; Discussion given at MCRD, Chow Hall, 20 Oct. 2017. In a letter Woody wrote the author on 25 June 2017, he copied many pages out of James Hallas' book *Uncommon Valor* which has a short biography of Woody in it and marked up several sections. In big cursive writing, next to the bazooka man firing at the pillbox, Woody writes "*False.*" Woody further derides Hallas writing "Reading this author's version…makes me almost sick and for sure angry. There is so much assumption in these pages and there is more that is not factual, I can hardly believe a person would take the liberty to write." When I discussed this with Hallas, he then produced the two articles Woody had given in 1956 and 1966 that show he wrote up his biography of Woody using these interviews *Woody actually gave*. When Hallas was confronted with what Woody wrote, Hallas said "What is the old geezer talking about. I quote his first-person responses to what journalists have been questioning him about throughout the years. I take offense of him not looking up my sources first before he attacks my writing" (Interview James 10 July 2017).

3077    Crowl & Isely, 267.

3078    NASLPR, Woody Williams, Statement Cpl. Schlager.

3079    Interview Gary Fischer, 11 Jan. 2020.

3080    MCHDQV, Dashiell File, #288, Cookson; NASLPR, Marquys Kenneth Cookson, Silver Star Citation. PFC Donald Graves heard about numerous people crawling on top of pillboxes and attacking them from the top, but they always did so staying as low as possible. Interview Don Graves, 11 Jan. 2020.

3081    Interview John Renstrom, 22 Jan. 2020. As a child, Renstrom had been shown how dangerous gasoline fumes were by his local firemen and, obviously, never forgot the lessons of his youth.

3082    Interview John Renstrom, 22 Jan. 2020; https://www.cob.org/services/safety/education/pages/gasoline.aspx; Tim Sharp, "How Hot is the Sun?," Space.com, 19 Oct. 2017; http://coolcosmos.ipac.caltech.edu/ask/7-How-hot-is-the-Sun-.

3083    https://astronomy.stackexchange.com/questions/11969/what-would-happen-if-we-stepped-on-the-sun

3084    Interview John Renstrom, 22-23 Jan. 2020. This happened other times on Iwo when an IJA "ammunition dump hidden" underground was ignited by an attack by flames—in this case a flamethrower. When this particular ridge exploded, it shook the entire ridge and "smoke poured out of the cave's rear entrance 60 yards away." Henri, et al., 108.

3085    Steve Tally, "Farmers Should Use Extra Caution with Gasoline," Purdue News Service, 4 June 1999; Interview John Renstrom, 22 Jan. 2020. The city of Bellington Washington put out a report that a gallon of vaporized gasoline can have the equivalent explosive power of 20 sticks of dynamite. So, Renstrom put anywhere between 1,000 to 5,000 "sticks of dynamite by pouring 55 gallons of gasoline into this bunker network.

3086    Interview Gene Rybak, 11 Jan. 2020.

3087    Interview Don Graves, 11 Jan. 2020.

3088    NACPM, RG 127 Container 14, Dashiell, Folders 1-2, #267, 3-4; Francois, 94.

3089    NACPM, RG 127 Container 14, Dashiell, Folders 1-2, #267, 1.

3090    NASLPR, Frederick Knowles Dashiell, Letter of Commendation by Erskine for Dashiell, 17 June 1945.

3091    MCHDHQ, Dashiell, "Combat Correspondent Dispatch," 17 May 1945, 1; MCHDHQ, Beck,

"Biography of Donald Beck," Dec. 2002, 16; NASLPR, File Woody Williams, Endorsement Letter Colonel Withers for Woody Williams MOH, 27 April 1945.

3092   NASLPR, File Woody Williams, Endorsement Letter Col. Withers for Woody's MOH, 27 April 1945, 2.

3093   Francois, 94; Interview Woody, 21 July 2016; WWII Museum Oral History Project, 3 Feb. 2011.

3094   Interview Woody, 21 July 2016.

3095   Toll, *Pacific Crucible*, 162.

3096   Mears, 98.

3097   https://www.youtube.com/watch?v=4_V6h6EJh9c; Documentary film "Desmond Doss: The Conscientious Objector (The Real Story)," 16 Nov. 2016; https://www.youtube.com/watch?v=1FvoY2DFAAs

3098   Interview George Waters, 21 Oct. 2019. Woody has also often mention that he and Waters were best friends at boot camp when the record shows they were in different classes at different times. Moreover, Woody has claimed Waters was there at the CP helping him gear up on the day he took out the pillboxes when the record shows Waters was in a different company and assigned to a different sector of the battle line than Woody.

3099   Interview Woody, 22 July 2016. In Draft 21 for this book (16 Feb. 2018), which Woody read and approved, Woody was having me write: "Waters made sure all weapons would be ready for Woody" on the day he destroyed the pillboxes (p. 367). Showing how much Woody adheres to this draft (a draft written before I knew many of the problems with Woody's entire narrative), he uses it in the legal documents he has had his lawyers create to attack me and earlier presses. See United States District Court for the Southern District of West Virginia Huntington Division, Hershel Woodrow "Woody" Williams, Plaintiff v. Bryan Mark Rigg, and John Doe Publishing Company, Civil Action Case 3:19-cv-00423, Document 11, Filed 1 Nov. 2019, 5.

3100   Interview Woody, 21 July 2019; Hallas, *Iwo Jima*, 137. In Draft 21, which Woody read and approved, it is written: "Woody remembers [Waters'] ring was not worn on his middle finger because Vernon's middle two fingers were grown together" (p. 645 fn 1651).

3101   NASLPR, Eleanor Pyles, From Dir. of Personnel, Marine Corps to The Com. Off., Cherry Point, 9 June 1945, Subj: Change of Marital Status, Case of Private Eleanor Pyles & From Com. Off. Aviation Women's Reserve Squadron-20 to Marine Corps Commandant, Subj: Discharge, case of Private Eleanor G. Pyles, 1. (c) The Medical Officer's letter certifying pregnancy, 14 July 1945.

3102   NASLPR, Eleanor G. Pyles, Service Record, Discharged at AWRS-20, Cherry Point, NC, Under Honorable Conditions, Future Address, 34; World War II Young American Patriots, 1941-45, WV, Marion County, 442; https://www.nolo.com/legal-encyclopedia/eligibility-va-benefits-depending-type-military-discharge.html. The *Stolen Valor Act of 2013* (Pub.L. 113–12; H.R. 258) is a U.S. federal law that the 113th U.S Congress passed in 2013. "The law amends the federal criminal code to make it a crime for" anyone to declare they have "served in the military, embellish their rank or fraudulently claim having received" a valorous award or medal specified in the Act, with the intent of securing "money, property, or other tangible benefit by convincing another that he or she received the award." This federal law is a revised version of a previous statute from 2005 struck down by the Supreme Court of the United States in 2012 after the case of United States v. Alvarez." With this law, individuals who lie about their military service or are imposters wearing medals or uniforms they never earned can be prosecuted under the law. Under these definitions, Lefty is a classic "Stolen Valor Case" according to lawyers Geoffrey Harper, Steve Stoghill and Tom Hancock, historian and USMC Colonel Charles Neimeyer and JAG officer Colonel Charles Jones. This is also the case since Lefty said he had the Bronze Star with V from Iwo Jima and a Purple Heart form Iwo Jima. See the article by Al Zdon, "A Marine in the Pacific," Minnesota Legionnaire, May 2015.

3103   NASLPR, Eleanor G. Pyles, Service Record, Discharged at AWRS-20, Cherry Point, NC, Under

Honorable Conditions, Future Address, 34; World War II Young American Patriots, 1941-45, WV, Marion County, 442; https://www.nolo.com/legal-encyclopedia/eligibility-va-benefits-depending-type-military-discharge.html.

3104    Hoffman, *Chesty*, xiii. To possibly explain what Woody has done, see the following article: Meera Senthilingam, "People Lie to Seem More Honest, Study Finds," CNN, 30 Jan. 2020.

3105    My ancestor, John Linton (8th great-grandfather), a friend of William Penn, left England in 1699 for Pennsylvania after being disinherited by his father, Sir Roger Lynton, for having converted to Quakerism and embraced a religion disloyal to the crown. See Herman LeRoy Collins, *Philadelphia, A Story of Progress*, NY, 1941; Albert J. Brown, *Clinton County Ohio: Its People, Industries and Institutions*, Indianapolis, IN, 1915, 621.

3106    George M. Marsden, *The Twilight of the American Enlightenment: The 1950's and the Crisis of Liberal Belief*, New York, 2014, xxiii-xxiv; Morris U. Schappes (ed.), *A Documentary History of the Jews in the United States 1654-1875*, New York, 1950, 80.

3107    Interview Barry Beck, 25 June 2019.

3108    Dan Barker, *Godless: How an Evangelical Preacher Became One of America's Leading Atheists*, Berkeley, 2008, 213.

3109    Christopher Hitchens, *The Portable Atheist*, New York, 2007, "Introduction" by Christopher Hitchens, xiii.

3110    Kem Stone, "*The Struggle of Sisyphus: Absurdity and Ethics in the Work of Albert Camus—Moralist Camus – The Plague*," July 2006, http://www.kemstone.com/Nonfiction/Philosophy/Thesis/plague.htm. Kem Stone is a pseudonym and not the author's real name. For excellent lectures of Camus, see Robert C. Solomon's lectures on him with the Teaching Company entitled "*No Excuses: Existentialism and the Meaning of Life*," Lecture 5, Disc 3, 2000.

3111    Sam Harris, *The End of Faith: Religion, Terror, and the Future of Reason*, NY, 2004, 44.

3112    Numbers 31: "1) The Lord said to Moses, 2) "Take vengeance on the Midianites for the Israelites….3) So Moses said to the people, "Arm some of your men to go to war against the Midianites and to carry out the Lord's vengeance on them….5) So twelve thousand men armed for battle…were supplied from the clans of Israel. 6). Moses sent them into battle…7). They fought against Midian, as the Lord commanded Moses, and killed every man…9). The Israelites captured the Midianite women and children…14). Moses was angry with the officers of the army [saying]…15). "Have you allowed all the women to live?" he asked them. 16). "They were the ones who followed Balaam's advice and the means of turning the Israelites away from the Lord…17). "Now kill all the boys. And kill every woman who has slept with a man, 18). but save for yourselves every girl who has never slept with a man." [The soldiers did as commanded]…32) The plunder remaining from the spoils that the soldiers took was…32,000 women who had never slept with a man. (NIV—New International Version, pp. 163-4).

3113    "The slaughter of Midianite women and children is rationalized in pious tradition as a justified punishment for the scheme 'to seduce the sons of Israel to unchastity and then to idolatry,' a scheme that the rabbis attributed to the despised Balaam." Jonathan Kirsch, *Moses: A Life*, NY, 1998, 323. While there is no historical evidence for the Israelites conquering the ancient lands of Canaan, most ultra-religious Jews and strong Christians still believe it actually happened and thus it is cited here. For a source that discusses the lack of historical evidence for events in the Torah, see H. H. Ben-Sasson (ed.), *A History of the Jewish People*, A. Malamat, "Origins and the Formative Period," Cambridge, 1976, 43, 52-9; Theodor H. Gaster, *Myth, Legend, and Custom in the Old Testament*, NY, 1969, 411; James L. Kugel, *How to Read the Bible: A Guide to Scripture Then and Now*, NY, 2007, 376-85.

3114    Robert Melson, *Revolution and Genocide: On the Origins of the Armenian Genocide and the Holocaust*, University of Chicago, 1996.

3115    Hitler Youth songs referred to Hitler as their "redeemer." Kertzer, 202. In China, people often called Mao their "Savior." Interview with 10th Degree Grand Kung Fu Master Johnny Lee, 13 Feb. 2015.

3116    Jonathan Steinberg, *All or Nothing: The Axis and the Holocaust 1941-1943*, NY, 1991, 244.

3117    Jung Chang and Jon Halliday, *Mao: The Unknown Story*, London, 2006, 3; Ian Kershaw, *The Hitler Myth: Image and Reality in the Third Reich*, Oxford, 1990, 105-20. Hitler was often referred to as Germany's Messiah by lay people and ministers. As historian Ian Kershaw wrote "Once he had become Chancellor, Hitler's language became pronouncedly 'messianic' in tone, and his public addresses were frequently replete with religious symbolism." Kershaw, *The Hitler Myth*, 107. In reference to 70 million, this is a high number put on the genocide by Chang and Halliday. The National Geographic Jan. 2006 edition on genocide numbers places China's extermination under Mao reaching the number of 30 million. Regardless of what the final figure is, Mao killed a lot of people and due to Chinese society not having placed much value on demographic studies like the Germans, the number killed under Mao is probably higher than most think. Nonetheless, the actual number is difficult to ascertain.

3118    Steinberg, *All or Nothing*, 244.

3119    Hitchens, *god is not Great*, 202-3.

3120    Dawkins, 249.

3121    Wallance, 201.

3122    William B. Breuer, *Sea Wolf: A Biography of John D. Bulkeley, USN*, Novato, CA, 1989, 10-1.

3123    Harris, 67.

3124    Stone, "*The Struggle of Sisyphus: Absurdity and Ethics in the Work of Albert Camus.*

3125    Michael Berenbaum, *The World Must Know: The History of the Holocaust as told in the United States Holocaust Memorial Museum*, Baltimore, 2007, xxi.

3126    Plato, *The Trial and Death of Socrates: Apology*, Cambridge, 1975, 39. Joseph Campbell states it even more dramatically when he said "The person who thinks he has found the ultimate truth is wrong. There is an often-quoted verse in Sanskrit, which appears in the Chinese *Tao-te Ching* as well: 'He who thinks he knows, doesn't know. He who knows that he doesn't know, knows. For in this context, to know is not to know. And not to know is to know.'" Joseph Campbell, *The Power of Myth with Bill Moyers*, NY, 1991, 65.

3127    Interview with Saburo Sakai by Colin Heaton, Nov. 1990.

3128    Tanaka, 1.

3129    "Another Attempt to Deny Japan's History," NY Times, 2 Jan. 2013; Justin McCurry, "Shinzo Abe, an Outspoken Nationalist, Takes Reins at Japan's LDP, Risking Tensions with China, South Korea," Global Post, 28 Sept. 2012; Norimitsu Onishi, "Abe Rejects Japan's Files on War Sex, NYTimes, 22 July 2018; "Japan PM Abe Demands Apology for South Korean Comments on Emperor Akihito," The Straits Times, 12 Feb. 2019; Tanaka, xv, xxii, xvii.

3130    Hitchens (ed.), *The Portable Atheist*, George Eliot, "Evangelical Teaching," 76; Tanaka, xi.

3131    Driscoll, 266-77; Richard Samuels, "Kishi and Corruption: An Anatomy of the 955 System," Working Paper No. 83, Dec. 2001, Japan Policy Research Institute; Bright, 517.

3132    Crowl & Isely, 500-1.

3133    Letter Shindo to IJAA, DD Dec. 2018.

3134    "Nationalist 'Japan Conference' Building Its Clout: Ten Days after the Meeting, Abe Officially Addressed the Issue of Revising the Pacifist Constitution," Korea JoongAng Daily, 3 May 2013; "Japanese Minister Yoshitaka Shindo Visits Yasukuni Shrine Provoking China's Ire," South China Morning Post, 1 Jan. 2014; https://www.revolvy.com/page/Yoshitaka-Shind%C5%8D

3135    "Behavioral Principle of the Right Wing Organization 'Japan Conference' Supporting the Abe Administration: Interview with Mr. Sueno, author of 'Japanese Conference Research'", Diamond Online, 20 May 2016; Josh Gelernter, "Japan Reverts to Fascism," National Review, 16 July 2016.

3136    Norihiro Kato, "Tea Party Politics in Japan," NY Times, 12 Sept. 2014; Jennifer E. Robertson, *Politics and Pitfalls of Japan Ethnography: Reflexivity, Responsibility, and Anthropological Ethics*, NY,

2009, 66; Emma Chanlett-Avery, Mark E. Manyin, Rebecca M. Nelson, Brock R. Williams and Taishu Yamakawa, *Japan-U.S. Relations: Issues for Congress*, Congressional Research Service, 16 Feb. 2017, 8-10; Tanaka, xxi-xxii, 257; JNMS, Vol. 1, No. 1, Sheng, "The Seven Why-Questions in the Writing of History," 10. Article 11 of the Japanese Constitution clearly states: "Japan accepts the judgements of the International Military Tribunal for the Far East and of other Allied War Crimes Courts both within and outside Japan, and will carry out the sentences imposed thereby upon Japanese nationals imprisoned in Japan." Wakabayashi (ed.), *The Nanking Atrocity, 1937-38*, Wakabayashi, "The Messiness of Historical Reality," 8.

3137    JNMS, Vol. 1, No. 1, Sheng, "The Seven Wh-Questions in the Writing of History," 3.

3138    TSHACNCDNM, Vol. 10, 267, Statement Keenan, Chief Counsel, Inter. Military Tribunal Far East, 4 June 1945, 5.

3139    Takashi Yoshida, 5.

3140    Ibid., 5.

3141    JNMS, Vol. 1, No. 1, Sheng, "The Seven Wh-Questions in the Writing of History," 3. See also *Hong Kong's War Crimes Trials*, Daqun, "Foreword," viii.

3142    Toland, Vol. II, 1078.

3143    Weiner (ed.), *Race, Ethnicity and Migration in Modern Japan*, Weiner, "Introduction," 2.

3144    The Carryall, Iwo Jima, Feb. 19, 1945—Oct. 15, 1945, 133rd NCB Navy Seabee Souvenir Newsletter, Reprint Ken Bingham, Vol. III, No. 8, Gulffort MS, 2011, 29.

3145    Scott, 515.

# Index

First Air Fleet (IJN), 206, 208

Carrier Strike Group I (USN), 303

1st Marine Brigade, 183, 194, 284, 285

1st Marine Division (1st MarDiv), 50, 194

First Mobile Fleet (Japanese), 199, 205, 206, 207, 208, 229

1st Provisional Marine Brigade, 284, 285

2nd Infantry Division (USA), 191

2nd Marine Division (2nd MarDiv), 183, 199, 200, 229

2nd Mixed Brigade (IJA), 133, 518

III Amphibious Corps, 170, 243, 251

3rd Army (USA), 191

3rd Marine Division (3d MarDiv) ("The Fighting Third"), xxxvi, xxxix, 6, 55, 147, 159-160, 161, 167, 183, 196, 197, 199, 209, 253, 257, 258, 260, 264, 265, 268, 283, 284, 285, 295, 307, 327, 344-345, 351, 381, 383, 387, 388, 390, 395, 402, 440, 445, 451, 457, 469, 477, 481, 482, 483, 495, 496, 505, 507, 509, 511, 523, 528, 531, 532, 559, 561, 562, 563, 567, 570, 587, 595, 614, 615, 629, 630, 641, 703, 705, 717, 747, 779

    Baseball League, 630, 631

    Boxing federation, 630

    Casualties on Iwo Jima, 493-494

    Cemetery, 559, 561

    Dedication of the Cemetery, 561

    Educational Programs, 629-630

    Orchestra, 630

    Replacements, 491-499

    Training its men to hate the enemy, 160-161

3rd Marines (Regiment) (USMC), 269, 498, 499

    Replacement controversy, 498-499

3rd Service Battalion (USMC), 469

3rd Tank Battalion (USMC), 309

4th Brigade (USA), 191

4th Marine Division (4th MarDiv), 136, 183, 199, 213, 214, 217, 229, 230, 245, 362, 365, 367, 371, 376, 381, 392, 395, 482, 483, 578, 627, 645, 647, 657, 662, 705, 722

    Casualties, 578

4th Marines (Regiment) (USMC), 55

Fifth Air Force (USAAF), 146

V Amphibious Corps, 170, 199, 217, 243, 301, 355, 499, 507, 595, 642, 645

    Size of Corps, 199, 355

5th Division (IJA), 202

Fifth Fleet (USN), 204, 206, 207, 329, 352, 488

5th Marine Division (5th MarDiv), xlii, 159, 345, 351, 365, 367, 376, 403, 482, 483, 528, 531, 532, 546, 549, 557-558, 559, 564, 587, 664, 705

    Iwo Jima Cemetery Dedication, 546-549, 559

Sixth Army (USA), 609

6th Bombardment Squadron (USAAF), 352

6th Division (IJA), 106, 107

6th Expeditionary Force (IJA), 241

6th Marines (Regiment) (USMC), 202, 561

7th Calvary Regiment (IJA), 314

7th Fleet (USN), 553

Seventh War Zone (Chinese Nationalist), 316

8th Marines (Regiment) (USMC), 202

9th Army (German), 638

9th Engineers (IJA), 322

9th Marines (Regiment) (USMC), xxiv, 271, 378, 463, 491, 499, 502, 588

10th Army (IJA), 99, 106

11th Battle Fleet (IJN), xlix

12th Army Group (Chinese Nationalist Army), 316

Cruiser Division TWELVE, 609

12th Independent Anti-Tank Battalion (IJA), 452

14th Army (British), 208

16th Division (IJA), 106

18th Division (IJA), 316

Twentieth Army Air Force (USAAF), 543

XXI Bomber Command (USAAF), 543

21st Marines (Regiment) (USMC), 147, 162-163, 252, 260, 265, 266, 269, 271, 272, 273, 294, 306, 309, 361, 385, 388, 391, 392, 395, 396, 398, 403, 409, 410, 411, 445, 455, 457, 463, 478, 485, 491, 493, 494, 496, 499, 502, 510, 517, 528, 535, 537, 560, 568, 570, 571, 581, 587, 614, 626, 630, 645, 705, 747, 754

    Casualties on Iwo Jima, 493-494, 560, 587

    Positions of 21st's battalions during the *Banzai* attack on Guam, 25-26 July 1944, (see Gallery 2, Photo 21, page 14)

21st Regimental Newsletter, 568, 571, 577, 580, 593, 705

23rd Army (IJA), 123-133, 315, 316, 317, 318, 513, 638

    The establishment of brothels, 131

    Size, 316

23rd Marines (Regiment) (USMC), 238-239, 379, 395, 510, 627

24th Marines (Regiment) (USMC), 393, 408, 465, 492, 493, 645, 647

    Casualties on Iwo Jima, 492-493

XIV U.S. Army Corps, 142

25th Marines (Regiment) (USMC), 230, 662

26th Marines (Regiment) (USMC), 409, 664

26th Tank Regiment (IJA), 133

27th Air Flotilla (IJN), 329

27th Infantry Division (USA), 183, 199, 284

28th Infantry Unit (IJA), 129

28th Marines (Regiment) (USMC), 403, 407, 465, 495, 532, 533, 587

29th Bombardment Group (USAAF), 352

29th Infantry Division (IJA), 241

31st Army (IJA),

32nd Replacement Battalion (USMC), 145

38th Division (IJA), 129

38th Regiment (IJA), 255, 315, 316

43rd Division (IJA), 202

44th Infantry Regiment (IJA), 133

47 *Rōnin*, 521

48th Independent Mixed Brigade (IJA), 255

    Task Force 51, 357

    Task Force 53,

    Task Force 58, 184, 352

76th Infantry Border Garrison Regiment (IJA), 1333

77th Infantry Division (USA), 183, 203, 244, 284

81st Infantry Division (USA), 340, 341

104th Division (IJA), 316

109th Division (IJA), 133

129th Artillery (USA), 624

145th Regiment (IJA), 455

147th Infantry (USA), 535, 550, 551

    Casualties on Iwo Jima, 551

148th General Hospital (Saipan), 601

*Sturmregiment* (Storm or Assault Regiment) 195, 638

299th Regiment (IJA), 129

322nd Regiment (USA), 340, 341

A-Day (invasion of the Philippines), 170

Abe, Prime Minister Shinzo, 123, 136, 138, 735, 736, 772, 774, 775, 851 fn 511

Racists and Xenophobic Affiliations, 773-775

Abbot, Cecilia, First Lady of Texas Mrs., 692

*Ace* (Fighter-Pilot), 649

Acts (Bible), 688

Adair, Admiral Charles (USN), 543-544

Adelup Point (Guam), 252

Aden Harbor (Yemen), 645

Adorno, Theodor, xxxiii

Aeschylus (Greek writer), xxix

Afghanistan, 702, 709

African Americans ("Negroes"), 148, 310, 537, 538, 539, 547, 733-734

*Afrika Korps* (Nazi Desert Troops—Africa Corps), 90

Agat Beach (Guam), 244, 252 (see Gallery 2, Photo 16, page 11)

Agaña (Hagåtña—Guam's capital), 177, 178, 244, 252 (see Gallery 2, Photo 16, page 11)

Agaña Bay, 262, 283 (see Gallery 2, Photo 16, page 11)

Agriculture Institute (Hopei Medical College), 114

Aguon, Edward L. G., 181, 282

Aiea Heights (Hawaii), 601

Air Raids, 512-513

Airfield No. 1 (Iwo Jima), 375, 376, 392, 393, 397, 409, 455, 456, 473, 488, 500, 561 (see Gallery 3, Photo 11, page 8)

Airfield No. 2 (Iwo Jima), 396, 463, 473, 474, 477, 478, 484, 537, 641 (see Gallery 3, Photo 11, page 8)

Aitape Beach (Papua New Guinea), 117

Akikusa, Tsuruji, xli, 357, 408, 516, 517, 523, 734 (see Gallery 1, photo 6, page 2; Gallery 4, Photo 9, page 5)

*Akashi* (Japanese name for Guam's capital Agaña), 178

*Akō* vendetta (legend of the 47 rōnin), 521

Al Gray Research Center (Marine Corps University), 779

Al-Qaida (Islamic Fascists), 769

Alabama (USA), 190

Alabama National Guard, 190

Alabama Polytechnic Institute, 190

Alaska (USA), 194, 323, 606, 846 fn 422

St. Albans Naval Hospital (New York), 665

Aleutians Islands (Alaska), 394

Alexander the Great, 341

Allen, Colonel John (USMC), 778

Allied Tribunal (Tokyo War Crimes Trials), 141

Allies, xix, xlvii, liii, 51, 78, 89, 92, 142, 168, 170, 183, 199, 208, 226, 297, 302, 303, 337, 381, 459, 571, 574, 638, 695, 712, 717, 719, 723, 768

Allison, Colonel Fred (USMC), 782

Alkek, Major Dr. David (USAF), 12

*Amagiri* (Japanese destroyer), 669

Amano, Tetsuro (Japanese Consul-General in Houston, Texas), 119

*Amaterasu Omikami*, 72, 80, 81

America's Production of War Materials, 696

America's Total KIAs in WWII, 703

American Church Mission (Nanking), 111

American Colonies, 764

American First Committee, xlvii, 3

American Legation in Beijing, 196

American Military University, xxii, 778

American Revolution, 71

Ammon-Zeus, 341

Amphibious Corps, 194

Amphibious Warfare, 159, 189, 190, 191, 192-198, 865 fn 894

Amphitheater (Iwo Jima), 457

Amtracs ("alligators" and LVTs-Landing Vehicle Tracked), 7, 194-195, 197, 248, 251, 252, 253, 308, 358, 373, 382, 865 fn 894, 887 fn 1569

Anderson U.S. Air Force Base (Guam), xlii, xliii

Anderson, Captain Carl E. "Squeaky" (USN), 393, 394

Anderson, Sergeant Charles C. Jr. (USMC), 373, 380

Anderson, Captain Charles C. Sr. (USN), 373

Andrew Sisters (musicians), 462

Andrews, Harold, xli

Angaur Island (Battle of), 340, 341, 609, 703, 706

Anti-American Japanese Propaganda, (see Gallery 1, Photos 13-14, page 5)

Anti-Comintern Pact (1936), 85, 843 fn 361

Anti-Japanese American Propaganda, (see Gallery 1, Photos 15-19, 32-33, pages 6-7, 16; Gallery 2, Photos 1, 3-4, page 1)

Anti-Semitism, 548-549

Apra Harbor (Guam), 252, 403 (see Gallery 2, Photos 14-15, page 10)

Araki, Kamikaze Haruo, 304

Araki, War Minister and General Sadao (IJA), 80, 81

Araki, Shigeko, 304

Archway Books, xxxii

Ares (the Greek god of war), 341

Argentina, 35, 694

Arlington National Cemetery, xxi, 634

Armageddon, 346

Armenian Genocide, 767

Armstrong, Corporal Connie (USMC), 238, 239

Army Air Academy (IJA), 298

Army Group Center (Nazi), 208

Army War College (USA),

Arnold, General Hap (USA), 169, 170, 554, 617

The Art of War, 301

Arthur, Aaron (archivist), 782

Article 11 of the San Francisco Peace Treaty, 774

Aryan (Nazi Race Myth), 79, 81, 85

Asaka, Lieutenant General and Prince Yasuhiko (IJA), 106

Asan Beach (Guam), 244, 253 (see Gallery 2, Photos 11, 16, 21 pages 8, 11, 14)

Asan Point (Guam), 252 (see Gallery 2, Photos 16, 21, pages 11, 21)

Asan River (Guam), 254 (see Gallery 2, Photo 21, page 14)

Ashihei, Hino (IJA), 49

Ashworth, Commander Frederick L. (USN), 555, 721

Asia (Orient), 31, 35, 83, 90, 94, 123,133, 184, 186, 187, 318, 694, 697, 712, 718

Asian Co-Prosperity Sphere, 90, 94, 109, 177, 518, 700

Asians, 90, 718, 733, 763

Aslito Airfield (Saipan), 229 Map of the Tinian Battle 1944, (see Gallery 2, Photo 10, page 7)

Astley, Sir Jacob,

Atheism (of servicemen), 339, 340, 341, 391, 392

Atlantic Monthly, 369

Atlantic Ocean, 4, 668, 694

Atomic Bombs, xxxvi, xxxviii, 120, 123, 240, 361, 369, 488, 532, 555, 556, 562, 564, 616-617, 684, 707, 709, 710, 713, 714, 715, 716, 717, 718, 719, 721, 722, 723, 724, 765, 769, 935 fn 3001
    End the war, 715-716
    Number killed in the two bombings, 120
    Preparation for follow on atomic bombs to the Hiroshima and Nagasaki bombings, 713

Attu Island (WWII battle (Alaska)), 298, 323, 336, 561, 846 fn 422

University of Auburn, 190

Auschwitz Nazi Extermination Camp (Krakow, Poland), 67, 100, 119, 719

Australia, xlvii, lii, 24, 51, 88, 89, 90, 108, 124, 164, 190, 694

Austria, 23

Awards and Decorations Board (USN), see Navy Awards and Decorations Board

AWOL, 664

Axis Powers, 5, 6, 31, 47, 86, 91, 184-185, 186, 458, 638, 694, 733, 768

Azuma, Shirō (IJA), 99

B-24 Bomber (Liberator), 345, 359, 690

B-29 Bomber (Superfortress), xli, xliii, 169, 170, 171, 225, 240, 289, 296, 304, 306, 345, 352, 362, 450, 488, 489, 511, 514, 527, 544, 552, 553, 554, 555, 568, 673, 710
    Bombing of Kobe, 527
    Bombing of Tokyo, 511-512

Bagration (Operation), 208

Bainbridge (Maryland), 632

Baka bombs, 389

Baker, Norman (USMC), (see Gallery 1, Photo 12, page 6)

Baltic States, 4

BAMS ("Broad-Ass Marines"—Nickname for Female Marines of WWII), 734

Band of Brothers, xxxvi

Banzai (suicidal mass wave of charging Japanese), 210, 303, 321, 379, 381, 416, 434, 462, 478, 535, 538, 699, 708

Banzai (on Guam), xxxvi, xxxix, lv, lvi, 262, 263-274, 278, 284, 290, 295, 413, 491, 690, 743, 747 (see Gallery 2, Photo 21, page 14)

Banzai (on Iwo Jima), 510, 533, 537, 538, 539, 540, 541

Banzai (on Saipan), 210, 211, 212, 290

Banzai (on Tinian), 230, 234-235, 236, 238

Banzai Ridge (Guam), 254, 262, 272, 290 (see Gallery 2, Photos 18-23, pages 12-15)

Baoding (China), 114

Barboursville (West Virginia), 689, 692

Barbary Pirates, 44

Barker, Dan (writer), 766

Barlow, PFC George (USMC), 645, 646, 647, 648

Baseball, 21, 456, 519-520, 630, 631

The Basic School (TBS—USMC), xxii, 778

Basilone, Gunnery Sergeant John, xxiv, 266-267, 376, 377, 378, 380, 462
    Death, 377, 378
    Medal of Honor, 266, 267, 376
    Navy Cross, 378

Bataan (Battle of), 50, 52, 150, 661

Bataan Death March, 49, 50, 51, 139, 161, 512

Bates, Professor Miner Searle, 111

Battalion Landing Team, 252

Battle of the Bulge, 685

Battle of the Philippine Sea, see Philippine Sea Battle or The Great Mariana Turkey Shoot

Bauer, Kimberly, 781

Bazooka, 233, 234, 237, 256, 278, 291, 305, 417, 437, 473, 653, 654, 655, 656, 665, 720, 749-751, 898 fn 1898, 924 fn 2710, 924 fn 2710, 938 fn 3072

Baylor University, 666

Bechtold, Elmer (USMC), 532

Beck, Barry (USMC), 158, 307, 581-582, 672, 692, 742, 765, 780

Beck, Captain Donald (USMC), xxiii, 148, 149, 157, 158, 163, 172, 173, 249-250, 260, 268, 269, 270, 272, 282, 283, 284, 291, 293, 294, 295, 306, 307, 309, 337, 338, 380, 381, 409, 412, 413, 414, 415, 418, 420, 421, 447, 448, 450, 451, 453, 462, 494, 495, 507, 511, 536, 557, 569, 572, 573, 577, 578, 579, 580, 581, 586, 587, 588, 593, 595, 598, 600, 615, 639, 650, 651, 653, 656, 657, 659, 671, 672, 673, 674, 675, 692, 693, 701, 702, 707, 724, 725, 726, 734, 741, 742, 747, 750, 754, 759, 763, 765, 775, 780, 784, 892 fn 1750, 895 fn 1815, 895 fn 1823 (see Gallery 2, Photo 5, page 1)
    Beck's reasoning to start the MOH process for Woody, 577-578, 671
    Beck's recommendation for Woody, 578
    Bronze Star with V, 450
    Exploration of why Beck never mentioned or contacted Woody after the war, 757-758
    Purple Heart, 293, 413
    Silver Star, 413

Beck, Mark (USMC), 157

Beck, Ruth, 157-158, 172, 173, 293, 294, 337, 587, 741

Beijing (Peking), 77, 114, 136, 196, 314, 769

Belarus, 208

Belgium, 23, 154, 302

Belleau Wood (WWI battle), 44, 57, 191, 196, 561

Bellevue Hospital (New York City), 374

Bellevue-Stratford (Philadelphia), 540

Belton, Major F. (USMC), 541

Berenbaum, Michael (historian), xvii, xxxiii, 770, 780, 845-846 fn 415

Berlin (German capital), 137, 208, 337

Berman, Paul, 142

Bernays, Edward L., 567, 677

Bernstein, George (USMC), 533-534, 673

Bernstein, Sammy (USMC), 559-560

Besser, Joe (actor), 537

Bethesda Naval Hospital (Maryland), 665

*Bhagavadgita*, 714

The Bible, 1, 19, 58, 160, 283, 339, 355, 635, 685, 777, 830 fn 68, 902 fn 2015

Billings, Corporal Dale (USMC), 430

Bingham, Ken (amateur historian), 779

Bishop, Jim, 379

Bismarck Islands, 146

Black Marines, see "Negroes"

Blackman, Lieutenant General Robert R. Jr. (USMC), 691, 692

Blandy, Rear Admiral William H.P. (USN), 350, 609, 658

Blechman, Esther, 570

Blechman, Rabbi Nathan, 402, 404, 570, 571

Blechman, Sergeant Solomon Israel (USMC), 402, 403, 404, 570, 893 fns 1766-1767
    Bronze Star with V, 403, 404
    Fort Bliss (Texas), 313

*Blitz* (USMC), 571

*Blitzkrieg* (Nazi), 203, 571

Blood Transfusions, 486

Blowflies, see Flies

Blue Network, 405

Body Bearers, 261

Boetticher, Lieutenant General von (German), 184

Bolivar, Rear Admiral Babette "Bette" (USN), xliii

Bonin Islands, liv, 183, 331, 554

Boot Camp (Marine Corps), 77-102

Borah, Senator William A. (Idaho), 4

Bornholz, Corporal Warren (USMC), 443, 444, 506, 549, 579, 581, 626, 724, 726, 784, 898-899 fn 1912

Bougainville (Battle of), 39, 144, 145, 146, 147, 164, 172, 253, 309, 338

Boy Scouts, 48, 288

Boyington, Major Gregory "Pappy" (USMC), 607, 608, 609, 649, 650, 677, 678, 739, 937 fn 3044
    Medal of Honor, 607, 610
    Silver Star, 607

Bradley, Corpsman John (USMC), 651, 924 fn 2694
    Misleads people about his role with the Flag Raising on Mt. Suribachi, 651, 924 fn 2694
    Navy Cross, 651

Bradley, General of the Army Omar (USA), 737

Brazil, 694

British (Britain), 132, 141, 149, 172, 202, 316, 317, 318, 638, 667, 697, 733

British Expeditionary Force (1940), 23

Broderick, George (USMC), xlii, 827 fn 31, 914 fn 2431

Brodie, Bernard (military strategist), 699-700

Brown, Vice Admiral Charles R. (USN), 303

Brown, PFC Francis S. (USMC), 659

Brown, Leonard (USA), 690-691

Brown, PFC Thomas N. (USMC), 446

Browning Automatic Rifle (BAR), 149, 153, 417, 418, 424, 445, 511, 639, 704

Buchan, Dr. William H. (USN), 271

Buddha (*Shakyamuni*), 80, 81, 143, 679

Buddhism, 30, 68, 75, 81,83, 84, 389, 488, 543, 574, 844 fn 391

    Monks in the IJA, 80, 105, 849 fn 474

    Hell (Underworld), 543

    *Sunzu* River, 488

    Support for Fascism and Hirohito, 68-69, 74, 75, 87

Bulgaria, 29, 184

Bulkeley, Vice Admiral John D. (USN), 769

*Bundesarchiv-Militärarchiv* (Freiburg i. Br., Germany), xvii

Burgess, Ann (psychologist), 125

Burke, Captain Arleigh (USN), 206

Burke, Edmund (philosopher and political scientist), cover page

Burkhalter, Gunnery Sergeant Durell (USMC), 387, 388, 485

Burleigh, Michael (historian), 543

Burma (Myanmar), lii, 35, 89, 226, 568, 722

Burrell, Lieutenant Colonel and historian Robert (USMC), xvii, xxxiv, 131, 546, 562, 613, 779-780

Bush, President George H. W. (USN), 50

Bush, Corpsman Robert (USN), 628

    Medal of Honor, 628

*Bushido* ("Way of the Warrior"), 68, 75, 78, 331, 461, 521

Butterfield, Colonel J. B. (USMC), 640

C-2 Explosives, 150, 308, 429, 430, 442, 476, 704

C-47 Transport Plane, 486, 619

C-Rations, 529

Caddy, PFC William R. (USMC), 647

    Medal of Honor, 647

Cadillac Deville, lv

California (USA), 1, 35, 88, 143, 144, 554, 569, 669

Caligula (Roman Emperor), 341

Calley, Lieutenant William Jr. (USA), 827 fn 20

Calvo, Governor Eddie (Guam), xliii

Cambodia, 304

Cambridge University (England), xvii, 778

Mount Cameron (Hong Kong), 132

Camp Jacques Farm (California USMC), 143-144

Campbell, Bob, 363

Campbell, Joseph (writer & philosopher), 762

Camus, Albert (writer & philosopher), 766, 770

Canada (Canadian), 24, 132, 316, 542, 694

Candidates' Class (what is today called Officer Candidates Class (OCS)), 402, 411

Canton (China), 131, 202, 719, 769

Caroline Islands, 76, 183

Carolinians, 200, 226

Carry, Corporal (USMC), 248

Carter, Lieutenant James D. (USN), 658, 659

*Cartwheel* (Operation), 146

Cates, General Clifton B. (USMC), 135, 136, 152, 392, 486, 493, 658, 662, 681, 698, 722

    Analysis of German and Japanese soldiers, 708

    No tolerance with Combat-Fatigue cases, 486

Catholic, 58, 73, 124, 175, 341, 488, 547, 549, 550, 733

    Last Rites, 467, 488

Caucasian (ethnicity), lv, 30, 82, 84, 310, 734

    Fights with Blacks in the U.S. Military, 310

CBS Radio, 511

Cedar Park (Texas), 692

Cemeteries, xxi, xxv, 67, 258, 290, 467, 473, 518, 546, 548, 550, 557, 558, 559, 560, 561, 563, 564, 689, 690, 691, 692, 744, 777, 833 fn 109

Centcom, xxiv, xxv, 129, 562, 777

Chaen, Corporal Kunimori (IJA), 496

    *Jôbun*, 496

    *Kanjo*, 496

Chamberlain, Prime Minister Neville, 768

Chambers, 2nd Lieutenant Howard F. (USMC), xvii, xxxvii, 409, 410, 411, 412, 414, 415, 418, 420, 421, 447, 449, 450, 451, 453, 462, 485, 486, 528, 564, 577, 579, 580, 581, 582, 586, 587, 590, 595, 596, 597, 598, 599, 601, 602, 603, 604, 613, 626, 636, 637, 639, 643, 650, 651, 652, 654, 656, 657, 671, 672, 673, 674, 677, 678, 693, 701,703, 707, 724, 725, 726, 727, 734, 750, 754, 755-756, 758, 760, 763, 775, 784, 895 fn 1823

    Purple Heart, 411, 602

    Refused to endorse Woody for the MOH, 411, 602, 755-756

    Silver Star, 411, 450, 602, 643, 657 (see Gallery 3, Photo 14, page 11)

    Wounds suffered on Iwo Jima, 485

Chamorro Elders, 286, 287

Chamorros (native citizens of Guam), 175, 176, 177, 178, 179, 180, 181, 242, 243, 249, 260, 279, 280, 282, 285, 286, 288

Chamorros (living on Saipan), 199, 200, 226, 227

Chapei (China), 512

Chaplains (Naval), 4, 159, 247, 283, 338, 339, 354, 360, 373, 393, 394, 467, 468, 546, 547-551, 559, 570, 698, 777, 780

*Charleston Gazette*, 577, 583, 588, 594, 656, 657

Charlotte (North Carolina), 148

Chateau Thiery (WWI battle), 191, 196, 561

Chemical Warfare (Japanese), 367-369

Chemical Warfare Service (U.S. Army Engineer Test Board), 154

Cherokee (Native American), 637

Cherry Point (NC) (USMC base), 157

Chiang Kai-Shek, *Generalissimo*, l, 77, 95, 105, 316, 318, 694, 832, 880 fn 94, 1354

Chiba Prefectures (Japan), 324

*Chicago Tribune*, 4, 65

Chichi Jima (Japan), 50, 133, 535, 554

Chikao, Yasuno, 117

Chikuhei, Nanjima, 79

China, xx, xxxv, xlvii, xlviii, l, lvii, 4, 23, 31, 35, 71, 72, 75, 76, 77, 78, 83, 88, 89, 92, 93, 95, 105, 116, 118, 124, 133-134, 135, 136, 137, 141, 172, 179, 186, 187, 226, 253, 313, 314, 460, 694, 697, 698, 722, 773, 775

China Marines, l, 55, 136, 252, 253, 295, 698

Chinese War Crime Trials in Nanking, 107, 127-128

Chokoho Incident, 856 fn 630

Chongqing (China), 512

Chonito Cliff (Guam), 253, 257

Chou Kings (China), 83

Christian, PFC (Photographer) (USMC), 363, 433, 439, 516

Christianity, 4, 58, 148, 149, 534, 548
    Japanese persecution of, 49

Christians, 60, 74, 283, 339-340, 341, 767

Christopher, PFC Merrill L. (USMC), 725
    Bronze Star with V, 725

Chungking (China), 93

Churchill, Prime Minister Winston, xix, xli, 8, 31, 88, 191, 316, 697, 716, 717

*Chuyo* (Japanese aircraft carrier), 662

Cirillo, Roger (historian), 613, 677

Citino, Robert (historian), 781

Civil War (American 1861-65), 11, 241, 401, 668

Civil War (China), xlvi, 77-78

Civilian Conservation Corps (CCC), 24, 25, 26, 27, 29, 31, 40, 163

Clancy, Tom (writer), 743

Clevenger (USMC), 266, 272, 747, 875 fn 1184

Clinton, President Bill, 661

Coast Guard, 610, 701

Cocaine (Japanese selling of), 31

Cohen, Philip (USMC), 305

William E. Colby Award, 761

Colby, Jack (USMC), (see Gallery 1, Photo 12, page 6)

Cold War, 122

Cole, Sergeant Darrell S. (USMC), 626, 627, 639, 644
    Medal of Honor, 626, 627, 639, 644-645
    Navy Cross, 639, 644-645

Cole, Margaret, 645

Collier, Peter, 577, 583

Coltrane, John R. (USMC), xli, xlii

Colonial System (American), 733

Colonial Wars (American colonies), 401

Columbia (Louisiana), 196

Columbia Network, 405

*Column of Strength* (memorial in San Francisco for Japanese "Comfort Women"), 120

Comanche (Native American), 172

Combat Fatigue (see also Shell-Shock), 393, 394, 407, 458, 459, 486, 493, 496, 506, 551, 560, 650, 702, 729, 730, 741, 814

"Comfort Women" (Japanese Sex Slaves or *ianfu*), 120, 130, 131, 179, 313, 328, 534, 542, 732, 772, 882 fn 1421

Commonwealth (British), 92, 126, 131, 191, 316, 695, 697

Communism, 30, 77-78, 85, 122, 718, 843 fn 361

Communists, l, 24, 77-78, 126, 141

Concentration Camps (Imperial Japan), 178, 280, 286, 863 fn 837

Concentration Camps (Nazi), 67, 100, 119, 167, 719, 770, 861 fn 782

Confederates (U.S. Civil War), 11, 190, 196, 241, 620

Confederation of A-Bomb and H-Bomb Sufferers (Japan), 122

Confucists, 74, 80, 83

Congress (United States), xlvi, xlviii, 23, 484, 616, 735

Conn Transportation Company (Fairmont, West Virginia), 33

Connecticut (USA), 142

Connellsville (Pennsylvania), 621

Conner, Lieutenant Garlin Murl (USA), 668
    Medal of Honor, 668

Conolly, Read Admiral Richard "Close-in" (USN), 170, 244, 246

Conrad, Joseph (writer), 115, 263

Cook, Haruko (historian), 217

Cook, Theodore (historian), 217

Cooke, Sergeant Bob (USMC), 371

Cookson, PFC Marquys Kenneth (USMC), 415, 652, 674, 705, 751, 753, 924 fn 2697
    Silver Star, 652

Cooper, Gary, 18

Cooper's Rock State Park (West Virginia), 25

Coral Sea (Battle of), 8

Cornelius, Staff Sergeant M. A. (Photographer) (USMC), 466

*Coronet* (Operation), 571

Corpsmen (Naval), 258, 259, 260, 261, 267, 270-271, 374, 378, 417, 503, 505, 586, 622, 636, 651, 658-660, 699, 741

Corregidor (Battle of), 51, 661

Correspondents (Marine or USMC), 398, 399, 400, 892 fn 1756

Corsair Fighter Plane (Vought F4U), 720
    "The Courage Division" (IJA), 133

Corum, James (historian), 661

Crabbe, PFC Ellery B. "Bud" (USMC), 266, 292, 305, 629, 780

Craigie, British Ambassador to Japan Sir Robert, 91

Creech, Chaplain D. G. (USN), 777

*Croix de Guerre*, 191

Crosby, Bing (singer), 22

Crowl, Philip A. (historian--USMC), xxxiii, 197, 452, 653, 699, 773

Crowley, Professor James B. (Yale University), xvii, 717, 777

Cruiser Division Twelve, 609

Crusades, 767

Cuba, 193

Culver Military Academy, 402

Cunningham, Captain Winfield S. (USN), 668

Curtis, Alan (actor), 537

Curtis, Joe (USMC), 572

Cushman, Lieutenant Colonel and later 25th USMC Commandant Robert E. Jr., xlviii, l, 271, 273, 274, 308, 378, 871-872 fn 1083, 904 fn 2081, 917 fn 2502
    Navy Cross, 273

Cuthriell, Chaplain Warren (IJN), 546, 549

Czechoslovakia, 23

D-Day (Iwo Jima), 357, 362, 374, 375

D-Day (Normandy), 170, 301-302

D-Day (Saipan), 170, 199

Dachau Death Train, 167

Dachau Nazi Concentration Camp, 167, 719

Dahmer, Jeffrey (serial killer), 128

*Daihatsu* amphibious landing craft (Japanese), 317

Dallas (Texas), xliii

Daly, First Sergeant Dan (USMC), 663
    Medals of Honor, 663
    Navy Cross, 663

Dante (Durante degli Alighieri), 67, 391, 420, 472

Daqun, Honorable Justice Liu (Judge of the Appeals Chamber of the International Criminal Tribunals for the Former Yugoslavia and Rwanda Genocide Trials), 137-138

Darwin (Australia), 90

Dashiell, Dorothy, 394-395

Dashiell, Fred (Freddy), 148, 593, 917 fn 2506 (see Gallery 3, Photo 30, page 15)

Dashiell, Sergeant Frederick K. "Dick" (USMC), 55, 148, 245, 328, 329, 332, 385, 394, 395, 398, 399, 400, 401, 402, 403, 411-412, 415, 417, 433, 448, 449, 455, 463, 464, 484, 491, 526, 527, 533, 557, 560, 568, 577, 580, 581, 582, 583, 588, 589-594, 614, 616, 630, 631, 657, 672, 674, 676, 677, 701, 742, 748, 749, 754, 755
    Article about Woody, May 1945, 589-593
    Doubted Woody's story, 402, 589, 593, 917 fn 2506
    Letter of Commendation, 399, 754 (see Gallery 3, Photo 12, page 9)
    Nicknames, 400

Dashiell, James, 401

Dashiell, John, 401

Dashiell, Professor John Frederick, 394-395

Dashiell, Rev. John Thomas, 401

Dashiell, Robert Brooke, 402

Dashiell, Vivian, 148

*Daughters of World War II*, xlii

Davis, Liselotte M. (Yale University), xvii

Day of Commemorating the End of the War (Japanese event), 122

DDT (*dichlorodiphenyltrichloroethane*), 467, 477, 536, 558, 559, 902 fn 2010

De Ome, Corporal F. E., 361, 410, 494

Declaration of the United Nations (1942), 140

Democratic Party (U.S.), 694

Demolitions (Demolition Men), 150, 291, 392, 415, 417, 418, 423, 442, 506, 529, 541, 593, 652, 704, 705, 706, 747

Denfeld, Vice Admiral Louis E. (USN), 623, 661

Desertion, 729

Destroyers for Bases, 694

*Detachment* (Operation), 338

Deuteronomy (Bible), 18

Devereux, Colonel James P.S. (USMC), 668

The Diet (Japanese), 78, 735

Dietrich, Captain Neil K. (USN), 611, 612

DiMaggio, Joe (baseball player), 519

Dimitrowski (Russia),

*Dinah Might* (B-29), 488, 489

Distinguished Service Medal, 51, 196, 356, 562, 608, 635

Divine, Robert A. (historian), 26

Divine Wind Attack Corps, see *Kamikaze*s.

Division of Operations and Training (USMC), 194

Dmitrowski (Russia), 638

Dobbins, Bill (USMC), 373

Dodds, PFC R. R. (USMC), 436

Dogs (for military use), see War Dogs

Doi, Colonel Teihichi (IJA), 317

Doihara, Lieutenant General Kenji (*aka* "Lawrence of Manchuria") (IJA), 30

Dominican Republic, 190-191

Domitian (Roman Emperor), 341

Donatich, John (Yale University Press), 827 fn 20
    Refuses to support authors and the freedom of speech, 827 fn 20

Doolittle Raid (1942), 8, 120, 832 fn 94

Dorman, Douglas (USMC), 690

Dorsey, Jimmy (singer), 462

Dorsey, Tommy (singer), 462

Doss, Medic Desmond (USA), 284, 757
    Medal of Honor, 284, 757

Douglas-Mansfield Act of 1952, 738

Dowell, Private Cam (USA), 519

Dower, John W. (historian), 122, 277

*Downfall* (Operation), 571, 617

*Dragoon* (Operation), 302

Drake, Mack (USMC), 266

Drake University, 580

Drea, Edward (historian), 168, 697

Dreyfuss (photographer) (USMC), 366

Drill Instructors (*aka* DI--Marine Corps), 18, 40, 41, 42, 44, 45, 47, 54, 57, 58, 426, 665, 837 fn 206

Driscoll, Mark (historian), 777

Duenas, Father Jesus Baza, 181

DUKWs (2½ ton amphibious trucks, called "ducks"), 362

Dunford, General Joseph (USMC), xliii, 691, (see Gallery 1, Photo 5, page 2)

Dunkirk (Battle of), 23

Dust Bowl (U.S. Depression), 17, 24

Dutch East Indies (Indonesia), lii, 35, 89, 92, 205

E. T. Vincent Cemetery (Quiet Dell, West Virginia), 17

Eastern Kentucky Veteran Cemetery, 692

Eastwood, Clint (actor and director), xxx, xliv, 133, 701

Eben Emael (Belgium WWII fort), 154

Eden, British Foreign Secretary Anthony, 126

Ecclesiastes (Bible), 679

Edgehill (Battle of),

Edson, Brigadier General Mike (USMC), 197, 335, 637, 643, 649
    Medal of Honor, 637, 649

Edson's Ridge (battle on Guadalcanal), 145

Eftang, PFC George F. (USMC), 283

Eisenhower, General of the Army Dwight D. (USA), 208, 661-662, 720

Elder, First Sergeant William R. "Hose Nose" (USMC), 414, 415, 417, 446, 485, 568, 579, 618, 754
    Purple Heart, 415
    St. Elizabeth's (Washington D.C.), 665

Elliot, Billy, xxxii

Camp Elliott (USMC), 147
Ellis, Major Earl H. (USMC), 76-77
Ellis, Harvey (USMC), 645
Emerson, Ralph Waldo (transcendentalist), 341
Emperor (Japanese), 49, 58, 68-75, 79-81, 83-84, 94, 122, 137-138,189, 203, 209-211, 225, 254-255, 276, 297, 304, 311, 318-320, 325, 335, 341-342, 389, 390, 396, 452-453, 496, 515, 519, 521-523, 542, 544-545, 551, 562, 572, 574, 683, 703-704, 709-710, 717, 841 fn 311, 843 fn 365
Emperor's Tokyo Division, See Imperial Guards
*Encyclical* (Catholic), 73, 548
*Encyclopedia of Genocide*, 118
Endo, Major General Saburo (IJA), 298
England, xlix, 4, 31, 118
Eniwetok Atoll, 39, 245
*Enola Gay* (B-29), 240, 555, 713
Ephesians (Bible), 339
Erskine, Lieutenant General Graves Blanchard "Blood and Guts" (USMC), xxiii, xxxvii, xxxix, 30, 195, 196, 197, 200, 202, 219, 224, 229, 276, 306, 307, 308, 323, 357, 358, 360, 378, 392, 399, 401,451, 453, 462, 463, 481, 482, 483, 484, 497, 498, 511, 517, 530-531, 532, 533, 535, 540, 551, 561, 562, 563, 564, 588, 595, 603, 618, 619, 629, 642, 643, 682, 693, 701, 706, 707, 708, 754, 763, 775, 784, 865 fn 897
    Analysis of German and Japanese soldiers, 708
    Chief-of-Staff of Amphibious Warfare, 197, 323
    Comments on Flamethrowers, 150
    Creator of "Amphibious Blitzkrieg," 195-201
    Dedication of the 3d MarDiv Cemetery, 561
    Developer of Amphibious Warfare Tactics, 195, 196
    Distinguished Service Medal, 562-563
    Divisional Board of Awards, 588, 595
    Educational Programs for his Division, 629-630
    Fights with Turner, 358, 359
    Legion of Merit, 197
    Plans the invasion of Saipan, 199
    Plans the invasion of Tinian, 229
    Praise for his Marines, 451
    Purple Heart Medal, 196, 561
    Responsibility for the main thrust through Iwo Jima, 482-483, 484
    Shaping the battle on Saipan, 210
    Silver Star Medal, 196, 561
    View on killing the Japanese, 276
Europe, 208, 345, 567, 661, 685, 694, 695, 697, 712, 733, 768
Executions of enemy soldiers by U.S. Servicemen, 167, 221, 235, 275
Exeter Academy, see Phillips Exeter Academy
Exodus (Bible), 18, 161
Fagan, Captain Francis (USMC), 641, 642, 643, 674, 761
    Medal of Honor process, 641-643
    Navy Crosses, 641-643
Fairmont (West Virginia), 16, 34, 621, 628, 689, 691
Fairmont Army Reserve Training Center, 692

Fairmont Hotel (San Francisco), 179
Farley, James Oelke (Park Ranger & historian), lvi, 784
Farnum (Photographer) (USMC), 439
Farrell, Don (historian), 779
Faulkner, Bill (USMC), 164
FDR, see Franklin Delano Roosevelt
Feast of the Immaculate Conception, 176
Fellers, Brigadier General Bonner (USA), 701
Ferneyhough, (Photographer) (USMC), 522
*Fidelis Historia*, xxxii
Field Sanitation, 369, 465, 466, 467, 468, 469, 471, 472, 473, 477, 558
Filip, Private Robert C. (USMC), 34, 646, 648, 650
    Navy Cross, 646, 648
Filipinos, 89
Finch, Barbara (journalist), 487
Finland, 4
First Amendment (U.S. Constitution), xxxi, xxxiii, 764-765
First Sino-Japanese War, see Sino-Japanese War (First)
Fischer, PFC Charles (USMC), 390-391, 443, 444, 506, 569, 570, 579, 581, 626, 724, 726, 784, 898-899 fn 1912, 899 fn 1913
Fisher, Radio Operator David (USAAF), 673
*Flags of Our Fathers* (2006), 701
Flamethrower (portable—man carried), 149-150, 164, 210, 277, 279, 291, 305, 385, 392, 414, 417, 418, 423, 424, 425, 426, 427, 428, 429, 430, 432-449, 506, 507, 511, 571-572, 586, 614, 652, 653, 654, 703, 704, 705, 706, 707, 708, 720, 721, 724, 747, 897 fn 1905 (see Gallery 2, Photo 1, page 1; Gallery 3, Photo 15, page 12)
    American development, 154-155
    Distributions within battalions and regiments, 414
    Fatal attributes besides fire of flamethrowers, 443-444
    German development, 150, 154
    High casualties of, 151
    Japanese use of, 150
    Nomenclature and specifications, 150-152
    Production numbers during WWII, 706
    Tactics, 152-154
Flamethrower (cannons), 251
Flamethrower Tanks (*Zippo* or *Satan* Tanks), 375, 476, 477, 507, 703 see Tanks
Fleet Marine Force, Pacific (FMFPAC), 355, 356, 525, 595, 596, 597, 598, 599, 600, 601, 602, 607, 619, 639, 640, 697, 779
Fletcher, Admiral Frank (USN), 8, 206, 610
Flies (on the battlefield), 369, 466, 467, 489, 521
Flores, Palacios, 280
Flying Tigers, 89
Fonte Hill (Guam), 262 (see Gallery 2, Photos 16, 22, pages 11, 15)
Football (American), 400, 401, 410, 411, 666
*Forager* (Operation), 167, 170, 183, 721, 871 fn 1056
Ford Model A, 15
Ford Model T, 15

Formosa (Taiwan), 75, 88

Forrestal, Secretary of the Navy James, 378, 405, 489, 562, 595, 596, 597, 601, 602-603, 604, 605, 610, 611, 612, 615, 636, 651, 670, 672
  Highest authority for medals for the U.S. Navy and United States Marine Corps, 604-605

Fort Bliss (Texas), 313

Fort Eustis (Virginia), 579, 602, 671

Fort Knox, 680

Fort Worth Christian School (Fort Worth, Texas) (FWC), xvii

Fox, D. (Photographer) (USMC), 469

France, 4, 23, 24, 44, 191, 194, 302

Francis, Sergeant R. (USMC), 427

Francois, Professor and Journalist Bill (also WWII paratrooper veteran) (USA), 442, 447, 577, 580, 583, 588, 653, 748, 754, 755, 916 fn 2481

Frank, Benis (historian), 399

Frank, Richard (historian), xvii, xxx, xxxii, 420, 553, 577, 715, 718, 779

Frankel, Staff Sergeant Bernice (USMC), 734

Frankl, Viktor, 256, 460, 728, 740, 741

Franklin, Benjamin, 764

Franklin, Staff Sergeant Joe (USMC), 412

Freedom of Speech, xxxii, xxxiii, 90, 560, 737, 765, 773, 811, 827 fn 20, 914 fn 2410

French, 733

French Indochina (Cambodia, Laos and Vietnam),

French Revolution, 71

Frey, Coach Edward (Exeter), xvii

Friendly Fire (fratricide), 261, 408, 465

Frogmen, see Underwater Demolition Teams

Fuji, Major Hajime (IJA), 303

Fussell, Paul (USA), 715

*Gaizin (Gaijin)*, 84, 85, 773, 844 fn 397

Gallipoli (WWI battle), 191

Galloway, Staff Sergeant J. F. (Photographer) (USMC), 441

Garapan (Saipan's capital), 201, 202 (see Gallery 2, Photo 7, page 4)

Gardner, Troy (USMC), 645

Garnett, Senior Corpsman Pharmacist's Mate First Class (USN), 260, 261, 270, 271, 272, 273, 291
  Bronze Star with V, 271

Gas Warfare (Japanese), 367, 368, 637, 722, 723

Gavutu-Tanambogo, (Battle of), 47

Gehrig, Lou (baseball player), 519

Geiger, Lieutenant General Roy S. (USMC), xxxvii, 170, 243, 251, 253, 285, 461, 479, 480, 481, 586, 587, 595, 596, 597, 598, 599, 600, 601, 602, 602-603, 604, 605, 606, 609, 610, 612, 613, 614, 616, 619, 641, 642, 644, 647, 651, 653, 657, 672, 678, 747, 756
  Distinguished Service Medal, 285
  Refuses to endorse Woody's MOH, 595, 596, 598, 599, 600, 601, 602, 603, 604, 606, 612, 613, 614, 616, 748

Geishas, 534

Gelernter, Josh (journalist), 774

Gendarmerie (Japanese), 136

General Agreement on Tariffs and Trade (GATT), 733

General Staff Officer (IJA), 203

Genesis (Bible), 19, 473

Geneva Convention (1929), 48, 166, 367, 531, 829 fn 51

Genocide, 138, 733, 763, 767

*Genroku Akō*, 521

German Military, 35 see *Wehrmacht*

Germans, 4, 44, 60, 68, 119, 137, 138, 142, 150, 172, 184, 186, 191, 202, 302, 561, 563, 576, 577, 637, 638, 663, 668, 685, 689, 733
  Fighting ability compared to the Japanese, 708

Germany, lvii, 4, 9, 14, 23, 29, 68, 76, 86, 88, 118, 119-120, 121, 159, 199, 302, 547, 567, 573, 574, 576, 694, 696, 697, 706, 721, 768
  East Germany, 718
  West Germany, 718

Germany's Supreme Army Command (*Oberkommando des Heeres*), 78

Gettysburg (Civil War Battle), 379, 473

Ghettos (Nazi), 769

GI Bill, 630

Gibson, Hoot (actor), 22

Gilbert Islands, 146, 298, 394

Gillon, Steven M. (historian), 671

Ginling College (China), 111

Ginther, James (USMC archivist), 782

Gittelsohn, Rabbi Roland (Naval Chaplain), 4, 159, 338, 339, 354, 355, 393, 394, 467, 546, 547, 548, 549, 550, 560, 561, 564, 702, 780, 836 fn 173 (see Gallery 3, Photo 18, page 14)
  Dealing with racist priests and Protestant ministers (Chaplains), 548-550
  Famous speech, "The Purest Democracy," 546-547, 548, 550, 559
  Protestant Chaplains defend Gittlesohn, 547-550

Global Combatting Terrorism Network US Special Operations Command, 129

God (Judeo-Christian), 4, 58, 231, 251, 287, 339, 341, 354, 355, 391, 444, 489, 570, 687, 699, 713, 769, 770
  Intervening to prevent the Holocaust 769
  The role of God in human affairs, 770

Gods (Japanese), 58, 319, 320, 341, 408, 490, 527, 708, 717, 767

Goebbels, Nazi Propaganda Minister Joseph, 82

*Gokashô*, 325

*The Gold Star Families Memorial Foundation*, xxxi, 690, 726, 742 (see Gallery 4, Photo 7, page 4)
  History of the Gold Star, 690

*Golden Girls*, 734

Golden Rule, 19

Golden Gate Bridge (California), 159

Gonorrhea, 54

Good Samaritan (The New Testament), 19

Goodman, Benny (jazz musician), 462

Graham, Ed, xli, 487 (see Gallery 1, Photo 12, page 6)

*The Grapes of Wrath* (1939), 24

"Grasshopper" Observation Planes, 486

Graves, Private Donald "Don" (USMC), 34, 423, 495, 496, 497, 498, 501, 502, 533, 582, 583, 586, 587, 594, 652, 672, 745, 753, 761, 780, 836 fn 173, 895 fn 1815, 938 fn 3080 (see Gallery 3, Photo 19, page 14)

  Problems with Woody Williams' narrative, 501-502, 582, 583, 586, 594, 672, 753

Gray, 29th USMC Commandant and General Alfred M. "Al" Jr., xix, xxii, xxiii, xxxiv, 538, 778

Gray, Sergeant Ross Franklin (USMC), 657-658

  Medal of Honor, 657-658

Great Britain, xlvii, xlviii, xlix, lii, 4, 24, 76, 88, 89, 91, 186, 694, 695

Great Depression, xlvii, 13, 14, 17, 19, 23, 24, 163, 401, 567

  Unemployment, 13

*Great Duty* (*Taigi*), 80

"The Great Marianas Turkey Shoot," xxxvi, 205, 207-208, 609

The Great War, see World War I (WWI)

The Greater East Asia Co-Prosperity Sphere" (*Dai Tō Kyōei Ken*), 80, 109

Green Beach (Guam), 252, 254,

Gregory (South Dakota), 163

Grew, Ambassador Joseph C., 161

Griggs, Corporal Billie (USMC), xli, 497, 498, 702, (see Gallery 1, Photo 12, page 6)

  Letter of Commendation, 497, 498

Groves, Major General Leslie (USA), 555

Guadalcanal (Battle of and island), 8, 47, 50, 51, 53, 54, 145, 146, 147, 148, 149, 150, 158, 163-164, 167, 183, 189, 190, 203, 242, 297, 357, 376, 477, 506, 563, 606, 608, 643, 694, 698, 707, 720, 724, 747

*Guale* (people of Guadalcanal), 148-149

Guam (Battle of or Liberation of), xxii, xxxvi, xxxix, liv, lvi, 39, 147, 169, 170, 229, 241-285, 287, 288, 289, 290, 291, 323, 336, 337, 338, 342, 343, 348, 352, 383, 402, 402-403, 411, 431, 486, 491, 499, 506, 563, 564, 570, 609, 650, 695, 698, 703, 721, 725, 765, 779, 780, 872 fn 1089

  Casualties, 290

  Economic cost of conquering Guam, 287

  Green Beach, 252, 254

  Korean Forced Laborers, 241

  Marines help civilians, 292

  Reunion 2015, xli, lvi, 779

  Reunion 2018, 655

  Seabees, 721

  Size of Japanese garrison, 241

Guam (island), xxxv, xliii, lii, 35, 49, 89, 90, 116, 118, 138, 158, 171, 175, 176, 177, 179, 180, 181, 182, 190, 199, 200, 206, 207, 208, 223, 225, 230, 295, 296, 297, 299, 301, 305, 307, 309, 310, 489, 587, 628, 629, 712, 744, 748, 754, 757, 760

  Number of deaths of citizens while under Japanese rule, 182

Guam Pacific Battlefield National Park, lvi

Guangdong Province (China), 129, 130, 131, 133

Guangzhou (China—also known as Canton), 131, 512

Guderian, General Heinz (German), 477

Gudmundsson, Major Brian (USMC), 778

Gypsies, 141

*Gyokusai* (honorable death/defeat), 209, 707-708

*Hachiman Daimyōjin* (one of the Shinto gods of war), 341

*Hacksaw Ridge* (2016 movie), 284

*Hagakure* (a popular *Bushido* text), 78

Hagåtña (Guam's capital), 177

Hague Convention (1907), 97, 98, 166, 829 fn 51

Haha Jima (Japan), 555

Haiti, 193, 196, 253

Hallas, James (journalist), 580, 643

Hallowell, Ned (Harvard University), 740

Halsey, Fleet Admiral William "Bull" (USN), 1, 8, 146, 147, 303, 607, 677, 725

Halwanger (USMC), 296

Hamas, Captain John (USMC), 667

Hamilton, Hanse (USMC), 780

Hammel, Eric (amateur historian), 704

Hammond, Bruce, (see Gallery 1, Photo 10, page 4)

Hammond, Ivan, (see Gallery 1, Photo 12, page 6)

Hankow (China), 719

Hanmou, General Yu (Chinese Nationalist Army), 316

Hane, Mikiso (historian), 183

Hanley, Flight Engineer Fiske (USAAF), xli, 289, 345, 511, 620 (see Gallery 1, Photo 4, page 2)

*Hand Grenade Hill* (Iwo Jima), 456

*Hannibal Lecters* (from the 1991 horror film *Silence of the Lambs*), 115

*Hara-Saku* (Operation), 316

Harada, Zen Master Sōgaku Daiun, 83, 574, 575

Harboard, Brigadier General James (USA), 191

Hardesty, Senator C. Howard (WV), 628

*Hari-kari* (Japanese ritual suicide), 138

Harper, Geoffrey (J.D.), xxxiii

Harrell, Sergeant William G. (USMC), 623

Harris, PFC Albert J. (USMC), 207

Harris, Sam (writer and neuroscientist), 342, 769

Harrison, Joyce (writer), xxxii

Harvard University, 313, 670, 740

Harvey, Corporal T. Fred (USMC), 172, 641-642, (see Gallery 1, Photo 2, page 1)

Hashimoto, Japanese pseudo-scientist, 79, 81

Hashimoto, Osaka Major Tōru, 120

Hassan, Thomas, xvii

Hattori, Colonel Takushiro (IJA), 203

  Operations Section Chief of the Senior Officers' Section of the IJA's General Staff Office, 203

Hawaii (United States), xliii, lii, 2, 8, 29, 55, 92, 176, 183, 203, 292, 564, 601, 619, 620

Hawkins, Colonel Mant (USMC), 129

Hayashi, Hirofumi (historian), 119

Hayauchi, Captain Masao (IJN), 452, 453

  *Jōbun*, 452-453

  *Kanjo*, 452-453

Hayden, Coast Guard Correspondent Vic (USCG), 339

Hayler, Rear Admiral Robert Ward (USN), 603-612, 640, 642, 644, 645, 651, 668, 678, 748, 749
    Double awarding of medals, 607
    Failure with U.S. torpedo development, 605-606
    Highest authority in the military besides the Secretary of the Navy to grant valorous awards, 604, 605, 609, 610
    Inspector of Ordnance (Naval Torpedo Station), 605
    Navy Commendation Medal, 605
    Navy Crosses, 603, 606, 607
    Silver Star, 607
Haynes, Bonnie, xxxi, 825 fn 5
Haynes, Major General Fred (USMC), 825 fn 5
Hays, Emmitt (USMC), 275
*Heart of Darkness*, 115
Heaton, Colin (historian), 780
"Hell Ships," 178
Hellcat fighter plane (F6F), 207, 405
Helldivers, 208
Helsely, George Jacob Jr., 11
Helton, Sergeant and Flamethrower Wayne B. (USMC), 425
Hemphill, Gunnery Sergeant Albert D. (USMC), 269, 278, 291, 295, 387, 494, 674, 701, 784
    Purple Heart, 278
    Silver Star, 278
Henderson Field (Guadalcanal), 164, 376
Hengel, Marian (Lefty's girlfriend), 463
Henning, Second Lieutenant Charles "Mike" "Ox" (USMC), 412-413, 450, 480, 481, 485, 517, 557, 579, 615, 616, 674, 742, 754
    Silver Star, 413, 480-481, 615
*Henry V* (Shakespeare), 520, 743
Hensely, Robert (USMC), 469
Hepburn, Admiral Arthur J. (USN), 610
Hermes Typewriter, 399
Herring, Lieutenant Junior Grade Rufus (USN), 348, 349
    Medal of Honor, 348, 349
*Hey Rookie* (1944 movie), 537
Heydrich, *SS*-General Reinhard, xxxv
Higgins, Andrew Jackson, 192
Higgins Boat (LCVP or Landing Craft, Vehicle, Personnel), 7, 171, 190, 192, 193, 195, 197, 247, 248, 251, 252, 253, 291, 308, 362, 363, 373, 384, 390-391, 549, 557, 720, 778
    General of the Army Eisenhower's assessment of its importance during WWII, 720
    General Smith's assessment of its importance during WWII, 720
Hill 283 (Iwo Jima), 457, 458
Hill 331 (Iwo Jima), 505
Hill 362A (Iwo Jima), 583, 900 fn 1962
Hill 362B (Iwo Jima), 900 fn 1962
Hill 362C (Iwo Jima), 457, 500, 505, 506, 509, 900 fn 1962
Hill 382 (Iwo Jima), 457, 627
Himmler, Head of the SS Heinrich, xxxv
Hiranuma, Japanese Prime Minister, 78

Hirohito, Japanese Emperor (*Shōwa* Emperor), xxiii, xxxviii, xlvii, 3, 4, 31, 68, 69, 74, 82, 87, 90, 91, 94, 101, 105, 116, 118, 119, 121, 132, 134, 136, 137, 138, 185, 186, 190, 195, 205, 206, 209, 225, 255, 299, 303, 310, 314, 318, 325, 338, 508, 556, 569, 576, 694, 697, 703, 707, 709, 710, 711, 712, 722, 723, 732, 763, 767, 768, 769, 774, 775, 841 fn 306, 934 fn 2964
    Absolute ruler, 68-70, 74
    Atomic bombs convince him to surrender, 709-713, 717
    Buddhist Monarch, 69
    Declaration of War against the US, 93-94, 185
    Gas warfare, 369
    High Shinto Priest, 69
    Human God, 72, 83, 341, 712, 769
    Imperial rescripts (*Tokuhō*), 73, 93-94
    Lack of remorse for WWII atrocities and crimes, 121-122
    Massive grave-monument to honor Hirohito, (see Gallery 1, Photo 27, page 13)
    Military ranks, 88
    Military orders and power, 68, 72, 77, 88, 841 fn 311
    Racism, 77
    Radicalizing of, 77
    Religious Upbringing, 77
    Renounces his godhood, 703, 717
    Saipan, 203
    Study habits, 70
    War atrocities, 77, 116, 142, 369
Hirota, Prime Minister Kōki, 137
Hiroshima (Japan), 120, 488, 684, 712, 713, 715
    Estimated Casualties from the atomic bomb dropped there, 120, 713
Hispanic, 733
Hitchens, Christopher (journalist), 47
Hitler, Adolf (*Der Führer*), xxxv, xlvii, xlviii, l, lii, 4, 9, 14, 24, 31, 35, 68, 70, 81, 82, 91, 110, 118, 121, 136-138, 186, 195, 299, 302, 314, 548, 567, 573, 694, 696, 697, 703, 718, 722, 732, 767, 768, 769, 831 fn 78, 843 fn 361, 922 fn 2641
    Declaration of War against the US, lii, 4, 5
    Fortress Europe, 208
    Hitler was a Catholic, 548
    *Mein Kampf*, 81, 85, 548, 768
    View of Christianity, 81
    View of Jesus, 81
    Views on the Japanese, 85
*Hitler's Jewish Soldiers*, 761
*HMS Bee* (gunboat), xlix
*HMS, Cricket* (gunboat), xlix
*HMS Ladybird* (gunboat), xlix
*HMS Prince of Wales* (battleship), 88
*HMS Repulse* (battlecruiser), 88
*HMS Scarab* (gunboat), xlix
Hoffman, Colonel Jon T. Hoffman (historian) (USMC), xvii, xxx, xxxii, xxxiv, 34, 35, 594, 613, 636-637, 762, 782, 922 fn 2641

Hogaboom, Colonel Robert "Bobby" (USMC), 619
Hoge, Corporal Marlin (USMC), 471
Holiday, Billie, 463
Holiday, Samuel Tom "Sam," xli, xlii, 367, 734 (see Gallery 1, Photo 8, page 3)
Hollaender, Colonel Walter (German), 638, 639, 761, 922 fn 2641
    Ritterkreuz, 638, 639
Hollywood, 651
Holmes, Wilfred J. "Jasper" (USN), 693
Holocaust (Japanese mass murder in Asia and the Pacific), See Japan's Holocaust
Holocaust (Nazi Extermination of the Jews), 9, 11, 119-120, 134, 142, 167, 769, 780
Holocaust Memorial (Berlin, Germany), 67
Holocaust Museum (Washington D.C.), 67
Homer (writer), 743
Homosexuality, 54, 141, 767, 839 fn 261
Hong, Lieutenant General Sa-ick (IJA), 242
Hong Kong, 89, 124-133, 135, 150, 217, 315, 316, 317, 534, 618, 638, 667, 719, 772, 775, 800, 880 fn 1352, 1354
Hong Kong Island, 126, 127, 317, 318
Hong Kong POW camps, 129, 131, 217
Hong Kong Red Cross Hospital, 126
Honk Kong War Trials, 128, 129
Honolulu (Hawaii), 244, 601
Hoolihan, Private Paul B. (USMC), 568, 577, 580, 583, 748, 749
Hopei Medical College (China), 114
Horie, Major Yoshitaka (IJA), 322
Horii, Major General Tomitarō (IJA), 177
Horses (for military use), 314, 880 fn 1343
Hōshō (Japanese air craft carrier), 77
Hotaling, Chaplain E. Gage (USN), 468
Hotel Hill (Oak Hill, West Virginia), 631
Hough, Major Frank (USMC), 274
House of Commons (British), 126
Howard, A. F. (USMC) (Deputy Commander to Lt. General Geiger), 598, 600
Hudson, Colonel Lewis C. (USMC), 662-663
Hull, Secretary of State Cordell, 6, 92
Hulitzsky (USMC), 414
Hume, David (Scottish Philosopher), cover page
Humor (Battlefield), 474, 475
Hungary, 29
The Hunt for Red October, 743
Huntington (West Virginia), 689
Hyaline Membrane Disease, 12
Hyman, Paula E. (Yale University), xvii
Hytken, Captain Frank (USMC JAG officer), 779
Ichigaya (Japan), 203
Ichimaru, Rear Admiral Tochinosuke (IJN), 321, 528
Idaho, 4
Ienaga, Saburō (historian), 31, 118, 712
Igata, Corporal Takayoshi (IJA), 200
Ikeda, Colonel Masuo (IJA), 455, 530-531, 532
Imperial General Headquarters (Japan), 321
Imperial Guards, 318

Imperial Headquarters, 116, 134, 189, 297, 298, 321, 346, 534, 535, 542, 544
    Refuses to follow international law, 134
Imperial Japanese Army (IJA), xxii, xxix, 103, 104, 109, 115, 132-133, 160, 843 fn 365
    47mm mobile cannon, 452
    Admission of crimes, 133-134
    Amphibious operations, 317, 318
    Army regulations, 77
    Attempts to curtail atrocities, 315
    Atrocities, 79, 86, 95, 100, 101, 102, 103, 104, 114, 115, 123-132
    Baby Killers, 109, 110, 112
    Beheadings, 89, 90, 94, 103, 104, 110, 111, 117, 180, 181, 182
    Bombings of cities and civilians, 512
    Brutalization of its troops, 84
    Burning of cities, 95
    Bushido, 78
    Cannibalism, 115
    Casualties in China, 694
    Comfort Women, 120, 130-131, 179
    Conflicts with the Navy, 698
    Destructions of Chinese farms, 109
    Economic Conquest, 89, 135
    Fatality Rate, 139-140
    Fighting capability compared to the Germans, 708
    Killing of Civilians, 95-101, 102, 104, 109, 110, 111, 112, 114, 125, 128, 134, 176, 177, 178, 179, 180, 181, 279, 280, 282, 285
    Killing of POWs, 95-101, 113, 115, 116, 124, 126, 129, 131, 133-134, 141, 178, 667
    Killing their own citizens, 213, 214, 239
    Killing their own soldiers, 138-139, 515-516, 520
    Korean Laborers, 133
    Lack of flexible command, 699
    Looting, 127, 134
    Marianas strategy, 203
    Night attacks, 462
    Pedophiles, 108, 120
    Policies on surrendering, 138
    Problems with alcohol, 264, 273, 274, 533
    Problems with supply, 331
    Psychology on fighting, 277
    Rapes, 95, 99, 106, 108-109, 110, 113, 114, 124, 125, 133-134, 178-179 see Rapes
    Sodomizing Chinese Men and Teenage Boys (Male Rape), 110.
    Soldiers committing suicide, 138-139, 271, 274, 276, 283, 289, 408, 515-516 see also Jiketsu
    Spiritual duties, 80
    Use of chemical/gas warfare, 368, 369
    Use of horses, 314
    View on orders and obedience, 78
    War Crimes, 134-135, 142

War in China, 77, 92, 94-95, 109, 133, 134, 135, 136, 141, 314

Imperial Japanese Navy (IJN), xlix, lii, 50, 53, 76, 169, 205, 239, 302, 304, 322, 331, 462
    Atrocities, xlix, 48-49, 179
    Conflicts with the army, 698
    Fights with the IJA, 542, 699
    Mariana Islands, 205, 206, 207
    Pearl Harbor, lii

Imperial Rescript (*Tokuhō*), 73, 78, 84, 93, 94, 712, 717, 844 fn 396

Imperial War Ministry, 134, 314, 315

Imphal, 208

Inchon Landing (Korea), 737

India, 842 fn 358

Indian (Commonwealth troops under the British), 316

Indianapolis (Indiana), 592

Indochina, 35, 304, 846 fn 419

Indonesia (Dutch East Indies), 89

Ink Spots (music group), 462

Inoguchi, Captain Rikibei (IJN), 304

Inouye, Captain Samaji (IJN), 509

International Criminal Tribunals for the Former Yugoslavia and Rwanda Genocide Trials, 138

International Law, 133

International Monetary Fund, 733

International Date Line, 176

International Safety Zone (Nanking), 101, 110, 111, 126

Internment Camps (Saipan), 226-227

Iraq, 702

Isaiah (Bible), 409, 732

Isely, Jeter A. (historian), xxxiii, 197, 452, 653, 699, 773

Ishii, General Shirō (IJA),

ISIS (Islamic Fascists), 769

Islamic Fascists, 763, 769

Isolationists, l, li. 3, 4, 19, 24, 694, 721

Israel, 67

Israelites (Bible), 767

Itagaki, Head of War Ministry and Lieutenant General Seishirō (IJA), 134
    Admission that all IJA troops were guilty of rape and murder, 134

Italian Armed Forces, 184

Italians, 119, 186, 733

Italy, xlvii, 4, 5, 23, 29, 31, 88, 194, 547, 695

Ito, Takeo (Commander of the 28th Infantry Unit IJA), 129

Iwabuchi, Rear Admiral Sanji (IJN), 518, 519, 540

Iwasaki, Ryuji (Deput Consul for Houston, TX), 121, 780

Iwo Jima (Battle of), xx-xxi, xxii, xxiv, xxxv, xxxvi, liv, lvi, 7, 133, 165, 299, 310, 315, 319, 320, 321, 322, 323, 324, 331, 342-343, 344, 350, 352, 353, 355, 356, 357, 358, 359, 360-410-445, 486, 489-556, 573, 610, 655, 698, 703, 707, 725, 735, 743, 744, 765, 775 (see Gallery 1, Photo 9, page 4; Gallery 2, Photo 24, page 16; Gallery 3, Photo 11, page 8)
    1968 American-Japanese Agreement, xxx, xliv
    70th Commemoration (2015), xli, liv, lv
    Airfield No. 1, 375, 376, 392, 393, 397, 409, 455, 456, 473, 488, 500, 561 (see Gallery 3, Photo 11, page 8)
    Airfield No. 2, 396, 463, 473, 474, 477, 478, 484, 537, 641 (see Gallery 3, Photo 11, page 8)
    Airfield No. 3, 484 (see Gallery 3, Photo 11, page 8)
    American bombardment of, 345
    Artillery tactics, 500, 528, 906 fn 2152
    Casualties, xxi, 379, 463, 487, 491, 492, 493, 494, 510, 635, 702
    Controversy about pre-landing bombardment, 350-351
    DNA Testing of Missing-in-Action Marines are not being conducted by the Japanese Government, xxx-xxxi
    Flag raising of "Old Glory" on Mt. Suribachi, 404, 405, 406
    Flight crews rescued by the seizing of the island, 488, 489
    *Kamikaze*s launched from, 347
    Japanese bombers attack Marine lines, 464
    Japanese defensives positions (Map), (see Gallery 2, Photo 24, page 16)
    Japanese infiltration tactics, 379, 392, 462
    Japanese Soldiers' Fighting Song, 330
    Japanese manpower, 329 (see Gallery 2, Photo 24, page 16)
    Justification for taking the island, 553-556
    Landing beaches, 364, 365 (see Gallery 3, Photos 4-5, 11, pages 4, 8)
    Map of Japanese defensive lines, (see Gallery 2, Photo 24, page 16)
    Marine Corps success on Iwo helped save it from disbandment after WWII, 737
    Mine fields, 354, 382
    Missing-in-Action Marines are not being looked for by the Japanese Government, xxx, xxxi, xliv
    Number of medals awarded to the participants, 635
    Pillboxes, 327, 410-430
    Red Beach, 376
    Reunion 2015, xli, liv, lv, 731, 734, 736, 779
    Reunion 2018, 655, 779
    Seabees, 698
    Size of American invasion force, 709
    Size of garrison, 327, 353, 369, 709
    Tunnels, 327, 490
    Types and numbers of heavy weapons, 353, 354
    U.S. military estimation of the size of Japanese force, 380, 381
    Yellow Beach, 1, 393
    Women on the island, 533-534, 882 fn 1421

Iwo Jima (island), xvii, xliii, xliv, liv, 138, 304, 773-774
    Japanese Grave Marker, (see Gallery 1, Photo 11, page 5)
    U.S. and Japanese agreements on the use thereof after 1968, xxx, xliv

Iwo Jima American Association (IJAA), xxx, 675, 734-735, 736, 773-774
Iwo To (modern Japanese spelling of Iwo Jima), xliv
Jackson, PFC Harry L. (USMC), 645-646, 648, 650
    Silver Star, 646, 648
Jacobson, Private First Class Douglas (USMC), 391, 623, 627
    Medal of Honor, 627
Jacques Farm, Camp (Southern California), 143, 144
James J. Peters VA Medical center (Bronx, New York), 570
Japan (*Nippon*), lvii, 4, 5, 23
    American aide to after WWII, 733
    Ancestor Worship, 138
    Drug trade in China, 30-1
    Gender Equality Law (1999), 774
    The Greater East Asia Co-Prosperity Sphere, 94, 109, 134, 177
    Surrender, 710-712, 717
    Training children to attack invading Americans, 709
    Violation of international treaties, 98
    War production, 696
    Willingness to sacrifice its entire population in the war (*ichioku gyokusai*), 574-575, 707-708
    Woman's Suffrage, 733
Japan's "Holocaust," xxxix, xlix, 9, 86-87, 117, 118, 142, 718-719, 733, 763-764, 769, 772, 773, 777, 845 fn 407, 846 fn 431, 851 fn 513
    Definition, 86-87, 118, 769, 845 fn 407, 846 fn 431
    Justification for naming, 118, 134
    Knowledge of in the world, 135, 136, 141
    Numbers, 86, 117, 118, 119-120, 133-134, 718, 850-851 fns 499-501, 505
Japan's Attacks on U.S. Territory
    Alaska, 298, 323
    California, 35
    Guam, xxxv, xliii, lii, 35, 49, 176-177
    Hawaii, lii liii, 1, 2, 3, 4, 6, 8, 23, 24, 29, 34, 35, 52
    Oregon, 35
    Wake Island, lii, 35, 48-49, 52
Japanese,
    Racism, 79, 87, 161-162 see *Yamato* Race
    *Hakko Ichiu*, 87
    View themselves, 79,83, 87 see *Yamato* Race
    Viewed by the West, 82, 85, 161-162
Japanese Americans, 661
Japanese Atrocities, 9, 47, 75-76, 93, 109, 110, 116, 133, 718-719, 773, 846 fn 431, 857 fn 645 (see Gallery 1, Photos 22-25, pages 9-11) see Japan's Holocaust
    Bataan Death March, 49, 138, 161
    Baoding, 114
    Beijing (Peking), 114, 135
    Cannibalism, 115
    Canton, 131
    Chekiang, 832 fn 94

Chemical Warfare, 202, 367-369
China campaigns, 77-78, 133
"Comfort Women" (Sex Slaves), 120, 130, 179
Concentration Camps (Guam), 178, 179, 180, 182, 280, 286, 863 fn 837
Desecration of bodies, 50, 115, 124
Guadalcanal, 50, 164-165
Guam, 178, 179, 180, 181, 182, 279, 280, 282, 285
Guangdong Providence, 129-130
Guangzhou, 131
"Hell Ships", 178
Hong Kong, 124-133, 315-316, 853 fn 564
    see *Rape of Hong Kong*
Jinan, 128
Manchuria, 75, 76, 77, 79, 88, 89, 95, 123, 133
Manila, 130, 142, 518-519, 849 fn 477, 853 fn 564 see *Rape of Manila*
Medical Experiments on Humans, 130, 179
Mufushan, 101
Nanking (*Nanjing*), xlix, 65, 66, 86, 93, 95, 99, 100, 101, 108-109, 110, 133-134, 135, 141, 848 fn 463 (see Gallery 1, Photos 22-25, pages 9-11) see *Rape of Nanking*
Nit'ang, 101
Paoting, 114
Phillipines, 140, 142 see *Rape of Manila*
Pingdingshan, 123
POWs, 49-50, 51, 65, 67, 94-98, 141
Puerto Princesa Prison on Palawan Island (Philippines), 51, 52
Rape Catholic Nuns, 124
Resettlement Programs, 126
Roosevelt's denunciation of, 135
Saipan, 200
Sex Crimes, 124-125, 131-133, see also "Comfort Women" and Rape
Shanghai, 94, 95, 100, 101
Shenzhen, 131
Singapore, 49
Slave Labor, 134
Soochow, 101
Starvation, 95, 118, 126, 131, 202, 280
Suzhou, 101
Wake Island, 48-49, 667
War Crimes, 134-135, 142
Wuhu, 101
Wuxi, 101
Japanese Civilians, 574
Japanese Constitution, xxxiii, 942, fn 3136
Japanese Government,
    Denial of Atrocities, xxx, 118-119, 120, 121, 132, 136, 774, 775, 937 fn 3036
    Falsification of history, 118-119, 120, 136, 774, 775, 937 fn 3036
    Obstruction of Historical Research, 118-119
    Reparations (refusal to pay), 120
    Suppression of Freedom of Speech, 120

Japanese Military Academy, 242

Jarosz, Shayne, 882 fn 1421

Jefferson, Thomas, xxxi, xxxiii, 764

Jerusalem (Israel), 67

Jesus Christ, 40, 80, 81, 687, 688

Jez, Leo (USMC), 478

Jewish Religion, 58, 733

Jews, 141, 167, 339, 341, 402, 403, 547, 548, 549, 640, 688, 734, 765, 766, 769
  Number of Jewish Marines at Iwo Jima, 549
  *Shema*, 467, 902 fn 2015

Jig-Day (Invasion of Tinian), 229

*Jiketsu* (literally self-determination, but in this context, suicide other than disembowelment), 139, 274, 276, 408, 471, 515, 516, 517, 535, 536, 565

Jim Crow Laws (Racial Segregation of the South), 734

*Jôbun*, 452, 496, 639

Johnson, Secretary of Defense Louis A., 737

Johnston Atoll, 619

Johnston, Philip, 366

Jones, Colonel Charles A. (USMC JAG officer), 613, 784 (see Gallery 2, Photo 14, page 10)

Jones, *Atlantic Monthly* Correspondent Edgar, 369

Jones, Gene (Photographer) (USMC), 398, 456, 462-463

Jones, Milton Emory, 401

Josephy, Sergeant Alvin (USMC), 474

Joyce, James (writer), 391

Kaenko, Corporal Hirke (IJA), 496

Kakehashi, Kumiko, 539, 543, 828 fn 37

*Kami* (Shinto spirits), 319

*Kamikaze* Runners, 373, 374

*Kamikaze*s, 140, 242, 289, 302, 303, 304, 305, 347, 389, 444, 574, 575, 576, 609, 699 708, 717, 720, 726, 858 fn 672, 915 fn 2462

Kanagawa Prefecture (Japan), 324

Kaenko, Corporal Hirke (IJA), 496
  *Jôbun*, 496
  *Kanjo*, 496

Kaneko, Japanese Minister of the Privy Council Kentarō, 199

*Kanji*, 83

*Kanjo*, 452, 496

Kansas (USA), 489

Kanto Plain (Japan), 170

Kase, Hideaki, 774

Katō, Admiral Kanji (IJN), 73

*Katori Junja* Shinto Shrine, 389

Katsuichi, Honda (journalist), 119

Kauffman, Staff Sergeant Kauffman (Photographer) (USMC), 441

Kavieng (New Ireland), 167

Kawai, Chiyoko, 534

KDKA (AM Radio station in Pittsburgh), 19

Keenan, Chief of Counsel for the International Military Tribunal for the Far East Joseph B., 718-719, 774

Kelly Jr., A. (Photographer) (USMC), 558

*Kempeitai* (Japanese notorious Secret Police), 126, 299, 545

Kennedy, Presdient John F. (USN), 12, 669, 670, 671, 681
  Navy and Marine Corps Medal, 670

Kennedy Sr., Ambassador Joseph P., 669, 670, 671

Kennedy, Patrick Bouvier, 12

Kennedy, Professor Paul (Yale University), xvii, 696, 721, 777

Kennedy, Master Sergeant Walter J. (USMC), 667

Kenney, Lieutenant General George (USA), 146

Kenshi, Takayama (IJA), 106

Kenyon, Colonel Howard N. "Red" (USMC), 588, 641, 642, 643, 917 fn 2502
  Navy Cross, 588

Keys, *London Daily Express* Journalist Henry, 519

Kharbine (Manchuria), 89

Lake Khasan (Battle of), 133

Kido, Lord Keeper of the Privy Seal Kōichi, 3, 868 fn 999

Kilgore, Eric (archivist), 782

Killen, Corporal Walter (USMC), 471

Kimmel, Admiral Husband E. (USN), xlvii

Kinami Iwo, 389

King, Dan (historian), 329, 369, 780

King, Fleet Admiral Ernest J. (USN), 8, 169, 192, 193, 206, 586, 611, 615, 616, 640, 678, 695, 721

King, Stephen (writer), 743

*King John* (Shakespeare), 525

King's Park (Hong Kong), 126

Kingman, Rear Admiral (USN), 609

Kinkaseki Mining Company, 135

Kinlacheeney, Navajo Code Talker Paul, 370

Donel C. Kinnard Memorial State Veterans Cemetery, 689

Kipling, Rudyard (writer), xii-xiv, 82

Kishi, Nobusuke ("Hirohito's devil"), 736, 772

Kiska Island (Alaska), 561, 606, 846 fn 422

Kitano Point (Iwo Jima), 528, 531 (see Gallery 3, Photo 11, page 8)

Knatchbull-Hugessen, Ambassador Sir Hugh, 829 fn 44

Knowles, Harvard V. (Exeter), xvii

Knox, Frank, 695

Kobe (Japan), 513

Kohima, 208

Koiso, Prime Minister Kuniaki, 535

Kolombangara Island (Battle of), 607

Konoe, Japanese Prime Minister and Prince Fumimaro, 30, 338

Konoye, Japanese Prime Minister, 85

Koo, Wellington, 512

Korea, 75, 83, 88, 133-134, 241-242, 737, 773, 775
  North Korea, 718
  South Korea, 718

Koreans, 199, 217, 226, 477, 515

Korean Forced Laborers (*Senjin or Chōkō*), 133, 135, 179, 238, 241-242, 477, 515

Korean Sex Slaves, See "Comfort Women"

Korean War, 737

Kotohito, Field Marshal Prince Kan'in (IJA), 134

Kowloon (Hong Kong), 126, 316

Krakow (Poland), 67

Kress, Tech Sergeant (Photographer) (USMC), 475

*Kristallnacht* (Night of the Broken Glass—Nazi Pogrom), 769

Krulak, 31st USMC Commandant and General Charles
      C., xvii. xix, xxi, xxx, xxxiv, xxxviii, 192, 557, 602,
      603, 611, 678, 693, 725, 726, 778, 864 fn 885
      Believes Woody should not have received the
          MOH, 678, 725, 726
Krulak, Lieutenant General Victor H. Krulak (USMC),
      44, 192, 194-195, 197, 549, 745, 778, 864 fns
      884-885
Krupp (German company), 76
Kula Gulf (Battle of), 606
*Kuomintang* (Chinese Nationalist Forces), 77, 105
Kuribayashi, General Tadamichi (IJA), xx. xxii, xxiii,
      xxix, xxx, xxxiv, xxxv, xxxviii, xxxix, 50, 90, 101,
      123-134, 138, 140, 217, 301, 310, 311, 312, 317,
      318, 319, 320, 321, 322, 323, 324, 325, 326,
      327, 328, 329, 330, 331, 332, 333, 334, 335,
      336, 346, 347, 350, 353, 366, 369, 370, 371,
      372, 380, 381, 390, 411, 416, 418, 449, 450,
      452, 463, 465, 482, 489, 490, 496, 510, 511,
      513, 527, 534, 538, 539-540, 541, 542, 544,
      555, 618, 638, 682, 683, 684, 701, 707, 732,
      735, 763, 765, 771, 773, 774, 783, 853 fn 551,
      854 fns 588, 590, 879 fn 1333, 881 fn 1395 (see
      Gallery 3, Photo 6, page 5)
      Amphibious assaults against Hong Kong
          Island, 317
      Battle doctrines, 140, 542
      Chief-of-Staff of the 23rd Army, 123-132, 315,
          316
      "Comfort Women," 131, 313, 328, 534
      "Courageous Battle Vows" for Iwo Jima, 331-
          332, 334
      Crimes, 123-133, 315-316, 318
      Dealings with the navy, 329
      Death, 540-543
      Engineers for Iwo Jima, 322, 323, 325, 327, 328,
          353
      Handling of the Rape Epidemic in Hong
          Kong, 131-132
      Harvard University, 313
      Horse husbandry (IJA), 314, 880, fn 1343
      Human Trafficking, 132
      Imports of Japanese prostitutes for the 23rd
          Army in China, 131
      Introduces the Kamikaze suicide doctrine on
          Iwo, 140-141
      Iwo Jima preparation/training, 324-325, 329,
          330, 332-334, 336, 707
      Japanese prostitutes, 131-132, 534
      Justification for defending Iwo Jima, 321-322,
          334, 335
      Main Ordnance, Military Service Department
          (War Ministry), 314
      Meeting with Hirohito, 318
      Micromanager, 321
      Military attaché (Canada), 313
      Preparations for the attack on Hong Kong, 315
      Promotion to general, 534, 544

      Recruitment for Comfort Women for the 23rd
          Army, 131
      Religious beliefs, 319, 320, 323, 331-333,
          490, 527, 534, 539-540
      Remount Administrative Section, Military
          Service Bureau (War Ministry), 314
      Sailors under his command disobey him, 510-511
      *Samurai* Tradition (Matsushiro clan), 313, 314, 319
      Takes command of Iwo Jima, 318-319, 320, 321
      Time in the US, 313
      Treatment of POWs, 132,135
      Use of indirect fire, 354
      View on attacking the U.S., 186, 314-315
      View on Americans, 328, 335
Kuribayashi, Taro, 336-337
Kuribayashi, Yoshi, 318, 330, 336, 337, 544, 682
Kursk (Battle of), 638
Kurusu, Japanese Ambassador to the U.S. Saburō, 6, 29
Kwajalein Atoll, (Battle of), 39, 203, 244, 561, 735
Kyushu (Japan), (see Gallery 4, Photo 2, page 2)
*Kwantung* Army (Japanese), 178, 202
Kyushu (Japan), 571
La Belle, PFC James Dennis (USMC), 647, 648
          Medal of Honor, 647, 648
La Porte (Texas), 346
LaCroix, Corporal Gedeon (USMC), 514, 515
Lambda Chi Alpha Fraternity, 410
Landchow (China), 512
Langlotz, Angela (J.D.), 779
Laos, 304
Lauriello, John, xli, 339, 370, 458, 461, 502, 503, 527,
          540, 593, 780 (see Gallery 1, Photo 3, page 1)
          Doubts some of Woody's claims, 593
Lawless, George, 577, 588, 657, 748
*LCI* (Landing Craft Infantry), 195, 197, 245, 251, 347,
          348, 349
*LCI-(G)s*, 251, 347
*LCI (G) 441*, 347
*LCI (G) 449*, 347, 348, 349
*LCI (G) 466*, 347
*LCI (G) 473*, 347
*LCI (G) 474*, 347
LCVP (Landing Craft, Vehicle, Personnel), see Higgins boat
League of Nations, 31, 77, 369, 512
Leahy, Fleet Admiral William (USN), 7, 8, 76, 77, 169,
          175, 625, 695, 745, 917 fn 2502, 934 fn 2954
*Lebensraum*, 77
Lee, Darol Eugene "Lefty" (USMC), xli, 164, 165, 253,
          258, 259, 260, 280, 282, 291, 292, 293, 384,
          390, 394, 396, 459, 486, 506, 564, 650, 692,
          728, 729, 730, 741, 742, 759, 780, 873 fn 1118,
          923 fns 2688-2689 (Gallery 2, Photos 14, 17,
          pages 10, 12; Gallery 4, Photo 8, page 5)
          Lied about his narrative, 729-730
          Misled People about Woody's Exploits, 650
          Purple Heart, 729
          Shell-Shock, 394, 730
          Stolen-Valor, 650, 939-940 fn 3102

Lee, Chaplain John P. (USN), 283

Legionnaires of Fairmont Post (WV), 628

Lejeune (USMC Camp), 157

Lejeune, USMC Commandant John Archer (USMC), 77

LeMay, General Curtis (USAAF), 122, 226, 306, 450, 511, 527, 563-564, 716, 717-718, 868 fn 999

Lend Lease, xlviii, 694

Leonard, Lieutenant Blake "Dusty" (USMC), 508
    Silver Star, 508

Leppert, Laura, xlii, 784, 825 fn 5, (see Gallery 4, Photo 9, page 5)

*Letters from Iwo Jima* (2006 movie), xxx, 701

Levi, Primo (Holocaust survivor), 119

Leviticus (Bible), 830 fn 68

Lewis, Park Ranger Kina Doreen (Guam), 178-179, 781

Lewis, Captain Robert A. (USAAF), 713

Leyte Gulf (Battle of), 223, 302, 304, 381, 609

LGBT Rights, 774

Liberal Democratic Party (LDP), xxx, 90, 133, 137, 684, 735, 772

Liberman, Lieutenant Sigmund L. (USA), 167
    Bronze Star with V, 167
    Purple Heart, 167

Libya, 44

Lindbergh, Charles, xlviii, 3, 668
    Medal of Honor, 668

Lindsley, PFC Bud (Photographer) (USMC), 377, 468

Litter Bearers, 261

*Little Boy* (atomic bomb), 240, 722

Longfellow, Henry Wadsworth (poet), 169

Lord Russell of Liverpool, 118

Los Angeles (California), xliii

Los Angeles Airport (LAX), xliii

Louisiana (USA), 196, 684

Louisiana State Guard, 196

Louisiana State University (LSU), 196

*LSM* (Landing Ships, Medium Transports), 452

*LST* (Landing Ship, Tank), 195, 196, 197, 339, 389

*LST 477*, 389

Lucas, Sergeant Huron J. (USMC), 388

Lucas, Lieutenant J. G. (USMC), 379

Lucas, Lieutenant Jim (USMC), 233-234

Lucas, Jacklyn "Jack," 34, 602, 623, 624, 626, 628, 645, 648, 650, 663-664, 665, 739, 745, 836 fn 173
    Medal of Honor, 602, 623, 624, 626, 628, 645, 648, 665
    *Luftwaffe*, 576, 685
    Lummus, 1st Lieutenant Jack (USMC), 666
    Medal of Honor, 666

Lunsford, Ernest (USMC), 371, 372, 380

Luxembourg, 23

Luzon (Battle of), 138-139, 722

LVTs (Landing Vehicle Tracked), see amtracs

Lynch, PFC Jessica (USA), 669, 671, 677
    Bronze Star with V, 669
    POW Medal, 669
    Purple Heart, 669

*MS St. Louis* (refugee ship), 769

*M. S. Weltevreden* (US Troop Ship), 144

MacArthur, General of the Army Douglas (USA), xix, xxxiii, 1, 122, 130, 146, 158, 183, 189, 208, 302, 310, 337, 519, 608, 617-618, 661-662, 667, 669, 695, 698, 701
    Medal of Honor, 661

MacMullen, Ramsay (Yale University), xvii

Madison, James, xxxiii, 764, 842 fn 337

Maecham, Charles (USMC), 268

Magee, Reverend John, 111

Maginot Line (France), 203

Mahan, Rear Admiral Alfred Thayer (USN), 205

*Mahayana* Buddhism, 84

Makin Island, 39, 47, 298

Mako Japanese Naval Base, 690

Malaya, lii, 89, 694

Mamaroneck (New York), 402

*The Man Who Shot Liberty Valance* (1962 movie), 635

*The Man's Magazine*, 577, 588, 594, 653, 656, 748 (see Gallery 4, Photo 3, page 2)

*Manchester Guardian*, 104

*Manchukuo* (Japanese Puppet Government in Manchuria), 30, 88, 773, 835, 842 fn 157, 358 see Manchuria

Manchuria, 75, 76, 77, 88, 89, 95, 123, 133, 202, 229, 241, 718, 723, 835 fn 157 see *Manchukuo*
    Citizens kill themselves as the Soviets invade in 1945, 710
    Japan's *Lebensraum*, 77, 710, 842, fn 358
    Size of Japanese garrison force, 710

Mandel, Ann R., 12

Mandel, Captain Dr. Lee (USN), 303, 780

Manenggon Japanese Concentration Camp, 178 (see Gallery 2, Photo 16, page 11)

Mangan Quarry (Guam), 254

Manhattan (New York), 240, 463, 555, 721

Manhattan Project, 170

Manifest Destiny (Japan's), 90

Manila (Philippines capital), 1, 130, 142, 518-519, 719

Manila's POW camp (Japanese), 242

Mao Zedong (Tse-Tung), 77, 767
    Slaughter of 70 million people, 767

Maoists, 767

Map of the "*Banzai*" attack on Guam, 25-26 July 1944, (see Gallery 2, Photo 21, page 14)

Map of the Guam Battle 1944, (see Gallery 2, Photo 16, page 11)

Map of the Iwo Jima Battle 1945, (see Gallery 3, Photos 11, 13, pages 8, 10)

Map of Japanese defensive positions on Iwo Jima, (see Gallery 2, Photo 24, page 16)

Map of MacArthur's and Nimitz's twin offenses in the Pacific, (see Gallery 1, Photo 31, page 16)

Map of the Marianas Campaign 1944, (see Gallery 2, Photo 6, page 3)

Map of the Saipan Battle 1944, (see Gallery 2, Photo 7, page 4)

Map of the Tinian Battle 1944, (see Gallery 2, Photo 10, page 7)

Marco Polo Bridge (Lugouqia Bridge) Incident (1937), 77

Marianas Islands, xxxvi, liv, 76, 167, 169, 170, 175, 183, 206, 207, 208, 209, 223, 243, 246, 297, 304, 319, 323, 394, 486, 489, 554, 709 (see Gallery 2, Photo 6, page 3)

Mariedas, Lieutenant James P. (USMC), 380

Marine Corps Enlistment and Voluntary Enlistment Policies, 34-35

Marine Corps Height Requirements (WWII), 34

Marine Corps Heritage Foundation, 691

Marine Corps Historical Division, xxx, 782

Marine Corps League, xlii

Marine Corps Maneuver units, 700

Marine Corps Recruit Depot (MCRD) in San Diego (Marine Corps Boot Camp), lvii, 37, 39-64
    Sex education, 54-55 (see Gallery 1, Photos 21-21, page 8)

Marine Corps University, 779

Marine Rifle Creed, 59

Marion Counties (West Virginia), 628, 689, 759

Marpi Mountain (Saipan), 210

Marpi Point (Saipan), 210, 213, 214, 215, 216, 219, 223, 238, 533 (see Gallery 2, Photos 7-9, pages 4-6)

Marpo Point (Tinian), 238, 534 (see Gallery 2, Photos 10, 13, pages 7, 9)

*Mars* (the Roman god of war), 341

Marshall, General of the Army George C. Marshall (USA), 8, 367, 661, 723
    Analysis of Using Gas Warfare, 723

Marshall Islands, 8, 76, 244, 245, 298, 394

Marshall Plan, 733

Marston Mat, 476, 477

Martin, 1st Lieutenant Harry L. (USMC), 537, 538, 539
    Medal of Honor, 537, 538, 539

Marxism, 767

Maryland (USA), 632

Massachusetts Institute of Technology (MIT), 605

Masters, Brigade Major John (British), 452

Masunaga, Buddhist Priest Reihō, 574

Matheney, PFC Cecil (USMC), 456

Matsui, General Iwane (IJA), 99, 101, 106, 107, 108, 111, 119, 133-134, 137, 314, 618, 718, 732, 774, 848 fn 459

Matsuoka, Japanese Foreign Minister Yōsuke, 85, 91

Matsushiro Clan (*Samurai*), 313

Matthew (New Testament Book), 161, 341

Matthews, Private Allen R. (USMC), 250

Maynard, Ken, 22

McCarthy, Captain Joseph J. (USMC), 623
    Medal of Honor, 623

McCormick, Colonel Robert, 4

McDougal, Chris, 781

*McGuffey's Readers*, 17

McKinney (Texas), 692

McKitterick, 2nd Lieutenant Gertrude "Trudy" (USMC), 734

McMillin, Captain George J. (USN), 177

McNair, General Lesley J. (USA), 37

Meadowdale (West Virginia), 621

Mears, Dwight (historian), 674

"Meatgrinder" (Iwo Jima), 457, 487

Medal of Honor (Highest Medal for valor in the United States Military), xx, xxii, xxiii, xxiv, xxxi, xxxvii, xxxviii, xliv, xlv, xlvi, xlviii, liii, lv, lvi, 34, 452, 496, 568, 577-605, 607, 608, 610, 615, 616-628, 635-666
    History of the MOH, xlvi, liv
    Number of Marines who received the MOH during WWII, 635, 757
    Number of Marines who received the MOH posthumously from WWII, 635
    Number of MOH Recipients from Iwo Jima, 635-636
    Number of recipients in the history of the United States, liv, lv, 635
    Revoking the Medal, 668

Mediterranean, 696

Meiji, Japanese Emperor, 68-69

Meiji Constitution, 68, 71

Meiji Restoration, 71

*Mein Kampf*, 81, 85, 768

Melville, Herman (novelist), xliii, 391, 523, 731-732

Memphis (Tennessee), 662

Memorial Day, 11

*The Memorial Hall of the Victims in Nanjing Massacre by Japanese Invaders*, 65-67

Mercury Chloride (used by Japanese medics to kill their wounded), 139

Meredith, Ruby (Woody's fiancée), 33, 155, 156, 157, 158, 171-172, 173, 275, 292, 434, 460, 621, 622, 631-632, 724, 743, 759-760 (see Gallery 4, Photos 1, 12, pages 1, 7)

Meredith, Reverend Thomas Grafton "Tom," 631

Messerschmidt, Manfred (historian), 825 fn 7

Mexicans, 90, 608, 637, 734

Mexico, 44

Meyer, Dakota (USMC), 668

Micronesia, 175

Midianites (Bible), 767

Midway Island (Battle of), 8, 9, 189, 190, 206, 297, 298

Military Academy (IJA), 242, 683

Miller, Ann, 537

Miller, Glenn (jazz musician), 296, 462

Millett, Colonel Allan (USMC) (historian), 57, 170, 737-738, 739, 869 fn 1007, 872 fn 1089, 930 fn 2829

Milligan, PFC Patrick J. (USMC), 388

Milton, John (writer), 391

Mines, 154, 205, 354, 382, 477, 544, 599, 623, 866 fn 920
    Terra Cotta, 382

Ministry of Foreign Affairs at the Japanese Embassy, 773, 783, 848 fn 463, 851 fn 511

Minnesota (USA), 730

*Mischlinge* (Nazi term for partial-Jews), 761, 922 fn 2641

Missionaries (Christian), 149, 848 fn 465

Missionaries (Shinto), 73, 841 fn 327

Mississippi River, 38

Missouri (USA), 568

Missouri National Guard, 624

Mitscher, Admiral Marc Andrew "Pete (USN), 206
Mittlebau-Dora Nazi Concentration Camp, 167, 861 fn 782
Mix, Tom (actor), 22
*Moby Dick* (novel), 731-732
Model, General Walter (German), 638
Moltke, Field Marshal Helmuth von, 275
Mongolia, 83
Mongols, 513
Montana (USA), 4, 25, 26, 29, 680, 734, 758
Montfort, Günther, xvii
Montezuma (Mexico), 44
Morgan, Michele, 537
Morgan, Martin (historian), 118
Morgantown (West Virginia), 25, 621
Morison, Samuel Eliot (historian), 351, 695-696
Moriyasa, Murase, 99, 101
Mormon Religion (The Church of Jesus Christ of Latter Day Saints), 73, 215, 230-231, 235, 733
Morotai (Indonesia), 302
Moscow (Russia), 24
Motoshima, Nagasaki Mayor Hitoshi, 122
Motoyama (Iwo Jima), 327, 483, 491 (see Gallery 3, Photo 11, page 8)
Mount Cameron (Hong Kong), 132
Mount Marpi (Saipan), 210
Mount Suribachi (Iwo Jima), xxiv, xliii, 350, 363, 366, 379, 403, 404, 408, 537, 702, 734 (see Gallery 1, Photo 9, page 4; Gallery 3, Photo 11, page 8)
    Famous Flag Raising (controversy), 651, 702
Mountcastle, Brigadier General and historian John (USA), 704-705
*MS St. Louis*, 769
Mufushan (China), 101
Mukai, Sub-Lieutenant Toshiaki (IJA), 104, 848 fn 469
Mukden Incident (18 Sept. 1931), 30
Mueller, Robert "Bob" (USMC), 583
Mulligan, James "Big Moon" (USMC), 465
Mulstay, Sergeant (Photographer) (USMC), 470
Munich Agreement (September 1938), 768
Murray, Major Clay M. (USMC), 396-397, 413, 462, 506
*Musashi* (Japanese battleship), 205
Muslim Fanatics, 767, 769
Mussolini, *il Duce* Benito, xlvii, 4, 5, 31, 82, 548, 567, 696, 722
Mutual Network, 405
Mutsuhito (Meiji Emperor), 71
Mutō, Lieutenant General Akira (IJA), 130
My Lai Massacre, 827 fn 20
Myers, Private James (USMC), 496, 497
Myō Kai, 30
Nagano, Justice Minister and Army Chief-of-Staff Shigeto (JDF), 119
Nagano Prefecture (Japan), 312
Nagasaki (Japan), 120, 121, 488, 712, 714
Estimated killed from the atomic bomb dropped there, 120
Nagatomi, Hakudo (IJA), 110
Nagoya (Japan), 513
Nagumo, Vice Admiral Chūichi (IJN), 184, 185, 223, 830 fn 55

Nakagawa, Colonel Kunio (IJA), 325
Nakajima, Lieutenant General Kesago (IJA), 106
Nakamura, 2nd Lieutenant Sadao (IJA), 452
    *Jôbun*, 452
    *Kanjo*, 452
Nakamura, Captain (IJA), 91
Nakane, Colonel Kaneji (IJA), 541-542, 543
*Namu Hachiman Daibosatsu* (one Shinto god of war), 408
Nanking (Chinese city, also spelled Nanjing), xlviii, 65, 67, 95, 96, 98, 100-101, 104, 105, 106, 107, 111, 126, 135, 314, 511, 513, 719 see *Rape of Nanking*
Nanking Garrison Force (Nationalist), 95
Nanpo Shoto, 331
Napalm, 229-230, 376, 433
Napoleon, Bonaparte, 416, 464, 614
Napoleon Complex, 81
Naro, Corporal William J. (USMC), 448, 517, 579, 590, 592, 616, 671, 754
National Museum of the Marine Corps, 691, 702
National Museum of the Pacific War, 781
National Park Service Center of War in the War in the Pacific at Guam, 781
Nationalist China, 77, 92, 95, 126, 316, 831 fn 55
Native American ("Indians"), 90, 172, 637, 733-734
Navajo Code Talkers, xli, xlii, 366, 367, 370, 734
Naval Academy (Annapolis, Maryland, USA), 357, 402, 605, 681, 745
Naval Torpedo Station (Rhode Island), 605
Naval Training Center Bainbridge (Maryland), 632
Naval War College (U.S.), 193
Navy Board of Awards and Decorations (sometimes referred to Navy Department Board of Decorations and Medals), 583, 603, 604, 606, 607, 608, 609, 614, 615, 619, 642, 657, 726
    Power and Responsibilities, 604
Navy Cross, xxiii, 51, 378, 510, 587, 603, 649, 662, 666, 725, 917 fn 2502
    Number awarded to men who served on Iwo Jima, 635
Navy Seals, 674, 675
Nazi-Soviet Non-Aggression Pact, l, 24, 834 fn 133
Nazi Germany, xxxvi, xlvii, xlviii, 5, 31, 77, 78, 88, 513, 549, 693, 733, 766
Nazis, 9, 35, 91, 119-120, 134, 154, 161, 208, 573, 576, 685, 697, 717, 723, 732, 769, 770
Nazi Genocide, 733
Nazi Military, 135
"Negroes", (American Blacks), 148, 537, 538, 539, 547, 548, 733-734
Neimeyer, Charles (USMC), xxx, xxxii, 674, 779
Nelson Mullins Riley & Scarborough, LLP, xxxiii
Nenninger, Tim (archivist), 782
Nero (Roman Emperor), 341
Netherlands, lii, 23, 88, 302
Neuman, Private Paul J. (USMC), 659
New Britain (Battle of), 39
New Caledonia, 144, 145, 146
New Deal (FDR's Social Reforms), 17, 23, 25
New Georgia (Battle of), 606
New Guinea, 146, 167, 183

New Hampshire (USA), 163
New Ireland, 167
New Mexico Army National Guard, 738
New Orleans (Louisiana), 684
New York (USA), 402, 665
New York City (New York), xliii, lvi, 240, 374
New York Giants (NFL), 666
*New York Times*, 418
New Zealand, 51, 88, 694
Newchwang (Manchuria), 79
Newport (Rhode Island), 193, 765
NFL (National Football League), 666, 669
NHK (Japanese Broadcasting Corporation), 3
Nicaragua, 196
Nietzsche, Friedrich (philosopher), 460
*Niho Budôkan* Hall (Tokyo), 122
*Nihonjin* (Japanese "Race"), 79
Nimitz, Fleet Admiral Chester (USN), xix, xxxv, xxxvi,
    xxxvii, lvii, 7, 8, 146, 158, 166, 197, 206, 208,
    243, 275, 286, 322, 344, 345, 356, 367, 369,
    403, 525-526, 554, 555, 586, 587, 599, 603,
    606, 608, 612, 614, 616, 617, 619, 636, 640,
    645, 658, 662, 672, 678, 698, 721, 722, 723,
    728, 748, 864 fn 871
        Analysis of Using Gas Warfare, 723
        Refuses to endorse Woody's MOH, 599, 603,
            606, 612, 616, 619, 642, 672, 678, 748
Nimitz Hill (Guam), 261, 295
*Nippon Kaigi* (a sort of Japanese "Holocaust" de-
    nial group but literally translates to "Japan
    Conference"), xxx, 133, 774, 775, 937 fn 3036
        Denies Gay Rights, 774
        Denies the *Rape of Nanking* happened, 774
        Denies the Tokyo War Trials after WWII were
            legitimate, 774
        Denies Women's Rights, 774
*Nippon Times*, 775
*Nisei* (second generation Japanese Americans), 530
Nishi, Colonel Baron Takeichi (IJA), 134
Nit'ang (China), 101
Nitta, Japanese Chief of Police on Saipan, 200
Noda, Sub-Lieutenant Tsuyoshi (IJA), 104, 848 fn 469
Nolte, PFC William C. Nolte (USMC), 392, 415, 895
    fn 1815
Nomura, Japanese Ambassador to the U.S. Kichisaburō, 6
Norfolk (Virginia), 193
Normandy (France, D-Day), xxxvi, 170, 184, 195, 208,
    302, 323, 345
North Africa, xxxvi, 90, 194, 345, 661, 694
North Atlantic Treaty Organization (NATO), 7333
North Carolina (USA), 157, 283
Nosarzewski, PFC Johnny (USMC), 614, 615, 705
        Navy Cross, 614
*Nostra Actate*, 548
Novak, Alex, xxxi
Numbers (Bible), 767
Nuremburg Trials, 141
Nuttall, Private Stanely H. (USMC), 660
O'Brien, Phillips Payson (historian), 169

Oahu (Hawaii), 380
Oak Hill (West Virginia), 631
Oath of Service (U.S. Military), 38
Obama, President Barack, 122, 123
Obata, Lieutenant General Hideyoshi (IJA), 284, 285
Oda, Sergeant Shezuo (IJA), 543
*Odin* (one of the Norse gods of war and revenge), 341
*Odyssey* (Homer), 743
Officer's Training Course (USMC), 411
Ogata, Colonel Kiyochi (IJA), 239
Oil (Petroleum), 88, 135, 184
        US Oil Embargo, 92
Okamura, Lieutenant General Yasuji (IJA), 134
Okinawa, 72, 183, 299, 303, 305, 402, 514, 555, 570,
        573, 574, 576, 610, 698, 706, 716, 722, 723
        Civilian casualties, 709
        Size of Japanese garrison force, 706
        Size of the U.S. invasion force, 706
Oklahoma (United States), 17, 637
Oldendorf, Vice Admiral Jesse B. (USN), 609
Olson, Barbara, 703, (see Gallery 3, Photo 21, page 16)
Olson, Lynne (historian), 184
Olson, Colonel Robert "Bob" (USA), 602, 612-613,
        672 (see Gallery 3, Photo 21, page 16)
Olympics (1932), 134
*Olympic* (Operation), 571, 718 (see Gallery 4, Photo
    2, page 2)
        Projected American casualties, 709
        Projected Japanese casualties, 709
Omagari, Lieutenant Satoru (IJA), 487, 520-521, 522, 523
Omata, Yukio (IJA), 96
*Omiyajima* (Japanese name for Guam), 178
Ona (West Virginia), lv
Onishi, Vice Admiral Takijiro (IJN), 574
Ormsby, Major Murray (Judge at the Hong Kong War
        Crimes Trial), 129
Operation *Bagration*, 208
Operation *Cartwheel* (Pacific Campaign), 146
Operation *Coronet*, 571
Operation *Detachment*, 338
Operation *Downfall*, 571, 617
Operation *Dragoon*, 302
Operation *Forager* (Marianas Islands), lv, 167, 170, 183,
        721, 871 fn 1056
Operation *Hara-Saku*, 316
Operation *Ketsu-Go*, 577
Operation *Olympic*, 571, 718 (see Gallery 4, Photo 2,
        page 2)
        Projected American casualties, 709
        Projected Japanese casualties, 709
Operation *Overlord*, 302, 345
Operation *Stevedore* (Guam's Liberation), 170, 179, 252
Operation *Torch*, 345
Operation *Zitadelle*, 638
Opium, 30, 71, 139
Opium Wars (1840s & 1850s), 71
Oppenheimer, J. Robert, 714
"Orange Plan", 77
Orel (Russia), 638

Oregon (USA), 35, 554

Orient, see Asia

Orote Peninsula (Gaum), 244, 252, 273 (see Gallery 2, Photo 16, page 11)

Orwell, George (writer), 85

Osaka (Japan), 513

Osama bin Laden, 674

Osborn, Company Sergeant Major John Robert (Canadian), 638
      Victoria Cross, 638

Oshima, Ambassador Hiroshi, 831 fn 78

Outrigger Hotel (Guam), xli

Ottoman Empire, 767

*Overlord* (Operation), 302, 345

Oya, Captain Goichi (IJN), 239

Ozawa, Vice Admiral Jisaburō (IJN), 206, 207, 208

P-51 Fighter Plane (Mustang), xli, 339, 540, 545, 546, 554

Pacific Fleet Board of Awards, 599

Pacific Ocean, 88

Paco Railroad Station (the Philippines), 637

Palau, 39, 609, 703

Pan American Airways, 175, 667

Pan-Asianism, 90, 177, 186

Paoting (China), 114

Papua (New Guinea), 8, 116, 117, 150, 189, 190

Paradise Valley (Saipan), 210

Park, President Chung-hee (Korea), 242

Parks, Larry, 537

Parris Island (South Carolina—Marine Corps Boot Camp), 38, 402

Parshall, Jonathan (historian), 784

Pascal, Blaise (philosopher), 768

Pasewark, Sergeant William "Bill" (USMC), xli, 469 (see Gallery 1, Photo 12, page 6)

Patch, Nathaniel (archivist), 782

Pattiwal, H., 117

PB4Y2 airplane, 552

Pea Ridge Methodist Church (West Virginia), 739

Pearl Harbor (Battle of in Hawaii), lii liii, 1, 2, 3, 4, 6, 8, 23, 24, 29, 34, 35, 52, 88, 90, 91, 93, 123, 136, 176, 184, 187, 192, 193, 223, 226, 315, 316, 349, 380, 512, 527, 694, 695, 697, 700
      Casualties, 380

Peck, Major General DeWitt (USMC), 570

Pedophiles (Japanese soldiers), 108, 120

Peleliu (Battle of), 324, 325, 381, 444, 698, 706
      Camp Pendleton (California), 587

Pennsylvania (USA), 32, 410, 602, 621, 734

Pepin, Gunnery Sergeant Lawrence Henry (USMC), 39-40, 54-55, 56, 58, 839 fn 261
      Homosexuality, 54, 56, 839 fn 261

Perry, Commodore Matthew (USN), 71

Pershing, General John J. (USA), 191
      James J. Peters VA Medical Center, 570

Pharaohs (Egyptian), 341

Philadelphia (Pennsylvania), 35, 602

Philippines (Battle of and island), lii, 1, 9, 35, 39, 49, 57, 89, 90, 116, 118, 140, 146, 150, 175, 183, 193, 194, 243, 302, 304, 337, 488, 518-519, 567-568, 609, 637, 695
      Japanese casualties, 208

Philippine Sea Battle, 205, 207, 209, 223, 609

Phillips, John, xxix

Phillips Exeter Academy (Exeter, New Hampshire), xvii, xxix, 778

Pickett's Charge (Civil War), 473

Pierce, Corpsman Francis J. (USN), 658-661
      Medal of Honor, 661
      Navy Cross, 661
      Silver Star, 661

Pingdingshan Incident (1932), 123

Pitner, Nathan (USMC), 415, 485-486
      Bronze Star with V, 486
      Purple Heart (1), 486
      Purple Heart (2), 486

Pittsburgh (Pennsylvania), 19

*The Plague*, 766, 770

Plato (philosopher), 761

Poison Gas, 202, 367, 369, 467, 721

Poland, 67, 769

Pole-Charges, 150, 308, 430

Polychronis, Andrew (Exeter), xvii

Pope Pius the XI, 548

Pope Pius the XII ("Hitler's Pope"), 548-549
      Anti-Semitism, 548-549
      Catholic Racism, 548-549

Portugal, 184

Port Arthur (China), 76

Portland (Oregon), 554

Potsdam Declaration, 571, 723

POWs, 49-50, 51, 65, 67, 95-98, 99, 120, 126, 129, 132, 134, 138, 141, 161, 162, 178, 217, 382, 477, 512, 514, 515, 517, 523, 539, 619, 620, 667, 718, 732
      American Handling of, 530, 531, 539
      Japanese Treatment of, 732

Powers, Brigadier General Thomas (USAAF), 513

Prange, Gordan W. (historian), 693

Prentky, Robert (psychologist), 125

Princeton University, xxxiii, 653, 699

Prisoner-of-War (POW), 49-50, 51, 65, 67, 95-98, 99, 120, 126, 129, 132, 134, 138, 141, 161, 162, 178, 217, 382, 477, 512, 514, 515, 517, 523, 539, 619, 620, 667, 718, 732
      American Handling of, 530, 531, 539
      Japanese Treatment of, 732

*Prior Restraint*, xxxi

Prock, Cam (FWC), xvii

Prodigal Son (The New Testament), 19

Protestant (Christian denomination), 58, 175, 467, 546, 547, 549, 733, 764

Prostitution, 55, 131-132

*Protocols of the Elders of Zion*, 828-829 fn 41

Proverbs (Bible), 618, 635

Pruitt, Private Albert C. (USA), 167

Prussian, 189

Pryor, Don (radio personality), 405

Psychoneurosis, see Shell-Shock

PTSD (post-traumatic stress disorder), 394, 420, 602, 702, 729, 730-731

Public Law 416, 82d Congress, 738

Puerto Princesa Prison on Palawan Island (Philippines), 51, 52

Pulitzer Prize, 403

Puller, Brigadier General Chesty (USMC), 594, 649, 650, 923 fn 2686
  Navy Crosses, 649

Punxsutawney (Pennsylvania), 410

Purple Beach (Iwo Jima), 535

Purple Heart, 54, 167, 196, 622, 649, 709

Pyles, Corporal Eleanor Genevieve (USMC) (Woody's girlfriend), 33, 155, 156, 157, 734, 759-760, 835-836 fn 170
  Good Conduct Medal, 759

Quiet Dell (West Virginia), 11, 16, 32 622, 689

Quinine, 135

*Ra* (Egyptian Sun God), 341

Rabaul, 146, 183

Rabe, John, 110, 111

*The Rabbi Saved by Hitler's Soldiers*, 761

Radio (i.e. Columbia, Mutual and Blue networks), 405

Radio Tokyo, 285

Raeder, Admiral Erich (German), lii

Ragliano, Al (USMC), 414

*Raiders* (Marine Corps), 145, 146, 335

Ramesses II (Egyptian Pharaoh), 341

Rape (and other Crimes against Women by Japan), 70, 95, 99, 106, 108, 109, 110, 114, 120, 124, 125, 130-131, 133-134, 136, 141, 178, 179, 518, 519, 710, 825 fn 7, 849 fn 477, 854 fn 565, 880 fn 1356
  Definition of Japanese Rape, 125
  Rape of children, 120

*The Rape of Guam*, 178-179

*The Rape of Hong Kong*, 124-133, 178

*The Rape of Manila*, 130, 142, 178, 518-519

*The Rape of Nanking*, xlix, 65, 66, 67, 85, 93, 95, 100-101, 108, 109, 111, 120, 121, 136, 142, 178, 314-315, 519, 769, 774, 848 fn 463 (see Gallery 1, Photos 22-25, pages 9-11)

Reagan, President Ronald, 37

Red Army, 208

Red Cross, 126, 279-280

Reed, Jim (USMC), 212, 219, 220, 701

Reese, Private John (USA), 637
  Medal of Honor, 637

Reffer Ships, 184

Refuges, 769

Regimental Combat Team (RCT), 147, 587

Regnery Books, xxxi

Reharing, M., 117

The Third *Reich* (Nazi), 136

*Reichstag*, lii, 4

Reisch, David, xxxii

Religious Racism, 546-549, 688, 767

Renstrom, Gunnery Sergeant Keith A. (USMC), 167, 214, 215, 216, 220, 221, 222, 224, 227, 230, 231, 232, 233, 234, 235, 236, 237, 375, 485, 506, 655, 662, 663, 674, 700, 701, 752-753, 760, 775, 784, 870 fn 1029, 938 fn 3081 (see Gallery 2, Photo 11, page 8)
  Bronze Star with V (1), 236, 662
  Bronze Star with V (2), 485
  Did not believe Woody's narrative, 655
  Navy Cross recommendation, 236, 662, 663
  Silver Star, 662

Reparations (German), 120, 852 fn 522

Reparations (Japanese), 120, 855 fn 522

Replacements, 464, 491-499, 560-561, 579

Republican Party (USA), xlviii, 694

*Rescued from the Reich*, 761

Revolutionary War (1776-1783), 11, 401

Reynolds, Donna (FWC), xvii

Reynolds, General Russel (USA) (Judge at the War Crimes Trial in Manila), 130

Rhode Island (USA), 193, 764

Richie, Alexandra, xxix

Ridlon, Captain Walter J. (USMC), 659, 660, 661, 760
  Medal of Honor, 661
  Navy Cross, 661

*Rifleman* (Marine Corps),

Rigg, Bryan Mark, xxii, xxv, xxxix, 634 (see Gallery 1, Photos 5-7, pages 2-3; Gallery 2, Photos 11-12, page 8; Gallery 2, Photos 15, 17, 20, pages 10, 12-13; Gallery 3, Photos 2, 15-16, 19-21, pages 2, 12-16; Gallery 4, Photos 5, 7, pages 3-4)

Rigg, Doris, 703

Rigg, 1st Lieutenant Frank (USA), 340, 341, 703
  Purple Heart, 340

Rigg, Ian, xvii, 634 (see Gallery 2, Photo 20, page 13)

Rigg, Justin, xvii, xlii (see Gallery 1, Photos 2-6, 8, pages 1-3; Gallery 2, Photos 15, 17, 20, pages 10, 12, 13; Gallery 3, Photo 18, page 14)

Rigg, Leona née Parr, 35

Rigg, Mark, 35

Rigg, Sophia, xvii, 65, 67, 781 (see Gallery 1, Photo 26, page 12)

Riggs, "Negro" Boxer for the 3d MarDiv, 630

Riggs, Professor Reverend Charles H., 111

Risshō University, 83

*Ritterkreuz* (Knight's Cross), 638, 639, 761
  Number awarded during WWII, 639

*RMS Titanic*, 345

"Rock and Shoals" Naval Legal Codes, 458, 459, 729

Rockey, Major General Keller (USMC), 664

Rocky Mountains (USA), 25

Rodriquez, Private Cleto "Chico" (USA), 637
  Medal of Honor, 637

Rogers, Brigadier General William W. (USMC), 381

Roi-Namur (WWII battle), 323, 735

Roman Catholic Church (Vatican), 549

Romans (Bible), 688

Romania, 29

Rome (Italy), 301

Rommel, Field Marshal Erwin (German), 90

Roosevelt, President Franklin Delano (FDR), xix, xlvii, xlviii, l, 3, 4, 6, 8, 13, 23, 25, 48, 88, 136, 189,

287, 367, 369, 376, 379, 484, 560, 567, 568,
    605, 617, 694, 698, 717, 832 fn 94, 914 fn 2410
    1933 Inauguration, 13
    Death, 568
    Declaration of the United Nations (1942), 141
    Denunciation of Japan's atrocities in China, 135
    U.S. Navy Assistant Secretary, 8
Roosevelt, Eleanor (First Lady), 276, 721
Roselle, 2nd Lieutenant Benjamin Jr. (USMC), 372, 380
Rosenthal, Associated Press photographer Joe, xxiv,
    403, 406, 634
Ross, Bill D. (USMC), 133, 400
Route 73 (West Virginia), 621
Rowan & Littlefield Publishers, xxxii
Ruff-O'Hearne, Jeanne, 121
Ruhl, PFC Donald J. (USMC), 626, 646-648, 650
    Medal of Honor, 626, 646-648
    Lord Russell of Liverpool, 118
Rubber, 135
Russia, 75, 76, 78, 91, 92, 172, 341, 547, 638, 697,
    723, 769, 773
Russians, 141, 337, 368, 710, 723
    Rape of Japanese citizens, 710
Russo-Japanese War, 76, 536
Ruth, Babe (baseball player), 519
Rwanda Genocide, 138
Rybak, Corie, (see Gallery 3, Photo 16, page 13)
Rybak, Lieutenant Colonel Gene (USMC), 504, 506, 507,
    655, 753, 784 (see Gallery 3, Photo 16, page 13)
Rybak, Pattie, (see Gallery 3, Photo 16, page 13)
Rybak, Zack, 503
Rybakiewicz, Joseph Anthony (USMC), 165, 412, 415,
    502, 503, 504, 506, 507, 510, 511, 530, 557, 655,
    674, 676, 677, 705, 753, 760, 784, 895 fn 1815
    Bronze Star with V, 504, 506, 510, 676
    Friendship with Vernon Waters, 502-503
    Problems with Woody's MOH, 504, 655
    Purple Heart, 504, 506
Ryukyu Islands, 72
Sachsenhausen Nazi Concentration Camp, 770
Saddam Hospital (Nasiriya), 669
Sakai, Hatsuyo, 708
Sakai, Sub-Lieutenant Saburō (IJN), 78, 84, 708, 720,
    771, 772
Sakai, General Takashi (IJA), 127, 128, 129, 131, 132,
    315, 316, 618, 718, 774
Sakhalin, 75
Saipan (Battle of), xxxvi, xxxix, 39, 167, 169, 170,
    183, 190, 197, 199, 200, 201, 202, 203, 204,
    206, 209, 223, 224, 226, 229, 244, 245, 249,
    255, 287, 290, 296, 299, 323, 336, 337, 344,
    348, 356, 357, 358, 362, 369, 372, 380, 381,
    402, 482, 485, 518, 561, 563, 573, 574, 609,
    655, 680, 695, 698, 701, 720, 735 (see Gallery 2,
    Photo 7, page 4)
    Casualties, 224
    Civilian casualties, 709
    Civilians flee the battle, 211
    Civilians kill themselves, 212, 213, 214, 215,
        216, 574
    Internment camps, 226
    Japanese soldiers kill their own civilians, 213,
        214
    Marines help civilians, 211, 218, 219, 220,
        221, 222, 224, 225
    Reunion 2018, 655
    Seabees, 721
    Size of Japanese garrison, 199
    Total civilian deaths, 217
Saitō, Tatsushi (archivist), 782
Saito, Lieutenant Toshio (IJN), 49, 209
Saitō, General Yoshitsugu (IJA), 202, 207, 208, 212,
    219, 222, 224
Salem (West Virginia), 568
Salerno (Italy), 254
Sammons, Jeffrey L. (Yale University), xvii
Samurai (Japanese), xx, 71, 132, 175, 235, 269, 312,
    313, 314, 318, 319, 449, 478, 521, 535, 542, 683
San Diego (California), 38, 147, 506, 554
San Francisco (California), xliii, 120, 179, 601, 602, 620
Sanburg, Carl (historian), 313
Sands of Iwo Jima (1949 movie), xxiv, 701
Sandusky (Ohio), 605
Sano, Major General (IJA), 315
Satan Tanks, see Flamethrower Tanks/Zippo Tanks
Satchel Charges, 150
Scapa Flow (UK), 605
Schlager, PFC Alexander (USMC), 306, 424, 426, 427,
    429, 442, 449, 462, 485, 579, 581, 582, 590,
    592, 616, 653-654, 671, 724, 742, 748, 751, 754,
    898 fn 1907, 924 fns 2700-2701
    Displeasure with Woody, 426
    Upset by Woody's interview in Man's
        Magazine, 426, 653-654, 748 (see Gallery
        4, Photo 3, page 2)
Schlager, Bill, 306, 742
Schlager, Lawrence, 654
Schmidt, Major General Harry "The Dutchman"
    (USMC), 229, 350, 355, 356, 380, 383, 395, 403,
    477, 482, 491, 507, 525, 595, 642, 643, 645, 658
    Distinguished Service Medal, 356
Schneersohn, Rebbe Joseph Isaac (leader of Lubavitch
    Judaism and a Religious Fanatic), 761
Schnell, Judith, xxxii
Schwartz, Corporal (Photographer) (USMC), 471
Schwartz, Private George B. (USMC), 438, 491, 579,
    590, 591, 616, 650, 671, 724, 754
Scotland, 605
Scott, James M. (historian), 519, 777
Scott, Lieutenant Commander John A. (USN), 662
    Navy Cross, 662
SCR-625 (mine detector), 382
Scrivens, Private Thomas E. (USMC), 457, 458, 459,
    460, 505, 727-729, 730, 756, 900 fn 1969
Sea of Japan, 76
Seabees (Naval construction battalions), 240, 296, 375, 475,
    489, 515, 538, 564, 565, 630, 698-699, 721, 779
    Accomplishments during WWII, 721
    Numbers during WWII, 721

Second Sino Japanese War (in China, it is known as War Against Japanese Aggression or the War of Resistance Against Japan), xlviii

Selzer, Dr. Richard, 777

Senda, Major General Sadasue (IJA), 133-134, 518

*Senjinkum* (Battle Ethics), 222-223

*Senninbari* (Thousand-stich belts), 342

*Seppuku* (ritualistic suicide by disembowelment, also known as *hara-kiri*), 138, 139, 515, 519, 542

*Sergeant York* (1941 movie), 18

Sermon on the Mount (The New Testament), 19

Shakespeare, William (playwright and actor), 313, 520, 525, 743

Shandong Province (China), 77

Shanghai (China), xlix, 55, 94, 95, 100, 101, 104, 136, 192, 512, 79

Shanghai Expeditionary Army (IJA), 142

Shank, 1st Lieutenant and Dr. Lawton Ely (USA), 666-668
    Medal of Honor, 666-668
    Navy Cross, 666, 668

Shanxi Province (China), 81

Sharon (Pennsylvania), 32

*Shawshank Redemption* (1994 movie), 743

Shell-Shock (see also Combat Fatigue), 394, 457, 458, 459, 460, 486, 685
    Numbers of on Iwo Jima, 741

Sheng, Zhang (historian), 775

Shenzhen (China), 131

Sherrod, *Time and Life* Correspondent Robert, 222, 346, 374, 396, 428, 429, 489

Shi, Jinghua, 781

Shigematsu, Major General Kiyoshi (IJA), 255

Shimada, Admiral Shigetarō (IJN), lii

Shimizu, Ryūzan, 83

Shintarō, Uno (IJA), 84, 110-111

Shindo, Yoshitaka, xxx, 90, 133, 137-138, 321-322, 557, 682-683, 684, 693, 735-737, 771, 773, 774, 775, 783, 937 fn 3036 (see also Gallery 1, Photo 1, page 1; Gallery 4, Photos 10, 13, pages 6, 8)
    China's government denounces behavior of, 137
    Claims of ignorance about his grandfather Kuribayashi's crimes, 133
    Disregard for the American dead on Iwo Jima who are still there, xxx-xxxi
    Explains Japan's reasons for making war, 736-737
    Fight with the author, 771-773
    Historians Crowl and Isely warn about the memory of his grandfather (Kuribayashi) starting a new nationalism, 773
    *Nippon Kaigi* membership, xxx, 774, 936 fn 3036
    Political bullying, xxx, 771-774
    Racist, 774
    Refusal to commit to DNA testing of MIA Marines on Iwo Jima, xxx-xxxi
    Recovers the dead on Iwo Jima, xxx-xxxi, 825-826 fn 8
    Religious beliefs, 319-320, 879 fn 1340

*Samurai* tradition, 313, 314
    Threats to shut down the island to Americans, xxx, 773-774
    Use of Kuribayashi's image in political posters, 133, 773
    Worshiping at the graves of War Criminals, 136, 137
    Xenophobic, fascist and anti-historical associations, xxx, 132, 774

Shinto Religion, 68, 71, 73, 80-81, 83, 90, 319, 341, 342, 389, 408, 488, 520-523, 527, 543, 774
    Contemporary death rituals on Iwo Jima, xxxi
    Fascist doctrines, 68, 80-81, 87
    Hell (underworld), 543
    Glorifying War Criminals, 136, 137
    Proselytizing efforts, 73
    *Tama*, 543

Shintoists, 73, 518, 763, 768

Ship Building Program (U.S.), 7, 187
    Aircraft carriers, 187

Shively, Allen (USMC), 268

*Shoguns*, 71

*Shohatsu* amphibious landing craft (Japanese), 317

Shoemaker, Sergeant and Flamethrower Leonard J. (USMC), 427

Showalter, Dennis (historian), 128

Siffleet, Australian Commando Leonard, 117

Sigler, PFC Franklin E. (USMC), 628, 665, 666
    Medal of Honor, 665, 666

Sigma Alpha Epsilon Fraternity (University of South Dakota), 163

*Silence of the Lambs* (1991 movie), 115

Silver Star, xxii, 196, 649, 662, 725
    Number awarded to men who served on Iwo Jima, 635

Simon, PFC Leo (USMC), 259, 260

Simon & Schuster, xxxii

Simpson, Corporal R. G. (Photographer) (USMC), 446

Singapore, 51, 89

Sino-Japanese War (First), 75

Sino-Japanese War (Second), 81, 92, 177, 315

Skaggs, Joseph, xxxii

*Skat*, 148

Skinner, James "Jim", xli, xlii, xliv, 167 (see Gallery 1, Photo 10, page 4)

Sledge, Eugene (USMC), 150, 152, 339-340, 378, 391, 392, 473, 525

Slim, General Bill (British), 208, 452, 459

Smith, General Holland M. "Howlin Mad" Smith (USMC), xxi, 29, 170, 190, 191, 192, 193, 194, 195, 197, 199, 207, 210, 224, 229, 243, 244, 253, 307, 310, 323, 328, 329, 343, 350, 355, 356, 357, 358, 360, 372, 374, 401, 405, 406, 407, 408, 450, 451, 461, 462, 481, 482, 483, 484, 497, 498, 525, 529, 546, 551, 602, 614, 640, 641, 645, 646, 652, 658, 664, 697, 707, 708, 720, 723, 741, 778, 864 fns 884-885, 865 fn 891
    Analysis of German and Japanese soldiers, 708

Analysis of using gas warfare, 723
Authority to award medals, 640-641
Controversy over not using the 3rd Marines at Iwo Jima, 498-499
*Croix de Guerre* Medal, 191
Invasion of Saipan, 200
Religious beliefs, 355
Smith, Lieutenant General Norman (USMC), 825 fn 5
Smith, Major General Ralph (USA), 284, 482, 483
Smith, Steven B. (Yale University), xvii, 842 fn 337
Smoak, Lieutenant Colonel Eustace (USMC), 484
Snow, Philip (historian), 128, 315, 316
Snowden, Lieutenant General Lawrence (USMC), xli, 734-735 (see Gallery 4, Photo 10, page 6)
Purple Heart, 735
Socrates (Greek philosopher), 770
Solomon Islands, 8, 144-145, 146, 190, 607
Soochow (China), 101
Sorenson, Maceline (Vernon Waters' girlfriend), 569
South Dakota (USA), 163, 734
Southwell, Sergeant Lyman D. (USMC), 413, 416, 417, 426, 427, 431, 461, 462, 517, 579, 581, 582, 589, 590, 591, 616, 671, 724, 897 fn 1868
Purple Heart, 517
Silver Star, 416, 460
Souvenirs of Japanese body parts and supplies taken by Marines, 164-166, 275, 276, 517-518 , 536
Soviet Union (U.S.S.R.), xlvii, 4, 91, 116, 133, 186, 459, 638, 697, 709-710, 723
Invasion of Manchuria (1945), 709-710, 723
U.S. support of, 186
Spain, 199
Spanish American War (1898), 175, 401
Spanish Flu Epidemic (1918-19), 14-15
Spector, Ronald (historian), 351, 553
Speer, Nazi Armament Minister Albert, 85
Spence, Jonathan (historian), 100
Spigot Mortars (Japanese), 354, 377, 378, 485, 729
Spruance, Admiral Raymond (USN), 8, 206, 207, 238, 243, 329, 352
*SS*-Deathhead (Nazi), 136, 167
St. Albans Naval Hospital (New York), 665
St. Elizabeth's Hospital (Washington D.C.), 665
St. Mihiel (WWI battle), 191, 196, 561
St. Stephen's College (Hong Kong), 124, 667
Stackpole Books, xxxii
Stalin, Joseph, xix, l, 82, 186, 710, 718, 723, 767
Standard Oil, xlviii
Stanley Village (Hong Kong), 317
Starpoint School (Texas Christian University), xvii
Stein, Tony (USMC), 34, 602, 626-627, 639-640, 643, 688, 775
Jewish Immigrant from Austria, 640
Medal of Honor, 602, 626-627, 639-640
Navy Cross, 640
Steinbeck, John (writer), 24
Steinberg, Professor Jonathan (University of Pennsylvania), xvii, 767-768

Steinhagen, Captain P. W. (USN), 607
*Stevedore* (Operation), 170, 179, 252
Stewart, Master Technical Sergeant Jess L. (USMC), 667
Stewart, Mary (TCU), xvii
Stich, John (Honorary Japanese Consul Dallas, TX), 780
Stilwell, Colonel Joseph "Vinegar Joe" (USA), xlviii, xlix, 96, 114
Stimson, Henry, 695, 709
Stock Market Crash (October 1929), 13
Stockham, Gunnery Sergeant Fred (USMC), 637
Medal of Honor, 637
Stoghill, Steven (J.D.), xxxiii, 939 fn 3102
Stone, Kem, 770
Strickland, PFC Wesley (USMC), 424, 581, 754
Stubbs, Major Paul (USMC), 602, 784
*Styx* (Greek mythological river), 488
Submarines (German), 694
Submarines (Japanese), 722
Submarines (USN), 205, 206, 208, 369, 544, 710
Suenaga, Colonel Tsunetaro (IJA), 255
Sugimoto, Lieutenant Colonel Gorō (IJA), 80, 81
Sugiyama, IJA Minister Hajime (IJA), 544
Suicide (Japanese Citizens), 212, 213, 214, 215, 216, 218, 238, 239
Sun Goddess (*Amaterasu Omikami*), 79, 80-81, 319
Sun-Tzu (military philosopher), 301
Super Bowl LII, lvi
Surigao Strait (Battle of), 609
Suzuki, Prime Minister Kantarō, 575, 576, 577
Synder, John A. (USMC), 647
Syria, 709
*Taiyo* (Japanese aircraft carrier), 662
Taiwan, 75, 88
Takaishi, Colonel Tadashi (IJA), 543
Takamasa, Iba, 773, 783
Takashina, Lieutenant General (IJA), 241, 254, 255, 284
Takeda, Colonel Hideyuki (IJA), 249
*Takemikazuchi-no-kami* (one of the Shinto gods of war), 341
Taliban (Islamic Fascists), 769
Talmud (Jewish Oral Bible), 1,
Tamuning (Guam), xli
Tanaka, Terumi, 122
Tanaka, Ryossaburo (Commander of 299th Regiment IJA), 129
Tanaka, Yuki (historian), 119, 700, 771, 825 fn 7
Tang, General Shengzhi (Chinese Nationalist), 95, 847 fn 453
Tani, Lieutenant General Hisao (IJA), 106, 107
Tanks, 143, 144, 203, 231, 232, 233, 234, 237, 278, 279, 283, 284, 317, 354, 362, 373, 374, 375, 380, 393, 408, 452, 461, 476, 477, 508, 535, 638, 668, 696, 703
Biggest tank battle of the Pacific War (Saipan), 203
Flamethrower tanks (*Zippos* or *Satan*s), 375, 476, 477, 720
Japanese attacks against Shermans, 452

Japanese tank attack on Tinian, 231-234, 237

Sherman tanks, 362, 373, 374, 375, 380, 393, 452, 476

Tank maneuvers on Guam, 278, 279, 283, 284

Taoists, 74, 83

Tarawa (Battle of), 39, 53, 203, 298, 357, 372, 402

Tassafaronga (Sea Battle of), 606

Tehran Conference (1943), 186

Ten Commandments (Bible), 18

Mt. Tenjo (Guam), 261

*Terror and Liberalism*, 142

Texas (USA), 163, 346, 637, 692

Texas Christian University (TCU), xvii

Texas Veterans Memorial Park (McKinney), 692

Thailand, 89, 846 fn 422

Thanksgiving (Holiday), 632

Thiaucounrt (WWI battle), 563

Thiel College, 410

Third *Reich*, 573, 767

*This is Your Life*, 757

Thomas, Norman, 4

*Thor* (one of the Norse gods of war and revenge), 341

Thoreau, Henry David (transcendentalist), 460-461

Thornton, Lieutenant Commander Michael E. (USN), 675

Tibbets, Colonel Paul (USAAF), 555

Tientsin, 314, 512, 769

Tillman, Pat (USA), 669, 670, 671

Silver Star, 669

Timperley, Harold John (journalist), 104, 108

Tinian (Battle of), xxxvi, 39, 167, 169, 170, 190, 197, 225, 227, 229, 230, 231-240, 287, 296, 299, 323, 336, 348, 372, 485, 555, 561, 573, 574, 609, 655, 662, 663, 695, 701, 735, 869 fn 1007

Casualties, 238

Civilian casualties, 709

Civilians commit suicide, 238, 239

Japanese soldiers killing their own civilians, 239

Marines help civilians, 238, 239, 574

Reunion 2018, 655

Size of Japanese garrison, 229

Tinian Town, 229, 230

Tischler, 2nd Lieutenant Richard, 284, 527-528, 549, 569, 570, 572, 573, 616, 674, 727, 742, 759, 784

Letter of Commendation, (see Gallery 3, Photo 17, page 13)

Tōgō, Admiral Heihachirō (IJN), 76

Tojo, Prime Minister and General Hideki (IJA), 5, 31, 32, 82, 85, 91, 101, 137, 161, 225, 254-255, 318, 319, 323, 490, 618, 698, 718, 732, 774

Conflicts with Yamamoto, 698

Declaration to the Japanese people about war, 91

Tojo's Line (The Mariana Islands), 203, 204, 249

*Tokuhō*, see Imperial Rescript

Tokugawa Government (Japan), 71, 184

Tokunaga, Colonel Isao (IJA), 129, 132, 217

Tokyo (Japan's capital), liii, 8, 91, 92, 122, 170, 189, 203, 296, 306, 310, 315, 324, 511, 513, 527, 532, 534, 544, 698, 723, 736, 771

Massive bombing of the city on 9 March 1945, 511-512

Tokyo Bay (Japan), 1,

Tokyo International Military Tribunal, 108, 512, 774

Tokyo Prefecture, liv

Tokyo Radio, 51

Toland, John (historian), 303

Toll, Ian W. (historian), 757

Tommasi, Al (USMC), 165

*Top Secret* (B-29), 555

*Torch* (Operation), 345

Torpedoes (U.S.), 605-606, 720

Toshiharu, Takahashi (IJA), 346

Toth, Brigadier General Andrew (USAF), xliii

Toth, Charlie (USMC), 341

*The Tower of Triumph* (Hong Kong), 132

Treaty of Versailles, 31

Tripartite Pact (1940), 4, 85, 88, 92

Tripoli (Libya), 44

Tripp, Corporal Alan B. (USMC), 440, 442, 446, 517, 557, 579, 580, 590, 616, 671, 724, 726, 754, 924 fns 2700-2701

Letter of Commendation, 440, 517

Truk (Japanese island base), 183, 322

Truman, President Harry S., xxxi, 122, 562, 568, 571, 585, 603, 610-611, 617, 622, 623, 624, 625, 650, 682, 713, 715, 716, 717, 734, 737

Truman tries to disband the Marine Corps, 737

Trump, President Donald, lvi, 675, 689, 691

Tsar, 76

Tsuneishi, Keiichi (historian), 119

Tsushima (Battle of), 76, 205, 206

Tuchman, Barbara (historian), l

Tuck, Leo (USN), xli

Tulagi (Battle of), 47, 53, 54

Turnage, Major General Allen (USMC), 159, 253, 307

Navy Cross, 253

Turner, Henry Ashby Jr. (Yale University), xvii, xxxiii

Turner, Admiral Richmond Kelly (USN), 8, 199, 329, 357, 358, 381, 405, 406, 407, 408, 499

Controversy over not using the 3rd Marines, 499

Fights with Erskine, 358

Fights with Smith, 360

Nicknames, 357, 358, 360

Tutankhamun (Egyptian Pharaoh), 341

*Two Tickets to London* (1943 movie), 537

U.S. Armed Forces, 8, 184, 287

Racial Integration of in 1948, 734

Size during WWII, 733, 831-832 fn 90

U.S. Army, 24, 27, 154, 209, 284, 313, 323, 345, 555, 563, 618, 663, 666, 669, 695

U.S. Army War College, 313

U.S. Constitution, xxxi, xxxiii, 71

U.S. Embargos (against Japan), 92

U.S. Military,
> Great Depression's effect on, 13
> Height requirement before and after WWII, 34-35
> Size of Armed Forces, 8
U.S. Office of Strategic Services, 136
U.S. Pacific Fleet, lii, 7, 29, 35, 88, 92, 194, 297, 695, 697
Uenuma, Dr. Masao, 777
Ukraine, 208
Ulithi Atoll (Battle of), 609
ULTRA (Intelligence gathering group), 662
Umi Yukabai, 298
Underhill, Major General James L. (USMC), 640
Underwater Demolition Teams (UDTs of Frogmen), 246, 347, 365
UNESCO (United Educational, Scientific and Cultural Organization), 121
Unit 731, 132, 202
Unit 8604, 131
United Airlines, xli
United Kingdom (U.K.), 31
United States of Soviet Russia (U.S.S.R.), 91, 186, 694
United States Marine Corps,
> Focus on Amphibious Warfare, 189-195
> Size before and during WWII, 193
University of Alabama, 190
University of Michigan, 401
University of North Carolina, 400, 401
University of South Dakota, 163
University Press of Kansas, xxxii
Uo, Mariko (SIMUL), 773, 846 fn 428
USO, 246, 621
*USNS Woody Williams*, 689, 702, (see Gallery 4, Photo 3, page 3)
*USS Arizona* (battleship), 2, 722
*USS Bayfield PA-33* (Troopship), 394, 814
*USS Bismarck Sea* (escort carrier), 389
*USS Bunker Hill* (aircraft carrier), 576
*USS California* (armored cruiser), 76
*USS Cole* (guided-missile destroyer), 645
*USS Dashiell* (DD-659), 402
*USS Deuel* (Troopship), 354, 664
*USS Drexler* (destroyer), 303
*USS DuPage APA-41* (Troopship), 403
*USS Eldorado* (Troopship), 360, 381, 405
*USS Essex* (aircraft carrier), 208
*USS Finbak* (submarine), 50
*USS Gunston Hall* (LSD-5), xli
*USS Honolulu* (cruiser), 606
*USS Hornet* (aircraft carrier), 303
*USS Indianapolis* (battle-cruiser), 286, 488, 722
*USS Iwo Jima LHD-7*, 702
*USS Kalinin Bay* (escort carrier), 303
*USS Keokuk* (anti-torpedo net tender), 389
*USS Indianapolis* (heavy cruiser), 286, 488, 722
*USS Lexington* (aircraft carrier), 207
*USS Lunga Point* (escort carrier), 389
*USS M.S. Weltevreden* (Troopship), 144

*USS Matsonia* (Troopship), 601
*USS Missouri* (battleship), 617, 712
*USS Nevada* (battleship), 349
*USS Oklahoma* (battleship), 2
*USS Panay* (gunboat), xlviii, xlix, l
*USS Penguin* (minesweeper), 177
*USS Pennsylvania* (battleship), 242
*USS Pensacola* (cruiser), 347
*USS President Adams APA-19* (Troopship), 338
*USS Rinsdale* (Troopship), 641
*USS Rixey* (Troopship), 146
*USS Saratoga* (aircraft carrier), 389
*USS Solace* (hospital ship), 259
*USS Tang* (submarine), 605-606
*USS Tullibee* (submarine), 605-606
*USS Terror* (CM 5 (minelayer)), 349
*USS Texas* (battleship), 345, 346 (see Gallery 3, Photos 1-3, pages 1-3)
*USS Wayne* (Troopship), 244
*USS Zaurak* (Troopship), 557
Valbracht, Chaplain (USN), 546
Van Horn Moseley, Brigadier General George (USA), 313
Van Stockum, Lieutenant Colonel Ronald R. (USMC), 262, 264, 268, 269, 273, 274, 413, 780
Van Wagner, Kathryn (USN), 165-166
Vandegrift, Commandant and General Alexander Archer (USMC), xxxvii, 47-48, 50, 51, 194, 195, 243, 246, 286, 355, 401, 403, 453, 463, 570, 603, 609, 610, 611, 612, 614, 622, 625, 626, 637, 642, 649, 664, 678, 681, 682, 728
> Distinguished Service Medal, 608
> Medal of Honor, 608
> Navy Cross, 608
> Speaks with Woody, 625-626
> Triple Awarding of Medals for the same act, 608
Vandegrift, Lieutenant Colonel Alexander Jr. (USMC), 625
Vandenlinde, Bob, 689
> Silver Star, 689
Vatican, 548-549
Vautrin, Professor Minnie, 111 (see Gallery 1, Photo 26, page 12)
Vasey, Rear Admiral Lloyd "Joe" (USN), 123
The Vatican, 548
VE Day, 573
Veterans' Hospital No. 81, 570
VJ Day, xxiii, 296, 616
Venereal diseases, 54-55, 839 fn 264
Veterans Administration (VA), 18, 679, 688-689
Veterans Day Parade in New York City (2019), lvi
Vichy France (Fascist France), 8, 88
Victoria (Hong Kong), 128, 318
Victoria Barracks (Hong Kong), 318
Victoria Cross, 452, 638, 639
Victoria Harbor (Hong Kong), 317
Victoria, Brian (Buddhist priest and historian), 75
Viet Cong, 620
Vietnam, 118, 304, 688

Vietnam War, xliv, 620, 688

Virginia (United States), 11, 193, 579, 602, 634, 671

Volcano Islands (Bonin Islands), liv

*Das Volk* (Nazi definition), 81

Voltaire, 65

Von Boetticher, Lieutenant General Fredrich (German), l

Von Clausewitz, Military Philosopher Carl, 92, 329, 334, 357, 370, 372, 375, 413, 436, 450-451, 464, 498, 560, 713, 720

Von Moltke, Field Marshal Helmuth, 275

Voorhees, Captain Melvin Robert (USMC), 572

W-Day (Invasion of Guam), 170, 199, 243, 246, 247, 250

Wagner, Lieutenant Commander Richard (USN), 611, 612

Wainwright, Major General Jonathan Mayhew IV (USA), 661, 662
    Medal of Honor, 661

Wakabayashi, Bob Tadashi (historian), 828-829 fn 41

Wake Island, (Battle of), lii, 35, 48-49, 52, 53, 57, 150, 666, 667

Walsh, Lieutenant William P. (USA), 167, 647
    Medal of Honor, 647

Wanchai district (Hong Kong), 132

War of 1812, 241

War Bond Campaigns (United States), 82

War Crimes, 134-135, 142

War Crimes Affidavits (Guam), 178-179

The War Department (U.S.), 297

War Dogs (Guam), 289, 390

War Fighting Lab (TBS—USMC), 778

War Ministry (IJA), 314

Warfield, Thomas G. (USN), 669

Warner, Sergeant John (USMC), 136

Washington, George, 764, 765

Washington D.C., 67, 186, 356, 603, 609, 610, 620, 628, 631, 665, 689, 702

Water Purification Engineers, 487

Waters, George (USMC), 680, 758

Waters, Kathryn, 680, 681

Waters, Verna, 680

Waters, Corporal Vernon (USMC), xvii, 257, 258, 291, 423, 478, 479, 480, 481, 500, 501, 502, 506, 527, 560, 569, 582, 595, 614, 674, 677, 680, 725, 727, 734, 742, 758-759, 761, 763, 775, 784, 836 fn 173, 896 fn 1832, 906 fn 2156, 906 fn 2160
    Missing Ring, 670-671
    Silver Star, 478-479, 480, 481, 582, 595

Watson, Private Wilson D. (USMC), 445, 602, 628, 635, 643, 644
    Medal of Honor, 445, 644

Wayne, John (actor), xxiv, 701

Weathers, Captain N. A. (USMC), 571

Weber, David R. (Exeter), xvii, 784

*Wehrmacht* (German Military), 24, 167, 184, 302, 314, 580, 638, 639, 708, 781,825 fn 7, 922 fn 2641

Weinberg, Gerhard (historian), 29, 553

Weiner, Michael (historian), 775

Weitz, Mark (Legal Scholar), 826 fn 11

Claims Woody's lawsuit is frivolous, 826 fn 11

Welch, PFC Warren (USMC), 510
    Navy Cross, 510

Wells, Sandy (journalist), 583

Wenneker, Rear Admiral Paul (Nazi), liii, 91

West, First Sergeant Chester H. (USA), 689
    Medal of Honor, 689

West Point (United States Military Academy), 681

West Virginia (United States), lv, lvii, 11, 12, 17, 18, 20, 25, 38, 171, 258, 306, 343, 568, 621, 628, 631, 632, 689, 691, 725, 734

Western Union Telegrams, 33

Wheeler, Senator Burton (Montana), 4, 29

White, William Allen, 695

The White House, 186, 611, 612, 613, 622-623, 629, 677

"White Race" (Japan's hatred for), xlviii, 30, 90, 132, 177, 316

Whitehall (Montana), 25

Wiedhahn, Colonel Warren H. (USMC), 675

Wiesel, Elie (Holocaust survivor and writer), 11

Wiesenthal, Simon, 761

Wildcat Fighter Plane (Grumman F4), 720

Willey, Wendell D. (USN), 839 fn 261

Willhite, Corporal Leighton (USMC), xli, 373, 374, 378, 380, 507, 508, 780 (see Gallery 1, Photo 12, page 6)
    Bronze Star with V, 508

Wilkie, Wendell, xlviii, 695

Williams, Brady, 744 (see Gallery 4, Photo 11, page 6)

Williams, Dr. & Captain Brian (USAF), 641

Williams, Gerald, 14, 15, 21, 685

Williams, Hershel Woodrow "Woody," xvii, xx, xxix, xxxiv, xxxv, xxxvii, xxxviii, xli, xliv, xlv, liii, lv, 14, 15, 138, 142, 147, 164, 223, 245, 251, 256, 257, 258, 261, 262, 263, 264, 275, 276, 278, 279, 289, 291, 293, 296, 297, 299, 301, 305, 308, 338, 343, 348, 360, 361, 382, 383, 384, 387, 393, 397, 399,403, 404, 405, 407, 409, 410, 412-413, 414, 415, 416, 417, 418, 419, 427-451, 453, 455, 457, 458, 459, 460, 462, 473, 476, 477, 478, 480, 484, 485, 490, 491, 493, 494, 495, 500, 501, 502, 504, 505, 506, 509, 514, 515, 520, 525, 526, 529, 541, 550, 552, 560, 567, 569, 571, 573, 626-633, 635, 636, 650, 653, 658, 671, 672-692, 693, 701, 705, 706, 707, 717, 724, 725, 726, 727-732, 733, 734, 742-762, 763, 765, 766, 775, 784-785, 895 fn 1815, 895 fn 1823, 896 fn 1851, 897 fn 1877, 898 fn 1907, 905 fn 2113, 916 fn 2481, 929 fns 2801-2802, 930 fn 2829, 936 fn 3015, 938 fn 3076 (see Gallery 1, Photos 10, 12, pages 4, 6; Gallery 2, Photos 14-15, page 10; Gallery 3, Photo 7, page 5; Gallery 4, Photos 1, 5, 7-9, 12, pages 1, 3-5, 7)
    Admission he would have killed fellow Marine Thomas E. Scrivens had Scrivens not obeyed him, 459-460
    Assault on fellow Marine, 457, 458, 459, 460
    Attack on the large pillbox with vent, 656,

751-754 (see Gallery 3, Photos 7-10, 13, pages 5-7, 10)

Attempts to make the author sign an unfair contract, xxxi

Awards and honors, 689

"*Banzai*" on Guam, 266-267, 269-270

Birth (problems with), 11-13

Boot camp, 39-64, 838 fn 221

Woody Williams bridge, (see Gallery 4, Photo 6, page 4)

Cancer survivor, lv

Childhood, 11, 15, 16, 20-1, 22, 23

Civilian Conservation Corps, xxxix, 24, 25

Commentary on shell-shock or PTSD, 730-731

Comments about Japanese citizens committing suicide, 216, 217

Controversial MOH, 577-616, 618-622, 629, 631

Conversion (Christian), 687

Dashiell's article about Woody's MOH actions, 589-593

Death of siblings (Controversy of), 14-15, 17, 744, 833 fn 109 (see Gallery 4, Photo 11, page 6)

Demands money for his MOH story, xxxi

Destroying military documents, 145

Difficult to interview, 742-743

Education after the war, 679-680

Enlistment, 34, 35, 37, 744-745, 836 fn 173

Explanation of why he almost killed a fellow Marine, PFC Thomas E. Scrivens, 728

Failure to become a Marine Corps officer, 688

Failure to reach out to the Marines who died protecting him, 726

Failure to reach out to his officer, Captain Beck, 726

Falsifying information about himself, 12-3, 14-15

Families of Woody's comrades denounce what he did with his lawsuit against the author and publishers, 742

Feelings about combat, 250, 460

Feelings about invading Japan, 572

Feelings about the flag raising on Mt. Suribachi, 404-405

Financial Benefits of having the MOH, 681

Flamethrower Training, 149-153, 291, 305, 724

GCT Score, 666, 688, 926 fn 2756

Girlfriend Eleanor Pyles, 33, 155, 156, 157, 158, 759-760, 835-836 fn 170

Girlfriend Ruby Meredith, 33, 155, 156, 158, 275, 292, 434, 621-622, 739-741, 743, 759-760, 920 fn 2600 (see Gallery 4, Photo 1, page 1)

Gold Star Families Project, xxxi, 690, 726 (see Gallery 4, Photo 7, page 4)

History of the Gold Star, 690

Handling the fame with the MOH, lvi

Height problems with the Marine Corps, 33-

34, 744-746, 836 fn 173

High School dropout, 25

Interaction with citizens of Guam, 292

IQ, 666

Knowledge of Japanese War Crimes, 136

Lack of understanding of WWII strategy, 159

Last Marine MOH recipient from WWII to be alive in 2020, lv

Lawsuit against author, xxxi, xxxii, 742, 811-824, 826 fn 11, 826 fn 14

Lawsuits against publishers, xxxii

Loosing track of time during battle, 525

Love letters to Ruby, 155, 156, 158, 171-172, 292 (see Gallery 2, Photo 2, page 2)

Marriage to Ruby, 631

Medal of Honor, 577-584, 585, 586-616, 618-622, 629, 631

Actions for MOH on Iwo Jima, (see Gallery 3, Photo 13, page 10)

Final version of the MOH citation, 585, 613, 623-626

Commandant Charles C. Krulak believes Woody should not have received the MOH, 678, 725, 929 fn 2810, 936 fn 3013

Never received the Marine Corps of 1945's endorsement, xxxvii, 595-603, 611-613, 619, 651, 653, 672

Sample citation for the MOH for Woody (First Draft), 584

Medal of Honor Society's Chaplain, 687

Military assignments, 147

Military service, 738

Military training, 143, 144, 149-150, 159, 160, 171

Napoleon Complex, 744

Nicknames, lv, 263, 306

Non-Profit organization *The Gold Star Families Memorial Foundation*, xxxi

Overweight at enlistment, 37-38, 736

Patrols, 289

Possible "Stolen Valor," 928 fn 2797

Problems with his story of a small *Banzai* attacking him on Iwo, 432, 581-582, 587, 754-755, 897 fn 1882, 916 fn 2486

Problems with the bazooka man on Iwo Jima, 655-656, 749-751, 898 fn 1898, 924 fn 2707, 938 fn 3076

Problems surrounding his birth, 12-13, 744

Problems with boot camp class rank, 747, 840 fn 287

Problems with his comrade Clevenger on Guam, 747, 875 fn 1184

Problems with when he entered and left the CCC, 835 fn 166

Problems with his enlistment, 34, 35, 37, 744-746, 836 fn 173

Problems counting his KIAs, 657

Problems surrounding his Medal of Honor actions, 421, 504, 568, 611, 612, 613, 725, 726, 747-749, 816-817, 917 fn 2509

Problems with narrative, xxxi, xxxvii, 12-17, 33-34, 37-38, 613, 651, 653, 654, 655-656, 672,673, 725, 742-761, 899 fn 1923, 899 fn 1929

Problems with standing on a pillbox to destroy it, 753-754

Problems about the deaths of his siblings, 14-15, 17, 744, 833 fn 109 (see Gallery 4, Photo 11, page 6)

Problems about Tripp's actions on the day Woody earned the MOH, 580-581, 898 fn 1904, 924 fns 2700-2701

Problems with the story about his "friend" Vernon Waters' death, 500-502, 680, 758-759, 872 fn 1110, 896 fn 1832, 897 fn 1891, 906 fn 2156, 907 fn 2169, 930 fn 2824, 939 fns 3098-3100

Problems with the story about Vernon Waters' ring, 680, 681, 758-759

PTSD, 684, 686, 731

*Public Figure,* xxxii

Purple Heart, 505

Reasons why there are so many problems with Woody's narrative, 743-744, 760-761

Receiving the Medal of Honor, 577-605, 610-616, 618-626, 629, 631

Religious beliefs, 635, 685, 687-688, 921 fn 2629, 931 fn 2848

Religions conversion, 687, 731, 739

Religious racism, 688

Religious upbringing, 18, 58, 339, 383

Ruby's death, 739-741

Should have received a Silver Star or Navy Cross, 725

Siblings' deaths, 14-15, 17, 744, 833 fn 109 (see Gallery 4, Photo 11, page 6)

Struggles with height, 33-34

Struggles with weight, 37-38

Suing author's publishers, xxxi

Taking out pillboxes, 417-451

Training as an NCO, 295

Threatening to sue author's publishers, xxxi

Threatens to kill fellow Marine Thomas E. Scrivens, 457, 458, 727-729

Troubled Youth, 24-5

U.S. Army National Guard, 738

Unconstitutional actions, xxxi

Uses Stolen-Valor person Lefty Lee to support his exploits on Iwo Jima, 650, 923 fns 2688-2689

Using demolitions, 529, 541

VA work, 738

View of Hirohito, 121

View of the Japanese, 138, 165, 239-240, 275-276, 382, 461, 526, 736

View on killing, 18, 275-276

View of dead Marines, 469

What can be proven about his MOH actions, 583

What the MOH means to Woody, 692

Work with the VA, 679-680

Wounds, 503

Williams, Lila, 744 (see Gallery 4, Photo 11, page 6)

Williams, Lloyd Dennis "June" Jr., 14, 15, 21, 22, 685

Williams, Lloyd Dennis Sr., 15, 17, 687

Williams, Lurenna "Rennie", 15, 17, 20, 37, 541, 592, 622, 744

Williams, Lieutenant Colonel Marlowe C. (USMC), 267-268, 269, 270, 272, 293, 294, 396, 462, 506, 704, 705

  Bronze Star with V, 396

  Silver Star, 272

Williams, Robert L. (USMC), 686, 687

Williams, Ruby, 680, 686, 739, 740, 741

Williams, Samuel, 744 (see Gallery 4, Photo 11, page 6)

Willis, Corpsman John H. (USN), 586

  Medal of Honor, 586

Wilson, President Woodrow, 8, 15

Winston & Strawn, LLP, xxxiii

*Windstalkers* (2002 movie), 367

Withers, Colonel Harnoll J. (USMC), 308, 309, 481, 490, 528, 557, 581, 587, 594, 595, 616, 754, 755

  Silver Star (1), 310

  Silver Star (2), 481

Wlach, Corporal John (USMC), 455-456, 457, 462, 760

Wong, H. S. "Newsreel", 94

"Woody" Williams Armed Forces Reserve Center (Fairmont, West Virginia), 689, 691

World Bank, 733

World War I (WWI), xlviii, 11, 14, 19, 23, 31, 76, 88, 173, 191, 196, 253, 367, 401, 509, 561, 568, 605, 624, 637, 722

  Gas victims, 368

  Gas warfare, 368, 722

World War II Museum (New Orleans, LA), 684, 730, 742, 755, 781

Wuhu (China), 101

Wuxi (China), 101

Wyoming (USA), 306

Yad Vashem (Jerusalem, Israel), 67

Yale University, xvii, xxxiii, 100, 696, 717, 721, 777

Yamada, Radioman (IJA),

Yamamoto, Fleet Admiral Eisuke (IJN) 320

Yamamoto, Fleet Admiral Isoroku (IJN), xlviii, lii, liii, 4, 8, 9, 29, 82, 185, 186, 190, 313, 341, 698, 830 fn 55, 832 fn 94, 863 fn 855, 864 fn 871, 879 fn 1333

  Conflicts with Tojo, 698

Yamashita, General Tomoyuki (IJA), 130, 518, 618

*Yamato* (Japanese battleship), 205

*Yamato* Race (the Japanese Sun clan), 79, 84, 85, 87, 772

Yameda, Yoshio (IJA), 531

Yangtze River, xlviii, xlix, 65, 100

Yasukuni Shrine (Shinto), 108, 136, 137, 249, 518, 534, 774 (see Gallery 1, Photos 28-30, pages 14-15)

Yasuo, Ueno (IJA), 531

Yasunaga, Colonel Yukio (JDF), 120, 783

Yellin, Captain Jerry (USAAF), xli, xlii, 339, 489, 540, 546, 780, 913 fn 2394 (see Gallery 1, Photo 7, 12, pages 3, 6)

Yellow Beach (Iwo Jima), 348, 393

Yemen, 645

Yizhend, Liu (historian), 774

Yokohama (Japan), 1

Yonai, Admiral Mitsumasa, 186

York, Sergeant Alvin (USA), 627
    Medal of Honor, 627

Yoshida, Yutaka (historian), xlix, 119

Yoshimi, Yoshiaki (historian), 119

Yoshimura, Osaka Major Horfunmi, 120

Yost, Dr. Hershel, 15

Young, Private Gerald H. (USMC), 506

Young, James (writer), 85

*Yushukan* War Museum (Tokyo), 119

Zedong, Chairman Mao, 77, 767

Zen Buddhism, 83-84, 105, 488, 768 see also Buddhism

Zen Buddhists, 83-84, 574, 763, 768

Zero Japanese fighter plane ("Zeke"), 207, 393, 720

Zinni, General Anthony (USMC), xvii, xix, xxiv, xxv, xxxiv, 562, 777-778

*Zippos* (flamethrower tanks), See Flamethrower Tanks

*Zitadelle* (Citadel), 638

Zurlinden, Lieutenant and Correspondent Cyril P. (USMC), 372